The Elements

Element	Symbol	Atomic number	Molar mass/g mol^{-1}
Actinium	Ac	89	227.03
Aluminum	Al	13	26.98
Americium	Am	95	241.06
Antimony	Sb	51	121.75
Argon	Ar	18	39.95
Arsenic	As	33	74.92
Astatine	At	85	210.
Barium	Ba	56	137.34
Berkelium	Bk	97	249.08
Beryllium	Be	4	9.01
Bismuth	Bi	83	208.98
Boron	B	5	10.81
Bromine	Br	35	79.91
Cadmium	Cd	48	112.40
Calcium	Ca	20	40.08
Californium	Cf	98	251.08
Carbon	C	6	12.01
Cerium	Ce	58	140.12
Cesium	Cs	55	132.91
Chlorine	Cl	17	35.45
Chromium	Cr	24	52.01
Cobalt	Co	27	58.93
Copper	Cu	29	63.54
Curium	Cm	96	247.07
Dysprosium	Dy	66	162.50
Einsteinium	Es	99	254.09
Erbium	Er	68	167.26
Europium	Eu	63	151.96
Fermium	Fm	100	257.10
Fluorine	F	9	19.00
Francium	Fr	87	223.
Gadolinium	Gd	64	157.25
Gallium	Ga	31	69.72
Germanium	Ge	32	72.59
Gold	Au	79	196.97

Element	Symbol	Atomic number	Molar mass/g mol^{-1}
Hafnium	Hf	72	178.49
Helium	He	2	4.00
Holmium	Ho	67	164.93
Hydrogen	H	1	1.008
Indium	In	49	114.82
Iodine	I	53	126.90
Iridium	Ir	77	192.2
Iron	Fe	26	55.85
Krypton	Kr	36	83.80
Lanthanum	La	57	138.91
Lawrencium	Lr	103	257.
Lead	Pb	82	207.19
Lithium	Li	3	6.94
Lutetium	Lu	71	174.97
Magnesium	Mg	12	24.31
Manganese	Mn	25	54.94
Mendelevium	Md	101	258.10
Mercury	Hg	80	200.59
Molybdenum	Mo	42	95.94
Neodymium	Nd	60	144.24
Neon	Ne	10	20.18
Neptunium	Np	93	237.05
Nickel	Ni	28	58.71
Niobium	Nb	41	92.91
Nitrogen	N	7	14.01
Nobelium	No	102	255.
Osmium	Os	76	190.2
Oxygen	O	8	16.00
Palladium	Pd	46	106.4
Phosphorus	P	15	30.97
Platinum	Pt	78	195.09
Plutonium	Pu	94	239.05
Polonium	Po	84	210.
Potassium	K	19	39.10
Praseodymium	Pr	59	140.91

Element	Symbol	Atomic number	Molar mass/g mol^{-1}
Promethium	Pm	61	146.92
Protactinium	Pa	91	231.04
Radium	Ra	88	226.03
Radon	Rn	86	222.
Rhenium	Re	75	186.2
Rhodium	Rh	45	102.91
Rubidium	Rb	37	85.47
Ruthenium	Ru	44	101.07
Samarium	Sm	62	150.35
Scandium	Sc	21	44.96
Selenium	Se	34	78.96
Silicon	Si	14	28.09
Silver	Ag	47	107.87
Sodium	Na	11	22.99
Strontium	Sr	38	87.62
Sulfur	S	16	32.06
Tantalum	Ta	73	180.95
Technetium	Tc	43	98.91
Tellurium	Te	52	127.60
Terbium	Tb	65	158.92
Thallium	Tl	81	204.37
Thorium	Th	90	232.04
Thulium	Tm	69	168.93
Tin	Sn	5[illegible]	[illegible]
Titanium	Ti	2[illegible]	[illegible]
Tungsten	W	7[illegible]	[illegible]
Uranium	U	9[illegible]	[illegible]
Vanadium	V	2[illegible]	[illegible]
Xenon	Xe	5[illegible]	[illegible]
Ytterbium	Yb	7[illegible]	[illegible]
Yttrium	Y	3[illegible]	[illegible]
Zinc	Zn	3[illegible]	[illegible]
Zirconium	Zr	4[illegible]	[illegible]

Inorganic Chemistry

Inorganic Chemistry

DUWARD F. SHRIVER

Morrison Professor of Chemistry
Northwestern University, Evanston, Illinois

P. W. ATKINS

University Lecturer and Fellow of
Lincoln College, Oxford

COOPER H. LANGFORD

Professor of Chemistry
Concordia University, Montreal

Oxford Melbourne Tokyo
OXFORD UNIVERSITY PRESS

Oxford University Press, Walton Street, Oxford OX2 6DP
Oxford New York Toronto
Delhi Bombay Calcutta Madras Karachi
Petaling Jaya Singapore Hong Kong Tokyo
Nairobi Dar es Salaam Cape Town
Melbourne Auckland
and associated companies in
Berlin Ibadan

Oxford is a trade mark of Oxford University Press

First published 1990
Reprinted (with corrections) 1990

British Library Cataloguing in Publication Data
Shriver, D. F. (Duward Felix), 1934–
Inorganic chemistry.
1. Inorganic chemistry
I. Title II. Atkins, P. W. (Peter William), 1940– III. Langford, C. H. 546
ISBN 0–19–855232–7
ISBN 0–19–855231–9 (pbk)

Library of Congress Cataloging in Publication Data
Shriver, D. F. (Duward F.), 1934–
Inorganic chemistry/Duward F. Shriver, P. W. Atkins, and Cooper H. Langford.
Includes bibliographical references.
1. Chemistry, Inorganic. I. Atkins, P. W. (Peter William), 1940–
II. Langford, Cooper Harold, 1934– . III. Title.
QD151.5.S57 1990 546—dc20 89–22964 CIP
ISBN 0–19–855232–7
ISBN 0–19–855231–9 (pbk)

Set by The Universities Press (Belfast) Ltd
Printed in Great Britain by Butler and Tanner, Frome, Somerset

Preface

Inorganic chemistry is a vast and thriving discipline that deals with over a hundred elements from many points of view and touches all other branches of science. One reason for the vigorous research activity that has been characteristic of inorganic chemistry since the 1950s is the development of techniques that allow us to extract detailed information about inorganic compounds and their reactions. In addition, powerful theoretical models have been developed that help to systematize and rationalize the enormous number of facts that are now known. We say models rather than model because the subject is so diverse that no one model is useful throughout inorganic chemistry, and a part of the fascination of the subject (to the practicing chemist, if not initially to the student) is the challenge of deciding how to approach an explanation.

In addition to these intellectual attractions, inorganic chemistry has considerable practical impact. Eight of the top ten industrial chemicals, by annual tonnage, are inorganic. Inorganic chemistry is also essential to the formulation and improvement of semiconductors, light guides, nonlinear optical devices, superconductors, advanced ceramic materials, and pharmaceuticals. In common with organic chemists, inorganic chemists are constantly trying to synthesize forms of matter that have not existed before (perhaps anywhere in the universe) and to add to the number of materials at our disposal. Inorganic chemistry is also essential to an understanding of the composition of the cosmos and our more immediate environment, and we introduce these subjects where we consider it appropriate.

The vigor and diversity of modern inorganic chemistry presents a challenge to instructors and textbook writers, who must decide what to emphasize and what merely to mention. In any discipline a consensus on priorities develops slowly, and in inorganic chemistry agreement is only beginning to form on the core topics that are relevant to it. The problem is particularly acute in the United States and Canada where little time is available for inorganic chemistry in the curriculum.

We started this project with some definite ideas about our priorities and approach to the subject. These ideas were refined as the writing progressed, and were often modified and sometimes discarded in response to classroom experience and upon advice from reviewers. However, we always intended to produce a text that could be used not only as a thorough introduction to inorganic chemistry but also in a variety of balanced short courses. Our aim from the start was to be selective and interpretative across the full range of the subject; we never intended to be comprehensive. We take the view that students need a strong sense of the range of syntheses, reactions, structures, and explanations that are encountered in inorganic chemistry and that they should be prepared to appreciate new, unexpected results. Thus the book begins with an account of structure, bonding, periodic trends, and concepts relating to mechanism, and these ideas are used throughout later chapters to interpret and systematize the facts of inorganic chemistry. We have taken particular care to present the bonding and mechanistic models that are commonly used and widely accepted and have tried to develop a sense of when it is appropriate to apply one approach rather than another.

We have aimed for a style that is clear and interesting to students but have not hesitated to introduce sophisticated ideas where they are central to the subject. Among the special features are worked examples and exercises throughout the text, which provide points of emphasis and illustrate how a concept is deployed in practice. The three-dimensional aspect of inorganic structures is important, and we have provided a large number of illustrations and structural drawings. We also give a sense of the continuing development of the subject by including information on the timing and motivation of discoveries and the existence of some unsolved problems. We also attempt to convey the impact of inorganic chemistry on our economic and physical well-being.

We have had to make a number of decisions about notation and units. In general we have adhered to IUPAC recommendations and have used SI units almost everywhere. About the only exception (the outcome of a lot of argument between the authors, resolved ultimately by horse trading) is that all bond lengths are given in angstroms, for these units are still dominant in the current literature. The vexing question of the numbering of the periodic table has been dealt with by public compromise: we use both 1 to 18 and (for the main groups) I to VIII. Thus we refer to the oxygen group as Group VI/16 but to the chromium group simply as Group 6.

This text should be ample for a one-year course in inorganic chemistry in the United States and Canada and as a general introductory text for all three years of courses like those in the United Kingdom. The text is also suitable for a one-semester course using Parts 1 and 2, together with material selected from later chapters. Instructors of one-semester courses should note that each chapter in Part 3 has an introductory section that will stand on its own.

Most of the references that we give are to the review literature. The scientific literature also contains many articles of a semi-popular style that provide entertaining discussions of the history and applications of inorganic chemistry. We occasionally cite these to convey some of the less formal aspects of the subject. General references that can supplement this text are the two single-volume compendia of inorganic chemistry: *Chemistry of the elements,* by N. N. Greenwood and A. Earnshaw (Pergamon Press, Oxford, 1984) and *Advanced inorganic chemistry,* by F. A. Cotton and G. Wilkinson (5th edn, Wiley, New York, 1988). Three useful multivolume compendia are *Comprehensive inorganic chemistry,* edited by J. C. Bailar, H. J. Emeleus, R. Nyholm, and A. F. Trotman-Dickenson (Pergamon Press, Oxford, 1973), *Comprehensive organometallic chemistry,* edited by G. Wilkinson, F. G. A. Stone, and E. W. Abel (Pergamon Press, Oxford, 1982), and *Comprehensive coordination chemistry,* edited by G. Wilkinson, R. D. Gillard, and J. McCleverty (Pergamon Press, Oxford, 1987). A mine of specific information on inorganic chemistry is provided by the *Gmelin handbook of inorganic chemistry* (Springer-Verlag, Berlin).

In refining the text we have benefited from incisive comments by students in the United States and Canada and from colleagues around the world. We are particularly indebted to Professor S. H. Strauss (Colorado State University) who in the course of compiling his accompanying *Guide to solutions* helped us considerably with the main text. Professors R. J. P. Williams and M. L.

H. Green (University of Oxford) and F. Basolo (Northwestern University) also gave very helpful advice at various stages. Professor D. A. Phillips (Wabash College) provided valuable suggestions on the level and approach for both text and exercises. We are grateful to Dr R. Speel (University of Oxford) for reading the entire page proofs and helping us to catch a number of stupidities before they were made public. Our joint publishers were particularly helpful at all stages of the project, and we are very grateful to them for their understanding attitude and their support.

Evanston
Oxford
Montreal
September 1989

D. F. S.
P. W. A.
C. H. L.

Acknowledgements

We wish to acknowledge most sincerely everyone who gave us so much advice at all stages in the preparation of the text. In particular we wish to record our thanks to:

Dr D. M. Adams, University of Leicester
Professor A. L. Allred, Northwestern University
Professor M. Aresta, University of Bari
Professor R. J. Balahura, University of Guelph
Professor J. M. Berg, Johns Hopkins University
Professor P. H. Bird, Concordia University
Dr P. L. Bogdan, Northwestern University
Professor B. Bosnich, University of Chicago
Professor M. I. Bruce, University of Adelaide
Professor J. Burdett, University of Chicago
Dr J. Burgess, University of Leicester
Professor R. L. Burwell, Jr., Northwestern University
Dr A. K. Cheetham, University of Oxford
Professor S. S. H. Y. Ching, Connecticut College
Dr C. B. Cooper III, Varian Associates
Professor A. H. Cowley, University of Texas at Austin
Dr P. A. Cox, University of Oxford
Professor P. Day, University of Oxford
Dr M. A. Drezdzon, Amoco Research Center
Professor R. D. Ernst, University of Utah
Professor E. A. V. Ebsworth, University of Edinburgh
Dr J. Emsley, King's College, London
Professor J. H. Espenson, Iowa State University
Dr J. Evans, University of Southampton
Professor J. E. Finholt, Carleton College
Professor C. D. Garner, University of Manchester
Professor H. B. Gray, California Institute of Technology
Professor M. Greenblatt, State University of New Jersey
Professor J. Halpern, University of Chicago
Professor D. T. Haworth, Marquette University
Dr W. Henderson, Northwestern University
Dr H. A. O. Hill, University of Oxford
Professor B. R. Hollebone, Carleton University
Professor J. A. Ibers, Northwestern University
Professor B. R. James, University of British Columbia
Dr B. F. G. Johnson, University of Cambridge
Dr J. W. Johnson, Exxon
Dr G. C. Joy, Universal Oil Products
Professor R. D. W. Kemmitt, University of Leicester
Professor S. F. A. Kettle, University of East Anglia
Professor J. A. Kornblatt, Concordia University
Professor M. M. Lerner, Oregon State University

Dr W. Levason, University of Southampton
Professor G. Long, University of Missouri–Rolla
Professor C. M. Lukehart, Vanderbilt University
Professor J. A. McLeverty, University of Birmingham
Professor K. W. Morse, Utah State University
Professor T. A. O'Donnell, University of Melbourne
Professor T. V. O'Halloran, Northwestern University
Professor R. T. Paine, University of New Mexico
Dr B. Pant, Concordia University
Dr A. Parkins, King's College, London
Dr C. S. G. Phillips, University of Oxford
Professor K. R. Poeppelmeier, Northwestern University
Professor L. Que, University of Minnesota
Dr D. W. H. Rankin, University of Edinburgh
Dr M. Sabat, University of Virginia
Professor M. J. Sailor, University of California, San Diego
Professor C. K. Schauer, University of North Carolina
Professor H. Schmidbaur, Technischen Universität München
Professor N. Serpone, Concordia University
Professor J. R. Shapley, University of Illinois
Mr D. N. Shriver, Evanston
Professor A. M. Stacy, University of California, Berkeley
Dr C. Tessier-Youngs, Case Western University
Professor J. M. Thomas, Royal Institution of Great Britain
Professor R. West, University of Wisconsin
Professor A. Wold, Brown University
Professor H.-C. zur Loye, Massachusetts Institute of Technology.

Copyright acknowledgements

Fig. 4B.2, J. M. Manoli, C. Potash, J. M. Begeaut, and W. P. Griffith, *J. chem. Soc. Dalton,* 192 (1980). **Fig. 10.B1,** Vacuum Atmospheres, Inc., Hawthorne, USA. **Fig. 10.5** M. J. Fink, M. J. Michalczyk, K. J. Haller, R. West, and J. Michl, *Organometallics,* **3,** 793 (1984). **Figs 13.11** and **13.12,** J. D. Corbett, *Acc. chem. Res.,* **14,** 239 (1981). **Fig. 17.14,** G. A. Samorjai, *Chemistry in two dimensions.* Cornell University Press, Ithaca (1981). **Fig. 18.4,** S. Iijima, *J. Solid State Chem.,* **14,** 52 (1975). **Fig. 18.20,** C. Kittel, *Introduction to solid state physics,* (6th edn). John Wiley, New York (1986). **Figs 18.34** and **18.36,** T. Hubanks and R. Hoffmann, *J. Amer. chem. Soc.,* **105,** 1152 (1983). **Fig. 15.14,** J. B. Goddard and F. Basolo, *Inorg. Chem.,* 936 (1968). **Fig. 15.15,** R. J. Guschl, R. Stewart, and T. L. Brown, *Inorg. Chem.,* **13,** 959 (1959). **Fig. 19.7,** L. Stryer, *Biochemistry.* W. H. Freeman & Co., New York (1989).

Contents

Part 1 · Structure

1 Atomic structure

The origin and distribution of the elements 4
Atomic structure and chemical periodicity 8
Many-electron atoms 21
Further reading 35
Exercises 35
Problems 36

2 Molecular structure

Electron pair bonds 38
Molecular symmetry 50
Molecular orbitals of diatomic molecules 58
Further reading 70
Exercises 71
Problems 72

3 Polyatomic molecules and solids

Molecular orbitals of polyatomic molecules 74
The molecular orbital theory of solids 92
Further information: symmetry-adapted orbitals 102
Further reading 104
Exercises 105
Problems 106

4 The structures of solids

Crystal structure 108
Metals 111
Ionic solids 115
Further information: thermodynamics 135
Further reading 139
Exercises 140
Problems 141

Part 2 · Reactions

5 Brønsted acids and bases

Brønsted acidity 146
Periodic trends in Brønsted acidity 156
Polyoxo compound formation 162
Further reading 166
Exercises 167
Problems 167

6 Lewis acids and bases

The Lewis definitions of acids and bases 170
The strengths of Lewis acids and bases 172
Representative Lewis acids 181
Heterogeneous acid–base reactions 186
Further reading 188
Exercises 188
Problems 189

7 *d*-Metal complexes

Structures and symmetries of complexes 192
Bonding in complexes 205
Reactions of complexes 217
Further reading 225
Exercises 225
Problems 226

8 Oxidation and reduction

Extraction of the elements 229
Reduction potentials 235
Redox stability in water 242
The diagrammatic presentation of potential data 246
Further reading 260
Exercises 260
Problems 261

Part 3 · *s*- and *p*-block elements

9 Hydrogen and its compounds

The element 266
Classification and structure of compounds 272
The reactivity of hydrogen compounds 279
The electron-deficient hydrides of the boron group 283
Electron-precise hydrides of the carbon group 287
Electron-rich compounds of Groups 15/V to 17/VII 289
Further information: nuclear magnetic resonance 292
Further reading 296
Exercises 297
Problems 298

10 Main group organometallics

Classification and structure 300
Ionic and electron-deficient compounds of Groups 1, 2, and 12 310
Electron-deficient compounds of the boron group 315
Electron-precise compounds of the carbon group 319
Electron-rich compounds of the nitrogen group 326
Further reading 328
Exercises 329
Problems 330

11 The boron and carbon groups

General properties of the elements 332
The boron group 334
The carbon group 342
Silicates and aluminosilicates 347
Boron cluster compounds and carboranes 353
Borides, carbides, and silicides 361
Electronic properties 364
Further reading 367
Exercises 367
Problems 368

12 The nitrogen and oxygen groups

The elements 370
The nitrogen group 371
The oxygen group 386
p-Block ring and cluster compounds 397
Further reading 400
Exercises 400
Problems 401

13 The halogens and the noble gases

The elements 403
Polyhalogen compounds 407
Compounds of halogens and oxygen 412
Metal halides 420
The noble gases 424
Further reading 428
Exercises 428
Problems 429

Part 4 · *d*- and *f*-block elements

14 Bonding and spectra of complexes

The electronic spectra of atoms 434
The electronic spectra of complexes 441
Lower symmetry and binuclear complexes 456
Further information: the addition of angular momenta 462
Further reading 463
Exercises 463
Problems 464

15 Reaction mechanisms of *d*-block complexes

Ligand substitution reactions 466
Substitution in square-planar complexes 471
Substitution in octahedral complexes 477
The mechanisms of redox reactions 486
Photochemical reactions 492
Further reading 494
Exercises 495
Problems 496

16 *d*- and *f*-block organometallics

Bonding 500
d-Block carbonyls 504
Other organometallics 517
Metal–metal bonding and metal clusters 528
Further reading 536
Exercises 537
Problems 538

Part 5 · Interdisciplinary topics

17 Catalysis

General principles 542
Homogeneous catalysis 546
Heterogeneous catalysis 558
Further reading 570
Exercises 570
Problems 571

18 Structure and properties of solids

Some general principles 573
Prototypical oxides and fluorides 580
Prototypical sulfides and related compounds 593
Further information: phase diagrams 599
Further reading 601
Exercises 602
Problems 602

19 Bioinorganic chemistry

Pumps and transport proteins 607
Enzymes exploiting acid catalysis 615
Redox catalysis 620
Further reading 631
Exercises 631
Problems 632

Appendices

A1. Electron configurations of atoms 635
A2. Ionization energies 637
A3. Electronegativities 640
A4. Standard potentials 642
A5. Character tables 664
A6. Symmetry-adapted orbitals 671
A7. Tanabe–Sugano diagrams 682

Answers to exercises 685

Formula index 693

General index 699

Part 1

Structure

The four chapters of this part lay the foundations of bonding theory and deal in turn with atoms, molecules, and solids. The emphasis in this part is on how theories can account for the rich variety of structures that are found. In due course we shall see that the same ideas also provide a framework for accounting for the reactions of substances.

Atomic structure

1

In this chapter we describe how the chemical elements have been formed, and why they are found with their characteristic abundances in the universe and on earth. We also review the periodic table, which is the most important device for organizing the properties of the 100 or so elements that are now known. The bulk of the chapter then describes the atomic properties of the elements and shows how they can be rationalized in terms of the behavior of electrons in atoms. To do this to an adequate level, we need to introduce some of the concepts of quantum mechanics, and so to a considerable extent this chapter is not typical of the rest of the text, where a much more qualitative approach is generally sufficient.

In the course of the chapter we meet some of the parameters that characterize the properties of atoms, and which are used to systematize the chemical properties of the elements and to organize inorganic chemistry. This parametrization of properties—effectively, bundling together a number of influences into a single concept—will be a recurring theme throughout the text.

The origin and distribution of the elements

1.1 The cosmic distribution of the elements

1.2 The distribution of the elements on earth

Atomic structure and chemical periodicity

1.3 Some principles of quantum mechanics

1.4 Atomic orbitals

Many-electron atoms

1.5 The building-up principle

1.6 Atomic parameters

Further reading

Exercises

Problems

The observation that the universe is expanding has led to the idea that 18 billion years ago all observable matter was concentrated into a pointlike region which exploded in the event called the Big Bang. With initial temperatures immediately after the Big Bang thought to be about 10^9 K, the fundamental particles produced in the explosion had too much kinetic energy to bind together in the forms we know today. However, the universe cooled as it expanded, the particles moved more slowly, and soon began to adhere under the influence of a variety of forces. In particular, the strong force—a short-range but powerful force between protons, neutrons, and each other—bound particles together into nuclei. Similarly, the electromagnetic force—a long-range force between electric charges—bound electrons to nuclei to form atoms. The properties of the only subatomic particles that we need to consider are summarized in Table 1.1.

Table 1.1. Subatomic particles of relevance to chemistry

Particle	Symbol	Mass	Charge*	Spin
Electron	e^-	9.11×10^{-31} kg (m_e) 5.486×10^{-4} u†	−1	$\frac{1}{2}$
Proton	p	1.67×10^{-27} kg (m_p) 1.0073 u	+1	$\frac{1}{2}$
Neutron	n	1.67×10^{-27} kg 1.0087 u	0	$\frac{1}{2}$
Photon	γ	0	0	1
Neutrino	ν	0	0	$\frac{1}{2}$
α particle	α	[$^4_2He^{2+}$ nucleus]	+2	0
β particle	β	[e^- ejected from nucleus]	−1	$\frac{1}{2}$
γ photon	γ	[electromagnetic radiation from nucleus]	0	1

* Multiple of elementary charge, 1.602×10^{-19} C.
† Atomic mass unit, 1 u = 1.6605×10^{-27} kg.

The origin and distribution of the elements

If current views are correct, by about 2 h after the start of the universe the temperature had fallen so much that most of the matter was in the form of 89 percent H atoms and 11 percent He atoms. In one sense, not much has happened since then: as Fig. 1.1 shows, H and He remain the most abundant elements in the universe. However, nuclear events have formed a wide assortment of other elements and have immeasurably enriched the variety of matter in the universe.

1.1 The cosmic distribution of the elements

The abundances of elements are summarized in Fig. 1.1. These data are only approximate because the elements are not uniformly distributed. One problem is that there are concentrations of certain elements in different regions of the surface of the earth. Another is that the earth is not uniform from its center to its crust. Moreover, the compositions of other planets are not the same as that of the earth. In fact, **terrestrial abundances**, the

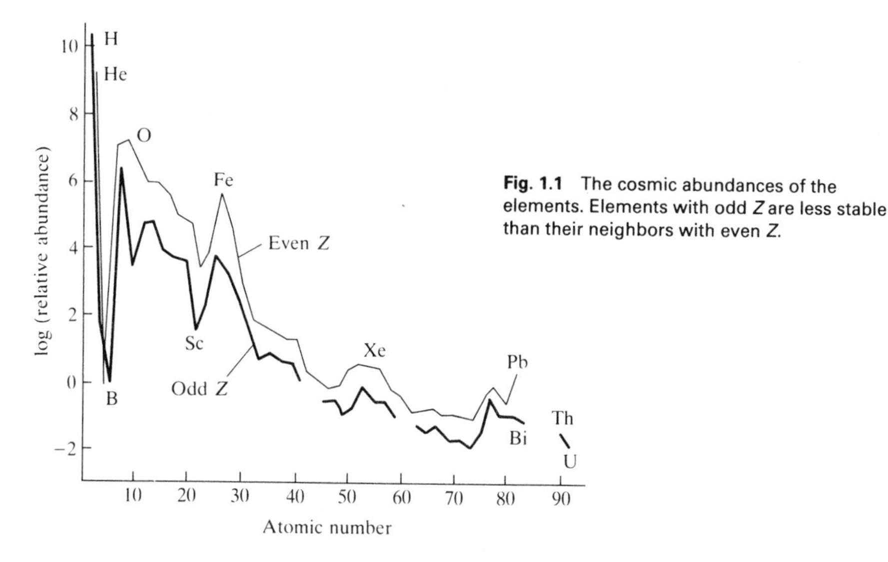

Fig. 1.1 The cosmic abundances of the elements. Elements with odd Z are less stable than their neighbors with even Z.

abundances of the elements on earth, often bear little relation to **cosmic abundances**, their abundances universe-wide. Nevertheless, there is a broad consensus that the information in the illustration is plausible. That being so, we can draw a number of conclusions.

In the first place, for reasons related to the stabilities of nuclei and the properties of the strong force, cosmic abundances alternate between elements of odd and even atomic number, with elements of even atomic number generally more dominant. Thus Fe ($Z = 26$) is more abundant than either Mn ($Z = 25$) or Co ($Z = 27$). Secondly, the overall trend is for cosmic abundance to decline with increasing atomic number. Thus, the heavier halogens (such as Br and I) are less abundant than the lighter halogens (F and Cl). There are also striking anomalies in the distribution. For instance, the very light elements Li, Be, and B are far less abundant than might be expected by extrapolation from the heavier elements. In contrast, Fe is anomalously abundant. Any theory of nucleosynthesis must account for this pattern.

Nucleosynthesis of light elements

The condensation of clouds of H and He atoms is thought to have led to the formation of the earliest stars. The collapse of these stars under the influence of gravity gave rise to high temperatures and densities within them, and nuclear fusion reactions began as nuclei merged together. Energy is released when light nuclei fuse to form elements with atomic number up to 26 (Fe), and these elements were formed in stellar interiors. Such elements are the 'ash' of the nuclear fusion events referred to as 'nuclear burning'. The burning reactions (which should not be confused with chemical combustion) involved H and He nuclei and a complicated fusion cycle catalyzed by C nuclei.

The low abundance of Li, Be, and B is consistent with the events taking place during the phase in the life of a star when its energy is provided by He burning. In this **helium cycle**, ^{8}Be is formed by collisions between α particles (^{4}He nuclei), but it then reacts with more α particles:

$$^{8}Be + \alpha \rightarrow {}^{12}C + \gamma$$

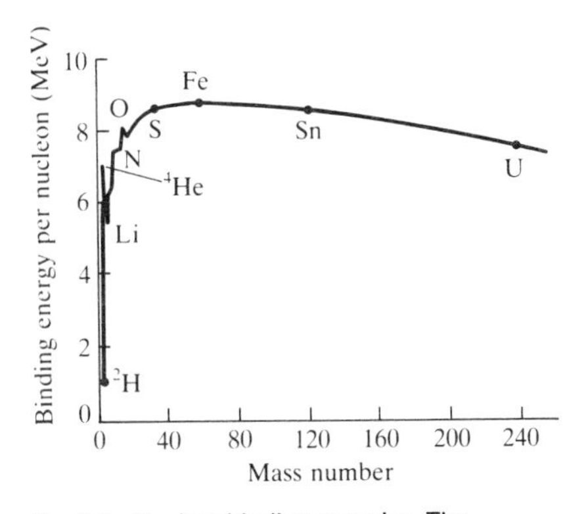

Fig. 1.2 Nuclear binding energies. The greater the binding energy, the more stable the nucleus. The most stable nucleus is ^{56}Fe.

The helium cycle stage of stellar evolution does not result in the formation of Li, Be, and B as stable end products. The origin of these three elements is still uncertain, but they may result from events in which C, N, and O nuclei are fragmented by collisions with high-energy particles, the process called 'spallation'. Elements can also be produced by nuclear reactions such as:

$$^{14}\text{N} + \text{n} \rightarrow {}^{14}\text{C} + \text{p}$$

(n is a neutron and p a proton.) This reaction still continues in the atmosphere under the impact of cosmic rays and contributes to the steady-state concentration of ^{14}C on earth.

The high cosmic abundance of Fe is consistent with its having the most stable of all nuclei (Fig. 1.2). This stability can be assessed from its binding energy, the difference in energy between the nucleus itself and the same numbers of individual protons and neutrons. The binding energy of ^{56}Fe, for example, is the difference in energy between the ^{56}Fe nucleus and 26 protons and 30 neutrons. Figure 1.2 shows the binding energy (expressed as an energy per nucleon) for all the elements, and we see that Fe occurs at the maximum of the curve.

Nucleosynthesis of heavy elements

Elements heavier than Fe are produced by a variety of processes. These include the capture of neutrons produced from the decay of lighter elements in reactions such as

$$^{23}\text{Ne} + \alpha \rightarrow {}^{26}\text{Mg} + \text{n}$$

Under conditions of intense neutron flux, as in a supernova (the explosion of a star), a given nucleus may capture several neutrons and become a progressively heavier isotope. However, there comes a point when it will eject an electron from the nucleus as a β particle, thereby increasing its atomic number by 1, and hence becoming a new element. An example is

$$\begin{aligned} {}^{98}\text{Mo} + \text{n} \rightarrow {}^{99}&\text{Mo} + \gamma \\ &\downarrow \\ {}^{99}&\text{Tc} + \beta + \nu \end{aligned}$$

(ν is a neutrino.) The daughter nucleus, the product of a nuclear reaction (^{99}Tc in this example), can absorb another neutron, and the process can continue, gradually building up the heavier elements.

1.2 The distribution of the elements on earth

Inorganic chemistry grew up among, and still largely deals with, the distribution of the elements as we find them on earth. In considering the terrestrial distribution of the elements, it proves useful to bear in mind the concept of volatility as it is used in geochemistry. Thus an element may be either volatile itself (like the noble gases) or form compounds that are volatile under the conditions that prevailed during the era following condensation of the earth. Some members of the two classes are as follows:

Volatiles: C, halogens, Hg, Cd, Ga, Ge, Pb, Bi, Zn
Nonvolatiles: Fe, Co, Ni, Si, Be, Ca

It is reasonable to suppose that the less volatile elements (in the geochemical sense) have a terrestrial distribution similar to the overall cosmic distribution. We can verify that by studying meteorites, for these are also products of the processes of formation of the solar system but have not experienced the heating episodes characteristic of the earth. It turns out that the more volatile elements (in the geochemical sense again) are indeed greatly depleted on earth relative to their cosmic abundance. This is explained by their loss during the high-temperature eras accompanying the formation of the earth.

Condensation processes

Figure 1.3 shows a widely accepted summary of the earth's composition. The elements with terrestrial abundances similar to their cosmic abundances are marked 'early condensates'. This name indicates that they appear in minerals that condensed at high temperatures and suffered little loss. A relatively depleted group (compared with the cosmic abundance) consists of the moderately volatile elements, which include Ag, Zn, Ge, Sn, and F. A highly depleted group of volatile elements include Cd, Hg, Pb, and the halogens other than F.

The early condensates include metallic Fe, which condenses at about 1500 K (with 12.5 percent Ni), diopside ($CaMgSi_2O_6$) condensing at about 1450 K, and anorthite ($CaAl_2Si_2O_8$) condensing at about 1350 K. As a result, Fe, O, Mg, and Si together make up more than 90 percent of the earth. The elements S, Ni, Al, and Ca account for another 6 to 7 percent.

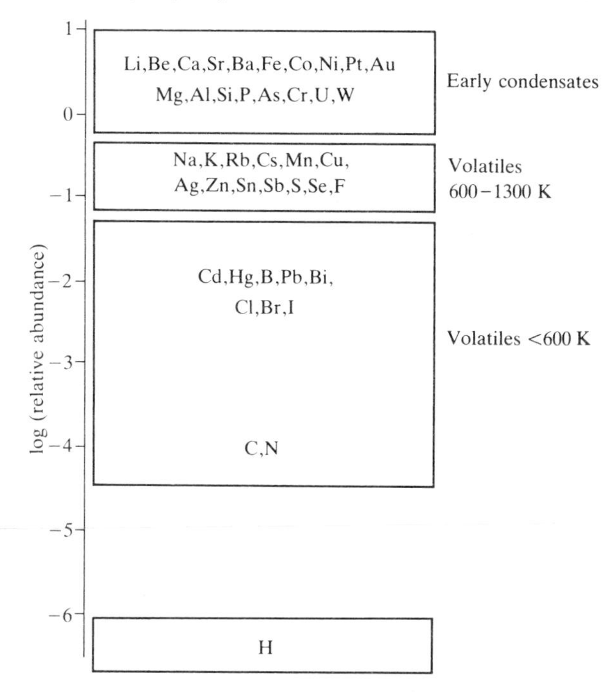

Fig. 1.3 The estimated abundances of the elements on earth (not merely the crust). Note that the early condensates are the most abundant and the volatiles that vaporize below 600 K are the least abundant.

Zonal distribution

The earth is divided into five main zones: the atmosphere, the hydrosphere, the crust, the mantle, and the core. The outermost three zones are permeated by the biosphere, the zone of life. The three inner zones are studied by seismography and are distinguished by properties that include density and elasticity.

The constant h is **Planck's constant**, a fundamental constant with the value 6.626×10^{-34} J s.

The de Broglie relation shows that the greater the momentum of the particle, the smaller the wavelength of its wavefunction. The relation has been confirmed by showing that electrons accelerated to a particular velocity by a potential difference display diffraction characteristics (a typical wave property) in agreement with the wavelength that the de Broglie relation predicts. An **electron diffraction** experiment is a practical application of the wave properties of electrons. An accelerating potential difference of about 20 kV is used to accelerate electrons to a speed at which they have a wavelength of about 0.1 Å, and the beam is passed through a gaseous sample. The wavelength is comparable to the bond lengths of molecules (which are typically about 1 Å) and the diffraction pattern that results can be analyzed to obtain the details of the structure of the molecules.

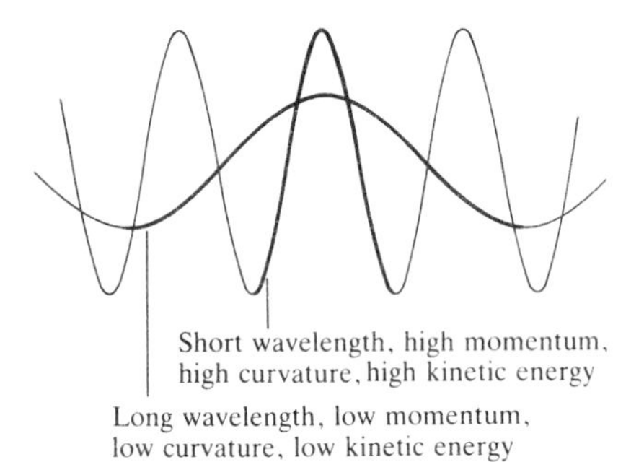

Fig. 1.5 The relation between curvature of the wavefunction and kinetic energy.

Since the kinetic energy of an electron is proportional to the square of its momentum (specifically, $E_K = p^2/2m_e$), the kinetic energy increases as the wavelength of the electron's wavefunction decreases. Another way of expressing this is to note that the smaller the wavelength, the more sharply ψ is curved. Therefore, all we need do in order to judge whether an electron has a high kinetic energy is to check whether its wavefunction changes rapidly from point to point. If it does, the kinetic energy is high (Fig. 1.5). If ψ varies only slowly from point to point, the kinetic energy is low.

The sharpness with which any mathematical function curves (including the sharpness of a wavefunction), is measured by its second derivative $d^2\psi/dx^2$. If this quantity is large, corresponding to ψ being sharply curved, the kinetic energy is high. This is consistent with the de Broglie relation, since a sharply curved wavefunction is also one with a small wavelength (Fig. 1.5). After a few more preliminaries, we shall see how to use this feature.

The uncertainty principle

A second major result of quantum mechanics is that it is impossible to specify the exact position and momentum of a particle simultaneously. This conclusion is expressed quantitatively by the **uncertainty principle**, which was deduced by Werner Heisenberg in 1926. This principle states that if the uncertainty in the position of a particle is Δx and the uncertainty in its momentum is Δp, then the product of these two uncertainties must satisfy the relation

$$\Delta p \Delta x \geq \tfrac{1}{2}\hbar \qquad (2)$$

The fundamental constant $\hbar$ stands for $h/2\pi$:

$$\hbar = 1.052 \times 10^{-34} \text{ J s}$$

We see that as the uncertainty Δx decreases, the uncertainty Δp must increase, and vice versa, so as to keep their product at least as big as $\tfrac{1}{2}\hbar$.

One of the important implications of the uncertainty principle for chemistry is that it denies the possibility of ascribing electrons to particular orbits around nuclei like planets traveling around the sun. For an orbit to be a meaningful concept, it would be necessary to specify the position and momentum of an electron at each instant. The uncertainty principle, however, tells us that this is impossible. The concept of orbit works for planets only because we do not try to measure their positions and momenta precisely enough for the uncertainty principle to be relevant.

Quantization

One of the most significant features of quantum mechanics is that it limits the energies of particles—such as electrons in atoms—to discrete values. This is in striking contrast to classical mechanics, where energies can take any value whatever the object. According to classical mechanics, for instance, a pendulum can swing with any energy (within reason). The restriction of any property to discrete values is called **quantization**; the restriction of energy to discrete values is **energy quantization** (certain other properties are also restricted to discrete values, as we see later).

The permitted energies of a system, such as an H atom, are called its **energy levels**. These levels, and the detailed form of the wavefunction, are found by solving the **Schrödinger equation**. This equation is one of the fundamental equations of quantum mechanics, and its formulation by Erwin Schrödinger in 1926 marks the birth of modern quantum theory. The Schrödinger equation is so central to the discussion of electrons, atoms, and molecules, that although we shall not need to find its solutions explicitly, it is at least appropriate to see what it looks like. For a particle of mass m moving in a one-dimensional region of space where its potential energy is V, the equation is

$$-\frac{\hbar^2}{2m}\frac{\mathrm{d}^2\psi}{\mathrm{d}x^2} + V\psi = E\psi \tag{3}$$

Although the Schrödinger equation is probably best regarded as a fundamental postulate, some justification of its form comes from noting that the first term, the one proportional to $\mathrm{d}^2\psi/\mathrm{d}x^2$, is essentially the kinetic energy of the particle. Therefore, the equation simply expresses in quantum mechanical terms the fact that the total energy E is the sum of the kinetic energy and the potential energy V.

Energy quantization

When the Schrödinger equation is solved for atoms and molecules, it is found that physically meaningful wavefunctions exist only for certain values of the energy E. That is, energy quantization is a direct consequence of the Schrödinger equation.

The simplest system that shows quantization is the 'particle in a box', a particle of mass m confined to a one-dimensional region of length L between two impenetrable walls. The wavefunctions are sine waves that fit into the region between the walls, so the only permitted wavelengths (Fig. 1.6) are

$$\lambda = \frac{2L}{n}, \qquad n = 1, 2, \ldots$$

Waves of this wavelength have the form

$$\psi = \sin\frac{2\pi x}{\lambda} = \sin\frac{n\pi x}{L}$$

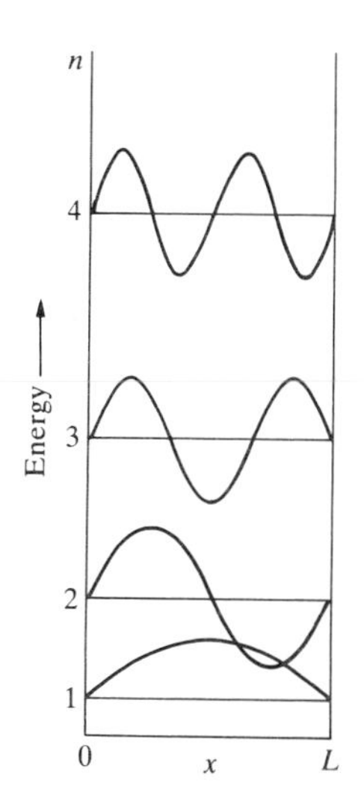

Fig. 1.6 The first four wavefunctions of a particle in a box have wavelengths $2L$, L, $\frac{2}{3}L$, and $\frac{1}{2}L$. In general, only waves of wavelength $2L/n$ fit between the walls.

These are the wavefunctions of the particle in this system. It follows from the de Broglie relation that the only permitted linear momenta of the particle are

$$p = \frac{h}{\lambda} = \frac{nh}{2L}, \qquad n = 1, 2, \ldots$$

Therefore, since the kinetic energy of the particle (the only kind of energy it has) is related to the linear momentum p by

$$E = \frac{p^2}{2m}$$

the only permitted energies of the particle are

$$E = \frac{n^2h^2}{8mL^2}, \qquad n = 1, 2, \ldots$$

The number n is called a **quantum number**, and once we know its value, we can calculate the energy. One feature to note is that in this one-dimensional system, we need *one* quantum number to specify the energy and the wavefunction. The wavefunctions have the same form as the harmonics of a vibrating string. The lowest level, $n = 1$, has no nodes within the box, and progressively higher energy levels have more nodes. We shall see that this relation between the number of nodes and the energy also applies to the wavefunctions of electrons in atoms and molecules.

A second important example is that of a particle undergoing **harmonic motion**, motion in which the particle is subjected to a restoring force proportional to its displacement, $F = -kx$. In one dimension the wavefunctions and energies are specified by a single quantum number v, and

$$E = (v + \tfrac{1}{2})\hbar\omega, \qquad v = 0, 1, 2, \ldots \tag{4}$$

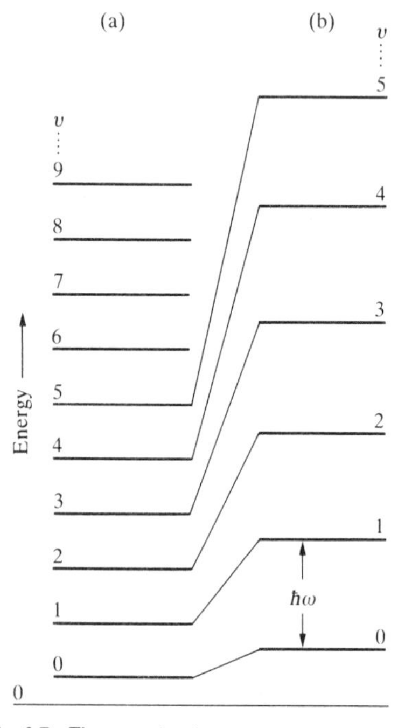

Fig. 1.7 The quantized energy levels of a harmonic oscillator with (a) low force constant and (b) high force constant.

In this expression $\omega = (k/m)^{\frac{1}{2}}$, where m is the mass of the particle and k is the **force constant**. The ladder of uniformly spaced energy levels, with separation $\hbar\omega$, is shown in Fig. 1.7. If we think of the oscillator as being atoms joined by a chemical bond, the separation between the quantized levels becomes greater as the stiffness of the bond increases (as k increases). It also becomes greater as the masses of the atoms decrease. For future reference, note that the minimum possible energy of a harmonic oscillator (when $v = 0$) is $\frac{1}{2}\hbar\omega$. This irremovable energy is called the **zero-point energy** of the oscillator.

A third very important example of energy quantization is that of the energy of an electron in a **hydrogenic atom**, a one-electron atom or ion of atomic number Z (with $Z = 1$ for hydrogen itself, $Z = 2$ for He^+, and so on). In this case the permitted energies are

$$E = \frac{-Z^2 m_e e^4}{32\pi^2\varepsilon_0^2\hbar^2} \times \frac{1}{n^2}, \qquad n = 1, 2, \ldots \tag{5}$$

where e is the fundamental charge, m_e is the mass of the electron, and ε_0 is permittivity of free space, a fundamental constant with the value $8.85 \times 10^{-12}\ J^{-1}\ C^2\ m^{-1}$. The permitted energy levels are shown in Fig. 1.8. Before we discuss the shapes of the wavefunctions corresponding to these levels, we shall consider two general points about the quantum theory of atoms: the transitions that occur between their levels and the interpretation of their wavefunctions.

Transitions

An experimental confirmation of energy quantization is the fact that discrete frequencies of light are absorbed and emitted by atoms. This is readily explained on the grounds that light of frequency ν consists of a stream of particles called **photons**, each one of which has an energy $h\nu$. The higher the frequency of the light, the more energetic each of its photons; the greater the intensity of the light, the greater the number of photons. Thus, when an atom absorbs a photon of frequency ν its energy increases by $h\nu$ (Fig. 1.9). Similarly, if it emits a photon of frequency ν, the energy of the atom decreases by $h\nu$. These remarks are summarized by the **Bohr frequency condition**, which states that if the magnitude of the change in energy of the atom is denoted by ΔE, the frequency of the light absorbed or emitted must satisfy

$$\Delta E = h\nu \qquad (6)$$

Since atomic energies are quantized, only certain changes of energy ΔE are possible, so only certain values of ν will occur in the light emitted or absorbed by atoms.

Since the frequency and wavelength of light are related by

$$\lambda = \frac{c}{\nu}$$

transitions between widely separated energy levels emit (or absorb) high frequency (short wavelength) light: transitions between closely spaced energy levels emit (or absorb) low frequency (long wavelength) light. The relation between color and wavelength is given in Table 1.2.

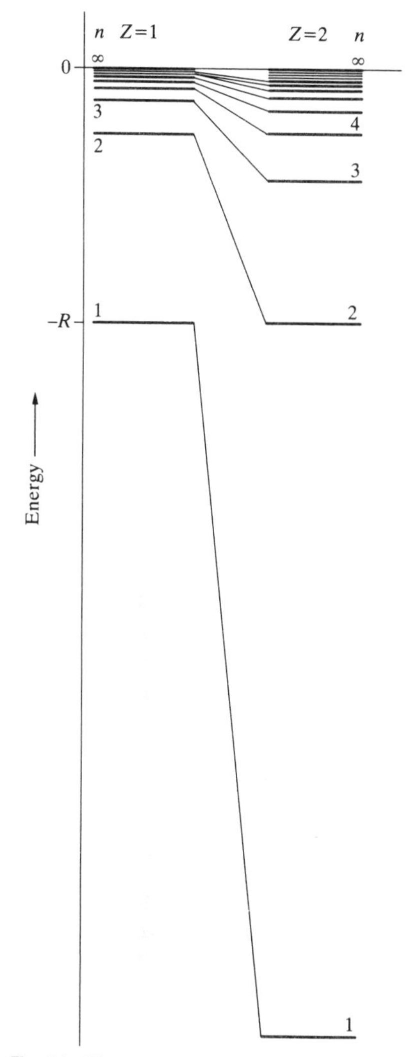

Fig. 1.8 The quantized energy levels of an H atom and an He atom.

Table 1.2. Color, frequency, and wavelength of electromagnetic radiation

Color	Frequency $\nu/10^{14}$ Hz	Wavelength λ/nm	Energy per photon $h\nu/10^{-19}$ J	E/eV
X-rays and γ-rays	10^3 and above	3 and below	660 and above	420 and above
Ultraviolet radiation	10	300	6.6	4.1
Visible light				
Violet	7.1	420	4.7	2.9
Blue	6.4	470	4.2	2.6
Green	5.7	530	3.7	2.3
Yellow	5.2	580	3.4	2.1
Orange	4.8	620	3.2	2.0
Red	4.3	700	2.8	1.8
Infrared radiation	3.0	1000	2.0	1.3
Microwaves and radiowaves	3×10^{-3} and below	1 mm and above	2.0×10^{-3} and below	1.3×10^{-3} and below

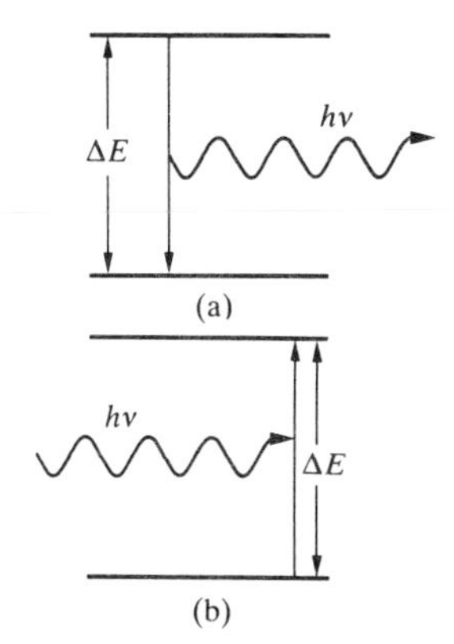

Fig. 1.9 (a) When an atom loses an energy ΔE it emits a photon of frequency ν that satisfies $h\nu = \Delta E$. (b) When it absorbs a photon of frequency ν, its energy increases by $\Delta E = h\nu$.

Since the properties of light, particularly its frequency, are an important source of information on energy levels, it is common to express the energies of the levels themselves in terms of the quantities and units normally used for discussing the properties of radiation. Of these the most common is the **wavenumber** $\tilde{\nu}$:

$$\tilde{\nu} = \frac{\nu}{c}$$

The dimensions of wavenumber are 1/length and the unit employed is normally cm^{-1}. The wavenumber in cm^{-1} can be pictured as the number of wavelengths of the radiation that fit into 1 cm, so the shorter the wavelength (the higher the frequency) the higher the wavenumber. Wavenumbers of light in the visible region are typically about $2 \times 10^4\,\mathrm{cm}^{-1}$; those in the ultraviolet are closer to $10^5\,\mathrm{cm}^{-1}$ or so. In terms of the wavenumber,

$$\Delta E = hc\tilde{\nu}$$

The Born interpretation

De Broglie did not know what he meant by there being a 'wave'—the wavefunction ψ, in modern terms—'associated' with a particle. This was clarified by the German physicist Max Born, who proposed that the square ψ^2 of the wavefunction at some point[2] is proportional to the probability of finding the particle there. Thus quantum mechanics does away with the classical concept of a *precise* path and leaves us with the possibility of making predictions only about the *probability* of finding a particle somewhere. For our purposes we are interested in what can be said about the locations of electrons in atoms, and for the rest of the chapter we consider them.

The precise interpretation of ψ is expressed in terms of the **volume element** $\mathrm{d}\tau$, an infinitesimal region of space, such as a minute region of an atom. If we wish to know the probability that an electron will be found in a volume element $\mathrm{d}\tau$ at a specified point, we evaluate ψ^2 at that point and multiply it by the size of the volume element, obtaining $\psi^2\,\mathrm{d}\tau$. The total probability of finding the electron *somewhere* in the universe is the sum (integral) of the probabilities of finding it in all the volume elements into which the universe can be pictured as being divided. However, since we also know that the probability of it being somewhere is 1 (that is, it is certainly somewhere), we can conclude that the wavefunction must satisfy the relation

$$\int \psi^2\,\mathrm{d}\tau = 1$$

Wavefunctions that satisfy this relation are said to be **normalized**; the Born interpretation is valid only for normalized wavefunctions.

[2] If the wavefunction is complex, in the sense of having real and imaginary parts, the probability is proportional to the square modulus $\psi^*\psi$ where ψ^* is the complex conjugate of ψ. For simplicity, we shall usually assume that ψ is real and write all formulas accordingly. However, in some cases we meet, the wavefunction is complex. We shall signal these in the text.

1.4 Atomic orbitals

The possible wavefunctions of an electron in a hydrogenic (one-electron) atom are called **atomic orbitals**. They are obtained by setting the potential energy in the Schrödinger equation proportional to $-Ze^2/r$, which is the Coulombic potential energy for the interaction of a charge $-e$ (the electron) and Ze (the nucleus of atomic number Z) separated by a distance r. The following section reviews the results of solving the Schrödinger equation using this potential energy.

Atomic quantum numbers

Since the electron in a hydrogenic atom is free to move in three-dimensional space, each hydrogenic atomic orbital is defined by three quantum numbers: n, l, and m_l. The **principal quantum number** n has already been introduced; it specifies the energy of the orbital through eqn 5. In a hydrogenic atom, all orbitals with the same value of n correspond to the same energy and hence are said to be **degenerate**. The principal quantum number therefore defines a series of **shells** of the atom, or orbitals with the same value of n and hence (in a hydrogenic atom) with the same energy.

The orbitals belonging to each shell (that is, orbitals of a given value of n) are classified into **subshells**. Each subshell of a shell is distinguished by a quantum number l called the **orbital angular momentum quantum number**. For a given principal quantum number n, the quantum number l can have the values

$$l = 0, 1, \ldots, n-1$$

giving n different values in all. Thus, the shell with $n = 2$ consists of two subshells of orbitals. It is common to refer to each subshell by a letter:

l:	0	1	2	3	4	...
	s	p	d	f	g	...

It follows that there is only one subshell (an s subshell) in the shell with $n = 1$, two subshells in the shell with $n = 2$ (the s and p subshells), three in the shell with $n = 3$ (the s, p, and d subshells), and so on. For most purposes in chemistry we need consider only s, p, d, and f subshells.

A subshell with quantum number l consists of $2l + 1$ individual orbitals. These are distinguished by the **magnetic quantum number** m_l which can take the $2l + 1$ values

$$m_l = l, l-1, l-2, \ldots, -l$$

Thus a d subshell of an atom consists of five individual atomic orbitals which we could distinguish with the quantum number

$$m_l = 2, 1, 0, -1, -2$$

There is only one orbital in an s subshell, the one with $m_l = 0$: this is called an ***s* orbital**. There are three orbitals in a p subshell, with quantum numbers $m_l = +1$, 0, and -1. These are called ***p* orbitals**. The five orbitals of a d subshell are called ***d* orbitals**, and so on.

Example 1.1: *Counting orbitals*

How many orbitals are there in a shell with quantum number n?

Answer. Although this can be answered algebraically, the simplest procedure is to draw up a table for the first few values of n and then to infer the general result. We use the rules that each shell consists of n subshells and that each subshell consists of $2l+1$ orbitals. For $n=1$ to 3 we can write:

Shell	Subshell	Number of orbitals	Total
$n=1$	$l=0$	1	1
$n=2$	$l=0$	1	4
	$l=1$	3	
$n=3$	$l=0$	1	9
	$l=1$	3	
	$l=2$	5	

The sequence clearly suggests that in general the number is n^2, as is in fact the case.

Exercise. Give the number of orbitals in each of the s, p, d, f, and g subshells of a shell with $n=5$.

Orbital angular momentum

In a spherical system (such as an atom) the **angular momentum**, the momentum of a particle traveling around the central nucleus, is quantized. Its magnitude is limited to the values

$$\text{angular momentum} = \{l(l+1)\}^{1/2}\hbar$$

This expression shows that an electron in an s orbital (with $l=0$) has zero angular momentum around the nucleus. As l increases on going from subshell to subshell, the angular momentum rises too, and an electron in a d orbital has a higher orbital angular momentum than an electron in a p orbital.

Angular momentum is often represented by a vector with a length proportional to the magnitude of the momentum and a direction that denotes the orientation of the motion. Specifically, the length of the vector is $\{l(l+1)\}^{1/2}$ units and its projection on to the z axis is m_l units (Fig. 1.10). The restriction of the orientation of the angular momentum to certain values is yet another example of quantization and is called **space quantization**.

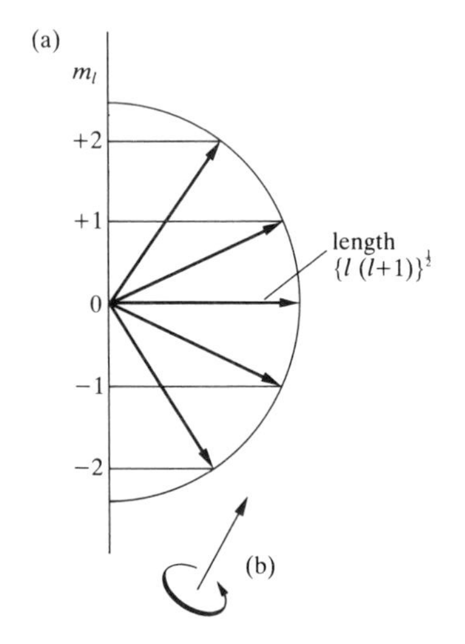

Fig. 1.10 (a) The significance of the quantum numbers l and m_l (for $l=2$). The former gives the magnitude of the orbital angular momentum, as represented here by the length of a vector. The latter specifies the component of momentum around the z axis as represented by the $2l+1$ orientations of the vector. (b) The relation between the vector (for $m_l=+2$ in this case) and the direction of the motion.

Electron spin

Two more quantum numbers are needed to define the state of an electron in a hydrogenic atom completely (in addition to the values of n, l, and m_l). These additional quantum numbers relate to the intrinsic angular momentum of an electron, its **spin**. This evocative name suggests that an electron can be considered as having an angular momentum arising from an intrinsic spinning motion, rather like the daily rotation of a planet about its polar axis as it travels in its annual orbit around the sun. However, spin is a purely quantum mechanical property and differs considerably from its classical namesake.

Although (like orbital angular momentum) the spin angular momentum is specified by a quantum number, in this case s, and has magnitude $\{s(s+1)\}^{1/2}\hbar$, the value of s is *invariably* $\frac{1}{2}$. Electron spin is similar to electron charge and mass: it has a value characteristic of the particle and cannot be changed. Like orbital angular momentum, spin angular momentum can have only certain orientations relative to a chosen axis, but in its case, only two orientations are allowed. They are distinguished by the quantum number m_s, which can take the values $+\frac{1}{2}$ and $-\frac{1}{2}$ and no others. These two spin states of the electron are normally represented by the two arrows ↑ ('spin-up') and ↓ ('spin-down'). Since s is fixed, it is common to say that the state of an electron in a hydrogenic atom is specified by four quantum numbers, namely n, l, m_l, and m_s.

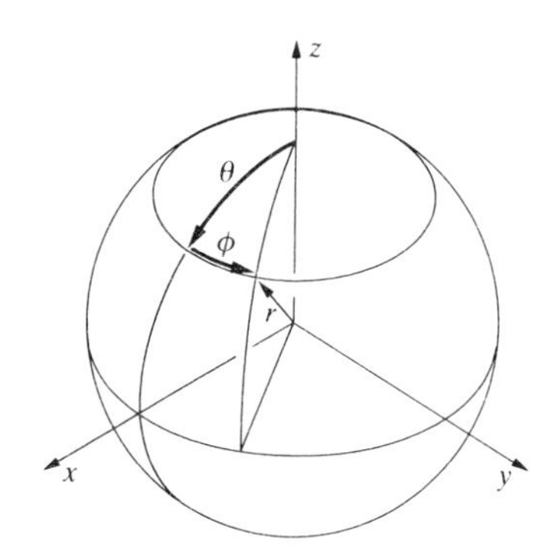

Fig. 1.11 The spherical polar coordinates: r is the radius, θ the colatitude, and ϕ the azimuth.

The shapes of hydrogenic orbitals

The mathematical expressions for some of the hydrogenic orbitals are shown in Table 1.3. Since the Coulomb potential of the nucleus is spherically symmetrical, it is convenient to express the orbitals in terms of the spherical polar coordinates r, θ, and ϕ defined in Fig. 1.11. In these coordinates the orbitals all have the form

$$\psi_{n,l,m_l}(r, \theta, \phi) = R_{n,l}(r) Y_{l,m_l}(\theta, \phi)$$

The expression may look somewhat daunting, but it expresses the simple idea that a hydrogenic orbital can be written as the product of a function R of the radius and a function Y of the angular coordinates. The **radial wavefunction** R determines the radial extent of the orbitals and the **angular wavefunction** Y expresses its angular shape. Most of the time we shall use pictorial representations and not the expressions themselves. The variations of some of the orbitals with distance r from the nucleus are shown in Fig. 1.12. The 1s orbital, for example, is a wavefunction that decays exponen-

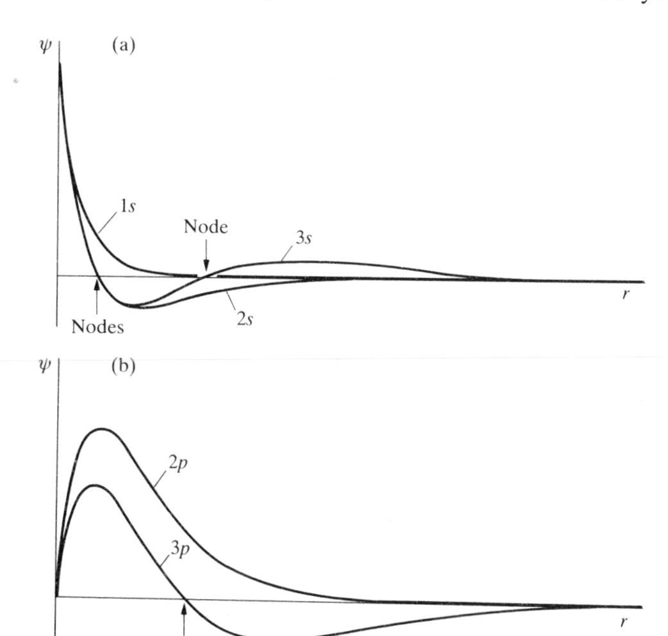

Fig. 1.12 The amplitude of (a) s orbitals and (b) p orbitals plotted against distance from the nucleus. All atomic orbitals decay exponentially with distance at large distances from the nucleus.

Table 1.3. Hydrogenic orbitals

(a) Radial wavefunctions

$$R_{n,l}(r) = f(r)\left(\frac{Z}{a_0}\right)^{3/2} e^{-\rho/2}$$

with a_0 the Bohr radius (0.53 Å) and $\rho = 2Zr/na_0$

n	l	$f(r)$
1	0	2
2	0	$(1/2\sqrt{2})(2-\rho)$
2	1	$(1/2\sqrt{6})\rho$
3	0	$(1/9\sqrt{3})(6-6\rho+\rho^2)$
3	1	$(1/9\sqrt{6})(4-\rho)\rho$
3	2	$(1/9\sqrt{30})\rho^2$

(b) Angular wavefunctions

$$Y_{l,m_l}(\theta, \phi) = \left(\frac{1}{4\pi}\right)^{1/2} y(\theta, \phi)$$

l	m_l	$y(\theta, \phi)$
0	0	1
1	0	$3^{1/2}\cos\theta$
1	± 1	$\mp(\frac{3}{2})^{1/2}\sin\theta\, e^{\pm i\phi}$
2	0	$(\frac{5}{4})^{1/2}(3\cos^2\theta - 1)$
2	± 1	$\mp(\frac{15}{4})^{1/2}\cos\theta\sin\theta\, e^{\pm i\phi}$
2	± 2	$(\frac{15}{8})^{1/2}\sin^2\theta\, e^{\pm 2i\phi}$

tially with distance from the nucleus. All orbitals decay exponentially at sufficiently great distances, but some oscillate before beginning to decay.

We have shown the mathematical expressions in Table 1.3 because they reveal some features that are worth noting. For example, since the 2*s* radial wavefunction ($R_{2,0}$) is proportional to the factor $2-\rho$, where ρ is proportional to r as specified in Table 1.3, we know at once that the orbital vanishes when $\rho=2$, corresponding to $r=2a_0/Z$ where a_0 is the **Bohr radius**:

$$a_0=\frac{4\pi\varepsilon_0\hbar}{m_e e^2}=5.29\times10^{-11}\ \text{m}\quad(0.529\ \text{Å})$$

That is, the 2*s* orbital has a **radial node**, a spherical surface on which $\psi=0$, that surrounds the nucleus at $r=2a_0/Z$. Similarly, the 3*s* orbital (with $n=3$ and $l=0$) has radial nodes at the two solutions of the quadratic equation $6-6\rho+\rho^2=0$.

Another feature to note from Table 1.3 is that *s* orbitals (those with $l=0$) do not vanish at the nucleus (at $\rho=0$) but all other orbitals do vanish there. For instance, the 3*d* orbital (with $n=3$ and $l=2$) is proportional to ρ^2, which is zero when $\rho=0$. This feature will turn out to be a substantial component of the explanation of the periodic table.

We see from the expressions in Table 1.3 for the angular wavefunction Y_{l,m_l} that orbitals with $l=0$ do not depend on the angular coordinates (in each case Y is a constant, $1/2\pi^{1/2}$). In other words, *s* orbitals are spherically symmetrical in the sense that there is the same probability of finding the electron at a given distance from the nucleus whatever the direction.

Because the force that binds the electron is centered on the nucleus, it is often of interest—for instance when we are judging how tightly the electron is bound—to know the probability of finding an electron at a given distance from the nucleus irrespective of its direction. The volume element of interest is then a thin spherical shell of radius r and thickness dr that surrounds the nucleus. The volume of such a shell is its surface area $4\pi r^2$ multiplied by its thickness dr, or $4\pi r^2\,\mathrm{d}r$. It then follows from the Born interpretation, that if a spherically symmetrical wavefunction has the value ψ at the distance r, then the probability of the electron being somewhere in the shell is $4\pi r^2\psi^2\,\mathrm{d}r$. This is often written $P\,\mathrm{d}r$, where $P=4\pi r^2\psi^2$ is called the **radial distribution function**. If we know the value of P at some radius r (which we know once we know ψ), we can state the probability of finding the electron somewhere in a shell of thickness dr at that radius simply by multiplying P by dr.

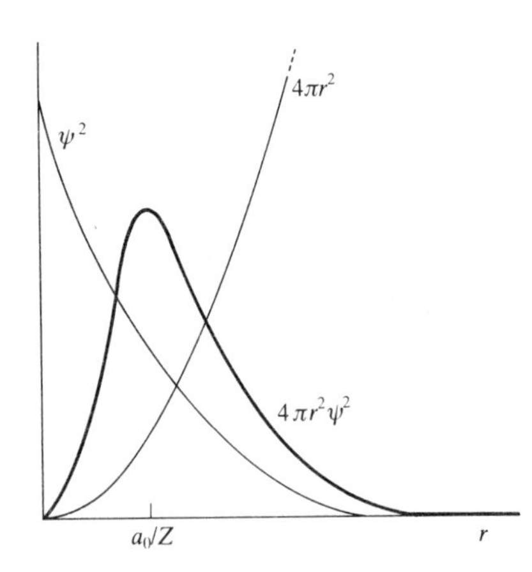

Fig. 1.13 The radial distribution function of a hydrogenic 1*s* orbital. The product of $4\pi r^2$ (which increases as r increases) and ψ^2 (which decreases exponentially) passes through a maximum at $r=a_0/Z$.

Since a 1*s* orbital decreases exponentially with distance and r^2 increases, the radial distribution function of the orbital goes through a maximum (Fig. 1.13). Therefore, there is a distance at which the electron is most likely to be found. In general, this distance decreases as the nuclear charge increases (because the electron is attracted more strongly to the nucleus). It increases as n increases, because the higher the energy, the more likely it is that the electron will be found far from the nucleus. The most probable distance of an electron from the nucleus in the ground state of a hydrogenic atom is at the point where the radial distribution function goes through a maximum. As we show in the following example, this occurs at

$$r=\frac{a_0}{Z}$$

We see that the most probable radius decreases as the atomic number increases.

Example 1.2: *Calculating the most probable distance of an electron from the nucleus*

Given that the wavefunction for a 1*s* orbital in a hydrogenic atom is proportional to $\psi = e^{-Zr/a_0}$, calculate the most probable distance of the electron from the nucleus.

Answer. The probability of finding the electron on a shell of radius r is proportional to

$$P = r^2 \times e^{-2Zr/a_0}$$

This has a maximum where $dP/dr = 0$; that is, where

$$\left(2r - \frac{2r^2 Z}{a_0}\right) e^{-2Zr/a_0} = 0$$

This is satisfied by $r = a_0/Z$.

Exercise. Repeat the calculation for a hydrogenic 2*s* orbital, which is proportional to $(2 - Zr/a_0)e^{-Zr/2a_0}$.

A *p* orbital has a marked angular dependence (Fig. 1.14). In the most common graphical representation, the shapes of the three *p* orbitals are identical, but lie parallel to each of the three cartesian axes. This is the origin of the labels p_x, p_y, and p_z, which are alternatives to the m_l values. We shall refer to p_x, p_y, and p_z as the **real forms** of the *p* orbitals (because the wavefunctions are real functions). The orbitals labeled with m_l are the **complex forms** (because the wavefunctions are complex).[3] The two forms differ in that the complex form of the orbitals with a definite value of m_l are *traveling waves* corresponding to the electron circulating in a definite sense around the nucleus. The real forms are *standing waves*, and do not correspond to a definite sense of circulation. We need the complex forms when we are discussing a free atom, for the electrons are then free to circulate. We need the real forms when the electron is concentrated in a definite region of space, such as in a chemical bond. In most applications, we shall use only the real forms of the *p* orbitals, and represent them by diagrams like those in Fig. 1.14. As in that illustration, we shall show the positive and negative amplitudes of the wavefunction by different shading: positive amplitude will be shown by light gray and negative by dark gray.

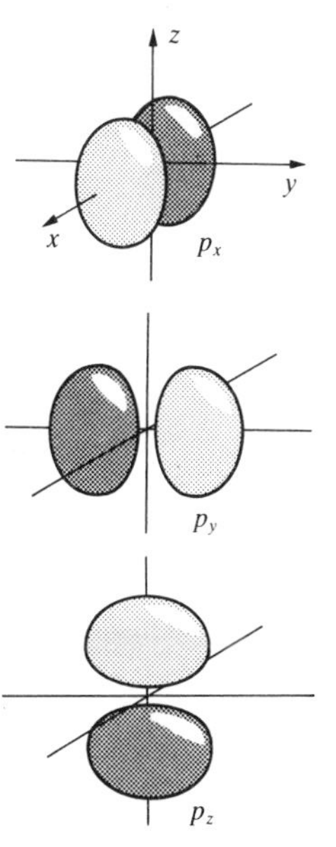

Fig. 1.14 The boundary surfaces of *p* orbitals. Each orbital has one nodal plane running through the nucleus. For example, the nodal plane of the p_z orbital is the *xy* plane. The lightly shaded lobe has a positive amplitude, the darker shaded one is negative.

The forms and labels we use for the *d* and *f* orbitals are shown in Figs. 1.15 and 1.16 respectively. The shapes and phases shown there are the (real) standing wave versions of the orbitals; these are the only ones we

[3] For example, an orbital with $m_l = +1$ is proportional to $e^{i\phi}$ and one with $m_l = -1$ is proportional to $e^{-i\phi}$, and both are complex. The linear combinations

$$\frac{e^{i\phi} + e^{-i\phi}}{2} = \cos\phi \quad \text{and} \quad \frac{e^{i\phi} - e^{-i\phi}}{2i} = \sin\phi$$

are both real and correspond to the p_x and p_y orbitals respectively. We see, therefore, that the complex forms correspond to definite orientations of the orbital angular momentum, but that the real forms are mixtures—superpositions—of these orientations. It is an important feature of quantum theory that linear combinations of solutions of the Schrödinger equation with the same energy are equally valid solutions.

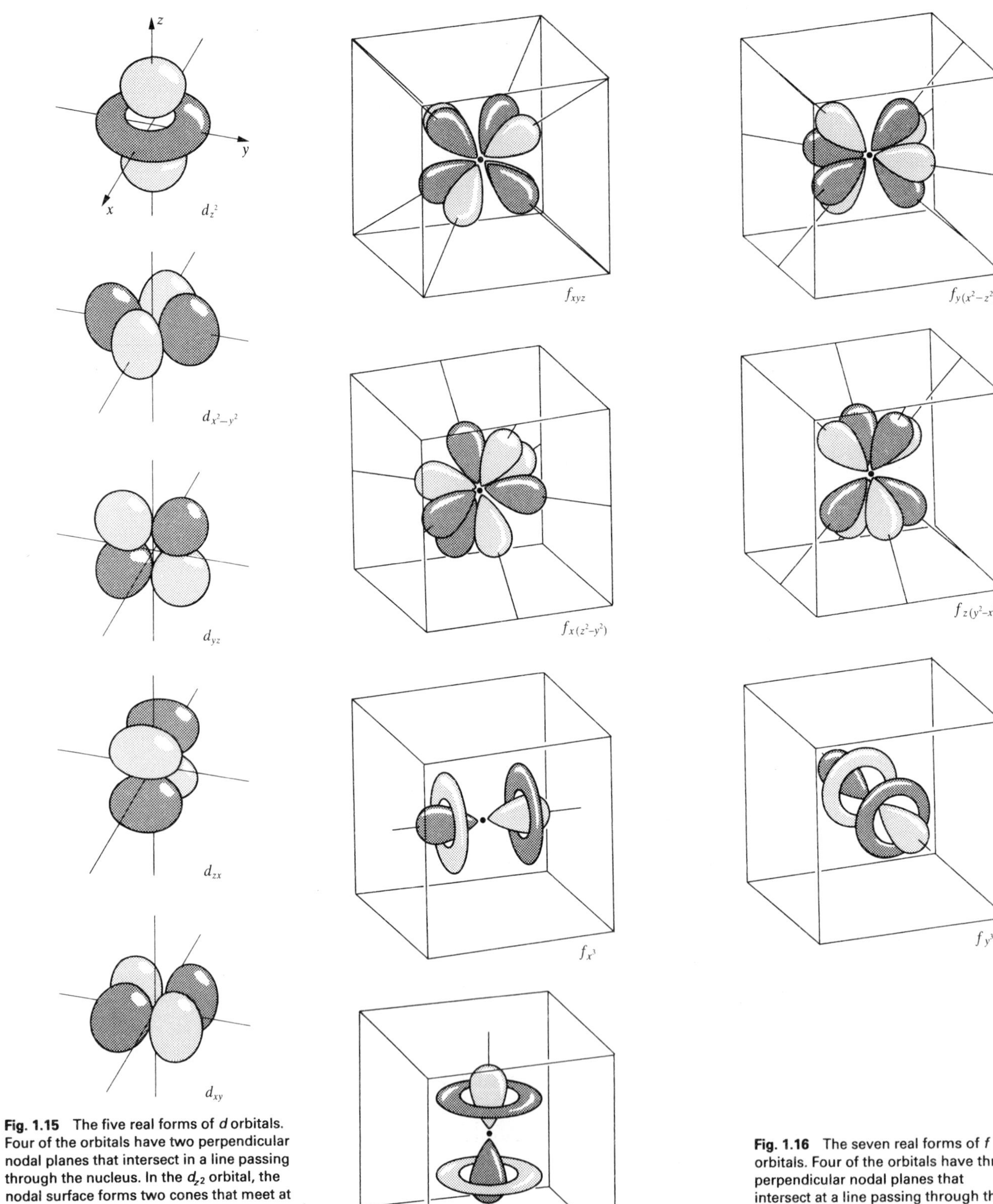

Fig. 1.15 The five real forms of *d* orbitals. Four of the orbitals have two perpendicular nodal planes that intersect in a line passing through the nucleus. In the d_{z^2} orbital, the nodal surface forms two cones that meet at the nucleus.

Fig. 1.16 The seven real forms of *f* orbitals. Four of the orbitals have three perpendicular nodal planes that intersect at a line passing through the nucleus. The other three have a double-cone nodal surface like the d_{z^2} orbital, and an additional nodal plane that bisects it.

need to consider explicitly. The d orbitals each have four lobes and two nodal planes (with the exception of the d_{z^2} orbital). The f orbitals each have six lobes and three nodal planes (with the exception of the f_{x^3}, f_{y^3}, and f_{z^3} orbitals).

Many-electron atoms

By a **many-electron atom** we mean an atom with more than one electron; so even He, with two electrons, is a many-electron atom. The exact solution of the Schrödinger equation for an atom with N electrons is a very complicated wavefunction with a value that depends simultaneously on the coordinates of all the electrons. Mathematically we would write this wavefunction $\Psi(\boldsymbol{r}_1, \boldsymbol{r}_2, \ldots, \boldsymbol{r}_N)$, or Ψ for short, showing that it depends on the coordinates $\boldsymbol{r}_i$ of all N electrons. However, this is far too complicated an approach for our purposes. An assumption widely made in chemistry is that each electron moves independently of the others in the field of the nucleus that is modified by a negative charge representing the average effect of all the other electrons in the atom. In this **orbital approximation**, we picture each electron in a many-electron atom as occupying a hydrogenic atomic orbital like those we have discussed already. In mathematical terms, this is expressed by saying that the true wavefunction can be approximated as a product of N hydrogenic wavefunctions:

$$\Psi = \psi(\boldsymbol{r}_1)\psi(\boldsymbol{r}_2) \ldots \psi(\boldsymbol{r}_N)$$

This expression implies that electron 1 is described by the wavefunction $\psi(\boldsymbol{r}_1)$, electron 2 by the wavefunction $\psi(\boldsymbol{r}_2)$, and so on.

1.5 The building-up principle

The building-up principle (or the *Aufbau* principle) is a procedure for arriving at the lowest-energy electron **configuration** of many-electron atoms, a list of occupied orbitals in the **ground state**, the state of lowest energy of an atom or ion.

The first step in the procedure is to establish the order of energies of the atomic orbitals in the atom. This is approximately the order of the hydrogenic orbitals, but with important modifications arising from electron–electron repulsions, which we describe in a moment. Then, with the energy levels established, the electrons needed to build the atom are pictured as entering the orbitals, starting at the one with lowest energy. However, the number of electrons that can occupy any orbital is limited by a fundamental feature of nature known as the **Pauli exclusion principle**:

> No more than two electrons may occupy a single orbital and, if two do occupy a single orbital, their spins must be paired.

By 'paired' we mean that one electron spin must be ↑ and the other ↓; the pair is denoted ⇅. Another way of expressing the principle is to note that an electron in an atom is described by four variable quantum numbers (we exclude s, which is fixed at $\frac{1}{2}$). Hence the exclusion principle amounts to saying that no two electrons can have the same four quantum numbers (n, l, m_l, and m_s).

Penetration and shielding

As a preliminary to discussing the building-up principle, we need to consider how the orbital energies differ from those in hydrogenic atoms.

An electron in an orbital experiences a Coulombic attraction to the nucleus and a Coulombic repulsion from all the other electrons present in the atom. We make the approximation of supposing that any given electron experiences a single **central field**, the sum of the field of the nucleus and the *average* field of all the other electrons treated as a point negative charge centered on the nucleus. This point negative charge reduces the nuclear charge from its true value Ze to an **effective nuclear charge** $Z_{\mathrm{eff}}e$ for a specific electron in the atom, those closer to the nucleus on average experiencing higher values of Z_{eff} than those that on average are further away. This reduction is called **shielding**, and the **shielding parameter** σ is a correction to the true nuclear charge:

$$Z_{\mathrm{eff}} = Z - \sigma$$

Once we know the effective nuclear charge we can write approximate forms of the atomic orbitals and begin to make estimates of their extent and other properties. This was first done by J. C. Slater, who devised a set of rules for estimating the value of Z_{eff} for an electron in any atom, and using the value to write down an approximate atomic orbital. The rules themselves have been superseded by more accurate calculated values, which are listed in Table 1.4.

The important feature revealed by Table 1.4 is that an *ns* electron in the valence shell is generally less shielded than an *np* electron. Note, for instance, that a 2*s* electron in F experiences an effective nuclear charge of $5.13e$ whereas the 2*p* electron experiences $5.10e$. That is, a valence *ns* electron is more strongly bound to the nucleus than is an *np* electron. The reason for the difference can be traced to the effect called **penetration**. Figure 1.17 shows that an *ns* electron has a higher probability of being found close to the nucleus than an *np* electron because an *ns* orbital has a nonzero amplitude at the nucleus but an *np* electron has a node there. Since an *ns* orbital approaches the nucleus more closely, it is less completely

Table 1.4. Effective nuclear charges Z_{eff}

	H							**He**
Z	1							2
1*s*	1.00							1.69
	Li	**Be**	**B**	**C**	**N**	**O**	**F**	**Ne**
Z	3	4	5	6	7	8	9	10
1*s*	2.69	3.68	4.68	5.67	6.66	7.66	8.65	9.64
2*s*	1.28	1.91	2.58	3.22	3.85	4.49	5.13	5.76
2*p*			2.42	3.14	3.83	4.45	5.10	5.76
	Na	**Mg**	**Al**	**Si**	**P**	**S**	**Cl**	**Ar**
Z	11	12	13	14	15	16	17	18
1*s*	10.63	11.61	12.59	13.57	14.56	15.54	16.52	17.51
2*s*	6.57	7.39	8.21	9.02	9.82	10.63	11.43	12.23
2*p*	6.80	7.83	8.96	9.94	10.96	11.98	12.99	14.01
3*s*	2.51	3.31	4.12	4.90	5.64	6.37	7.07	7.76
3*p*			4.07	4.29	4.89	5.48	6.12	6.76

Source: E. Clementi and D. L. Raimondi, *Atomic screening constants from SCF functions.* IBM Research Note NJ-27 (1963).

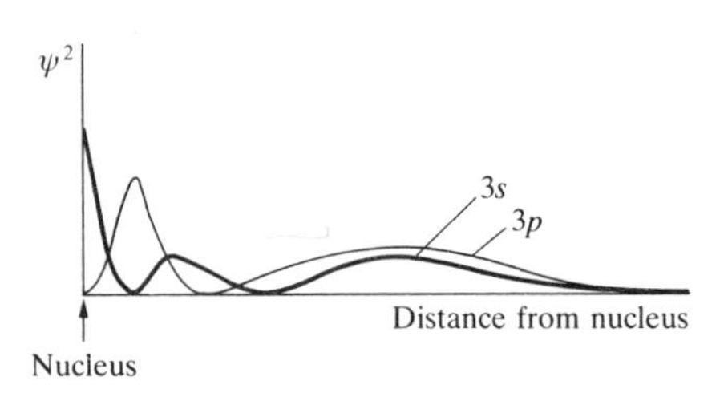

Fig. 1.17 The penetration of a 3*s* electron through the inner core is greater than that of a 3*p* electron since the latter has a node at the nucleus.

shielded by electrons in inner shells and hence experiences a higher effective nuclear charge. A similar difference arises between the *p* and *d* electrons of the valence shell, and a typical order is $ns < np < nd < nf$ because *s* orbitals are the most penetrating and *f* orbitals are the least.

A typical energy level diagram for neutral atoms is shown in Fig. 1.18. As will be seen, the effects are quite subtle, and the ordering of the orbitals depends strongly on the number of electrons present in the atom. For example, the effects of penetration are very pronounced for 4*s* electrons in K and Ca, and in these elements the 4*s* orbitals lie lower in energy than the 3*d* orbitals. However, from Sc onwards in the neutral atom the 3*d* orbitals lie lower than the 4*s* orbitals. The energies of the orbitals through the periodic table are shown in Fig. 1.19. In the elements from Ga onward, the 3*d* orbital energies fall well below that of the 4*s* orbitals and the outermost electrons are unambiguously those of the 4*s* and 4*p* subshells.

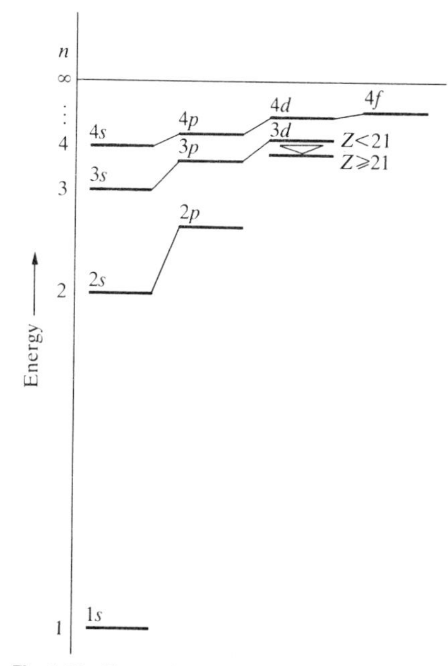

Fig. 1.18 Energy levels of many-electron atoms (specific case). This is the diagram used in the building-up principle, with up to two electrons being allowed to occupy each orbital. Note the change in relative energy of the 4*s* and 3*d* orbitals at $Z = 21$.

Ground state electron configurations

In the building-up principle, orbitals of the neutral atoms are treated as being occupied in the order

$$1s \quad 2s \quad 2p \quad 3s \quad 3p \quad 4s \quad 3d \quad 4p \ldots$$

Allowing each orbital to accommodate two electrons (so a *p* subshell can accommodate up to six electrons and a *d* subshell up to ten), leads to the ground state configuration of any atom. The configurations of the first five

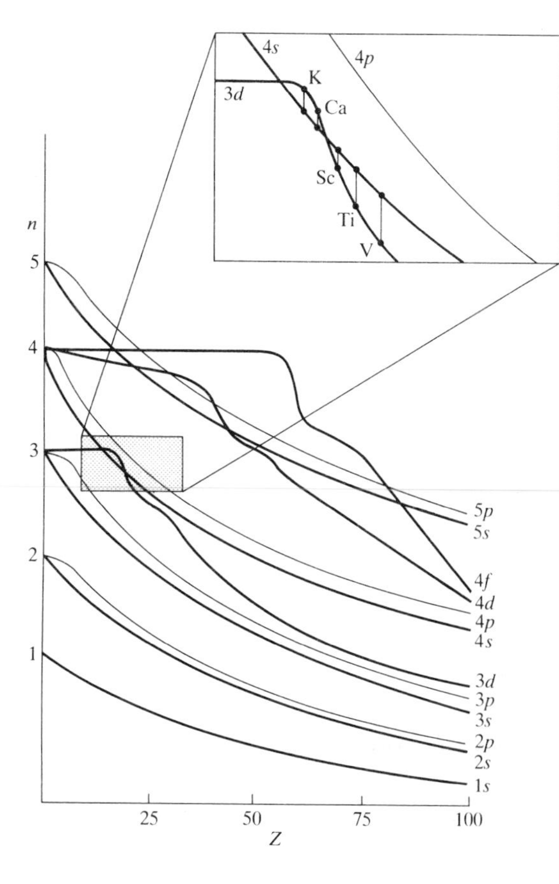

Fig. 1.19 Energy levels of many-electron atoms in the periodic table. The inset shows a magnified view of the order near $Z = 20$.

elements, for instance, are

H	He	Li	Be	B
$1s^1$	$1s^2$	$1s^22s^1$	$1s^22s^2$	$1s^22s^22p^1$

When more than one orbital is available for occupation, such as when the $2p$ orbitals are starting to be filled in B and C, we adopt one of **Hund's rules**:

> When more than one orbital is available for occupation, electrons occupy separate orbitals and do so with parallel spins (↑↑).

In this arrangement, the two electrons have a lower energy than when they occupy the same orbitals with antiparallel spins (↑↓). The difference is related to the quantum mechanical effect called **spin correlation**, the tendency of two electrons with parallel spins to stay apart from one another and hence to repel each other less. It follows that the configuration of O is $1s^22s^22p_x^22p_y^12p_z^1$ (it is arbitrary which of the p orbitals is occupied first: we have adopted the order p_x, p_y, p_z) and that, in a more condensed notation, the configurations of the atoms C through Ne are

C $1s^22s^22p^2$ N $1s^22s^22p^3$ O $1s^22s^22p^4$

F $1s^22s^22p^5$ Ne $1s^22s^22p^6$

The $1s^22s^22p^6$ configuration is a **closed shell**, a shell containing its full complement of electrons, and is denoted [Ne]. The $1s^2$ configuration of He is also closed, and is denoted [He]. Lithium can therefore be denoted [He]$2s^1$, showing more clearly that it consists of a single $2s$ electron around a helium-like core. Fluorine is denoted [He]$2s^22p^5$, showing that it is one electron short of a neon-like closed shell. The outermost electrons surrounding a closed shell are called the **valence electrons**.

Example 1.3: *Accounting for trends in shielding*

Suggest a reason why the increase in Z_{eff} (Table 1.4) for a $2p$ electron is smaller between N and O than between C and N.

Answer. The valence configurations of the three atoms are

C $2s^22p^2$ N $2s^22p^3$ O $2s^22p^4$

On going from C to N, the additional electron occupies an empty $2p$ orbital. On going from N to O, the additional electron must occupy a $2p$ orbital that is already occupied by one electron. It therefore experiences a stronger repulsion and the increase in atomic number is more completely canceled than between C and N.

Exercise. Account for the larger increase in effective nuclear charge for a $2p$ electron on going from B to C compared with a $2s$ electron on going from Li to Be.

The configuration of Na is obtained by adding one more electron to a neon-like set of orbitals, and is [Ne]$3s^1$, showing that it consists of a single electron outside a closed shell. Now the same sequence begins again, with the $3s$ and $3p$ orbitals complete at Ar, with configuration [Ne]$3s^23p^6$, denoted [Ar]. Since the $3d$ orbitals are so much higher in energy, this configuration is effectively a closed shell. Moreover, the $4s$ orbital is next in line for occupation, so the configuration of K is analogous to that of Na, with a single electron outside a closed shell: it is specifically $[Ar]4s^1$. The next electron, for Ca, also enters the 4*s* orbital, giving $[Ar]4s^2$, resembling Mg. However, next in line for occupation are the 3*d* orbitals, so there is now a change in the sequence of occupation.

In the *d* block, the *d* orbitals are in the course of occupation according to the formal rules of the building-up principle. However, the energy levels in Fig. 1.18 and 1.19 are for *individual* atomic orbitals and do not fully take into account interelectronic repulsions. For most of the *d* block, the outcome of the repulsions is that the ground state configurations are of the form $3d^n4s^2$, with the 4*s* orbitals fully occupied despite individual 3*d* orbitals being lower in energy. An additional complication is that in some cases a lower total energy may be obtained by forming a half-filled or filled *d* subshell at the expense of an *s* electron, so close to the center of the block the ground configuration is likely to be d^5s^1 and not d^4s^2 (as for Cr), and close to the right of the block the configuration is likely to be $d^{10}s^1$ rather than d^9s^2 (as for Cu). A similar complication occurs in the *f* block, where the *f* orbitals are being occupied. Thus the electron configuration of Gd is $[Xe]4f^75d^16s^2$.

The complication of the orbital energy not being a true guide to the total energy disappears when the 3*d* orbital energies fall well below that of the 4*s* orbitals, for then the balance is less subtle and the orbital energies themselves are a better guide to the overall energy of the atom. This is the case in the positive ions of the *d*-block elements, where the removal of electrons reduces the complicating effects of electron–electron repulsions, and all cations have d^n configurations. Similarly, in the *f* block, the ions have f^n configurations, such as $[Xe]4f^5$ for Sm^{3+}. In later chapters (starting in Chapter 7) we shall see the great significance of the configurations of the *d*-metal ions, because the subtle modulation of their energies is the basis of important properties of their complexes: for the purposes of chemistry, the configurations of these ions are more important than those of the neutral atoms. The complication also disappears in the *p* block, where the valence electrons are in *ns* and *np* orbitals and the $(n-1)d$ electrons are so low in energy that they do not contribute substantially to bonding.

Example 1.4: *Deriving an electron configuration*

Give the ground state electron configurations of the Ti atom and the Ti^{3+} ion.

Answer. For the atom, we add $Z = 22$ electrons in the order specified above, with no more than two electrons in any one orbital. This results in the configuration

$$\text{Ti} \quad 1s^22s^22p^63s^23p^64s^23d^2, \quad \text{or} \quad [\text{Ar}]4s^23d^2$$

with the two 3*d* electrons in different orbitals with parallel spins. However, since the 3*d* orbitals lie below the 4*s* orbitals once we are past Ca, it is more revealing to reverse their order. The configuration is therefore reported as $[Ar]3d^24s^2$. The configuration of the cation is obtained by removing the *s* electrons and as many *d* electrons as required (in that order). We must remove three electrons in all, two *s* electrons and one *d* electron. The configuration of Ti^{3+} is therefore $[Ar]3d^1$.

Exercise. Give the electron configurations of Ni and Ni^{2+}.

A list of ground state electron configurations of all the elements is given in Appendix 1.

1.6 Atomic parameters

Each period of the periodic table corresponds to the completion of the *s* and *p* subshells, the period number being the value of the principal quantum number *n* of the shell being completed.

The group numbers are closely related to the number of electrons in the valence shell, but the precise relation depends on the group number *G* (and the numbering system adopted). In the 1–18 numbering system recommended by IUPAC:

Block	Number of valence electrons
s, d	G
p	$G - 10$

In the roman numeral system, the group number is equal to the number of *s* and *p* valence electrons for the *s* and *p* blocks. Thus, selenium belongs to Group 16 and is in the *p* block; hence it has $16 - 10 = 6$ valence electrons. Thallium belongs to Group III, so it has 3 valence *s* and *p* electrons.

Metallic and ionic radii

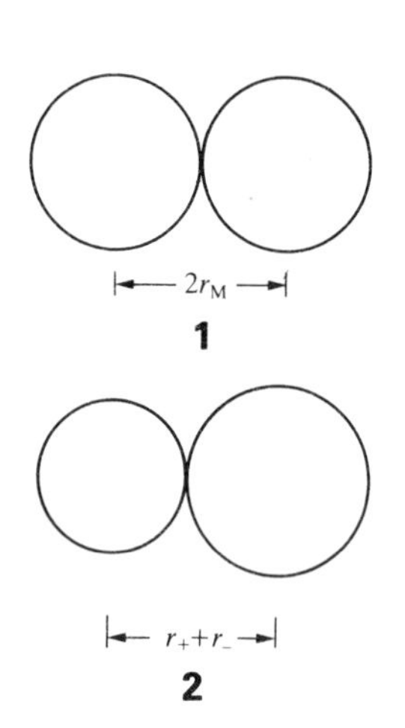

1

2

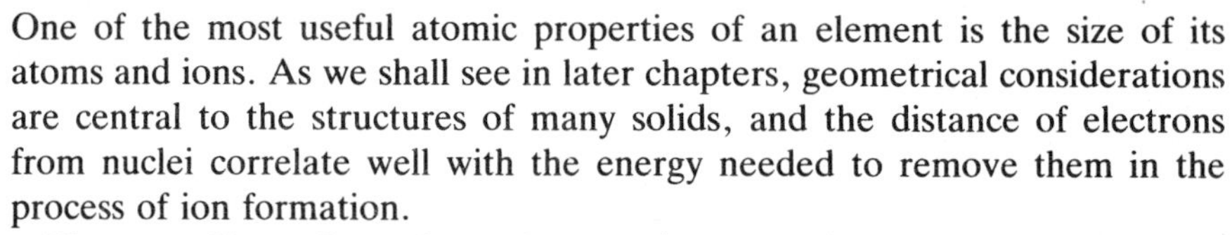

One of the most useful atomic properties of an element is the size of its atoms and ions. As we shall see in later chapters, geometrical considerations are central to the structures of many solids, and the distance of electrons from nuclei correlate well with the energy needed to remove them in the process of ion formation.

The **metallic radius** of an element (Table 1.5) is defined as half the distance between the centers of nearest neighbor atoms in the solid metal (**1**, but see Section 4.5 for a refinement of this definition). The **ionic radius** (Table 1.6) is defined as the distance between the centers of the cation and anion, apportioned on the basis that the radius of the O^{2-} ion is 1.40 Å (**2**; as before, see Section 4.5 for a refinement of this definition too). For instance, the ionic radius of Mg^{2+} is obtained by subtracting 1.40 Å from the internuclear separation of neighboring Mg^{2+} and O^{2-} ions in MgO.

Table 1.5. Metallic radii (in Å)*

Li	**Be**													
1.57	1.12													
Na	**Mg**											**Al**		
1.91	1.60											1.43		
K	**Ca**	**Sc**	**Ti**	**V**	**Cr**	**Mn**	**Fe**	**Co**	**Ni**	**Cu**	**Zn**	**Ga**		
2.35	1.97	1.64	1.47	1.35	1.29	1.27	1.26	1.25	1.25	1.28	1.37	1.53		
Rb	**Sr**	**Y**	**Zr**	**Nb**	**Mo**	**Tc**	**Ru**	**Rh**	**Pd**	**Ag**	**Cd**	**In**	**Sn**	
2.50	2.15	1.82	1.60	1.47	1.40	1.35	1.34	1.34	1.37	1.44	1.52	1.67	1.58	
Cs	**Ba**	**Lu**	**Hf**	**Ta**	**W**	**Re**	**Os**	**Ir**	**Pt**	**Au**	**Hg**	**Tl**	**Pb**	**Bi**
2.72	2.24	1.72	1.59	1.47	1.41	1.37	1.35	1.36	1.39	1.44	1.55	1.71	1.75	1.82

* The values refer to coordination number 12 (see Section 4.3). *Source:* A. F. Wells, *Structural inorganic chemistry,* 5th edn. Clarendon Press, Oxford (1984).

Table 1.6. Ionic radii (in Å)*

Li^{+} (4) 0.59	Be^{2+} (4) 0.27	B^{3+} (4) 0.12	C	N^{3-} 1.71	O^{2-} (6) 1.40	F^{-} (6) 1.33
Na^{+} (6) 1.02	Mg^{2+} (6) 0.72	Al^{3+} (6) 0.53	Si	P^{3-} 2.12	S^{2-} (6) 1.84	Cl^{-} (6) 1.81
K^{+} (6) 1.38	Ca^{2+} (6) 1.00	Ga^{3+} (6) 0.62	Ge	As^{3-} 2.22	Se^{2-} (6) 1.98	Br^{-} (6) 1.96
Rb^{+} (6) 1.49	Sr^{2+} (6) 1.16	In^{3+} (6) 0.79	Sn	Sb^{3-}	Te^{2-} (6) 2.21	I^{-} (6) 2.20
Cs^{+} (6) 1.70	Ba^{2+} (6) 1.36	Tl^{3+} (6) 0.88	Pb	Bi^{3-}	Po^{2-}	At^{-}

* Numbers in parentheses are the coordination numbers of the ions. Values for ions without a coordination number stated are estimates.
Source: R. D. Shannon and C. T. Prewitt, *Acta Crystallogr.*, **B25**, 925 (1969).

The metallic and ionic radii of the elements show a variation that is an important component of many explanations in inorganic chemistry. As successive electrons are added to an atom on going across a period, they enter orbitals of the same shell (we disregard the *d*-block metals for the time being). However, the effective nuclear charge increases across the period, draws in the electrons, and results in a more compact atom (Table 1.5 and Fig. 1.20). The general decrease in metallic and ionic radii across a period should be remembered.

At the end of each period, the next electron must enter a shell with a higher principal quantum number. This new shell surrounds the completed inner core of the atom, and hence results in an atom or ion of greater radius than in the preceding period. This is illustrated for the radii of Li and K. The trend to remember is that the radii of atoms and ions increase on descending a group.

Period 6 shows an interesting and important modification to this otherwise general trend. We see from Fig. 1.20 that the metallic radii in the

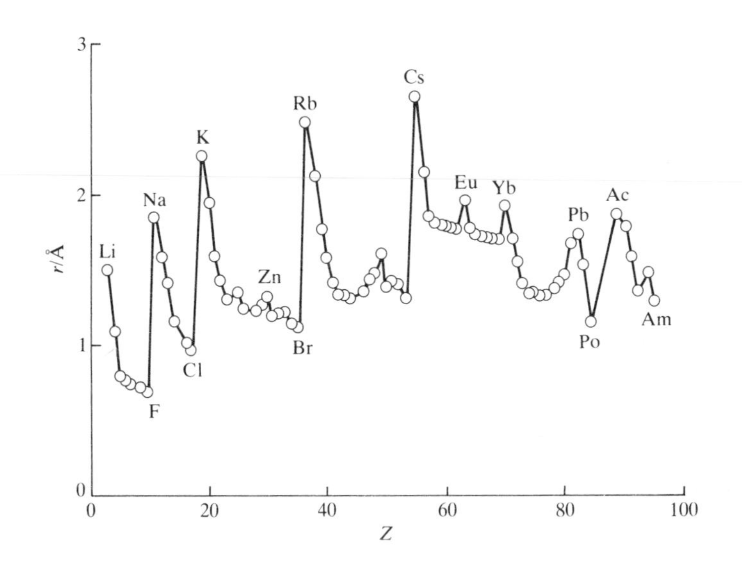

Fig. 1.20 The variation of radius through the periodic table. Note the contraction of radii following the lanthanides in Period 6. We have included nonmetallic elements in the figure using radii derived by taking half the distance between like atoms in a molecule.

third row of the d block are very similar to those in the second row, not significantly larger. The reduction of radii below what extrapolation might suggest is called the **lanthanide contraction**. The name points to its origin. The elements in the third row of the d block are preceded by the 14 elements of the f block, in which the $4f$ orbitals are being occupied. These orbitals have poor shielding properties, and the repulsions between electrons being added on crossing the f block fail to compensate for the increasing nuclear charge. The dominating effect of the latter is to draw in all the electrons, and to result in a more compact atom.

The existence of the lanthanide contraction should make us alert for its analogs in the d block. Indeed, there is also a contraction along a series of d metals, which arises from the poor shielding by the d electrons. This is quite noticeable between Sc (metallic radius 1.64 Å) and Ni (1.25 Å) in the first row of the block, and between Y (1.82 Å) and Ru (1.34 Å) in the second row. In due course we shall see that contractions like these play an important role in controlling the properties of the compounds these elements form.

Ionization energy

The ease with which an electron can be removed from an atom is measured by its **ionization energy** I, the minimum energy needed to remove an electron from a gas-phase atom:

$$A(g) \rightarrow A^+(g) + e^-(g)$$

The **first ionization energy** I_1 signifies the ionization of the least tightly bound electron from the neutral atom, the **second ionization energy** I_2 is the ionization of the resulting cation, and so on. Ionization energies are conveniently expressed in electronvolts, eV, where 1 eV is the energy acquired by an electron when it falls through a potential difference of 1 V. Since this is equal to $e \times 1$ V, it is easy to deduce that

$$1\ \text{eV} = 96.49\ \text{kJ mol}^{-1}$$

The ionization energy of the H atom is 13.6 eV, so to remove an electron from an H atom is equivalent to moving the electron through 13.6 V. In thermodynamic calculations it is more convenient to use the **ionization enthalpy**, the standard enthalpy of the process written above. The ionization enthalpy is RT larger than the ionization energy (because 2 mol of gas particles replace 1 mol when ionization occurs). Because RT is only 2.5 kJ mol^{-1} at room temperature, the difference between ionization energy and enthalpy can often be ignored. In this book we express ionization energies in electronvolts and ionization enthalpies in kJ mol^{-1}.

First ionization energies vary systematically through the periodic table (Table 1.7 and Fig. 1.21), being smallest at the lower left (near Cs) and greatest at the upper right (near He). Their variation follows the pattern of effective nuclear charge already mentioned in connection with the building-up principle, but there are some subtle modulations arising from the effect of electron–electron repulsions within the same subshell.

One feature that we can explain is the observation that the first ionization energy of B is smaller than that of Be despite its higher nuclear charge. This is readily explained by noting that on going to B, the outermost electron

Table 1.7. First and second (and some higher) ionization energies of the elements (in eV*)

H							**He**
13.60							24.58
							54.40
Li	**Be**	**B**	**C**	**N**	**O**	**F**	**Ne**
5.39	9.32	8.30	11.26	14.54	13.61	17.42	21.56
75.62	18.21	25.15					
	153.85	37.92					
		259.30					
Na	**Mg**	**Al**	**Si**	**P**	**S**	**Cl**	**Ar**
5.14	7.64	5.98	8.15	11.0	10.36	13.01	15.76
47.29	15.03	18.82					
	80.12	28.44					
		153.77					
K	**Ca**	**Ga**	**Ge**	**As**	**Se**	**Br**	**Kr**
4.34	6.11	6.00	8.13	10	9.75	11.84	14.00
31.81	11.87						
	51.21						
Rb	**Sr**	**In**	**Sn**	**Sb**	**Te**	**I**	**Xe**
4.18	5.69	5.79	7.33	8.64	9.01	10.44	12.13
27.5	11.03						
Cs	**Ba**	**Tl**	**Pb**	**Bi**	**Po**	**At**	**Rn**
3.89	5.21	6.11	7.42	8			10.74
25.1	10.00						

* To convert to kJ mol^{-1}, multiply by 96.485. See Appendix 2 for a longer list.
Source: C. E. Moore, *Atomic energy levels.* NBS Circular 467 (1949).

occupies a $2p$ orbital and hence is less strongly bound than if it had entered a $2s$ orbital. As a result, the value of I_1 falls back to a lower value. The decrease between N and O has a slightly different explanation. The valence configurations of the two atoms are

$$\mathrm{N}\quad 2s^2 2p_x^1 2p_y^1 2p_z^1 \qquad \mathrm{O}\quad 2s^2 2p_x^2 2p_y^1 2p_z^1$$

We see that in O, two electrons are present in a single $2p$ orbital. They are so close together that they repel each other strongly, and this offsets the greater nuclear charge.

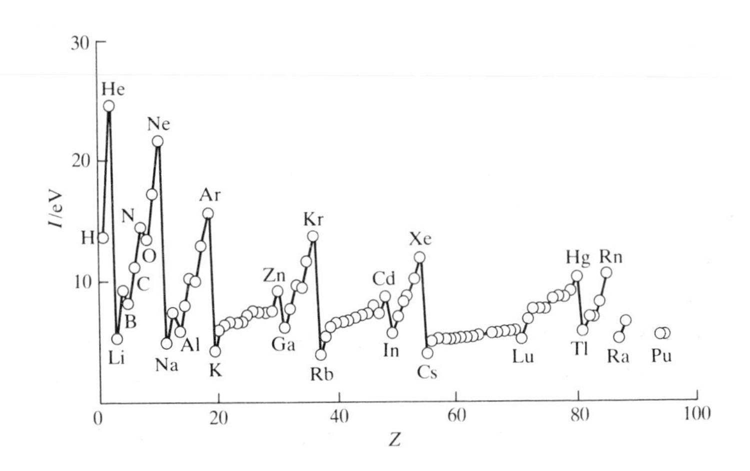

Fig. 1.21 The variation of first ionization energies through the periodic table.

Example 1.5: *Accounting for a variation in ionization energy*

Account for the decrease in first ionization energy between P and S.

Answer. The valence configurations of the two atoms are

$$\text{P} \quad 3s^2 3p_x^1 3p_y^1 3p_z^1 \qquad \text{S} \quad 3s^2 3p_x^2 3p_y^1 3p_z^1.$$

As for N and O, in the S configuration, two electrons are present in a $3p$ orbital. They are so close together that they repel each other strongly, and this offsets the greater nuclear charge.

Exercise. Account for the decrease in first ionization energy between F and Cl.

For F and Ne the next electrons also occupy half-filled orbitals, and continue the trend from O, their higher values indicated by the high value of Z_{eff}. The value of I_1 falls back sharply from Ne to Na as the outermost electron must occupy the next shell with an increased value of n.

Electron affinity

The **electron-gain enthalpy** ΔH_{eg} is the enthalpy change when a gas-phase atom gains an electron:

$$\mathrm{A}(g) + \mathrm{e}^-(g) \rightarrow \mathrm{A}^-(g)$$

Electron gain may be either exothermic or endothermic (Table 1.8). Although the electron-gain enthalpy is the thermodynamically appropriate term, much of inorganic chemistry is discussed in terms of a closely related property, the **electron affinity** A_e of an element, the negative of its electron gain enthalpy. If the electron affinity of an element A is higher than that of another element B, electron gain is more exothermic for A than for B. As in the case of ionization energy, we can distinguish between the electron-gain enthalpy and the electron-gain energy; the former we report in kJ mol^{-1} and the latter in eV.

Table 1.8. Electron affinities of the main group elements (in eV*)

H 0.754							**He** −0.5
Li 0.618	**Be** −0.5	**B** 0.277	**C** 1.263	**N** −0.07	**O** 1.461 −8.75	**F** 3.399	**Ne** −1.2
Na 0.548	**Mg** −0.4	**Al** 0.441	**Si** 1.385	**P** 0.747	**S** 2.077 −5.51	**Cl** 3.617	**Ar** −1.0
K 0.502	**Ca** −0.3	**Ga** 0.30	**Ge** 1.2	**As** 0.81	**Se** 2.021	**Br** 3.365	**Kr** −1.0
Rb 0.486	**Sr** −0.3	**In** 0.3	**Sn** 1.2	**Sb** 1.07	**Te** 1.971	**I** 3.059	**Xe** −0.8

* To convert to kJ mol^{-1}, multiply by 96.485. The first values refer to the formation of the ion X^- from the neutral atom X; the second values to the formation of X^{2-} from X^-.

Source: H. Hotop and W. C. Lineberger, *J. Phys. Chem. Ref. Data*, **14**, 731 (1985).

An element has a high electron affinity if the additional electron can enter a shell where it experiences a strong effective nuclear charge. This is the case for elements toward the top right of the periodic table, as we have already explained. Therefore, elements close to F can be expected to have the highest electron affinities. The second electron-gain enthalpy, the enthalpy change for the attachment of a second electron to an initially neutral atom, is invariably endothermic since the electron repulsion outweighs the nuclear attraction.

Example 1.6: *Accounting for the variation in electron affinity*

Account for the large decrease in electron affinity between Li and Be despite the increase in nuclear charge.

Answer. The electron configurations of the two atoms are [He]$2s^1$ and [He]$2s^2$. The additional electron enters the 2*s* orbital of Li, but it must enter the 2*p* orbital of Be, and hence is much less tightly bound. In fact, the nuclear charge is so well shielded in Be that the electron gain is endothermic.

Exercise. Account for the decrease in electron affinity between C and N.

The Mulliken electronegativity

As we shall see in Chapter 2, the discussion of electron distributions in molecules often involves the question of the relative tendencies of each atom to acquire control of shared electrons. This *molecular* issue is widely discussed using an *atomic* property called the **electronegativity** χ, which is defined, somewhat loosely, as the power of an atom to attract electrons to itself when it is part of a molecule (Table 1.9).

The concept of electronegativity was introduced by Linus Pauling using an analysis of molecular energies. However, the American chemist and physicist Robert Mulliken proposed an alternative approach that expresses the property as a more directly atomic parameter. He observed that the issue of electron-sharing in molecules is one of an atom becoming a cation (by electron loss) or an anion (by electron gain), and proposed a definition of the **absolute electronegativity** as the average value of the ionization

Table 1.9. Pauling electronegativities

H 2.20							**He**
Li 0.98	**Be** 1.57	**B** 2.04	**C** 2.55	**N** 3.04	**O** 3.44	**F** 3.98	**Ne**
Na 0.93	**Mg** 1.31	**Al** 1.61	**Si** 1.90	**P** 2.19	**S** 2.58	**Cl** 3.16	**Ar**
K 0.82	**Ca** 1.00	**Ga** 1.81	**Ge** 2.01	**As** 2.18	**Se** 2.55	**Br** 2.96	**Kr** 3.0
Rb 0.82	**Sr** 0.95	**In** 1.78	**Sn** 1.96	**Sb** 2.05	**Te** 2.10	**I** 2.66	**Xe** 2.6
Cs 0.79	**Ba** 0.89	**Tl** 2.04	**Pb** 2.33	**Bi** 2.02			

See Appendix 3 for a longer list and comparison with other scales.

Source: A. L. Allred, *J. Inorg. Nucl. Chem.*, **17**, 215 (1961); L. C. Allen and J. E. Huheey, *ibid.*, **42**, 1523 (1980).

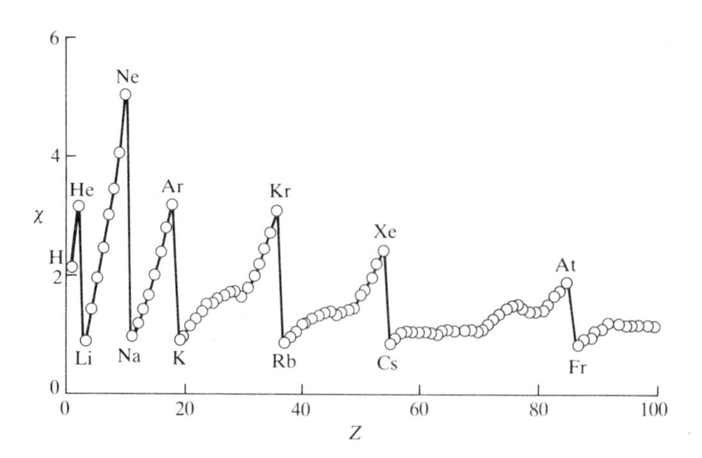

Fig. 1.22 The variation of electronegativity in the periodic table.

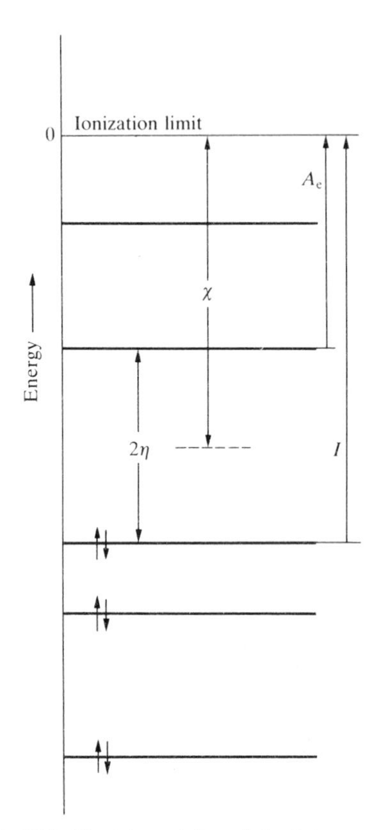

Fig. 1.23 The interpretation of absolute electronegativity and absolute hardness in terms of atomic energy levels.

energy and the electron affinity of the atom:

$$\chi = \tfrac{1}{2}(I + A_e)$$

The ionization energy and electron affinity in this definition relate to a specified state of an atom, which is called its **valence state**.[4] The valence state is the state of the atom as it can be supposed to be when it is part of a molecule. Hence, some calculation is involved, because the ionization energy and electron affinity to be used in calculating χ are mixtures of their values for the contributing spectroscopic states. We need not go into the calculation, but the resulting values are given in Appendix 3 and compared there with the Pauling values. The two scales are very much in line, but there is some controversy over the best way of expressing the relationship between them. One reasonably accurate conversion between the two is

$$\chi_P = 1.35\chi_M^{1/2} - 1.37$$

where χ_P denotes a Pauling electronegativity and χ_M a Mulliken electronegativity.

The variation of χ through the periodic table is shown in Fig. 1.22. Since the elements near F are the ones with both high ionization energies and appreciable electron affinities, these elements have the highest electronegativities. The absolute electronegativity can be interpreted in terms of atomic energy levels (Fig. 1.23) and in particular the location of the **frontier orbitals**, the uppermost filled and lowest unfilled orbitals. Both I and A_e, and hence the electronegativity, are large if the two frontier orbitals are low in energy. Such an atom will release an electron only with reluctance and will tend to acquire one from another atom if its valence shell is incomplete.

The Allred–Rochow electronegativity

The force exerted by the effective nuclear charge at the periphery of an atom has been adopted as another measure of the atom's electronegativity. A spherical ion of radius r and charge number z behaves like a point charge located at its center. The coulombic potential at its surface, the **ionic potential** of the ion, is proportional to z/r and the force the ion exerts on

[4] A helpful review of modern values, and their relation to the Pauling scale, has been given by S. G. Bratsch in *J. chem. Educ.*, **65**, 34 (1988).

charges in its vicinity is proportional to z/r^2. The distance dependence of the coulombic force is the origin of the Allred–Rochow definition of electronegativity, which sets χ proportional to Z_{eff}/r^2, using for r the covalent radius of the atom in Å:

$$\chi = \frac{0.3590 Z_{eff}}{(r/\text{Å})^2} + 0.744$$

With the Allred–Rochow scale we gain another insight into the variation of χ through the periodic table since we already know how Z_{eff} and r vary from element to element. Elements with the highest electronegativity can now be seen to be those with the highest effective nuclear charge and the smallest radii: these lie close to F.

Hardness and softness

The *difference* between the ionization energy of the neutral atom and its anion (Fig. 1.23) is a measure of the **hardness** η of the element:

$$\eta = \tfrac{1}{2}(I - A_e)$$

This is half the separation between the two frontier orbitals, so the hardness is large when the two levels are widely separated in energy. The hardness is low (the atom is 'soft') when the frontier orbitals are close together.[5] In Section 6.3 we meet hardness again and see the motivation for introducing it into chemistry.

The hardest atoms are those with high ionization energies and low electron affinities. If the ionization energy is much larger than the electron affinity, as is often the case, hardness is correlated with high ionization energy. Hence, the hardest atoms and ions are the small atoms and ions near F. The softest atoms and ions are the ones with low ionization energies and electron affinities. These are the atoms and ions of the heavier alkali metals and the heavier halogens. The light atoms of a group are generally hard and the heavier atoms soft.

Perturbation theory

The significance of the hardness is that it indicates the responsiveness of the atom to the presence of electric fields, particularly those arising from neighboring atoms and ions. It is closely related to the older concept of **polarizability** α, the ease with which an atom or ion can be distorted by an electric field. The justification for this remark lies in a technique of quantum mechanics called **perturbation theory**, which was devised in order to describe the distortions that systems undergo when they are subjected to a **perturbation**, an external influence.

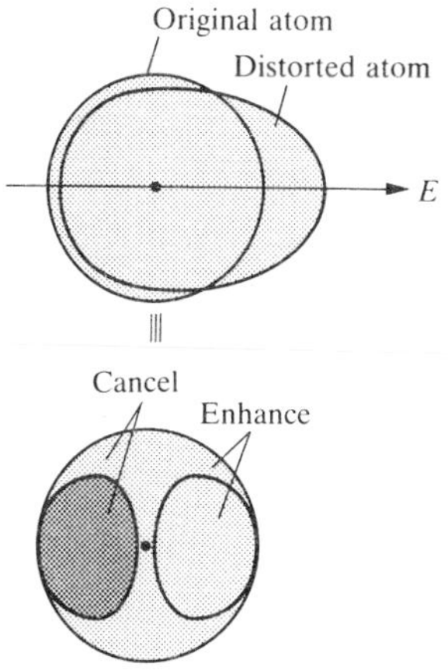

The aim of perturbation theory is to reproduce the distortions of an electron distribution by mixing together wavefunctions of the original, unperturbed, system. An example is that of an electric field applied to an H atom in its ground state. The 1*s* orbital bulges towards the positively charged electrode (Fig. 1.24) because it is energetically favorable for the electron to be on that side of the nucleus.

The shape of the distorted orbital can be reproduced by writing it as a

Fig. 1.24 The distortion of a 1*s* orbital caused by an electric field and its approximation as a superposition of *s* and *p* orbitals.

[5] An instructive review of hardness and tables of values is given in R. G. Pearson, *Inorg. Chem.*, **27**, 734 (1988).

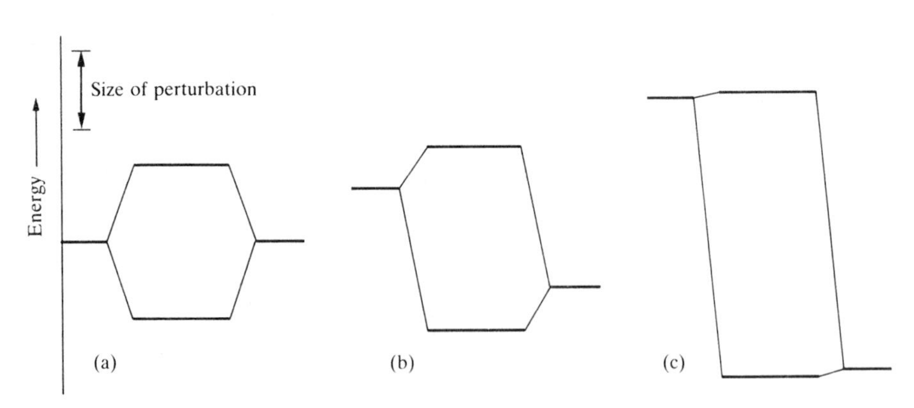

Fig. 1.25 The shifts in energies of the states of a system when it is subjected to a perturbation. (a) Levels initially degenerate, giving a large shift in position. (b) When the levels differ in energy, the shift is small; (c) the shift is very small when they differ by an energy that is very large compared with the strength of the perturbation.

superposition—a sum—of the original 1*s* orbital and a small amount of the $2p_z$ orbital (if the field lies along z). The positive lobe of the *p* orbital adds to the *s* orbital on one side and the negative lobe partly cancels the *s* orbital on the other side. That is, the perturbed atom is described, approximately at least, by the distorted wavefunction

$$\psi = \phi_{1s} + \lambda\phi_{2p}$$

where λ is a coefficient that is found from quantum mechanics. The larger $|\lambda|$, the greater the distortion of the atom. Specifically, λ^2 (or $\lambda^*\lambda$ if λ is complex) is proportional to the probability of finding the electron in the 2*p* orbital given that we know it to be in the 1*s* orbital when the perturbation is absent. The distortion can in fact be reproduced exactly by mixing in the appropriate amount of all possible hydrogen orbitals, but this simple two-component approximation is adequate for illustration.

Quantum mechanics gives a recipe[6] for the value of λ:

$$\lambda = \frac{\text{energy of the perturbation}}{\text{energy separation of the states mixed by the perturbation}}$$

This shows that the stronger the perturbation, the larger the coefficient λ and hence the larger the distortion. However, even a strong perturbation may produce only a weak distortion if the energy-level separation is large, for then the denominator is large.

In other words, a system is stiff and unresponsive (hard) if its energy levels are widely separated. A system with energy levels that are almost degenerate is highly responsive to a perturbation (soft) even if the perturbation is weak since then the denominator is small and λ may be large. Similar arguments are widely used for semiquantitative statements about bond formation.

Perturbation theory also gives an expression for the shifts in the energy levels of an atom or molecule when a perturbation is present. When a perturbation is applied, the two levels move apart, one going to lower energy and the other to higher energy (Fig. 1.25). Suppose for the moment

[6] The general formula for the coefficient λ_i of the orbital ψ_i of energy E_i that should be mixed into the original orbital ψ_0 of energy E_0 is

$$\lambda_i = \frac{-\int \psi_i^* H' \psi_0 \, d\tau}{E_i - E_0}$$

where H' is the perturbation hamiltonian, the difference between the actual hamiltonian and the hamiltonian of the original, unperturbed system. See P. W. Atkins, *Molecular quantum mechanics*. Oxford University Press (1983), Chapter 8.

we consider an atom with a pair of degenerate energy levels (this might be the case if we were considering an excited state of an H atom in which the electron occupies a 2*s* orbital, which is degenerate with the $2p_z$ orbital). Then the two levels move apart to the full extent of the perturbation. For instance, if the perturbation has energy 1 eV, one level decreases in energy by 1 eV and the other rises by the same amount. When the two levels are not degenerate initially, they move apart to a smaller extent than the full perturbation. The actual expression for the energy change has the form[7]

$$\Delta E = \frac{(\text{energy of the perturbation})^2}{\text{energy separation of the states mixed by the perturbation}}$$

If the levels are initially widely separated, the denominator is large, and the levels change only slightly in energy (just as there is only a small change in wavefunction). The closer the energy levels initially, the greater they shift when a perturbation is present.

One of the most important perturbations on an atom in chemistry is that arising from the presence of another atom nearby. The changes in wavefunction and energies that then result are responsible for the adhesion of one atom to another. This opens up an approach to chemical bonding, the topic we consider next.

Further reading

P. A. Cox, *The elements: their origin, abundance, and distribution.* Oxford University Press (1989). A concise and readable account.

P. Henderson, *Inorganic geochemistry.* Pergamon Press, Oxford (1982). An introductory survey of geochemistry, including the origin of the elements.

J. E. Fergusson, *Inorganic chemistry and the earth.* Pergamon Press, Oxford (1982). A similar span to the text above, but from a more chemical perspective.

R. J. Puddephatt and P. K. Monaghan, *The periodic table of the elements.* Oxford University Press (1986). A short survey of the structure of the periodic table and the trends in properties of the elements.

J. Emsley, *The elements.* Oxford University Press (1989). A very useful collection of data and information in a handy format.

P. W. Atkins, *Physical chemistry.* Oxford University Press and W. H. Freeman & Co., New York (1990). Chapters 11 and 12 give an introduction to the principles of atomic structure.

[7] Specifically, the lowering in energy of ψ_0 is from E_0 to E_0', where

$$E_0' = E_0 + \sum_n \frac{(\int \psi_0^* H' \psi_n \, d\tau)^2}{E_0 - E_n}$$

The sum is over all n other than $n = 0$.

Exercises

1.1 Without consulting reference material, draw the form of the periodic table with the numbers of the groups and the periods and identify the *s*, *p*, and *d* blocks. Identify as many elements as you can.

1.2 Contrast the relative terrestrial and cosmic abundances of (a) H and (b) He. Account for the difference.

1.3 Why is Fe an abundant element on earth, and why are the elements with higher atomic numbers increasingly rare?

1.4 What potential difference is needed to accelerate an electron from rest to a velocity at which it has a wavelength of 0.25 Å?

1.5 What is the ratio of the energy of a ground state He^+ ion to that of a Be^{3+} ion?

1.6 The ionization energy of H is 13.6 eV. What is the difference in energy between the $n = 1$ and $n = 6$ levels?

1.7 Define (a) degeneracy of energy levels, (b) principal quantum number. What is the relation of the possible angular momentum quantum numbers to the principal quantum number?

1.8 Using sketches for 2*s* and 2*p* orbitals, distinguish between (a) the radial wavefunction, (b) the radial distribution function, and (c) the angular wavefunction.

1.9 Explain the term penetration, and apply the concept to the rationalization of the trend in energy of the 4*s* orbital relative to that of the 3*d* orbital for the elements in Period 4.

1.10 Compare the first ionization energy of Li with that calculated for ionization of a 2*s* electron from H and a 2*s* electron from Li^{2+}. Account for the location of the experimental value between these two values in terms of penetration and shielding.

1.11 The second ionization energies of some Period 4 elements are

Ca	Sc	Ti	V	Cr	Mn
11.87	12.80	13.58	14.15	16.50	15.64 eV

Identify the orbital from which ionization occurs and account for the trend in values.

1.12 Give the ground state electron configurations of (a) C, (b) F, (c) Ca, (d) Ga^{3+}, (e) Bi, (f) Pb^{2+}, (g) Nd, and (h) U.

1.13 Give the ground state electron configurations of (a) Sc, (b) V^{3+}, (c) Mn^{2+}, (d) Cr^{2+}, (e) Co^{3+}, (f) Cr^{6+}, (g) Cu, and (h) Gd^{3+}.

1.14 Account for the trends, element by element, across Period 3 in (a) ionization energy, (b) electron affinity, and (c) electronegativity.

1.15 From the data in Tables 1.7 and 1.8, contrast the hardness of Na with that of Na^+. What atomic characteristics account for this difference?

1.16 Using the data in Tables 1.7 and 1.8, discuss the correlation of size with electronegativity and hardness.

Problems

1.1 Show that an atom with the configuration ns^2np^6 is spherically symmetrical. Is the same true of an atom with the configuration ns^2np^3?

1.2 According to the Born interpretation, the probability of finding an electron in a volume element $d\tau$ is proportional to $\psi^*\psi\, d\tau$. (a) What is the most probable location of an electron in an H atom in its ground state? (b) What is its most probable distance from the nucleus, and why is this different? (c) What is the most probable distance of a 2*s* electron from the nucleus?

1.3 The ionization energies of Rb and Ag are respectively 4.18 eV and 7.57 eV. Calculate the ionization energies of an H atom with its electron in the same orbitals as in these two atoms and account for the differences in values.

1.4 Taking into account the processes of nucleosynthesis and probable events during the formation of the earth, suggest an explanation of the following mass percentage composition data of the earth's crust: 48 percent O, 26 percent Si, 8.5 percent Al, 5 percent Fe, 3.5 percent Ca, 2.8 percent Na, 2.5 percent K, and 2.0 percent Mg. (The last value is controversial; magnesium is sometimes quoted as following calcium in abundance.)

1.5 When 58.4 nm radiation from a He discharge lamp is directed on a sample of Kr, electrons are ejected with a velocity of $1.59 \times 10^6\ m\ s^{-1}$. The same radiation ejects electrons from Rb at $2.45 \times 10^6\ m\ s^{-1}$. What are the ionization energies (in eV) of the two elements?

Molecular structure

2

In this chapter we explore the reasons why atoms bond together into molecules with definite shapes and numbers of neighbors. We review the early approach to molecular structure devised by Lewis and see how his ideas may be extended to account for the shapes of simple molecules. Then we refine these ideas. First, we make more precise what we mean by the symmetry of a molecule and provide a foundation for analyzing the structures, properties, and spectra of molecules. Then we develop a discussion of the electronic structures of diatomic molecules from the same point of view as we adopted for atoms, where electrons occupy orbitals. Now, though, the orbitals spread over both atoms in the molecule. We shall see that this approach accounts for many features of Lewis's approach. Moreover, it is readily extended to polyatomic molecules, as we see in the following chapter.

Electron pair bonds
2.1 Lewis structures: a review
2.2 Molecular shape and VSEPR theory
Molecular symmetry
2.3 An introduction to symmetry analysis
2.4 Applications of symmetry
Molecular orbitals of diatomic molecules
2.5 An introduction to the theory
2.6 Homonuclear diatomic molecules
2.7 Heteronuclear diatomic molecules
2.8 Bond properties
Further reading
Exercises
Problems

Simple molecules are discrete groups of atoms in which each atom is linked to a definite number of partners by chemical bonds. These bonds have characteristic properties, such as lengths and strengths that, to a good approximation, can be transferred from molecule to molecule. A theory of molecular structure must be able to explain what determines how many bonds an atom may form—such as why S often forms four bonds but sometimes six—and why the local properties of a bond are approximately the same despite changes in other parts of the molecule. The theory should also account for the characteristic shapes of molecules, including why SO_4^{2-} is tetrahedral and why H_2O is angular with an HOH angle of 104°.

Numerical solution of the Schrödinger equation can provide these explanations in some cases, but for most inorganic molecules rigorous calculations are not feasible. However, we can often identify correlations with well-chosen qualitative concepts supported by approximate theories, and in this way rationalize the rich diversity of facts that is so characteristic of the subject. In this respect, organic chemistry and much of biochemistry are similar to each other but different from inorganic chemistry, which needs to deal with a wider range of properties and phenomena. Thus, inorganic chemistry demands a more diverse and sometimes more sophisticated range of bonding models than other fields.

Electron-pair bonds

The earliest successful model of a chemical bond was proposed by Gilbert Lewis in the opening decades of this century. By 1916 he had identified a chemical bond as a shared electron pair, but he had no idea why a pair and not some other number of electrons should be responsible for bond formation. In this section we review Lewis's elementary (but powerful) theory of the electron-pair bond, the main features of which we assume to be familiar. In the next section, we examine the quantum mechanical version of bonding theory and see an explanation of some of the concepts he adopted.

2.1 Lewis structures: a review

Lewis supposed that each pair of electrons involved in bonding lies between and is shared by two neighboring atoms. This **covalent bond**, the shared electron pair, is denoted A—B. Double (A═B) and triple (A≡B) bonds consist of two and three shared pairs of electrons respectively. Unshared pairs of valence electrons on atoms are called **lone pairs**. Although lone pairs do not contribute directly to the bonding, they do influence the shape of a molecule and its chemical properties.

The octet rule

Lewis found that he could account for the existence of a wide range of stable molecules by supposing that each atom continued to acquire shares in electrons until its valence shell had eight electrons. This is the **octet rule**. As we have seen, a closed-shell, noble gas configuration is achieved when eight electrons occupy the *s* and *p* subshells of the valence shell. The H atom completes a 'duplet', the configuration of its neighbor He.

The octet rule provides a simple way of constructing **Lewis structures**, which are diagrams that show the pattern of bonds in a molecule. In most cases we can construct a Lewis structure in three steps.

1. *Decide how many electrons are to be included in the structure by adding together all the valence electrons provided by the atoms.*

Each atom provides all its valence electrons (thus, H provides one electron and O provides six). Each negative charge on an ion corresponds to one more electron; each positive charge corresponds to one electron less.

2. *Write the chemical symbols of the atoms in the arrangement that shows which are bonded together.*

In most cases we know the arrangement or can make an informed guess. For example, the less electronegative element is usually the central atom of a molecule, as in CO_2, SO_4^{2-}, and PCl_5.

3. *Distribute the electrons in pairs so that there is one pair of electrons between each pair of atoms bonded together, and then supply electron pairs (to form multiple bonds or lone pairs) until each atom has an octet.*

All bonding pairs are then represented by a single line. If we are dealing with a polyatomic ion, the charge is assumed to be possessed by the ion as a whole, not by a particular individual atom. However, for some applications, it is convenient to ascribe a **formal charge** (F.C.) to each atom using the definition

$$\begin{aligned}\text{F.C.} = {} & \text{Number of valence electrons in atom} \\ & - \text{Number of lone pair electrons} \\ & - \tfrac{1}{2} \times \text{Number of shared electrons}\end{aligned}$$

Formal charge is helpful for choosing between more than one possible Lewis structure, since it is generally the case that the lowest energy structure is the one with the smallest formal charges on the atoms. Thus, of the two structures

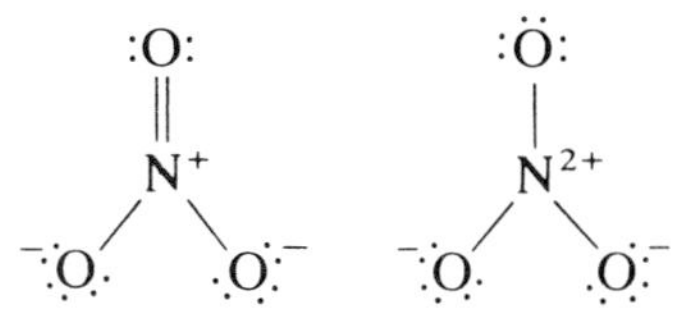

```
       :O:                :Ö:
       ||                  |
       N+                 N2+
      /  \               /   \
  ⁻:O.    .O:⁻       ⁻:O.     .O:⁻
```

we select the first and reject the second.

Example 2.1: ***Writing a Lewis structure***

Write a Lewis structure for the BF_4^- ion.

Answer. The atoms supply $3 + (4 \times 7) = 31$ valence electrons; the single negative charge of the ion reflects the presence of an additional electron. We must therefore accommodate 32 dots in 16 pairs around the 5 atoms. One solution is

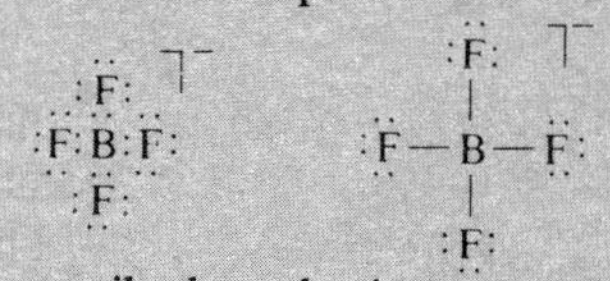

```
     :F: ⁻              :F: ⁻
   :F:B:F:          :F—B—F:
     :F:                :F:
```

The negative charge is ascribed to the ion as a whole, not to an individual atom.

Exercise. Write a Lewis octet structure for the PCl_3 molecule.

Table 2.1. Lewis structures of some common molecules

Molecule	Lewis structures*		
H_2	H—H		
N_2, CO	:N≡N:	:C≡O:	
O_3, SO_2, NO_2^-	:Ö—Ö=Ö: (angular)	:Ö—S̈=Ö: (angular)	[:Ö—N̈=Ö:]$^-$ (angular)
NH_3, SO_3^{2-}	H—N̈(—H)—H	[:Ö—S̈(=Ö)—Ö:]$^{2-}$	
PO_4^{3-}, SO_4^{2-}, ClO_4^-	[:Ö—P(=Ö)(—Ö:)—Ö:]$^{3-}$	[:Ö—S(=Ö)(=Ö)—Ö:]$^{2-}$	[Ö=Cl(=Ö)(=Ö)—Ö:]$^-$

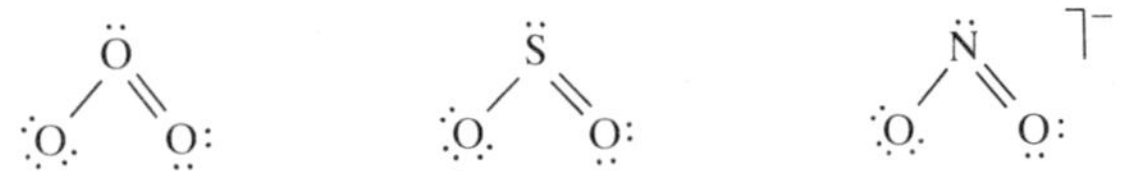

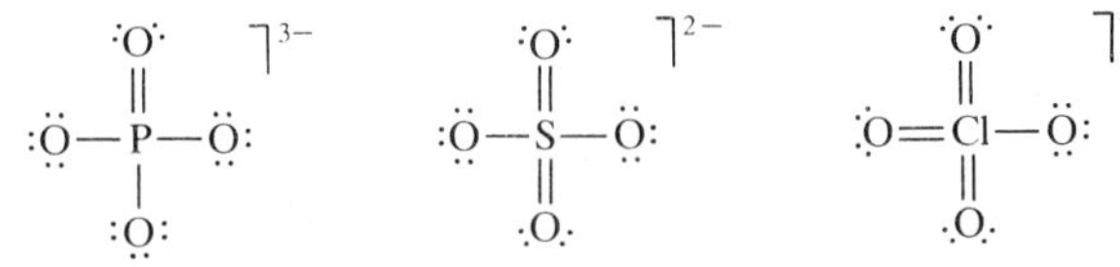

* Only representative resonance structures are given. Shapes are indicated only for diatomic and triatomic molecules.

Table 2.1 gives some examples of Lewis structures for common molecules and molecular ions. It is important to remember that Lewis structures do not convey the shape of the molecule, but only the pattern and numbers of bonds. The BF_4^- ion, for example, is a regular tetrahedron, not the planar structure depicted in Example 2.1. However, in simple cases (particularly for planar molecules), an attempt is often made to show the relative locations of the atoms, and the linear CO_2 molecule and the angular O_3 molecule are written

Ö=C=Ö   :Ö—Ö=Ö: (angular)

Resonance

In many cases a single Lewis structure is an inadequate description of the molecule. Thus, the Lewis structure of ozone given above suggests incorrectly that one OO bond is different from the other, whereas in fact they have identical lengths (1.28 Å) intermediate between those of typical single O—O and double O=O bonds (1.48 Å and 1.21 Å respectively). This deficiency is repaired by introducing the concept of **resonance**, in which the actual structure of the molecule is taken to be a blend of all the feasible Lewis structures corresponding to a given atomic arrangement. Resonance is indicated by a double-headed arrow:

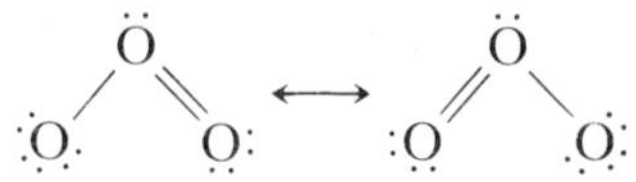

:Ö—Ö=Ö: ⟷ :Ö=Ö—Ö:

Resonance should be pictured as a *blending* of structures, not a flickering alternation between them. In quantum mechanical terms, the electron distribution of each structure is represented by a wavefunction, and the actual wavefunction ψ of the molecule is the superposition of the individual wavefunctions for each structure:

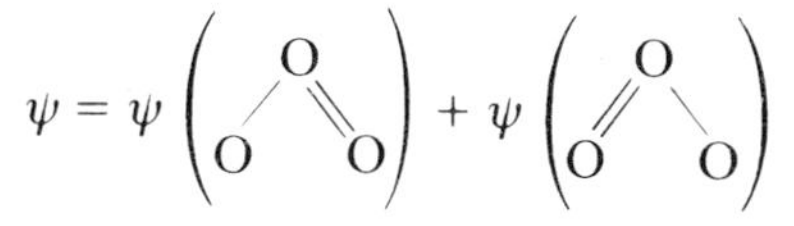

$$\psi = \psi\left(\text{O}\diagup\overset{\text{O}}{}\diagdown\!\!\diagdown\text{O}\right) + \psi\left(\text{O}\diagup\!\!\diagup\overset{\text{O}}{}\diagdown\text{O}\right)$$

The first of the wavefunctions on the right-hand side, for instance, represents an electron distribution in which there is a double bond between the left-hand pair of atoms in O_3. Writing the wavefunction as a superposition with equal contributions from each structure signifies that the probabilities of finding two electron pairs between the left-hand pair of O atoms is the same as finding two between the right-hand pair.

Resonance both averages the bond characteristics and results in a lowering of the energy of the molecule below that of any single contributing structure. The energy of the O_3 resonance hybrid, for instance, is lower than that of either individual structure alone.

Resonance is most important when there are several structures of identical energy that can be written to describe the molecule, as in the case of O_3. In this case, all the structures of the same energy contribute equally to the overall structure. However, structures with different energies may also contribute, but in general the higher the energy of a Lewis structure the smaller the contribution it makes to the overall structure. The BF_3 molecule, for instance is a resonance superposition of the structures

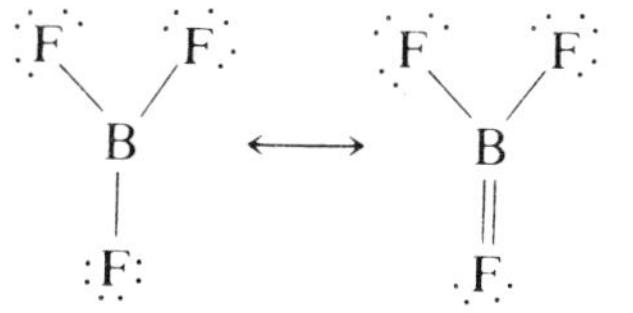

:F: :F: B — :F: ⟷ :F: :F: B = F

as well as several others, but the first of the two structures dominates (it has smaller formal charges than the other structure) and BF_3 is regarded as having single B—F bonds. In contrast, of the two analogous structures of the NO_3^- ion

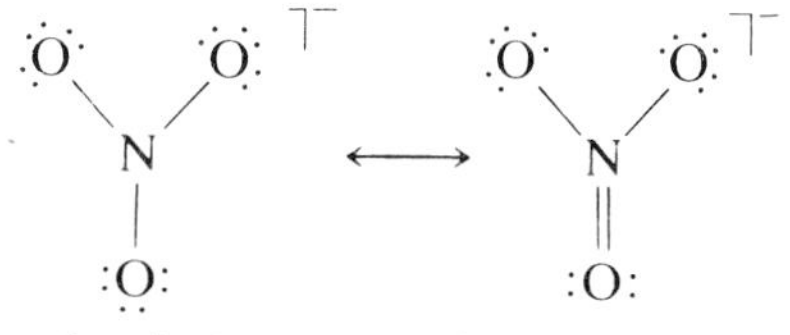

:O: :O: N — :O: ⟷ :O: :O: N = O

the second (and its analogs) dominates (as we saw earlier, it has the smaller formal charges), and we treat the ion as having partial double-bonded character.

Example 2.2: *Writing resonance structures*

Write the resonance structures that make the greatest contribution to the structure of the nitrate ion.

Answer. One Lewis structure is given above (the second of the pair). However, the double bond may be in any of three locations to give structures that have identical energies. Therefore, all three structures contribute equally to the overall resonance hybrid (see margin right).

Exercise. Write resonance structures for the NO_2^- ion.

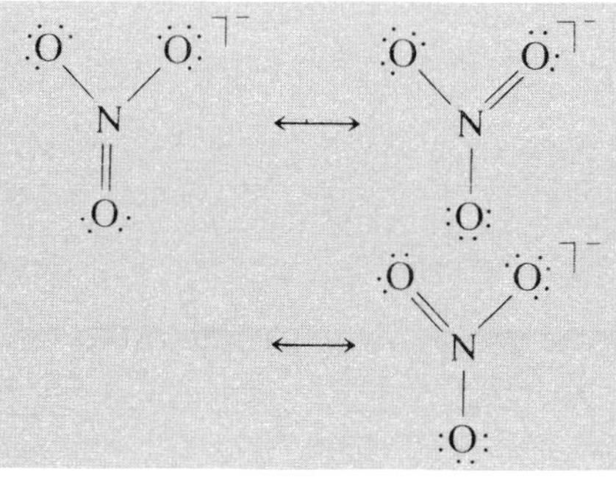

Hypervalence

Although Period 2 elements obey the octet rule quite well, Period 3 and subsequent elements show deviations from it. For example, we can account for the existence of PCl_5 in Lewis terms only if the P atom has 10 electrons in its valence shell, one pair for each P—Cl bond. Similarly SF_6, a very stable molecule, must have 12 electrons around the S atom if each F atom is to be bound to the central S atom by an electron pair. Compounds of this kind, for which the Lewis structures demand more than an octet of electrons around at least one atom, are called **hypervalent compounds**.

Example 2.3: *Writing a Lewis structure for a hypervalent compound*

The SO bond length in SO_4^{2-} is intermediate between the values expected for S—O and S=O bonds. This suggests that the ion is a resonance hybrid of Lewis structures with both bond types. Write a Lewis structure for SO_4^{2-} in which the S atom forms S=O bonds and has six shared electron pairs.

Answer. We need to accommodate $6 + (4 \times 6) + 2 = 32$ electrons. If we allow for sulfur–oxygen double bonds, one possible structure is

:Ö:]²⁻
Ö::S::Ö denoted Ö=S=Ö with :Ö: above and below (single bonds)

The actual structure of the sulfate ion (which is tetrahedral) is a resonance hybrid of structures like this and others in which the octet is not expanded.

Exercise. Write a Lewis structure for the perxenate ion (XeO_6^{4-}) in which the Xe atom has eight electron pairs.

The occurrence of hypervalence in the elements of Periods 3 to 6 has been explained by noting that atoms of those elements have low-lying unfilled *d* orbitals available to accommodate the additional electrons. Thus P can accommodate more than eight electrons if it uses its vacant 3*d* orbitals. In PCl_5, for instance, at least one 3*d* orbital must be used. The rarity of hypervalence in Period 2 is then ascribed to the absence of 2*d* orbitals in these elements.[1] However, a more compelling reason for the rarity of hypervalence in Period 2 may be the difficulty of packing more than four atoms around a single small central atom, and may have little to do with the availability of *d* orbitals. What scant evidence there is (from computations of the kind described later) suggests that there are hypervalent compounds in which *d* orbitals are not used to any significant amount. In Section 3.2 we shall see how to account for hypervalent compounds without making use of *d* orbitals.

2.2 Molecular shape and VSEPR theory

Molecular shapes can be determined experimentally using X-ray diffraction (Box 4.1 of Chapter 4), electron diffraction, and spectroscopy, particularly

[1] Five- and six-coordinate B and C are known in carboranes and metal carbides (Chapter 11), metal clusters (Chapter 16), and in other molecules. See, for instance, J. C. Martin, *Science,* **221,** 509 (1983).

Table 2.2. The description of molecular shapes

Description of shape	Shape	Examples
Linear	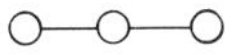	HCN, CO_2
Angular		H_2O, O_3, NO_2^-
Trigonal planar	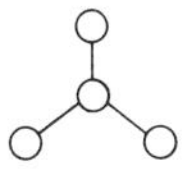	BF_3, SO_3, NO_3^-, CO_3^{2-}
Trigonal pyramidal	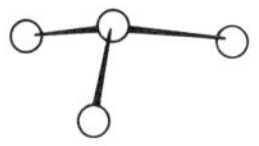	NH_3, SO_3^{2-}
Tetrahedral	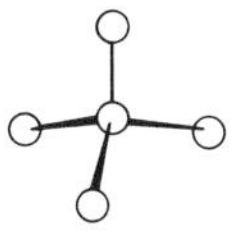	CH_4, SO_4^{2-}, NSF_3
Square planar	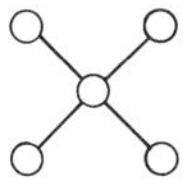	XeF_4
Square pyramidal	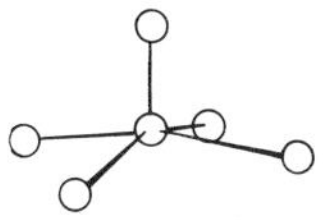	$Sb(Ph)_5$
Trigonal bipyramidal	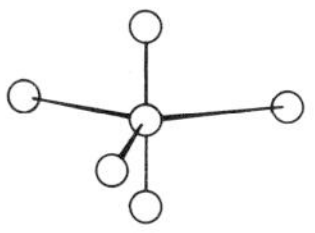	$PCl_5(g)$, SOF_4
Octahedral	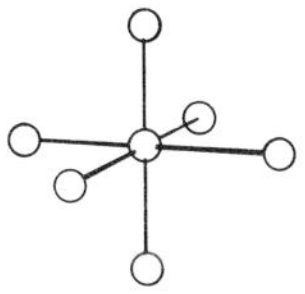	SF_6, PCl_6^-, $IO(OH)_5$

infrared and Raman spectra. They can be reported, in simple cases, by stating whether the molecule is linear, tetrahedral, and so on (Table 2.2) and giving the bond lengths. Individual bond angles are reported for molecules in which the angles are not implied by the symmetry of the shape. Thus, SF_6 is octahedral and all its bond angles are necessarily 90°; NH_3 is pyramidal, and although its symmetry requires all three HNH angles to be the same, we need to specify that the angles are in fact 107° because this is not implied by the symmetry.

Even when the shape does not correspond to one of the symmetrical geometrical figures, one may be a good starting point for the description of the molecule. The 'sawhorse' shape of the SF_4 molecule, for example, is closely related to a trigonal bipyramid. Where the molecule is less symmetrical than one of the shapes shown in Table 2.2, or where it is more complex, it is necessary to give the individual coordinates of the atoms.

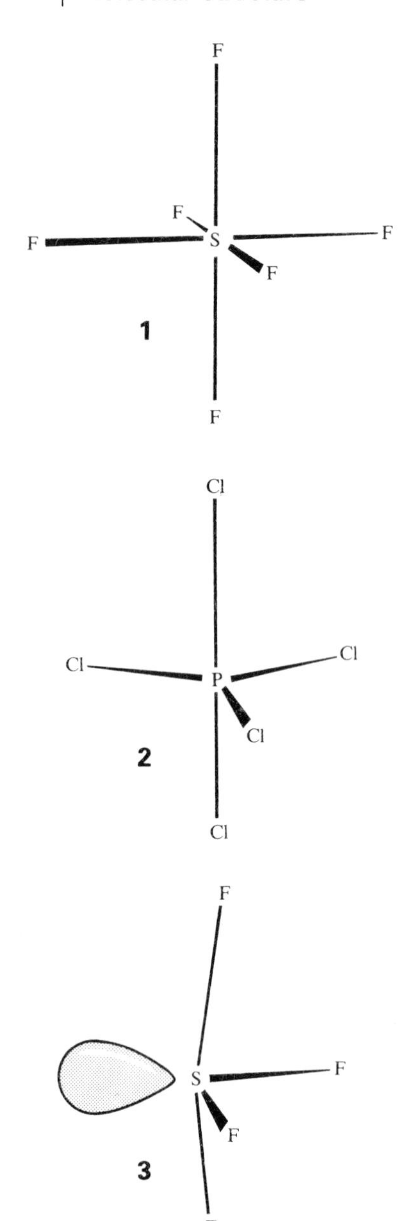

Predicting shapes: VSEPR theory

The **valence shell electron pair repulsion theory** (VSEPR theory) is a simple extension of Lewis's ideas which is surprisingly successful for predicting the shapes of polyatomic molecules. The theory stems from suggestions made by Nevil Sidgwick and Herbert Powell in 1940 and extended and put into a more modern context by Ronald Gillespie and Ronald Nyholm. When using VSEPR theory we need to remember that the *name* of the shape of a molecule (such as linear or tetrahedral) is determined by the geometrical arrangement of the atoms disregarding the location of any lone pairs. The NH_3 molecule, for example, is pyramidal, not tetrahedral; the NH_4^+ ion is tetrahedral.

The primary assumption of VSEPR theory is that electron pairs repel each other and so take up positions as far apart as possible. Thus, three electron pairs lie at the corners of an equilateral triangle, four pairs lie at the corners of a regular tetrahedron, and so on (Table 2.3). The SF_6 molecule, with six electron pairs around the S atom, is therefore predicted to be octahedral (**1**) and the PCl_5 molecule, with five pairs, is predicted to be trigonal bipyramidal (**2**). An alternative, slightly less favorable structure is the square pyramid, and there are several examples of this structure too, often slightly distorted (**3**).

The NH_3 molecule has four electron pairs that are disposed tetrahedrally, but as one of them is a lone pair the molecule itself is classified as trigonal pyramidal. The apex of the pyramid is occupied by the lone pair. Similarly, H_2O has a tetrahedral arrangement of its electron pairs, but as two of the pairs are lone pairs, the molecule is reported as angular.

According to VSEPR theory, a multiple bond is treated as though it were a single electron pair, and its two or three electron pairs are treated as a single 'superpair'. The linear structure of O=C=O, for example, is predicted on the grounds that the C atom effectively has two superpairs. This rule does away with any worry about which resonance structure to consider. Thus, an octet all-single-bonded Lewis structure of SO_4^{2-} and one with two double bonds and an expanded octet are both predicted to be tetrahedral.

The basic shapes for seven electron pairs are less readily predicted than others, partly because so many different conformations correspond to similar energies. VSEPR theory is not reliable for these fluxionally mobile compounds. The Te atom in $[TeCl_6]^{2-}$, for example, has one lone pair and is attached to six neighbors, yet it appears to be a regular octahedron. Lone pairs are stereochemically less influential when they belong to heavy *p*-block elements.

Table 2.3. The basic arrangement of electron pairs according to VSEPR theory

Number of electron pairs	Arrangement
2	Linear
3	Trigonal planar
4	Tetrahedral
5	Trigonal bipyramidal
6	Octahedral

Modifications of the basic shape

Once the basic shape of a molecule has been identified, minor adjustments are made to the shape to take into account the differences in electrostatic repulsion between bonding pairs and lone pairs. These repulsions are generally assumed to lie in the following order:

lone pair/lone pair repulsions
are stronger than
lone pair/bonding pair repulsions
are stronger than
bonding pair/bonding pair repulsions

In elementary accounts, the greater repelling effect of a lone pair is explained by supposing that the lone pair is on average closer to the nucleus than a bonding pair, and therefore repels other electron pairs more strongly. However, the true origin of the difference is obscure.[2] An additional detail about this order of repulsions is that, given the choice between an axial and an equatorial site for a lone pair in a trigonal bipyramidal array, it occupies the equatorial site since it is then repelled less by the bonding pairs.

Example 2.4: *Using VSEPR theory to predict shapes*

Predict the shape of the SF_4 molecule.

Answer. First, we write the Lewis structure of SF_4:

```
        :F:
         |
   :F—S:
       /  \
    :F.   .F:
```

This structure has five electron pairs around the central atom, which adopt a trigonal bipyramidal arrangement (Table 2.3). Repulsion from the one lone pair is minimized if it occupies an equatorial site, when it interacts with the two axial bonding pairs strongly, rather than being axial, when it would interact strongly with three equatorial bonding pairs. The S—F bonds then bend away from the lone pair (**3**).

Exercise. Predict the shape of XeF_4.

The angle between the O—H bonds in H_2O decreases slightly from its tetrahedral value (109.5°) as the two lone pairs move apart. This is in agreement with the observed HOH bond angle, 104°. A similar effect accounts for the HNH bond angle of 107° in NH_3. It also accords neatly with the regular tetrahedral shape of NH_4^+, in which all four bonds are equivalent.

Hybridization

We saw in Chapter 1 that electrons occupy atomic orbitals, such as *s* and *p* orbitals, that point in directions seemingly unrelated to the bonding directions predicted by VSEPR theory. However, we can bring the Lewis and VSEPR descriptions of molecules into line with the orbital picture. One way to do this is to use the quantum mechanical result that eight electrons in one *s* and three *p* orbitals (the configuration s^2p^6) have a distribution in space that is the same as that obtained by treating the electrons as occupying four equivalent **hybrid orbitals** (Fig. 2.1) directed towards the corners of a regular tetrahedron. Each of these hybrid orbitals is a mixture of the original *s* and *p* orbitals:

Fig. 2.1 A set of four sp^3 hybrid orbitals.

[2] The physical basis of VSEPR theory, and a more succinct formulation, is given by R. F. W. Bader, R. J. Gillespie, and P. J. MacDougall, *J. Am. chem. Soc.*, **110**, 7329 (1988).

$$h_1 = s + p_x + p_y + p_z \qquad h_3 = s - p_x + p_y - p_z$$
$$h_2 = s - p_x - p_y + p_z \qquad h_4 = s + p_x - p_y - p_z$$

Each hybrid is a combination of atomic orbitals, and an electron that occupies any one of them has partial s and p orbital character. Since each hybrid is composed of one s orbital and three p orbitals, the four hybrid orbitals are known as **sp^3-hybrids**. A hybrid description is equivalent to an unhybridized orbital description in the sense that the charge distribution of the configuration s^2p^6 is the same as the description $h_1^2h_2^2h_3^2h_4^2$.

The distinctive shapes of the hybrids can be thought of as arising from the interference of the atomic orbitals, the addition or cancellation of the amplitudes of the wavefunctions. Thus, in h_1 the positive regions of the s and p orbitals all reinforce each other but the positive s orbital largely cancels the negative lobes of the three p orbitals. This results in an orbital with a strong amplitude in one direction but only a small amplitude in the opposite direction, as shown in two dimensions in Fig. 2.2a. A similar interference between the orbitals in the combinations with minus signs gives enhancements and cancellations in different directions (Fig. 2.2b and c), and the four different combinations of signs for the orbitals in three dimensions give the four different lobes of the hybrid orbitals.

Hybrid orbitals can be used to recreate an orbital version of any of the distributions predicted by VSEPR theory. Since we cannot simply 'lose' orbitals, N atomic orbitals combine to give N hybrid orbitals. If we want to recreate a trigonal planar arrangement of three electron pairs, we form three **sp^2-hybrids** from an s orbital and two p orbitals lying in the plane (as in Fig. 2.2). Therefore, if we are seeking to describe the electron distribution in a molecule that VSEPR theory predicts to be trigonal planar, a reasonable first approximation to the electron wavefunctions is to suppose that each pair occupies an sp^2-hybrid orbital.

Where more than four electron pairs must be accommodated (in expanded octet molecules) we must include d orbitals in the hybridization. For five electron pairs, we need five hybrid orbitals and therefore must use one d orbital in addition to the s and p orbitals. As Table 2.4 shows, sp^3d hybridization leads to five hybrids that point toward the corners of a trigonal

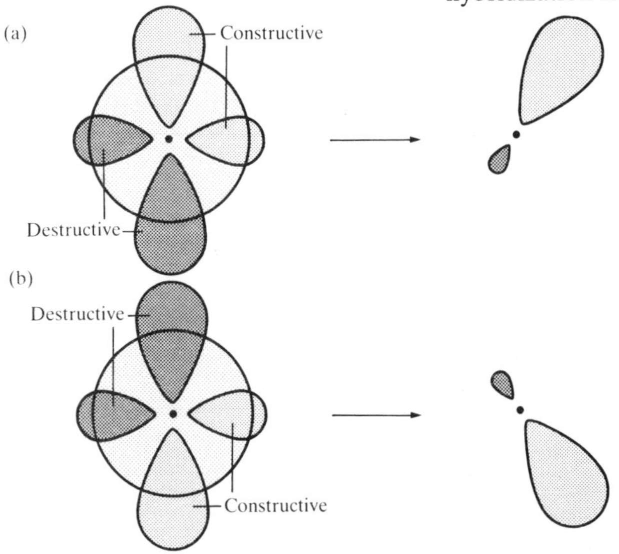

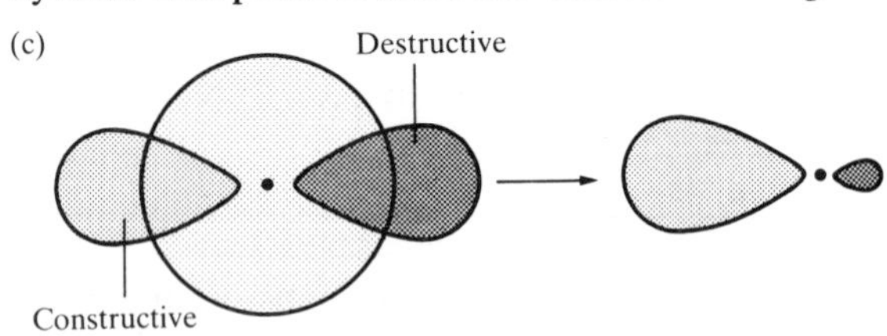

Fig. 2.2 Hybridization in two dimensions. The regions of atomic orbitals with the same phase (sign, represented by shades of gray) add to give a lobe in a particular direction, but cancel in other directions. (a), (b), and (c) show the hybrids obtained by using different signs in the three combinations of three atomic orbitals.

Table 2.4. Some hybridization schemes

Coordination number	Arrangement	Composition
2	Linear	*sp*, *pd*, *sd*
	Angular	*sd*
3	Trigonal planar	sp^2, p^2d
	Unsymmetrical planar	*spd*
	Trigonal pyramidal	pd^2
4	Tetrahedral	sp^3, sd^3
	Irregular tetrahedral	spd^2, p^3d, pd^3
	Square planar	p^2d^2, sp^2d
5	Trigonal bipyramidal	sp^3d, spd^3
	Tetragonal pyramidal	sp^2d^2, sd^4, pd^4, p^3d^2
	Pentagonal planar	p^2d^3
6	Octahedral	sp^3d^2
	Trigonal prismatic	spd^4, pd^5
	Trigonal antiprismatic	p^3d^3

Source: H. Eyring, J. Walter, and G. E. Kimball, *Quantum chemistry*. Wiley, New York (1944).

bipyramid. Similarly, where we needed to accommodate six electron pairs in a regular octahedral arrangement, we need two *d* orbitals: the six sp^3d^2 hybrids point in the required directions.

The role of hybridization must be kept in perspective. It cannot be used to predict molecular shapes; it is a way of creating localized orbitals that reproduce the observed shapes of molecules.

Shape and timescale

One problem with the experimental determination of the shape of a molecule is that different techniques sometimes lead to apparently conflicting conclusions. For example, pentacarbonyliron ($Fe(CO)_5$, **4**) has infrared and Raman spectra consistent with a trigonal bipyramidal structure with distinct axial and equatorial CO groups. However, when the molecule is studied by ^{13}C-NMR, a single resonance line is observed, indicating that the five CO groups are identical.

4

The conflict is resolved by realizing that the molecule is **fluxional**, or able to change from one conformation to another, and that each technique observes the molecule on a characteristic timescale. In the infrared and Raman experiments, the interaction of the incident photon with the molecule is almost instantaneous, and the results are an almost instantaneous snapshot of its structure. The timescale of an NMR experiment is much longer (we shall say more precisely what this means in a moment), and the spectrum represents an average of the conformations of the molecule. Hence, in an NMR experiment, the axial and equatorial CO groups look identical.

We can decide the timescale of a spectroscopic determination of structure by considering the quantum mechanical effect called **lifetime broadening**. According to quantum mechanics, the energy of a state becomes less well defined as the lifetime of the state decreases. A state has a precisely determined energy only if it persists for ever, and the shorter the lifetime, the wider the range of energies of the state (Fig. 2.3). The relation between

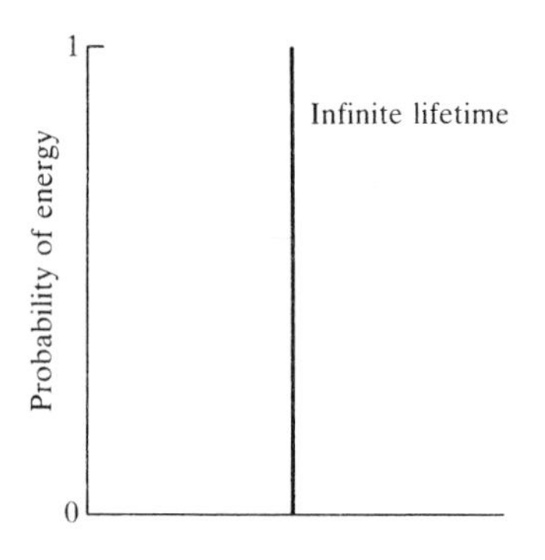

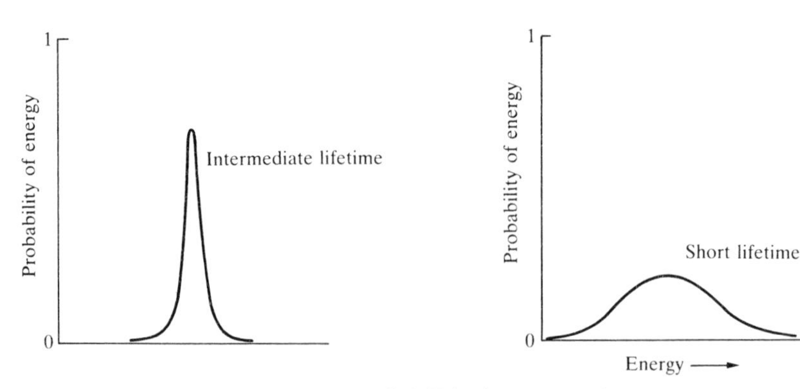

Fig. 2.3 The relation between lifetime of a state and its energy. A state with an infinite lifetime has a precisely defined energy; a state that has a short lifetime has an energy that may be anywhere within a range.

the lifetime τ of a state and the spread ΔE in its energy is

$$\tau \Delta E \approx \frac{h}{2\pi} \qquad (1)$$

In terms of the equivalent frequency spread (using $\Delta E = h\Delta\nu$),

$$\tau \Delta \nu = \frac{1}{2\pi} \qquad (2)$$

Thus, if a state lasts for only 1 ps, its frequency may lie anywhere in a range of width 160 GHz (1 GHz $= 10^9$ Hz).[3]

Suppose that a static molecule has two inequivalent sites for an atom (such as the axial and equatorial sites for the C atoms in $Fe(CO)_5$). These atoms give rise to two distinct NMR signals at frequencies ν_1 and ν_2 that are characteristic of the two sites. However, if the molecule can rearrange in such a way that the atoms in the sites are interchanged, each resonance frequency will span a range $\Delta\nu$ given in eqn 2, with τ the residence time of an atom in one of the sites. The distinction between the two sites is lost, and the molecule consequently appears to have higher symmetry, when

$$\Delta\nu \approx \nu_2 - \nu_1$$

because the frequencies of the two transitions then become indistinguishable. By combining this condition with eqn 2, we see that the sites become indistinguishable when

$$\tau \approx \frac{1}{2\pi(\nu_2 - \nu_1)} \qquad (3)$$

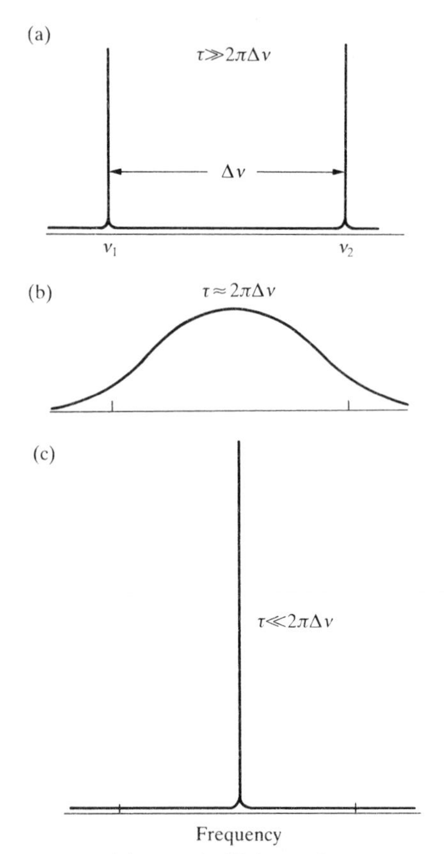

Fig. 2.4 (a) When the lifetime τ of a conformation of a molecule is infinite, the spectrum shows narrow absorptions. (b) As the lifetimes shorten, the absorptions broaden and merge. (c) When the lifetime is very short, a single, sharp absorption is obtained at the mean of the two frequencies.

The effect is illustrated in Fig. 2.4. If the distinct sites survive for a lifetime shorter than the τ given by eqn 3, we cannot distinguish between the two absorptions and the experiment gives the average of the two. If the interchange is so slow that each site survives for longer than the τ given by eqn 3, we can distinguish the two absorptions.

It is generally the case that the higher the frequency of the technique, the more precise the molecular coordinates it can be used to obtain. This is largely because a high-frequency technique (like infrared spectroscopy) is

[3] The lifetime broadening effect, and the equations that describe it, are reminiscent of the uncertainty principle for position and momentum (Section 1.3). However, they are derived in different ways and are best regarded as unrelated. See P. W. Atkins, *Molecular quantum mechanics*. Oxford University Press (1983), p. 96.

Table 2.5. Approximate timescales of common structural techniques

Technique	Timescale/s
Electron diffraction*	10^{-20}
Neutron diffraction*	10^{-18}
X-ray diffraction*	10^{-18}
Ultraviolet spectroscopy	10^{-15}
Visible spectroscopy	10^{-14}
Infrared spectroscopy	10^{-13}
Vibrational Raman spectroscopy	10^{-13}
Electron paramagnetic resonance	10^{-4} to 10^{-8}
Nuclear magnetic resonance	10^{-1} to 10^{-9}
Mössbauer spectroscopy	10^{-7}
Chemical separation of isomers	10^{2} and longer

* In a number of cases, particularly diffraction, the observation times are so long (typically hours) that the image obtained is an envelope spanning all atomic locations, and the theoretical timescales are not particularly relevant.

generally accompanied by large frequency differences between different groups, and so fluxional lifetimes must be very short before they damage the resolution of the experiment. Table 2.5 summarizes the typical timescales of techniques commonly used in inorganic chemistry. If the difference in resonance frequency between sites for an NMR experiment is 100 Hz, only lifetimes longer than 2 ms will allow the conformations to be distinguished. In contrast, the difference in vibrational frequencies between sites might be 10^{12} Hz (corresponding to a difference in wavenumber of 300 cm^{-1}), so they can be distinguished even if they survive for little more than 0.2 ps. Hence, vibrational spectroscopy shows an almost instantaneous conformation of a fluxional molecule.

Example 2.5: *Judging the timescale of a technique*

In a certain fluxional molecule, two groups that are interchanged by the conversion have vibrational absorptions at 1650 cm^{-1} and 1655 cm^{-1}. Calculate the minimum lifetime of the states for which the separate absorptions could be distinguished.

Answer. The absorptions are indistinguishable if the lifetimes are less than $1/2\pi\Delta\nu$, or in terms of wavenumber, less than

$$\tau = \frac{1}{2\pi \times c\Delta\tilde{\nu}}$$

$$= \frac{1}{2\pi \times 2.998 \times 10^{10}\ \mathrm{cm\ s^{-1}} \times 5\ \mathrm{cm^{-1}}}$$

$$= 1 \times 10^{-12}\ \mathrm{s}$$

1 ps is time for the molecule to undergo only about 10 oscillations in each conformation.

Exercise. Two groups in a molecule gave NMR absorptions with shifts separated by 1.10 ppm at 500 MHz, corresponding to a frequency separation of 550 Hz. What is the minimum lifetime for distinguishing them?

Molecular symmetry

One aspect of the shape of a molecule is its 'symmetry' (we define the technical meaning of this term in a moment) and the systematic treatment of symmetry uses **group theory**. This is a rich and powerful subject, but we will confine our use of it at this stage to classifying molecules and drawing some general conclusions about their properties.

2.3 An introduction to symmetry analysis

Our initial aim is to define the symmetry of molecules much more precisely than we have done so far, and to provide a notational scheme that conveys their symmetry. In subsequent chapters we extend the material presented here to applications in bonding and spectroscopy, and it will become clear that symmetry analysis is one of the most pervasive techniques in inorganic chemistry.

Symmetry operations and elements

A fundamental concept of group theory is the **symmetry operation**. This is an action, such as a rotation through a certain angle, that leaves the molecules apparently unchanged. An example is the rotation of an H_2O molecule by 180° (but not any smaller angle) around the bisector of the HOH angle. Associated with each symmetry operation there is a **symmetry element**; this is a point, a line, or a plane with respect to which the symmetry operation is performed. The most important symmetry operations and their corresponding elements are listed in Table 2.6. All these operations leave at least one point of the molecule unmoved, just as a rotation of a sphere leaves its center unmoved. Hence they are the operations of 'point-group symmetry'. The **identity operation** E leaves the whole molecule unchanged.

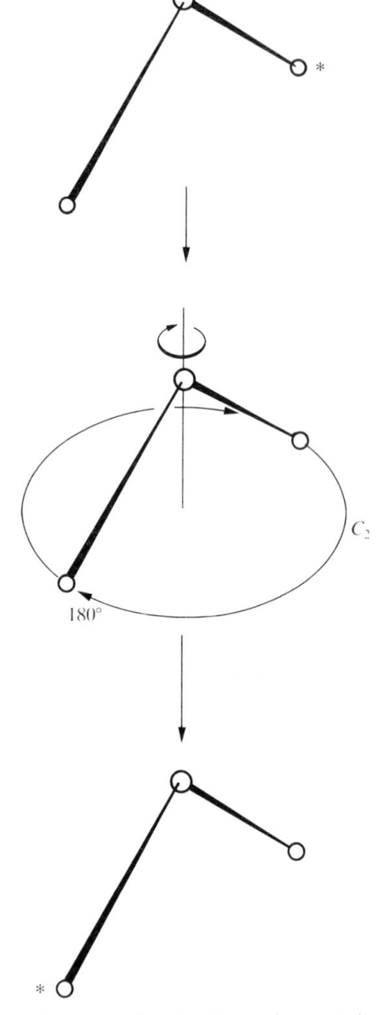

Fig. 2.5 An H_2O molecule may be rotated through any angle about the bisector of the HOH bond angle, but only a rotation of 180°, C_2, leaves it apparently unchanged.

The rotation of an H_2O molecule by 180° ($n = 2$) around a line bisecting the HOH angle is a symmetry operation, so the H_2O molecule possesses a 'twofold' rotation axis C_2 (Fig. 2.5). In general, an ***n*-fold rotation** is a symmetry operation if the molecule appears unchanged after rotation by $360°/n$. The corresponding symmetry element is a line, the ***n*-fold rotation axis C_n**, about which the rotation is performed. The triangular pyramidal NH_3 molecule has a threefold rotation axes, denoted C_3, but there are two operations associated with it, one a rotation by 120° and the other a rotation

Table 2.6. Important symmetry operations and symmetry elements

Symmetry element	Symmetry operation	Symbol
	Identity*	E
n-Fold symmetry axis	Rotation by $2\pi/n$	C_n
Mirror plane	Reflection	σ
Center of inversion	Inversion	i
n-Fold axis of improper rotation†	Rotation by $2\pi/n$ followed by reflection perpendicular to rotation axis	S_n

* The symmetry element can be thought of as the molecule as a whole.
† Note the equivalences $S_1 = \sigma$ and $S_2 = i$.

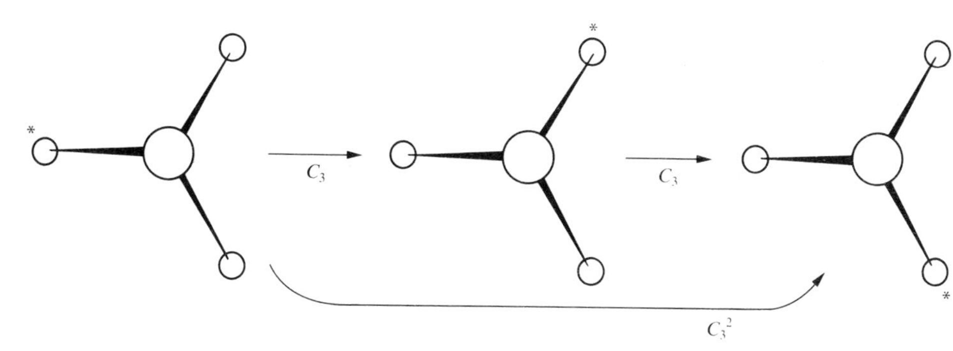

Fig. 2.6 A threefold rotation and the corresponding C_3 axis in NH_3. There are two rotations associated with this axis, one through 120° (C_3) and the other through 240° (C_3^2).

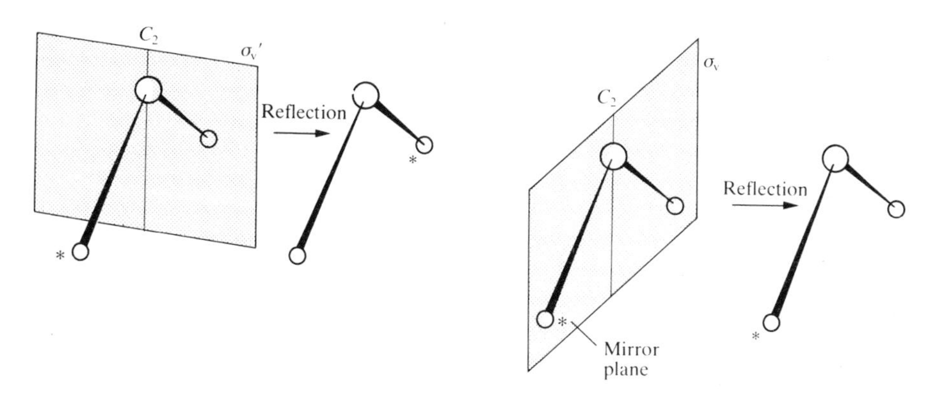

Fig. 2.7 The two vertical mirror planes σ_v and σ_v' in H_2O and the corresponding operations. Both planes cut through the C_2 axis.

through twice this angle (Fig. 2.6). The two operations are denoted C_3 and C_3^2. We do not need to consider C_3^3, a rotation through $3 \times 120° = 360°$, since it is equivalent to the identity.

The **reflection** of an H_2O molecule in either of the two planes shown in Fig. 2.7 is a symmetry operation; the corresponding symmetry element is a **mirror plane** σ. The H_2O molecule has two mirror planes which intersect at the bisector of the HOH angle. Because the planes are vertical (in the sense of being parallel to the rotational axis of the molecule), they are labeled σ_v and σ_v'. The C_6H_6 molecule has a mirror plane σ_h in the plane of the molecule. The h signifies that the plane is 'horizontal' in the sense that the principal rotational axis of the molecule is perpendicular to it. This molecule also has two more sets of three mirror planes that intersect the sixfold axis (Fig. 2.8). In such cases, the members of one set are called 'vertical' and the members of the other are called 'dihedral'. The symmetry elements (and the associated operations) are denoted σ_v and σ_d respectively.

The **inversion** operation consists of imagining that each point of the molecule is taken through a single center and projected an equal distance on the other side (Fig. 2.9). The symmetry element is the **center of inversion** i, the point through which the projection is made. An N_2 molecule has a center of inversion midway between the two nitrogen nuclei. The H_2O molecule does not possess this element, but a benzene molecule and a sulfur hexafluoride molecule both do. In due course we shall see the importance of recognizing that an octahedron has a center of inversion but that a tetrahedron does not.

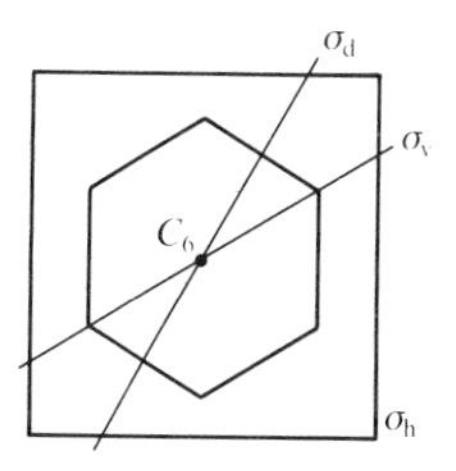

Fig. 2.8 Some of the symmetry elements of the benzene ring. There is one horizontal reflection plane (σ_h) and two sets of vertical reflection plane (σ_v and σ_d); one example of each is shown.

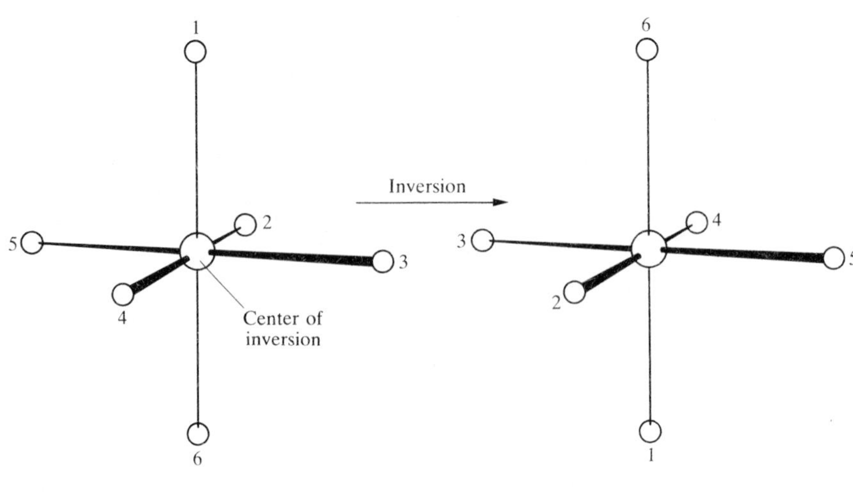

Fig. 2.9 The inversion operation and the center of inversion i in SF_6.

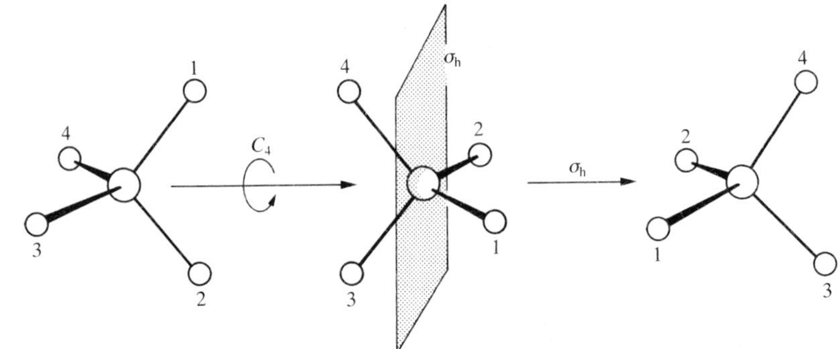

Fig. 2.10 A fourfold axis of improper rotation S_4 in the CH_4 molecule.

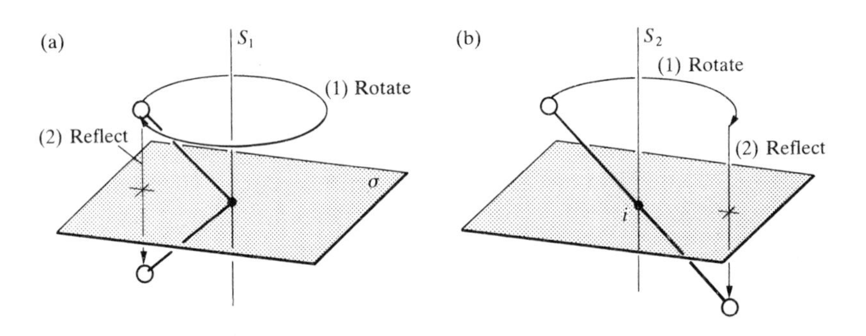

Fig. 2.11 (a) An S_1 axis is equivalent to a mirror plane and (b) an S_2 axis is equivalent to a center of inversion.

The operation of **improper rotation** is composite: it consists of a rotation followed by a reflection in the plane perpendicular to the axis of rotation. Figure 2.10 shows a fourfold improper rotation of a tetrahedral CH_4 molecule: in this case, the operation consists of a 90° rotation about an axis bisecting two HCH bond angles followed by a reflection through a plane perpendicular to the rotation axis. Neither the operation C_4 nor the reflection σ_h alone is a symmetry operation for CH_4, but their product $C_4 \times \sigma_h$ is a symmetry operation, the improper rotation S_4. The symmetry element, the **improper rotation axis** S_n (S_4 in the example), is the corresponding combination of an n-fold rotational axis and a perpendicular mirror plane. The only S_n axes that exist are those for $n = 1$ and $n = 2, 4, 6, \ldots$. Moreover, S_1 is equivalent to a horizontal reflection σ_h and S_2 is equivalent to the inversion i (Fig. 2.11).

Example 2.6: *Identifying a symmetry element*

Which conformation of a CH_3CH_3 molecule has an S_6 axis?

Answer. We need to find a conformation that leaves the molecule looking the same after a 60% rotation followed by a reflection in a plane perpendicular to that axis. The conformation and axis are shown in (**5**); this is the 'staggered' conformation of the molecule, and also the one of lowest energy.

Exercise. Identify a C_3 axis of an NH_4^+ ion. How many of these axes are there in the ion?

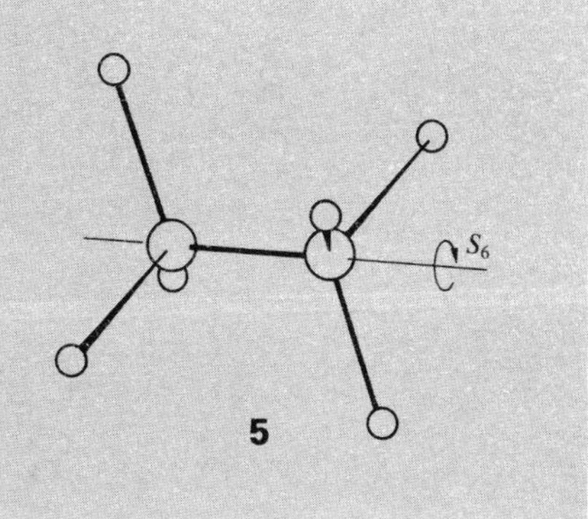

The point groups of molecules

The symmetry elements possessed by molecules determine the **point group** to which they belong. Assigning a molecule to a particular group depends on making a list of symmetry elements it possesses and then comparing it with the list that is characteristic of each point group. For example, if a molecule has only the identity element (CHBrClF is an example), we list its elements as E and look for the group that has *only* this one. In fact, the group called C_1 is the group with only the element E, so the CHBrClF (**6**) molecule belongs to that group. The molecule CH_2BrCl belongs to a slightly richer group: it has the elements E (all groups have that element) and a mirror plane. The group of elements E, σ is called C_s, so the CH_2BrCl molecule belongs to that group. This process can be continued, with molecules being assigned to the group that matches the symmetry elements they possess. Some of the more common groups and their names are listed in Table 2.7. Assigning a molecule to its group depends on listing the symmetry elements it possesses and then referring to the table. However, it is often easier to work through the tree in Fig. 2.12 and to arrive at the correct point group by answering the questions at each decision point on the chart. (Note that we need to be careful to distinguish the names of the groups, C_2 and so on, from the symbols for the symmetry elements, such as C_2, and the corresponding operations, also C_2. The context will always make it clear what interpretation is intended.)

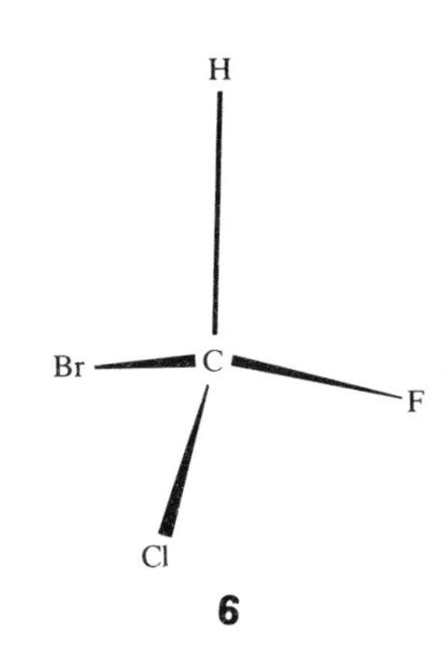

Table 2.7. The composition of some common groups

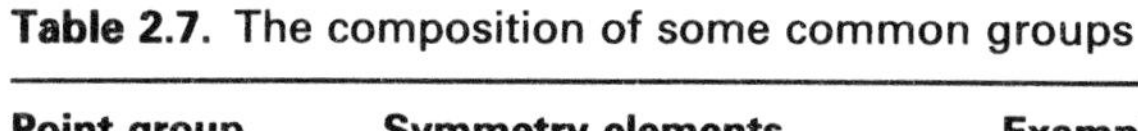

Point group	Symmetry elements	Examples
C_1	E	SiBrClFI
C_2	E, C_2	H_2O_2
C_s	E, σ	NHF_2
C_{2v}	$E, C_2, 2\sigma_v$	H_2O, SO_2Cl_2
C_{3v}	$E, C_3, 3\sigma_v$	NH_3, PCl_3, $POCl_3$
$C_{\infty v}$	E, C_∞	CO, HCl, OCS
D_{2h}	$E, 3C_2, 2\sigma_v, \sigma_h, i$	N_2O_4, B_2H_6
D_{3h}	$E, C_3, 3C_2, 3\sigma_v, \sigma_h$	BF_3, PCl_5
D_{4h}	$E, C_4, C_2, i, S_4, \sigma_h, \ldots$	XeF_4, *trans*-MA_4B_2
$D_{\infty h}$	$E, C_\infty, \infty\sigma_v, \ldots$	H_2, CO_2, C_2H_2
T_d	$E, 3C_2, 4C_3, 6\sigma, 3S_4$	CH_4, $SiCl_4$
O_h	$E, 6C_2, 4C_3, 4S_6, 3S_4, i \ldots$	SF_6

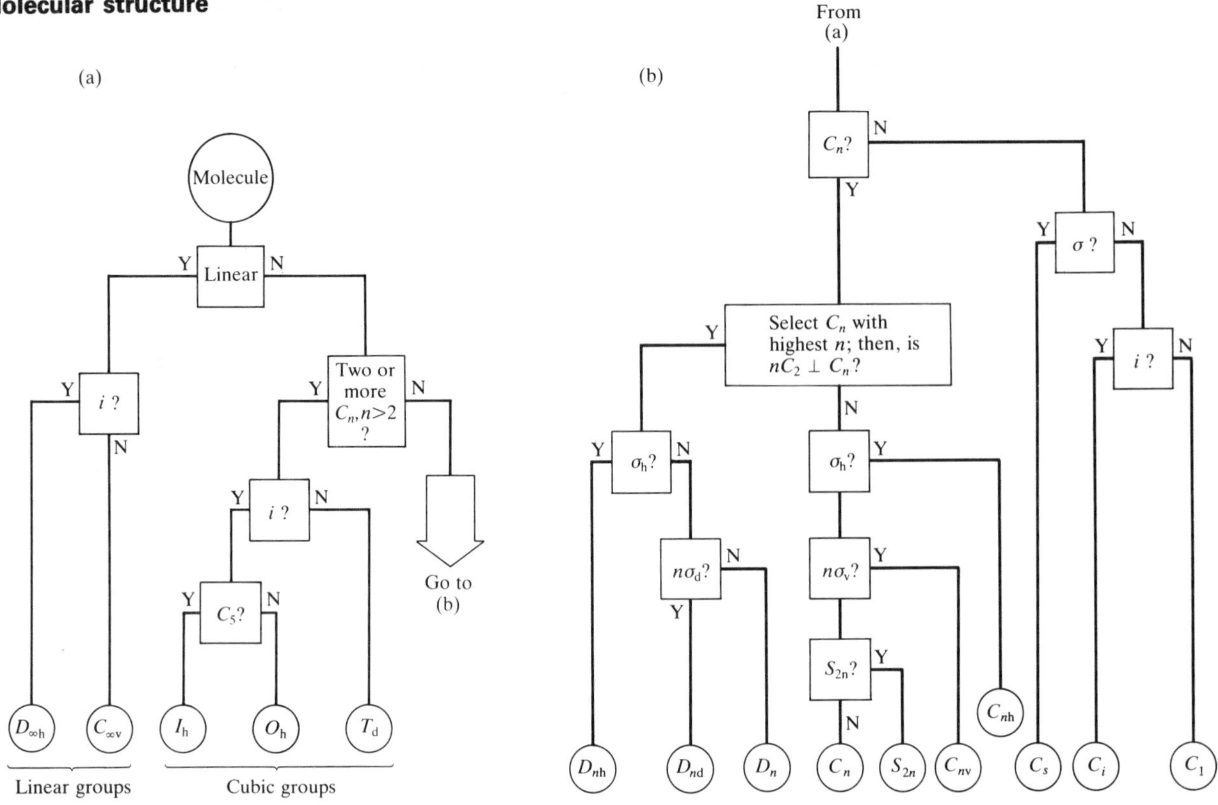

Fig. 2.12 The decision tree for identifying a molecular point group. After passing through part (a), go to part (b) if necessary.

Example 2.7: *Identifying the point group of a molecule*

To what point groups do H_2O and NH_3 belong?

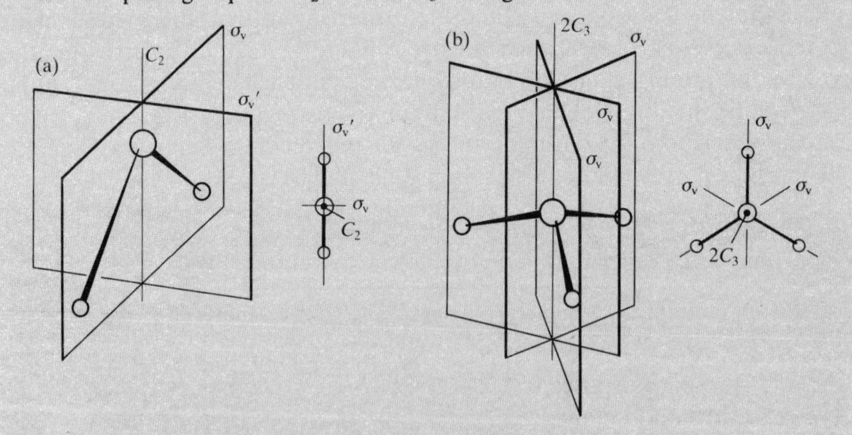

Fig. 2.13 The symmetry elements of (a) H_2O and (b) NH_3. The diagrams on the right are views from above and summarize the diagrams to their left.

Answer. The symmetry elements are shown in Fig. 2.13.

(a) H_2O possesses the identity (E), a twofold rotation axis (C_2), and two vertical mirror planes ($2\sigma_v$). The set of elements (E, C_2, $2\sigma_v$) corresponds to the group C_{2v}.

(b) NH_3 possesses the identity (E), a threefold axes (C_3). and three vertical mirror planes ($3\sigma_v$). The set of elements (E, C_3, $3\sigma_v$) corresponds to the group C_{3v}.

Exercise. Identify the point groups of (a) BF_3, a trigonal planar molecule and (b) the tetrahedral SO_4^{2-} ion.

Linear molecules with a center of symmetry (H_2, CO_2, HC≡CH, **7**) belong to the point group $D_{\infty h}$. A molecule that is linear but has no center of symmetry (HCl, OCS, NNO, **8**) belongs to the point group $C_{\infty v}$. Tetrahedral (T_d) and octahedral (O_h) molecules (Fig. 2.14), which are of great importance in coordination chemistry, have more than one principal axis of symmetry: a tetrahedral CH_4 molecule, for instance, has four C_3 axes, one along each C—H bond. A closely related group, the icosahedral group I_h characteristic of the icosahedron (Fig. 2.14c), is important for boron compounds. The three groups are quite easy to recognize: a regular tetrahedron has four equilateral triangles for its faces, an octahedron has eight, and an icosahedron has 20.

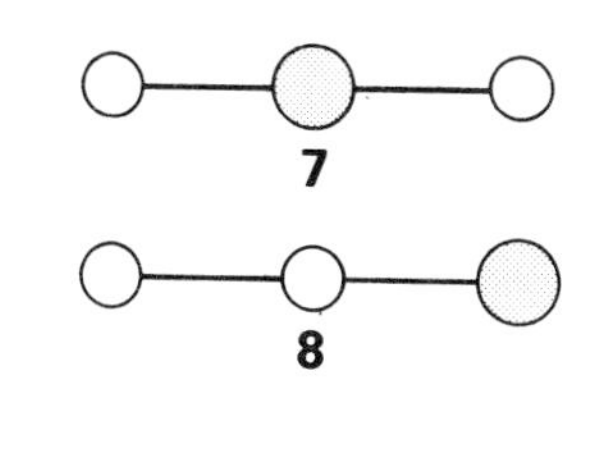

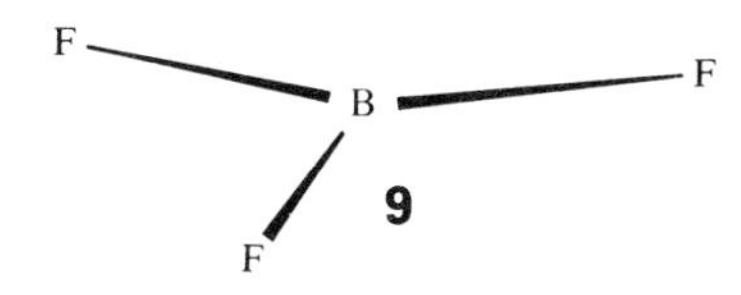

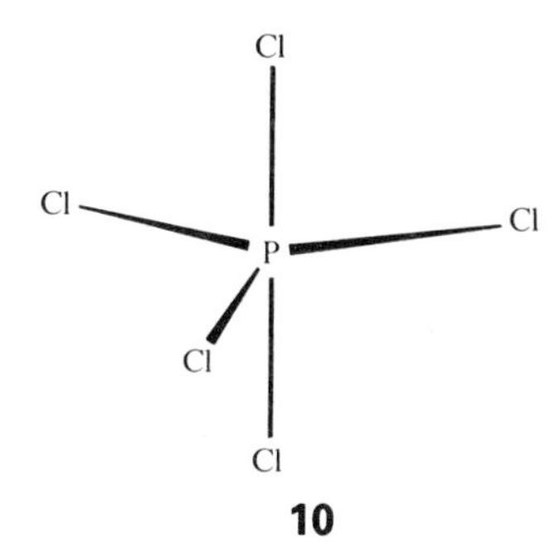

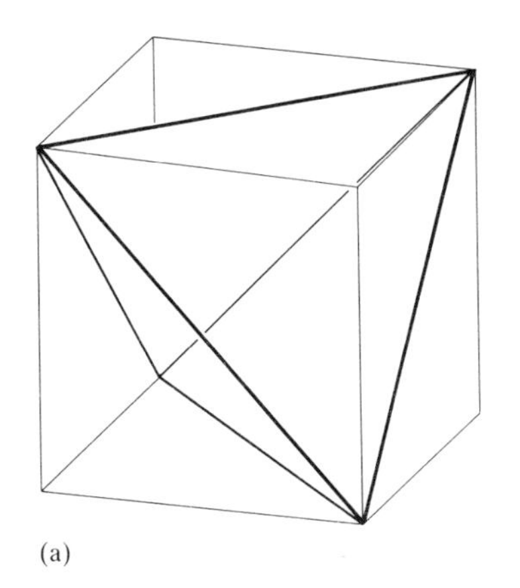

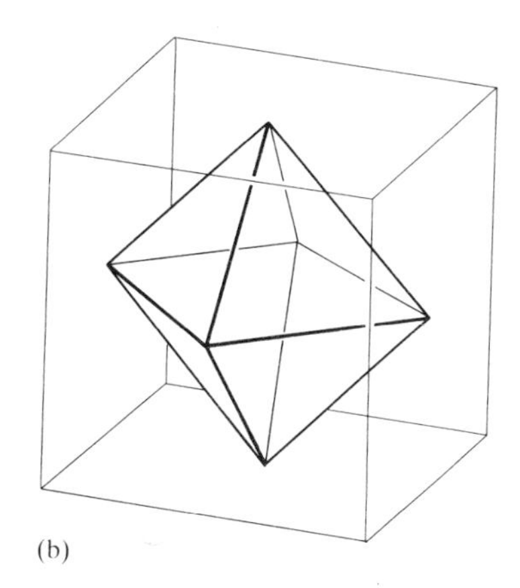

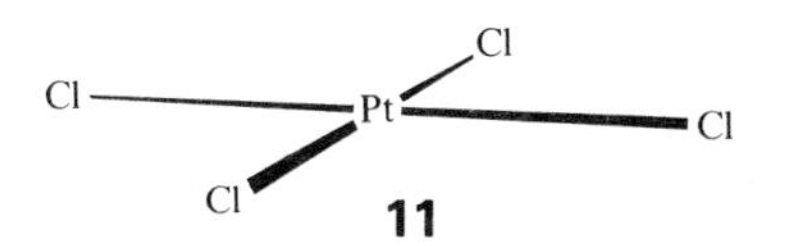

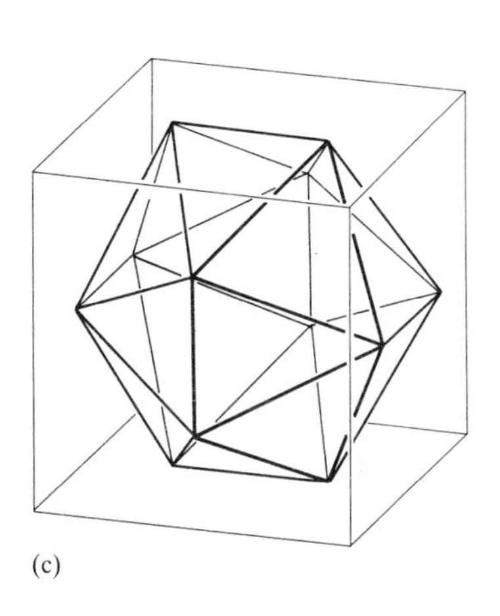

Fig. 2.14 Shapes with the symmetries of the groups (a) T_d, the tetrahedron, (b) O_h, the octahedron, and (c) I_h, the icosahedron. They are all closely related to the symmetries of a cube.

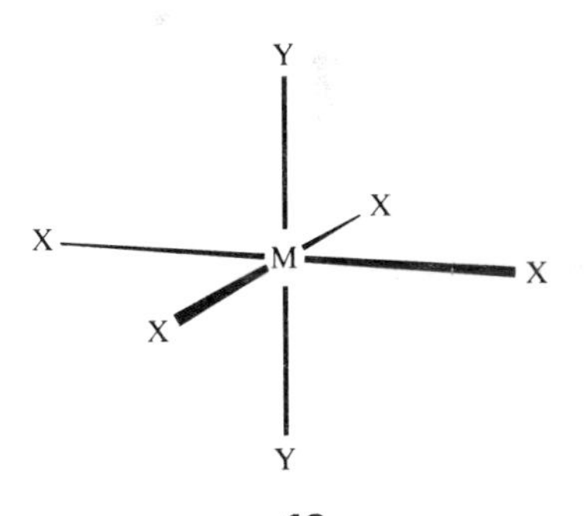

For our present purposes, the most important groups are C_{2v} and C_{3v} and the tetrahedral and octahedral groups. We shall also sometimes encounter $C_{\infty v}$ and $D_{\infty h}$ for linear molecules, D_{3h} for trigonal planar molecules (such as BF_3, **9**) and trigonal bipyramidal molecules (such as PCl_5, **10**), and D_{4h} for square-planar molecules (**11**) and 'octahedral' molecules with two substituents opposite each other, as in (**12**). This last example shows that the point-group classification of a molecule is more precise than the casual use of the term 'octahedral' or 'tetrahedral'. For instance, a molecule may loosely be called octahedral even if it has six different groups attached to the central atom. However, it only belongs to the octahedral point group O_h if all six groups are identical (**13**).

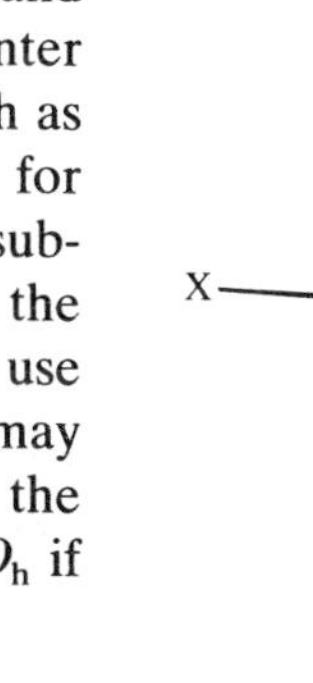

2.4 Applications of symmetry

A simple application of the group classification of molecules is the prediction of whether a compound is polar or chiral. In fact, in most cases we do not need to use the sledgehammer of group theory to decide whether a molecule has these characteristics, but these examples will give an impression of the approach that can be adopted when the result is less obvious.

Polar molecules

A **polar molecule** is a molecule with a permanent electric dipole moment. A molecule cannot have a permanent electric dipole perpendicular to any mirror plane: this must be so, because a dipole perpendicular to a mirror plane would require equal and opposite charges either side of that plane (Fig. 2.15a), in which case the plane would not be a symmetry element. Similarly, a molecule cannot have a permanent electric dipole perpendicular to any axis of symmetry (Fig. 2.15b). Such a dipole would require opposite charges on either side of the axis, which is inconsistent with the axis being a symmetry element.

Since a symmetry axis rules out a dipole in a perpendicular direction, it follows that any molecule that has both a C_n axis and a perpendicular C_2 axis or σ_h plane cannot have a dipole in *any* direction (Fig. 2.15c). Any molecule belonging to a D point group is of this kind, so any such molecule must be nonpolar; BF_3 (D_{3h}) is therefore nonpolar. Likewise, molecules belonging to the tetrahedral, octahedral, and icosahedral groups must be nonpolar; hence SF_6 (O_h) and CCl_4 (T_d) are nonpolar.

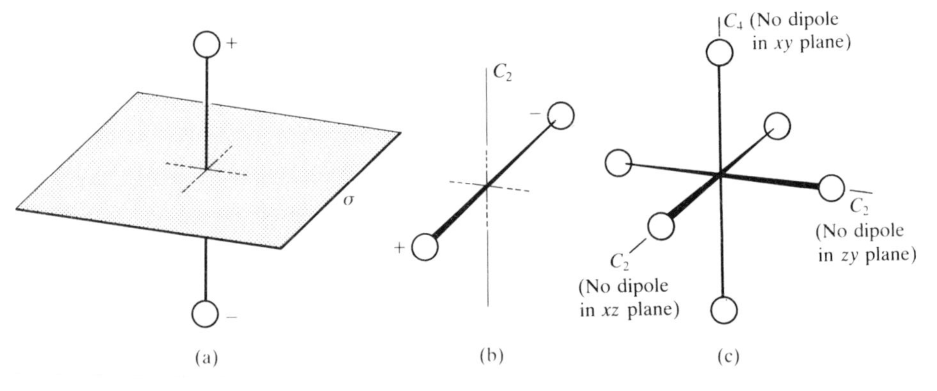

Fig. 2.15 (a) The presence of a mirror plane rules out a dipole in the direction shown here. (b) The presence of a symmetry axis rules out any dipole in the perpendicular plane. (c) If both a C_n symmetry axis and a perpendicular C_2 axis are present, a dipole is ruled out in all directions.

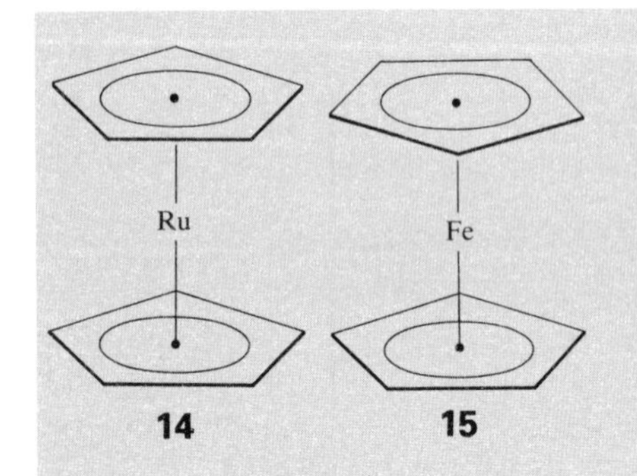

Example 2.8: *Judging whether a molecule can be polar*

The ruthenocene molecule (**14**) is a pentagonal prism with the Ru atom sandwiched between two C_5H_5 rings. Is it polar?

Answer. We decide whether the point group is D or cubic, because in neither case can it have a permanent electric dipole. Reference to the chart in Fig. 2.12 shows that a pentagonal prism belongs to the point group D_{5h}. Therefore, the molecule must be nonpolar.

Exercise. A conformation of the ferrocene molecule (**15**) that lies 4 kJ mol^{-1} above the lowest energy conformation is a pentagonal antiprism. Is it polar?

Chiral molecules

A **chiral molecule** (from the Greek word for hand) is a molecule that is distinguishable from its mirror image in the same way that left and right hands are distinguishable. Chiral molecules are optically active and rotate the plane of polarized light. A chiral molecule and its mirror image partner are called **enantiomers** (from the Greek word for 'both').

The group-theoretical criterion of whether or not a molecule is superimposable on its mirror image is that it should not have an improper-rotation axis S_n; if it has an S_n axis, then it is not chiral. However, we must be alert for improper-rotation axes that are present in disguise. Thus, we have seen (Fig. 2.11) that a mirror plane alone is an S_1 axis (S_1) and a center of inversion is equivalent to an S_2 axis. Therefore, molecules with either a mirror plane or a center of inversion have improper-rotation axes and cannot be chiral.

Molecules without a center of inversion and a mirror plane (and hence with neither S_1 nor S_2 axes) are usually chiral, but it is important to verify that a higher-order improper-rotation axis is not present. For instance, the quaternary ammonium ion (**16**) has neither a mirror plane (S_1) nor an inversion center (S_2), but it does have an S_4 axis and so it is not chiral.

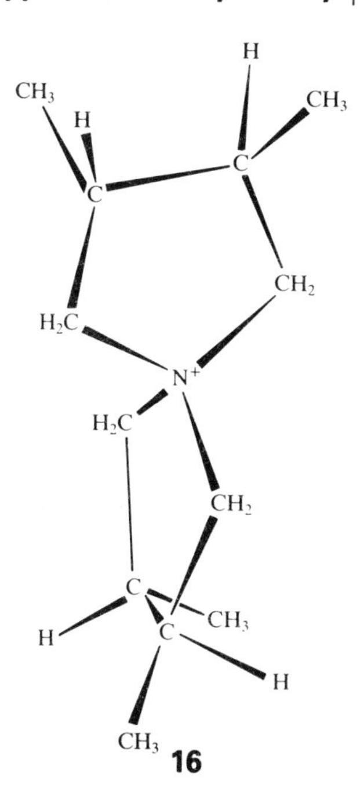

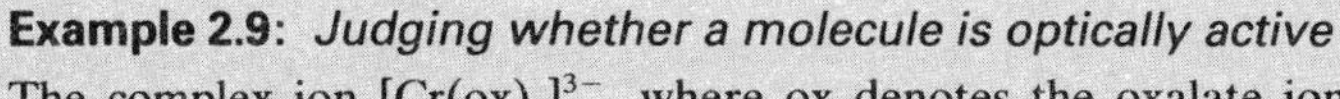
Example 2.9: *Judging whether a molecule is optically active*

The complex ion $[Cr(ox)_3]^{3-}$, where ox denotes the oxalate ion $[O_2CCO_2]^{2-}$ has the structure shown as (**17**). Is it chiral and hence optically active?

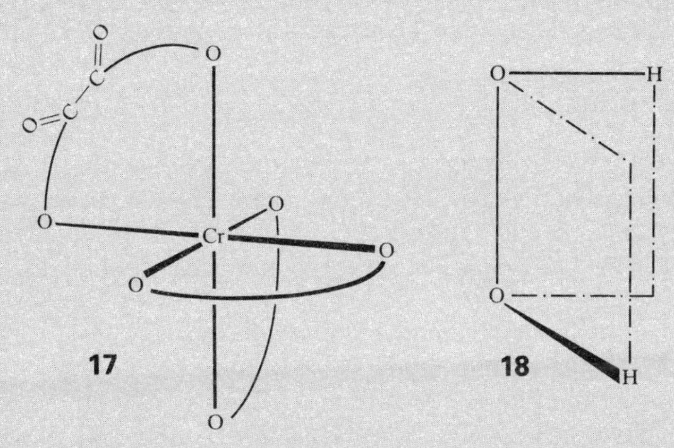

Answer. We begin by identifying the point group. Working through the chart in Fig. 2.12 shows that the ion belongs to the point group D_3. This group consists of the elements (E, C_3, $3C_2$) and hence does not contain an improper-rotation axis. It is chiral and hence may be optically active.

Exercise. Is the skew form of H_2O_2 (**18**) chiral?

We leave this preliminary introduction to group theory at this stage, but take it up again in the next chapter. So far we have seen how it sharpens our use of terms like 'symmetrical' and 'octahedral' and how it enables us to classify molecules according to certain aspects of their shapes. In later chapters we shall see that it also provides powerful procedures for constructing and analyzing the orbitals that electrons occupy in molecules, which is the subject we begin in the next section.

Molecular orbitals of diatomic molecules

We shall now generalize the *atomic* orbital description of atoms in a very natural way to **molecular orbitals**. These describe how electrons spread over all the atoms in a molecule and bind them together into a stable unit.

2.5 An introduction to the theory

We begin by considering **homonuclear diatomic molecules**, which are molecules consisting of two identical atoms, as in N_2, O_2, and F_2. The same ideas apply to diatomic ions, such as the superoxide ion, O_2^-, and the Hg(I) ion, Hg_2^{2+}. They are also readily extended to **heteronuclear diatomic molecules**, or molecules built from two different atoms and, as we see in Chapter 3, to polyatomic molecules and solids composed of huge numbers of atoms and ions. In parts of this section we shall include molecular fragments in the discussion, such as an SF fragment of the SF_6 molecule or an OO fragment of H_2O_2, since a number of the concepts we introduce for diatomic molecules also apply to pairs of atoms bound together as parts of larger molecules.

The approximation of the theory

As in the description of the electronic structures of atoms, we set out by making the **orbital approximation**. That is, we assume that a reasonable approximation (at least initially) is that the actual wavefunction Ψ of the N electrons in the molecule can be written as a product of N one-electron wavefunctions ψ:

$$\Psi(\boldsymbol{r}_1, \boldsymbol{r}_2, \ldots, \boldsymbol{r}_N) = \psi(\boldsymbol{r}_1)\psi(\boldsymbol{r}_2) \ldots \psi(\boldsymbol{r}_N)$$

This implies that electron 1 is described by the wavefunction $\psi(\boldsymbol{r}_1)$, electron 2 by the wavefunction $\psi(\boldsymbol{r}_2)$, and so on. These one-electron wavefunctions are the molecular orbitals of the theory. As for atoms, the square of a one-electron wavefunction gives the probability distribution for that electron in the molecule.

The next approximation is motivated by noticing that when an electron is close to the nucleus of one atom, its wavefunction closely resembles an atomic orbital of that atom. For instance, when an electron is close to the nucleus of a hydrogen atom, its wavefunction is like a 1*s* orbital of that atom. Therefore we may suspect that we can construct a reasonable first approximation to the molecular orbital by adding together atomic orbitals contributed by each atom. This is called the **linear combination of atomic orbitals** (LCAO) approximation, a 'linear combination' being simply a sum with various weighting coefficients. In the most elementary form of molecular orbital theory, only the valence shell atomic orbitals are used to form molecular orbitals. Thus the molecular orbitals of H_2 are approximated by adding together two hydrogen 1*s* orbitals, one from each atom:

$$\psi = c_{\mathrm{A}}\phi_{1s}(\mathrm{A}) + c_{\mathrm{B}}\phi_{1s}(\mathrm{B})$$

In this case the **basis set**, the atomic orbitals from which the molecular orbital is built, consists of two hydrogen 1*s* orbitals, one on atom A and the other on atom B. The coefficients c in the linear combination show the extent to which each atomic orbital contributes to the molecular orbital:

the greater the value of c^2, the greater the contribution of that orbital to the molecular orbital.

We shall show below that the linear combination that models the lowest energy exact solution of the Schrödinger equation has equal contributions from each 1*s* orbital ($c_A^2 = c_B^2$). As a result, electrons in this orbital are equally likely to be found near each nucleus. Specifically, the coefficients are $c_A = c_B = 1$, and

$$\psi_+ = \phi_{1s}(A) + \phi_{1s}(B) \qquad (4a)$$

(We disregard normalization throughout the chapter.[4]) We shall also show that the combination that models the next higher energy orbital also has equal contributions from each 1*s* orbital ($c_A^2 = c_B^2$) but that the coefficients have opposite signs ($c_A = +1$, $c_B = -1$):

$$\psi_- = \phi_{1s}(A) - \phi_{1s}(B) \qquad (4b)$$

The relative signs of the coefficients in molecular orbitals play a very important role in determining the energies of the orbitals, for, as we shall now explain, they tell us whether atomic orbitals interfere constructively or destructively in different regions of the molecule, and hence lead to an accumulation or reduction of electron density in those regions.

Two further preliminary points should be noted. One is that we see from this discussion that *two* molecular orbitals may be constructed from *two* atomic orbitals. In due course, we shall see that if we start out with a basis set of N atomic orbitals, then we can construct N molecular orbitals from them. Secondly, as in atoms, each molecular orbital may be occupied by two electrons with paired spins. This aspect of the Pauli principle will soon be seen to play a major role in determining the electronic structures of molecules, just as it does in atoms.

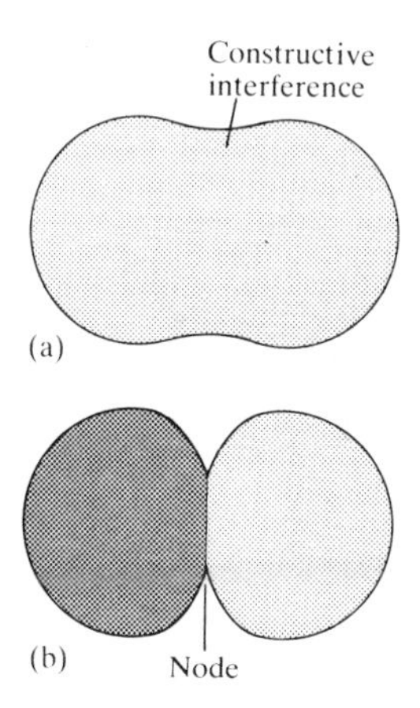

Fig. 2.16 (a) The enhancement of electron density in the internuclear region arising from the constructive interference between the atomic orbitals on neighboring atoms. (b) The destructive interference that leads to a node when the overlapping orbitals have opposite phases.

Bonding and antibonding orbitals

The orbital ψ_+ is a **bonding orbital**. It is so called because the energy of the molecule is lowered if it is occupied by electrons. In elementary discussions of the chemical bond, the bonding character of ψ_+ is ascribed to the constructive interference between the two atomic orbitals and the enhanced amplitude this interference causes between the two nuclei (Fig. 2.16a). An electron that occupies ψ_+ has an enhanced probability of being found in the internuclear region, and can interact strongly with both nuclei. Hence **orbital overlap**, the spreading of one orbital into the region occupied by another, leading to enhanced probability of electrons being found in the internuclear region, is taken to be the origin of the strength of bonds.[5]

[4] The wavefunction ψ_+ (like all wavefunctions) must have the property that

$$\int \psi_+^2 \, d\tau = 1$$

if the Born interpretation is to be applicable. This is achieved by multiplying ψ_+ by a constant N called the 'normalization constant'. For clarity in the nonmathematical presentation that we are adopting, we omit N, because the form of the orbitals is then much easier to see.

[5] For a straightforward, readable discussion of conventional wisdom on the origin of the strengths of bonds according to molecular orbital theory, see R. L. DeKock and H. B. Gray, *Chemical structure and bonding*. Benjamin-Cummings, Menlo Park (1980).

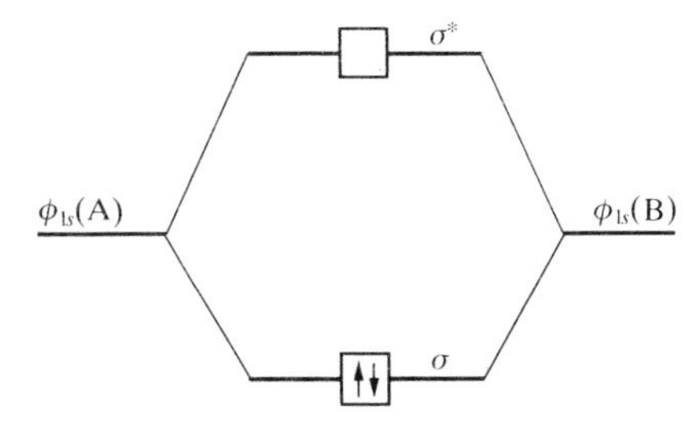

Fig. 2.17 The molecular orbital energy level diagram for H_2 and analogous molecules.

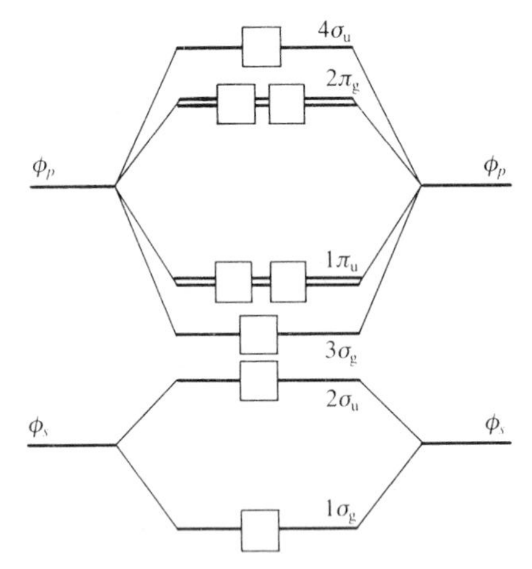

Fig. 2.18 The molecular orbital energy level diagram for the Period 2 homonuclear diatomic molecules without taking into account *sp* mixing.

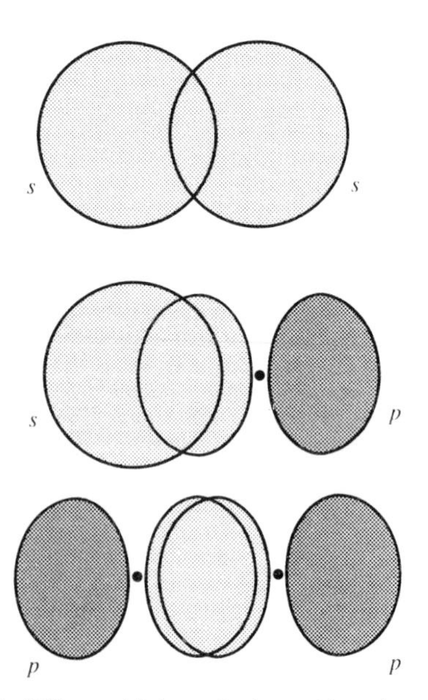

Fig. 2.19 σ orbitals can be formed in various ways, which include (a) *s*, *s* overlap, (b) *s*, *p* overlap, and (c) *p*, *p* overlap, with the *p* orbitals directed along the internuclear axis in each case.

The orbital ψ_- is an **antibonding orbital**. It is so called because if it is occupied, the energy of the molecule is higher than for the two separated atoms. The greater energy of an electron in this orbital reflects the destructive interference between the two atomic orbitals, which cancels their amplitudes and gives rise to a **nodal plane** between the two nuclei (Fig. 2.16b). Electrons that occupy ψ_- are largely excluded from the internuclear region and are forced to occupy less favorable locations. We shall see later that it is generally true that the energy of a molecular orbital in a polyatomic molecule is greater the more internuclear nodes it has, for this reflects an increasingly complete exclusion of electrons from the regions between nuclei.

The Pauli exclusion principle limits the number of electrons that can occupy any molecular orbital to no more than two, and those two electrons must be paired ($\uparrow\downarrow$). The Pauli principle is the origin of the importance of the electron pair in bond formation: two electrons is the maximum number that can occupy an orbital that contributes to the stability of the molecule. The H_2 molecule, for example, is stable because two electrons can occupy the orbital ψ_+ and both can contribute to the lowering of its energy (Fig. 2.17). One electron is less effective (but H_2^+ is known as a transient gas-phase ion) and three are also less effective because the third electron would have to occupy the antibonding orbital ψ_- and so destabilize the molecule. With four electrons, when two can occupy ψ_+ and two can occupy ψ_-, the antibonding effect of the electrons in ψ_- completely overcomes the bonding effect of the pair in ψ_+, and there is no net bonding. Hence a four-electron molecule with only 1*s* orbitals available for bond formation, such as He_2, cannot be expected to be stable.

2.6 Homonuclear diatomic molecules

The **minimal basis set** is the smallest set of atomic orbitals from which useful molecular orbitals can be built. In Period 2 diatomic molecules, the minimal basis set consists of the one valence *s* and three valence *p* orbitals on each atom, giving eight atomic orbitals in all. As in hybridization, where we saw that *N* orbitals gave *N* hybrids, we cannot simply 'lose' orbitals, and quantum theory shows that *N* atomic orbitals give rise to *N* molecular orbitals. We shall now show how the minimal basis set of eight valence-shell atomic orbitals is used to construct eight molecular orbitals. Then we shall use the Pauli principle to predict the ground-state electron configurations of the molecules.

The orbitals

The energies of the atomic orbitals are shown on either side of Fig. 2.18. We can form **σ orbitals** by allowing overlap between atomic orbitals that have cylindrical symmetry around the internuclear axis; this axis is conventionally labeled *z*. The notation σ is the analog of *s* for atomic orbitals, and signifies that the orbital has cylindrical symmetry (and so, when viewed along the internuclear axis, looks like an *s* orbital). Both orbitals shown in Fig. 2.16 are σ orbitals. Orbitals that can form σ orbitals include the 2*s* and $2p_z$ orbitals on the two atoms, as we show in Fig. 2.19. As a first approximation, these can be divided into two sets: the 2*s* orbitals overlap to give bonding and antibonding σ orbitals, and so do the two $2p_z$

orbitals. This gives the σ energy levels shown in Fig. 2.18; we label the σ orbitals in order 1σ, 2σ, and so on, starting with the lowest energy. The remaining two $2p$ orbitals on each atom, which have a nodal plane through the x axis, overlap to give **π orbitals** (Fig. 2.20). Bonding and antibonding π orbitals can be formed from the overlap of the $2p_x$ and the $2p_y$ overlap (separately) and give rise to the two pairs of doubly-degenerate energy levels shown in Fig. 2.18.

We have drawn Fig. 2.18 on the assumption that the s and p_z orbitals contribute to separate σ orbitals. However, since they all have cylindrical symmetry with respect to the internuclear axis, in actuality they may mix to an extent that depends on their relative energies. When this mixing is taken into account, the energies of the σ orbitals shown in Fig. 2.18 are modified, and in some homonuclear diatomic molecules (specifically for B_2 through N_2), the arrangement shown in Fig. 2.21 is a better representation. The relative energies of the σ and π orbitals depend on the molecule, and varies across Period 2 as shown in Fig. 2.22.

In the case of homonuclear diatomics, it is convenient (particularly for spectroscopic discussions) to signify the symmetry of the molecular orbitals with respect to their inversion through the center of symmetry of the molecule. The orbital is designated g (for *gerade,* even) if it is identical under inversion and u (for *ungerade,* odd) if it changes sign. Thus a bonding σ orbital is g and an antibonding σ orbital is u (**19**). On the other hand, a bonding π orbital is u and an antibonding π orbital is g (**20**). We have included these labels in Figs. 2.18 and 2.21.

If we were considering complexes containing two neighboring d-metal atoms, as in $[Cl_4ReReCl_4]^{2-}$, we should also allow for the possibility of forming bonds from d orbitals. The d_{z^2} orbital has cylindrical symmetry with

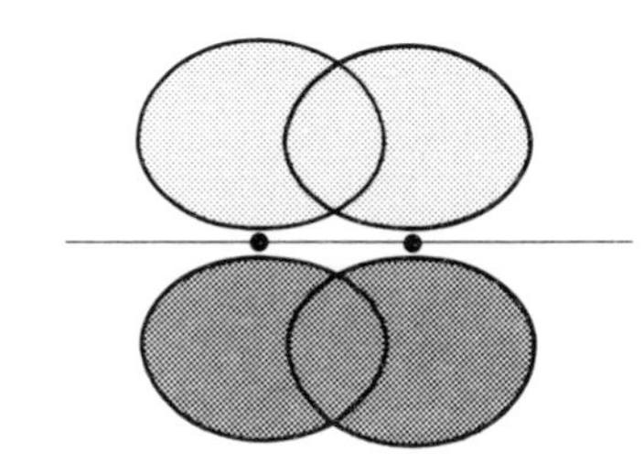

Fig. 2.20 Two p orbitals can overlap to form a π orbital.

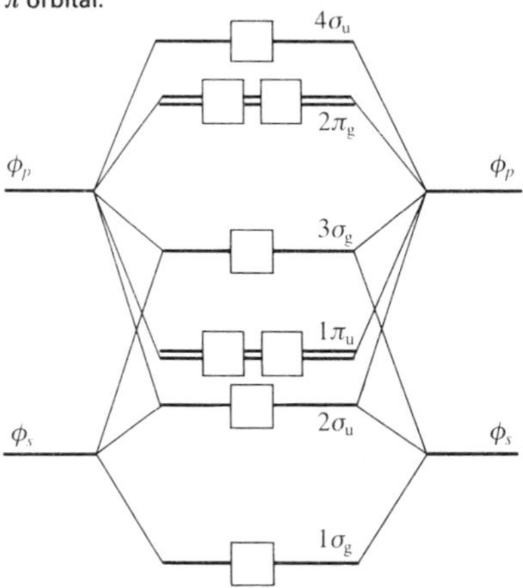

Fig. 2.21 The alternative ordering of orbital energies found in some homonuclear diatomic molecules.

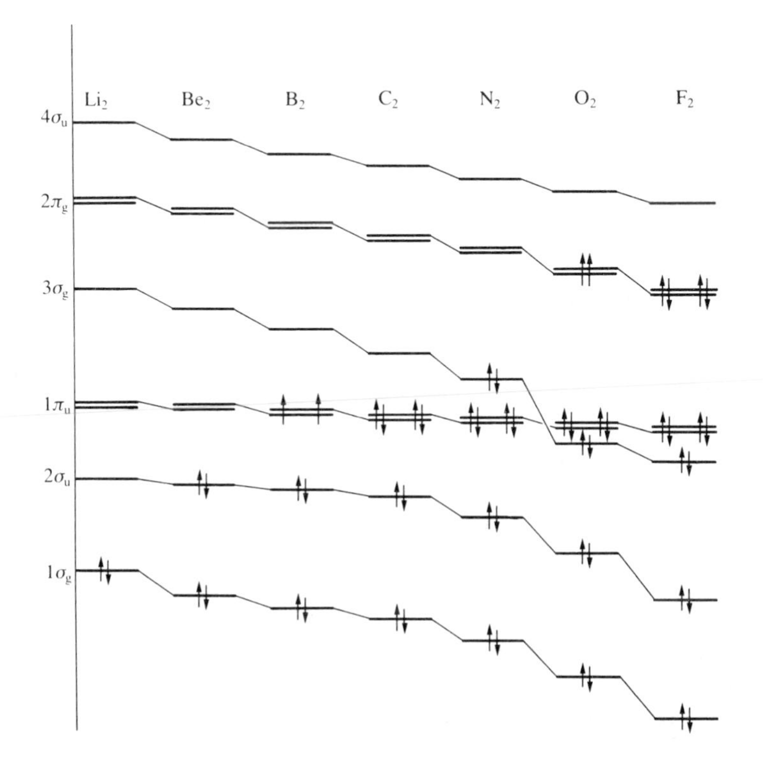

Fig. 2.22 The variation in orbital energies for Period 2 homonuclear diatomic molecules as far as F_2.

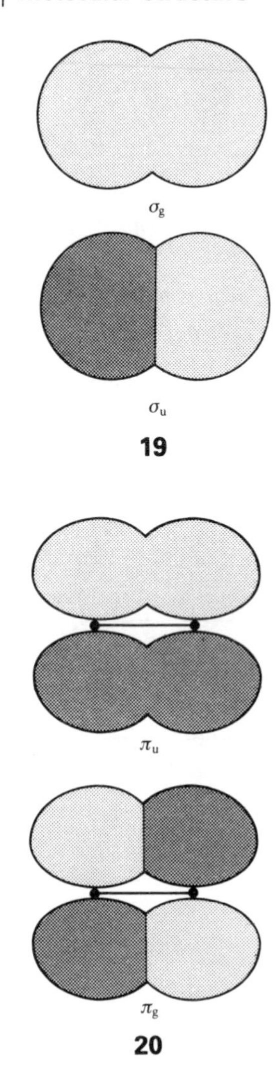

respect to the internuclear (z) axis, and hence can contribute to the σ orbitals that are formed from s and p_z orbitals. The d_{yz} and d_{zx} orbitals both look like p orbitals when viewed along the axis, and hence can contribute to the π orbitals formed from p_x and p_y. The new feature is $d_{x^2-y^2}$ and d_{xy}, which have no counterpart in the orbitals we have discussed up to now. They can overlap with matching orbitals on the other atom to give rise to doubly-degenerate pairs of bonding and antibonding **δ orbitals** (Fig. 2.23). As we shall see in Chapter 14, δ orbitals are important when discussing bonds between metal atoms and lead to descriptions of complexes in terms of quadruple bonds, as in $[Cl_4Re{\equiv}ReCl_4]^{2-}$.

The building-up principle for molecules

We use the building-up principle in conjunction with the energy level diagram in the same way as for atoms. The order of occupation of the orbitals is the order of increasing energy as depicted in Fig. 2.22. Each orbital can accommodate up to two paired electrons. If more than one orbital is available for occupation (because they happen to have identical energies), the orbitals are occupied separately. In that case, the electrons in the half-filled orbitals adopt parallel spins.

With very few exceptions, these rules lead to the actual ground state configuration of the Period 2 diatomics. For example, the electron configuration of N_2, with ten valence electrons, is

$$N_2 \quad 1\sigma_g^2 2\sigma_u^2 1\pi_u^4 3\sigma_g^2$$

Molecular orbital configurations are written like those for atoms: the orbitals are listed in order of increasing energy, and the number of electrons in each one is indicated by a superscript.

Example 2.10: *Writing electron configurations of diatomic molecules*

Give the ground-state electron configurations of the oxygen molecule, O_2, the superoxide ion, O_2^-, and the peroxide ion, O_2^{2-}.

Answer. The O_2 molecule has 12 valence electrons. The first 10 recreate the N_2 configuration except for the reversal of the order of the $1\pi_u$ and $3\sigma_g$ orbitals (see Fig. 2.22). Next in line for occupation are the doubly-degenerate $2\pi_g$ orbitals. The last two electrons enter these orbitals separately, and have parallel spins. The configuration is therefore

$$O_2 \quad 1\sigma_g^2 2\sigma_u^2 3\sigma_g^2 1\pi_u^4 2\pi_g^2$$

The O_2 molecule is interesting because the lowest energy configuration has two unpaired electrons in different π orbitals. Hence, O_2 oxygen is paramagnetic (tends to move into a magnetic field). The next two electrons can be accommodated in the $2\pi_g$ orbitals, giving

$$O_2^- \quad 1\sigma_g^2 2\sigma_u^2 3\sigma_g^2 1\pi_u^4 2\pi_g^3$$
$$O_2^{2-} \quad 1\sigma_g^2 2\sigma_u^2 3\sigma_g^2 1\pi_u^4 2\pi_g^4$$

(We are assuming that the orbital order does not change; this might not be the case.)

Exercise. Write the valence electron configuration for S_2^{2-} and Cl_2^-.

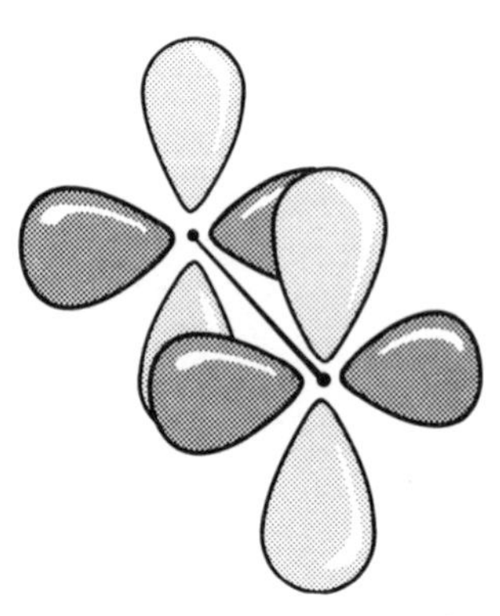

Fig. 2.23 The formation of δ orbitals from d-orbital overlap.

The **highest occupied molecular orbital** (the HOMO) is the molecular orbital that is occupied last according to the building-up principle. The

lowest unoccupied molecular orbital (the LUMO) is the next higher molecular orbital. Jointly, the two are the **frontier orbitals** of the molecules. In Fig. 2.22, the HOMO of F_2 is $2\pi_g$ and its LUMO is $4\sigma_u$; for N_2 the HOMO is $3\sigma_g$ and the LUMO is $2\pi_g$. We shall increasingly see that the frontier orbitals play special roles in structural and kinetic studies.

A final basic feature of molecular structure is the **molecular potential energy curve**, the plot of the energy of the molecule as the internuclear separation is changed (Fig. 2.24). The curve shown in the illustration is the typical shape for the ground state of a diatomic molecule. The energy of the molecule—the combined effect of all the electrons in the molecule—decreases as the two atoms approach and bonds begin to form. However, below a certain separation the energy begins to rise again as the bonding electrons cannot accumulate sufficiently in the internuclear region to overcome the internuclear repulsion. The minimum of the curve is the **equilibrium bond length** R_e of the molecule. The depth of the minimum is the **dissociation energy** D of the molecule, and the deeper the minimum the more strongly the atoms are bonded together. The steepness of the well shows how rapidly the energy of the molecule rises as the bond is stretched or compressed, and hence governs the force constant of the bond and (in combination with the masses of the atoms) the vibrational frequency of the molecule. The more confining the potential well, the greater the force constant and (for given masses) the higher the vibrational frequency.

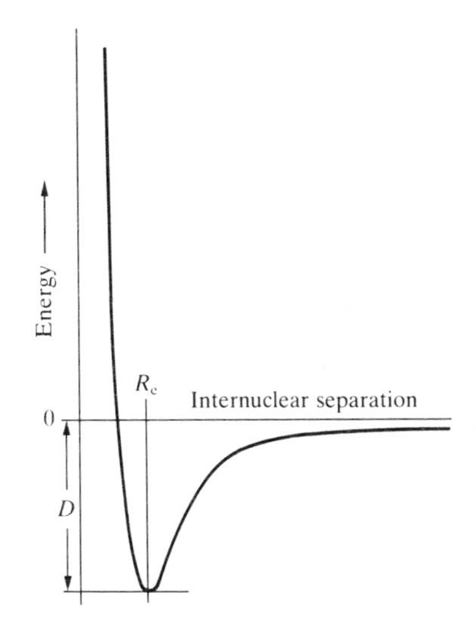

Fig. 2.24 A molecular potential energy curve showing how the total energy of a molecule varies as the internuclear separation changes.

Photoelectron spectroscopy

Photoelectron spectroscopy provides a very direct experimental portrayal of the features of molecular structure predicted by molecular orbital theory. In **ultraviolet photoelectron spectroscopy** (UV-PES), photon absorption results in the ejection of electrons from the molecule. In this technique, a sample is irradiated with 'hard' (high frequency) ultraviolet radiation (typically 21.2 eV radiation) and the kinetic energies of the **photoelectrons**—the ejected electrons—are measured. Since a photon of frequency ν has energy $h\nu$, if it expels an electron with ionization energy I from the molecule, the kinetic energy E_K of the photoelectron will be

$$E_K = h\nu - I \tag{5}$$

The lower in energy the electron lies initially (that is, the more tightly it is bound), the greater its ionization energy, and hence the lower its kinetic energy if ejected (Fig. 2.25). Since the peaks in a photoelectron spectrum correspond to the various kinetic energies of photoelectrons ejected from different energy levels of the molecule, the spectrum gives a vivid portrayal (and in many cases an experimental confirmation) of the molecular orbital energy level scheme for the molecule.

The UV-photoelectron spectrum of N_2 is shown in Fig. 2.26. We see that the photoelectrons have a series of discrete ionization energies close to 15.6 eV, 16.7 eV, and 18.8 eV. This strongly suggests a shell structure for the arrangement of electrons in the molecule. All the values are close to the ionization energy of the atom (14.5 eV). Since that energy corresponds to the removal of a valence electron, this suggests that when a molecule forms, the valence electrons arrange themselves into the molecular analog of shells with slightly different strengths of binding: these can be

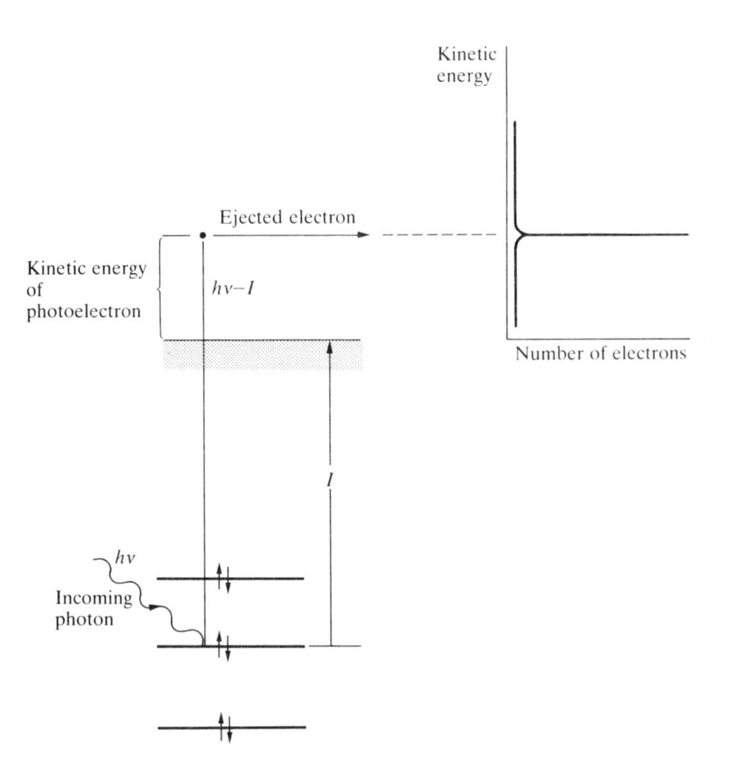

Fig. 2.25 The photoelectron experiment. The incoming photon has energy $h\nu$; when it ejects an electron from an orbital of ionization energy I, the electron acquires a kinetic energy $h\nu - I$. The spectrometer detects the numbers of photoelectrons with different kinetic energies.

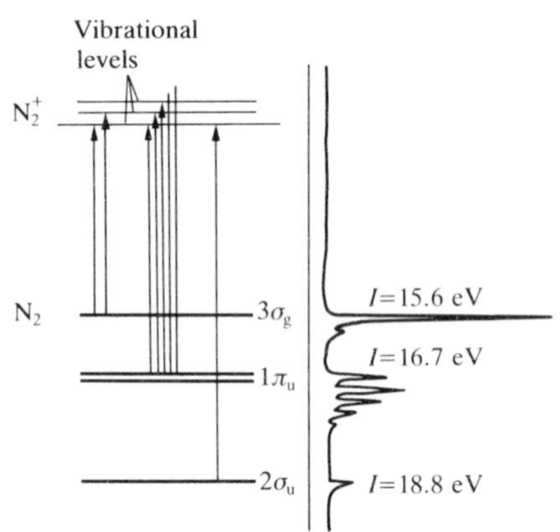

Fig. 2.26 The UV photoelectron spectrum of N_2. The fine structure in the spectrum arises from the excitation of vibrations in the cation formed by photoejection.

identified with the σ and π orbitals that we have discussed already. The lines corresponding to the lowest ionization energy (close to 15.6 eV) are photoelectrons ejected from the HOMO, which is the $3\sigma_g$ orbital. The lines at 16.7 eV are electrons ejected from the lower-energy 1π orbitals. The only other orbital accessible with 21.2 eV photons is the $2\sigma_u$ orbital, and it is responsible for the line at 18.8 eV.

The detailed structure of each group of lines in a UV-photoelectron spectrum is a result of the ionized molecule being left in a vibrationally excited state after it is ionized. Some of the energy of the incoming photon is used to excite the vibration, so less energy is available for the kinetic energy of the ejected electron. Each quantum of vibration that is excited results in a corresponding reduction in kinetic energy; hence the photoelectron appears with a series of kinetic energies separated by the energy of each vibrational excitation that has occurred.

The vibrational structure in a photoelectron spectrum can be very helpful in assigning the origin of a line. For example, extensive vibrational structure occurs when the ejected electron comes from an orbital in which it exerts a strong force on the nuclei, for the loss of the electron has a pronounced effect on the forces experienced by the nuclei. If the orbital is neutral in this sense, photoejection leaves the force field largely undisturbed and there is little vibrational structure.

2.7 Heteronuclear diatomic molecules

The molecular orbitals of heteronuclear diatomic molecules differ from those of homonuclear diatomic molecules by having unequal contributions from each atomic orbital. Each molecular orbital has the form

$$\psi = c_A\phi(A) + c_B\phi(B)$$

as in homonuclear molecules, but the coefficients are no longer equal in magnitude. If c_A^2 is larger than c_B^2, the orbital is composed principally from $\phi(A)$ and an electron that occupies the molecular orbital is more likely to be found near atom A than atom B. The opposite is true for a molecular orbital in which c_B^2 is larger than c_A^2.

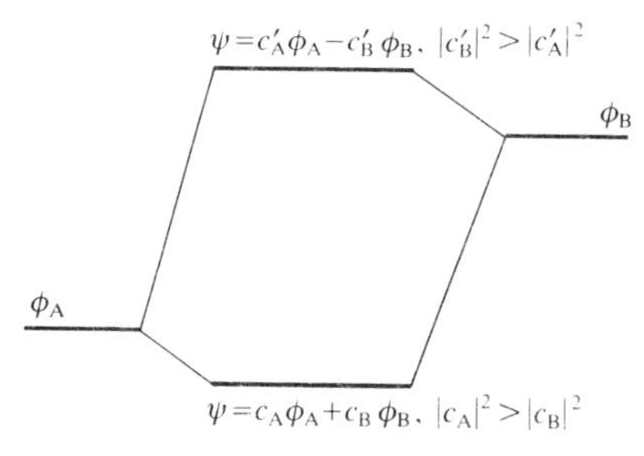

Fig. 2.27 When two atomic orbitals with different energies overlap, the lower molecular orbital is primarily composed of the lower atomic orbital, and vice versa. Moreover, the shift in energies of the two levels is less than if they had had the same energy in the atoms.

Molecular orbitals between different atoms

In general, the atomic orbital with lower energy (the one contributed by the more electronegative atom) has the larger coefficient in the bonding molecular orbital (Fig. 2.27). This is because the bonding electrons are then likely to be found close to that atom and hence be in an energetically favorable location. The same atomic orbital has the smaller coefficient in the antibonding combination, so antibonding electrons are more likely to be close to the atom with the higher energy atomic orbitals (the less electronegative atom).

A second difference from homonuclear diatomic molecules stems from the energy mismatch between the two sets of orbitals on the two different atoms. We saw in Section 1.6 (Fig. 1.25) that two levels interacted less strongly as their energies diverged. This implies that the energy lowering as a result of the overlap of atomic orbitals on different atoms in a heteronuclear molecule is less pronounced than in a homonuclear molecule, when the orbitals have the same energies. However, this does not necessarily mean that AB bonds are weaker than AA bonds, since other factors (orbital size, closeness of approach) are also important. CO, for example, which is isoelectronic with N_2, has a higher bond enthalpy (1070 kJ mol^{-1}).

Hydrogen fluoride

As an illustration of these points, we can consider a simple heteronuclear diatomic molecule, HF. The valence orbitals available for molecular orbital formation are the 1*s* orbital of H and the 2*s* and 2*p* orbitals of F; there are $1+7=8$ valence electrons to accommodate in the molecular orbitals.

The σ orbitals for HF can be constructed by allowing an H 1*s* orbital to overlap the 2*s* and $2p_z$ orbitals of F (*z* being the internuclear axis). These three atomic orbitals combine to give three σ molecular orbitals of the form

$$\psi = c_1\phi_{1s}(\mathrm{H}) + c_2\phi_{2s}(\mathrm{F}) + c_3\phi_{2p_z}(\mathrm{F}).$$

This leaves the $2p_x$ and $2p_y$ orbitals on the F atom unaffected as they have π symmetry and there are no valence H orbitals of that symmetry. The π orbitals are therefore **nonbonding orbitals**, neither bonding nor antibonding in character and confined to a single atom.

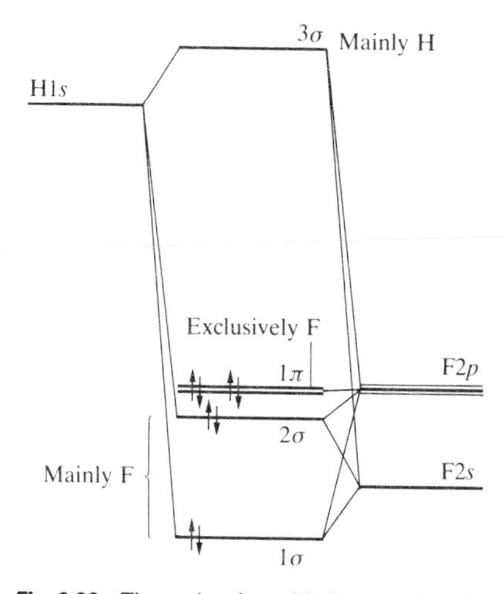

Fig. 2.28 The molecular orbital energy level diagram for HF. The relative positions of the atomic orbitals reflects the ionization energies of the atoms.

The resulting energy level diagram is shown in Fig. 2.28. The 1σ bonding orbital is predominantly F 2*s* orbital in character since that orbital lies so low in energy and it plays the major role in the bonding orbitals it forms. The 2σ orbital is largely nonbonding and confined mainly to the F atom. The bulk of its density lies on the opposite side of the F atom to the H atom, so it barely participates in bonding. The 3σ orbital is antibonding, and principally H 1*s* in character, since this atomic orbital has a relatively high energy and hence contributes predominantly to the high energy antibonding molecular orbital.

Two of the eight electrons enter the 1σ orbital, forming a bond between

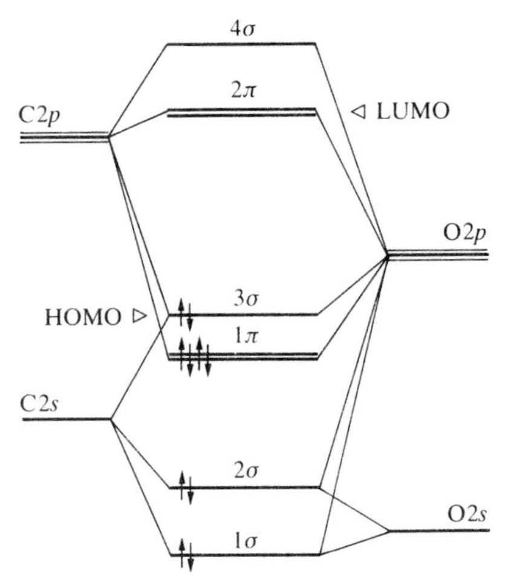

Fig. 2.29 The molecular orbital energy level diagram of CO.

the two atoms. Six more enter the 2σ and 1π orbitals; these are largely nonbonding and confined mainly to the F atom. This accounts for all the electrons, so the configuration of the molecule is $1\sigma^2 2\sigma^2 1\pi^4$. One important feature to note is that all the electrons occupy orbitals that are predominantly on the F atom. It follows that we can expect the HF molecule to be polar, with a partial negative charge on the F atom; this is consistent with the observed dipole moment of 1.91 D.[6]

Carbon monoxide

The molecular orbital energy level diagram for carbon monoxide (and the isoelectronic CN^- ion) is a somewhat more complicated example than HF because both atoms have $2s$ and $2p$ orbitals that can participate in the formation of σ and π orbitals. The energy level diagram is shown in Fig. 2.29. The ground configuration is

$$\text{CO} \quad 1\sigma^2 2\sigma^2 1\pi^4 3\sigma^2$$

The HOMO in CO is 3σ, which is a largely nonbonding lone pair on the C atom. The LUMO is the doubly degenerate pair of antibonding π orbitals, with mainly C $2p$ orbital character. This combination of frontier orbitals is very significant, and we shall see that it is one reason why metal carbonyls are such a characteristic feature of d element chemistry (Chapter 16).

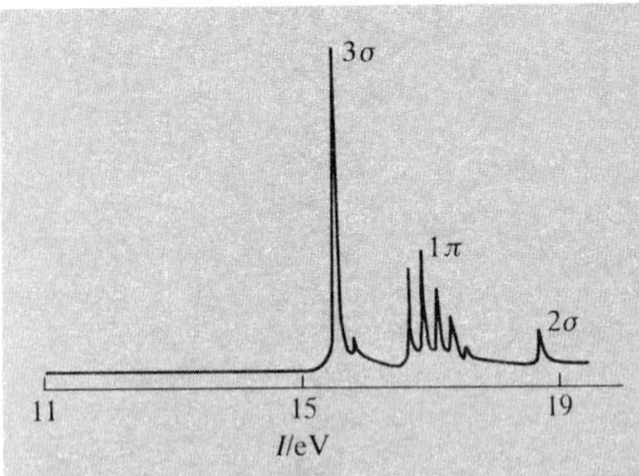

Fig. 2.30 The UV photoelectron spectrum of CO.

Example 2.11: *Interpreting a photoelectron spectrum*

Assign the lines in the photoelectron spectrum of CO shown in Fig. 2.30.

Answer. We can argue by analogy with the spectrum of the isoelectronic N_2 molecule. The lowest ionization energy (14 eV) is for the most weakly-bound electrons, those in the 3σ orbitals. The band at 17 eV are for the next lower electrons, those in the 1π orbitals. The band at 20 eV are electrons from the deeper 2σ orbital.

Exercise. Predict the photoelectron spectrum of the SO molecule.

Although the difference in electronegativities between C and O is large, the experimental value of the electric dipole moment of CO (0.1 D) is small; moreover, the negative end of the dipole is on the C atom despite its being the less electronegative atom. These difficulties stem from the fact that the lone pairs and bonding pairs have such a complex distribution. It is wrong to conclude that because the bonding electrons are mainly on the O atom, O must be the negative end of the dipole, since this ignores the balancing effect of the lone pair on the C atom. Electronegativity arguments are particularly unreliable when antibonding orbitals are occupied in the molecule.

[6] The SI unit for reporting dipole moments, the product of charge in coulombs and distance in meters, is C m. However, it proves more convenient to adopt the (non-SI) Debye unit D, which was originally defined in terms of electrostatic units. All we need is the conversion

$$1\,\text{D} = 3.336 \times 10^{-30}\,\text{C m}$$

A dipole consisting of charges e and $-e$ separated by 1 Å has a dipole moment of 4.8 D. In most cases, dipole moments are somewhat smaller since they arise from partial charges; typical values are close to 1 D.

2.8 Bond properties

We now bring the discussion full circle, and show that although seemingly very different, molecular orbital theory does elucidate many features of the Lewis description of molecules. We have already seen the origin of the importance of the electron pair: two electrons is the maximum number that can occupy a bonding orbital and hence contribute to a chemical bond. We now extend this concept by introducing the 'bond order'.

Bond order

The **bond order** (B.O.) is an attempt to judge the number of bonds between pairs of atoms in a molecule and to provide a link with the Lewis description. The bond order in effect identifies a shared electron pair as counting as a 'bond' and an electron pair in an antibonding orbital as an 'antibond' between two atoms. More precisely, it is defined as

$$\text{B.O.} = \tfrac{1}{2} \times (\text{number of electrons in bonding orbitals} - \text{number of electrons in antibonding orbitals})$$

For example, for N_2, $1\sigma_g^2 2\sigma_u^2 1\pi_u^4 3\sigma_g^2$ since σ_g and π_u orbitals are bonding and σ_u are antibonding,

$$\text{B.O.} = \tfrac{1}{2} \times (2 + 4 + 2 - 2) = 3$$

The bond order of 3 corresponds to a triply-bonded molecule, which is in line with the Lewis structure N≡N. The high bond order is reflected in the high enthalpy change accompanying dissociation of the molecule, +946 kJ mol^{-1}, one of the highest for any molecule. The bond order of the isoelectronic CO molecule is also 3, in accord with the Lewis structure C≡O.

Electron loss from N_2 leads to the formation of a molecular cation N_2^+, in which the bond order is reduced to 2.5. This is accompanied by a corresponding decrease in bond strength (to 855 kJ mol^{-1}) and increase of the bond length from 1.09 Å for N_2 to 1.12 Å for N_2^+. The bond order of F_2 is 1. This is consistent with the Lewis structure F—F and describing the molecule as having a single bond.

The definition of bond order allows for the possibility that an orbital is only singly occupied. The bond order in O_2^-, for example, is 1.5, since three electrons occupy the $2\pi_g$ antibonding orbitals. Isoelectronic molecules and ions have the same bond order, so F_2 and O_2^{2-} both have bond order 1, and N_2, CO, and NO^+ all have bond order 3.

Bond lengths

As we have remarked, the equilibrium bond length R_e of the molecule is the internuclear separation at the minimum of the molecular potential energy curve (Fig. 2.24). This is not quite the same as the experimentally observed bond length, which is an average over the bond lengths explored during a molecular vibration. However, we shall ignore the difference, which is negligibly small for most purposes. Bond lengths may be measured spectroscopically (for instance, from the rotational structure of gas-phase spectra) or by X-ray diffraction of solids and electron diffraction of gases. Some typical bond lengths (which are of the order of 1 to 2 Å) are shown in Table 2.8.

Table 2.8. Bond lengths

	R_e/Å
H_2^+	1.06
H_2	0.74
HF	0.92
HCl	1.27
HBr	1.41
HI	1.60
N_2	1.09
O_2	1.21
F_2	1.44
Cl_2	1.99
I_2	2.67

Source: G. Herzberg, *Spectra of diatomic molecules*. Van Nostrand, Princeton, NJ, (1950).

Table 2.9. Covalent radii* (r_{cov}/Å)

H	0.37						
C	0.77 (1) 0.67 (2) 0.60 (3)	**N**	0.74 (1) 0.65 (2)	**O**	0.74 (1) 0.57 (2)	**F**	0.72
Si	1.18	**P**	1.10	**S**	1.04 (1) 0.95 (2)	**Cl**	0.99
Ge	1.22	**As**	1.21	**Se**	1.17	**Br**	1.14
		Sb	1.41	**Te**	1.37	**I**	1.33

*Values are for single bonds except where otherwise stated (in parentheses).

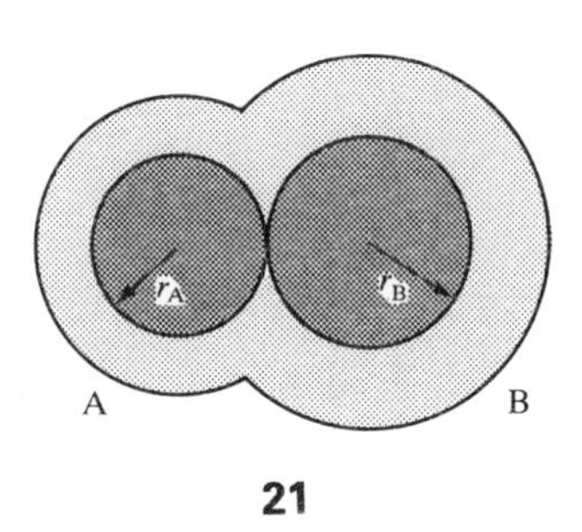

21

In general, the greater the atomic radii of the atoms, the greater the lengths of the bonds they form. To a reasonable first approximation, bond lengths can be expressed as the sum of characteristic contributions from each atom, which is called its **covalent radius** (**21**); some values are given in Table 2.9. With covalent radii available, we can predict, for example, that the bond length in HCl is 0.37 Å + 0.99 Å = 1.36 Å (the experimental value is 1.28 Å). Covalent radii vary through the periodic table in much the same way as metallic and ionic radii (Section 1.6), and for the same reasons, and are smallest close to F.

Bond strength

We have also seen that the depth of the minimum of the molecular potential energy curve below the energy of the separated atoms is the bond dissociation energy D (Fig. 2.24). In fact, this is not quite true, since we should allow for the zero-point vibrational energy of the bond, a characteristic that we deal with later. The greater the bond dissociation energy, the greater the strength of binding between the atoms forming the bond. This is an important feature when we come to consider the types of reaction that a molecule can undergo, for an atom that is tightly bound has a high activation energy for reaction.

When making use of bond strengths in thermodynamic cycles, it is often more convenient to use the **bond dissociation enthalpy**, the enthalpy change for the dissociation

$$\text{A—B}(g) \rightarrow \text{A}(g) + \text{B}(g)$$

The bond dissociation enthalpy differs from the bond energy by about RT, or 2.5 kJ mol^{-1} at room temperature. In line with our convention about ionization energies and enthalpies, we quote bond energies in eV and bond enthalpies in kJ mol^{-1}.

Bond dissociation enthalpies can be transferred from one molecule to another, at least to a good approximation. That is, a O—H bond, for example, has much the same strength whatever the identity of the rest of the molecule. The bond dissociation enthalpy of a polyatomic molecule is derived from the complete atomization of the molecule into gas-phase atoms,

$$H_2O(g) \rightarrow 2H(g) + O(g) \qquad \Delta H^{\ominus} = +920\ \text{kJ}$$

and the **mean bonding enthalpy** B is this quantity divided by the number of

Table 2.10. Mean bond enthalpies* (B/kJ mol^{-1})

	H	C	N	O	F	Cl	Br	I	S	P	Si
H	436										
C	412	348 (1) 612 (2) 518 (a)									
N	388	305 (1) 613 (2) 890 (3)	163 (1) 409 (2) 945 (3)								
O	463	360 (1) 743 (2)	157	146 (1) 497 (2)							
F	565	484	270	185	155						
Cl	431	338	200	203	254	242					
Br	366	276				219	193				
I	299	238				210	178	151			
S	338	259			496	250	212		264		
P	322									200	
Si	318			466							226

* Values are for single bonds except where otherwise stated (in parentheses). (a) Denotes aromatic.

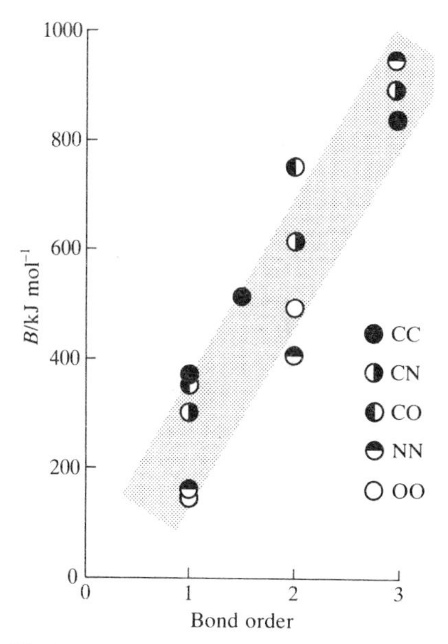

Fig. 2.31 Bond order/bond strength correlation.

bonds dissociated. In this example, the mean O—H bond enthalpy is one-half the atomization enthalpy, or 460 kJ mol^{-1}, since two O—H bonds are dissociated. We can use mean bond enthalpies to make rough estimates of reaction enthalpies when enthalpies of formation are unavailable. Some values are given in Table 2.10.

Example 2.12: *Making estimates using mean bond enthalpies*

Estimate the reaction enthalpy for the production of SF_6 from SF_4.

Answer. The reaction is

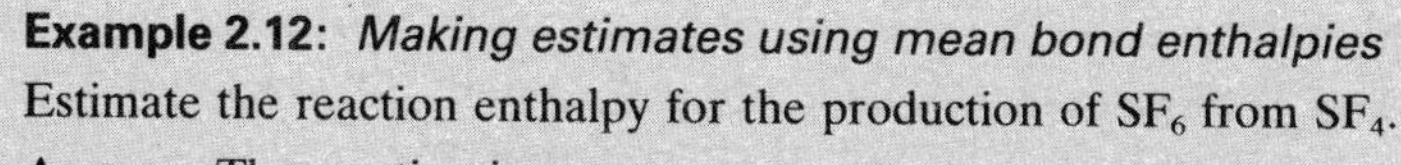

$$SF_4(g) + F_2(g) \rightarrow SF_6(g)$$

One F—F bond must be broken (energy investment 155 kJ) and two S—F bonds are formed (energy released 2 × 496 kJ). The net enthalpy change is therefore

$$\Delta H = +155\ \text{kJ} - 2 \times 496\ \text{kJ} = -837\ \text{kJ}$$

Hence, the reaction is strongly exothermic.

Exercise. Estimate the enthalpy of formation of H_2S from S_8 (a cyclic molecule) and H_2.

Bond correlations

The strengths and lengths of bonds correlate quite well with each other and with the bond order. Thus, the bond enthalpy between a given pair of atoms increases with bond order (Fig. 2.31) and the bond length decreases (Fig. 2.32) as the bond order increases.

The sensitivity of the bond length to the bond order varies with the elements. In Period 2 it is relatively weak in CC bonds, with the result that a C═C double bond is less than twice as strong as a C—C single bond. This

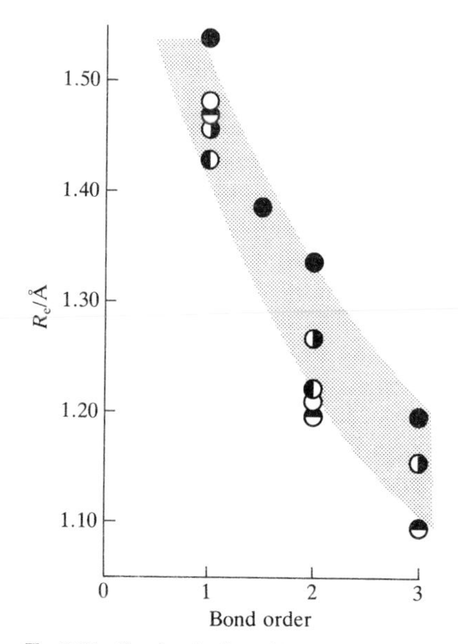

Fig. 2.32 Bond order/bond length correlation. The code for the points is the same as in Fig. 2.31.

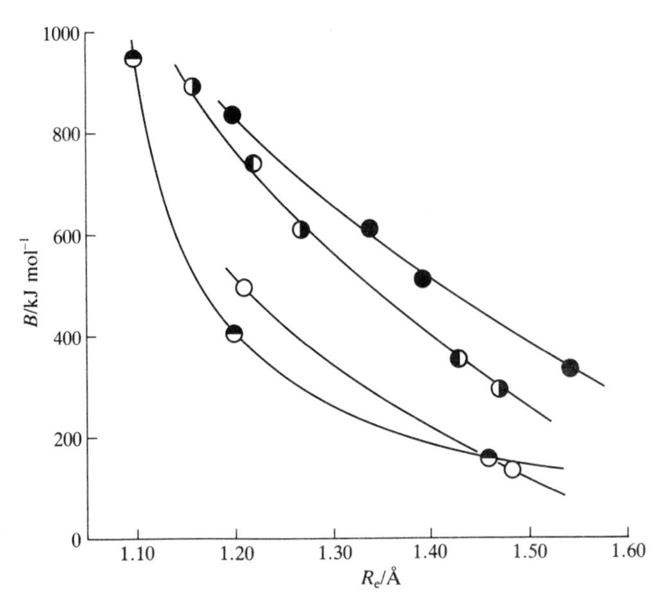

Fig. 2.33 Bond strength/bond length correlation. The code for the points is the same as in Fig. 2.31.

difference has profound consequences in organic chemistry, particularly for the reactions of unsaturated compounds. It implies that it is energetically favorable (but kinetically slow in the absence of a catalyst) for ethylene and acetylene to polymerize, a process which forms C—C single bonds at the expense of the appropriate numbers of multiple bonds.

Familiarity with carbon's properties must not be extrapolated without caution to the bonds between other elements. An N═N double bond ($409\ \mathrm{kJ\ mol^{-1}}$) is more than twice as strong as an N—N single bond ($163\ \mathrm{kJ\ mol^{-1}}$), and an N≡N triple bond ($945\ \mathrm{kJ\ mol^{-1}}$) is more than five times as strong. It is on account of this trend that NN multiply bonded compounds are stable relative to polymers or three-dimensional compounds having only single bonds. The same is not true of phosphorus, where the P—P, P═P, and P≡P bond enthalpies are $200\ \mathrm{kJ\ mol^{-1}}$, $310\ \mathrm{kJ\ mol^{-1}}$, and $481\ \mathrm{kJ\ mol^{-1}}$ respectively. In this case, single bonds are stable relative to the matching number of multiple bonds, and white phosphorus exists as singly-bonded tetrahedral P_4 molecules, and not as P≡P molecules.

The two correlations with bond order taken together imply that, for a given pair of elements, the bond enthalpy increases as bond length decreases (Fig. 2.33). This is a useful feature to bear in mind when considering the stabilities of molecules, since bond lengths may be readily available from independent sources.

Further reading

Bonding

The first three books are introductions to the treatment of molecules in terms of molecular orbitals:

R. L. DeKock and H. B. Gray, *Chemical structure and bonding*. Benjamin-Cummings, Menlo Park (1980).

J. N. Murrell, S. F. A. Kettle, and J. M. Tedder, *The chemical bond*. Wiley, New York (1978).

R. McWeeney, *Coulson's Valence*. Oxford University Press (1983).

R. J. Gillespie, *Molecular geometry*. Van Nostrand-Reinhold, New York (1972). This book describes VSEPR theory in some detail.

Symmetry

These books are introductions to symmetry; the first is a more pictorial presentation than the other two.

S. F. A. Kettle, *Symmetry and structure*. Wiley, New York (1985).
F. A. Cotton, *Chemical applications of group theory*. Wiley, New York (1971).
D. C. Harris and M. D. Bertolucci, *Symmetry and spectroscopy*. Oxford University Press (1978).

Structural techniques

E. A. V. Ebsworth, D. W. H. Rankin, and S. Cradock, *Structural methods in inorganic chemistry*. Blackwell Scientific, Oxford (1987). This is a general book that covers the principles of spectroscopic techniques such as NMR and vibrational spectroscopy as well as X-ray diffraction.
K. Nakamoto, *Infrared and Raman spectra of inorganic and coordination compounds*. Wiley, New York (1986). A more specialized text which introduces vibrational spectroscopy and provides a very useful collection of data.

Exercises

2.1 Write Lewis structures for (a) $[GeCl_3]^-$, (b) SeF_4, (c) $[FCO_2]^-$, (d) CO_3^{2-}, (e) $[AlCl_4]^-$, (f) NOF. Where more than one resonance structure is important, give examples of all major contributors. Determine the geometry about the central atom using VSEPR theory.

2.2 Write Lewis structures for (a) XeF_4 (b) PF_5 (c) BrF_3 (d) $TeCl_4$ (e) ICl_2^- and determine the geometry about the central atom using VSEPR theory.

2.3 Arrange the following techniques in order of increasing sensitivity to lifetime broadening: X-ray photoelectron spectroscopy, infrared vibrational spectroscopy, NMR spectroscopy, visible and UV electronic spectroscopy.

2.4 Using the covalent radii in Table 2.9, calculate the bond lengths in (a) CCl_4 (1.77 Å), (b) $SiCl_4$ (2.01 Å), (c) $GeCl_4$ (2.10 Å). (The values in parentheses are experimental values.)

2.5 Draw sketches to identify the following symmetry elements: (a) a C_3 axis and a σ_v plane in the NH_3 molecule; (b) a C_4 axis and a σ_h plane in the square planar $[PtCl_4]^{2-}$ ion.

2.6 Which of the following molecules and ions has (1) a center of inversion, (2) an S_4 axis: (a) CO_2, (b) C_2H_2, (c) BF_3, (d) SO_4^{2-}?

2.7 Determine the symmetry elements and assign the point group of (a) NH_2Cl, (b) CO_3^{2-}, (c) SiF_4, (d) H—C≡N, (e) SiFClBrI, (f) BrF_4^-.

2.8 Determine the symmetry elements of (a) an s orbital, (b) a p orbital, (c) a d_{xy} orbital, and (d) a d_{z^2} orbital.

2.9 State the symmetry elements that imply that a molecule is nonpolar. Using symmetry criteria, determine for each of the species in Exercise 2.7 whether it is polar.

2.10 State the symmetry criteria for chirality and determine whether any of the species in Exercise 2.7 can be optically active.

2.11 With the aid of Fig. 2.22, write the electron configurations of (a) Be_2, (b) B_2, (c) C_2^-, and (d) F_2^+.

2.12 Using molecular orbital diagrams, determine the number of unpaired electrons in (a) O_2^-, (b) O_2^+, (c) BN, and (d) NO^-

2.13 By analogy with Figs. 2.21 and 2.29, determine the bond order of (a) S_2, (b) Cl_2, and (c) NO^- and compare it with the bond order determined from Lewis structures.

2.14 A molecule shows two NMR proton signals separated by 2.00 ppm at 90 MHz. What is the maximum rate at which the two might be interconverting?

2.15 What is the expected effect on bond order and bond distance of the following ionization processes?

(a) $O_2 \rightarrow O_2^+ + e^-$

(b) $N_2 + e^- \rightarrow N_2^-$

(c) $NO \rightarrow NO^+ + e^-$

2.16 Using data from Table 2.10, calculate the standard enthalpy of the reaction

$$2H_2(g) + O_2(g) \rightarrow 2H_2O(g)$$

The experimental value is -484 kJ. What is the error in using the bond enthalpy approximation?

2.17 The common forms of N and P are $N_2(g)$ and $P_4(s)$ respectively. Do the single-bond and multiple-bond enthalpies suggest a reason?

Problems

2.1 Using concepts from Chapter 1, particularly the effects of penetration and shielding, account for the variation of single bond covalent radii with position in the periodic table.

2.2 Develop an argument based on bond enthalpies for the importance of Si—O bonds in substances common in the earth's crust in preference to Si—Si or Si—H bonds. How and why does this differ from the behavior of C?

2.3 Consider a molecule IF_3O_2 (with I as the central atom). How many isomers are possible? Which is likely to be most stable? Assign point group designations to each isomer.

2.4 When an atom of helium absorbs a photon to form the excited configuration $1s^12s^1$ (here called He*) a weak bond forms to give the diatomic molecule He—He*. Construct a molecular orbital description of the bonding in the molecule.

Polyatomic molecules and solids

3

The concepts introduced in Chapter 2 also apply to polyatomic molecules too, the only difference being that the molecular orbitals spread over all the atoms in the molecule. We shall see that we can achieve a great simplification in the classification, description, and construction of these orbitals by exploiting the symmetry of the molecular framework, and we shall describe the results of this process. As in Chapters 1 and 2, we continue to look for ways of identifying the major factors that influence the shapes and structures of molecules, and introduce the concept of correlation diagrams and isolobality, both of which are valuable in the interpretation of inorganic chemistry. One major class of compounds that we do not deal with here is that of the complexes formed by metal ions. They are treated in Chapter 7 after we have introduced some additional concepts.

In the final part of the chapter we allow the number of atoms in the aggregate to grow to infinity. This lets us extend the principles of molecular orbitals to the description of the electronic structures of solids. We consider metals and semiconductors, both of which owe their ability to conduct electricity to the molecular orbitals that extend throughout the solid.

Molecular orbitals of polyatomic molecules
3.1 The construction of molecular orbitals
3.2 Polyatomic molecules in general
3.3 The symmetries of orbitals
3.4 Characteristics of molecular orbitals
The molecular orbital theory of solids
3.5 Molecular orbital bands
3.6 Semiconduction
3.7 Superconduction
Further information: symmetry-adapted orbitals
Further reading
Exercises
Problems

We can use molecular orbital theory to discuss in a uniform manner the electronic structures of triatomic molecules, finite groups of atoms, and solids. In each case the molecular orbitals resemble those of diatomic molecules, the only important difference being that we build the orbitals from a more extensive basis set of atomic orbitals. As we remarked in Chapter 2, from N atomic orbitals it is possible to construct N molecular orbitals.

Molecular orbitals of polyatomic molecules

First, we see how to extend the ideas introduced in Chapter 2 to polyatomic molecules. We shall see that the distinction between bonding and antibonding orbitals, which in a diatomic molecule is related to the presence or absence of an internuclear node, can be generalized to the point where we can judge the relative energies of polyatomic molecular orbitals from their numbers of nodes. Another analogy with diatomic molecules that we shall develop is the use of symmetry to construct and label orbitals. This amplifies the point made in Chapter 2, that we cannot form a molecular orbital by combining an *s* orbital on one atom with a perpendicular *p* orbital on the other atom. The symmetry aspects are less obvious in polyatomic molecules, but nonetheless they can be found and prove very useful. Labeling polyatomic molecules according to their symmetries is an extension of what we have already done for diatomic molecules, where we used the labels σ and π, and g and u: these are in fact symmetry labels. We shall meet the analogs of these labels for polyatomic molecules: they are more elaborate because they must convey more information, but the idea behind them is exactly the same.

3.1 The construction of molecular orbitals

Just as the simplest diatomic molecules H_2^+ and H_2 led us into the structures of all diatomic molecules, so we can use the simplest (but transient) polyatomic species H_3^+ and H_3 to identify some of the central features of all polyatomics. Although these species might seem far removed from real inorganic chemistry, we shall see in fact that the orbitals we are about to derive occur widely, as in NH_3 and, in a more disguised form, in other molecules, such as BF_3. Since a major question in connection with polyatomic molecules is their shapes, we shall begin by considering two possibilities for H_3 and H_3^+, linear and triangular, and look for reasons why one shape may have a lower energy than the other.

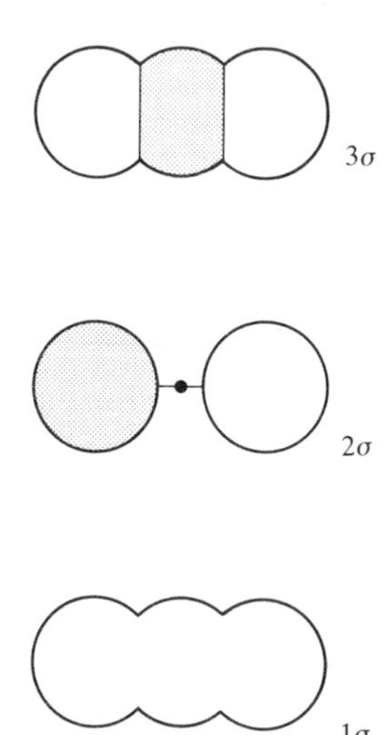

Fig. 3.1 The three molecular orbitals of the linear H_3 molecule formed from overlap of the H 1*s* orbitals. Throughout this chapter, white and gray or different shades of gray denote opposite signs of the wavefunction.

The linear H_3 molecule

We saw in Chapter 2 that the LCAO approximation allows us to write the two orbitals in H_2 as

$$1\sigma = \phi_{1s}(A) + \phi_{1s}(B) \qquad 2\sigma = \phi_{1s}(A) - \phi_{1s}(B)$$

(We have changed the notation slightly, from ψ_+ to 1σ and from ψ_- to 2σ since this will make it easier to generalize.) The corresponding orbitals in the linear H_3 molecule are combinations of all three atomic orbitals to give three molecular orbitals (Fig. 3.1). As in Chapter 2, we shall omit normalization constants since the form of the orbitals is then much clearer; where we judge it appropriate, we give the normalized forms in footnotes.

A specific calculation shows that the most strongly bonding of the three combinations is

$$1\sigma = \phi_{1s}(A) + \phi_{1s}(B) + \phi_{1s}(C)$$

However, it is easy to appreciate that this orbital does have a low energy because it is bonding between H_A and H_B and between H_B and H_C. It is also bonding between H_A and H_C, but as these are so far apart, this is less important. 1σ is a σ orbital because it has cylindrical symmetry around the molecular axis. The next higher orbital is also a σ orbital:

$$2\sigma = \phi_{1s}(A) - \phi_{1s}(C)$$

This has no contribution from the central atom, and since the outer two orbitals are so far apart there is negligible interaction between them. Therefore, it is called a **nonbonding** orbital, and neither lowers nor raises the energy of the molecule when it is occupied. The third orbital that can be formed from the basis set is another σ orbital:

$$3\sigma = \phi_{1s}(A) - 2\phi_{1s}(B) + \phi_{1s}(C)$$

This is antibonding between both neighboring pairs of atoms and therefore has the highest energy of the three orbitals we have formed.[1]

The three orbitals 1σ, 2σ, and 3σ introduce one feature that should be kept in mind: the energy of molecular orbitals generally increases as the number of nodes between neighboring atoms increases. The physical reason is that electrons are progressively excluded from the internuclear regions as the number of nodes increases.

The triangular H_3 molecule

Now we suppose that H_3 is an equilateral triangle. Its three molecular orbitals have the same form as in the linear molecule, but the ones we have called 2σ and 3σ are now **degenerate**; that is, they have the same energy. It is not immediately obvious that 2σ and 3σ are degenerate, but group theory can be used to show that the two are in fact rigorously degenerate. We shall not go into the symmetry analysis involved, but simply show that the degeneracy is physically plausible. In the triangular molecule, A and C are next to each other (Fig. 3.2), so 2σ is antibonding between them and hence higher in energy than in the linear molecule. On the other hand, the 3σ orbital is bonding between A and C (but still antibonding between A and B and between C and B), so its energy is lower than before. The two orbitals therefore move together in energy as the linear molecule is converted into the triangular molecule, and when all three H—H separations are the same the energies are equal.

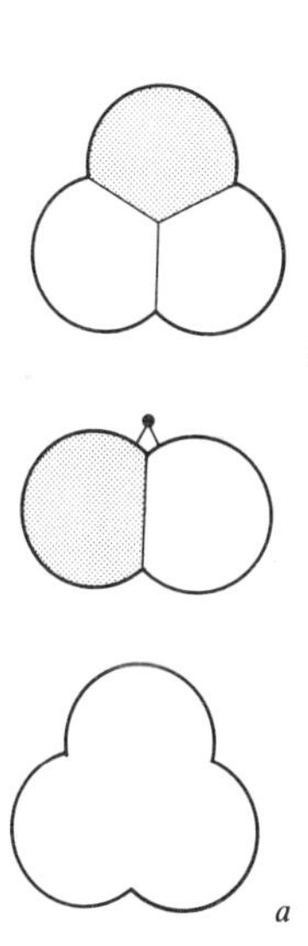

Fig. 3.2 The three molecular orbitals of the equilateral triangular H_3 molecule formed from overlap of the H 1*s* orbitals.

[1] The normalized form of these three orbitals, if we ignore overlap, are

$$1\sigma = \frac{1}{\sqrt{3}}\{\phi_{1s}(A) + \phi_{1s}(B) + \phi_{1s}(C)\}$$

$$2\sigma = \frac{1}{\sqrt{2}}\{\phi_{1s}(A) - \phi_{1s}(C)\}$$

$$3\sigma = \frac{1}{\sqrt{6}}\{\phi_{1s}(A) - 2\phi_{1s}(B) + \phi_{1s}(C)\}$$

In the triangular molecule it is no longer strictly appropriate to use the label σ because that refers to a molecule with cylindrical $C_{\infty v}$ symmetry. However, it is often convenient to continue to use the notation σ and π when we are concentrating on the form that an orbital takes between any two neighboring atoms. When using these labels for nonlinear polyatomic molecules, we must always remember that they refer to the *local* symmetry of the orbital, its symmetry with respect to one particular atom–atom axis, and not the molecule as a whole.

The correct procedure for labeling orbitals in polyatomic molecules according to their symmetries is to note that the triangular H_3 molecule belongs to the group D_{3h}, and to label it appropriately. We shall explain in the next section how this is done: for our present purposes all we need know is the following:

a denotes a nondegenerate orbital,
e denotes a doubly degenerate orbital,
t denotes a triply degenerate orbital.

Subscripts are sometimes added to these letters, as in a_1, e_g, and t_2, since it is sometimes necessary to distinguish different *a*, *e*, and *t* orbitals according to a more detailed analysis of their symmetries. The letter *b* also denotes a nondegenerate orbital in certain cases. All this will be explained later in the chapter.

The orbital we have called 1σ is nondegenerate and hence has the label *a* (specifically, using the full notation of the D_{3h} group, it is a_1'). The two orbitals 2σ and 3σ are doubly degenerate and so must be labeled *e* (their full label in the group is e'). Therefore, the orbitals are

$$a_1' = \phi_{1s}(A) + \phi_{1s}(B) + \phi_{1s}(C)$$

$$e' = \begin{cases} \phi_{1s}(A) - \phi_{1s}(C) \\ \phi_{1s}(A) - 2\phi_{1s}(B) + \phi_{1s}(C) \end{cases}$$

We can imagine bending a linear molecule into an equilateral triangle, and following the energies of the σ orbitals of the linear molecule as they become the a_1' and e' orbitals of the triangular molecule. The resulting diagram (Fig. 3.3) is called a **correlation diagram**. We shall see that diagrams like this play an important role in understanding the shapes, spectra, and reactions of polyatomic molecules.

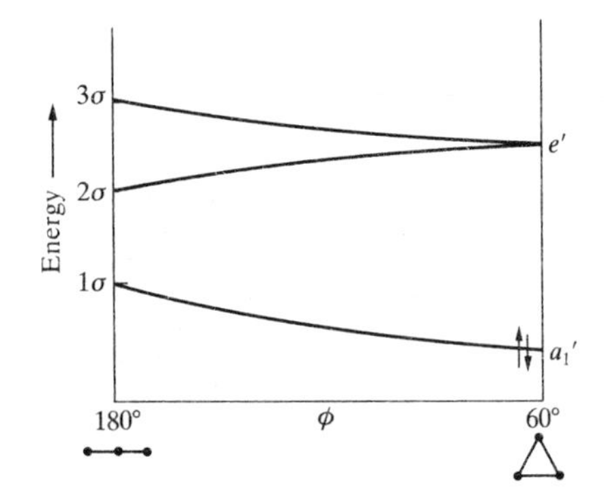

Fig. 3.3 The orbital correlation diagram for the H_3 molecule shows how the energies of the orbitals change as the linear molecule is converted into the equilateral triangular molecule.

Electron configurations in H_3^+

Now that we have the form of the molecular orbitals and know their relative energies, we can deduce the electron configuration of H_3^+ in each geometry.

The two electrons in H_3^+ both occupy the lowest energy orbital. If the molecule is linear, the resulting configuration will be $1\sigma^2$; if triangular, $a_1'^2$. We can decide which geometry has the lower energy only by detailed calculation. However, the fact that a_1' is bonding between all three atoms but 1σ is bonding only between A and B and B and C suggests that the lowest energy is obtained for the triangular arrangement. This is in fact the case, and spectroscopic data indicate that H_3^+ is an equilateral triangular species with configuration $a_1'^2$.

The important point, as can be seen from Fig. 3.2, is that the a_1' orbital spreads equally over all the atoms. The same is true of the 1σ orbital in the linear molecule. Therefore, the two electrons form a bond that binds the entire cluster together. In other words, molecular orbitals are **delocalized orbitals**, in the sense that their bonding or antibonding influence is spread over several atoms and not localized between two atoms.

Molecular orbitals for chains and rings of atoms

It is quite easy to extend these ideas to build up the molecular orbitals for chains and rings of more than three atoms. In order to be as useful as possible for what is to follow, we shall consider the π orbitals that may be formed by combining p orbitals on atoms of an element E. In many cases E will be carbon, but various other p-block elements also form chains and rings. We shall see later that we can treat some large molecules as formed by combining the orbitals of a central metal atom with the orbitals we are about to construct, as in the compound bis(benzene)chromium.

The linear chains in Fig. 3.4 represent the π orbitals formed from p orbitals projecting above and below the plane of the paper. An open circle represents a positive lobe above the page, and a shaded circle represents a negative lobe. We have given a rough idea of the sizes of the coefficients by the sizes of the circles. The molecular orbitals are arranged in order of increasing energy, with the lowest energy at the bottom. The regular pattern of nodal planes should be noted: they resemble the harmonics of a vibrating string. The same diagrams apply to angular molecules; thus we can use the diagram for E_3 to describe the π orbitals of the V-shaped allyl group ($CH_2{=}CH{-}CH_2 \leftrightarrow CH_2{-}CH{=}CH_2$) or the nitrite ion.

Figure 3.5 shows the analogous diagrams for cyclic systems. A new feature is that some of their orbitals are doubly degenerate (as in triangular H_3). The degeneracy of the orbitals in cyclic E_4 is probably easier to accept than in E_3 because the two degenerate orbitals are changed into each other by a simple 90° rotation of the molecule. As with chain systems, the greater the number of nodes, the greater the antibonding character and hence the higher the energy of the orbital.

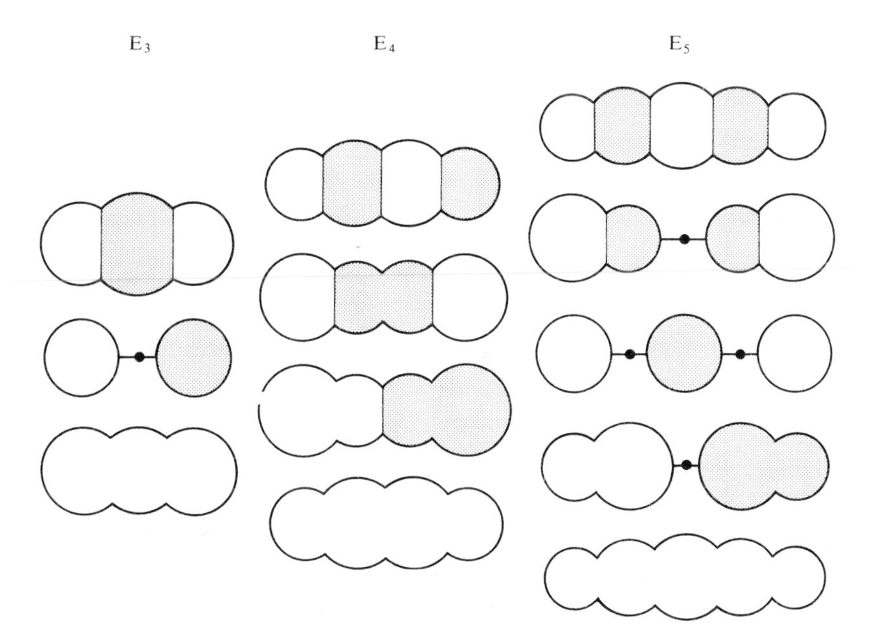

Fig. 3.4 Nodal properties of π orbitals in linear chains of atoms. Opposite signs of the orbitals are represented by the presence or absence of shading and the sizes of the atomic orbitals indicate the size of the coefficients. An open circle represents a positive lobe above the page, a shaded circle represents a negative lobe.

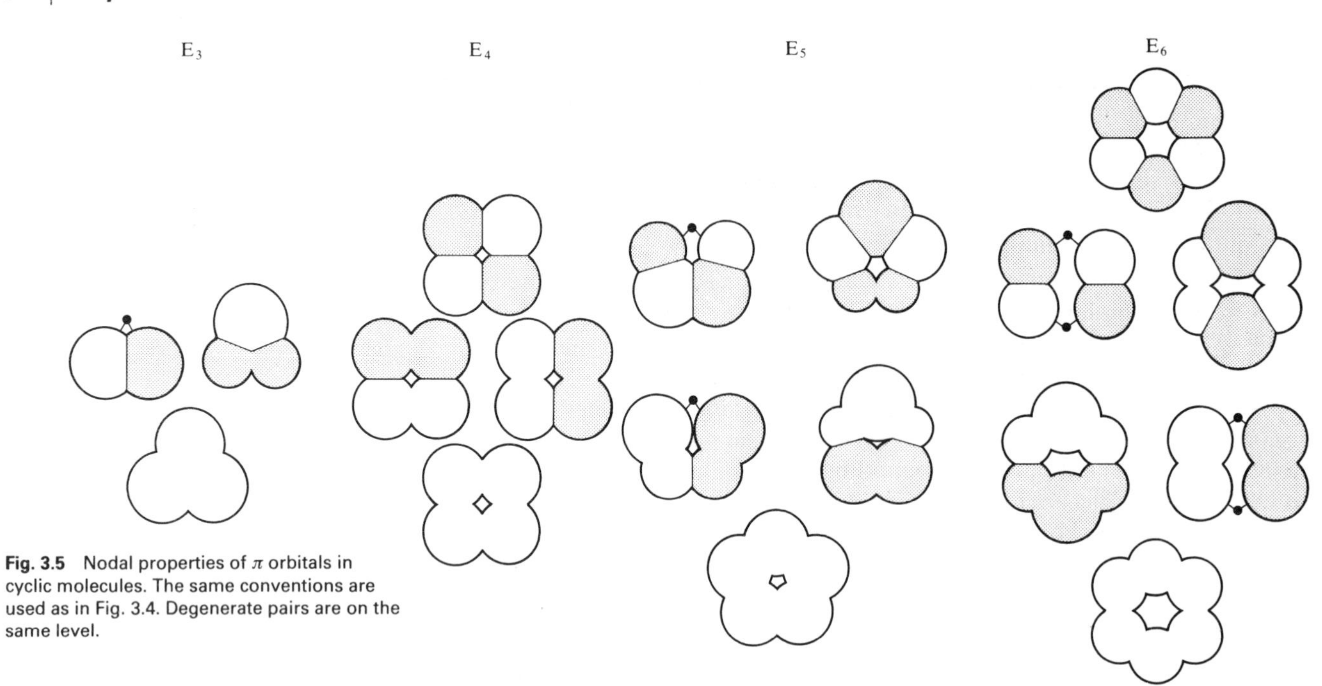

Fig. 3.5 Nodal properties of π orbitals in cyclic molecules. The same conventions are used as in Fig. 3.4. Degenerate pairs are on the same level.

Example 3.1: *Constructing molecular orbitals for polyatomic molecules*

Draw the linear combinations of p_z orbitals that form molecular orbitals in a six-membered linear chain.

Answer. A simple approach is to draw the sine waves representing the first six vibrations of a string (Fig. 3.6a), and then to replace the curves by a circle on each atom (Fig. 3.6b). The size of the circle is proportional to the amplitude of the wave at the atom in question, and the sign is that of the wave.

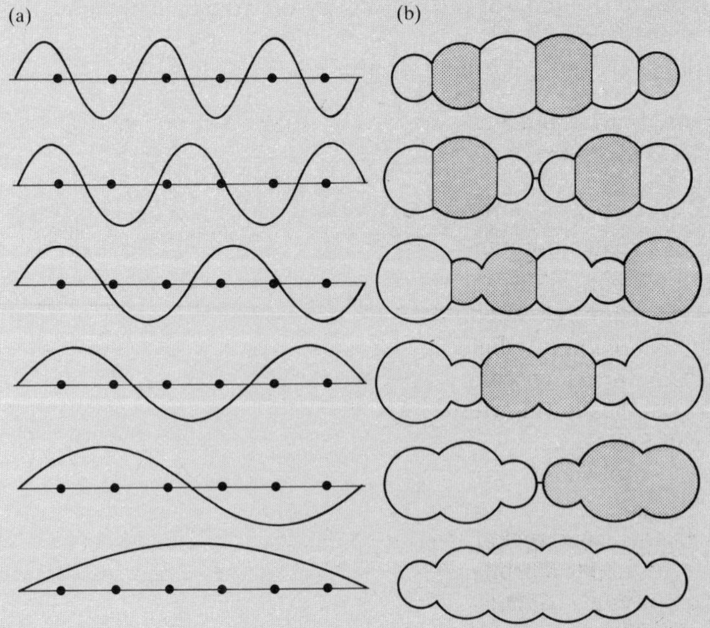

Fig. 3.6 (a) The first six harmonics of a stretched string and (b) the molecular orbitals they suggest.

Exercise. Construct the π molecular orbitals formed by an E_8 ring (we shall see later that such a ring occurs in uranocene).

3.2 Polyatomic molecules in general

We shall now introduce the molecular orbitals of some polyatomic molecules and generalize the discussion to include three-dimensional systems. First, we shall describe the result of a molecular orbital analysis of three important cases. The aim is to see what the theory has to say about the electronic structure of the molecules and in particular the origin of the bonding in them and their stabilities. Once we have become familiar with the appearance of molecular orbitals and the manner in which they are formed by combining atomic orbitals, we return to some simple cases and show how a symmetry analysis can be used to suggest the forms of orbitals. By the end of the chapter we shall be able to use the combinations shown in Appendix 6 to construct approximate molecular orbitals for a wide range of simple molecules. That is, in this chapter, we construct a kind of phrase book of terms for discussing the molecular orbitals of polyatomic molecules without going as far as presenting the details of the grammar of the language.

In Appendix 6 we summarize in a similar pictorial manner the composition of the orbitals for a number of common molecular shapes. The orbitals shown there apply to *all* molecules of a given symmetry type irrespective of the details of the energies of the contributing atomic orbitals. Later we shall see that the patterns can be used to *construct* molecular orbitals. For the time being, though, we shall use them only as a guide to labeling the orbitals.

The formation of molecular orbitals

The features we have introduced by considering H_3 are present in all the molecules we consider from now on. In each case, we can write the molecular orbital as a sum of *all* the orbitals on all the atoms in the molecule:

$$\psi = \sum_i c_i \phi_i \qquad (1)$$

where the ϕ_i are atomic orbitals and the index i runs over all the atoms in the molecule. From N atomic orbitals we can construct N molecular orbitals. Then:

1. The greater the number of nodes in a molecular orbital the greater the antibonding character and the higher the orbital energy.
2. Interactions between non-nearest neighbor atoms are weakly bonding (lower the energy slightly) if the orbital lobes on these atoms have the same sign (interfere constructively) but are weakly antibonding if the signs are opposite (interfere destructively).

To these two points we can add one made in Chapter 2:

3. Orbitals constructed from lower energy atomic orbitals lie lower in energy (so atomic s orbitals produce lower energy molecular orbitals than atomic p orbitals of the same shell).

In NH_3, for example, each molecular orbital is the combination of seven atomic orbitals: the three H 1s orbitals, the N 2s orbital, and the three N 2p

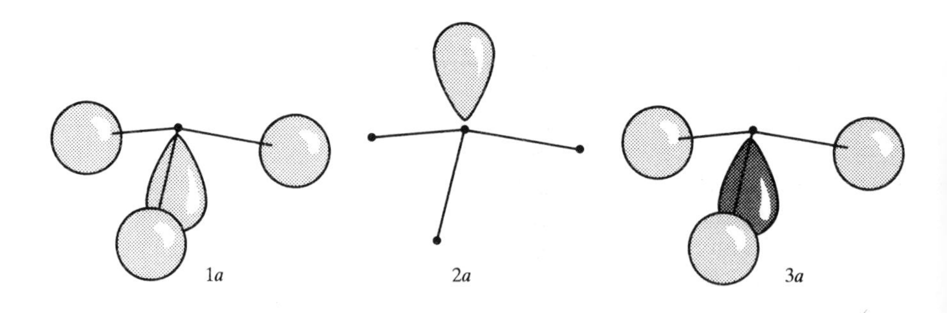

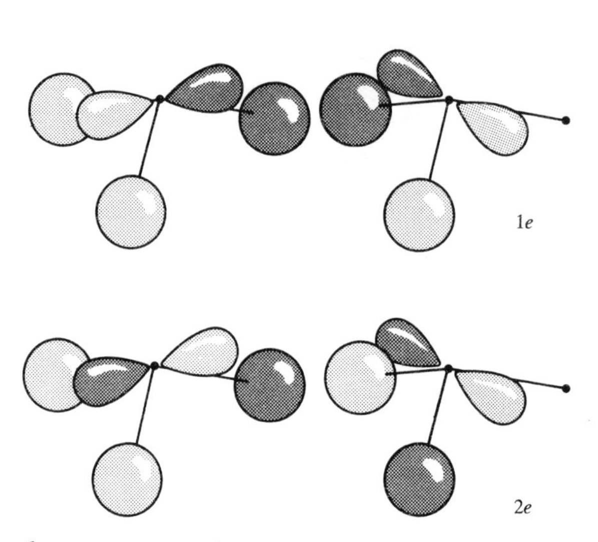

Fig. 3.7 A representation of the orbitals in NH_3.

orbitals. From these seven atomic orbitals we can construct seven molecular orbitals. The form of the orbitals (which is obtained using group theory) is shown in Fig. 3.7; the resulting energy level diagram, which we can construct approximately by assessing the numbers of nodes and the energies of the contributing atomic orbitals (and, in principle, obtain by explicit calculation from the Schrödinger equation), is shown in Fig. 3.8. Three of the orbitals are nondegenerate and are labeled a: one is strongly bonding, one almost nonbonding, and the third strongly antibonding. The remaining four orbitals fall into two doubly degenerate sets, one bonding and the other antibonding. They are therefore designated $1e$ and $2e$ respectively.

There are eight valence electrons to accommodate in NH_3 to give the lowest energy configuration (five from N and one from each H atom). They enter the molecular orbitals in increasing order of energy, starting with the orbital of lowest energy and taking note of the Pauli exclusion principle, that no more than two electrons can occupy any one orbital. The first two electrons enter $1a$ and fill it. The next four enter the doubly degenerate $1e$ orbitals, and fill them. The last two enter the $2a$ orbital, which calculations show is almost nonbonding. The resulting overall ground state electron configuration is therefore $1a^2 1e^4 2a^2$. No antibonding orbitals are occupied, so the molecule is a stable species (in the sense of having a lower energy than the separated atoms). The conventional description of NH_3 as a molecule with a lone pair is also mirrored in the configuration, for the HOMO is $2a$, which is largely confined to the N atom and makes only a small contribution to the bonding.

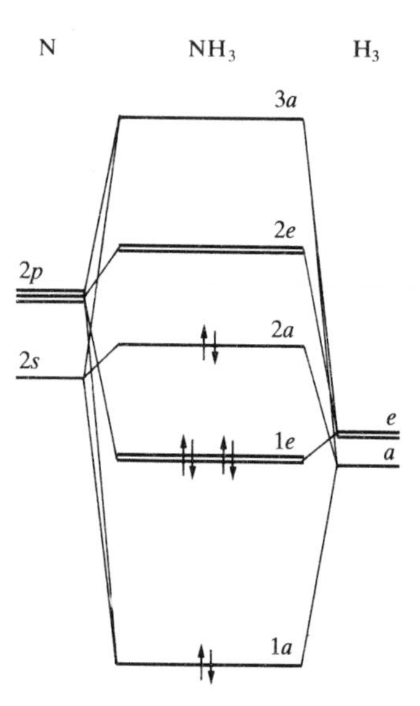

Fig. 3.8 The molecular orbital energy level diagram of NH_3 when it has its observed bond angle (107°) and bond lengths.

Hypervalence

A slightly more complicated example, but one that makes an important point, is the octahedral molecule SF_6. A simple basis set that is adequate to illustrate the point we want to make consists of the valence shell *s* and *p* orbitals of an S atom and one *p* orbital of each of the six F atoms and pointing toward the S atom. From these 10 atomic orbitals we can construct 10 molecular orbitals. Calculations indicate that four of the 10 orbitals are bonding and four are antibonding; the two remaining orbitals are nonbonding (Fig. 3.9). The labels of the orbitals[2] are

Antibonding:	$2a$ (nondegenerate)
	$2t$ (triply degenerate)
Nonbonding:	e (doubly degenerate)
Bonding:	$1t$ (triply degenerate)
	$1a$ (nondegenerate)

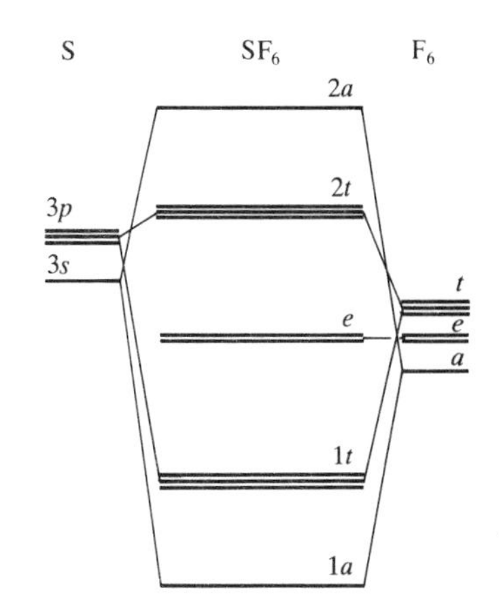

Fig. 3.9 The molecular orbital energy level diagram of SF_6.

There are 12 electrons to accommodate: the first two can enter $1a$ and the next six can enter $1t$. The remaining four just fill the nonbonding pair of orbitals, resulting in $1a^2 1t^6 e^4$. As we see, none of the antibonding orbitals is occupied. The theory, therefore, accounts for the formation of SF_6: with four bonding orbitals and two nonbonding orbitals occupied, the average S—F bond order is $\frac{2}{3}$.

The important point is that the bonding in the hypervalent SF_6 molecule can be explained without using S *d* orbitals to expand the octet. This does not mean that *d* orbitals cannot participate in the bonding, but they are not *necessary* for bonding six F atoms to the central S atom. Molecular orbital theory takes hypervalence into its stride by having available plenty of orbitals, not all of which are antibonding. Therefore the question of when hypervalence can occur appears to depend on factors other than *d*-orbital availability.

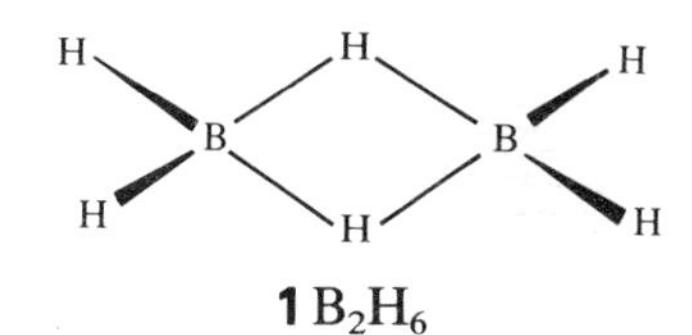

1 B_2H_6

Electron deficiency

The formation of molecular orbitals by combining several atomic orbitals accounts effortlessly for the existence of **electron-deficient compounds**, which are compounds for which there are insufficient electrons for a Lewis structure. We can see this most easily with diborane (B_2H_6, **1**). In this molecule, the eight atoms contribute a total of 14 valence orbitals for the construction of 14 molecular orbitals (four from each B atom, making eight, and one each from the six H atoms). About seven of these molecular orbitals will be bonding or nonbonding, which is more than enough to accommodate the 12 valence electrons provided by the atoms. There should therefore be nothing mysterious about the existence of this molecule.

A slightly different viewpoint is achieved by concentrating on the two BHB fragments and the three molecular orbitals that can be formed by combining the atomic orbitals contributed by the three atoms (Fig. 3.10).[3] The bonding orbital spanning these three atoms can accommodate two electrons and pull the molecule together. Since the bridging bond is formed

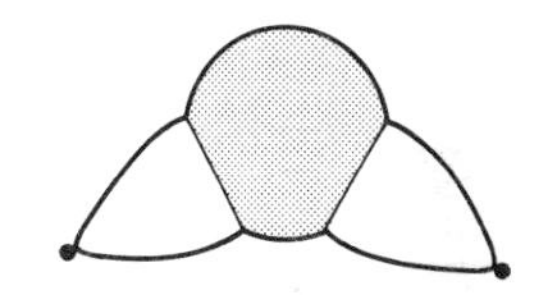

Fig. 3.10 The molecular orbitals formed by two B orbitals and one H orbital on an atom lying between them, as in B_2H_6. Two electrons occupy the bonding combination and hold all three atoms together.

[2] For a broad idea of their shapes, refer to Appendix 6 and see the patterns of orbitals corresponding to each symmetry type.

[3] This is the first of several examples where we construct molecular orbitals from the orbitals of molecular fragments. This approach is a halfway stage between the atomic orbitals and the final molecular orbitals, and is very useful for drawing analogies between molecules.

from two electrons in a molecular orbital built from orbitals on three atoms, it is called a **three-center, two-electron bond** and denoted (3c,2e). That two electrons bind two pairs of atoms (BH and HB) is immaterial, but it is an insurmountable difficulty for the Lewis approach. Electron deficiency is in fact well developed not only in boron (where it was first clearly recognized), but also in carbocations and a variety of other classes of compounds that we encounter later in the text. The electronic structure of metals may be regarded as another example of its occurrence.

3.3 The symmetries of orbitals

We shall now see in more detail the significance of the labels used for the orbitals, and gain more insight into the construction of molecular orbitals. We shall proceed informally and pictorially to give a flavor of group theory, but not the specific calculations involved. The objective here is to show how to identify the symmetry label of an orbital from a drawing like those in Appendix 6 and, conversely, to appreciate what a symmetry label signifies. The arguments later in the book are all based on simply 'reading' molecular orbital diagrams qualitatively. If you are interested in the details of the calculations on which the illustrations are based, and in particular the construction of combinations of specific symmetry, you will find further guidance in the *Further information* section at the end of this chapter.

Character tables and symmetry labels

Molecular orbitals of diatomic molecules (and linear polyatomic molecules) are labeled σ, π, and so on. These labels refer to the symmetry of the orbitals with respect to rotations around the principal symmetry axis of the molecule (or, locally, to the axis of a bond). Thus, a σ orbital does not change sign under any rotation, a π orbital changes sign when rotated by 180°, and so on (Fig. 3.11). We can generalize this labeling of orbitals according to their behavior under rotations and extend it to nonlinear polyatomic molecules, where there are reflections and inversions to take into account as well as rotations. Moreover, just as we can assign the labels σ and π to individual *atomic* orbitals in a linear molecule (and speak of a p_z orbital, for instance, as having σ symmetry), so we can also often ascribe symmetry labels to individual atomic orbitals in polyatomic molecules. This is important, because orbitals that have different symmetry types do not contribute to the same molecular orbital. Symmetry labels may also be ascribed to certain linear combinations of atomic orbitals, such as the combination $\phi_{1s}(A) + \phi_{1s}(B) + \phi_{1s}(C)$ of the three H atomic orbitals in the

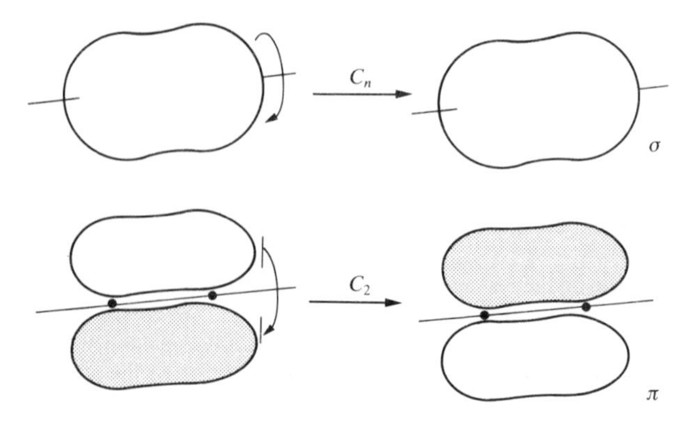

Fig. 3.11 The σ and π classification of orbitals is based on their symmetry with respect to rotations around an axis: a σ orbital is unchanged by any rotation; a π orbital changes sign when rotated by 180°.

Table 3.1. The C_{3v} character table

	E	$2C_3$	$3\sigma_v$	
A_1	1	1	1	s, p_z, d_{z^2}
A_2	1	1	−1	
E	2	−1	0	$(p_x, p_y), (d_{xz}, d_{zy}), (d_{x^2-y^2}, d_{yx})$*

* (a, b) denotes a degenerate pair of orbitals; the characters in the table refer to the symmetry of the pair jointly.

NH_3 molecule. The motivation to bear in mind is that, through a symmetry analysis of orbitals and groups of orbitals, we can find the general form of molecular orbitals as combinations of units associated with fragments of the molecules.

The labels a, a_1, e, e_g, and so on reflect the behavior of the orbitals under all the symmetry operations of the relevant molecular point group. The particular label is assigned by referring to the **character table** of the group, a table that characterizes the different symmetry types possible in a point group. Thus, when we assign the labels σ and π, we use

	C_2	(i.e. rotation by 180°)
σ	1	(i.e. no change of sign)
π	−1	(i.e. change of sign)

This is a fragment of a character table, the +1 entry showing that an orbital remains the same and the −1 that it changes sign under the operation C_2.

The entries in a complete character table are derived using the formal techniques of group theory and are called **characters**, χ. These numbers characterize the essential features of each symmetry type in a way that will be illustrated as we go on.[4] We shall show this by considering the C_{3v} character table (Table 3.1); character tables for other groups are given in Appendix 5.

The columns in the table are labeled with the symmetry operations of the group. The rows, which summarize the characteristic symmetries of orbitals, are labeled with the **symmetry type** (the analogs of σ and π). By convention, the symmetry types are given upper-case letters (such as A_1 and E) but the orbitals to which they apply are labeled with the lower-case italic equivalents (so an orbital of symmetry type A_1 is called an a_1 orbital). Note that we must be careful to distinguish the identity element E (italic, a column heading) from the symmetry label E (roman, a row label).

The entry in the column headed by the identity operation E gives the degeneracy of the orbitals. Thus, in a C_{3v} molecule, any orbital with a symmetry label a_1 or a_2 must be nondegenerate and have a character of 1 in the column headed E. Conversely, if we know that we are dealing with a nondegenerate orbital in a C_{3v} molecule, its symmetry type must be either A_1 or A_2 and the orbital will be labeled either a_1 or a_2. Similarly *any* doubly degenerate pair of orbitals in C_{3v} must be labeled e and have a character 2 in the column labeled E. Since there are no characters with the value 3 in the column headed E, we know at a glance there can be no triply degenerate orbitals in a C_{3v} molecule.

[4] For an account of the generation and use of character tables without too much mathematical background, see P. W. Atkins, *Physical chemistry*, 4th edn. Oxford University Press and W. H. Freeman & Co.(1990).

Example 3.2: *Using a character table to judge degeneracy*

Can a trigonal planar molecule such as BF_3 have triply degenerate orbitals?

Answer. Trigonal planar molecules belong to the point group D_{3h}. Reference to the character table for this group (Appendix 5) shows that the maximum degeneracy is 2, since no character exceeds 2 in the column headed E. Therefore, the orbitals cannot be triply degenerate.

Exercise. The SF_6 molecule is octahedral. What is the maximum possible degree of degeneracy of its orbitals?

For the A and B symmetry types, the entries under the elements other than the identity E show the behavior of an orbital or a set of orbitals under the corresponding operations:

Character	*Significance*
1	the orbital is unchanged
−1	the orbital changes sign
0	the orbital undergoes a more complicated change

By comparing the changes that occur to an orbital under each operation and comparing the resulting 1, −1, or 0 with the entries in a row of the character table for the point group, we can identify the symmetry label of the orbital.

In the case of degenerate orbitals, the characters in a row of the table are the *sums* of the characters summarizing the behavior of the individual orbitals. Thus, if one member of a pair remains unchanged under an operation but the other changes sign, the entry is reported as $\chi = 1 - 1 = 0$.

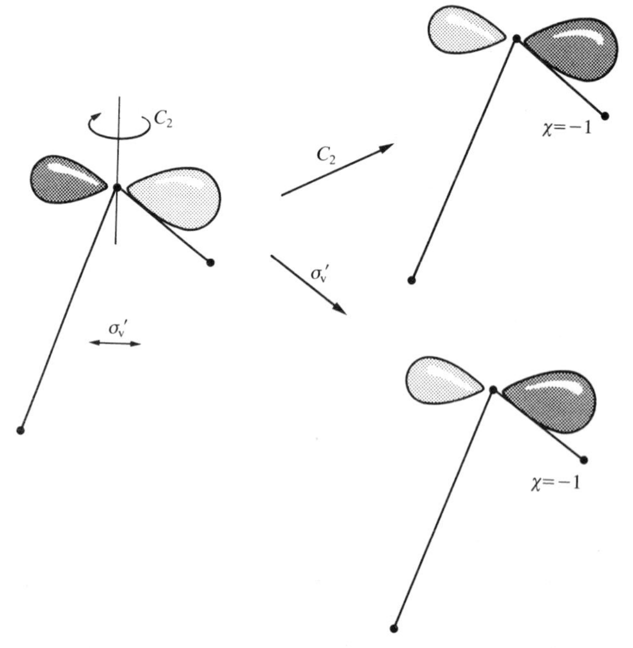

Fig. 3.12 In a C_{2v} molecule such as H_2O, the $2p_x$ orbital on the central atom changes under a C_2 rotation, which signifies that it is a b orbital. That it also changes sign under the reflection σ_v' identifies it as b_1.

We now see why the first column tells us the degeneracy: it is the sum of 1 for each orbital in the degenerate set (because each orbital remains unchanged under the identity operation), so a nondegenerate orbital gives $\chi = 1$, a doubly degenerate orbital gives $\chi = 1 + 1 = 2$, and so on.

As an example, consider a $2p_x$ orbital on the O atom of H_2O. This is a C_{2v} molecule, so we know by referring to the C_{2v} character table that the labels available for the orbitals are a_1, a_2, b_1, and b_2. We can decide the label for $2p_x$ by noting that under a 180° rotation (C_2) it changes sign (Fig. 3.12), so it must be either b_1 or b_2 since only these have character -1 under C_2. This orbital also changes sign under the reflection σ'_v, which identifies it as b_1. As we shall see, any molecular orbital built from this atomic orbital will also be a b_1 orbital. Similarly, $2p_y$ changes sign under C_2 but not under σ'_v and so contributes to b_2 orbitals.

A slightly more complicated example is the symmetry classification of the combination $\phi_1 = \phi(A) + \phi(B) + \phi(C)$ of the three H 1s orbitals in the C_{3v} molecule NH_3 (Fig. 3.13). Since ϕ_1 is nondegenerate, it is either a_1 or a_2. It remains unchanged under a C_3 rotation and under any of the vertical reflections, so its characters are

E	$2C_3$	$3\sigma_v$
1	1	1

Comparison with the character table shows that ϕ_1 is of symmetry type A_1 and therefore that it contributes to a_1 molecular orbitals in NH_3.

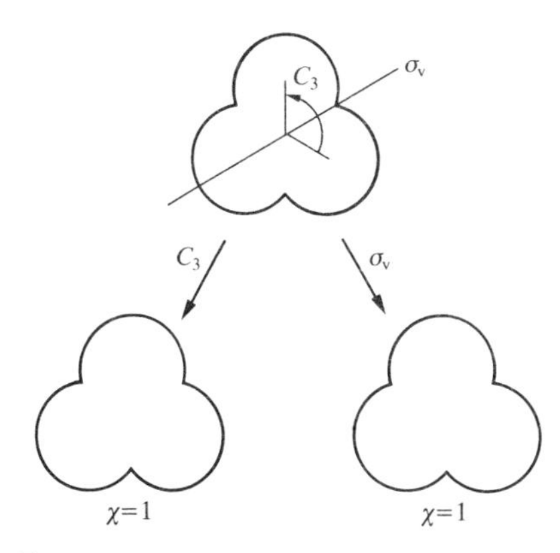

Fig. 3.13 The combination $\phi_1 = \phi(A) + \phi(B) + \phi(C)$ of the three H 1s orbitals in the C_{3v} molecule NH_3 remains unchanged under a C_3 rotation and under any of the vertical reflections.

Example 3.3: *Identifying the symmetry type of orbitals*

Identify the symmetry type of the orbital

$$\psi = \phi - \phi'$$

in a C_{2v} NO_2 molecule, where ϕ is a $2p_x$ orbital on one O atom and ϕ' that on the other.

Answer. The combination is shown in Fig. 3.14. Under a C_2 rotation ψ changes into itself, implying a character of 1. Under the reflection σ_v both orbitals change sign, so $\psi \rightarrow -\psi$, implying a character of -1. Under σ'_v, ψ also changes sign, so the character for this operation is also -1. The characters are therefore

E	C_2	σ_v	σ'_v
1	1	-1	-1

These match the characters of the A_2 symmetry label, so ψ can contribute to an a_2 orbital.

Exercise. Identify the symmetry type of the combination $\phi_{1s}(A) - \phi_{1s}(B) + \phi_{1s}(C) - \phi_{1s}(D)$ in a square-planar array of H atoms.

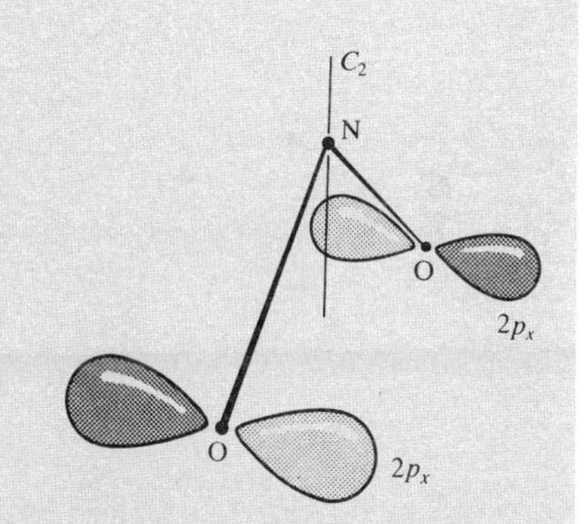

Fig. 3.14 The combination of O $2p_x$ orbitals referred to in Example 3.3.

Symmetry-adapted orbitals

We have already seen an illustration of the important conclusion deduced from group theory that molecular orbitals must be constructed from atomic orbitals, or combinations of atomic orbitals, with the same symmetry label. Thus, in a linear molecule with the z axis as the bond axis an s orbital and a p_z orbital both have σ symmetry, and may combine to form molecular

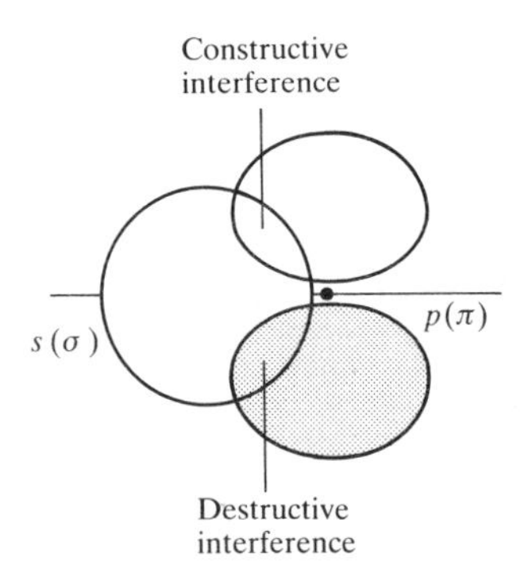

Fig. 3.15 An *s* orbital (with σ symmetry) has zero net overlap with a *p* orbital (with π symmetry) because the constructive interference between the parts of the atomic orbitals with the same sign exactly matches the destructive interference between the parts with opposite signs.

orbitals. On the other hand, the *s* orbital and the p_x orbitals have different symmetries (σ and π respectively) and hence cannot contribute to the same molecular orbital (Fig. 3.15). As can be seen from the illustration, the contribution from the region of constructive interference is canceled by the contribution from the region of destructive interference.

We can appreciate how the restriction works in a nonlinear molecule by considering NH_3 again. We have just seen that in NH_3 the combination ϕ_1 has a_1 symmetry. The N 2*s* and N $2p_z$ orbitals also have the same set of characters, as shown explicitly to the right of the character table for C_{3v} (Table 3.1), and so these atomic orbitals also have a_1 symmetry. Since they have the same symmetry as the ϕ_1 combination, they can all contribute to molecular orbitals of a_1 symmetry. All three of the resulting **symmetry-adapted orbitals**, or combinations of orbitals with a specific symmetry, have the form

$$a_1 = c_{2s}\phi_{2s}(\mathrm{N}) + c_{2p}\phi_{2p_z}(\mathrm{N}) + c_{\mathrm{H}}\{\phi(\mathrm{A}) + \phi(\mathrm{B}) + \phi(\mathrm{C})\}$$

Only three such linear combinations are possible (because the H combination $\phi(\mathrm{A}) + \phi(\mathrm{B}) + \phi(\mathrm{C})$ counts as a single orbital), and are labeled $1a_1$, $2a_1$, and $3a_1$ in order of increasing energy (the order of increasing number of nodes).

We have seen that the combinations ϕ_2 and ϕ_3 have E symmetry in C_{3v}. The character table shows that the same is true of the N $2p_x$ and N $2p_y$ orbitals, and this is confirmed by noting that jointly the two 2*p* orbitals behave exactly like ϕ_1 and ϕ_2 (Fig. 3.16). It follows that ϕ_2 and ϕ_3 can combine with these two *p* orbitals and form doubly degenerate bonding and antibonding *e* orbitals of the molecule. These two orbitals have the form

$$e = \begin{cases} c_1\phi_{2p_x}(\mathrm{N}) + c_2\phi_2 \\ c_1\phi_{2p_y}(\mathrm{N}) + c_2\phi_3 \end{cases}$$

We label the bonding pair 1*e* and the antibonding pair 2*e*.

Example 3.4: *Identifying symmetry-adapted orbitals*

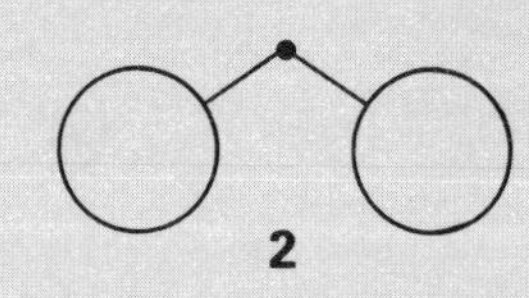

2

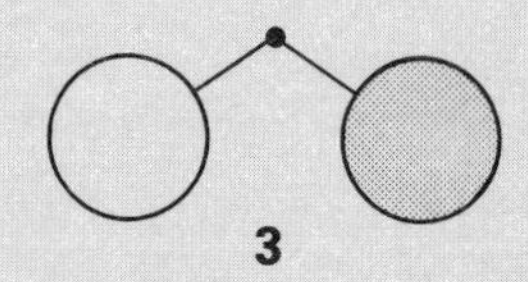

3

The H 1*s* orbitals in H_2O (point group C_{2v}) form two linear combinations $\phi_+ = \phi(1) + \phi(2)$ (**2**) and $\phi_- = \phi(1) - \phi(2)$ (**3**). What symmetry labels do they have? With what O orbitals will they overlap to form molecular orbitals?

Answer. Under C_2, ϕ_+ does not change sign but ϕ_- does; their characters are +1 and −1 respectively. Under the reflections, ϕ_+ does not change sign; ϕ_- changes sign under σ_v, so its character is −1 for this operation. The characters are therefore

	E	C_2	σ_v	σ_v'
ϕ_+	1	1	1	1
ϕ_-	1	−1	−1	1

This table identifies the orbitals as a_1 and b_2 respectively. According to the right of the character table, the O 2*s* and O $2p_z$ orbitals also have a_1 symmetry; O $2p_y$ has b_2 symmetry. The linear combinations that can be formed are therefore

$$a_1 = c_1\phi_{2s}(\mathrm{O}) + c_2\phi_{2p_z}(\mathrm{O}) + c_3\phi_+$$

$$b_2 = c_1\phi_{2p_y}(\mathrm{O}) + c_2\phi_-$$

Exercise. What is the symmetry label of the combination $\phi = \phi(\mathrm{A}) + \phi(\mathrm{B}) + \phi(\mathrm{C}) + \phi(\mathrm{D})$ in CH_4, where A, B, C, and D designate the H atoms?

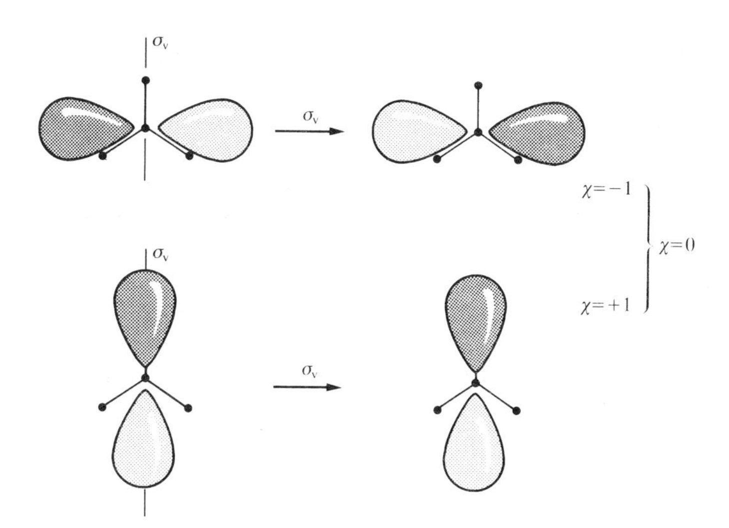

Fig. 3.16 An N $2p_x$ orbital in NH_3 changes sign under a σ_v reflection but a $2p_y$ orbital is left unchanged. Hence the degenerate pair jointly has character 0 for this operation. The plane of the paper is the *xy* plane.

Group theory has nothing to say about the energies of orbitals (other than to identify degeneracies); to calculate the energies we must resort to quantum mechanics, and to assess them experimentally we must turn to techniques such as photoelectron spectroscopy. In simple cases, though, we can use the general rules set out previously (Section 3.2) to judge the relative energies of the orbitals. The $1a_1$ orbital, being composed of the low-lying N 2*s* orbital, will lie lowest in energy and its antibonding partner, $3a_1$, will probably lie highest. The *e* bonding orbital is next after $1a_1$, and is followed by the $2a_1$ orbital, which is largely nonbonding. This leads to the energy level scheme that was shown in Fig. 3.8.

3.4 Characteristics of molecular orbitals

We can now bring a number of features of structure, bonding, and reactivity into focus using molecular orbital theory.

The Walsh diagram of an H_2X molecule

First, we deal with the connection between the molecular approach and the VSEPR theory as a means of accounting for molecular shapes. The latter focuses on a single factor, the repulsion between electron pairs. However, the shape adopted by a polyatomic molecule actually depends on many factors, including electron–nucleus attractions, electron–electron repulsions, and electron kinetic energies. The daunting task of assessing the various contributions to the total energy from orbitals of different symmetry types was greatly simplified by an approach devised by A. D. Walsh in a classic series of papers published in 1953. It is the simplest analysis available that shows the connection between molecular orbital theory and molecular geometry. It provides an instructive alternative to VSEPR theory since it relates molecular shape to the occupation of molecular orbitals and does not consider electron-pair repulsions explicitly. Once again we see how in inorganic chemistry we seek to identify various dominating influences, and how sometimes it is possible to do so in more than one way. There is no simple theory of molecular shape that can take all the relevant factors into account at once.

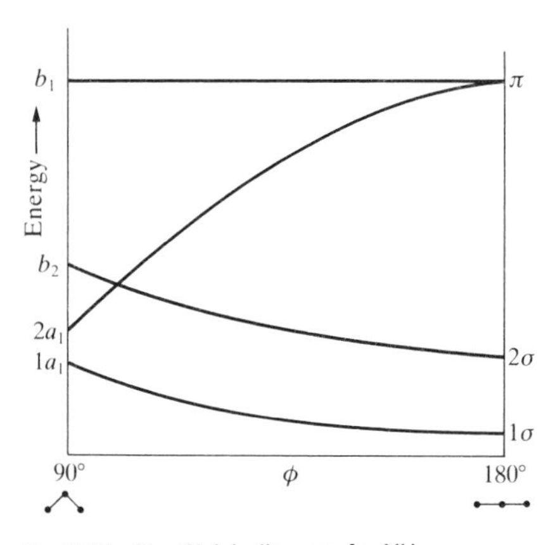

Fig. 3.17 The Walsh diagram for XH_2 molecules.

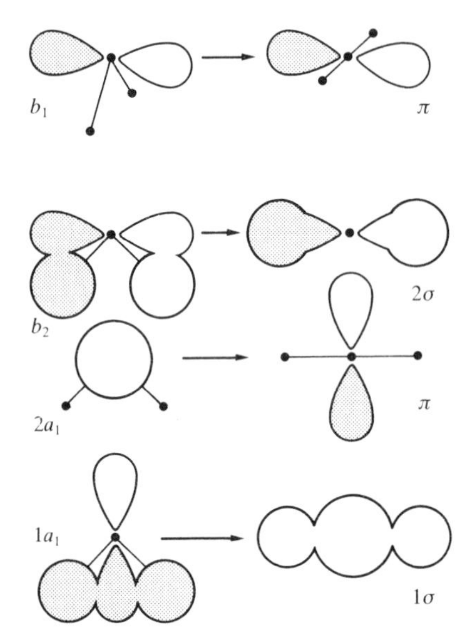

Fig. 3.18 The composition of the molecular orbitals of an XH_2 molecule at the two extremes of the correlation diagram shown in Fig. 3.17.

Walsh's approach to the discussion of the shape of an H_2X triatomic molecule (such as BeH_2 and H_2O) is illustrated in Fig. 3.17, which is an example of a **Walsh diagram**. This is a diagram showing the variation of orbital energy with molecular geometry. It is constructed by considering how the composition and energy of each molecular orbital changes as the bond angle changes from 90° to 180°. A Walsh diagram is in fact just a more elaborate version of the correlation diagram that we illustrated for H_3^+ earlier in the chapter.

The orbitals we consider in the angular C_{2v} molecule (Example 3.4) are

$$a_1 = c_1\phi_{2s} + c_2\phi_{2p_z} + c_3\phi_+$$

$$a_2 \quad \text{does not exist}$$

$$b_1 = \phi_{2p_x}$$

$$b_2 = c_1\phi_{2p_y} + c_2\phi_-$$

As explained before, there are three a_1 orbitals and two b_2 orbitals; they are shown on the left of Fig. 3.18. In the linear $D_{\infty h}$ molecule, the molecular orbitals are

$$\sigma_g = c_1\phi_{2s} + c_2\{\phi(A) + \phi(B)\}$$

$$\pi_g \quad \text{does not exist}$$

$$\pi_u = \phi_{2p_x} \text{ and } \phi_{2p_z}$$

$$\sigma_u = c_1\phi_{2p_y} + c_2\{\phi(A) - \phi(B)\}$$

The lowest energy orbital in 90° H_2X is the one labeled $1a_1$, which is built from the overlap of the X $2p_z$ orbital with the $\phi(A) + \phi(B)$ combination of H $1s$ orbitals. As the bond angle changes to 180° the two H $1s$ orbitals overlap less but the contribution of the X $2s$ orbital increases. In the 180° molecule, X $2s$ is the only contribution from the X atom to the $1a_1$ orbital (Fig. 3.18). The replacement of X $2p_z$ by X $2s$ lowers the energy of the orbital. The energy of the $1b_2$ orbital is also lowered because the adverse H—H overlap decreases and the H $1s$ orbitals move into a better position for overlap with the X $2p_y$ orbital. The biggest change occurs for the $2a_1$ orbital. It is a pure X $2s$ orbital in the 90° molecule, but correlates with a pure X $2p_z$ orbital in the 180° molecule. Hence, it shows a steep rise in energy as the bond angle increases. The $1b_1$ orbital is a nonbonding X $2p$ orbital perpendicular to the molecular plane in the 90° molecule and remains nonbonding in the linear molecule. Hence, its energy barely changes with angle.

The principal feature that determines whether or not the molecule is angular is whether the $2a_1$ orbital is occupied. This is the orbital that has considerable X $2s$ character in the angular molecule but not in the linear molecule. Hence, a lower energy is achieved if, when it is occupied, the molecule is angular. The shape adopted by an H_2X molecule therefore depends on the number of electrons that occupy the orbitals.

The simplest XH_2 molecule in Period 2 is the transient gas-phase BeH_2 molecule (BeH_2 normally exists as a polymeric solid), in which there are four valence electrons. These four electrons occupy the lowest two molecular orbitals: if the lower energy is achieved with the molecule angular, then that will be its shape. We can decide whether the molecule is angular or not by accommodating the electrons in the lowest two orbitals corresponding to an *arbitrary* bond angle in Fig. 3.17. We then note that the HOMO decreases in energy if we proceed to the right of the diagram and that the lowest total energy is obtained when the molecule is linear. Hence, BeH_2 is predicted to be linear and to have the configuration $1\sigma_g^2 1\sigma_u^2$. In CH_2, which has two more electrons than BeH_2, three of the molecular orbitals must be occupied. In this case, the lowest energy is achieved if the molecule is angular and has configuration $1a_1^2 2a_1^2 1b_2^2$. In general, any XH_2 molecule with from five to eight valence electrons is predicted to be angular. The observed bond angles are

BeH_2	BH_2	CH_2	NH_2	OH_2
180	131	136	103	105

While both VSEPR theory and the Walsh approach can account qualitatively for these results, neither provides a reliable way of estimating the angles quantitatively.

Example 3.5: *Using a Walsh diagram to predict a shape*

Predict the shape of the H_2O molecule on the basis of a Walsh diagram for an XH_2 molecule.

Answer. We choose an intermediate bond angle along the horizontal axis of the XH_2 diagram in Fig. 3.17 and accommodate eight electrons. The resulting configuration is $1a_1^2 2a_1^2 1b_2^2 b_1^2$. The $2a_1$ orbital is occupied, so we expect the nonlinear molecule to be more stable than the linear.

Exercise. Is any Period 3 XH_2 molecule expected to be linear? If so, which?

Walsh applied his approach to molecules other than compounds of hydrogen, but the correlation diagrams soon become very complicated. Nevertheless, this represents a valuable complement to VSEPR theory since it traces the influences on molecular shapes of the occupation of orbitals spreading over the entire molecule and concentrates less on localized repulsions between pairs of electrons. Correlation diagrams like those introduced by Walsh are frequently encountered in contemporary discussions of the shapes of complex molecules, and we shall see a number of examples in later chapters. They illustrate how inorganic chemists can sometimes identify and weigh competing influences by considering two extreme cases (such as linear and 90° XH_2 molecules), and then rationalize the fact that the state of a molecule is a compromise intermediate between the two extremes.

Localization

A striking feature of the Lewis approach to chemical bonding is its accord with chemical instinct, for it identifies something that can be called 'an A—B bond'. Both O—H bonds in H_2O, for instance, are treated as localized, equivalent, structures, since each is an electron pair shared between O and H. This feature appears to be absent from molecular orbital theory, for molecular orbitals are delocalized and pairs of electrons in them bind all the atoms together, not just a specific pair. The concept of an A—B bond as existing independently of other bonds in the molecule, and of being transferable from one molecule to another, appears to have been lost. However, we shall now show that the molecular orbital description is

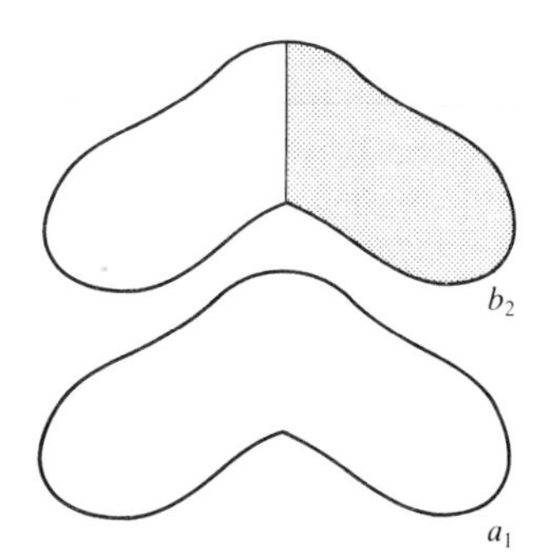

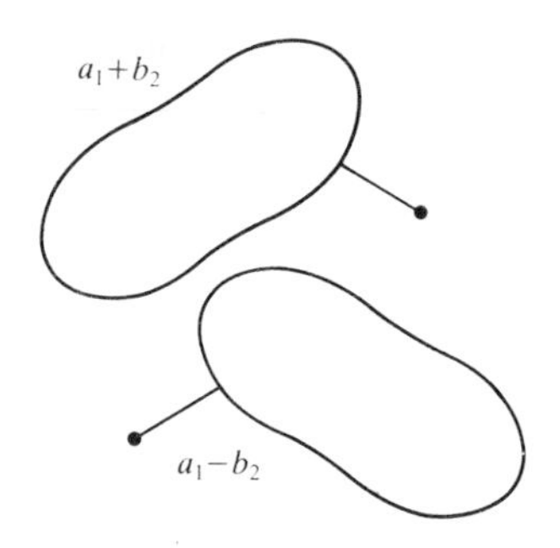

Fig. 3.19 The two occupied $1a_1$ and $1b_2$ orbitals of the H_2O molecule and their sum $1a_1 + 1b_2$ and difference $1a_1 - 1b_2$. In each case we form a localized orbital between a pair of atoms.

mathematically almost equivalent to a localized description of the overall electron distribution.

The demonstration hinges on the fact that quantum theory shows that we can take linear combinations of orbitals and still arrive at the same overall electron density distribution and energy of a molecule. We encountered an example of this when we dealt with hybridization, where we considered linear combinations of atomic orbitals on a single atom. In that case, the hybrid orbitals gave the same overall electron distribution as the original unhybridized orbitals. The same is true of molecular orbitals: we can take combinations and obtain the same *overall* electron distribution, but with individual orbitals that are distinctly different.

Consider the H_2O molecule. The two occupied orbitals of the delocalized description, $1a_1$ and $1b_2$, are shown in Fig. 3.19. If we form the sum $1a_1 + 1b_2$, the negative half of $1b_2$ cancels half the $1a_1$ orbital, leaving a localized orbital between O and the other H. Likewise, when we form the difference $1a_1 - 1b_2$, the other half of the $1a_1$ orbital is canceled, leaving a localized orbital between the other pair of atoms. That is, by taking sums and differences of delocalized orbitals, localized orbitals are created (and vice versa). Since these are two equivalent ways of describing the same overall electron population, one cannot be said to be better than the other. Hence it is reasonably well justified to use the localized description of a molecule when chemical evidence suggests that it is appropriate.

Table 3.2. A general indication of the properties when localized and delocalized descriptions are appropriate

Localized appropriate	Delocalized appropriate
Bond strengths	Electronic spectra
Force constants	Photoionization
Bond lengths	Electron attachment
Brønsted acidity	Magnetism
VSEPR description of molecular geometry	Walsh description of molecular geometry
	Reduction potentials

Table 3.2 suggests when it is appropriate to select a delocalized description or a localized description. The delocalized description is needed for dealing with global properties of the entire molecule. Such properties include electronic spectra (UV and visible transitions), photoionization spectra, ionization and electron attachment energies, and redox potentials. In contrast, ideas of bond strength, bond length, bond force constant, and some aspects of reactions (such as acid–base neutralizations) depend upon the characteristics of a fragment of the total molecule. For these properties, the localized description is best because it focuses attention on the distribution of electrons in and around a particular bond.

Isolobal analogies

A concept that is found throughout chemistry is the substitution of one fragment for another in a molecule. Thus, we may view $N(CH_3)_3$ as derived from NH_3 by substitution of a CH_3 fragment for each H atom. In current terminology, we speak of the structurally analogous fragments as being **isolobal** and express the relationship by the symbol $\leftarrow\!\!\!\!\!\!\!\!\;\;\circ\!\!\!\rightarrow$.

Two fragments are isolobal if their uppermost orbitals have the same symmetry (such as the σ symmetry of the H 1*s* and the C sp^3 orbital), similar energies, and the same electron occupation (one in each case in H 1*s* and C sp^3). In many of the applications of isolobality, we can adopt a simple localized orbital viewpoint. This allows us to identify families of isolobal fragments, such as

Once we have done so, we can anticipate that we may be able to form molecules such as $H_3C{-}Br$ and $(OC)_5Mn{-}CH_3$ by analogy with H—H.

In this localized orbital viewpoint we can also identify some isolobal fragments with two available singly occupied orbitals:

Some three-orbital isolobal fragments can also be identified:

The existence of these families suggests that we should expect to encounter molecules such as *cyclo*-C_4H_8, $O(CH_3)_2$, and $N(CH_3)_3$. These are all known. The complexes $Co_4(CO)_{12}$ (**4**) and $Co_3(CO)_9CH$ (**5**) are two more examples. However, isolobal analogies must be used with care, for they may also tempt us to postulate the existence of $(OC)_5Mn{-}O{-}Mn(CO)_5$ and $(OC)_5Fe{=}Fe(CO)_5$, but both are unknown. Isolobal analogies—like molecular orbitals themselves—provide useful correlations and hints, but they are no substitute for chemical knowledge.

The localized orbitals in the diagrams given above are not the most appropriate starting point for the construction of a full molecular orbital description. In this context we should translate the isolobal analogy

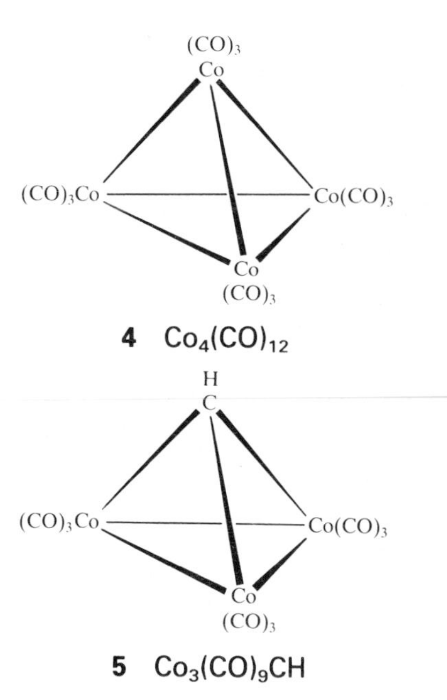

4 $Co_4(CO)_{12}$

5 $Co_3(CO)_9CH$

into diagrams that show the HOMOs and LUMOs of the two fragments:

The molecular orbital theory of solids

The molecular orbital theory of small molecules can be extended to account for the properties of solids, which are aggregations of a virtually infinite number of atoms. If for the moment we concentrate on one particular kind of solid, metals, we must be able to explain their characteristic luster, their good electrical and thermal conductivity, and their malleability. All these properties can be explained in terms of the ability of the atoms to contribute electrons to a common sea. The luster and electrical conductivity stem from the mobility of these electrons, either in response to the oscillating electric field of an incident ray of light or to a potential difference. The high thermal conductivity is also a consequence of electron mobility, since an electron can collide with a vibrating atom, pick up its energy, and transfer it to another atom elsewhere in the solid. The ease with which metals can be mechanically deformed is another aspect of electron mobility, since the electron sea can quickly readjust to a deformation of the solid and continue to bind the atoms together.

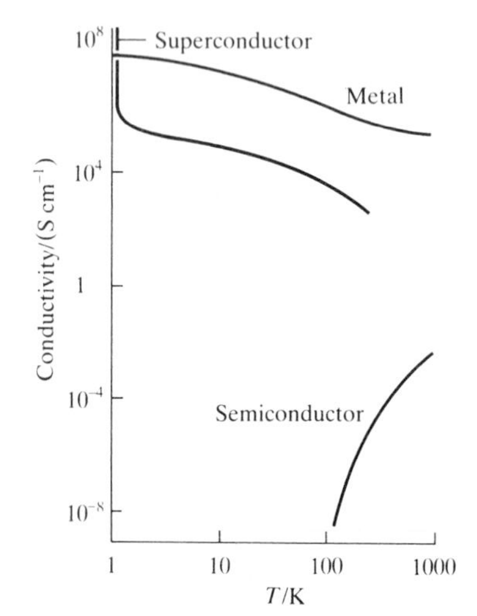

Fig. 3.20 The variation of the electrical conductivity of a substance with temperature enables us to classify it as a metal or a semiconductor.

Electronic conduction is also a characteristic of the solids known as **semiconductors**. The criterion for distinguishing between a metal and a semiconductor is the temperature-dependence of the electric conductivity:[5] the conductivity of a metal *decreases* with temperature; that of a semiconductor *increases* (Fig. 3.20). It is also generally the case (but not the criterion for distinguishing them) that conductivities of metals at room temperature are higher than those of semiconductors. An **insulator** is a substance with a very low electrical conductivity. However, when that conductivity can be measured, it is found to increase with temperature, like that of a semiconductor. For some purposes, therefore, it is possible to disregard the classification 'insulator' and to treat all solids as either metals or semiconductors. Superconductors are a special class of materials that have zero electrical resistance.

[5] The resistance R of a sample is measured in ohm, Ω. The inverse of the resistance is called the conductance G and is measured in siemens S, where $1\,\mathrm{S} = 1\,\Omega^{-1}$. The resistance of a sample increases with its length l and decreases with its cross-sectional area A, and we write

$$R = \frac{\rho l}{A}$$

where ρ is the resistivity of the substance. The units of resistivity are Ω m. The conductivity σ is the reciprocal of the resistivity, and its units are S m^{-1} or S cm^{-1}.

Example 3.6: *Classifying substances according to their electrical character*

The following table gives the electrical conductivities of bismuth at three different temperatures. What type of material is bismuth?

273 K	373 K	573 K
9.1×10^5 S m^{-1}	6.4×10^5 S m^{-1}	7.8×10^5 S m^{-1}

Answer. The conductivity decreases between the first two temperatures, showing that bismuth has metallic conductivity. The increase when the temperature is raised to 573 K might suggest that bismuth then becomes a semiconductor; however, we have to be alert for phase transitions that might upset the reasoning. Bismuth actually melts at 544 K, so the third value refers to the liquid, not the solid.

Exercise. Up to 125 K the conductivity of VO increases sharply with increasing temperature, and reaches 1×10^{-4} S m^{-1}. At about 125 K the conductivity rises abruptly to 1×10^{2} S m^{-1} and then declines slowly to about 5×10^{1} S m^{-1} near 400 K. What behavior is illustrated by (a) the low-temperature and (b) the high-temperature form of VO?

3.5 Molecular orbital bands

The central idea underlying the description of the electronic structure of solids is that the valence electrons donated by the atoms spread through the entire structure. This is expressed more formally by making a simple extension of molecular orbital theory in which the solid is treated like an indefinitely large molecule. (In solid state physics, this approach is called the 'tight-binding approximation'.) The description in terms of delocalized electrons is also valid for nonmetallic solids (such as ionic solids) but it often seems more natural for metals. We shall therefore begin by showing how metals are described in terms of molecular orbitals. Then we shall show that the same principles can be applied, but with a different outcome, to ionic and molecular solids.

Band formation by orbital overlap

The overlap of a large number of atomic orbitals leads to molecular orbitals that are closely spaced in energy and so form a virtually continuous **band** that covers a range of energies (Fig. 3.21). On an energy diagram, bands are often separated by **band gaps**, which are values of the energy for which there are no orbitals.

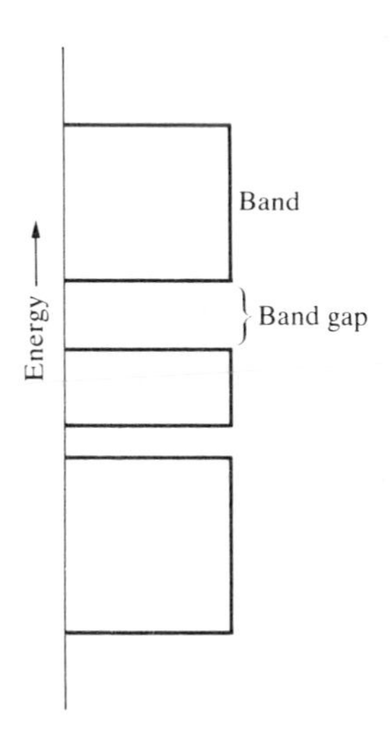

Fig. 3.21 The electronic structure of a solid is characterized by a series of bands of orbitals which are separated by gaps in the energy for which there are no orbitals.

The formation of bands can be understood by considering a line of atoms, and supposing that each atom has an s orbital that overlaps the s orbitals on its immediate neighbors (Figs. 3.22 and 3.23). When the line consists of only two atoms, there is a bonding and an antibonding molecular orbital. When a third atom joins them, there are three orbitals, the middle one being nonbonding. As more atoms are added, each one contributes an atomic orbital, and hence one more molecular orbital is formed.

When there are N atoms in the line there are N molecular orbitals. The total width, which remains finite even as N approaches infinity (Fig. 3.23), depends on the strength of the interaction between neighboring atoms. Since the width remains finite, it follows that the separation of the levels

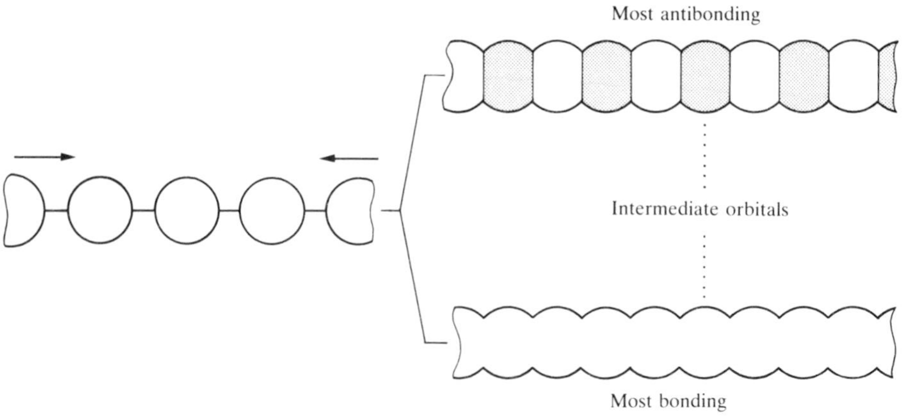

Fig. 3.22 A band can be thought of as formed by bringing up atoms successively to form a line of atoms. N atomic orbitals give rise to N molecular orbitals.

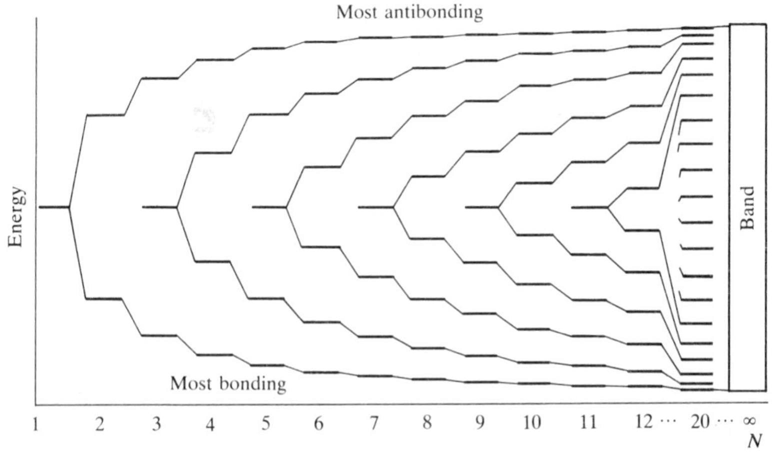

Fig. 3.23 The energies of the orbitals that are formed when N atoms are brought up to form a line.

within it approaches zero as N increases. That is, the band consists of a countable number but near continuum of energy levels. The orbital of lowest energy is fully bonding between all nearest neighbors. The uppermost orbital has a node between each nearest neighbor and is fully antibonding between any atom and its two immediate neighbors. Intermediate orbitals have intermediate bonding and antibonding character. As for molecules, the greater the number of nodes, the higher the energy.

The band just described is built from s orbitals and is called an **s band**. If there are p orbitals available, a ***p* band** can be constructed from their overlap as shown in Fig. 3.24. Since p orbitals lie higher in energy than s orbitals of the valence shell there is a gap between the s band and the p band (Fig. 3.25) provided the band width is not too great. The ***d* band** is similarly constructed from the overlap of d orbitals.

The Fermi level

Electrons occupy the orbitals in the bands in accord with the building-up principle. If each atom supplies one s electron, then at $T = 0$ the lowest $\frac{1}{2}N$ are occupied. The highest occupied orbital at $T = 0$ is called the **Fermi level**; it lies near the center of the band (Fig. 3.26). At temperatures above absolute zero, the population P of the orbitals is given by the **Fermi–Dirac distribution**, which is a version of the Boltzmann distribution that takes into

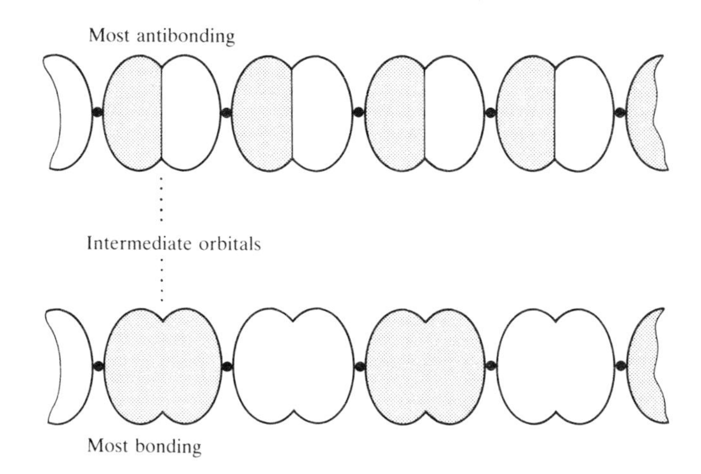

Fig. 3.24 An example of a p band in a one-dimensional solid.

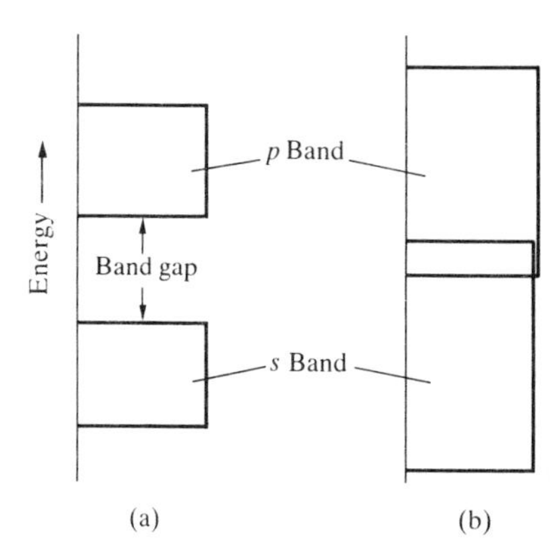

Fig. 3.25 (a) The s and p bands of a solid and the gap between them. Whether there is in fact a gap depends on the separation of the s and p orbitals of the atoms and the strength of interaction between the atoms. (b) If the interaction is strong, the bands are wide and may overlap.

account the effect of the Pauli exclusion principle:

$$P = \frac{1}{e^{(E-E_F)/kT} + 1} \quad (2)$$

E_F is the **Fermi energy**, the energy of the level for which $P = \frac{1}{2}$. The Fermi energy depends on the temperature and at $T = 0$ is equal to the energy of the Fermi level. The shape of the Fermi–Dirac distribution is shown in Fig. 3.27. For energies well above the Fermi energy, the 1 in the denominator can be neglected, and the populations resemble a Boltzmann distribution, decaying exponentially with increasing energy:

$$P \approx e^{-(E-E_F)/kT} \quad (3)$$

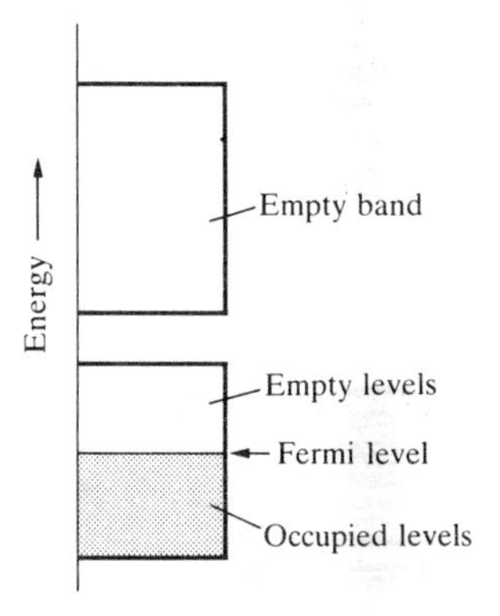

Fig. 3.26 If each of the N atoms supplies one s electron, then at $T = 0$ the lowest $\frac{1}{2}N$ are occupied, and the Fermi level lies near the center of the band.

When the band is not completely full, the electrons close to the Fermi surface can easily be promoted to nearby empty levels. As a result, they are mobile, and can move relatively freely through the solid. The substance is an electronic conductor. One way of seeing that this is so is to think of the individual orbitals in a band as being standing waves. As we saw when discussing the real and complex representations of atomic p orbitals (Section 1.4), standing waves can be regarded as superpositions of traveling waves corresponding to motion in opposite directions. In the absence of a potential difference, the two directions of travel are degenerate and are equally populated up to the Fermi level (Fig. 3.28a). However, when a potential difference is applied, electrons traveling in one direction have a different energy from those traveling in the opposite direction, and the two sets of orbitals are no longer equally populated (Fig. 3.28b). Consequently, there are now more electrons traveling in one direction that in the other, and an electric current flows through the solid.

We have seen that the criterion of metallic conduction is the decrease of conductivity with increasing temperature. This is the opposite of what we might expect if the conductivity were governed by the Boltzmann distribution of electrons. The competing effect can be identified once we recognize that the ability of an electron to travel smoothly through the solid in a

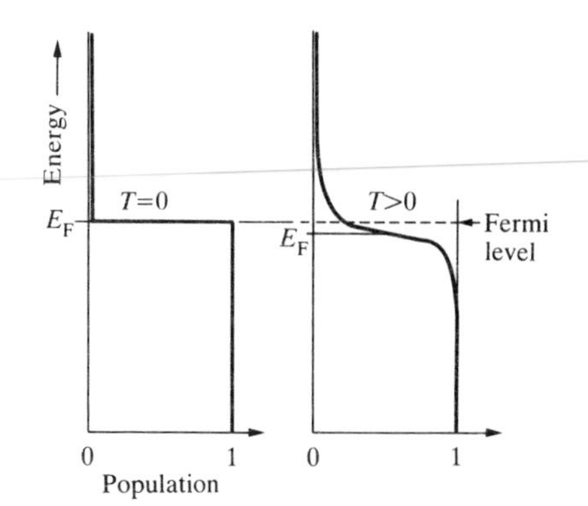

Fig. 3.27 The shape of the Fermi distribution (a) at $T = 0$ and (b) at $T > 0$. The population decays exponentially well above the Fermi level.

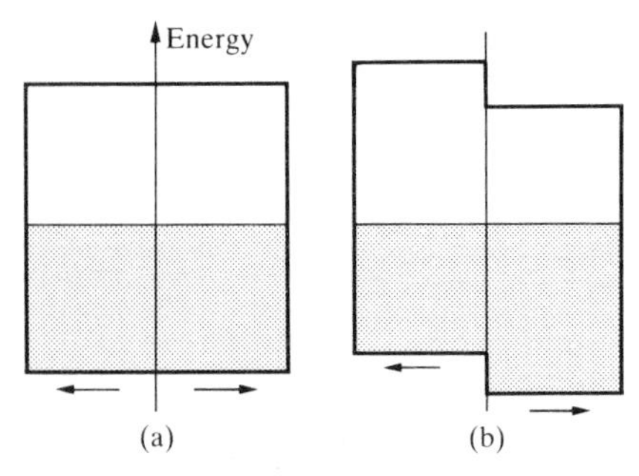

Fig. 3.28 Another way of representing the bands is to draw the orbitals for motion to the right and motion to the left separately. (a) In the absence of a potential difference applied across the metal, the corresponding orbitals are degenerate. However, when a field is applied, (b) one 'half band' has a lower energy than the other. It is more heavily populated, and the net result is a flow of electrons.

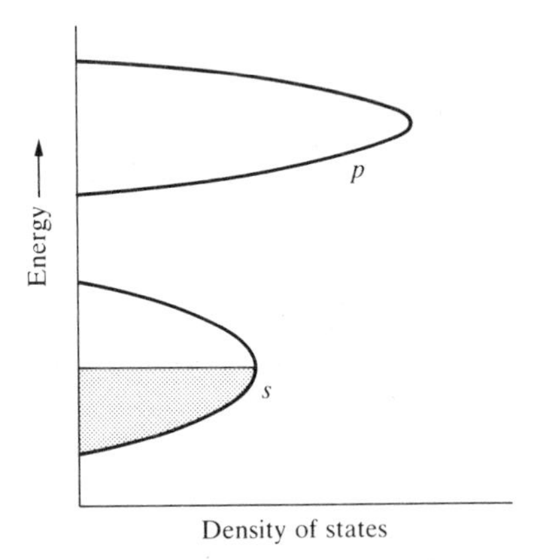

Fig. 3.29 A typical density of states in a metal.

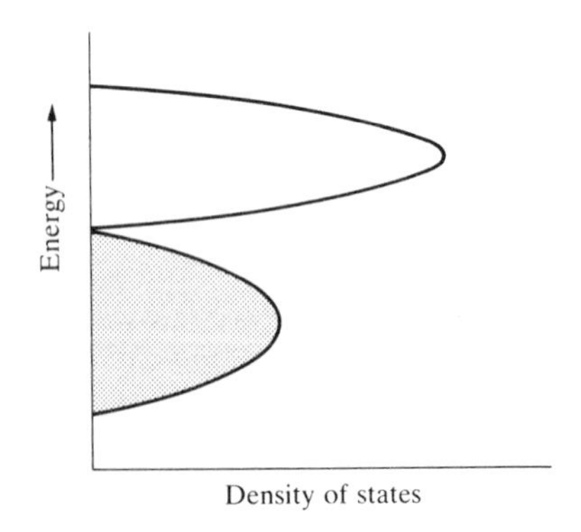

Fig. 3.30 The density of states typical of a semimetal.

conduction band depends on the uniformity of the arrangement of the atoms. An atom vibrating vigorously at a site is equivalent to an impurity that disrupts the orbitals. This reduces its ability to conduct from one edge of the solid to the other, so the conductivity of the solid is less than at $T = 0$. If we think of the electron as described by a wave propagating through the solid, we would say that it was 'scattered' by the impurity—the atomic vibration. This **carrier scattering** increases with increasing temperature as the lattice vibrations increase and accounts for the observed inverse temperature dependence of the conductivity of metals.

Densities of states

The number of energy levels per unit energy increment is called the **density of states** ρ. It is not uniform across a band, meaning that the energy levels are packed together more closely at some energies than at others. This is apparent even in one dimension, for the center of the band is relatively sparse in orbitals compared with its edges (Fig. 3.23). In three dimensions the variation of density of states is more like that shown in Fig. 3.29, with the greatest density of states near the center of the band and relatively sparse at the edges. The reason for this behavior can be traced to the number of ways of producing a particular linear combination of atomic orbitals. There is only one way of forming a fully bonding molecular orbital (the lower edge of the band) and only one way of forming a fully antibonding orbital (the upper edge). However, there are many ways (in a three-dimensional array of atoms) of forming a molecular orbital with an energy in the interior of the band.

The density of states is zero in the band gap—there are no energy levels there. However, in certain special cases a full band and an empty band might coincide in energy,[6] but with a zero density of states at their conjunction (Fig. 3.30). Solids of this kind are called **semimetals**. Since they have only a few electrons that can act as carriers, semimetals are characterized by a low metallic conductivity. One important example is graphite, which is a semimetal in directions parallel to the sheets of carbon atoms. The use of the term semimetal in this context should be carefully distinguished from another unrelated usage, in which it denotes a *chemical* character intermediate between that of metal and nonmetal; for this character we use the term **metalloid**.

Photoelectron and X-ray analysis of bands

Evidence for the existence of bands and a map of their densities of states can be obtained experimentally with photoelectron spectroscopy in much the same way as for discrete molecules. The densities of states of discrete molecules consist of a series of widely separated narrow spikes, each spike corresponding to the energy of a discrete molecular orbital. These spikes appear in the photoelectron spectrum as photoelectrons with discrete ionization energies.

Analogous information can be inferred for solids from **X-ray emission bands**. In this technique, electrons are ejected from the inner closed shells

[6] The coincident orbitals at the top and bottom of the two bands are not identical wavefunctions since they differ in wavelength, a feature not shown in the illustration. That is, the electrons that occupy orbitals at the top and bottom of the bands have the same energy but different linear momenta.

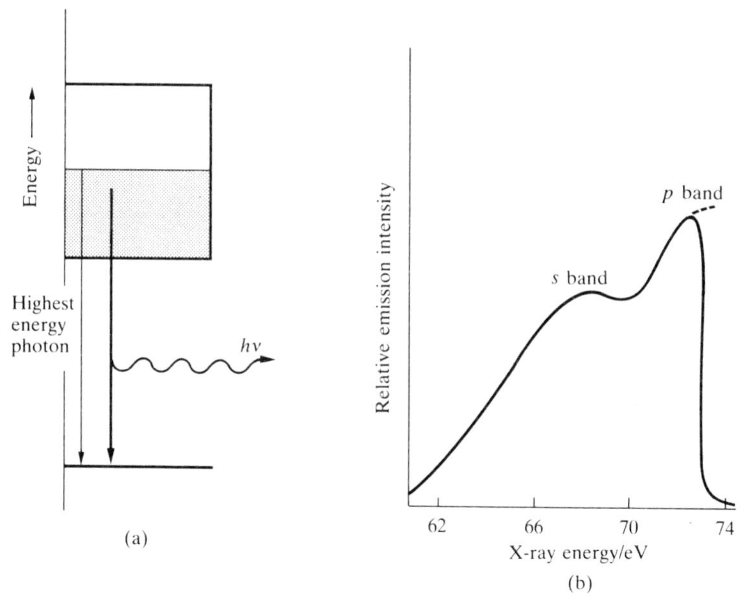

Fig. 3.31 (a) The formation of an X-ray emission band and (b) a typical example (aluminum).

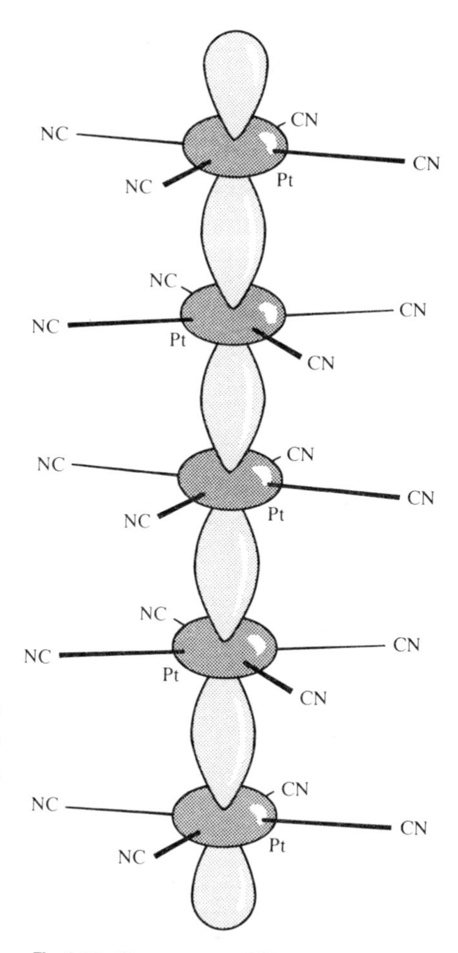

Fig. 3.32 The structure of KCP, $K_2Pt(CN)_4Br_{0.3}{\cdot}3H_2O$.

of the atoms by electron bombardment, and X-rays are emitted as electrons from the valence bands fall into the vacancy (Fig. 3.31). In general, a range of X-ray frequencies is produced, corresponding to electrons originating from any of the occupied states of the valence bands. The intensity profile of the emission band depends on the density of states in the band, and so the latter can be deduced. However, the correspondence between the intensity and the density of states is not straightforward. The X-ray emission intensity, like any electromagnetic radiation emission, also depends on selection rules and transition probabilities (and hence on the details of the wavefunctions of the initial and final states).

Special features of one-dimensional solids

In recent years, a series of solids have been investigated that have a partially filled band formed by a linear chain of metal atoms. An example is $K_2Pt(CN)_4Br_{0.3}{\cdot}3H_2O$, commonly called 'KCP' (Fig. 3.32). These 'one-dimensional solids' are not quite as simple as has been implied by the discussion so far, for a theorem due to Rudolph Peierls states that at $T = 0$, *no* one-dimensional solid is a metal!

The origin of the theorem can be traced to a hidden assumption. We have supposed that the atoms lie in a line with a regular separation. However, the actual spacing in a one-dimensional solid (and any solid) is determined by the distribution of the electrons, not the other way round, and there is no guarantee that the state of lowest energy is a solid with a regular lattice spacing. In fact, in a one-dimensional solid at $T = 0$, there always exists a distortion, a **Peierls distortion**, which leads to a lower energy than in the perfectly regular solid.

An idea of the origin and effect of a Peierls distortion can be obtained by considering a one-dimensional solid of N atoms and N valence electrons (Fig. 3.33). Such a line of atoms distorts to one that has long and short alternating bonds. Although the longer bond is energetically unfavorable,

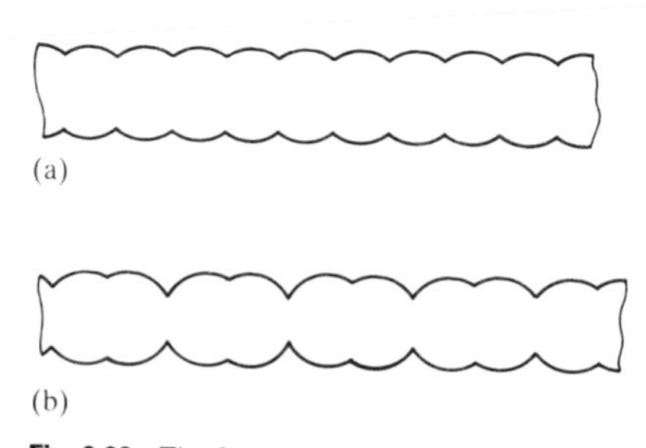

Fig. 3.33 The formation of a Peierls distortion: the energy of the line of atoms with alternating bond lengths is lower than that of the uniformly spaced atoms.

the strength of the short bond more than compensates for the weakness of the long bond, and the net effect is a lowering of energy below that of the regular solid. Now, instead of the electrons near the Fermi level being free to move through the solid, they are trapped between the longer-bonded atoms. The Peierls distortion introduces a band gap in the center of the original conduction band, and the filled orbitals are separated from the empty orbitals. Hence, the distortion results in semiconductor properties.

The conduction band in KCP is a *d* band formed principally by overlap of Pt $5d_{z^2}$ orbitals. The small proportion of Br, which exists as Br^- in the solid, removes a small number of electrons from this otherwise full *d* band, hence turning it into a conduction band. Indeed, at room temperature, KCP has a lustrous bronze color with its highest conductivity along the axis of the Pt chain. However, below 150 K the conductivity drops sharply. This can be explained by the onset of a Peierls distortion, which traps the uppermost electrons in regions where the Pt—Pt distances are short and separated by long Pt—Pt bond length regions. That is, at low temperatures the distortion of the lattice introduces a band gap at the Fermi level. At higher temperatures the motion of the atoms averages the distortion to zero, the separation is regular (on average), the gap is absent, and the solid is metallic.

Insulators

If enough electrons are present to fill a band completely, and there is a considerable energy gap before an empty orbital becomes available (Fig. 3.34), the substance is an insulator. In an NaCl crystal, for instance, the N Cl^- ions are nearly in contact and their 3*s* and 3*p* valence orbitals overlap to form a narrow band of $4N$ levels. The Na^+ ions have small but nonzero overlap and also form a band. The electronegativity of chlorine is so much greater than that of sodium that the chlorine band lies well below the sodium band, and the band gap is about 7 eV. A total of $8N$ electrons are to be accommodated (7 from each chlorine atom, one from each sodium atom). These enter the lower chlorine band, fill it, and leave the sodium band empty. Since $kT \approx 0.03$ eV at room temperature, electrons cannot easily be promoted into empty orbitals.

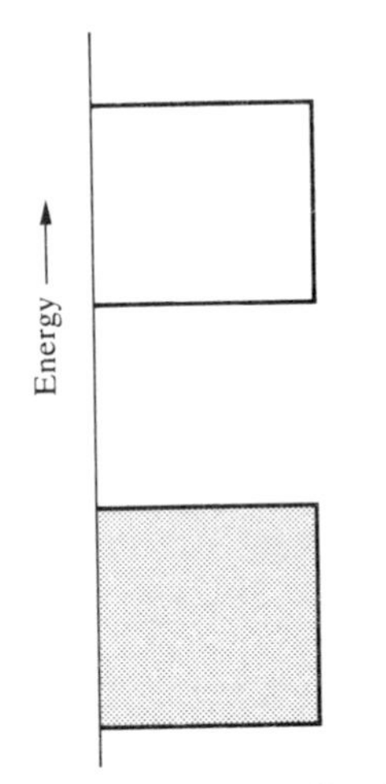

Fig. 3.34 The structure of a typical insulator: there is a significant gap between the filled and empty bands.

We normally think of an ionic or molecular solid as consisting of discrete ions or molecules, yet according to the picture we have just described it appears that they should be regarded as having a band structure. The two pictures can be reconciled in much the same way as we showed that a delocalized bonding description of discrete molecules is virtually equivalent to a description in terms of localized bonds. In this case, it is possible to show that a *full* band is equivalent to a sum of localized electron densities. In NaCl, a full band built from Cl orbitals is equivalent to a collection of discrete Cl^- ions. As with molecules, the delocalized band picture is needed for description of spectra where processes involve one electron at a time such as photoelectron spectra and X-ray spectra.

3.6 Semiconduction

The characteristic physical property of a semiconductor is that its electrical conductivity increases strongly with temperature. At room temperature, conductivities of semiconductors are typically intermediate between those of metals and insulators (in the region of 10^3 S cm^{-1}). The dividing line between insulators and semiconductors is a matter of the size of the band gap, and even diamond is now being considered as a promising semiconductor material. The actual value of the conductivity is an unreliable criterion because, as its temperature is increased, a given substance may have a low, intermediate, or high conductivity. The band gap and conductivity that are taken as indicating semiconduction rather than insulation depend on the application being considered.

Intrinsic semiconductors

In an **intrinsic semiconductor**, the band gap is so small that the Fermi distribution results in some electrons populating the empty upper band (Fig. 3.35). This introduces negative carriers into the upper level and positive holes into the lower and the solid is conducting. The upper band is broader than the lower, more localized band. A semiconductor at room temperature is generally much less strongly conducting than a conventional metal because only electrons and holes in the exponentially decaying tail of the Fermi distribution are active as charge carriers. The strong temperature dependence follows from the exponential Boltzmann-like temperature dependence of the electron population in the upper band.

We can anticipate that the conductivity of a semiconductor will show an Arrhenius-like temperature dependence of the form

$$\sigma = Ae^{-E_a/kT} \quad (4)$$

on account of the exponential dependence of the population of charge carriers. But how is the activation energy related to the band gap E_g? We can answer this by deciding how $E - E_F$, which appears in the high-temperature form of the Fermi–Dirac distribution (eqn 3) depends on E_g. This hinges on knowing the location of the Fermi energy (the energy at which $P = \frac{1}{2}$).

In a simple picture of the band structure, the Fermi energy is approximately half-way between the upper and lower bands (Fig. 3.36), and the

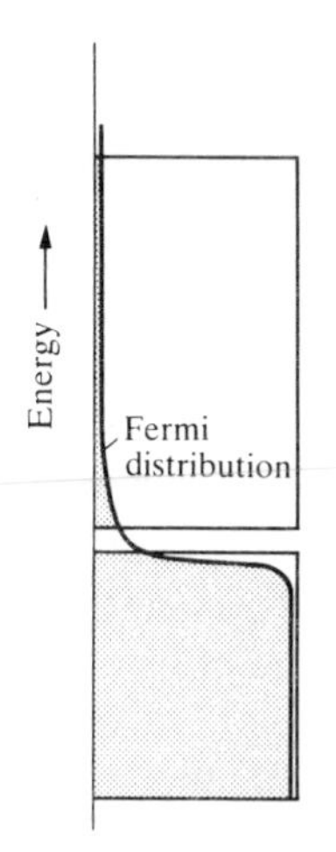

Fig. 3.35 In an intrinsic semiconductor, the band gap is so small that the Fermi distribution results in some electrons populating the empty upper band.

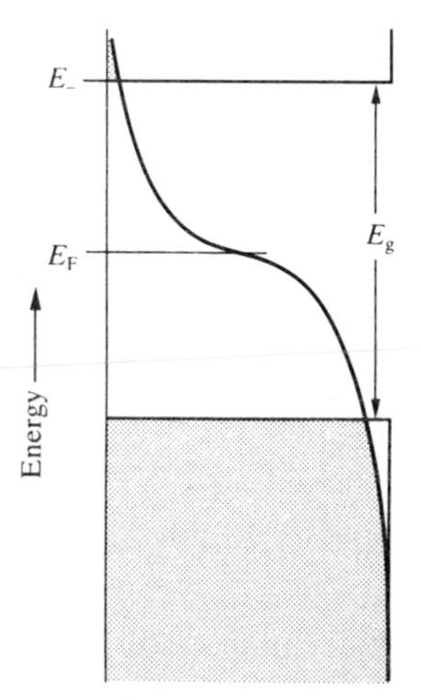

Fig. 3.36 The relation between the Fermi distribution and the band gap.

energy of the lowest level of the upper band E_- is related to E_F and E_g by

$$E_- - E_F \approx \tfrac{1}{2}E_g$$

It follows that the number of charge carriers in the upper band, and hence the conductivity, is approximately proportional to

$$n \propto e^{-E_g/2kT}$$

Hence, the temperature dependence of the conductivity of a semiconductor can be expected to be Arrhenius-like with an activation energy equal to half the band gap, $E_a \approx \tfrac{1}{2}E_g$. This is found to be the case in practice.

Example 3.7: *Determining the band gap*

The conductance G of a sample of germanium varied with temperature as indicated below. Estimate the value of E_g.

T/K	312	354	420
G/S	0.0847	0.429	2.86

Answer. From eqn 4 we see that the analysis is similar to that used to obtain the activation energy of a chemical reaction. Since the conductance is proportional to the conductivity σ, we can write

$$\sigma = bG = Ae^{-E_a/kT}$$

Taking logarithms gives

$$\ln b + \ln G = -E_a/kT$$

Therefore, a plot of $\ln G$ against $1/T$ should yield a straight line of slope $-E_a/k$. The data give a slope of -4.26×10^3, and since $k = 8.614$ eV K^{-1}, we find $E_a = 0.367$ eV. Since $E_g = 2E_a$, this gives $E_g = 0.73$ eV.

Exercise. What is the conductance of the sample at 370 K?

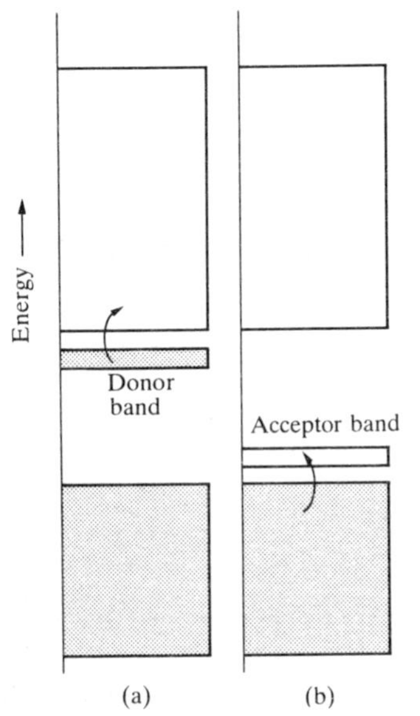

Fig. 3.37 The band structure in (a) an n-type semiconductor and (b) a p-type semiconductor.

Extrinsic semiconductors

The number of electron carriers can be increased if atoms with more electrons than the parent element can be introduced in the process called **doping**. Remarkably low levels of dopant concentration are needed—only about one atom per 10^9 atoms of the host material—so it is essential to achieve very high purity.

If As atoms are introduced into a silicon crystal, then one additional electron will be available for each **dopant** atom that is substituted. Note that the doping is *substitutional* in the sense that the dopant atom takes the place of an Si atom. If the donor atoms, the As atoms, are far apart from each other, their electrons will be localized and the **donor band** will be very narrow (Fig. 3.37a). Commonly the energy of the filled dopant band is close to that of the empty band of the lattice. For $T > 0$, some of its electrons will be thermally promoted into the empty conduction band. In other words, thermal excitation will lead to the transfer of an As electron on to a neighboring Si atom from where it will be able to migrate through the lattice in the molecular orbitals formed by Si–Si overlap. This gives rise to **n-type semiconductivity**, the n indicating that the charge carriers are negative electrons.

An alternative substitutional procedure is to dope the silicon with atoms of an element with fewer electrons per atom, such as gallium. A dopant atom of this kind effectively introduces holes into the solid. More formally, the dopant atoms form a very narrow, empty **acceptor band** that lies above the full Si band (Fig. 3.37b). At $T = 0$ the acceptor is empty, but at higher temperatures it can accept thermally excited electrons and hence withdraw them from the Si valence band. By doing so, it introduces holes into the latter and hence allows the remaining electrons to be mobile. Since the charge carriers are now effectively positive holes in the lower band, this type of semiconductivity is called **p-type semiconductivity**.

Several *d*-metal oxides, including ZnO and Fe_2O_3, are n-type semiconductors. In this case the property is due to nonstoichiometry and a small deficit of O atoms. The electrons that should occupy the localized O atomic orbitals (giving a very narrow oxide band, essentially localized individual ions) occupy a previously empty conduction band formed by the metal orbitals. The conductivity decreases when the solid is heated in oxygen because the deficit of O atoms is replaced. As these O atoms are replaced, electrons are withdrawn from the conduction band.

p-Type semiconductivity is observed for some *d*-metal chalcogenides and halides with lower oxidation numbers, including Cu_2O, FeO, FeS, and CuI. In these nonstoichiometric compounds a deficit of electrons is equivalent to the oxidation of some of the metal atoms, leaving a hole in the band formed by the ions of lower oxidation number. When these compounds are heated in oxygen, the conductivity increases because more holes are formed in the metal ion band as oxidation progresses.

3.7 Superconduction

A **superconductor** is a substance that conducts electricity without resistance. Until 1987, the only known superconductors (which included metals, some oxides, and some halides) needed to be cooled to below about 20 K before they became superconducting. However, in 1987 the first 'high-temperature' superconductors were discovered; their superconduction is well established at 120 K and spasmodic reports of even higher temperatures have appeared. We will not consider these high-temperature materials at this stage (they are discussed in Chapter 18), but sketch the ideas behind the mechanism of low-temperature superconduction.

The central concept of low-temperature superconduction is the existence of a **Cooper pair**, a pair of electrons that exists on account of their interaction indirectly through vibrational displacements of the atoms in the lattice. Thus, if one electron is in a particular region of a solid, the nuclei there move toward it to give a distorted local structure (Fig. 3.38). Since that local distortion is rich in positive charge, it is favorable for a second electron to join the first. Hence, there is a virtual attraction between the two electrons, and they move together as a pair. The local distortion can be easily disrupted by thermal motion of the ions, so the virtual attraction occurs only at very low temperatures.

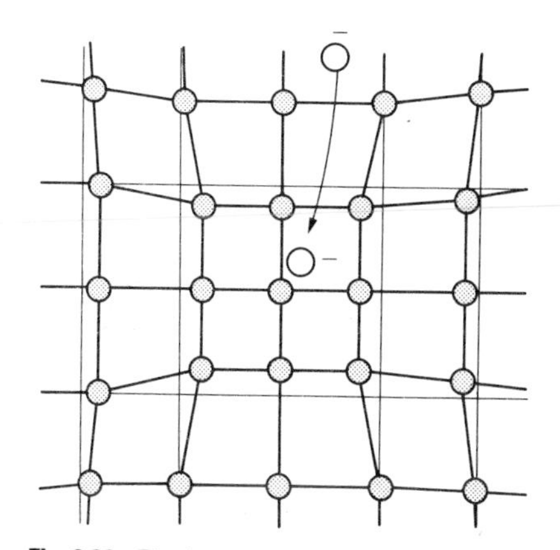

Fig. 3.38 The formation of a Cooper pair. One electron distorts the crystal lattice, and the second electron has a lower energy if it goes to that region. This effectively binds the two electrons into a pair.

A Cooper pair undergoes less scattering than an individual electron as it travels through the solid, since the distortion caused by one electron can attract back the other electron should it be scattered out of its path in a collision. This has been likened to the difference between the motion of a herd of cattle, with members of the herd that are deflected from their path by boulders in their way, and a team of cattle yoked together, which will travel forward largely regardless of obstacles. Since the Cooper pair is stable against scattering, it can carry charge freely through the solid, and hence give rise to superconduction.

Further information: symmetry-adapted orbitals

In this section we shall take the symmetry analysis of orbitals one stage further than in the text. Our aim is to show how to use character tables to construct linear combinations of orbitals of a specific symmetry. The justification for the procedure we are about to describe is based on the mathematics of group theory, and we do not intend to go that far. These remarks are best regarded as a bridge between what we have asserted in the text and what is proved in the references we give to group theory in our suggestions for further reading.

The central concept we need to generate symmetry adapted orbitals from a set of atomic orbitals is that of a **projection operator** P. This operator is so called because it 'projects out' from the basis set the linear combinations we require (as we shall see in action shortly). A projection operator is defined as follows:

$$P = \frac{1}{h}\sum_{O} \chi(O)O$$

where h is the order of the group (the number of operations in the group), $\chi(O)$ is the character of the operation O, and O is the operation itself. That is, P is a weighted sum of the symmetry operations of the group, the weighting coefficients being the characters of the operations for the symmetry type we are seeking to generate.

	E	C_2
A	1	1
B	1	−1

As an example, consider the group C_2, which consists of the identity operation E and the two-fold rotation C_2 and hence is of order 2. Its character table is shown in the margin. The projection operator is the sum

$$P = \tfrac{1}{2}\{\chi(E)E + \chi(C_2)C_2\}$$

where $\chi(E)$ is the character of the identity operation and $\chi(C_2)$ that of the rotation. If we want to project out the linear combination of orbitals of symmetry type a we use the characters $\chi(E) = 1$ and $\chi(C_2) = 1$ and obtain

$$P = \tfrac{1}{2}(E + C_2)$$

If we wanted the orbitals of symmetry type b (the only other type in this group), we use $\chi(E) = 1$ and $\chi(C_2) = -1$ and obtain

$$P = \tfrac{1}{2}(E - C_2)$$

Next, we see how P acts. The simplest approach is to consider a basis set composed of an s orbital on each atom and write them $\phi_s(\mathrm{A})$ and $\phi_s(\mathrm{B})$. In order to work out how P acts, we need to know how the basis is affected by the operations of the group. This is straightforward in the group C_2 because E leaves each orbital unchanged and C_2 interchanges them:

$$E\phi_s(\mathrm{A}) = \phi_s(\mathrm{A}) \qquad E\phi_s(\mathrm{B}) = \phi_s(\mathrm{B})$$

$$C_2\phi_s(\mathrm{A}) = \phi_s(\mathrm{B}) \qquad C_2\phi_s(\mathrm{B}) = \phi_s(\mathrm{A})$$

It follows that when P acts on either orbital, it generates a sum of the two orbitals. For example, if we take the projection operator for symmetry type a and apply it to $\phi_s(\mathrm{A})$, we get

$$P\phi_s(\mathrm{A}) = \tfrac{1}{2}(E + C_2)\phi_s(\mathrm{A}) = \tfrac{1}{2}\{\phi_s(\mathrm{A}) + \phi_s(\mathrm{B})\}$$

The same linear combination is obtained if we use $\phi_s(\mathrm{B})$ in place of $\phi_s(\mathrm{A})$, and so that gives no new information. We have now projected the combination of a symmetry from the basis, and so can write

$$a = \tfrac{1}{2}\{\phi_s(\mathrm{A}) + \phi_s(\mathrm{B})\}$$

The same procedure, but with the projection operator for B, projects out the symmetry-adapted combination of that symmetry:

$$P\phi_s(\mathrm{A}) = \tfrac{1}{2}(E - C_2)\phi_s(\mathrm{A}) = \tfrac{1}{2}\{\phi_s(\mathrm{A}) - \phi_s(\mathrm{B})\}$$

Hence we can write

$$b = \tfrac{1}{2}\{\phi_s(\mathrm{A}) - \phi_s(\mathrm{B})\}$$

and have now obtained all the symmetry-adapted linear combinations that it is possible to form from a basis set of two orbitals.

Exactly the same procedure may be used for other groups and other bases. However, we run into a complication when we try to generate degenerate orbitals, such as the two e orbitals in the group C_3 (or C_{3v} and D_{3h}) and their analogs in other groups. We remarked in the text that the characters in the text are the *sums* of the characters for the individual components of a degenerate set. Therefore, when we apply the projection operator P we produce sums of the two symmetry-adapted linear combinations that we are trying to generate.

The simplest example that makes this clear is that of a basis set of three s orbitals in the group C_3 (of order 3), for which the character table is given in the margin. We have to remember that the column headed $2C_3$ relates to the *two* symmetry operations C_3 (rotation by 120°) and C_3^2 (rotation by 240°). The projection operator is therefore

	E	$2C_3$
A	1	1
E	2	−1

$$P = \tfrac{1}{3}\{\chi(E)E + \chi(C_3)C_3 + \chi(C_3^2)C_3^2\}$$

It follows that when we want to generate the combination of a symmetry we use

$$P = \tfrac{1}{3}\{E + C_3 + C_3^2\}$$

and when we want the doubly-degenerate e combination we use

$$P = \tfrac{1}{3}\{2E - C_3 - C_3^2\}$$

As in the C_2 example, we need to know the effect of the individual operations of the group. The identity E leaves each one unchanged and

$$C_3\phi_s(\mathrm{A}) = \phi_s(\mathrm{B}) \qquad C_3\phi_s(\mathrm{B}) = \phi_s(\mathrm{C}) \qquad C_3\phi_s(\mathrm{C}) = \phi_s(\mathrm{A})$$

$$C_3^2\phi_s(\mathrm{A}) = \phi_s(\mathrm{C}) \qquad C_3^2\phi_s(\mathrm{B}) = \phi_s(\mathrm{A}) \qquad C_3^2\phi_s(\mathrm{C}) = \phi_s(\mathrm{B})$$

There is no problem when we try to generate the a combination, because when we apply P to $\phi_s(\mathrm{A})$ we get

$$P\phi_s(\mathrm{A}) = \tfrac{1}{3}\{E + C_3 + C_3^2\}\phi_s(\mathrm{A}) = \tfrac{1}{3}\{\phi_s(\mathrm{A}) + \phi_s(\mathrm{B}) + \phi_s(\mathrm{C})\}$$

and we get the same linear combination when we apply P to the other two orbitals. Hence

$$a = \tfrac{1}{3}\{\phi_s(\mathrm{A}) + \phi_s(\mathrm{B}) + \phi_s(\mathrm{C})\}$$

This is the form of the a orbital we used for H_3 in the text. However, when we apply the projection operator for the e orbitals, we get a different combination in each case:

$$P\phi_s(A) = \tfrac{1}{3}\{2E - C_3 - C_3^2\}\phi_s(A) = \tfrac{1}{3}\{2\phi_s(A) - \phi_s(B) - \phi_s(C)\}$$
$$P\phi_s(B) = \tfrac{1}{3}\{2E - C_3 - C_3^2\}\phi_s(B) = \tfrac{1}{3}\{2\phi_s(B) - \phi_s(C) - \phi_s(A)\}$$
$$P\phi_s(C) = \tfrac{1}{3}\{2E - C_3 - C_3^2\}\phi_s(C) = \tfrac{1}{3}\{2\phi_s(C) - \phi_s(A) - \phi_s(B)\}$$

These three combinations are not independent of each other because the sum of the last two is

$$\tfrac{1}{2} \times \tfrac{1}{3}\{2\phi_s(B) - \phi_s(C) - \phi_s(A)\} + \tfrac{1}{2} \times \tfrac{1}{3}\{2\phi_s(C) - \phi_s(A) - \phi_s(B)\} = -\tfrac{1}{3}\{2\phi_s(A) - \phi_s(B) - \phi_s(C)\}$$

which, apart from the overall sign, is the same as the first. However, the difference of the two is

$$\tfrac{1}{2} \times \tfrac{1}{3}\{2\phi_s(B) - \phi_s(C) - \phi_s(A)\} - \tfrac{1}{2} \times \tfrac{1}{3}\{2\phi_s(C) - \phi_s(A) - \phi_s(B)\} = \tfrac{1}{2}\{\phi_s(B) - \phi_s(C)\}$$

and this combination is different from the first. We see that the use of characters has resulted in the generation of the sum of two linear combinations, only one of which is independent of a combination that has been generated previously. The two linearly independent symmetry-adapted combinations of E symmetry are therefore

$$\tfrac{1}{3}\{2\phi_s(A) - \phi_s(B) - \phi_s(C)\} \quad \text{and} \quad \tfrac{1}{2}\{\phi_s(B) - \phi_s(C)\}$$

which are the combinations we used in the text when discussing H_3 and related molecules.

The complication that arises when we consider degenerate orbitals using character tables can be avoided by more advanced group-theoretical methods. It should not be allowed to obscure the fact that projection operators and character tables jointly provide a very powerful technique for generating linear combinations of atomic orbitals of a given symmetry. They were used to generate the orbitals quoted in the text and illustrated in Appendix 6.

Further reading

Structure and bonding

The first two books are simple introductions to chemical bonding:

R. L. DeKock and H. B. Gray, *Chemical structure and bonding*. Benjamin-Cummings, Menlo Park (1980).

J. N. Murrell, S. F. A. Kettle, and J. M. Tedder, *The chemical bond*. Wiley, New York (1978).

More thorough discussions are given in the following:

R. McWeeney, *Coulson's Valence*. Oxford University Press (1979).

J. K. Burdett, *Molecular shapes: theoretical models of inorganic stereochemistry*. Wiley, New York (1980).

T. A. Albright, J. K. Burdett, and M.-H. Wangbo, *Orbital interactions in chemistry*. Wiley, New York (1985). This text has a thorough discussion of isolobality.

Point groups

Four reasonably elementary accounts of the use of point groups and character tables in chemistry are:

S. F. A. Kettle, *Symmetry and structure.* Wiley, New York (1985).
B. E. Douglas and C. A. Hollingsworth, *Symmetry in bonding and spectra.* Academic Press, New York (1985).
D. C. Harris and M. D. Bertolucci, *Symmetry and spectroscopy.* Oxford University Press (1978).
F. A. Cotton, *Chemical applications of group theory.* Wiley, New York (1990).

Solids

The following three texts give further details on the solid state concepts that we have introduced, and do so at about the same level:

P. A. Cox, *The electronic structure and chemistry of solids.* Oxford University Press (1987).
M. F. C. Ladd, *Structure and bonding in solid state chemistry.* Wiley, New York (1979).
M. H. B. Stiddard, *The elementary language of solid state physics.* Academic Press, New York (1975).

Exercises

3.1 How many independent linear combinations are possible for four 1*s* orbitals evenly spaced in a line? Draw pictures of the linear combinations of H 1*s* orbitals for the hypothetical linear H_4 molecule. From a consideration of the number of nonbonding and antibonding interactions, arrange these molecular orbitals in order of increasing energy.

3.2 Determine the symmetry group appropriate to the pyramidal SO_3^{2-} ion. What is the maximum degeneracy of a molecular orbital in this ion? If the sulfur orbitals are 3*s* and 3*p*, which of these can contribute to molecular orbitals of this maximum degeneracy?

3.3 Determine the point group of the PF_5 molecule. (Use VSEPR, if necessary, to assign geometry.) What is the maximum degeneracy of its molecular orbitals? Which P 3*p* orbitals contribute to a molecular orbital of this degeneracy?

3.4 In a metal complex MH_4L_2 ($L =$ a ligand), four H atoms form a square planar array around the central atom. Draw the symmetry-adapted combinations of H 1*s* orbitals. Determine the point group for the complex and assign the symmetry label of each of these symmetry-adapted orbitals. Which metal *d* orbitals have the correct symmetry to form molecular orbitals with these H linear combinations?

3.5 Construct a qualitative molecular orbital diagram for the hypothetical $[H{-}F{-}H]^+$ linear ion. Use as a basis set the 1*s* orbitals on the two H atoms and the 2*s* and 2*p* orbitals on F. First construct linear combinations of H orbitals then choose the F orbitals of appropriate symmetry to construct molecular orbitals.

3.6 In Chapter 2 we defined bond order. (a) Find the average N—H bond order in NH_3 by calculating the net number of bonds and dividing by the number of N—H links. (b) Find the average bond order in SF_6.

3.7 From the relative atomic orbital and molecular orbital energies depicted in Fig. 3.9, describe the character as mainly F or mainly S for the frontier orbitals *e* (the HOMO) and 2*t* (the LUMO) in SF_6. Explain your reasoning.

3.8 Using the concept of isolobality, give (a) the hydrogen–nitrogen molecule or molecular fragment that is isolobal with CH_3^-, (b) the hydrogen–boron molecule or molecular fragment that is isolobal with the O atom, (c) a nitrogen-containing species that is isolobal with $[Mn(CO)_5]^-$.

3.9 (a) Draw a simple band picture to distinguish a metal from a semiconductor. (b) Explain how the temperature dependence of the electrical conductivity can be used to distinguish a metal from a semiconductor. (c) Can the temperature dependence of the conductivity be used to distinguish an insulator from a semiconductor?

3.10 Decide whether the following are likely to be n-type or p-type semiconductors: (a) As-doped Ge, (b) Ga-doped Ge, (c) Si-doped Ge.

3.11 The promotion of an electron from the valence band into the conduction band in pure TiO_2 by light absorption requires a wavelength of less than 350 nm. Calculate the energy gap in eV between the valence and conduction bands.

3.12 When TiO_2 is heated in hydrogen, a blue color develops, indicating light absorption in the red. Does reduction of Ti(IV) to Ti(III) correspond to n-doping or p-doping?

3.13 A semiconductor with advantages for fast circuits is made from gallium arsenide. If GaAs is doped with Se, is it n-doped or p-doped?

3.14 CdS is used as a photoconductor in light-meters. The band gap is about 2.4 eV. What is the greatest wavelength of light that can promote an electron from the valence band to the conduction band in CdS?

3.15 The band gap of Si as determined from optical absorption spectra is 1.14 eV. Calculate the ratio of the conductivities at 373 K and 273 K.

Problems

3.1 Group theory is often used by chemists as an aid in the interpretation of infrared spectra. For example, there are four N—H bonds in NH_4^+ and therefore four stretching modes possible. There is the possibility that several vibrational modes occur at the same frequency, and hence are degenerate. A quick glance at the character table will tell if degeneracy is possible. (a) In the case of the tetrahedral NH_4^+ ion, is it necessary to consider the possibility of degeneracies? (b) Are degeneracies possible in any of the vibrational frequencies of $NH_2D_2^+$?

3.2 Figure 3.39 shows the bonding energy levels for CH_3^+. What is the point group assumed for CH_3^+ in this illustration? What H 1*s* orbital linear combination participates in a_1'? What C orbitals contribute to a_1'? What H linear combinations participate in the bonding e' pair? What C orbitals are e'? Which C orbitals are e''? Is any H linear combination e''? Now add two H 1*s* orbitals on the z axis (above and below the plane), modify the linear combinations of each symmetry type accordingly, and construct a new a'' linear combination. Are there bonding and nonbonding (or only weakly antibonding orbitals) that can accommodate 10 electrons and allow the C to become hypervalent? (*Hint*: Refer to Appendices 5 and 6).

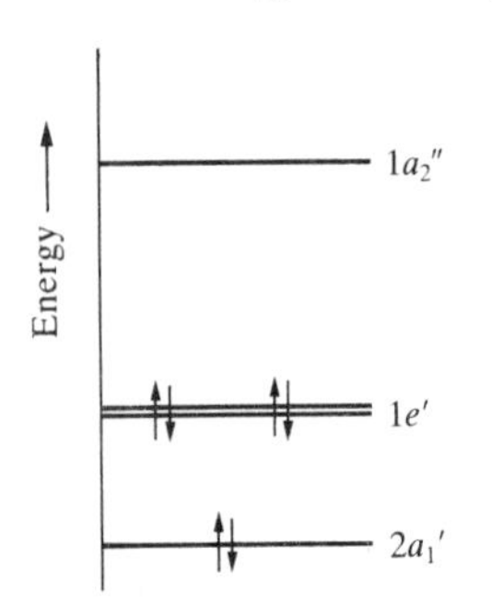

Fig. 3.39 The molecular orbital energy level diagram for CH_3^+.

3.3 Construct a Walsh diagram correlating the orbitals of a square-planar H_4 (D_{4h}) molecule with those of a linear H_4 ($D_{\infty h}$) molecule.

3.4 Construct a Walsh diagram correlating the orbitals of a trigonal planar XH_3 (D_{3h}) molecule with those of a trigonal pyramidal XH_3 (C_{3v}) molecule.

3.5 The 'K-shell' X-ray emission and absorption spectra of aluminum oxide are illustrated in Fig. 3.40. It is so called because emission arises from valence band electrons falling into a vacancy in the 'K-shell', an alternative name for the $n = 1$ shell, produced by bombardment. Absorption arises from promotion of an electron from the K-shell into the conduction band. What is the band gap in Al_2O_3? Is Al_2O_3 an insulator or a semiconductor? Are energy levels dense near the band edges or near the band centers? Which peak gives an idea of the distribution of levels derived mainly from O orbitals?

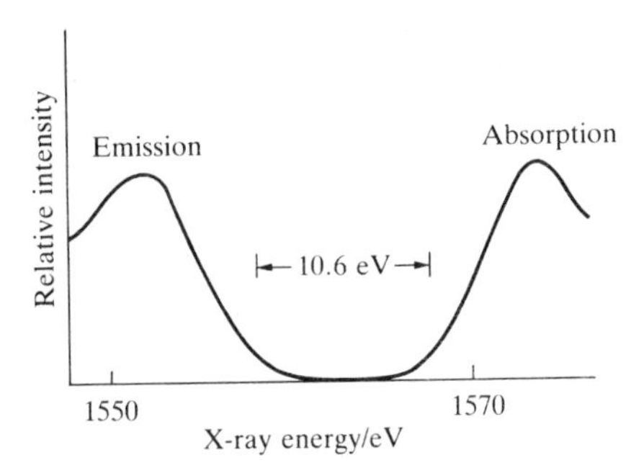

Fig. 3.40 The X-ray emission and absorption spectra of Al_2O_3.

The structures of solids

4

In this chapter we survey the patterns adopted by atoms and ions in simple solids and look for reasons why one arrangement may be preferred to another. We begin by seeing that the simplest model, in which atoms are represented by spheres and the structure is the outcome of stacking them together densely, is a good description of many metals and a very useful starting point for the discussion of alloys and ionic solids. We then see that the structures of many ionic solids can be expressed in terms of a few commonly occurring structures, which in a number of cases can also be related to the packing of spheres. We see that some of these structures are more likely when the bonding in the solid has a partially covalent character.

In the final part of the chapter we look for atomic parameters that help us to rationalize why one particular structure is obtained rather than another. One parameter is the relative radii of the ions, which brings in the material we introduced in Chapter 1. Then, by considering the energetics of crystal formation, we identify another parameter which is relevant when ionic interactions are dominant. In later chapters we shall see that this electrostatic parameter plays a role whenever charge–charge interactions are important. In the closing sections of this chapter, we see that the parameters we have introduced—the ionic radii and the electrostatic parameter—help us to account for a number of chemical properties that involve the formation and decomposition of solids.

Crystal structure
4.1 Crystal lattices
4.2 The packing of spheres
Metals
4.3 Metallic elements
4.4 Alloys
Ionic solids
4.5 Characteristic structures of ionic solids
4.6 The rationalization of structures
4.7 Lattice enthalpies
4.8 Consequences of lattice enthalpies
Further information: thermodynamics
Further reading
Exercises
Problems

The thermodynamically most stable stacking patterns adopted by atoms and ions in solids are the arrangements that correspond to minimum Gibbs free energy at the temperature and pressure in question. The Gibbs free energy is difficult to assess in general, but in the case of ionic solids it is possible to analyze the contributions in terms of the coulombic interactions between the ions and to rationalize trends in the stabilities and solubilities of some inorganic compounds. In the course of doing so, we shall also see how to use similar arguments to suggest synthetic procedures and account for some of the reactions that solids undergo.

The forces that favor one structure over another are often finely balanced. As a result, many crystalline solids are **polymorphic**, or exist with different crystal forms, and undergo phase transitions to different structures as the temperature or pressure is changed. Polymorphism is a property of all kinds of solids, and examples include the black and white allotropes of phosphorus and the calcite and aragonite forms of calcium carbonate.

Crystal structure

Our first task is to develop the concepts that we need for describing the structures of crystals and the methods used to generate simple models of the stacking arrangements.

4.1 Crystal lattices

The structures of crystalline solids are best discussed in terms of the **unit cell**. This is a component of the crystal that reproduces the whole when stacked together repeatedly (Fig. 4.1), the stacked cells being related by pure translations. That is, all the cells in the crystal are related to each other by displacements without rotation, reflection, or inversion. We have a wide range of choices in selecting a unit cell, as the two-dimensional example in Fig. 4.1 shows, but it is generally preferable to choose a cell that shows the full symmetry of the arrangement of the atoms. Thus, in Fig. 4.1, the unit cell (a) is preferred to (b).

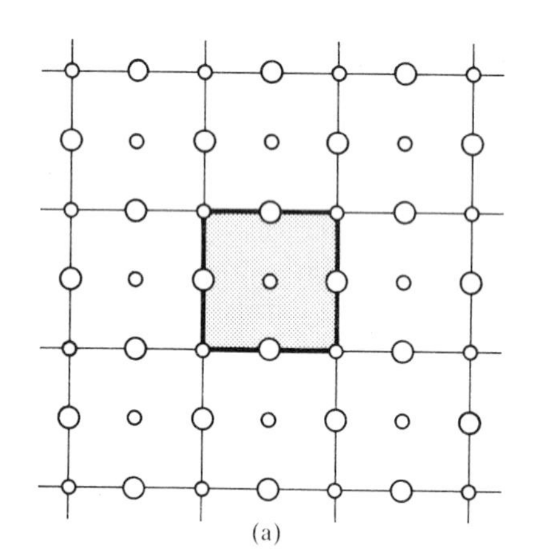

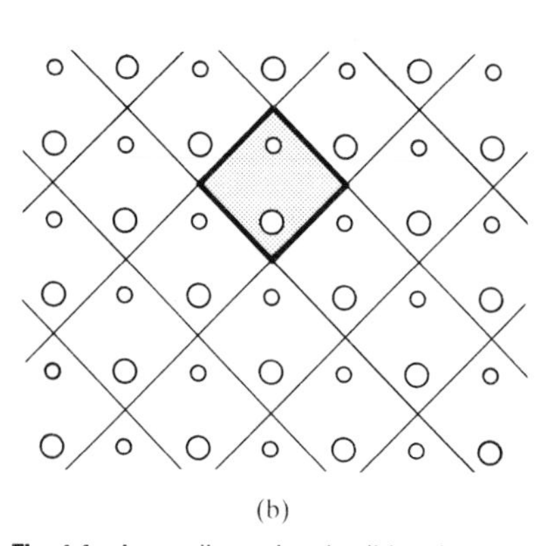

Fig. 4.1 A two-dimensional solid and two choices of unit cell. The entire crystal is reproduced by translational displacements of either unit cell, but (a) is generally preferred because it displays the maximum symmetry of the structure, but (b) does not.

The pattern of atoms, ions, or molecules in a crystal is depicted by an array of points called the **lattice**. The lattice points do not necessarily lie at the centers of atoms, but denote some common position of an **asymmetric unit**, the atom, ion, molecule, or group of molecules from which the actual crystal is built. Each point in the lattice shown in Fig. 4.2 implies the location of an M^+ and X^- pair of ions (the asymmetric unit). However, the point may be located on the cation, on the anion, or at an arbitrary point relative to either of them. The relation of the point to the asymmetric unit is arbitrary, but once chosen it is the same throughout the crystal.

The unit cell is formed by joining the lattice points with straight lines. This may also be done in an arbitrary manner, so long as pure translational repetition of the unit cell reconstructs the entire lattice.

4.2 The packing of spheres

The structures of many solids can be described in terms of the stacking of spheres that represent the atoms or ions. Metals are particularly simple in this respect because (for elements) all the atoms are identical and we can treat the solid as though it consists of spheres of identical size. In many

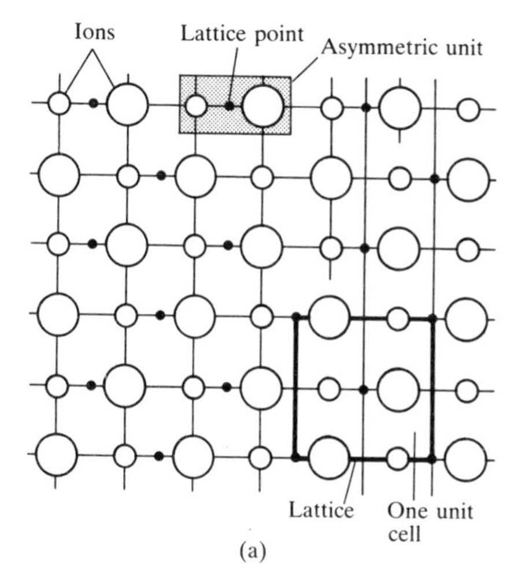

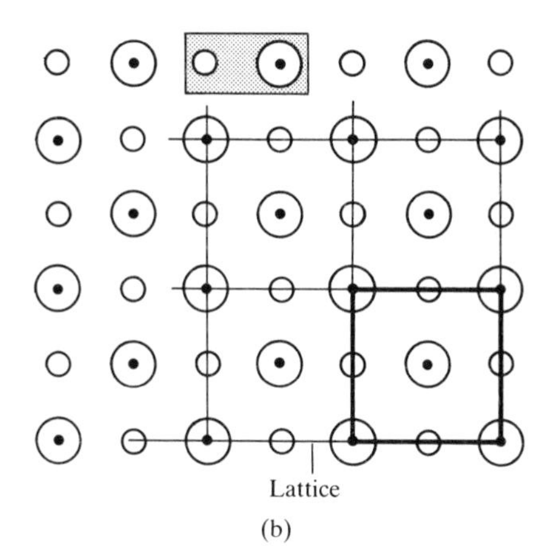

Fig. 4.2 Each lattice point represents the location of an asymmetric unit, in this case an Na^+Cl^- formula unit. The relation of the point to the asymmetric unit is arbitrary (but, once chosen, fixed throughout the crystal). (a) The lattice point is chosen to lie between the Na^+ and Cl^- ions. (b) An equally valid choice places the lattice point on a Cl^- ion.

cases, the atoms are free to pack together as closely as geometry allows: this occurs if there are no specific bonding forces that favor particular local arrangements. That is, metals are often **close-packed structures**—structures having least waste of space—with each sphere having its geometrically maximal number of neighbors. Close-packed structures are also useful starting points for the discussion of substances other than metals, and we shall introduce them in a general way.

The **coordination number** (C.N.) of a lattice is the number of nearest neighbors around any given atom or ion. The coordination number is often large (typically eight or twelve) for metals, intermediate for ionic solids (typically six), and low for molecular solids (typically four). This variation is reflected to some extent in the densities of the three types of solid, with the densest materials (metals) being most closely packed and having high coordination numbers.

The close-packing of spheres

Close-packed structures of identical spheres are often pictured as formed by laying close-packed layers on top of each other. The structure is started by placing a sphere in the indentation between two touching spheres, so making a triangle (**1**). The layer is then formed by continuing this process of laying spheres together in the indentations between those already in place. A complete close-packed layer consists of spheres in contact (Fig. 4.3) with each sphere having six nearest neighbors in the plane. This is shown by the shaded spheres.

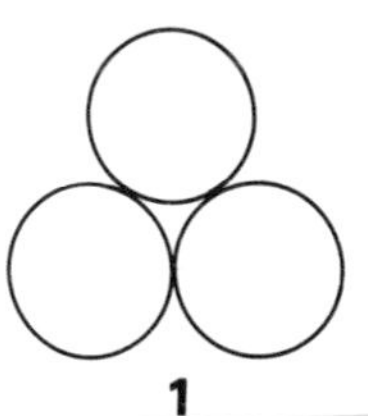

1

We form the second layer by placing spheres in the dips of the first layer. The third layer can be laid in either of two ways and can give rise to either of two **polytypes**. These are structures that are the same in two dimensions (in this case, in the planes) but different in the third. The coordination number is 12 for each polytype.

In one polytype, the spheres of the third layer lie directly above the spheres of the first. This ABAB . . . pattern of layers gives a lattice with a hexagonal unit cell, and hence is said to be **hexagonally close-packed** (hcp, Fig. 4.4a). In the other polytype, the spheres of the third layer are placed above the *gaps* in the first layer. Thus the second layer covers half the holes

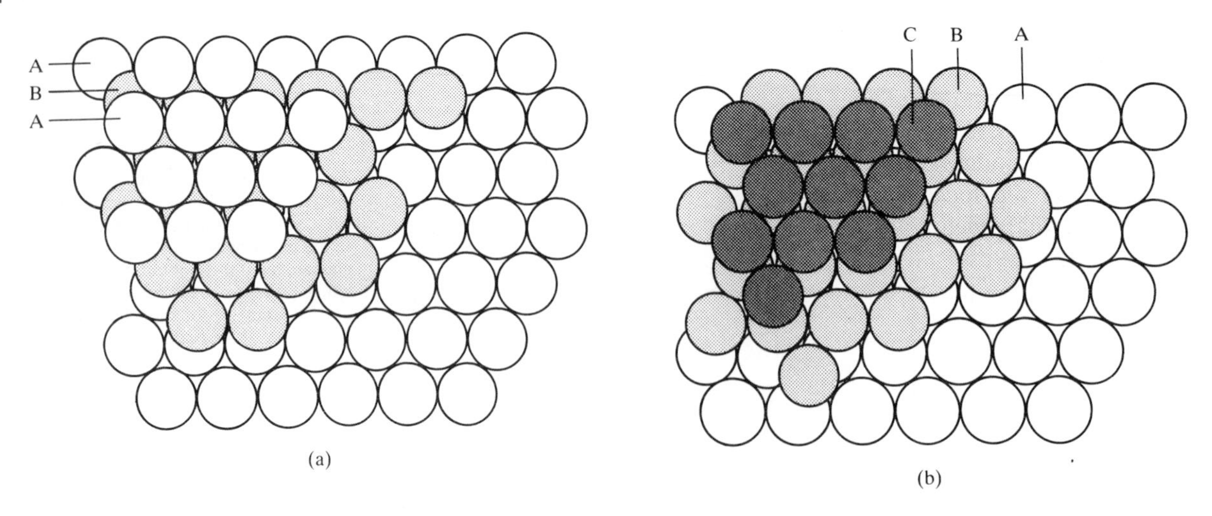

Fig. 4.3 The formation of two close-packed polytypes. (a) The third layer reproduces the first, giving an ABA structure. (b) The third layer lies above the gaps in the first layer, giving an ABC structure. The different shadings denote the different layers of identical spheres.

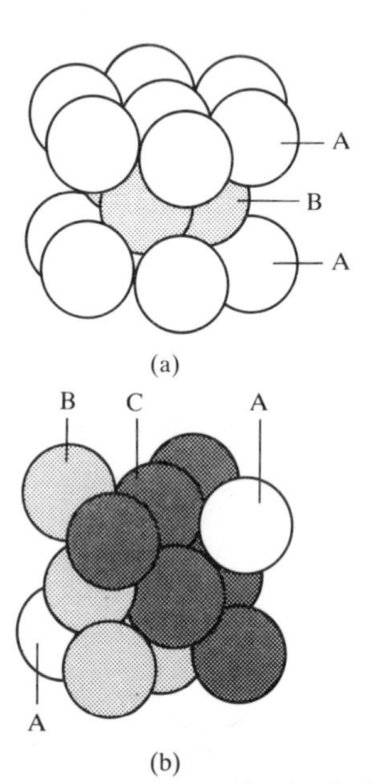

Fig. 4.4 (a) The hexagonal (hcp) unit of the ABAB . . . close-packed solid and (b) the cubic (fcc) unit of the ABCABC . . . polytype. The tints of the spheres correspond to the layers shown in Fig. 4.3.

in the first layer and the third layer lies above the remaining holes. This results in an ABCABC . . . pattern, and corresponds to a lattice with a face-centered cubic unit cell (Fig. 4.4b). Hence it is **cubic close-packed** (ccp) or, more specifically, **face-centered cubic** (fcc).

Holes in close-packed structures

A feature of a close-packed structure is the existence of two types of **hole**, or unoccupied space. (The space represented by the holes is not empty in a real solid since electron density does not end as abruptly as the hard-sphere model suggests.) These holes are important because many structures, including those of some alloys and many ionic compounds, can be regarded as formed from a close-packed arrangement in which additional atoms or ions occupy the holes.

One type of hole is an **octahedral hole** (shaded in Fig. 4.5a); it lies between two oppositely directed planar triangles of spheres in adjoining layers. If there are N atoms in the crystal, there are N octahedral holes. These holes are distributed in an fcc lattice as shown in Fig. 4.6a; this illustration also shows that the hole has local octahedral symmetry in the sense that it is surrounded by six nearest neighbor lattice points arranged octahedrally. If each sphere has radius r, each octahedral hole can accommodate another atom with a radius no larger than $0.414r$ (that is, $2^{1/2} - 1$ times r).

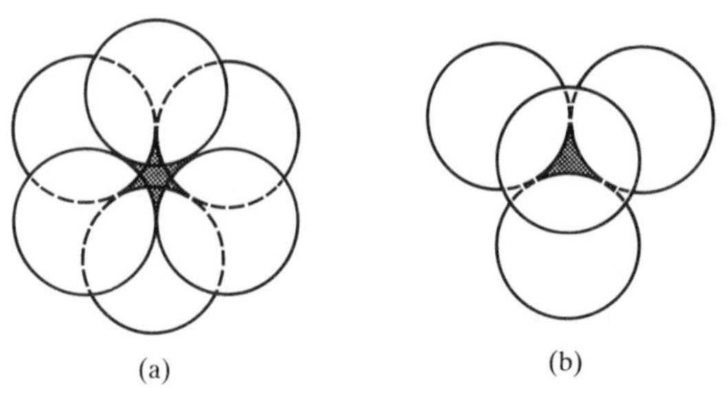

Fig. 4.5 (a) An octahedral hole in the cleft between six spheres. (b) A tetrahedral hole in the cleft between four spheres in a close-packed lattice.

A **tetrahedral hole** (T, shaded in Fig. 4.5b) is formed by a planar triangle of touching spheres that is capped by a single sphere lying in the dip between them. The apex of the tetrahedron may be directed up (T) or down (T′) in the crystal. There are N tetrahedral holes of each type ($2N$ in all), and they can accommodate another atom of radius no greater than $0.225r$. Figure 4.6b shows the location of tetrahedral holes in an fcc lattice; we also see from it that each hole has four nearest neighbor lattice points arranged tetrahedrally. Larger spheres may be accommodated in the holes if the close-packing of the parent structure is relaxed.

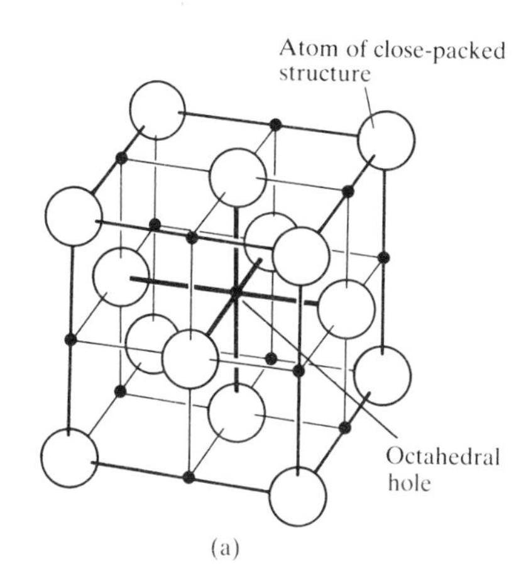

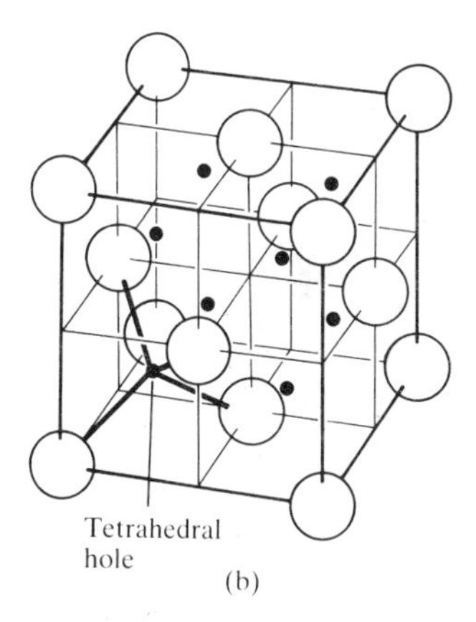

Fig. 4.6 The locations of (a) octahedral holes and (b) tetrahedral holes relative to the atoms in an fcc structure. The holes take names from the disposition of atoms around them.

Metals

Many metals are close packed. Since their atoms have only a weak tendency toward covalency, and a consequent weak directional character to their bonding, they often achieve the maximum number of nearest neighbors. One consequence of this close packing is that metals often have high densities. Indeed, the elements deep in the d block, near Ir and Os, include the densest solids known.

4.3 Metallic elements

Metallic radii

An informal definition of the **metallic radius** of an element is that it is half the distance between the nuclei of neighboring atoms in the solid at room temperature and pressure. However, it is found that this distance depends on the coordination number of the lattice, and generally the higher the coordination number, the greater the distance and therefore the greater the metallic radius.

Many metals and ionic compounds are polymorphic, in the sense that a change in temperature or pressure leads to a change in structure. In an extensive study of internuclear separations in a wide variety of polymorphic elements and alloys, Goldschmidt found that the apparent radius of an atom increases with increasing coordination number. He found that the average relative ratios are as follows:

Coordination number	Relative radius
12	1
8	0.97
6	0.96
4	0.88

Since, when comparing different elements, it is desirable to put them all on the same footing, it is common to adjust the empirical internuclear separation to the value that would be expected if the element were in fact close-packed (with a coordination number of 12). Thus, the empirical metallic radius of Na is 1.85 Å, but that is for a structure in which the coordination number is eight. We therefore multiply this radius by $1/0.97 = 1.03$ and obtain 1.91 Å as the radius that Na would have if it were close-packed.

The 'corrected' Goldschmidt values were listed in Table 1.5 and were the values we used when we discussed the periodicity of atomic size (Section

1.6). The essential features of that discussion to bear in mind now are that metallic radii generally increase down a group and decrease across a period. As we also remarked in Section 1.6, metallic radii reveal the presence of the lanthanide contraction in Period 6, with radii that are smaller than simple extrapolation from earlier periods would suggest. As we explained there, this contraction can be traced to the poor shielding effect of *f* electrons.

Close-packed metals

Which close-packed polytype—hcp or fcc—a metal adopts (if either) depends on the details of the elements, the interaction of atoms with second-nearest neighbors, and the residual effects of some directional character in their atomic orbitals. A rough guide to the type of structure that can be expected is that metals with more valence electrons per valence orbital tend to be cubic and those with fewer tend to be hexagonal.

We should also not lose sight of the fact that a close-packed structure need not be either of the regular ABAB . . . or ABCABC . . . polytypes, for these are only extreme, but common, polytypes. An infinite range of polytypes can in fact occur since the planes may stack in a more complex manner. For example, above 500°C, Co is fcc, but it undergoes a transition when cooled, and the structure that results is a randomly stacked set (ABACBABABC . . .) of close-packed layers. In some samples of Co (and of SiC too), the polytypism is not random, for the sequence of planes repeats after several hundred layers. It is hard to account for this behavior in terms of valence forces. The long-range repeat may be a consequence of a spiral growth of the crystal that requires several hundred turns before a stacking pattern is repeated.

Structures that are not close-packed

Not all metals are close-packed, and some other packing patterns use space nearly as efficiently. Even metals that are close-packed often undergo a phase transition to a less closely packed structure when they are heated and their atoms undergo large-amplitude vibrations.

One common structure is the **body-centered cubic** (cubic-I or bcc) structure in which there is a lattice point at the center of a cube with lattice points at each corner (Fig. 4.7). Metals with this structure have a coordination number of eight. Although this is less close-packed than the ccp and hcp structures (for which the coordination number is 12), the difference is not very great because the central atom has six second-nearest neighbors only 15 percent further away. This leaves 32 percent of the space unfilled compared with 26 percent in the close-packed structures.

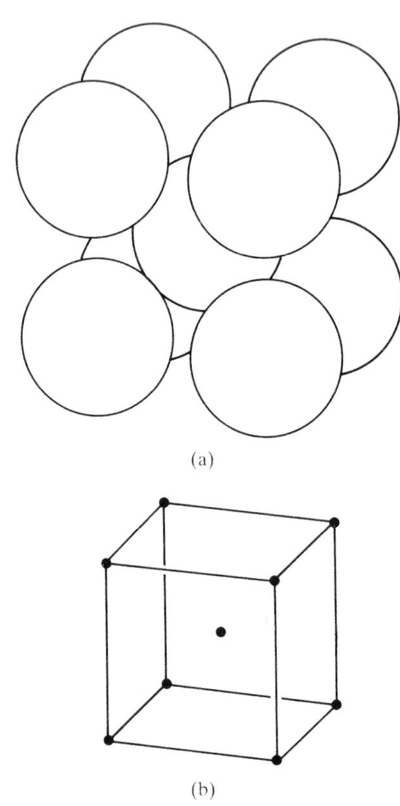

Fig. 4.7 The body-centered cubic unit cell and its lattice-point representation.

The least common metallic structure is the **primitive cubic** (cubic-P) structure (Fig. 4.8), in which atoms occupy lattice points located at the corners of a cube. The coordination number is only 6. One form of Po (α-Po) is the only example of this structure among the elements under normal conditions. Solid Hg, however, has a closely related structure. It is obtained from the simple cubic arrangement by stretching the cube along one of its body diagonals.

Metals that have structures more complex than those described so far can sometimes be regarded as slightly distorted versions of simple structures. Zn and Cd, for instance, have almost hcp structures, but the three neighbors above and below the close-packed plane are slightly further away than in pure hcp. This suggests a degree of covalent bonding between the atoms in the plane: the bonding draws these atoms together and in doing so, squeezes out the atoms of the neighboring layers.

Polymorphism of metals

Polymorphic phases of metals are generally (but not always systematically) labeled α, β, γ, . . . with increasing temperature. Some metals revert to a low-temperature form at higher temperatures. Iron, for example, is polymorphic: α-Fe, which is bcc, is stable up to 906°C, γ-Fe, which is fcc, is stable up to 1401°C, and then α-Fe is stable again up to the melting point at 1530°C. β-Fe, which is hcp, is formed at high pressures.

The room-temperature polymorph of Sn is *white tin* (β-Sn): it undergoes a transition to *gray tin* (α-Sn) below 14.2°C but the conversion occurs at an appreciable rate only after prolonged exposure to a much lower temperature. Gray tin has a diamond-like structure (Fig. 4.9). The structure of white tin is unusual in that each atom has four nearest neighbors which are more distant than in gray tin, as would be expected for the high-temperature form. However, white tin is the appreciably denser polymorph ($7.31\ \mathrm{g\,cm^{-3}}$ compared with $5.75\ \mathrm{g\,cm^{-3}}$ for gray tin). The explanation is that in white tin the second-nearest neighbors are closer than in gray tin, so overall the solid is more compact. A further point of interest, which shows that crystal structure can influence chemical properties, is that when Sn dissolves in concentrated hydrochloric acid, white tin forms Sn(II) chloride whereas gray tin forms Sn(IV) chloride.

The bcc structure is common at high temperatures for metals that are close-packed at low temperatures: in simple terms, the increased atomic vibrations demand a less close-packed structure. For many metals (among them Ca, Ti, and Mn), the transition temperature is above room temperature. For others (among them Li and Na), the transition temperature is below room temperature. The fact that a bcc structure is favored by a small number of valence electrons per orbital suggests that a dense electron sea is needed to draw cations together into the close-packed arrangement and that the alkali metals do not have enough valence electrons for this to be achieved.

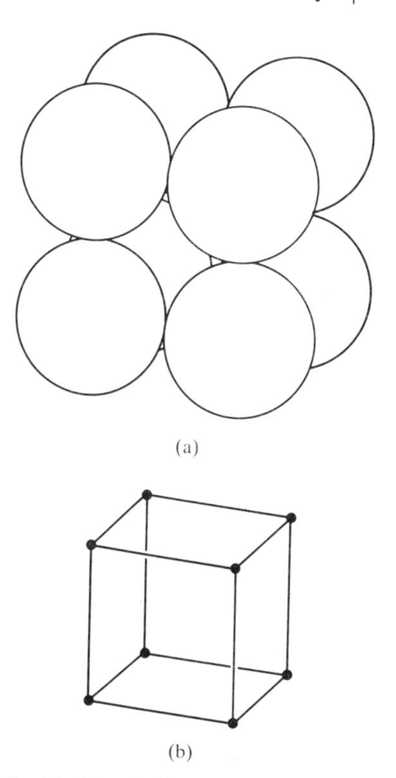

Fig. 4.8 The primitive cubic unit cell and its lattice-point representation.

4.4 Alloys

An **alloy** is a blend of metals prepared by mixing the components when molten and then cooling the mixture. Alloys may be homogeneous solid solutions, in which the atoms of one metal are distributed randomly among the atoms of the other, or they may be compounds with a definite composition and internal structure. Solid solutions are sometimes classified as either **substitutional** or **interstitial**. Substitutional solid solutions are those in which atoms of the solute metal occupy some of the locations of the solvent metal atoms (Fig. 4.10a). Interstitial solid solutions are those in which the solute atoms occupy the gaps between the solvent atoms (Fig. 4.10b). However, this distinction is not particularly real, since interstitial atoms often lie in a definite lattice (Fig. 4.10c), and hence can be regarded as substitutional. A better viewpoint is that a new structure is formed, and its relation to the original lattice may be largely coincidental.

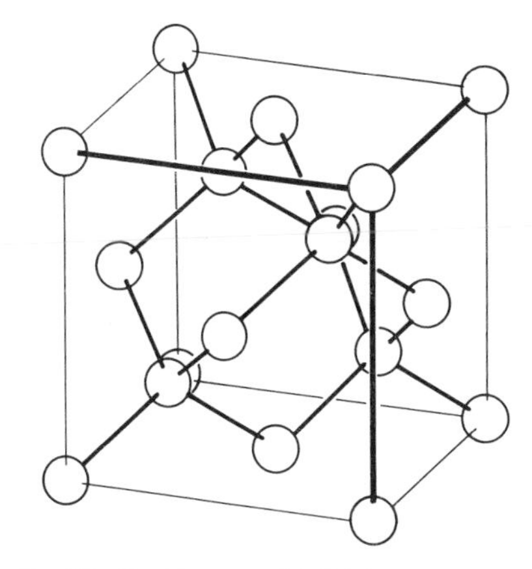

Fig. 4.9 The structure of α-Sn. Note its relation to the fcc unit cell in Fig. 4.6b, with an additional Sn atom in half the tetrahedral holes. This structure is also that of diamond, silicon, and germanium.

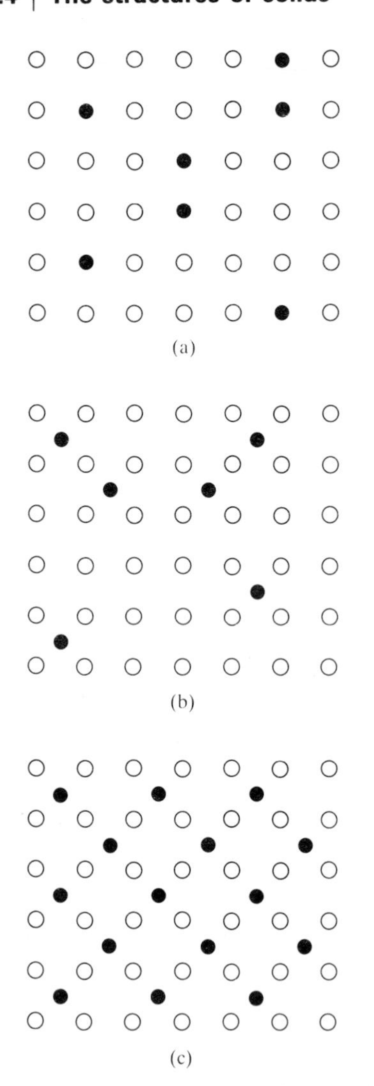

Fig. 4.10 (a) Substitutional and (b) interstitial alloys. (c) In some cases an interstitial alloy may be regarded as a substitutional alloy derived from another lattice.

Substitutional alloys

Substitutional solid solutions are formed only by atoms with metallic radii that are within about 15 percent of each other. Moreover, the crystal structures of the two metals should be the same, for this indicates that the directional forces between the atoms are compatible. Finally, the two metals should have similar electropositive character, for otherwise compound formation would be more likely. Thus, although Na and K are chemically similar and have bcc structures, the metallic radius of an Na atom (1.86 Å) is 18 percent smaller than that of a K atom (2.26 Å), and the two solid metals are immiscible. On the other hand, Cu and Ni, two neighbors late in the *d* block, have similar electropositive character, similar crystal structures (both fcc), and similar metallic radii (Ni 1.25 Å, Cu 1.28 Å, 2.3 percent different), and form a continuous series of solid solutions, ranging from pure Ni to pure Cu. The other neighbor of Cu, Zn, has a similar chemical character and metallic radius (1.37 Å, 7 percent larger), but it is hcp, not fcc. In this instance, Zn and Cu are partially miscible as solids and form solid solutions over a limited concentration range.

Interstitial alloys

Interstitial solid solutions are formed by solute nonmetals that are small enough to inhabit the interstices in the solvent structure. The small atoms enter the host solid with preservation of the crystal structure of the original metal, and often result in nonstoichiometric substances.

The largest solute atom that can enter a close-packed solid without distorting the structure appreciably is one that just fits an octahedral hole, which as we have seen has radius $0.414r$. Hence, on a purely geometrical basis, and with no reconstruction of the crystal structure, in order to accommodate H, B, C, or N atoms, the metal atoms must have radii no smaller than 0.90 Å, 1.95 Å, 1.88 Å, and 1.80 Å respectively. However, since the late *d*-block members of the first row do form an extensive series of interstitial solid solutions with B, C, and N despite their having radii no larger than 1.60 Å, we must conclude that specific metal/nonmetal bonding occurs. Hence, these substances are best regarded as examples of compounds of the nonmetals, and they are treated as such in Chapters 11 and 12.

Intermetallic compounds

When some liquid mixtures of metals are cooled, they form phases with a definite structure, which is often not related to the parent structure. These phases are called **intermetallic compounds**. They include β-brass (CuZn) and compounds of composition $MgZn_2$, Cu_3Au, and Na_5Zn_{21}.

The **Zintl phases** (which are named for E. Zintl, who first characterized them) are intermetallic compounds formed by strongly electropositive elements (the alkali metals and alkaline earth metals) and a less electropositive metal (typically from the early *d* block or the early *p* block). Examples of Zintl phases include NaTl, Mg_2Sn, $CaZn_2$, and LiZn. Their electronic structure is equivalent to the transfer of the valence electrons from the more electropositive element to the less electropositive element. As a result, the compositions sometimes reflect the conventional valencies of the metals (as in Mg_2Sn, but not in $CaZn_2$).

The transfer of an electron from Na to Tl in NaTl results in an atom that is isoelectronic with C, and in fact the Tl atoms are found to lie in a diamond lattice. Just as diamond has full bands, so too does NaTl, and it is a colorless nonmetallic solid. In contrast, LiZn does not have enough electrons in the diamondlike Zn bands, and is colored and a metallic conductor. Extensive transfer of electrons from Ca to the Zn bands also occurs in $CaZn_2$, and in this case the Zn atoms form a graphite-like layer structure of Zn hexagons with Ca^{2+} ions between them.

Ionic solids

We now consider **ionic solids**, a class of solids that can be modeled, at least to a first approximation, as an assembly of oppositely charged spheres. As we did for metals, we start by describing the common structures that are encountered. After that, we shall see how to rationalize the structures in terms of the energetics of crystal formation. The structures we shall describe were obtained using X-ray diffraction (Box 4.1) and were among the first inorganic solids to be examined in this way.

4.5 Characteristic structures of ionic solids

The ionic structures we describe in this section are prototypes of a wide range of solids. For instance, although the rock-salt structure takes its name from a mineral form of NaCl, it is characteristic of numerous other solids (Table 4.1). Many of the structures we describe can be regarded as derived from arrays in which the anions (sometimes the cations) stack together in fcc or hcp patterns and the counter ions occupy the octahedral and tetrahedral holes in the lattice. Throughout the following discussion, it will be helpful to refer back to Fig. 4.6 to see how the structure being described is related to the hole patterns shown there. The close-packed layer usually needs to expand in order to accommodate the counter ions, but this is often a minor perturbation of the anion arrangement. Hence, the close-packed structure is often a good starting point for the discussion of ionic structures.

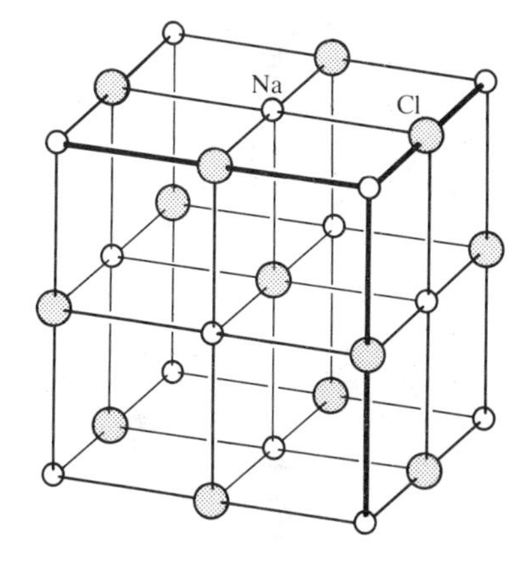

Fig. 4.11 The rock-salt structure. Note its relation to the fcc structure in Fig. 4.6a, with an anion in each octahedral hole. Alternatively, treat the lattice as the location of anions, in which case the cations occupy the octahedral holes.

The rock-salt structure

The rock-salt structure (Fig. 4.11) may be regarded as based on an fcc array of bulky Cl^- anions in which the cations occupy the *N* octahedral holes. We

Table 4.1. Compounds with particular crystal structures

Crystal structure	Example*
Antifluorite	K_2O, K_2S, Li_2O, Na_2O, Na_2Se, Na_2S
Cesium chloride	**CsCl**, CaS, TlSb, CsCN, CuZn
Fluorite	**CaF_2**, UO_2, $BaCl_2$, HgF_2, PbO_2
Nickel arsenide	**NiAs**, NiS, FeS, PtSn, CoS
Perovskite	**$CaTiO_3$**, $BaTiO_3$, $SrTiO_3$
Rock-salt	**NaCl**, LiCl, KBr, RbI, AgCl, AgBr, MgO, CaO, TiO, FeO, NiO, SnAs, UC, ScN
Rutile	**TiO_2**, MnO_2, SnO_2, WO_2, MgF_2, NiF_2
Sphalerite (zinc blende)	**ZnS**, CuCl, CdS, HgS, GaP, InAs
Wurtzite	**ZnS**, ZnO, BeO, MnS, AgI, AlN, SiC, NH_4F

* The substance in bold type is the one that gives its name to the structure.

Box 4.1 Single crystal X-ray diffraction

X-ray diffraction is the most widely used and least ambiguous method for the precise determination of the positions of atoms in molecules and solids. X-ray structure determinations play a much more prominent role in inorganic chemistry than in organic chemistry because inorganic molecules and solids are structurally more diverse. Thus structural inferences from spectroscopic data frequently suffice for organic molecules, but spectroscopy is less successful for the unambiguous characterization of new inorganic compounds. Additionally, the bonding in inorganic molecules is more varied than in organic molecules, so inorganic chemists depend on bond distance and bond angle information to infer the nature of bonds.

A typical X-ray diffractometer (Fig. 4B.1) consists of an X-ray source with a fixed wavelength, a mount for a single crystal of the compound being investigated, and an X-ray detector. The positions of the detector and the crystal, which is typically as small as 0.2 mm on a side, are controlled by a computer. For certain orientations of the crystal relative to the X-ray beam the crystal diffracts the X-rays at a fixed angle and the intensity is measured when the detector is placed in the direction of this diffracted beam. Under computer control, the detector is scanned through each reflection while the intensities are recorded and stored. It is common to collect data on the intensities and positions of over 1000 reflections, and to obtain more than 10 observed reflections for each structural parameter to be determined (positions of atoms and a range of locations associated with their thermal motion). A trial structure is chosen, either by means of a 'direct methods' program or by hints from the diffraction data together with a knowledge of physically reasonable arrangements of atoms. This structural model is refined by systematic shifts in the atom positions until satisfactory agreement between observed and calculated X-ray diffraction intensities is obtained.

The pictorial display of an X-ray structure often has the appearance of Fig. 4B.2. This type of computer-generated drawing is called an **ORTEP diagram** (an acronym for Oak Ridge Thermal Ellipsoid Program). The ORTEP diagram depicts the bond distances and bond angles. In addition, the atoms are displayed as ellipsoids that indicate the amplitude of their thermal motion. Since the restoring force for bond angle bending is generally less than for bond stretching, the ellipsoids are generally elongated in directions perpendicular to the bonds, as clearly shown by the N atoms of the CN^- ligands in Fig. 4B.2.

[References: A. K. Cheetham, Chapter 2 in *Solid state chemistry techniques* (eds A. K. Cheetham and P. Day). Oxford University Press (1987); M. F. C. Ladd and R. A. Palmer, *Structure determination by X-ray crystallography*. Plenum, New York (1985); J. P. Glusker and K. N. Trueblood, *Crystal structure analysis: a primer*. Oxford University Press (1985).]

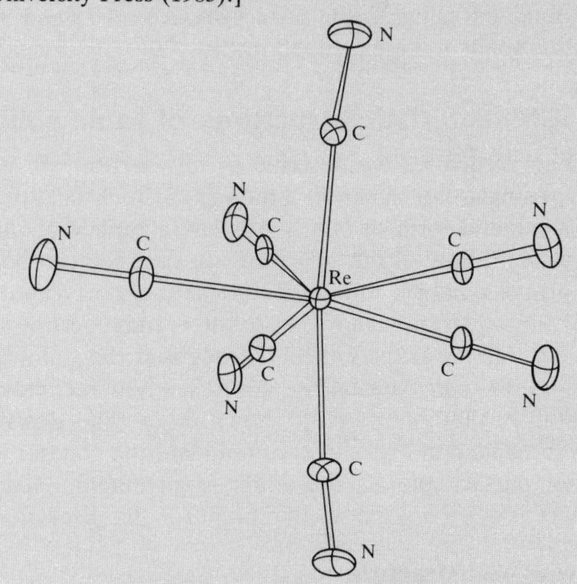

Fig. 4B.2 An ORTEP diagram of $[Re(CN)_7]^{4-}$ in $K_4[Re(CN)_7]{\cdot}2H_2O$.

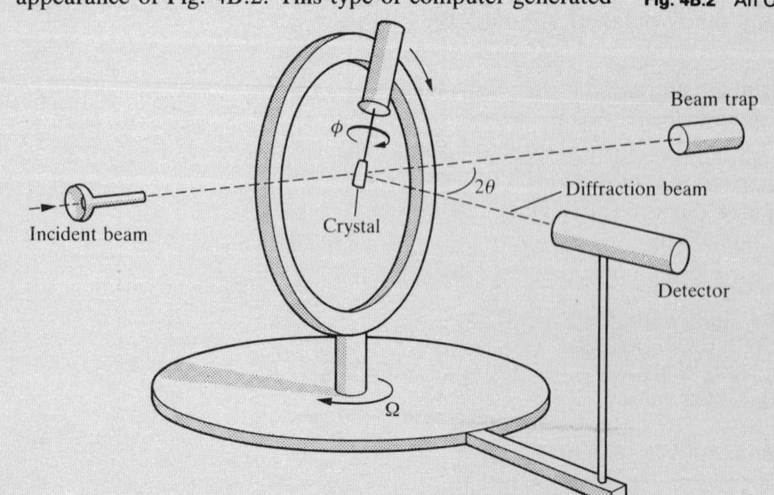

Fig. 4B.1 A schematic diagram of an X-ray diffractometer.

can see from the diagram that each ion is surrounded by an octahedron of six counterions. The coordination number of each type of ion is therefore 6, and the structure is said to have '(6, 6)-coordination'. In this notation, the first number in parentheses is the coordination number of the cation, and the second number is the coordination number of the anion.

The six nearest neighbors of the central ion are at the centers of the faces of the unit cell and are all ions of opposite charge. The twelve second-nearest neighbors, those next further away, are at the centers of the edges of the cell, and are all ions of the same charge as the central ion. The eight third-nearest neighbors are at the corners of the unit cell, and have opposite charge to the central ion.

When assessing the number of ions of each type in a unit cell we must take into account that any ions that are not fully inside the cell are shared by neighboring cells. Thus, an ion on a face is shared by two cells and contributes $\frac{1}{2}$ to the cell in question. An ion on an edge is shared by four cells and hence contributes only $\frac{1}{4}$. An ion on a corner is shared by eight cells that share the corner, and so it contributes only $\frac{1}{8}$. In the unit cell shown in Fig. 4.11, there are four Na^+ ions and four Cl^- ions. Hence, each unit cell contains four NaCl formula units.

Cesium chloride

Much less common than the rock-salt structure is the cesium chloride structure, which is shown by CsCl, CsBr, and CsI as well as some other compounds formed of ions of similar radii to these, including NH_4Cl (Table 4.1). The cesium chloride structure (Fig. 4.12) has a cubic unit cell with each lattice point occupied by a halide anion and a metal cation at the cell center (or vice versa). The coordination number of both types of ion is eight, for they are of such comparable radii that this energetically highly favorable (8, 8) form of packing, with numerous counter ions adjacent to a given ion, is feasible.

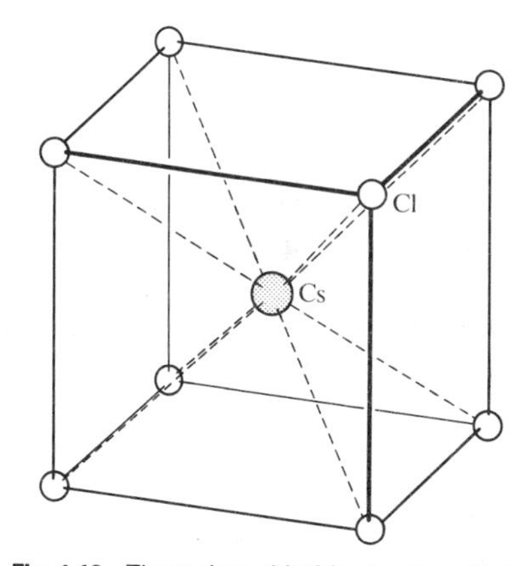

Fig. 4.12 The cesium chloride structure. Note that the corner ions, which are shared by eight cells, are surrounded by eight nearest-neighbor center atoms.

Sphalerite

The sphalerite (ZnS) or zinc blende structure (Fig. 4.13) is based on an expanded fcc anion lattice, but now the cations occupy one type of tetrahedral hole. Since each ion is surrounded by four neighbors, the structure has (4, 4)-coordination.

Example 4.1: *Counting the number of ions in a unit cell*

How many ions are there in the unit cell shown in Fig. 4.13?

Answer. An ion fully inside the cell belongs entirely to that cell and counts as 1. One on the face is shared by two adjacent unit cells and counts $\frac{1}{2}$. One on an edge is shared by four cells and so counts $\frac{1}{4}$. An atom at a corner position is shared by eight cells and so counts $\frac{1}{8}$. In the sphalerite structure, the count is as follows:

Inside:	4	counts 4
Face:	6	counts 3
Edge:	0	
Corner:	8	counts 1
Total:		8

Note that there are four cations and four anions.

Exercise. Count the ions in the unit cell in Fig. 4.11.

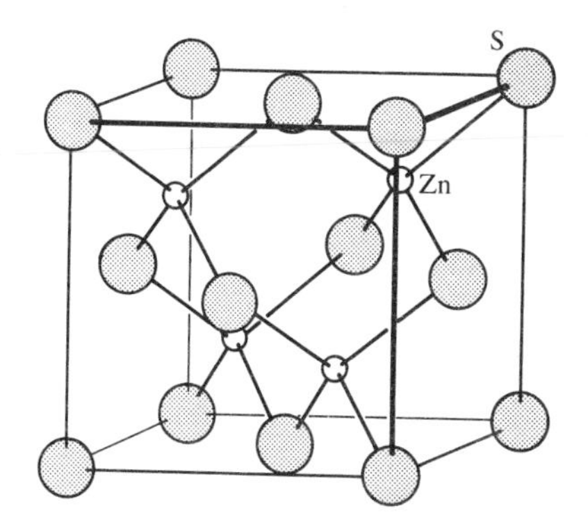

Fig. 4.13 The sphalerite (zinc blende) structure. Note its relation to the fcc structure in Fig. 4.6b, with half the tetrahedral holes occupied by Zn^{2+} ions.

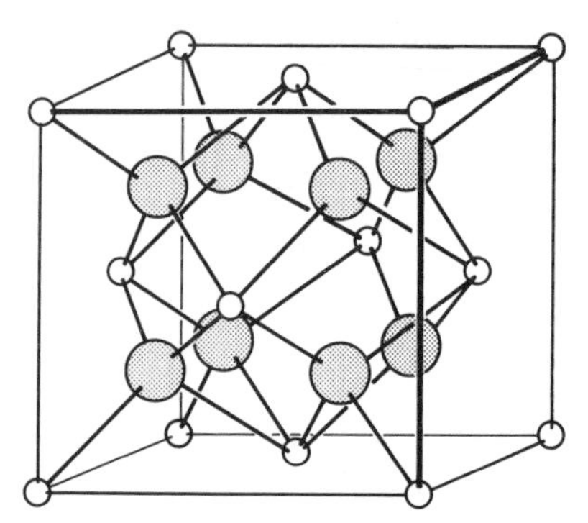

Fig. 4.14 The antifluorite and fluorite structures. These too are related to the fcc structure, but all the tetrahedral holes are occupied (Fig. 4.6b).

Fluorite and antifluorite

The last of the three simple structures based on the expanded fcc lattice of anions is the antifluorite structure (Fig. 4.14), which is shown by some alkali metal oxides, including K_2O. In it, the cations (which are twice as numerous as the anions) occupy both types of tetrahedral hole. (Recall that if there are N atoms, there are $2N$ tetrahedral holes.) The fluorite structure, the inverse of the antifluorite, takes its name from its exemplar fluorite (CaF_2). In it, the Ca^{2+} cations lie in an expanded fcc array and the F^- anions in the two types of tetrahedral hole.

In the antifluorite structure, the cations in their tetrahedral holes have four nearest neighbors. The anion site is surrounded by a cubic array of cations; hence, the local site symmetry is cubic. The coordination of the lattice is therefore (4, 8), which is consistent with there being twice as many cations as anions. In the fluorite structure, the coordination is the opposite of this, namely (8, 4).

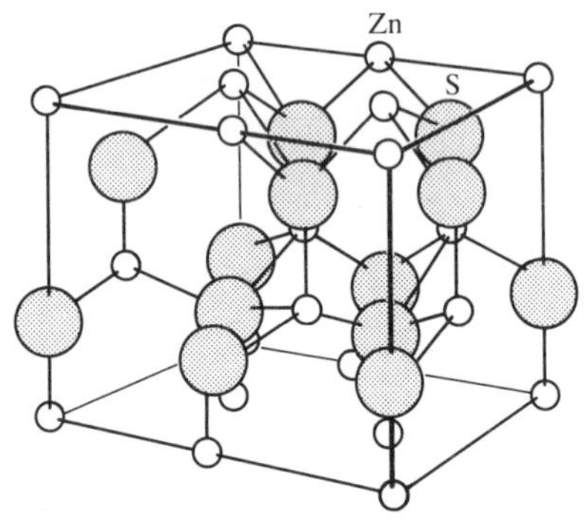

Fig. 4.15 The wurtzite structure, which is derived from the hcp structure (Fig. 4.4a).

Wurtzite

The mineral wurtzite (Fig. 4.15) is another polymorph of zinc sulfide. It differs structurally from sphalerite in being derived from an expanded hcp anion array rather than an fcc array, but as in sphalerite the cations occupy one type of tetrahedral hole. This structure, which has (4, 4)-coordination, is also that of ZnO, AgI, and one polymorph of SiC as well as several other compounds (Table 4.1). In comparing wurtzite and sphalerite, the local symmetries of the cations and anions are identical toward their nearest neighbors but differ at the second-nearest neighbors.

Nickel arsenide

Nickel arsenide (NiAs, Fig. 4.16) is also based on an expanded, distorted hcp anion array, but the Ni cations now occupy the octahedral holes and the As ions lie in a trigonal prismatic arrangement. It is also the structure of NiS, FeS, and a number of other sulfides. The structure is typical of MX compounds containing soft cations in combination with soft anions, which suggests that the NiAs structure is favored by covalence.

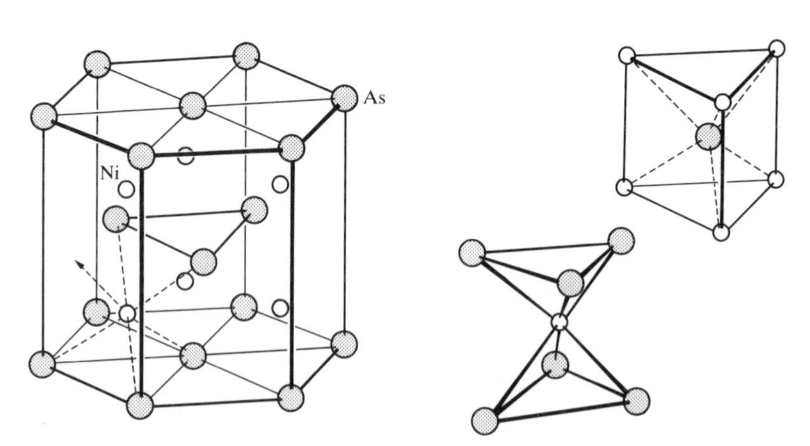

Fig. 4.16 The nickel-arsenide structure, another structure derived from the hcp structure (Fig. 4.4a). Note the prismatic and trigonal antiprismatic local symmetries of the As and Ni atoms respectively.

Rutile

Rutile (Fig. 4.17) is a mineral form of titanium dioxide, TiO_2. It is also an example of an hcp anion lattice, but now the cations occupy only half the octahedral holes. This results in a tetragonal structure that reflects the strong tendency of Ti to acquire octahedral coordination. The structure consists of TiO_6 octahedra, with the oxide ions shared by neighboring Ti ions. Each Ti ion is surrounded by six O ions and each O ion is surrounded by three Ti ions; hence the structure has (6, 3)-coordination. The principal ore of tin, cassiterite (SnO_2), has the rutile structure, as do a number of fluorides (Table 4.1).

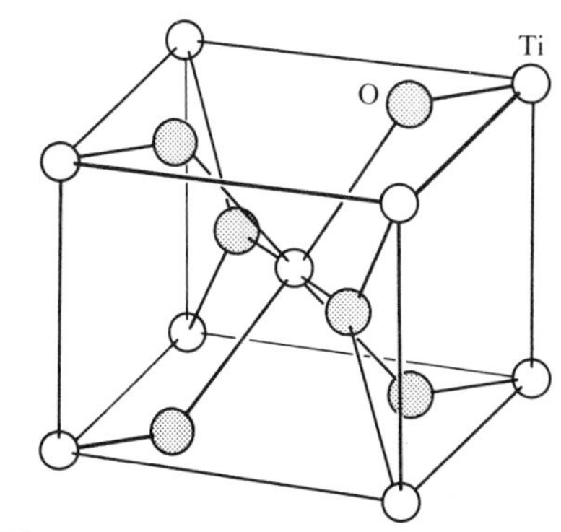

Fig. 4.17 The rutile structure. Rutile itself is one polymorph of TiO_2.

Perovskite

The mineral perovskite $CaTiO_3$ has a structure that is the prototype of many ABX_3 solids (Table 4.1), particularly oxides. In its ideal form (Fig. 4.18), the perovskite structure is cubic with the A atoms surrounded by 12 O atoms and the B atoms surrounded by 6 O atoms. The sum of the charges on the A and B ions must be 6, but that can be achieved in several ways ($A^{2+}B^{4+}$ and $A^{3+}B^{3+}$ among them), including the possibility of mixed oxides of formula $A(B_{\frac{1}{2}}B'_{\frac{1}{2}})O_3$, as in $La(Ni_{\frac{1}{2}}Ir_{\frac{1}{2}})O_3$. The perovskite structure is closely related to the materials that show interesting electrical properties, including piezoelectricity, ferroelectricity, and high-temperature superconductivity. We shall explore these properties in Chapter 18.

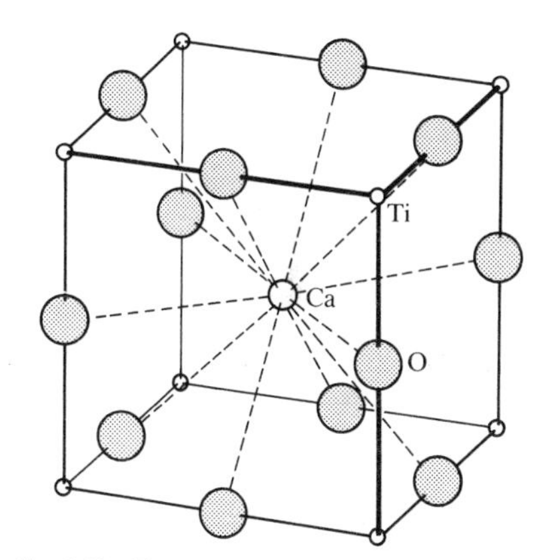

Fig. 4.18 The perovskite structure. Perovskite itself is $CaTiO_3$. In general, Ca is A and Ti is B in an ABX_3 solid.

4.6 The rationalization of structures

We noted in Chapter 3 that all solids can be treated theoretically using the delocalized orbitals of the band picture. However, when those bands are full, the localized bonding description is equivalent and is frequently more convenient. In the case of ionic solids, the filled band is often very well approximated by electrons in the orbitals of the anions and the empty band represents the vacancies in the cation orbitals, and valence electrons have been transferred to form the cations and anions. We can treat the energetics of such solids very simply in terms of the **ionic model** in which a collection of charged spheres interact coulombically. Departures from the predictions of this model are likely to be common, since many solids are more covalent than ionic. Even conventional 'good' ionic solids, such as the alkali metal halides, have significant covalent character. Nevertheless, the ionic model provides an attractively simple scheme—and a parameter—for correlating many properties.

Ionic radii

A difficulty that confronts us at the outset is the meaning of the term **ionic radius**. When a solid is formed from a single element we can take half the internuclear separation of nearest neighbors as a measure of their radii. However, in an ionic solid, nearest neighbors are different species, and the early workers in the field (notably Bragg, Pauling, and Goldschmidt) were confronted with the problem of apportioning the separation between the individual ions.[1] The most direct way to solve the problem is to make an assumption about the radius of one ion, and then to use that value to compile a set of self-consistent values for all other ions. The O^{2-} ion has the advantage of being found in combination with a wide range of elements and being reasonably hard (that is, unpolarizable, so its size does not vary much

[1] A helpful survey of the history of the problem of defining and measuring ionic radii has been given by R. D. Shannon and C. T. Prewitt, *Acta Crystallogr.*, **B25**, 925 (1969).

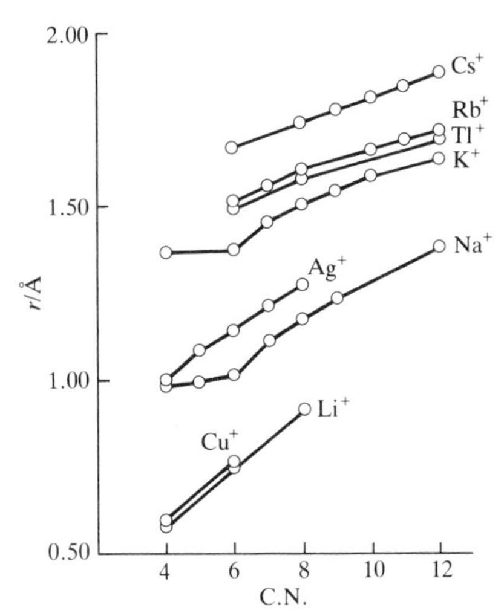

Fig. 4.19 The variation of ionic radius with coordination number.

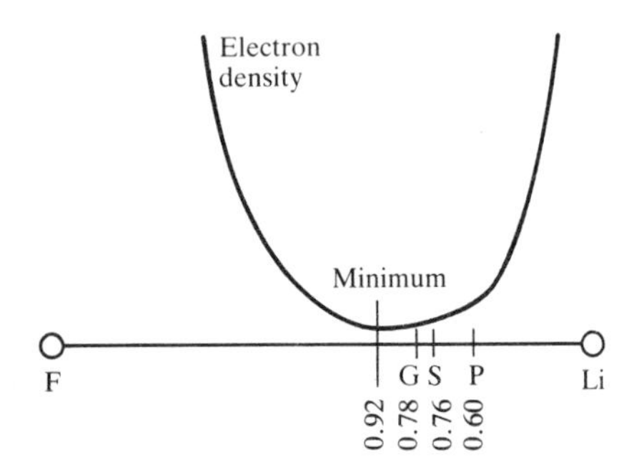

Fig. 4.20 The variation in electron density along the Li–F axis in LiF. The point P indicates the Pauling radii of the ions, G the original (1927) Goldschmidt radii, and S the Shannon radii.

as the identity of the accompanying cation is changed). In a number of compilations, therefore, the values are based on $r(O^{2-}) = 1.40$ Å. This value not only produces a consistent set of radii but also satisfies a number of theoretical criteria proposed by Linus Pauling. However, be warned that this value is by no means sacrosanct: the values compiled by Goldschmidt were based on $r(O^{2-}) = 1.32$ Å, and J. C. Slater even published a set of radii that were self-consistent but used the value $r(O^{2-}) = 0.60$ Å.

The lesson at this stage is that for certain purposes (such as predicting the sizes of unit cells) ionic radii can be helpful, but they are reliable only if they are all based on the same fundamental choice (such as the value 1.40 Å for O^{2-}). It is very dangerous to mix values from different sources.

An additional complication that was first noted by Goldschmidt is that ionic radii vary with the coordination number of the ions. As illustrated in Fig. 4.19, we see that the radius increases with coordination number (as happens with the metallic radius too). Hence, when comparing ionic radii, we should compare like with like, and use values for a single coordination number (typically six).

The problems of the early workers have only partly been resolved by developments in X-ray diffraction. Thus, it is now possible to measure the electron density between two neighboring ions, locate the minimum, and identify that with the boundary between the two ions. However, as we see from Fig. 4.20, the electron density passes through a very broad minimum, and its exact location may be very sensitive to the identities of the two neighbors. That being so, and in the general spirit of inorganic chemistry, it is still probably more useful to parametrize the sizes of ions in a self-consistent manner than to seek detailed quantum mechanical values of individual radii in certain combinations. Very extensive lists of self-consistent values, which have been compiled by analyzing X-ray data on thousands of compounds, particularly oxides and fluorides, are now available, and some are given in Table 4.2.

Table 4.2. Ionic radii (in Å)*

Li^+	**Be^{2+}**	**B^{3+}**		**O^{2-}**	**F^-**
0.59 (4)	0.27 (4)	0.12 (4)		1.35 (2)	1.28 (2)
0.76 (6)				1.38 (4)	1.31 (4)
Na^+	**Mg^{2+}**	**Al^{3+}**		1.40 (6)	1.33 (6)
0.99 (4)	0.49 (4)	0.39 (4)		1.42 (8)	
1.02 (6)	0.72 (6)	0.53 (6)			
1.16 (8)	0.89 (8)				
K^+	**Ca^{2+}**	**Ga^{3+}**			
1.38 (6)	1.00 (6)	0.62 (6)			
1.51 (8)	1.12 (8)				
1.59 (10)	1.28 (10)				
1.60 (12)	1.35 (12)				
Rb^+	**Sr^{2+}**	**In^{3+}**	**Sn^{2+}**	**Sn^{4+}**	
1.49 (6)	1.16 (6)	0.79 (6)		0.69 (6)	
1.60 (8)	1.25 (8)	0.92 (8)	1.22 (8)		
1.73 (12)	1.44 (12)				
Cs^+					
1.67 (6)					
1.74 (8)					
1.88 (12)					

* Numbers in parentheses are the coordination numbers of the ions.
Source: R. D. Shannon and C. T. Prewitt, *Acta Crystallogr.*, **B25**, 925 (1969). The radius of the NH_4^+ ion is approximately 1.46 Å.

The general trends shown in Table 4.2 for ionic radii are the same as for metallic radii. Thus, ionic radii increase on going down a group:

$$Li^+ < Na^+ < K^+ < Rb^+ < Cs^+$$

but with the lanthanide contraction (Section 1.6) restricting the increase among the heaviest ions. The radii of ions of the same charge decrease across a period:

$$Ca^{2+} > Mn^{2+} > Zn^{2+}$$

When an ion can occur in environments with different coordination numbers, its radius increases as the coordination number increases:

$$4 < 6 < 8 < 10 < 12$$

If an element can exist with several different oxidation numbers, for a given coordination number its ionic radius decreases with increasing positive charge:

$$Fe^{2+} > Fe^{3+}$$

Since a positive charge indicates a reduced number of electrons, and hence a more dominant nuclear attraction, cations are usually smaller than anions.

The radius ratio

To a limited extent it is possible to rationalize the structures of ionic compounds in terms of the **radius ratio** ρ of the ions, the ratio of the radius of the smaller ion ($r_<$) to that of the larger ($r_>$):

$$\rho = \frac{r_<}{r_>}$$

In most cases, $r_<$ is the cation radius r_+ and $r_>$ is the anion radius r_-, so then

$$\rho = \frac{r_+}{r_-}$$

The minimum radius ratio that can tolerate a given coordination number can be calculated by considering the geometrical problem of packing together spheres of different sizes. The results are listed in the margin. If the radius ratio falls below the minimum given, ions of opposite charge will not be in contact and ions of like charge will touch. According to a simple electrostatic argument, the lower coordination number, in which the contact of oppositely charged ions is restored, then becomes favorable. As the ionic radius of the M^+ ion increases, more anions can pack around it, as we have seen for CsCl with (8, 8) coordination compared with NaCl and its (6, 6) coordination.

In some cases, the radius ratio can be used to predict the structure that a compound is likely to adopt. The relations are least reliable for the simplest structures, such as the alkali metal halides and the alkaline earth metal oxides. They are most reliable for complex fluorides and oxides and for the (very ionic) oxoanion salts. A major part of the reason for the unreliability of the rules is the breakdown of the ionic model.

Coordination number	Radius ratio	Diagram
8	0.732	**2**
6	0.414	**3**
4	0.225	**4**
3	0.155	**5**

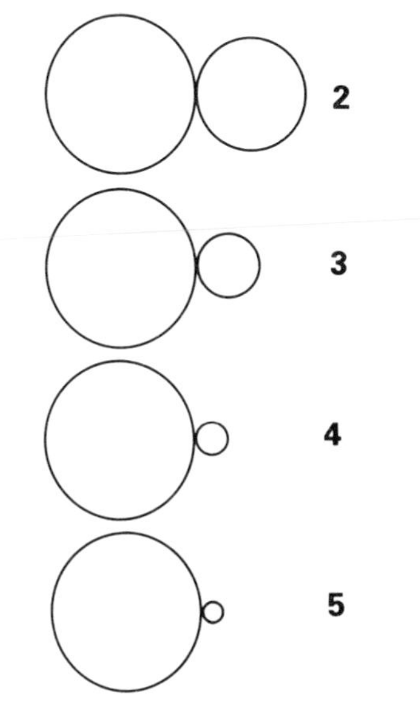

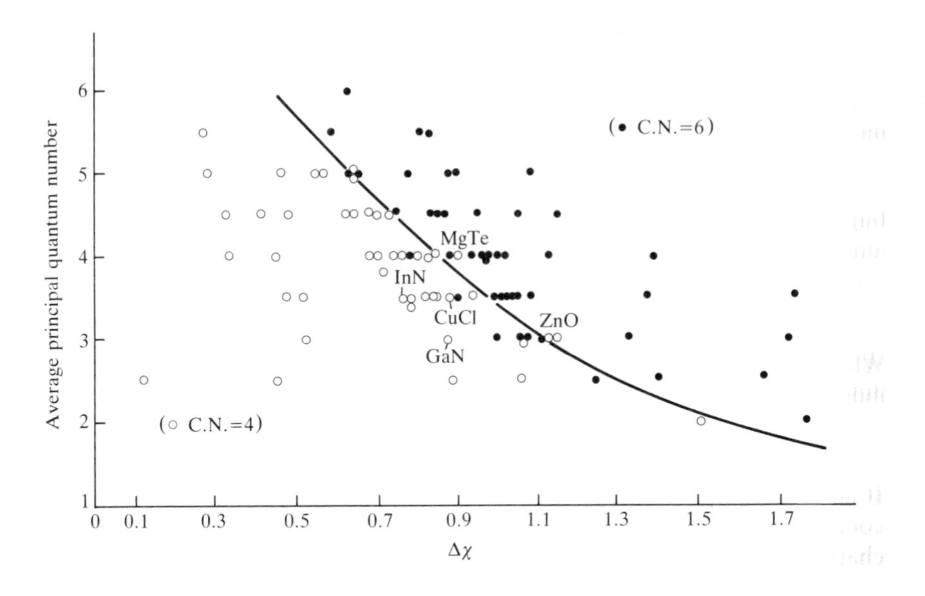

Fig. 4.21 A structure map for compounds of formula MX. A point is defined by the electronegativity difference between anion and cation $\Delta\chi$ and the average principal quantum number n. Its location in the map indicates the coordination number expected for that pair of properties.

Structure maps

Even if we accept that the use of radius ratios is inadequate, we can still make progress in the rationalization of structures by collecting enough information empirically and looking for a pattern in it. This is the thought behind the compilation of **structure maps**. These are empirically compiled maps, like the one in Fig. 4.21, that show how the type of crystal structure depends on the electronegativity difference $\Delta\chi$ between the elements involved.[2] This dependence on $\Delta\chi$ points to the map being some kind of representation of the ionic or covalent character in the bonds in the solid. The role of the average principal quantum number is to indicate the general size of the ions, since the larger the average n (the higher the period number), the larger the ions.

Figure 4.21 is an example of a structure map for MX compounds. We see that the structures we have been discussing fall in quite distinct regions of the map. Elements with large $\Delta\chi$ have (6,6)-coordination, such as is found in rock salt and rutile structures; elements with small $\Delta\chi$ (and hence where there is the expectation of covalence) have a lower coordination number. In terms of a structure map representation, GaN is in a more covalent region of Fig. 4.21 than ZnO because $\Delta\chi$ is appreciably smaller.

Example 4.2: *Using a structure map*

What type of coordination environment should be expected for MgS?

Answer. The electronegativities of Mg and S are 1.3 and 2.6 respectively, so $\Delta\chi = 1.3$. The average principal quantum number is 3 (since both Mg and S are in Period 3). The point $\Delta\chi = 1.3$, $n = 3$ lies just in the C.N. = 6 region of the structure map. This is consistent with the observed rock salt structure of MgS.

Exercise. Predict the coordination environment of RbCl.

[2] E. Moser and W. B. Pearson, *Acta Crystollgr.*, **12,** 1015 (1959).

4.7 Lattice enthalpies

The criterion of stability of a crystal lattice under conditions of constant temperature and pressure, and hence the thermodynamic criterion of why one structure is adopted rather than another, is the Gibbs free energy of lattice formation:

$$M^+(g) + X^-(g) \rightarrow MX(s) \qquad \Delta G^\ominus = \Delta H^\ominus - T\Delta S^\ominus$$

If $\Delta G^\ominus$ is more negative for the formation of a structure X rather than Y, then the transition from Y to X is spontaneous under the prevailing conditions, and we can expect the solid to be found with that structure. We shall draw on a number of concepts from thermodynamics in this and the following chapters, and a review of the essential material is in the *Further information* section at the end of this chapter.

The process of lattice formation from the gas of ions is so exothermic that at and near room temperature the contribution of the entropy may be neglected (this is rigorously true at $T = 0$). Hence, discussions of lattice energetics normally focus, initially at least, on the lattice enthalpy. That being so, we look for the structure that is formed most exothermically and identify it as the thermodynamically stable form.

Lattice enthalpy

The **lattice enthalpy** ΔH_L is the standard enthalpy change accompanying the formation of a gas of ions from the solid:

$$MX(s) \rightarrow M^+(g) + X^-(g) \qquad \Delta H_L$$

Since lattice disruption is endothermic, lattice enthalpies are always positive. If we neglect entropy considerations, we expect the stable crystal structure to be the one with the greatest lattice enthalpy under the prevailing conditions.

We can find lattice enthalpies from enthalpy data using a thermodynamic **Born–Haber cycle**, such as that shown in Fig. 4.22. The aim is to form a closed path in which the lattice formation is one stage between the elements and the solid compound, with the enthalpy of formation of the compound completing the cycle. As in Fig. 4.22, the lattice enthalpy is incorporated by using a path that leads to the formation of the gas of ions (by atomization and ionization of the elements). The value of the lattice enthalpy—the only unknown in a well-chosen cycle—is found from the requirement that the sum of the enthalpy changes around a complete cycle is zero (since enthalpy is a state property):

$$\Delta H(\text{atomization}) + \Delta H(\text{ionization}) + \Delta H(\text{lattice formation}) + \Delta H(\text{decomposition to elements}) = 0$$

This is the same as

$$\Delta H(\text{atomization}) + \Delta H(\text{ionization}) - \Delta H_L - \Delta H_f^\ominus = 0$$

where $\Delta H_f^\ominus$ is the standard enthalpy of formation of the compound (the reverse of its decomposition). Therefore,

$$\Delta H_L = \Delta H(\text{atomization}) + \Delta H(\text{ionization}) - \Delta H_f^\ominus$$

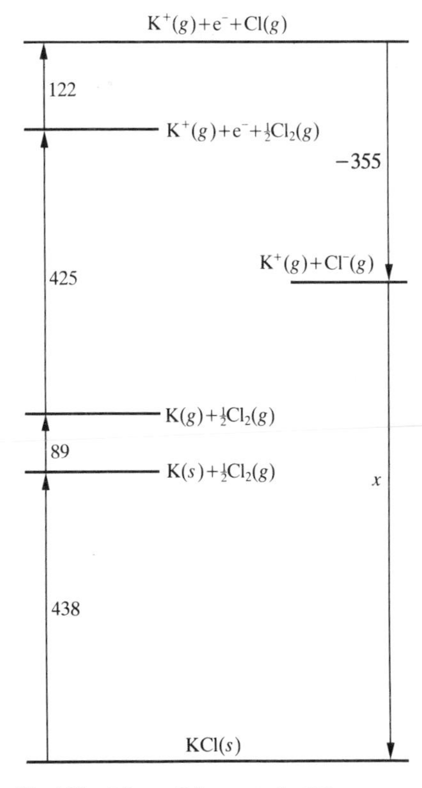

Fig. 4.22 A Born–Haber cycle for KCl.

Example 4.3: *Using a Born–Haber cycle to determine a lattice enthalpy*

Calculate the lattice enthalpy of KCl(*s*) using a Born–Haber cycle and the following information:

	$\Delta H^{\ominus}/(\text{kJ mol}^{-1})$
Sublimation of K(*s*)	+89
Ionization of K(*g*)	+425
Dissociation of $Cl_2(g)$	+244
Electron attachment to Cl(*g*)	−355
Formation of KCl(*s*)	−438

Answer. The required cycle is shown in Fig. 4.22. The first step is the sublimation of solid K:

$$K(s) \rightarrow K(g) \qquad \Delta H^{\ominus}/(\text{kJ mol}^{-1}) \quad +89$$

followed by its ionization:

$$K(g) \rightarrow K^+(g) + e^-(g) \qquad +425$$

Chloride ions are formed by dissociation of Cl_2:

$$\tfrac{1}{2}Cl_2(g) \rightarrow Cl(g) \qquad +122$$

followed by electron gain:

$$Cl(g) + e^-(g) \rightarrow Cl^-(g) \qquad -355$$

The solid is now formed:

$$K^+(g) + Cl^-(g) \rightarrow KCl(s) \qquad x$$

and the cycle is completed by decomposing KCl(*s*) into its elements:

$$KCl(s) \rightarrow K(s) + \tfrac{1}{2}Cl_2(g) \qquad +438$$

(the reverse of its formation). The sum of the enthalpies is $x + 719\ \text{kJ mol}^{-1}$; however, the sum must be equal to zero, so $x = -719\ \text{kJ mol}^{-1}$. The lattice enthalpy is the negative of this enthalpy change, $\Delta H_L = +719\ \text{kJ mol}^{-1}$.

Exercise. Calculate the lattice enthalpy of magnesium bromide from the following data:

	$\Delta H^{\ominus}/(\text{kJ mol}^{-1})$
Sublimation of Mg(*s*)	+148
Ionization of Mg(*g*)	+2187
Vaporization of $Br_2(l)$	+31
Dissociation of $Br_2(g)$	+193
Electron attachment to Br(*g*)	−331
Formation of $MgBr_2(s)$	−524

Once the lattice enthalpy is known, it can be used to judge the character of the bonding in the solid. If the value calculated on the assumption that the lattice consists of ions interacting coulombically is in good agreement with the measured value, we may adopt a largely ionic model of the compound. A discrepancy indicates a degree of covalence. The following sections describe one approach to the electrostatic calculation.

Coulombic contributions to lattice enthalpies

The total coulombic potential energy of a crystal is the sum of the individual coulomb potential energy terms

$$V_{AB} = \frac{(z_A e) \times (z_B e)}{4\pi\varepsilon_0 r_{AB}}$$

for ions of charge numbers z_A and z_B separated by a distance r_{AB} (ε_0 is the vacuum permittivity, 8.85×10^{-12} C^2 J^{-1} m^{-1}). The sum over all the pairs of ions in the solid may be carried out for any crystal structure, although in practice it converges very slowly because nearest neighbors contribute a large negative term, second-nearest neighbors an only slightly weaker positive term, and so on. The overall result is that the attraction between the cations and anions predominates and yields a favorable (negative) contribution to the energy of the solid.

For example, in a uniformly spaced one-dimensional line of alternating cations and anions with $z_A = +z$ and $z_B = -z$, the interaction of one ion with all the other is proportional to

$$-\frac{2z^2}{d} + \frac{2z^2}{2d} - \frac{2z^2}{3d} + \frac{2z^2}{4d} - \ldots = \frac{-2z^2}{d}(1 - \tfrac{1}{2} + \tfrac{1}{3} - \tfrac{1}{4} + \ldots)$$

$$= \frac{-2z^2}{d} \ln 2$$

(The factor of 2 comes from the fact that the same ions occur on both sides of the central ion.) As in this case, it is found that, apart from the explicit appearance of the ion charge numbers, the value of the sum depends only on the type of the lattice and on a single scale parameter which may be taken as the separation of the centers of nearest neighbors, d. We may write

$$V = \frac{-e^2}{4\pi\varepsilon_0} \times \frac{z^2}{d} \times 2 \ln 2$$

The first factor is a collection of fundamental constants. The second is specific to the identities of the ions and to the scale of the lattice. The third term, $2 \ln 2 = 1.386$, characterizes the symmetry of the lattice (in this case a straight line) and is the simplest example of a **Madelung constant** $\mathscr{A}$. In simple solids, the Madelung constant is specific to the crystal type and independent of the interionic distances.

In general, the total potential energy per mole of formula units in an arbitrary crystal structure is

$$V_m = N_A \frac{e^2}{4\pi\varepsilon_0}\left(\frac{z_A z_B}{d}\right) \times \mathscr{A}$$

Some computed values of the Madelung constant for a variety of lattices are given in Table 4.3. The general trend is for the values to increase with coordination number. This reflects the fact that a large contribution comes from nearest neighbors, and these are more numerous when the coordination number is large. The values for the rock-salt structure (with a coordination number of six) and cesium-chloride structure (with a coordination number of eight) illustrate the trend. However, this does not necessarily mean that the interactions are stronger in the cesium chloride structure, since the potential energy also depends on the scale of the lattice. Thus d may be so large in lattices with ions big enough to adopt eightfold coordination that it overcomes the small increase in Madelung constant and results in a less favorable potential energy.

Table 4.3. Madelung constants*

Structural type	$\mathscr{A}$
Cesium chloride	1.763
Fluorite	2.519
Rock-salt	1.748
Rutile	2.408
Sphalerite	1.638
Wurtzite	1.641

* The values given are for the geometric factor $\mathscr{A}$ described in the text (Section 4.7). Some sources cite values that include the charge numbers of the ions (so, for instance, the value for CaF_2 is quoted as 5.039), and it is necessary to verify the definition before using them.

Example 4.4: *Analyzing the coulombic contribution to the lattice enthalpy*

Which has the more favorable electrostatic energy, KCl with a rock-salt structure or KCl with a cesium chloride structure?

Answer. The coulombic energy is proportional to the Madelung constant and inversely proportional to the scale of the lattice. Therefore, we should compare the values of $\mathscr{A}/d$ for the two substances. The Madelung constants for the two structures are

$$\mathscr{A}(\text{rock salt}) = 1.748 \qquad \mathscr{A}(\text{cesium chloride}) = 1.763$$

Notice that the coordination numbers of the lattices are different and use the appropriate radii. The ionic radii are $r(K^+, \text{C.N.} = 6) = 1.38$ Å, $r(K^+, \text{C.N.} = 8) = 1.51$ Å, $r(Cl^-) = 1.81$ Å, so

$$d(\text{rock salt}) = 1.38\ \text{Å} + 1.81\ \text{Å} = 3.19\ \text{Å}$$

$$d(\text{cesium chloride}) = 1.51\ \text{Å} + 1.81\ \text{Å} = 3.32\ \text{Å}$$

The ratio of the potential energies is

$$\frac{V(\text{rock salt})}{V(\text{cesium chloride})} = \frac{\mathscr{A}(\text{rock salt})}{\mathscr{A}(\text{cesium chloride})} \times \frac{d(\text{cesium chloride})}{d(\text{rock salt})}$$

$$= \frac{1.748}{1.763} \times \frac{3.32\ \text{Å}}{3.19\ \text{Å}}$$

$$= 1.03$$

That is, the coulombic attraction in the rock-salt structure is marginally more favorable than in the cesium-chloride structure. This brings us close to what must be the explanation of the radius-ratio rules.

Exercise. Repeat the analysis for CsCl in the two structures.

Another positive contribution to the lattice enthalpy arises from the **van der Waals attractions** between the ions and molecules, the weak intermolecular interactions that are responsible for the formation of condensed phases of neutral species. The dominant contribution of this kind is often the **dispersion interaction** (the 'London interaction'), the interaction between fluctuations in the instantaneous electric dipoles of the molecules. The potential energy of this interaction is inversely proportional to the sixth power of the separation, so in the crystal as a whole we can expect it to vary as the sixth power of the scale of the lattice:

$$V_m = \frac{-N_A C}{d^6}$$

C is a constant that depends on the substance. For ions of low polarizability, this contribution is only about 1 percent of the coulombic contribution and we shall ignore it in lattice enthalpy calculations of ionic solids.

Repulsions arising from overlap

When two closed-shell ions are in contact, another contribution to the total energy is the repulsion arising from the overlap of their electron distributions.[3] Since orbitals decay exponentially towards zero at large distances from the nucleus, and repulsive interactions depend on the overlap of orbitals, it is plausible that their contribution to the potential energy has the form

$$V = +N_A C' e^{-d/d^*}$$

where C' and d^* are constants. The parameters can be estimated from measurements of compressibilities, since these reflect the increase in potential energy that occurs when ions are pressed together by an applied force. Although the values of d^* measured in this way actually span a range, it is often found that setting $d^* = 0.345$ Å gives reasonable agreement with experiment. We shall see in a moment that C' cancels and need not be known.

The total energy for an arbitrary scale size d is the sum of the coulombic and overlap contributions:

$$V = N_A \frac{e^2}{4\pi\varepsilon_0}\left(\frac{z_A z_B}{d}\right) \times \mathscr{A} + N_A C' e^{-d/d^*}$$

At $T = 0$ (when there is no thermal motion, and hence no nuclear kinetic energy), the ions adopt separations such that the attractive interactions are balanced by the repulsive interactions and V is a minimum. The minimum potential energy is obtained where $dV/dd = 0$, which occurs when

$$N_A C' e^{-d/d^*} = -N_A \frac{e^2}{4\pi\varepsilon_0} z_A z_B \times \frac{\mathscr{A} d^*}{d^2}$$

Substitution of this relation into the preceding one results in the **Born–Mayer equation**:

$$V = \frac{N_A z_A z_B e^2}{4\pi\varepsilon_0 d}\left(1 - \frac{d^*}{d}\right)\mathscr{A}$$

The experimental scale of the lattice may now be used for d. Since the potential energy of the ions is zero when they are dispersed as a gas, the negative of V may be identified as the lattice enthalpy (strictly, its value at $T = 0$).

As remarked previously, the agreement between the experimental lattice enthalpy and the value calculated on this ionic model of the solid is a measure of the extent to which the solid is ionic. Reasonable agreement signifies that the compound is ionic; poor agreement signifies that there is a significant degree of covalence. Some data are given in Table 4.4. We see that alkali metal halides give quite good agreement, the best with the hard halide ions (F^-) and the worst with the soft halide ions (I^-). The worst agreement is for soft cation–soft anion combinations, such as CuI, AgI, and TlI. These compounds are more likely to be substantially covalent, and hence not conform to the ionic model on which the Born–Mayer equation is based.

[3] The repulsive interactions between ions of like charge are essential to the stability of the solid, for Earnshaw's theorem (which was also a source of concern for early models of the atom) implies that coulombic forces alone cannot result in a stable, static structure. A more obvious point is that without repulsive interactions, the cations and anions would coalesce.

Table 4.4. Measured and calculated lattice enthalpies

Compound*	$\Delta H_L/(\text{kJ mol}^{-1})$ (calc)	(expt)	calc/expt (percent)
LiF^a	1033	1037	99.6
$LiCl^a$	845	852	99.2
$LiBr^a$	798	815	97.9
LiI^a	740	761	97.2
NaF^a	915	926	98.8
$NaCl^a$	778	786	99.0
$NaBr^a$	739	752	98.3
NaI^a	692	705	98.2
KF^a	813	821	99.0
KCl^a	709	717	98.9
KBr^a	680	689	98.7
KI^a	640	649	98.6
RbF^a	778	789	98.6
$RbCl^a$	686	695	98.7
$RbBr^a$	659	668	98.7
RbI^a	622	632	98.4
CsF^a	748	750	99.7
$CsCl^b$	652	676	96.4
$CsBr^b$	632	654	96.6
CsI^b	601	620	96.9
CuCl	904	993	91.0
CuBr	870	976	89.1
CuI	833	963	86.5
AgF	920	969	94.9
AgCl	833	912	91.3
AgBr	816	900	90.7
AgI	778	886	87.8
TiCl	686	748	91.7
TiBr	665	732	90.8
TiI	636	707	90.0

* *a* = NaCl; *b* = CsCl structures.
Source: D. Cubicciotti, *J. chem. Phys.*, **31**, 1646 (1959). The calculated values use a more complete ionic model that includes terms beyond those in the Born–Mayer equation.

Example 4.5: *Judging whether a solid is ionic*

Is AgCl an ionic solid?

Answer. The experimental lattice enthalpy (Table 4.4) is 912 kJ mol^{-1}. We judge whether the solid is ionic by testing whether the value calculated from the ionic model is in reasonable agreement with this value. AgCl has a rock-salt structure, so $\mathcal{A} = 1.748$. The sum of the ionic radii is 3.07 Å; we take $d^* = 0.345$ Å, as suggested in the text. The calculated lattice enthalpy is therefore

$$\Delta H_L = \frac{N_A e^2}{4\pi\varepsilon_0} \times \frac{1}{3.07\ \text{Å}} \times \left(1 - \frac{0.345}{3.07}\right) \times 1.748$$

For calculations of this kind, a useful quantity is

$$\frac{N_A e^2}{4\pi\varepsilon_0} = \frac{6.022 \times 10^{23}\ \text{mol}^{-1} \times (1.602 \times 10^{-19}\ \text{C})^2}{4\pi \times 8.854 \times 10^{-12}\ \text{C}^2\ \text{J}^{-1}\ \text{m}^{-1}}$$

$$= 1.39\ \text{MJ Å mol}^{-1}$$

Therefore,

$$\Delta H_L = 1.39\ \text{MJ Å mol}^{-1} \times \frac{1}{3.07\ \text{Å}} \times 0.89 \times 1.748$$

$$= 700\ \text{kJ mol}^{-1}$$

Since this is considerably different from the experimental value, we infer that the ionic model is poor for AgCl and that the solid has a substantial degree of covalence. Note that the calculation is much less sophisticated than that used to compile Table 4.4.

Exercise. Is MgO properly regarded as an ionic solid (experimental value: 3850 kJ mol^{-1})?

The Kapustinskii equation

The Russian chemist A. F. Kapustinskii noticed that if the Madelung constants for a number of structures are divided by the number of ions per formula unit, approximately the same value is obtained for them all. Moreover, he also noticed that the order of values so obtained increases with the coordination number. Therefore, since as we have seen, the ionic radius also increases with coordination number, the variation in $\mathcal{A}/nd$ from one structure to another can be expected to be quite small. This led him to propose that there exists a hypothetical rock-salt structure that is energetically equivalent to the true structure of any ionic solid. That being so, the lattice enthalpy can be calculated by using the Madelung constant and the appropriate ionic radii for (6, 6)-coordination. The resulting expression is called the **Kapustinskii equation**:

$$\Delta H_L = \frac{-nz_+z_-}{d}\left(1 - \frac{d^*}{d}\right)K$$

with $d = r_+ + r_-$, $K = 1.21$ MJ Å mol^{-1}, and n the number of ions in each formula unit.

Example 4.6: *Using the Kapustinskii equation*

Estimate the lattice enthalpy of KNO_3.

Answer. In order to use the Kapustinskii equation we need the number of ions per formula unit ($n = 2$), their charge numbers ($z_+ = +1$, $z_- = -1$), and the sum of their thermochemical radii (1.38 Å + 1.89 Å = 3.27 Å). Then, with $d^* = 0.345$ Å,

$$\Delta H_L = \frac{2}{3.27} \times \left(1 - \frac{0.345\ \text{Å}}{3.27\ \text{Å}}\right) \times 1.21\ \text{MJ mol}^{-1}$$

$$= 662\ \text{kJ mol}^{-1}$$

Exercise. Evaluate the lattice enthalpy of $CaSO_4$.

Table 4.5. The thermochemical radii of ions, r/Å

Main-group elements							
BeF_4^{2-}	BF_4^-	CO_3^{2-}	NO_3^-				
2.45	2.28	1.85	1.89				
			PO_4^{3-}	SO_4^{2-}	ClO_4^-	CrO_4^{2-}	MnO_4^-
			2.38	2.30	2.36	2.30	2.40
			AsO_4^{3-}	SeO_4^{2-}		MoO_4^{2-}	
			2.48	2.43		2.54	
			SbO_4^{3-}	TeO_4^{2-}	IO_4^-		
			2.60	2.54	2.49		
Complex ions							
		$[TiCl_6]^{2-}$	$[IrCl_6]^{2-}$	$[SiF_6]^{2-}$	$[GeCl_6]^{2-}$		
		2.48	2.54	1.94	2.43		
		$[TiBr_6]^{2-}$	$[PtCl_6]^{2-}$	$[GeF_6]^{2-}$	$[SnCl_6]^{2-}$		
		2.61	2.59	2.01	2.47		
		$[ZrCl_6]^{2-}$			$[PbCl_6]^{2-}$		
		2.47			2.48		

Source: A. F. Kapustinskii; *Q. Rev. Chem. Soc.*, **20**, 203 (1956).

The Kapustinskii equation allows us to ascribe a meaning to the 'radii' of nonspherical molecular ions, for their values can be adjusted until the calculated value of the lattice enthalpy matches that obtained experimentally from the Born–Haber cycle. The self-consistent parameters obtained in this way are called **thermochemical radii** (Table 4.5). They may be used to estimate the stabilities of lattices, and hence the formation enthalpies, of a wide range of compounds.

4.8 Consequences of lattice enthalpies

The Born–Mayer equation shows that for a given lattice type (a given value of $\mathcal{A}$), the lattice enthalpy increases with increasing ion charge numbers (as $z_A z_B$). The lattice enthalpy also increases as the scale d of the lattice decreases. Energies that vary as the **electrostatic parameter** ξ, where

$$\xi = \frac{z^2}{d}$$

z being an ion charge number and d a scale length, are widely adopted in inorganic chemistry, as indicative that an ionic model is appropriate.[4] In Chapter 7, we shall see how ξ applies to some complex formation equilibria.

Thermal stabilities of ionic solids

Thermal stabilities of solid inorganic compounds can be discussed in terms of their Gibbs free energies of decomposition into specified products. In many cases it is sufficient to consider only the reaction enthalpy, for the reaction entropy is either negligible or almost constant for the substances being compared. Moreover, as we shall see, most of these calculations center on differences of lattice enthalpy between the solid reactants and solid products. These in turn may be discussed, at least in a general way, in terms of the Kapustinskii equation. Hence the variation in stability may be expressed in terms of the variation of the electrostatic parameter ξ.

As an illustration of what is involved, consider the thermal stability of carbonates:

$$MCO_3(s) \rightarrow MO(s) + CO_2(g)$$

The experimental observation is that the **decomposition temperature**, the temperature at which $\Delta G^\ominus$ becomes negative and the reaction favorable, increases with cation radius. Calcium carbonate, for instance, decomposes at a higher temperature than magnesium carbonate.

From the thermodynamic relation

$$\Delta G^\ominus = \Delta H^\ominus - T\Delta S^\ominus$$

it follows that the decomposition temperature is reached when

$$T = \frac{\Delta H^\ominus}{\Delta S^\ominus}$$

[4] The correlation of properties with ξ is a very useful guide for many properties, but we should not take it as proof of the dominance of specifically charge–charge interactions. Often there is a correlation of a property with a wide range of parameters such as z^2/r^2 or z/r; indeed, a correlation can often be found with almost any expression with charge number in the numerator and radius in the denominator.

The decomposition entropy is almost constant for all carbonates because it is dominated by the formation of gaseous carbon dioxide in each case. Hence, the higher the reaction enthalpy, the higher the decomposition temperature (Table 4.6).

Thermal stability of large cation–large anion combinations

We can use the relation between decomposition temperature and reaction enthalpy to demonstrate that *large cations stabilize large anions*. The key to the explanation is obtained by considering the difference between the lattice enthalpy of a salt (such as MCO_3) and its decomposition product (MO). The reaction enthalpy of the decomposition can be written

$$\begin{aligned}\Delta H^{\ominus} &= C - \Delta H_L(\mathrm{MO}) + \Delta H_L(\mathrm{MCO_3}) \\ &= C - L\end{aligned}$$

where C is a constant (independent of the metal M) relating to the formation of the gas CO_2. Therefore

$$L = \Delta H_L(\mathrm{MO}) - \Delta H_L(\mathrm{MCO_3})$$

reflects the change in the stability of the lattice when the oxide is formed. In general, L is positive (since the lattice enthalpy of the oxide is greater than that of the carbonate), and hence subtracts from C. The decomposition is endothermic, but the greater the increase in lattice enthalpy in the decomposition (the larger L), the less endothermic it is. Hence, the greater the increase in lattice enthalpy on forming MO from MCO_3, the lower the decomposition temperature.

First, we give the pictorial explanation. We see from Fig. 4.23 that the relative change in scale of the lattice is large when the salt of a small cation and a large anion becomes an oxide. The change in scale is relatively small when the salt has a large cation initially. As the illustration shows in an exaggerated way, when the cation is very big, the change in size of the anion barely affects the scale of the lattice. Therefore, the lattice enthalpy difference, which depends on ξ and hence is inversely proportional to the lattice scale, is more favorable to decomposition in the former case than in the latter. It is therefore more favorable for the salt of the small cation to decompose, and its decomposition temperature is lower.

We can obtain the quantitative dependence of the reaction enthalpy on the identity of the cation by expressing the lattice enthalpies in L in terms of

Table 4.6. Decomposition data on carbonates*

	Mg	Ca	Sr	Ba
$\Delta G^{\ominus}/(\mathrm{kJ\,mol^{-1}})$	+48.3	+130.4	+183.8	+218.1
$\Delta H^{\ominus}/(\mathrm{kJ\,mol^{-1}})$	+100.6	+178.3	+234.6	+269.3
$\Delta S^{\ominus}/(\mathrm{J\,K^{-1}\,mol^{-1}})$	+175.0	+160.6	+171.0	+172.1
$\theta/°\mathrm{C}$	300	840	1100	1300

* Data are for the reaction

$$\mathrm{MCO_3}(s) \rightarrow \mathrm{MO}(s) + \mathrm{CO_2}(g) \qquad \Delta X^{\ominus}(298\,\mathrm{K})$$

θ is the temperature required to reach 1 bar pressure of CO_2, and has been estimated from the 298 K reaction enthalpy and entropy.
Source: Gibbs function data calculated from NBS Thermochemical tables.

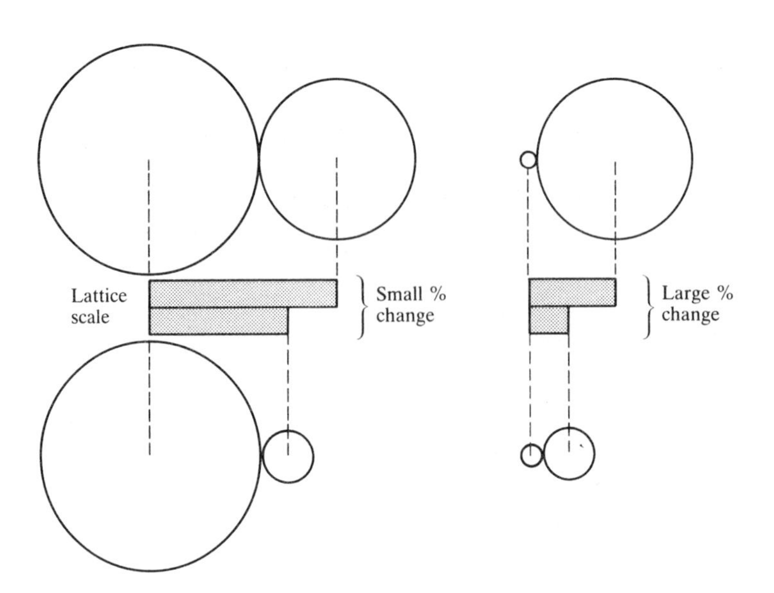

Fig. 4.23 A greatly exaggerated representation of the change in lattice parameter for cations of different sizes. (a) When the anion changes size (when CO_3^{2-} decomposes to O^{2-} and CO_2, for instance), and the cation is large, the lattice scale changes by a relatively small amount. (b) If the cation is small, the *relative* change in lattice scale is large, and decomposition is thermodynamically more favorable.

the Kapustinskii equation, which gives

$$L \propto \frac{1}{r(M^+) + r(O^{2-})} - \frac{1}{r(M^+) + r(CO_3^{2-})}$$

Since the CO_3^{2-} ion is larger than the O^{2-} ion, L is positive, which reduces the overall endothermic character of the decomposition. That is, it is energetically favorable for the bulky CO_3^{2-} ion to be replaced by a small O^{2-} ion. If $r(M^+)$ is large, however, the difference in the two denominators is not great (if $r(M^+)$ were infinite, we could ignore the anion radii in the denominators, and the two terms would cancel). Hence, the greatest difference and the maximum contribution to exothermicity arises when $r(M^+)$ is small. As we saw pictorially, the decomposition reaction becomes more endothermic the greater the ionic radius of M.

The role of charge

The contribution of L depends on the charge of the cations as

$$L \propto z(M^+)\left\{\frac{1}{r(M^+) + r(O^{2-})} - \frac{1}{r(M^+) + r(CO_3^{2-})}\right\}$$

When the charge number is large, the exothermic contribution from L is much more important and decompositions occur at lower temperature. Thus, alkaline earth carbonates (M^{2+}) are less stable than the corresponding alkali metal carbonates (M^+).

Example 4.7: *The dependence of stability on ionic radius*

Present an argument to account for the fact that, when they burn in oxygen, Li forms the oxide Li_2O but Na forms the peroxide Na_2O_2.

Answer. Since the small Li^+ ion results in Li_2O having a more favorable lattice enthalpy than Na_2O, the decomposition reaction $M_2O_2 \rightarrow M_2O + \frac{1}{2}O_2$ is thermodynamically more favorable for Li_2O_2 than Na_2O_2.

Exercise. Predict the order of decomposition temperatures for alkaline earth sulfates in the reaction $MSO_4(s) \rightarrow MO(s) + SO_3(g)$.

The stability of high oxidation numbers in the presence of small anions

A similar argument can be used to account for the greater ability of fluorine, compared with the other halogens, to stabilize the high oxidation numbers of metals. Thus, the only known halides of Ag(II), Co(III), and Mn(IV) are the fluorides. Another sign of the decrease in stability of the heavier halides of metals with high oxidation numbers is that the iodides of Cu(II) and Fe(III) decompose on standing at room temperature.

To explain these observations, we consider the redox reaction

$$MX + \tfrac{1}{2}X_2 \rightarrow MX_2$$

Our aim is to show why this reaction is most favored for $X = F$. If we ignore entropy contributions, this implies that we must show that the decomposition is most exothermic for fluorine.

One contribution to the reaction enthalpy is the conversion of $\frac{1}{2}X_2$ to X^-. The conversion of $\frac{1}{2}F_2$ to F^- is more favored than for the conversion of $\frac{1}{2}Cl_2$ to Cl^- because, although the electron affinity of F is lower than that of Cl (Table 1.8), the F—F bond enthalpy is lower than that of Cl—Cl.

The lattice enthalpies, however, play the major role. From the Kapustinskii equation, the difference in lattice energy between MX and MX_2 is proportional to

$$\frac{1}{r(M^+) + r(X^-)} - \frac{2}{r(M^{2+}) + r(X^-)}$$

This is a negative quantity, reflecting the exothermicity of the process in which MX is converted into MX_2, a solid with a more highly charged cation. As $r(X^-)$ increases, though, this difference becomes small. We can see this algebraically by supposing that the anion is so large that the cation radii can be ignored; then the difference is $-1/r(X^-)$, which becomes smaller as the anion becomes larger. Hence, both the lattice energy and the X^- formation enthalpy lead to a less exothermic reaction as the halogen changes from F to I. So long as entropy factors are similar, this corresponds to an increase in thermodynamic stability of MX relative to MX_2 on going from $X = F$ to $X = I$ down the group.

Solubility

Lattice enthalpies play a role in solubilities, but one that is much more difficult to analyze than for reactions. The solubility of an ionic compound depends on the Gibbs free energy of the process

$$MX(s) \rightarrow M^+(aq) + X^-(aq)$$

in which the interactions responsible for the lattice enthalpy are replaced by hydration (and by solvation in general). Since there are much less sharp differences between the lattice enthalpy and hydration enthalpy than there are between the lattice enthalpies of different compounds, the two terms almost balance. This leaves the value of $\Delta G^{\ominus}$ dominated by the entropy change that accompanies dissolving. On that, the Kapustinskii equation is silent.

However, a rule that is quite widely obeyed is that salts of ions with widely different radii are generally soluble. Conversely, the least soluble salts are those of ions with similar radii. That is, in general, *asymmetry in size favors solubility*. The experimental basis of this rule is shown by the data in Fig. 4.24 where we see that there is a correlation between the enthalpy of solution of a salt and the difference of the hydration enthalpies of the two ions. If the cation has a larger hydration enthalpy than its anion partner (reflecting the asymmetry in their sizes), or vice versa, the enthalpy of solution of the salt is exothermic (reflecting the favorable solubility equilibrium).

These features can be explained using the ionic model. We can write the lattice enthalpy, as before, as

$$\Delta H_{\mathrm{L}}^{\ominus} = \frac{f_1}{r_+ + r_-}$$

The hydration enthalpy is the sum of individual ion contributions:

$$\Delta H_{\mathrm{H}}^{\ominus} = \frac{f_2}{r_+} + \frac{f_3}{r_-}$$

where f_1, f_2, and f_3 are constants. If the radius of one ion is small, the term in the hydration enthalpy for that ion will be large. However, the contribution to lattice enthalpy is joint, and one small ion cannot make the denominator of the expression small by itself. Thus one small ion can contribute to solvation in a way it cannot contribute to lattice energy.

Two familiar series of salts illustrate these trends. In gravimetric analysis, Ba^{2+} is used to precipitate SO_4^{2-} quantitatively, and the solubilities of the alkaline earth metal sulfates decrease from Mg to Ba. In contrast, the solubility of the alkaline earth metal hydroxides increases from Mg to Ba: $Mg(OH)_2$ is the sparingly soluble 'milk of magnesia' but $Ba(OH)_2$ is used as a soluble hydroxide for preparation of solutions of OH^-. The first case shows that a large anion requires a large cation for precipitation. The

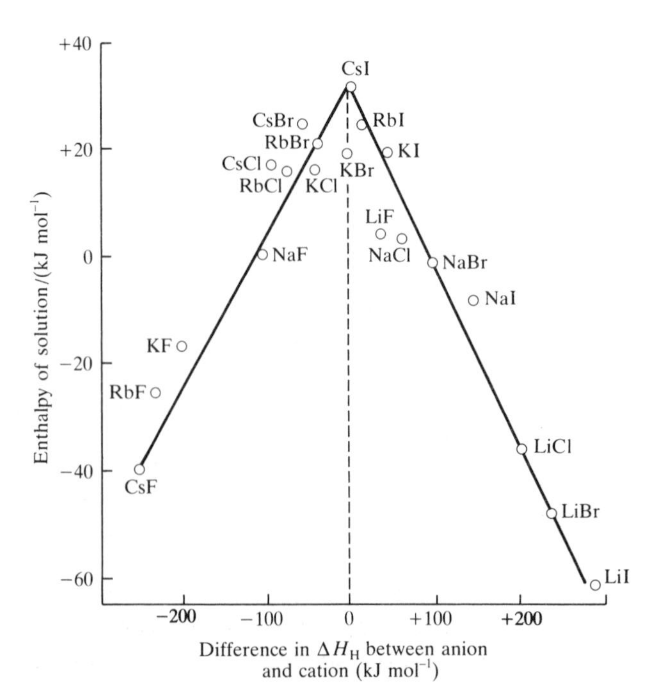

Fig. 4.24 The correlation between enthalpies of solution of halides and the differences between the hydration enthalpies of the ions. Dissolving is most exothermic when the difference is large.

second case shows that a small anion requires a small cation for precipitation.

A parallel analysis can be carried out for charge type. Ions of different charge type tend to produce more soluble salts. The least soluble salts tend to be those having ions of the same charge type.

Further information: thermodynamics

Thermodynamics is an extensive and subtle subject, but most of what is needed for everyday purposes in inorganic chemistry can be expressed by a handful of equations and the concepts they represent.

Consequences of the First Law of thermodynamics

The principal concept of the First Law of thermodynamics that is required in chemistry is the property called **enthalpy**, H. The change in the enthalpy of a system is equal to the energy transferred to the system as heat at constant pressure:

$$\Delta H = q_p \qquad \text{(A1)}$$

Once we know the change in enthalpy that accompanies a reaction we can predict the heat output that accompanies that reaction when it occurs at constant pressure. Alternatively, we can predict the energy that must be supplied to it as heat in order for it to occur. Reactions that are accompanied by a decrease in enthalpy (and hence release heat into their surroundings) are called **exothermic reactions**. They are far more common than **endothermic reactions**, which are reactions that are accompanied by an increase in enthalpy and hence absorb heat as they occur.

The enthalpy change that accompanies a reaction is often reported for a specific set of conditions, when it is known as the **standard reaction enthalpy**, $\Delta H^{\ominus}$. The value of $\Delta H^{\ominus}$ is specifically the enthalpy change that accompanies the conversion of the pure (unmixed) reactants changing to pure products, all at 1 bar pressure (1 bar $= 10^5$ Pa) at the specified temperature. Data are commonly reported for 298.15 K (25.00°C), and unless we state otherwise, all the $\Delta H^{\ominus}$ values that we report in this text are for that temperature. One very important reaction enthalpy is the **standard enthalpy of formation**, $\Delta H_f^{\ominus}$, which is the standard reaction enthalpy that accompanies the formation of a compound from its elements. For example, the standard enthalpy of formation of $H_2O(l)$ is the standard enthalpy change for the reaction

$$H_2(g) + \tfrac{1}{2}O_2(g) \rightarrow H_2O(l) \qquad \Delta H_f^{\ominus}(H_2O, l) = -285.83\ \text{kJ mol}^{-1}$$

By definition, the standard enthalpy of formation of an element is zero at all temperatures.

The central characteristic of the enthalpy is that it is a **state property**. That is, it is a property that depends only on the current state of the system and is independent of how that state was prepared. One implication is that the enthalpy change around a **cycle**, a sequence of reactions that ends at its starting point, is zero. This is the justification for the analysis of lattice enthalpies using a Born–Haber cycle described in Section 4.7. A special case arises when the cycle consists of a single reaction and its reverse, for then it implies that the reverse of a reaction is the negative of the enthalpy change for the forward reaction:

$$N_2(g) + 3H_2(g) \rightarrow 2NH_3(g) \qquad \Delta H^{\ominus} = -92.22\ \text{kJ mol}^{-1}$$

$$2NH_3(g) \rightarrow N_2(g) + 3H_2(g) \qquad \Delta H^{\ominus} = +92.22\ \text{kJ mol}^{-1}$$

The second important implication of H being a state function is that the same change in enthalpy occurs whatever the sequence of reactions between the reactants and the products. In particular, the same change in enthalpy occurs even if the reaction sequence includes *hypothetical* steps in which we imagine the reactants being dismantled into their elements and then these elements being reassembled into the products:

$$\text{reactants} \rightarrow \text{elements} \rightarrow \text{products} \qquad \Delta H^{\ominus}$$

It follows that a standard reaction enthalpy can be constructed by combining the enthalpies of formation for all the reactants and products:

$$\Delta H^{\ominus} = \sum n\Delta H_f^{\ominus}(\text{products}) - \sum n\Delta H_f^{\ominus}(\text{reactants}) \qquad \text{(A2)}$$

where the n are the stoichiometric coefficients in the reaction of interest. For this expression to be valid, the enthalpies of formation must be those for the temperature of interest.

Consequences of the Second Law of thermodynamics

The Second Law of thermodynamics introduces the **entropy** S of the system. This is also a state property, and can be interpreted as a measure of the disorder of a system: the greater the entropy, the greater the disorder. Thus, the entropy of a substance increases as it is heated and when it undergoes a phase transition from a solid to a liquid or from a liquid to a gas. It will be useful to bear in mind that the molar entropy of a substance is generally much larger when it is a gas than when it is either a solid or a liquid. Entropies may either be calculated from spectroscopic data or measured calorimetrically by determining heat capacities down to temperatures close to $T = 0$, and values are reported as the **standard molar entropy** $S^{\ominus}$ of the substance (*not* its entropy of formation). Once we know the standard molar entropies of the substances involved in a reaction we can calculate the **standard entropy of reaction** $\Delta S^{\ominus}$ using

$$\Delta S^{\ominus} = \sum nS^{\ominus}(\text{products}) - \sum nS^{\ominus}(\text{reactants}) \qquad \text{(A3)}$$

Although this resembles eqn A2, we must remember that the entropies of elements are not zero and contribute to the sums.

The Second Law is a statement about the direction of **spontaneous processes**, which are processes that may occur without needing to be driven:

A spontaneous reaction is accompanied by an increase in entropy:

$$\Delta S(\text{total}) > 0$$

where $\Delta S(\text{total})$ is the change in the entropy of the system and its surroundings.

More informally, we can regard the Second Law as summarizing the observation that processes occur in the direction that leaves the universe in a state of greater disorder either because atoms are more widely dispersed (as in the spontaneous expansion of a gas) or because the energy is dispersed more chaotically (as when thermal motion is generated in the environment). It is very important to appreciate that the Second Law is a statement about the *total* entropy of the system and its surroundings. Thus, a process that leads to a *decrease* in disorder of a system (as in the formation of liquid water from gaseous hydrogen and oxygen) may be spontaneous so long as a greater increase in entropy occurs in the surroundings.

It might seem a cumbersome problem to keep track of changes in the surroundings, but this is in fact very simple, for under conditions of constant pressure and temperature, the change in entropy of the surroundings $\Delta S'$ is simply

$$\Delta S' = \frac{-\Delta H}{T}$$

where ΔH is the enthalpy change of the system. The total entropy change of the system and its surroundings is therefore

$$\Delta S(\text{total}) = \Delta S + \Delta S' = \Delta S - \frac{\Delta H}{T}$$

The beautiful feature of this result is that we can judge whether a process (in chemistry, a reaction) is spontaneous from properties (ΔH and ΔS) relating to the system alone.

This aspect of the approach is normally taken one stage further by introducing a single property of the system, the **Gibbs free energy** G, which is defined through

$$\Delta G = \Delta H - T\Delta S \qquad \text{(A4)}$$

Since

$$\Delta G = -T\Delta S(\text{total}) \qquad \text{(A5)}$$

the presence of the negative sign implies that spontaneous changes are accompanied by a *decrease* in Gibbs free energy at constant temperature and pressure. Moreover, we can see from eqn A4 that a negative ΔG is expected when ΔH is negative (for exothermic reactions) and ΔS is positive (for reactions that create disorder in the system), and the latter plays a more important role when the temperature is high (because T multiplies ΔS). However, it should not be forgotten that *both* contributions to ΔG are in fact related to the total entropy change.

The Gibbs free energy is a state property, and may be tabulated and manipulated like the enthalpy and entropy. Thus, we can define the **standard Gibbs free energy of reaction** $\Delta G^{\ominus}$ as the change in G when the reactants in their standard states (pure, unmixed, and under 1 bar pressure) change into products in their standard states. A special case is the **standard Gibbs free energy of formation** $\Delta G_f^{\ominus}$, the standard reaction Gibbs free energy for the formation of the compound from its elements, as in

$$H_2(g) + \tfrac{1}{2}O_2(g) \rightarrow H_2O(l) \qquad \Delta G_f^{\ominus}(H_2O, l) = -237.13\ \text{kJ mol}^{-1}$$

As for the reaction enthalpy, the standard Gibbs free energy of a reaction may be calculated by combining the formation values:

$$\Delta G^{\ominus} = \sum n\Delta G_{\mathrm{f}}^{\ominus}(\text{products}) - \sum n\Delta G_{\mathrm{f}}^{\ominus}(\text{reactants}) \qquad \text{(A6)}$$

The thermodynamics of equilibrium

The central focus of chemical thermodynamics is on the conditions that correspond to equilibrium. This is a condition when a system has no spontaneous tendency to change in either direction. In thermodynamic terms, this condition is the one for which the total entropy is at a maximum and hence $\Delta S(\text{total})$ does not increase for any infinitesimal change in the abundance of products or reactants. At equilibrium, therefore, $\Delta S(\text{total}) = 0$ for any change, and therefore at constant temperature and pressure from eqn A5,

$$\Delta G = 0 \quad \text{at equilibrium}$$

The change in Gibbs free energy that occurs when a reaction occurs in such a huge abundance of reactants and products that the mixture remains effectively constant in composition is given by

$$\Delta G = \Delta G^{\ominus} + RT \ln Q \qquad \text{(A7a)}$$

where Q is the **reaction quotient**:

$$a\mathrm{A} + b\mathrm{B} \rightarrow c\mathrm{C} + d\mathrm{D} \qquad Q = \frac{[\mathrm{C}]^c[\mathrm{D}]^d}{[\mathrm{A}]^a[\mathrm{B}]^b} \qquad \text{(A7b)}$$

(In formal thermodynamics, and in practice when ions are involved in the reaction and departures from ideality are important, the molar concentrations are replaced by activities.) In effect, the $\Delta G^{\ominus}$ in eqn 7 takes care of the change in Gibbs free energy that arises from the conversion of the identities of the substances that are present, and the term $RT \ln Q$ takes into account the change that arises from the mixing of the substances (and the greater entropy that mixing implies).

At equilibrium, $\Delta G = 0$. The reaction quotient when the reaction mixture has its equilibrium composition is denoted K and is called the **equilibrium constant** K of the reaction. Making these two substitutions in eqn A7 results in

$$0 = \Delta G^{\ominus} + RT \ln K$$

Therefore, the relation between the equilibrium constant and the standard Gibbs free energy of a reaction is

$$\Delta G^{\ominus} = -RT \ln K \qquad \text{(A8)}$$

This is one of the most important equations in chemical thermodynamics, for through it and eqn A6 we can make predictions about the equilibrium compositions of reaction mixtures.

The central feature of eqn A8 to bear in mind is that if $\Delta G^{\ominus} < 0$, the equilibrium constant K is greater than 1 and products are favored. If $\Delta G^{\ominus} > 0$, $K < 1$ and reactants are favored.

The response of equilibrium to the conditions

We can express the response of the equilibrium composition of a system to changes in the pressure and temperature by considering how $\Delta G^\ominus$ responds and then using eqn A8.

The standard Gibbs free energy of reaction is defined for a *fixed* pressure of 1 bar, and so $\Delta G^\ominus$ does not change when the pressure at which the reaction is carried out is changed. Therefore K does not change when the pressure is changed, and the equilibrium constant is independent of the applied pressure.

The standard Gibbs free energy of reaction does change when the temperature is changed. The simplest approach is to suppose that the whole of the variation stems from the T in eqn A4 and that the reaction enthalpy and entropy are constant. Then for two temperatures T and T',

$$\Delta G^\ominus = \Delta H^\ominus - T\Delta S^\ominus \qquad \Delta G^{\ominus\prime} = \Delta H^\ominus - T'\Delta S^\ominus$$

It then follows that

$$\ln K' - \ln K = \frac{\Delta G^\ominus}{RT} - \frac{\Delta G^{\ominus\prime}}{RT'} = \frac{\Delta H^\ominus}{R}\left(\frac{1}{T} - \frac{1}{T'}\right) \tag{A9}$$

This is the **van't Hoff equation**. We see that if the reaction is exothermic ($\Delta H^\ominus < 0$) and the temperature of the equilibrium mixture is raised ($T' > T$, so the temperature factor is positive), the right-hand side is negative. This implies that $K' < K$ and the formation of products is disfavored, in accord with the predictions of Le Chatelier's principle.

If we wish to concentrate on the temperature dependence of the reaction free energy itself (as distinct from K), we use the result from thermodynamics that

$$\frac{dG^\ominus}{dT} = -S^\ominus$$

Therefore, if the entropy of the substance is large (as for a gas), its Gibbs free energy responds strongly to changes in the temperature. This equation applies to all the substances participating in a reaction, so

$$\frac{d\Delta G^\ominus}{dT} = -\Delta S^\ominus \tag{A10}$$

Therefore, if the reaction produces gases (and $\Delta S^\ominus$ is strongly positive), the reaction free energy decreases sharply with increasing temperature. This aspect of the behavior of reaction free energies is taken up in Chapter 8.

Further reading

Some introductory texts on solid-state inorganic chemistry at about the level of this text are:

M. F. C. Ladd, *Structure and bonding in solid state chemistry*. Wiley, New York (1979).

D. M. Adams, *Inorganic solids*. Wiley, New York (1974).

M. H. B. Stiddard, *The elementary language of solid state physics*. Academic Press, New York (1975).

A. R. West, *Solid state chemistry and its applications*. Wiley, New York (1984).

P. A. Cox, *The electronic structure and chemistry of solids*. Oxford University Press (1987).

The standard reference book, which surveys the structures of a huge number of elements and compounds, is

A. F. Wells, *Structural inorganic chemistry*. Clarendon Press, Oxford (1984).

Two very useful, thoughtful introductory texts on the application of thermodynamic arguments to inorganic chemistry are:

W. E. Dasent, *Inorganic energetics*. Cambridge University Press (1982).

D. A. Johnson, *Some thermodynamic aspects of inorganic chemistry*. Cambridge University Press (1982).

Exercises

4.1 Which of the following schemes for ordering close-packed planes are not ways of generating close-packed lattices? (a) ABCABC..., (b) ABAC..., (c) ABBA..., (d) ABCBC... (e) ABABC... (f) ABCCBA...

4.2 Draw one layer of close-packed spheres. On this layer mark the positions of the centers of the B layer atoms using the symbol ⊗ and, with the symbol ○ mark the positions of the centers of the C layer atoms of an fcc lattice.

4.3 Show that the largest sphere that can fit into an octahedral hole in a close-packed lattice has a radius equal to $2^{1/2} - 1 = 0.414$ times that of the spheres composing the lattice. (Hint: Consider the four circles obtained by cutting a plane through the hole and the surrounding square of spheres.)

4.4 The interstitial alloy WC has the rock-salt structure. Describe it in terms of the holes in a close-packed structure.

4.5 β-Brass, which has a composition close to CuZn, has a cesium chloride structure. Is this alloy substitutional or interstitial?

4.6 Sketch the *s* and *p* blocks of the periodic table and mark the boxes for the elements that contribute monatomic cations and anions to solids that are described well by the ionic model. Identify the elements.

4.7 In a rock-salt structure: (a) What is the coordination number of the anion and cation? (b) How many Na^+ ions occupy second-nearest neighbor locations of an Na^+ ion? (c) Pick out the closest-packed plane of Cl^- ions. (Hint: this hexagonal plane will be perpendicular to a threefold axis.)

4.8 In a cesium chloride structure: (a) What is the coordination number of the anion and cation? (b) How many Cs^+ ions occupy second-nearest neighbor locations of a Cs^+ ion?

4.9 How many Cs^+ ions and how many Cl^- ions are in the CsCl unit cell? How many Zn^{2+} ions and how many S^{2-} ions are in the sphalerite unit cell?

4.10 Imagine the construction of an MX_2 lattice from the bcc CsCl lattice by removal of half of the Cs^+ ions to leave tetrahedral coordination around each Cl^-. What is this MX_2 structure?

4.11 Given the following data for the length of a side of the unit cell for compounds that crystallize in the NaCl structure, determine the cation radii: MgSe (5.45 Å), CaSe (5.91 Å), SrSe (6.23 Å), BaSe (6.62 Å). (To determine the Se^{2-} radius, assume that the Se^{2-} ions are in contact in MgSe.)

4.12 Using the structure map in Fig. 4.21, predict the coordination numbers of the cations and anions in (a) LiF, (b) RbBr, (c) SrS, and (d) BeO. Which of these compounds is likely to be well described by an ionic model?

4.13 What are the coordination numbers of the cations and anions in the sphalerite and wurtzite structures?

4.14 Confirm that in rutile (Fig. 4.17) the stoichiometry is consistent with the structure.

4.15 Figure 4.18 shows the perovskite structure of $CaTiO_3$. Confirm that the stoichiometry is consistent with the structure.

4.16 Calculate the enthalpy of formation of the hypothetical compound KF_2 assuming a CaF_2 structure. Use the Born–Meyer equation to obtain the lattice enthalpy and estimate the radius of K^{2+} by extrapolation of trends in Table 4.2. Ionization enthalpies and electron gain enthalpies are given in Tables 1.7 and 1.8. What factor prohibits the formation of this compound despite the favorable lattice enthalpy?

4.17 The coulombic attraction of nearest-neighbor cations and anions accounts for the bulk of the lattice energy of an ionic compound. With this fact in mind, estimate the order of increasing lattice energy for the following solids, all of which crystallize in the rock-salt structure: (a) MgO, (b) NaCl, (c) LiF. Give your reasoning.

4.18 Which of the following pairs of isostructural compounds are likely to be more stable with respect to thermal decomposition? Give your reasoning. (a) $MgCO_3$, $CaCO_3$

(decomposition products $MO + CO_2$). (b) CsI_3, $N(CH_3)_4I_3$ (both compounds contain I_3^-; decomposition products $MI + I_2$ and the radius of $N(CH_3)_4^+$ is much bigger than that of Cs^+).

4.19 Which of LiCl and KCl would you expect to be more soluble in water. Explain your answer.

4.20 Recommend a cation for the quantitative precipitation of carbonate ion in water.

4.21 You have just synthesized a complex ion of the formula $[CoCl(NH_3)_5]^{2+}$ in aqueous solution. To obtain a crystalline sample, should you try precipitation with sulfate or chloride?

Problems

4.1 In the structure of MoS_2, the S atoms are arranged in close-packed layers which repeat themselves in the sequence AAA . . . The Mo atoms occupy holes with a coordination number of 6. Show that each Mo atom is surrounded by a trigonal prism of S atoms.

4.2 Show that the maximum fraction of the available volume occupied by hard spheres on various lattices is (a) simple cubic: 0.52, (b) bcc: 0.68, (c) fcc: 0.74.

4.3 X-ray diffraction experiments indicate that NH_4NO_3 has the CsCl structure at 150°C. An NO_3^- ion can replace a Cl^- ion and preserve the symmetry of the lattice only if it has at least the symmetry of the Cl^- ion site or behaves as if it did. (a) What is the local point-group of the site it occupies? (b) What is the point-group symmetry of NO_3^-? (c) How might the high-temperature results be explained?

4.4 The common oxidation number for an alkaline earth metal is +2. Aided by the Born–Meyer equation for lattice enthalpy and a Born–Haber cycle, show that CaCl is an exothermic compound. Use a suitable analogy to estimate an ionic radius for Ca^+. The sublimation enthalpy of Ca(*s*) is 176 kJ mol^{-1}. Show that an explanation for the non-existence of CaCl can be found in the enthalpy change for the reaction

$$2CaCl(s) \rightarrow Ca(s) + CaCl_2(s)$$

4.5 LiF is the least soluble of the alkali metal halides. AgI is the least soluble of the silver halides. Can these facts be reconciled by the ionic model? (Note: the ionic radius of Ag^+ is close to that of K^+.) If not, how are they to be understood?

Part 2

Reactions

In this part we turn our attention from bonding and structural aspects of inorganic chemistry to the reactions that inorganic compounds undergo. It is possible to classify the great majority of reactions into three classes, namely acid–base, redox, and radical, and here we focus on the first two. Much of the discussion in the first two chapters is devoted to an understanding of what it means to call a substance an acid or a base and to the factors that influence the equilibrium position of the reactions between them. In Chapter 7 we see how to build on these concepts to discuss the structures and properties of *d*-metal complexes as a special case of the interaction between acids and bases. We consider redox reactions in the final chapter of this part, Chapter 8, and hence provide a basis for understanding most of the reactions of inorganic chemistry. As well as considering equilibria, we need to take note of the rates of reactions, and throughout this part we shall see how kinetic considerations may affect the predictions expected thermodynamically.

Brønsted acids and bases

5

The focus of this chapter is on substances that can participate in proton transfer reactions. We look for ways of rationalizing the different strengths with which substances can donate or accept protons, and to do so we use some of the parameters that we have been introducing, such as the electronegativity and the electrostatic parameter. As we shall see, the solvent plays a very important role in the transfer, and we describe how the intrinsic properties of a substance may be disentangled from the role of the solvent. The concepts we describe can account for a wide range of chemical reactions, and we conclude the chapter with a survey of how acidity may influence the formation of polymers of oxo ions. This is an important process in geochemistry, for it controls mineral formation, and in biochemistry, where it controls energy exchange.

Brønsted acidity
5.1 The proton in water
5.2 Acid equilibria in water
5.3 Factors governing acid strength
5.4 Solvent leveling
Periodic trends in Brønsted acidity
5.5 Periodic trends in aqua acid strength
5.6 Simple oxoacids
5.7 Anhydrous oxides
Polyoxo compound formation
5.8 Polymerization of aqua ions to polycations
5.9 Polyoxoanions of early *d*-block metals
5.10 Nonmetal polyoxoanions
Further reading
Exercises
Problems

The chemical manufactured in greatest mass annually is sulfuric acid; in third and fourth places in annual production (in 1989) are the bases lime and ammonia. These figures emphasize the central role of acids and bases in industrial chemistry. Acids and bases are equally important in laboratory chemistry, where many of the synthetic and analytical reactions that are used routinely and in the development of new materials are acid–base neutralizations of one kind or another.

In this chapter, we discuss the essential features of acids and bases in terms of proton transfer between molecules. In Chapter 6 we take a more general view, and see that an even wider range of reactions can be treated from a single viewpoint.

Brønsted acidity

One essential aspect of the behavior of acids and bases was identified in 1923 by Johannes Brønsted in Denmark and Thomas Lowry in England. They concentrated on the transfer of protons from one substance to another, and suggested that any substance that acts as a proton donor should be classified as an **acid**, and any substance that acts as a proton acceptor should be classified as a **base**. Substances that act in this way are called **Brønsted acids** and **Brønsted bases** respectively.

An example of a Brønsted acid is HF, which can donate a proton to another molecule, such as H_2O when it dissolves in water:

$$HF(g) + H_2O(l) \rightarrow H_3O^+(aq) + F^-(aq)$$

An example of a Brønsted base is NH_3, since it can accept a proton from an acid:

$$HF(aq) + NH_3(aq) \rightarrow NH_4^+(aq) + F^-(aq)$$

The definitions make no reference to the environment in which proton transfer occurs, and so apply equally to the gas phase and to solution in any solvent.

Water is an example of a substance that can act as both a Brønsted acid and a Brønsted base. Thus, it acts as an acid toward NH_3:

$$H_2O(l) + NH_3(aq) \rightarrow OH^-(aq) + NH_4^+(aq)$$

and as a base toward H_2S:

$$H_2S(aq) + H_2O(l) \rightarrow H_3O^+(aq) + HS^-(aq)$$

5.1 The proton in water

Since the most important environment for acid–base reactions is an aqueous solution, we set the stage by characterizing the behavior of the proton in water.

The hydronium ion

1.01
100 to 120° variable

1 H_3O^+

When an acid donates a proton to a water molecule, the product is the **hydronium ion**, H_3O^+. Protonation of all common solvent molecules is exothermic, and the free hydrogen ion is not encountered in solution.

The dimensions of the hydronium ion (**1**) are taken from the crystal structure of $H_3O^+ClO_4^-$ in which it is the cation. However, the structure H_3O^+ is almost certainly an oversimplified description of the ion in water, for it participates in extensive hydrogen bonding. An analysis of the rates of proton transfer by Manfred Eigen and his colleagues at Gottingen (as part of their pioneering study of very fast reactions using temperature-jump and ultrasonic relaxation techniques) indicates that the hydronium ion in water is best represented as $H_9O_4^+$ (**2**).

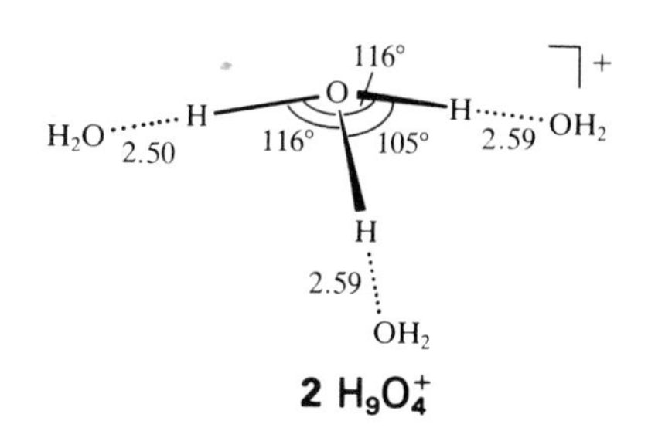

2 $H_9O_4^+$

The mobility of hydrogen ions in water

The formula $H_9O_4^+$ does not convey the whole sense of the structure of the ion. We have to bear in mind that $H_9O_4^+$ is not a fixed unit in solution, for a proton can transfer rapidly from one H_2O molecule to another. Early evidence for the rapidity of this transfer came from the observation that the mobility of the proton in water (as measured from its electric conductivity) is about three times that of typical ions. The explanation, which is called the **Grotthus mechanism**, is that the migration of a proton is in fact not an actual movement of the ion through the solvent but a cooperative rearrangement of the atoms: the proton jumps from one O atom to the next along a hydrogen bond (Fig. 5.1), the receiving molecule becomes the cation, and one of its protons can now migrate to another neighbor in the same way. The migration is therefore a cooperative process that takes place through a network of several hydrogen-bonded H_2O molecules.

Eigen's work shows that a proton remains at one end of an O—H $\cdots$ O hydrogen bond for a half life of 1 to 4 ps (1 ps $= 10^{-12}$ s) before it jumps and forms O $\cdots$ H—O. The O—H bond is not the only example of a very labile (short-lived) structure: fast proton transfers from one electronegative atom (O, N, Cl, . . .) to another are common. Table 5.1 shows some rate constants for such proton transfers. In each case, the reaction is overall second-order and first-order in the donor and acceptor concentrations, and the ratio of the forward to the reverse rate constants is the equilibrium constant for the transfer reaction. The rate of the favorable reaction is diffusion-controlled in each case; that is, the reaction is limited by the rate at which the acid and the base diffuse together through the solvent. The rate of the unfavorable reaction is limited by an activation barrier with a height equal to the (endothermic) enthalpy of reaction. That is, the reaction is as fast as possible in both directions.

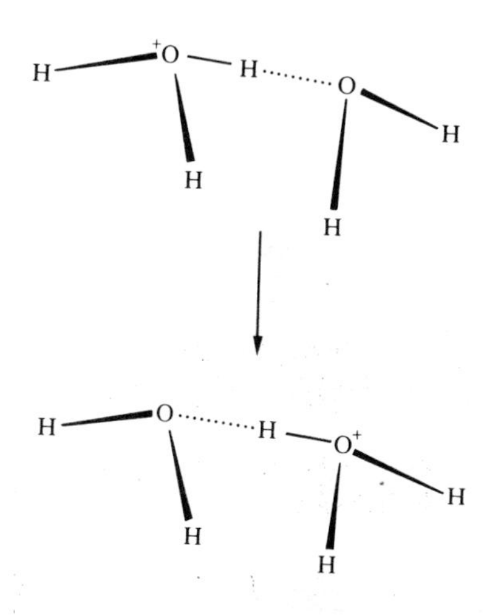

Fig. 5.1 The Grotthus mechanism for the effective migration of H^+ in water by rearrangement of covalent and hydrogen bonds.

Table 5.1. Rate constants* for aqueous proton transfer reactions at 25°C

Reaction	$k_{\rightarrow}/(M^{-1}\,s^{-1})$	$k_{\leftarrow}/(M^{-1}\,s^{-1})$
$H_3O^+ + OH^- \rightleftharpoons H_2O + H_2O$	1.4×10^{11}	2.5×10^{-5}
$H_3O^+ + SO_4^{2-} \rightleftharpoons HSO_4^- + H_2O$	1×10^{11}	7×10^{7}
$H_3O^+ + NH_3 \rightleftharpoons NH_4^+ + H_2O$	4.3×10^{10}	8.4×10^{5}
$OH^- + NH_4^+ \rightleftharpoons H_2O + NH_3$	3.4×10^{10}	6×10^{5}

* $k_{\rightarrow}$ is the rate constant for the forward reaction and $k_{\leftarrow}$ is that for the reverse reaction.

5.2 Acid equilibria in water

The dynamic equilibria

$$HF(aq) + H_2O(l) \rightleftharpoons H_3O^+(aq) + F^-(aq)$$
$$H_2O(l) + NH_3(aq) \rightleftharpoons OH^-(aq) + NH_4^+(aq)$$

give a more complete description of the behavior of the acid HF and the base NH_3 in water. Brønsted acid–base chemistry is mainly a matter of rapidly attained proton-transfer equilibria, and we concentrate on this aspect.

Conjugate acids and bases

In the first of these two equilibria, H_3O^+ is a Brønsted acid in the reverse reaction because it donates a proton to F^-. The latter anion, being a proton acceptor, is a Brønsted base. Likewise, in the reverse reaction in the second equilibrium, NH_4^+ is a Brønsted acid because it donates a proton to OH^-. The fact that OH^- accepts a proton identifies it as a Brønsted base.

We can express the symmetry of the forward and reverse reactions, both of which depend on the transfer of a proton from an acid to a base, by writing the general **Brønsted equilibrium** as

$$\text{Acid}_1 + \text{Base}_2 \rightleftharpoons \text{Base}_1 + \text{Acid}_2$$

Base_1 is called the **conjugate base** of Acid_1, and Acid_2 is the **conjugate acid** of Base_2. Thus, H_3O^+ is the conjugate acid of H_2O and F^- the conjugate base of HF. Likewise, OH^- is the conjugate base of H_2O, HS^- the conjugate base of H_2S, and NH_4^+ the conjugate acid of NH_3. By 'acid' and 'base' we normally mean the reactants in the chemical equation as written; the products are their conjugates. However, there is no fundamental distinction between an acid and a conjugate acid or a base and a conjugate base: a conjugate acid is just another acid and a conjugate base is just another base. Thus, HF is the conjugate acid of the base F^- and NH_3 is the conjugate base of the acid NH_4^+.

The classification of acid and base may depend on the reaction in which the substance is a participant, and a substance classified as a base in one reaction may be able to act as an acid in another (and vice versa). An example is HS^-, the conjugate base of the acid H_2S; it is an acid in its own right because it can donate its remaining proton:

$$HS^-(aq) + H_2O(l) \rightleftharpoons S^{2-}(aq) + H_3O^+(aq)$$

Thus, the conjugate base of HS^- is the sulfide ion S^{2-}. Similarly, although OH^- is the conjugate base of H_2O, it can donate the remaining proton and hence act as an acid. The conjugate base of the acid OH^- is the oxide ion O^{2-}.

Example 5.1: *Identifying acids and bases*

Identify the Brønsted acid and its conjugate base in (a) $HSO_4^- + OH^- \rightleftharpoons SO_4^{2-} + H_2O$, (b) $PO_4^{3-} + H_2O \leftrightharpoons HPO_4^{2-} + OH^-$, (c) $H_2Fe(CO)_4 + CH_3OH \leftrightharpoons [FeH(CO)_4]^- + CH_3OH_2^+$.

Answer. In (a), the hydrogen sulfate ion (HSO_4^-) transfers a proton to hydroxide; it is therefore the acid and the SO_4^{2-} ion produced is its conjugate base. In (b), H_2O transfers a proton to the phosphate ion acting as a base; thus H_2O is the acid and the OH^- ion is its conjugate base. In (c), $H_2Fe(CO)_4$ is the acid that transfers a proton to methanol its conjugate base is $[FeH(CO)_4]^-$.

Exercise. Identify the acid, base, conjugate acid, and conjugate base in (a) $HNO_3 + H_2O \rightleftharpoons H_3O^+ + NO_3^-$, (b) $CO_3^{2-} + H_2O \rightleftharpoons HCO_3^- + OH^-$, (c) $NH_3 + H_2S \rightleftharpoons NH_4^+ + HS^-$.

The strengths of Brønsted acids

We express the strength of a Brønsted acid HA in aqueous solution by the **acidity constant** (or **acid ionization constant**, K_a), the equilibrium constant for proton transfer between acid and water:

$$HA(aq) + H_2O(l) \rightleftharpoons A^-(aq) + H_3O^+(aq) \qquad K_a = \frac{a(H_3O^+)a(A^-)}{a(HA)a(H_2O)}$$

$a(X)$ is the activity of X, its effective thermodynamic concentration, in the solution at equilibrium. Since the activity of pure water (as for any pure liquid at 1 bar) is 1, and is close to 1 for dilute solutions, the approximation is normally made that $a(H_2O) = 1$. When thermodynamic precision is not required, as in approximate descriptive work or when concentrations are very low (less than about 1 mM), activities of solutes are often replaced by molar concentrations [X]. We shall make both approximations throughout this text, and use

$$K_a = \frac{[H_3O^+][A^-]}{[HA]}$$

The proton transfer from water acting as a Brønsted acid is described by the water **autoprotolysis constant** K_w:

$$H_2O(l) + H_2O(l) \rightleftharpoons H_3O^+(aq) + OH^-(aq) \qquad K_w = [H_3O^+][OH^-]$$

Since molar concentrations and acidity constants span many orders of magnitude, it proves convenient to report them as their logarithms (to the base 10) using

$$pH = -\log [H_3O^+] \qquad pK_a = -\log K_a \qquad pK_w = -\log K_w$$

At 25°C, $pK_w = 14.00$. Another advantage of using logarithms is the existence of the simple relation between the standard Gibbs free energy of the reaction and its equilibrium constant:

$$\Delta G^\ominus = -RT \ln K = 2.303RT\, pK \qquad (1)$$

At 25°C,

$$\Delta G^\ominus = 5.71 \times pK \text{ kJ mol}^{-1}$$

The acidity constants of some common acids are given in Table 5.2. The range of over 24 orders of magnitude in K_a may seem enormous, but in fact it corresponds to a variation of only about 340 kJ mol^{-1} in $\Delta G^\ominus$ for the proton transfer reaction. Substances with negative pK_a values (corresponding to $K_a > 1$ and usually to $K_a \gg 1$) are classified as **strong acids**, in the sense that the proton transfer equilibrium lies in favor of donation to water. Substances with positive pK_a values (corresponding to $K_a < 1$) are classified as **weak acids**; for them, proton transfer equilibrium lies in favor of nonionized HA. The conjugate bases of strong acids are **weak bases** since it is thermodynamically unfavorable for them to accept a proton from H_3O^+. A general feature to keep in mind, therefore, is that the weaker the acid, the stronger its conjugate base.

Table 5.2. Acidity constants for aqueous solution at 25°C

Acid	HA	A^-	K_a	pK_a
Hydriodic	HI	I^-	10^{11}	−11
Perchloric	$HClO_4$	ClO_4^-	10^{10}	−10
Hydrobromic	HBr	Br^-	10^9	−9
Hydrochloric	HCl	Cl^-	10^7	−7
Sulfuric	H_2SO_4	HSO_4^-	10^2	−2
Hydronium ion	H_3O^+	H_2O	1	0.0
Sulfurous	H_2SO_3	HSO_3^-	1.5×10^{-2}	1.81
Hydrogen sulfate ion	HSO_4^-	SO_4^-	1.2×10^{-2}	1.92
Phosphoric	H_3PO_4	$H_2PO_4^-$	7.5×10^{-3}	2.12
Hydrofluoric	HF	F^-	3.5×10^{-4}	3.45
Pyridinium ion	$HC_5H_5N^+$	C_5H_5N	5.6×10^{-6}	5.25
Carbonic	H_2CO_3	HCO_3^-	4.3×10^{-7}	6.37
Hydrogen sulfide	H_2S	HS^-	9.1×10^{-8}	7.04
Ammonium ion	NH_4^+	NH_3	5.6×10^{-10}	9.25
Hydrocyanic	HCN	CN^-	4.9×10^{-10}	9.31
Hydrogen carbonate ion	HCO_3^-	CO_3^{2-}	4.8×10^{-11}	10.32
Hydrogen arsenate ion	$HAsO_4^{2-}$	AsO_4^{3-}	3.0×10^{-12}	11.53
Hydrogen sulfide ion	HS^-	S^{2-}	1.1×10^{-12}	11.96
Hydrogen phosphate ion	HPO_4^{2-}	PO_4^{3-}	2.2×10^{-13}	12.67

Polyprotic acids

The successive acidity constants of **polyprotic acids** (substances that can donate more than one proton) are defined analogously. For a diprotic acid H_2A, such as H_2S for example, there are two successive proton donations and two acidity constants:

$$H_2A(aq) + H_2O(l) \rightleftharpoons HA^-(aq) + H_3O^+(aq) \qquad K_{a1} = \frac{[H_3O^+][HA^-]}{[H_2A]}$$

$$HA^-(aq) + H_2O(l) \rightleftharpoons A^{2-}(aq) + H_3O^+(aq) \qquad K_{a2} = \frac{[H_3O^+][A^{2-}]}{[HA^-]}$$

The second acidity constant, K_{a2}, is almost always smaller than K_{a1} (and hence pK_{a2} is generally larger than pK_{a1}). In the case of oxoacids, pK_{a2} is typically about 5 units larger than pK_{a1}, corresponding to a decrease in equilibrium constant by a factor of about 10^5, a decrease that is largely independent of the identity of the central atom. The large nonspecific decrease is consistent with an electrostatic model of the acid in which, in the second ionization, a proton must separate from a center with one more unit of negative charge than in the first ionization. Since additional electrostatic work must be done to remove the positively charged proton, the ionization is less favorable.

5.3 Factors governing acid strength

Aqueous solution pK_a values exhibit complex trends through the periodic table, as may be judged by noting the pK_a values of even such an apparently homogeneous group as the hydrohalic acids:

	HF	HCl	HBr	HI
pK_a	+3	−7	−9	−11

In order to make sense of them, it is useful to consider proton transfer in the gas phase first and then to consider the effects of the solvent.

Gas-phase acidities

The simplest reaction of a hydrogen ion is its attachment to a base B in the gas phase. This reaction is characterized by the **proton-gain enthalpy** $\Delta H_p^\ominus$, the standard enthalpy of the reaction

$$B(g) + H^+(g) \rightarrow BH^+(g) \qquad \Delta H_p^\ominus(B)$$

The enthalpy of protonation is sometimes reported in terms of the **proton affinity** A_p, which is the negative of the proton gain enthalpy. When $\Delta H_p^\ominus$ is negative, corresponding to exothermic proton attachment, we speak of the proton affinity as being high. If proton gain is only weakly exothermic, we say that the proton affinity of the base (B or A^-) is low.

Proton transfer from HA to B in the gas phase,

$$HA(g) + B(g) \rightarrow HB^+(g) + A^-(g) \qquad \Delta H^\ominus$$

is the difference of proton gain by B:

$$B(g) + H^+(g) \rightarrow HB^+(g) \qquad \Delta H_p^\ominus(B)$$

and proton gain by A^-:

$$A^-(g) + H^+(g) \rightarrow HA(g) \qquad \Delta H_p^\ominus(A^-)$$

The enthalpy of proton transfer is therefore the difference of the two proton-gain enthalpies

$$\Delta H^\ominus = \Delta H_p^\ominus(B) - \Delta H_p^\ominus(A^-)$$

To discuss the strength of HA as an acid, we really need to know the Gibbs free energy, not the enthalpy, of this reaction. However, since the entropies of gas-phase species are all quite similar, the change in entropy is small in the transfer reaction and, to a good approximation, $\Delta G^\ominus \approx \Delta H^\ominus$. Hence, as a first approximation, we can discuss trends by considering the enthalpy alone.

The study of proton affinities in the gas phase has been made possible by developments in mass spectrometry, including high-pressure mass spectrometry, flowing afterglow spectrometry, and ion cyclotron resonance spectrometry.[1] Some of the experimental values obtained are given in Table 5.3. All protonation enthalpies are negative, corresponding to an exothermic proton gain and positive proton affinities. The higher the proton affinity of A^-, the weaker the gas-phase acidity of HA and the higher its gas-phase pK_a. Thus, we see from the table that HF has a lower gas-phase acidity than HCl, which in turn is lower than that of HI.

Gas-phase binary acids

The gas-phase acidities of the *p*-block binary acids increase across a period and down a group. Thus HF is a stronger acid than H_2O and HI is the strongest of the hydrogen halides.

[1] The techniques of obtaining gas phase energies have developed spectacularly in fairly recent years. A good account is found in M. T. Bowers (ed.), *Gas phase ion chemistry*. Academic Press, New York (1978). See especially the chapter by J. E. Bartmess and R. T. McIver, Jr., *The gas phase acidity scale*.

Table 5.3. Gas phase and solution proton affinities* of bases

Conjugate acid	Base	A_p/kJ mol^{-1}	A'_p/kJ mol^{-1}
HF	F^-	1553	1150
HCl	Cl^-	1393	1090
HBr	Br^-	1353	1079
HI	I^-	1314	1068
CH_4	CH_3^-	1741	≈1380
NH_3	NH_2^-	1670	1351
PH_3	PH_2^-	1548	1283
H_2O	OH^-	1634	1188
HCN	CN^-	1476	1183
H_3O^+	H_2O	723	1130
NH_4^+	NH_3	865	1182
$C_5H_5NH^+$	C_5H_5N	936	1160

* $A_p = -\Delta H_p^\ominus$ is the gas-phase proton affinity; A'_p the effective proton affinity for the base in water.
Source: J. E. Bartmess and R. J. McIver, in *Gas phase ion chemistry*, M. T. Bowers (ed.), vol 2, Academic Press, New York (1979).

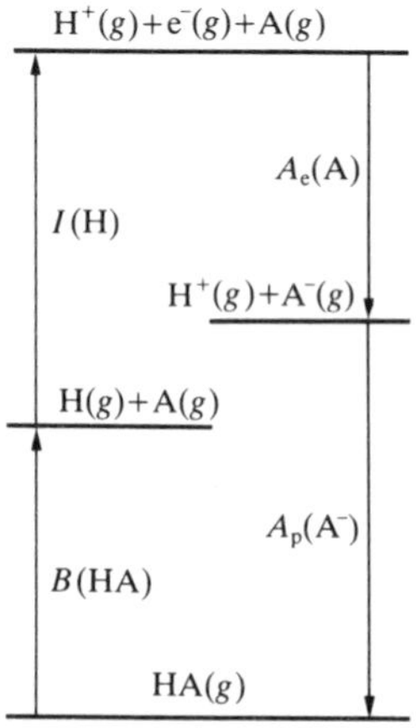

$H^+(g)+e^-(g)+A(g)$
$A_e(A)$
$I(H)$
$H^+(g)+A^-(g)$
$H(g)+A(g)$
$A_p(A^-)$
$B(HA)$
$HA(g)$

Fig. 5.2 The thermodynamic cycle for the analysis of gas-phase acidity and basicity. Proton affinities (A_p) are the negative of proton gain enthalpies ($\Delta H_p^\ominus$).

Some insight into the origin of these trends is obtained by treating proton gain as the outcome of the steps shown in Fig. 5.2:

$A^-(g) \rightarrow A(g) + e^-(g)$	$A_e(A)$, the electron affinity of A,
$H^+(g) + e^-(g) \rightarrow H(g)$	$-I(H)$, the negative of the ionization energy of H,
$H(g) + A(g) \rightarrow HA(g)$	$-B(HA)$, the negative of the H—A bond enthalpy,
$A^-(g) + H^+(g) \rightarrow HA(g)$	$\Delta H_p^\ominus = A_e(A) - I(H) - B(HA)$

That is, the proton affinity of A^-, the negative of $\Delta H_p^\ominus$, is

$$A_p(A^-) = B(HA) + I(H) - A_e(A)$$

The dominant variation in proton affinity across a period arises from the trend in electron affinity of A, which increases from left to right and hence reduces the proton affinity of A^-. Thus, the gas-phase acidity of HA increases as the electron affinity of A increases. Hence it also increases as the electronegativity of A increases. The dominant effect on descending a group is the decrease in the HA bond dissociation enthalpy, which lowers the proton affinity of A^-. The overall result of both effects is therefore a decrease in proton affinity from the upper left to the lower right in the *p* block. Hence, there is an increase in gas-phase acidity in the same direction. These trends can be seen in the data in Table 5.3, which show that A_p decreases down the halogens from F to I and decreases from left to right across Period 2 from N to F.

The enthalpy of proton transfer to water

The enthalpy of proton transfer to an H_2O molecule in the gas phase is obtained by specializing to $B = H_2O$ in the general expression and considering

$$\Delta H^{\ominus} = \Delta H_p^{\ominus}(H_2O) - \Delta H_p^{\ominus}(A^-)$$
$$= A_p(A^-) - A_p(H_2O)$$

Since mass spectrometry gives $A_p(H_2O) = 723\ kJ\ mol^{-1}$, only acids with conjugate bases having proton affinities smaller than $723\ kJ\ mol^{-1}$ will have negative (exothermic) proton transfer enthalpies to an H_2O molecule, and hence undergo transfer. However, there are no such acids in Table 5.3; the few examples that could be found include cations like N_2H^+ and H_2I^+, which have very low proton affinities.

Water molecules may also act as gas-phase acids toward suitable bases:

$$H_2O(g) + B(g) \rightarrow HB(g) + OH^-(g)$$
$$\Delta H^{\ominus} = \Delta H_p^{\ominus}(B) - \Delta H_p^{\ominus}(OH^-)$$
$$= A_p(OH^-) - A_p(B)$$

However, since $A_p(OH^-) = 1634\ kJ\ mol^{-1}$, a large value, this reaction is exothermic (and hence favorable if entropy differences are ignored) only for bases that have proton affinities in excess of $1634\ kJ\ mol^{-1}$. Ions for which this is the case include the very strong bases H^- and CH_3^-.

Solvation

Gas-phase acidities suggest that very few species are reactive toward water. This surprising conclusion, however, is a result of omitting the effects of solvation. Hydration radically alters the gas-phase proton transfer enthalpy, but it is difficult to take into account partly because more than one H_2O molecule is involved. Nevertheless, it is possible to establish a scale of **effective proton affinities** A_p' obtained from mass spectroscopic measurements on the attachment of a proton to a cluster of water molecules in the gas phase:

$$H^+(g) + (H_2O)_n(g) \rightarrow H^+(H_2O)_n(g) \qquad \Delta H_p^{\ominus} = -1130\ kJ\ mol^{-1}$$

The proton transfer in aqueous solution may be expressed as the difference of two values as before:

$$HA(aq) + H_2O(l) \rightarrow H_3O^+(aq) + A^-(aq) \qquad \Delta H^{\ominus} = A_p'(A^-) - A_p'(H_2O)$$

using the values of A_p'.

The effective proton affinity of I^- in water is $1068\ kJ\ mol^{-1}$: this is smaller than its gas-phase value because the I^- ion is stabilized by hydration. It is also now smaller than the effective proton affinity of water, so proton transfer to water is exothermic. This is consistent with the fact that HI is a strong acid in water. All the halide ions except F^- have effective proton affinities smaller than that of H_2O, which is consistent with all the hydrohalic acids except HF being strong acids in water. Solvation, we see, is the key contribution to the familiar energetics of proton transfer.

Factors influencing the solvation enthalpy

The solvation of a gas-phase ion is always strongly exothermic. The magnitude of the enthalpy of solvation $\Delta H_S^{\ominus}$ ($\Delta H_H^{\ominus}$, the enthalpy of hydration, if the solvent is water) depends on the radius of the ions, the relative permittivity (the dielectric constant) of the solvent, and the possibility of specific bonding (especially hydrogen bonding) between the ions and the solvent. We now consider how the effects of solvation can be rationalized in terms of a useful but approximate electrostatic theory derived on the assumption that a solvent is a continuous dielectric medium.

The Gibbs free energy of solvation of an ion in a solvent of relative permittivity ε_r can be estimated from the **Born equation**:

$$\Delta G^{\ominus} = \frac{-Nz^2e^2}{8\pi\varepsilon_0 r}\left(1 - \frac{1}{\varepsilon_r}\right)$$

The dominant quantities in this equation are the charge number z, the ionic radius r, and the relative permittivity ε_r of the solvent (ε_0 is the vacuum permittivity, a fundamental constant). We should recognize the factor z^2/r as the electrostatic parameter ξ (which is likely to occur whenever ionic interactions are dominant). We can therefore write the Born equation as

$$\Delta G^{\ominus} = \frac{-Ne^2}{8\pi\varepsilon_0} \times \xi \times \left(1 - \frac{1}{\varepsilon_r}\right)$$

which emphasizes its dependence on the properties of the ion (the factor ξ) and the properties of the solvent. Since the free energy of solvation is proportional to ξ, small ions of high charge number are strongly stabilized in polar solvents. The Born equation also shows that the larger the relative permittivity, the more negative the value of $\Delta G^{\ominus}$. This is an important consideration for water, for which $\varepsilon_r \approx 80$, compared with nonpolar solvents for which ε_r may be as low as 2.

A large negative free energy of solvation favors the formation of ions in solution as compared with the gas phase. The physical reason is the stabilization of the ion by the favorable interaction between its charge and the polar solvent molecules. This stabilizes anionic bases A^- relative to the neutral (or less negatively charged) parent acid HA, and hence the acid HA is strengthened by a polar solvent. Cationic acids (such as NH_4^+) are stabilized by solvation and hence their acidity is reduced by a polar solvent.

The proton affinity of a neutral base B is enhanced by solvation because the conjugate acid BH^+ is stabilized. As a general rule, proton transfer leading to charge separation is favored by polar solvents. Proton transfer that leads to charge reduction is favored by less polar media.

A further factor in **protic solvents**, those capable of participating in proton transfer, is hydrogen bonding. Thus water has a greater stabilizing effect on small, highly electronegative ions than the Born equation predicts, particularly for F^-, OH^-, and Cl^-, to which it can act as a hydrogen-bond donor. Since H_2O has lone pairs on O, it can also be a hydrogen-bond acceptor. Acidic ions such as NH_4^+ are stabilized by hydrogen bonding and have a correspondingly reduced acidity. An example is the increased acidity of HCl in CH_3OH, which can stabilize Cl^- by the formation of $Cl^- \cdots H$—OCH_3, in comparison with HCl in dimethylformamide $(CH_3)_2NCHO$, which has a similar relative permittivity but does not have hydrogen-bond donor properties.

5.4 Solvent leveling

Discrimination in water

Any acid stronger than H_3O^+ in water donates a proton to H_2O and forms H_3O^+. Consequently, no acid stronger than H_3O^+ can survive in water. No experiment conducted in water can tell us which is the stronger acid of HCl and HBr because both react essentially completely to give H_3O^+. Water is said to have a **leveling effect** that brings all stronger acids down to the acidity of H_3O^+. Since the effective proton affinity of H_2O in water is 1130 kJ mol^{-1}, all acids with conjugate bases having an effective proton affinity smaller than 1130 kJ mol^{-1} are leveled in water.

An analogous limit can be found for bases in water. Any base strong enough to react completely with water to give the OH^- ion will be leveled; hence OH^- is the strongest base that can exist in water. The gas-phase proton affinity of OH^- is 1634 kJ mol^{-1} and hydration of the OH^- ion reduces this to 1188 kJ mol^{-1} in water. Any base A^- with an effective proton affinity in water significantly greater than 1188 kJ mol^{-1} will be completely converted to the conjugate acid HA, producing OH^- in the process. For this reason, we cannot study NH_2^- or CH_3^- in water by dissolving their salts because both generate OH^- ions quantitatively and are fully protonated to NH_3 and CH_4.

The range of acidity that can be studied in water lies *approximately* between the effective proton affinities of 1130 kJ mol^{-1} and 1188 kJ mol^{-1}, a range of 58 kJ mol^{-1}: if the proton affinity of A^- is less than 1130 kJ mol^{-1}, HA will be converted completely to A^- and all the protons will be present as H_3O^+. If the proton affinity of A^- is greater than 1188 kJ mol^{-1}, the solute will exist only as HA at the expense of forming OH^- ions from water.

It is interesting to express the acidity range as an equilibrium constant. The contribution of the reaction entropy is important in solution, and we must use $\Delta G^{\ominus} = +81$ kJ mol^{-1} in place of $\Delta H^{\ominus} = +58$ kJ mol^{-1} in

$$\Delta G^{\ominus} = 2.303RT\, \mathrm{p}K$$

This gives $\mathrm{p}K = 14$, which is the value of $\mathrm{p}K_w$. The window of unleveled strengths, which in terms of proton affinities spans 58 kJ mol^{-1}, can be interpreted as the value of $\mathrm{p}K_w$.

Discrimination in nonaqueous solvents

For any solvent, the range over which acid and base strengths can be discriminated is given by its autoprotolysis constant. For water, the range is 14. For liquid ammonia

$$NH_3(l) + NH_3(l) \rightleftharpoons NH_4^+(am) + NH_2^-(am) \qquad \mathrm{p}K_{am} = 33$$

so the range of discrimination is considerably wider. It is quite easy to understand the physical basis of why this is so. Since the proton affinity of NH_2^- is distinctly greater than that of OH^-, strong bases that are leveled in water will not be leveled in ammonia. However, the proton affinity of NH_3 is distinctly greater than that of H_2O, so acids that are weak in water may be leveled in ammonia. Hence the basic solvent ammonia has a window for measurement of acid strengths that is widened and shifted toward weaker acids.

The reverse is also true. Going from water to a more strongly acid solvent, such as CH_3COOH, moves the range of distinguishable acids toward stronger acids. The proton affinity of CH_3COOH is less than that of water, so acids that transfer protons to water completely do not do so completely to CH_3COOH. Since $CH_3CO_2^-$ has a lower proton affinity than OH^-, weaker bases may convert CH_3COOH quantitatively to $CH_3CO_2^-$.

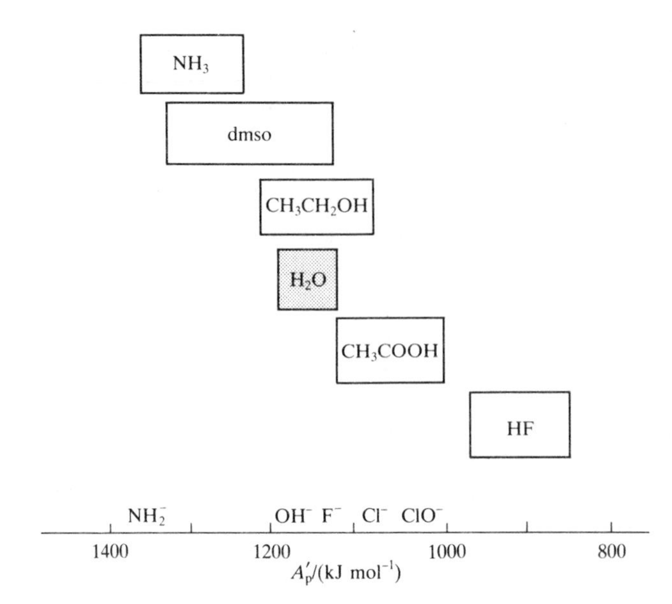

Fig. 5.3 The acid–base discrimination windows of a variety of solvents shown as effective proton affinities (A'_p). In each case, the width of the window is proportional to the autoprotolysis constant pK of the solvent. Several common bases are shown with their effective proton affinities to illustrate the range of the scale.

To determine the acidity constants of strong acids, such as HCl, we need data from acidic solvents, such as acetic acid. To study strong bases, such as NH_2^-, we need data from basic solvents, such as liquid ammonia. The discrimination windows of a number of solvents are shown in Fig. 5.3. The window for dimethylsulfoxide (dmso) is wide because pK for the reaction

$$CH_3\overset{\overset{\displaystyle O}{\|}}{S}CH_3 + CH_3\overset{\overset{\displaystyle O}{\|}}{S}CH_3 \rightleftharpoons CH_3\overset{\overset{\displaystyle O}{\|}}{S}CH_2^- + CH_3\overset{\overset{\displaystyle OH^+}{\|}}{S}CH_3 \quad pK = 37$$

is very large. Consequently, dmso can be used to study a wide range of acids (from H_2SO_4 to PH_3).

Periodic trends in Brønsted acidity

Now that we have seen the principles of Brønsted acidity, we can explore how it manifests itself in compounds formed by elements in different regions of the periodic table. The most important acids in water are those that donate protons from a hydroxyl group (—OH): a donatable proton of this kind is called an **acid proton** to distinguish it from other protons that may be present in the molecule.

There are three classes of hydroxyl group acids to consider:

1. **Aqua acids**, in which the acid proton is on a water molecule coordinated to a central metal ion.

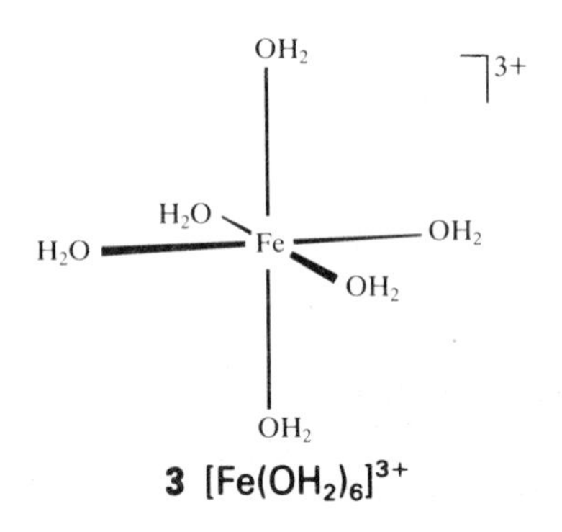

3 $[Fe(OH_2)_6]^{3+}$

An example is $[Fe(OH_2)_6]^{3+}$ (**3**). Aqua acid proton transfer equilibrium is a special case of a Brønsted equilibrium:

$$E(OH_2)(aq) + H_2O(l) \rightleftharpoons [E(OH)]^-(aq) + H_3O^+(aq)$$

as in

$$[Fe(OH_2)_6]^{3+}(aq) + H_2O(l) \rightarrow [Fe(OH_2)_5(OH)]^{2+}(aq) + H_3O^+(aq)$$

2. **Hydroxoacids**, in which the acid proton is on a hydroxyl group without a neighboring oxo group (=O).

An example is $Si(OH)_4$ (**4**), which is important in the formation of minerals.

3. **Oxoacids**, in which the acid proton is on a hydroxyl group with an oxo group attached to the same atom.

Sulfuric acid, H_2SO_4 (**5**), is an example of an oxoacid.

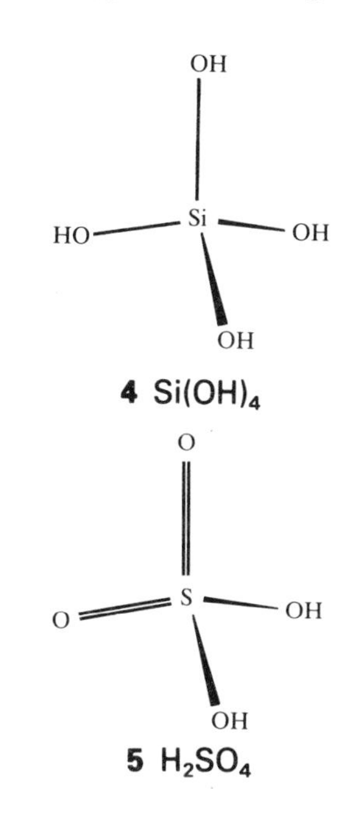

4 $Si(OH)_4$

5 H_2SO_4

We can think of the three classes of acid as successive stages in the deprotonation of an aqua acid:

$$\underset{\text{aqua acid}}{H_2O{-}E{-}OH_2} \xrightarrow{-2H^+} \underset{\text{hydroxo acid}}{HO{-}E{-}OH]^{2+}} \xrightarrow{-H^+} \underset{\text{oxoacid}}{HO{-}E{=}O]^{3+}}$$

An example of these successive stages is provided by a *d*-block metal in an intermediate oxidation state, such as Ru(IV):

$$[\text{Ru}(\text{OH}_2)_2]^{4+} \underset{+2H^+}{\overset{-2H^+}{\rightleftharpoons}} [\text{Ru}(\text{OH})_2]^{2+} \underset{+H^+}{\overset{-H^+}{\rightleftharpoons}} [\text{Ru}(\text{O})(\text{OH})]^{+}$$

Aqua acids are characteristic of central atoms with low oxidation numbers and of elements in the *s* block and *d* block and of metals on the left of the *p* block. Oxoacids are found where the central element has a high oxidation number. In addition, an element from the right of the *p* block with one of its intermediate oxidation numbers may also produce an oxoacid ($HClO_2$ is an example).

5.5 Periodic trends in aqua acid strength

The variation of the strengths of aqua acids through the periodic table can be rationalized to some extent in terms of an ionic model, in which the metal cation is represented by a sphere carrying z units of positive charge.

The gas-phase pK_a is proportional[2] to the work of removing a proton to infinity from a distance equal to the sum of the ionic radius, r, and the diameter of a water molecule, d. Because the higher the cation charge and the smaller its radius the more easily the proton is lost, the model predicts that the acidity should increase with increasing z and with decreasing ionic radius, very roughly as the electrostatic parameter $\xi = z^2/(r+d)$. The trends this model predicts for the gas phase will also apply in solution if the effects of solvation are reasonably constant.

The scope of the ionic model can be judged from Fig. 5.4. Elements that form ionic solids (principally those from the *s* block) have pK_a values that are quite well described by the ionic model. Several *d*-block ions (such as Fe^{2+} and Cr^{3+}) lie reasonably near the curve, but deviate to higher acid strength (lower pK_a). The deviation indicates that the metal ions repel the departing proton more strongly than is predicted by the ionic model. This can be rationalized by supposing that the cation charge is not fully localized on the central ion but is delocalized over the ligands and hence is closer to

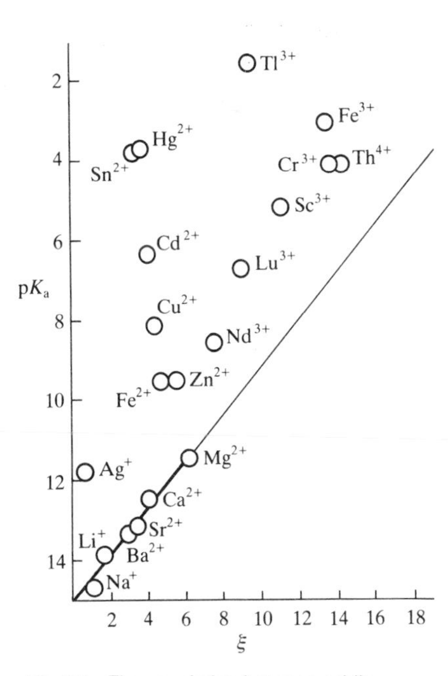

Fig. 5.4 The correlation between acidity constant and electrostatic parameter ξ of aqua ions. Note that only hard ions with low charge follow the correlation; all others are more acidic than the correlation suggests.

[2] This comes from $pK_a \propto \Delta G^{\ominus}$ coupled with the identification of ΔG and electrical work, so $pK_a \propto w_{\text{electrical}}$.

the departing proton, which is equivalent to attributing covalence to the E—O bond.

The strengths of the aqua acids exceed the predictions of the ionic model by a large margin for the *d*-block and *p*-block metals (such as at Cu^{2+} and Sn^{2+}). This is largely because covalent bonding is more important than ionic bonding, and the ionic model is unrealistic.

Example 5.2: *Accounting for trends in aqua acid strength*

Account for the trend in acidity $[Fe(OH_2)_6]^{2+} < [Al(OH_2)_6]^{3+} < [Fe(OH_2)_6]^{3+} \approx [Hg(OH_2)_n]^{2+}$.

Answer. The weakest acid is the Fe^{2+} complex, on account of its relatively large radius and low charge. The increase of charge to +3 increases the acid strength. The greater acidity of Al^{3+} can be explained by its smaller radius. The anomalous ion in the series is the Hg^{2+} complex. This reflects the failure of an ionic model, for in this complex there is considerable transfer of positive charge to oxygen.

Exercise. Arrange the following ions in order of increasing acidity: $[Na(OH_2)_n]^+$, $[Sc(OH_2)_6]^{3+}$, $[Mn(OH_2)_6]^{2+}$, $[Ni(OH_2)_6]^{2+}$.

5.6 Simple oxoacids

The simplest oxoacids are the **mononuclear acids**, which contain one atom of the parent element. They include H_2CO_3, HNO_3, H_3PO_4, and H_2SO_4. These oxoacids are formed by the electronegative elements at the upper right of the periodic table and by other elements with high oxidation numbers. Table 5.4 summarizes some of their structures. One interesting feature in that table is the occurrence of the planar H_2CO_3, and HNO_3 molecules but not their analogs in later periods. This is explained by noting that π bonding is more important in the Period 2 elements, so the atoms are more likely to be constrained to lie in a plane.

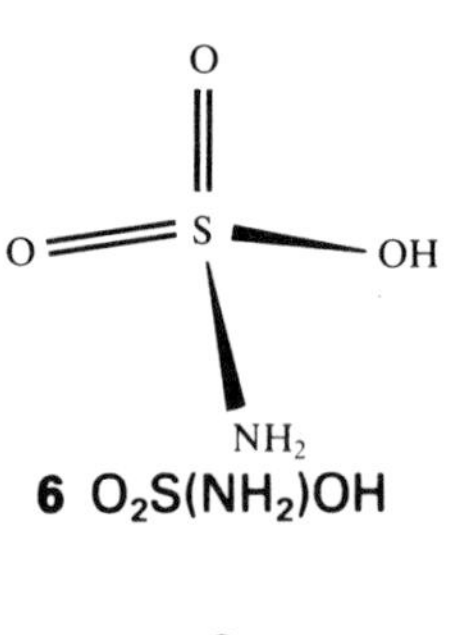

6 $O_2S(NH_2)OH$

Substituted oxoacids

One or more hydroxyl groups of the oxoacids may be replaced by other groups to give a series of substituted oxoacids. These include fluorosulfuric acid, $O_2SF(OH)$, and aminosulfuric acid, $O_2S(NH_2)OH$ (**6**). Since F is highly electronegative, it withdraws electrons from the central S atom and confers on it a higher effective positive charge. In contrast, the less electronegative NH_2 group can donate electron density to S by π bonding. This reduces the positive charge of the central atom and weakens the acid. Another electron donor substituent is CH_3—, as in methylsulfonic acid CH_3SO_3H, but its effect is smaller because CH_3 has no lone pair.

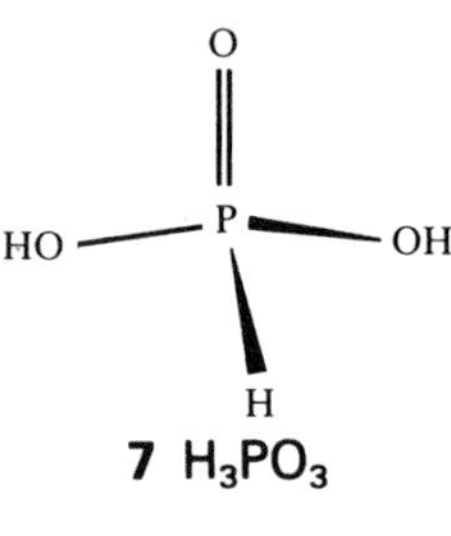

7 H_3PO_3

A trap for the unwary is the substitution of an H atom for an entire —OH group, as in phosphorous acid H_3PO_3. Phosphorous acid is in fact a diprotic acid, for the substitution results in a P–H bond (**7**) and a nonacidic proton. This is confirmed by NMR and Raman spectra, and the formula is best written as $OPH(OH)_2$. Substitution of an oxo group (as distinct from a hydroxyl group) can also occur. An important example is thiosulfuric acid, $OS_2(OH)_2$ (**8**), in which an S atom replaces an O atom of sulfuric acid.

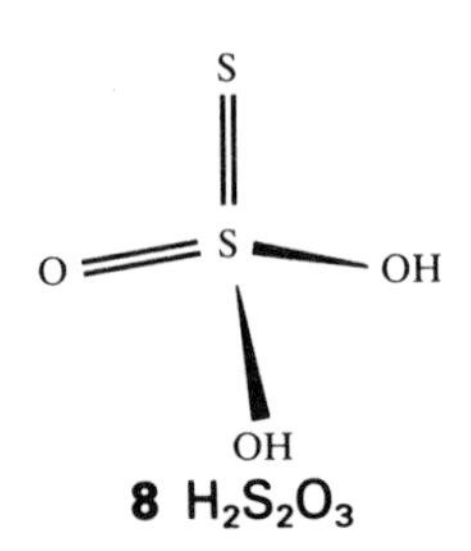

8 $H_2S_2O_3$

Table 5.4. The structures and acidity constants of oxoacids*

$p=0$	$p=1$	$p=2$	$p=3$
HO–Cl 7.2	$OC(OH)_2$ 3.6	$O_2N(OH)$ −1.4	
$Si(OH)_4$ 10	$OP(OH)_3$ 2.1, 7.4, 12.7 $OCl(OH)$ 2.0	$O_2S(OH)_2$ −2.0, 1.9	$O_3Cl(OH)$ −10
$Te(OH)_6$ 7.8, 11.2	$OPH(OH)_2$ 1.8, 6.6 $OI(OH)_5$ 1.6, 7.0	$O_2Cl(OH)$ −1.0	
	$OAs(OH)_3$ 2.3, 6.9, 11.5 $OSe(OH)_2$ 2.6, 8.0		

* Numbers are successive pK_a values.

Pauling's rules

The strengths of mononuclear oxo acids can be predicted using two rules devised by Linus Pauling:

> 1. For the oxoacid $O_pE(OH)_q$, $pK_a \approx 8 - 5p$.

Neutral hydroxoacids with $p = 0$ have $pK_a \approx 8$, acids with one oxo group have $pK_a \approx 3$, and acids with two oxo groups have $pK_a \approx -2$.

> 2. The successive pK_a values for polyprotic acids (those with $q > 1$), increase by five units for each successive proton transfer.

Sulfuric acid, $O_2S(OH)_2$, has $p = 2$ and $q = 2$, and $pK_{a1} \approx -2$, $pK_{a2} \approx +3$.

The success of these simple rules may be gauged by inspection of Table 5.4, in which acids are grouped according to p, the number of oxo groups. That the estimates are good to about ±1 is pleasantly surprising. The variation in strengths down a group is not large, and the complex and perhaps canceling effects of changing structures allow the rules to work moderately well. The more important variation across the periodic table from left to right and the effect of change of oxidation number are taken into account by the number of oxo groups characteristic of the neutral acids. In Group 15/V, the oxidation number +5 requires one oxo group (as in $OP(OH)_3$) whereas in Group 16/VI the oxidation number +6 requires two (as in $O_2S(OH)_2$).

Structural anomalies

Perhaps the most interesting use of the Pauling rules is to detect structural anomalies. For example, carbonic acid, $OC(OH)_2$, is commonly reported as having $pK_{a1} = 6.4$ but the rules predict $pK_{a1} = 3$. The reduced acidity is the result of treating the concentration of dissolved CO_2 as if it were all H_2CO_3. However, the equilibrium

$$CO_2(aq) + H_2O(l) \rightleftharpoons OC(OH)_2(aq)$$

favors $OC(OH)_2$ only to the extent of about 1 percent, so the actual concentration of acid is much less than the concentration of dissolved CO_2. When this is taken into account, the true pK_{a1} of H_2CO_3 is about 3.6, as the Pauling rules predict.

The value $pK_{a1} = 1.8$ reported for sulfurous acid, H_2SO_3, suggests another anomaly, this time acting in the opposite direction. In fact, spectroscopic studies have failed to detect the molecule $OS(OH)_2$ in solution, and the equilibrium constant for

$$SO_2(aq) + H_2O(l) \rightleftharpoons H_2SO_3(aq)$$

is less than 10^{-9}. The equilibria of dissolved SO_2 are complex, and a simple analysis is inappropriate. The ions that have been detected include HSO_3^- and $S_2O_5^{2-}$ and there is evidence for an S—H bond in the solid salts of the hydrogen sulfite ion.[3]

The pK_a values of aqueous CO_2 and SO_2 call attention to an important point: not all nonmetal oxides react fully with water to form acids. Carbon monoxide is another example: although it is formally the anhydride of formic acid, HCOOH, it does not in fact react with water at room temperature to give the acid. The same is true of some metal oxides: OsO_4, for example, can exist as dissolved neutral molecules.

Example 5.3: *Using Pauling's rules*

Predict the formulas that are consistent with the following pK_a values: H_3PO_4, 2.12; H_3PO_3, 1.80; H_3PO_2, 2.0.

Answer. All three values are in the range with which Pauling's first rule associates with one oxo group. This suggests the formulas; $(HO)_3PO$, $(HO)_2HPO$, and $(HO)H_2PO$. That is, the second and the third are derived from the first by replacement of —OH by —H bound to P.

Exercise. Predict the pK_a values of (a) H_3PO_4, (b) $H_2PO_4^-$, (c) HPO_4^{2-}. Experimental values are given in Table 5.4.

5.7 Anhydrous oxides

We have treated oxoacids as derived from the deprotonation of parent aqua acids. It is also useful to take the opposite viewpoint and to consider aqua and oxoacids as derived by hydration of the oxides of the central element.

[3] Elegant solution spectroscopic studies of the sulfite problem have been done by R. E. Connick and his collaborators. See *Inorg. Chem.*, **21,** 103 (1982) and **25,** 2414 (1986).

This emphasizes the acid and base properties of oxides and their correlation with the location of the element in the periodic table.

Acidic and basic oxides

An **acidic oxide** is one which, on dissolution in water, binds an H_2O molecule and releases a proton to the surrounding solvent:

$$CO_2(g) + H_2O(l) \rightarrow [OC(OH)_2](aq)$$
$$[OC(OH)_2](aq) + H_2O(l) \rightleftharpoons [O_2C(OH)]^-(aq) + H_3O^+(aq)$$

An equivalent interpretation is that an acidic oxide is one that reacts with an aqueous base (an alkali):

$$CO_2(g) + OH^-(aq) \rightarrow [O_2C(OH)]^-(aq)$$

A **basic oxide** is one to which a proton is transferred when it dissolves in water:

$$CaO(s) + H_2O(l) \rightarrow Ca^{2+}(aq) + 2OH^-(aq)$$

The equivalent interpretation in this case is that a basic oxide is one that reacts with an acid:

$$CaO(s) + 2H^+(aq) \rightarrow Ca^{2+}(aq) + H_2O(l)$$

Since acidic and basic oxide character often correlates with other chemical properties, a wide range of properties can in fact be predicted from a knowledge of the character of oxides. In a number of cases the correlations follow from the basic oxides being largely ionic and of acidic oxides being largely covalent.

Amphoterism

An **amphoteric oxide** (from the Greek word meaning both) is an oxide that reacts with both acids and bases. Thus, Al_2O_3 reacts with acids and alkalis:

$$Al_2O_3(s) + 6H_3O^+(aq) + 3H_2O(l) \rightarrow 2[Al(OH_2)_6]^{3+}(aq)$$
$$Al_2O_3(s) + 2OH^-(aq) + 3H_2O(l) \rightarrow 2[Al(OH)_4]^-(aq)$$

Amphoterism is observed for the lighter elements of Groups 2 and 13/III, as illustrated by BeO, Al_2O_3, and Ga_2O_3. It is also observed for some of the *d*-block elements, as exemplified by TiO_2 and V_2O_5, and some of the heavier elements of Groups 14/IV and 15/V such as SnO_2, As_2O_5, and Sb_2O_5. The location of the amphoteric oxides in the periodic table for elements having their characteristic group oxidation number (that is, G in the *s* block and $18 - G$ in the *p* block, where G is the group number) is shown in Fig. 5.5. They lie on the frontier between acidic and basic oxides, and hence serve as an important guide to the character of an element.

An important issue in the *d* block is the oxidation number necessary for amphoterism. Figure 5.6 shows the oxidation number for which an element in the first row of the block has an amphoteric oxide. We see that on the left of the block, from Ti to Mn and perhaps Fe, oxidation number +4 is amphoteric (with higher values acidic and lower values basic). On the right of the block, amphoterism occurs at lower oxidation numbers, +3 for Co and Ni, and +2 for Cu and Zn.

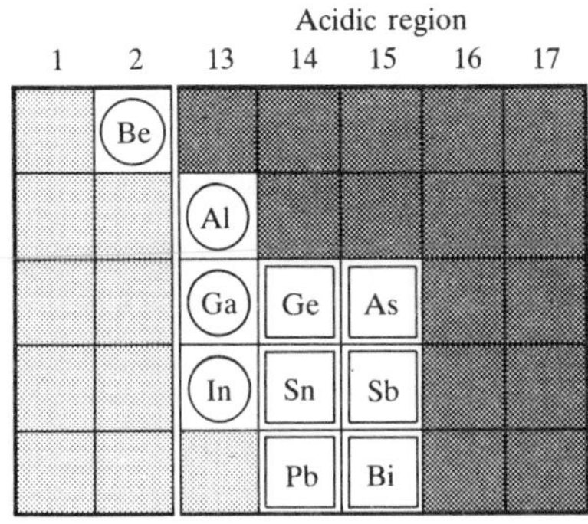

Fig. 5.5 The elements in circles have amphoteric oxides even in their highest oxidation states. The elements in boxes have acidic oxides in their maximum oxidation states and amphoteric oxides at lower oxidation numbers.

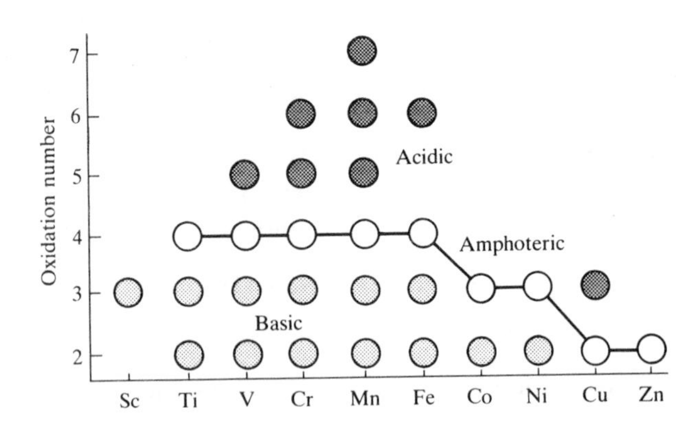

Fig. 5.6 The influence of oxidation number on the acid–base character of oxides of the elements in the first row of the *d* block. Oxidation numbers connected by the line are amphoteric.

Example 5.4: *Using oxide acidity in qualitative analysis*

In the traditional scheme of qualitative analysis, a solution of metal ions is oxidized and then aqueous ammonia is added to raise the pH. The ions Fe^{3+}, Ce^{3+}, Al^{3+}, Cr^{3+}, and V^{3+} precipitate as hydroxides. Addition of H_2O_2 and NaOH redissolves Al, Cr, and V. Discuss these steps in terms of the acidities of oxides.

Answer. When the oxidation number is +3, all the metal oxides are sufficiently basic to be insoluble in a pH$\approx$10 solution. Al(III) is amphoteric and redissolves in strong base to give aluminate ions, $[Al(OH)_4]^-$. V(III) and Cr(III) are oxidized by H_2O_2 to give vanadate ions, $[VO_4]^{2-}$, and chromate ions, $[CrO_4]^{2-}$, which are acidic.

Exercise. If Ti(IV) ions were present in the sample, how would they behave?

Polyoxo compound formation

One of the most important aspects of the reactivity of acids containing the O—H group is the formation of condensation polymers. Polycation formation reduces the average positive charge per central atom by the removal of a proton:

$$2[Al(OH_2)_6]^{3+}(aq) \rightarrow [(H_2O)_5Al{-}OH{-}Al(OH_2)_5]^{5+}(aq) + H_3O^+(aq)$$

Polyanion formation reduces the average negative charge per central atom:

$$2[CrO_4]^{2-}(aq) + 2H_3O^+(aq) \rightarrow [O_3Cr{-}O{-}CrO_3]^{2-}(aq) + 3H_2O(l)$$

The importance of polymers like these can be judged by the fact that they account for most of the mass of oxygen in the earth's crust, because they include almost all silicate minerals. They also include the phosphate polymers used for energy storage in living cells. The silicates are so important that they are treated separately (Chapter 11).

5.8 Polymerization of aqua ions to polycations

As the pH of the solution is increased, the aqua ions of metals that have basic or amphoteric oxides generally undergo polymerization and precipitation. One application is to the separation of metal ions, since the precipitation occurs quantitatively at different pH characteristic of each metal.

Only the most basic metals (of Groups 1 and 2) have no important solution species beyond the aqua ion. However, the solution chemistry becomes very rich as the amphoteric region of the periodic table is approached. The two most common examples are polymers formed by Fe(III) and Al(III), both of which are abundant in the earth's crust. In acid solutions, both form octahedral hexa-aqua ions, $[Al(OH_2)_6]^{3+}$ and $[Fe(OH_2)_6]^{3+}$. In solutions of pH > 4, both precipitate as gelatinous hydrous oxides:

$$[Fe(OH_2)_6]^{3+}(aq) + nH_2O(l) \rightarrow Fe(OH)_3 \cdot nH_2O(s) + 3H_3O^+(aq)$$
$$[Al(OH_2)_6]^{3+}(aq) + nH_2O(l) \rightarrow Al(OH)_3 \cdot nH_2O(s) + 3H_3O^+(aq)$$

The precipitated polymers, which are often of colloidal dimensions, slowly crystallize to stable mineral forms.

Aluminum(III) and iron(III) behave differently at intermediate pH, the region between the existence of aqua ions and precipitation. Relatively few Fe species have been characterized, but ones that have include the two monomers (**9, 10**), a dimer (**11**), and a polymer containing about 90 Fe atoms. In contrast, Al(III) forms a series of discrete polymeric cations in which the monomer unit consists of a central Al^{3+} ion surrounded tetrahedrally by four O atoms (**12**). A 'simple' polymeric cation of this kind is $[AlO_4(Al(OH)_2)_{12}]^{7+}$, with a charge of 0.54 per Al atom. An impression of the structure is given in Fig. 5.7, where the AlO_6 octahedra are represented as octahedral blocks packed around the central tetrahedron; the larger Fe^{3+} ion does not fit so well in such a structure.

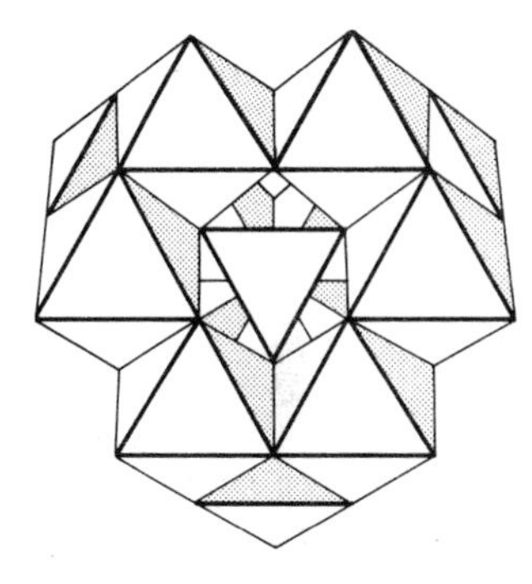

Fig. 5.7 The structure of the $[AlO_4(Al(OH)_2)_{12}]^{7+}$ ion, with AlO_6 groups represented by octahedra around the central tetrahedron which represents the AlO_4 unit.

The structures of the Al(III) polymers are built up by the addition of AlO_6 octahedra to the edges of the tetrahedron and the extensive network, neatly packed in three dimensions, contrasts to the long polymers of the

9 $[Fe(OH_2)_5OH]^{2+}$

10 $[Fe(OH_2)_4(OH)_2]^+$

11 $[Fe_2(OH_2)_8(OH)_2]^{4+}$

12 $[AlO_4]^{5-}$

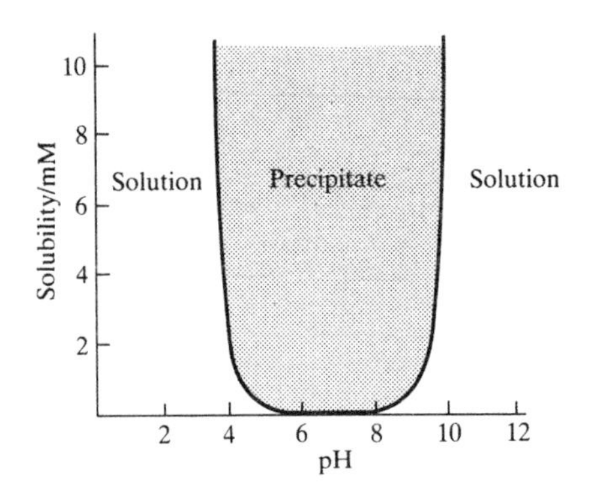

Fig. 5.8 The variation of the solubility of Al_2O_3 with pH, expressed as total concentration of Al (*s*). In the extreme acid region the Al is present as $[Al(OH_2)_6]^{3+}$. In the extreme basic region it is present as $[Al(OH)_4]^-$.

Fe(III) case. Aluminum polycations and similar ions are used in water treatment to precipitate anions (such as F^-) that are present as pollutants in effluents from Al refining plants.[4]

As the pH is increased, H^+ ions are picked off these polycations and their charge is reduced. The point at which the net charge is zero is known as the 'point of zero charge'. Since both Fe(III) and Al(III) form amphoteric oxides, increasing the pH to a sufficiently high value can lead to the redissolution of their oxides as anions (Fig. 5.8).

5.9 Polyoxoanions of early *d*-block metals

Polyoxoanions form from the oxoacids of the early *d*-block elements with high oxidation numbers. This polymerization is important for V(V), Mo(VI), W(VI), and (to a lesser extent) Nb(V), Ta(V), and Cr(VI).

Isopolyanions

A solution formed by dissolving the amphoteric oxide V_2O_5 in a strongly basic solution is colorless, and the dominant species is the tetrahedral $[VO_4]^{3-}$ ion (the analog of the colorless PO_4^{3-} ion). As the pH is increased, the solution passes through a series of deeper colors from orange to red. This indicates a complicated series of condensations and hydrolyses that yield ions including (in an order from basic to acidic solution) $[V_2O_7]^{4-}$, $[V_3O_9]^{3-}$, $[V_4O_{12}]^{4-}$, $[HV_{10}O_{28}]^{5-}$, and $[H_2V_{10}O_{28}]^{4-}$. We should notice the successive reduction of anionic charge per V atom as the polyanions grow. The strongly acidic solution is pale yellow and contains the hydrated $[VO_2]^+$ ion (**13**).

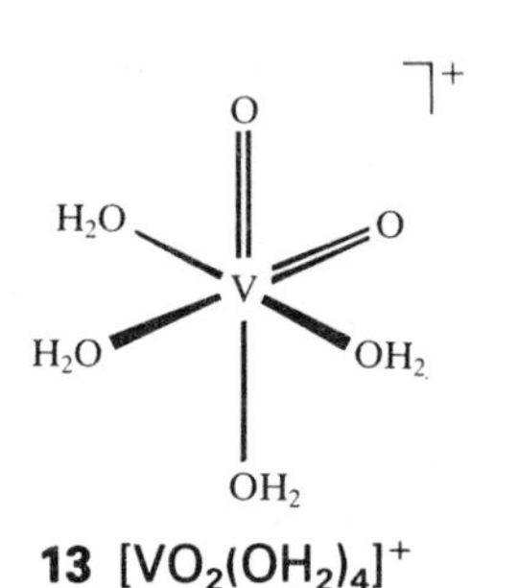

13 $[VO_2(OH_2)_4]^+$

Similar polyoxoanions are formed by the other elements in the early part of the *d* block. Those with only one type of metal atom are known as **isopolyanions** or **isopolymetallates**. This distinguishes them from the **heteropolyanions** in which there are two different central elements. The latter include $[PMo_{12}O_{40}]^{3-}$, which has PO_4 inside a basket of MoO_6 octahedra. This ion is used in colorimetric analysis of phosphate, for on mild reduction it gives a strong blue color. Fresh insights into the nature of these polyanions have been derived from recent NMR studies of solutions and X-ray analysis of the solids. This work has been stimulated by interest in polymetallates as catalysts for the selective air oxidation of organic molecules.

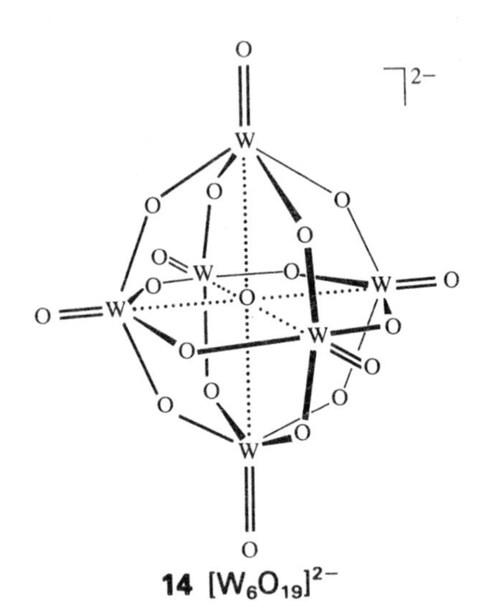

14 $[W_6O_{19}]^{2-}$

Structural analogies

Polyoxoanions exhibit large cluster structures related to solid lattices, with analogies that extend to subtle features of both structure and bonding. We shall consider the ion $[W_6O_{19}]^{2-}$ (**14**) as a representative of the class.

The O atoms in $[W_6O_{19}]^{2-}$ form a close-packed array which is a fragment of a cubic close-packed structure (Section 4.5) and the W atoms occupy its octahedral holes. That is, the $[W_6O_{19}]^{2-}$ ion is like a fragment of a rock-salt structure (Fig. 4.11). Since the tungsten has its highest oxidation number,

[4] Some of the factors have been discussed by N. Parthasathey, J. Buffle, and W. Haerdi in *Canad. J. Chem.*, **64**, 24 (1986). They describe how coprecipitation by hydrous oxides depends on the different polycation structures that form under different conditions of addition of aluminum sulfate (alum) to the effluent and the changing initial effluent pH.

there are no *d* electrons available to form metal–metal bonds, and the unit is held together only by W–O–W bonds.

Three distinct oxygen resonances are observed in the ^{17}O-NMR spectrum, which indicates that there are three distinct oxygen sites. X-ray analysis shows that around each W atom there is one short W=O bond (1.69 Å), four longer bonds to bridging O atoms (1.92 Å) and one long weak bond to an O atom (2.33 Å). The structural unit is (**15**). Six of these units can be assembled into $[W_6O_{19}]^{2-}$ as shown in (**14**). The structure can be considered to be a *neutral* W_6O_{18} fragment of the solid oxide with an O^{2-} ion trapped at its center. The O^{2-} ion is weakly bonded to the six surrounding W atoms of the $[O{=}WO_4]^{4-}$ fragments. From this point of view, many other W(VI) and Mo(VI) polyoxoanions can be interpreted as based on M_nO_{3n} neutral cages surrounding anionic subunits.

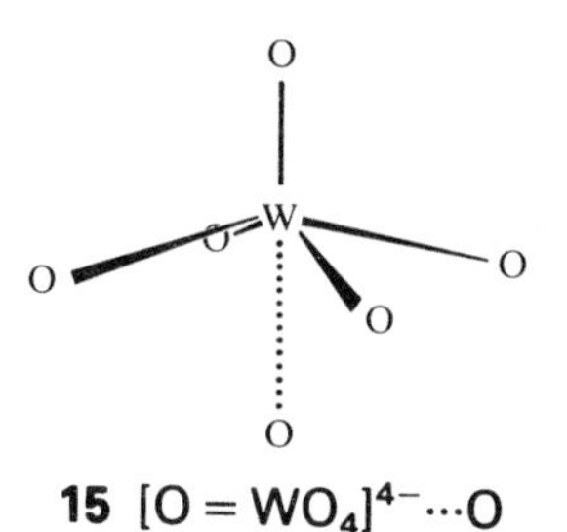

15 $[O=WO_4]^{4-}\cdots O$

A second class of structures is represented by the anion $[CH_2Mo_4O_{15}H]^{3-}$, which has a twofold axis of symmetry. It can be regarded as based on a close-packed ABA arrangement of O atoms with Mo atoms occupying octahedral holes. In fact, the structure is slightly distorted: the structural unit is (**16**) and that of the ion is (**17**). A large family of ions related to this structure also exists. Since a variety of these discrete polyanions can be prepared containing organic and organometallic fragments, they have been used to try to model the reactive sites of heterogeneous oxide catalysts for selective air oxidation of organic molecules.[5]

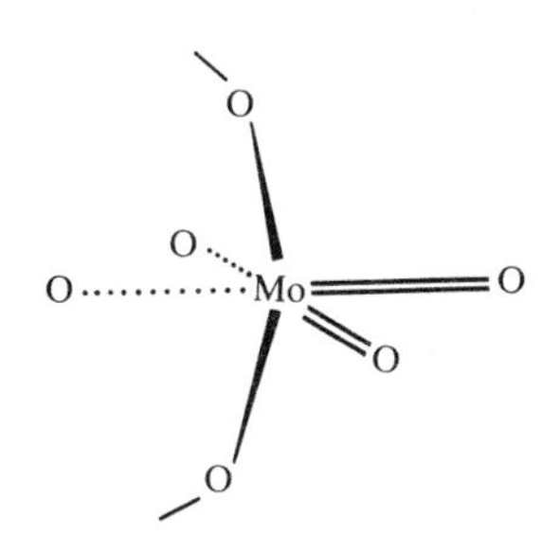

16 $O\cdots[O_2MoO_2]\cdots O$

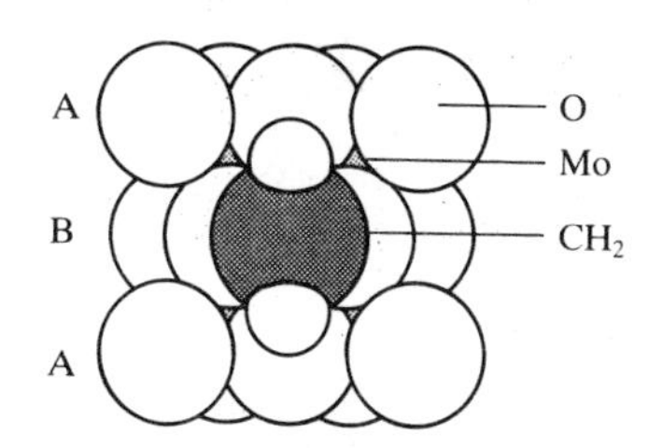

17 $[CH_2Mo_4O_{15}H]^{3-}$

5.10 Nonmetal polyoxoanions

The polymerization of the anions of the nonmetal oxoanions is different from that of the *d*-metal analogs, and the common species in solution are rings and chains. As we have remarked, the silicates are very important examples of polymeric oxoanions, and we discuss them separately in Chapter 11. In this section we illustrate some features of polyoxoanions using phosphates as examples.

The simplest polymerization reaction, starting with the *orthophosphate ion*, PO_4^{3-}, is

$$2PO_4^{3-} + 2H^+ \longrightarrow \left\{ \begin{matrix} & O & & O & \\ & \| & & \| & \\ O- & P & -O- & P & -O \\ & | & & | & \\ & O & & O & \end{matrix} \right\}^{4-} + H_2O$$

Elimination of water consumes protons and reduces the average charge number of each P atom to −2. The *diphosphate ion,* $[P_2O_7]^{4-}$, can be drawn as (**18**) if each phosphate group is represented as a tetrahedron with the corners representing the location of the four O atoms.

18 $[P_2O_7]^{4-}$

Phosphoric acid can be prepared by hydrolysis of the solid P(V) oxide, P_4O_{10}. An initial step using a limited amount of water produces the *metaphosphate ion* (**19**) with the anion formula $[P_4O_{12}]^{4-}$. However, chromatographic analysis shows that this is only the simplest reaction among

19 $[P_4O_{12}]^{4-}$

[5] A review of the chemistry of these ions, exhibiting the fascinating interplay between NMR and X-ray studies for the determination of structures and analysis of reactions, is found in V. W. Day and W. G. Klemperer, *Science,* **228,** 533 (1985).

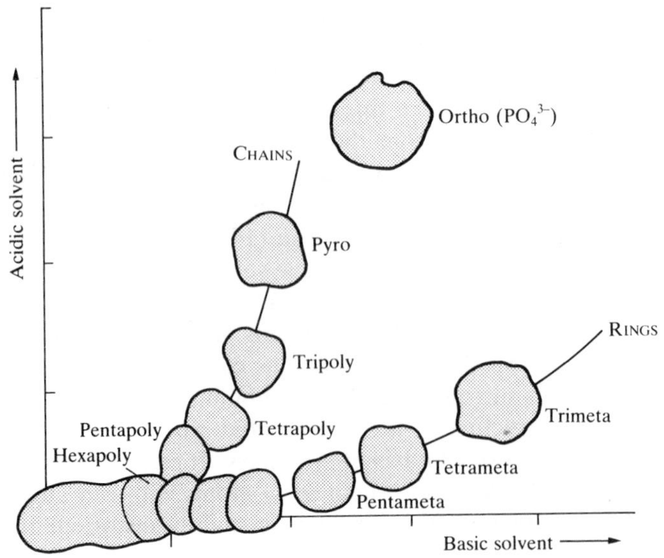

Fig. 5.9 Two-dimensional paper chromatogram of a complex mixture of phosphates formed by condensation reactions. The sample spot was placed at the lower left, basic solvent separation was used first, followed by acid solvent, which separates open chain from cyclic phosphates.

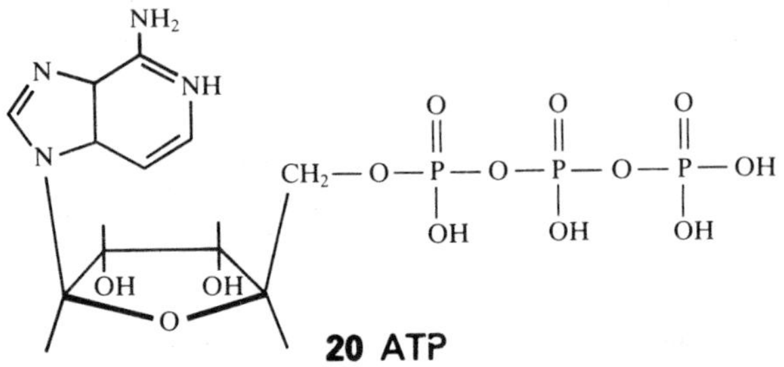

20 ATP

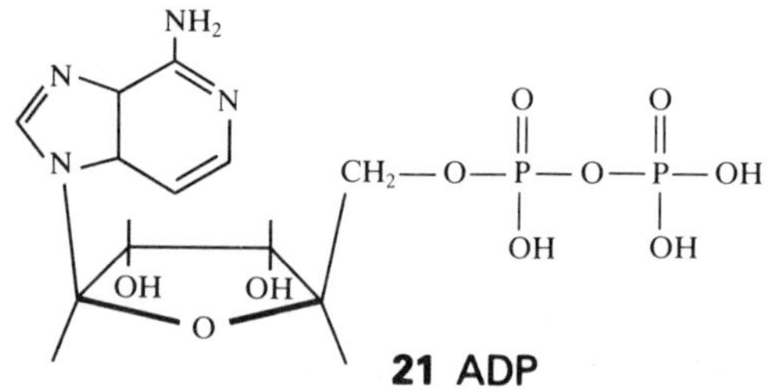

21 ADP

many, and the separation of products from the hydrolysis of P(V) oxide by column chromatography reveals the presence of chain species with one to nine P atoms. Higher polymers are also present and can be removed from the column only by hydrolysis. A two-dimensional paper chromatogram is shown in Fig. 5.9. The upper spot sequence corresponds to linear polymers and the lower to rings. P_n-chain polymers with $n = 10$ to 50 can be isolated as mixed amorphous glasses analogous to those formed by silicates and borates.

The biological importance of polyphosphates was mentioned at the outset. At physiological pH values (close to 7.4), the P—O—P bond is unstable with respect to hydrolysis. The key to energy exchange in metabolism is the hydrolysis of ATP (**20**) to ADP (**21**):

$$ATP^{4-} + 2H_2O \rightarrow ADP^{3-} + HPO_4^{2-} + H_3O^+ \qquad \Delta G^{\ominus} = -41\ \text{kJ} \quad \text{at} \quad \text{pH} = 7.4$$

Energy storage in metabolism depends on the subtle construction of pathways to make ATP from ADP.

Further reading

R. P. Bell, *The proton in chemistry*. Cornell University Press, Ithaca (1969). A classic discussion of Brønsted acidity with many examples drawn from organic chemistry.

J. E. Bartmess and R. T. McIver, Chapter 11 in *Gas phase ion chemistry*, M. T. Bowers (ed). Academic Press, New York (1978). The source of the information on the measurement of gas phase acidities used in this chapter.
C. F. Baes, Jr. and R. E. Messmer, *The hydrolysis of cations*. Wiley-Interscience, New York (1976). A survey of the acidities and the polymerization of aqua ions.
W. Stumm and J. J. Morgan, *Aquatic chemistry*. Wiley-Interscience, New York (1981). The standard text on the chemistry of natural waters.
M. T. Pope, *Heteropoly and isopoly oxometallates*. Springer-Verlag, Berlin (1983). A summary of the chemistry of early *d*-block polyions.

Exercises

5.1 Sketch an outline of the *s* and *p* blocks of the periodic table and indicate on it the elements that form (a) strongly acidic oxides and (b) strongly basic oxides, and (c) show the regions for which amphoterism is common.

5.2 Identify the conjugate bases corresponding to the following acids: $[Co(NH_3)_5(OH_2)]^{3+}$, HSO_4^-, CH_3OH, $H_2PO_4^-$, $Si(OH)_4$, HS^-.

5.3 Identify the conjugate acids of the bases C_5H_5N (pyridine), HPO_4^{2-}, O^{2-}, CH_3COOH, $[Co(CO)_4]^-$, CN^-.

5.4 List the bases HS^-, F^-, I^-, NH_2^- in order of increasing proton affinity.

5.5 List the acids H_2O, HCl, HI, CH_4 in order of increasing gas-phase acidity.

5.6 Calculate the gas-phase proton transfer enthalpy for the reaction $NH_4^+(g) + Cl^-(g) \rightarrow NH_3(g) + HCl(g)$. What factor makes this reaction less favorable in aqueous solution?

5.7 Aided by Fig. 5.3 (taking solvent leveling into account), identify which bases from the following lists are (a) too strong to be studied experimentally; (b) too weak to be studied experimentally; (c) of directly measureable base strength.
(i) CO_3^{2-}, O^{2-}, ClO_4^-, and NO_3^- in water
(ii) HSO_4^-, NO_3^-, and ClO_4^- in H_2SO_4.

5.8 The aqueous solution pK_a values for HOCN, H_2NCN, and CH_3CN are approximately 4, 10.5, and 20 (estimated) respectively. Explain the trend in these —CN derivatives of binary acids and compare them with H_2O, NH_3, and CH_4. Is —CN electron donating or withdrawing?

5.9 The pK_a value of $HAsO_4^{2-}$ is 11.5. Is this value consistent with the two Pauling rules?

5.10 Draw the structures and indicate the charges of the tetraoxoanions of X = Si, P, S, and Cl. Summarize and account for the trends in the pK_a values of their neutral parent oxoacids.

5.11 Which of the following pairs is the stronger acid? Give reasons for your choice. (a) $[Fe(OH_2)_6]^{3+}$ or $[Fe(OH_2)_6]^{2+}$, (b) $[Al(OH_2)_6]^{3+}$ or $[Ga(OH_2)_6]^{3+}$, (c) $Si(OH)_4$ or $Ge(OH)_4$, (d) $HClO_3$ or $HClO_4$, (e) H_2CrO_4 or $HMnO_4$, (f) H_3PO_4 or H_2SO_4.

5.12 Arrange the oxides Al_2O_3, B_2O_3, BaO, CO_2, Cl_2O_7, SO_3 in order from the most acidic through amphoteric to the most basic.

5.13 Arrange the acids HSO_4^-, H_3O^+, H_4SiO_4, CH_3GeH_3, NH_3, HSO_3F in order of increasing acid strength.

5.14 Na^+ and Ag^+ have similar ionic radii. Which aqua ion is the stronger acid? Why?

5.15 Which of the elements Al, As, Cu, Mo, Si, B, Ti form oxide polyanions and which form oxide polycations?

5.16 When a pair of aqua cations forms an M—O—M bridge with the elimination of water, what is the general rule for the change in charge per M atom on the ion?

5.17 Write a balanced equation for the formation of $P_4O_{12}^{4-}$ from PO_4^{3-}. Write a balanced equation for the dimerization of $[Fe(OH_2)_6]^{3+}$ to give $[(H_2O)_4Fe(OH)_2Fe(OH_2)_4]^{4+}$.

5.18 Write balanced equations for the main reaction occurring when (a) H_3PO_4 and Na_2HPO_4 and (b) CO_2 and $CaCO_3$ are mixed in aqueous media.

5.19 Identify the point groups of $[W_6O_{19}]^{2-}$ (**14**) and $[Fe_2(OH)_2(OH_2)_8]^{4+}$ (**11**). Neglect H atom positions.

Problems

5.1 The proton affinity (A_p) of CF_3 is 1572 kJ mol^{-1}. The corresponding value for $(C_2H_5)_3P$ is 982 kJ mol^{-1}. Calculate the enthalpy of transfer of a proton from HCF_3 to $(C_2H_5)_3P$ in the gas phase. Assuming entropy changes are negligible, calculate an equilibrium constant for this reaction. List the factors that are important when the reaction is transferred from the gas phase to a dipolar aprotic solvent. Why must the solvent be a very weak acid if this equilibrium is to be studied?

5.2 Methanol (CH_3OH) and dimethylformamide are solvents of similar relative permittivity (dielectric constant) but methanol has the added ability to act as a hydrogen-bonding donor. List the factors that influence the relative acidity of

HX (X = Cl, Br, I) in each of the two solvents. Account for the greater differentiation among the three in methanol.

5.3 A standard trick in analytical chemistry for improving the detection of the equivalence point in titrations of weak bases with strong acids is to use acetic acid as a solvent. Explain the basis of this approach.

5.4 In the gas phase, the base strength of amines increases regularly along the series $NH_3 < CH_3NH_2 < (CH_3)_2NH < (CH_3)_3N$. Consider the role of steric effects and electron-donating ability of CH_3 in determining this order. In aqueous solution, the order is reversed. What solvation effect is likely to be responsible?

5.5 The unstable species $Si(OH)_4$ is a weaker acid than H_2CO_3. Write balanced equations to show how dissolving a solid $M_2SiO_4(s)$ can lead to a reduction in the pressure of $CO_2(g)$ over an aqueous solution. Explain why silicates in ocean sediments might limit the increase of CO_2 in the atmosphere.

5.6 The precipitation of $Fe(OH)_3$ discussed in the chapter is used to clarify waste waters since the gelatinous hydrous oxide is very efficient at coprecipitation of some contaminants and entrapment of others. The solubility product of $Fe(OH)_3$ may be written as: $K_{sp} = [Fe(OH_2)_6^{3+}][OH^-]^3 \approx 10^{-38}$. Since the autoprotolysis constant of water links $[H_3O^+]$ to $[OH^-]$ by $K_w = [H_3O^+][OH^-] = 10^{-14}$, we can rewrite the solubility product by substitution as $[Fe^{3+}(aq)]/[H^+]^3 = 10^4$. (a) Balance the reaction for precipitation of $Fe(OH)_3(s)$ when Fe(III) nitrate is added to water. (b) If 5.6 kg of $Fe(NO_3)_3 \cdot 9H_2O$ is added to 100 L of water, what is the final pH and $[Fe(OH_2)_6]^{3+}$, neglecting other forms of dissolved Fe(III)?

5.7 The frequency of the symmetrical M—O breathing vibration of the octahedral aqua ions, $[M(OH_2)_6]^{2+}$ increases along the series, $Ca^{2+} < Mn^{2+} < Ni^{2+}$. How does this trend relate to the trend in acidity?

Lewis acids and bases

6

In this chapter we move on from the Brønsted theory of acids and bases to a more general viewpoint encompassing a wider range of substances and reactions. As in the case of Brønsted acids, the 'Lewis acids' that we shall now introduce show a number of periodic trends. However, the concept of a Lewis acid is so general that we lose the simplicity of a single acidity scale (for a given solvent) analogous to the pK_a scale for Brønsted acids. We describe two ways of coping with this problem. One involves classifying Lewis acids qualitatively as 'hard' or 'soft'. The other approach, which is more quantitative, is based on fitting thermochemical data to a set of parameters characteristic of each acid or base. As with Brønsted acidities, we shall see that Lewis acidities are also strongly affected by solvent effects. We conclude the chapter by considering polyoxoanions of aluminum and silicon, as we did in Chapter 5. However, with the concepts we develop here, we can approach their properties from a new viewpoint.

The Lewis definitions of acids and bases
6.1 Examples of Lewis acids and bases
6.2 The fundamental types of reaction
The strengths of Lewis acids and bases
6.3 The contributions to Lewis acid and base strengths
6.4 Solvent influences
6.5 Thermodynamic correlations
Representative Lewis acids
6.6 Boron and carbon group acids
6.7 Nitrogen and oxygen group acids
6.8 Halogen acids
Heterogeneous acid–base reactions
6.9 Surface acidity
6.10 Solid and molten acids in industrial processes
Further reading
Exercises
Problems

The Brønsted concept of acids and bases focuses on the transfer of a proton between molecules. While more general than the theories of acids and bases that preceded it, it still fails to take into account reactions between substances that show similar features but in which no protons are transferred. This deficiency was remedied by a more general concept of acidity introduced in the same year as the Brønsted concepts (1923) by G. N. Lewis, but which became influential only in the 1930s.

The Lewis concepts unify a wide range of chemistry. Moreover, whereas we can observe the binding of a free proton to a Brønsted base only in the gas phase using specialized equipment, the interaction of a free molecular Lewis acid and a Lewis base can often be studied under everyday laboratory circumstances.

The Lewis definitions of acids and bases

A **Lewis acid** is a substance that acts as an electron pair acceptor. A **Lewis base** is a substance that acts as an electron pair donor. We denote a Lewis acid by A and a Lewis base by :B, often omitting any other lone pairs that may be present. The fundamental reaction of Lewis acids and bases is the formation of a **complex** A—B, in which A and :B bond together by sharing the electron pair supplied by the base. It is appropriate to mention that the terms acidity and basicity are reserved for discussions of the equilibrium position of reactions. When electron donation is involved in a kinetic process determining the rate of reaction, the term base is replaced by the term **nucleophile**. Correspondingly, an electron acceptor, acting to determine a rate, is called an **electrophile**.

6.1 Examples of Lewis acids and bases

The proton is a Lewis acid because it can attach to an electron pair. It follows that any Brønsted acid, since it provides protons, exhibits Lewis acidicity too.[1] All Brønsted bases are Lewis bases, since a proton acceptor is also an electron pair donor. However, since the proton is not a part of the definition, a wider range of substances can be classified as Lewis acids and bases than can be classified in the Brønsted scheme.

We meet many examples of Lewis acids later, but we should be alert for the following possibilities:

1. A metal cation can bond to an electron pair supplied by the base.

An example is the hydration of Cu^{2+}, where the O atom lone pairs of H_2O (acting as a Lewis base) attach to the central cation. The cation is therefore the Lewis acid.

2. A molecule with an incomplete octet can complete its octet by accepting an electron pair.

A prime example is $B(CH_3)_3$, which can accept the lone pair of NH_3 and other donors; hence, $B(CH_3)_3$ is a Lewis acid.

[1] The Brønsted acid HA is the complex formed by the Lewis acid H^+ with the Lewis base A^-. This is why we said that a Brønsted acid 'exhibits' Lewis acidity rather than saying that a Brønsted acid *is* a Lewis acid.

3. A molecule or ion with a complete octet can rearrange its valence electrons and accept an additional electron pair.

CO_2 acts as a Lewis acid when it forms HCO_3^- by accepting an electron pair from an O atom in an OH^- ion.

4. A molecule or ion may be able to expand its octet (or simply be large enough) to accept an electron pair.

An example is the formation of the complex $[SiF_6]^{2-}$ when two F^- ions (the Lewis bases) bonds to SiF_4 (the acid).

5. A closed-shell molecule may be able to use one of its antibonding molecular orbitals to accommodate an incoming electron pair.

An example of this behavior is the ability of tetracyanoethylene (TCNE) to accept a lone pair into its π^* orbital, and hence to act as an acid.

Example 6.1: *Identifying Lewis acids and bases*

Identify the Lewis acids and bases in the reactions (a) $BrF_3 + F^- \rightarrow [BrF_4]^-$, (b) $(CH_3)_2CO{:} + I_2 \rightarrow (CH_3)_2CO{—}I_2$, (c) $KH + H_2O \rightarrow KOH + H_2$.

Answer. (a) The acid BrF_3 adds the base $:F^-$. (b) Acetone (propanone) is acting as a base, donating a lone pair of electrons from O into an empty antibonding orbital of the I_2 molecule, which therefore is acting as an acid. (c) The ionic hydride complex KH provides the base H^- to displace the acid H^+ from water to give H_2 along with KOH which in the solid state can be viewed as the base OH^- combined with the very weak acid K^+.

Exercise. Identify the acids and bases in the reactions (a) $FeCl_3 + Cl^- \rightarrow [FeCl_4]^-$, (b) $I^- + I_2 \rightarrow I_3^-$, (c) $[{:}SnCl_3]^- + (CO)_5MnCl \rightarrow (CO)_5Mn{—}SnCl_3 + Cl^-$.

6.2 The fundamental types of reaction

The simplest Lewis acid–base reaction in the gas phase is complex (or sometimes 'adduct') formation

$$A + {:}B \rightarrow A{—}B$$

Figure 6.1 shows the interaction of orbitals responsible for the bonding. The stability of the complex stems from the fact that the new, lower-energy bonding orbital that is formed is populated by the two electrons supplied by the base, the antibonding orbital is left unoccupied, and the net result is a lowering of energy. Three examples are

Acid	Base		Complex
BF_3 +	$:NH_3$	→	$F_3B{—}NH_3$
SO_3 +	$:OEt_2$	→	$O_3S{—}OEt_2$
$SnCl_2$ +	$:py$	→	$Cl_2Sn{—}py$

(py is pyridine, C_5H_5N.) All three reactions involve Lewis acids and bases that are independently stable in the gas phase or in solvents that do not form complexes with them, and hence are easily studied experimentally.

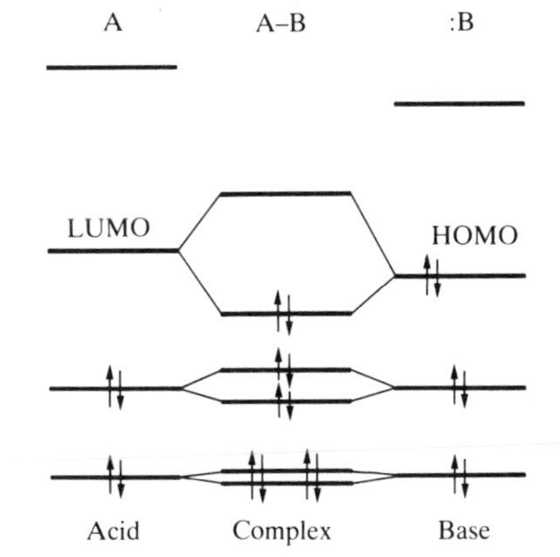

Fig. 6.1 Localized molecular orbital representation of the interaction between frontier orbitals in the formation of a complex between a Lewis acid A and a Lewis base :B.

Displacement reactions

A **displacement** of one Lewis base by another is a reaction of the form

$$B{-}A + {:}B' \rightarrow {:}B + A{-}B'$$

An example is

$$Et_2O{-}BF_3 + {:}NC_5H_5 \rightarrow Et_2O + F_3B{-}NC_5H_5$$

All Brønsted proton transfer reactions are of this type, as in

$$HS^-(aq) + H_2O(l) \rightarrow S^{2-}(aq) + H_3O^+(aq)$$

In this reaction, the Lewis base H_2O displaces the Lewis base S^{2-} from its complex with the acid H^+. Displacement of one acid by another is also possible, as in the reaction

$$\underset{A}{BF_3} + \underset{B-A'}{py{-}SnCl_2} \rightarrow \underset{B-A}{py{-}BF_3} + \underset{A'}{SnCl_2}$$

Metathesis reactions

A **metathesis reaction** (the name comes from the Greek word for 'interchange') is the interchange of partners:

$$A{-}B + A'{-}B' \rightarrow A{-}B' + A'{-}B$$

Metathesis is also called 'double replacement'. In a metathesis, the displacement of the base :B by :B′ is assisted by the extraction of :B by the acid A′. An example is the reaction

$$Et_3Si{-}I + AgBr \rightarrow Et_3Si{-}Br + AgI$$

In this reaction the base Br^- displaces the base I^-, the latter being extracted from its complex Et_3Si^+ by the acid Ag^+.

Solvents as acids and bases

Most solvents are either electron-pair acceptors or donors and hence are either Lewis acids or bases. It follows that a Lewis displacement reaction often occurs when a solute dissolves in a solvent, and that the subsequent reactions of the solution are also usually either displacements or metatheses. For example, when SbF_5 dissolves in BrF_3, the following displacement reaction occurs:

$$SbF_5 + BrF_3 \rightarrow [SbF_6]^- + [BrF_2]^+$$

In the reaction, the strong Lewis acid SbF_5 displaces the weaker Lewis acid BrF_2^+ from its complex with the Lewis base F^-. A more familiar example of the solvent as participant in a reaction is found in Brønsted theory. In this theory, the acid (H^+) is always regarded as complexed with the solvent, as in H_3O^+ if the solvent is water, and reactions are treated as the transfer of the acid, the proton, from a basic solvent molecule to another base.

The strengths of Lewis acids and bases

The proton (H^+) was the key ion in the discussion of Brønsted acid and base strengths. In the Lewis theory, we must allow for a greater variety of electron-pair acceptors and hence a greater variety of factors that influence strength. Nevertheless, it turns out that we can discuss some general trends by focusing on a few central aspects. Since the elementary acid–base reaction is

$$A + :B \rightleftharpoons A\text{—}B$$

the strength of the acid A can be expressed thermodynamically in terms of the equilibrium constant (or Gibbs free energy; we reviewed the relation between them in the *Further information* section of Chapter 4) for this formation reaction:

$$K_f = \frac{[AB]}{[A][:B]} \qquad \Delta G^{\ominus} = -RT \ln K_f$$

We can then set up a scale of strengths by choosing a common acid A and arranging the bases :B in order of their K_f values (or pK_f values, where $pK_f = -\log K_f$). It is important to remember, though, that different reference bases might yield different scales. When the reference acid is H^+, $K_f = 1/K_a$, where K_a is the acidity constant defined in Chapter 5.

6.3 The contributions to Lewis acid and base strengths

Four contributions are largely responsible for the magnitude of $\Delta G^{\ominus}$. One is the dependence of the strength of the A—B bond. Another is the rearrangement of the substituents of the acid and base that may be necessary to permit formation of the complex. The third is steric interaction between substituents on the acid and the base. Finally, in solution, we must take into account the solvation of the acid, the base, and the complex.

Electronic considerations: hard and soft acids and bases

In Chapter 5 we encountered the trends in basicity that are found when we consider the single acid H^+. We saw there that strong bases are found among compounds in which the electron-pair donor atom comes from the upper right of the *p* block. When we consider more general acids, we find that in many cases (among them Al^{3+}, Cr^{3+}, and BF_3) there are excellent correlations between the order of affinity toward bases obtained with them and the order obtained when H^+ is used as the acid. Thus the value of pK_f for the formation of complexes of carboxylate ions and Sc^{3+} ions is proportional to the pK_a of the carboxylic acids.

However, not all acids behave like H^+. If we are to deal with the interactions of acids and bases containing elements from throughout the periodic table, we need to consider at least two main classes of substance. These are the 'hard' and 'soft' acids and bases. This classification was introduced by R. G. Pearson and is a generalization—and a more evocative renaming—of the distinction between two types of behavior which originally were named simply 'class A' and 'class B' by Arland, Chatt, and Davies. The two classes are identified by the opposite order of strengths (as measured by K_f) with which they form complexes with halide ion bases:

Class A bond in the order $I^- < Br^- < Cl^- < F^-$
Class B bond in the order $F^- < Cl^- < Br^- < I^-$

A representative class A acid is the Al^{3+} ion; one of class B is Hg^{2+}. In the case of Al^{3+}, the binding strength increases as the electrostatic

Table 6.1. The classification of Lewis acids and bases†

	Hard	Borderline	Soft
Acids	H^+, Li^+, Na^+, K^+ Be^{2+}, Mg^{2+}, Ca^{2+} Cr^{3+}, SO_3, BF_3	Fe^{2+}, Co^{2+}, Ni^{2+} Cu^{2+}, Zn^{2+}, Pb^{2+} SO_2, BBr_3	Cu^+, Ag^+, Au^+, Tl^+, Hg^+ Pd^{2+}, Cd^{2+}, Pt^{2+}, Hg^{2+} BH_3
Bases	F^-, OH^-, H_2O, NH_3 CO_3^{2-}, NO_3^-, O^{2-} SO_4^{2-}, PO_4^{3-}, ClO_4^-	NO_2^-, SO_3^{2-}, Br^- N_3^-, N_2 C_6H_5N, SCN^- (*N*)	H^-, R^-, CN^-, CO, I^- SCN^- (*S*), R_3P, C_6H_6 R_2S

† The italicized element is the site of attachment to which the classification refers.

parameter ξ $(=z^2/r)$ increases, which is consistent with an ionic model of the bonding. In the case of Hg^{2+}, the binding strength increases with increasing polarizability (the responsiveness of the electron distributions to perturbation). This suggests that class A acids are cations that form complexes in which simple coulombic interactions are dominant, and that class B acids are cations that form more complexes in which covalent bonding is significant.

A similar classification applies to neutral molecular acids and bases. For example, the Lewis acid phenol forms a more stable complex with $(C_2H_5)_2O$: than with $(C_2H_5)_2S$:. This is analogous to the preference of Al^{3+} for F^- over Cl^-. In contrast, the Lewis acid I_2 forms a more stable complex with $(C_2H_5)_2S$:. Since it resembles Al^{3+}, phenol is class A, whereas I_2, which resembles Hg^{2+}, is class B.

The two classes of Lewis acid are now usually called 'hard' (the old class A) and 'soft' (the old class B). The **hard acids** preferentially bind lighter basic atoms within a group:

$$F^- \gg Cl^- > Br^- > I^- \qquad R_2O \gg R_2S \qquad R_3N \gg R_3P$$

The **soft acids** show the opposite trend down each group:

$$F^- \ll Cl^- < Br^- < I^- \qquad R_2O \ll R_2S \qquad R_3N \ll R_3P$$

The definitions of hardness imply the following rules:

Hard acids tend to bind to hard bases.
Soft acids tend to bind to soft bases.

When a series of acids and bases is analyzed with these rules in mind, it is possible to identify the classification summarized in Table 6.1.

The interpretation of hardness

Hard acids and bases are generally best described in terms of ionic interactions. Soft acids and bases are more polarizable than hard acids and bases and are more richly covalent.

We can interpret molecular hardness and softness in terms of frontier orbitals in much the same way as we did for atomic hardness (Section 1.6). When the frontier orbital separation is small (Fig. 6.2, see also Fig. 1.23), the electron distribution is easily rearranged by a perturbation and the molecule is soft. When the separation is large, the electron distribution

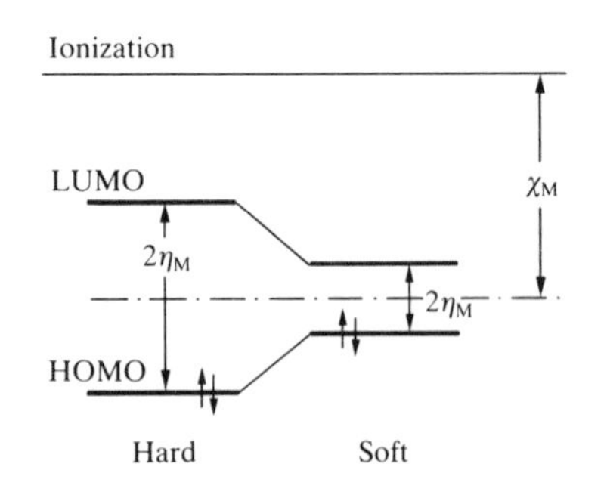

Fig. 6.2 The relation between the frontier orbital separation in a molecule and the molecular hardness η_M. Both molecules have the same absolute molecular electronegativity χ_M.

resists rearrangement even when the perturbation is moderately strong. A strong hard acid does not have a low-lying LUMO. A strong hard base has a low energy, strongly bound HOMO. Since the electronic structures are barely perturbable, their interaction is primarily electrostatic. In contrast, the LUMO and HOMO of a soft acid–base pair rearrange substantially to give a covalent bond.

Chemical consequences of hardness

The concepts of hardness and softness help to rationalize a great deal of inorganic chemistry. For example, they are relevant to the structure of the earth, for the tendency of soft acids to prefer soft bases and hard acids to prefer hard bases explains two of the categories of the Goldschmidt classification (Section 1.2). The lithophile elements, which are found primarily in the crust in association with the hard base O^{2-} in silicate minerals, include the hard acids Li, Na, Ti, Al, and Cr. The chalcophile elements, which include the soft acids Ag, Zn, Cd, Pb, Sb, Bi, are found principally in combination with the soft base S^{2-} (or selenide and telluride) minerals.

Example 6.2: *Explaining the Goldschmidt classification*

The common ores of Ni, Zn, and Cu are sulfides. In contrast, Al is obtained from the oxide and Ca from the carbonate. Can this be explained in terms of hardness?

Answer. Both O^{2-} and CO_3^{2-} are hard bases; S^{2-} is a soft base. Ni, Zn, and Cu are considerably softer acids than Al or Ca. Hence the hard–hard and soft–soft rule accounts for the sorting observed.

Exercise. Which of the metals Cd, Rb, Cr, Pb, Sr, Pd might be expected to be found in aluminosilicate minerals and which in sulfides?

Hardness is also useful for choosing preparative conditions and predicting the directions of reactions. Elements with high oxidation numbers generally provide hard acid centers and hence are stabilized by hard bases (F^- and O^{2-}). For example, Ag(II) is found as $[AgF_4]^{2-}$, Fe(VI) as $[FeO_4]^{2-}$, and Os(VIII) as OsO_4. In contrast, elements with low oxidation numbers act as soft centers. Metals with low oxidation numbers may be obtained using soft bases such as CO. Thus, Cr(0) is obtained with CO as $Cr(CO)_6$ or with benzene as $Cr(C_6H_6)_2$.

We can use hardness as a guide to the outcome of metathesis reactions. Thus, R_3Si^+ is a hard acid, and salts of the form AgX′ convert R_3SiX to R_3SiX' if X′ is harder than X. This is because the harder base X′ prefers the hard Si atom center whereas the softer base X prefers the softer acid Ag^+. An illustration of these preferences is that the sequence

$$R_3SiNC \xrightarrow{AgCl} R_3SiCl \xrightarrow{AgF} R_3SiF$$

is thermodynamically favorable.

Another type of illustration concerns the way in which the thiocyanate ion SCN^- functions as an **ambidentate base**, a molecule or ion that can donate an electron pair from more than one atom. The SCN^- ion is a base by virtue of both the harder N atom and the softer S atom; the ion binds to the hard

Si atom through N. However, with a soft acid, such as a metal ion with a low oxidation number, the ion bonds through S. Pt(II), for example, forms Pt—SCN.

Example 6.3: *Using the concept of hardness for synthesis*

Propose routes for preparation of (a) B_2H_6 from $NaBH_4$ and (b) R_2Hg from a Grignard reagent.

Answer. (a) Boron trifluoride is a hard acid; F^- is a hard base and H^- is a soft base. Therefore, the number of hard–hard interactions is increased in the redistribution reaction

$$3NaBH_4 + 4BF_3 \rightarrow 3NaBF_4 + 2B_2H_6$$

(b) The carbanionic base R^- is softer than the hard Cl^- ion; Mg^{2+} is a harder acid than the Hg^{2+} ion. Thus the reaction yields the soft–soft, hard–hard combinations

$$HgCl_2 + 2RMgCl(\text{Grignard}) \rightarrow R_2Hg + 2MgCl_2$$

Exercise. Propose a synthesis of (a) C_2H_2 from CaC_2 and (b) $P(CH_3)_3$ from PCl_3.

Although the concepts of hardness and softness are useful, it must be appreciated that they are qualitative and that they focus on electronic factors. That being so, we must develop an appreciation of their limitations and reliability in different chemical contexts. We shall acquire this understanding as we work through the examples given in the remainder of the text. In particular, we must keep structural changes, steric factors, and solvation in mind.

Structural change and steric factors

So far, we have treated the orbitals of acids and bases as unchanging when the complex forms. Commonly, however, addition of the base to the acid requires rearrangements of geometry at one or both of the participants, and this rearrangement often makes a major contribution to the Gibbs free energy of the interaction. An example is the conversion of BF_3 from trigonal planar to pyramidal when it forms a complex with an amine base. The formation of the complex may be considered as taking place in two steps. First, the acid changes its geometry. This directs the acceptor orbital (the acid's LUMO) toward the incoming amine lone pair (the HOMO of the base) which then forms a σ bond. Related examples include the conversion of CS_2 from linear to angular when $:PR_3$ adds to it, and the conversion of the tetrahedral SiF_4 molecule to the octahedral $[SiF_6]^{2-}$ ion upon addition of $2F^-$.

Figure 6.3 shows the orbital and energy diagrams for each step when a base adds to BF_3. The orbital levels of the original trigonal planar BF_3 acid are shown on the left. Since its LUMO is a π antibonding orbital which is partly delocalized over the three F atoms, it does not have good overlap with the HOMO of the amine donor. The energy of the trigonal pyramidal molecule is greater on account of loss of the π bonding, but the complex formation is energetically favorable overall because the distortion produces

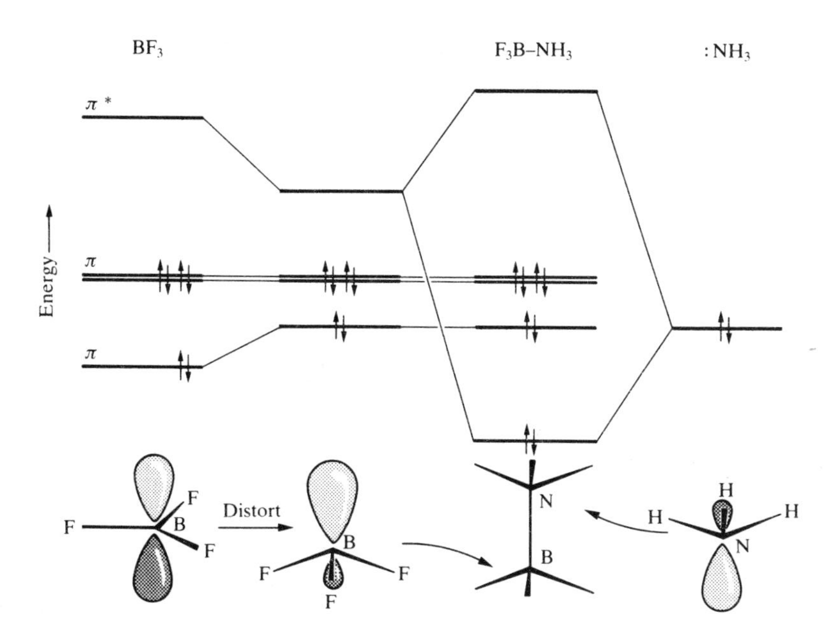

Fig. 6.3 When a planar BF_3 molecule is distorted to a trigonal pyramid, the LUMO falls to a lower energy as it hybridizes and acquires *s*-orbital character. The modified LUMO can interact strongly with the NH_3 HOMO (far right), resulting in a strong B—N σ bond.

a LUMO that is ideally adapted to overlap with the HOMO of the base. Hence, the investment of energy can be more than recovered.

It is possible to identify steric effects on Lewis acidity in the reactions of substituted pyridines with trimethylboron. The reaction enthalpy when pyridine forms a complex with $B(CH_3)_3$ is $-71\ \mathrm{kJ\ mol^{-1}}$. If a *meta* methyl group is substituted into the pyridine (**1**), the electron donor effect of methyl stabilizes the complex and the reaction enthalpy is $3\ \mathrm{kJ\ mol^{-1}}$ more exothermic. However, moving the methyl group to the *ortho* position (**2**) results in a reaction enthalpy which is *less* exothermic by $29\ \mathrm{kJ\ mol^{-1}}$. The same substitution barely changes the basicity of the pyridine toward a small acid, the proton. The destabilization is therefore attributed to crowding between the *ortho*-methyl of the base and the methyl groups of the acid.

A further factor is that the alkyl groups on the B atom move closer together when it forms a complex and changes from trigonal planar to pyramidal (**3a** → **3b**). This suggests that if its substituents are branched, then the B atom should be less strongly acidic, which is observed in practice.

1

2

3a **3b**

Example 6.4: *Identifying steric effects*

Predict which of the following combinations will form the stronger complex: (a) $B(CH_3)_3$ or $B(i\text{-}C_3H_7)_3$ with NH_3 and (b) bicyclic $P(C_2H_4)_3CH$ or $P(C_2H_5)_3$ with BCl_3.

Answer. (a) Of the two Lewis acids, $B(CH_3)_3$ contains less bulky substituents and as a result has less steric crowding in the complex. (b) The bicyclic phosphine should form the stronger complex because the CH_2 groups are pinned back away from the complexation center.

Exercise. Which of 2-methyltetrahydrofuran and 3-methyltetrahydrofuran should form the stronger complex with BF_3?

6.4 Solvent influences

Large solvent effects can often be treated in terms of the acidity and basicity of the solvents themselves. The exceptions are acid–base reactions that change the net charge and so are sensitive to the permittivity (dielectric constant) of the medium. We saw how to treat effects of permittivity in Chapter 5 and these considerations also apply to Lewis acid–base equilibria: reactions that increase net charge are favored by an increase in solvent permittivity; those that decrease net charge are disfavored by high solvent permittivity.

Solvent acidity and basicity

Solvents with basic character are common. Most of the well known polar solvents, including water, alcohols, ethers, amines, dimethylsulfoxide (dmso, $(CH_3)_2SO$), dimethylformamide, and acetonitrile (CH_3CN), are hard Lewis bases. Dmso is an interesting example of an ambidentate solvent that is hard on account of its O donor atom and soft on account of its S donor atom. Reactions of acids and bases in these solvents are generally displacements:

$$O_2S\text{—(dmso)} + :NH_3 \rightarrow O_2S\text{—}NH_3 + \text{dmso}$$

In Chapter 5 we introduced one specific type of solvent acidity without actually using the term acidity: that was hydrogen bonding. We can now see that hydrogen bonding is complex formation between A—H (the Lewis acid) and :B (the Lewis base) to give the complex conventionally denoted A—H · · · B. We pointed out that what we now call the hard bases N, O, and F had their Brønsted basicity reduced by hydrogen bond formation to hydroxylic and amine solvents. These solvents act as the A—H molecule and hence play the role of a Lewis acid. Such acidic solvents must be displaced if proton transfer to the base is to occur:

$$\underset{\text{A—B}}{\text{H—O—H}\cdots NH_3} + \underset{\text{A}'\text{—B}'}{H_3O^+} \rightarrow \underset{\text{A}'\text{—B}}{NH_4^+} + \underset{\text{A—B}'}{2H_2O}$$

Liquid SO_2 is a good acidic solvent for dissolving benzene, and a variety of solvents—including dimethylsulfoxide and benzene—have sufficiently low-lying LUMOs to function as weakly acidic solvents. This ability can enhance the solubility of basic solutes. However, the same solvents frequently display more important base properties; for instance, the I_2 molecule can act as an acceptor toward a benzene molecule.

Only the saturated hydrocarbons among common solvents lack significant Lewis acid or base character. Unsaturated hydrocarbons, however, may act as acids or bases by using their π or π^* orbitals as frontier orbitals. Electronegatively substituted alkanes such as haloalkanes (e.g. $CHCl_3$) are slightly acidic at the hydrogen atom.

Solvation parameters

We can express the basicity of solvents quantitatively using the reaction enthalpy for the formation of a complex with a reference acid. Victor Gutmann selected the strong Lewis acid $SbCl_5$ in 1,2-dichloroethane for this role, the relevant reaction being

$$SbCl_5 + :B \rightarrow Cl_5Sb\text{—B} \qquad \Delta H^{\ominus}$$

Table 6.2. Donor and acceptor numbers and relative permittivities (dielectric constants) at 25°C

Solvent	D.N.	A.N.	ε_r
Acetic acid		52.9	6.2
Acetone	17.0	12.5	20.7
Benzene	0.1	8.2	2.3
Carbon tetrachloride		8.6	2.2
Diethyl ether	19.2	3.9	4.3
Dimethylsulfoxide	29.8	19.3	45
Ethanol	19.0	37.1	24.3
Pyridine	33.1	14.2	12.3
Tetrahydrofuran	20.0	8.0	7.3
Water	18	54.8	81.7

Source: V. Guttman, *Coordination chemistry in nonaqueous solutions.* Springer-Verlag, Berlin (1968).

The negative of $\Delta H^{\ominus}$ (for historical reasons, in units of kcal mol^{-1}) is called the **donor number** (D.N.) of the solvent. Some representative values are collected in Table 6.2: the higher the donor number, the stronger the Lewis base.

A corresponding parameter to measure solvent acidity, the **acceptor number** (A.N.), has been introduced. In this case triethylphosphine oxide, $(C_2H_5)_3PO$:, is used as a reference base and the NMR chemical shift of ^{31}P is measured for the base dissolved in the pure solvent. The scale is set at 0 for the base in hexane and at 100 in $SbCl_5$. On this arbitrary basis, numbers similar in magnitude to donor numbers are obtained and some are shown in Table 6.2. As for bases, the higher the acceptor number, the stronger the Lewis acid.

Example 6.5: *Predicting solvent effects*

Which of the solvents (denoted Sol) would more favor the reaction as written? What factor in solvent behavior is most important (e.g. hardness, D.N., A.N., ε_r)? (a) $AgClO_4(s) + n\text{Sol} \rightarrow \text{AgSol}_n^+ + ClO_4^-$ where Sol may be dmso or CH_3OH, (b) $F_3B\text{—}NH_3 \rightarrow BF_3 + NH_3$ in CH_3OCH_3 or CH_3OH, (c) $CH_3Cl + I^- \rightarrow CH_3I + Cl^-$ in dmf or CH_3OH, and (d) $CH_3I + NH_3 \rightarrow CH_3NH_3^+ + I^-$ in CH_3CN or CH_3OCH_3.

Answer. (a) Dmso: although donor numbers and permittivities are similar, since Ag^+ is a soft acid, when it bonds through the soft S atom, dmso is more favorable than the hard base CH_3OH. The very weak base ClO_4^- is insensitive to solvent changes. (b) CH_3OH: both solvents are similarly basic toward the acid BF_3 and charge is not important, but CH_3OH can function as an acid (by hydrogen bond donation) toward NH_3. This is indicated by its higher acceptor number. (c) CH_3OH: both solvents have similar basicity, but CH_3OH is a stronger hard acid (by hydrogen bonding) toward Cl^-, a hard base. This strength is also indicated by its larger acceptor number. (d) CH_3CN: charge separation is important. Both solvents are moderately strong bases and weak acids (comparable D.N. and A.N.). The higher relative permittivity of CH_3CN is the important factor.

Exercise. Select the better solvent for the following processes: (a) $[CoCl(NH_3)_5]^{2+} + \text{Sol} \rightarrow [Co(NH_3)_5\text{Sol}]^{3+} + Cl^-$, Sol = CH_3OH or H_2O; (b) $AgClO_4(s) \rightarrow Ag^+(s) + ClO_4^-(s)$, Sol = benzene or cyclohexane.

6.5 Thermodynamic correlations

It is possible to go beyond the qualitative (but theoretically motivated) discussion that we have given so far and to correlate the strengths of acid–base interactions using empirical parameters. The simplest case where this may be done is for the formation of complexes in the gas phase. However, since many such reactions go to completion, it is difficult to measure the equilibrium constant and hence to obtain reliable values of $\Delta G^{\ominus}$. Nevertheless, enthalpy data are available and provide a good indication of trends in complex formation for reactions in which entropy changes are similar.

We shall consider gas-phase complex formation reactions of the type

$$A(g) + {:}B(g) \rightarrow A\text{—}B(g)$$

Many examples have been studied, among them

$$SO_3 + {:}NH_3 \rightarrow O_3S\text{—}NH_3$$

$$I_2 + {:}OEt_2 \rightarrow I_2\text{—}OEt_2$$

$$(CH_3)_3Al + {:}py \rightarrow (CH_3)_3Al\text{—}py$$

$$Cl_2Sn + {:}N(CH_3)_3 \rightarrow Cl_2Sn\text{—}N(CH_3)_3$$

It has been found that their standard reaction enthalpies can be expressed by the **Drago–Wayland equation**:

$$-\Delta H^{\ominus} = E_A E_B + C_A C_B \text{ kJ mol}^{-1}$$

The parameters E and C were introduced with the idea that they represented 'electrostatic' and 'covalent' factors, but in fact they must accommodate all factors except solvation. The compounds for which the parameters are listed in Table 6.3 satisfy the equation with an error of less

Table 6.3. Drago–Wayland parameters for some acids and bases*

	E	*C*
Acids		
Antimony pentachloride	15.1	10.5
Boron trifluoride	20.2	3.31
Iodine	2.05	2.05
Iodine monochloride	10.4	1.70
Phenol	8.86	0.904
Sulfur dioxide	1.88	1.65
Trichloromethane	6.18	0.325
Trimethylboron	12.6	3.48
Bases		
Ammonia	2.78	7.08
Benzene	0.23	2.9
Methylamine	1.30	5.88
p-Dioxane	2.23	4.87
Pyridine	1.17	6.40
Trimethylphosphine	0.84	6.55

* E and C parameters are often reported to give ΔH in kcal mol^{-1}; we have multiplied both by $\sqrt{4.184}$ to obtain ΔH in kJ mol^{-1}.

than ±3 kJ mol^{-1}, as do a much larger number of examples discussed in the reading suggested at the end of the chapter.

Example 6.6: *Using the Drago–Wayland equation*

Predict the enthalpies of complex formation of (a) $(CH_3)_3B—NH_3$, (b) $O_2S—C_6H_6$.

Answer. (a) From Table 6.3, for $B(CH_3)_3$, $E_A = 12.6$ and $C_A = 3.48$ and for NH_3 $E_B = 2.78$ and $C_B = 7.08$. Hence,

$$-\Delta H^\ominus = 12.6 \times 2.78 + 3.48 \times 7.08 \text{ kJ mol}^{-1}$$
$$= 59.7 \text{ kJ mol}^{-1}$$

The experimental value is -57.5 kJ mol^{-1}. (b) Likewise, for SO_2 $E_A = 1.88$ and $C_A = 1.65$ and for C_6H_6 $E_B = 0.23$ and $C_B = 2.9$. Hence,

$$-\Delta H^\ominus = 1.88 \times 0.23 + 1.65 \times 2.9 \text{ kJ mol}^{-1}$$
$$= 5.2 \text{ kJ mol}^{-1}$$

The experimental value is -4.2 kJ mol^{-1}.

Exercise. Estimate the enthalpies of formation of the complexes (a) ICl-(*p*-dioxane) [-31.4 kJ mol^{-1}], (b) $(CH_3)_3B—NH_2(CH_3)$ [-73.7 kJ mol^{-1}]. For comparison, experimental values are in brackets.

The Drago–Wayland equation is very successful and useful. In addition to giving the enthalpies of complex formation for over 1500 acid–base pairs, these enthalpies can be combined to calculate the enthalpies of displacement and metathesis reactions. Moreover, the equation is useful for reactions of acids and bases in nonpolar, nondonor solvents as well as the gas phase. The major limitation is that it is restricted to substances that can be conveniently studied in the gas phase or 'inert' solvents. This means that it is limited in the main to neutral molecules.

Unfortunately a careful study of the trends in the parameters has not yet yielded to any simple theoretical interpretation. Some trends that have already been discussed can be recognized in the parameters, but attempts to interpret E and C as electrostatic and covalent factors have failed. Moreover, it has been found that the equation is least successful for highly exothermic complexes (those for which $\Delta H^\ominus$ is strongly negative). This suggests that it cannot cope with cases where there is a large perturbation of the acid by the base or *vice versa*.

Representative Lewis acids

All the Brønsted acids and bases mentioned in Chapter 5 are also Lewis acids and bases. We have already considered a wide range of bases in the Brønsted and Lewis senses, among the most important being the halide ions, X^-, the amines, the phosphines, water, alcohols, ethers, and sulfides. We have said less about acids in general, so in this section we survey *p*-block acids. The important difference in viewpoint that we should keep in mind is that Lewis acidity focuses on the bonding of the base to the acid. Brønsted acidity focuses on the *transfer* of the proton, not individual complex formation.

The metal cations of the *s* and *d* blocks of the periodic table are all Lewis acids. A characteristic feature of these ions is their ability to bind several bases simultaneously and to form complexes such as $[Co(NH_3)_6]^{2+}$ and the aqua ions. Because of the great variety of structures and reactions, we shall deal with this major topic separately in Chapter 7. Here we concentrate on the unification that can be achieved in the description of the reactions of *p*-block elements.

6.6 Boron and carbon group acids

The halides of B and Al are among the most familiar Lewis acids. The planar BX_3 and AlX_3 molecules have a *p* orbital perpendicular to the plane (**4**) and can accept a lone pair from a Lewis base:

$$X_3B + :N(CH_3)_3 \rightarrow X_3B\text{—}N(CH_3)_3$$

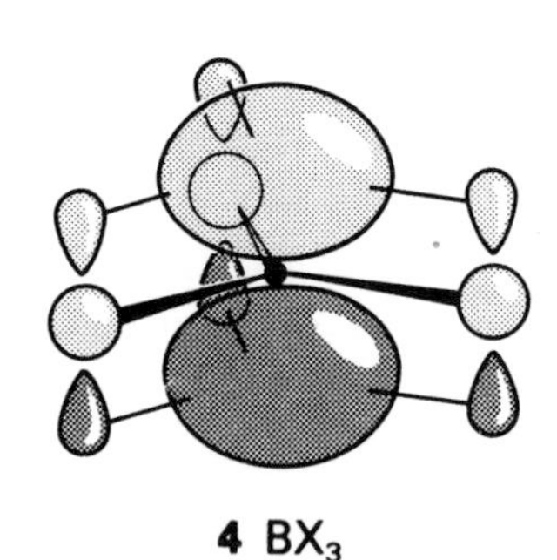

4 BX_3

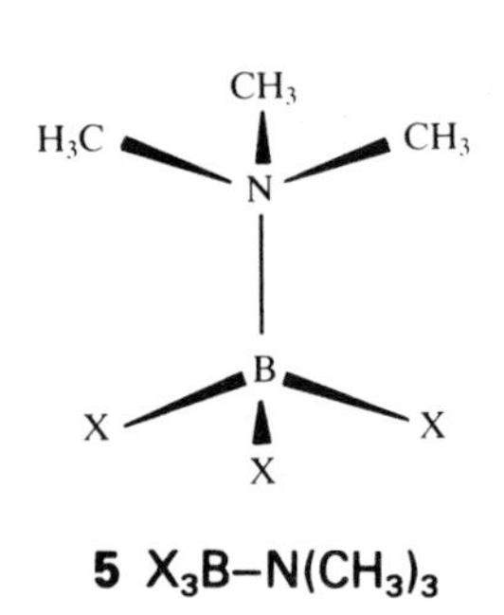

5 $X_3B\text{–}N(CH_3)_3$

As we have already remarked, the acid molecule becomes pyramidal as the complex (**5**) is formed (Fig. 6.3).

Boron halides

The order of stability of complexes with BX_3 is

$$BF_3 < BCl_3 < BBr_3$$

This is opposite to the order expected on the basis of the relative electronegativities of the halogens, which suggest that F, the most electron withdrawing, ought to leave the B atom in BF_3 most electron deficient and hence most acidic. The paradox is resolved by noting that the halogen atoms in the BX_3 molecule can form π bonds with the empty B *p* orbital (**4**) and that these π bonds must be broken in order to free the acceptor orbital for complex formation. In a sense, complex formation is a subtle type of displacement reaction in which the incoming base, the amine, must displace the halogens from their role as bases by virtue of their π bonding. The small F atom is the best able to form π bonds with the B *p* orbital. (It is a general rule that p—p π bonding is strongest for Period 2 elements.) Thus, the BF_3 molecule has the strongest π bond to be broken when the amine attacks and forms a complex.

Boron trifluoride is widely used as an industrial catalyst. Its role there is to extract bases bound to carbon and hence to generate carbocations:

$$R_3C\text{—}X + BF_3 \rightarrow R_3C^+ + [XBF_3]^-$$

It is a gas, but it dissolves in diethyl ether to give a solution which is convenient to use. This dissolving is also an aspect of Lewis acidity, for the BF_3 molecule forms a complex with the lone pair on the O atom of a solvent molecule.

An important derivative of boron trifluoride is the tetrafluoroborate anion $[BF_4]^-$. Since this ion is the complex of the acid BF_3 with the base F^-, it retains very little basic character and consequently has very little tendency to coordinate further. Moreover, with four bonds, the B atom is fully coordinated and $[BF_4]^-$ is an inert anion.

Aluminum halides

In contrast to the boron halides, the aluminum halides are dimers in the gas phase; the chloride, for example, has molecular formula Al_2Cl_6 in the vapor (**6**). This tendency to dimerize illustrates the reduced tendency of p orbitals to participate in π bonding in Period 3 as compared with Period 2. The molecule is a 'self acid–base complex', since each Al atom acts as an acid toward a Cl atom bonded to the other Al atom.

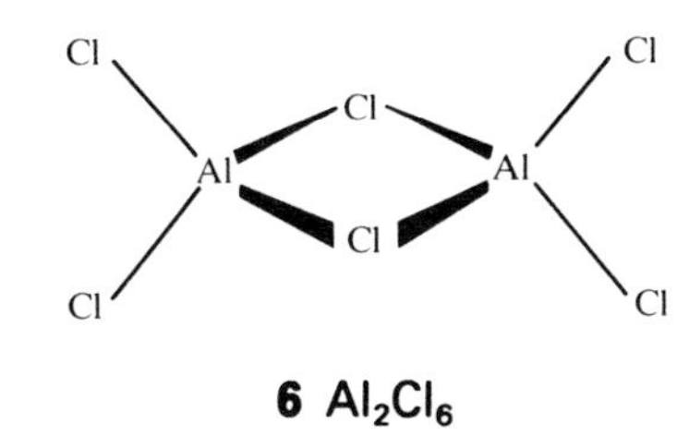

6 Al_2Cl_6

Aluminum chloride is widely used as a Lewis acid catalyst in organic reactions. The classic examples are Friedel–Crafts alkylation (the attachment of R— to an aromatic ring) and acylation (the attachment of R—CO—):

$$2CH_3COCl + Al_2Cl_6 \rightarrow 2CH_3CO^+ + 2[AlCl_4]^-$$

$$CH_3C(=O)^+ + C_6H_6 \rightarrow CH_3C(=O)\text{—}C_6H_5 + H^+(\text{solvated})$$

Silicon and tin complexes

Unlike carbon, Si can expand its octet (or it is simply large enough) and become hypervalent:

$$SiF_4(g) + 2HF(aq) \rightarrow [SiF_6]^{2-}(aq) + 2H^+(aq)$$

Since the very hard base F^-, aided by a proton, can displace the very hard base O^{2-} from Si, hydrofluoric acid is corrosive toward glass (SiO_2). The trend of acidity for SiX_4,

$$SiI_4 < SiBr_4 < SiCl_4 < SiF_4$$

is the reverse of that for BX_3 and follows the increase in the electron-withdrawing power of the halogen from I to F. Coordination number 6, as in $[SiF_6]^-$, is not the only state of coordination for Si above 4. For example, $PhSi(OC_6H_4O)_2$ is a five-coordinate trigonal bipyramid (**7**).

7

Tin(II) chloride is both a Lewis acid and a base. As an acid, it forms $[SnCl_3]^-$ with a Cl^- ion. The complex retains a lone pair and it is sometimes more revealing to write it as $:SnCl_3^-$. It acts as a base to give metal–metal bonds, such as in the complex $(CO)_5Mn\text{—}SnCl_3$ cited above. Compounds like this are currently the focus of much attention in inorganic chemistry, as we see later in the text.

6.7 Nitrogen and oxygen group acids

The heavier elements of the nitrogen group (Group 15/V) form some of the most important Lewis acids, SbF_5 being among the most widely studied. This Lewis acid can be used to produce some of the strongest Brønsted acids, as in the reaction.

$$2HF + SbF_5 \rightarrow [H_2F]^+ + [SbF_6]^-$$

A **superacid**, a mixture so called because it can protonate almost all organic compounds, can be produced by dissolving SbF_5 in a mixture of HSO_3F and SO_3. The simplest of the many reactions occurring in this mixture is

$$SbF_5 + 2HSO_3F \rightarrow [H_2SO_3F]^+ + [FSO_2O\text{—}SbF_5]^-$$

where the protonated fluorosulfuric acid acts as the powerful Brønsted acid.

Sulfur dioxide is both a Lewis acid and a Lewis base. Its acidity is weak and conventional:

$$O_2S: + :NR_3 \longrightarrow O_2S{-}NR_3$$

The SO_2 molecule acts as an ambidentate base because it can donate either its S or its O lone pairs to an appropriate acid. It can also act as a Lewis acid at the S atom. Its complex with the strong hard acid SbF_5 is formed by donation from the hard O atom base; with the soft donor Ir(I) the soft S atom acts as an acceptor (**8**). Sulfur bonding is also common for borderline cations in the *d* block.

8 $[IrCl(CO)(PPh_3)_2SO_2]$

Sulfur trioxide is a strong Lewis acid and a very weak (O donor) Lewis base. Its acidity is illustrated by the reactions

$$SO_3 + NR_3 \longrightarrow O{=}S(O)_2{-}NR_3$$

$$SO_3 + RSH \longrightarrow O{=}S(OH)(O){-}SR$$

A classic aspect of the acidity of SO_3 is its highly exothermic reaction with water in the formation of sulfuric acid. The problem of having to remove large quantities of heat from the reactor used for the commercial production of the acid is solved using another aspect of its Lewis acidity. Prior to dilution it is dissolved in sulfuric acid to form the complex mixture known as oleum. This reaction is in fact an example of complex formation:

$$SO_3 + H_2SO_4 \rightarrow H_2S_2O_7$$

followed by

$$H_2S_2O_7 + H_2O \rightarrow 2H_2SO_4$$

6.8 Halogen acids

Lewis acidity is expressed in an interesting and subtle way by the dihalogen molecules, especially I_2 and Br_2. I_2 is violet in the gas phase and in $CHCl_3$ solution; in water, acetone, or ethanol solution it is brown. The color changes because the complex formed from the interaction of the lone pair on O atoms in these solvents with a low-lying σ^* orbital of the dihalogen has a strong optical absorption.

Complexes can form between molecules that lack obvious valence-shell acceptor sites, as between I_2 and H_2O.[2] However, once we examine the

[2] The terms 'donor-acceptor complex' and 'charge-transfer complex' were at one time used to denote these complexes. However, the distinction between these and the more familiar Lewis acid–base complexes is arbitrary and in the current literature the terms are used interchangeably.

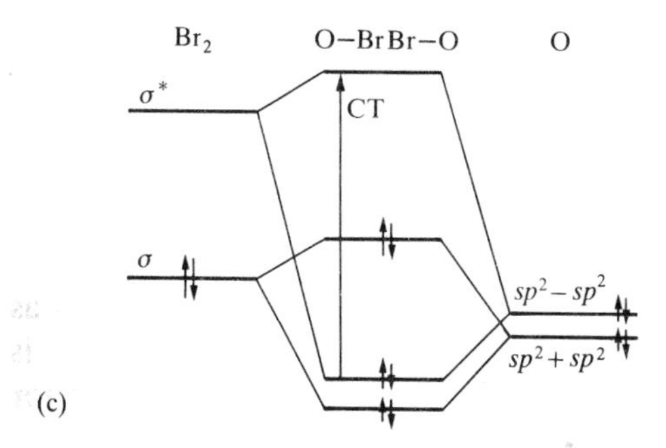

Fig. 6.4 (a) The structure of $(CH_3)_2COBr_2$ as shown by X-ray diffraction. (b) The orbital overlap responsible for complex formation. (c) A partial molecular orbital energy level diagram for the interaction of the σ and σ^* orbitals of Br_2 with the appropriate symmetry-adapted combinations of sp^2 orbitals on the two O atoms. The donor–acceptor charge-transfer band in the near UV is labeled CT.

interaction in more detail (such as by noting the role of empty antibonding orbitals, which all molecules possess) we see that they are complexes in the normal Lewis sense. The dihalogens also show their acidity by forming complexes using the π electrons of benzene (**9**).

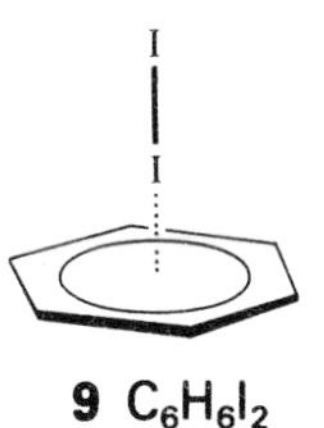

9 $C_6H_6I_2$

The structure of the complex of Br_2 with acetone is shown in Fig. 6.4. The figure also shows the transition responsible for the new 'donor–acceptor band' in the spectrum. The orbital from which this transition originates is predominantly the lone pair orbital of the base, the ketone. The receiving orbital is predominantly the LUMO of the dihalogen acid. Thus, to a first approximation, the transition promotes an electron from the base to the acid. It is therefore called a **charge-transfer transition**. Such transitions frequently initiate photochemical ionization reactions in suitably polar solvents.

The triiodide ion (I_3^-) is an example of a complex between a halogen acid (I_2) and a halogen base (I^-). One of the applications of its formation is to render molecular iodine soluble in water so that it can be used as a titration reagent:

$$I_2(s) + I^-(aq) \rightarrow I_3^-(aq) \qquad K = 725$$

The triiodide ion is one example of the polyhalide ions that are described more fully in Section 13.4.

Heterogeneous acid–base reactions

Some of the most important reactions involving the Lewis acidity of inorganic compounds occur at solid surfaces. For example, **surface acids**, which are solids with a high surface area and Lewis acid sites, are used as catalysts in the petrochemical industry for the isomerization and alkylation of aromatic compounds. The surfaces of many materials that are important in the chemistry of soil and natural waters also have Lewis acid sites.[3]

6.9 Surface acidity

Alumina and aluminosilicates

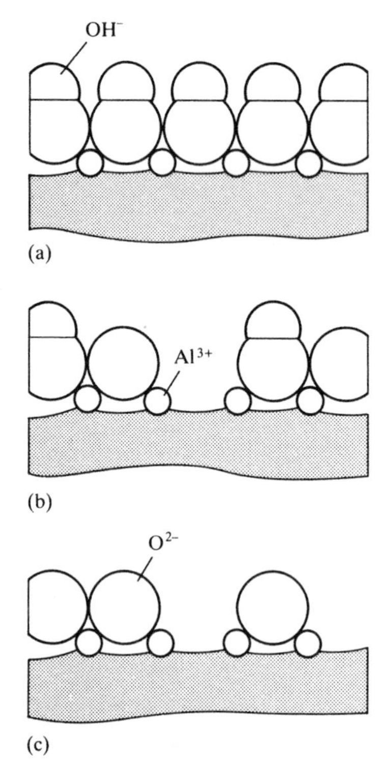

Fig. 6.5 Schematic representation of the surface of γ-alumina viewed in cross-section. The gray area represents the atoms below the surface layers. (a) The hydroxylate alumina surface is covered by OH groups that act as mild Brønsted acids. (b) Heating to 400°C produces partially dehydroxylated alumina in which Lewis acid (Al^{3+} ions) and base sites (O^{2-} ions) and Brønsted acid sites coexist. (c) At 900°C, dehydroxylated alumina is formed, with Lewis acid and base sites.

Alumina (Al_2O_3) is a surface acid on account of the high oxidation number (+3) of Al (Fig. 6.5). When freshly precipated hydrous aluminum oxide is heated above 150°C, dehydration begins and the surface undergoes reactions such as

$$\begin{array}{ccc} OH & OH & OH \\ | & | & | \\ Al & Al & Al \\ \hline \end{array} \longrightarrow \begin{array}{ccc} & O & OH \\ Al & \diagup\diagdown \; Al & Al \\ \hline \end{array} + H_2O$$

These create Al^{3+} ions at the surface, which act as Lewis acids, as well as unprotonated O^{2-} ions, which act as Lewis bases. This acid–base pair is always produced together, but the Lewis acid sites are more important for surface catalysis.

Aluminosilicates, in contrast to alumina, can display strong Brønsted acidity. The formation of the surface may be thought of as the condensation of $Si(OH)_4$ units with $(H_2O)Al(OH)_3$ units:

$$\begin{array}{cc} OH & OH_2 \\ | & | \\ Si & Al \\ \hline \end{array} \longrightarrow \begin{array}{cc} \multicolumn{2}{c}{OH} \\ Si & Al \\ \hline \end{array} + H_2O$$

This produces a strong Brønsted acid at the site where the proton is retained to balance the smaller positive charge of the Al^{3+} ion.

The characters and strengths of the surface acids can be studied by methods similar to those used to study species in solution, in that surface acid sites may be titrated with solutions of bases. It is also possible to use the IR spectra of the adsorbed molecules to study reactive surface sites. For example, the Lewis sites on alumina surfaces are revealed by the infrared spectrum of adsorbed pyridine, which bears a strong resemblance to the infrared spectrum of simple Lewis complexes, such as $C_5H_5N—BH_3$ (Table 6.4).

[3] An interesting account is found in G. Sposito, *The surface chemistry of soils.* Oxford University Press (1984).

Table 6.4. Assignment of C—C vibrational maxima (in cm^{-1}) for pyridine and dimethylpyridine adsorbed on two representative aluminosilicate surfaces*

Sample 1 surface	Assignment	Sample 2 surface
1575 (m)	B_1 mode of PyAl	1448 (s)
1595 (vs)	A_1 mode of PyAl	1490 (s)
1627 (s)	B_1 mode of PyH	1550 (m)
1655 (s)	A_1 mode of PyH	1590 (vs)

* Brønsted sites are denoted PyH; Lewis by PyAl. Intensities: medium (m); strong (s); very strong (vs).

Organic dehydration reactions can occur on surface acids. These include the conversion of alcohols to ethers and alkenes:

$$\underset{/////////}{\overset{|}{OH}} + C_2H_5OH \longrightarrow \underset{///////}{\overset{|}{OC_2H_5}} + H_2O$$

$$\underset{////////////}{\overset{|}{OC_2H_5}\quad \overset{|}{OC_2H_5}} \longrightarrow C_2H_5\text{-O-}C_2H_5 + \underset{//////}{O}$$

$$\underset{//////}{\overset{|}{OC_2H_5}} \longrightarrow H_2C{=}CH_2 + \underset{///////}{\overset{|}{OH}}$$

Acid catalysis of alkene isomerization and initiation of cationic alkene polymerization can occur if a surface Brønsted acid protonates a C═C double bond to form a carbocation:

$$\underset{///////////}{Si{-}OH{-}Al} + CR_2{=}CR_2 \rightarrow R_2CH\overset{+}{C}R_2 + \underset{///////////}{Si{-}O^{-}{-}Al}$$

Silica surfaces

Silica surfaces do not readily produce Lewis acid sites since —OH groups remain tenaciously attached at the surface of SiO_2 derivatives, and Brønsted acidity is dominant. The Brønsted acidity of silica surfaces themselves is only moderate (and comparable to that of acetic acid). However, as we have already remarked, aluminosilicates display *strong* Brønsted acidity.

Surface reactions carried out using the Brønsted acid sites of silica gels are used to prepare thin coatings of a wide variety of organic groups using surface modification reactions such as

$$\underset{/////////}{\overset{|}{Si}\text{-OH}} + HOSiR_3 \longrightarrow \underset{///////////}{\overset{|}{Si}\text{-O-}SiR_3} + H_2O$$

$$\underset{////////}{\overset{|}{Si}\text{-OH}} + ClSiR_3 \longrightarrow \underset{///////////}{\overset{|}{Si}\text{-O-}SiR_3} + HCl$$

Thus silica gel surfaces can be modified to have affinities for specific classes of molecules. This greatly expands the range of stationary phases that can be used for chromatography. The surface O—H groups on glass can be modified similarly, and glassware treated in this manner is sometimes used in the laboratory when proton-sensitive compounds are being studied.

6.10 Solid and molten acids in industrial processes

The Lewis viewpoint systematizes many reactions in solids and molten salt solutions. It recognizes that reactions often involve the transfer of a basic anion (typically O^{2-}, S^{2-}, or Cl^-) from one cationic acid center to another. For example, the reaction of CaO with SiO_2 to give the Ca^{2+} salt of the polyanion $[SiO_3^{2-}]_n$ can be regarded as the transfer of the base O^{2-} ion from the weak acid Ca^{2+} to the stronger acid 'Si^{4+}':

$$\mathrm{CaO} + \underset{|}{\mathrm{Si}}\text{–O–}\underset{|}{\mathrm{Si}} \longrightarrow \underset{|}{\mathrm{Si}}\text{–O–Ca}^{2+}\text{–O–}\underset{|}{\mathrm{Si}}$$

////////////   //////////////////////

This reaction is a model for slag formation, which removes silicates from the molten iron phase during reduction of iron ores in a blast furnace, and the floating of slag on top of iron is a microcosm of the core/mantle/crust division of the earth. Similar molten salt and solid reactions are involved in the formation of glass and ceramics. In these, alkali metal oxides or hydroxides transfer a basic O^{2-} ion to the acid silicate center.

Further reading

Two general treatments of acids and bases are:

R. S. Drago and N. A. Matwiyoff, *Acids and bases.* Heath, Boston (1968).

W. B. Jensen, *The Lewis acid–base concepts.* Wiley, New York (1980).

More specialized books are:

R. G. Pearson in Chapter 1, A. Scott (ed.), in *Survey of progress in chemistry,* 1. Academic Press, New York (1969). This is an account of the hard and soft classification from the originator of the terminology.

V. Gutmann, *Coordination chemistry in nonaqueous solutions.* Springer-Verlag, Berlin (1968). This monograph analyzes the role of solvents in detail.

Exercises

6.1 Sketch a diagram of the *p* block of the periodic table. Identify as many elements as you can that form Lewis acids in one of their lower oxidation states.

6.2 For each of the following processes, identify the acids and bases involved and characterize the processes as complex formation or acid–base displacement. Identify the species that exhibit Brønsted acidity as well as Lewis acidity.

(a) $SO_3 + H_2O \rightarrow HSO_4^- + H^+$
(b) $CH_3[B_{12}] + Hg^{2+} \rightarrow [B_{12}]^+ + CH_3Hg^+$; $[B_{12}]$ designates the Co-porphyrin, vitamin B_{12}
(c) $KCl + SnCl_2 \rightarrow K^+ + [SnCl_3]^-$
(d) $AsF_3(g) + SbF_5(l) \rightarrow [AsF_2]^+[SbF_6]^-(s)$
(e) Ethanol dissolves readily in pyridine.

6.3 Select the compound on each line with the named characteristic and state the reason for your choice.

(a) Strongest Lewis acid:

BF_3	BCl_3	BBr_3
$BeCl_2$	BCl_3	
$B(n\text{-}Bu)_3$	$B(t\text{-}Bu)_3$	

(b) Most basic toward $B(CH_3)_3$

Me_3N	Et_3N
$(2\text{-}CH_3)C_5H_4N$	$(4\text{-}CH_3)C_5H_4N$

6.4 Using hard–soft concepts, which of the following reactions have an equilibrium constant greater than 1? Unless otherwise stated, assume gas phase or hydrocarbon solution and 25°C.

(a) $R_3PBBr_3 + R_3NBF_3 \rightleftharpoons R_3PBF_3 + R_3NBBr_3$

(b) $SO_2 + (C_6H_5)_3PHOC(CH_3)_3 \rightleftharpoons (C_6H_5)_3PSO_2 + HOC(CH_3)_3$

(c) $CH_3HgI + HCl \rightleftharpoons CH_3HgCl + HI$

(d) $[AgCl_2]^-(aq) + 2CN^-(aq) \rightleftharpoons [Ag(CN)_2]^-(aq) + 2Cl^-(aq)$

6.5 $(CH_3)_2N—PF_2$ has two basic atoms, P and N. One is bound to B in a complex with BH_3, the other to B in a complex with BF_3. Decide which is which and state your reason.

6.6 Enthalpies of reaction of trimethylboron with NH_3, CH_3NH_2, $(CH_3)_2NH$, and $(CH_3)_3N$ are −58, −74, −81, and −74 kJ mol^{-1} respectively. Why is trimethylamine out of line?

6.7 With the aid of the E and C values in Table 6.3, discuss the relative basicity of (a) acetone and dimethylsulfoxide, (b) dimethylsulfide and dimethylsulfoxide. Comment on a possible ambiguity for dmso.

6.8 By exploiting the patterns of Lewis acid–base behavior as in Example 6.3, give balanced chemical equations for the preparation of the following compounds:

(a) $KB(C_6H_5)_4$ using $K(C_6H_5)$
(b) KPF_6 using PF_5
(c) $(CH_3)_4Sn$ using CH_3I
(d) $(SiH_3)_2O$ from SiH_3I
(e) KSO_2F from SO_2
(f) BF_3 using AsF_3

6.9 Give the equation for the dissolution of SiO_2 glass by HF and interpret the reaction in terms of Lewis and Brønsted acid–base concepts.

6.10 Al_2S_3 gives off a foul odor when it becomes damp. Write a balanced chemical equation for the reaction and discuss it in terms of acid–base concepts.

6.11 Describe the solvent properties which would (a) favor displacement of Cl^- by I^- from an acid center, (b) favor basicity toward an acid center of R_3As over R_3N, (c) favor acidity of Ag^+ over Al^{3+}, (d) promote the reaction $2FeCl_3 + ZnCl_2 \rightarrow Zn^{2+} + 2[FeCl_4]^-$. In each case, suggest a specific solvent which might be suitable.

6.12 Why are strongly acid solvents (e.g. SbF_5/HSO_3F) used in the preparation of esoteric cations like I_2^+ and Se_8^+ whereas strongly basic solvents are needed to stabilize anionic species such as S_4^{2-} and Pb_9^{4-}?

6.13 Describe the acidic or basic characteristics required of a solvent to allow (a) amphoteric behavior of AsF_5 and (b) high solubility of a BF_3. Explain why diethyl ether is a good solvent for BF_3.

6.14 The Lewis acid $AlCl_3$ assists the acylation of benzene as described in Section 6.6. Propose a mechanism for a similar reaction on an alumina surface.

6.15 Comment on the fact that the only important ore of Hg is cinnabar, HgS, while Zn occurs in nature as sulphides, silicates, carbonates, and oxides.

6.16 Water has a considerably higher boiling point than ammonia; it is also often a better solvent than ammonia for salts but a poorer solvent for nonpolar organic compounds. Is there any connection between these statements?

6.17 The f-block elements are found as M(III) lithophiles in silicate minerals. What does this indicate about their hardness?

6.18 Consider the reaction forming metasilicates from carbonates;

$$CaCO_3(s) + SiO_2(s) \rightarrow [CaSiO_3]_n + CO_2(g)$$

Identify the stronger acid between SiO_2 and CO_2.

6.19 The ores of Ti, Ta, and Nb may be brought into solution near 800°C using sodium disulfate. A simplified version of the reaction is

$$TiO_2 + Na_2S_2O_7 \rightarrow Na_2SO_4 + TiO(SO_4)$$

Identify the acids and bases.

6.20 Sketch the shapes of AsF_5 and its complex with F^- (Use VSEPR if necessary) and identify their point groups. What is the point group of $X_3B—N(CH_3)_3$ (**5**) and $Al_2Cl_6(g)$ (**6**)?

Problems

6.1 An electrically conducting solution is produced when $AlCl_3$ is dissolved in the basic polar solvent CH_3CN. Give formulas for the conducting species and describe their formation using Lewis acid-base concepts.

6.2 The complex anion $[FeCl_4]^-$ is yellow whereas $[Fe_2Cl_6]$ is reddish. Dissolution of 0.1 mol $FeCl_3(s)$ in 1 L of either $POCl_3$ or $PO(OR)_3$ produces a reddish solution which turns yellow on dilution. Titration of red solutions in $POCl_3$ with Et_4NCl solutions leads to a sharp color change (from red to yellow) at 1:1 mole ratio of $FeCl_3/Et_4NCl$. Vibrational spectra suggest that oxochloride solvents form adducts with typical Lewis acids *via* coordination of oxygen. Compare the following two sets of reactions as possible explanations of the observations:

(a) $Fe_2Cl_6 + 2POCl_3 \rightleftharpoons 2[FeCl_4]^- + 2[POCl_2]^+$
$POCl_2^+ + Et_4NCl \rightarrow Et_4N^+ + POCl_3$

(b) $Fe_2Cl_6 + 4POCl_3 \rightleftharpoons [FeCl_2(OPCl_3)_4]^+ + [FeCl_4]^-$

Both equilibria are shifted to products by dilution.

6.3 In the traditional scheme for the separation of metal ions from solution that is the basis of qualitative analysis, ions of Au, As, Sb, and Sn precipitate as sulfides but 'redissolve' on addition of excess ammonium polysulfide. In contrast, ions of Cu, Pb, Hg, Bi, and Cd precipitate as sulfides but do not redissolve. In the language of Chapter 5, the first group is 'amphoteric' for reactions involving SH^- in place of OH^-. The second group is less acidic. Locate the amphoteric boundary in the periodic table for sulfides implied by this information. Compare this with the amphoteric boundary for hydrous oxides in Fig. 5.5. Does this agree with describing S^{2-} as a softer base than O^{2-}?

6.4 SO_2 and $SOCl_2$ can undergo an exchange of radioactively labeled S. The exchange is catalyzed by Cl^- and $SbCl_5$. Suggest mechanisms for these two exchange reactions with the first step being the formation of an appropriate complex.

6.5 Using the symmetry-adapted linear combinations shown in Appendix 6, construct the molecular orbital diagram for AsF_5. Compare this to the diagram for SF_6 in Section 3.2, since SF_6 is isoelectronic with the complex AsF_6^-. Discuss the orbital changes that occur when AsF_5 adds a base.

d-Metal complexes

7

We now consider a broad class of compounds in which a central metal atom acting as a Lewis acid forms a complex with several Lewis bases. This chapter differs from the others in this part because we emphasize some new aspects of structure and bonding as well as the reactions of complexes. In this way we can present a general survey of the field known as *coordination chemistry*. Furthermore, for the first time we introduce details of the mechanisms by which chemical reactions occur, and this new theme will grow in importance later in the book.

We shall describe the structures of complexes in terms of *ligand field theory,* which develops the molecular orbital theory of polyatomic molecules we introduced in Chapter 3 by making use of the high symmetry of complexes. Ligand field theory provides an excellent example of how a single parameter—we shall come to know it as the *ligand field splitting*—can be used to correlate a wide range of properties, including structure, spectra, magnetic properties, and some aspects of thermochemistry.

Metal complexes range from hydrated metal ions to complicated metalloenzymes. We concentrate in this introductory chapter on complexes formed by the *d*-block elements, but many of the same ideas apply to complexes of other metals too. Some of the ideas we present are developed further in Chapter 14, and that chapter could be tackled immediately after this.

Structures and symmetries of complexes
7.1 Constitution and isomerism
7.2 Types of ligands and nomenclature
7.3 Chiral complexes
Bonding in complexes
7.4 Molecular orbital theory of octahedral complexes
7.5 Correlation of theory with experiment
7.6 Complexes with lower symmetry
Reactions of complexes
7.7 Coordination equilibria
7.8 Rates and mechanisms of ligand substitution
Further reading
Exercises
Problems

The credit for elucidating the principal features of the structures of *d*-metal complexes belongs to the Swiss chemist Alfred Werner (1866–1919) whose training was in organic stereochemistry. Werner combined the interpretation of optical and geometrical isomerism, patterns of reactions, and conductance data in work that remains a model of how to use chemical evidence effectively and imaginatively.

The structures of *d*-metal complexes can now be studied in many ways. When single crystals of the compound can be grown, X-ray diffraction (Box 4.1) gives precise shapes, bond distances, and angles. It is also possible to infer the geometries of complexes with long lifetimes in solution (such as the classic complexes of Co(III), Cr(III), and Pt(II) and many 4*d* and 5*d* organometallic compounds) by Werner's method of analyzing patterns of reactions. NMR (a technique we review in the *Further information* section of Chapter 9) can be used to study complexes with lifetimes longer than microseconds. Very short-lived complexes, those with lifetimes comparable to diffusional encounters in solution (a few nanoseconds), can be studied by vibrational and electronic spectroscopy.

Structures and symmetries of complexes

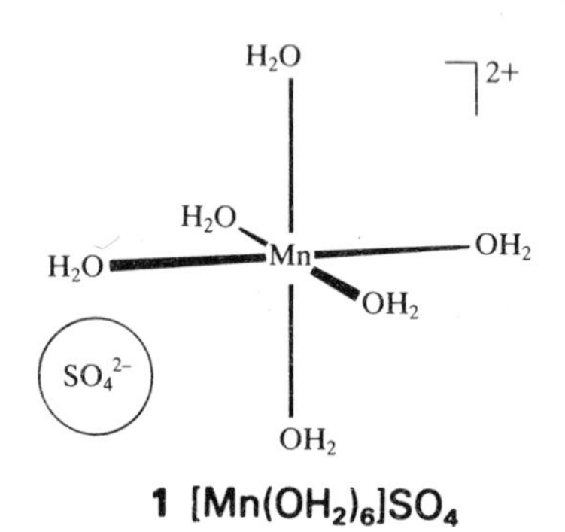

1 $[Mn(OH_2)_6]SO_4$

The **coordination sphere** of a complex consists of the ligands directly attached to the metal center, and the number of ligands in the coordination sphere is called the **coordination number** (C.N.) of the complex. The wide range of coordination numbers that can occur, from 1 to 12, distinguishes metal complexes from the simpler Lewis complexes described in Chapter 6 and is the origin of their structural richness and chemical diversity.

Although we shall concentrate on the first coordination sphere of the metal ion and its associated ligands, we should keep in mind that complex cations can associate electrostatically with anionic ligands without displacement of the ligands already present (**1**). The product of this association is called an **outer-sphere complex**. For $[M(OH_2)_6]^{2+}$ and SO_4^{2-} ions, the concentration of outer-sphere complex $[M(OH_2)_6]^{2+}SO_4^{2-}$ exceeds that of the **inner-sphere complex** $[M(OH_2)_5SO_4]$ in which the SO_4^{2-} ligand is attached to the metal ion. It is worth remembering that most methods of measuring complex formation equilibria do not distinguish outer-sphere from inner-sphere complexation but simply detect the sum of all bound ligands. Outer-sphere complexation is commonly observed between multipositive cations and multinegative anions.

7.1 Constitution and isomerism

Table 7.1 gives examples of complexes with each of the common coordination numbers, their shapes, and their point group classifications. In this section we describe some examples of each class, indicating how some are prepared and their structures identified. One feature of complexes that we shall meet almost immediately is the existence of a rich variety of isomers. A particularly common type of isomerism is **geometrical isomerism**, in which the same ligands differ in their spatial arrangement around the metal center.

Table 7.1. A brief catalog of some d^n complexes

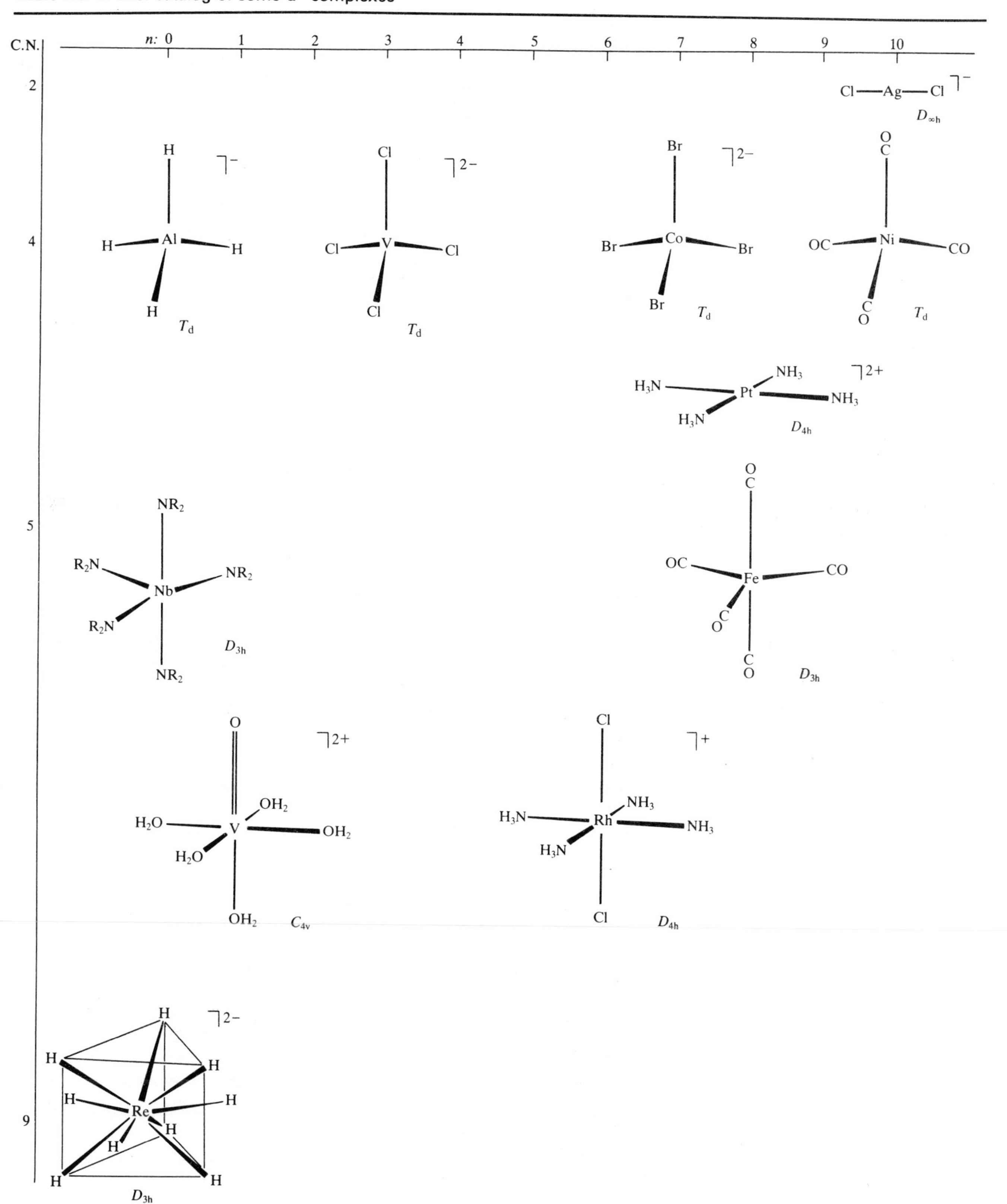

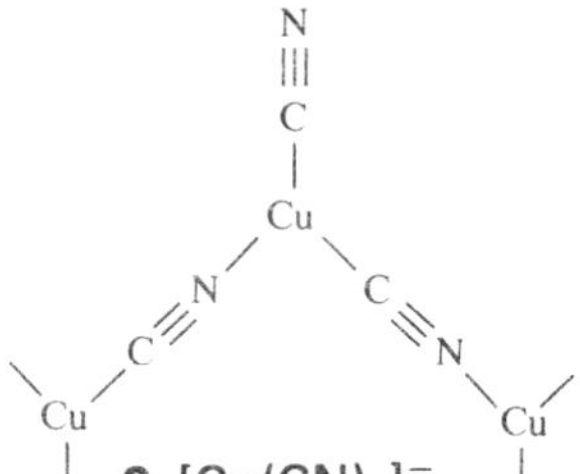

2 $[Cu(CN)_2]^-$

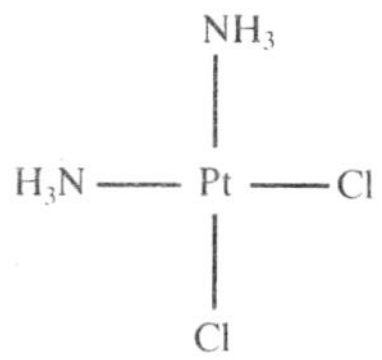

3a *cis*-$[PtCl_2(NH_3)_2]$

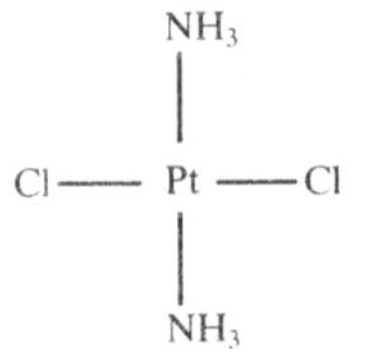

3b *trans*-$[PtCl_2(NH_3)_2]$

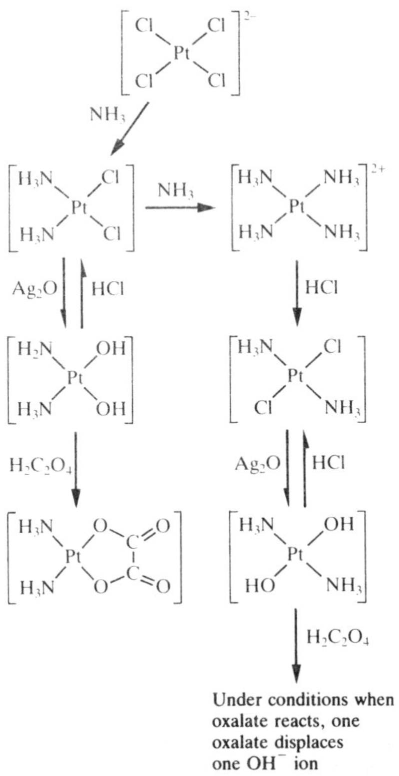

Under conditions when oxalate reacts, one oxalate displaces one OH^- ion

Fig. 7.1 The preparation of *cis*- and *trans*-diamminedichloro-platinum(II) and a chemical method for distinguishing the isomers.

Low coordination numbers

Complexes with C.N. = 1 (ML) and 2 (ML_2) are found in the gas phase at high temperatures but are rare under ordinary circumstances. The complex $[AgCl_2]^-$, which is responsible for the dissolution of AgCl in aqueous solutions containing excess Cl^- ions, is one example. The toxic complex $Hg(CH_3)_2$, which is formed by the action of the methylating enzymes of microorganisms on $Hg^{2+}(aq)$, is another.

Many two-coordinate complexes readily gain additional ligands to form four-coordinate complexes. In some cases, the empirical formula of a solid suggesting two-coordination conceals a polymer with a higher coordination number. This emphasizes that there is not always an obvious correlation between a formula and a structure. The salt $K[Cu(CN)_2]$ in the solid state, for example, contains a chain-like anion (**2**) with three-coordinate Cu atoms. Three-coordination is rare among *d*-metal complexes, and ML_3 compounds are usually chains or networks with a higher coordination number and shared ligands.

Four-coordination

Four-coordination is common. Tetrahedral complexes are favored over higher coordination numbers if the central atom is small or the ligands large (such as Cl^-, Br^-, and I^-), for then steric effects override the energy advantage of forming more metal–ligand bonds. Four-coordinate *s* and *p* block complexes with no lone pairs are almost always tetrahedral, as in the examples $[BeCl_4]^{2-}$, $[BF_4]^-$, $[ZnCl_4]^{2-}$, and $[SnCl_4]$. Tetrahedral complexes are common for oxoanions of metal atoms in high oxidation states and for halide complexes of M^{2+} ions in the first row of the *d* block. Some examples are $[VO_4]^{3-}$, $[MnO_4]^-$, $[FeCl_4]^-$, $[CoCl_4]^{2-}$, $[Ni(CO)_4]$, and $[OsO_4]$.

Werner studied[1] a series of four-coordinate Pt(II) complexes formed by the reactions of $PtCl_2$ with NH_3 and HCl. Since he was able to isolate *two* nonelectrolytes of formula $[PtCl_2(NH_3)_2]$, they cannot be tetrahedral and are, in fact, square-planar geometrical isomers. The one with like ligands on adjacent corners of the square is the *cis* isomer (**3a**) and the one with like

Example 7.1: *Identifying isomers from chemical evidence*

Using the reactions in Fig. 7.1, show how the *cis* and *trans* geometries may be assigned.

Answer. The *cis* isomer reacts with Ag_2O to lose Cl^-. The resulting hydroxo product adds one oxalic acid molecule ($H_2C_2O_4$) across the neighboring positions. The *trans* isomer loses Cl^- but the product cannot displace the two OH groups with only one $H_2C_2O_4$ molecule. A reasonable explanation is that the $H_2C_2O_4$ molecule cannot reach across the square plane to bridge two *trans* positions. This is supported by X-ray crystallography.

Exercise. The two square-planar isomers of $[PtBrCl(PR_3)_2]$ (where —PR_3 is a trialkylphosphine group) have different phosphorus NMR spectra. One (A) shows a single ^{31}P group of lines, the other (B) shows two distinct ^{31}P resonances each similar to the single resonance region of A. Which is *cis* and which is *trans*?

[1] G. B. Kauffman gives a fascinating account of the history of structural coordination chemistry in *Inorganic coordination compounds*. Wiley, New York (1981).

ligands opposite is the *trans* isomer (**3b**). Geometrical isomerism such as this is far from being of only academic interest: Pt complexes are used in cancer chemotherapy, and it is found that only *cis*-Pt(II) complexes can bind to the bases of DNA and be effective. Square-planar structures are observed for complexes of the d^8 metal ions Rh^+, Ir^+, Pd^{2+}, Pt^{2+}, Au^{3+}, and sometimes Ni^{2+}.

For simple ligands, only d^8 configurations favor square-planar over tetrahedral geometry. Examples of square-planar complexes in the first row of the *d* block generally have ligands that can form π bonds with the metal by accepting electrons from it. These include $[Ni(CN)_4]^{2-}$ and complexes such as (**4**). The four-coordinate d^8 complexes of the elements belonging to the second and third rows of the *d* block are almost invariably square planar.

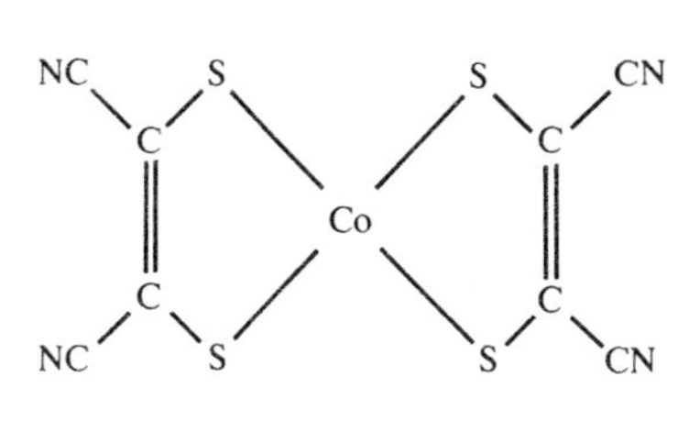

4 $[Co(C_2S_2(CN)_2)_2]$

Five-coordination

The geometry of five-coordination is a delicate balance between trigonal bipyramidal and square pyramidal, one being converted into the other by a simple distortion (Fig. 7.2). The conversion of the trigonal bipyramidal complex in the illustration to the trigonal bipyramidal isomer brings a pair of ligands from the equatorial to the axial positions and vice versa. This transposition is called a **Berry pseudorotation**. The delicacy of the energy balances involved is underlined by the fact that $[Ni(CN)_5]^{3-}$ can exist with both geometries in the same crystal. The structures of five-coordinate complexes as determined by X-ray diffraction studies of single crystals are controlled by subtle factors, and shapes intermediate between the ideal forms are common.[2]

Square pyramidal five-coordination is found among the biologically important porphyrins, where the ligand ring enforces a square-planar structure and a fifth ligand attaches above the plane. Structure (**5**) shows the active center of myoglobin, the oxygen transport protein; the lifting of the Fe atom above the plane is important to its function, as we see in Chapter 19. One effective method for inducing five-coordination is to use a ligand that contains an atom able to bind at an axial location of a trigonal bipyramid, with the remainder of the molecule extending an arm that reaches down to each of the three equatorial positions (**6**).

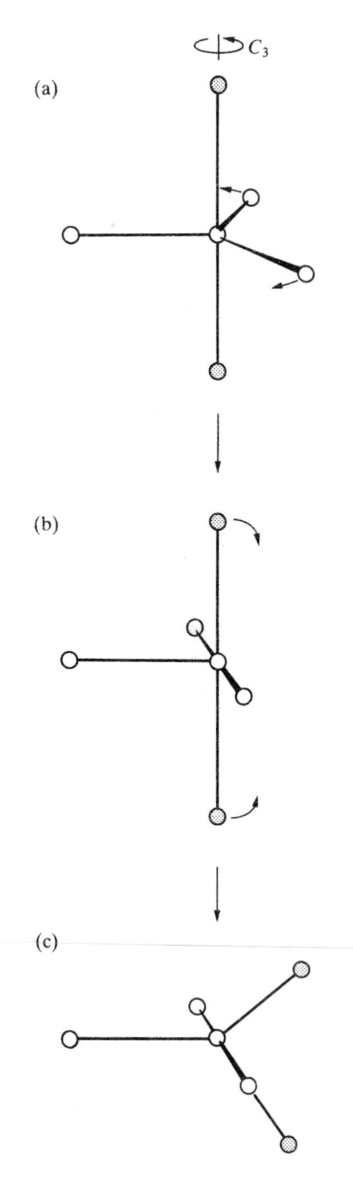

Fig. 7.2 A Berry pseudorotation in which a trigonal bipyramidal complex (top) distorts into a square pyramidal isomer and then becomes trigonal bipyramidal again, but with two initially equatorial groups now axial.

Six-coordination

Six-coordination is the most important for configurations ranging from d^1 to d^9 in the *d* block, but is much less common for complexes formed by elements in the *f* block. A few examples representative of a wide range of six-coordinate complexes that can occur are $[Sc(OH_2)_6]^{3+}$, $[Cr(NH_3)_6]^{3+}$, $[Mo(CO)_6]$, $[Fe(CN)_6]^{4-}$, and $[RhCl_6]^{3-}$. Even some halides of the *f*-block elements can display six-coordination. Full octahedral symmetry (O_h) of the coordination sphere, with all six ligands the same, is common. However, for

[2] E. L. Muetterties and L. J. Guggenberger, *J. Amer. chem. Soc.* **96**, 1748 (1974), report an elegant series of five-coordinate structures designed to show a smooth transition from ideal trigonal bipyramid (TBP) to ideal square pyramidal (SP). The compounds in question were (TBP)$[CdCl_5]^-$, $[P(C_6H_5)_5]$, $[Co(C_6H_7NO)_5]^{2+}$, $[Ni(CN)_5]^{3-}$, $[Nb(NC_5H_{10})_5]$, $[Sb(C_6H_5)_5]$ (SP).

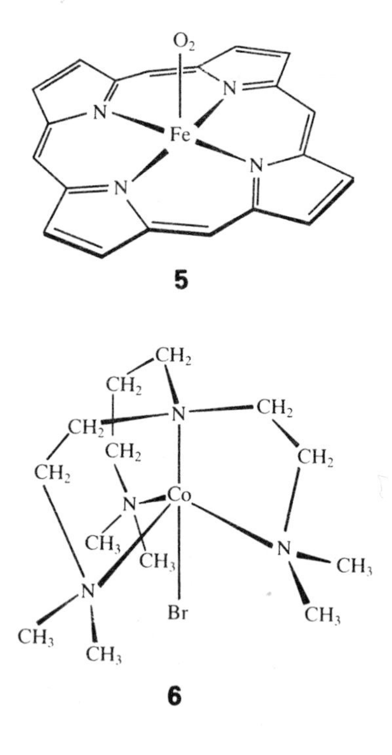

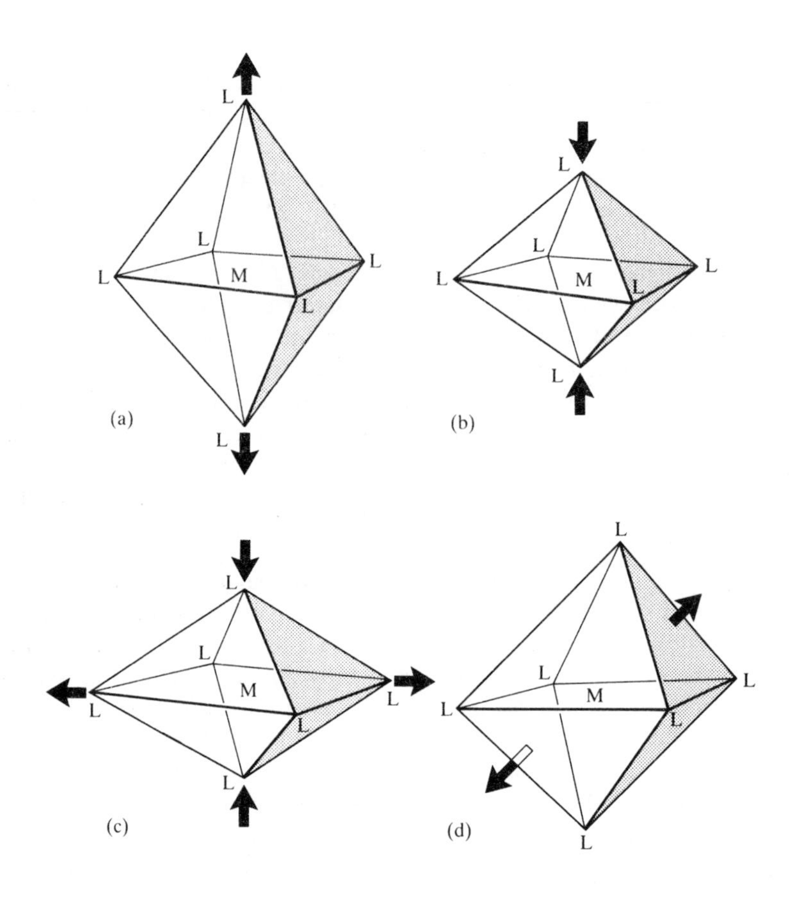

Fig. 7.3 (a) and (b) Tetragonal (D_{4h}) distortions of a regular octahedron, (c) rhombic (D_{2h}), and (d) trigonal (D_{3d}) distortions. The last can lead to a trigonal prism (D_{3h}) by a further 60° rotation of the faces containing the arrows.

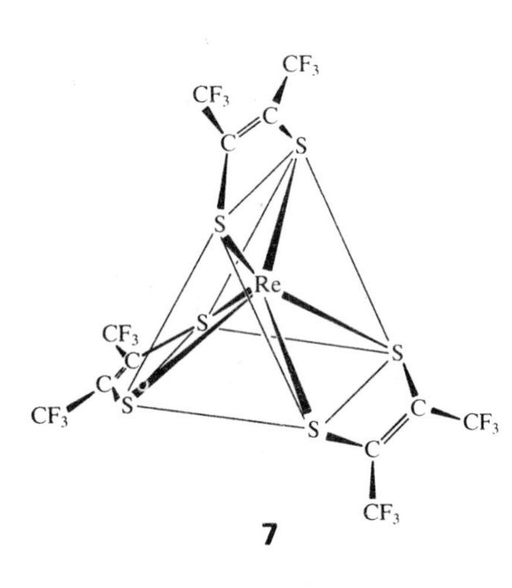

the d^9 configuration (particularly Cu^{2+} complexes) significant distortions from O_h symmetry occur even though all ligands may be identical.

The regular octahedron is important because it is also the starting point for discussions of complexes of lower symmetry, such as those shown in Fig. 7.3. The simplest distortion from O_h symmetry is tetragonal (D_{4h}), and occurs when two ligands are significantly different from the other four. Rhombic (D_{2h}) and trigonal (D_{3d}) distortions are also common. Trigonal distortion gives rise to a large family of structures that are intermediate between regular octahedral and trigonal prismatic (D_{3h}). The trigonal prism itself is rare, but was first found in MoS_2 and WS_2; it is also the shape of several complexes of formula $[M(S_2C_2R_2)_3]$ (**7**).

Geometrical isomerism in six-coordination

Geometrical isomerism is similar in six-coordinate complexes and square-planar complexes. For example, the two X ligands of an ML_4X_2 complex may be placed on adjacent octahedral positions to give a *cis* isomer or on opposite positions to give a *trans* isomer. The *trans* isomer is D_{4h} and the presence of the different ligands is equivalent to a tetragonal distortion. The *cis* isomer has C_{2v} symmetry (see **8** and **9**).

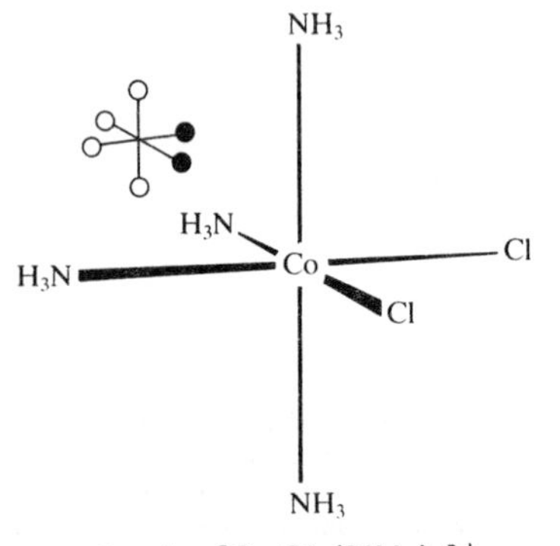

8 *cis*–$[CoCl_2(NH_3)_4]^+$

The preparation of specific isomers often requires considerable ingenuity, and it is important to acquire a sense of what is involved in studying isomerism in the laboratory. Thus, although air oxidation of Co(II) solutions containing ammonia produces six-coordinate Co(III) complexes, the number of structures obtained in this way is large because the most stable Co(II) complex, $[Co(NH_3)_6]^{2+}$, is only slowly oxidized to $[Co(NH_3)_6]^{3+}$. Bubbling air through a solution of a Co(II) salt and $(NH_4)_2CO_3$ yields $[CoCO_3(NH_3)_4]^+$ with the CO_3^{2-} ligand occupying two coordination positions; this complex can be used as a starting point for preparing others. For instance, on reaction with strong acids, the CO_3^{2-} ligand is replaced; if concentrated HCl is used, the violet *cis*-$[CoCl_2(NH_3)_4]Cl$ (**8**) can be isolated. In contrast, reaction with a mixture of HCl and H_2SO_4 gives the bright green *trans*-$[CoCl_2(NH_3)_4]Cl$ isomer (**9**).

One of the products of the oxidation of Co(II) in the presence of nitrite ions and ammonia is the yellow non-electrolyte $[Co(NO_2)_3(NH_3)_3]$. There are two ways of arranging the ligands in this complex. In one (**10a**), two ligands are *trans* to each other with the third in between; this is designated the ***mer*** isomer (for meridional). The second (**10b**), where all three ligands surround one face of the octahedron, is the ***fac*** isomer (for facial).

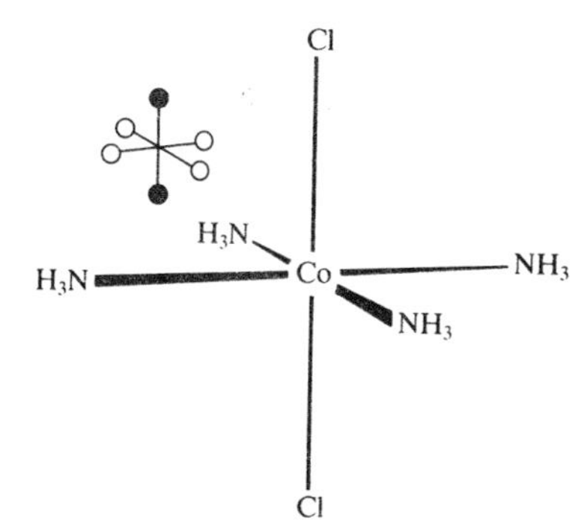

9 *trans*–$[CoCl_2(NH_3)_4]^+$

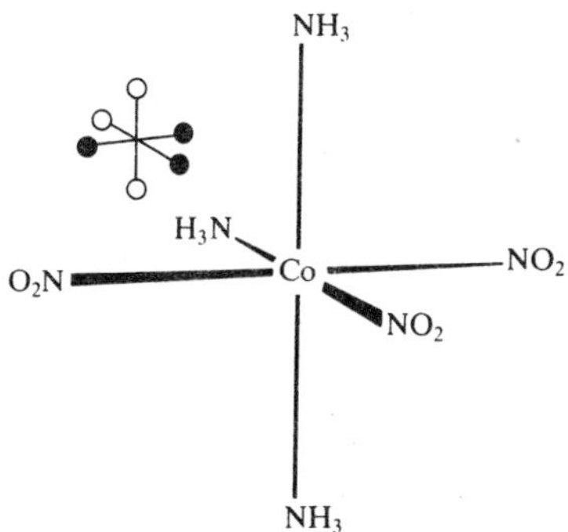

10a *mer*–$[Co(NO_2)_3(NH_3)_3]$

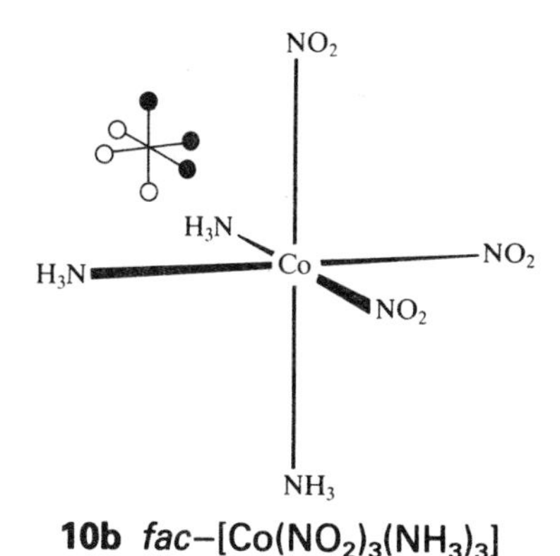

10b *fac*–$[Co(NO_2)_3(NH_3)_3]$

Example 7.2: *Identifying types of isomerism*

When the four-coordinate square-planar complex $[IrCl(PPh_3)_3]$ (where PPh_3 = triphenylphosphine) reacts with Cl_2, the six-coordinate product $[IrCl_3(PPh_3)_3]$ is formed by a reaction known as 'oxidative addition' (Section 15.13). What isomers of the product are possible?

Answer. Structures (**11a**) and (**11b**) show the arrangement of the three Cl^- ions in the *fac* and *mer* isomers respectively.

11a *fac*–$[IrCl_3(PPh_3)_3]$

11b *mer*–$[IrCl_3(PPh_3)_3]$

Exercise. When the anion of the amino acid glycine $H_2NCH_2CO_2^-$ (gly^-) is used to dissolve Co(III) oxide, both the N and an O atom of gly^- coordinate and two Co(III) nonelectrolyte *mer* and *fac* isomers of $[Co(gly)_3]$ are formed. Sketch them.

High coordination numbers

Seven-coordination is common for heavier *d*-metals with high oxidation numbers. It resembles five-coordination in the similarity in energy of its various geometries. Limiting forms include the pentagonal bipyramid and capped octahedra and trigonal bipyramids with the seventh ligand on one face. There are a number of intermediate structures and interconversions are relatively facile. Examples include $[Mo(CNR)_7]^{2+}$, $[ZrF_7]^{3-}$, $[TaCl_4(PR_3)_3]$, and $[ReOCl_6]^{2-}$ from the *d* block and $[Gd_2S_3]$ from the *f* block. A method to force seven-coordination rather than six on the lighter elements is to synthesize a ring of five donor atoms (**12**) which then occupy the equatorial positions, leaving the axial positions free for two more ligands.

12

Stereochemical nonrigidity is also shown in eight-coordination, where the same complex may be square antiprismatic (**13**) in one crystal but dodecahedral (**14**) in another.

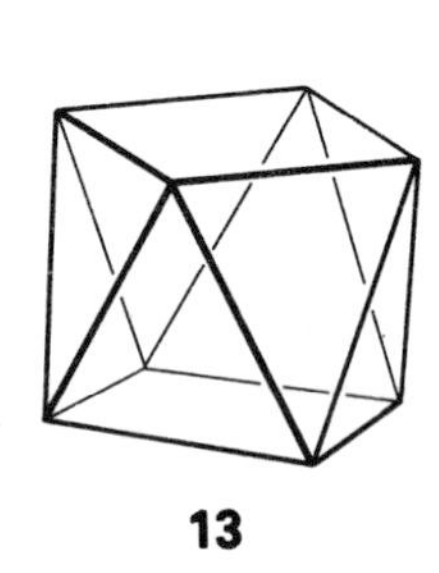

13

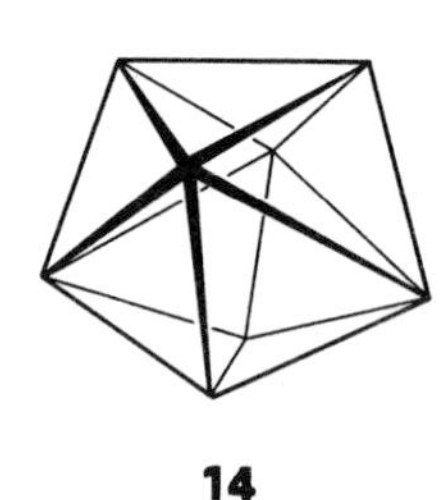

14

Nine-coordination is important in the structures of *f*-block elements, for their relatively large ions can act as host to a large number of ligands. Examples include $[Nd(OH_2)_9]^{3+}$ and the MCl_3 solids, with M ranging from La to Gd. An example of nine-coordination in the *d* block is $[ReH_9]^{2-}$ (Table 7.1), which has small enough ligands for this coordination number to be feasible. Coordination numbers 10 and 12 are rare: they are encountered in the *f* block but are rare in the *d* block. Examples include $[Ce(NO_3)_6]^{2-}$ (**15**), which is formed in the reaction of Ce(IV) salts with HNO_3 where each ligand is bonded to the metal atom by two O atoms, and the ten-coordinate complex $[Th(ox)_4(OH_2)_2]^{4-}$, in which each oxalate ion provides two O atoms.

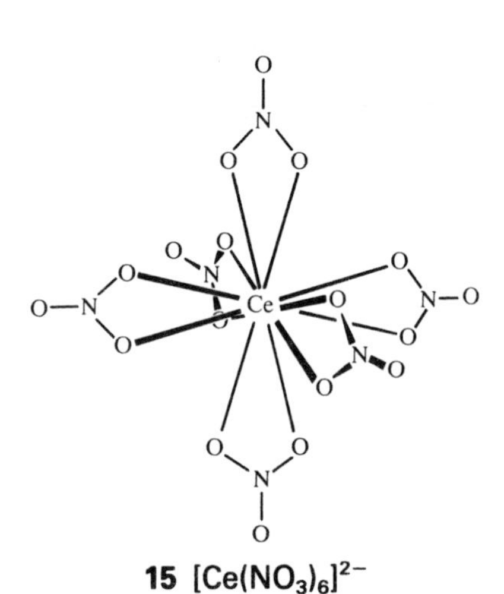

15 $[Ce(NO_3)_6]^{2-}$

Polymetallic complexes

A considerable amount of attention has been given recently to the synthesis of **polymetallic complexes**, which are complexes that contain more than one metal atom. In some cases, the metal atoms are held together by bridging ligands, in others there are direct metal–metal bonds, and in yet others there are both types of link. The term **cluster** is usually reserved for polymetallic complexes in which there are direct metal–metal bonds. Some of these polymetallic complexes are illustrated in Fig. 7.4.

Polymetallic complexes arise in the first *d* series through the formation of hydroxo and oxo bridges, as we saw in Chapter 5 when discussing the polymerization of aqua cations (Section 5.8). The Cu complex in Fig. 7.4a, with acetate ion bridges, is an analogous example. Some biologically important Fe-based mediators of electron transfer are built from similar S-bridged species. A synthetic analog of these biological complexes is the compound shown in Fig. 7.4b. In Chapter 16 we shall see the great importance of CO as a bridging ligand in complexes of the type illustrated in Fig. 7.4c.

In the second and third *d* series, clusters with metal–metal bonds become prominent. A very simple example from classical chemistry is the Hg(I) cation Hg_2^{2+}, and complexes derived from it, such as Hg_2Cl_2. Metal–metal

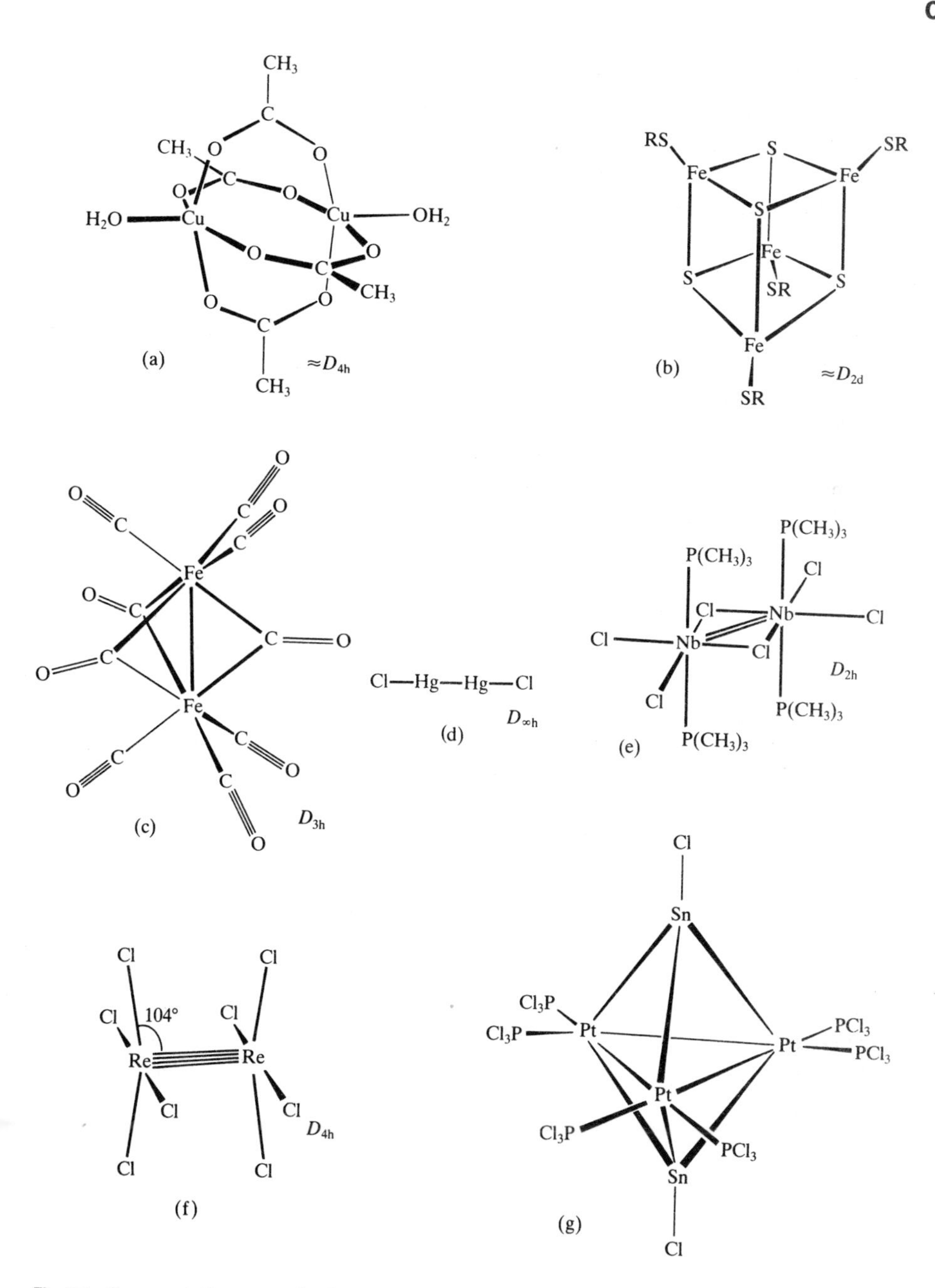

Fig. 7.4 Representative types of polymetallic complex. (a) The copper(II) acetate dimer. (b) A synthetic Fe—S complex which models biochemically important electron-transfer agents. (c) A carbonyl with bridging CO ligands (the Fe—Fe distance is short enough for Fe—Fe bonding, but there is controversy about its importance). (d) Mercury(I) chloride, with an Hg—Hg bond. (e) An Nb complex with a Nb═Nb double bond. (f) $[Re_2Cl_8]^{2-}$, the first stable complex containing a quadruple metal–metal bond. (g) A complex with both Pt—Pt and Pt—Sn bonds.

bonded complexes range from the d^2 Nb(II) complex with a Nb═Nb double bond (Fig. 7.4e) to the d^7 green aqua ion Rh_2^{4+}; in this case, the d^7 configuration is stabilized by Rh—Rh bonding. The Re(IV) chlorine complex shown in Fig. 7.4f is regarded as having a Re≣Re quadruple bond. A final example (at this stage; there will be more in Chapter 16) is the cluster shown in Fig. 7.4g, which has both Pt—Pt and Pt—Sn bonds.

7.2 Types of ligands and nomenclature

We shall outline here a few key ideas of nomenclature and introduce a number of common ligands. We begin by considering complexes that contain only **monodentate ligands**, ligands that have only one point of attachment to the metal atom, as distinct from **polydentate ligands** which have more than one.

Monodentate ligands

Complexes are named with their ligands (Table 7.2) in alphabetical order. The ligand names are followed by the name of the metal with either its oxidation number in parentheses, as in Co(II), or with the overall charge on the complex specified in parentheses, as in hexacyanocobaltate(3^-). We add the suffix -ate to the name of the metal (sometimes in its Latin form) if the complex is an anion, as in hexacyanoferrate(II) for $[Fe(CN)_6]^{4-}$.

We should write the formula of a complex inside square brackets whether it is charged or not; however, in casual usage, neutral complexes are often written without brackets, as in $Ni(CO)_4$. The metal symbol is given first, then the anionic ligands, and finally the neutral ligands. This order is sometimes varied in order to clarify which ligand is involved in a reaction. Polyatomic ligand formulas are sometimes written in an unfamiliar sequence

Table 7.2. Typical ligands and their names

Name	Formula	Abbreviation	Classification*
Acetylacetonato	$CH_3COCHCOCH_3^-$	acac	B(O)
Ammine	NH_3		M(N)
2,2-Bipyridine	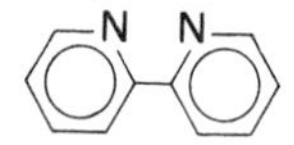 (two linked pyridine rings: N N)	bipy	B(N)
Carbonato	CO_3^{2-}		M(O), B(O)
Chloro	Cl^-		M(Cl)
Diethylenetriamine	$NH(C_2H_4NH_2)_2$	dien	T(N)
Ethylenediamine	$H_2NCH_2CH_2NH_2$	en	B(N)
Ethylenediaminetetraacetato	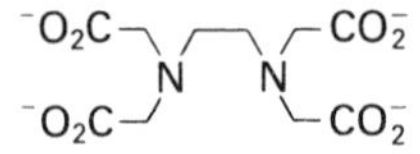 $(^-O_2CCH_2)_2NCH_2CH_2N(CH_2CO_2^-)_2$	edta	H(N, O)
Glycinato	$NH_2CH_2CO_2^-$	gly	B(N, O)
Maleonitriledithiolato	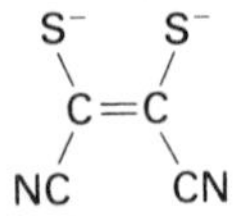 $(^-S)(NC)C{=}C(CN)(S^-)$	mnt	B(S)
Nitrilotriacetato	$N(CH_2CO_2^-)_3$	nta	Q(N, O)
Oxalato	$C_2O_4^{2-}$	ox	B(O)
Tetraazacyclotetradecane	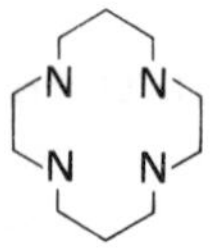 (14-membered ring with four N)	cyclam	Q(N)
Triethylenetetramine	$N(C_2H_4NH_2)_3$	tren	Q(N)

* M: monodentate, B: bidentate, T: tridentate, Q: quadridentate, H: hexadentate.

(as for OH_2 in $[Fe(OH_2)_6]^{2+}$) to emphasize the Lewis acid–base character of their bonding to the metal atom.

The number of occurrences of a ligand in a complex is indicated by the prefixes mono-, di-, tri-, and tetra-. Where confusion with the names of ligands is likely (as with ethylenediamine) we use the alternative prefixes bis-, tris-, and tetrakis-, with the ligand name in parentheses. For example, dichloro- or dimethylamino- are clear but bis(methylamine) designates more clearly that there is only one methyl group on each N. Ligands that bridge two metal centers are denoted by μ in front of the name of the relevant ligand. The number of metal atoms in a bridged complex is indicated by the prefixes di-, tri-, etc.

Ambidentate ligands

As we saw in connection with Lewis bases in general (Section 6.3), bases with alternative donor atoms are called **ambidentate**. An example is the thiocyanate ion (NCS^-) which can attach to Cr(III) either by the hard N atom, to give isothiocyanato complexes, or by the softer S atom, to give thiocyanato complexes. Compounds with S—Cr bonds can form in fast reactions in which Cr(II) attacks a Co(III)—NCS complex, as in

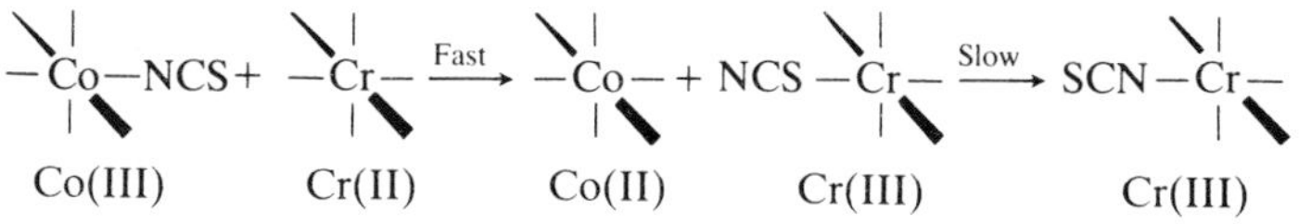

$$-Co-NCS + -Cr- \xrightarrow{\text{Fast}} -Co- + NCS-Cr- \xrightarrow{\text{Slow}} SCN-Cr-$$

Co(III) Cr(II) Co(II) Cr(III) Cr(III)

These Cr(III)—SCN thiocyanato complexes are unstable and rearrange to give the Cr(III)—NCS isothiocyanato complex. Two more examples of ambidentate ligands are $—NO_2^-$ and $—ONO^-$, and $—SO_3^{2-}$ and $—OSO_2^{2-}$.

The existence of ambidentate character gives rise to the possibility of **linkage isomerism**, in which the same ligand is linked through alternative atoms. This type of isomerism accounts for the red and yellow isomers of the formula $[CoNO_2(NH_3)_5]^{2+}$. The red compound has a Co—O link (**16a**) and is called a nitrito complex. The yellow isomer, which forms from the unstable red form on standing, has a Co—N link (**16b**) and is a nitro complex.

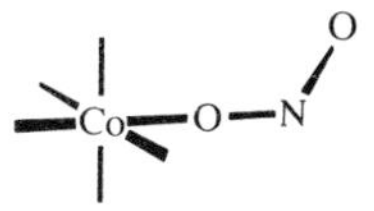

16a

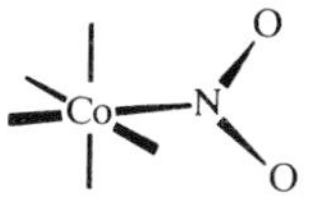

16b

Example 7.3: *Naming complexes*

Name (a) $[Cr(edta)]^-$, (b) *trans*-$[PtCl_2(NH_3)_4]^{2+}$, (c) $[Ni(CO)_3(py)]$.

Answer. (a) This complex uses hexadentate edta as the sole ligand. The four negative charges of the ligand make an anion of Cr(III) with a single negative charge, namely ethylenediaminetetraacetatochromate(III). (b) The complex is a cation despite its two anionic ligands. Thus, the Pt oxidation number is +4. Following the alphabetical order rules, the name is *trans*-tetraamminedichloroplatinum(IV). (c) All the ligands are neutral, so the oxidation number of Ni is 0; py is pyridine. The name of the complex is tricarbonylpyridinenickel(0).

Exercise. Give formulas corresponding to the following names: (a) *cis*-diaquadichloroplatinum(II); (b) diamminetetrakis(isothiocyanato)chromate-(III); (c) tris(ethylenediamine)rhodium(III).

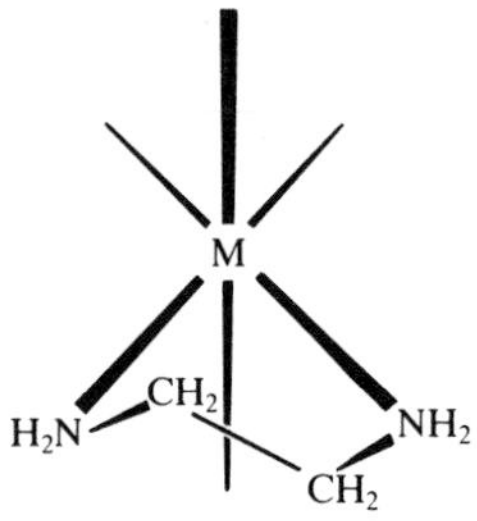

Chelating ligands

Polydentate ligands are often **chelating** (from the Greek for claw) in that they can form a ring that includes the metal atom. The resulting complex is called a **chelate**. An example is the bidentate ligand ethylenediamine ($NH_2CH_2CH_2NH_2$, en), which forms a five-membered ring (**17**). A more complex example is the hexadentate ligand ethylenediaminetetraacetic acid (edta) which is used to trap metal ions (**18**).

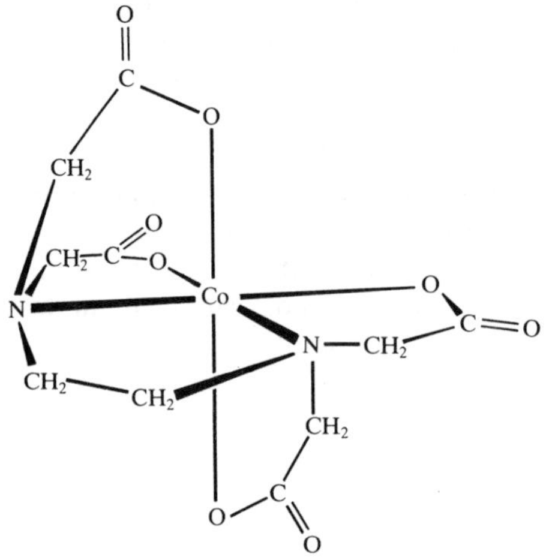

17

18 $[Co(edta)]^-$

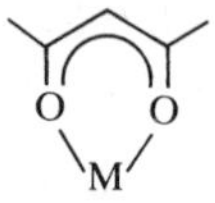

19

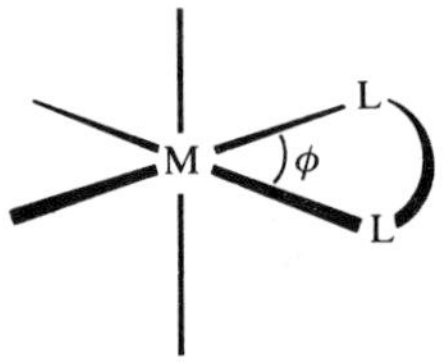

20

Ring stereochemistry

The L—M—L angle in an octahedral complex is 90°. In a chelate formed from saturated C and N centers, such as (**17**), the five-membered ring can fold into a conformation that preserves the tetrahedral angles within the ligand and yet still achieve the L—M—L 90° angle. Six-membered rings are reasonably stable and may be favored if electron delocalization (by conjugation of single and double bonds) can then occur. β-Diketones, for example, coordinate as the anions of their enols in six-membered ring structures (**19**). Amino acids that can form five- or six-membered rings also chelate readily.

The degree of strain in a chelating ligand can be expressed in terms of the **bite angle** ϕ (**20**), the L—M—L angle in the chelate ring. The small bite angle for ligands with short donor atom separations is one of the main causes of distortion from octahedral toward trigonal symmetry in six-coordinate complexes (**21**). However, steric factors do not explain everything. Dithiolene ligands (like those in **7**) can produce complexes that are almost trigonal prismatic even though the donor separation is large. In this case it is thought that S—S bonding straps the donor atoms together and hence favors a twisting of the octahedron.

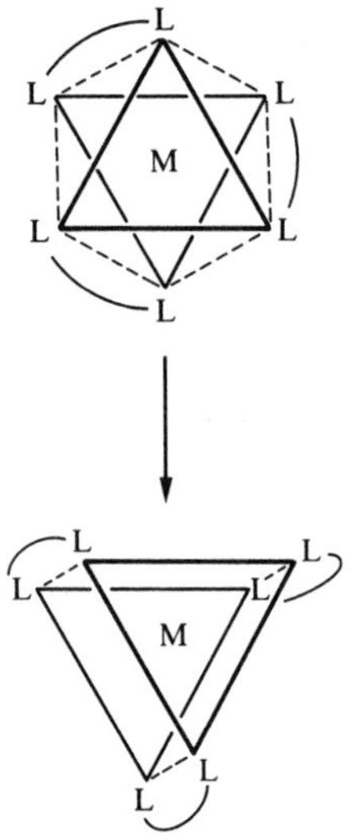

A special case of the chelation occurs with a **macrocyclic ligand**, a polydentate ligand in which several donor atoms form a large ring even before the complex is formed. An example is the porphyrin ring (**22**), which is found (in modified forms) complexed with Fe at the oxygen-binding site

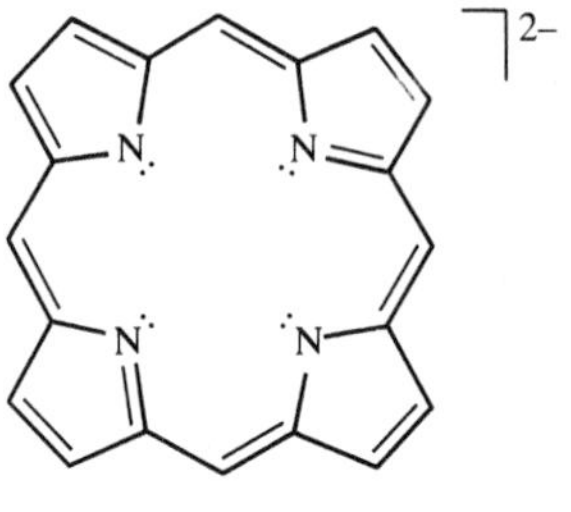

21

22 Porphyrin

of hemoglobin and with Mg in chlorophyll. A number of electron-transfer enzymes are Fe porphyrins. A saturated analog of the porphyrin ligand is tetraazacyclotetradecane (called familiarly 'cyclam', Table 7.2).

Template synthesis

A metal ion such as Ni(II) can be used to assemble a group of ligands which then undergo a condensation reaction to form a macrocyclic ligand. (In a condensation reaction a bond is formed between two molecules, and a small molecule—often H_2O—is eliminated.) This trick, which is called **template synthesis**, can be applied to produce a surprising variety of macrocyclic ligands. The most generally useful condensation reaction is the Schiff's base condensation of an amine and a ketone, of which an example is

$$(CH_3)_2C{=}O + H_2NCH_3 \rightarrow (CH_3)_2C{=}NCH_3 + H_2O$$

(A Schiff's base, an *anil*, is the product $R_2C{=}N{-}R'$.) A good example of a template synthesis is the reaction that follows coordination of one butadione molecule and two 1,2-aminoethylthiol molecules to Ni(II). This complex then condenses with 1,2-di(bromomethyl)benzene to form a macrocycle:

$$CH_3COCOCH_3 + 2NH_2CH_2CH_2SH + Ni^{2+} \longrightarrow$$

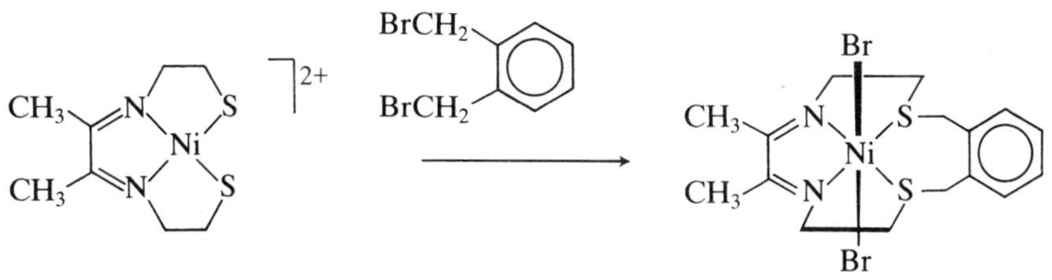

7.3 Chiral complexes

A **chiral** complex is a complex that is not superimposable on its own mirror image. The formal criterion of chirality (Section 2.4) is the absence of an axis of improper rotation (S_n, an n-fold axis in combination with a horizontal mirror plane). The presence of such a symmetry element is implied by the presence of either a mirror plane (which is equivalent to an S_1 axis) or a center of inversion (which is equivalent to an S_2 axis), and if either of these elements is present the complex is achiral. As remarked in Section 2.4, we must also be alert for higher-order axes of improper rotation (particularly S_4), since the presence of any S_n axis implies achirality.

Optical isomerism

The existence of a pair of distinct chiral isomeric complexes that are each other's mirror image (like a right and left hand), and which have lifetimes that are long enough for them to be separable, is called **optical isomerism**. The two mirror-image isomers jointly make up an **enantiomeric pair**. Optical isomers are so called because they are optically active, in the sense that one enantiomer rotates the plane of polarized light in one direction and the other rotates it through an equal angle in the opposite direction.

With Co(III), for example, ethylenediamine forms a violet and green pair of complexes, the *cis* and *trans* isomers of dichlorobis(ethylenediamine)cobalt(III), $[CoCl_2(en)_2]^+$. (It also forms the yellow complex tris(ethylenediamine)cobalt(III) ion $[Co(en)_3]^{3+}$.) As can be seen from Fig. 7.5, the *cis* isomer of the bis complex has a nonsuperimposable mirror image. It is therefore chiral and hence (since the complexes are long-lived) optically active. The *trans* isomer of the bis complex has a mirror plane; it is achiral and optically inactive.

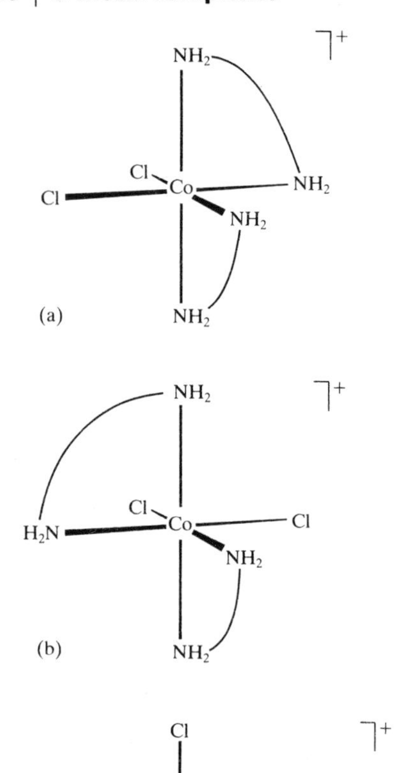

Fig. 7.5 (a) and (b) Enantiomers of *cis*-$[CoCl_2(en)_2]^+$ and (c) the nonchiral *trans* isomer. The curves represent the —CH_2—CH_2— bridges in the en ligands. The C atoms of the en ligand do not lie in the mirror plane, but the conformation fluctuates rapidly.

Example 7.4: *Detecting chirality*

Which of the complexes (a) $[Cr(edta)]^-$, (b) $[Ru(bipy)_3]^{2+}$, (c) $[PtCl(dien)]^+$ are chiral?

Answer. The complexes are shown schematically in (**23**) and (**24**). Neither (**23**) nor (**24**) has a mirror plane or a center of inversion; so both are chiral (they also have no higher S_n axis); (**25**) has a plane of symmetry and hence is achiral. (Although the CH_2 groups in a dien ligand are not in the mirror plane, they fluctuate rapidly above and below it.)

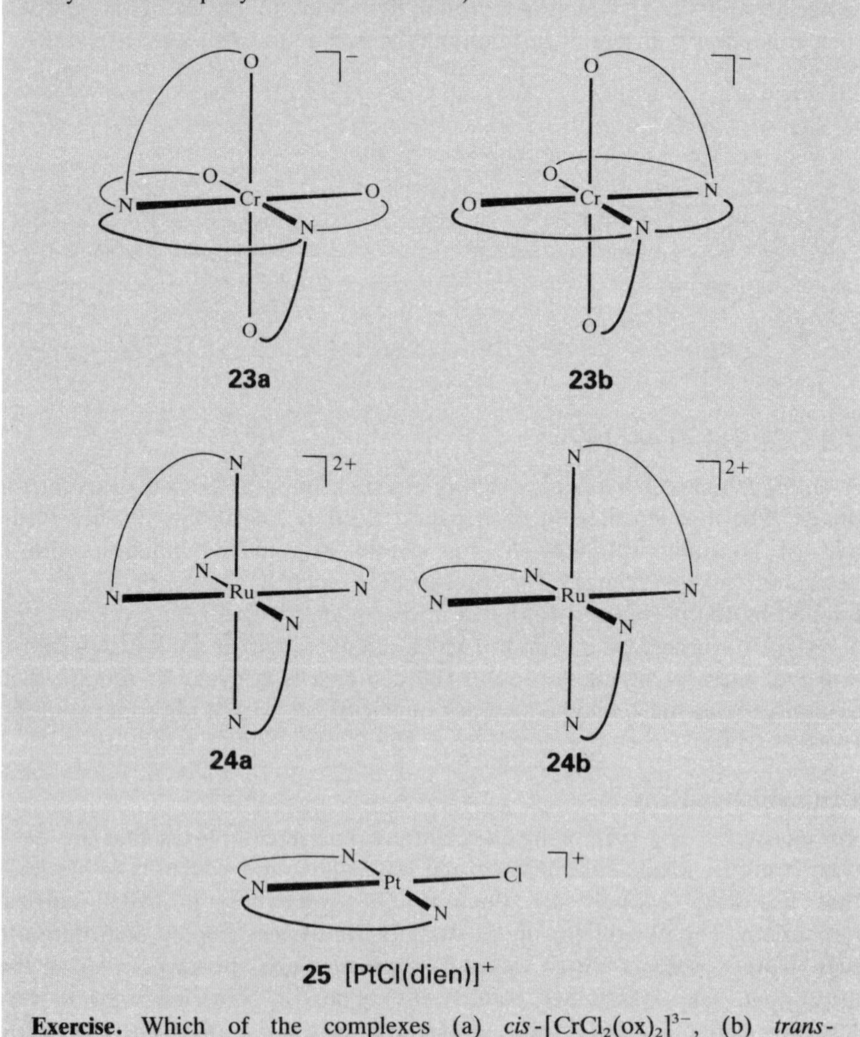

25 $[PtCl(dien)]^+$

Exercise. Which of the complexes (a) *cis*-$[CrCl_2(ox)_2]^{3-}$, (b) *trans*-$[CrCl_2(ox)_2]^{3-}$, (c) *cis*-$[RhH(CO)(PR_3)_2]$ are chiral?

The absolute configurations of chiral complexes are described by imagining a view looking along a threefold rotation axis of a regular octahedron (Fig. 7.6) and noting the handedness of the helix formed by the ligands. Left rotation of the helix is then designated Λ and right rotation Δ. The designation of the absolute configuration must be distinguished from the experimentally determined direction in which an isomer rotates polarized light: some Δ compounds rotate in one direction, others rotate in the opposite direction and the direction may change with wavelength. The isomer that rotates to the right (when viewed into the oncoming beam) at a specified wavelength is called *d*- or (+)- and the one rotating the plane to the left is called *l*- or (−)-.

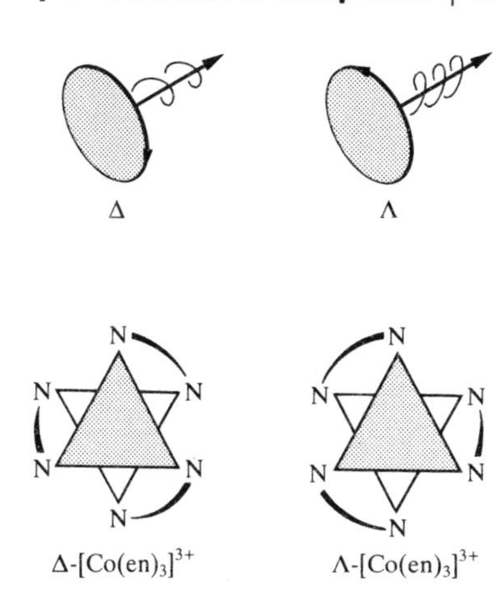

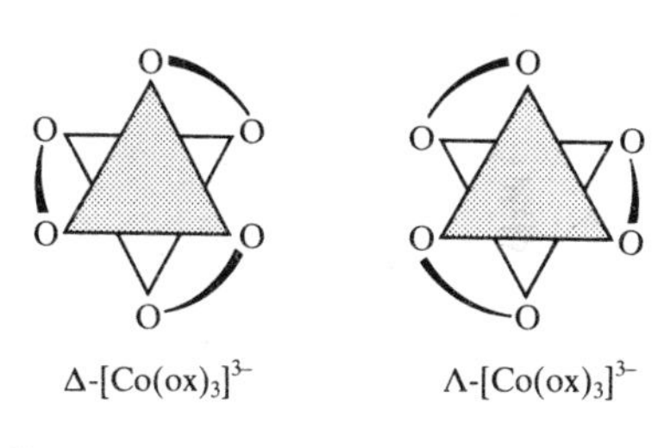

Fig. 7.6 Absolute configurations of $[M(L—L)_3]$ complexes; Δ is a right-hand screw and Λ is a left-hand screw, as is indicated by the diagrams at the top of the figure by the direction that a screw would turn when being driven in the direction shown.

Resolution of enantiomers

Optical activity is the only manifestation of chirality when a single chiral center exists in a compound. However, as soon as more than one chiral center is present, other physical properties—such as solubility and melting points—are affected. One method of resolving a pair of enantiomers into the individual isomers is therefore to prepare **diastereomers**. As far as we need be concerned, these are isomeric compounds that contain two chiral centers, one being of the same absolute configuration in both compounds and the other being enantiomeric between the two compounds. An example is the salt of an enantiomeric pair with an optically pure second species, and hence of composition $(\Delta-A)(\Delta-B)$ and $(\Lambda-A)(\Delta-B)$. Since diastereomers differ in physical properties, they are separable by conventional techniques. For instance, the resolution of Δ-$[Co(en)_3]^{3+}$ from Λ-$[Co(en)_3]^{3+}$ can be achieved by forming salts with the naturally chiral anion Δ-bromocamphorsulfonate. The solubility of (Δ-$[Co(en)_3]$)(Δ-Br-camphorsulfonate)$_3$ differs from that of (Λ-$[Co(en)_3]$)(Δ-Br-camphorsulfonate)$_3$. The diastereomers can be separated by fractional crystallization and then the $[Co(en)_3]^{3+}$ isomers isolated by conversion to chlorides.

Bonding in complexes

One feature that puzzled Werner and his contemporaries was how the metal ion manages to bind so many ligands. This is no longer a puzzle from the viewpoint of molecular orbital theory: with several metal *d* orbitals available in the valence shell, enough delocalized molecular orbitals can be built to accommodate all the electrons necessary to bind the ligands.

7.4 Molecular orbital theory of octahedral complexes

A *d*-metal atom can utilize up to nine orbitals (one *s*, three *p*, and five *d*) to form molecular orbitals. For coordination number N (and considering only N appropriately orientated ligand orbitals), N bonding and N antibonding orbitals can be formed, with $9-N$ *d* orbitals remaining unused as nonbonding orbitals. Thus a complex with N metal–ligand bonds has the molecular

orbital scheme

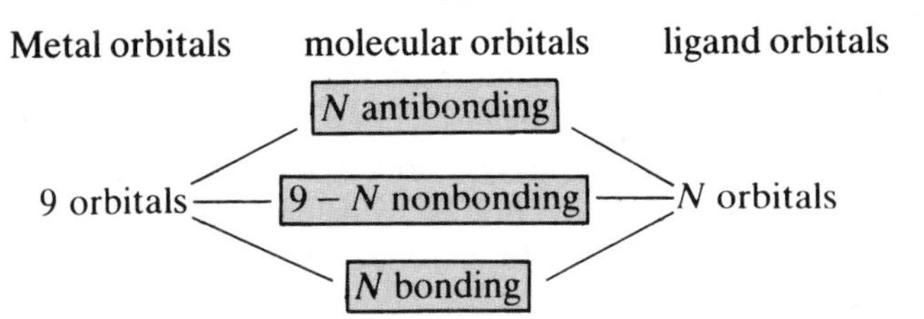

(In the case of six-coordination, this scheme corresponds to six bonding orbitals, three nonbonding orbitals, and six antibonding orbitals.)

Filling the N bonding and $9-N$ nonbonding orbitals (but none of the antibonding orbitals) requires nine electron pairs. This conclusion suggests an **18-electron rule** optimum for d-block complexes as the analog of the Lewis octet rule for main-group elements. The rule is satisfied by many of the stable diamagnetic (all electrons paired) d-block complexes. These include d^6 complexes, such as those of Co(III) amines, metal carbonyls, and the vast majority of organometallic compounds (Chapter 16). However, exceptions to the 18-electron rule are much more numerous than deviations from the octet rule for the s and p blocks, since antibonding orbitals are often readily accessible and vacancies in the nonbonding orbitals do little to destabilize the complex.

σ Bonding

We begin the systematic discussion of the electronic structures of complexes by imagining an octahedral complex in which each ligand has a single valence orbital directed toward the central metal atom with local σ symmetry around the M—L axis. Examples include the isolobal NH_3 molecule and F^- ion.

In an octahedral (O_h) environment, the metal orbitals divide by symmetry into four sets (Fig. 7.7 and Appendix 6):

Metal orbital	Symmetry label
s	a_{1g} (non-degenerate)
p_x, p_y, p_z	t_{1u} (triply degenerate)
d_{xy}, d_{yz}, d_{zx}	t_{2g} (triply degenerate)
$d_{x^2-y^2}, d_{z^2}$	e_g (doubly degenerate)

Group theory shows that $d_{x^2-y^2}$ and d_{z^2} form a doubly degenerate pair even though this is not obvious from the diagrams.

We can form six symmetry-adapted linear combinations of the six ligand orbitals. These can be taken from Appendix 6 and are shown in Fig. 7.7 together with the metal orbitals of corresponding symmetry. One (unnormalized) ligand combination is a nondegenerate a_{1g} orbital:

$$a_{1g} \quad \sigma_1+\sigma_2+\sigma_3+\sigma_4+\sigma_5+\sigma_6$$

where the σ_i denote ligand σ orbitals. Three form a triply degenerate t_{1u} set:

$$t_{1u} \quad \sigma_1-\sigma_3, \qquad \sigma_2-\sigma_4, \qquad \sigma_5-\sigma_6$$

The remaining two form a doubly degenerate e_g pair:

$$e_g \quad \sigma_1-\sigma_2+\sigma_3-\sigma_4, \qquad 2\sigma_6+2\sigma_5-\sigma_1-\sigma_2-\sigma_3-\sigma_4$$

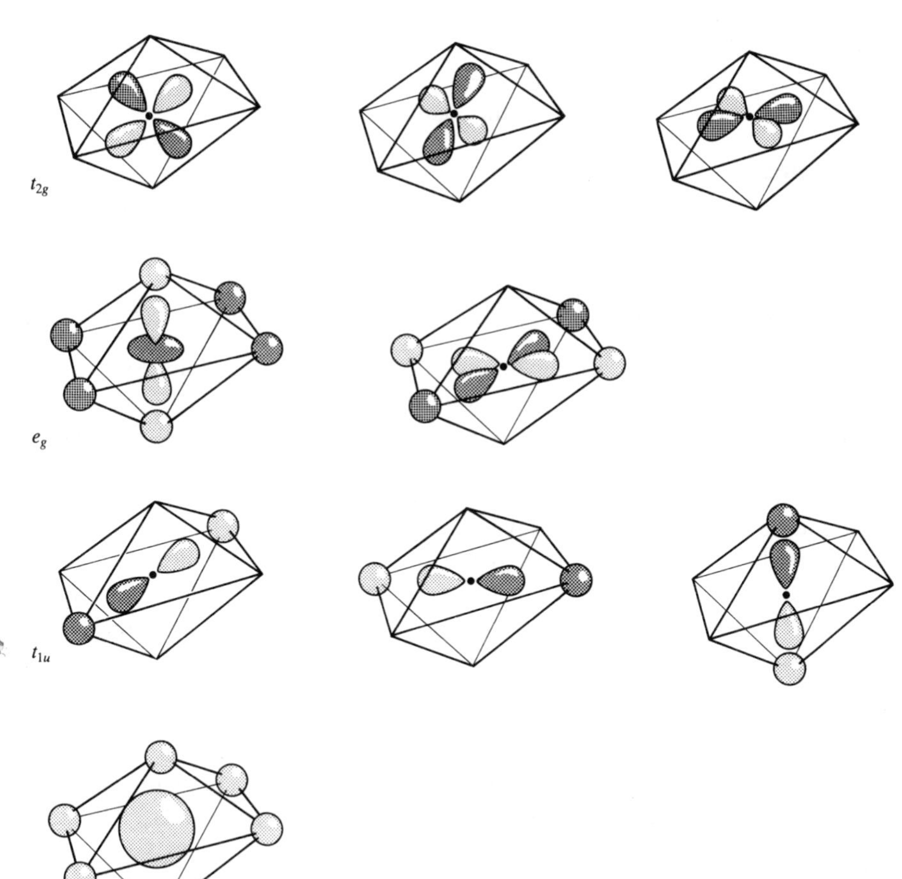

Fig. 7.7 Symmetry-adapted combinations of ligand σ orbitals (represented here by spheres) in an octahedral complex. For symmetry adapted orbitals in other point groups, see Appendix 6.

There is no combination of ligand σ orbitals that has the symmetry of the metal t_{2g} orbitals, so the latter do not participate in σ bonding.[3]

Molecular orbital calculations (adjusted to agree with experimental data) result in the molecular orbital energy level diagram shown in Fig. 7.8. For NH_3, F^-, and most other ligands, the ligand σ orbitals are derived from atomic valence orbitals that lie well below the metal *d* orbitals in energy. As a result, the six bonding molecular orbitals of the complex are mainly ligand in character. Their energies increase as the number of nodes increases. These six bonding orbitals can accommodate the 12 electrons provided by the six ligand lone pairs. The number of electrons to accommodate in addition to these depends on the number of *d* electrons supplied by the central metal ion.

The frontier orbitals of the complex are the nonbonding t_{2g} orbitals (the HOMO, purely metal in character) and the antibonding e_g orbitals (the

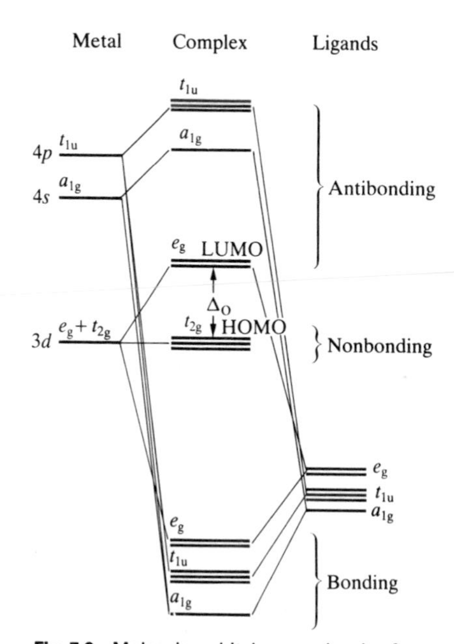

Fig. 7.8 Molecular orbital energy levels of a typical octahedral complex.

[3] The normalized forms of the orbitals (with overlap neglected) are:

$$a_{1g} \quad (\tfrac{1}{6})^{1/2}(\sigma_1 + \sigma_2 + \sigma_3 + \sigma_4 + \sigma_5 + \sigma_6)$$

$$t_{1u} \quad (\tfrac{1}{2})^{1/2}(\sigma_1 - \sigma_3), \qquad (\tfrac{1}{2})^{1/2}(\sigma_2 - \sigma_4), \qquad (\tfrac{1}{2})^{1/2}(\sigma_5 - \sigma_6)$$

$$e_g \quad (\tfrac{1}{4})^{1/2}(\sigma_1 - \sigma_2 + \sigma_3 - \sigma_4) \qquad (\tfrac{1}{12})^{1/2}(2\sigma_6 + 2\sigma_5 - \sigma_1 - \sigma_2 - \sigma_3 - \sigma_4).$$

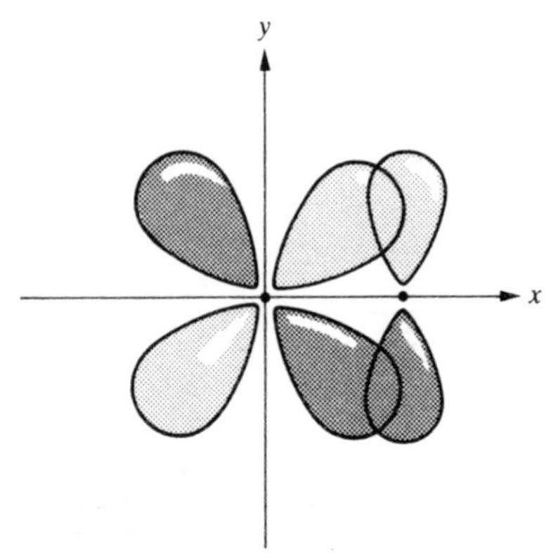

Fig. 7.9 The π overlap that may occur between a ligand *p* orbital perpendicular to the M–L axis and a metal d_{xy} orbital.

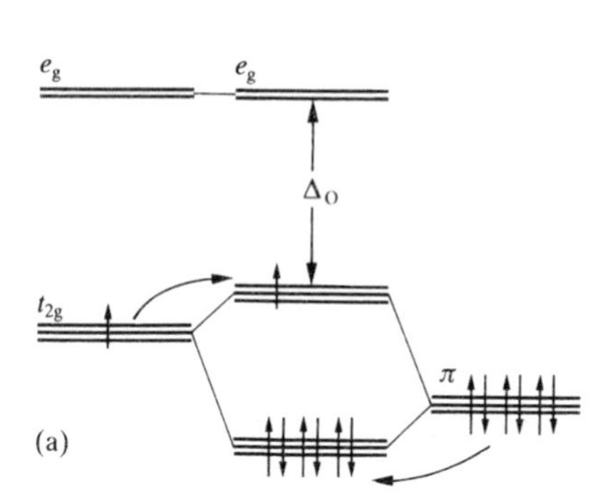

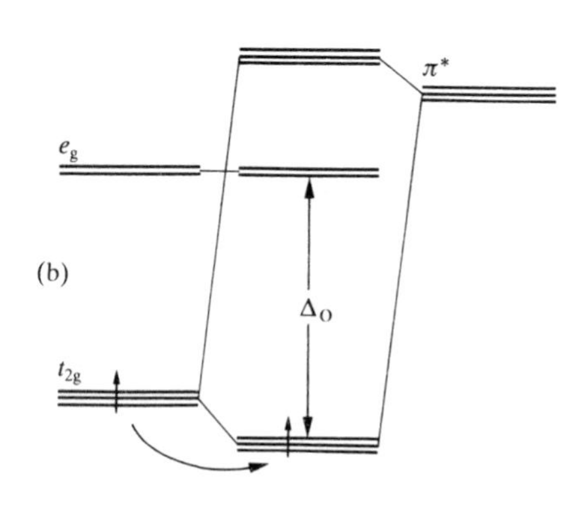

Fig. 7.10 The effect of π bonding on the ligand field splitting. (a) Ligands that act as π bases decrease Δ_O. (b) Ligands that act as π acids increase Δ_O.

LUMO, largely metal in character). Thus, the octahedral **ligand field splitting** Δ_O, the HOMO–LUMO separation, is approximately the splitting of the metal *d* orbitals caused by the ligands.

π Bonding

If the ligands have orbitals with local π symmetry with respect to the M—L axis (as two *p* orbitals of a halide ligand do), they may form π orbitals with the t_{2g} metal orbitals (Fig. 7.9). The effect of this π bonding on the value of Δ_O depends on whether the π ligand orbitals act as electron donors or acceptors.

Lewis **π-base ligands** are ligands with *filled*, locally π symmetry orbitals and no low-energy vacant *p* or π^* orbitals; they include Cl^-, I^-, and H_2O. Their full π orbitals lie lower in energy than the *d* orbitals of the metal. When they form molecular orbitals with the t_{2g} *d* orbitals, the bonding combination lies lower than the ligand orbitals (Fig. 7.10a) and the antibonding combination lies above the *d* orbitals. The electrons supplied by the ligand lone pairs (the π lone pairs, that is) occupy and fill the bonding combination, leaving the *d* electrons to occupy the antibonding t_{2g} orbitals. The net effect is the *reduction* in the ligand field splitting as the t_{2g} HOMO orbitals become antibonding.

Lewis **π-acid ligands** are ligands that, in addition to full π orbitals, have accessible *empty* orbitals of locally π symmetry, which are typically antibonding orbitals *within* the ligand. If they are low enough in energy and have sufficient overlap with the metal t_{2g} orbitals, their π character may be stronger than the π donor character of the full ligand *p* orbitals. The most important example of such a ligand is CO. The π^* orbital of CO has its largest amplitude on the C atom and it is of the correct symmetry for overlap with the t_{2g} orbitals of the metal. In contrast, the full bonding π orbital of CO is largely localized on the O atom (because that is the more electronegative atom), and is not available for overlap with the metal orbitals.

The antibonding character of the orbitals within a π-acid ligand is sufficient to place them higher in energy than the metal *d* orbitals. Thus, when they form molecular orbitals, the t_{2g} orbitals of mainly metal *d* orbital character acquire bonding character (Fig. 7.10b). The net result is that Δ_O is increased. (We can determine whether or not a ligand is a net π donor or acceptor toward a particular central metal atom by experiment, as we shall describe later when we introduce the spectrochemical series.)

Ligand field splitting parameters

The ligand field splitting Δ_O may be calculated (with difficulty) or estimated from experiments that measure the HOMO–LUMO separation. Its value gives a good indication of bonding trends in the absence of π bonding, and the stronger the M—L σ bond, the greater the ligand field splitting. However, when π bonding is important, Δ_O must be interpreted more cautiously since it is a composite of competing trends in σ and π bonding, and one parameter cannot summarize two effects.

It is conventional to take as the zero of energy for the frontier orbitals the average energy of the t_{2g} and e_g orbitals. However, it must not be forgotten that the *d* orbitals of successive elements in a period have progressively lower energy (on account of the increasing nuclear charge). Hence σ bonds

Table 7.3. Ligand field stabilization energies*

d^n	Example	Octahedral				Tetrahedral	
		Strong field		Weak field			
		N	LFSE	*N*	LFSE	*N*	LFSE
d^0	Ca^{2+}, Sc^{3+}	0	0	0	0	0	0
d^1	Ti^{3+}	1	0.4	1	0.4	1	0.6
d^2	V^{3+}	2	0.8	2	0.8	2	1.2
d^3	Cr^{3+}, V^{2+}	3	1.2	3	1.2	3	0.8
d^4	Cr^{2+}, Mn^{3+}	2	1.6	4	0.6	4	0.4
d^5	Mn^{2+}, Fe^{3+}	1	2.0	5	0	5	0
d^6	Fe^{2+}, Co^{3+}	0	2.4	4	0.4	4	0.6
d^7	Co^{2+}	1	1.8†	3	0.8†	3	1.2
d^8	Ni^{2+}	2	1.2†	2	1.2†	2	0.8
d^9	Cu^{2+}	1	0.6†	1	0.6†	1	0.4
d^{10}	Cu^{+}, Zn^{2+}	0	0	0	0	0	0

* *N* is the number of unpaired electrons; LFSE is in units of Δ_O or Δ_T; the calculated relation is $\Delta_T \approx 0.45\,\Delta_O$.
† If undistorted.

become stronger across a period as the metal *d* orbitals approach the ligand σ orbitals in energy and the energy of the complex is lowered. Changes in energy that stem from the difference in population of the nonbonding and antibonding orbitals are modifications of this trend. The overall energy lowering is not normally shown explicitly, and the e_g and t_{2g} orbital energies are normally shown relative to a zero of energy located at the mean of their actual energies.

Since there are three t_{2g} and two e_g orbitals, the t_{2g} orbitals lie $\frac{2}{5}\Delta_O$ below the average energy and the e_g orbitals lie $\frac{3}{5}\Delta_O$ above the average. Therefore, with the zero of energy chosen as the average energy, the t_{2g} orbitals have an energy of $-0.4\Delta_O$ and the e_g orbitals have an energy of $+0.6\Delta_O$. Hence the net energy of a $t_{2g}^x e_g^y$ configuration relative to the average energy of the orbitals is

$$(-0.4x + 0.6y)\Delta_O$$

This is called the **ligand field stabilization energy** (LFSE). Some values are given in Table 7.3.

Ground-state electron configurations

We use the molecular orbital energy level diagram in Fig. 7.11 as a framework for the building-up principle. The first three metal *d* electrons of a d^n complex occupy separate t_{2g} nonbonding orbitals, and do so with parallel spins. The next electron may enter one of the t_{2g} orbitals and pair with the electron already there; but if it does so, it experiences a strong coulombic repulsion, which is called the **pairing energy** *P*. Alternatively, it may occupy one of the antibonding e_g levels and, although avoiding the pairing penalty, have an energy higher by the amount Δ_O.

The ions Ti^{2+}, V^{2+}, and Cr^{2+} have electron configurations d^2, d^3, and d^4 respectively. In the first two ions, the *d* electrons may occupy the lower t_{2g} orbitals and the complexes are stabilized by $2 \times 0.4\Delta_O = 0.8\Delta_O$ and $3 \times 0.4\Delta_O = 1.2\Delta_O$. Two possibilities arise with the d^4 Cr^{2+} ion. If the fourth electron occupies a t_{2g} orbital, the LFSE is $1.6\Delta_O$. However, this can be

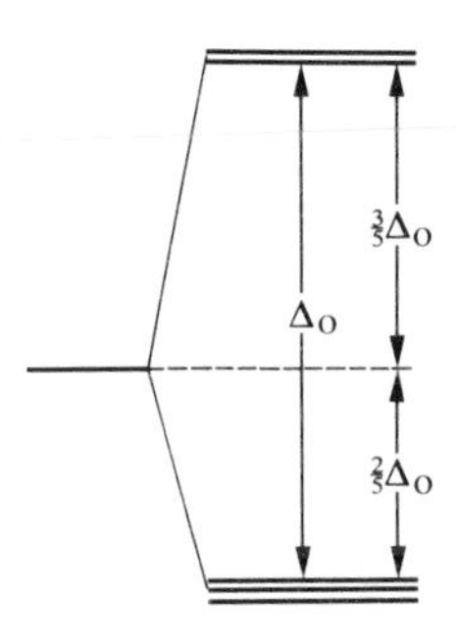

Fig. 7.11 The orbital energy level diagram used in the application of the building-up principle in a ligand field analysis. The ligand field splitting Δ_O is the outcome of both σ and π bonding effects.

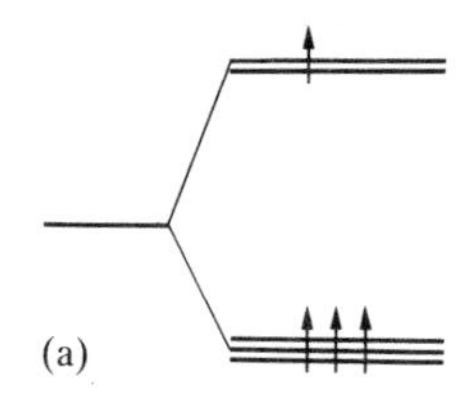

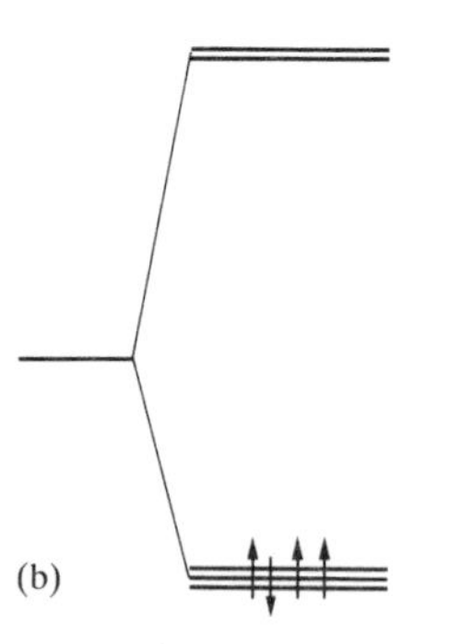

Fig. 7.12 (a) The weak-field, high-spin $t_{2g}^3e_g^1$ and (b) strong-field, low-spin t_{2g}^4 configurations of a d^4 complex.

achieved only by allowing two electrons to occupy one orbital, which is energetically unfavorable to an extent P. The net stabilization is therefore $1.6\Delta_O - P$. Alternatively, the fourth electron may occupy an e_g orbital without incurring the pairing energy. In this case the net stabilization is $3 \times 0.4\Delta_O - 0.6\Delta_O = 0.6\Delta_O$. Which configuration adopted depends on which of $1.60\Delta_O - P$ and $0.60\Delta_O$ is the larger.

Weak-field and strong-field limits

The occupation scheme adopted depends on the relative sizes of Δ_O and P. If $\Delta_O < P$, which is called the **weak-field case** (Fig. 7.12a), occupation of the upper orbital is more favorable because the electron repulsion is minimized; the configuration adopted is then $t_{2g}^3e_g^1$. If $\Delta_O > P$, which is called the **strong-field case** (Fig. 7.12b), pairing is more favorable despite the repulsion because it is energetically expensive to occupy the upper orbitals; the configuration adopted is then t_{2g}^4. For example, $[Cr(OH_2)_6]^{2+}$ has the ground-state configuration $t_{2g}^3e_g^1$ (Fig. 7.12a) whereas $[Cr(CN)_6]^{4-}$ has the configuration t_{2g}^4 (Fig. 7.12b). The fact that $[RuCl_6]^{2-}$ is t_{2g}^4 (Fig. 7.12b) is evidence that heavier metals, because of their better overlap, have large Δ_O and hence favor strong-field cases.

In the weak-field case, the lowest energy is achieved if all the spins are parallel, which for four electrons gives a total spin of 2. In the strong-field case, two of the four electrons must be paired, leaving two unpaired and a net spin of 1. The difference in total spin in the two cases gives rise to the classification of complexes as **high spin** and **low spin**. The terms high spin and weak field should not be used interchangeably, nor should low spin and strong field. It is not necessarily the case that a strong field gives a low spin complex (or a weak field case a high spin complex). For example, $[Cr(NH_3)_6]^{3+}$ is a strong field complex, but with a d^3 configuration it must be high spin. The distinction between high spin and low spin can be made experimentally (as we shall describe) but the distinction between strong field and weak field depends on a model of the bonding.

A similar discussion applies to d^5, d^6, and d^7 octahedral complexes. Table 7.3 summarizes the configurations, the total spin, and the ligand field stabilization energies in each case.

7.5 Correlation of theory with experiment

We have emphasized previously that in inorganic chemistry trends are often expressed in terms of one or two parameters. Perhaps the best example of this procedure is found in ligand field theory, for we shall now see how the single parameter Δ_O helps to correlate magnetic, spectroscopic, and thermodynamic properties of complexes. As far as the determination of Δ_O is concerned, two steps are usually required: we determine the ground state configuration using magnetic measurements; then we estimate Δ_O using spectroscopy.

Spectroscopic measurements

The optical absorption of complexes can often be ascribed to transitions that depend on the ligand field splitting of the frontier orbitals. This is generally small compared with typical orbital separations in p-block molecules.

Table 7.4. Ligand field splittings Δ_O of ML_6 complexes*

Ions		Ligands				
		Cl^-	H_2O	NH_3	en	CN^-
d^3	Cr^{3+}	13.7	17.4	21.5	21.9	26.6
d^2	Mn^{2+}	7.5	8.5		10.1	30
d^5	Fe^{3+}	11.0	14.3			(35)
d^6	Fe^{2+}		10.4			(32.8)
	Co^{3+}		(20.7)	(22.9)	(23.2)	(34.8)
	Rh^{3+}	(20.4)	(27.0)	(34.0)	(34.6)	(45.5)
d^8	Ni^{2+}	7.5	8.5	10.8	11.5	

* Values are in multiples of $1000\ cm^{-1}$; entries in parentheses are for high spin complexes.
Source: H. B. Gray, *Electrons and chemical bonding*. Benjamin, Menlo Park (1965).

Indeed, Δ_O is often small enough for visible light to promote an electron and hence give rise to the wide range of colors that are so characteristic of these compounds.

Figure 7.13 shows the optical absorption spectrum of $[Ti(OH_2)_6]^{3+}$ with its first absorption maximum at $20\,300\ cm^{-1}$, which is assigned to the transition $e_g \leftarrow t_{2g}$. (In keeping with spectroscopic notation, the higher-energy orbital is shown first.) We can identify $20\,300\ cm^{-1}$ as Δ_O for the complex and note that it converts to $243\ kJ\ mol^{-1}$, or about $41\ kJ\ mol^{-1}$ per ligand. It is much more complicated to obtain values of Δ_O for complexes with more than one d electron. This is because the energy of a transition depends not only on orbital energies (which we wish to know) but also on the repulsion energies between the several electrons present. The subject is treated more fully in Chapter 14, and the results from the analyses described there have been used to obtain the values of Δ_O in Table 7.4.

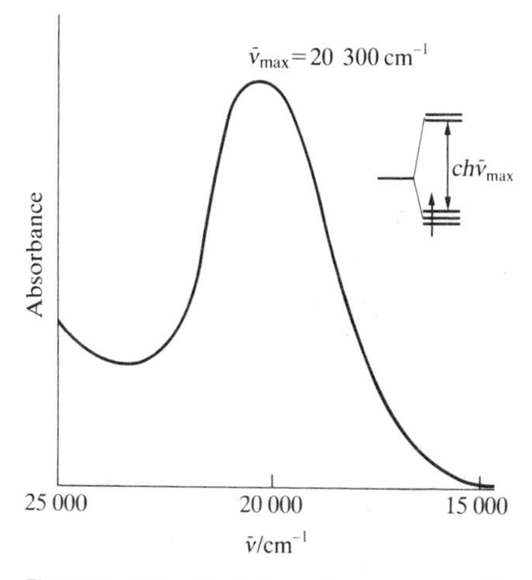

Fig. 7.13 The optical absorption spectrum of $[Ti(H_2O)_6]^{3+}$.

The spectrochemical series

The observed values of Δ_O show interesting and useful regularities with variation of both ligand and metal. As to the ligand, it has been found that the order of the values of Δ_O is approximately the same for all metal ions and can be expressed by the **spectrochemical series**, in which ligands are arranged in order of their ligand field strength:

$$I^- < Br^- < S^{2-} < \mathit{S}CN^- < Cl^- < NO_3^- < F^- < C_2O_4^{2-} < H_2O < \mathit{N}CS^- < CH_3CN < NH_3 < en < bipy < phen < \mathit{N}O_2^- < PPh_3 < CN^- < CO$$

H^- and CH_3^- also occupy a position among the strongest-field ligands in this series. The italicized atom indicates the atom that forms the link.

This empirical ordering can be readily interpreted by molecular orbital theory. Two factors are the energy difference and overlap of ligand orbitals with the metal d orbitals: these determine the strength of the σ bonds and of the energy of the antibonding e_g orbitals. However, other factors must play a part, for low values of Δ_O are associated with π base ligands and the reduction they cause in Δ_O. The middle region of the series includes many ligands with little tendency to form π bonds (such as NH_3). High values of Δ_O are associated with π acid ligands such as CO and the increase they cause in Δ_O. In fact, the overall ordering of the spectrochemical series of ligands may be interpreted in broad terms as

$$\pi \text{ bases} < \text{weak } \pi \text{ bases} < \text{no } \pi \text{ effects} < \pi \text{ acids}$$

Representative ligands that match these classes are

$$I^- < Br^- < Cl^- < F^- < H_2O < NH_3 < PR_3 < CO$$

π bases   weak π bases   no π effects π acids

Notable exceptions where the σ bonding effect is clearly important include CN^-, which is stronger than its limited π acidity implies, and H^-, which is very strong.

The values of Δ_O also depend in a systematic way on the metal, and it is not in general possible to say that a particular ligand is 'high field' or 'low field' without also considering the metal involved. The most important metal trends to keep in mind are that Δ_O increases both with increasing oxidation number and on descending a group. The latter trend reflects the improved σ bonding of the more expanded 4*d* and 5*d* orbitals compared with the compact 3*d* orbitals. The spectrochemical series for metal ions is (in part)

$$Mn^{2+} < V^{2+} < Co^{2+} < Fe^{2+} < Ni^{2+} < Fe^{3+} < Co^{3+} < Mn^{4+} < Mo^{3+} < Rh^{3+} < Ru^{3+} < Pd^{4+} < Ir^{3+} < Pt^{4+}$$

The extremes of the metal and ligand series give a reasonably reliable way of predicting whether a particular combination of metal and ligand is weak field or strong field. Thus, if the metal is to the right of its series, and the ligand is to the right of its series, we can be reasonably confident that the combination will be a strong field complex. If they are both to the left of their series, we can expect the complex to be weak field. Where intermediate members of both series are combined, it may be necessary to look up detailed information to determine the class of the complex. It is worth bearing in mind that H_2O and the halide ions tend to give weak field complexes with 3*d* metals, that NH_3 frequently produces strong field complexes, and that the CN^- ion almost always does so.

Magnetic measurements

Complexes are classified as **diamagnetic** if they tend to move out of a magnetic field and **paramagnetic** if they tend to move into a magnetic field. The two can be distinguished by hanging a sample of the substance between the poles of an electromagnet and seeing whether it appears to weigh less or more when the field is on. Paramagnetism stems principally from the presence of unpaired electron spins.

In a free atom or ion, both the orbital and the spin angular momenta give rise to a magnetic moment. When the atom or ion is part of a complex, its orbital angular momentum may be eliminated—the technical term is **quenched**—as a result of the interactions of the electrons with their environment. The electron spin, however, survives, and gives rise to **spin-only paramagnetism**. Quenching is more important for *d* electrons than for *f* electrons, which are deeply buried in atoms and often experience their environment only weakly. Hence, whereas *d*-metal complexes can often be discussed as though quenching has occurred completely, that is not generally true of *f*-block complexes.

Since the ground electron configurations of d-metal complexes have characteristic numbers of unpaired electron spins, a principal use of magnetic susceptibility measurements is to identify these configurations. The unquenched contribution of orbital angular momentum can usually be ignored in a first approximation. More detailed magnetic investigations, particularly the observation of the variation of susceptibility with temperature, are used to evaluate theoretical predictions about electron distributions.

The spin contribution

The magnetic moment μ of a complex with total spin quantum number S is

$$\mu = 2\{S(S+1)\}^{1/2}\mu_B$$

where μ_B is the **Bohr magneton**:

$$\mu_B = \frac{e\hbar}{2m_e} = 9.274 \times 10^{-24}\ \mathrm{J\,T^{-1}}$$

Since each unpaired electron has a spin quantum number of $\frac{1}{2}$, $S = \frac{1}{2}n$, where n is the number of unpaired electrons; therefore

$$\mu = \{n(n+2)\}^{1/2}\mu_B$$

Measuring the magnetic moment of a complex therefore provides a way of counting the number of unpaired electrons present.

The spin-only magnetic moments for some configurations are listed in Table 7.5 and compared there with experimental values for a number of complexes from the first row of the d block. For most 3d and some 4d complexes, experimental values lie reasonably close to spin-only predictions, so it becomes possible to identify correctly the number of unpaired electrons in a ground state of one of these configurations. The kind of information the measurements provide is illustrated by considering $[Fe(OH_2)_6]^{3+}$. Compounds containing hexaaqua Fe(III) are paramagnetic with a magnetic moment of 5.3μ_B. As shown in Table 7.5, this is reasonably close to the value for five unpaired electrons and a high-spin $t_{2g}^3e_g^2$ configuration.

Table 7.5. Calculated spin-only magnetic moments

Ion	n	S	μ/μ_B Calculated	Experiment
Ti^{3+}	1	$\frac{1}{2}$	1.73	1.7–1.8
V^{3+}	2	1	2.83	2.7–2.9
Cr^{3+}	3	$\frac{3}{2}$	3.87	3.8
Mn^{3+}	4	2	4.90	4.8–4.9
Fe^{3+}	5	$\frac{5}{2}$	5.92	5.3

Example 7.5: *Inferring an electron configuration from a magnetic moment*

The magnetic moment of an octahedral Co(II) complex is 4.0μ_B. What is its electron configuration?

Answer. A Co(II) complex is d^7. The two possible configurations are $t_{2g}^5e_g^2$ (high spin) with three unpaired electrons or $t_{2g}^6e_g^1$ (low spin) with one unpaired electron. The spin-only magnetic moments (Table 7.5) are 3.87μ_B and 1.73μ_B respectively. Therefore, the only consistent assignment is the high spin $t_{2g}^5e_g^2$.

Exercise. The magnetic moment of the complex $[Mn(NCS)_6]^{4-}$ is 6.06μ_B. What is its electron configuration?

Orbital contributions

The potassium salt of $[Fe(CN)_6]^{3-}$ has $\mu = 2.3\mu_B$, which is between the spin-only values for one and two unpaired electrons (1.7μ_B and 2.8μ_B

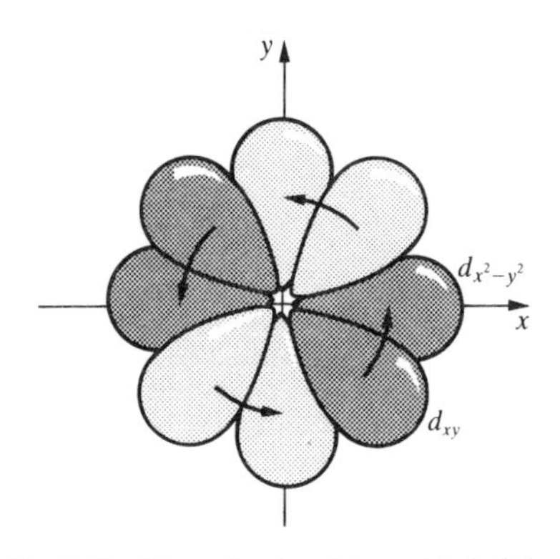

Fig. 7.14 If there is a low-lying orbital of the correct symmetry, the applied field may induce the circulation of the electrons in a complex and hence generate orbital angular momentum. This shows the way that circulation may arise when the field is applied perpendicular to the *xy* plane (and perpendicular to this page).

respectively). This suggests that in this complex, the spin-only assumption has failed and that an orbital magnetic contribution is important.

For orbital angular momentum to contribute, and hence for the paramagnetism to differ significantly from the spin-only value, there must be an orbital similar in energy to that of the orbitals occupied by the unpaired spins. If that is so, the electrons can make use of it to circulate through the framework of the complex and hence generate angular momentum and a magnetic moment (Fig. 7.14). Therefore, we can use departures from spin-only values of magnetic moments to assess energy separations between orbitals and hence to infer degrees of distortion or the value of Δ_O.

More specifically, an orbital must be available that is related by rotational symmetry to the occupied orbital (as a d_{xy} orbital is related to $d_{x^2-y^2}$ by a 45° rotation about the z axis, Fig. 7.14) and should not contain an electron with the same spin as the first electron. These conditions are fulfilled whenever any two of the three t_{2g} (d_{xy}, d_{xz}, d_{yz}) orbitals contain an odd number of electrons. Departure from spin-only values is generally large for $3d^1$ and $3d^2$ and low-spin $3d^4$ and $3d^5$ complexes. In other $3d$ complexes the orbital angular momentum contributions of the electrons cancel each other and spin-only values are reliable. The magnetic properties of $4d$, $5d$, and f-block complexes are more difficult to systematize, and we postpone their discussion until Chapter 14. However, one common characteristic of the heavier d-metal complexes is that they tend to form low-spin complexes.

Thermochemical correlations

Molecular orbital theory predicts that complexes should become more stable as the energies of the d orbitals fall toward that of the ligand orbitals and σ bonds strengthen. This greater stability should be reflected in more exothermic complex formation across the period. However, we should not expect a linear variation because ligand field stabilization energies vary across the period in a wavelike manner. As Table 7.3 shows, it increases from d^1 to d^3, decreases again to d^5 then rises to d^8. (As we see in the next section, d^9 is a special case.)

Figure 7.15 shows this pattern in the case of the hydration enthalpies ΔH_H

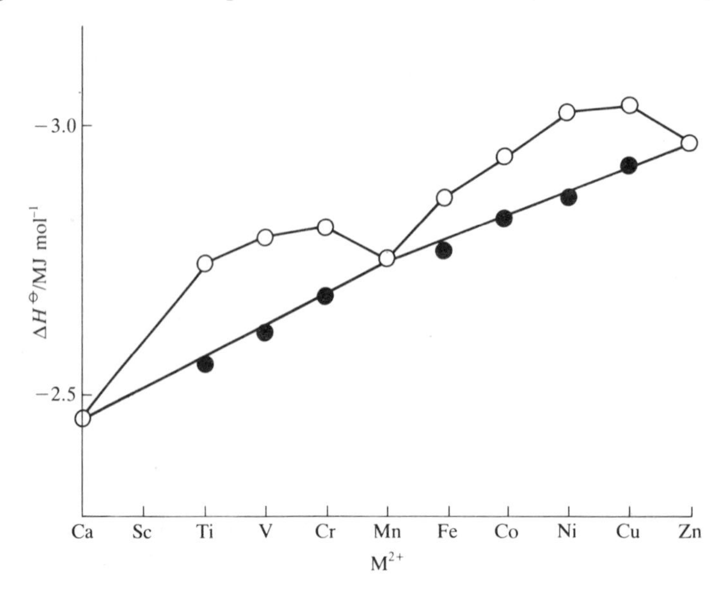

Fig. 7.15 The hydration enthalpy of M^{2+} ions of the first row of the *d* block. The straight line shows the trend when the ligand field stabilization energy has been subtracted from the observed values.

of the M^{2+} ions of the first row of the d block, the enthalpy change accompanying

$$M^{2+}(g) + 6H_2O(l) \rightarrow [M(OH_2)_6]^{2+}(aq)$$

The straight line in the illustration shows the general trend and the open circles mark the experimental points. The filled circles have been calculated by subtracting the high-spin LFSE from ΔH_H using the spectroscopic values of Δ_O in Table 7.4.

Example 7.6: *Using the LFSE to account for properties*

The oxides of formula MO, which all have octahedral coordination of the metal ions, have the following lattice enthalpies:

CaO	TiO	VO	MnO
3460	3878	3913	3810 kJ mol^{-1}

Account for the trends in terms of the LFSE.

Answer. The general trend across the d block is the increase from CaO (d^0) to MnO (d^5), both of which have an LFSE of zero. Since O^{2-} is a weak-field ligand, TiO (d^2) has an LFSE of $0.8\Delta_O$ and VO (d^3) has an LFSE of $1.2\Delta_O$ (Table 7.3). If we suppose that Δ_O is fairly constant across the series, the contribution of LFSE to lattice enthalpies is in the same order as the experimental values of the lattice enthalpy.

Exercise. Account for the variation in lattice enthalpy of the octahedral fluorides MnF_2 (2780 kJ mol^{-1}), FeF_2 (2926 kJ mol^{-1}), CoF_2 (2976 kJ mol^{-1}), NiF_2 (3060 kJ mol^{-1}), and ZnF_2 (2985 kJ mol^{-1}).

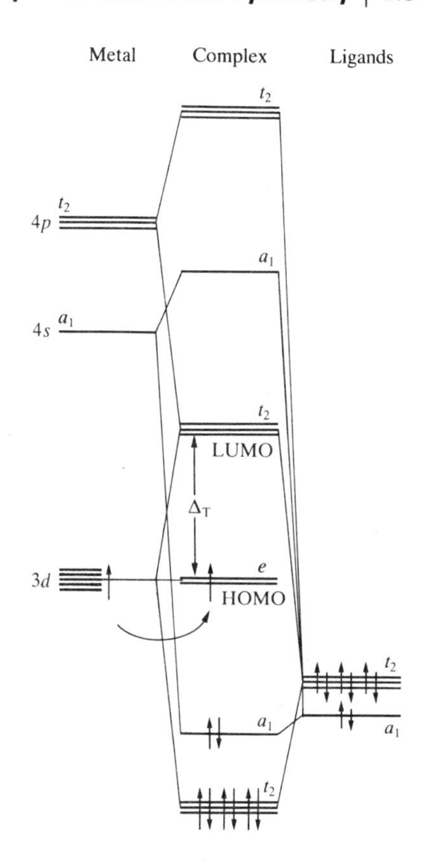

Fig. 7.16 The molecular orbital energy level diagram for tetrahedral ML_4 complexes. The symmetry-adapted combinations of ligand orbitals are given in Appendix 6.

7.6 Complexes with lower symmetry

Tetrahedral complexes

The principal qualitative difference between octahedral and tetrahedral complexes results from reversal of the order of the frontier orbitals: in octahedral complexes the triply degenerate t_2 orbitals are nonbonding and lie below the doubly degenerate antibonding e orbitals. As is shown in Appendix 6, in a tetrahedral environment the e pair of d orbitals has no matching combination of ligand orbitals, and hence these two orbitals remain nonbonding in the complex. The three t_2 orbitals, however, do have ligand partners, and form bonding and antibonding combinations (Fig. 7.16). A secondary difference is that the ligand field splitting Δ_T in a tetrahedral complex is less than Δ_O, as might be expected for complexes with fewer ligands (in fact, $\Delta_T < \frac{1}{2}\Delta_O$, which is harder to explain). Hence, only weak-field tetrahedral complexes are common and are the only ones we need consider. Some values of Δ_T are collected in Table 7.6. Among $3d^n$ systems, tetrahedral complexes are especially important for d^7 Co(II).

Table 7.6. Values of Δ_T for representative tetrahedral complexes

Complex	Δ_T/cm^{-1}
VCl_4	9010
$[CoCl_4]^{2-}$	3300
$[CoBr_4]^{2-}$	2900
$[CoI_4]^{2-}$	2700
$[Co(NCS)_4]^{2-}$	4700

Tetragonal and square-planar complexes

The regular octahedron is a useful starting point for the discussion of six-coordinate complexes with distorted geometries. A tetragonal distortion, corresponding to extension along the z axis and compression on the x and y axes, reduces the antibonding character of the $e_g(d_{z^2})$ orbital and hence

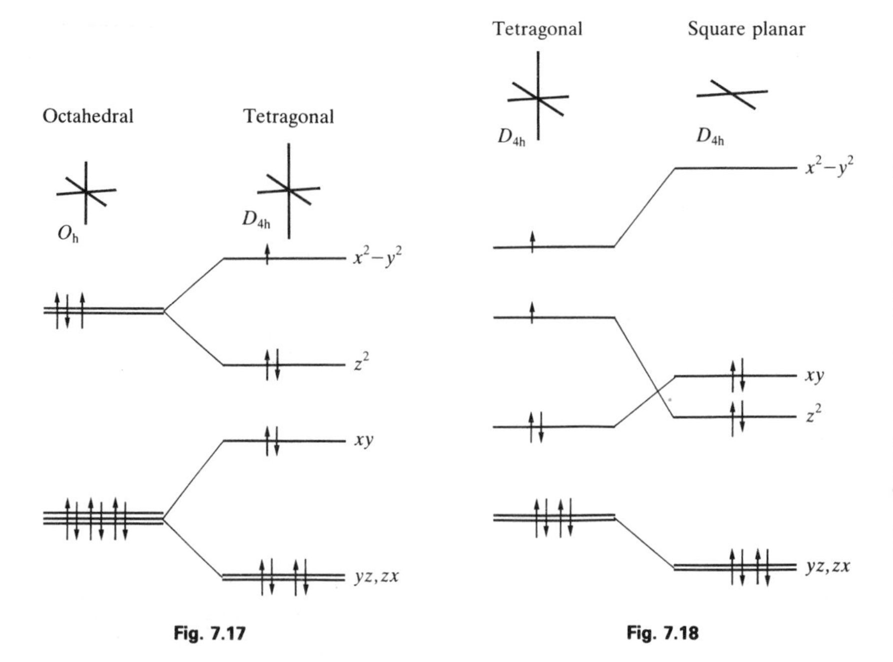

Fig. 7.17 A frontier orbital energy level diagram showing how a tetragonal distortion (compression along x and y and extension along z) affects the energy levels. The electron occupation is for a d^9 complex.

Fig. 7.18 The change in energy levels of a complex as it distorts from tetragonal to square planar. A d^8 configuration is shown.

lowers its energy (Fig. 7.17), but it increases the energy of the $e_g(d_{x^2-y^2})$ orbital. Therefore, if one, two, or three electrons occupy the e_g orbitals (as in the d^7, d^8, and d^9 complexes) a tetragonal distortion may be energetically advantageous. For example, in a d^9 complex ($t_{2g}^6e_g^3$) such a distortion leaves two electrons stabilized and one destabilized. As a result, Cu^{2+} complexes usually depart considerably from octahedral symmetry and are more stable than pure-octahedral ligand field stabilization predicts. Low-spin d^7 complexes (e_g^1) may show a similar distortion, but this case is rare.

The distortion of d^8 ($t_{2g}^6e_g^2$) complexes may be large enough to encourage the two e_g electrons to pair in the d_{z^2} orbital. This distortion actually often goes as far as the total loss of the ligands on the z axis and the formation of square-planar complexes, such as those found for Pt(II), Pd(II), Au(III), and $[Ni(CN)_4]^{2-}$. The frontier orbital region of the molecular orbital diagrams of these complexes is shown in Fig. 7.18. In these, all electrons are paired because the full advantage of filling the d_{z^2} orbital must be exploited to compensate for the loss of two ligands.

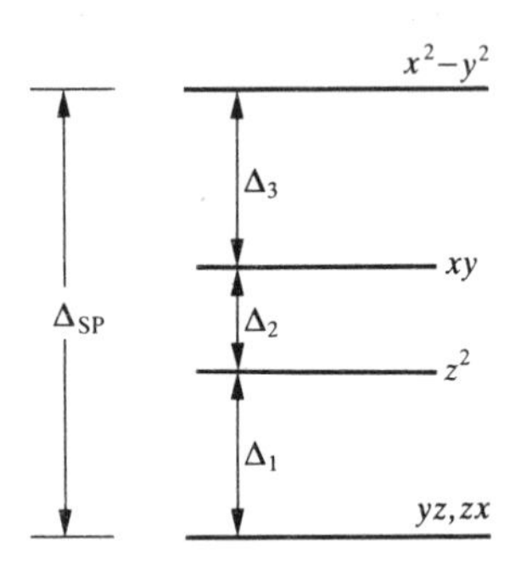

Fig. 7.19 The orbital splitting parameters for a square-planar complex.

The sum of the three distinct orbital splittings shown for a square-planar complex in Fig. 7.19 is denoted Δ_{SP}. This sum is greater than Δ_O; simple theory predicts that $\Delta_{SP} = 1.3\Delta_O$ for complexes of the same metal and ligands.

The Jahn–Teller theorem

The tetragonal distortions just described represent specific examples of the consequences of the **Jahn–Teller theorem**:

> If the ground electronic configuration of a nonlinear molecule is degenerate, the molecule will distort so as to remove the degeneracy and become more stable.

An octahedral d^9 complex is degenerate because the odd electron can occupy either the $d_{x^2-y^2}$ orbital or the d_{z^2} orbital. A tetragonal distortion lowers the energy of the $d_{z^2}^1$ configuration . Distortion to square planar, with the electrons paired in d_{z^2}, stabilizes the molecule, and can be regarded as an extreme response to the demands of the Jahn–Teller theorem. Note that the theorem does not refer to the degeneracies of the orbitals themselves but to the degeneracies of the configurations that arise when they are occupied.

The Jahn–Teller theorem only identifies an unstable geometry; it does not predict the preferred distortion. The examples we have cited involve the elongation of two *trans* bonds and the compression of the four that lie in a plane. The alternative distortion, compression along an axis and elongation in a plane, would also remove the degeneracy. Which occurs in practice is a matter of energetics, not symmetry. However, since the axial elongation weakens only two bonds whereas elongation in the plane would weaken four, axial elongation is more common than axial compression.

Apparent exceptions to the Jahn–Teller theorem are sometimes encountered. Some Cu(II) complexes, for instance, seem to be undistorted. This symmetry may, however, only be apparent, for it could be a result of the timescale of the measurement and the detection of a time-average of fluxional structures. For example, the electron paramagnetic resonance (EPR, Section 14.8) signal from $[Cu(OH_2)_6]^{2+}$ in a $[Zn(OH_2)_6][SiF_6]$ host crystal at room temperature detects the unpaired electron as a single isotropic line, which suggests an octahedral environment. However, if a distortion alternates among the three equivalent axes faster than the difference in their resonance frequencies, the observed spectral lines will lie at the average of the locations of the three lines of the distorted species. Since resonance frequency differences are of the order of megahertz in EPR spectroscopy, an average spectrum is observed if a distortion survives for less than a microsecond.

The hopping of a distortion from one orientation to another is called the **dynamic Jahn–Teller effect**, and its rate depends on the temperature. In the EPR spectrum of $[Cu(OH_2)_6]^{2+}$ a static distortion (more precisely, one effectively stationary on the timescale of the resonance experiment) occurs when the temperature is below 20 K, and the distinct signals from the different environments are observed.

Reactions of complexes

The reactions of d-metal complexes are almost always studied in solution. The solvent molecules compete for the central metal ion, so the formation of a complex with another ligand is a **substitution reaction**, a reaction in which an incoming group displaces a ligand (a solvent molecule) already present. The incoming group is called the **entering group** and the displaced ligand is the **leaving group**. We normally denote the leaving group as X and the entering group as Y. Then a substitution reaction is the Lewis displacement reaction

$$M—X + Y \rightarrow M—Y + X$$

Both the thermodynamics and the kinetics of complex formation lead to an understanding of the reactions of complexes, and we shall introduce both aspects here. We give a more thorough account of the kinetics of reactions in Chapter 15.

7.7 Coordination equilibria

A specific example of a coordination equilibrium is the reaction of Fe(III) with NCS^- to give the complex $[FeNCS(OH_2)_5]^{2+}$, which is closely related to the analytically useful red complex used to detect either Fe or NCS^-:

$$[Fe(OH_2)_6]^{3+} + NCS^- \rightleftharpoons [FeNCS(OH_2)_5]^{2+} + H_2O \qquad K_f = \frac{[(FeNCS)^{2+}]}{[Fe^{3+}][NCS^-]}$$

The equilibrium constant K_f is the **formation constant** of the complex. The concentration of H_2O does not appear because it is taken to be constant in dilute solution and is absorbed into K_f. A ligand for which K_f is large is one that binds more tightly than H_2O. A ligand for which K_f is small may not be a weak ligand in an absolute sense, but merely weaker than H_2O.

The discussion of stabilities is more involved when more than one ligand may be replaced. In the series from $[Ni(OH_2)_6]^{2+}$ to $[Ni(NH_3)_6]^{2+}$, for instance, there are as many as six steps even if *cis–trans* isomerism is ignored. For the general case of the complex ML_n, the **stepwise formation constants** are

$$M + L \rightleftharpoons ML \qquad K_1 = \frac{[ML]}{[M][L]}$$

$$ML + L \rightleftharpoons ML_2 \qquad K_2 = \frac{[ML_2]}{[ML][L]}$$

$$ML_2 + L \rightleftharpoons ML_3 \qquad K_3 = \frac{[ML_3]}{[ML_2][L]}$$

$$\vdots \qquad \vdots \qquad \vdots$$

$$ML_{n-1} + L \rightleftharpoons ML_n \qquad K_n = \frac{[ML_n]}{[ML_{n-1}][L]}$$

These are the constants to consider when seeking to understand the relations between structure and reactivity. When we want to calculate the concentration of the final product (the complex ML_n) we use the **overall formation constant** β_n:

$$\beta_n = \frac{[ML_n]}{[M][L]^n}$$

The overall formation constant is the product of the stepwise constants:

$$\beta_n = K_1 K_2 K_3 \cdots K_n$$

The inverse of K_f, the **dissociation constant** K_d, is also sometimes useful:

$$ML \rightleftharpoons M + L \qquad K_d = \frac{[M][L]}{[ML]}$$

K_d has the same form as K_a for acids, which facilitates comparisons between metal complexes and Brønsted acids. K_d and K_a can be tabulated together if the proton is considered to be simply another cation.

Trends in successive formation constants

It is commonly observed that stepwise formation constants lie in the order

$$K_1 > K_2 > K_3 \cdots > K_n$$

This general trend can be explained quite simply in terms of the numbers of ligands present and the number of opportunities for reaction. We can find a simple explanation by considering the decrease in the number of the ligand H_2O molecules available for replacement in the formation step, as in

$$M(OH_2)_5L + L \rightarrow M(OH_2)_4L_2 + H_2O$$

compared with

$$M(OH_2)_4L_2 + L \rightarrow M(OH_2)_3L_3 + H_2O$$

This reduces the number of H_2O ligands available for replacement as n increases. Conversely, the increase in the number of bound L groups increases the importance of the reverse of these reactions as n increases. Therefore, as long as the reaction enthalpy is largely unaffected, the equilibrium constants lie progressively in favor of the reactants as n increases. That such a simple explanation is more or less correct is illustrated by data for the successive complexes in the series from $[Ni(OH_2)_6]^{2+}$ to $[Ni(NH_3)_6]^{2+}$ (Table 7.7). The enthalpy changes for the six successive steps vary by less than 2 kJ mol^{-1} from 16.7 to 18.0 kJ mol^{-1}.

A reversal of the relation $K_n < K_{n+1}$ is usually an indication of a major change in the structure and bonding at the metal center as more ligands are added. An example is that the tris(bipyridine) complex of Fe(II) is strikingly stable compared with the bis complex. This can be correlated with the change from a weak field $t_{2g}^4e_g^2$ to a strong field t_{2g}^6 configuration. A contrasting example is the anomalously low value of K_3/K_2 (approximately $\frac{1}{7}$) for the halogeno complexes of Hg(II). This decrease is too large to be explained statistically and suggests the onset of four-coordination:

$$[HgX_2](aq) + X^-(aq) + H_2O(l) \rightarrow [Hg(OH_2)X_3]^-(aq)$$

Table 7.7. Formation constants for Ni(II) ammines

n	pK_f	K_n/K_{n-1}	
		Experimental	**Statistical***
1	−2.72		
2	−2.17	0.28	0.42
3	−1.66	0.31	0.53
4	−1.12	0.29	0.56
5	−0.67	0.35	0.53
6	−0.03	0.2	0.42

* Based on ratios of numbers of ligands available for replacement, with the reaction enthalpy assumed constant.

Example 7.7: *Interpreting irregular successive formation constants*

The formation of Cd complexes with Br^- exhibit the successive equilibrium constants $K_1 = 1.56$, $K_2 = 0.54$, $K_3 = 0.06$, $K_4 = 0.37$. Suggest an explanation of why K_4 is larger than K_3.

Answer. The anomaly suggests a structural change. Aqua complexes are usually six-coordinate whereas halo complexes are commonly tetrahedral. The reaction of the complex with three Br^- groups to add the fourth is

$$[CdBr_3(H_2O)_3]^-(aq) + Br^-(aq) \rightarrow [CdBr_4]^{2-}(aq) + 3H_2O(l)$$

This step is entropically favored because of the release of three molecules of water from the relatively restricted coordination sphere environment. The result is an increase in K.

Exercise. A square-planar four-coordinate Fe(II) porphyrin complex may add two further ligands axially. The maximally coordinate complex $[FePL_2]$ is low spin whereas $[FePL]$ is high spin. Account for an increase of the second formation constant for L with respect to the first.

The chelate effect

When K_1 for a bidentate chelate ligand, such as ethylenediamine, is compared with the value of β_2 for the corresponding diligand complex (a diammine), it is found that the former is generally larger:

$$[Cu(OH_2)_6]^{2+} + en \rightleftharpoons [Cu(OH_2)_4(en)]^{2+} + 2H_2O$$

$$\log K_1 = 10.6 \qquad \Delta H^{\ominus} = -54\ \text{kJ mol}^{-1} \qquad \Delta S^{\ominus} = +23\ \text{J K}^{-1}\ \text{mol}^{-1}$$

$$[Cu(OH_2)_6]^{2+} + 2NH_3 \rightleftharpoons [Cu(OH_2)_4(NH_3)_2]^{2+} + 2H_2O$$

$$\log \beta_2 = 7.7 \qquad \Delta H^{\ominus} = -46\ \text{kJ mol}^{-1} \qquad \Delta S^{\ominus} = -8.4\ \text{J K}^{-1}\ \text{mol}^{-1}$$

(CH₃)₂N N(CH₃)₂

26

27

28 bipy

29 phen

Essentially the same two Cu—N bonds are formed in each case, yet the formation of the chelate is distinctly more favorable. The **chelate effect** is this greater stability of chelated complexes compared with their nonchelated analogs.

We can trace the chelate effect primarily to differences in reaction entropies between chelated and nonchelated complexes in dilute solutions. The chelation reaction results in an increase in the number of independent molecules in solution but the reaction with monodentate ligands produces no net change. The former therefore has the more positive entropy change and hence is the more favorable process. The entropy changes measured in dilute solution support this interpretation.

The chelate effect is of great practical importance. The majority of reagents used in complexometric titrations in analytical chemistry are multidentate chelates like edta. Most metal binding sites in biomolecules are chelating ligands. When a formation constant is measured as $10^{12}\ \text{M}^{-1}$ to $10^{25}\ \text{M}^{-1}$ it is generally a sign that the chelate effect is in operation.

Steric effects

Steric effects also have an important influence on formation constants. They are particularly important in chelate formation since ring completion may be difficult geometrically. Chelate rings with five members are generally the most stable, as we explained in Section 7.2. Six-membered rings are reasonably stable and may be favored if electron delocalization can occur.

An interesting trend is observed when the basic site on the ligand is crowded but the cavity remains large enough to accept a proton. The formation constants for the metal complexes correlate well with the Brønsted basicity if steric factors are considered. Figure 7.20, for instance, shows correlations for substituted pyridine bases. All lines have the same slope but there is a reduction of stability for the metal complexes for each blocking R next to the donor N. The extreme example of such steric effects is a ligand called engagingly 'Proton Sponge' (**26**). This compound can strongly bind a proton but is sterically blocked from binding any other cations.

Diimine ligands (**27**), such as bipyridine (**28**) and phenanthroline (**29**), are constrained to form five-membered rings. The great stability of their complexes is probably a result of their ability to act as π acids as well as σ bases, and to form π bonds with full metal d orbitals and their vacant ring π^* orbitals. This is favored by electron population in the metal t_{2g} orbitals, which allows the metal to act as a π base and donate to the ligand rings. An example is the complex $[Ru(bipy)_3]^{2+}$ (**30**). The small bite angle of these ligands distorts the complex from octahedral symmetry.

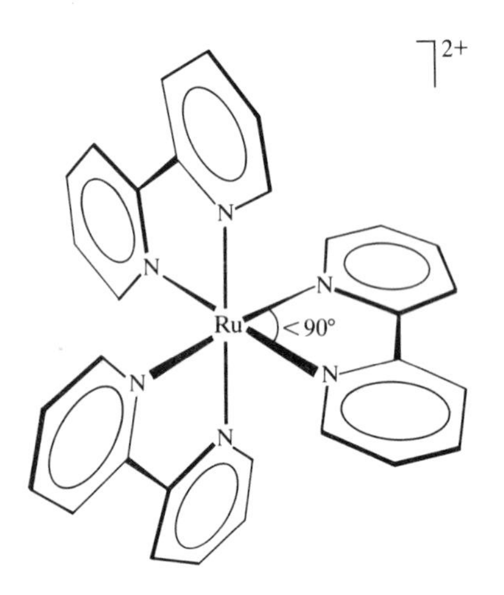

30 $[Ru(bipy)_3]^{2+}$

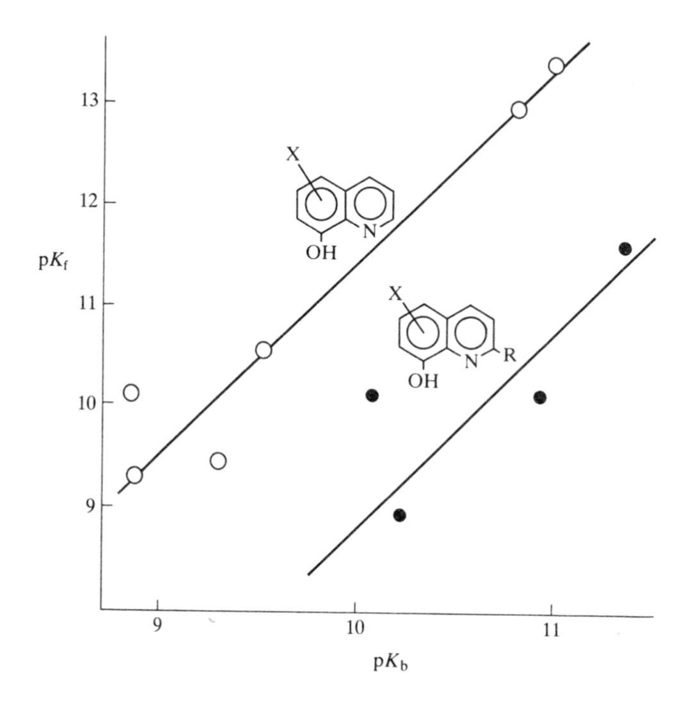

Fig. 7.20 The correlation of formation constant pK_f (for complexation with Cu^{2+}) and Brønsted basicity pK_b for two series of substituted 8-quinoline ligands. The filled circles are for more sterically hindered ligands.

The role of hardness

Figure 7.20 suggests that there may be a quantitative relationship between Brønsted and Lewis basicity. However, the proton is a hard acid, which is not true of all metal ions. One of the major trends in stability constants with variation of the ligand is the one based on the hardness of the ligand as a Lewis base. For hard acid centers, the trend in stability constants follows the hardness of the donor atom:

$$N > P > As \qquad O > S \qquad F^- > Cl^- > Br^- > I^-$$

In contrast, soft acids show the opposite trends in formation constants:

$$As > P > N \qquad S > O \qquad I^- > Br^- > Cl^- > F^-$$

Elements late in the *d* block with low oxidation numbers are soft, but the magnitudes of the formation constants of *s*-block and most of the lighter *d*-block metal ions in combination with O, N, F, and Cl donor ligands can be rationalized by a simple electrostatic argument. For these hard ligands, the largest values of K_f are found for metal ions of high oxidation number, such as Fe^{3+} and Al^{3+}; the smaller the ionic radius (as for Al^{3+} in this pair), the larger the value. In other words, there is a correlation with ξ. In accord with this analysis, Be^{2+} and Mg^{2+} have large values of K_f because their radii are small. For hard ligands, only two main types of deviation from the trends predicted by correlation of K_f with ξ occur. The first can be explained by correction for ligand field stabilization energy in the first *d* series. Larger deviations begin to occur as the soft heavier *d* ions and ions of the *p* block are approached and covalence begins to dominate.

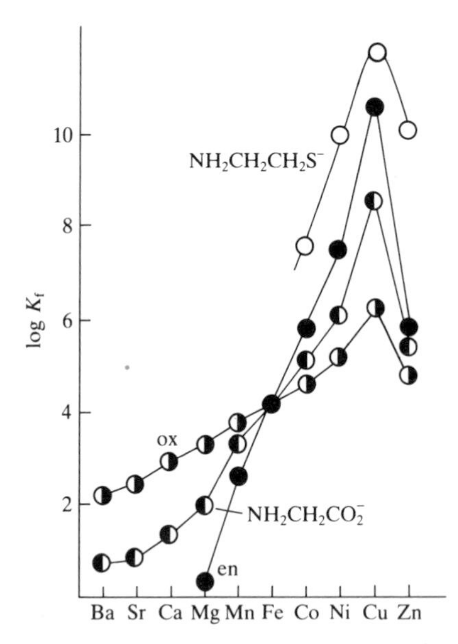

Fig. 7.21 The variation of formation constants for the M^{2+} ions of the Irving–Williams series.

The Irving–Williams series

Figure 7.21 is obtained when log K_f is plotted along the first *d* series of divalent metal ions (M^{2+}). This variation is summarized by the **Irving–Williams series** for the order of formation constants. For divalent cations:

$$Ba^{2+} < Sr^{2+} < Ca^{2+} < Mg^{2+} < Mn^{2+} < Fe^{2+} < Co^{2+} < Ni^{2+} < Cu^{2+} > Zn^{2+}$$

The order is relatively insensitive to the choice of ligands.

In the main, the Irving–Williams series reflects electrostatic effects. However, beyond Mn^{2+} there is a sharp increase in the value of K_f for Fe(II), d^6; Co(II), d^7; Ni(II), d^8; and Cu(II), d^9. These ions have additional stabilizations that are proportional to the ligand field stabilization energies shown for d^6 to d^9 in Table 7.3. However, there is one important exception: the stability of Cu(II) complexes is greater than that for Ni(II) despite the fact that Cu(II) has an additional antibonding e_g electron. This is a consequence of the stabilizing influence of the Jahn–Teller distortion, which enhances the value of K_f. In the tetragonally distorted complex, there is strong binding of the four ligands in the plane. Note that the remaining distorted axial positions are more weakly bound.

7.8 Rates and mechanisms of ligand substitution

Rates of reaction are as important as equilibria in coordination chemistry. The numerous isomers of the ammines of Co(III) and Pt(II), which were so important to the development of the subject, could not have been isolated if ligand substitutions and interconversion of the isomers had been fast.

Lability and inertness

Complexes that survive for long periods (at least a minute) are called **inert**. Complexes that undergo more rapid equilibration are called **labile**. Octahedral complexes of the first *d* series show an interesting correlation with electron configuration: strong-field d^3 and d^6 complexes (such as Cr(III) and Co(III) complexes respectively) are generally inert; all others are generally labile.

Figure 7.22 shows the characteristic lifetimes of octahedral complexes of the important aqua metal ions. We see a range of lifetimes starting at about 1 ns, which is approximately the time it takes for a molecule to diffuse one molecular diameter in solution. At the other end of the scale are lifetimes in

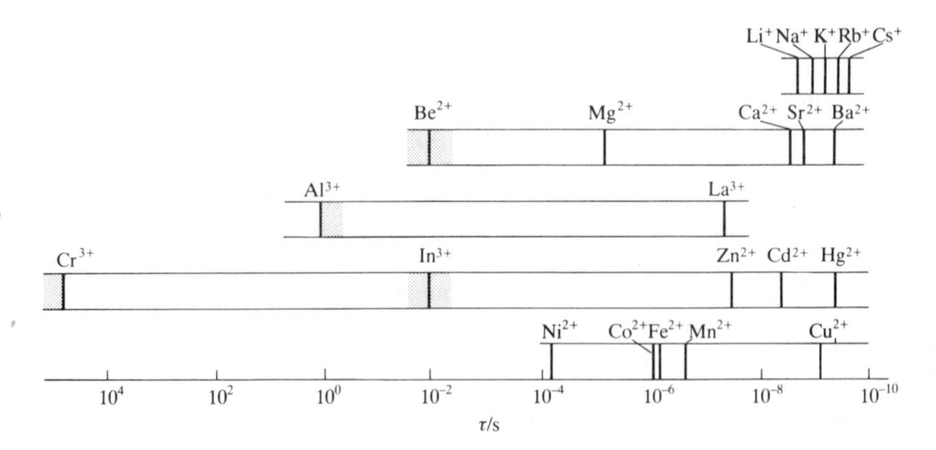

Fig. 7.22 Characteristic lifetimes for exchange of water molecules in hexaaqua complexes.

years. Even so, the illustration does not show the longest times that could be considered, which are comparable to geological eras.

There are a number of generalizations that help us to anticipate the lability of the complexes we are likely to meet. Thus, all complexes of *s*-block ions except the smallest (Be^{2+} and Mg^{2+}) are very labile. Across the first *d* series, complexes of M(II) ions are moderately labile with distorted Cu(II) complexes among the most labile. Complexes of M(III) ions are distinctly less labile. The M(III) ions of the *f* block are all extremely labile. In the first *d* series, the least labile M(II) and M(III) ions are those with greatest LFSE. Among the second and third *d* series complexes, inertness is quite common, which reflects the strengths of the metal–ligand bonding.

Nucleophilicity

Since the formation of any complex in solution is a substitution reaction, we need a preliminary idea of the mechanisms of these reactions in order to understand lability and inertness.

The activation barrier in a substitution reaction is determined by two main factors. One is the energy required to break the bond between the metal ion and the leaving group. The other is the energy released as the metal ion forms a bond to the entering group. The latter assistance is the kinetic equivalent of basicity and is called **nucleophilicity** (an affinity for positive centers). Just as relative basicity is measured by comparing *equilibrium* constants for acid–base reactions, nucleophilicity is measured by comparing *rate* constants for substitution reactions, and the faster the reaction with an entering group, the greater its nucleophilicity.

The kinetic analog of acidity is **electrophilicity** (an affinity for negative centers): the faster the reaction of a Lewis acid with an entering group, the greater the electrophilicity of the acid. It turns out that the most telling information about mechanisms of ligand substitution comes from studies of the variation of the ligands and observing their effect on the rate of substitution reactions, and we concentrate on this aspect.

Associative reactions

Reactions of square-planar complexes of Pt(II) are first order in the complex, first order in the entering group, and second order overall. The rate constants vary widely with the choice of the nucleophile. For example, *trans*-$[PtCl_2(py)_2]$ reacts with various entering groups (Y) to give *trans*-$[PtClY(py)_2]$ and the rate constant (in units of $M^{-1}\,s^{-1}$) varies from 4.7×10^{-4} for $Y = NH_3$, through 3.7×10^{-3} for Br^- and 0.107 for I^-, to 6.00 for thiourea, a factor of 10^4. The corresponding change in the rate constants of reactions in which the leaving group is changed from Cl^- to I^- is a factor of only 3.5. In general, effects of entering groups exceed those of leaving groups. The sensitivity of the reaction to the entering nucleophile in contrast to its modest insensitivity to the leaving group indicates that the nucleophile is the major factor determining the overall activation energy. This behavior is expected in an **associative substitution reaction**, a reaction in which the reactant passes through an activated complex with an increased coordination number. The observation that *cis* or *trans* isomers substitute with retention of the original stereochemistry suggests that the activated complex of square planar complexes is approximately a trigonal bipyramid:

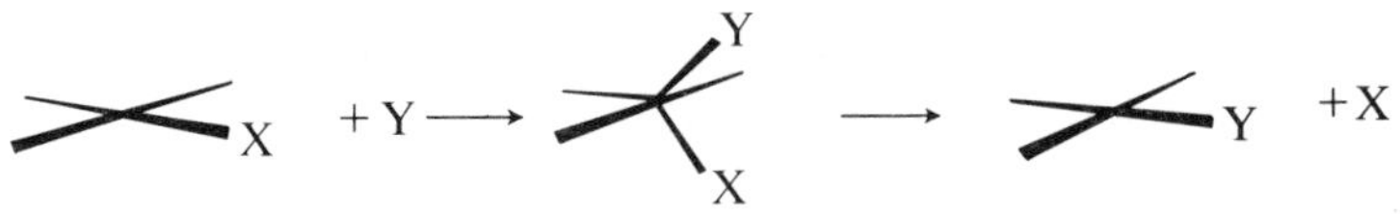

X +Y ⟶ Y X ⟶ Y +X

It is difficult to generalize about rates of associative substitutions because both the initial complex and the nucleophile must be specified.[4] However, it can be said that among the d^8 systems with similar entering and leaving groups, most are comparatively inert. Typically, lability decreases in the series

$$\mathrm{Ni(II)} > \mathrm{Pd(II)} \gg \mathrm{Pt(II)} \approx \mathrm{Rh(I)}$$

Dissociative substitutions

Octahedral Ni(II) complex formation reactions are representative of reactions of octahedral complexes. They are first order in $Ni(aq)^{2+}$, first order in the entering group, and second order overall. However, in contrast to the square-planar case, there is only a comparatively small change in the rate constant as the entering group is varied: for example, there is only a factor of 10 difference between the rate constants for the formation of $[Ni(OH_2)_4SO_4]$ and $[Ni(OH_2)_4(phen)]^{2+}$ (phen is 1,10-phenanthroline). The higher rate constant of SO_4^{2-} substitution arises from the electrostatic attraction between Ni^{2+} and SO_4^{2-} ions, which increases the probability of their encountering each other in solution.

The principal factor determining the activation energy for the replacement of a ligand in such an octahedral complex is the breaking of the bond between the metal and the leaving group. The entering group plays only a minor role in the substitution reaction. Reactions of this kind are called **dissociative substitution reactions**; they are reactions in which the activated complex has a lower coordination number than the reactant. The pathway can be written as follows:

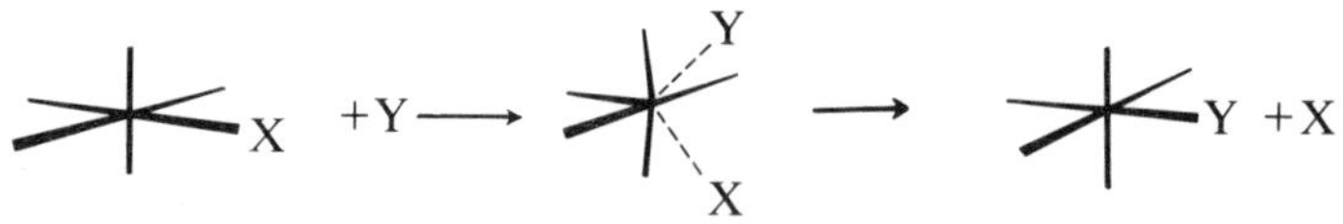

X +Y ⟶ Y X ⟶ Y +X

In practice, the dissociative character is identified from the insensitivity of the rate to variation of the entering group. Judged in this way, most octahedral substitutions seem more dissociative than associative. This is helpful because the lability of an octahedral coordination compound can be described without specifying the entering group: lability is a property of the complex itself. Among the aqua metal ions shown in Fig. 7.22, the ones that form 'weak' bonds on account of their relatively low charge and large ionic radius are more labile than those with higher charge and smaller radius. This correlation is consistent with a dissociative mechanism of substitution, which involves the breaking of a bond.

That there is a correlation of inertness with LFSE has a complicated

[4] One of the most interesting generalizations that can be made is that the ligand *trans* to the leaving group plays an important role in determining reactivity, a role comparable to the entering nucleophile. This '*trans* effect' is discussed in Section 15.3.

explanation, but the fact itself is worth remembering. The strong-field d^6 Co(III) and d^3 Cr(III) complexes are inert. The weak-field d^3 and d^8 V(II) and Ni(II) complexes are distinctly less labile than other high-spin complexes. All four have large LFSE. We explore this feature further in Chapter 15.

Further reading

Three introductory texts that cover much of the material in this chapter are:

R. DeKock and H. B. Gray, *Chemical structure and bonding*. Benjamin-Cummings, Menlo Park (1980).

F. Basolo and R. Johnson, *Coordination chemistry*. Science Reviews, Northwood (1987).

B. Figgis, *Introduction to ligand fields*. Wiley, New York (1966).

Two sources of detailed information on stability and lability are:

A. E. Martell and R. M. Smith, *Critical stability constants*. Plenum Press, New York. The four volumes are arranged according to ligand: **1** *Amino acids*, **2** *Amines*, **3** *Other organic ligands*, **4** *Inorganic ligands*.

M. L. Tobe, *Inorganic reaction mechanisms*. Thames and Nelson, London (1972).

A useful general reference is:

G. Wilkinson, R. D. Gillard, and J. McCleverty (eds.), *Comprehensive coordination chemistry*. Pergamon Press, Oxford (1987). Volume 1 of this seven-volume set provides chapters on history, coordination numbers and geometry, ligand field theory, and reactions.

Exercises

7.1 Preferably without using reference material, write out the *d*-block elements in the form of the periodic table and indicate (a) the *d*-electron count for their M^{2+} ions, (b) the elements that form (i) tetrahedral, (ii) square-planar tetrahalo complexes of formula $[MX_4]^{2-}$, (c) trends in the stability constants for complex formation between hard ligands and M^{2+} ions in the first row of the block.

7.2 Draw the structures of complexes that contain the ligand (a) en, (b) ox, (c) tren, and (d) edta.

7.3 Name and draw structures of the octahedral complex ions (a) *cis*-$[CrCl_2(NH_3)_4]^+$, (b) *trans*-$[Cr(NCS)_4(NH_3)_2]^-$, and (c) $[Co(C_2O_4)(en)_2]^+$. Is the oxalato complex *cis* or *trans*?

7.4 Write the formulas of (a) pentaamminechlorocobalt(III) chloride; (b) hexaaquairon(III) nitrate, (c) *cis*-dichlorobisethylenediammineruthenium(II) ion, (d) μ-hydroxobis[pentaamminechromium(III)] chloride.

7.5 Draw all possible isomers of (a) octahedral $[Ru(NH_3)_4Cl_2]$, (b) square-planar $[Ir(PR_3)_2H(CO)]$, (c) tetrahedral $[CoCl_3(OH_2)]^-$, and (d) octahedral $[CoCl_2(en)(NH_3)_2]^+$.

7.6 Which of the following complexes are chiral? (a) $[Cr(ox)_3]^{3-}$, (b) *cis*-$[PtCl_2(en)]$, (c) *cis*-$[RhCl_2(NH_3)_4]^+$, (d) $[Ru(bipy)_3]^{2+}$, (e) $[Co(edta)]^-$, (f) *fac*-$[Co(NO_2)_3(dien)]$, (g) *mer*-$[Co(NO_2)_3(dien)]$. Draw the enantiomers of the complexes identified as chiral and identify the plane of symmetry in structures identified as achiral.

7.7 A pink solid has the empirical formula $CoCl_3{\cdot}5NH_3{\cdot}H_2O$. A solution of this salt is also pink and rapidly gives 3 mol AgCl on titration with $AgNO_3$ solution. When the pink solid is heated, it loses 1 mol H_2O to give a purple solid with the same ratio of NH_3:Cl:Co (recall that Co(III) complexes are inert). Deduce the structures of the two octahedral complexes and draw and name them.

7.8 The hydrated chromium chloride that is available commercially has the overall composition $CrCl_3{\cdot}6H_2O$; it is blue and produces a solution with a molar electrical conductivity similar to that of $[Co(NH_3)_6]Cl_3$. In contrast, $CrCl_3{\cdot}5H_2O$ is green and has a lower molar conductivity in solution. If a dilute acidified solution of the green complex is allowed to stand for several hours, it turns blue. Interpret these observations with structural diagrams.

7.9 A complex $[PtCl_2(NH_3)_2]$ was identified as the *trans* isomer. It reacts slowly with solid Ag_2O to produce $[Pt(NH_3)_2(OH_2)_2]^{2+}$. This complex does not react with ethylenediamine to give a complex incorporating one molecule of en. Name and draw the structure of the diaqua complex.

7.10 A complex with empirical formula $PtCl_2{\cdot}2NH_3$ is an insoluble solid which, when ground with $AgNO_3$, gives a solution containing $[Pt(NH_3)_4](NO_3)_2$ and a solid phase identified as $Ag_2[PtCl_4]$. Give the structures and name of the parent $PtCl_2{\cdot}2NH_3$ compound.

7.11 Phosphine and arsine analogs of $[PtCl_2(NH_3)_2]$ were prepared in 1934 by Jensen. He reported zero dipole moments for the β isomers, where the β designation represents the product of a synthetic route analogous to that of the ammines. Give the structures of the complexes.

7.12 Which of the following complexes obey the 18-electron rule? (a) $[Cu(NH_3)_4]^{2+}$, (b) $[Fe(CN)_6]^{4-}$, (c) $[Fe(CN)_6]^{3-}$, (d) $[Cr(NH_3)_6]^{3+}$, (e) $[Cr(CO)_6]$, (f) $[Fe(CO)_5]$.

7.13 In *trans*-$[W(CO)_4(PR_3)_2]$, the alkylphosphine ligands are placed on the z axis of the coordinate system. Sketch the symmetry-adapted linear combination of σ orbitals from the two P atoms that can combine with the metal d_{z^2} orbital. Show the bonding and antibonding orbitals that may be formed.

7.14 Determine the number of unpaired d electrons, and the ligand field stabilization energy as a multiple of Δ_O or Δ_T for each of the following complexes using the spectrochemical series to decide which are strong field and which weak field. (a) $[Co(NH_3)_6]^{3+}$, (b) $[Fe(OH_2)_6]^{2+}$, (c) $[Fe(CN)_6]^{4-}$, (d) $[Cr(NH_3)_6]^{3+}$, (e) $[W(CO)_6]$, (f) tetrahedral $[FeCl_4]^-$, and tetrahedral $[Ni(CO)_4]$.

7.15 Both H^- and $(C_6H_5)_3P$ are ligands of similar field strength high in the spectrochemical series. Recalling that phosphines display π acidity, is π-acid character required for strong-field behavior? What factors account for the strength of each ligand?

7.16 Comment on the interpretation of the following series of oxide lattice enthalpies (in kJ mol^{-1}), all of which have the rock-salt structure. Do not overlook the overall trend across the d series.

CaO(3461) TiO(3879) VO(3912) MnO(3808)
FeO(3921) CoO(3988) NiO(4071)

7.17 Bearing in mind the Jahn–Teller theorem, predict the structure of $[Cr(OH_2)_6]^{2+}$.

7.18 The spectrum of d^1 $Ti^{3+}(aq)$ is attributed to a single electronic transition $e_g \leftarrow t_{2g}$. The band shown in Fig. 7.13 is not symmetrical and suggests that more than one state is involved. Is it possible to explain this observation using the Jahn–Teller theorem?

7.19 Take into account proton basicity and using the correlation with proton basicity and the chelate effect, predict which complex of Ni^{2+} with each pair of ligands is the more stable: (a) CH_3OH or CH_3NH_2, (b) $(CH_3)_2CHCH_2NH_2$ or $NH_2CH(CH_3)CH_2NH_2$, (c) NH_3 or NF_3, (d) three $NH_2CH_2CH_2NH_2$ or two $NH_2CH_2CH_2NHCH_2CH_2NH_2$.

7.20 The ligands F^- and O^{2-} are used to stabilize complexes with high oxidation number. Phosphines and CO are especially common in the preparation of complexes with low metal oxidation number. Account for these generalizations in terms of Lewis base type (see Table 6.1).

7.21 In analytical textbooks, a major class of titrations based on complex formation are called 'chelatometric' rather than simply 'complexometric'. Why are they so called and why is chelation important?

7.22 Why is it not possible to separate and isolate the isomers *cis*-$[CuCl_2(NH_3)_4]$ and *trans*-$[CuCl_2(NH_3)_4]$. Is it likely to be possible to separate and isolate the isomers *cis*-$[IrCl_2(NH_3)_4]^+$ and *trans*-$[IrCl_2(NH_3)_4]^+$?

7.23 The rate constants for the formation of $[CoX(NH_3)_5]^{2+}$ from $[Co(NH_3)_5(OH_2)]^{3+}$ for $X = Cl^-$, Br^-, N_3^-, and SCN^- differ by no more than a factor of two. What is the mechanism of the substitution?

7.24 If a substitution process is associative, why may it be difficult to characterize an aqua ion as labile or inert?

Problems

7.1 Air oxidation of Co(II) carbonate and aqueous NH_4Cl gives a pink chloride salt with a ratio of $4NH_3{:}Co$. On addition of HCl to a solution of this salt, a gas is rapidly evolved and the solution slowly turns violet on heating. Complete evaporation of the violet solution yields $CoCl_3{\cdot}4NH_3$. When this is heated in concentrated HCl, a green salt can be isolated which analyzes as $CoCl_3{\cdot}4NH_3{\cdot}HCl$. Write balanced equations for all the transformations occurring after the air oxidation. Give as much information as possible concerning the isomerism occurring and give the basis of your reasoning. If you know that the form of $[CoCl_2(en)_2]^+$ that is resolvable into enantiomers is violet, is that helpful?

7.2 By considering the splitting of the octahedral orbitals as the symmetry is lowered, draw the symmetry-adapted linear combinations and the molecular orbital energy level diagram for σ bonding in a *trans*-$[MX_2L_4]$ complex. Assume that the ligand X is lower in the spectrochemical series than L.

7.3 Draw the appropriate symmetry adapted linear combinations and the molecular orbital diagram for σ bonding in a square-planar complex. The group is D_{4h}. Take note of the small overlap of the ligand with the d_{z^2} orbital. What is the effect of π bonding?

7.4 The equilibrium constants for the successive reactions of ethylenediamine with Co^{2+}, Ni^{2+}, and Cu^{2+} are as follows:

$$[M(OH_2)_6]^{2+} + en \rightleftharpoons [M(en)(OH_2)_4]^{2+} + 2H_2O \qquad K_1$$
$$[M(en)(OH_2)_4]^{2+} + en \rightleftharpoons [M(en)_2(OH_2)_2]^{2+} + H_2O \qquad K_2$$
$$[M(en)_2(OH_2)_2]^{2+} + en \rightleftharpoons [M(en)_3]^{2+} + 2H_2O \qquad K_3$$

Ion	log K_1	log K_2	log K_3
Co^{2+}	5.89	4.83	3.10
Ni^{2+}	7.52	6.28	4.26
Cu^{2+}	10.55	9.05	−1.0

Discuss whether these data support the generalizations in the text about successive formation constants and the Irving–Williams series. How do you account for the very low value of K_3 for Cu^{2+}?

7.5 When Co(II) salts are oxidized by air in a solution containing NH_3 and $NaNO_2$, a yellow solid, $[Co(NO_2)_3(NH_3)_3]$, can be isolated. In solution it displays a conductivity small enough to be attributed to impurities only. Werner treated it with HCl to give a complex which, after a series of further reactions, he identified as *trans*-$[CoCl_2(NH_3)_3(OH_2)]^+$. It required an entirely different route to prepare *cis*-$[CoCl_2(NH_3)_3(OH_2)]^+$. Is the yellow substance *fac* or *mer*? What assumption must you make in order to arrive at a conclusion?

8 Oxidation and reduction

Extraction of the elements
8.1 The reduction of oxide ores
8.2 Elements extracted by oxidation
Reduction potentials
8.3 Redox half-reactions
8.4 Kinetic factors
Redox stability in water
8.5 Reactions with water
8.6 Disproportionation
8.7 Oxidation by atmospheric oxygen
The diagrammatic presentation of potential data
8.8 Latimer diagrams
8.9 Frost diagrams
8.10 pH dependence
8.11 Effect of complex formation on potentials
8.12 Trends in stabilities of the oxidation states of the metals
Further reading
Exercises
Problems

We close Part 2 with a discussion of the third major class of chemical reactions in which electrons are effectively transferred from one substance to another and oxidation and reduction takes place. An understanding of oxidation and reduction grew out of the early work on the extraction of elements from their natural sources, and this remains of considerable technological importance. The thermodynamic analysis of such reactions is well established, and we start with that. Kinetic considerations are of minor importance in these reactions because the high temperatures generally employed result in rapid reaction. However, when we turn to oxidation and reduction reactions in solution at about room temperature, kinetic considerations are important. In these cases we must decide first whether the reaction is thermodynamically feasible, and then whether it is kinetically practicable.

Oxidation and reduction in solution are open to study by electrochemical techniques. As a result, much of the data relating to the reactions is available in the form of electric potentials, and we shall describe how such information is used. We shall introduce several diagrammatic techniques that summarize thermodynamic relationships in a compact form. One application that we shall illustrate is the use of these diagrams to summarize trends in the stabilities of the oxidation states of the *d*- and *f*-block elements. Their application to the *p*-block elements will be illustrated more fully in Part 4.

A very large class of reactions can be regarded as occurring by the loss of electrons from one substance and their gain by another. Since electron gain is called **reduction** and electron loss is **oxidation**, the joint process is called a **redox reaction**. The substance that supplies electrons is the **reducing agent** and the substance that removes electrons is the **oxidizing agent**. These remarks show that redox reactions are similar to proton transfer reactions, but instead of a proton being transferred from the Brønsted acid to the Brønsted base, one or more electrons are transferred (perhaps with accompanying atoms) from the reducing agent to the oxidizing agent.

Extraction of the elements

The original definition of reduction was a reaction in which an oxide is converted to an element. Similarly, oxidation originally meant the reverse reaction, in which an element combined with oxygen to produce the oxide. Both terms have been generalized to yield the electron-transfer definitions, but these special cases are still the basis of a major part of chemical industry. In particular, reduction is an important technique for the conversion of ores to metals.

8.1 The reduction of oxide ores

Oxygen has been a component of the atmosphere since photosynthesis became a dominant process over a billion years ago, and many metals are found as their oxides. After about 4000 BC, copper could be extracted from its ores at temperatures attainable in primitive hearths, and the process of **smelting** was discovered, in which ores are reduced by heating the melt with a reducing agent such as carbon. Since many important ores of these easily reduced metals are sulfides, the smelting was often preceded by conversion of the sulfide to oxide by 'roasting' in air.

It was not until nearly 1000 BC, with the beginning of the Iron Age, that the higher temperatures needed to achieve the reduction of less readily reduced elements, such as iron, could be attained. Carbon remained the dominant reducing agent until the end of the nineteenth century, and metals that needed higher temperatures for their production could not be produced, even though their ores were reasonably abundant.

The technological breakthrough that resulted in the conversion of aluminum from a rarity to a major construction metal was the introduction of electrolysis. The availability of electric power also expanded the scope of carbon reduction, for electric furnaces can reach much higher temperatures than carbon-combustion furnaces such as the blast furnace. Thus, magnesium is also a twentieth-century metal even though the **Pidgeon process**, the electrothermal reduction of the oxide, uses carbon as the reducing agent.

Ellingham diagrams

Despite the fact that redox reactions do not always reach equilibrium, thermodynamics can at least be used to identify which reactions are feasible. The criterion of feasibility is that, under the prevailing conditions (and at constant temperature and pressure), the Gibbs free energy of reaction ΔG is negative (*Further information,* Chapter 4). It is usually sufficient to consider the *standard* Gibbs free energy $\Delta G^{\ominus}$ because it is related to the equilibrium constant through

$$\Delta G^{\ominus} = -RT \ln K$$

Hence a negative value of $\Delta G^{\ominus}$ corresponds to $K > 1$ and therefore to a 'favorable' reaction. Reaction rates are also relevant, but at high temperatures reactions are often fast, and we can normally assume that any thermodynamically permissible process can occur.

The free energies of metal oxide reductions depend on temperature: that is reflected in the progressive achievement of different metals through history as higher temperatures became accessible. This dependence is shown on an **Ellingham diagram**, a plot of the standard Gibbs free energy of formation of oxides against temperature (Fig. 8.1). An Ellingham diagram provides a way of treating an overall redox reaction as the difference of one of the following oxidations:

$$2C(s) + O_2(g) \rightarrow 2CO(g) \qquad \Delta G^{\ominus}(C, CO)$$
$$2CO(g) + O_2(g) \rightarrow 2CO_2(g) \qquad \Delta G^{\ominus}(CO, CO_2)$$
$$C(s) + O_2(g) \rightarrow CO_2(g) \qquad \Delta G^{\ominus}(C, CO_2)$$

and the oxidation

$$\frac{2}{x}M(s \text{ or } l) + O_2(g) \rightarrow \frac{2}{x}MO_x(s) \quad \Delta G^{\ominus}(M)$$

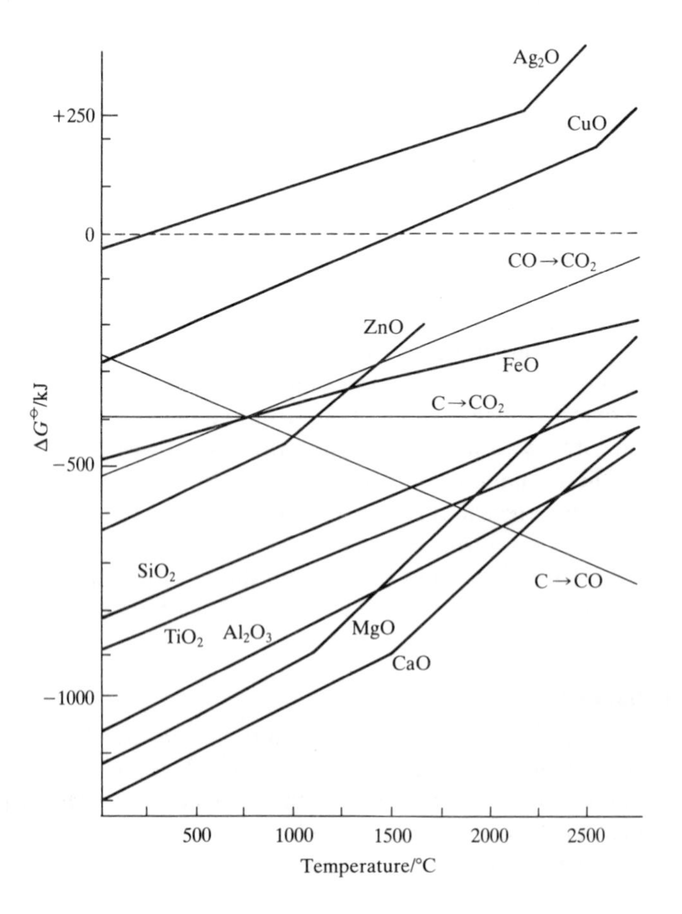

Fig. 8.1 An Ellingham diagram for the reduction of metal oxides. The standard free energies are for the formation of the oxides from the metal and for the three carbon oxidations quoted in the text.

where M is the metal of interest. All these reactions are written for 1 mol O_2, so they can be combined easily. An Ellingham diagram shows the variation of the $\Delta G^\ominus$ of each individual reaction with temperature.

We can understand the slopes of the lines in the diagram in terms of the reaction entropy $\Delta S^\ominus$ and the relation $d\Delta G^\ominus/dT = -\Delta S^\ominus$. This shows that the greater the reaction entropy, the greater the slope of $\Delta G^\ominus$. If there is a net consumption of gaseous reactants in the oxidation reaction, the reaction entropy is negative and the slope is positive. This is the case with metal oxide formation and the oxidation of CO to CO_2. The oxidation of C(*s*) to CO(*g*) causes a net increase in the number of gaseous molecules and hence has a positive reaction entropy: in this case, the slope of the line is negative.

If $\Delta G^\ominus(\mathrm{M})$ itself becomes positive at a particular temperature, the reverse reaction, thermal decomposition of the oxide, is allowed without a reducing agent being needed. Figure 8.1 shows, for instance, that Ag_2O is unstable with respect to Ag and O_2 above 250°C. For most metal oxides, a reducing agent is required, and we should consider the overall reaction obtained by subtracting the metal oxidation from one of the carbon oxidations, as in

$$\mathrm{C}(s) + \mathrm{O_2}(g) \rightarrow \mathrm{CO_2}(g) \qquad \Delta G^\ominus(\mathrm{C, CO_2})$$

$$\frac{2}{x}\mathrm{M}(l) + \mathrm{O_2}(g) \rightarrow \frac{2}{x}\mathrm{MO}_x(s) \qquad \Delta G^\ominus(\mathrm{M})$$

$$\mathrm{C}(s) + \frac{2}{x}\mathrm{MO}_x(s) \rightarrow \frac{2}{x}\mathrm{M}(l) + \mathrm{CO_2}(g) \qquad \Delta G^\ominus = \Delta G^\ominus(\mathrm{C, CO_2}) - \Delta G^\ominus(\mathrm{M})$$

In general, the Gibbs energy for any combination is

$$\Delta G^\ominus = \Delta G^\ominus(\mathrm{C}) - \Delta G^\ominus(\mathrm{M})$$

where $\Delta G^\ominus(\mathrm{C})$ is one of the carbon oxidations. It follows that metal oxide reduction is thermodynamically favorable for temperatures at which the line for the metal oxide is above any one of the lines for carbon oxidation, for then $\Delta G^\ominus$ for the metal oxide reduction by carbon is negative.

Example 8.1: *Using the Ellingham diagram*

What is the lowest temperature at which ZnO can be reduced to the metal by carbon? What is the overall reaction at this temperature.

Answer. The ZnO line in Fig. 8.1 rises above the C, CO oxidation at approximately 950°C, so above this temperature the reaction is thermodynamically feasible. The contributing reactions are:

$$2\mathrm{C}(s) + \mathrm{O_2}(g) \rightarrow 2\mathrm{CO}(g)$$

$$2\mathrm{Zn}(l) + \mathrm{O_2}(g) \rightarrow 2\mathrm{ZnO}(s)$$

Overall (subtracting):

$$2\mathrm{C}(s) + 2\mathrm{ZnO}(s) \rightarrow 2\mathrm{Zn}(l) + 2\mathrm{CO}(g)$$

or simply

$$\mathrm{C}(s) + \mathrm{ZnO}(s) \rightarrow \mathrm{Zn}(l) + \mathrm{CO}(g)$$

Exercise. What is the minimum temperature for reduction of MgO by C? Give the overall reaction.

Similar principles apply to other types of reductions. For instance, an Ellingham diagram can be used to explore whether a metal M′ can be used as a reducing agent for the oxide of another metal M. In this case, we note from the diagram whether $\Delta G^{\ominus}(\mathrm{M})$ lies above $\Delta G^{\ominus}(\mathrm{M'})$, for M′ is now taking the place of C. When

$$\Delta G^{\ominus} = \Delta G^{\ominus}(\mathrm{M'}) - \Delta G^{\ominus}(\mathrm{M})$$

is negative, the reaction

$$\mathrm{MO + M' \rightarrow M + M'O}$$

(and it analogs for MO_2 and so on) is feasible. For example, since in Fig. 8.1 the line for Si is above the line for Mg at temperatures greater than 2200°C, Mg may be used to reduce SiO_2 below that temperature. This reaction has in fact been employed to produce low-grade Si.

Chemical reductions

Industrial processes for achieving the reductive extraction of metals show a greater variety than the thermodynamic analysis might suggest. We can see this by considering three important examples of low, moderate, and extreme difficulties of reduction.

The least difficult reductions include that of Cu ores. Roasting and smelting are still widely used examples of the **pyrometallurgical extraction** of Cu, the extraction of a metal utilizing strong heating. However, recent techniques seek to avoid the major environmental problems caused by the production of the large quantity of SO_2 that accompanies roasting. One promising development is the **hydrometallurgical extraction** of Cu, its extraction by reduction of aqueous solutions of its ions (with H_2 or scrap iron). In this process, the oxide or sulfide ores are dissolved in acid with the aid of O_2 to give a solution of Cu^{2+} ions (as the sulfate). The reduction itself is

$$\mathrm{Cu^{2+}}(aq) + \mathrm{H_2}(g) \rightarrow \mathrm{Cu}(s) + 2\mathrm{H^+}(aq)$$

As well as being environmentally relatively benign, this process also allows economic exploitation of lower-grade ores.

The extraction of Fe is of intermediate difficulty. In a blast furnace (Fig. 8.2), which is still the major source of the element, the charge of Fe ores (Fe_2O_3, Fe_3O_4), coke, and limestone is heated with a blast of hot air. As the exothermic reactions proceed, a composition and temperature gradient is set up in the furnace. Below 700 to 800°C, CO reduces the ores to FeO. The Ellingham diagram in Fig. 8.1 indicates that reduction to Fe by CO can occur above about 600°C and that direct reduction, in which the reducing agent is C, can occur near 1000°C. In this region, limestone also decomposes to CaO and CO_2.

More difficult than the extraction of either Cu or Fe is the extraction of Si from its oxide. Silicon of 96 to 99 percent purity is prepared by reduction of quartzite or sand (SiO_2) with high purity coke. The Ellingham diagram shows that the reduction is feasible only at temperatures in excess of about 1500°C. This is achieved in an electric arc furnace in the presence of excess

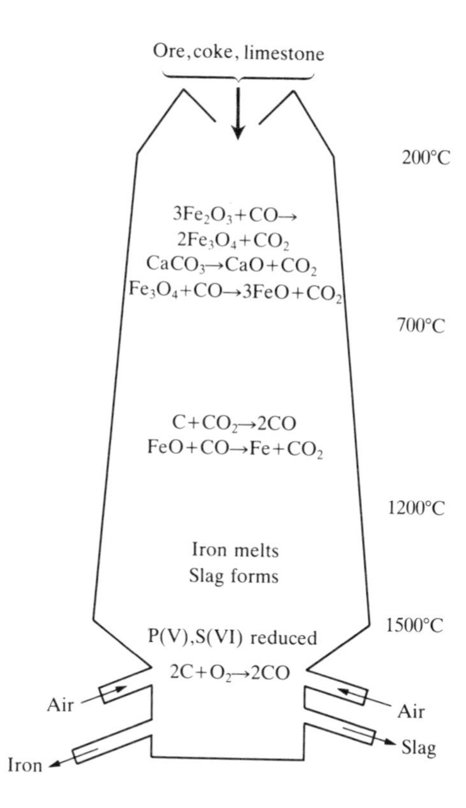

Fig. 8.2 A schematic diagram of a blast furnace showing the typical composition and temperature profile.

silica (to prevent the accumulation of SiC):

$$SiO_2 + 2C \xrightarrow{1500°C} Si + 2CO$$

$$2SiC + SiO_2 \rightarrow 3Si + 2CO$$

Very pure Si for semiconductor applications is made by preparing volatile compounds, such as $SiCl_4$. These are purified by exhaustive fractional distillation and then reduced to the element with pure hydrogen. The resulting semiconductor-grade Si is melted and grown into single crystals by pulling the growing crystal slowly from the melt.

Electrolytic reduction

The Ellingham diagram shows that the direct reduction of Al_2O_3 with C becomes feasible only above 2000°C, which is uneconomically expensive. However, the reduction can be brought about electrolytically, and all modern production uses the **Hall–Héroult process**, which was invented in 1886 independently by Charles Hall in the United States and Paul Héroult in France.

Electrolytic reduction can be regarded as a technique for driving a reduction by coupling it (through electrodes and external circuitry) to a reaction or physical process with a more negative ΔG. The free energy available from the external source can be assessed from the potential difference E it produces across the electrodes using the thermodynamic relation

$$\Delta G = -nFE$$

where n is the number of moles of electrons transferred and F is Faraday's constant ($F = 96.5\ \text{kC mol}^{-1}$). Hence, since the total Gibbs energy change of the coupled internal and external process is

$$\Delta G + \Delta G(\text{external process}) = \Delta G - nFE_{ext}$$

if the potential difference of the external source exceeds

$$E_{ext} = \frac{\Delta G}{nF}$$

the reduction is thermodynamically feasible, for then the overall process occurs with a decrease in free energy.

Example 8.2: *Estimating the potential difference needed for electrolysis*

Estimate the minimum potential difference needed to reduce Al_2O_3 at 500°C.

Answer. The reaction free energy for the decomposition

$$\tfrac{2}{3}Al_2O_3 \rightarrow \tfrac{4}{3}Al + O_2$$

can be taken from the Ellingham diagram in Fig. 8.1 where the free energies are given per mole of O_2. For the reaction as written, $\Delta G^{\ominus} = +960\ \text{kJ}$. Since 4 mol e^- are transferred (because in the reaction as written, $\tfrac{4}{3}$ mol Al^{3+} are reduced to the metal), we take $n = 4$ mol. Therefore,

$$E_{ext} = \frac{960\ \text{kJ}}{4\ \text{mol} \times 96.49\ \text{kC mol}^{-1}} = 2.5\ \text{V}$$

That is, at 500°C, a potential difference of at least 2.5 V must be applied to the oxide to bring about reduction.

Exercise. Estimate the potential difference needed to reduce TiO_2 electrolytically to the metal.

In practice, the bauxite ore may be considered a mixture of the acidic oxide SiO_2 and the amphoteric oxides Al_2O_3 and Fe_2O_3 (with some TiO_2 too). The Al_2O_3 is extracted with aqueous NaOH. This separates the Al and Si oxides from the Fe (which needs a more concentrated alkali for significant reaction). The neutralization of the solution with CO_2 precipitates $Al(OH)_3$, leaving silicates in solution. The $Al(OH)_3$ is then dissolved in fused cryolite (Na_3AlF_6) and the melt is reduced electrolytically at a cathode. Since the commercial process requires 4.5 V with a current density of about $1\ \text{A cm}^{-2}$, plants use massive quantities of electricity. Consequently, Al is often produced where electricity is cheap (in Quebec for instance) rather than where bauxite is mined (in Jamaica, for example).

8.2 Elements extracted by oxidation

The halogens are the most important elements extracted by oxidation. The standard reaction free energy for the oxidation of Cl^- ions in water:

$$2Cl^-(aq) + 2H_2O(l) \rightarrow 2OH^-(aq) + H_2(g) + Cl_2(g) \qquad \Delta G^{\ominus} = +422\ \text{kJ}$$

is strongly positive, which suggests that electrolysis is required. The minimum potential difference that can bring it about is 2.2 V (since $n = 2$ mol for the reaction as written).

It may seem that there is a problem with the competing reaction

$$2H_2O(l) \rightarrow 2H_2(g) + O_2(g) \qquad \Delta G^{\ominus} = +414\ \text{kJ}$$

which is feasible when the applied potential difference is only 1.2 V (since $n = 4$ mol). The question underlines the importance of kinetic factors. The rate of oxidation of water (in effect the electrolytic preparation of O_2) is extremely slow at potentials where it first becomes favorable thermodynamically. This is expressed by saying that the reduction has a high **overpotential** η, where η is the potential difference *in addition* to the equilibrium value needed to achieve a significant rate of reaction. Consequently, the electrolysis of brine produces Cl_2, H_2, and aqueous NaOH, but not much O_2. One problem faced by the industry is therefore to try to develop uses for both Cl_2 and NaOH in a market balance as closely adjusted to the stoichiometry of the reaction as possible.

Oxygen, not fluorine, is produced if aqueous fluoride solutions are electrolyzed. Therefore, F_2 is prepared by the electrolysis of an anhydrous HF/KF mixture, which is conducting and melts at 72°C. The more readily oxidizable halogens, Br_2 and I_2, are obtained by chemical oxidation of the aqueous halides with Cl_2.

Since O_2 is available from fractional distillation of air, chemical methods are not necessary (but might become necessary if we colonize the other planets). Sulfur is an interesting mixed case. Elemental S is either mined or produced by oxidation of the H_2S obtained from 'sour' natural gas by trapping in ethanolamine $HOCH_2CH_2NH_2$. The oxidation is accomplished by the **Claus process**, which exploits the following three reactions:

$$2H_2S + O_2 \xrightarrow{\text{Low temperature}} 2S + 2H_2O$$

$$2H_2S + 3O_2 \rightarrow 2SO_2 + 2H_2O$$

$$2H_2S + SO_2 \xrightarrow{\text{Oxide catalyst, 300°C}} 3S + 2H_2O$$

The oxide catalyst used in the third step is typically Fe_2O_3 or Al_2O_3. Since emission of sulfur gases to the atmosphere is a major source of pollution, the Claus process is environmentally valuable.

The only important metals to be obtained by oxidation are the ones that occur native (that is, as the element). Gold is an example, since it is difficult to separate the granules of metal in low-grade ores by simple 'panning'. Dissolving the Au depends on promoting its oxidation, which is favored by complexation with CN^- ions, when it forms $[Au(CN)_2]^-$ ions. This complex can be reduced to the metal by reaction with a reactive metal, such as Zn:

$$2[Au(CN)_2]^-(aq) + Zn(s) \rightarrow 2Au(s) + [Zn(CN)_4]^{2-}(aq)$$

Reduction potentials

One of the points we have made is that one reaction may be used to drive another if the overall ΔG is negative. We have also seen that free energies may be expressed as a potential difference. We can combine these two remarks and obtain a very useful way of discussing redox reactions.

8.3 Redox half-reactions

It is convenient to think of a redox reaction as the sum of two **half-reactions**. In a reduction half-reaction, a substance gains electrons, as in

$$2H^+(aq) + 2e^- \rightarrow H_2(g)$$

In an oxidation half-reaction, a substance loses electrons, as in

$$Zn(s) \rightarrow Zn^{2+}(aq) + 2e^-$$

We do not ascribe a state to the electrons, since the division into half-reactions is only conceptual and need not correspond to an actual physical separation of the two processes. The oxidized and reduced species in a half-reaction form a redox **couple**. A couple is written with the oxidized species before the reduced species, as in H^+/H_2 and Zn^{2+}/Zn; in general, we denote a redox couple Ox/Red.

It is usually desirable to adopt the convention of writing *all* half-reactions as reductions. (This is reminiscent of discussing Brønsted bases in terms of the properties of their conjugate acids, so that all substances are discussed as acids.) Since an oxidation is the reverse of a reduction, we can write the second of these two half-reactions as

$$Zn^{2+}(aq) + 2e^- \rightarrow Zn(s)$$

The overall reaction is then the *difference* of the two reduction half-reactions with the number of electrons matching. This is similar to our discussion of the Ellingham diagram, in which all the reactions were oxidations, and the overall reaction was a difference of reactions with matching numbers of O_2 molecules.

Standard electrode potentials

Each reduction half-reaction can be thought of as having its own $\Delta G^{\ominus}$, the standard Gibbs free energy of the overall reaction being their difference. The overall reaction is favorable in the direction that corresponds to a negative value of this $\Delta G^{\ominus}$.

Since reduction half-reactions must always occur in pairs in any actual reaction, only the difference in their standard free energies has experimental significance. Therefore, we can chose one half-reaction as having $\Delta G^{\ominus} = 0$, and report all other values relative to it. By convention, the specially chosen half-reaction is the reduction of hydrogen ions:

$$2H^+(aq) + 2e^- \rightarrow H_2(g) \qquad \Delta G^{\ominus} = 0$$

With this choice, the reduction of Zn^{2+} ions is found by noting (experimentally) that for the overall reaction

$$Zn^{2+}(aq) + H_2(g) \rightarrow Zn(s) + 2H^+ \qquad \Delta G^{\ominus} = +147\ \text{kJ mol}^{-1}$$

Then, since the H^+ reduction half-reaction makes zero contribution, it follows that

$$Zn^{2+}(aq) + 2e^- \rightarrow Zn(s) \qquad \Delta G^{\ominus} = +147\ \text{kJ mol}^{-1}$$

Gibbs free energies of overall reactions may be measured electrochemically by setting up a galvanic cell (Fig. 8.3). We then measure the potential

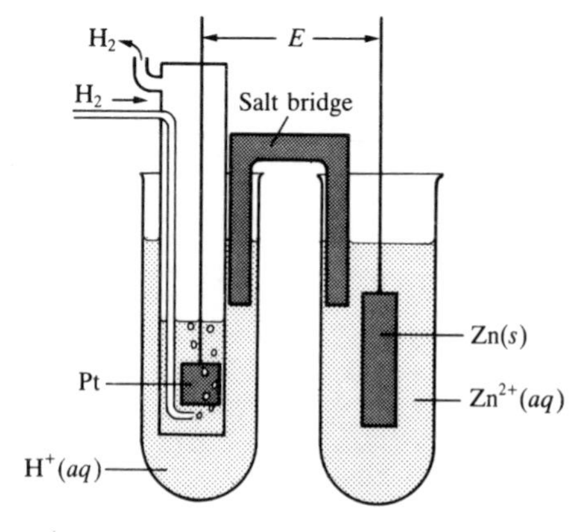

Fig. 8.3 A schematic diagram of a galvanic cell. The standard cell potential $E^{\ominus}$ is the potential difference when the cell is not generating current and all the substances are in their standard states.

difference between its electrodes[1] and, if desired, convert the potential difference using $\Delta G = -nFE$. Tabulated values—normally for standard conditions[2]—are usually kept in the units in which they were measured, namely volts (V).

The potential that corresponds to the $\Delta G^\ominus$ of a half-reaction is written $E^\ominus$ and is called the **standard reduction potential**. Since $\Delta G^\ominus$ for the reduction of H^+ is arbitrarily set at zero, the standard reduction potential of the H^+/H_2 couple is also zero (at all temperatures):

$$2H^+(aq) + 2e^- \rightarrow H_2(g) \qquad E^\ominus(H^+, H_2) = 0$$

Similarly, for the Zn^{2+}/Zn couple, for which $n = 2$ mol,

$$Zn^{2+}(aq) + 2e^- \rightarrow Zn(s) \qquad E^\ominus(Zn^{2+}, Zn) = -0.76\text{ V}$$

The value of $E^\ominus$ for an overall reaction is the difference of the two standard potentials of the reduction half-reactions into which it can be divided:

$$2H^+(aq) + 2e^- \rightarrow H_2(g) \qquad E^\ominus = 0$$

$$Zn^{2+}(aq) + 2e^- \rightarrow Zn(s) \qquad E^\ominus = -0.76\text{ V}$$

Difference:

$$2H^+(aq) + Zn(s) \rightarrow Zn^{2+}(aq) + H_2(g) \qquad E^\ominus = +0.76\text{ V}$$

The consequence of the negative sign in $\Delta G^\ominus = -nFE^\ominus$ is that the rule that a reaction is favorable if $\Delta G^\ominus < 0$ is replaced by the rule that the reaction is favorable if $E^\ominus > 0$. Since $E^\ominus > 0$ for the reaction above, we know that Zn has a thermodynamic tendency to reduce H^+ ions under standard conditions. The same is true for any couple with a negative standard reduction potential.

The electrochemical series

A negative standard reduction potential signifies a couple in which the reduced species (the Zn in Zn^{2+}/Zn) is a reducing agent for H^+ ions under standard conditions in aqueous solution. That is, if $E^\ominus(\text{Ox, Red}) < 0$, the substance 'Red' is a strong enough reducing agent to reduce H^+ ions.

A short list of $E^\ominus$ values at 25°C is given in Table 8.1. The list is arranged in the order of the **electrochemical series**:

Strongly positive couples	Ox/Red	Ox strongly oxidizing
	⋮	
Strongly negative couples	Ox/Red	Red strongly reducing

The reduced member of a couple is able thermodynamically (not always kinetically) to reduce the oxidized member of any couple that lies above it in the series.

[1] In practice, we must ensure that the cell is acting reversibly in a thermodynamic sense, which means that the potential difference must be measured with negligible current flowing.

[2] Standard conditions are all substances at 1 bar pressure and unit activity. For our purposes, there is negligible difference in using 1 atm in place of 1 bar. It is helpful to bear in mind that for reactions involving H^+ ions, standard conditions correspond to pH = 0, approximately 1 M acid.

Table 8.1. Selected standard reduction potentials at 25°C

Couple	$E^{\ominus}$/V
$F_2(g) + 2e^- \rightarrow 2F^-(aq)$	+2.87
$Ce^{4+}(aq) + e^- \rightarrow Ce^{3+}(aq)$	+1.76
$MnO_4^-(aq) + 8H^+(aq) + 5e^- \rightarrow Mn^{2+}(aq) + 4H_2O(l)$	+1.51
$Cl_2(g) + 2e^- \rightarrow 2Cl^-(aq)$	+1.36
$[IrCl_6]^{2-}(aq) + e^- \rightarrow [IrCl_6]^{3-}(aq)$	+0.87
$Fe^{3+}(aq) + e^- \rightarrow Fe^{2+}(aq)$	+0.77
$[PtCl_4]^{2-}(aq) + 2e^- \rightarrow Pt(s) + 4Cl^-(aq)$	+0.60
$I_3^-(aq) + 2e^- \rightarrow 3I^-(aq)$	+0.54
$[Fe(CN)_6]^{3-}(aq) + e^- \rightarrow [Fe(CN)_6]^{4-}(aq)$	+0.36
$AgCl(s) + e^- \rightarrow Ag(s) + Cl^-(aq)$	+0.22
$2H^+(aq) + 2e^- \rightarrow H_2(g)$	0
$AgI(s) + e^- \rightarrow Ag(s) + I^-(aq)$	−0.15
$Zn^{2+}(aq) + 2e^- \rightarrow Zn(s)$	−0.76
$Al^{3+}(aq) + 3e^- \rightarrow Al(s)$	−1.68
$Ca^{2+}(aq) + 2e^- \rightarrow Ca(s)$	−2.84
$Li^+(aq) + e^- \rightarrow Li(s)$	−3.04

Example 8.3: *Using the electrochemical series*

Among the couples in Table 8.1 is the permanganate ion, MnO_4^-, the common analytical reagent used in redox titrations of iron. Which of the following ions can permanganate oxidize in acidic solution: Fe^{2+}, Cl^-, Ce^{3+}?

Answer. The standard reduction potential of MnO_4^- to Mn^{2+} in acidic solution is +1.51 V. The potentials for the listed ions are +0.77, +1.36, and +1.72 V respectively. Permanganate ion is a sufficiently strong oxidizing agent in acidic solution to oxidize the first three ions, which have less positive reduction potentials. It cannot oxidize Ce^{3+}, which has a more positive reduction potential.

Exercise. Another common analytical oxidizing agent is an acidic solution of dichromate ions ($Cr_2O_7^{2-}$), for which $E^{\ominus} = +1.38$ V. Is it useful for titration of Fe^{2+}? Could there be a side reaction when Cl^- is present?

The Nernst equation

Standard reduction potentials are signposts of spontaneous change under *standard* conditions. In order to judge the tendency of a reaction to run in a particular direction under *nonstandard* conditions, we need to know the sign and value of ΔG under the relevant conditions. For this, we use the thermodynamic result (*Further information*, Chapter 4) that

$$\Delta G = \Delta G^{\ominus} + RT \ln Q$$

where Q is the **reaction quotient**

$$a\mathrm{Ox_A} + b\mathrm{Red_B} \rightarrow a'\mathrm{Red_A} + b'\mathrm{Ox_B} \qquad Q = \frac{[\mathrm{Red_A}]^{a'}[\mathrm{Ox_B}]^{b'}}{[\mathrm{Ox_A}]^{a}[\mathrm{Red_B}]^{b}}$$

The reaction is spontaneous under the stated conditions if ΔG is negative. This criterion can be expressed in terms of potentials by substituting

$\Delta G = -nFE$ and $\Delta G^{\ominus} = -nFE^{\ominus}$, which gives

$$E = E^{\ominus} - \frac{RT}{nF}\ln Q \qquad (1)$$

This is called the **Nernst equation**. The reaction is spontaneous if E is positive. Since at equilibrium, $E = 0$ and $Q = K$, we have

$$E^{\ominus} = \frac{RT}{nF}\ln K \qquad (2)$$

If we regard E as the difference of two potentials, just as $E^{\ominus}$ is the difference of two standard potentials, we can write the potential of each couple as

$$E = E^{\ominus} - \frac{RT}{nF}\ln Q \qquad Q = \frac{[\mathrm{Red}]^{a'}}{[\mathrm{Ox}]^{a}}$$

A useful numerical form of this expression is that at 25°C,

$$E = E^{\ominus} - \frac{0.059\ \mathrm{V}}{n}\log Q$$

where $\log Q$ is $\log_{10} Q$. It follows that each increase of an order of magnitude in concentration ratio ($Q \rightarrow 10Q$) corresponds to a $(59/n)$ mV decrease in potential.

Example 8.4: *Using the Nernst equation*

What is the dependence of the reduction potential of the H^+/H_2 couple on pH when the hydrogen pressure is 1 bar and the temperature is 25°C?

Answer. The reduction half-reaction is

$$2H^+(aq) + 2e^- \rightarrow H_2(g)$$

The Nernst equation for the couple is therefore

$$E(H^+, H_2) = E^{\ominus} - \frac{0.059\ \mathrm{V}}{2}\log\frac{p(H_2)}{[H^+]^2}$$
$$= 0 + 0.059\ \mathrm{V} \times \log[H^+]$$

(because $E^{\ominus} = 0$ and $\log p = 0$ when $p = 1$ bar). Hence

$$E(H^+, H_2) = -59\ \mathrm{mV} \times \mathrm{pH}$$

That is, the reduction potential becomes more negative by 59 mV for each unit increase in pH.

Exercise. What is the potential for reduction of MnO_4^- to Mn^{2+} in neutral (pH = 7) solution?

8.4 Kinetic factors

Overpotential

A negative reduction potential for a metal ion indicates that the corresponding metal is thermodynamically capable of reducing H^+ ions or any more positive couple under standard conditions in aqueous solution; it does not assure us that a mechanistic pathway exists for this reduction to be realized. There are no fully general rules that predict when reactions are likely to be fast, for the factors that control them are diverse (as we see in Chapter 15). However, a useful rule of thumb (but one with some notable exceptions) is that reactions with potentials that are positive by more than approximately 0.6 V for a one-electron transfer are likely to proceed rapidly. All metal ions having $E^{\ominus} < -0.6$ V are therefore likely to reduce H^+ ions at an appreciable rate under standard conditions.

The existence of the overpotential explains why some metals reduce acids and not water itself. These metals (which include Fe and Zn) have negative reduction potentials, but their potentials are not low enough to achieve the necessary overpotential in neutral solution (at pH = 7). The difference $E(H^+, H_2) - E(Fe^{2+}, Fe)$ can be increased if $E(H^+, H_2)$ is made more positive, which can be done by lowering the pH from 7 toward a more acidic value. When the difference exceeds the overpotential, the metal brings about reduction at an appreciable rate.

Example 8.5: *Taking overpotential into account*

Is Fe likely to be rapidly oxidized to $Fe^{2+}(aq)$ in solution at pH = 7 and 25°C?

Answer. We need to judge the value of the potential difference of the couples under the prevailing conditions. The Nernst equation for the reaction

$$Fe(s) + 2H^+(aq) \rightarrow Fe^{2+}(aq) + H_2(g) \qquad n = 2$$

$$Q = \frac{p(H_2)[Fe^{2+}]}{[H^+]^2}$$

at 25°C is

$$E = E^{\ominus} - \tfrac{1}{2} \times 0.0591\ \text{V} \log Q$$

Substituting $p(H_2) = 1$ bar and pH = 7 (i.e. $[H^+] = 10^{-7}$ M) and $E^{\ominus} = +0.47$ V (from the standard potentials of Fe^{2+}, Fe and H^+, H_2), gives

$$E = 0.47\ \text{V} - 0.0295\ \text{V} \times \log [Fe^{2+}]/10^{-14}$$

If the concentration of Fe^{2+} approaches 1 M,

$$E = 0.47\ \text{V} - 0.0295\ \text{V} \times 14 = +0.05\ \text{V}$$

Although E is positive, it is smaller than the overpotential typically required for a significant reaction rate. This suggests that the reaction will be slow on a laboratory time scale.

Exercise. If an exposed surface is maintained, Mg can undergo rapid oxidation in water at pH = 7, 25°C. What is the value of E for this reaction?

Electron transfer

In some cases, the overpotential can be interpreted mechanistically, particularly when the reaction is in homogeneous solution. In these cases we can gain some insight into the factors that control reaction rates.

The transfer of one electron often takes place by **outer-sphere electron transfer**, a process in which no change occurs in the coordination sphere of the redox center (the atom undergoing a change of oxidation number). This transfer can be fast, and it is found that logarithms of rates are often proportional to differences of standard potentials: the more favorable the equilibrium, the faster the reaction. The quantitative form of the dependence is discussed in Section 15.12.

The next simplest case is an **inner-sphere** electron transfer, in which a change does occur in the coordination sphere of the redox center. A classic example is the reduction of $[CoCl(NH_3)_5]^{2+}$ by $Cr^{2+}(aq)$:

$$[CoCl(NH_3)_5]^{2+}(aq) + [Cr(OH_2)_6]^{2+}(aq) + 5H_2O(l) \rightarrow [Co(OH_2)_6]^{2+}(aq) + [CrCl(OH_2)_5]^{2+}(aq) + 5NH_3(aq)$$

Since it is known that replacement of Cl^- at the nonlabile Co(III) center is slow, the rearrangement must accompany reduction to the more labile Co(II) complex. It is also frequently found in this kind of reaction that the more favorable the potential, the faster the reaction.

Noncomplementary redox reactions, those in which the change in oxidation numbers of the oxidizing and reducing agents are unequal, are often slow. This is because the reaction cannot be accomplished in a single electron-transfer step but must take place in several. The reaction may be slow if any one of the steps is slow, or has such an unfavorable equilibrium constant that the concentration of an essential intermediate is low. The second case can be pictured as follows. The reaction

$$2Fe^{3+}(aq) + Tl^{+}(aq) \rightarrow 2Fe^{2+}(aq) + Tl^{3+}(aq)$$

might require a rapid but thermodynamically unfavorable ($K \ll 1$) initial step

$$Fe^{3+}(aq) + Tl^{+}(aq) \rightarrow Fe^{2+}(aq) + Tl^{2+}(aq) \qquad E^{\ominus} = -1.4\ V$$

Then, since the formation of the intermediate Tl^{2+} ion is highly unfavorable, the next step

$$Fe^{3+}(aq) + Tl^{2+}(aq) \rightarrow Fe^{2+}(aq) + Tl^{3+}(aq)$$

is slowed for lack of reactant.

Atom transfer

Many redox reactions occur by atom transfer, typically O or H, and hence an inner-sphere mechanism. For example, a radioisotope tracer study of the oxidation of NO_2^- ion by hypochlorous acid, HOCl, for which the overall reaction is

$$NO_2^-(aq) + {}^{18}OCl^-(aq) \rightarrow NO_2{}^{18}O^-(aq) + Cl^-(aq)$$

shows that the reaction proceeds by attack of NO_2^- on the O atom of OCl^-, with displacement of Cl^- (**1**).

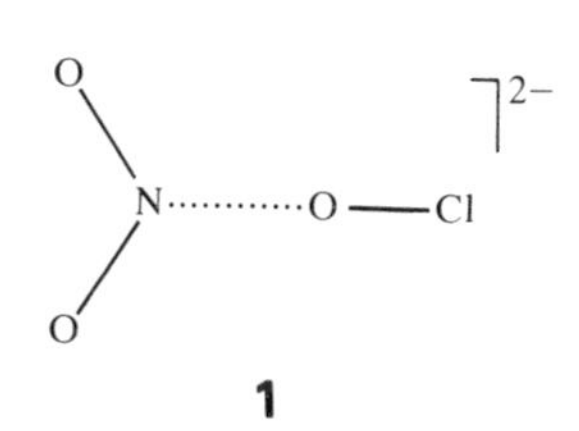

1

It is found that the rate of reaction varies strongly with the oxidation number of the central atom in an oxoanion, and that the lower the oxidation number, the faster the reaction. For example, rates of oxoanion reduction lie in the order

$$ClO_4^- < ClO_3^- < ClO_2^- < ClO^-$$
$$ClO_4^- < SO_4^{2-} < HPO_4^{2-}$$

The size of the central atom is also important, and the larger the central atom the greater the rate:

$$ClO_3^- < BrO_3^- < IO_3^-$$

Iodate reactions are fast, and equilibrate rapidly enough to be useful in titrations.

One final and very useful empirical rule is that the formation and decomposition of the common diatomic molecules O_2, N_2, and H_2 have complex mechanisms and are usually slow. We have already discussed (in the context of overpotential) the slow kinetics of reactions involving O_2 and H_2.

Redox stability in water

An ion or molecule in solution may be destroyed by a redox reaction with any of the other species present. Therefore, when assessing the stability of a species in solution we must bear in mind all possible reactants: the solvent, other solutes, the solute itself (disproportionation), and dissolved O_2.

8.5 Reactions with water

Water may act as an oxidizing agent, being reduced to H_2 in the process. It may also act as a reducing agent, when it is oxidized to O_2. Species that are thermodynamically stable in water must have potentials lying between the limits defined by these two processes.

Oxidation by water

The reaction of metals with water or aqueous acid is actually the oxidation of the metal by water or hydrogen ions, since the overall reaction is one of the following types:

$$M(s) + 2H_2O(l) \rightarrow M^{2+}(aq) + H_2(g) + 2OH^-(aq)$$

$$M(s) + 2H^+(aq) \rightarrow M^{2+}(aq) + H_2(g)$$

M can be any *s*-block metal other than Be, and a first *d*-series metal from Group 4 through at least Group 7 (Ti, V, Cr, Mn). A number of other metals undergo similar reactions but with different numbers of electrons being transferred. An example from Group 3 is

$$2Sc(s) + 6H^+(aq) \rightarrow 2Sc^{3+}(aq) + 3H_2(g)$$

When the standard potential for the reduction of a metal ion to the metal is negative, the metal should undergo oxidation in 1 M acid with the evolution of H_2. However, the role of the overpotential must always be considered.

Although reaction of Mg and Al with moist air is thermodynamically feasible, both metals can be used for years in the presence of water and oxygen. They survive because they are **passivated**, protected against reaction, by an impervious film of oxide. MgO and Al_2O_3 are both very stable, and form a protective skin on the metal beneath. A similar passivation occurs with Fe, Cu, and Zn. The process of 'anodizing' a metal, in which the metal is made an anode in an electrolytic cell, is one in which partial oxidation produces a smooth, hard passivating film on its surface.

Reduction by water

Water can act as a reducing agent through the half-reaction

$$2H_2O(l) \rightarrow 4H^+(aq) + O_2(g) + 4e^-$$

This is the reverse of the reduction half-reaction

$$O_2(g) + 4H^+(aq) + 4e^- \rightarrow 2H_2O(l) \qquad E^\ominus = +1.23\ \text{V}$$

Reduction by water and, equivalently, oxidation of water are of considerable industrial and biochemical interest. For interest, the process is driven in the electrolytic generation of H_2 and O_2 and is involved in photosynthesis.

The strongly positive value $E^\ominus = +1.23$ V shows that acidified water is a poor reducing agent except toward strong oxidizing agents. An example of the latter is $Co^{3+}(aq)$, for which $E^\ominus(Co^{3+}, Co^{2+}) = +1.82$ V. It is reduced by water with the evolution of O_2:

$$4Co^{3+}(aq) + 2H_2O(l) \rightarrow 4Co^{2+}(aq) + O_2(g) + 4H^+(aq) \qquad E^\ominus = +0.59\ \text{V}$$

This $E^\ominus$ is very close to the notional overpotential needed for a significant reaction rate. Since H^+ ions are produced in the reaction, changing from acidic to neutral or basic solution favors the oxidation, because lowering the concentration of H^+ ions favors the products.

Only a few oxidizing agents (Ag^{2+} is another example) can oxidize water rapidly enough to give appreciable rates of O_2 evolution. This is because overpotentials are too high for many reagents that are thermodynamically capable of acting. Indeed, standard potentials greater than +1.23 V occur for several redox couples that are regularly used in aqueous solution, including Ce^{4+}/Ce^{3+}, the acidified dichromate ion couple $Cr_2O_7^{2-}/Cr^{3+}$, and the acidified permanganate couple MnO_4^-/Mn^{2+}. The origin of the overpotential is the activation barrier to the transfer of four electrons and the formation of an oxygen–oxygen double bond. It remains a challenge to inorganic chemists to find good catalysts for O_2 evolution: some progress has been made using coordination compounds of Ru. Existing catalysts include the relatively poorly understood coatings that are used in the anodes of cells used for the commercial electrolysis of water. They also include the even more mysterious enzyme system that is found in the O_2 evolution apparatus of the plant photosynthetic center. This system is based on Mn with four identifiable levels of oxidation; although Nature is elegant and efficient, it is also complex!

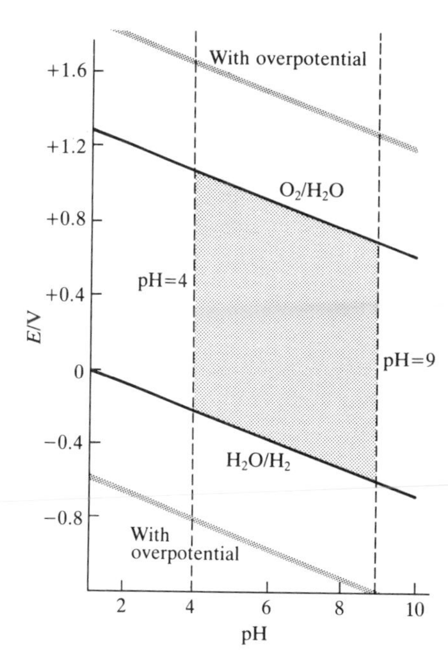

Fig. 8.4 The stability field of water. The vertical axis is the reduction potentials of redox couples in water: those above the upper line can oxidize water; those below the lower line can reduce water. The gray lines are the boundaries when overpotential is taken into account, and the broken vertical lines represent the normal pH range for natural waters. Hence, the gray area is the stability field for natural waters.

The stability field of water

A reducing agent that can reduce water to H_2 sufficiently rapidly, or an oxidizing agent that can oxidize water to O_2 sufficiently rapidly, cannot survive in aqueous solution. This is the analog of the Brønsted leveling effect (Section 5.4) for acid and bases. It is conventional in redox chemistry to refer not to the leveling effect but to the **stability field** of water (Fig. 8.4). A stability field consists of the values of reduction potential and pH for which water is thermodynamically stable against oxidation and reduction.

We can identify the boundaries of the stability field by finding the dependence of E on pH for the relevant half-reactions. The Nernst equation for the O_2, H^+/H_2O couple is

$$O_2(g) + 4H^+(aq) + 4e^- \rightarrow 2H_2O(l) \qquad n = 4 \qquad Q = \frac{1}{p(O_2)[H^+]^4}$$

so

$$E = E^{\ominus} + \frac{RT}{4F} \ln p(O_2)[H^+]^4$$

If the partial pressure of oxygen is 1 bar, at 25°C this becomes

$$E = 1.23\ \text{V} - 0.059\ \text{V} \times \text{pH}$$

Any substance with a *higher* potential than this can be reduced by water, with the production of O_2. Hence, this expression defines the upper boundary of the stability field. The reduction of water to H_2 occurs by the half-reaction

$$2H^+(aq) + 2e^- \rightarrow H_2(g) \qquad n = 2 \qquad Q = \frac{p(H_2)}{[H^+]^2}$$

so

$$E = E^{\ominus} - \frac{RT}{2F} \ln \frac{p(H_2)}{[H^+]^2}$$

If the partial pressure of hydrogen is about 1 bar and the temperature 25°C, this becomes

$$E = -0.059\ \text{V} \times \text{pH}$$

Any substance with a potential that is *lower* than this is thermodynamically capable of reducing water to H_2, so the line gives the lower boundary of the stability field.

Couples that are thermodynamically stable in water lie between the limits defined by the sloping lines in Fig. 8.4. A couple that lies outside the field is unstable. The couple consists either of too strong a reducing agent (below the H_2 production line) or too strong an oxidizing agent (above the O_2 production line). The stability field in 'natural' water is represented by the addition of two vertical lines at pH = 4 and pH = 9 which mark the limits on pH commonly found in lakes and streams. Diagrams like this are widely used in geochemistry, as we shall see in Section 8.10.

8.6 Disproportionation

Since $E^{\ominus}(Cu^+, Cu) = +0.52$ V and $E^{\ominus}(Cu^{2+}, Cu^+) = +0.16$ V, and both lie within the stability field of water, Cu^+ ions neither oxidize nor reduce water. Despite this, Cu(I) is not stable in aqueous solution because it can undergo **disproportionation**, a reaction in which the oxidation number of an element is simultaneously raised and lowered:

$$Cu^+(aq) + Cu^+(aq) \rightarrow Cu^{2+}(aq) + Cu(s)$$

This is thermodynamically feasible since $E^{\ominus} = 0.52\ \text{V} - 0.16\ \text{V} = +0.36$ V. We can obtain a more quantitative picture of the equilibrium position by employing eqn (2) in the form

$$0.36\ \text{V} = \frac{0.059\ \text{V}}{n} \log K$$

Since one electron is transferred in the reaction, $K = 1.3 \times 10^6$.

Hypochlorous acid is also subject to disproportionation:

$$5HOCl(aq) \rightarrow 2Cl_2(g) + ClO_3^-(aq) + 2H_2O(l) + H^+(aq)$$

This reaction is the difference of the following two half-reactions:

$$4HOCl(aq) + 4H^+(aq) + 4e^- \rightarrow 2Cl_2(g) + 4H_2O(l) \qquad E^\ominus = +1.63\text{ V}$$
$$ClO_3^-(aq) + 5H^+(aq) + 4e^- \rightarrow HOCl(aq) + 2H_2O(l) \qquad E^\ominus = +1.43\text{ V}$$

So overall $E^\ominus = 1.63\text{ V} - 1.43\text{ V} = +0.20\text{ V}$, implying that $K = 3 \times 10^{13}$ and reaction is favorable.

Example 8.6: *Assessing the importance of disproportionation*

Show that manganate(VI) ions are unstable with respect to disproportionation into Mn(VII) and Mn(II) in acidic aqueous solution.

Answer. The overall reaction must involve reduction of one Mn(VI) to Mn(II) balanced by four Mn(VI) oxidizing to Mn(VII). The equation is therefore

$$5MnO_4^{2-}(aq) + 8H^+(aq) \rightarrow 4MnO_4^-(aq) + Mn^{2+}(aq) + 4H_2O(l)$$

This reaction can be regarded as the difference of the reduction half-reactions

$$MnO_4^{2-}(aq) + 8H^+(aq) + 4e^- \rightarrow Mn^{2+}(aq) + 4H_2O(l) \qquad E^\ominus = +1.75\text{ V}$$
$$4MnO_4^-(aq) + 4e^- \rightarrow 4MnO_4^{2-}(aq) \qquad E^\ominus = +0.56\text{ V}$$

The difference of the standard potentials is +1.19 V and the disproportionation is essentially complete. A practical consequence is that high concentrations of MnO_4^{2-} ions cannot be obtained in acidic solution but must be prepared in a basic solution.

Exercise. The standard reduction potentials for Fe^{2+}, Fe and Fe^{3+}, Fe^{2+} are −0.41 V and +0.77 V respectively. Can Fe^{2+} disproportionate under standard conditions?

The reverse of disproportionation is **comproportionation**. In comproportionation, two substances with the same element having different oxidation numbers form a product in which the element has an intermediate oxidation number. An example is

$$Ag^{2+}(aq) + Ag(s) \rightarrow 2Ag^+(aq) \qquad E^\ominus = +1.18\text{ V}$$

The large positive potential indicates that Ag(II) and Ag(0) are completely converted to Ag(I) in aqueous solution.

8.7 Oxidation by atmospheric oxygen

When a solution is exposed to air, for example because it is contained in an open beaker, the possibility of reaction between the solutes and dissolved O_2 must be considered. As an example, consider a solution containing Fe^{2+}. The standard Fe^{3+}/Fe^{2+} reduction potential is +0.77 V, which suggests that Fe^{2+} should survive in water. Moreover, the oxidation of metallic Fe should not proceed beyond Fe(II), since further oxidation to Fe(III) is unfavorable (by 0.77 V) under standard conditions. However, the picture changes considerably in the presence of O_2. In fact, Fe(III) is the most common form of Fe in the earth's crust, and most Fe in sediments that have been deposited from aqueous environments is present as Fe(III).

The potential for

$$4Fe^{2+}(aq) + O_2(g) + 4H^+(aq) \rightarrow 4Fe^{3+}(aq) + 2H_2O(l) \qquad E^\ominus = +0.44\ \text{V}$$

shows that the oxidation of Fe^{2+} by O_2 is thermodynamically favorable. However, +0.44 V does not provide a large overpotential for rapid reaction, and atmospheric oxidation of Fe(II) in aqueous solution is slow in the absence of catalysts. As a result, it is possible to use Fe^{2+} aqueous solutions in laboratory procedures without elaborate precautions.

Example 8.7: *Judging the importance of atmospheric oxidation*

The oxidation of copper-clad roofs to a characteristic green color is another example of atmospheric oxidation in a damp environment. Estimate the potential for oxidation of Cu by O_2.

Answer. The half-reactions are

$$O_2(g) + 4H^+(aq) + 4e^- \rightarrow 2H_2O \qquad E^\ominus = +1.23\ \text{V}$$

$$Cu^{2+}(aq) + 2e^- \rightarrow Cu(s) \qquad E^\ominus = +0.34\ \text{V}$$

The difference is $E^\ominus = +0.89$ V, so atmospheric oxidation is favorable in the reaction

$$2Cu(s) + O_2(g) + 4H^+(aq) \rightarrow 2Cu^{2+}(aq) + 2H_2O(l)$$

Nevertheless, copper-clad roofs do last. This is because their familiar green surface is a passive layer of an almost impenetrable hydrated copper carbonate and sulfate formed from oxidation in the presence of atmospheric CO_2 and SO_2.

Exercise. The reduction potential for reduction of sulfate ($SO_4^{2-}(aq)$ to $SO_2(aq)$) is +0.16 V according to the reaction

$$SO_4^{2-}(aq) + 4H^+(aq) + 2e^- \rightarrow SO_2(aq) + 2H_2O(l)$$

What is the possible fate of SO_2 emitted into fog or clouds?

The diagrammatic presentation of potential data

There are several useful diagrammatic summaries of the relative thermodynamic stabilities of a series of compounds in which one element has several different oxidation numbers.

8.8 Latimer diagrams

The simplest type of diagram was introduced by Wendell Latimer, one of the pioneers in the application of thermodynamics to solution inorganic chemistry. A **Latimer diagram** uses the notation

$$\text{Ox} \xrightarrow{E^\ominus/\text{V}} \text{Red}$$

The diagram for chlorine in acid solution, for instance, is

$$\underset{+7}{ClO_4^-} \xrightarrow{+1.20} \underset{+5}{ClO_3^-} \xrightarrow{+1.18} \underset{+3}{ClO_2^-} \xrightarrow{+1.70} \underset{+1}{HClO} \xrightarrow{+1.63} \underset{0}{Cl_2} \xrightarrow{+1.36} \underset{-1}{Cl^-}$$

As in this example, oxidation numbers are occasionally written under (or over) the species. The notation

$$ClO_4^- \xrightarrow{+1.20} ClO_3^-$$

denotes

$$ClO_4^-(aq) + 2H^+(aq) + 2e^- \rightarrow ClO_3^-(aq) + H_2O(l) \qquad E^\ominus = +1.20\ V$$

and

$$HClO \xrightarrow{+1.63} Cl_2$$

denotes

$$2HClO(aq) + 2H^+(aq) + 2e^- \rightarrow Cl_2(g) + 2H_2O(l) \qquad E^\ominus = +1.63\ V$$

As we can see from these examples, a Latimer diagram summarizes a great deal of information in a compact form and (as we shall explain) shows the relationships between the various species in a particularly clear manner. The potential data in Appendix 4 are presented in the form of Latimer diagrams because they are compact and their arrangement by element makes it easy to locate the information required.

A Latimer diagram contains sufficient information to derive the standard potentials of nonadjacent couples. The connection is obtained from the relation $\Delta G^\ominus = -nFE^\ominus$, and the fact that the overall $\Delta G^\ominus$ for two successive steps is the sum of the individual values. Therefore, we convert the individual $E^\ominus$ values to $\Delta G^\ominus$ values by multiplying by the relevant $-nF$ factor, add them together, and then convert the sum back to $E^\ominus$ for the nonadjacent couple by dividing by $-nF$ for the overall electron transfer. Since the factors $-F$ cancel in this procedure, in practice we form

$$E_{13} = \frac{n_1 E_{12}^\ominus + n_2 E_{23}^\ominus}{n_1 + n_2}$$

Example 8.8: *Extracting $E^\ominus$ for nonadjacent oxidation numbers*

Use the Latimer diagram to calculate the value of $E^\ominus$ for reduction of HClO to Cl^- in acidic aqueous solution.

Answer. In this case the steps are

$$HClO(aq) + H^+(aq) + e^- \rightarrow \tfrac{1}{2}Cl_2(g) + H_2O(l) \qquad E^\ominus(HClO, Cl_2) = +1.63\ V$$

$$\tfrac{1}{2}Cl_2(g) + e^- \rightarrow Cl^-(aq) \qquad E^\ominus(Cl_2, Cl^-) = +1.36\ V$$

The potential for ClO^- to Cl^- is

$$\frac{E^\ominus(HClO, Cl_2) + E^\ominus(Cl_2, Cl^-)}{2} = \frac{1.63\ V + 1.36\ V}{2}$$

$$= 1.50\ V$$

Exercise. Calculate $E^\ominus$ for the reduction of $HClO_3$ to HClO in aqueous solution.

A Latimer diagram shows the species for which disproportionation is likely: a species can disproportionate into its neighbors if the potential on the left is lower than the one on the right. Thus, H_2O_2 has a tendency to disproportionate into O_2 and H_2O under acid conditions:

$$O_2 \xrightarrow{+0.70} H_2O_2 \xrightarrow{+1.76} H_2O$$

We can test this by considering the two half-reactions:

$$H_2O_2(aq) + 2H^+(aq) + 2e^- \rightarrow 2H_2O(l) \qquad E^{\ominus} = +1.76\ \text{V}$$
$$O_2(g) + 2H^+(aq) + 2e^- \rightarrow H_2O_2(aq) \qquad E^{\ominus} = +0.70\ \text{V}$$

and forming the difference:

$$2H_2O_2(aq) \rightarrow 2H_2O(l) + O_2(g) \qquad E^{\ominus} = +1.06\ \text{V}$$

Since $E^{\ominus} > 0$, the disproportion is spontaneous.

8.9 Frost diagrams

A **Frost diagram** for an element X is a plot of $nE^{\ominus}$ for the couple X(N)/X(0) against its oxidation number N (Fig. 8.5). The qualitative feature of Frost diagrams to remember is that the steeper the line joining two points in a diagram, the higher the potential of the corresponding couple. It follows that we can make thermodynamic predictions about the reaction between any two couples by comparing the slopes of the corresponding lines. The oxidizing agent in the couple with the more positive slope (the more positive $E^{\ominus}$) is liable to undergo reduction. The reducing agent of the couple with the less positive slope (the less positive $E^{\ominus}$) is liable to undergo oxidation.

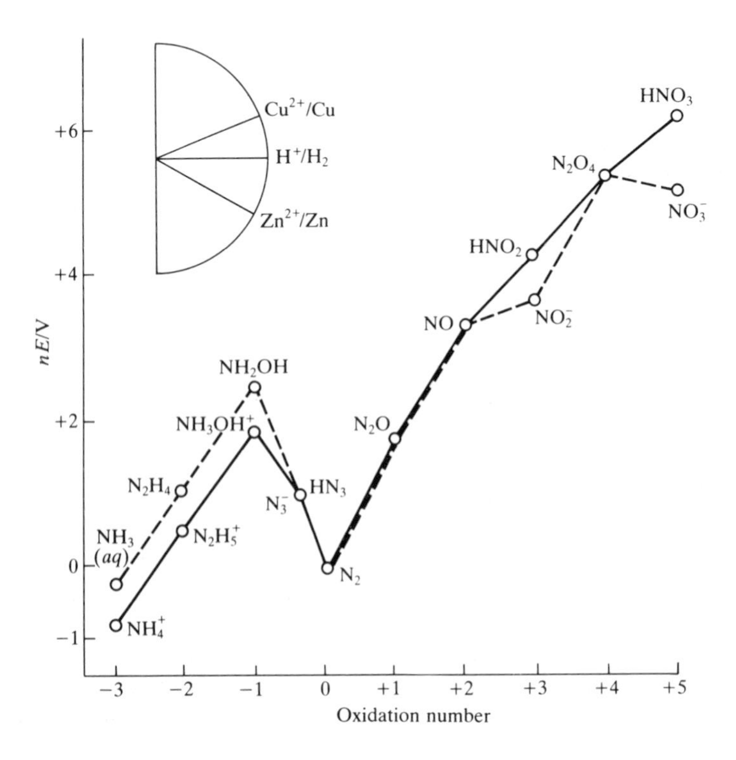

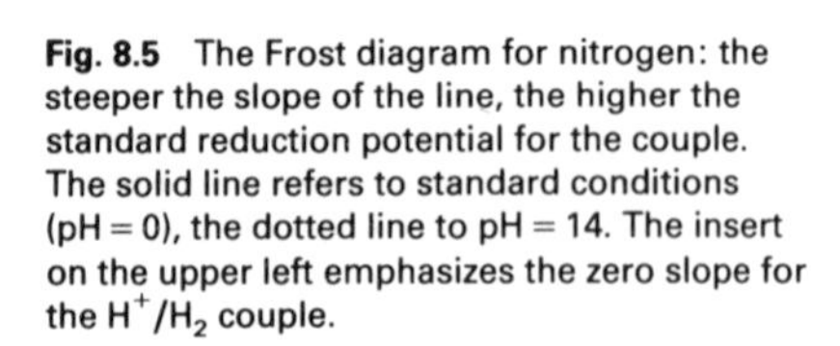

Fig. 8.5 The Frost diagram for nitrogen: the steeper the slope of the line, the higher the standard reduction potential for the couple. The solid line refers to standard conditions (pH = 0), the dotted line to pH = 14. The insert on the upper left emphasizes the zero slope for the H^+/H_2 couple.

Example 8.9: *Constructing a Frost diagram*

Construct a Frost diagram for O from the Latimer diagram

$$O_2 \xrightarrow{+0.70} H_2O_2 \xrightarrow{+1.76} H_2O \quad (O_2 \to H_2O: +1.23)$$

Answer. The oxidation numbers of O are 0, −1, and −2 in the three species. For the change of oxidation number from 0 to −1 (O_2 to H_2O_2), $E^\ominus = +0.70$ V and $n = -1$, so $nE^\ominus = -0.70$ V. Since the oxidation number of O in H_2O is −2, and $E^\ominus$ for the formation of H_2O is +1.23 V, $nE^\ominus = -2.46$ V. These results are plotted in Fig. 8.6. As a check, look at the slope of the line joining H_2O_2 and H_2O. At the point corresponding to oxidation number −1, $nE^\ominus = -0.70$ V and at oxidation number −2 it is −2.46 V, a difference of −1.76 V. The change in oxidation number from H_2O_2 to H_2O is −1. Therefore, $E^\ominus$ for the couple H_2O_2 is $(-1.76\text{ V})/(-1) = +1.76$ V, in accord with the Latimer diagram.

Exercise. Construct a Frost diagram from the Latimer diagram for Tl:

$$Tl^{3+} \xrightarrow{+1.25} Tl^{+} \xrightarrow{-0.34} Tl \quad (Tl^{3+} \to Tl: +0.72)$$

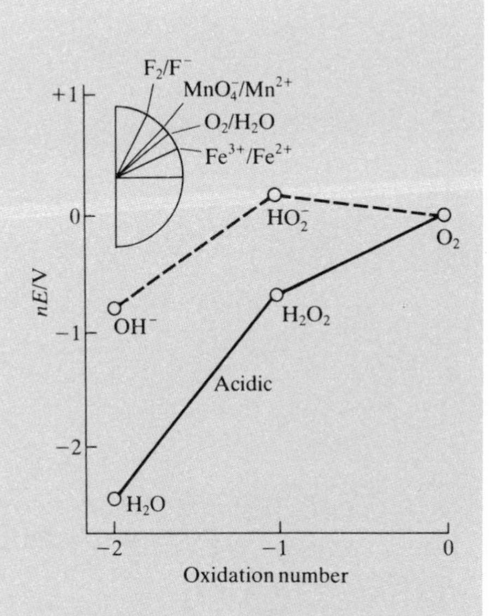

Fig. 8.6 The Frost diagram for O in acidic solution (full line) and basic solution (broken line).

The utility of Frost diagrams is best appreciated by using them. For example, the steep slope connecting HNO_3 to lower oxidation numbers in Fig. 8.5 shows that HNO_3 is a good oxidizing agent under standard conditions. By comparing the slope of the Cu^{2+}/Cu couple in the upper left corner of Fig. 8.5 with the nitrogen couples, we see that HNO_3/NO has a more positive slope than Cu^{2+}/Cu; hence, HNO_3 can oxidize Cu to Cu^{2+}. The diagram indicates that the slope remains steep all the way to N_2, implying that N_2 should be the product if excess Cu is present. However, we must not forget that there may be kinetic limits to these thermodynamic predictions, and that kinetic barriers are particularly common in *p*-block redox chemistry. In this case, N_2 is not formed rapidly, and NO is commonly the gas evolved when Cu is heated with dilute HNO_3.

An ion or molecule in a Frost diagram is unstable with respect to disproportionation if it lies above the line connecting two adjacent species. This is illustrated in **2**, where we show geometrically that the average free energy of the two terminal species lies below that for a substance of intermediate oxidation number, implying that the latter is unstable with respect to disproportionation. A specific example is NH_2OH in Fig. 8.5, which is unstable with respect to disproportionation into NH_3 and N_2. However, it is one of the intricacies of nitrogen chemistry that slow formation of N_2 often prevents its production; in practice, NH_2OH is resistant to disproportionation.

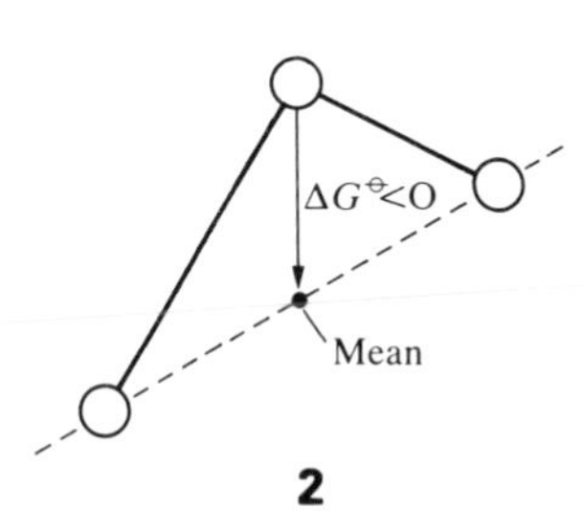

2

A substance that lies below the line connecting its neighbors in a Frost diagram is more stable than they are because its average free energy is lower (**3**). Accordingly, the comproportionation reaction is thermodynamically favorable. The N in NH_4NO_3, for instance, has two ions with oxidation numbers −3 and +5. Since N_2O, in which the oxidation number of N has the intermediate value +1, lies below the line joining NH_4^+ to NO_3^-, their

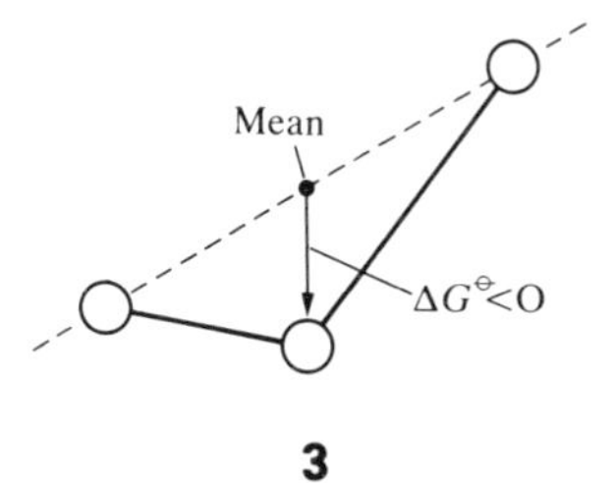

3

comproportionation is possible:

$$NH_4^+ + NO_3^- \rightarrow N_2O + 2H_2O$$

The reaction is not normally fast, but if initiated can become explosive and in fact is often used in place of dynamite for blasting rocks.

When the points for three substances lie approximately on a straight line, no one species will be the exclusive product. The three reactions

$$NO(g) + NO_2(g) + H_2O(l) \rightarrow 2HNO_2(aq) \quad \text{(rapid)}$$

$$3HNO_2(aq) \rightarrow HNO_3(aq) + 2NO(g) + H_2O(l) \quad \text{(rapid)}$$

$$2NO_2(g) + H_2O(l) \rightarrow HNO_3(aq) + HNO_2(aq)$$

(slow and normally rate determining)

are examples of this behavior. They are all important in the industrial synthesis of HNO_3 by the oxidation of NH_3.

Frost diagrams give a quick impression of the chemical properties of an element. We shall use them frequently in this context in the following chapters to convey the sense of trends in the redox properties of the members of a group. For numerical calculations, however, Latimer diagrams and other explicit tabulations of standard potentials are more convenient.

Example 8.10: *Using a Frost diagram to judge the stability of ions*

Figure 8.7 shows a Frost diagram for Mn. (a) Comment on the stability of Mn^{3+}. (b) What is the oxidation number of Mn in the product when MnO_4^- is used as an oxidizing agent in aqueous acid?

Answer. (a) Mn^{3+} lies above the line joining Mn^{2+} to MnO_2. It should disproportionate. Since this is a one-electron transfer, it is likely to be reasonably rapid. (b) With the exception of solid MnO_2, all Mn oxidation numbers between $HMnO_4$ and Mn^{2+} lie above the line connecting the two and should, therefore, disproportionate to give Mn^{2+} and MnO_4^-, so Mn^{2+} is expected. MnO_2 is also a strong oxidizing agent since it has a slope to Mn^{2+} greater than $HMnO_4$ to H_2MnO_4. It follows that most species that reduce MnO_4^- also reduce MnO_2. The minimum at Mn^{2+} represents the common product in acidic solution.

Exercise. Could O_2 serve as an oxidizing agent to oxidize Mn^{2+} to Mn^{3+} in acidic solution? Could H_2O_2 serve?

8.10 pH dependence

So far, we have discussed Latimer and Frost diagrams primarily for aqueous solution at pH = 0, but we can equally well construct them for other conditions. As in Appendix 4, Latimer diagrams are commonly presented for both pH = 0 and pH = 14.

Conditional Latimer diagrams

'Basic Latimer diagrams' are expressed in terms of reduction potentials at pH = 14 (pOH = 0). The potentials are denoted $E_B^\ominus$ and the dotted line in the Frost diagram (Fig. 8.5) is based on them. The important difference is the stabilization of NO_2^- against disproportionation, since its point no longer

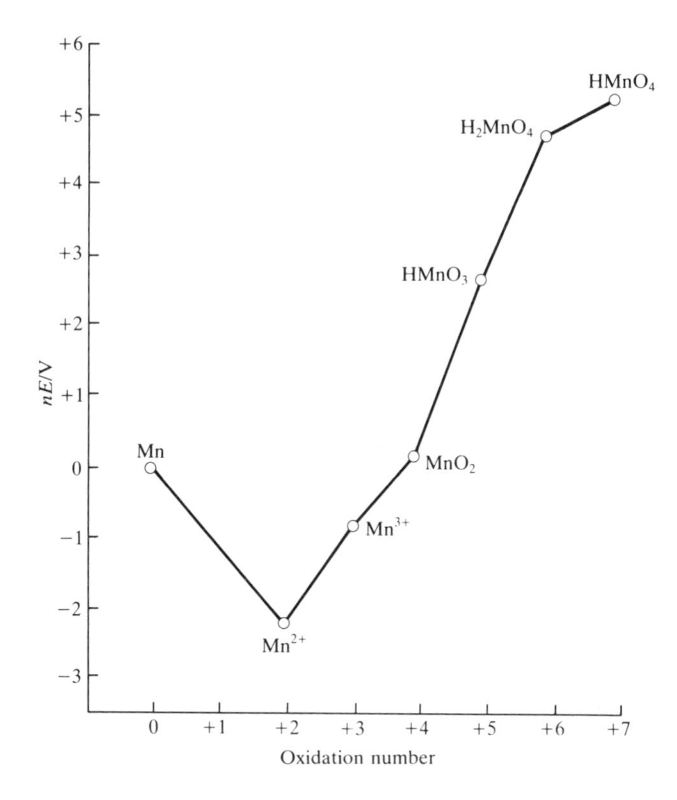

Fig. 8.7 The Frost diagram for Mn in acidic solution.

lies above the line connecting its neighbors. The practical outcome is that metal nitrites can be isolated whereas HNO_2 cannot.

In some cases, there are marked differences between strongly acidic and basic solutions, as for the phosphorus oxoanions:

$$E^{\ominus} \quad H_3PO_4 \xrightarrow{-0.28} H_3PO_3 \xrightarrow{-0.50} H_3PO_2 \xrightarrow{-0.51} P \xrightarrow{-0.06} PH_3$$

$$E^{\ominus}_{B} \quad PO_4^{3-} \xrightarrow{-1.12} HPO_3^{2-} \xrightarrow{-1.56} H_2PO_2^{2-} \xrightarrow{-2.05} P \xrightarrow{-0.89} PH_3$$

This example illustrates an important general point about oxoanions. When their reduction requires removal of oxygen, it consumes H^+ ions, and all oxoanions are stronger oxidizing agents in acidic than in basic solution.

Reduction potentials in neutral solution (pH = 7) are denoted $E^{\ominus}_{W}$. These are particularly useful in biochemical discussions since cell fluids are buffered near pH = 7.00. Basic or neutral potentials can be used for construction of Frost diagrams and both acid and base conditions may be plotted on the same diagram, as in Figs. 8.5 and 8.6.

Example 8.11: *Using conditional diagrams*

KNO_2 is stable in basic solution, but when the solution is acidified, a gas is evolved which turns brown on exposure to air. What is the reaction?

Answer. Figure 8.5 shows that the NO_2^- ion in basic solution lies below the line joining NO to NO_3^-. It resists disproportionation. On acidification, the HNO_2 point rises and the line between NO, HNO_2, and N_2O_4 implies that all three species are present at equilibrium. The brown gas is NO_2 formed from the reaction of NO evolved from the solution with air. In solution, the species of oxidation number +4 tends to disproportionate. The escape of NO from the solution prevents its disproportionation to N_2O and HNO_2.

Exercise. By reference to Fig. 8.5, compare the strength of NO_3^- as an oxidizing agent in acidic and basic solution.

Pourbaix diagrams

A **Pourbaix diagram** (Fig. 8.8) is a diagram that is used for discussing the general relations between redox activity and Brønsted acidity. The regions in the diagram indicate the conditions of pH and potential under which each species is stable.

We can see how the diagrams are constructed by considering some of the reactions involved. The reduction half-reaction

$$Fe^{3+}(aq) + e^- \rightarrow Fe^{2+}(aq) \qquad E^{\ominus} = +0.77\ \text{V}$$

does not involve H^+ ions, and so its potential is independent of pH, giving a horizontal line on the diagram. If the environment contains a couple with a reduction potential above this line (a more positive, more oxidizing potential), the oxidized species, Fe^{3+}, will be the stable species. Hence, the horizontal line is a type of phase boundary that separates the regions where Fe^{3+} and Fe^{2+} are stable.

Another reaction to consider is

$$2Fe^{3+}(aq) + 3H_2O(l) \rightarrow Fe_2O_3(s) + 6H^+(aq)$$

This is not a redox reaction, and the regions of stability of $Fe^{3+}(aq)$ and $Fe_2O_3(s)$ are independent of any redox couples that may also be present. However, it does depend on pH, with the $Fe^{3+}(aq)$ ion stable at low pH and the Fe_2O_3 stable at high pH. Hence, the regions where each species is stable are separated by a vertical line at some pH that depends on the strength of Fe_2O_3 as a base.

A third reaction we need to consider is

$$Fe_2O_3(s) + 6H^+(aq) + 2e^- \rightarrow 2Fe^{2+}(aq) + 3H_2O(l)$$

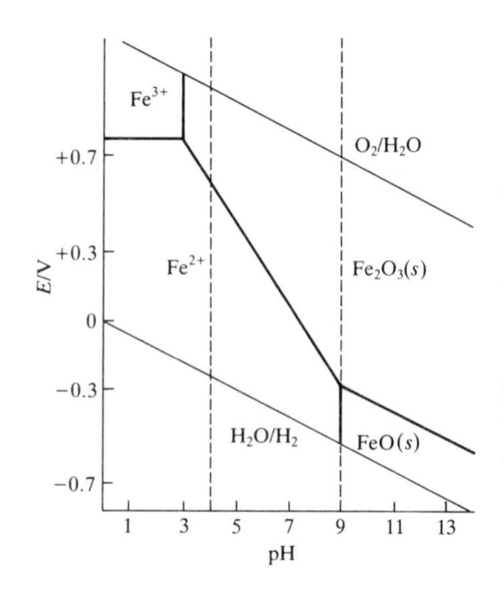

Fig. 8.8 A simplified Pourbaix diagram for some important naturally occurring compounds of Fe. The broken vertical lines represent the normal pH range in natural waters.

This reaction depends on the pH, as is shown explicitly by writing the Nernst equation for its potential:

$$E = E^{\ominus} - \frac{0.059\ \text{V}}{2}\log\frac{[\text{Fe}^{2+}]^2}{[\text{H}^+]^6}$$
$$= E^{\ominus} - 0.059\ \text{V}\log[\text{Fe}^{2+}] - 0.177\ \text{V}\times\text{pH}$$

This shows that the potential falls linearly as the pH increases, as is shown in Fig. 8.8. (The line has been drawn for a solution concentration of 10^{-5} M Fe^{2+}.) The region of potential and pH that lie above this line corresponds to conditions in which the oxidized species (the Fe_2O_3) is stable; those below it correspond to conditions in which the reduced species (the Fe^{2+}) is stable. We see that a stronger oxidizing agent is required to oxidize and precipitate $Fe^{2+}(aq)$ in acidic media than in basic media.

The vertical line at pH = 9 divides the regions at which either the reactants or the products are stable in the reaction

$$\text{Fe}^{2+}(aq) + \text{H}_2\text{O}(l) \rightarrow \text{FeO}(s) + 2\text{H}^+(aq)$$

This is not a redox reaction, and the vertical line shows that $FeO(s)$ is favored when the pH exceeds 9. Another line separates the regions in which FeO and Fe_2O_3 are stable:

$$\text{Fe}_2\text{O}_3(s) + 2\text{H}^+(aq) + 2\text{e}^- \rightarrow 2\text{FeO}(s) + \text{H}_2\text{O}(l)$$

The potential of this reduction depends on the pH, but when we write the Nernst equation

$$E = E^{\ominus} - \frac{RT}{2F}\ln\frac{1}{[\text{H}^+]^2}$$
$$= E^{\ominus} - 0.136\ \text{V pH}$$

we see that it has a less steep slope than for the Fe_2O_3/Fe^{2+} couple (since the number of H^+ ions involved is smaller).

Finally, we have drawn on the diagram the two sloping lines that act as boundaries for the stability field of water (Fig. 8.4). As we saw earlier, any couples present with potentials more positive than the upper line will oxidize water to O_2, and any with potentials more negative than the lower line will reduce it to H_2. All the couples that we need consider when discussing the iron redox reactions therefore lie within the stability field.

Natural waters

We can rationalize the chemistry of natural waters in terms of Pourbaix diagrams of the kind we have just constructed. Thus, where fresh water is in contact with the atmosphere, it is saturated with O_2 and many species may be oxidized by this powerful oxidizing agent ($E^{\ominus} = +1.23$ V). More fully reduced forms are found in the absence of O_2, especially where there is organic matter to act as a reducing agent. The major acid system that controls the pH of the medium is the CO_2–H_2CO_3–HCO_3^-–CO_3^{2-} diprotic system, where atmospheric CO_2 provides the acid and dissolved

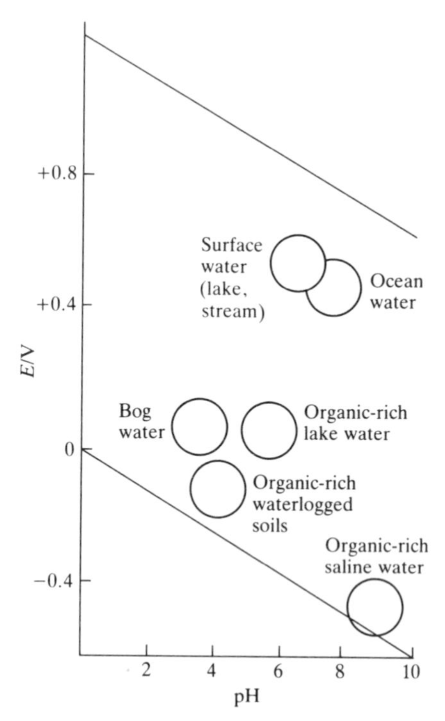

Fig. 8.9 The stability field of water showing regions typical of various natural waters.

carbonate minerals can provide the base. Biological activity is also important, for respiration consumes O_2 and releases CO_2. This acidic oxide reduces the pH and hence makes the reduction potential more negative. The reverse process, photosynthesis, consumes CO_2 and releases O_2. This consumption of acid raises the pH and makes the reduction potential less negative. The condition of typical natural waters—their pH and the potentials of the redox couples they contain—is summarized in Fig. 8.9.

From Fig. 8.8 we see that Fe^{3+} can exist in water if the environment is oxidizing; hence, where O_2 is plentiful and the pH is low (below 4), Fe will be present as Fe^{3+}. Since few natural waters are so acidic, Fe^{3+} is very unlikely to be found in them. The Fe in insoluble Fe_2O_3 can enter solution as Fe^{2+} if it is reduced, which occurs when the condition of the water lies below the sloping boundary in the diagram. We should observe that as the pH rises, Fe^{2+} can form only if there are strong reducing couples present, and its formation is very unlikely in oxygen-rich water. If we compare Figs. 8.8 and 8.9, we see that Fe will be reduced and dissolved as Fe^{2+} in bog waters and organic-rich waterlogged soils (pH near 4.5 and E near +0.03 V and −0.1 V respectively).

It is instructive to analyze a Pourbaix diagram in conjunction with an understanding of the physical processes that occur in water. As an example, consider a lake where the temperature gradient tends to prevent vertical mixing. At the surface, the water is fully oxygenated and the Fe must be present in particles of the insoluble Fe_2O_3; these particles will tend to settle. At depth, the O_2 content is low. If the organic content or other sources of reducing agents are sufficient, the oxide will be reduced and iron will dissolve as Fe^{2+}. The Fe(II) ions will then diffuse toward the surface where they encounter O_2 and are oxidized to insoluble Fe_2O_3 again.

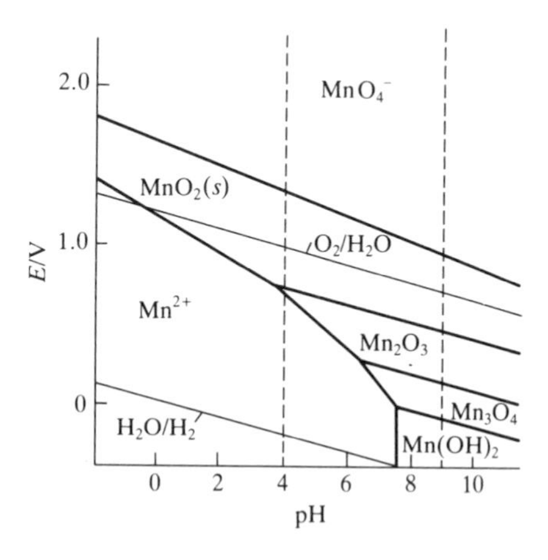

Fig. 8.10 A section of the Pourbaix diagram for Mn. The broken vertical lines represent the normal pH range in natural waters.

Example 8.12: *Using a Pourbaix diagram*

Figure 8.10 is part of a Pourbaix diagram for Mn. Identify the environment in which the solid MnO_2 or its corresponding hydrous oxides are important.

Answer. MnO_2 is thermodynamically favored only if $E > +0.6$ V. Under mildly reducing conditions, the stable species is $Mn^{2+}(aq)$. Thus, MnO_2 is important in well-aerated waters near the air–water boundary where E approaches the value for the O_2/H_2O couple.

Exercise. Use Figs. 8.8 and 8.9 to evaluate the possibility of finding $Fe_2O_3(s)$ in a waterlogged soil.

8.11 Effect of complex formation on potentials

The formation of metal complexes affects reduction potentials. For example, the extraction of Au by air oxidation (Section 8.2) depends on the complexation of Au(I) by CN^-. Another example can be seen in Table 8.1, where two reduction potentials are given for Fe(III) to Fe(II):

$$[Fe(OH_2)_6]^{3+}(aq) + e^- \rightarrow [Fe(OH_2)_6]^{2+}(aq) \qquad E^\ominus = +0.77\ \text{V}$$
$$[Fe(CN)_6]^{3-}(aq) + e^- \rightarrow [Fe(CN)_6]^{4-}(aq) \qquad E^\ominus = +0.36\ \text{V}$$

To analyze this difference, we can treat the reduction of the cyano complex as the outcome of the following sequence:

$$\begin{array}{ccc} [Fe(CN)_6]^{3-}(aq) & \xrightarrow[-6CN^-]{+6H_2O \; [1]} & [Fe(OH_2)_6]^{3+}(aq) \\ \downarrow [4]\; e^- & & \downarrow [2]\; e^- \\ [Fe(CN)_6]^{4-}(aq) & \xleftarrow[-6H_2O]{+6CN^- \; [3]} & [Fe(OH_2)_6]^{2+}(aq) \end{array}$$

Overall ([1] + [2] + [3]),

$$[Fe(CN)_6]^{3-}(aq) + e^- \rightarrow [Fe(CN)_6]^{4-}(aq)$$

The more favorable potential for reduction of $[Fe(H_2O)_6]^{3+}$ compared with $[Fe(CN)_6]^{3-}$ indicates that steps [1] and [3] contribute unfavorably. As we noted in Chapter 7, the Fe(III) complex is more stable than the Fe(II) complex, so the overall reaction is disfavored by conversion of $[Fe(CN)_6]^{3-}$ to $[Fe(CN)_6]^{4-}$.

The central lesson of this example is that complexation alters potentials. The formation of a more stable complex when the metal has the higher oxidation number favors oxidation and makes the reduction potential more negative. The formation of a more stable complex when the metal has the lower oxidation number favors reduction and the potential becomes more positive.

Example 8.13: *Assessing the effect of complexation on potential*

In general, CN^- forms more stable complexes than Br^-. Which complex, $[Ni(CN)_4]^{2-}$ or $[NiBr_4]^{2-}$, is expected to have the more positive potential for reduction to Ni(*s*)?

Answer. The reduction half-reaction is

$$[NiX_4]^{2-}(aq) + 2e^- \rightarrow Ni(s) + 4X^-(aq)$$

The more stable the complex, the higher its resistance to reductive decomposition. Since $[Ni(CN)_4]^{2-}$ is more stable, it is more difficult to reduce than $[NiBr_4]^{2-}$, and therefore will have the more negative reduction potential.

Exercise. Which couple has the higher reduction potential, $[Ni(en)_3]^{2+}/Ni(s)$ or $[Ni(NH_3)_6]^{2+}/Ni(s)$?

8.12 Trends in stabilities of the oxidation states of the metals

Except for the *s* block and aluminum in the *p* block, the metallic elements display a rich variety of oxidation numbers. Our main emphasis here will be on the aqua ions (aqua complexes) and oxides in contact with water. We should bear in mind that changes in the ligands around an ion in solution or the occurrence of one of these metals in solid compounds other than simple oxides may alter the trends we describe. However, most of the general trends, with small modification of detail, are followed by oxides, fluorides, chlorides, and other hard ligand systems. They provide an excellent framework for systematizing the descriptive chemistry of the metal ions.

The *d* block

The maximum oxidation number of the elements in Groups 3 through 8 is equal to the group number (which equals the number of valence electrons). This group oxidation number is most stable for elements on the left of the first *d* series, an example being scandium, which is stable in water only as Sc(III). The trend in stability of the group oxidation numbers is further illustrated by the Frost diagram in Fig. 8.11. The group oxidation numbers for Sc, Ti, and V fall in the lower part of the diagram, indicating that they are relatively stable. In contrast, the group oxidation numbers of the Group 6 and 7 metals Cr and Mn fall in the upper part of the diagram, indicating that they are susceptible to reduction. The maximum group oxidation number is never achieved in Groups 8 through 12 of Period 4 (with Fe, Co, Ni, Cu, and Zn). Chlorides can be prepared for the earliest metals with their group oxidation number ($ScCl_3$ and $TiCl_4$) but the much stronger oxidizing agent F is necessary to achieve the group oxidation number for V (Group 5) and Cr (Group 6), which form VF_5 and CrF_6. Beyond Group 6 in

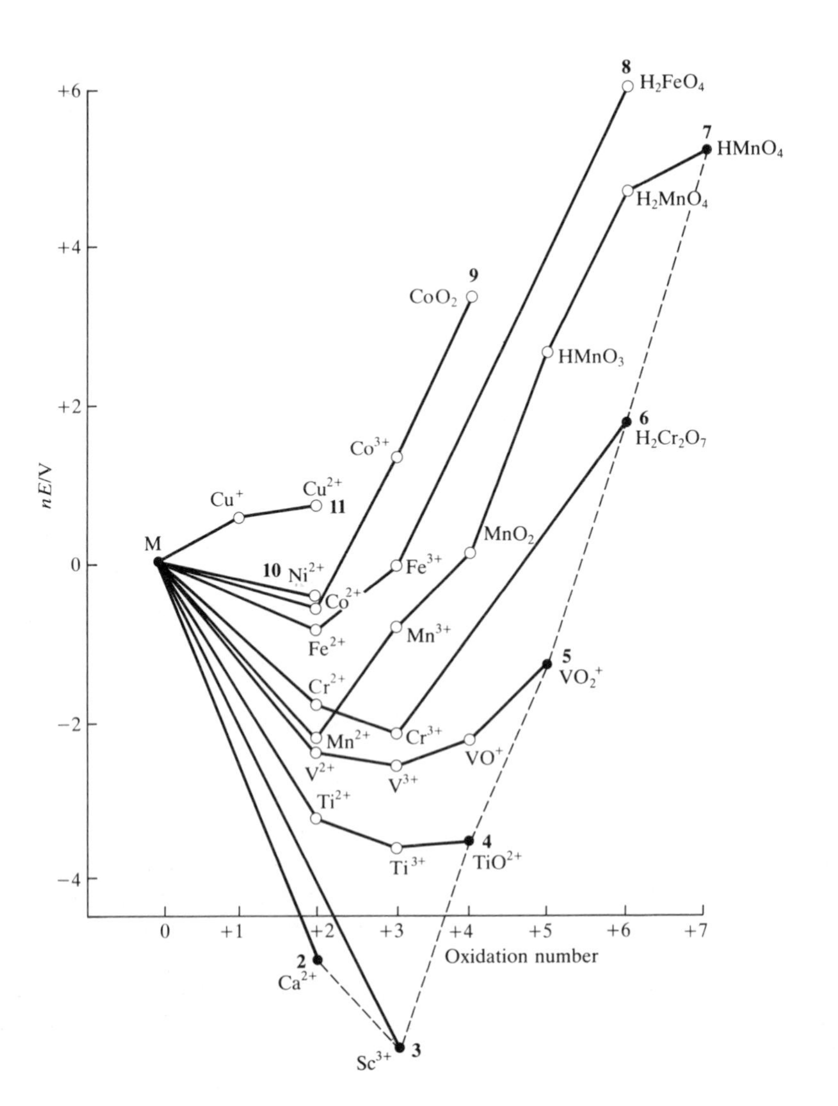

Fig. 8.11 A Frost diagram for the first series of *d*-block elements in acidic solution. The bold numbers designate the group numbers and the broken line connects species in their group oxidation states.

Period 4, even F cannot bring out the group oxidation number. The lack of MnF_7 and FeF_8 may be the result of both steric factors and the difficulty of achieving high oxidation states.

In Period 4, the oxidation number +2 becomes increasingly stable from left to right in the period. For example, among the early members of this series, M(II) is susceptible to oxidation by water, Ti^{2+}, V^{2+}, and Cr^{2+} all being unstable with respect to oxidation by H^+:

$$2V^{2+}(aq) + 2H^+(aq) \rightarrow 2V^{3+}(aq) + H_2(g) \qquad E^\ominus = +0.26\ \mathrm{V}$$

Beyond Cr (for Mn^{2+}, Fe^{2+}, Co^{2+}, Ni^{2+}, and Cu^{2+}), we find that the +2 state is stable with respect to reaction with water. In fact, for Ni and Cu the only aqua ions are M(II). The same is nearly true for Co because Co(III) is rapidly reduced to Co(II) by water:

$$4Co^{3+}(aq) + 2H_2O(l) \rightarrow 4Co^{2+}(aq) + O_2(g) + 4H^+(aq) \qquad E^\ominus = +0.69\ \mathrm{V}$$

In Groups 4 through 10, the highest oxidation number becomes more stable going down a group. This is illustrated for the chromium group in Fig. 8.12. Note that Cr(VI) is well above Mo(VI) and W(VI), indicating that the maximum oxidation number is more stable for the heavier members. The diagram also illustrates the common observation that the biggest change in stability of the maximum oxidation state is between the first and second rows of the *d* block. Thus Cr(VI) in the form of dichromate is a strong oxidizing agent in acid solution, whereas MoO_3 and WO_3 are not. The Frost diagrams for Mo and W are quite flat (Fig. 8.12), and as a consequence Mo and W do not exhibit the pronounced stability of the +3 oxidation number so characteristic of Cr.

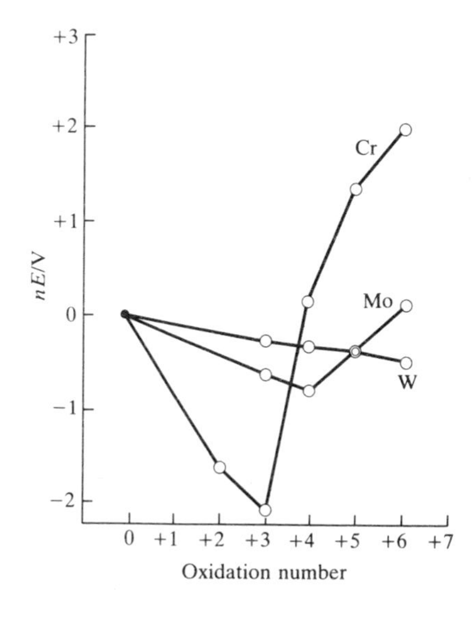

Fig. 8.12 A Frost diagram for the Cr group in the *d* block (Group 6) in acidic solution.

Groups 11 and 12 and the *p* block

The metals in Groups 12 through 15/V display trends in oxidation states that contrast strongly with those we have just described. Zn and Ga, in Groups 12 and 13/III, are much more readily oxidized than Cu (Group 11), so the noble character (resistance to oxidation) that has been developing across the *d* block is largely absent in Group 12. This may be seen by comparing the reduction potential for Zn (−0.76 V) with that for Cu (+0.34 V).

Although in the *d* block the high oxidation states are favored for heavy members of a group, the opposite is true in the *p* block. Figure 8.13 shows that the group oxidation state is preferred for Ga and not for Tl, and the latter favors an oxidation number that is two less than the group number. This tendency to prefer the lower oxidation state is the **inert pair effect**. It is a consequence of the greater energy separation of the *s* and *p* orbitals of heavy atoms, which makes the former less available for bond formation. Thus Tl(I), Pb(II) and Bi(III) are the preferred oxidation states of these three elements. The elements are strong oxidizing agents when they have their group oxidation numbers and are present as Tl(III), Pb(IV), and Bi(V).

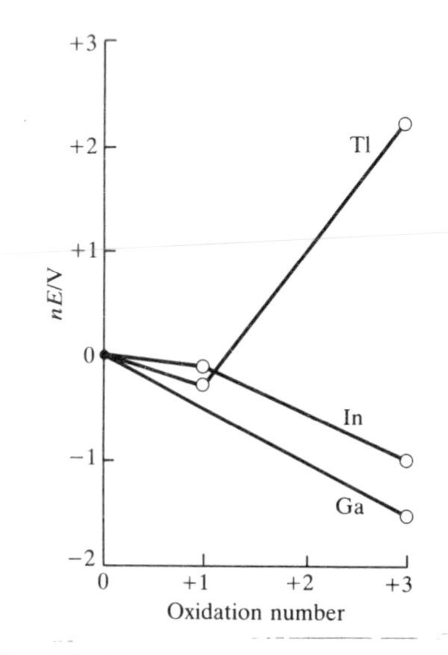

Fig. 8.13 A Frost diagram for the early *p*-block metals.

Table 8.2. Some oxidation states and potentials of the lanthanides

Z	Species and $E^{\ominus}$/V						
58	$Ce^{4+}(f^0)$	+1.72	$Ce^{3+}(f^1)$	−2.34			Ce
59			$Pr^{3+}(f^2)$	−2.35			Pr
60			$Nd^{3+}(f^3)$	−2.6	$Nd^{2+}(f^4)$	−2.2	Nd
62			$Sm^{3+}(f^5)$	−1.55	$Sm^{2+}(f^6)$	−2.67	Sm
63			$Eu^{3+}(f^6)$	−0.35	$Eu^{2+}(f^7)$	−2.80	Eu
66			$Dy^{3+}(f^9)$	−2.5	$Dy^{2+}(f^{10})$	−2.2	Dy
69			$Tm^{3+}(f^{12})$	−2.3	$Tm^{2+}(f^{13})$	−2.3	Tm
70			$Yb^{3+}(f^{13})$	−1.05	$Yb^{2+}(f^{14})$	−2.8	Yb

Lanthanides

The lanthanides[3] La (lanthanum) through Yb (ytterbium) favor the +3 oxidation number with a uniformity unprecedented in the periodic table. There is no simple explanation for this uniformity because the other properties of the elements that may be thought relevant vary significantly. For example, the radii of the M^{3+} ions contract greatly, from 1.18 Å for La^{3+} to 0.85 Å for Lu^{3+} (lutetium), and this 28 percent decrease in radius leads to a great increase in the hydration enthalpy across the series. Detailed analysis shows that there is fortuitous cancellation of the various terms for sublimation, solvation, ionization in the Born–Haber cycle for aqua ion formation and that it is only a coincidence that the potential for the reduction of La^{3+} to the metal is −2.38 V and that for Lu^{3+} at the other end of the block has almost the same value, −2.30 V.

Superimposed on this uniformity there are some atypical oxidation states (Table 8.2) that are most prevalent when the ion can attain an empty (f^0), half-filled (f^7), or filled (f^{14}) subshell. This is analogous to the extra stability of the empty, half-full, and full d subshell in the d block (Section 1.5). Thus Ce^{3+}, which is an f^1 ion, can be oxidized to the f^0 ion Ce^{4+}, which is a strong and useful oxidizing agent. The next most common of the atypical oxidation numbers is Eu^{2+}, which is an f^7 ion, and readily reduces water.

Actinides

A common oxidation number for the actinides, which for the present discussion will include Ac (actinium) through Lr (lawrencium) is +3. Unlike the lanthanides, however, the early members of the actinide series occur

[3] There is an ongoing controversy over whether the lanthanides should include 14 elements from La through Yb or be displaced one position to the right, Ce through Lu. We will include 15 elements in this discussion, La to Lu. Similar controversy extends to the actinides. See W. B. Jensen, *J. chem. Educ.*, **59**, 634 (1982).

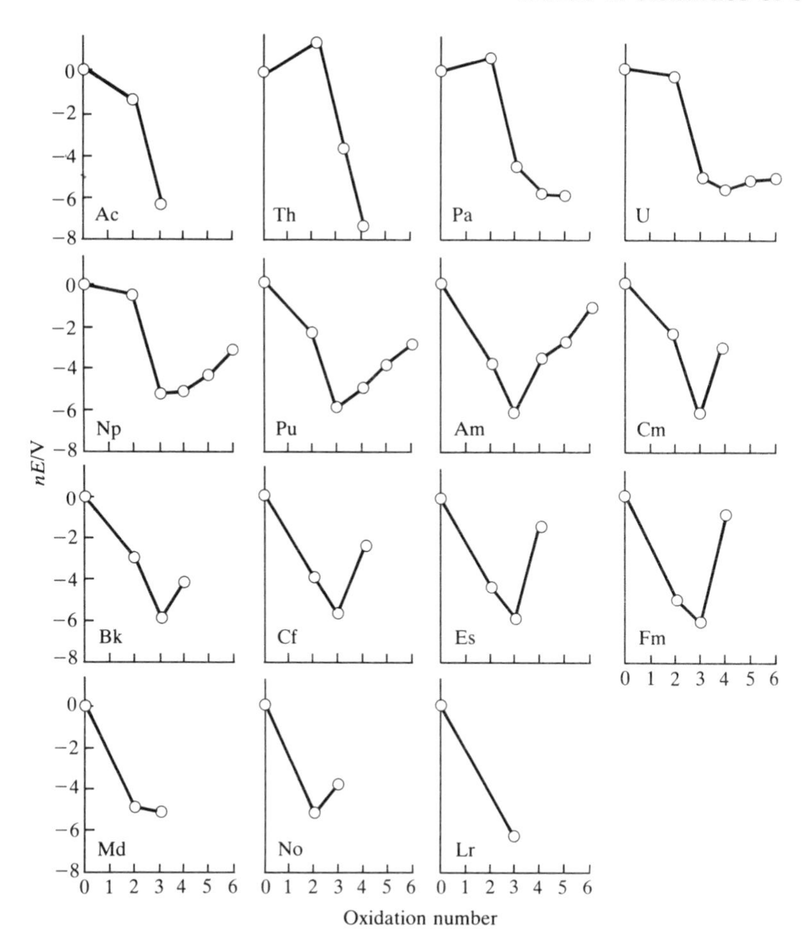

Fig. 8.14 Frost diagrams for some of the actinides.

with a rich variety of oxidation numbers. The Frost diagrams (Fig. 8.14) shows that oxidation numbers higher than +3 are preferred for the first elements, namely Th (thorium), Pa (protactinium), U (uranium), and Np (neptunium). The linear or nearly linear MO_2^+ and MO_2^{2+} ions are the dominant aqua species for oxidation numbers +5 and +6. Figure 8.14 shows that with increasing atomic number, M(III) becomes progressively more stable relative to higher oxidation states, and it is the dominant oxidation number for elements 96 to 100, namely Cm (curium), Bk (berkelium), Cf (californium), and Es (einsteinium). These elements therefore resemble the lanthanides.

Example 8.14: *Judging qualitative trends in redox stability*

On the basis of trends in the *d* block, suggest suitable M(II) aqua ions for use as reducing agents in analytical chemistry.

Answer. Since the group oxidation number is favored for the early *d*-block elements, appropriate reducing agents include $Ti^{2+}(aq)$, $V^{2+}(aq)$, and $Cr^{2+}(aq)$. The ions $Mn^{2+}(aq)$ and $Fe^{2+}(aq)$ are only weak reducing agents. The ions $Co^{2+}(aq)$, $Ni^{2+}(aq)$, and $Cu^{2+}(aq)$ are not oxidizable in water.

Exercise. Identify the lanthanide species with its oxidation number that is used (a) as a strong oxidizing agent and (b) as a good reducing agent.

Further reading

A. J. Bard, R. Parsons, and R. Jordan, *Standard potentials in aqueous solution.* M. Dekker, New York (1985). The most recent critical collection of potential data. Comments on rates are also included.

J. O'M. Bockris and A. K. N. Reddy, *Modern electrochemistry.* Plenum Press, New York (1970). A comprehensive account of electrochemical phenomena.

M. Pourbaix, *Atlas of electrochemical equilibria in aqueous solution.* Pergamon Press, Oxford (1966). The source of Pourbaix diagrams for many elements.

R. M. Garrels and C. L. Christ, *Solutions, minerals, and equilibria.* Harper and Row, New York (1965). A standard text on aqueous geochemistry, which makes use of Pourbaix diagrams throughout and analyzes the implications of chemical equilibria in geology.

W. Stumm and J. J. Morgan, *Aquatic chemistry.* Wiley, New York (1981). The standard reference on natural water chemistry.

D. C. Harris, *Introduction to analytical chemistry.* W. H. Freeman, New York (1986). Chapters 14 and 16 are good introductions to electrochemical cells and electrochemical measurements.

The following two books give introductions to and information on inorganic reaction mechanisms:

D. Katokis and G. Gordon, *Mechanisms of inorganic reactions.* Wiley, New York (1987).

J. Atwood, *Inorganic and organometallic reaction mechanisms.* Brooks-Cole, Monterey (1985).

Exercises

8.1 Consult the Ellingham diagram (Fig. 8.1) and determine if there are conditions under which Al might be expected to reduce MgO. Comment on these conditions.

8.2 Using standard potential data from Appendix 4 as a guide, write balanced equations for the reactions that might be expected for each of the following species in aerated aqueous acid. If the species is stable, write 'no reaction'. (a) Cr^{2+}, (b) Fe^{2+}, (c) Cl^-, (d) HOCl, (e) Zn(*s*).

8.3 Write the Nernst equation for (a) the reduction of O_2:

$$O_2(g) + 4H^+(aq) + 4e^- \rightarrow 2H_2O(l)$$

(b) the reduction of $Fe_2O_3(s)$:

$$Fe_2O_3(s) + 6H^+(aq) + 6e^- \rightarrow 2Fe(s) + 3H_2O(l)$$

and in each case express the formula in terms of pH. What is the potential for the O_2 reduction at pH = 7 and $p(O_2) = 0.20$ bar (the partial pressure of oxygen in air)?

8.4 Given the following potentials in basic solution:

$$CrO_4^{2-}(aq) + 4H_2O(l) + 3e^- \rightarrow Cr(OH)_3(s) + 5OH^-(aq) \qquad E^\ominus = -0.11\text{ V}$$

$$[Cu(NH_3)_2]^+(aq) + e^- \rightarrow Cu(s) + 2NH_3(aq) \qquad E^\ominus = -0.10\text{ V}$$

calculate $E^\ominus$, $\Delta G^\ominus$, and K for the reduction of CrO_4^{2-} and $[Cu(NH_3)_2]^+$ by H_2 in basic solution. Comment on why $\Delta G^\ominus$ and K are so different in the two cases despite the values of $E^\ominus$ being so similar.

8.5 Answer the following questions using the Frost diagram in Fig. 8.15. (a) What are the consequences of dissolving Cl_2

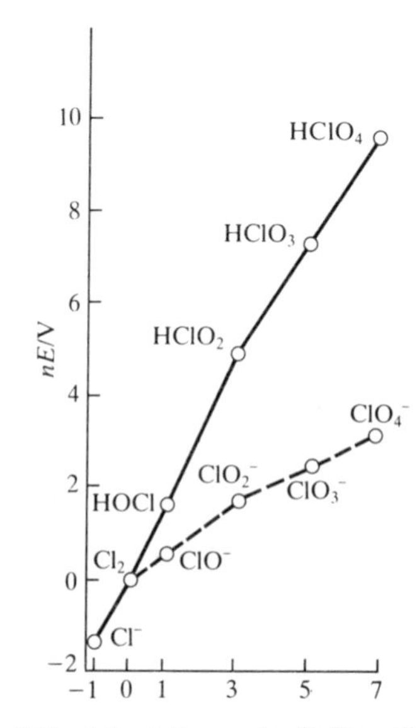

Fig. 8.15 A Frost diagram for Cl. The solid line refers to standard conditions (pH = 0) and the broken line to pH = 14.

in aqueous basic solution? (b) What are the consequences of dissolving Cl_2 in aqueous acid? (c) Is the failure of $HClO_3$ to disproportionate in acidic aqueous solution a thermodynamic or a kinetic phenomenon?

8.6 Write equations for the main net reaction which you would predict in the following experiments: (a) N_2O is bubbled into aqueous NaOH solution, (b) Zn metal is added to aqueous NaI_3, (c) I_2 is added to excess acidic aqueous $HClO_3$.

8.7 Characterize the condition of acidity or basicity which would most favor the following transformation in aqueous solution: (a) $Mn^{2+} \rightarrow MnO_4^-$, (b) $ClO_4^- \rightarrow ClO_3^-$, (c) $H_2O_2 \rightarrow O_2$, (d) $I_2 \rightarrow 2I^-$.

8.8 Comment on the likelihood that the following reactions occur by a simple outer-sphere electron transfer, simple atom transfer, or a multistep mechanism: (a) $HOI(aq) + I^-(aq) \rightarrow I_2(aq) + OH^-(aq)$, (b) $[Co(phen)_3]^{3+}(aq) + [Cr(bipy)_3]^{2+}(aq) \rightarrow [Co(phen)_3]^{2+}(aq) + [Cr(bipy)_3]^{3+}(aq)$, (c) $IO_3^-(aq) + 8I^-(aq) + 6H^+(aq) \rightarrow 3I_3^-(aq) + 3H_2O(l)$.

8.9 Use the Latimer diagram for chlorine to determine the potential for reduction of ClO_4^- to Cl_2 acidic solution. Write a balanced equation for this half-reaction.

8.10 Convert the values of reduction potentials for $[Fe(CN)_6]^{3-}$, Fe^{3+}, and Cl_2 from the data in Table 8.1 into a series of vectors of the type shown in the upper left of Fig. 8.5.

8.11 Use Fig. 8.9 to find the approximate potential of an aerated lake at $pH = 6$. With this information and Figs. 8.8 and 8.10, or from Appendix 4, predict the species at equilibrium for the elements (a) Fe, (b) Mn, (c) S.

8.12 Fe^{2+} and H_2S are important species at the bottom of a lake where O_2 is scarce. If the $pH = 6$, what is the maximum value of E characterizing the environment?

8.13 Edta forms stable complexes with hard acid centers. How will complexation with edta affect the reduction of M^{2+} to the metal in the first *d* series?

8.14 Sketch the first series of the *d* block, including the symbols of the elements. Indicate those elements for which the group oxidation number is stable by 's'; those for which the group oxidation number can be reached but is a powerful oxidizing agent by 'o'; and those for which the group oxidation number is not achieved by 'n'.

8.15 State the trend in the stability of the group oxidation number on descending a group in the periodic table. Illustrate the trend using standard potentials in acid solution for the reduction to the metal of the maximum oxidation number Cr versus W species in Group 6. Compare and contrast the case of Al versus Tl in the *p* block.

8.16 From the Frost diagrams in Fig. 8.14, pick out the most stable positive oxidation states of the actinides and contrast this with the most stable states of the lanthanides.

Problems

8.1 Using standard potential data, suggest why permanganate is not a suitable oxidizing agent for the quantitative estimation of Fe^{2+} in the presence of HCl but becomes so if sufficient Mn^{2+} and phosphate ion are added to the solution. (Hint: Phosphate complexes Fe^{3+}, thereby stabilizing it.)

8.2 By a consideration of data for Cr, Mo, and W, G. W. Seaborg has estimated that element 106 (Unh) would have the Latimer diagram in acid solution given below. Construct a Frost diagram and comment on the extrapolation to Unh of the properties summarized in Fig. 8.14, with particular attention to the stabilities of oxidation numbers +3 and +6.

$$UnhO_3 \xrightarrow{0.5} Unh_2O_5 \xrightarrow{-0.2} UnhO_2 \xrightarrow{-0.7} Unh^{3+} \xrightarrow{0.0} Unh$$

8.3 Refer to the data in Problem 8.2. Predict the results of the following experiments with unnilhexium: (a) Unh is placed in 3 M $HCl(aq)$, (b) $UnhCl_3$ is dissolved in acidic aqueous Fe^{2+}, (c) $UnhO_2$ and Unh are contact with each other and acid solution, (d) Unh is treated with hot concentrated HNO_3.

8.4 R. A. Binstead and T. J. Meyer have described (*J. Amer. chem. Soc.*, **109**, 3287 (1987)) the reduction of $[Ru^{(IV)}O(bipy)_2(py)]^{2+}$ to $[Ru^{(III)}(OH)(bipy)_2(py)]^{2+}$. The study revealed that the rate of the reaction is strongly influenced by the change of the solvent from H_2O to D_2O. What does this suggest about the mechanism of the reaction? Comment on the relation of this result to the cases of simple electron transfer and simple atom transfer considered in the chapter.

8.5 It is often found that O_2 is a slow oxidizing agent. Suggest a mechanistic explanation that takes into account the two reduction potentials

$$O_2(g) + 4H^+(aq) + 4e^- \rightarrow 2H_2O(l) \qquad E^{\ominus} = +1.23\ V$$
$$O_2(g) + 2H^+(aq) + 2e^- \rightarrow H_2O_2(aq) \qquad E^{\ominus} = +0.70\ V$$

Part 3

s- and *p*-block elements

In this part we take a more detailed look at the properties of individual elements and their compounds. We shall not try to be comprehensive, which would be overwhelming, but aim to identify the most characteristic properties and reactions and to consider the more important compounds. In many cases we shall be able to rationalize the behavior we describe in terms of the models of bonding and structure and the classification of reactions developed in the preceding chapters. We shall also discuss new areas of inorganic chemistry where explanations are still being developed. It is important to appreciate, however, that the explanations widely presented in inorganic chemistry vary from sound deductions drawn from thermodynamics to speculations about bonding and mechanism. Some of these explanations will be overthrown as the subject advances; the facts, for the most part, will survive. The principal rationalizing theme of these chapters will be the periodicity of the properties of the elements; however, many fruitful research projects have stemmed from noticing an anomaly in a periodic trend, and we shall develop this point too.

Hydrogen and its compounds

9

This chapter deals with hydrogen. Despite its simplicity, hydrogen has a very rich chemistry, and it has in fact been at the forefront of much of the discussion so far, particularly in the discussion of acids and bases. Here we summarize some of the properties of the element and its binary compounds. The latter range from solid ionic compounds to molecular compounds that may be electron-deficient, electron-precise, or electron-rich (in the sense that we shall explain). We shall see that we can understand many of the reactions of these compounds in terms of their ability to supply an H^- anion or an H^+ cation, and that we can sometimes anticipate when one tendency is likely to dominate. We conclude the chapter with a brief survey of one of the striking characteristics of certain compounds of hydrogen, their ability to participate in the formation of hydrogen bonds.

The element
9.1 Nuclear properties
9.2 Hydrogen atoms and ions
9.3 Dihydrogen
9.4 The production of dihydrogen

Classification and structure of compounds
9.5 The saline hydrides
9.6 Metallic hydrides
9.7 Binary molecular compounds

The reactivity of hydrogen compounds
9.8 Stability and synthesis
9.9 Reactions

The electron-deficient hydrides of the boron group
9.10 Diborane
9.11 The tetrahydroborate ion
9.12 The hydrides of aluminum and gallium

Electron-precise hydrides of the carbon group
9.13 Silanes
9.14 Germane, stannane, and plumbane

Electron-rich compounds of Groups 15/V to 17/VII
9.15 Formation
9.16 Hydrogen bonding
9.17 Brønsted acidity and Lewis basicity

Further information: nuclear magnetic resonance

Further reading

Exercises

Problems

Hydrogen is the most abundant element in the universe. It is also one of the most important elements in the earth's crust, where it is found in minerals, in the oceans, and in all living things. Since the H atom has only one electron, it might be thought that the element's chemical properties will be mundane, but this is far from the case. Hydrogen has richly varied chemical properties, for despite its single electron it can bond to more than one atom simultaneously. Moreover, it ranges in character from a strong Lewis base (as the hydride ion H^-) to a strong Lewis acid (as the hydrogen cation H^+, the proton).

The element

9.1 Nuclear properties

There are three isotopes of hydrogen, with mass numbers 1, 2 (deuterium, ^2H or D), and 3 (tritium, ^3H or T). The justification for giving these isotopes their own names is the difference in their masses, which make their chemical and physical differences readily observable.

The lightest isotope, ^1H—which is very occasionally called protium—is by far the most abundant. Deuterium has variable natural abundance with an average value of about 0.016 percent. Neither ^1H nor ^2H is radioactive, but tritium is:

$$^3H \rightarrow {}^3_2He + \beta \qquad t_{\frac{1}{2}} = 12.4\ y$$

Tritium's abundance of 1 in 10^{21} H atoms in surface water reflects a steady state between its production by bombardment of cosmic rays on the upper atmosphere and its loss by radioactive decay. However, the low natural abundance of the isotope is augmented artificially because it is manufactured on an undisclosed scale for use in thermonuclear weapons. This synthesis uses the neutrons from a fission reactor with Li as a target:

$$^1_0n + {}^6_3Li \rightarrow {}^3_1H + {}^4_2He$$

Deuterium and tritium are used as **tracers**, or isotopes that can be followed through a series of reactions, in the study of reaction mechanisms. Thus we can trace the transfer of hydrogen between different molecules by following the isotopic composition of the products and any intermediates using NMR, IR, or mass spectroscopy. Tritium is sometimes the preferred tracer because it can be detected by its radioactivity, which is a far more sensitive probe than spectroscopy.

Isotope effects

The physical and chemical properties of isotopically substituted molecules are usually very similar. For example, the bond enthalpies, vapor pressures, and Lewis acidities of $^{10}BF_3$ and $^{11}BF_3$ are nearly identical. However, the same is not true when D is substituted for H, for the mass of the substituted atom is doubled. Table 9.1, for instance, shows that the differences in boiling points and bond enthalpies are easily measurable for H_2 and D_2. Although less striking, the bond enthalpy for E—H and E—D bonds (where E stands for an element) are also measurably different (Table 9.1). The difference in boiling point between H_2O and D_2O reflects the greater

Table 9.1. The effect of deuteration on physical properties

	H_2	D_2	H_2O	D_2O
Normal boiling point/°C	−252.8	−249.7	100.00	101.42
Mean bond enthalpy/(kJ mol^{-1})	436.0	443.3	463.5	470.9

strength of the O · · · D—O hydrogen bond compared with the O · · · H—O bond.

Much of the difference between E—D and E—H bonds (and hydrogen bonds) arises from differences in their zero-point vibrational energies. We know from the expression for the vibrational energy levels of a harmonic oscillator (eqn 4 of Section 1.3) that the minimum permitted vibrational energy is obtained with the vibrational quantum number $v = 0$, when

$$E_0 = \tfrac{1}{2}\hbar\omega \quad \text{with } \omega = \left(\frac{k}{\mu}\right)^{\frac{1}{2}}$$

k is the force constant of the bond and μ is the **reduced mass**:

$$\frac{1}{\mu} = \frac{1}{m_A} + \frac{1}{m_B}$$

(m_A and m_B are the masses of the two atoms.) For two identical atoms of mass m, $\mu = \frac{1}{2}m$. The D_2 molecule has a larger reduced mass than H_2, and therefore a lower zero-point energy and a higher bond energy (Fig. 9.1). The difference in zero-point energy also accounts for the differences between the strengths of hydrogen bonds with and without deuteration.

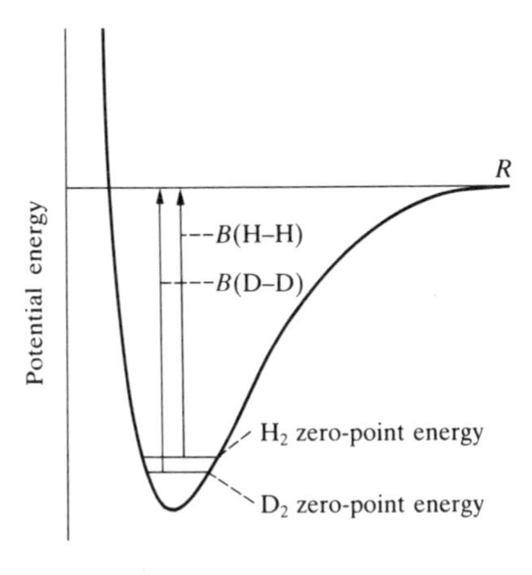

Fig. 9.1 Molecular potential energy curve for H_2 and D_2 and the differences in zero-point energy and, because of that, the differences in dissociation energy.

Reaction rates are also measurably different for processes in which E—H and E—D bonds are broken, made, or rearranged. The detection of this **kinetic isotope effect** can often help to support a proposed reaction mechanism. Kinetic isotope effects are frequently observed when an H atom is transferred from one atom to another in an activated complex. The effect on rate when this occurs is referred to as a **primary isotope effect**, and D transfer can be as much as ten times slower than H transfer. A substantial **secondary isotope effect** may occur when H is not transferred. For example, the rate of hydrolysis of the Cr—NCS link in *trans*-$[Cr(NCS)_4(NH_3)_2]^-$ is twice as fast when the ligands are ND_3 than when they are NH_3, even though no N—H bonds are broken in the reaction. This isotope effect is ascribed to the change in the strength of the N—H · · · O hydrogen bond between the complex and the solvent and the effect of this bonding on the ease with which the NCS^- ligand is able to depart from the Cr(III). The rate is greater for D than for H because N—D · · · O bonds are stronger than N—H · · · O bonds.

Isotopic substitution and IR spectra

Since the frequencies of IR transitions between vibrational levels depend on the masses of atoms, they are strongly influenced by substitution of D for H. The heavier isotope results in the lower frequency. Inorganic chemists often take advantage of this isotope effect to determine whether a particular IR transition involves significant motion of hydrogen. To a first approximation, the frequency of an E—H stretch is given by the equation for the isolated bond treated as a harmonic oscillator:

$$\nu = \frac{1}{2\pi}\left(\frac{k}{\mu}\right)^{1/2}$$

Since the force constants for E—D and E—H stretches are to a good approximation the same, their vibrational frequencies are approximately in the ratio

$$\frac{\nu_D}{\nu_H} = \left(\frac{\mu_H}{\mu_D}\right)^{1/2}$$

Furthermore, since the light nucleus dominates the reduced mass (set $m_E \gg m_H$ or m_D in the expression for μ), this ratio simplifies to

$$\frac{\nu_D}{\nu_H} = \left(\frac{m_H}{m_D}\right)^{1/2}$$

The same expression applies to the ratio of wavenumbers too.

Example 9.1: *Identifying an IR band by isotopic substitution*

For a dialkyl arsine R_2AsH, an IR band, thought to be due to an As—H stretch, is observed at $2080\,cm^{-1}$. For the deuterated compound R_2AsD the band is shifted to $1475\,cm^{-1}$. Do the data support the assignment of the band as an As—H stretch?

Answer. The expected wavenumber for As—D is $2080\,cm^{-1} \times 1/2^{1/2} = 1471\,cm^{-1}$, in reasonable agreement with the observed $1475\,cm^{-1}$. Thus, the assignment is confirmed.

Exercise. An IR band thought to be the Co—H stretch is observed at $1840\,cm^{-1}$ for $[Co(CN)_5(H)]^{3-}$. What is the expected wavenumber of the corresponding band in $[Co(CN)_5(D)]^{3-}$?

Nuclear spin and NMR spectra

Another important property of the hydrogen nucleus is its spin, which makes possible the technique of ^{1}H-NMR spectroscopy (see *Further information*). Although the nuclei of many different elements can be observed readily with modern NMR spectrometers, ^{1}H-NMR is still the most common spectroscopic method for the determination of the structures of molecules containing hydrogen.

9.2 Hydrogen atoms and ions

The H atom has a high ionization enthalpy ($+1310\,kJ\,mol^{-1}$) and a low electron affinity ($A_e = 77\,kJ\,mol^{-1}$). The absolute electronegativity calculated from these values is 2.3. This is similar to the electronegativities of B, C, and Si, so their E—H bonds are not very polar.

Atoms and monatomic ions

With *s*-block elements, hydrogen forms salt-like compounds in which the H atom acquires electron density and can be regarded as the **hydride ion** H^-. Conversely, when combined with the highly electronegative elements from the right of the periodic table, the E—H bond is best regarded as covalent and polar, with the H atom carrying the (very small) partial positive charge. In keeping with these trends, H is assigned oxidation number −1 when in combination with metals and +1 when in combination with nonmetals:

Oxidation number:	+1 −1	+3 −1	−3 +1	+1 −1
Compound:	NaH	AlH_3	NH_3	HCl

Although the oxidation number is not the charge on an atom, it does provide a guide to the variation of the polarities of hydrogen compounds through the periodic table.

When a high-voltage discharge is passed through hydrogen gas at low pressure, the molecules dissociate, ionize, and recombine to form a plasma that contains spectroscopically observable amounts of H, H_2^+, and H_3^+. The free hydrogen cation (H^+, the proton) has a very high charge/radius ratio and so it is not surprising to find that it is a very strong Lewis acid. In the gas phase it readily attaches to other molecules and atoms, including He, to form HeH^+. In condensed phases, H^+ is always found in combination with a Lewis base, and its ability to transfer between Lewis bases gives it the special role in chemistry that we explored in detail in Chapter 5.

Molecular cations

The molecular cations H_2^+ and H_3^+ have only a transitory existence in the gas phase and are unknown in solution. Spectroscopic data indicate that H_3^+ is an equilateral triangle (Fig. 9.2) and hence that it is the simplest example of a three-center two-electron bond (a 3c,2e bond) in which three nuclei are bound by only two electrons. This bond is represented by the three-spoke pattern shown in the illustration. As we saw in Section 3.1, there is nothing particularly puzzling about this form of bonding, since it is a natural outcome of the formation of molecular orbitals that are delocalized over the three atoms.

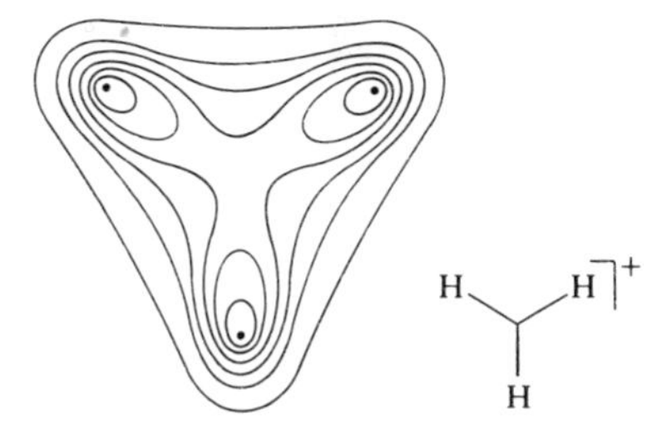

Fig. 9.2 Calculated electron density distribution in H_3^+ and the representation of the 3c,2e bond.

Tracer experiments provide evidence that H_3^+ may exist transitorily as a reaction intermediate. Thus, D_2 gas undergoes an exchange reaction with superacids such as HF/SbF_5, presumably through the triangular intermediate D_2H^+, which can split up to give HD or D_2:

$$D_2(g) + HF/SbF_5 \rightarrow \left[H{-}\!\!<^{D}_{D}\right]^+ [SbF_6]^- \rightarrow HD \text{ or } D_2 \;+\; D^+ \text{ or } H^+$$

9.3 Dihydrogen

The stable form of elemental hydrogen under normal conditions is **dihydrogen**, H_2, more informally and henceforth simply called 'hydrogen'. The H_2 molecule has a high bond enthalpy (436 kJ mol^{-1}) and a short bond length (0.74 Å). Since it has so few electrons, its intermolecular forces are weak, and the gas condenses to a liquid only when cooled to 20 K.

Hydrogen reacts slowly with most other elements, partly because of its high bond enthalpy and hence high activation energy for reaction. Under special conditions, however, the reactions are rapid. These include the activation of the molecule by homolytic dissociation on a metal surface or a metal complex, heterolytic dissociation by a surface or metal ion, or initiation of a radical-chain reaction.

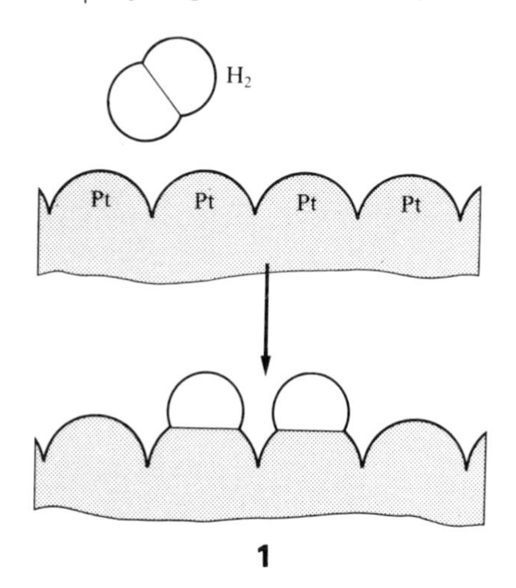

1

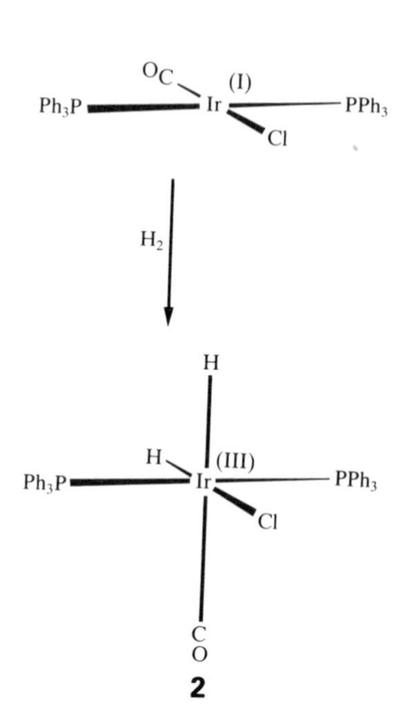

2

3

Homolytic dissociation by a metal surface or a metal complex

Two examples of homolytic dissociation are chemisorption on platinum (**1**) and coordination to the Ir atom in a complex (**2**). The former accounts for the use of finely divided platinum metal to catalyze the hydrogenation of alkenes and the release or reaction of H_2 in electrochemical cells. Thus the overpotential for H_2 formation is much smaller at a platinum electrode, on which a surface hydride can form, than on a mercury electrode, where the formation of a surface hydride is unfavorable.

Similarly, H_2 readily bonds to the Ir atom in complexes such as $[IrClCO(PPh_3)_2]$ to give a complex containing two hydrido (H^-) ligands:

$$\underset{\text{Ir(I)}}{[\text{IrClCO(PPh}_3)_2]} + \text{H}_2 \rightarrow \underset{\text{Ir(III)}}{[\text{IrCl(H)}_2\text{CO(PPh}_3)_2]}$$

This type of reaction is called **oxidative addition** because the metal center undergoes a formal increase in oxidation number when reactant adds to it. Oxidative addition is common with Rh(I), Ir(I), and Pt(0) complexes. When considering hydrido complexes, we should recall from Section 7.5 that H^- is a high-field ligand. Many *d*-block hydrido complexes are known, and they are discussed in more detail in Chapters 16 and 17.

It has recently been discovered that the H_2 molecule can coordinate to a metal atom without cleavage of the H—H bond.[1] The first such compound was $[W(CO)_3(H_2)(P^iPr_3)_2]$ (**3**; ^{i}Pr denotes isopropyl, —$CH(CH_3)_2$); several dozen compounds of this type are now known. The significance of this discovery is that it provides an example of a species intermediate between molecular H_2 and a dihydrido complex. Calculations indicate that the bonding consists of donation of electron density from the H—H bond into a vacant metal *d* orbital and a simultaneous back π bonding from a different *d* orbital into the empty σ^* orbital of H_2 (**4**). This bonding pattern is similar to that of CO and ethylene with metal atoms (Sections 7.4 and 16.6). The back-donation of electron density from the metal to a σ^* orbital on H_2 is consistent with the observation that the H—H bond is cleaved when the metal center is made electron-rich by highly basic ligands.

Heterolytic dissociation by a surface or a metal ion

An example of heterolytic dissociation is the reaction of H_2 with a ZnO surface, which appears to produce a Zn(II)-bound hydride and an O-bound proton:

$$\text{H}_2 + \underline{\text{Zn–O–Zn–O}} \longrightarrow \underline{\overset{\text{H}^-}{\text{Zn}}\text{–O–Zn–}\overset{\text{H}^+}{\text{O}}}$$

This reaction is thought to be an important step in the catalytic hydrogenation of carbon monoxide to methanol,

$$\text{CO}(g) + 2\text{H}_2(g) \xrightarrow{\text{Cu/ZnO}} \text{CH}_3\text{OH}(g)$$

[1] An informative account of this discovery and its implications is given by G. Kubas, *Acc. chem. Res.*, **21**, 120 (1988).

which is carried out on a very large scale worldwide. Another example is the sequence

$$H_2(g) + Cu^{2+}(aq) \rightarrow [CuH]^+(aq) + H^+(aq) \rightarrow Cu(s) + 2H^+(aq)$$

which is important in the hydrometallurgical reduction of Cu^{2+} (Section 8.1). The $[CuH]^+$ ion has only a transitory existence.

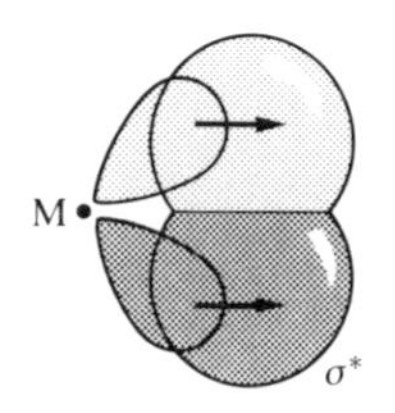

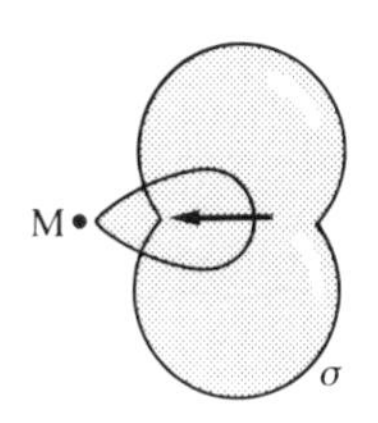

4

Radical-chain mechanisms

Chain mechanisms account for the thermally or photochemically initiated reactions between H_2 and the halogens:

Initiation: $Br_2 \xrightarrow{h\nu} 2Br\cdot$

Propagation: $Br\cdot + H_2 \rightarrow HBr + H\cdot$

$H\cdot + Br_2 \rightarrow HBr + Br\cdot$

The activation energy of radical attack is low because the loss of one bond occurs as the new bond is formed, so once initiated the production of HBr is very rapid. Chain termination occurs when the radicals recombine:

$$2H\cdot \rightarrow H_2 \qquad 2Br\cdot \rightarrow Br_2$$

Termination becomes more important toward the end of the reaction when the concentrations of H_2 and Br_2 are low.

Example 9.2: *Recognizing initiation steps in a radical reaction*

Suggest a mechanistic explanation for the observation that the addition of NO gas to a mixture of H_2 and Cl_2 leads to an explosion.

Answer. The NO, a radical, initiates a very fast reaction between H_2 and Cl_2 by generation of $Cl\cdot$ radicals:

$$\cdot NO + Cl_2 \rightarrow ClNO + Cl\cdot$$

The $Cl\cdot$ produced then reacts with H_2 in a radical chain mechanism in which the propagation steps are similar to those already given for Br_2:

$$H_2 + \cdot Cl \rightarrow HCl + \cdot H$$

$$\cdot H + Cl_2 \rightarrow HCl + \cdot Cl$$

Exercise. Suggest why, under some conditions, a radical chain reaction will slow down when a foreign radical is introduced.

9.4 The production of dihydrogen

Molecular hydrogen is not present in significant quantities in the earth's crust, but is produced in huge quantities to satisfy the needs of industry. The main commercial process is currently **steam reforming**, the catalyzed reaction of water and hydrocarbons (typically methane from natural gas) at high temperatures:

$$CH_4(g) + H_2O(g) \xrightarrow{1000^\circ C} CO(g) + 3H_2(g) \qquad (1)$$

A similar reaction, but with coke as the reducing agent, is sometimes called the 'water-gas reaction':

$$C(s) + H_2O(g) \xrightarrow{1000°C} CO(g) + H_2(g)$$

This reaction was once a primary source of H_2, and it may become important again when natural hydrocarbons are depleted. As with the reduction of metal oxides by CO (Section 8.1), the negative free energy of both reactions results in part from the strength of the bond in CO. The yield of H_2 can be increased by means of a subsequent **shift reaction**:

$$CO(g) + H_2O(g) \rightarrow CO_2(g) + H_2(g) \qquad (2)$$

Hydrogen is also produced as a by-product of the electrolysis of brine during the manufacture of Cl_2 and NaOH (Section 13.1), but electrolytic production will become the primary source only if cheap electricity becomes available. Another source is as a by-product of the dehydrogenation of alkanes to produce alkenes for the petrochemical industry and arenes for high-octane gasolines:

$$CH_3CH_3(g) \xrightarrow{\Delta} CH_2{=}CH_2(g) + H_2(g)$$

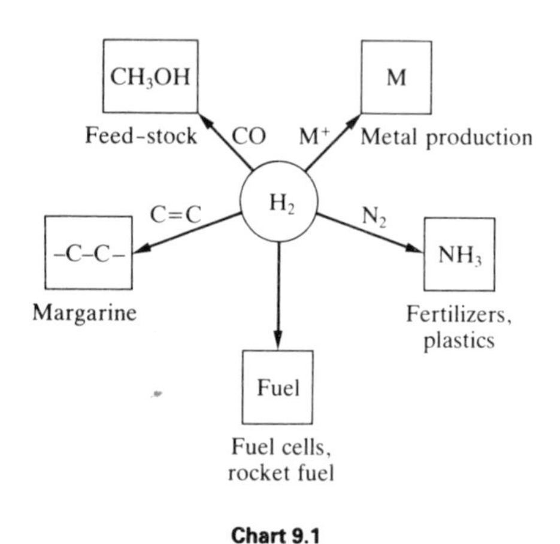

Chart 9.1

The hydrogen produced in one of these ways is normally used directly in an integrated chemical plant, such as an ammonia synthesis plant or a petroleum refinery (Chart 9.1).

Because of its low mass and high enthalpy of combustion, hydrogen is an excellent fuel for large rockets. Its more general use as a fuel has been analyzed seriously since the early 1970s when petroleum came into short supply. Strategies have been devised for a **hydrogen economy**, in which hydrogen is the primary fuel. One scenario is to collect solar energy with photovoltaic cells and employ the resulting electricity to electrolyze water. The hydrogen from this electrolysis would be a form of stored energy that could be burned as a fuel when needed or used as a feed-stock in chemical processes. The current prices of petroleum, natural gas, and coal are too low to make such a hydrogen economy viable, but the idea might have its day.[2] Quite aside from the eventual scarcity of fossil fuels, when hydrogen burns it produces water, which is not a 'greenhouse gas' like CO_2, so a hydrogen economy would not necessarily lead to the global warming that now seems to confront us. The availability of photochemically produced H_2 would also eliminate the CO_2 resulting from steam reforming and the shift reaction. Thus a first step toward a hydrogen economy might be the use of electrolytic hydrogen in the chemical and petrochemical industries.

Classification and structure of compounds

The hydrogen compounds of the highly electropositive *s*-block metals are nonvolatile, electrically nonconducting, crystalline solids. These properties as well as their structures lead to their classification as **saline hydrides** (or

[2] Research on the generation of hydrogen by advanced photochemical methods is described in M. Grätzel (ed.), *Energy resources through photochemistry and catalysis.* Academic Press, New York (1983).

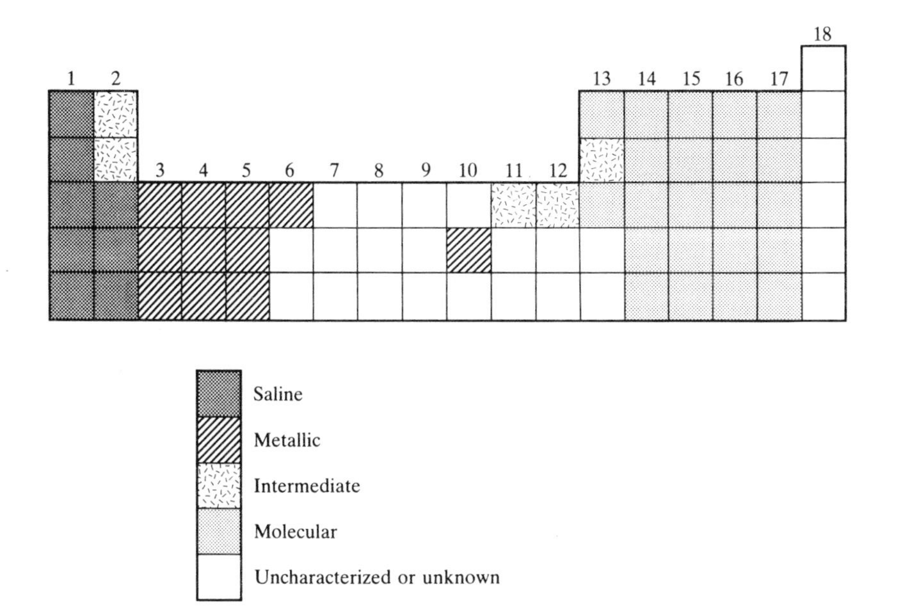

Fig. 9.3 Classification of the binary hydrogen compounds of the *s*-, *p*-, and *d*-block elements. Although some *d* elements, such as Fe and Ru, do not form binary hydrides, they do form metal complexes containing the hydride ligand.

'salt-like hydrides'). On the other hand, the *d*-block and *f*-block metals form hydrides that are often nonstoichiometric, electrically conducting solids, and are called **metallic hydrides**. Most of the *p*-block binary hydrogen compounds are volatile molecular compounds, and accordingly are classified as **molecular hydrides**. We show the distribution of these hydrides in the periodic table in Fig. 9.3. As so often in chemistry, the main point of this classification scheme is to emphasize the major trends in properties, and in fact there is a continuum of structural types. We shall see, for instance, that some elements (among them Be and Al) form hydrides that cannot be classified as either strictly saline or strictly molecular.

Proton-NMR is particularly useful for studying hydrides because it can detect the presence of H nuclei in a compound and be used to identify the nuclei nearby. Some typical proton chemical shifts are shown in Fig. 9.4. A hydrogen ligand attached to a late *d*-block element has an NMR signal that

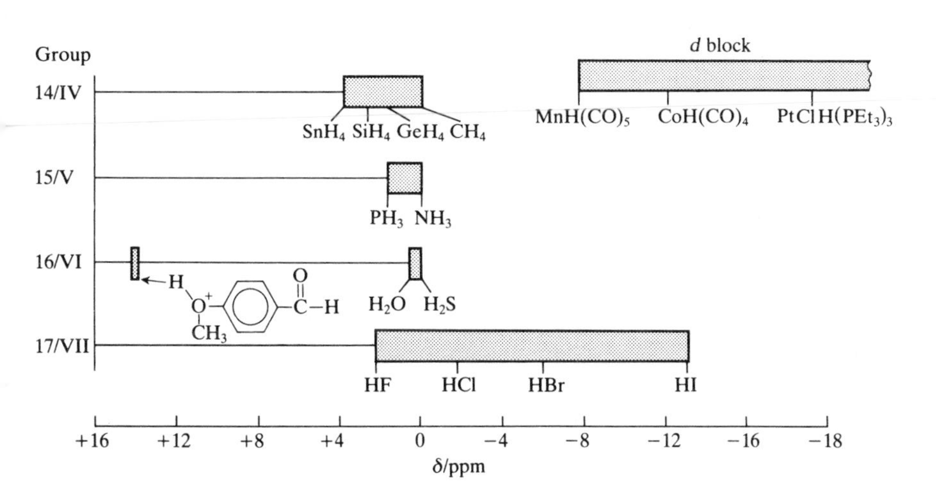

Fig. 9.4 Typical ^{1}H-NMR chemical shifts.

Table 9.2. Structures of *s*-block hydrides

Compound	Crystal structure
LiH, NaH, KH, RbH, CsH	Rock-salt
MgH_2	Rutile
CaH_2, SrH_2, BaH_2	Distorted $PbCl_2$

From: A. F. Wells, *Structural inorganic chemistry*. Oxford University Press (1984).

is much more shielded than the usual standard (tetramethylsilane). Theoretical considerations suggest that this shielding is not simply the consequence of high electron density around the proton of the hydride ligand. Only in the *s*-block hydrides is the electron density great enough for 'H^-' to be a reasonably accurate portrayal of the entity present in the compound. Strong acids have strongly deshielded ^{1}H-NMR signals (Fig. 9.4), and in this case the simple picture of depletion of electron density around the proton does appear to be adequate.

9.5 The saline hydrides

The Group 1 hydrides have rock-salt structures, and the Group 2 hydrides have crystal structures like those of some heavy metal halides (Table 9.2). Both observations are consistent with, but certainly do not prove, the substantial H^- ion character of these solids. The ionic radius of H^- determined from the dimensions of these structures varies quite widely, from 1.26 Å in LiH to 1.54 Å in CsH; its radius in the gas phase is calculated to be around 2.0 Å. This wide variability reflects the loose control that the single charge of the proton has on its two surrounding electrons and the resulting high compressibility of H^-.

The saline hydrides are insoluble in common nonaqueous solvents but they do dissolve in molten alkali halides. Electrolysis of this molten-salt solution produces hydrogen gas at the anode (the site of oxidation)

$$2H^-(melt) \rightarrow H_2(g) + 2e^-$$

thus providing chemical evidence for the existence of H^-. The reaction of saline hydrides with water is vigorous:

$$NaH(s) + H_2O(l) \rightarrow H_2(g) + NaOH(aq)$$

Indeed, the finely divided compound can ignite if it is left exposed to humid air. Such fires are difficult to extinguish because even CO_2 is reduced when it comes into contact with hot metal hydrides (water, of course, is useless, as it forms even more hydrogen); they may be blanketed with an inert solid, such as silica sand.

The absence of suitable solvents limits the use of saline hydrides as reagents, but this problem is partially overcome by the availability of commercial dispersions of finely divided NaH in oil.[3] This material is reactive by virtue of the dispersion's high surface area. On account of their

[3] Even more finely divided and reactive alkali metal hydrides can be prepared from the metal alkyl and hydrogen: P. A. A. Klusener, L. Brandsma, H. D. Verkruijsse, P. von Rague Schleyer, T. Friedl, and R. Pi, *Angew. Chem. Int. Ed. Engl.*, **25**, 465 (1986).

vigorous reaction with water, saline hydrides such as CaH_2 are sometimes used to remove the last traces of moisture from a gas or a solvent. Large amounts of water should not be removed in this manner because the strongly exothermic reaction evolves flammable hydrogen.

9.6 Metallic hydrides

Metallic hydrides are known for all *d*-block metals for Groups 3 through 5 and for the *f*-block metals (Fig. 9.5). However, the only hydrides in Group 6 are those of Cr, and no hydride phases are known for the unalloyed metals in Groups 7 through 9. This last region is sometimes referred to as the **hydride gap**. The *f*-block elements form hydrides with the limiting formulas MH_2 and MH_3, but large deviations from the ideal stoichiometric composition are generally found.

The hydrides of the Group 10 elements are especially important because Ni, Pd, and Pt metals are often used as hydrogenation catalysts (Section 17.7). However, somewhat surprisingly, at moderate pressures only Pd forms a stable bulk phase; its composition is PdH_x, with $x < 1$. Nickel forms hydride phases at very high pressures but Pt does not form any at all. Despite this, there is no doubt that hydrogen is chemisorbed on the surface of Ni and Pt to form surface hydrides, and Pt metal is the most versatile catalyst for hydrogenations (Section 17.7).

Most metallic hydrides are metallic conductors (hence their name) and have variable composition. For example, at 550°C a compound ZrH_x exists over a composition range from $ZrH_{1.30}$ to $ZrH_{1.75}$; it has the fluorite structure (Fig. 4.14) with a variable number of the anion sites unoccupied. We can understand their variable stoichiometry and metallic conductivity in terms of a model in which the band of delocalized orbitals responsible for

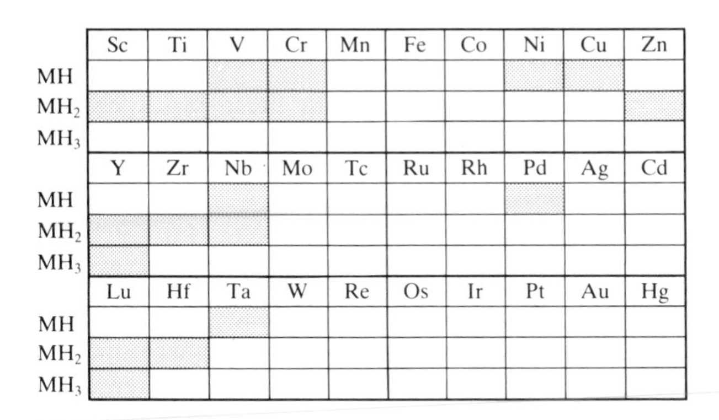

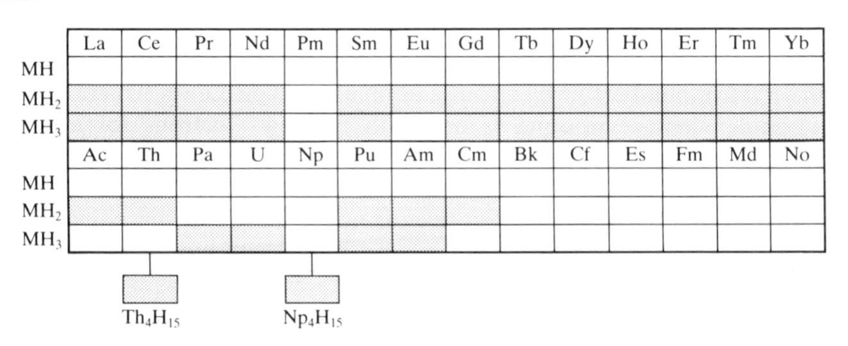

Fig. 9.5 Hydrides formed by *d*- and *f*-block elements. The formulas are limiting stoichiometries based on the structural type. In many cases they are not attained. For example, PdH_x never attains $x = 1$. (From G. G. Libowitz, *Solid state chemistry of the binary metal hydrides*. Benjamin (1965).)

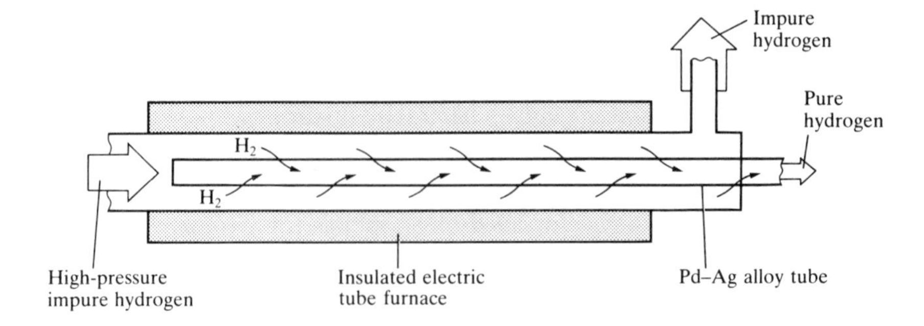

Fig. 9.6 Schematic diagram of a hydrogen purifier. Hydrogen diffuses through the Pd–Ag tube as H atoms but impurities do not.

the conductivity accommodates electrons supplied by arriving H atoms. The H^+ ions as well as the metal cations take up equilibrium positions in this electron sea. The conductivities of metallic hydrides typically vary with hydrogen content, and this can be correlated with the extent to which the conduction band is filled or emptied as hydrogen is added or removed. Thus, CeH_{2-x} is a metallic conductor, whereas CeH_3 (which has a full conduction band) is an insulator.

Another striking property of many metallic hydrides is that the H atoms diffuse rapidly through the solid at slightly elevated temperatures. This mobility is utilized in the ultrapurification of H_2 by diffusion through a Pd–Ag alloy tube (Fig. 9.6). The high mobility and variable composition also make the metallic hydrides potential hydrogen storage media. Thus the alloy $LaNi_5$ forms a hydride phase with a limiting composition $LaNi_5H_6$, and at its limiting composition it contains more hydrogen per unit volume than liquid H_2. A less expensive system with the composition $FeTiH_x$ ($x < 1.95$) is now commercially available for low-pressure hydrogen storage and it has been used in vehicle trials.

9.7 Binary molecular compounds

Hydrogen forms binary molecular compounds with the *p*-block elements. They include the familiar Period-2 compounds CH_4, NH_3, H_2O, and HF, as well as the corresponding compounds of the heavier elements in each group.

Nomenclature and classification

The systematic names of the molecular hydrogen compounds are usually formed from the name of the element and the suffix -ane. Common usage, however, often deviates from this recommendation (Table 9.3), as in the -ine of phosphine and the -ide of hydrogen sulfide and the hydrogen halides. The nonsystematic names 'ammonia' and 'water' are universally used rather than 'azane' and 'oxane'.

A useful classification of the molecular hydrogen compounds takes note of the relative numbers of electrons and bonds in their Lewis structures. An **electron-deficient compound** is one with too few electrons for a Lewis structure to be written with an octet around the central atom. An **electron-precise compound** is one with the correct number of electron pairs for bond formation with none left over as nonbonding electron pairs on the central atom. An **electron-rich compound** is one with more electron pairs than are needed for bond formation, the extra electron pairs being present as nonbonding electron pairs on the central atom.

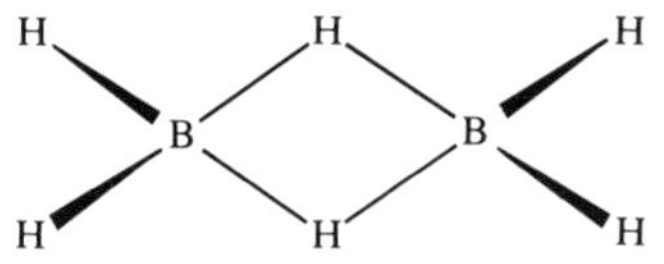

Table 9.3. Some molecular hydrogen compounds

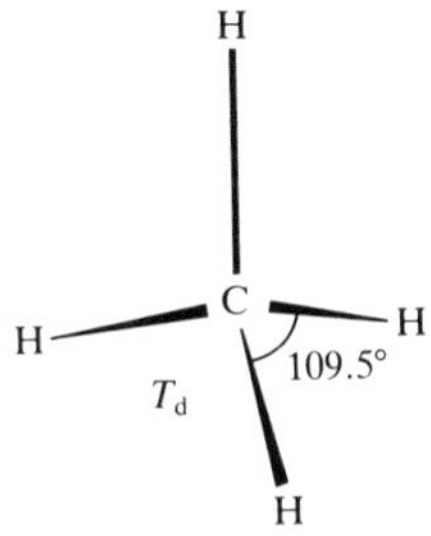

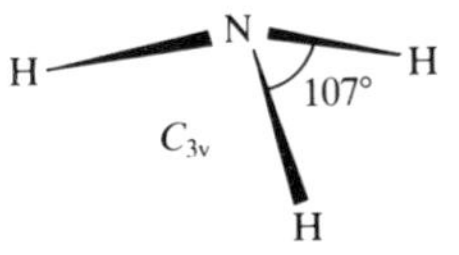

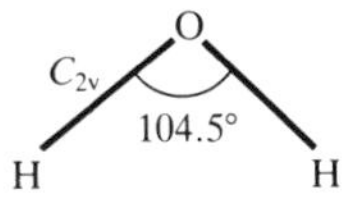

Group	Formula and name
13/III	B_2H_6, diborane
14/IV	CH_4, methane SiH_4, silane GeH_4, germane SnH_4, stannane
15/V	NH_3, ammonia PH_3, phosphine AsH_3, arsine SbH_3, stibine
16/VI	H_2O, water H_2S, hydrogen sulfide H_2Se, hydrogen selenide H_2Te, hydrogen telluride
17/VII	HF, hydrogen fluoride HCl, hydrogen chloride HBr, hydrogen bromide HI, hydrogen iodide

Diborane (B_2H_6) is a classic example of an electron-deficient compound (Section 3.2). Its Lewis structure would require at least 14 valence electrons to bind the eight atoms together, but the molecule in fact has only 12. A simple description of its structure is one in which a bridging H and two B atoms are joined by a 3c,2e bond (**5**): each B atom has an octet of electrons and each H atom has a pair of electrons, so the valence shells of all atoms are satisfied. That the 3c,2e hydrogen bridge bond is weaker than the terminal 2c,2e bonds is reflected in diborane's chemical properties.

Methane is an example of an electron-precise compound. Electron-precise binary compounds are formed by the elements in Group 14/IV (each of which has four valence electrons). The molecules are tetrahedral (**6**) with the expected increase in E—H bond length down the group.

Ammonia is an example of an electron-rich compound. Its four atoms could be bonded by only six valence electrons, whereas in fact it has eight, the extra two forming a nonbonding lone pair. A part of the reason for the existence of electron-rich compounds is that H has too low an electronegativity to draw out the highest oxidation number of Group 15/V to 17/VII elements. For example, in combination with H, P forms the electron-rich phosphine (PH_3) and not the electron-precise PH_5. With the more electronegative Cl phosphorus forms both PCl_3 and PCl_5.

General aspects of properties

The shapes of the electron-rich compounds can all be predicted by the VSEPR rules (Section 2.2). Thus, NH_3 is pyramidal (**7**), H_2O angular (**8**), and HF (necessarily) linear. However, the simple VSEPR rules do not indicate the considerable change in bond angle between NH_3 and its heavier analogs or between H_2O and its analogs in Group 16/VI. As noted in Table 9.4, the bond angles for the Period 2 molecules NH_3 and H_2O are slightly

Table 9.4. Bond angles (in degrees) for Group 15/V and 16/VI hydrogen compounds

NH_3	106.6	H_2O	104.5
PH_3	93.8	H_2S	92
AsH_3	91.8	H_2Se	91
SbH_3	91.3	H_2Te	89

Source: A. F. Wells, *Structural inorganic chemistry*. Oxford University Press (1984).

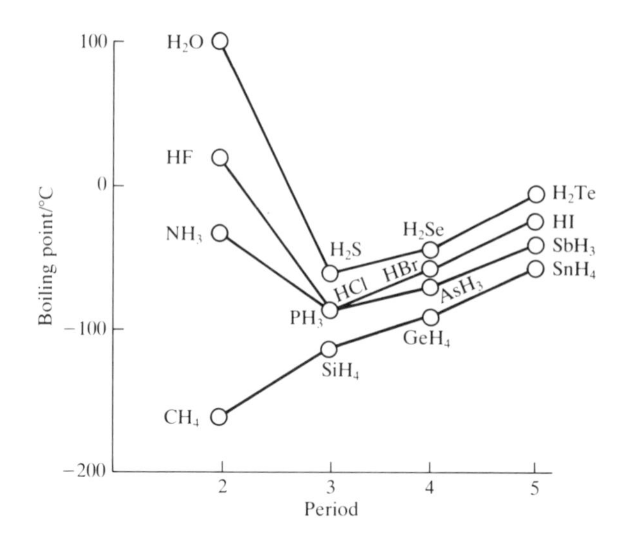

Fig. 9.7 Normal boiling points of *p*-block binary hydrogen compounds.

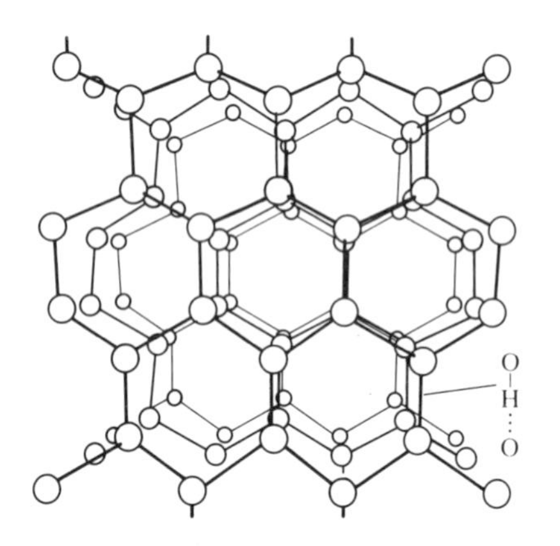

Fig. 9.8 The structure of ice. Spheres represent the oxygen atoms. Hydrogen atoms lie on the lines joining oxygen atoms.

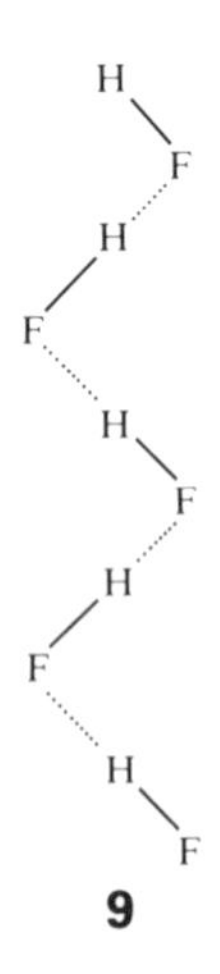

9

less than the tetrahedral angle, but for their heavier analogs the bond angle is as small as 90°.

An important consequence of the simultaneous presence of highly electronegative atoms (N, O, and F) and lone pairs in the electron-rich compounds is the possibility of forming hydrogen bonds. Some of the effects of hydrogen bonding include anomalously high boiling points (Fig. 9.7), the open network structure of ice (Fig. 9.8 and Section 9.16), and the existence of chains in solid HF (**9**) that survive partially even in the vapor. The heavier elements of these groups do not form strong hydrogen bonds: that H_2S, PH_3, and HCl are not associated is indicated by their boiling points, which are in line with that of their electron-precise analog silane (SiH_4).

Example 9.3: *Correlating the classification and properties of hydrogen compounds*

Classify the following compounds, and discuss their physical properties. $HfH_{1.5}$, PH_3, CsH, B_2H_6. For the molecular compounds specify their sub-classification (electron-deficient, electron-precise, or electron-rich).

Answer. The two hydrides $HfH_{1.5}$ and CsH will be solids. The former is an example of a metallic hydride, which are common for *d*- and *f*- block metals, and it exhibits significant electrical conductivity. CsH is a saline hydride typical of the *s*-block metals. It is an electrical insulator with the rock-salt crystal structure. The *p*-block molecular hydrides PH_3 and B_2H_6 have low molar mass and are therefore expected to be highly volatile. (They are in fact gases under normal conditions.) The Lewis structure indicates that PH_3 has a lone pair on P and that it is therefore an electron-rich molecular hydride. As explained in the text, diborane is electron deficient.

Exercise. Give examples of hydrogen compounds of elements in Groups 2, 11, and 16/VI; classify them, and discuss whether any of them are likely to participate in hydrogen bonding.

The reactivity of hydrogen compounds

9.8 Stability and synthesis

Stability

The standard free energies of formation of the hydrogen compounds of *s*- and *p*-block elements (Table 9.5) reveal a regular variation in their stabilities. With the possible exception of BeH_2, for which good data are unavailable, all the *s*-block hydrides are exoergic ($\Delta G_f^\ominus < 0$) and therefore thermodynamically stable with respect to their elements at room temperature. This trend is erratic in the boron group, in that only AlH_3 is exoergic. In all the other groups of the *p* block, the hydrogen compounds of the first members of the groups (CH_4, NH_3, H_2O, and HF) are stable and become progressively less stable down the group. This general trend in stability is largely a reflection of the decreasing E—H bond strength down a group in the *p* block (Fig. 9.9). The weak bonds formed by the heavier elements are often attributed to poor overlap with the diffuse *s* and *p* orbitals of the atoms of these elements. The heavier members become more stable on going from Group 14/IV across to the halogens. For example, SnH_4 is highly endoergic whereas HI is barely so.

Table 9.5. Standard Gibbs energy of formation, $\Delta G_f^\ominus/(\text{kJ mol}^{-1})$, of binary *s*- and *p*-block hydrogen compounds at 25°C

Period	Group 1 I	2 II	13 III	14 IV	15 V	16 VI	17 VII
2	LiH (*s*) −68.4	BeH_2 (*s*) (+20)	B_2H_6 (*g*) +86.7	CH_4(*g*) −50.7	NH_3(*g*) −16.5	H_2O (*l*) −237.1	HF (*g*) −273.2
3	NaH (*s*) −33.5	MgH_2 (*s*) −35.9	AlH_3 (*s*) (−1)	SiH_4 (*g*) +56.9	PH_3 (*g*) +13.4	H_2S (*g*) −33.6	HCl (*g*) −95.3
4	KH (*s*) (−36)	CaH_2 (*s*) −147.2	GaH_3 >0	GeH_4 (*g*) +113.4	AsH_3 (*g*) +68.9	H_2Se (*g*) +15.9	HBr (*g*) −53.5
5	RbH (*s*) (−30)	SrH_2 (*s*) (−141)		SnH_4 (*g*) +188.3	SbH_3 (*g*) +147.8	H_2Te (*g*) >0	HI (*g*) +1.7
6	CsH (*s*) (−32)	BaH_2 (*s*) (−140)					

Data from *J. Phys. Chem. Ref. Data* **11**, Supplement 2, (1982). Values in parenthesis are based on $\Delta H_f^\ominus$ data from this source and entropy contributions estimated by the method of W. M. Latimer, p. 359 of *Oxidation potentials*. Prentice-Hall, Englewood Cliffs, NJ, (1952).

Synthesis

The three common methods for synthesizing binary hydrogen compounds are direct combination of the elements, protonation, and metathesis.

(1) *Direct combination of the elements*:

$$2Li(l) + H_2(g) \xrightarrow{\Delta} 2LiH(s)$$

This method is used commercially for the synthesis of exoergic compounds, including NH_3 and the hydrides of Li, Na, and Ca. However, in some cases

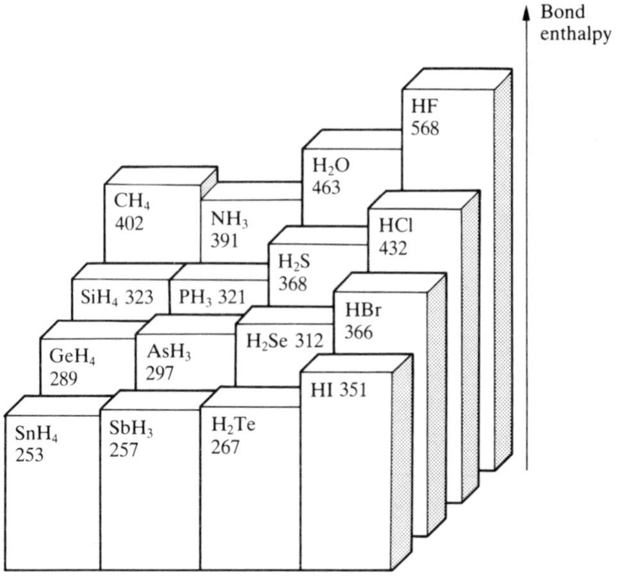

Fig. 9.9 Average bond enthalpies (in kJ mol^{-1} at 298 K) compiled from data in D. D. Wagman *et al., J. Phys. Chem. Ref. Data* **11**, Supplement 2 (1982).

forcing conditions (high pressure, high temperature, and a catalyst) are necessary to overcome the unfavorable kinetics. The high temperature used for the Li reaction is an example: it melts the metal and hence helps to break up the surface layer of hydride that would otherwise passivate it. This inconvenience is avoided in many laboratory preparations by adopting one of the following reactions, which may also be used to prepare endoergic compounds (ones that are thermodynamically unstable with respect to their elements and so cannot be formed from them directly).

(2) *Protonation of a Brønsted base*:

$$Li_3N(s) + 3H_2O(l) \rightarrow 3LiOH(aq) + NH_3(g)$$

The Brønsted base in this reaction is the nitride ion, N^{3-}. The starting materials are too expensive for the reaction to be suitable for the commercial production of NH_3, but it is very useful in the laboratory for the preparation of ND_3. The success of the reaction depends on the proton donor being stronger than the conjugate acid of the N^{3-} anion (NH_3 in this case). Water is a sufficiently strong acid to protonate the very strong base N^{3-}, but a stronger acid, such as H_2SO_4, is required to protonate the weak base Cl^-:

$$NaCl(s) + H_2SO_4(l) \rightarrow NaHSO_4(s) + HCl(g)$$

(3) *Metathesis* (*double replacement*) *of a halide or pseudohalide with a hydride*:

$$LiAlH_4 + SiCl_4 \rightarrow LiAlCl_4 + SiH_4 \qquad (3)$$

This reaction involves (at least formally) the exchange of Cl^- ions for H^- ions in the coordination sphere of the Si atom. Hydrides of the more electropositive elements (LiH, NaH, and $[AlH_4]^-$) are the most active H^- sources. The favorite sources are often the $[AlH_4]^-$ and $[BH_4]^-$ ions in salts such as $LiAlH_4$ and $NaBH_4$, which are soluble in ether solvents, such as $CH_3OC_2H_4OCH_3$, which solvate the alkali metal ion. Of these anion complexes, $[AlH_4]^-$ is much the strongest hydride donor (Section 9.12).

9.9 Reactions

Three types of reaction can lead to the scission of an E—H bond. They are similar to the processes already considered for H—H bond cleavage:

(1) *Heterolytic cleavage by hydride transfer*:

$$\text{E—H} \rightarrow \text{E}^+ + \text{:H}^-$$

(2) *Homolytic cleavage*:

$$\text{E—H} \rightarrow \text{E}\cdot + \text{H}\cdot$$

(3) *Heterolytic cleavage by proton transfer*:

$$\text{E—H} \rightarrow \text{:E}^- + \text{H}^+$$

A free H^- or H^+ ion is unlikely to be formed in reactions that take place in solution. Instead, the ion is transferred via a complex that involves a hydrogen bridge.

It is often not clear which of these three primary processes occurs in practice. For example, any of them can lead to the addition of EH across a double bond:

$$\text{E—H} + \text{H}_2\text{C}{=}\text{CH}_2 \rightarrow \text{H}_2\underset{|}{\overset{\text{E}}{\text{C}}}\text{—}\underset{|}{\overset{\text{H}}{\text{C}}}\text{H}_2$$

However, there are some patterns of behavior that at least suggest which one is likely to occur.

Heterolytic cleavage and hydridic character

Compounds with **hydridic character**—that is, compounds that can transfer an H^- ion and hence participate in Process 1—react vigorously with Brønsted acids, with the evolution of H_2:

$$\text{NaH}(s) + \text{H}_2\text{O}(l) \rightarrow \text{NaOH}(aq) + \text{H}_2(g)$$

A compound that takes part in this reaction with a weak proton donor (such as water, as in this example) is said to be **strongly hydridic**. If a compound requires a strong proton donor, it is classified as **weakly hydridic** (an example is germane, GeH_4). Hydridic character is most pronounced toward the left of a period where the element is most electropositive (in the *s*-block) and decreases rapidly after Group 13/III.

Hydrido complexes are readily formed with Lewis acids from the boron group:

$$2\text{LiH} + \text{B}_2\text{H}_6 \xrightarrow{\text{R}_2\text{O}} 2\text{LiBH}_4$$

These complexes are very useful because they take part in simple metathesis reactions, as in eqn 3.

Homolytic cleavage and radical character

Homolytic cleavage (Process 2) appears to occur readily for the hydrogen compounds of some *p*-block elements, especially the heavier elements. For example, the use of a radical initiator ($Q\cdot$ in the following equations, typically a peroxide) greatly facilitates the reaction of trialkylstannanes (R_3SnH) with haloalkanes, R—X. The overall reaction is

$$R_3SnH + R'X \rightarrow R'H + R_3SnX$$

and the proposed mechanism is

Initiation: $R_3Sn\text{—}H + Q\cdot \rightarrow R_3Sn\cdot + QH$

Propagation: $R_3Sn\cdot + XR' \rightarrow R_3SnX + R'\cdot$
(rate-determining halogen abstraction)
$R_3SnH + R'\cdot \rightarrow R'H + R_3Sn\cdot$

Termination: $R'\cdot + R'\cdot \rightarrow R'_2$
$R'\cdot + R_3Sn\cdot \rightarrow R_3SnR'$
$R_3Sn\cdot + R_3Sn\cdot \rightarrow R_3Sn\text{—}SnR_3$

The order of reactivity for various haloalkanes with alkylstannanes is

$$RF < RCl < RBr < RI$$

Thus fluoroalkanes do not react with R_3SnH, chloroalkanes require heat, photolysis or chemical radical initiators, and bromoalkanes and iodoalkanes react spontaneously at room temperature. This trend indicates that the rate-determining step is halogen abstraction. Similarly, the tendency toward radical reactions increases toward the heavier elements in each group, and Sn—H compounds are in general more prone to radical reactions than Si—H compounds. This trend mirrors the decrease in E—H bond strength down a group (Fig. 9.9) and the decrease in the stretching frequency.[4] In terms of wavenumbers:

H_2O	H_2S	H_2Se
3652 and 3756 cm^{-1}	2611 and 2684 cm^{-1}	2260 and 2350 cm^{-1}

As the E—H bond weakens, the molecular potential energy curve becomes shallower and hence the force constant becomes smaller. The force constant is greatest when the potential well has steeply inclined walls.

Heterolytic cleavage and protic character

Compounds reacting by deprotonation (Process 3) are said to show **protic behavior**: in other words, they are Brønsted acids. We saw in Section 5.2 that Brønsted acid strength increases from left to right across a period in the *p* block and down a group. One striking example of the former trend is the increase in acidity from CH_4 to HF.

[4] There are two EH_2 stretches for each molecule, a symmetric stretch in which both bonds stretch and contract in phase, and an antisymmetric stretch in which one bond contracts as the other lengthens.

Example 9.4: *Suggesting the nature of hydrogen transfer processes*

Suggest the most likely mechanism for hydrogen transfer in (a) the reaction of HF with NH_3 and (b) the reaction betweeen $[AlH_4]^-$ and BF_3 in ether.

Answer. The electronegativity difference between E and H in each case suggests that both are likely to involve heterolytic processes with HF acidic and $[AlH_4]^-$ hydridic. Free H^+ or H^- is unlikely to be involved. HF is likely to transfer a proton via a hydrogen-bonded complex. The $[AlH_4]^-$ ion is likely to transfer an H^- ion by a hydrogen-bridged complex.

Exercise. What is a likely mechanism for the reaction of Et_3PbH with CH_3Br?

The electron-deficient hydrides of the boron group

Boron hydrides were first prepared in a pure form by the German chemist Alfred Stock, who investigated them from 1912 to the early 1930s. Stock was motivated by the thought that boron, as carbon's neighbor in the periodic table, was likely to form an analogous series of compounds. He was served well by this analogy, since it led him to prepare and characterize six boron hydrides. His achievement was all the more noteworthy because he had to develop new techniques to deal with these highly reactive compounds, and in particular he needed to develop vacuum line techniques (Box 9.1). However, although the analogy was fruitful, we shall see that boron chemistry is unlike carbon chemistry in many interesting ways.

9.10 Diborane

Synthesis

Stock's entry into the world of boron hydrides was the protolysis of magnesium boride, but this has been superseded by syntheses that give better yields of specific compounds. Thus the simplest boron hydride, diborane B_2H_6, may be prepared by metathesis of a boron halide with either $LiAlH_4$ or $LiBH_4$ in ether:

$$3LiEH_4 + 4BF_3 \rightarrow 2B_2H_6 + 3LiEF_4 \qquad (E = B, Al)$$

Diborane is a gas and ignites on contact with air. It decomposes very slowly at room temperatures, forming higher boron hydrides and an involatile and insoluble yellow solid, the boron chemist's counterpart of the organic chemist's 'black tar'.

Stock prepared several members of two classes of boron hydrides with formulas B_nH_{n+4} and B_nH_{n+6}, the former class being the more stable. Examples include tetraborane(10) B_4H_{10}, pentaborane(9) B_5H_9, and pentaborane(11) B_5H_{11}. The nomenclature, in which we specify the number of B atoms by a prefix and give the number of H atoms in parentheses, should be noted. Thus, the systematic name for diborane is diborane(6); however, as there is no diborane(8), the simpler term diborane is almost always used.

All the boranes are colorless, diamagnetic substances. They range from gases (B_2 and B_4 hydrides), through volatile liquids (B_5 and B_6 hydrides), to the sublimable solid $B_{10}H_{14}$. Only the chemistry of diborane will be considered here; we discuss the higher boron hydrides in Chapter 11.

Box 9.1 Chemical vacuum lines

Chemical vacuum line techniques provide a method for handling gases or volatile liquids without exposure to the air. They enable a wide range of operations necessary for the preparation and characterization of air-sensitive compounds to be performed.

A vacuum line, such as the one illustrated in Fig. 9B.1, is evacuated by a high vacuum pumping system. Once the apparatus is evacuated, condensable gases can be moved from one site into another by low-temperature condensation. For example, a sample of gas from a storage bulb on the right might be condensed onto a solid reactant in the reaction vessel on the left. Products of different volatility can be separated by passing the vapors through U-traps. The first trap in the series is held at the highest temperature (perhaps −78°C, by means of a carbon dioxide slush bath surrounding the trap) to capture the least volatile component; the next trap might be held at −196°C by means of a dewar of liquid nitrogen; noncondensable gases such as H_2 might be pumped away.

Aside from the exclusion of air and ease of manipulating vapors, the vacuum line provides a closed system in which gaseous reactants or products can be measured quantitatively. For instance, a sample of gas in the bulb could be allowed to expand into the adjacent manometer where its pressure is measured. If the volume of the system has been calibrated, the amount of gas present can be calculated using the gas laws. The identity and purity of a compound can be determined on the vacuum line by measuring its vapor pressure at one or more fixed temperatures, and comparison with tabulated vapor pressures. NMR, IR, and other spectroscopic cells can also be attached to the vacuum line and loaded by condensation of a vapor.

Glass vacuum systems such as the one illustrated here are widely used for volatile hydrogen compounds, organometallic compounds, and halides. Since hydrogen fluoride and highly reactive fluorides corrode glass, they are generally handled in a vacuum line constructed from nickel tubing, with metal valves in place of the stopcocks, and electronic pressure transducer gauges in place of the mercury-filled manometers shown here.

[References: R. J. Angelici, *Synthesis and techniques in inorganic chemistry*. Saunders, Philadelphia (1977); D. F. Shriver and M. A. Drezdzon, *The manipulation of air-sensitive compounds*. Wiley, New York (1986).]

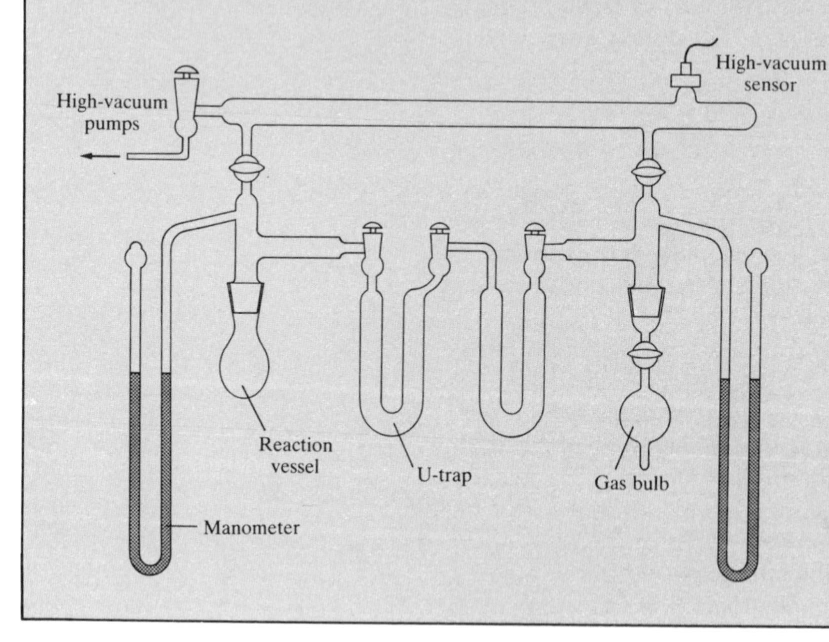

Fig. 9B.1 A simple chemical vacuum line suitable for the preparation, characterization, and quantitative measurement of condensable gases, particularly gases that are sensitive to air and moisture.

Oxidation

All the boron hydrides are flammable, and several of the lighter ones, including diborane, react spontaneously with air, often with explosive violence and a green flash (an emission from an excited state of the reaction intermediate BO). The final product of the reaction is the hydrated oxide:

$$B_2H_6(g) + 3O_2(g) \rightarrow 2B(OH)_3(s)$$

Air oxidation is fairly general for the p-block hydrides. Only HF and H_2O do not burn in air; some (as with B_2H_6) ignite spontaneously as soon as they come into contact with air.

The lighter boranes are readily hydrolyzed by water:

$$B_2H_6(g) + 6H_2O(l) \rightarrow 2B(OH)_3(aq) + 6H_2(g)$$

Coordination by H_2O acting as a Lewis base is an important step in the mechanism.

Lewis acidity

As implied by the mechanism of hydrolysis, diborane and many other light boron hydrides act as Lewis acids and are cleaved by reaction with Lewis bases. Two different cleavage patterns have been observed, namely 'symmetric cleavage' and 'unsymmetric cleavage'.

In **symmetric cleavage**, B_2H_6 is broken symmetrically into two BH_3 fragments, each of which forms a complex with a Lewis base:

$$B_2H_6 + 2N(CH_3)_3 \rightarrow 2H_3B{-}N(CH_3)_3$$

Many complexes of this kind exist. They are interesting partly because they are isoelectronic with hydrocarbons. For instance, the product of the reaction above is isoelectronic with 2,2-dimethylpropane (neopentane, $C(CH_3)_4$).

The direct reaction of B_2H_6 and NH_3 results in **unsymmetrical cleavage**, cleavage leading to an ionic product:

$$B_2H_6 + 2NH_3 \rightarrow [H_2B(NH_3)_2][BH_4]$$

Unsymmetrical cleavage of this kind is generally observed when B_2H_6 and a few other boron hydrides react with strong, sterically uncrowded bases at low temperatures.

Hydroboration

An important component of a synthetic organic chemist's repertoire of reactions is **hydroboration**, the addition of H—B across a multiple bond:

$$H_3B{-}OR_2 + H_2C{=}CH_2 \xrightarrow{R_2O} CH_3CH_2BH_2 + R_2O$$

From the viewpoint of an organic chemist, the C—B bond in the primary product of hydroboration is an intermediate stage in the formation of C—H or C—OH bonds, to which it can be converted. From the viewpoint of the inorganic chemist, the reaction is a convenient method for the preparation of wide variety of organoboranes. The hydroboration reaction is one of a class of reactions in which E—H adds across the multiple bond; hydrosilation (Section 9.13) is another example.

9.11 The tetrahydroborate ion

The simplest borohydride anion is the tetrahydroborate ion BH_4^-, which may be prepared by the reaction of B_2H_6 with LiH in ether. We can view this reaction as another example of the Lewis acidity of BH_3 toward the strong Lewis basicity of H^-. The BH_4^- ion is isoelectronic with CH_4 and NH_4^+, and the three show a smooth variation in chemical properties as the electronegativity of the central atom increases:

	BH_4^-	CH_4	NH_4^+
Character:	hydridic	—	protic

CH_4 is neither acidic nor basic under the conditions prevailing in aqueous solution.

Sodium tetrahydroborate, $NaBH_4$, more informally 'sodium borohydride', is a very useful laboratory and commercial reagent. It is used as a mild source of H^- ions, as a general reducing agent, and as a precursor for most boron–hydrogen compounds. Although it is thermodynamically unstable with respect to hydrolysis, the reaction is very slow at high pH.

Example 9.5: *Predicting the reactions of boron–hydrogen compounds*

By means of a chemical equation, indicate the products resulting from the interaction of equal amounts of $[HN(CH_3)_3]Cl$ with $LiBH_4$ in tetrahydrofuran (THF).

Answer. The interaction of the hydridic BH_4^- ion with the protic $[HN(CH_3)_3]^+$ ion will evolve hydrogen to produce trimethylamine and BH_3. In the absence of other Lewis bases, the BH_3 would coordinate to THF; however the stronger Lewis base trimethylamine is produced in the initial reactions, so the overall reaction will be:

$$[HN(CH_3)_3]Cl + LiBH_4 \rightarrow H_2 + H_3BN(CH_3)_3 + LiCl$$

The boron in the $H_3BN(CH_3)_3$ product has four tetrahedrally attached groups.

Exercise. Write plausible equations for the reaction of B_2H_6 with propylene in THF in the stoichiometric ratio 1:2.

9.12 The hydrides of aluminum and gallium

The hydrides of In and Tl are too unstable to be worth considering here. Pure gallium hydride has been prepared only recently,[5] but some derivatives have been known for some time. The range of binary aluminum hydrides is much more limited than those of boron. The metathesis of their halides with LiH leads to lithium tetrahydroaluminate, $LiAlH_4$, or the analogous tetrahydrogallate, $LiGaH_4$:

$$4LiH + ECl_3 \xrightarrow{R_2O} LiEH_4 + 3LiCl \qquad (E = Al, Ga)$$

The direct reaction of Li, Al, and H_2 leads to the formation of either $LiAlH_4$ or Li_3AlH_6 depending on the conditions of the reaction. Their formal analogy with halogen complexes such as $[AlCl_4]^-$ and $[AlF_6]^{3-}$ should be noted.

The $[AlH_4]^-$ and $[GaH_4]^-$ ions are tetrahedral, and are much more hydridic than $[BH_4]^-$. The latter property is consistent with the higher electronegativity of B compared with Al and Ga. Thus they are also much stronger reducing agents, and react violently with water. $LiAlH_4$ is commercially available and widely used when a strong hydridic reagent or a

[5] A. J. Downs, M. J. Goode, and C. R. Pulham, Gallane at last! Synthesis and properties of a binary gallium hydride, *J. Am. chem. Soc.*, **111**, 1936 (1989).

reducing agent stronger than the tetrahydroborate ion is needed. Under conditions of controlled protolysis, both ions lead to complexes of aluminum or gallium hydride:

$$LiAlH_4 + [(CH_3)_3NH]Cl \rightarrow (CH_3)_3N\text{—}AlH_3 + LiCl + H_2$$

In striking contrast to BH_3 complexes, these complexes will add a second molecule of base to form a 5-coordinate compound (**10**):

$$(CH_3)_3N\text{—}EH_3 + N(CH_3)_3 \rightarrow [(CH_3)_3N]_2EH_3 \qquad (E = Al, Ga)$$

This behavior is consistent with the trend for Period 3 and higher *p*-block elements to form 5- and 6-coordinate hypervalent compounds (Section 3.2).

Aluminum hydride (AlH_3) is a solid that is best regarded as saline, like the hydrides of the *s*-block metals. The alkylaluminum hydrides, such as $Al_2(C_2H_5)_4H_2$ (**11**), are well-known molecular compounds. Hydrides of this kind are used to couple alkenes, the initial step being the addition of Al—H across the C=C double bond, as in hydroboration.

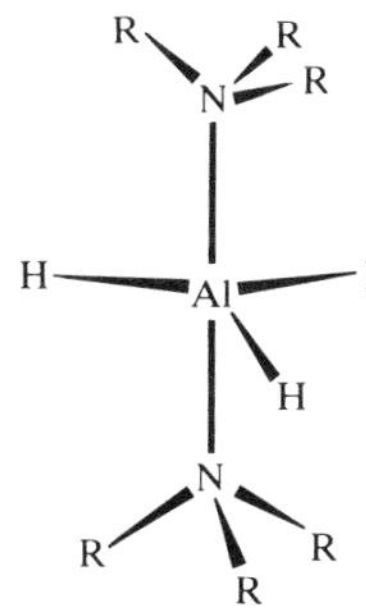

10

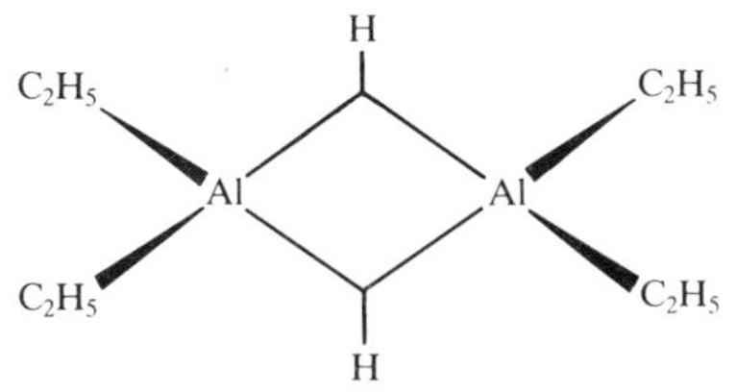

11

Electron-precise hydrides of the carbon group

The electron-precise hydrides of the carbon group (Group 14/IV) have no lone pairs of electrons and so are not Lewis bases. Their lack of lone pairs also means that they are not associated by hydrogen bonding. Moreover, the C atom in hydrocarbons has no vacant, energetically accessible orbitals, and so is not a Lewis acid. The numerous hydrocarbons are best regarded from the viewpoint of organic chemistry, and this section concentrates mainly on the **silanes**, the silicon hydrides. The extensive chemistry of metals and metalloids bound to hydrocarbon ligands is discussed in Chapters 10, 16, and 17.

9.13 Silanes

The silanes, with their greater number of electrons and stronger intermolecular forces, are less volatile than their hydrocarbon analogs: whereas propane (C_3H_8) boils at −44°C and is a gas under normal conditions, its analog trisilane (Si_3H_8) is a liquid and boils at 53°C. Their chemical properties are less fully documented than those of the alkanes and other hydrocarbons. This is partly because there have been few practical or theoretical reasons to prepare many of them, and their greater reactivity makes their investigation more demanding. Stable unsaturated analogs of alkenes, alkynes, and aromatics are unknown for the simple silanes.

Synthesis

Partly because he happened to use magnesium boride contaminated with magnesium silicide, Stock also found himself studying the silicon hydrides. He identified four Si analogs of the alkanes, specifically SiH_4, Si_2H_6, Si_3H_8, and Si_4H_{10}. Separation of these products by modern gas chromatography suggests that Si_4H_{10} in fact consists of a mixture of a linear and a branched isomer analogous to butane and 2-methylpropane.

The silanes are thermally less stable than the hydrocarbons. Thus, silanes are cracked at moderate temperatures, when they form a mixture of silane itself and higher silanes:

$$Si_2H_6 \xrightarrow{300°C} H_2 + SiH_4 + \text{higher silanes}$$

Complete decomposition occurs above about 500°C, when the silane decomposes to silicon and hydrogen. These **pyrolysis reactions**, reactions in which a compound is degraded by heat, have considerable technological utility because silane is used as a source of pure crystalline Si by the semiconductor industry:

$$SiH_4 \xrightarrow{500°C} Si(s, \text{crystalline}) + 2H_2$$

In this process a thin film of Si can be laid down on a heated substrate. A related reaction, the decomposition of silane in an electric discharge, yields amorphous Si:

$$SiH_4 \xrightarrow{\text{Electric discharge}} Si(s, \text{amorphous}) + 2H_2$$

Amorphous Si is used in photovoltaic devices, such as power sources for pocket calculators. 'Amorphous Si' is in fact a misnomer, since it contains a significant proportion of hydrogen, which IR spectroscopy shows to be bound by Si—H bonds. The Si—H bonds attach to the Si orbitals at sites where the jumbled structure does not permit four Si—Si bonds to form around an Si atom.

Stock's protolysis of a silicide is no longer of much importance in the preparation of Si—H bonds. Instead, the laboratory preparation is generally based on the metathesis of Si—Cl or Si—Br compounds with $LiAlH_4$:

$$4R_3SiCl + LiAlH_4 \xrightarrow{\text{Ether}} 4R_3SiH + LiAlCl_4$$

Although useful in the laboratory, this method is too expensive for the commercial production of silane. Instead, that begins with the reaction between less expensive reagents, such as HCl and either Si or iron silicide, which form trichlorosilane, $HSiCl_3$. When heated, the trichlorosilane forms silane and tetrachlorosilane:

$$4SiHCl_3 \xrightarrow{\Delta} SiH_4 + 3SiCl_4$$

Reactions

The silanes are generally more reactive than their hydrocarbon analogs, and ignite spontaneously on contact with air. They also react explosively with F_2, Cl_2, and Br_2. Silane itself is a reducing agent in solution. For example, when it is bubbled through an oxygen-free aqueous solution containing Fe^{3+}, reduction to Fe^{2+} results.

Si—H bonds are not readily hydrolyzed in neutral water, but the reaction is rapid in strong acid or in the presence of traces of base. Likewise, alcoholysis is accelerated by catalytic amounts of alkoxide:

$$SiH_4 + 4ROH \xrightarrow{OR^-} Si(OR)_4 + 4H_2$$

Kinetic studies indicate that the reaction proceeds through a structure in which OR^- attacks the Si atom while H_2 is being formed via a kind of H···H hydrogen bond between hydridic and protic hydrogen atoms (**12**).

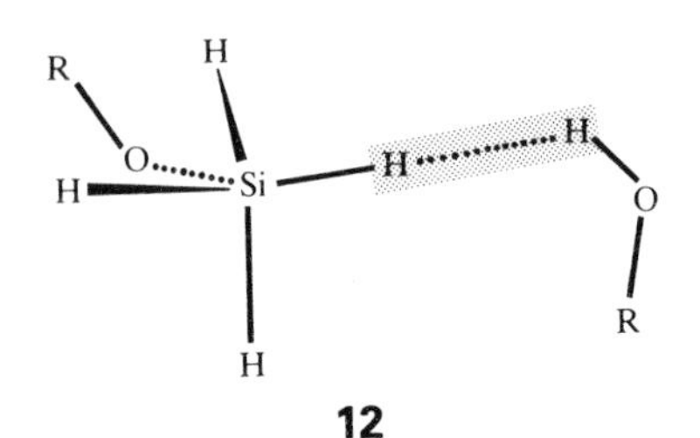

12

The silicon analog of hydroboration is **hydrosilation**, the addition of Si—H across the multiple bonds of alkenes and alkynes. This reaction, which is used in both industrial and laboratory syntheses, can be carried out under conditions (300°C or UV irradiation) that produce a radical intermediate. In practice it is usually performed under far milder conditions using a Pt complex as catalyst:

$$CH_2{=}CH_2 + SiH_4 \xrightarrow[\text{i-PrOH}]{H_2PtCl_6} CH_3CH_2SiH_3$$

The current view is that this reaction proceeds through an intermediate (**13**) in which both the alkene and silane are attached to the metal atom.

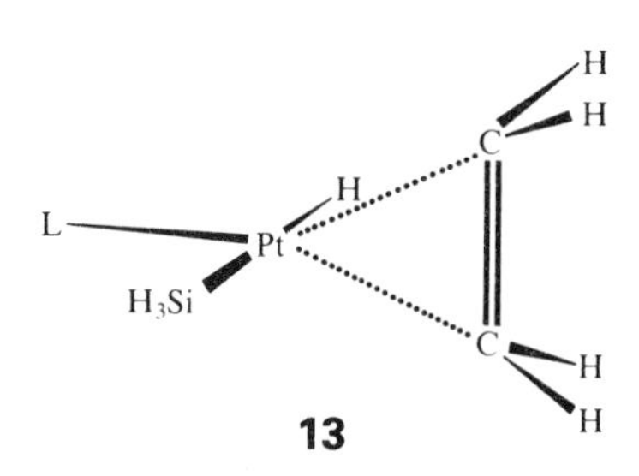

13

9.14 Germane, stannane, and plumbane

The decreasing stability of the hydrides on going down the group severely limits the chemical investigation of stannanes and plumbane. Germane (GeH_4) and stannane (SnH_4) can be synthesized by the reaction of the appropriate tetrachloride with $LiAlH_4$ in THF solution. It has been claimed that plumbane has been synthesized in trace amounts by the protolysis of a Mg—Pb alloy. The presence of alkyl or aryl groups stabilizes the hydrides of all three elements. For example, trimethylplumbane, $(CH_3)_3PbH$, begins to decompose at −30°C, but it manages to survive for several hours at room temperature.

Electron-rich compounds of Groups 15/V to 17/VII

The neutral binary hydrogen compounds in Groups 15/V to 17/VII all have lone pairs on the central atom. As we have already remarked, this is consistent with the angular structure of H_2O and its analogs and with the pyramidal structure of NH_3. The Lewis basicity and ability to take part in hydrogen bonding both stem from the presence of these lone pairs.

9.15 Formation

The synthesis of ammonia

Ammonia (NH_3) is produced in huge quantities worldwide for use as a fertilizer and as a primary source of nitrogen in the production of many chemicals. Virtually only one method, the **Haber process**, is used for the entire global production. This process is the direct combination of N_2 and H_2 over Fe as catalyst. The reaction is carried out at high temperature (around 450°C) to overcome the kinetic inertness of N_2, and at high pressure (around 270 atm) to overcome the thermodynamic effect of an unfavorable equilibrium constant at 450°C (Section 17.7). So novel and great were the chemical and engineering problems arising from the then (early twentieth century) uncharted area of large-scale high-pressure technology, that two Nobel Prizes were awarded. One went to Fritz Haber (in 1918), who developed the chemical process. The other went to Carl Bosch (in 1931), the chemical engineer who designed the first plants to realize Haber's process.

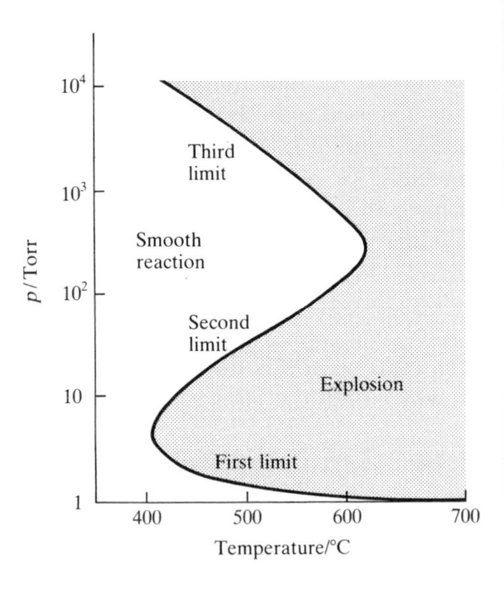

Fig. 9.10 The variation of the character of the $H_2 + O_2$ reaction with pressure and temperature.

The reaction of oxygen and hydrogen

The mechanism of reaction of hydrogen gas with oxygen gas to produce water has been extensively studied. A hint of its complexity is evident from the pressure dependence of the rate of the reaction (Fig. 9.10). This shows that at 550°C an increase in total pressure can turn an explosive reaction into a slow reaction and that a further pressure increase leads to an explosion again.

It is now recognized that the complexity of the reaction stems from the coexistence of a branched-chain mechanism as well as a simple chain propagation step. In the simple chain, a radical carrier (·OH) is consumed with the production of another carrier (·H),

$$\cdot OH + H_2 \rightarrow H_2O + \cdot H$$

In a branching chain, more than one radical carrier is produced when one radical reacts:

$$\cdot H + O_2 \rightarrow \cdot OH + \cdot O\cdot$$

$$\cdot O\cdot + H_2 \rightarrow \cdot OH + \cdot H$$

Under normal reaction conditions, the chain carriers are scavenged by collisions with the walls of the reactor or other chain-terminating gas-phase encounters. However, the branching steps allow the rate of carrier production to outstrip quenching, producing a cascade of radical carriers, an increase in rate, and an explosion. Hydrogen in air poses a serious explosion hazard because of the very wide range of hydrogen partial pressures for which its combustion is explosive.

Hydrogen halides

The hydrogen halides may be formed by direct reaction of the elements in a radical chain reaction, such as that for the reaction between Br_2 and H_2. With the lighter halogens (F_2 and Cl_2), the reaction is explosive under a wide range of conditions. However, all commercial HF and most HCl is synthesized by the protonation of halide ions:

$$CaF_2(s) + H_2SO_4(l) \rightarrow CaSO_4(s) + 2HF(g)$$

This reaction cannot be used for the preparation of HBr and HI because concentrated H_2SO_4 oxidizes the bromide and iodide ions to their elements.

9.16 Hydrogen bonding

A **hydrogen bond** consists of an H atom between atoms of more electronegative nonmetallic elements. This definition includes the widely recognized N—H···N and O—H···O hydrogen bonds but excludes the B—H—B bridges in boron hydrides. It also excludes the W—H—W links present in $[(OC)_6WHW(CO)_6]^-$, for tungsten is more electronegative than hydrogen on the Pauling scale, but it is a metal.

Structural and energetic aspects

Although hydrogen bonds are much weaker than conventional bonds (Table 9.6) they have important consequences for the properties of the electron-rich hydrogen compounds of Period 2, including their densities, viscosities, vapor pressures, and acid–base characters (Section 5.3). Hydrogen bonding is readily detected by the shift to lower wavenumber and broadening of E—H stretching bands in IR spectra (Fig. 9.11). The origin of the broadening appears to be the thermal disorder of the hydrogen-bonded system at ordinary temperatures, which gives rise to slightly different local environments for the covalent E—H bonds. Thermodynamic data on hydrogen-bonded systems reveal a wide range of hydrogen bond strengths (Table 9.6). The majority of hydrogen bonds are weak. In these cases, the H atom is not midway between the two nuclei, even when the heavier linked atoms are identical. For example, the $[ClHCl]^-$ ion is linear but the H atom is not midway between the Cl atoms. In contrast, the bifluoride ion, $[FHF]^-$, has a strong hydrogen bond and the H atom appears to be midway between the F atoms, and the F—F separation (2.26 Å) is significantly less than twice the van der Waals radius of the F atom (2 × 1.35 Å).

Very strong hydrogen bonds, such as that in $[FHF]^-$, are generally thought to have a single minimum in their potential energy curve, whereas weak hydrogen bonds are thought to have double minima (Fig. 9.12). The latter may be symmetric (as in Fig. 9.12b) or asymmetric, as in hydrogen bonds between water and an amine.

Table 9.6. Comparison of hydrogen bond enthalpies with the corresponding E—H covalent bond enthalpies (kJ mol^{-1})

	Hydrogen bond (···)	Covalent bond (—)
HS—H···SH_2	−7	−363
H_2N—H···NH_3	−17	−431
HO—H···OH_2	−22	−452
F—H···F—H	−29	−568
HO—H···Cl^-	−55	−432
F···H···F^-	−165	−568

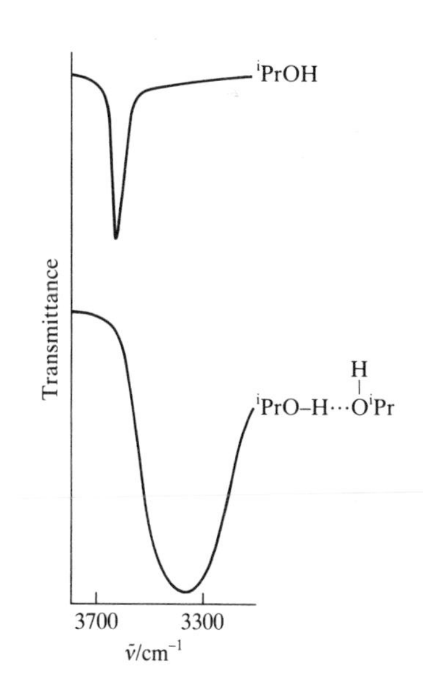

Fig. 9.11 In the upper curve, isopropanol (2-propanol) is present as unassociated molecule in dilute solution. In the lower curve the pure alcohol is associated through hydrogen bond. This lowers the frequency and broadens the O—H stretching absorption band. (From N. B. Colthup, L. H. Daly, and S. E. Wiberley, *Introduction to infrared and Raman spectroscopy*. Academic Press, New York (1975).)

Ice and clathrate hydrates

One of the most interesting manifestations of hydrogen bonding is the structure of ice and its analogs. There are seven different phases of ice above 2 kbar, but only one at lower pressures. The low-pressure phase crystallizes in a hexagonal unit cell with each O atom surrounded tetrahedrally by four others. These O atoms are held together by hydrogen bonds (Fig. 9.8) with O—H···O and O···H—O bonds largely randomly distributed through the solid. The resulting structure is quite open, which accounts for the density of ice being lower than that of water. When ice melts, the network of hydrogen bonds partially collapses.

Water can also form **clathrate hydrates**, consisting of hydrogen-bonded

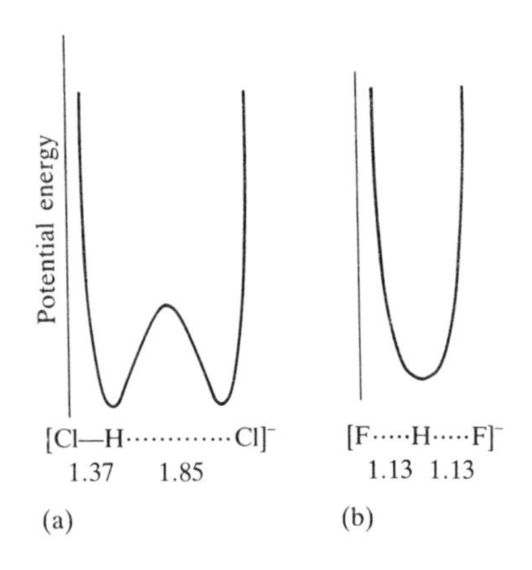

Fig. 9.12 The variation of the potential energy with the position of the proton between two atoms in a hydrogen bond. (a) The double-minimum potential characteristic of a weak hydrogen bond. (b) The single-minimum potential characteristic of many strong hydrogen bonds.

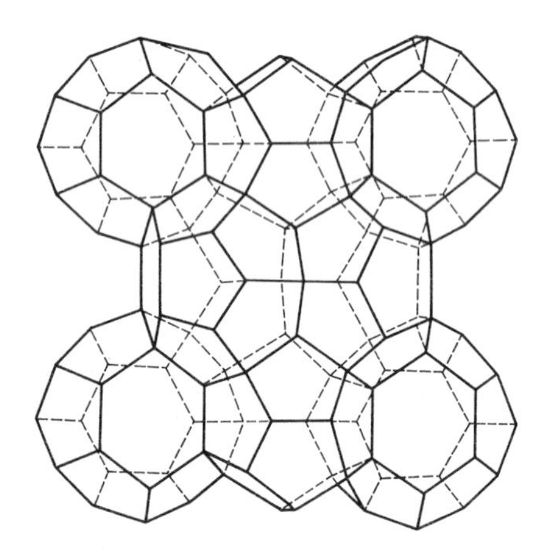

Fig. 9.13 The cages of water molecules in clathrate hydrates, such as $Cl_2{\cdot}(H_2O)_{7.25}$. Each intersection is the location of an O atom. Two of the central 14-hedra are about 80 percent occupied by Cl_2.

cages of water molecules surrounding foreign molecules or ions. One example is the clathrate hydrate with the composition $Cl_2{\cdot}(H_2O)_{7.25}$ (Fig. 9.13). In this structure, the cages with O atoms defining their corners consist of 14-faced polyhedra and 12-faced polyhedra in the ratio 3:2. The O atoms are held together by hydrogen bonds, and Cl_2 molecules occupy the 14-faced polyhedra. Similar clathrate hydrates are formed at elevated pressures and low temperatures with Ar, Xe, and CH_4, but in these cases all the polyhedra appear to be occupied. Aside from their interesting structures—which illustrate the organization that can be enforced by hydrogen bonding—clathrate hydrates are often used as models for the way in which water appears to become organized around nonpolar groups, such as those in proteins. Methane clathrate hydrates occur in the earth at high pressures, and it is estimated that huge quantities of natural gas are trapped in these formations.

Some ionic compounds form clathrate hydrates in which the anion is incorporated into the framework by hydrogen bonding. This type of clathrate is particularly common with the very strong hydrogen bond acceptors F^- and OH^-. One such example is $N(CH_3)_4F{\cdot}4H_2O$.

9.17 Brønsted acidity and Lewis basicity

The acid–base properties of the hydrogen compounds of Groups 15/V to 17/VII are very important and have already have been discussed (Section 5.3). However, HF has some special properties that set it apart from the other hydrogen halides.

The most striking property of HF is its reaction with glass and other silicates to form $[SiF_6]^{2-}$ (Section 6.8). Hydrogen fluoride is also an effective protein denaturing agent and can inflict deep and painful burns to the skin and eyes. It is a much stronger acid in the pure liquid state than in aqueous solution, for in the pure liquid its autoionization is favored by the formation of the bifluoride ion:

$$3HF(l) \rightarrow [H_2F]^+ + [HF_2]^- \qquad pK_{HF} = 11.1$$

The Lewis base character of the electron-rich Group 15/V to 17/VII compounds stems from their possession of lone pairs. However, their base strength decreases down each group, and although NH_3 and H_2O are common ligands in metal complexes, the binary hydrogen compounds of the elements in Periods 3 and 4 (PH_3, AsH_3, H_2S, and H_2Se) are encountered only very occasionally. More often, their interaction with *d*-block cations leads to deprotonation to form the metal phosphide, arsenide, and so on. The alkyl and aryl analogs, such as $P(CH_3)_3$ and $AsPh_3$, are more robust in the presence of *d*-block metal ions and—as we shall see in Chapter 16—many complexes of them are known.

Further information: nuclear magnetic resonance

Nuclear magnetic resonance (NMR) is the most powerful and widely used spectroscopic method for the determination of molecular structures in solution, pure liquids, and gases. In many cases, it provides information about shape and symmetry with greater certainty than is possible with other spectroscopic techniques, such as infrared and Raman spectroscopy. However, unlike X-ray diffraction (Box 4.1), NMR studies of molecules in solution generally do not provide detailed bond distance and angle information. NMR also provides information about the rate and nature of the interchange of ligands in fluxional molecules (Section 16.7).

NMR can be observed only for compounds containing elements with magnetic nuclei (those with nonzero nuclear spin). The sensitivity is dependent on several parameters, including the abundance of the isotope and the size of its nuclear magnetic moment. For example, ^{1}H with 99.98 percent natural abundance and a large magnetic moment, is easier to observe than ^{13}C, which has a smaller magnetic moment and only 1.1 percent natural abundance. With modern multinuclear NMR techniques it is easy to observe spectra for approximately 20 different nuclei, including many elements that are important in inorganic chemistry, such as ^{1}H, ^{7}Li, ^{11}B, ^{13}C, ^{15}N, ^{19}F, ^{23}Na, ^{27}Al, ^{29}Si, ^{31}P, ^{195}Pt, and ^{199}Hg. With more effort, useful spectra can also be obtained using many other nuclei.

Measurement

A nucleus with spin I can take up $2I+1$ distinct orientations relative to the direction of an applied magnetic field. Each orientation has a different energy, with the lowest level (marginally) the most highly populated. The energy of the transition between these nuclear spin states is measured by exciting nuclei in the sample with a radiofrequency pulse or pulse sequence and then observing the return of the nuclear magnetization back to equilibrium. After data processing (Fourier transformation), the data are displayed as an absorption spectrum (Fig. 9.14), with peaks at frequencies corresponding to transitions between the different nuclear energy levels.

Chemical shifts

The frequency of an NMR transition depends on the *local* magnetic field, B_{loc}, the nucleus experiences. This is the sum of the field of the magnet, B_{appl}, and an additional local field arising from the effect of the applied field on the molecule, which is written $-\sigma B_{appl}$:

$$B_{loc} = (1-\sigma)B_{appl}$$

σ, which expresses the role of the local chemical environment, is called the

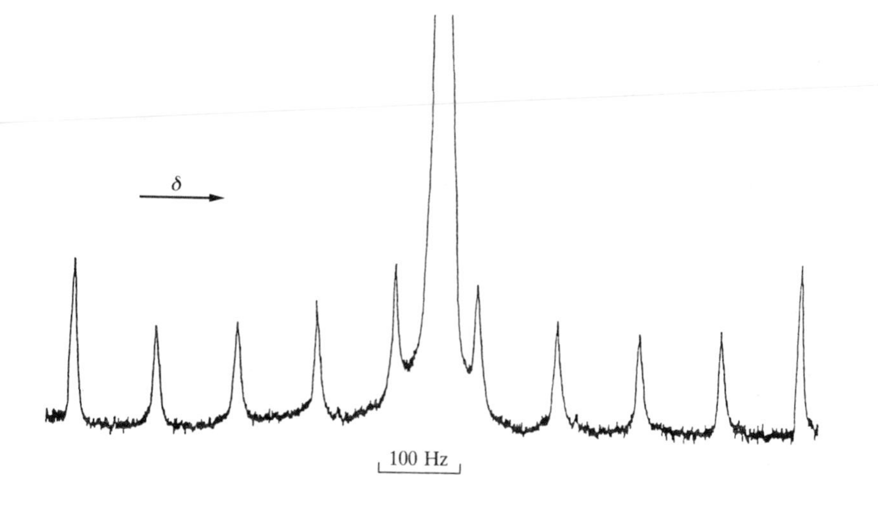

Fig. 9.14 The ^{1}H-NMR spectrum of GeH_4. (Source: E. A. V. Ebsworth, D. W. H. Rankin, and S. Cradock, *Structural methods in inorganic chemistry*. Blackwell, Oxford (1987).)

shielding constant. In practice, the position of an NMR signal is expressed as the **chemical shift** δ. This is defined in terms of the difference between the resonance frequency of nuclei in the sample and that of a reference compound:

$$\delta = \frac{\nu_{\text{sample}} - \nu_{\text{ref}}}{\nu_{\text{ref}}} \times 10^6$$

A common standard for ^1H, ^{13}C, or ^{29}Si NMR spectra is tetramethylsilane, $Si(CH_3)_4$ (TMS).

When δ for a signal is negative, the nucleus involved is said to be **shielded** relative to the standard. Conversely, a positive δ corresponds to nucleus that is **deshielded** with respect to the reference. An H atom bound to a closed-shell, low-oxidation-state, *d*-block element from Groups 6 through 10 (such as $HCo(CO)_4$) is generally found to be highly shielded whereas in an oxoacid, such as H_2SO_4, it is deshielded (in each case, relative to TMS). From these examples it might be supposed that the higher the electron density around a nucleus the greater its shielding. However, as several factors contribute to the shielding, a simple physical interpretation of chemical shifts in terms of electron density is generally not possible.

The chemical shifts of ^1H and other nuclei in various chemical environments are tabulated, and so empirical correlations can often be used to identify compounds or the element to which the resonant nucleus is bound. For example, the ^1H chemical shift in CH_4 is only $\delta = 0.1$ because the H nuclei are in an environment similar to that in TMS, but the ^1H chemical shift is $\delta = 3.1$ for H bonded to Ge in GeH_4. Chemical shifts are different for the same element in inequivalent positions within a molecule. Thus in ClF_3 the chemical shift of the equatorial ^{19}F nucleus is separated by $\delta = 120$ from that of the axial F nuclei.

Spin–spin coupling

Structural assignment is often helped by the observation of the **spin–spin coupling** of nuclei, which gives rise to a multiplet of lines in the spectrum. The strength of spin–spin coupling, which is reported as the **spin–spin coupling constant** J (in hertz, Hz), decreases rapidly with distance through chemical bonds, and in many cases is greatest when the two atoms are directly bonded to each other. For simple so-called **first-order spectra**, which are being considered here, the coupling constant is equal to the separation of adjacent lines in a multiplet. As can be seen in Fig. 9.14, $J(^1H\text{—}^{73}Ge) \approx 100$ Hz. The chemical shift is measured at the center of the multiplet.

The allowed transitions contributing to a multiplet all occur at the same frequency when the nuclei are related by symmetry. Thus, a single ^1H signal is observed for the CH_3I molecule because the three H nuclei are related to each other by a threefold axis. Similarly, in the spectrum of GeH_4 (Fig. 9.14) the single central line arises from the four equivalent H atoms in GeH_4 molecules that contain Ge isotopes of zero nuclear spin. This strong central line is flanked by 10 evenly spaced but less intense lines that arise from a small fraction of GeH_4 that contains ^{73}Ge, for which $I = \frac{9}{2}$. The properties of spin–spin coupling are such that a multiplet of $2I + 1$ lines results in the spectrum of a spin-$\frac{1}{2}$ nucleus when that nucleus (or a set of symmetry related spin-$\frac{1}{2}$ nuclei) is coupled to a nucleus of spin I. In the present case, the ^{1}H nuclei are coupled to the ^{73}Ge nucleus to yield a $2\times\frac{9}{2}+1=10$ line multiplet.

The coupling of the nuclear spins of different elements is called **heteronuclear coupling**: the Ge—H coupling discussed above is an example. **Homonuclear coupling** between nuclei of the same element is detectable when the nuclei are unrelated by the symmetry operations of the molecule, as in the ^{19}F-NMR spectrum of ClF_3 (Fig. 9.15). The signal ascribed to the two axial F nuclei is split into a doublet by the single equatorial F nucleus, and the latter is split into a triplet by the two axial F nuclei. Thus the pattern of ^{19}F resonances readily distinguishes this unsymmetrical structure from two more symmetric possibilities, trigonal planar and trigonal pyramidal, both of which would have equivalent F nuclei and hence a single ^{19}F resonance.

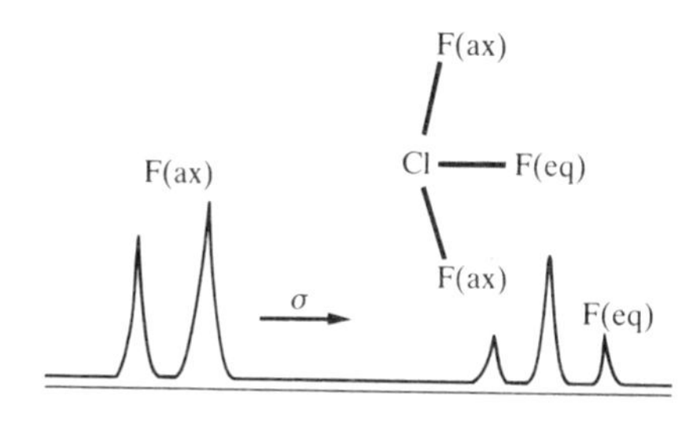

Fig. 9.15 The ^{19}F-NMR spectrum of ClF_3 (Source: R. S. Drago, *Physical methods in chemistry*. Saunders, Philadelphia (1977).)

The sizes of coupling constants are often related to the geometry of a molecule by noting empirical trends. In square-planar Pt(II) complexes, J(Pt—P) is sensitive to the group *trans* to a phosphine ligand and the value of J(Pt—P) increases in the following order of *trans* ligands:

$$PR_3 < H^- < R^- < NH_3 < Br^- \leq Cl^-.$$

For example, *cis*-$[PtCl_2(PEt_3)_2]$, where Cl^- is *trans* to P, has J(Pt—P) = 3.5 kHz, whereas *trans*-$[PtCl_2(PEt_3)_2]$, with P *trans* to P, has J(Pt—P) = 2.4 kHz. These systematics permit us to differentiate *cis* and *trans* isomers quite readily.

Intensities

Two aspects of line intensities are useful in the analysis of simple NMR spectra. The first is the intensity pattern within a multiplet. The second is the relative integrated intensity for signals of magnetically inequivalent nuclei.

The general rule for a multiplet that arises from coupling to spin-$\frac{1}{2}$ nuclei is that the ratios of the line intensities are given by Pascal's triangle:

Intensity ratio	Coupling to
1	0 other nuclei
1 1	1 spin-$\frac{1}{2}$ nucleus
1 2 1	2 spin-$\frac{1}{2}$ nuclei
1 3 3 1	3 spin-$\frac{1}{2}$ nuclei
1 4 6 4 1	4 spin-$\frac{1}{2}$ nuclei

This pattern is different from those generated by nuclei with higher spin moments. For example the ^{1}H NMR spectrum of HD is a triplet as a result of coupling with the ^{2}H nucleus ($I=1$, so $2I+1=3$), but the intensities of the components in this triplet are equal (Fig. 9.16).

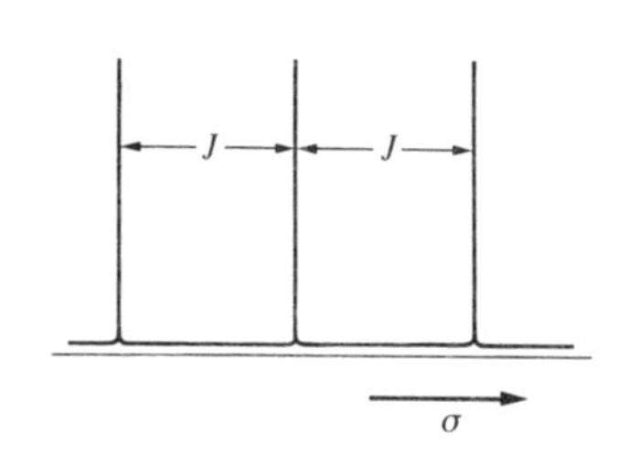

Fig. 9.16 Schematic ^{1}H-NMR spectrum of HD.

The integrated intensity under a signal arising from a set of symmetry-related nuclei is proportional to the number of nuclei in the set. As an example we return to ClF_3 (Fig. 9.15), where the relative integrated intensities are 2 (doublet) to 1 (triplet). This pattern is consistent with the structure and the splitting pattern, for it indicates the presence of two symmetry-related F nuclei and one magnetically inequivalent F nucleus.

Sometimes is it advantageous to eliminate spin–spin coupling by special

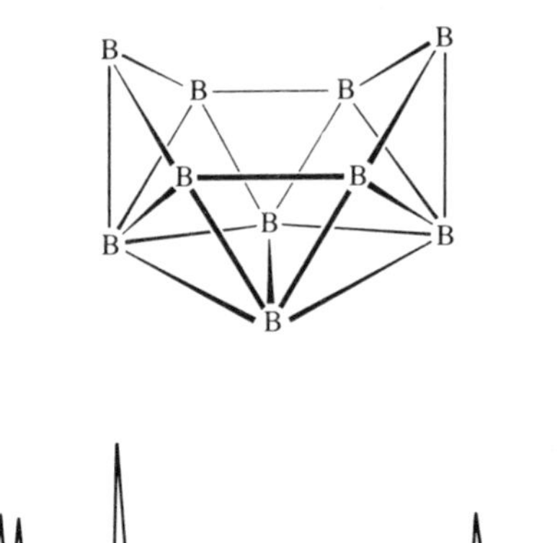

Fig. 9.17 The ^{11}B-NMR spectrum of $B_{10}H_{14}$. ^{11}B—^{1}H coupling has been suppressed and no B—B coupling is observed. The most intense peak belongs to the four equivalent B nuclei; the rest correspond to signals for each of the three equivalent pairs of B nuclei.

electronic techniques. Figure 9.17 shows the ^{11}B-NMR spectrum of $B_{10}H_{14}$ in solution, collected under conditions in which proton spin coupling has been eliminated. Each set of symmetry-related B nuclei gives rise to a signal that has an intensity approximately proportional to the number of B atoms it contains.

Solid-state NMR

Recent developments of far-reaching importance in inorganic chemistry have made possible the observation of high-resolution NMR spectra on solids. One of these high-resolution techniques, referred to as CPMAS-NMR, involves high-speed sample spinning at a 'magic angle' with respect to the field axis; the technique is a combination of magic angle spinning (MAS) with cross polarization (CP). Compounds containing ^{13}C, ^{31}P, and ^{29}Si and many other nuclei have been studied in the solid state using these techniques. An example is the use of ^{29}Si MAS-NMR spectra to infer the positions of Si atoms in natural and synthetic aluminosilicates.

Further reading

N. N. Greenwood and A. Earnshaw, *Chemistry of the elements*. Pergamon, Oxford (1984). Chapter 3 of this comprehensive text gives a good summary of hydrogen chemistry.

A. F. Wells, *Structural inorganic chemistry*. Oxford University Press (1984). Chapter 5 gives structural data and correlations relating to many hydrogen compounds, and Chapter 15 describes water and hydrates.

Kirk–Othmer encyclopedia of chemical technology. Wiley-Interscience, New York (1980). Vol. 12, p. 930 is a description of aspects of the industrial production of hydrogen. Metal hydrides are also covered on p. 772.

E. A. Evans, *Tritium*. Wiley, New York (1974).

K. M. Mackay, *Hydrogen compounds of the metallic elements*. Spon, London (1966).

E. L. Muetterties (ed.), *Boron hydride chemistry*. Academic Press, New York, (1975).

E. L. Muetterties, *Transition metal hydrides*. Marcel Dekker, New York (1971). This book deals mainly with *d*-block hydrido complexes but some broader aspects are covered too.

Two very good general accounts of hydrogen bonding are:

M. D. Joesten and L. J. Schaad, *Hydrogen bonding*. Marcel Dekker, New York (1974).

J. Emsley, Very strong hydrogen bonds. In *Chem. Soc. Rev.*, **9**, 91 (1980).

Nuclear magnetic resonance

E. A. V. Ebsworth, D. W. H. Rankin, and S. Cradock, *Structural methods in inorganic chemistry*. Blackwell, Oxford (1987). This is an excellent introduction to multinuclear NMR as practiced in inorganic chemistry.

J. K. M. Sanders and B. K. Hunter, *Modern NMR spectroscopy*. Oxford University Press (1987). A readable description of modern NMR techniques.

L. M. Jackman and F. A. Cotton (eds.), *Dynamic nuclear magnetic resonance spectroscopy*. Academic Press, New York (1975). The principles and examples of the use of NMR to follow fluxional processes in molecules.

J. Mason (ed.), *Multinuclear NMR spectroscopy*. Plenum Press, New York (1987). A description of NMR spectra for a large number of elements throughout the periodic table.

Exercises

9.1 Preferably without consulting reference material, construct the periodic table, identify the elements, and indicate (a) positions of salt-like, metallic, and molecular hydrides, (b) trends in $\Delta G_f^\ominus$ for the hydrogen compounds of the *p*-block elements, (c) the groups in which the molecular hydrides are electron deficient, electron precise, and electron rich.

9.2 Name and classify the following hydrogen compounds: (a) BaH_2, (b) SiH_4, (c) NH_3, (d) AsH_3, (e) $PdH_{0.9}$, (f) HI.

9.3 Pick out the compound from Exercise 9.2 that provides the most pronounced example of the following chemical characteristic and give balanced equations that illustrate each of the characteristics: (a) hydridic character, (b) Brønsted acidity, (c) Lewis basicity.

9.4 Divide the compounds in Exercise 9.2 into those that are solids, liquids, or gases at room temperature and pressure. Which of the solids are likely to be good electrical conductors?

9.5 Use Lewis structures and VSEPR theory to predict the structures of H_2Se, P_2H_4, and H_3O^+ and assign point groups. Assume a skew structure for P_2H_4.

9.6 Describe which of the following reactions would be likely to give the highest proportion of HD: (a) $H_2 + D_2$ equilibrated over a Pt surface, (b) $D_2O + NaH$, (c) electrolysis of HDO.

9.7 Phosphorus-31 is 100 percent abundant and has a nuclear spin of $\frac{1}{2}$. Sketch the ^{1}H-NMR and the ^{31}P-NMR spectra of PH_3 and $P(OCH_3)_3$.

9.8 Write balanced chemical equations for three major industrial preparations of H_2. Propose more convenient reactions for use in the laboratory.

9.9 Pick the compound in the following list that is most likely to undergo radical reactions with alkyl halides, and describe the reason for your choice: H_2O, NH_3, $(CH_3)_3SiH$, $(CH_3)_3SnH$.

9.10 Arrange H_2O, H_2S, and H_2Se in order of (a) increasing acidity, (b) increasing basicity toward a hard acid such as the proton.

9.11 Describe the three common methods for the synthesis of binary hydrogen compounds and illustrate each one with a balanced chemical equation.

9.12 Give balanced chemical equations for laboratory methods of synthesizing (a) H_2Se, (b) SiD_4, (c) $Ge(CH_3)_2H_2$ from $Ge(CH_3)_2Cl_2$, and (d) for the commercial synthesis of SiH_4 from Si and HCl.

9.13 Is B_2H_6 stable in air? If not, write the equation for the reaction. Describe a step-by-step procedure for transferring B_2H_6 quantitatively from a gas bulb at 200 Torr into a reaction vessel containing diethyl ether.

9.14 What is the trend in hydridic character of $[BH_4]^-$, $[AlH_4]^-$ and $[GaH_4]^-$? Which is the strongest reducing agent? Give the equations for the reaction of $[GaH_4]^-$ with excess 1 M HCl(*aq*).

9.15 Give a balanced chemical equation for the formation of pure silicon from crude silicon.

9.16 What are some important differences between Period 2 and Period 3 hydrogen compounds?

9.17 What type of compound is formed between water and Kr at low temperatures and elevated Kr pressure? Describe the structure in general terms.

9.18 Use the concepts introduced in Chapter 3 to draw the molecular orbitals and the molecular orbital energy level diagram of the bifluoride ion, HF_2^-.

9.19 Sketch the approximate potential energy surfaces for the hydrogen bond between H_2O and the Cl^- ion and contrast this with the potential energy surface for the hydrogen bond in $[FHF]^-$.

9.20 Describe the origin of the number of NMR absorption peaks and their intensities in the ^{31}P spectrum and the ^{19}F spectrum of PF_3 (Fig. 9.18).

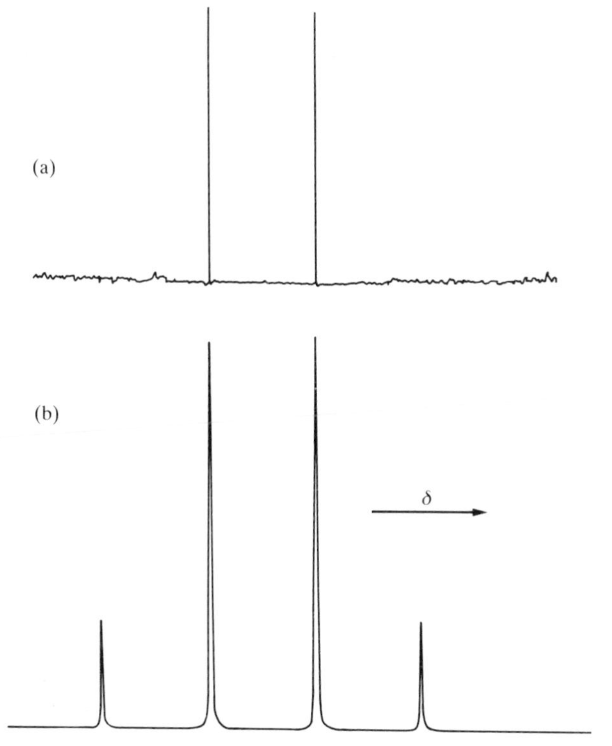

Fig. 9.18 NMR spectra of PF_3. (a) ^{19}F-NMR spectrum, (b) ^{31}P-NMR spectrum. (Source: S. Ching, Northwestern University, 1988.)

9.21 Sketch the ^{11}B-NMR spectra expected for $[HB(CH_3)_3]^-$ and for $[H_2B(CH_3)_2]^-$ assuming that only the protons directly bonded to B atoms are coupled with the ^{11}B nucleus.

9.22 Sketch the ^{1}H-NMR spectra of the BH_4^- ion. Naturally occurring boron is about 80 percent ^{11}B ($I = \frac{3}{2}$) and 20 percent ^{10}B ($I = 3$).

Problems

9.1 Sketch the molecular orbital energy level diagram for the HeH^+ molecule-ion. Why do you suppose that HeH^+ is unstable on contact with common solvents and surfaces even though the isolated molecule-ion has an acceptable electronic structure.

9.2 Borane exists as the molecule B_2H_6 and trimethylborane exists as a monomer $B(CH_3)_3$. In addition, the molecular formulas of the compounds of intermediate compositions are observed to be $B_2H_5(CH_3)$, $B_2H_4(CH_3)_2$, $B_2H_3(CH_3)_3$, and $B_2H_2(CH_3)_4$. Based on these facts, describe the probable structures and bonding in the latter series.

9.3 Observations on a complex in which HD replaces H_2 were used to show that H_2 is bound in $W(CO)_3(P^iPr_3)_2(H_2)$ without H—H bond rupture (G. Kubas *et al.*, *J. Am. chem. Soc.*, **106**, 451 (1984)). If the dihydrogen complex has an H—H stretch at 2695 cm^{-1}, what would be the expected wavenumber of the HD complex? What should be the pattern of the ^{1}H-NMR signal arising from coupling with D in the HD complex?

9.4 Hydrogen bonding can influence many reactions, including the rates of O_2 binding and dissociation from metalloproteins (G. D. Armstrong and A. G. Sykes, *Inorg. Chem.*, **25**, 3135 (1986)). Describe the nature of the evidence for (or against) hydrogen bonding with O_2 in the metalloproteins hemerythrin, myoglobin, and hemocyanin.

9.5 Summarize how deuterium tracer studies were used to provide evidence for the reversible hydroboration of benzene, and describe the proposed mechanism. (D. F. Gaines *et al.*, *Inorg. Chem.*, **24**, 621 (1985).)

9.6 Spectroscopic evidence has been obtained for the existence of $[Ir(C_5H_5)(H_3)(PR_3)]^+$, a complex in which one ligand is formally H_3^+. Devise a plausible molecular orbital scheme for the bonding in the complex, assuming that an angular H_3 unit occupies one coordination site and interacts with the e_g and t_{2g} orbitals of the metal.

Main group organometallics

10

Classification and structure
10.1 Comparison with hydrogen compounds
10.2 Structures
10.3 Synthesis
10.4 Stability
10.5 Reaction patterns
Ionic and electron-deficient compounds of Groups 1, 2, and 12
10.6 Alkali metals
10.7 Alkaline earth metals
10.8 The zinc group
Electron-deficient compounds of the boron group
10.9 Organoboron compounds
10.10 Organoaluminum compounds
10.11 Organometallic compounds of gallium, indium, and thallium
Electron-precise compounds of the carbon group
10.12 Organosilicon compounds
10.13 Organometallic compounds of germanium, tin, and lead
Electron-rich compounds of the nitrogen group
10.14 Noncatenated compounds of arsenic
10.15 Catenated singly and multiply bonded compounds
Further reading
Exercises
Problems

There are many similarities between the chemical properties of hydrogen compounds and the alkyl derivatives of the main-group elements. This is due in part to the similar electronegativities of hydrogen and carbon and the similar strengths and polarities of element–carbon and element–hydrogen bonds. Here we consider the organometallic compounds of the main-group elements and include the zinc group, which is very similar. The organometallics of the *d* and *f* blocks are treated in Chapter 16 and the role of organometallics in catalysis is treated in Chapter 17.

We shall see that most organometallics can be synthesized using one of four M—C bond-forming reactions. Moreover, many of their properties can be rationalized in terms of a small number of classes of reaction. One feature that we shall constantly emphasize is the important role of steric congestion around the central atom. This congestion is responsible for the ability of some organometallics to withstand hydrolysis, and has allowed the synthesis of compounds containing multiple bonds between heavy elements.

The origins of main-group organometallic chemistry are generally traced to the English chemist E. C. Frankland, who synthesized diethylzinc in 1849 and then proceeded to develop the use of organozinc reagents in organic synthesis. Around the turn of the century, the French chemist Victor Grignard developed a convenient synthesis of organomagnesium halides, and these more reactive 'Grignard reagents' largely displaced organozinc compounds as intermediates in organic syntheses.

The organic derivatives of Li, Mg, B, Al, and Si received considerable attention in the early part of this century. These and other developments led to several major industrial applications for organometallics, such as alkene polymerization catalysts and silicone polymers. More recently, a revival of research on *p*-block organometallics has produced new classes of compounds and has stimulated a better understanding of bonding in compounds of the *p*-block elements.

Classification and structure

A compound is regarded as organometallic if it contains at least one carbon–metal (C—M) bond. In this context the suffix 'metallic' is interpreted broadly to include metalloids such as B, Si, and As as well as true metals. For the purposes of this chapter, therefore, we shall take 'organometallic' to include compounds with B—C and Si—C bonds as well as Mg—C bonds. In general we shall speak of E—C bonds, where E is an element, but refer to M—C bonds when the element M is specifically an electropositive metal.

The *s*-block organometallics are commonly named as organic derivatives, as in methyllithium for CH_3Li. Compounds that have considerable ionic character are named as salts, as in sodium naphthalide for $Na[C_{10}H_8]$. The *p*-block organometallics are also often named as simple organic compounds, such as either trimethylboron or borontrimethyl for $B(CH_3)_3$. They may also be named as derivatives of their hydrogen counterparts, as in trimethylborane for $B(CH_3)_3$, tetramethylsilane for $Si(CH_3)_4$, and trimethylarsine for $As(CH_3)_3$.

10.1 Comparison with hydrogen compounds

The polarities and strengths of many E—C bonds are similar to those of E—H bonds. As a result, many structural and chemical similarities exist between alkyl compounds and their simple hydrogen analogs. This may be seen by comparing Fig. 10.1 and Fig. 9.3. Thus, Fig. 10.1 indicates that

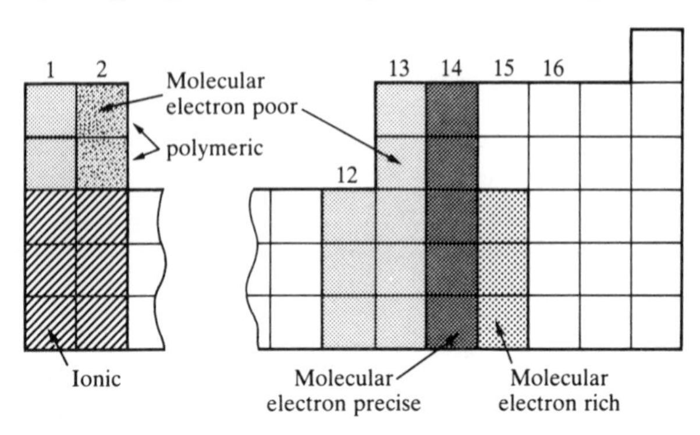

Fig. 10.1 Classification of methyl compounds of the metals and metalloids.

alkyl compounds of the elements in the lower left of the periodic table (such as KCH_3) are noticeably ionic, like the saline hydrides (such as KH). As with the hydrogen compounds, simple organometallics of the Group 14/IV and 15/V elements are molecular.

These analogies extend to the electron-deficient compounds. Thus, electron-deficient organometallics are often associated through alkyl or aryl bridges, the analog of the hydrogen bridges in B_2H_6. For instance, CH_3 groups bridge the two Al atoms in trimethylaluminum (**1**) and the Li atoms in methyllithium (**2**). These alkyl bridges are common in the organic compounds of light *s*-block elements (Li, Na, Be, and Mg) and for Al.

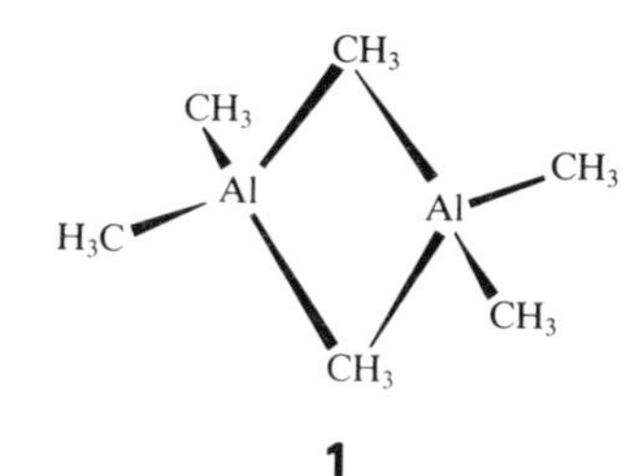

1

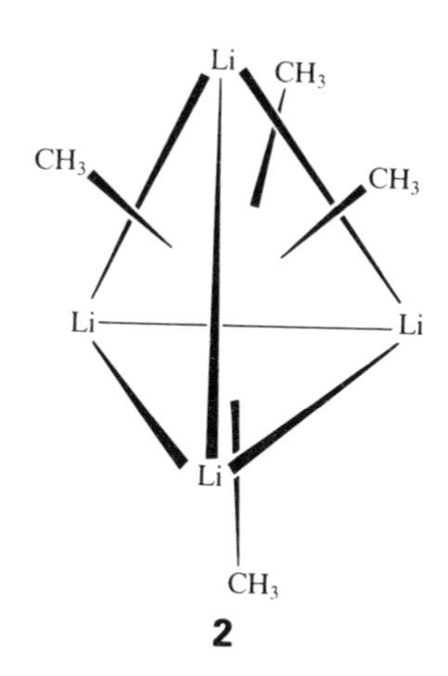

2

However, there are differences between organometallic and hydrogen compounds, which stem in part from the tendency of alkyl groups to avoid formation of highly ionic compounds. For example, methyllithium and methylsodium form molecular crystals (**2**) even though the M—C bonds are highly polar, and methyl derivatives of the heavier alkali metals crystallize in the nickel arsenide structure. As this structure is typical of soft-metal, soft-anion combinations (Section 4.5), it is indicative of compounds that have significant covalent character.

10.2 Structures

The great diversity of organic groups found in organometallic compounds tends to blur simple classification schemes. However, it is helpful to keep in mind two points. One is that the metal–carbon bonds of the *s*-block organometallics are all highly polar, with the possible exception of organoberyllium compounds. The second point is that this high polarity is largely absent from organometallics formed by the more electronegative metalloids, such as B, Si, Ge, and As.

s-Block organometallics

Molecular orbital theory provides a straightforward description of the structure of organometallics, even of those as unusual as methyllithium (**2**). Thus, a totally symmetric combination of three Li orbitals on each face of the Li_4 tetrahedron and one hybrid orbital from CH_3 gives an orbital that can accommodate one electron pair (Fig. 10.2), leading to a 4-center, 2-electron bond (a 4c,2e bond). Similarly, Be and Mg form many compounds with alkyl bridges that can be described in terms of 3c,2e bonds, as in the methyl-bridged polymers $[Be(CH_3)_2]_n$ and $[Mg(CH_3)_2]_n$ (**3**).

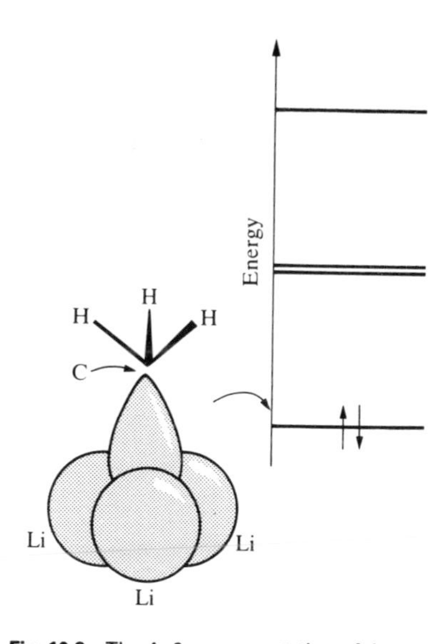

Fig. 10.2 The 4c,2e representation of the bonding between a face-bridging methyl group and the Li atoms in $Li_4(CH_3)_4$.

Compounds with greater ionic character are known in which the alkali metal is combined with cyclic or polycyclic arenes (in which a delocalized π^* orbital can accept an electron). One such compound, sodium naphthalide, $Na[C_{10}H_8]$, contains $[C_{10}H_8]^-$, a radical anion in which a single unpaired electron occupies an antibonding π orbital of naphthalene. Because of the delocalization of the charge, the effective radius of the anion is large and the coulombic interaction between cation and anion is weak. As a result, solids containing the anion salts are slightly dissociated in polar aprotic solvents, such as tetrahydrofuran or hexamethylphosphoramide, $((CH_3)_2N)_3PO$.

3 M = Be or Mg

The zinc group

In striking contrast to the ionic solid hydride ZnH_2, the much less electropositive metals of Group 12 form the molecular compounds

CH_3—M—CH_3

4 M = Zn,Cd,Hg

$Zn(CH_3)_2$, $Cd(CH_3)_2$, and $Hg(CH_3)_2$ (**4**). These compounds all consist of linear molecules that are not associated in the solid, liquid, or gas states or in hydrocarbon solution.

Since these Group-12 organometallics have only four valence electrons (two from the metal and two from the alkyl groups) we might expect them, like their Be and Mg analogs, to complete their valence shells by association through alkyl bridges. They do not do so, and localized 2c,2e bonds suffice for the description of the bonding. To find an explanation of this behavior we must look for reasons why the energy of the linear R—M—R arrangement is lower than that of a bridged structure. The answer may lie in the ability of the small, highly polarizing CH_3 group to induce *p* and *d* mixing, for these orbitals have similar energies. It may be able to do this more readily than it can mix the more widely separated *s* and *p* orbitals to achieve the sp^3 hybridization needed for the bridged structure.[1]

The boron group

Unlike BH_3, which exists as the dimer, trimethylboron is a monomeric molecular compound, probably on account of the steric repulsion that would result between methyl groups in the bridged structure. In support of this explanation, the bigger Al atoms in alkylaluminum compounds do allow dimerization: they form 3c,2e bonds composed of orbitals supplied by the two Al atoms and bridging CH_3 groups (**1**). However, that size is not the only factor is shown by the fact that $Ga(CH_3)_3$ is monomeric even though Ga is larger than Al.

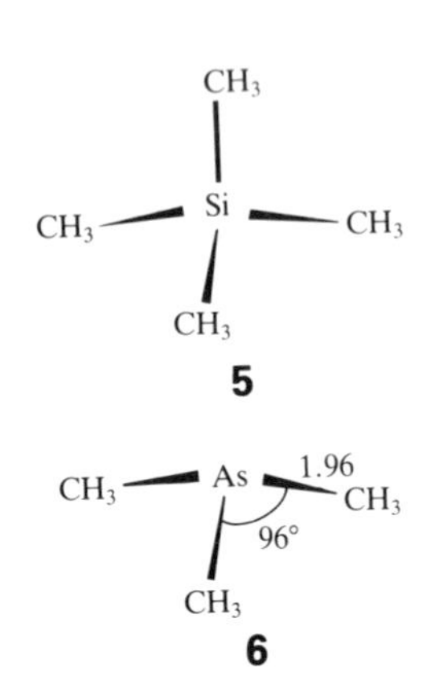

The carbon and nitrogen groups

The electron-precise compounds of the carbon group, such as tetramethylsilane (**5**), and the electron-rich compounds of the nitrogen group, such as trimethylarsine (**6**), are adequately described by conventional, localized 2c,2e center bonds, and their shapes can be rationalized by VSEPR theory. However, that the simplest form of VSEPR theory cannot account for everything is shown by the fact that C—E—C bond angles for trimethyl compounds of the heavy elements (As and Sb) are close to 90°. This is reminiscent of the bond angles of the corresponding hydrogen compounds, such as AsH_3 and SbH_3 (Table 9.4).

10.3 Synthesis

Most organometallic compounds can be synthesized using one of four M—C bond-forming reactions. These are reaction of a metal with an organic halide, transmetallation, metathesis, and addition.

Reaction of metals with haloalkanes and haloarenes

The net reaction of an electropositive metal M and a halogen-substituted hydrocarbon is

$$2M + RX \rightarrow MR + MX$$

[1] This is one of several models that have been proposed to account for the tendency of Zn(II), Cd(II), Hg(II), and Ag(I) to form linear complexes with certain ligands. For a theoretical assessment of these models, see J. A. Tossell and D. J. Vaughan, *Inorg. Chem.*, **20,** 3333 (1981).

where RX is a haloalkane or haloarene. For example, methyllithium is produced commercially by the reaction

$$2Li + CH_3Cl \rightarrow LiCH_3 + LiCl \qquad (1)$$

With other active metals, such as Mg, Al, and Zn, the reaction generally yields the organometal halide. A familiar example is the synthesis of a Grignard reagent, an alkylmagnesium halide:

$$Mg(s) + CH_3Br(soln) \xrightarrow{(C_2H_5)_2O} CH_3MgBr(soln) \qquad (2)$$

Transmetallation

In a **transmetallation** reaction, one metal takes the place of another, as in

$$M + M'R \rightarrow M' + MR$$

The metal M is more electropositive than M′ (which is typically Hg), so the reaction is analogous to a simple inorganic reaction in which a less electropositive metal is displaced from a compound by a more electropositive metal.

Since the metals of Groups 1, 2, and 13/III are all more electropositive than Hg, transmetallation can be carried out between them and dimethylmercury, as in

$$2Ga(l) + 3Hg(CH_3)_2(l) \rightarrow 3Hg(l) + 2Ga(CH_3)_3(l) \qquad (3)$$

The reaction is typically carried out in a sealed glass tube containing dimethylmercury and an excess of gallium, and kept at 60°C for about a day. During the reaction, three liquid phases (Hg, Ga, and the mixture of organometallics) are present, and as the reaction proceeds the lowest layer (Hg) increases in volume, the middle layer (Ga) decreases, and there is little noticeable change in the clear uppermost organometallic layer.

Metathesis

The metathesis (double replacement) of an organometallic MR and a binary halide EX is

$$MR + EX \rightarrow MX + ER$$

An approximate but instructive view of the reaction is that it takes place by the exchange of a formal carbanion (R^-) and a halide (X^-). Metathesis is an effective way of preparing a large number of organoelement compounds, with alkyllithium, magnesium, and aluminum compounds and the halides of the Group 13/III to 15/V elements the most common reagents:

$$Li_4(CH_3)_4 + SiCl_4 \rightarrow 4LiCl + Si(CH_3)_4 \qquad (4)$$

$$Al_2(CH_3)_6 + 2BF_3 \rightarrow 2AlF_3 + 2B(CH_3)_3 \qquad (5)$$

Since hydrocarbon groups form stronger covalent than ionic bonds, and the halogen forms strong ionic bonds with the electropositive metal, the alkyl or aryl group migrates from the less to the more electronegative element:

	MR				+		EX			$\rightarrow$ ER + MX
	M						E			
	Li	Mg	Al	Zn		Si	B	As	P	
χ	0.98	1.31	1.61	1.65		1.90	2.04	2.18	2.19	

Metatheses involving the *same* central element are often referred to as **redistribution reactions**. For example, silicon tetrachloride and tetramethylsilicon (tetramethylsilane) undergo redistribution when heated and produce a variety of chloromethylsilanes:

$$SiCl_4 + (CH_3)_4Si \rightarrow (CH_3)_3SiCl + (CH_3)_2SiCl_2 + \ldots$$

Example 10.1: *Classifying and predicting a potential M—C bond-forming reaction*

Determine the type of reaction that might occur between $Al_2(CH_3)_6$ and $GeCl_4$.

Answer. We are presented with an organometallic compound and a metal halide, so we should explore the possibility that these compounds can undergo metathesis. The organometallic compound has a metal (Al) that is more electropositive than the central atom in the halide (Ge); therefore the metathesis

$$3GeCl_4 + 2Al_2(CH_3)_6 \rightarrow 3Ge(CH_3)_4 + 4AlCl_3$$

should be thermodynamically favorable.

Exercise. What is the likely reaction of magnesium with dimethylmercury?

Addition of E—H to a multiple bond

The net outcome of an alkene addition reaction is

$$E{-}H + CH_2{=}CH_2 \rightarrow EH_2C{-}CH_3 \qquad (6)$$

This reaction is driven mainly by the high C—H bond strength relative to most E—H bond strengths, and occurs with a wide variety of E—H containing compounds. We have already encountered several examples in Chapter 9, the most important being:

Hydroboration (Section 9.10):

$$H_3BOR_2 + R'CH{=}CH_2 \xrightarrow{R_2O} R'CH_2CH_2BH_2 \qquad (7a)$$

Hydrosilation (Section 9.13):

$$SiH_4 + CH_2{=}CH_2 \xrightarrow[\text{i-PrOH}]{H_2PtCl_6} CH_3CH_2SiH_3 \qquad (7b)$$

In the hydroboration and hydrosilation of unsymmetrical alkenes, the more bulky B or Si group adds to the sterically less hindered C atom, and the smaller H atom adds to the more hindered C atom, as illustrated in eqn 7.

Table 10.1. Standard molar enthalpies of formation of gaseous methyl derivatives of the *p*-block elements ($\Delta H_f^{\ominus}$/kJ mol^{-1})

	Group			
Period	**12**	**13 III**	**14 IV**	**15 V**
2		$B(CH_3)_3$ −124	$C(CH_3)_4$ (−167)	$N(CH_3)_3$ (−24)
3		$Al(CH_3)_3$ −74	$Si(CH_3)_4$ −239	$P(CH_3)_3$ (−101)
4	$Zn(CH_3)_2$ +53	$Ga(CH_3)_3$ −45	$Ge(CH_3)_4$ (−71)	$As(CH_3)_3$ (+13)
5	$Cd(CH_3)_2$ +101	$In(CH_3)_3$?	$Sn(CH_3)_4$ (+21)	$Sb(CH_3)_3$ (+32)
6	$Hg(CH_3)_2$ +94	$Tl(CH_3)_3$ +	$Pb(CH_3)_4$ +136	$Bi(CH_3)_3$ (+194)

Sources: D. D. Wagman *et al., J. phys. chem. Ref. Data* **11,** Supplement 2 (1982). Data in parentheses are from M. E. O'Neill and K. Wade in *Comprehensive organometallic chemistry,* (ed. G. Wilkinson, F. G. A. Stone, and E. W. Abel). Pergamon, Oxford (1982).

10.4 Stability

As reliable Gibbs free energies of formation of organometallics are not widely available, it is necessary to resort to the enthalpies of formation in Table 10.1 for an approximate guide to the stability of the compounds with respect to their elements. As with hydrogen compounds, the formation of methyl compounds of the light elements of Groups 13/III to 15/V is exothermic (and so they are probably stable with respect to their elements). However, the stability decreases down the groups, and the heaviest elements (Tl, Pb, and Bi) form endothermic compounds. Their instability is a consequence in part of the decrease in M—C bond strength down a group (Table 10.2).

Table 10.2. Average bond enthalpies B/(kJ mol^{-1})

Group			
12 $E(CH_3)_2$	**13 III** $E(CH_3)_3$	**14 IV** $E(CH_3)_4$	**15 V** $E(CH_3)_3$
	B 365	C 358	N 314
	Al 274	Si 311	P 276
Zn 177	Ga 247	Ge 249	As 229
Cd 139	In 160	Sn 217	Sb 214
Hg 121	Tl	Pb 152	Bi 141

Source: M. E. O'Neill and K. Wade, *Comprehensive organometallic chemistry,* (ed. G. Wilkinson, F. G. A. Stone, and E. W. Abel). **1,** 1, Pergamon, Oxford (1982).

As with the analogous hydrogen compounds, the endothermic heavy metal organometallics are susceptible to homolytic cleavage of the M—C bond. For example, when $Pb(CH_3)_4$ is heated in the gas phase, it undergoes homolytic Pb—C bond cleavage to produce methyl radicals. Similarly, dimethylcadmium has been known to decompose explosively.

10.5 Reaction patterns

Oxidation

All organometallic compounds are potentially reducing agents; those of the electropositive elements are in fact strong reducing agents. It is important to be mindful of the potential fire hazard this character implies, for the compounds of the more electropositive elements ignite spontaneously on contact with air. The strong reducing character also presents a potential explosion hazard if the compounds come into contact with oxidizing agents.

All compounds of the electropositive elements that have unfilled valence orbitals, or that can dissociate to fragments with unfilled orbitals, are spontaneously flammable in air: these include $Li_4(CH_3)_4$, $Zn(CH_3)_2$, $B(CH_3)_3$, and $Al_2(CH_3)_6$. Accordingly, volatile compounds, such as alkylboron compounds, are handled in a vacuum line (Box 9.1) and inert atmosphere techniques are used for the others (Box 10.1). Compounds such as $Si(CH_3)_4$ and $Sn(CH_3)_4$, which do not have low-lying empty orbitals, require elevated temperatures to initiate combustion. Combustion takes place by radical chain processes (Section 9.9).

Box 10.1 Inert atmosphere techniques

Since they react so readily with oxygen, moisture, and carbon dioxide, many organometallics are handled in an inert atmosphere. The simplest of these techniques (conceptually, at least) is the **inert-atmosphere glove box** (Fig. 10.B1). This apparatus allows manipulations to be carried out in a large metal enclosure filled with N_2, Ar, or He. The enclosure has a window, and a tightly attached pair of rubber gloves are used to manipulate chemicals inside the box. Two other important components of the box are a source of pure inert gas and an antechamber—an airlock—which permits items to be brought into or out of the box while avoiding the influx of air. Since air slowly diffuses into the box through the rubber gloves, it is common to maintain the purity of atmosphere inside the chamber by a recirculating gas purifier or a constant flush of inert gas.

Another common method of handling air-sensitive compounds uses standard glassware constructed so that an inert atmosphere can be maintained inside it. Typically this apparatus is equipped with sidearms for pumping out the air and introducing inert gas (Fig. 10.B2). When apparatus containing air-sensitive compounds must be reconfigured, it is opened under a flush of inert gas as illustrated at the bottom of Fig. 10.B2. This type of apparatus can be employed for all the standard synthetic operations such as reactions in solution, filtration, and crystallization. The apparatus is often referred to as **Schlenk ware** in recognition of the German chemist Wilhelm Schlenk, who introduced this general design during his pioneering research in organometallic chemistry during the first quarter of the twentieth century.

A variation on the above method for transferring air-sensitive solution utilizes syringes or metal cannula (flexible, small diameter metal tubing). For these operations, a flask is fitted with a rubber serum bottle cap (a septum) which can be pierced with a syringe needle or sharpened cannula. Figure 10.B3 illustrates the transfer of a solution from one flask to the other by a cannula and a pressure differential between the two flasks.

[J. J. Eisch, *Organometallic synthesis*, Vol. 2. Academic Press, New York (1981).
A. L. Wayda and M. Y. Darensbourg (eds.), *Experimental organometallic chemistry: A practicum in synthesis and characterization*. ACS Symposium Series 357, American Chemical Society, Washington (1988).
D. F. Shriver and M. A. Drezdzon, *The manipulation of air-sensitive compounds*. Wiley, New York (1986).]

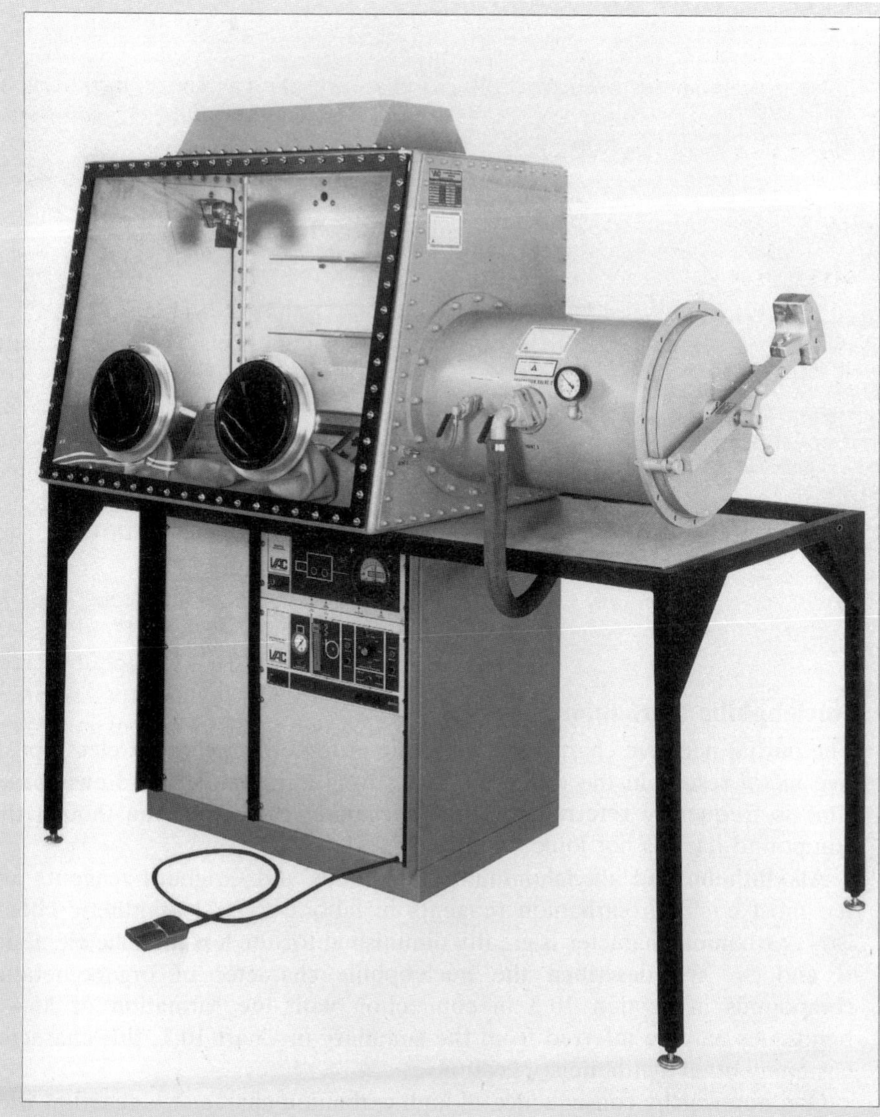

Fig. 10.B1 Inert-atmosphere glove box. The cylindrical airlock is on the right. The enclosure under the table contains a vacuum pump, which is used to evacuate the transfer lock before items are brought into the main chamber. This enclosure below the table also contains a circulation pump and columns, which remove H_2O and O_2 from the glove box atmosphere. (Reproduced by permission of Vacuum Atmospheres Corp., California.)

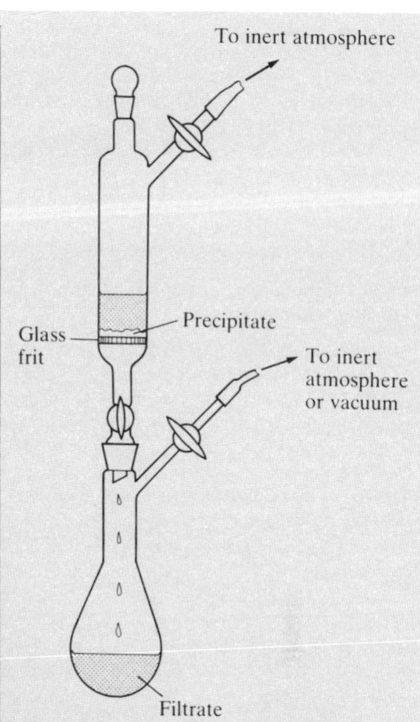

Fig. 10.B2 Typical inert-atmosphere Schlenk ware. In this example, a reaction mixture has been filtered in the upper part of the apparatus and the filtrate collected in the Schlenk tube illustrated in the lower part of the illustration.

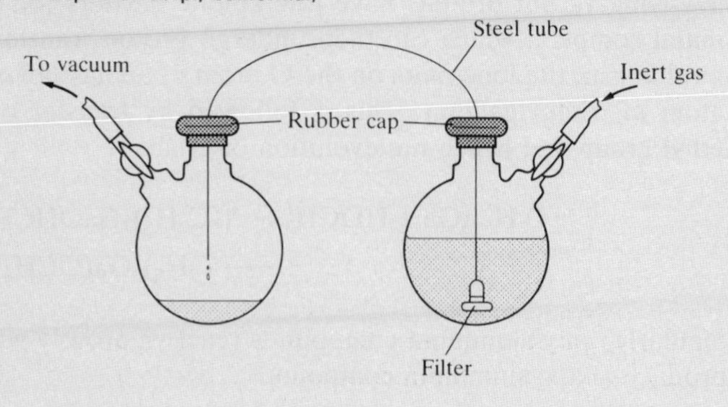

Fig. 10.B3 Filtration and liquid transfer by means of a pressure differential and stainless steel tube (cannula).

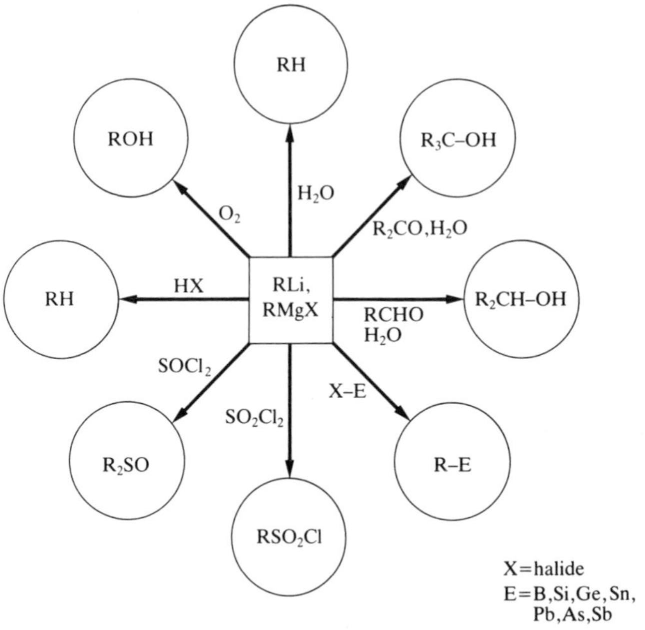

Chart 10.1 Some reactions of alkyllithium and of Grignard reagents.

RLi, RMgX
H_2O → RH
R_2CO, H_2O → R_3C–OH
RCHO, H_2O → R_2CH–OH
X–E → R–E
SO_2Cl_2 → RSO_2Cl
$SOCl_2$ → R_2SO
HX → RH
O_2 → ROH

X = halide
E = B, Si, Ge, Sn, Pb, As, Sb

Nucleophilic (carbanion) character

The partial negative charge of an organic group attached to an electropositive metal results in the group being a strong nucleophile and Lewis base. This is frequently referred to as its **carbanion character** even though the compound itself is not ionic.

Alkyllithium and alkylaluminum compounds and Grignard reagents are the most common carbanion reagents in laboratory-scale synthetic chemistry; carbanion character is greatly diminished for the less metallic elements B and Si. We described the nucleophilic character of organometallic compounds in Section 10.3 in connection with the formation of M—C bonds. As may be inferred from the summary in Chart 10.1, this character has many other synthetic applications.

One noteworthy consequence of high carbanion character is reaction with very weak Brønsted acids, including water and alcohols. The criteria for judging when this reaction can be rapid are similar to those for judging whether reaction with O_2 is likely. Thus, organometallic compounds with low-lying vacant orbitals have low kinetic barriers for the formation of an initial complex, which can then undergo proton transfer. According to this mechanism, the lone pairs on the O atom of an alcohol coordinate to the Ga atom in triethylgallium. This is followed by transfer of the proton to the ethyl group and hence the evolution of ethane:

$$(C_2H_5)_3Ga + HOCH_3 \rightarrow [(C_2H_5)_3GaOHCH_3]$$
$$\rightarrow (C_2H_5)_2Ga(OCH_3) + C_2H_6$$

Similarly, alkylaluminum compounds react vigorously with excess alcohol to produce alkoxyaluminum compounds:

$$Al_2(CH_3)_6 + 6C_2H_5OH \rightarrow 2Al(OC_2H_5)_3 + 6CH_4$$

Again, it appears that protolysis occurs by prior coordination:

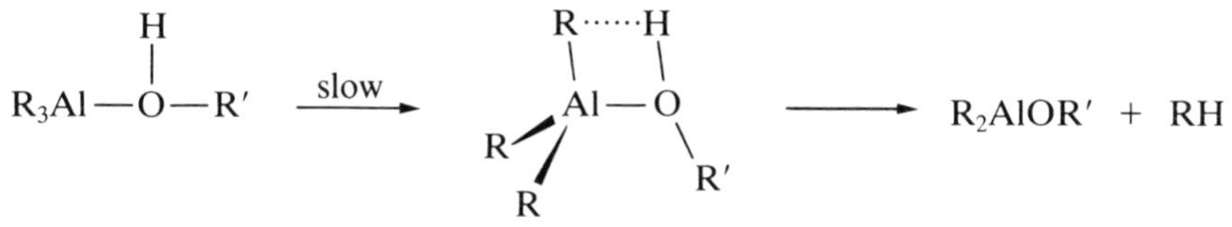

$$R_3Al\text{—}\overset{\displaystyle H}{\overset{|}{O}}\text{—}R' \xrightarrow{\text{slow}} \begin{matrix} R\cdots\cdots H \\ | \quad\quad | \\ R_2Al\text{—}O\text{—}R' \end{matrix} \longrightarrow R_2AlOR' + RH$$

In keeping with the lengthening of the O—H bond this mechanism implies, the reaction shows a significant isotope effect, with $k(\text{H—OCH}_3)/k(\text{T—OCH}_3) = 2.9$ when R is octyl.

Because the protolysis reaction liberates hydrogen with organometallic compounds, water cannot be used to extinguish a fire involving trialkylaluminum compounds or any other organometallic compounds of electropositive metals. Protolysis of electropositive organometallics with sterically hindered alcohols, such as *tert*-butanol, proceeds at a moderate rate and therefore provides a convenient way of destroying reactive organometallic wastes.

Lewis acidity

Electron-deficient organometallics are Lewis acids on account of the unoccupied orbitals on the metal. An illustration of this acidity is the synthesis of organometallic anions, such as the tetraphenylborate ion:

$$B(C_6H_5)_3 + Li(C_6H_5) \rightarrow Li[B(C_6H_5)_4] \tag{8}$$

This reaction may be viewed as the transfer of the strong base $C_6H_5^-$ from the weak Lewis acid Li^+ to the stronger acid B(III).

Organometallics that are bridged by organic groups can also serve as Lewis acids and undergo bridge cleavage. For example, $Al_2(CH_3)_6$ is cleaved by tertiary amines to form a simple Lewis acid–base complex:

$$Al_2(CH_3)_6 + 2N(C_2H_5)_3 \rightarrow 2(CH_3)_3AlN(C_2H_5)_3$$

This reaction illustrates once again the weakness of the 3c,2e Al—CH_3—Al bond. Solvents such as THF coordinate to the Li atoms in $Li_4(CH_3)_4$, but with such a mild base the metal cluster is not disrupted.

β-hydrogen elimination

A **β-hydrogen elimination** is a reaction in which the metal atom extracts an H atom from the next-nearest neighbor (β) atom, as in the reaction

$$M\text{—}CH_2\text{—}CH_3 \rightarrow M\text{—}H + H_2C\text{=}CH_2$$

It is the reverse of the addition of an M—H bond to an alkene (eqn 7), and under some conditions significant equilibrium concentrations of both reactants and products are observed. The reaction mechanism is believed to involve the formation of β-hydrogen bridges to an open coordination site on the metal:

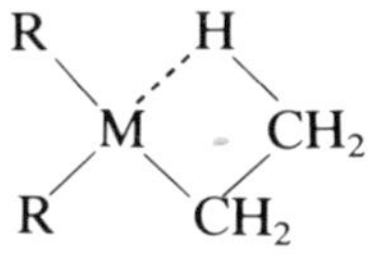

$$\begin{matrix} R \quad\quad H \quad\quad \\ \diagdown \; \cdot\cdot \diagup \;\; \diagdown \quad \\ M \quad\quad CH_2 \\ \diagup \quad \diagdown \;\; \diagup \quad \\ R \quad\quad CH_2 \quad \end{matrix}$$

It follows from this mechanism that we can expect atoms with low coordination numbers to undergo the reaction. Indeed, it occurs with trialkylaluminum compounds but not with tetraalkylgermanium compounds.

Example 10.2: *Predicting the products of thermal decomposition*

Summarize the thermal stabilities of (a) $Bi(CH_3)_3$ and (b) $Al_2(^iBu)_6$, and give equations for their decomposition.

Answer. (a) As with the other heavy *p*-block elements, BiC bonds are weak and readily undergo homolytic cleavage. The resulting methyl radicals will react with surrounding molecules or form ethane:

$$2Bi(CH_3)_3 \xrightarrow{\Delta} 2Bi(s) + 3H_3C{-}CH_3$$

(b) The $Al_2(^iBu)_6$ dimer readily dissociates. At elevated temperatures dissociation is followed by β-hydrogen elimination, which is common for organometallics that (1) have alkyl groups with β-hydrogens and (2) can form stable M—H bonds:

$$Al_2(^iBu)_6 \xrightarrow{\Delta} 2Al(^iBu)_3 \xrightarrow{\Delta\Delta} Al(^iBu)_2H + i\text{-}C_4H_8$$

Exercise. Describe the probable mode of thermal decomposition of $Pb(CH_3)_4$.

Ionic and electron-deficient compounds of Groups 1, 2, and 12

In this section and the following we discuss the chemistry of certain organometallic compounds in more detail. We shall concentrate on the organometallic compounds of Li, Mg, Zn, and Hg because of their interesting properties and utility in syntheses.

10.6 Alkali metals

Organolithium compounds

The synthetically useful organolithium compounds are available commercially as solutions. Methyllithium is generally handled in ether solution, but alkyllithium compounds with longer chains are soluble in hydrocarbons. Since the commercial preparation of alkyllithium compounds is by the reaction of the metal with the organic halide, they are often contaminated by the starting materials. However, to an organometallic chemist contamination can be interesting, and a complex of LiBr with phenyllithium has been characterized by X-ray diffraction.[2]

Although methyllithium exists as a tetrahedral cluster in the solid state and in solution, many of its higher homologs exist in solution as hexamers or equilibrium mixtures of aggregates ranging up to hexamers. These larger aggregates can be broken by strong Lewis bases, such as chelating amines. For example, the interaction of TMEDA[3] with phenyllithium produces a

[2] H. Hope and P. P. Power, *J. Am. chem. Soc.*, **105**, 5320 (1983).

[3] TMEDA is *N,N,N′,N′*-tetramethylethylenediamine. It is a favorite chelating ligand in *p*-block organometallic chemistry because it lacks the mildly acidic N—H bonds of ethylenediamine, and therefore does not undergo protolysis with carbanionic organometallics such as the alkyllithium compounds.

complex containing two Li atoms bridged by phenyl groups with each Li atom coordinated by the chelating diamine (**7**).

In addition to the common organolithium compounds with one Li atom per organic group, a variety of **polylithiated** organic molecules—organometallic compounds containing several Li atoms per molecule—are known.[4] The simplest of these is dilithiomethane, Li_2CH_2, which can be prepared by the pyrolysis of methyllithium. This compound crystallizes in a distorted antifluorite structure (Section 4.5), but the finer details of the orientation of the CH_2 groups are as yet unknown.

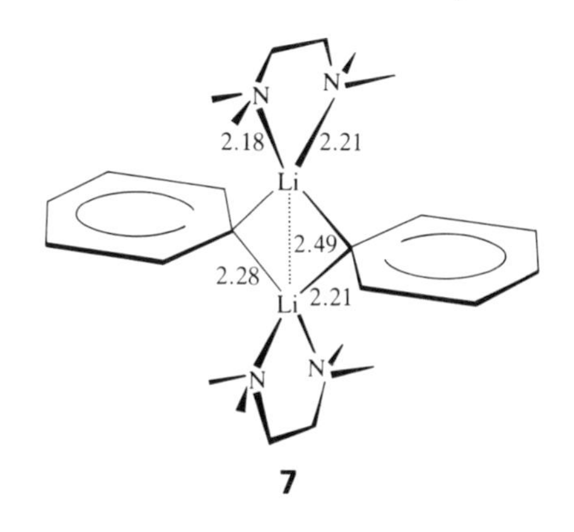

7

Radical anion salts

Sodium naphthalide is an example of an organometallic salt with a delocalized radical anion, the $C_{10}H_8^-$ ion. These are readily prepared by the direct interaction of the aromatic compound with an alkali metal. Thus, naphthalene dissolved in THF reacts with sodium metal to produce a dark green solution of sodium naphthalide:

$$Na + C_{10}H_8 \xrightarrow{THF} Na[C_{10}H_8]$$

EPR spectra show that the odd electron is delocalized in an antibonding orbital of $C_{10}H_8$.

The formation of the radical anion is more favorable when the π LUMO of the arene is low in energy. Simple molecular orbital theory predicts that the LUMO occurs at progressively lower energies on going from benzene to more extensively conjugated hydrocarbons. This is analogous to the lowering of energy levels of an electron in a box as the length of the box is increased (Section 1.3). The prediction is borne out by the reduction potentials of aromatic hydrocarbons (Table 10.3). For this reason, the radical anion of benzene cannot be formed in most solvents, but naphthalene and more extensively conjugated arenes readily yield alkali metal salts.

Table 10.3. Standard reduction potentials of some conjugated hydrocarbons (relative to Hg in 2-methoxyethanol)

Compound	$E^{\ominus}$/V
Benzene	>2.1
Biphenyl	2.08
Naphthalene	1.98
Phenanthrene	1.94
Anthracene	1.46

Sodium naphthalide and similar compounds are highly reactive reducing agents and are often preferred to sodium because—unlike sodium itself—they are readily soluble in ethers. The resulting homogeneous reaction is generally faster and easier to control than a heterogeneous reaction between one component in solution and pieces of sodium metal, which are often coated with unreactive oxide or with insoluble reaction products. As indicated in Table 10.3, an additional advantage of the radical anion reagents is that the reduction potential of the reagent can be tailored to a particular task by choice of the aromatic group.

Another route to delocalized anions is the reductive cleavage of acidic C—H bonds by an alkali metal or alkylmetallic compound. Thus, in the presence of a good coordinating ligand such as TMEDA, butyllithium reduces dihydronaphthalene to a diamagnetic dinegative anion:

$$2TMEDA + 2LiBu + C_{10}H_{10} \rightarrow [(TMEDA)Li]_2[C_{10}H_8] + 2BuH$$

TMEDA contributes to the favorable reaction free energy through its affinity for the Li^+ produced in the reaction.

[4] A description of X-ray structures of lithium compounds is given by W. N. Setzer and P. von R. Schleyer, *Adv. organomet. Chem.*, **24**, 353 (1985).

The planar aromatic cyclopentadienide ion $C_5H_5^-$ can be prepared by the reductive cleavage of the C—H in cyclopentadiene, using either sodium metal or NaH:

$$C_5H_6 + Na \xrightarrow{THF} Na[C_5H_5] + \tfrac{1}{2}H_2$$

Sodium cyclopentadienide readily undergoes metathesis with a variety of halides of *p*-block elements to produce either σ- or π-bonded cyclopentadienyl compounds. We shall discuss it later in the chapter. It is also used to synthesize a wide variety of *d*-block organometallic compounds, as we shall see in Chapter 16.

10.7 Alkaline earth metals

Organoberyllium and organomagnesium compounds have significant covalent character, whereas the analogous compounds of the heavier congeners are more ionic. However, as the latter have not yet been thoroughly investigated, this judgment might change.

A characteristic of organoberyllium and organomagnesium compounds is their tendency to adopt four-coordination; for Be, three-coordination is also observed. The dimethyl compounds, for instance, appear to be bridged species (**3**). That more bulky groups lead to decreased association is shown by the dimeric structure of diethylberyllium in benzene solution compared with the monomeric $Be(^tBu)_2$.

Formation and structure

Because Mg is much more electropositive than Be, ether-solvated dialkylberyllium compounds can be prepared by metathesis with a Grignard reagent:

$$BeCl_2 + 2RMgX + (C_2H_5)_2O \xrightarrow{(C_2H_5)_2O} BeR_2 \cdot O(C_2H_5)_2 + 2MgXCl$$

When an ether-free product is needed, transmetallation with dialkylmercury can be used because Be is more electropositive than Hg:

$$Be + Hg(CH_3)_2 \rightarrow Be(CH_3)_2 + Hg$$

Bis(cyclopentadienyl)beryllium is structurally very interesting because it has quite different structures in the gas and solid phases. It is readily prepared by metathesis of a beryllium halide with $Na[C_5H_5]$:

$$BeCl_2 + 2Na[C_5H_5] \rightarrow Be(C_5H_5)_2 + 2NaCl$$

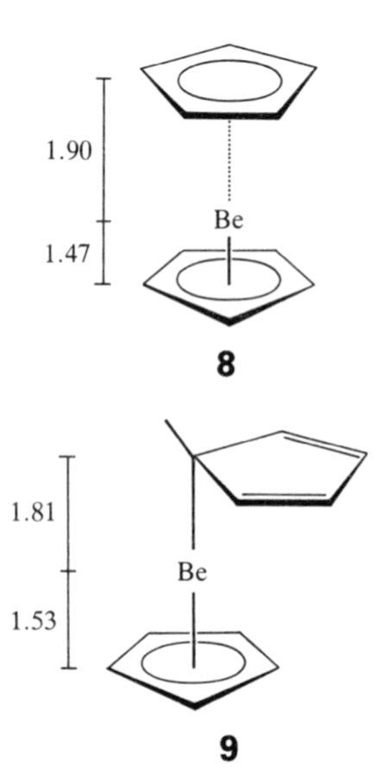

The gas-phase molecule appears to be an antisymmetrical sandwich (**8**) in which all five C atoms of each C_5H_5 group are within bonding distance of the Be atom. This fivefold **pentahapto** bonding is denoted η^5 to distinguish it from **monohapto** bonding, denoted η^1, in which only one C atom is within bonding reach of the metal atom. (The concept of hapticity, a word derived from the Greek for 'to fasten', is discussed in more detail in the introduction to Chapter 16.) X-ray crystal structure data demonstrate that in the solid state the compound has the mixed η^5-C_5H_5, η^1-C_5H_5 coordination (**9**). Recent electron diffraction data for the molecule in the gas phase are compatible with either structure **8** or **9**, but the latter fits the data better. This ambiguity illustrates the difficulty of obtaining precise structural data on large molecules in the gas phase.

Reactions

Simple organoberyllium compounds react with oxygen, water, and other weak Brønsted acids:

$$Be(CH_3)_2 + CH_3OH \rightarrow (CH_3)Be(OCH_3) + CH_4$$

Dimethylberyllium also readily forms complexes with Lewis bases:

$$Be(CH_3)_2 + N(CH_3)_3 \rightarrow (CH_3)_2BeN(CH_3)_3$$

$$\underset{\text{TMEDA}}{Be(CH_3)_2 + (CH_3)_2NCH_2CH_2N(CH_3)_2} \longrightarrow (CH_3)_2Be[N(CH_3)_2CH_2CH_2N(CH_3)_2]$$

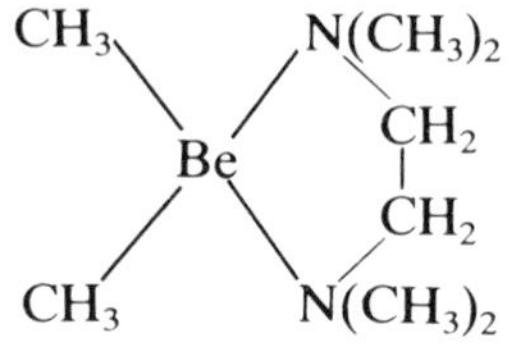

As Be compounds are highly toxic, and there appear to be no compelling reasons to use them as reagents, organoberyllium compounds do not have significant commercial applications. Their application as synthetic intermediates in the laboratory is confined to the synthesis of other Be compounds.

Grignard reagents

Organomagnesium compounds are familiar to any student of organic chemistry as useful carbanion reagents. The most common of these are the alkylmagnesium halides, or **Grignard reagents**, prepared by the reaction of an haloalkane with magnesium metal.[5] This reaction is carried out in ether, and because a coating of oxide on the magnesium acts as a kinetic barrier, a trace of iodine is often added to initiate the reaction.

From the early work carried out by Wilhelm Schlenk, it is known that redistribution equilibria occur in ether solution. The simplest of these, often called the **Schlenk equilibrium**, is

$$2RMgX \xrightarrow{(C_2H_5)_2O} MgR_2 + MgX_2$$

The addition of dioxane to the equilibrium mixture leads to the precipitation of a dioxane complex of the magnesium halide, $MgX_2 \cdot (C_4H_8O_2)$, and the filtrate can be evaporated to yield the dialkylmagnesium. More recent spectroscopic studies indicate a rather complex set of equilibria between alkylmagnesium halides in ether solution. In line with these observations, the species crystallized from ether solution include a wide variety of structures, such as monomeric 4-coordinate Mg complexes (**10**) and larger clusters (**11**).

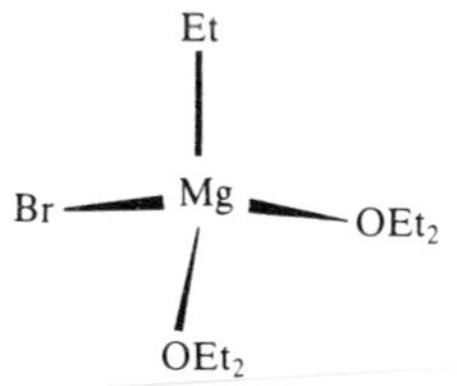

Because Grignard reagents are produced in ether solution, their use is limited to reactions in which that mildly Lewis basic solvent is not objectionable. The ether may form an unwanted complex with Lewis acid products, such as the formation of $(CH_3)_3BO(C_2H_5)_2$ in the reaction of BF_3 with CH_3MgBr in diethyl ether. In these cases, alkylaluminum or alkyllithium compounds are preferred because they can be used in hydrocarbon solution.

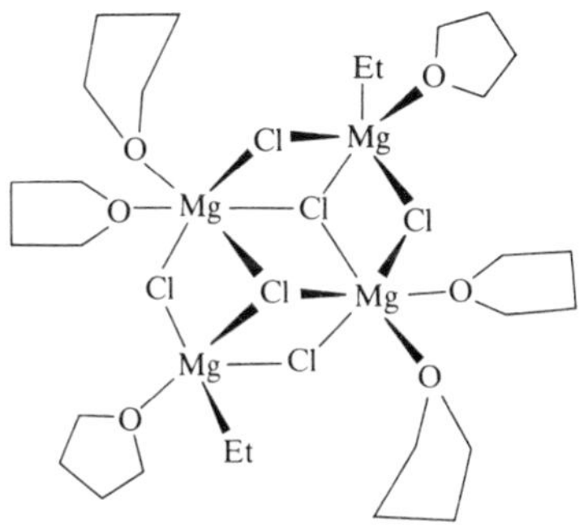

[5] A translation of Grignard's first full-length paper on this subject is available in *J. chem. Educ.*, **47**, 290 (1970).

10

11

10.8 The zinc group

As we mentioned earlier in the chapter, the dialkyl compounds of Zn, Cd, and Hg are remarkable for their lack of association through alkyl bridges. Another feature to bear in mind is that dialkylzinc compounds are only weak Lewis acids, organocadmium compounds are even weaker, and organomercury compounds do not act as Lewis acids except under special circumstances.

Organozinc and organocadmium compounds

A convenient synthesis of organometallic compounds of Zn is metathesis with alkylaluminum or alkyllithium compounds:

$$ZnCl_2 + Al_2(CH_3)_6 \rightarrow Zn(CH_3)_2 + [(CH_3)_2AlCl]_2$$

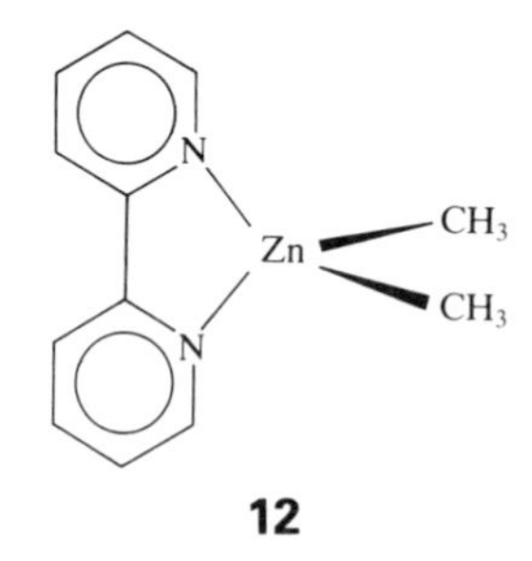

12

Alkylzinc compounds are pyrophoric and readily hydrolyzed, whereas alkylcadmium compounds react more slowly with air. Because of their mild Lewis acidity, dialkylzinc compounds form stable complexes with amines, especially chelating amines (**12**). The C—Zn bond has greater carbanionic character than the C—Cd bond. One property in which this is apparent is that an alkylzinc compound, but neither alkylcadmium nor alkylmercury compounds with their less polar M—C bonds, adds across a carbonyl C=O bond in a ketone:

$$Zn(CH_3)_2 + (CH_3)_2C{=}O \rightarrow (CH_3)_3C{-}O{-}ZnCH_3$$

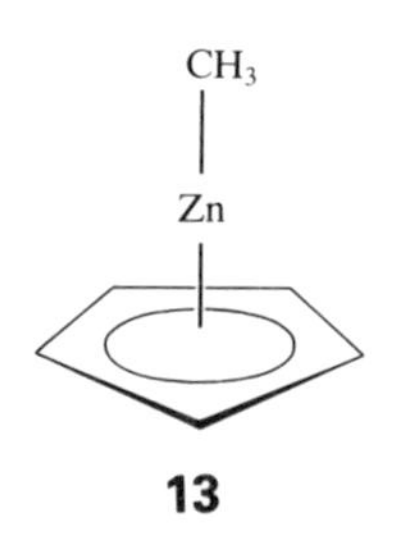

13

This reaction is similar to those of organolithium, organomagnesium, and organoaluminum compounds.

Once again, the cyclopentadienyl compounds are structurally unusual. Methyl(cyclopentadienyl)zinc is monomeric in the gas phase with a pentahapto C_5H_5 group (**13**). In the solid it is associated in a zigzag chain (**14**), each C_5H_5 group being pentahapto with respect to *two* Zn atoms.

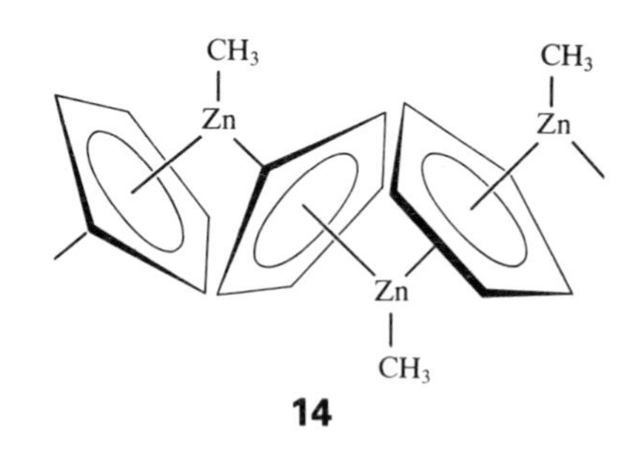

14

Organomercury compounds

Organomercury compounds are readily prepared by metathesis of Hg(II) halides with strong carbanion reagents, such as Grignard reagents or trialkylaluminum compounds:

$$2RMgX + HgX_2 \rightarrow HgR_2 + MgX_2$$

As we have already remarked, dialkylmercury compounds are versatile starting materials for the synthesis of many organometallics of more electropositive metals by transmetallation. However, because of the high toxicity of alkylmercury compounds, other syntheses are often preferred. In striking contrast to the high sensitivity of dimethylzinc to oxygen, dimethylmercury survives exposure to air.

The toxicity of Hg has been a matter of general public concern following the incidence of brain damage and death it caused among the inhabitants of Minamata, Japan. This incident arose because mercury from a factory was allowed to escape into a bay where it found its way into fish that were later eaten. Research since that time has shown that bacteria found in sediments are capable of methylating mercury, and that species such as $Hg(CH_3)_2$ and $[HgCH_3]^+$ enter the food chain because they readily penetrate cell walls.

The bacteria appear to produce $Hg(CH_3)_2$ as a means of eliminating toxic mercury ions through their cell walls and into the environment.[6]

Electron-deficient compounds of the boron group

The electron-deficient organometallics of Group 13/III are all molecular and are often named as derivatives of their simple hydrogen compounds. Thus trimethylboron is also known as trimethylborane. Of these compounds only the organoaluminum compounds are strongly associated through 3c,2e bonds.

10.9 Organoboron compounds

Trimethylboron is colorless, gaseous (b.p. −22°C), and monomeric. It is spontaneously flammable on contact with air but is not rapidly hydrolyzed by water. The alkylboranes can be synthesized by metathesis between boron halides and strong carbanion agents such as Grignard reagents or organoaluminum compounds:

$$BF_3 + 3CH_3MgBr \xrightarrow{Bu_2O} B(CH_3)_3 + 3MgBrF$$

(Dibutyl ether is used here rather than diethyl ether because it has a much lower vapor pressure than trimethylboron, and the weakly bonded complex separates into its components by distillation.)

Although the trialkyl and triaryl borons are mild Lewis acids, strong carbanion reagents lead to anions of the type $[BR_4]^-$ (as in eqn 8). The best known of these is the tetraphenylborate ion. This anion hydrolyzes very slowly in neutral or basic water and is a useful bulky anion in synthetic chemistry. Many other four-coordinate compounds of organoboranes are known. Incorporation of B into heterocycles is common, as in the reaction of 1,2-dihydroxybenzene with a trialkylboron at elevated temperatures:

$$C_6H_4(OH)_2 + BR_3 \rightarrow C_6H_4O_2B{-}R + 2RH$$

Organohaloboron compounds are more reactive than simple trialkylboron compounds. One preparation is by the reaction of boron trichloride and the stoichiometric amount of an alkylaluminum in a hydrocarbon solvent:

$$3BCl_3 + 6AlR_3 \rightarrow 3R_2BCl + 6AlR_2Cl$$

The products may be subjected to the full range of protolysis reactions with

[6] Mercury poisoning occurs because of the very high affinity of the soft Hg atom for sulfhydryl groups in enzymes. It was a serious problem for early scientists, including Newton in the eighteenth century and Alfred Stock in the early twentieth century, both of whom worked with Hg in poorly ventilated laboratories. Accounts of the biochemistry of organometallics are given by J. S. Thayer, *Organometallic compounds and living organisms*. Academic Press, New York (1984) and in P. J. Craig (ed.), *Organometallic compounds in the environment*. Wiley, New York (1986).

ROH, R_2NH, and other reagents, and to metatheses:

$$(CH_3)_2BCl + HOR \rightarrow (CH_3)_2BOR + HCl$$

$$(CH_3)_2BCl + (C_4H_9)Li \rightarrow (CH_3)_2B(C_4H_9) + LiCl$$

Among the many organoboranes, an especially interesting series contains B—N linkages. A BN fragment is isoelectronic with a CC fragment and B—N and C—C analogs generally have the same structures. However, despite this formal similarity, their chemical and physical properties are often quite different (Section 11.2).

10.10 Organoaluminum compounds

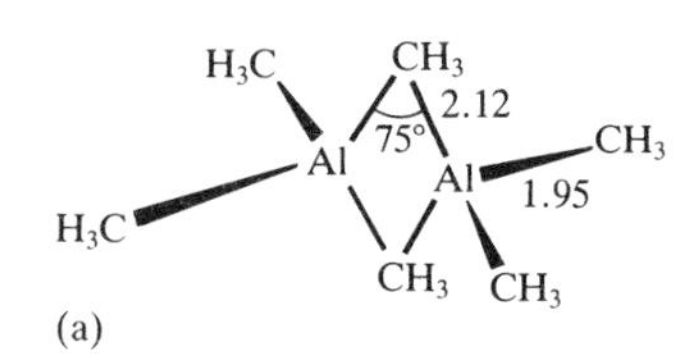

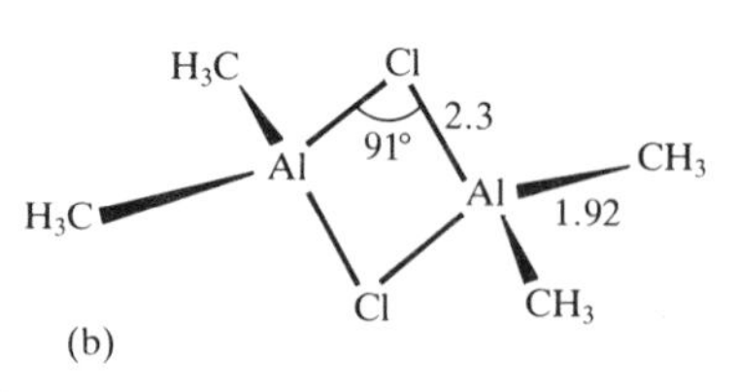

Fig. 10.3 Comparison of the bond angles between the 3c,2e bond for the Al_2CH_3 bridge and the conventional 2c,2e bonds for the Al_2Cl bridge.

Alkylaluminum compounds exist as bridged dimers (**1**) in which the sp^3 orbital of a CH_3 group is isolobal with a hydrogen 1*s* orbital in B_2H_6, and plays the same role in forming the 3c,2e bond. One of the distinguishing features of the methyl bridge bond is the small Al—C—Al angle, which is approximately 75°. A much larger angle is observed for halide bridges (Fig. 10.3), for the halogen atom can use more orbitals to form molecular orbitals.

Alkylaluminums dissociate slightly in the pure liquid to an extent that increases with the bulk of the alkyl group:

$$Al_2(CH_3)_6 \rightleftharpoons 2Al(CH_3)_3 \qquad K = 1.52 \times 10^{-8}$$

$$Al_2(C_4H_9)_6 \rightleftharpoons 2Al(C_4H_9)_3 \qquad K = 2.3 \times 10^{-4}$$

(The equilibrium constants refer to 25°C and are expressed in mole fractions.) With very bulky groups, dissociation is virtually complete: trimesitylaluminum, for instance, is a monomer.[7] These examples provide clear evidence for the powerful role of steric effects on the structures of the alkylaluminum compounds.

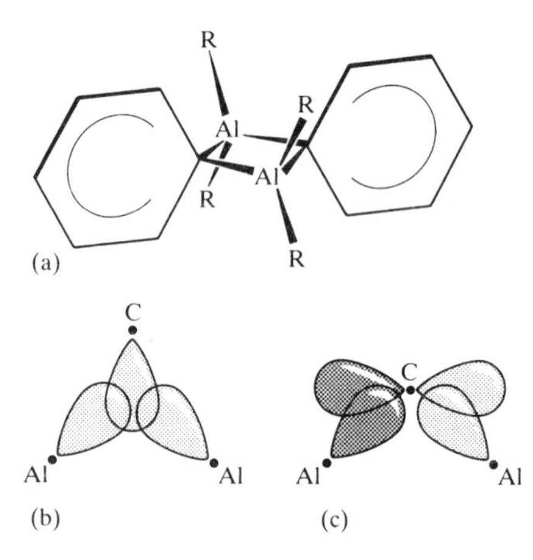

Fig. 10.4 Structure and bonding of the phenyl bridge in Ph_6Al_2. (a) Structure illustrating the perpendicular orientation of the bridging phenyl relative to the AlCAlC plane. (b) The 3c,2e bond formed by a symmetric combination of C and Al orbitals. (c) An additional interaction between the $p\pi$ orbital on C and an antisymmetric combination of Al orbitals.

Similarly, triphenylaluminum exists as a dimer with bridging η^1-phenyl groups lying in a plane perpendicular to the line joining the two Al atoms (Fig. 10.4). We can explain this structure by a model in which the bridge is a 3c,2e Al—C—Al bond and there is electron donation from the phenyl π orbitals to the Al atoms. The stronger bridge bonds (and greatest tendency to form bridges) in the alkylaluminum compounds are found for groups that have orbitals that are appropriately directed for them to form two conventional 2c,2e bonds: these include NR_2, OR, Cl, and Br. Phenyl and vinyl bridges between Al atoms are somewhat weaker and alkyl groups form the weakest bridges of all.

Alkylaluminum compounds have been extensively investigated on account of their use as alkene polymerization catalysts and chemical intermediates. They are relatively inexpensive carbanion reagents for the replacement of halogens by organic groups by metathesis. A general laboratory-scale preparation of trimethylaluminum is transmetallation with dimethylmercury:

$$2Al + 3Hg(CH_3)_2 \rightarrow Al_2(CH_3)_6 + 3Hg$$

CH3 CH3

CH3

[7] The mesityl group (left) is commonly used on account of its bulk. It plays a prominent role in the discussion of organosilicon chemistry in Section 10.12.

The commercial synthesis uses the reaction of Al metal with chloromethane to produce $(CH_3)_4Al_2Cl_2$, which is then reduced with Na:

$$3Al_2Cl_2(CH_3)_4 + 6Na \rightarrow 2Al_2(CH_3)_6 + 2Al + 6NaCl$$

The commercial synthesis of triethylaluminum and higher homologs is by the addition of H_2 and the appropriate alkene to aluminum metal at elevated pressures and temperatures:

$$2Al + 3H_2 + 6RHC{=}CH_2 \xrightarrow[200\text{–}50\ \text{atm}]{60\text{–}110^\circ\text{C}} Al_2(CH_2CH_2R)_6$$

It is probable that this reaction proceeds by the formation of a surface Al—H species that adds across the C═C bond of the alkene. The commercial use of alkylaluminum compounds would be very limited without this relatively economical synthesis.

As might be expected from the more electropositive nature of Al, the alkylaluminum compounds have much greater carbanionic character than the alkylboron compounds. As a result, alkylaluminum compounds are very sensitive to water and oxygen, and most are pyrophoric; the neat liquid and solutions must both be handled using inert-atmosphere techniques (Box 10.1). However, this reactivity can be turned to advantage, for the susceptibility of the Al—R bond to protolysis provides a simple method for the preparation of aluminum alkoxides and amides:

$$2AlR_3 + 2HOR' \longrightarrow \begin{array}{c} R_2Al{-}OR' \\ |\quad\ \ | \\ R'O{-}AlR_2 \end{array} \text{(and/or a trimer)} + 2RH$$

$$2AlR_3 + 2HNR'_2 \longrightarrow \begin{array}{c} R_2Al{-}NR'_2 \\ |\quad\ \ | \\ R'_2N{-}AlR_2 \end{array} \text{(and/or a trimer)} + 2RH$$

Alkylaluminums are mild Lewis acids and form complexes with ethers, amines, and anions.

β-Hydrogen elimination (Section 10.5) yields dialkylaluminum hydride when triethylaluminum and higher alkylaluminums are heated. Triisobutylaluminum has a strong tendency to undergo this reaction:

$$2Al(i\text{-}C_4H_9)_3 \longrightarrow (i\text{-}C_4H_9)_2Al\underset{H}{\overset{H}{\diamond}}Al(i\text{-}C_4H_9)_2 + 2(i\text{-}C_4H_8)$$

The general preference of hydride for the bridging position indicates that the 3c,2e bond is stronger for H than an alkyl group, probably because the small H atom can lie more readily between the atoms.

Example 10.3: ***Proposing structures for some boron and aluminum organometallics***

Based on your knowledge of the bonding in organoboron and organoaluminum compounds, propose structures for compounds having the empirical formulas (a) $B(^iPr)_3$, (b) $Al(C_2H_5)_2Ph$, (c) $Al(C_2H_5)_2P(CH_3)_2$.

Answer. (a) Triisopropylboron, as with all the simple alkylborons, will be monomeric and the B atom and the three C atoms bonded to it should lie in a plane. (b) If the attached groups have moderate bulk, the alkyl and arylaluminum compounds are associated into dimers through multicenter bonds. Recalling that the tendency toward bridge structures is

$$PR_2^- > X^- > H^- > Ph^- > R^-$$

(where R is alkyl), the structures will be

(b) 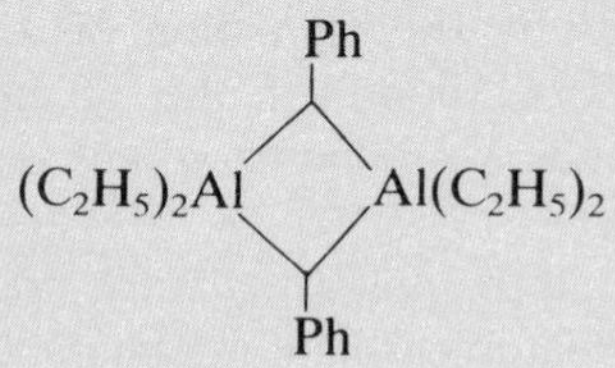 $(C_2H_5)_2Al(\mu\text{-Ph})_2Al(C_2H_5)_2$ — two Ph groups bridging the two Al atoms

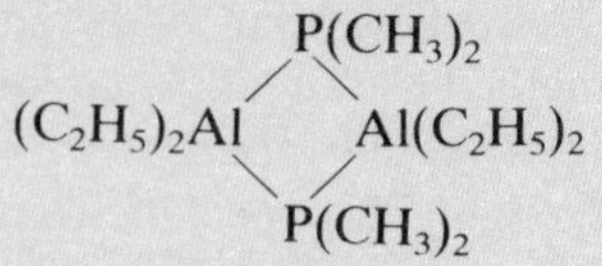

(c) $(C_2H_5)_2Al(\mu\text{-P(CH_3)_2})_2Al(C_2H_5)_2$ — two $P(CH_3)_2$ groups bridging the two Al atoms

Exercise. Propose a structure for $Al_2(^iBu)_4H_2$.

10.11 Organometallic compounds of gallium, indium, and thallium

There is a striking alternation of structures for the alkyl compounds of the first three elements in Group 13/III. As we have seen, the first member of the series, trimethylboron, is a monomer with three-coordinate B; the next, trimethylaluminum, is a dimer with four-coordinate Al; all the succeeding members (trimethylgallium, trimethylindium, and trimethylthallium) are monomeric in solution and in the gas phase. The decrease in carbanion character going down the group is evident in the progressive resistance to hydrolysis of the trialkyl compounds. In this connection, an interesting aspect of Tl chemistry is the tendency of $[CH_3TlCH_3]^+$ to exist as a linear isolated ion. A clue to the explanation is that the ion is isoelectronic and isostructural with CH_3HgCH_3.

Gallium

Trialkylgallium compounds can be synthesized by the reaction of alkyllithium compounds with $GaCl_3$ in a hydrocarbon solvent:

$$3Li_4(C_2H_5)_4 + 4GaCl_3 \rightarrow 12LiCl + 4Ga(C_2H_5)_3$$

Trialkylgallium compounds are mild Lewis acids, and so the corresponding metathesis in ether produces the complex $(C_2H_5)_2OGa(C_2H_5)_3$. Similarly, the use of excess Li reagent leads to the uptake of a fourth alkyl group by the Ga atom to form a salt:

$$Li_4(C_2H_5)_4 + GaCl_3 \rightarrow 3LiCl + Li[Ga(C_2H_5)_4]$$

Trialkylgallium compounds are pyrophoric and react with weak Brønsted acids such as water, alcohols, and thiols. In keeping with the decrease in carbanion character with the more electronegative metal atom, their reactivity toward Brønsted acids is somewhat less than that of alkylaluminum compounds:

$$n\text{Ga(CH}_3)_3 + n\text{H}_2\text{O} \rightarrow [(\text{CH}_3)_2\text{GaOH}]_n + n\text{CH}_4 \qquad n = 2, 3, \text{ or } 4$$

$$2\text{Ga(CH}_3)_3 + 2\text{ROH} \rightarrow [\text{Ga(CH}_3)_2\text{OR}]_2 + 2\text{CH}_4$$

The GaR_2 group is fairly resistant to protolysis, and complexes having the $Ga(CH_3)_2$ group (**15**) can be handled in aqueous solution.

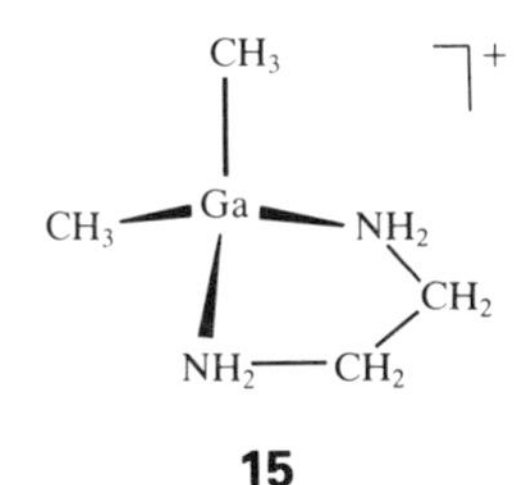

15

Indium and thallium

Alkylindium and alkylthallium compounds may be prepared by reactions analogous to those used to make the alkylgallium compounds. Trimethylindium is monomeric in the gas phase and in the solid the bond distances indicate that association is weak (if present at all).

A new aspect of organometallic chemistry becomes evident this deep in the boron group, for now the inert pair effect can play a role and give rise to stable indium(I) and thallium(I) organometallics. Examples of these compounds are $(\eta^5\text{-C}_5\text{H}_5)\text{In}$ and $(\eta^5\text{-C}_5\text{H}_5)\text{Tl}$, which exist as monomers in the gas phase but are associated as solids. Cyclopentadienylthallium is useful as a synthetic reagent in organometallic chemistry because it is not as highly reducing as $Na[C_5H_5]$ and the insolubility of TlCl provides an added driving force for metatheses. A drawback is that Tl is even more poisonous than Hg, so the disposal of the reaction by-products must be done with care.

Electron-precise compounds of the carbon group

In Group 14/IV we meet organometallics that are compounds formed by carbon with its own congeners. Since the electronegativity of C is similar to that of its congeners, the bonds they form to each other are not very polar. This low polarity in conjunction with the greater steric protection of the four-coordinate central atom appears to be responsible for the greater hydrolytic stability of the carbon-group organometallics compared with those of the boron group. The organometallic chemistry of Si, Sn, and Pb has been very extensively investigated, in part because of the variety of the applications of the compounds.

10.12 Organosilicon compounds

Structures and properties

In contrast to the M—C bonds of more electropositive elements such as Al, the Si—C bond is resistant both to hydrolysis and to air oxidation. Many oxo-bridged organosilicon compounds can be synthesized. Hexamethyldisiloxane $(CH_3)_3Si\text{—}O\text{—}Si(CH_3)_3$ is one example, and as with all simple organosilicon compounds, it is resistant to moisture and air. Two properties of these materials, the very weak Lewis basicity of the O atom and ready deformation of the Si—O—Si bond angle, have been rationalized by a model in which lone electron pairs on O are partially delocalized into vacant σ^* or d orbitals of Si (**16**). This delocalization reduces the directionality of

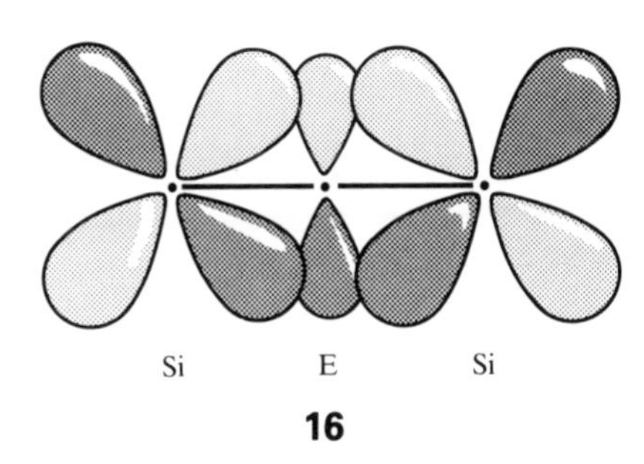

16

the Si—O single bond and hence makes the structure more flexible. This flexibility permits silicone elastomers to remain rubber-like down to very low temperatures. The delocalization also provides a means of decreasing the basicity of an attached O atom. The delocalization accounts for the properties of trisilylamine, $(SiH_3)_3N$, which is planar and very weakly basic.[8]

Another closely related observation is the relative ease of deprotonation of tetramethylsilane:

$$BuLi + Si(CH_3)_4 \rightarrow Li[CH_2Si(CH_3)_3] + BuH$$

The C atom is less basic than in simple hydrocarbons as a result of delocalization of electron density on to silicon.

Example 10.4: *Predicting the relative basicity of silyl ethers and amines*

Predict the relative Lewis basicity of (a) $(H_3Si)_2O$ and $(H_3C)_2O$ and (b) $(H_3Si)_3N$ and $(H_3C)_3N$.

Answer. If delocalization of the O or N lone pairs occurs as shown in (**16**), the silyl ether and silyl amine should be the weaker Lewis base in each pair. This prediction is borne out by the observation that whereas complexes of $B(CH_3)_3$ with dimethyl ether and trimethylamine are readily observed, their Si analogs do not form stable complexes.

Exercise. If π bonding between Si and the lone pairs of N is important, what difference in structure between $(H_3Si)_3N$ and $(H_3C)_3N$ do you expect?

Formation of Si—C bonds

A convenient way of linking alkyl groups to Group 14/IV elements is by metathesis of E—Cl using Grignard or alkyllithium reagents:

$$Li_4(CH_3)_4 + SiCl_4 \rightarrow 4LiCl + Si(CH_3)_4$$

Large quantities of methylchlorosilanes are needed in industry to synthesize silicone rubber and oils, and Grignard reagents are too expensive to be used on such a scale. It was to satisfy this need that Eugene Rochow (then—in the early 1940s—at General Electric in the USA) developed a process for their 'direct synthesis' from elemental silicon and an alkyl or aryl halide in the presence of copper as a catalyst:

$$Si + RX \xrightarrow{250\text{–}550^\circ C,\ Cu} R_nSiX_{4-n}$$

The conditions are usually adjusted to favor the formation of dimethyldichlorosilane, but other useful halosilanes are also produced. This relatively inexpensive direct process transformed silicone polymers from expensive laboratory curiosities to widely used materials.[9]

Redistribution reactions

Redistribution reactions are useful for the synthesis of a variety of Group 14/IV compounds. The reaction is readily performed in the laboratory and

[8] The interesting shapes of silyl molecules of N and O in the gas and solid phases have been reviewed by E. A. V. Ebsworth, *Acc. chem. Res.*, **20,** 295 (1987).

[9] A personal account of his discovery of the direct process is given by E. G. Rochow in his *Silicon and silicones.* Springer-Verlag, Berlin (1987).

is employed commercially to prepare halosilanes:

$$2SiCl_4 + Si(CH_3)_4 \rightarrow xSiCl(CH_3)_3 + ySiCl_2(CH_3)_2 + zSiCl_3(CH_3)$$

Traces of Lewis acids such as $AlCl_3$ are effective catalysts for this reaction. Since Si—Cl and Si—CH_3 bonds are being broken and then reformed, it might be expected that the product distribution from an initial 1:1 mixture of $SiCl_4$ and $Si(CH_3)_4$ would yield a statistical mixture of products:

$SiCl_4$:	$SiCl_3(CH_3)$:	$SiCl_2(CH_3)_2$:	$SiCl(CH_3)_3$:	$Si(CH_3)_4$
1	4	6	4	1

However, subtle aspects of the bonding and steric interactions often lead to nonstatistical mixtures.

Metatheses using organohalosilane, germane, and stannane compounds are very useful for laboratory-scale syntheses, especially for preparing compounds with mixed organic substituents:

$$4SiCl_3(C_6H_5) + 3Li_4(CH_3)_4 \rightarrow 4Si(CH_3)_3(C_6H_5) + 12LiCl$$

Studies of the kinetics of substitution reactions on silicon suggest that the mechanisms of these reactions are associative (Section 7.8). Thus the rate law is generally second-order overall:

$$\text{rate} = k[SiR_3X][Y]$$

and k depends on the identity of the entering group Y. That redistribution reactions on silicon and heavier members of the group are more facile than those on carbon also suggests that Si forms a 5-coordinate activated complex more readily than does C. This is in line with the existence of many more five-coordinate Si inorganic compounds than five-coordinate C compounds. (The latter are largely confined to carboranes and metal cluster compounds.)

Stereochemical investigations reveal that substitution reactions may proceed with either inversion or retention of configuration. The mechanism that accounts for inversion is very similar to that proposed for associative substitutions on C, with the entering and leaving groups occupying the axial positions in a trigonal bipyramidal activated complex:

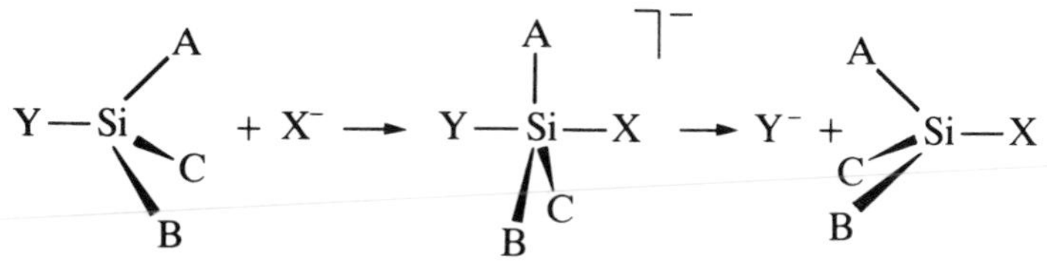

$$Y\text{—}SiABC + X^- \rightarrow [Y\text{—}SiABC\text{—}X]^- \rightarrow Y^- + ABCSi\text{—}X$$

Retention is common when Y is a poor leaving group, such as H^- or OR^-. This kind of reaction is believed to involve fluxionality (Section 2.2) of a five-coordinate intermediate that places Y on the pseudo threefold axis and *trans* to one of the original substituents:

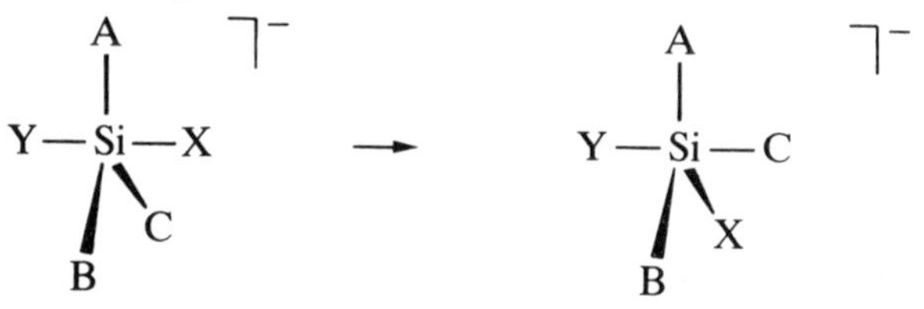

$$[Y\text{—}Si(A)(B)(C)\text{—}X]^- \rightarrow [Y\text{—}Si(A)(B)(X)\text{—}C]^-$$

The rearrangement is then followed by departure of the leaving group:

$$[\text{Y—Si(A)(B)(X)—C}]^- \rightarrow \text{Y—Si(A)(B)(X)} + \text{C}^-$$

The ability of the Si atom to form a five-coordinate intermediate with sufficiently long lifetime to rearrange results in retention of configuration in some nucleophilic displacement reactions. On the other hand, the smaller C atom, which has a much less stable five-coordinate activated complex, immediately undergoes inversion.

Protolysis of silyl halides

The protolysis of Si—X bonds is a convenient route to a myriad of compounds, particularly **siloxanes**, which are compounds with Si—O bonds. Thus Si—Cl bonds are susceptible to protolysis by H—O, H—N, and H—S bonds, but the Si—C bond is not. This difference accounts, for instance, for the formation of hexamethyldisiloxane in the reaction of water with chlorotrimethylsilane. The initial reaction is the hydrolysis of the Si—Cl bond:

$$2(CH_3)_3SiCl + 2H_2O \rightarrow 2[(CH_3)_3SiOH] + 2HCl$$

This is followed by a slower reaction, the elimination of water to form the Si—O—Si link:

$$2[(CH_3)_3SiOH] \rightarrow (CH_3)_3Si\text{—}O\text{—}Si(CH_3)_3 + H_2O$$

The condensation of the SiOH compound is analogous to the transformation of metal hydroxide complexes to polycations (Section 5.8) and to the polymerization of $Si(OH)_4$ in aqueous solution to produce silica gel. These reactions demonstrate the tendency of Si—OH groups to eliminate water.

When dimethylsilicon dichloride is hydrolyzed, both cyclic (**17**) and long-chain compounds are produced:

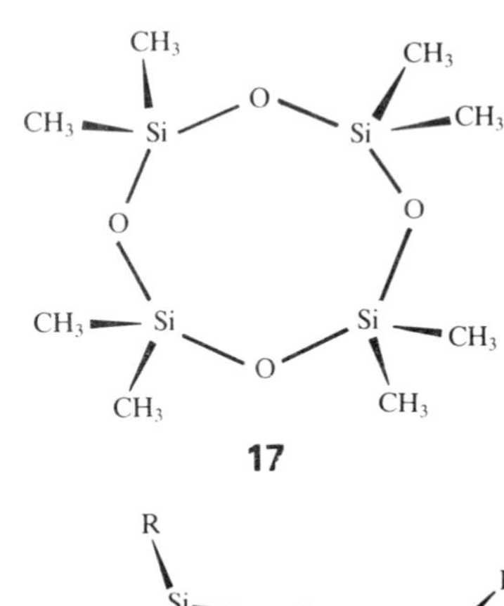

17

$$(CH_3)_2SiCl_2 + H_2O \rightarrow HO[Si(CH_3)_2O]_nH + [(CH_3)_2SiO]_4 + \text{etc}$$

When $RSiCl_3$ is used as a starting material and contains bulky organic groups, a variety of more elaborate structures are possible, including cage compounds (**18**). Similar reactions occur with ammonia and with primary and secondary amines to produce a variety of **silazanes**, compounds with Si—N bonds. For example, excess $(CH_3)_3SiCl$ reacts with NH_3 to produce $((CH_3)_3Si)_2NH$. In this case, the protolysis is incomplete because of the bulky groups around the Si atom; the less hindered $HSi(CH_3)_2Cl$ yields $((CH_3)_2HSi)_3N$.

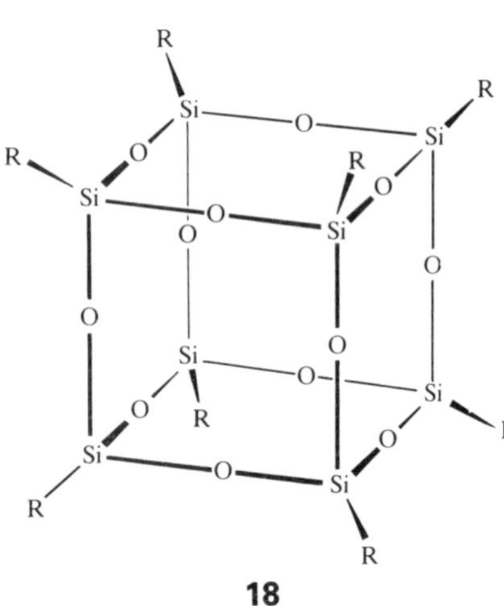

18

Siloxanes undergo redistribution reactions to form polymers, the **silicones**, in the presence of sulfuric acid as a catalyst:

$$n[(CH_3)_2SiO]_4 + (CH_3)_3SiOSi(CH_3)_3 \xrightarrow{H_2SO_4} (CH_3)_3SiO[Si(CH_3)_2O]_{4n}Si(CH_3)_3$$

In this reaction, the hexamethyldisiloxane provides the $Si(CH_3)_3$ end groups, so the higher the proportion of it present, the lower the molar mass of the resulting polymer. Silicones, including poly(dimethylsiloxane), are elastomers, and are produced on a large scale. Their useful properties include their flexibility at low temperatures, their ability to repel water, and their resistance to air oxidation. Additionally, their low toxicity leads to their use in medical and cosmetic implants.

Example 10.5: ***Contrasting the ease of hydrolytic cleavage of Al—CH_3 and Si—CH_3 bonds***

Give equations for the possible reactions of trimethylaluminum and tetramethylsilane with water, and identify the difference in bonding in the two compounds that accounts for the difference in behavior.

Answer. Organoaluminum compounds have considerable carbanion character, which correlates with the highly electropositive character of Al. This bonding description is consistent with the experimental observation that trimethylaluminum hydrolyzes rapidly on contact with water:

$$Al_2(CH_3)_6 + 6H_2O \rightarrow 2Al(OH)_3 + 6CH_4$$

As mentioned in Section 10.5, this reaction probably occurs by initial coordination of the O atom in H_2O to the Al atom of the alkylaluminum. In contrast, Si and C have similar electronegativities and they form covalent bonds of low polarity. Their low polarity and relative steric congestion prevents coordination prior to hydrolysis. This resistance to hydrolysis is illustrated by the use of silicone polymers as waterproofing agents on cloth or leather and the use of silicone cements to isolate electrical and electronic devices from water.

Exercise. Contrast the hydrolysis reactions of $Al_2(CH_3)_6$ and $Ga(CH_3)_3$.

Catenated compounds

Catenation means the formation of chains, such as $R_3Si—SiR_3$ and $R_3Si—SiR_2—SiR_3$. Considerable progress has been made in the synthesis of compounds containing the Si—Si bond, which is weaker than the C—C bond but not dramatically so. Open-chain, cyclic, bicyclic, and cage alkylsilicon compounds are known (**19**, **20**). These compounds are prepared by a reductive halide elimination reaction:

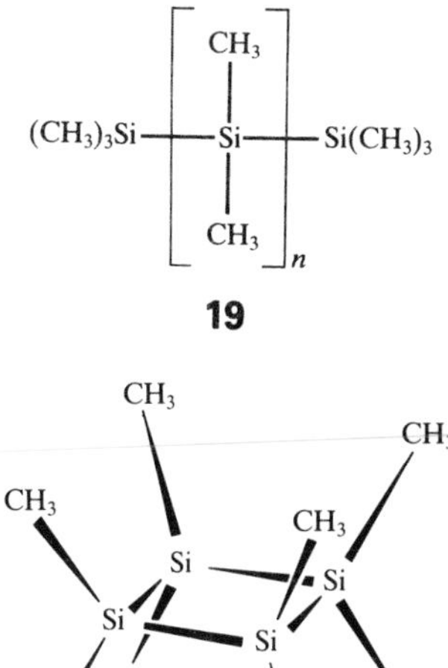

19

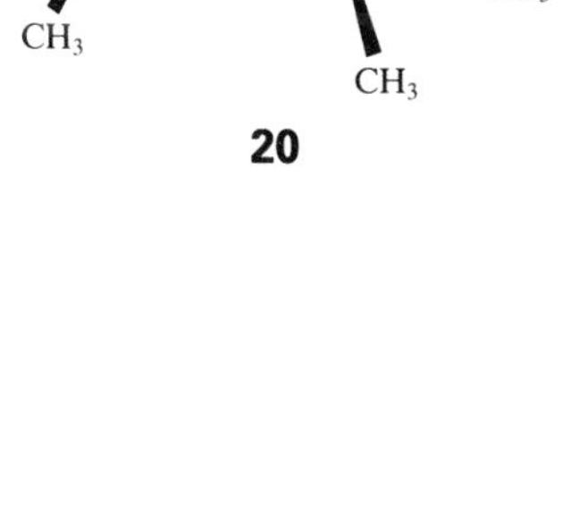

20

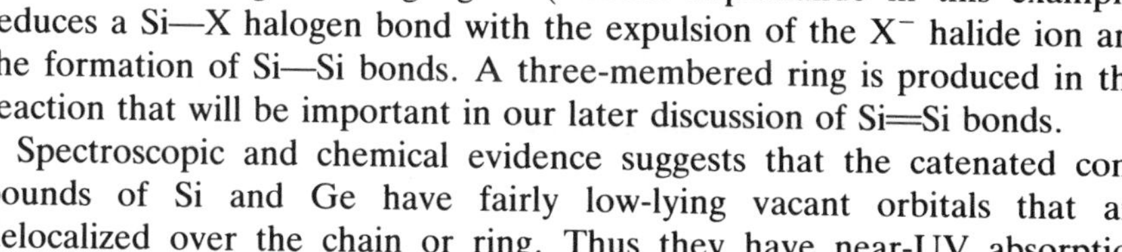

$$3(Xyl)_2SiCl_2 + 6Li[C_{10}H_8] \xrightarrow[-78°C]{(CH_3)OC_2H_4O(CH_3)} \underset{\diagdown\;\diagup \atop Si(Xyl)_2}{(Xyl)_2Si—Si(Xyl)_2} + 6LiCl + 6C_{10}H_8$$

In this equation, Xyl is the xylyl group (2,6-dimethylphenyl). In this type of reaction, a strong reducing agent (lithium naphthalide in this example) reduces a Si—X halogen bond with the expulsion of the X^- halide ion and the formation of Si—Si bonds. A three-membered ring is produced in this reaction that will be important in our later discussion of Si=Si bonds.

Spectroscopic and chemical evidence suggests that the catenated compounds of Si and Ge have fairly low-lying vacant orbitals that are delocalized over the chain or ring. Thus they have near-UV absorption bands that decrease in energy with increasing chain length. There is also evidence that this electronic transition is the promotion of an electron from a σ-bonding molecular orbital of the Si—Si or Ge—Ge backbone into an excited σ^*-orbital that is also delocalized along the Si—Si or Ge—Ge chain.

Another area of silicon chemistry that emphasizes the delocalized nature of the σ orbitals is the formation of cyclic permethylated silanes by controlled reduction. This produces a radical anion in which the odd electron occupies a delocalized Si—Si σ^* orbital:

$$Si_6(CH_3)_{12} + Na \xrightarrow{R_2O} Na^+[Si_6(CH_3)_{12}]^-$$

The EPR spectrum of the anion shows that the unpaired electron occurs with equal probability on all six Si atoms, proving that the electron is fully delocalized (at least on the timescale of this experiment, which is about 1 μs.)

Multiply-bonded compounds

All attempts to synthesize stable compounds having Si=C or Si=Si bonds were unsuccessful until recently. As is the case for carbon, two single Si—Si bonds are stronger than one Si=Si double bond, so there is a thermodynamic tendency for Si=C or Si=Si groups to couple together. Carbon C=C bonds are protected from coupling by the high activation energy of the reaction, but there is much less kinetic protection for multiple bonds involving Si or Ge.

A clue that multiply-bonded Si compounds might exist came from the detection of transient compounds such as $(CH_3)_2Si{=}CH_2$ in the gas phase and in low-temperature inert gas matrices. However, an extensive chemistry was not developed around this molecule because it rapidly forms dimers and polymers:

$$2(CH_3)_2Si{=}CH_2 \longrightarrow \begin{array}{c}(H_3C)_2Si{-}CH_2\\ \vert \quad\;\; \vert \\ H_2C{-}Si(CH_3)_2\end{array}$$

Success had to await the discovery that bulky substituents block dimerization and polymerization. Compounds containing Si=C bonds, the **silenes**, and Si=Si bonds, the **disilenes**, are now known. These new silenes and disilenes not only extend our understanding of chemical bonding but they are also useful synthetically.[10]

Synthesis, structure, and properties

Photochemical cleavage of Si—Si bonds, following excitation into an antibonding Si—Si σ orbital, provides a convenient route to disilenes. An example is the photolysis of a cyclic trisilane with bulky substituents:

$$2\begin{array}{c}(Xyl)_2Si{-}Si(Xyl)_2\\ \diagdown\;\diagup\\ Si(Xyl)_2\end{array} \xrightarrow{h\nu} 3(Xyl)_2Si{=}Si(Xyl)_2$$

The bulky groups on the Si atoms prevent cyclization and polymerization, and this leads to a stable disilene product.

[10] Silene chemistry is reviewed by A. G. Brook and K. M. Baines, *Adv. organomet. Chem.*, **25**, 1 (1986). A helpful survey of disilene chemistry is given by R. West in *Angew. Chem. Int. Ed. Engl.*, **26**, 1201 (1987).

X-ray structure determinations show the expected bond length decrease from Si—Si. For example, tetramesityldisilene (Fig. 10.5) has an Si═Si bond length of 2.16 Å, which is about 0.20 Å shorter than a typical Si—Si single bond. A significant difference between the disilenes and alkenes, however, is the greater ease of distorting the $R_2Si{=}SiR_2$ group from planarity. We shall see that this tendency is even more pronounced in heavier Group 14/IV analogs, and an explanation in terms of bonding will be given there.

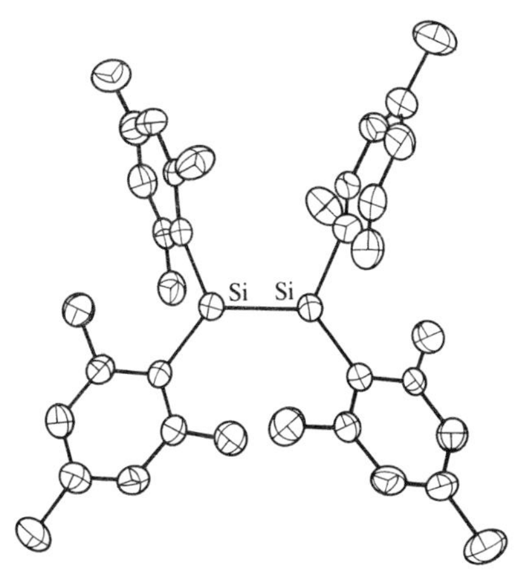

Fig. 10.5 An ORTEP diagram of the structure of $(Mes)_2Si{=}Si(Mes)_2$ as determined by X-ray single crystal diffraction. (From M. J. Fink, M. J. Michalczyk, K. J. Haller, R. West, and J. Michl, *Organometallics*, **3**, 793 (1984).) Note how the bulky mesityl groups shield the Si atoms.

The electronic absorption spectra of the disilenes contain a band in the visible or near-UV region, and the compounds are often brightly colored. This indicates that the π and π^* orbitals are closer in energy than in the alkenes, in which the corresponding absorption is well into the UV (Fig. 10.6).

The energies of the bonding and antibonding π orbitals can be inferred from oxidation and reduction potentials in solution. It is found that the oxidation of disilenes occurs at less positive potentials and reduction at less negative potentials than for corresponding alkenes. As depicted in Fig. 10.6, this indicates that the energies of π and π^* orbitals of the disilenes are between those of the corresponding alkenes. As a result, disilenes are better Lewis π bases (from the filled π orbital) and better Lewis π acids (into the π^* orbital).

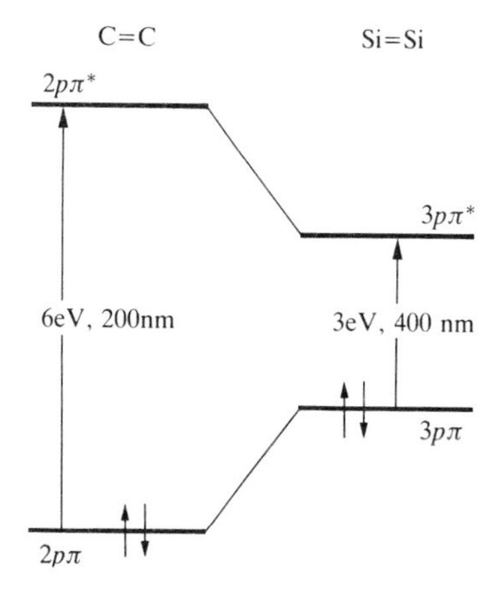

Fig. 10.6 Approximate energy levels for alkenes and disilenes. (From R. West, *Angew. Chem. Int. Ed. Engl.*, **26**, 1201 (1987).)

Molecules that are not sterically encumbered have access to the Si═Si bond in disilenes, and some of the reactions are very similar to those found for alkenes. For example hydrogen halides and halogens add across the Si═Si double bond:

$$R_2Si{=}SiR_2 + HX \longrightarrow \begin{array}{c} \mathrm{H}\quad\mathrm{X} \\ |\quad\;| \\ \mathrm{R{-}Si{-}Si{-}R} \\ |\quad\;| \\ \mathrm{R}\quad\mathrm{R} \end{array}$$

$$R_2Si{=}SiR_2 + X_2 \longrightarrow \begin{array}{c} \mathrm{X}\quad\mathrm{X} \\ |\quad\;| \\ \mathrm{R{-}Si{-}Si{-}R} \\ |\quad\;| \\ \mathrm{R}\quad\mathrm{R} \end{array}$$

where X = Cl or Br. More interesting, perhaps, are the contrasts with the chemistry of alkenes. Unlike alkenes, disilenes undergo addition of ROH across the Si═Si bond:

$$R_2Si{=}SiR_2 + R'OH \longrightarrow \begin{array}{c} \mathrm{R'O}\quad\mathrm{H} \\ |\quad\;| \\ \mathrm{R{-}Si{-}Si{-}R} \\ |\quad\;| \\ \mathrm{R}\quad\mathrm{R} \end{array}$$

They also undergo 2 + 2 additions with some alkynes:

$$R_2Si{=}SiR_2 + R'C{\equiv}CH \longrightarrow \begin{array}{c} \mathrm{R'}\qquad\quad\mathrm{H} \\ \mathrm{C{=}C} \\ |\quad\;| \\ \mathrm{R{-}Si{-}Si{-}R} \\ |\quad\;| \\ \mathrm{R}\quad\mathrm{R} \end{array}$$

10.13 Organometallic compounds of germanium, tin, and lead

Many of the reactions of organotin compounds (organostannanes) and organolead compounds (organoplumbanes) are similar to those already mentioned for Si and Ge. One major difference from organosilicon chemistry, however, is the existence (because of the inert-pair effect) of some Ge(II), Sn(II), and Pb(II) organometallics. Another contrast with Si—C chemistry results from the rapid decrease in E—C bond strength down the group (Table 10.2). Because of the latter, organolead compounds generally decompose above 100°C.

The gas-phase decomposition of alkyllead compounds occurs with the formation of alkyl radicals:

$$Pb(CH_3)_4(g) \xrightarrow{\Delta} Pb(s) + 4CH_3\cdot(g)$$

This decomposition was the reason why, for many years, tetramethyllead and tetraethyllead were added to gasoline to improve its octane rating. The alkyl radicals generated by the decomposition of the tetraalkyllead in the hot combustion chamber of the engine are radical chain terminators which reduce the tendency toward explosive combustion. The toxicity of lead, however, has prompted the elimination of organolead additives from gasoline in many parts of the world.

Organotin compounds find many different applications, ranging from stabilizers for poly(vinyl chloride) plastics to fungicides and antifouling paints for the hulls of boats. Recently, though, some of these applications have come under close scrutiny because they may harm benign and desirable organisms: organotin compounds kill not only barnacles but oysters too.

In addition to the simple monomeric organogermanium and organotin compounds, catenated compounds are known (**21** and **22**), and synthetic routes to these compounds resemble those used for their silicon analogs.

$(CH_3)_3Ge-[Ge(CH_3)_2]_n-Ge(CH_3)_3$

21

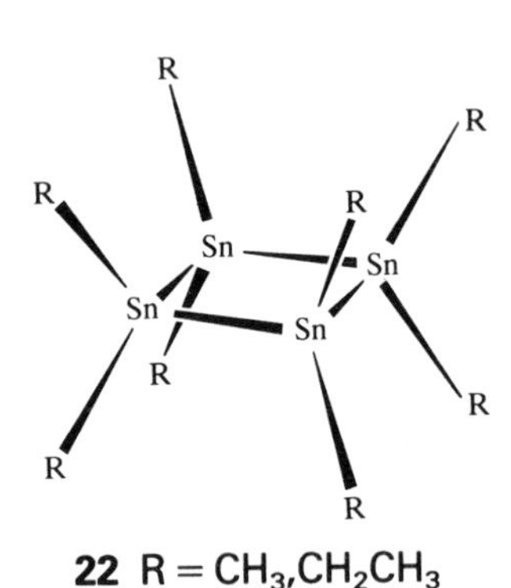

22 R = CH_3,CH_2CH_3

Germenes ($R_2Ge{=}CR_2$), stannenes ($R_2Sn{=}CR_2$), digermenes ($R_2Ge{=}GeR_2$), and distannenes ($R_2Sn{=}SnR_2$) have recently all been prepared. As with the corresponding Si compounds, it is necessary for the R groups to be bulky to prevent association. The digermenes and distannenes are distinctly nonplanar, and several bonding explanations have been proposed. In one of them, the nonplanarity is attributed to an unconventional pattern of multiple bonding, which is between the sp^2 σ orbital on one atom with a p π orbital on the other (**23**). The Ge—Ge and Sn—Sn multiple bonds are long, and dissociation has been observed in solution to form the carbene-like divalent compounds GeR_2 and SnR_2. As expected from the increased stability of the divalent state going down the group, the SnR_2 species are more stable than their GeR_2 counterparts. The divalent state is also seen in $Sn(\eta^5\text{-}C_5H_5)_2$ and $Pb(\eta^5\text{-}C_5H_5)_2$ which have angular structures in the gas phase (**24**), indicating that a stereochemically active lone pair may be present on the metal.

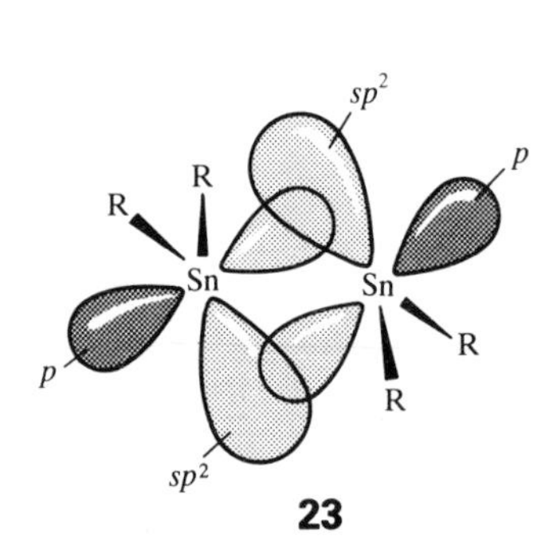

23

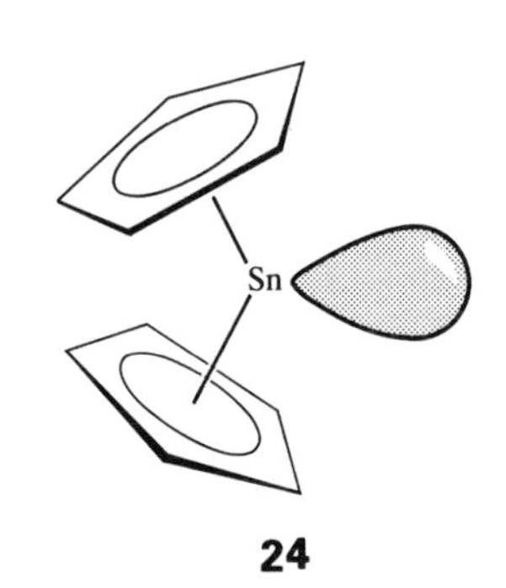

24

Electron-rich compounds of the nitrogen group

Since the first two members of the nitrogen group (N or P) are nonmetals, we do not consider them here. Recently, there has been significant work on the organic derivatives of the heavier elements in the group, and it has resulted in interesting new compounds with unusual bonding patterns and oxidation numbers as well as the common oxidation numbers +3 and +5 for the group (**6**, **25**). There are many similarities in the organic chemistry of these elements and we shall concentrate on the chemistry of one of them, arsenic.

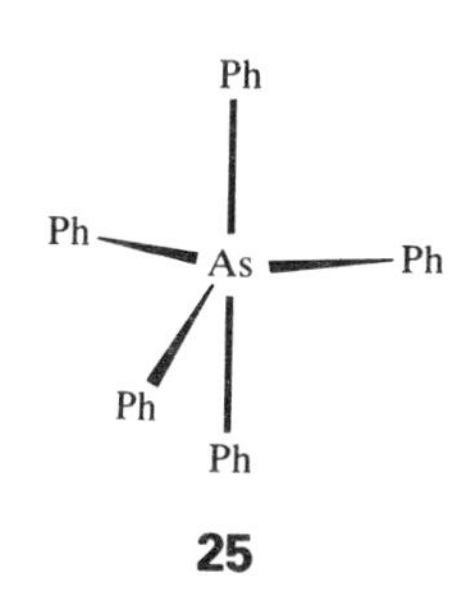

25

10.14 Noncatenated compounds of arsenic

The aryl and alkyl arsines, such as trimethylarsine (**6**), are employed as ligands in *d*-metal chemistry. As expected for a heavier element, the As(III) atom in these compounds is a weaker and softer base than is the P(III) atom in the corresponding phosphines. The trialkylarsines also act as nucleophiles toward haloalkanes to produce tetraalkylarsonium salts, which contain As(V):

$$As(CH_3)_3 + (CH_3)Br \rightarrow [As(CH_3)_4]Br \qquad (9)$$

This type of reaction cannot be used for the preparation of the tetraphenylarsonium ion, $[AsPh_4]^+$, because triphenylarsine is a much weaker nucleophile than trimethylarsine. Instead, the synthesis reaction

$$Ph_3As{=}O + PhMgBr \rightarrow [AsPh_4]Br + MgO$$

is used. This reaction may look unfamiliar, but we can view it as a metathesis in which the Ph^- anion replaces the formal O^{2-} ion attached to the As(V) atom and results in a compound in which the arsenic is formally As(V).

The tetraphenylarsonium (as well as tetraalkylammonium and tetraphenylphosphonium) cation is sometimes used in synthetic inorganic chemistry as a bulky cation to stabilize bulky complex anions. We shall see below that it is also a starting material in the preparation of other As(V)

Example 10.6: *Correlating oxidation numbers and stabilities*

Describe the stability of the alkyl compounds of the adjacent elements Ge and As with (a) their group oxidation number and (b) the group oxidation number less 2. Correlate these stabilities with the stability of the oxidation states for these elements.

Answer. The most common oxidation number for organogermanium compounds is +4, the group oxidation number. Only a few organogermanium(II) compounds are known. In contrast, As(V) compounds such as $As(CH_3)_5$ are unstable and the lower organoarsenic(III) compounds, such as $As(CH_3)_3$, are the most common and most stable. One way of explaining this trend is that on going to the right along a period, the elements become more electronegative, and it becomes progressively more difficult to bring out their high oxidation states. This is particularly so with a substituent, such as CH_3, with a low electronegativity. These trends in the stability of oxidation states follow the trends for simple inorganic compounds of germanium and arsenic.

Exercise. Compare formulas of the most stable hydrogen compounds of germanium and arsenic (Chapter 9) with those of their methyl compounds. Is the result understandable in terms of the relative electronegativity of C and H?

26

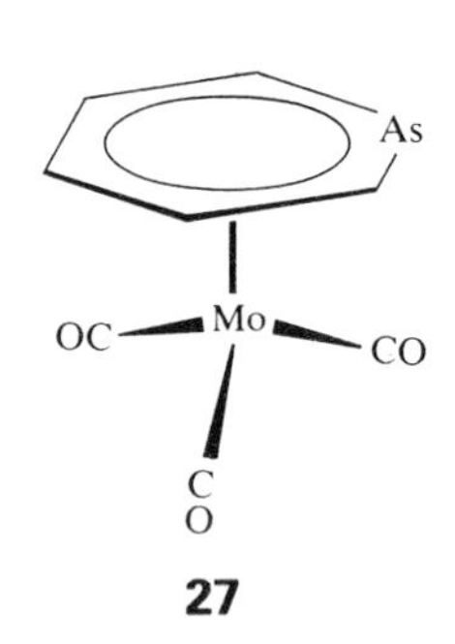

27

organometallics. One of the more interesting compounds of As(III) is arsabenzene (**26**), an analog of pyridine. It lacks the σ-basicity of pyridine but is known to form organometallic π complexes analogous to those formed by benzene (**27**).[11]

The action of phenyllithium on a tetraphenylarsonium salt produces the neutral compound pentaphenylarsenic, a compound (formally) of As(V):

$$[AsPh_4]Br + LiPh \rightarrow AsPh_5 + LiBr$$

Pentaphenylarsine is trigonal bipyramidal, as expected from VSEPR considerations. We have seen (Section 2.2) that a square pyramidal structure is often close in energy to the trigonally bipyramidal structure, and $SbPh_5$ is in fact square pyramidal. A similar reaction under carefully controlled conditions yields the unstable compound $As(CH_3)_5$.

10.15 Catenated singly and multiply bonded compounds

Tetramethyldiarsine, $(CH_3)_2AsAs(CH_3)_2$, a catenated compound, was one of the first organometallic compounds to be made. Cyclic arsines are also known, including $(PhAs)_6$. The synthesis of $(PhAs)_6$ is accomplished by abstracting iodine from phenylarsenic diiodide:

$$6PhAsI_2 + 12Hg \rightarrow (PhAs)_6 + 6Hg_2I_2$$

Among the other As—As bonded compounds is one containing an As_3 ring (**28**).

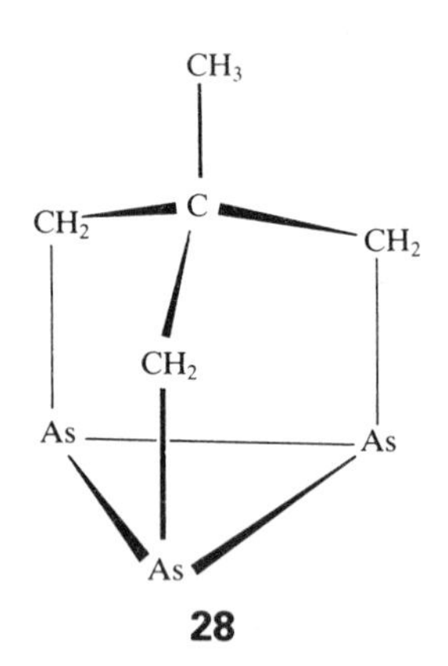

28

As with the disilenes, it has been found recently that bulky substituents make it possible to prevent the polymerization and cyclization of E=E groups, and this strategy leads to formal double-bonded systems of formula RE=ER, where E may be P, As, or Sb, including the compound shown as (**29**).

29 (E,E′) = (P,As),(P,Sb),(As,P)

Further reading

J. S. Thayer, *Organometallic chemistry, an overview*. VCH, Weinheim (1988); R. West and T. J. Barton, *J. chem. Educ.*, **57**, 165 and 334 (1980). These references provide accessible introductory material. The first surveys the organometallic compounds of the *s* and *p* blocks and their associated biological activity. The second pair of articles is a primer of organosilicon chemistry.

G. Wilkinson, F. G. A. Stone, and E. W. Abel (eds.), *Comprehensive organometallic chemistry*. Pergamon, Oxford (1982). Volumes 1 and 2 are devoted to the *s* and *p* blocks. Chapter 1, *Main group structure and bonding relationships* (**1**, 1) by M. E. O'Neill and K. Wade, provides a good introduction, and later chapters provide thorough coverage.

B. J. Aylett, *Organometallic compounds*. Chapman and Hall, London (1979). Group 14/IV and 15/V organometallics.

J. J. Eisch, *Organometallic syntheses* **2.** Academic Press, New York (1981). The volume provides a thorough description of methods and detailed procedures for the synthesis of important *p*-block organoelement compounds.

A. H. Cowley and N. C. Norman, The synthesis, properties and reactivities of stable compounds featuring double bonds between heavier group 14 and 15 elements. *Prog. Inorg. Chem.*, **34**, 1 (1986). The title implies the breadth of this review of the new discoveries in multiply bonded compounds.

[11] Other interesting chemical and structural properties of this series of heterocycles are described by A. J. Ashe, *Acc. chem. Res.*, **11**, 153 (1978).

Exercises

10.1 Explain why certain of the following compounds qualify as organometallic compounds while others do not: (a) $B(CH_3)_3$, (b) $B(OCH_3)_3$, (c) $(NaCH_3)_4$, (d) $SiCl_3(CH_3)$, (e) $N(CH_3)_3$, (f) sodium acetate, (g) $Na[B(C_6H_5)_4]$.

10.2 Without consulting reference material, construct the periodic table for the *s*- and *p*-block elements and the Group 12 elements. Head each group with the formula of a representative methyl compound and indicate (a) the positions of the ionic methyl compounds, (b) the positions of the electron-deficient, electron-precise, and electron-rich methyl compounds, (c) trends in $\Delta H_f^{\ominus}$ in the *p* block.

10.3 Write formulas for each of the following compounds and for nonionic compounds give an alternative name (i.e., as derivatives of their hydrogen compounds if they were named as organometallic compounds and vice versa): (a) trimethylbismuth, (b) tetraphenylsilane, (c) tetraphenylarsonium bromide, (d) potassium tetraphenylborate.

10.4 Name each of the following compounds, classify them, and sketch the structure of each molecular compound: (a) $SiH(C_2H_5)_3$, (b) $BCl(C_6H_5)_2$, (c) $Al_2Cl_2(C_6H_5)_4$, (d) $Li_4(C_2H_5)_4$, (e) $RbCH_3$.

10.5 Describe the tendency toward association through methyl bridges for the trimethyl compounds of B, Al, Ga, and In. Explain the difference between B and Al.

10.6 Sketch the structures of each of (a) methyllithium, (b) trimethylboron, (c) hexamethyldialuminum, (d) tetramethylsilane, (e) trimethylarsene.

10.7 For each of the following compounds, indicate those that may serve as (1) good carbanion nucleophile reagents, (2) mild Lewis acids, (3) mild Lewis bases at the central atom, (4) strong reducing agents. (A compound may have more than one of these properties.) (a) $Li_4(CH_3)_4$, (b) $Zn(CH_3)_2$, (c) $(CH_3)MgBr$, (d) $B(CH_3)_3$, (e) $Al_2(CH_3)_6$, (f) $Si(CH_3)_4$, (g) $As(CH_3)_3$.

10.8 For an appropriate compound from Exercise 10.7, give balanced chemical equations for: (a) the reactions of one of the carbanion reagents with $AsCl_3$, and with $SiPh_2Cl_2$, (b) the reaction of a Lewis acid with NH_3, (c) the reaction of a Lewis base with the Lewis acid $[HgCH_3][BF_4]$.

10.9 Give examples of each of the following reaction types and describe the factors that favor reaction in each case: (a) reaction of a metal with an organic halide, (b) transmetallation, (c) metathesis.

10.10 Determine which compound in (a) $Na[C_{10}H_8]$ and $Na[C_{14}H_{10}]$, (b) $Na[C_{10}H_8]$ and $Na_2[C_{10}H_8]$ (where $C_{10}H_8 =$ naphthalene, $C_{14}H_{10} =$ anthracene) is likely to be the stronger reducing agent and explain the physical basis for your answer.

10.11 Determine the potential reaction type, write a reasonable balanced chemical equation (or NR for no reaction) and explain the systematics of the reaction type that led you decide whether or not a reaction is likely:
(a) Ca with dimethylmercury
(b) Hg with diethylzinc
(c) Methyllithium with triphenylchlorosilane in ether
(d) Tetramethylsilane with zinc chloride in ether
(e) $HSi(CH_3)_3$ with ethylene in a solution of chloroplatinic acid in isopropyl alcohol.

10.12 Give balanced chemical equations that illustrate:
(a) Direct synthesis of $SiCl_2(CH_3)_2$
(b) Redistribution of $SiCl_2(CH_3)_2$

10.13 Using Si and a chloromethane as the primary starting materials, give equations and conditions for the synthesis of a poly(dimethylsiloxane).

10.14 Disregarding compounds with metal–metal bonds, describe the trends in the oxidation states for the organometallics of Groups 13/III through 15/V.

10.15 For the simple organometallic compounds, briefly describe the periodic trends in (a) metal–carbon bond enthalpies, (b) Lewis acidity, (c) hydrolytic stability for Groups 1, 2, 12, 13/III, and 14/IV.

10.16 Summarize the trend in each process listed below, giving your reasoning:
(a) The relative ease of pyrolysis of $Si(CH_3)_4$ and $Sn(CH_3)_4$ at 300°C.
(b) The relative Lewis acidity of $Li_4(CH_3)_4$, $B(CH_3)_3$, $Si(CH_3)_4$ and $Si(CH_3)Cl_3$.
(c) The relative Lewis basicity of $Si(CH_3)_4$ and $As(CH_3)_3$
(d) The tendency of $Li_4(CH_3)_4$ and $Hg(CH_3)_2$ to displace halide from $GeCl_4$.

10.17 Taking into account sensitivity to oxygen and moisture and the volatility of the compound, indicate the general techniques (vacuum line, inert atmosphere, Schlenk apparatus, or open flasks) that are most appropriate for handling the following organometallics: (a) $Li_4(CH_3)_4$ in ether, (b) trimethylboron, (c) tri-isobutylaluminum, (d) $AsPh_3$ (solid), (e) $(CH_3)_6SiOSi(CH_3)_3$ (liquid).

10.18 The 'saturated' cyclic compound $Si_6(CH_3)_{12}$ displays a near-UV absorption band and it can be reduced by Na to produce a cyclic radical anion $[Si_6(CH_3)_{12}]^-$. Interpret these phenomena in terms of the electronic structure of $Si_6(CH_3)_{12}$.

10.19 Give equations and conditions for the synthesis of $R_2Si{=}SiR_2$ from R_2SiCl_2 and indicate the type of R group that is necessary to yield a stable product.

Problems

10.1 The bulk price per mole of trimethylaluminum is an order of magnitude greater than that of triethylaluminum. Describe the methods of synthesis available for these two compounds and from an approximate consideration of the price of the starting materials explain the price differential of the compounds.

10.2 Based on general trends in atomic energy levels, explain the probable order of the π–π^* separation in $R_2Si{=}SiR_2$ and $R_2Ge{=}GeR_2$. Give specific examples of how this energy order might be reflected in the reactions of digermene in comparison with disilenes.

10.3 Although tetraalkyltin compounds are not Lewis acids, the alkyltin halides are. C. Yoder and coworkers have determined thermodynamic parameters of complex formation of triphenylphosphine oxide with a series of trialkyltin halides (*Organometallics,* **5,** 118 (1986)). Summarize the trends in enthalpies of complex formation and discuss reasonable explanations for these trends.

10.4 The synthesis of the first stable germene was reported by C. Couret *et al., J. Am. chem. Soc.,* **109,** 4411 (1987). Summarize the reactions involved in this synthesis and speculate on why each reaction is favorable. Describe the interaction of Lewis bases with this germene.

The boron and carbon groups

11

This chapter and the following two introduce the general chemical properties of the *p*-block elements. The *p* block is a very rich region of the periodic table, for its members show a much greater variation in properties than those of the *s* and *d* blocks, and range from metals such as aluminum to the highly electronegative nonmetals such as fluorine. A single viewpoint cannot cope adequately with this great diversity, and we shall see that we need to adjust our perspective as we travel across the block. For instance, as we travel toward the right we shall see that the number of oxidation states available to the elements increases, so redox properties become more important. This is in contrast to the elements on the upper left of the block (boron, carbon, and silicon) for which redox reactions are less important. However, these elements make up for lack of richness in their redox properties by their ability to bind to themselves to form chains, rings, and clusters. Compounds of the elements with oxygen are important throughout the block, and will frequently recur in the discussion.

General properties of the elements
11.1 Physical and chemical properties
11.2 Production
The boron group
11.3 Compounds of boron with the electronegative elements
11.4 Aluminum and gallium
11.5 Indium, thallium, and low oxidation state gallium
The carbon group
11.6 Carbon
11.7 Silicon and germanium
11.8 Tin and lead
Silicates and aluminosilicates
11.9 Extended silicate structures
11.10 Aluminosilicates
11.11 Molecular sieves
Boron cluster compounds and carboranes
11.12 Boron subhalides
11.13 Higher boron hydrides
11.14 Carboranes
Borides, carbides, and silicides
11.15 Borides
11.16 Carbides
11.17 Silicides
Electronic properties
11.18 Band gaps and conductivities
11.19 Optical properties
Further reading
Exercises
Problems

12	13	14	15
	B	C	N
	Al	Si	P
Zn	Ga	Ge	As
Cd	In	Sn	Sb
Hg	Tl	Pb	Bi
	III	IV	V

The elements of Groups 13/III and 14/IV have interesting and diverse physical and chemical properties as well as considerable importance to industry and in nature. Carbon, of course, plays a central role in organic chemistry, but it also forms many binary and organometallic compounds (which we discuss in Chapters 10 and 16). In combination with oxygen, carbon's congener silicon dominates inorganic natural products just as carbon in combination with hydrogen dominates organic natural products. The other elements of these two groups are also vital to modern high technology, particularly as semiconductors.

General properties of the elements

The elements of the two groups show a wide variation in abundance in crustal rocks, the oceans, and the atmosphere. Carbon, aluminum, and silicon are all abundant, but the low cosmic abundance of boron, like that of lithium and beryllium, reflects how these light elements are sidestepped in nucleosynthesis (Section 1.1). The low abundance of heavier members of both groups is in keeping with the progressive decrease in nuclear stability of the elements that follow iron (Section 1.2). Except for germanium, all the carbon-group elements are more abundant than adjacent members of the boron and nitrogen groups. This difference stems from the greater stability of nuclei that have even numbers of protons (and therefore even atomic number) compared with those having odd numbers.

11.1 Physical and chemical properties

The *p*-block elements show a wide range of physical properties (Table 11.1). This is particularly true in the two groups we consider here, for in them the lightest members are nonmetals and the heaviest are metals. The physical similarities are particularly strong between boron and its two nearby diagonal neighbors, silicon and germanium. For example, all three are hard, semiconducting solids. The occurrence of two or more significantly different polymorphs is a common characteristic of *p*-block elements, and is well illustrated by elemental boron and carbon (as we describe below). The only members of the two groups that crystallize in the close-packed structures typical of metals are aluminum, thallium, and lead.

The chemical properties of boron, carbon, silicon, and germanium are distinctly those of nonmetals. Their electronegativities are similar to hydrogen's and, as we saw in Chapters 9 and 10, they form many covalent hydrogen and alkyl compounds. Boron, aluminum, carbon, and silicon are strong **oxophiles** and **fluorophiles**, in the sense that they have high affinities for oxygen and fluorine respectively. We therefore class these four elements as 'hard'. Their oxophilic character is evident in the existence of an extensive series of oxoanions—the borates, aluminates, carbonates, and silicates. In contrast, thallium and lead have higher affinities for soft anions, such as I^- and S^{2-} ions, than for hard anions. We therefore classify these two elements as 'soft'.

Table 11.1 shows that, for most of the members of the two groups, the dominant oxidation number is the group oxidation number, which is +3 for Group 13/III and +4 for Group 14/IV. The major exceptions are thallium and lead, for which the most common oxidation number is 2 less than the

Table 11.1. Properties of the boron and carbon group elements

Element	I/kJ mol^{-1}	χ_P*	r_{cov}/Å †	r_{ion}/Å ‡	Appearance and properties	Common oxidation numbers§
Group 13/III						
B	899	2.04	0.88		Dark semiconductor	**3**
Al	578	1.61	1.26	0.54	Metal	**3**
Ga	579	1.81	1.26	0.62	Metal mp 30°C	1, **3**
In	558	1.78	1.44	0.80	Soft metal	1, **3**
Tl	589	2.04	1.47	0.89	Soft metal	**1**, 3
Group 14/IV						
C	1086	2.55	0.77		Hard insulator (diamond)	**4**
					Semimetal (graphite)	
Si	786	1.90	1.17	0.40	Hard semiconductor	**4**
Ge	760	2.01	1.22	0.53	Hard semiconductor	2, **4**
Sn	708	1.96	1.40	0.69	Metal	2, 4
Pb	715	2.33	1.46	0.92	Soft metal	**2**, 4

* Pauling values recalculated by A. L. Allred, *J. Inorg. Nucl. Chem.*, **17**, 215 (1961).

† Covalent radii from M. C. Ball and A. H. Norbury, *Physical data for inorganic chemists*. Longman, London (1974).

‡ Ionic radii from R. D. Shannon, *Acta Crystallogr.*, **A32**, 751 (1975), for elements with C.N. = 6 and in their maximum group oxidation states.

§ Most common oxidation number in bold type.

group maximum, being +1 for thallium and +2 for lead. This is a manifestation of the inert pair effect (Section 8.12).

11.2 Production

The chemically hard elements boron, aluminum, and silicon reveal their oxophilic character by their widespread occurrence in nature as oxides and oxoanions. Accordingly, these elements can be recovered only by using strongly reducing conditions, boron by reduction of its oxide with magnesium, silicon by electric-arc furnace reduction with carbon, and aluminum by cathodic reduction of a fused electrolyte (Section 8.1). Far more aluminum is produced than all the other elements in the two groups, and among all metals its production is second only to that of iron. We have already described the preparation of ultrapure semiconductor-grade silicon from $SiCl_4$ and hydrogen (Section 8.1).

As we saw when discussing Ellingham diagrams (Fig. 8.1), the oxides of the heavy elements in the two groups are easier to reduce than those of the light elements, so carbon reduction can be used to obtain them from their ores. For example, tin is commonly found as the mineral cassiterite, SnO_2, and is recovered by carbon reduction. The chemically soft heavier elements of the groups occur widely as sulfide ores, such as galena, PbS, the principal source of lead. The lead is recovered by roasting the sulfide in air to produce lead oxide

$$2PbS(s) + 3O_2(g) \xrightarrow{600°C} 2PbO(s) + 2SO_2(g)$$

That oxide is then reduced with carbon in a blast furnace, in which the overall reaction is

$$2PbO(s) + C(s) \xrightarrow{\Delta} 2Pb(l) + CO_2(g)$$

The cumulative toxicity of lead has resulted in its elimination from many consumer products, its greatest remaining use being in lead–acid batteries.

Many of the rarer elements of the two groups are recovered as by-products of more common metals. Gallium is obtained from the production of aluminum; germanium and thallium are recovered from zinc and lead smelting.

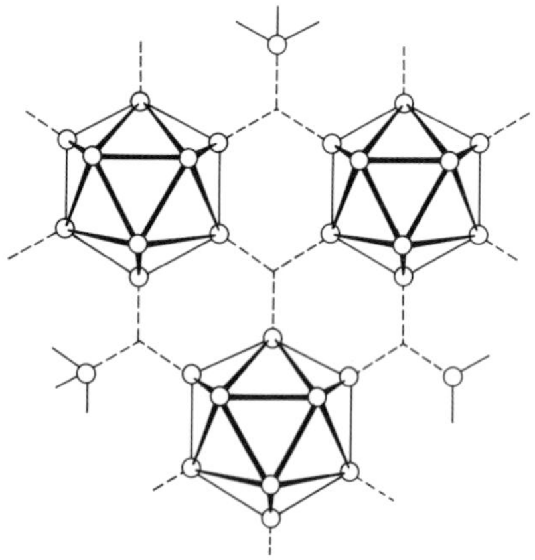

The boron group

Fig. 11.1 A section of α-rhombohedral boron perpendicular to the three-fold axis of the crystal. The dashed lines represent 3c,2e bonds connecting the B_{12} icosahedra. The B_{12} icosahedra are bonded by 2e,2c bonds to similar B_{12} units above and below this plane.

The boron group shows considerable structural diversity. Boron itself, for instance, exists in several hard and refractory polymorphs. The three solid phases for which crystal structures are available contain the icosahedral (20-faced) B_{12} unit (Fig. 11.1) as a building block. This icosahedral unit is a recurring motif in boron chemistry, and we shall meet it again in the structures of borides and boron hydrides.

All the other members of the group are metals. Like aluminum, gallium[1] is a good electrical conductor; it has a white silvery appearance and the other typical characteristics of a metal. The structure of gallium, however, is unlike that of any other metal in that the bonding in the solid is highly directional. Thus, each Ga atom has one nearest neighbor at 2.47 Å and six others between 2.70 and 2.79 Å away. Even more remarkably, experimental evidence suggests that Ga_2 units persist when the metal melts. The attendant physical properties of gallium are its very low melting point (30°C) and an unusually wide liquid range (of 2403°C).

The structures of the elements rapidly return to normal further down the group. Thus, indium has a slightly distorted fcc structure and thallium has a close-packed structure, both of which are typical of metals.

11.3 Compounds of boron with the electronegative elements

In this section we introduce the boron halides, which are very useful reagents and Lewis acid catalysts, and the numerous oxides and oxoanions. We shall also describe the boron nitrides. The latter are intriguing largely because the BN unit is isoelectronic with the CC unit, and hence compounds that contain it are analogs of the hydrocarbons.

Halides

All the boron trihalides except BI_3 may be prepared by direct reaction of the halogens with elemental boron. However, the preferred method for BF_3 is the reaction of B_2O_3 with CaF_2 in H_2SO_4, as the acid's high affinity for water helps to drive the reaction

$$B_2O_3(s) + 6HF(l) \rightarrow 2BF_3(g) + 3H_2O(l)$$

[1] Gallium, one of the elements predicted by Mendeleev, was discovered by the French chemist-spectroscopist Lecoq de Boisbaudran in 1875. Standard references state that he named gallium after the Latin name for France. There is the added possibility: *le coq* (the rooster) is *gallus* in Latin.

Table 11.2. Properties representative of the boron trihalides

Halide	mp/°C	bp/°C	$\Delta G_f^{\ominus}$/kJ mol^{-1} *
BX_3, X = F	−127	−100	−1112
Cl	−107	12	−339
Br	−46	91	−232
I	49	210	+21

* For the formation of the gaseous trihalide at 25°C.

to completion. Some interesting subhalides can also be prepared, as we shall describe in Section 11.12.

Boron trihalides consist of trigonal planar BX_3 molecules. Unlike the halides of the other elements in the group, they are unassociated in the gas, liquid, or solid states. However, halogen exchange does occur and it may well go through a transient dimer (**1**) like those characteristic of the gaseous aluminum halides. BF_3 and BCl_3 are gases, BBr_3 is a volatile liquid, and BI_3 is a solid (Table 11.2). This trend is consistent with the increase in strength of dispersion forces with the number of electrons in the molecules.

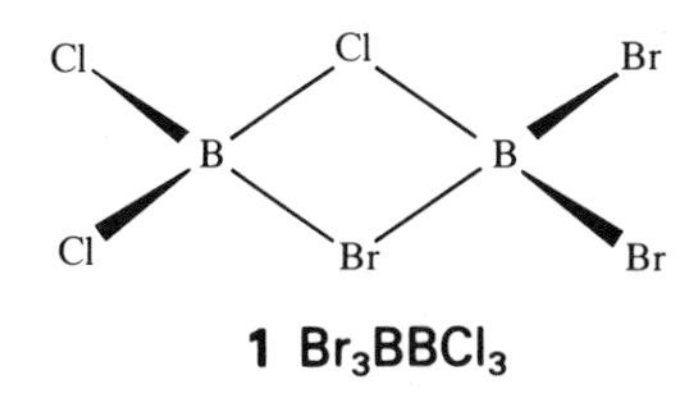

1 Br_3BBCl_3

The trihalides are Lewis acids. We have already drawn attention to the order of their strengths in this role, which is $BF_3 < BCl_3 \leq BBr_3$ and contrary to the order of electronegativity of the attached halogens (Section 6.6). This is thought to stem from greater X—B π bonding for the lighter, smaller halogens and the partial occupation of the *p* orbital on the B atom by electrons supplied by the halogen atoms. All the boron trihalides form simple Lewis complexes with suitable bases, as in the reaction

$$BF_3(g) + NH_3(g) \rightarrow F_3B\text{—}NH_3(s)$$

However, the chlorides, bromides, and iodides are susceptible to protolysis by mild proton sources such as water, alcohols, and even amines:

$$BCl_3(g) + 3H_2O(l) \rightarrow B(OH)_3(aq) + 3HCl(aq)$$

Example 11.1: *Predicting the products of reactions of the boron trihalides*

Predict the probable products of the following reactions, and write the balanced chemical equations.

(a) BF_3 and excess NaF in acidic aqueous solution.
(b) BCl_3 and excess NaCl in acidic aqueous solution.
(c) BBr_3 and excess $NH(CH_3)_2$ in a hydrocarbon solvent.

Answer. (a) The F^- ion is a hard and fairly strong base; BF_3 is a hard and strong Lewis acid with a high affinity for the F^- ion. Hence, the reaction should result in a complex:

$$BF_3(g) + F^-(aq) \rightarrow [BF_4]^-(aq)$$

Excess F^- and acid prevent the formation of hydrolysis products such as $[BF_3OH]^-$, which are formed at high pH.

(b) Unlike B—F bonds, which are only mildly susceptible to hydrolysis, the other boron–halogen bonds are vigorously hydrolyzed by water. We can anticipate that BCl_3 will hydrolyze rather than coordinate to Cl^-:

$$BCl_3(g) + 3H_2O(l) \rightarrow B(OH)_3(aq) + 3HCl(aq)$$

(c) BBr_3 will undergo protolysis with formation of a B—N bond:

$$BBr_3(g) + 3NH(CH_3)_2 \rightarrow B(N(CH_3)_2)_3 + 3HBr(g)$$

The HBr will protonate any excess dimethylamine.

Exercise. Write and justify balanced equations for plausible reactions between (a) BCl_3 and ethanol, (b) BCl_3 and pyridine in hydrocarbon solution, (c) BBr_3 and $F_3BN(CH_3)_3$.

The tetrafluoroborate anion $[BF_4]^-$, which is mentioned in the example, is a very weak Lewis base, and is used in preparative chemistry when a relatively large noncoordinating anion is needed. The other tetrahaloborate anions, $[BCl_4]^-$ and $[BBr_4]^-$, can be prepared in nonaqueous solvents. However, because of the ease with which B—Cl and B—Br bonds undergo hydrolysis, they are stable in neither water nor alcohols.

Oxides and oxo compounds

Boric acid $B(OH)_3$ is a very weak Brønsted acid in aqueous solution. However, the equilibria are more complicated than the simple Brønsted proton transfer reactions characteristic of the later *p*-block oxoacids. Boric acid is in fact primarily a *Lewis* acid, and the complex it forms with H_2O is the actual source of protons:

$$B(OH)_3(aq) + 2H_2O(l) \rightleftharpoons H_3O^+(aq) + [B(OH)_4]^-(aq) \qquad pK_a = 9.2$$

As is typical for many of the lighter elements in the two groups, there is a tendency for the anion to polymerize by condensation with the loss of H_2O. Thus, in concentrated neutral or basic solution, equilibria such as

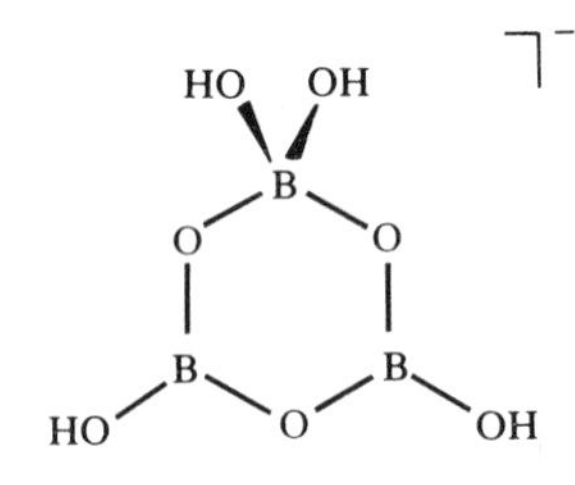

2 $[B_3O_3(OH)_4]^-$

$$3B(OH)_3(aq) \rightleftharpoons [B_3O_3(OH)_4]^-(aq) + H^+(aq) + 2H_2O(l) \qquad K = 1.4 \times 10^{-7}$$

occur to yield polynuclear anions (**2**).

The reaction of boric acid with an alcohol in the presence of sulfuric acid leads to the formation of simple **borate esters**, which are compounds of the form $B(OR)_3$:

$$B(OH)_3 + 3CH_3OH \xrightarrow{H_2SO_4} B(OCH_3)_3 + 3H_2O$$

3

Borate esters are much weaker Lewis acids then the boron trihalides, presumably because the O atom acts as an intramolecular π base, like the F atom in BF_3, and donates electron density to the *p* orbital of the B atom. Hence, judging from Lewis acidity, an O atom is more effective than an F atom as a π base toward boron. 1,2-Diols have a particularly strong tendency to form borate esters on account of the chelate effect, for they produce a cyclic borate ester (as in **3**). The chelate effect is also responsible for the decrease in pK_a from 9.2 to 5.2 when the hexose mannitol is added to boric acid.

4 $[B_3O_6]^{3-}$

As with silicates and aluminates, there are many polynuclear borates, and both cyclic and chain species are known. An example is the cyclic polyoxoborate anion $[B_3O_6]^{3-}$ (**4**), the conjugate base of (**2**). A notable feature of borate formation is the possibility of both three-coordinate B atoms, as in (**4**), and four-coordinate B atoms, as in $[B(OH)_4]^-$ and for one of the B atoms in (**2**). Polyborates form by sharing one O atom with a neighboring B atom, as in (**2**) and (**3**); structures in which two adjacent B atoms share two or three O atoms are unknown.

The rapid cooling of molten B_2O_3 or metal borates often leads to the formation of borate glasses. Although these glasses themselves have little technological significance, the borosilicate glasses (such as Pyrex) have low thermal expansivities and hence little tendency to crack when heated or cooled rapidly. Because of this property, borosilicate glass is used extensively for cooking ware and laboratory ware.

Compounds with nitrogen

The simplest binary compound of boron and nitrogen is easily synthesized by heating boron oxide with a nitrogen compound:

$$B_2O_3(l) + 2NH_3(g) \xrightarrow{1200^\circ C} 2BN(s) + 3H_2O(g)$$

The form of boron nitride this produces, and the thermodynamically stable phase under normal laboratory conditions, has a structure like that of graphite (Section 11.6). The planar sheets of alternating B and N atoms consist of edge-shared hexagons, and the B—N distance within the sheet (1.45 Å) is much shorter than the distance between sheets (3.33 Å, Fig. 11.2). The difference between the structures of graphite and boron nitride, however, lies in the register of the atoms of neighboring sheets. In boron nitride the hexagonal rings are stacked directly over each other, with B and N atoms alternating in successive layers; in graphite, the hexagons are staggered. Molecular orbital calculations suggest that the stacking in BN stems from a partial positive charge on B and partial negative charge on N. This is consistent with the greater electronegativity of N compared with B.

As with graphite, layered boron nitride is a slippery material that is used as a lubricant. Unlike graphite, it is a colorless electrical insulator, for there is a large energy gap between the filled and vacant π bands. As we saw in Section 3.5, this is to be expected in a heteroatomic solid. The size of the bandgap is consistent with the very small number of compounds formed by donation of electrons to the conduction band or electron capture from the valence band by atoms that intrude between the sheets (see the discussion of the 'intercalation' compounds of graphite in Section 11.6).

Layered boron nitride changes into a cubic phase (Fig. 11.3) at high pressures and temperatures (60 kbar and 2000°C). This phase is a hard crystalline analog of diamond, but as it has a lower lattice enthalpy, it has a slightly lower mechanical hardness (Fig. 11.4). Cubic boron nitride is manufactured and used as an abrasive for certain high-temperature applications in which diamond cannot be used because it forms carbides with the material being ground.

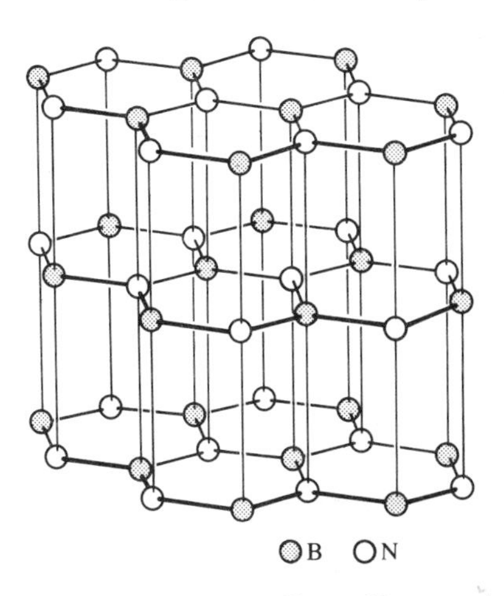

Fig. 11.2 The structure of layered hexagonal boron nitride. Note that the atoms are in register between layers.

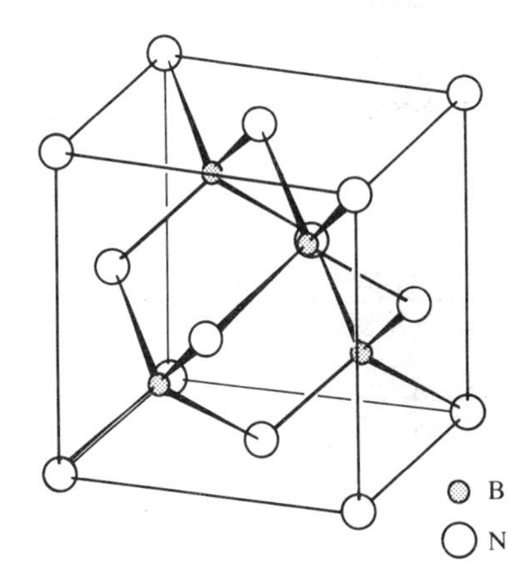

Fig. 11.3 The sphalerite structure of cubic boron nitride.

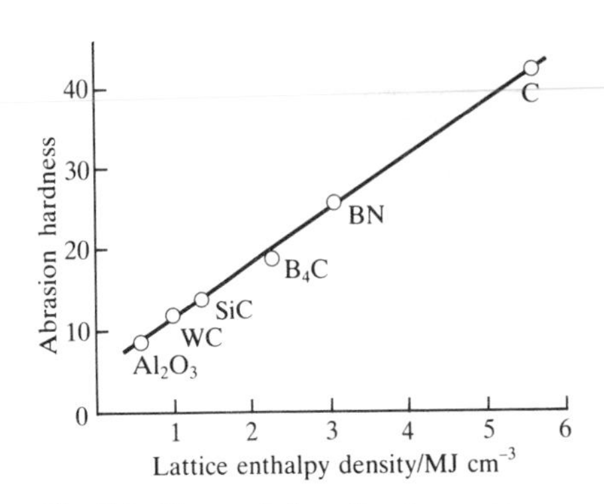

Fig. 11.4 The correlation of hardness with lattice enthalpy density. The point for carbon represents diamond and for boron nitrides it represents the diamond-like sphalerite structure.

Saturated BN compounds

Many analogs of saturated hydrocarbons can be synthesized by reaction between a nitrogen Lewis base and a boron Lewis acid:

$$\tfrac{1}{2}B_2H_6 + N(CH_3)_3 \rightarrow H_3B{-}N(CH_3)_3$$

These **amine-boranes** are isoelectronic with hydrocarbons, but their properties are significantly different. The most striking difference in physical properties is found between the two simplest analogs, ammoniaborane

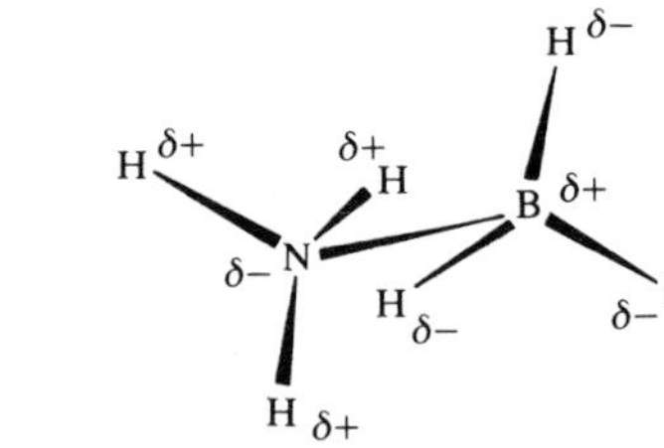

($H_3N—BH_3$) and ethane ($H_3C—CH_3$). The former is a solid at room temperature with a vapor pressure of a few torr, whereas ethane is a gas that condenses at −89°C. This difference can be traced to the difference in polarity of the two molecules: ethane is nonpolar, whereas ammoniaborane has a large dipole moment (5.2 D). The H atoms attached to the electronegative N atom have a partial positive charge and those on the less electronegative B atom have a partial negative charge (**5**).

Several B—N analogs of the amino acids have been prepared, including ammoniacarboxyborane, H_3NBH_2COOH, the analog of glycine, H_2NCH_2COOH. Some of these compounds display significant physiological activity, including tumor inhibition and reduction of serum cholesterol in certain cases.

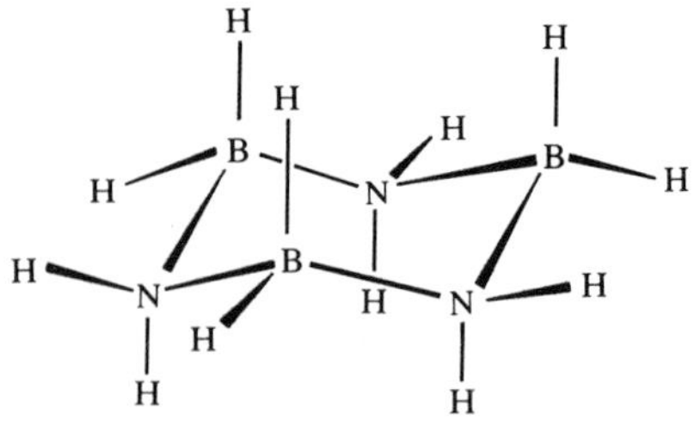

6 $N_3B_3H_{12}$

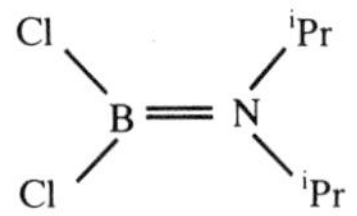

7 $Cl_2B=N(^iPr)_2$

Unsaturated BN compounds

Aminoborane, $H_2N{=}BH_2$, is isoelectronic with ethylene. It has only a transient existence in the gas phase since it readily forms cyclic ring compounds such as the cyclohexane analog (**6**). However, the aminoboranes do survive as monomers when the double bond is shielded from reaction by bulky alkyl groups on the N atom and by Cl atoms on the B atom (**7**). For instance, monomeric aminoboranes can be synthesized readily by the reaction of a dialkylamine, which is protic, and a boron halide:

$$R_2NH + BCl_3 \rightarrow R_2N{=}BCl_2 + HCl \qquad R = {}^iPr, \text{ xylyl}$$

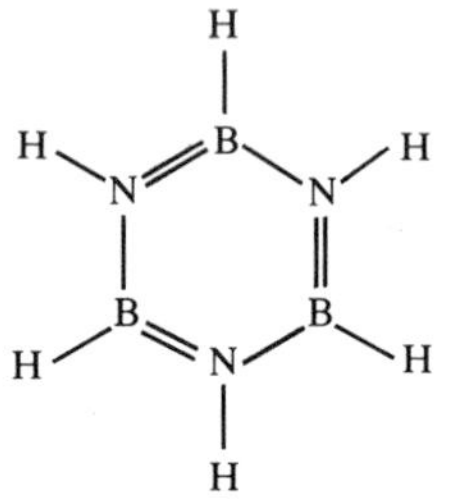

Apart from layered BN, the best-known unsaturated compound of boron and nitrogen is borazine, $H_3B_3N_3H_3$ (**8**), which is isoelectronic and isostructural with benzene. Borazine was first prepared in Stock's laboratory in 1926 by the reaction between diborane and ammonia. Many symmetrically trisubstituted derivatives can be made by more recent procedures that depend on the protolysis of B—Cl bonds of BCl_3 by an ammonium salt:

$$3NH_4Cl + 3BCl_3 \xrightarrow{\Delta,\ C_6H_5Cl} Cl_3B_3N_3H_3 + 9HCl$$

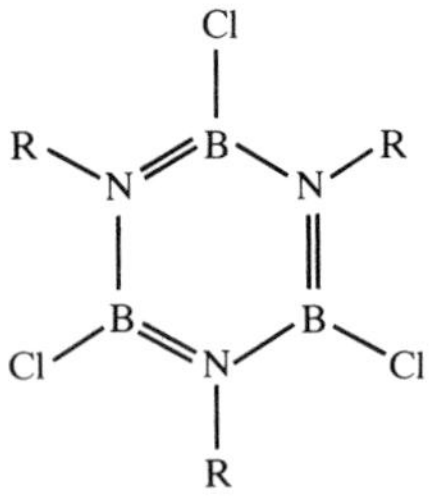

8 $H_3B_3N_3H_3$

A primary ammonium chloride yields *N*-alkyl substituted *B*-trichloroborazines (**9**).

9 $Cl_3B_3N_3R_3$

Although borazine is isoelectronic with benzene, there is little resemblance between them chemically. Once again, the difference in electronegativities of boron and nitrogen is influential, and B—Cl bonds are much more labile than C—Cl bonds in chlorobenzene because the π electrons are concentrated on the N atoms. This leaves the B atoms partially positive and open to electrophilic attack. A sign of the difference is that the reaction of chloroborazines with a Grignard reagent or hydride source results in the substitution of Cl by alkyl, aryl, or hydride groups. Another example of the difference is the ready addition of HCl to borazine to produce a trichlorocyclohexane analog (**10**):

$$3HCl + H_3B_3N_3H_3 \rightarrow Cl_3B_3H_3N_3H_6$$

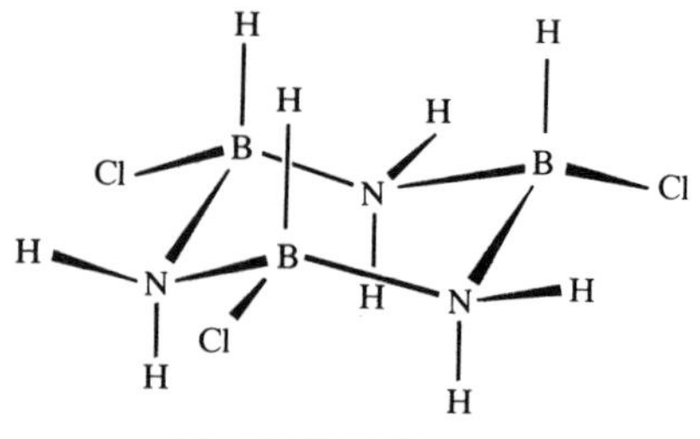

10 $Cl_3B_3N_3H_9$

The electrophile H^+ in this reaction attaches to the partially negative N atom and the nucleophile Cl^- attaches to the partially positive B atom.

Ultraviolet spectra indicate that the separation of the π and π^* orbitals in borazine is greater than in benzene. This recalls the much greater separation of the π and π^* bands in layered BN compared with graphite.

Example 11.2: *Preparing borazine compounds*

Give balanced chemical equations for the synthesis of borazine starting with NH_4Cl and other reagents of your choice.

Answer. The reaction of NH_4Cl with BCl_3 in refluxing chlorobenzene will yield the *B*-trichloroborazine:

$$3NH_4Cl + 3BCl_3 \rightarrow H_3N_3B_3Cl_3 + 9HCl$$

The Cl atoms in *B*-trichloroborazine can be displaced by hydride ions from reagents such as lithium borohydride, to yield borazine.

$$3LiBH_4 + H_3N_3B_3Cl_3 + 3THF \xrightarrow{THF} H_3N_3B_3H_3 + 3LiCl + 3THF{\cdot}BH_3$$

Exercise. Suggest a reaction or series of reactions for the preparation of N,N',N''-trimethyl-B,B',B''-trimethylborazine, starting with methylamine and boron trichloride.

11.4 Aluminum and gallium

Aluminum and gallium have similar characteristics when their oxidation number is +3. However, whereas no simple Al(I) compounds are known, simple Ga(I) compounds are known in nonaqueous solvents and in the solid state (Table 11.3). Gallium in these lower oxidation states has a lot in common with indium and thallium, and this aspect of its chemistry is best deferred to Section 11.5.

Halides

Although direct reaction of aluminum and gallium with a halogen yields the halides, both these electropositive metals react with HCl or HBr gas, and this is usually a more convenient route:

$$2Al(s) + 6HCl(g) \xrightarrow{100^\circ C} 2AlCl_3(s) + 3H_2(g)$$

In the laboratory, a 'hot tube' reactor (Fig. 11.5) is often used. The halides of both elements are available commercially, but it is common to synthesize them in the laboratory when a material free of hydrolysis products is

Table 11.3. Structures and properties of some aluminum and gallium halides

Halide		Structure	C.N.	mp/°C	bp/°C
AlX_3	X = F	Ionic	6	1291*	
	Cl	Layered	6	190*	
	Br	Molecular	4	97	263
	I	Molecular	4	191	360
GaX_3	X = F	FeF_3	6	1000	950*
	Cl	Molecular	4	78	210
	Br	Molecular	4	122	278
$GaCl_2$		$Ga[GaCl_4]$†		164	537

* Sublimes.
† Mixed oxidation state.

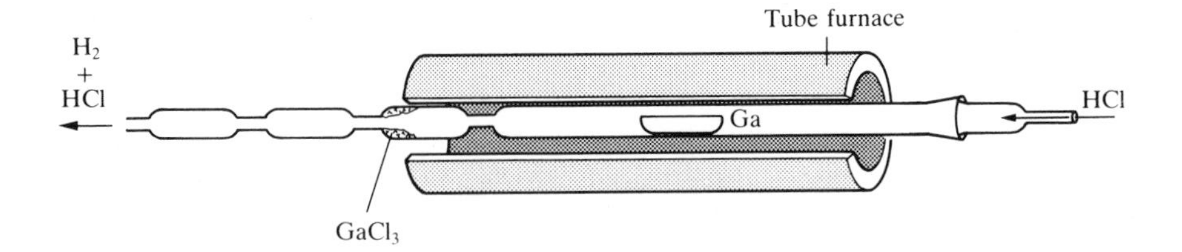

Fig. 11.5 Preparation of gallium trichloride in a hot tube. The molten gallium metal is contained in a small ceramic boat. Gallium trichloride vapor condenses into crystals on the cool walls of the tube downstream from the boat.

required. Halides of Ga(I) and Ga(II), such as $GaCl_2$, can be prepared by a comproportionation reaction in which the Ga(III) halide is heated with Ga metal:

$$2GaX_3 + Ga \xrightarrow{\Delta} 3GaX_2 \quad X = Cl, Br, \text{ or I but not F}$$

Because the F^- ion is so small, the fluorides AlF_3 and GaF_3 are hard solids that have much higher melting points and sublimation enthalpies than the other halides. Their high lattice enthalpies also result in their having very limited solubility in most solvents, and they do not act as Lewis acids to simple donor molecules. In contrast, the heavier halides are soluble in a wide variety of polar solvents and are excellent Lewis acids. However, despite their low reactivity toward most donors, AlF_3 and GaF_3 form salts of the type Na_3AlF_6 and Na_3GaF_6 which contain octahedral $[MF_6]^{3-}$ complex ions.

The relative Lewis acidities of the halides reflect the relative chemical hardnesses of the two elements. Thus, toward the hard O atom Lewis base of ethyl acetate, the Lewis acidity is weaker the softer the element:

$$BCl_3 > AlCl_3 > GaCl_3$$

In contrast, toward the soft S atom of dimethyl sulfide, the Lewis acidity is stronger the softer the element:

$$GaX_3 > AlX_3 > BX_3 \qquad (X = Cl \text{ or } Br)$$

In keeping with the general tendency toward higher coordination numbers for the heavier *p*-block elements, halides of aluminum and its heavier congeners may take on more than one Lewis base and become hypervalent:

$$AlCl_3 + N(CH_3)_3 \rightarrow Cl_3AlN(CH_3)_3$$

$$Cl_3AlN(CH_3)_3 + N(CH_3)_3 \rightarrow Cl_3Al(N(CH_3)_3)_2$$

Oxides

The stable form of Al_2O_3, **α-alumina**, is a very hard and refractory material. In its mineral form it is known as corundum and as a gemstone it is sapphire. The blue color of the latter arises from a charge-transfer transition from Fe^{2+} to Ti^{4+} ion impurities. The structure of α-alumina and gallia, Ga_2O_3, consists of an hcp array of O^{2-} ions with the metal ions occupying two-thirds of the octahedral holes in an ordered array. Ruby is α-alumina in which a few percent of the Al^{3+} is replaced by Cr^{3+}. The Cr(III) is red rather than the normal characteristic violet of $[Cr(OH_2)_6]^{3+}$ or Cr_2O_3 because when Cr^{3+} substitutes for the smaller Al^{3+} ion, the O ligands are compressed about Cr^{3+}. This compression increases the ligand field parameter Δ_O and shifts the first spin-allowed *d–d* band in the spectrum toward the blue.

As we discussed in Section 6.9, dehydration of aluminum hydroxide at temperatures below 900°C leads to the formation of **γ-alumina**, a metastable polycrystalline form with a very high surface area. Partly on account of its surface acid and base sites, this material is used as a solid phase in column chromatography and as a catalyst and catalyst support (Section 17.5).

11.5 Indium, thallium, and low oxidation state gallium

Halides

The monohalides GaX, InX, and TlX are known for X = Cl, Br, and I and TlF is known. Under ordinary conditions the Tl(I) halides are insulators, as is typical of ionic compounds. However, at high pressures a new phase is formed with a significant electrical conductivity that decreases with increasing temperature. This behavior signifies metallic conduction (Section 3.5).

An interesting point is that TlI_3 is not a halide of Tl(III) but of Tl(I), for structural data show that it contains the triiodide ion I_3^-. This compound forms when solutions of Tl^{3+} and I^- are mixed because Tl^{3+} can oxidize the I^- ion. A series of subhalides are also known for gallium, indium, and thallium in which M^+ and M^{3+} ions exist jointly. For instance, structural data indicate that the compound $GaCl_2$ is in fact $Ga(I)[Ga(III)Cl_4]$.

There is only a fine line between the formation of a mixed oxidation state ionic compound containing both M^+ and M^{3+} ions and the formation of a compound that contains M—M bonds. For example, mixing $Ga(I)[Ga(III)Cl_4]$ with a solution of $[N(CH_3)_4]Cl$ in a nonaqueous solvent yields the compound $[N(CH_3)_4]_2[Cl_3Ga—GaCl_3]$, in which the anion has an ethane-like structure with a Ga—Ga bond.

Example 11.3: *Proposing reactions of the boron group halides*

Propose balanced chemical equations (or indicate no reaction) for reactions between (a) $(C_2H_5)_3AsBCl_3$ and $(C_2H_5)_3NGaCl_3$ in toluene, (b) $(C_2H_5)_3NGaCl_3$ and GaF_3 in toluene, and (c) TlCl and NaI in H_2O.

Answer. (a) Ga(III) is softer than B(III); therefore, the soft–soft Ga—As and hard–hard B—N combinations are favored:

$$(C_2H_5)_3AsBCl_3 + (C_2H_5)_3NGaCl_3 \rightarrow (C_2H_5)_3AsGaCl_3 + (C_2H_5)_3NBCl_3$$

(b) NR, because GaF_3 has a very high lattice energy and so it is not a good Lewis acid. (c) Tl(I) is soft, so it combines with the softer I^- rather than Cl^-:

$$TlCl(s) + NaI(aq) \rightarrow TlI(s) + NaCl(aq)$$

like silver halides, Tl(I) halides have low solubility in water, so the reaction will probably proceed very slowly.

Exercise. Propose, with reasons, the chemical equation (or indicate no reaction) for reactions between (a) $(CH_3)_3NBF_3$ and $GaCl_3$ (hint: halide exchange is possible) and (b) $TlCl_3$ and formaldehyde (CH_2O) in acidic water (hint: formaldehyde is easily oxidized to CO_2 and H^+).

The carbon group

We discuss carbon in many contexts throughout this text, including organometallic compounds in Chapters 10 and 16 and catalysis in Chapter 17. In this section we shall focus attention on the more classically 'inorganic' properties of the element and its oxides.

All the elements of the group except lead have at least one solid phase with the diamond structure. That of tin, which is called **gray tin**, is not stable at room temperature; the more stable phase, **white tin**, has six nearest neighbors in a highly distorted octahedral array. This distortion is reminiscent of the unsymmetrical environment within tin's neighbors gallium and indium. As we remarked above, lead has a close-packed structure; like its neighbor thallium it is a soft metal.

11.6 Carbon

Diamond and graphite

Diamond and graphite, the two common crystalline forms of elemental carbon, are strikingly different. Diamond is an electrical insulator; graphite is a good conductor. Diamond is the hardest known substance and hence the ultimate abrasive; graphite is slippery, and frequently used as a lubricant. Because of its durability, clarity, and high refractive index, diamond is one of the most highly prized gem stones; graphite is soft and black with a slightly metallic luster, and is neither durable nor particularly attractive. The origin of these widely different properties can be traced to the very different structures and bonding in the two polymorphs.

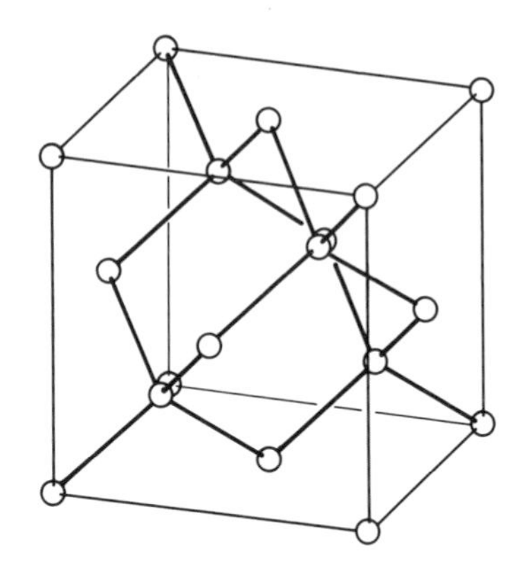

Fig. 11.6 The cubic diamond structure.

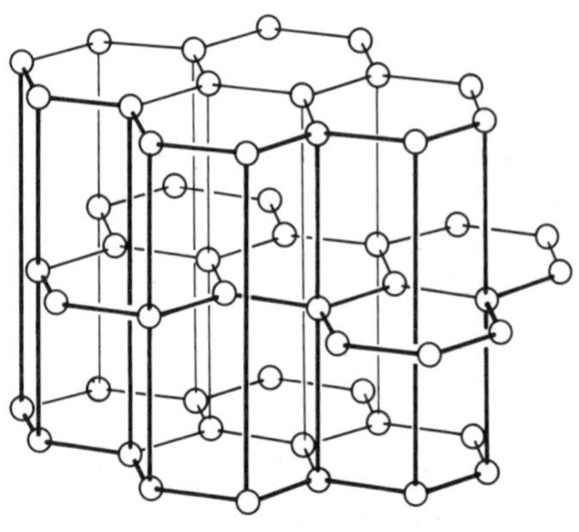

Fig. 11.7 The hexagonal graphite structure.

In diamond (Fig. 11.6), each C atom forms single bonds (of length 1.54Å) with four adjacent C atoms at the corners of a regular tetrahedron. The result is a tightly bonded, covalent, three-dimensional crystal. On the other hand, graphite (Fig. 11.7) consists of stacks of planar layers within which each C atom has three nearest neighbors at 1.42 Å. The σ bonds between neighbors within the sheets are formed from the overlap of sp^2 hybrids, and the remaining perpendicular p orbitals overlap to form π bonds that are delocalized over the plane. The planes themselves are widely separated from each other (at 3.35 Å), which indicates that there are only weak forces between them. These forces are sometimes, but not very appropriately, called **van der Waals forces** (because they are weak, like intermolecular forces), and the region between the planes is called the **van der Waals gap**. The ready cleavage of graphite parallel to the planes of atoms (which is enhanced by the presence of impurities) accounts for its slipperiness. Diamond can be cleaved, but this ancient craft requires considerable expertise since the forces in the crystal are more symmetrical.

The transformation of diamond to graphite at room temperature and pressure is spontaneous ($\Delta G^{\ominus} = -2.90\,\text{kJ mol}^{-1}$) but does not occur at an observable rate under ordinary conditions. Since diamond is the denser phase, it is favored by high pressures, and large quantities of diamond abrasive are manufactured commercially by a d-metal catalyzed high-temperature, high-pressure process. The d-metal (typically nickel) dissolves the graphite at high temperature and pressure (1800°C and 70 kbar), and the less soluble diamond phase crystallizes from it. The synthesis of gem-quality diamonds is possible but not yet economical.

Since the high-pressure synthesis of diamond is costly and cumbersome, a low-pressure process would be highly attractive. It has in fact been known for a long time that microscopic diamond crystals mixed with graphite can be formed by depositing C atoms on a hot surface. The C atoms are produced by the pyrolysis of methane, and the atomic hydrogen also produced in the pyrolysis plays an important role in favoring diamond over graphite.[2] One hypothesis is that the reaction of atomic hydrogen with the graphite to produce volatile hydrocarbons is faster than its reaction with diamond. The success of this mode of synthesis is important because then diamond films might find applications ranging from the hardening of surfaces subjected to wear to the construction of electronic devices.

The electrical conductivity and many of the chemical properties of graphite are closely related to the structure of its conjugated π bonds. Its conductivity perpendicular to the planes is low ($5\ \mathrm{S\,cm^{-1}}$ at 25°C) and increases with increasing temperature, signifying that graphite is a semiconductor in that direction. The electrical conductivity is much higher parallel to the planes ($3 \times 10^4\ \mathrm{S\,cm^{-1}}$ at 25°C) but decreases as the temperature is raised. This behavior indicates metallic conduction in that direction.[3] The anisotropy of the conductivity is consistent with a simple band model in which the mobile electrons are in a half-full π band extending over the sheets (Fig. 11.8).

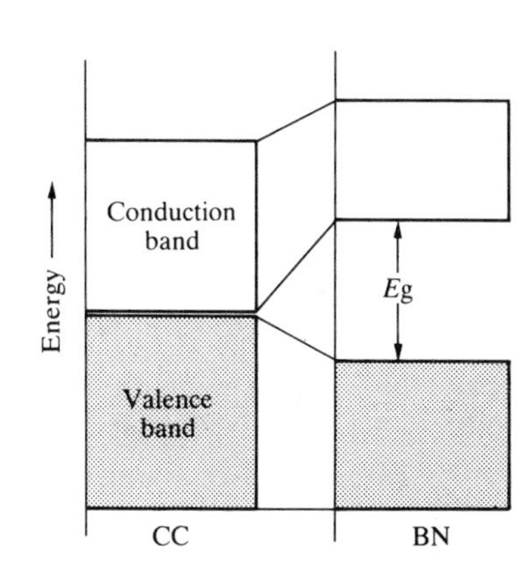

Fig. 11.8 A schematic diagram of the π valence and conduction bands for graphite and layered boron nitride.

A chemical consequence of the small separation of the HOMO and LUMO is that graphite may serve as either an electron donor or an electron acceptor toward atoms and ions that intercalate between its sheets. Thus, K atoms reduce graphite by donating their valence electron to the LUMO π band and the resulting K^+ ions penetrate between the layers. An example of an oxidation by removal of electrons from the HOMO band is the formation of the substances called **graphite bisulfates** by heating graphite with a mixture of sulfuric and nitric acids. In this reaction, electrons are removed from the full valence π band, and HSO_4^- ions penetrate between the sheets to give substances of approximate formula $(C_{24})^+SO_3(OH)^-$. In both cases, the change in the population of the conduction band leads to a modification of the electrical properties of the graphite.

Partially crystalline carbon

There are many forms of carbon that have a low degree of crystallinity. These **partially crystalline** materials have considerable commercial importance, for they include **carbon black**, **activated carbon**, and **carbon fibers**. Since single crystals suitable for complete X-ray analyses of these materials are not available, their structures are uncertain. However, what information there is suggests that their structures are similar to that of graphite, but the degree of crystallinity and shapes of the particles differ.

Carbon black is a very finely divided form of carbon. It is prepared (on a scale that exceeds 8×10^9 kg annually) by the combustion of hydrocarbons under oxygen-deficient conditions. Electron micrographs indicate that this form of carbon consists of a folded version of the graphite network (Fig. 11.9). The major uses of carbon black are as a pigment, in printer's ink (as

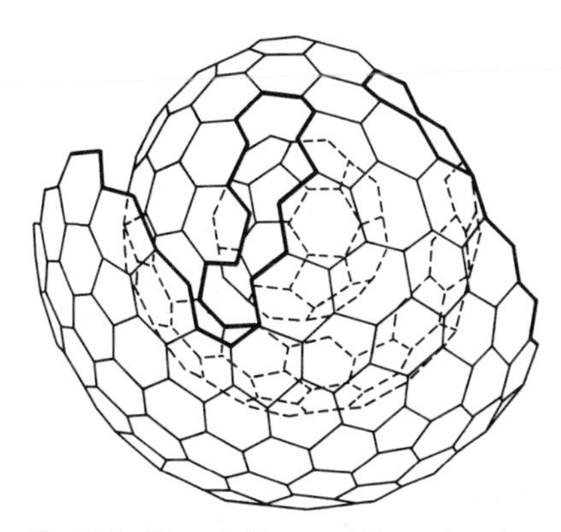

Fig. 11.9 The spiraling graphitic soot nucleus believed to form as a result of imperfect closure of a curved network as C atoms condense.

[2] J. C. Anderson and C. C. Hayman, Low-pressure metastable growth of diamond and "diamondlike" phases, in *Science*, **241**, 913 (1988).

[3] More precisely, graphite is a semimetal in that direction (Section 3.5).

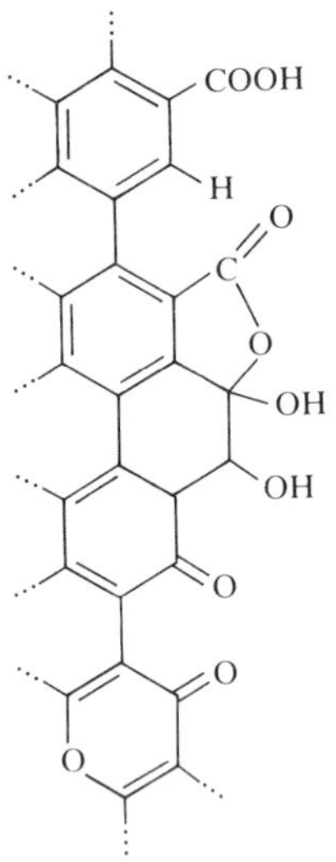

11

on this page), and as a filler for rubber goods, including automobile tires, where it greatly improves the strength and wear resistance of the rubber and helps to protect it from degradation by sunlight.

Activated carbon is prepared from the controlled pyrolysis of organic material, including coconut shells. It has a high surface area—in some cases exceeding $1000\,m^2\,g^{-1}$—which arises from the small particle size, and is therefore a very efficient adsorbent for molecules, including organic pollutants from drinking water, noxious gases from the air, and impurities from reaction mixtures. There is evidence that the parts of the surface defined by the edges of the hexagonal sheets are covered with oxidation products, including carboxyl and hydroxyl groups (**11**). This may account for some of the surface activity.

Carbon fibers are made by the controlled pyrolysis of asphalt fibers or synthetic fibers and are incorporated into a variety of high-strength plastic products, such as tennis rackets and aircraft components. The structure bears a resemblance to that of graphite, but in place of the extended sheets the layers consist of ribbons parallel to the axis of the fiber. The strong in-plane bonds (which resemble those in graphite) give the fiber its very high tensile strength.

Example 11.4: *Comparing bonding in diamond and boron*

A B atom in elemental boron is often bonded to five other B atoms but a C atom in diamond is bonded to four nearest neighbors. Suggest an explanation of this difference.

Answer. B and C both have four orbitals available for bonding (one s and three p). C has four valence electrons, one for each orbital, and it can therefore use all its electrons and orbitals in forming 2c,2e bonds with the four neighboring C atoms. B has one less electron; therefore, to use all its orbitals, it forms 3c,2e bonds. The formation of these three-center bonds brings another B atom into binding distance.

Exercise. Describe how the electronic structure of graphite is altered when it reacts with (a) potassium, (b) bromine.

Compounds with oxygen and sulfur

The two familiar oxides of carbon, CO and CO_2, have already been mentioned in several contexts; among its less familiar oxides is the gas carbon suboxide, O=C=C=C=O. Physical data on all three compounds are summarized in Table 11.4. In particular, it should be noted that the

Table 11.4. Properties of some oxides of carbon

Oxide	mp/°C	bp/°C	$\tilde{\nu}$(CO)/cm^{-1}	k(CO)/N m^{-1}	Bond length/Å	
					CC	CO
CO	−199	−191.5	2145	1860		1.13
OCO	−78*		2349, 1318	1550		1.16
OCCCO	−111	7	2200, 2290		1.28	1.16

* Sublimes; at 5 atm, CO_2 melts at −57°C.

C—O bond length of the monoxide is short and the force constant high; this is in accord with its triply bonded Lewis structure :C≡O: (Section 2.7).

The uses of CO that we have described elsewhere include the reduction of metal oxides in blast furnaces (Section 8.1) and the shift reaction (Section 9.4) for the production of H_2:

$$CO(g) + H_2O(g) \rightarrow CO_2(g) + H_2(g)$$

In Chapter 17, where we deal with catalysis, we shall describe its conversion to methanol, acetic acid, and aldehydes. The CO molecule has very low Brønsted basicity and negligible Lewis acidity toward neutral electron pair donors. Despite its weak Lewis acidity, CO is attacked by strong Lewis bases at high pressure and somewhat elevated temperatures. Thus, the reaction with hydroxide yields the formate ion:

$$CO(g) + OH^-(s) \rightarrow HCO_2^-(s)$$

Similarly, the reaction with methoxide ions (CH_3O^-) yields the acetate ion ($CH_3CO_2^-$).

Carbon monoxide is an excellent ligand toward metal atoms in low oxidation states, as we shall see in some detail in Chapter 16. Its well-known toxicity is an example of this behavior, for its binds to the Fe ion in hemoglobin, so excluding the attachment of O_2, and the victim suffocates. An interesting point is that H_3BCO can be prepared from B_2H_6 and CO at high pressures, in a rare example of CO coordinating to a simple Lewis acid. That a similar complex is not formed by BF_3 is a sign that the softer BH_3/soft CO combination is favorable whereas the harder BF_3/soft CO combination is not.

The C—O bond is longer and the stretching force constants smaller in CO_2 than in CO, which is consistent with there being a C=O double bond in place of the C≡O triple bond of the monoxide. Carbon dioxide is only a very weak Lewis acid, but it does add water to form carbonic acid H_2CO_3 (Section 5.6).

Although CO_2 is not directly toxic and is always present in the atmosphere, its increasing abundance there (through the combustion of fossil fuels and through deforestation) has given rise to a concern of a rather different kind. There is currently a worry that the increase in atmospheric CO_2 may cause global warming by the so-called **greenhouse effect**. In this effect, a molecule in the atmosphere permits the passage of visible light but absorbs strongly in the infrared, and so hinders the radiation of heat from the earth. In the past, nature has managed to stabilize the concentration of atmospheric CO_2 by precipitation of calcium carbonate in the deep oceans, but it seems that the rate of diffusion of CO_2 into the deep waters is too slow to compensate for the increased influx of CO_2 to the atmosphere.[4]

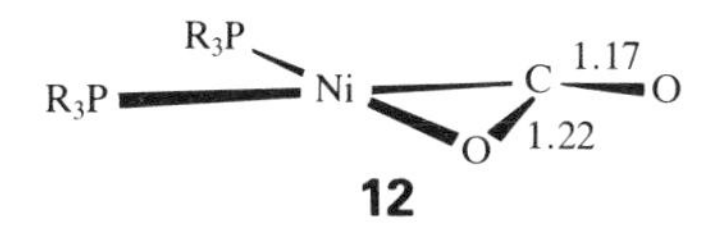

12

Much of the chemistry of CO_2 is based on its mild Lewis acidity, as in the formation of CO_3^{2-} ions in basic solution and its reaction with carbanion reagents to produce carboxylic acids. Metal complexes of both CO_2 (**12**) and CO_3^{2-} (**13**) are known, but they are far less important than the metal carbonyls. The neutral CO_2 molecule acts as a Lewis acid when it forms complexes with metal atoms in low oxidation states, and the bulk of the

13 $[Co(NH_3)_4CO_3]^+$

[4] R. A. Berner and A. C. Lasaga, Modeling the geochemical carbon cycle, in *Scientific American*, **258**(3), 74 (1988).

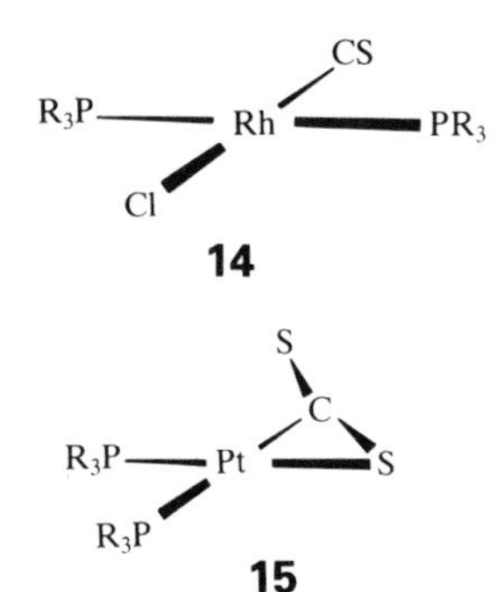

bonding occurs by electron donation from the metal atom into an antibonding π orbital of CO_2. On the other hand, the CO_3^{2-} ion coordinates as a Lewis base to metal atoms in positive oxidation states, and it can act as either a monodentate or a bidentate ligand.

The sulfur analogs of CO and CO_2, CS and CS_2, are known, but the former is an unstable transient molecule and the latter is endoergic ($\Delta G_f^{\ominus} = +65\ \text{kJ mol}^{-1}$). Some complexes of CS (**14**) and CS_2 (**15**) are known, and their structures are similar to those formed by CO and CO_2. In basic aqueous solution, CS_2 undergoes hydrolysis and yields a mixture of carbonate ions, CO_3^{2-}, and trithiocarbonate ions, CS_3^{2-}.

Compounds with nitrogen

Hydrogen cyanide, HCN, is produced in large amounts by the high temperature catalytic combination of methane and ammonia, and is used as an intermediate in the synthesis of common polymers such as poly(methyl methacrylate) and polyacrylonitrile. It is highly volatile (b.p. 26°C) and, like the CN^- ion, highly poisonous. In some respects the toxicity of the CN^- ion is similar to that of the isoelectronic CO molecule, because both form complexes with iron porphyrin molecules. However, whereas CO attaches to the Fe in hemoglobin and causes oxygen starvation, CN^- attaches to the Fe in cytochrome *c*, and blocks electron transfer.

Unlike the neutral ligand CO, the negatively charged CN^- ion is a strong Brønsted base ($pK_a = 9.4$) and a much poorer Lewis π acid. Its coordination chemistry is therefore mainly associated with metal ions in positive oxidation states.

11.7 Silicon and germanium

The full range of tetrahalides is known for silicon and germanium, and all of them are volatile molecular compounds. Lower down the group germanium shows signs of an inert pair in that it also forms nonvolatile dihalides. Among the silicon tetrahalides, the most important is the tetrachloride, which is prepared by direct reaction of the elements:

$$Si(s) + 2Cl_2(g) \rightarrow SiCl_4(l)$$

Much of the production of the tetrachloride is destined for purification by distillation followed by reduction with hydrogen to produce semiconductor-grade silicon:

$$SiCl_4(g) + 2H_2(g) \rightarrow Si(s) + 4HCl(g)$$

We described the preparation of 'amorphous silicon', which is better described as a solid silicon hydride SiH_x ($x \leq 0.5$), in Section 9.13.

Silicon and germanium halides are mild Lewis acids. They display this character when they add one or two ligands to yield complexes with a five-coordinate or six-coordinate central atom:

$$SiF_4(g) + 2F^-(aq) \rightarrow [SiF_6]^{2-}(aq)$$
$$GeCl_4(l) + N{\equiv}CCH_3(l) \rightarrow Cl_4GeN{\equiv}CCH_3(s)$$

11.8 Tin and lead

Dihalides and tetrahalides of tin are well known. The tetrachloride, bromide, and iodide are molecular compounds, but the tetrafluoride is an ionic solid since the small F^- ion permits a six-coordinate structure (this is reminiscent of the fluorides of the boron group). Lead tetrafluoride is also an ionic solid but, because of the inert pair effect, $PbCl_4$ is an unstable compound that decomposes into $PbCl_2$ and Cl_2 at room temperature. Lead tetrabromide and tetraiodide are unknown, and the dihalides dominate the halogen chemistry of lead. The arrangement of halogen atoms around the central metal atom in the dihalides of tin and lead often deviates from simple tetrahedral or octahedral coordination, and is attributed to the presence of a stereochemically active lone pair. The tendency to achieve the distorted structure is more pronounced with the small F^- ion, for the steric constraints are then less severe, and less distorted structures are observed with larger halides.

The oxides of lead are very interesting from both fundamental and technological standpoints. In the red form of PbO, the Pb(II) ions are four-coordinate (Fig. 11.10), but the O^{2-} ions around the Pb(II) lie in a square. As for the halides, this can be rationalized by the presence of a stereochemically active lone pair on the metal atom. Lead also forms oxides of mixed oxidation state. The best known is **red lead**, Pb_3O_4, which contains Pb(IV) in an octahedral environment and Pb(II) in an irregular 6-coordinate environment. The assignment of different oxidation numbers to the lead in these two sites is based on the shorter Pb—O distances for the atom identified as Pb(IV). The maroon form of lead(IV) oxide, PbO_2, crystallizes in the rutile structure. This oxide is a component of the cathode of a lead–acid battery.

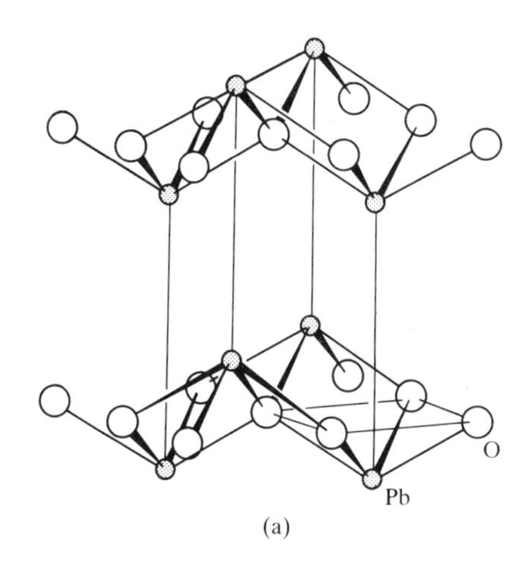

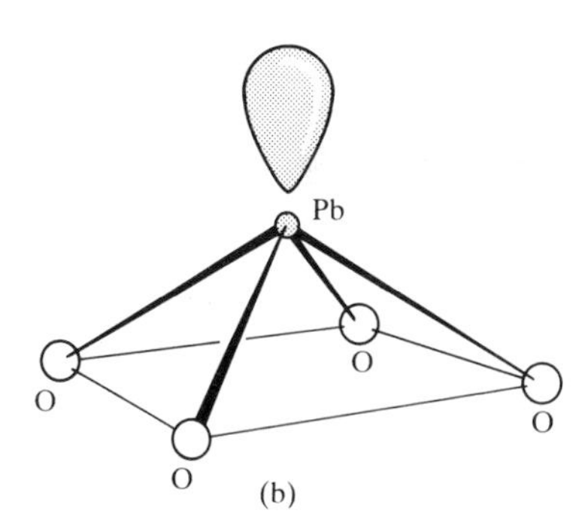

Fig. 11.10 (a) The structure of PbO. (b) The square pyramidal array of PbO showing the possible orientation of a stereochemically active lone pair of electrons.

The chemistry of the lead–acid battery is noteworthy because as well as being the most successful rechargeable battery, it illustrates the role of kinetics as well as thermodynamics in the operation of cells. In its fully charged state, the active material on the cathode is PbO_2 and at the anode it is Pb; the electrolyte is dilute sulfuric acid. One notable feature of this arrangement is that the lead-containing reactants and products at both electrodes are insoluble. When the cell is producing current, the reaction at the cathode is the reduction of Pb(IV) as PbO_2 to Pb(II), which in the presence of sulfuric acid is deposited on the electrode as insoluble $PbSO_4$:

$$PbO_2(s) + HSO_4^-(aq) + 3H^+(aq) + 2e^- \rightarrow PbSO_4(s) + 2H_2O(l)$$

At the anode, lead is oxidized to Pb(II), which is also deposited as the sulfate:

$$Pb(s) + SO_4^{2-}(aq) \rightarrow PbSO_4(s) + 2e^-$$

The overall reaction is

$$PbO_2(s) + 2HSO_4^-(aq) + 2H^+(aq) + Pb(s) \rightarrow 2PbSO_4(s) + 2H_2O(l)$$

The potential difference of about 2 V is remarkably high for a cell in which an aqueous electrolyte is used, because it exceeds by far the potential for the oxidation of water to O_2, which is 1.23 V. The success of this battery hinges on the high overpotentials (and hence low rates) of oxidation of H_2O on PbO_2 and of reduction of H_2O on lead.

Silicates and aluminosilicates

Although some examples of octahedrally coordinated Si are known, tetrahedral coordination is found in the great majority of the silicates. Their

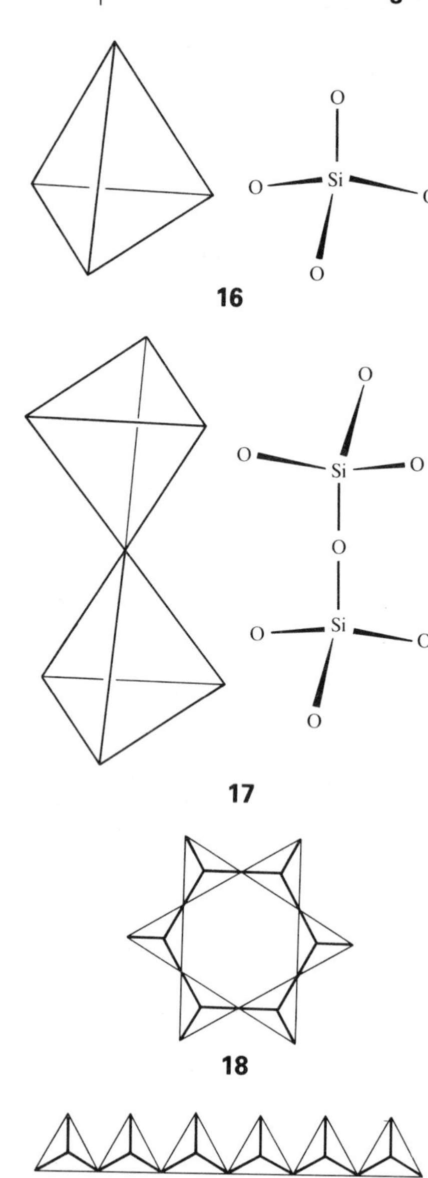

complicated structures are often easier to comprehend if we draw the SiO_4 unit as a tetrahedron with the Si atom at the center and O atoms at the vertices. The representation is often cut to the bone by drawing the SiO_4 unit as a simple tetrahedron with the atoms omitted. In general, these tetrahedra share vertices and (much more rarely) edges or faces. Each terminal O atom contributes -1 to the charge of the SiO_4 unit, but each shared O atom contributes zero. Thus orthosilicate (**16**) is $[SiO_4]^{4-}$, disilicate (**17**) is $[O_3SiOSiO_3]^{6-}$, and the SiO_2 unit of silica has no net charge because all O atoms are shared.

11.9 Extended silicate structures

With the above principles of charge balance in mind, it should be clear that an endless single-stranded chain or a ring of SiO_4 units, which has two shared O atoms for each Si atom, will have the formula and charge $[(SiO_3)^{2-}]_n$. An example of a compound containing such a cyclic metasilicate ion is the mineral beryl $Be_3Al_2Si_6O_{18}$, a major source of beryllium, which contains the $[Si_6O_{18}]^{12-}$ ion (**18**). Emerald is beryl in which Cr^{3+} ions are substituted for some Al^{3+} ions. A chain metasilicate (**19**) is present in the mineral jadeite, $NaAl(SiO_3)_2$, one of two different minerals sold as jade. In addition to other configurations for the single chain, there are double-chain silicates which include a family of minerals known commercially as 'asbestos'.[5]

Example 11.5: *Determining the charge on a cyclic silicate*

Determine the charge on the cyclic silicate anion $[Si_3O_9]^{n-}$.

Answer. The ion is a 6-membered ring with alternating Si and O atoms and 6 terminal O atoms, 2 on each Si atom. Since each terminal O atom contributes -1 to the charge, the overall charge is -6.

Exercise. Do the same for $[Si_4O_{12}]^{n-}$.

Silica and many silicates crystallize slowly, and by cooling the melt at an appropriate rate amorphous solids known as **glasses** can be obtained instead. In some respects these glasses resemble liquids. As with liquids, their structures are ordered over distances of only a few interatomic spacings (such as within a single SiO_4 tetrahedron). Unlike liquids, their viscosities are very high, and for many practical purposes they behave like solids.

The composition of silicate glasses has a strong influence on their physical properties. For example, fused quartz (amorphous SiO_2) softens at about 1600°C, borosilicate glass (which contains boron oxide, as we have already seen) softens at about 800°C, and crystal glass (which contains lead oxides, and is highly refracting on account of the high electron density of the Pb atoms) softens at even lower temperatures. The variation in softening point can be understood by appreciating that the Si—O—Si links in silicate glasses

[5] Asbestos has many commercially desirable qualities, but the fine asbestos particles are readily airborne and workers who handle the mineral are subject to a degeneration of the lung tissue which may manifest itself many years after the exposure. See L. Michaelis and S. S. Chissick (eds.), *Asbestos: Properties, applications, and hazards.* Wiley, New York (1979).

form the framework that imparts rigidity. When basic oxides such as Na_2O and CaO are incorporated (as in sodalime glass), they react with the SiO_2 melt and converts Si—O—Si links into terminal Si—O groups and hence lower its softening temperature (see also Chapter 18).

11.10 Aluminosilicates

As if the complexity of silicates were not enough, an even greater complexity is possible when Al atoms replace some of the Si atoms. The resulting **aluminosilicates** are largely responsible for the rich variety of the mineral world. We have already seen that in γ-alumina, Al^{3+} ions are present in both octahedral or tetrahedral holes. This versatility carries over into the aluminosilicates, where Al may substitute for Si in tetrahedral sites, enter an octahedral environment external to the silicate framework or, more rarely, occur with other coordination numbers. Since the aluminum occurs as Al(III), its presence in place of Si(IV) in an aluminosilicate renders the overall charge negative by one unit. An additional cation, such as H^+, Na^+, or $\frac{1}{2}Ca^{2+}$, is therefore required for each Al atom that replaces an Si atom. As we shall see, these additional cations have a profound effect on the properties of the materials.

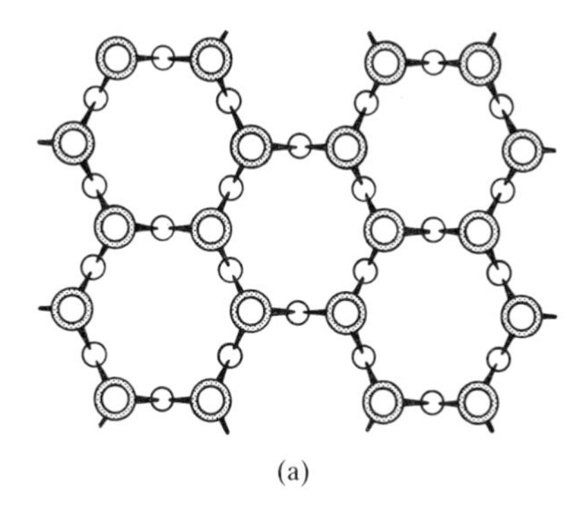

(a)

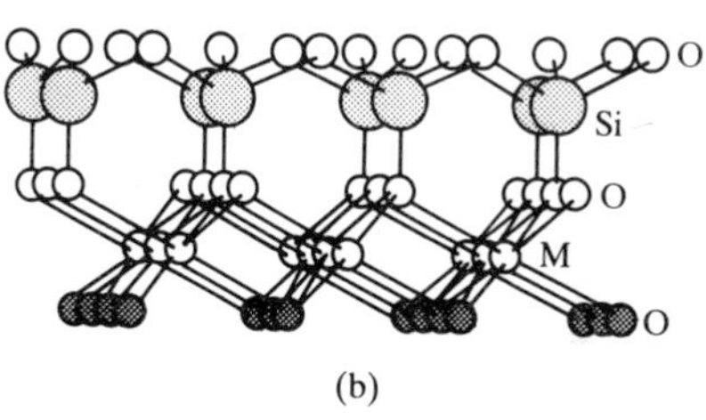

(b)

Fig. 11.11 (a) A net of SiO_4 tetrahedra with one O atom on top of each Si atom projecting toward the viewer. (b) Edge view of the above net, with one O/Si incorporated into a net of MO_6 octahedra. This structure is for the mineral chrysotile, for which M = Mg. When M is Al^{3+} and each of the O atoms on the bottom is replaced by an OH group this structure is close to that of the 1 : 1 clay mineral kaolinite.

Layered aluminosilicates

Many important minerals are varieties of layered aluminosilicates that also contain metals such as Mg, Li, and Fe: they include clays, talc, and various micas. In one class of layered aluminosilicate, the repeating unit consists of a silicate layer with the structure shown in Fig. 11.11. An example of a simple aluminosilicate of this type (simple, that is, in the sense of there being no additional elements) is the mineral kaolinite, $Al_2(OH)_4Si_2O_5$, which is used commercially as china clay. The electrically neutral layers are held together by rather weak hydrogen bonds, so the mineral readily cleaves and incorporates water between the layers.

A larger class of aluminosilicates has Al^{3+} ions sandwiched between silicate layers (Fig. 11.12). One such mineral is pyrophyllite, $Al_2(OH)_2Si_4O_{10}$. The mineral talc $Mg_3(OH)_2Si_4O_{10}$ is obtained when three Mg^{2+} ions replace two Al^{3+} ions in the octahedral sites. In talc (and in pyrophyllite) the repeating layers are neutral, and as a result talc readily cleaves between them. This accounts for its familiar slippery feel.

Muscovite mica, $KAl_2(OH)_2Si_3AlO_{10}$, has charged layers because one Al(III) substitutes for one Si(IV) in the pyrophylite structure. The resulting negative charge is compensated by a K^+ ion that lies between the repeating layers (Fig. 11.12). Because of this electrostatic cohesion, muscovite is not soft like talc, but it is readily cleaved into sheets. More highly charged layers with dipositive ions between the layers lead to greater hardness.

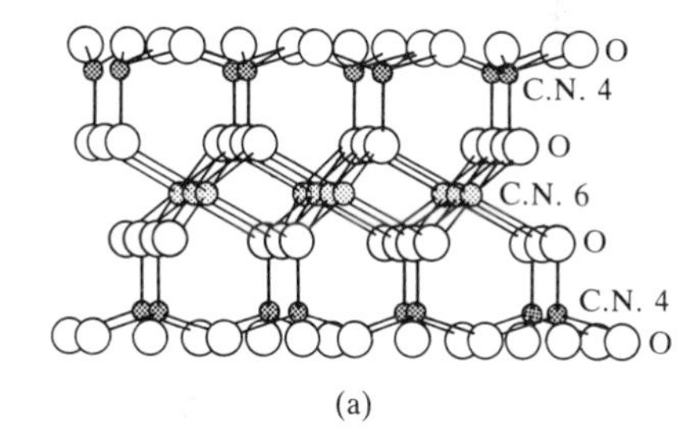

(a)

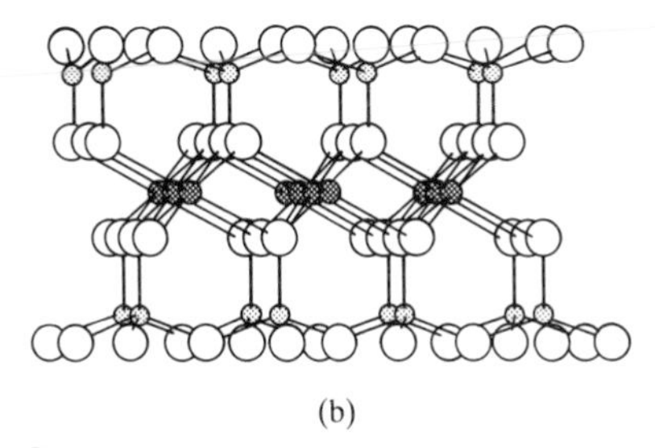

(b)

Fig. 11.12 (a) The structure of 2 : 1 clay minerals such as muscovite mica $KAl_2(OH)_2Si_3AlO_{10}$ where K^+ resides between the charged layers (exchangeable cation sites), Si^{4+} resides in sites of C.N. 4 and Al^{3+} in sites of C.N. 6. (b) In talc, Mg^{2+} resides in the octahedral sites and O atoms on the top and bottom are replaced by OH groups.

Three-dimensional aluminosilicates

There are many minerals based on a three-dimensional aluminosilicate framework. The **feldspars**, for instance, which are the most important class of rock-forming minerals (and include granite), belong to this group. The aluminosilicate frameworks of feldspars are built up by sharing all vertices of SiO_4 or AlO_4 tetrahedra. Cavities in the three-dimensional network then accommodate ions such as K^+ and Ba^{2+}. Two examples are the feldspars orthoclase, $KAlSi_3O_8$, and albite, $NaAlSi_3O_8$.

11.11 Molecular sieves

The **molecular sieves** are crystalline aluminosilicates having open structures with apertures of molecular dimensions. These substances represent a major triumph of solid state chemistry, for their synthesis and our understanding of their properties combine clever structure determination, imaginative synthetic chemistry, and important practical applications.

General description

The name 'molecular sieve' is very appropriate, because molecular sieves can capture molecules that are smaller than the aperture dimensions and so can be used to separate molecules of different sizes. A subclass of molecular sieves, the **zeolites**, have an aluminosilicate framework with cations trapped inside tunnels or cages. In addition to their function as molecular sieves, zeolites can exchange their ions for those in a surrounding solution.

Since the cages are defined by the crystal structure, they are highly regular and of precise size. Consequently, molecular sieves capture molecules with greater selectivity than do solids with high surface area, such as silica gel or activated carbon, where molecules may be caught in irregular voids between the small particles. Zeolites are also used for shape-selective heterogeneous catalysis. For example, the molecular sieve ZSM-5 is used to synthesize 1,2-dimethylbenzene (*o*-xylene) for use as an octane booster in gasoline. The other xylenes are not produced because the catalytic process is controlled by the size and shape of the zeolite cages and tunnels. This and other applications are summarized in Table 11.5 and discussed in Chapter 17.

One interesting aspect of zeolite chemistry is that large molecules can be synthesized from smaller molecules inside the zeolite cage. The result is like a ship in a bottle, because once assembled the large molecule is too large to escape. For example, Na^{+} ions in a Y-type zeolite may be replaced by Fe^{2+} ions (by ion exchange). The resulting Fe^{2+}-Y zeolite is heated with phthalonitrile, which diffuses into the zeolite and condenses around the Fe^{2+} to form iron phthalocyanine, which is too large to escape from the cage:

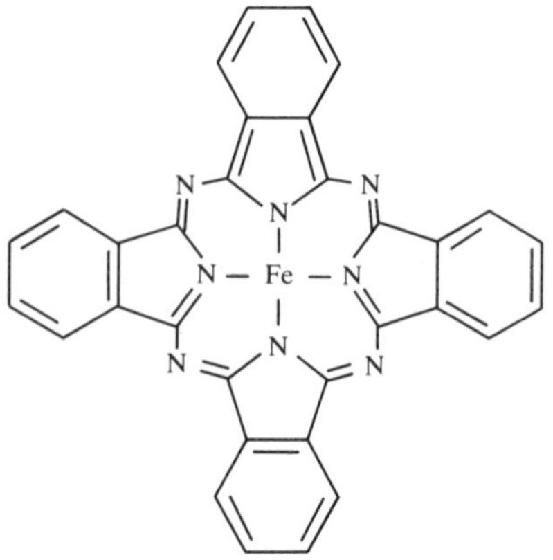

Synthesis

Synthetic procedures have added to the many naturally occurring zeolite varieties that have specific cage sizes and specific chemical properties within the cages. These synthetic zeolites are sometimes made at atmospheric pressure, but more often they are produced in a high-pressure autoclave.

Table 11.5. Some uses of zeolites

Function	Application
Ion exchange	Water softeners in washing detergents
Absorption of molecules	Selective gas separation in: industrial processes, gas chromatography, laboratory experiments
Solid acid and shape selectivity	Cracking of high-molecular-weight hydrocarbons for fuel and petrochemical intermediates. Alkylation and isomerization of aromatics for gasoline and polymer intermediates

Their open structures appear to form around hydrated cations or other large cations such as NR_4^+ ions introduced into the reaction mixture. For example, a synthesis may be performed by heating colloidal silica to 100 to 200°C in an autoclave with an aqueous solution of tetrapropylammonium hydroxide. The microcrystalline product, which has the typical composition $\{[N(C_3H_7)_4][OH]\}(SiO_2)_{48}$, is converted into the zeolite by burning away the C, H, and N of the quaternary ammonium cation at 500°C in air. Aluminosilicate zeolites are made by including high-surface-area alumina in the starting materials.

Framework representation

A wide range of zeolites has been prepared with varying cage and bottleneck sizes (Table 11.6). Their structures are based on approximately

Table 11.6. Composition and properties of some molecular sieves

Molecular sieve	Composition	Pore size/Å	Chemical properties
A	$Na_{12}[(AlO_2)_{12}(SiO_2)_{12}]\cdot xH_2O$	4	Absorbs small molecules; ion exchanger, hydrophilic
X	$Na_{86}[(AlO_2)_{86}(SiO_2)_{106}]\cdot xH_2O$	8	Absorbs medium-sized molecules; ion exchanger, hydrophilic
Chabazite	$Ca_2[(AlO_2)_4(SiO_2)_8]\cdot xH_2O$	4–5	Absorbs small molecules; ion exchanger, hydrophilic
ZSM-5	$Na_3[(AlO_2)_3(SiO_2)]\cdot xH_2O$	5.5	Moderately hydrophilic
ALPO-5	$AlPO_4\cdot xH_2O$	8	Moderately hydrophilic
Silicalite	SiO_2	6	Hydrophobic

tetrahedral MO_4 units, which in the great majority of cases are SiO_4 and AlO_4. Since the structures involve many such tetrahedral units, it is common practice to abandon the polyhedral representation in favor of one that emphasizes the position of the Si and Al atoms. In this scheme, the Si or Al atom lies at the intersection of four line segments (Fig. 11.13) and the O atom bridge lies on the line segment. This **framework representation** has the advantage of giving a clear impression of the shapes of the cages and channels in the zeolite.

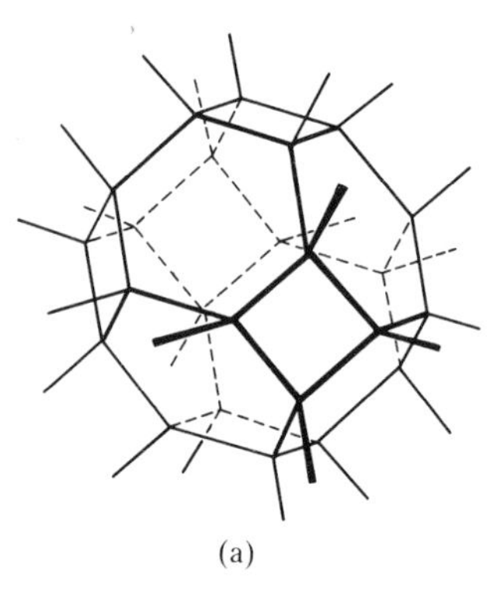
(a)

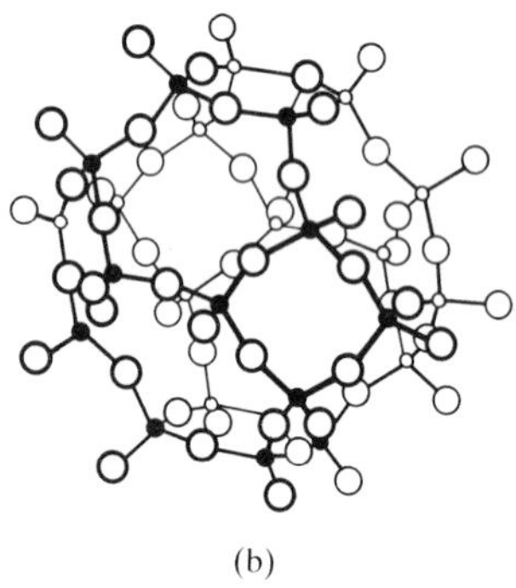
(b)

Fig. 11.13 (a) Framework representation of a truncated octahedron (truncation perpendicular to the fourfold axes of the octahedron). (b) Relation of Si and O atoms to the framework. Note that a Si atom is at each vertex of the truncated octahedron and an O atom is approximately along each edge.

Sodalite zeolites

A large and important class of zeolites is based on the **sodalite cage**. This is a truncated octahedron (Fig. 11.13) formed by slicing off each vertex. The truncation leaves a square face in the place of each vertex and the triangular faces of the octahedron are transformed into regular hexagons. 'Zeolite type A' is based on sodalite cages that are joined by O bridges between the square faces (Fig. 11.14). Eight such sodalite cages are linked in a cubic pattern with a large central cavity called an ***α* cage**. The α cages share octagonal faces, with an open diameter of 4.20 Å. Thus water or other small molecules can fill them and diffuse through octagonal faces. However, these faces are too small for molecules with van der Waals diameters larger than 4.20 Å, and they cannot diffuse.[6]

Example 11.6: *Analyzing the structure of the sodalite cage*

Identify the fourfold and sixfold axes in the truncated octahedral polyhedron used to describe the sodalite cage.

Answer. There is one fourfold axis running through each pair of opposite square faces. This gives a total of six fourfold axes. Similarly, a set of four sixfold axes run through opposite sixfold faces.

Exercise. How many Si and Al atoms are there in one sodalite cage?

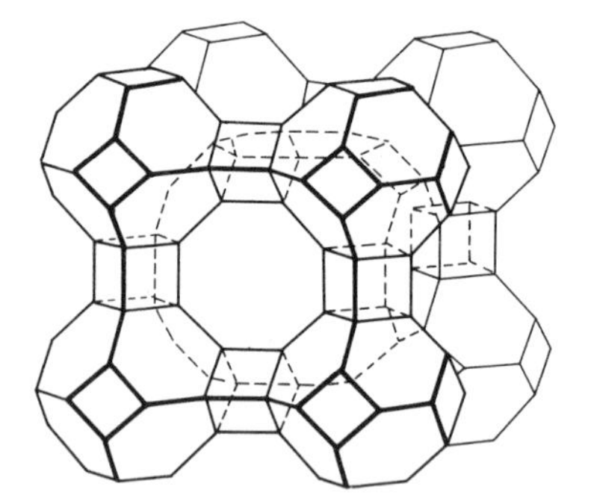

Fig. 11.14 Framework representation of a type-A Zeolite. Note the sodalite cages (truncated octahedra), the small cubic cages, and the central supercage.

Ion exchange properties and polarity

The charge on the aluminosilicate zeolite framework is neutralized by cations lying within the cages. In the type-A zeolite, Na^+ ions are present and the formula is $Na_{12}(AlO_2)_{12}(SiO_2)_{12}{\cdot}xH_2O$. Numerous other ions, including *d*-block cations and NH_4^+, can be introduced by ion exchange with aqueous solutions, and another of the major uses of zeolites is for ion exchange. In this connection, zeolites are widely used in laundry detergent to remove di- and trivalent ions that decrease the effectiveness of the surfactant. Zeolites have largely replaced polyphosphates because the latter, which are nutrients, find their way into natural waters and stimulate the growth of algae.

In addition to the control of properties by selecting a zeolite with the appropriate cage and bottleneck size, the zeolite can be chosen for its affinity for polar or nonpolar molecules according to its polarity (Table

[6] The combination of solid state ^{29}Si-NMR and ^{27}Al-NMR with high-resolution electron microscopy has revolutionized the structural chemistry of zeolites and other aluminosilicates. An interesting account is given by J. M. Thomas and C. R. A. Catlow, *Prog. Inorg. Chem.*, **35**, 1 (1987).

11.6). The aluminosilicate zeolites, which always contain charge-compensating ions, have high affinities for polar molecules such as H_2O and NH_3. In contrast, the nearly pure silica molecular sieves bear no net electric charge and are nonpolar to the point of being mildly hydrophobic. Another group of mildly hydrophobic zeolites is based on the aluminum phosphate frameworks, for $AlPO_4$ is isoelectronic with Si_2O_4 and the framework is similarly uncharged.

Boron cluster compounds and carboranes

11.12 Boron subhalides

Boron halides containing B—B bonds have been prepared. The best known of these have the formula B_2X_4, with X = F, Cl, and Br, and the tetrahedral cluster compound B_4Cl_4. The preferred route to B_2Cl_4 is to pass an electric discharge through BCl_3 gas in the presence of a Cl atom scavenger, such as mercury vapor. Spectroscopic data indicate that BCl is produced by electron impact on BCl_3:

$$BCl_3(g) \xrightarrow{\text{electron impact}} BCl(g) + 2Cl(g)$$

The Cl atoms are scavenged by mercury vapor and removed as Hg_2Cl_2, and the BCl is thought to combine with BCl_3 to yield B_2Cl_4.

Metathesis reactions can be used to make B_2X_4 derivatives from B_2Cl_4. The thermal stability of these derivatives increases with increasing tendency of the X group to form a π bond with B:

$$B_2Cl_4 < B_2F_4 < B_2(OR)_4 \ll B_2(NR_2)_4$$

For a long time it was thought that X groups with the lone pairs necessary for π bonding were essential for the stability of B_2X_4 compounds, but recently diboron compounds with alkyl groups have been prepared. When the alkyl groups are bulky, compounds that are stable at room temperature are obtained, such as $B_2(^tBu)_4$.

Diboron tetrachloride is a highly reactive volatile molecular liquid. The B_2Cl_4 molecules are planar (**20**) in the solid state but staggered (**21**) in the gas. This suggests that rotation about the B—B bond is quite easy, as is expected for a B—B single bond. One of the interesting reactions of B_2Cl_4 is its addition across a C═C double bond:

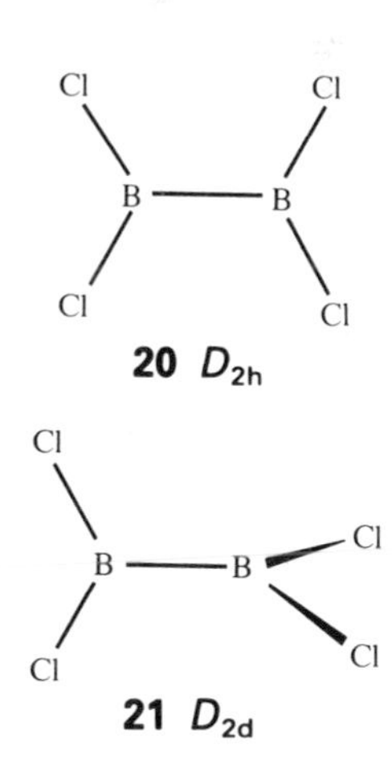

20 D_{2h}

21 D_{2d}

$$B_2Cl_4 + C_2H_4 \xrightarrow{\text{low temperature}} Cl_2BCH_2CH_2BCl_2$$

A secondary product in the synthesis of B_2Cl_4 is B_4Cl_4, a pale yellow solid composed of tetrahedral molecules. Higher boron halide clusters are also known, but they do not conform to the formulas and structures of the BH compounds discussed below. Although the inclusion of B—Cl π bonding in the molecular-orbital treatment helps to rationalize the structures of some of the boron subhalides, they are not yet fully understood.

11.13 Higher boron hydrides

We discussed the synthesis and chemistry of B_2H_6 and the existence of some higher boron hydrides in Section 9.10. In this section we describe the structures and properties of the cage-like boron hydrides. These include

Alfred Stock's series B_nH_{n+4} and B_nH_{n+6} (Section 9.10) as well as the more recently discovered $[B_nH_n]^{2-}$ closed polyhedra. The boron hydrides occur with a variety of shapes, some resembling nests and others (to the imaginative eye) butterflies. All the boron hydrides are electron deficient.

Bonding

The modern understanding of the bonding in boron hydrides derives from the work of Christopher Longuet-Higgins who, as an undergraduate at Oxford, published a seminal paper in which he introduced the concept of 3c,2e bonds. He later developed a fully delocalized molecular orbital treatment for boron polyhedra and predicted the stability of the icosahedral ion $[B_{12}H_{12}]^{2-}$, which was subsequently verified. William Lipscomb and his students in the USA used single-crystal X-ray diffraction to determine the structures of a large number of boron hydrides, and extended the concept of multicenter bonding to these more complex compounds.

The boron cluster compounds are best considered from the standpoint of fully delocalized molecular orbitals containing electrons that contribute to the stability of the entire molecule. However, it is sometimes fruitful to identify groups of three atoms and to regard them as bonded together by versions of the 3c,2e bonds of the kind that occur in diborane itself (**5** in Chapter 9). In the more complex boranes, the three centers of the 3c,2e bonds may be B—H—B bridge bonds, but they may also be bonds in which three B atoms lie at the corners of an equilateral triangle with their sp^3 hybrid orbitals overlapping at its center (**22**).

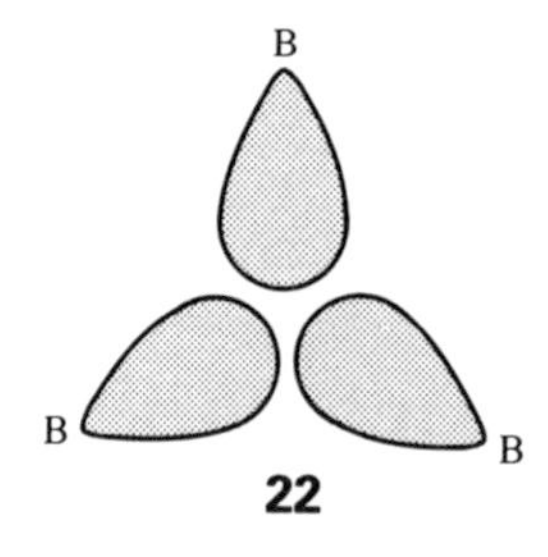

22

Classification and electron counting

A correlation between the number of electrons (counted in a specific way), the formula, and the shape of the molecule was established by the British chemist Kenneth Wade in the 1970s.[7] Wade's rules apply to a class of polyhedra called **deltahedra** (because they are made up of triangular faces) and can be used in two ways. For boranes and borohydrides, they enable us to predict the general shape of the molecule or anion from its formula. However, since the rules are also expressed in terms of the number of electrons that contribute to the framework of the species, we can extend them to other substances in which there are atoms other than boron, such as the carboranes and other *p*-block clusters. Here we concentrate on the boron compounds, where knowing the formula is sufficient for predicting the shape. However, so that we can cope with other clusters we shall show how to count the framework electrons too.

For boranes, the building block from which the deltahedron is constructed is assumed to be one B—H group. The electrons in the B—H bond are ignored in the counting procedure, but all others are included whether or not it is obvious that they help to hold the skeleton together. By the 'skeleton' we mean the framework of the cluster with each B—H group counted as a unit. If a B atom happens to carry two H atoms, only one of the B—H bonds is treated as a unit (this rather odd feature is just a part of Wade's rules: it works—counting electrons in this way provides a parameter

[7] Wade has given a very clear account of his rules in 'The key to cluster shapes', K. Wade, *Chem. Brit.*, **11**, 177 (1975).

that helps to correlate properties). For instance, in B_5H_{11}, one of the B atoms has two 'terminal' H atoms, but only one B—H bond is treated as a unit, and the other pair of electrons is treated as part of the skeleton. A B—H group makes two electrons available to the skeleton (the B atom provides three and the H atom one, but of these four, two are used for the B—H bond).

Example 11.7: *Counting skeletal electrons*

Count the number of skeletal electrons in B_4H_{10} (**23**).

Answer. Four B—H units contribute $4 \times 2 = 8$ electrons, and the six additional H atoms contribute a further six electrons, giving 14 in all. The resulting seven pairs are distributed as shown in (**24**): two are used for the additional terminal B—H bonds, four are used for the four B—H—B bridges, and one is used for the central B—B bond.

Exercise. How many skeletal electrons are present in B_5H_{11}?

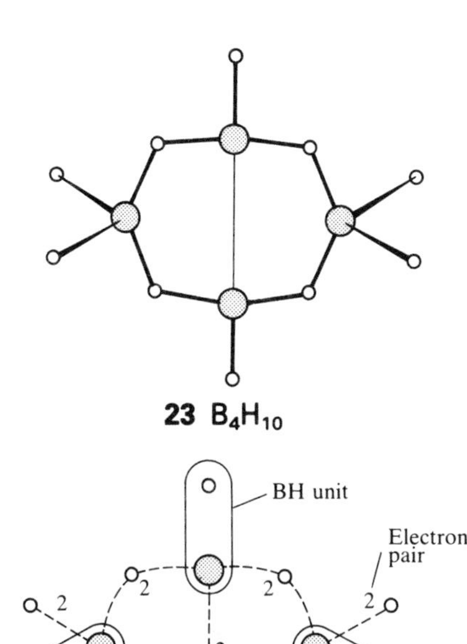

23 B_4H_{10}

24 B_4H_{10}

According to Wade's rules (Table 11.7), boranes of formula $[B_nH_n]^{2-}$ will be found to have the *closo* structure (from the Greek for 'cage') with a B atom at each corner of a closed deltahedron and no B—H—B bonds. Such structures have $n+1$ pairs of skeletal electrons. This series of anions is known for $n = 5$ to 12 and examples include the trigonal bipyramidal $[B_5H_5]^{2-}$ ion, the octahedral $[B_6H_6]^{2-}$ ion, and the icosahedral $[B_{12}H_{12}]^{2-}$ ion. The *closo*-hydroborates and their carborane analogs (Section 11.14) are often thermally stable and fairly unreactive.

Boranes of formula B_nH_{n+4} have the *nido* structure (the name is derived from the Latin for 'nest'). They can be viewed as a *closo*-borane that has lost one vertex but may have B—H—B bonds as well as B—B bonds. The compounds in this series contain $n+2$ pairs of skeletal electrons. An example is B_5H_9 (Table 11.8). In general, the thermal stability of the *nido*-boranes is intermediate between that of *closo*- and *arachno*-boranes.

The boranes of formula B_nH_{n+6} have the *arachno* structure (from the Greek for 'spider') and are *closo*-borane polyhedra less two vertices (and may have B—H—B bonds); they are so called because they resemble an untidy spider's web. The *arachno*-boranes have $n+3$ skeletal electron pairs. One example of an *arachno*-borane is pentaborane(11) (B_5H_{11}, Table 11.8). As with most *arachno*-boranes, pentaborane(11) is thermally unstable at room temperature and is highly reactive.

Table 11.7. Classification and electron count of boron hydrides

Type	Formula*	Skeletal electron pairs	Examples
Closo	$[B_nH_n]^{2-}$	$n+1$	$[B_5H_5]^{2-}$ to $[B_{12}H_{12}]^{2-}$
Nido	B_nH_{n+4}	$n+2$	B_2H_6, B_5H_9, B_6H_{10}
Arachno	B_nH_{n+6}	$n+3$	B_4H_{10}, B_5H_{11}
Hypho†	B_nH_{n+8}	$n+4$	None‡

* In some cases, protons can be removed; thus $[B_5H_8]^-$ is a *nido* analog of B_5H_9.
† The name comes from the Greek word for 'net'.
‡ Some derivatives are known.

Table 11.8. Structures of some *nido* and *arachno* boron hydrides

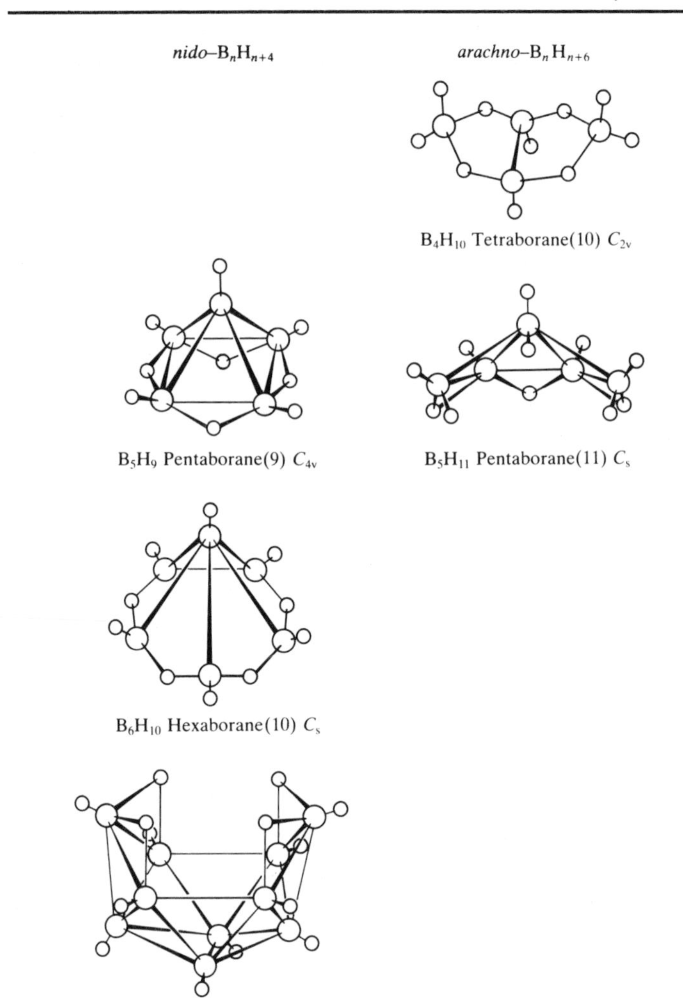

Example 11.8: *Using Wade's rules*

What is the structure of $[B_5H_5]^{2-}$?

Answer. The structural class of $[B_5H_5]^{2-}$ can be determined from the formula. In this case we note that the general formula is $[B_nH_n]^{2-}$, which is characteristic of a *closo*-borane. Alternatively, we can count the number of skeletal electron pairs and from that deduce the structural type. Assuming one B—H bond per B atom, there are five B—H units to take into account and therefore ten skeletal electrons plus two from the overall −2 charge. The total number of pairs is therefore $5 \times 1 + 1 = 6$, or $n + 1$, skeletal electron pairs. This number is characteristic of *closo* clusters. The closed polyhedron must contain triangular faces and five vertices; therefore a trigonal bipyramidal structure is indicated.

Exercise. How many framework electron pairs are present in B_5H_9 and to what structural category does it belong? Sketch the structure.

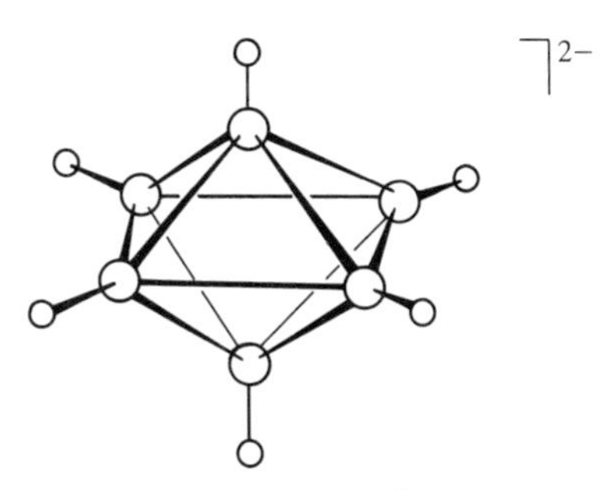

25 *closo*-$[B_6H_6]^{2-}$

A very useful structural correlation between *closo*, *nido*, and *arachno* compounds is based on the observation that clusters having the same numbers of skeletal electrons are related by removal of successive B—H groups and the addition of the appropriate numbers of electrons and H atoms. This type of process relates the octahedral *closo*-$[B_6H_6]^{2-}$ anion (**25**) to the square-pyramidal *nido*-B_5H_9 borane (**26**), and that in turn to the butterfly-like *arachno*-$[B_4H_{10}]$ molecule (**27**):

$$closo\text{-}[B_6H_6]^{2-} \xrightarrow{-BH,\,-2e,\,+4H} nido\text{-}B_5H_9 \xrightarrow{-BH,\,+2H} arachno\text{-}B_4H_{10}$$

Each structure has seven framework pairs. This correlation of cluster types is illustrated in more detail in Fig. 11.15. However, although this approach gives a good way of deducing the shapes of the molecules, it does not represent how they are interconverted chemically.

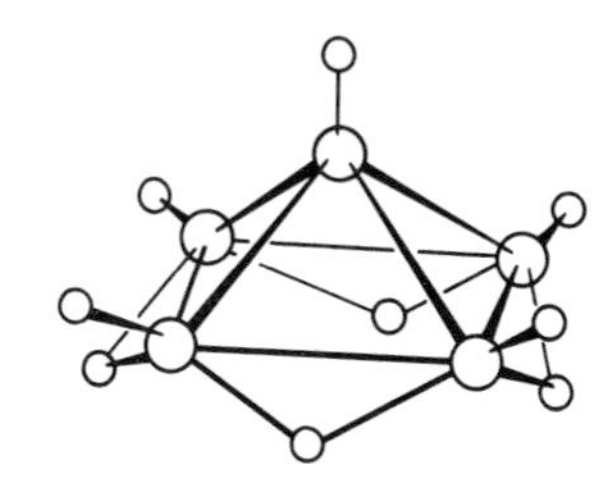

26 *nido*-B_5H_9

Molecular orbitals

Wade's rules have been justified by molecular orbital calculations. We shall illustrate the style of reasoning involved by considering the first of them (the $n+1$ rule). In particular we shall show that $[B_6H_6]^{2-}$ has a low energy if it has an octahedral *closo* structure, as predicted by the rules.

A B—H bond utilizes one electron and one orbital of the B atom, leaving three orbitals and two electrons for the skeletal bonding. One of these orbitals, which is called a **radial orbital**, can be considered to be a boron *sp* hybrid pointing toward the interior of the fragment (**28**). The remaining two boron *p* orbitals, the **tangential orbitals**, are perpendicular to the radial orbital (**29**). The shapes of the 18 symmetry adapted linear combinations of these 18 orbitals in an octahedral B_6H_6 cluster can be found in Appendix 6, and we show the ones with net bonding character in Fig. 11.16.

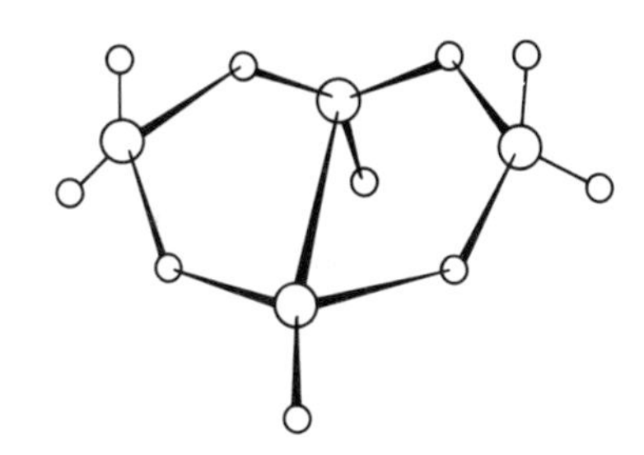

27 *arachno*-B_4H_{10}

The lowest energy orbital is totally symmetric (a_{1g}) and arises from in-phase contributions from all the radial orbitals. Calculations show that the next higher orbitals are the t_{1u} orbitals, each one of which is a combination of four tangential and two radial orbitals. Above these three degenerate orbitals lies another three t_{2g} orbitals, giving seven bonding orbitals in all. Hence, there are seven net bonding orbitals delocalized over the skeleton, and they are separated by a considerable gap from the remaining 11 largely antibonding orbitals (Fig. 11.17).

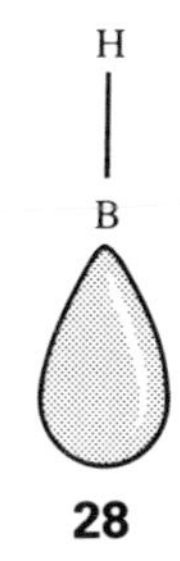

28

There are seven electron pairs to accommodate, one pair from each of the six B atoms and one pair from the overall −2 charge. These seven pairs can all enter and fill the seven bonding skeleton orbitals, and hence give rise to stable structure, in accord with the $n+1$ rule. Note that the unknown neutral octahedral B_6H_6 molecule would have too few electrons to fill the t_{2g} bonding orbitals and would be expected to be unstable.

11.14 Carboranes

Closely related to the polyhedral borohydrides is a large family of **carboranes**, which are clusters that contain both B and C atoms. Now we begin to see the full generality of Wade's approach, for BH^- is isoelectronic

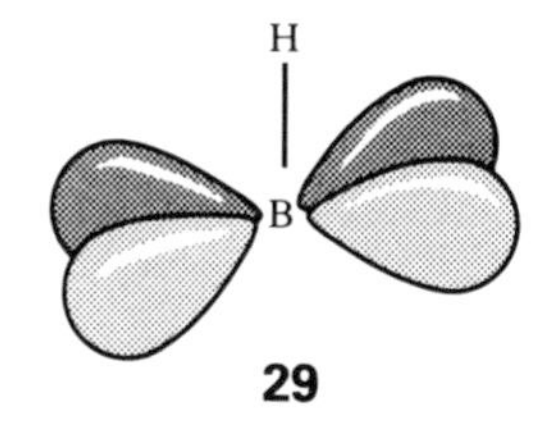

29

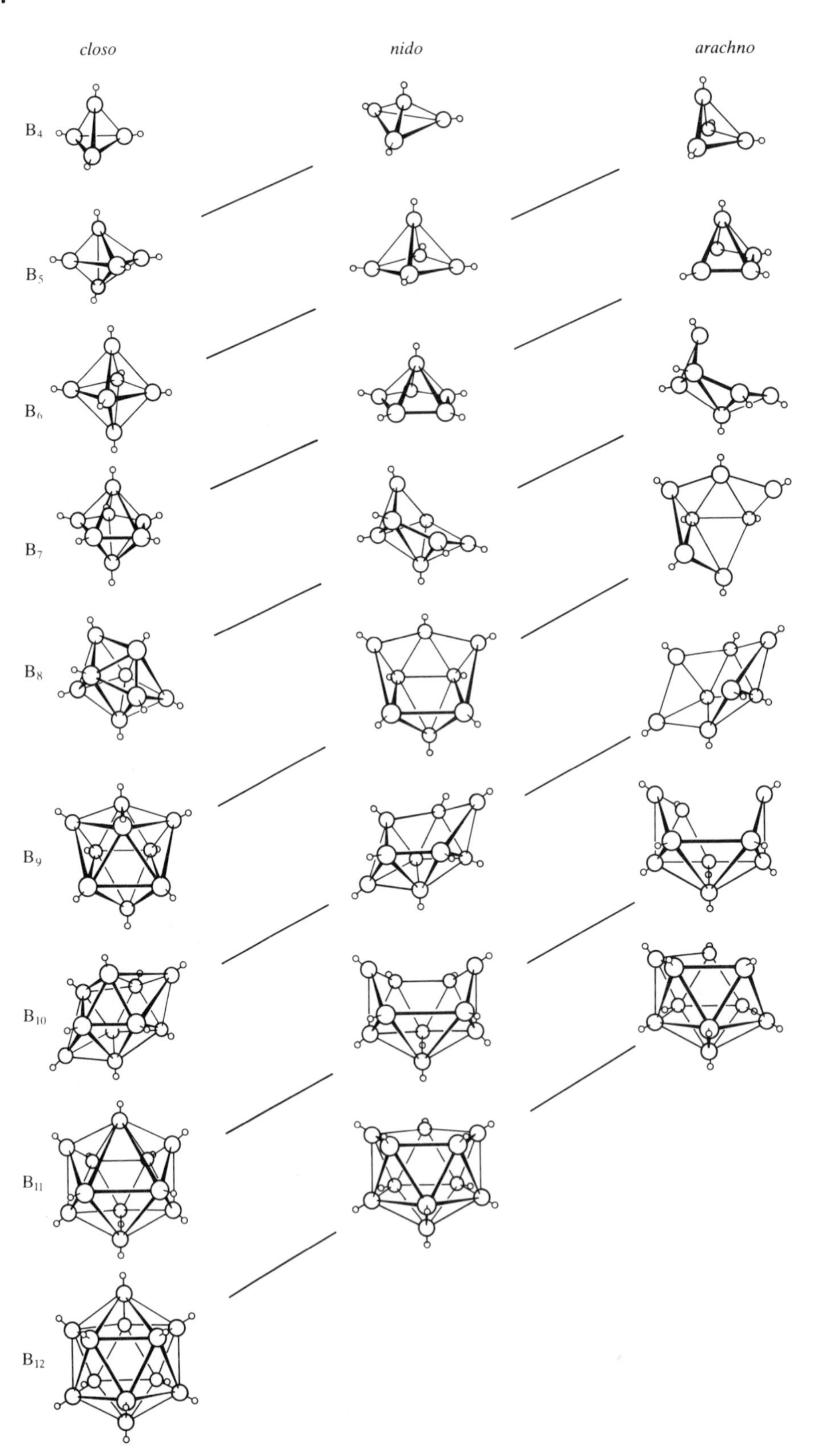

Fig. 11.15 Structural relations between *closo*, *nido*, and *arachno* boranes and heteroatom boranes. The diagonal lines connect species that have the same number of skeletal electrons. Hydrogen atoms beyond those in the B—H framework and charges have been omitted. (From R. W. Rudolph, *Acc. chem. Res.*, **9**, 446 (1976).)

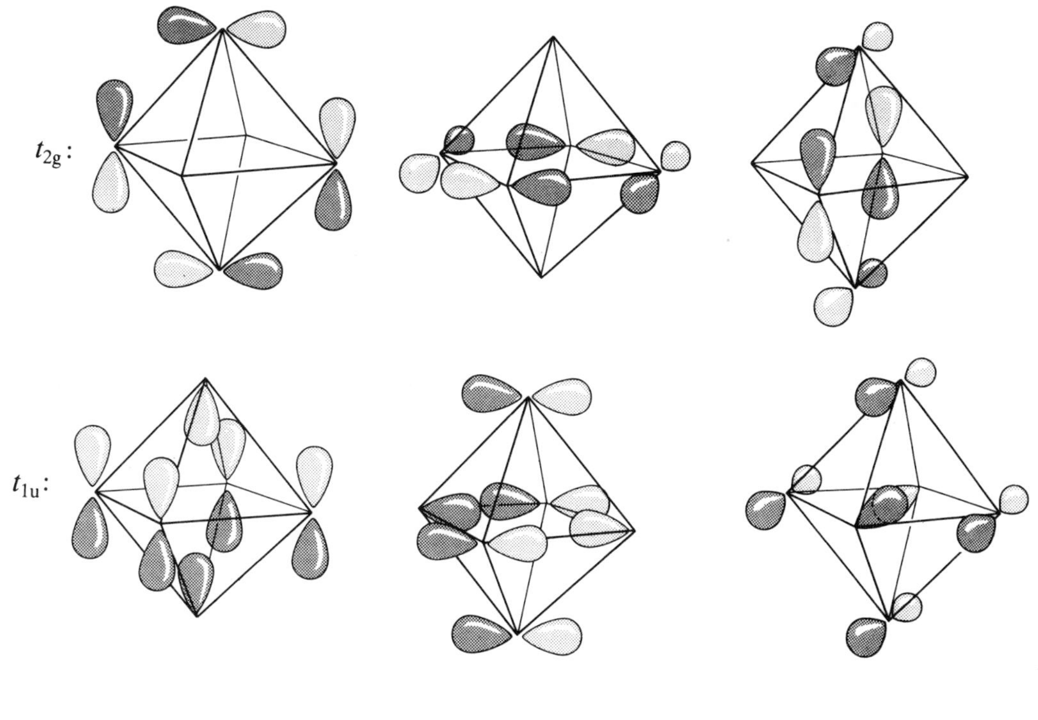

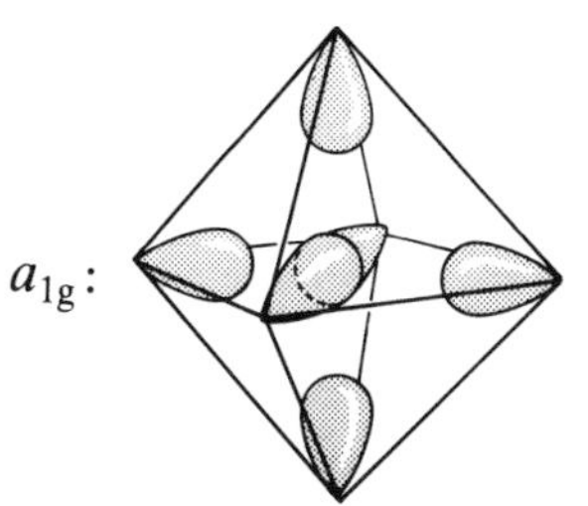

Fig. 11.16 Radial and tangential framework bonding molecular orbitals for $[B_6H_6]^{2-}$.

and isolobal with C—H (**30**), and we can expect the polyhedral borohydrides and carboranes to be related. Thus an analog of $[B_6H_6]^{2-}$ (**25**) is the neutral carborane $B_4C_2H_6$ (**31**).

Entry into the world of carboranes is facilitated in practice by the ease of preparing *closo*-1,2-$B_{10}C_2H_{12}$ (**32**) by the addition of acetylene to decaborane(14):

$$B_{10}H_{14} \xrightarrow{(1)\ Et_2S;\ (2)\ C_2H_2} B_{10}C_2H_{12} + 3H_2$$

The 1,2-compound produced in this reaction can be converted thermally to the 1,7- (**33**) and 1,12-isomers (**34**). The H atoms attached to the C atoms are very mildly acidic in the *closo*-$B_{10}C_2H_{12}$ clusters, so it is possible to lithiate these compounds with butyllithium:

$$B_{10}H_{10}C_2H_2 + 2LiC_4H_9 \rightarrow B_{10}H_{10}C_2Li_2 + 2C_4H_{10}$$

These dilithiocarboranes are good nucleophiles and thus open the way to the synthesis of a wide range of carborane derivatives. For example, reaction with CO_2 gives a dicarboxylic acid carborane:

$$B_{10}H_{10}C_2Li_2 + CO_2 \xrightarrow{(1)\ \text{react},\ (2)\ 2H_2O} B_{10}H_{10}C_2(COOH)_2$$

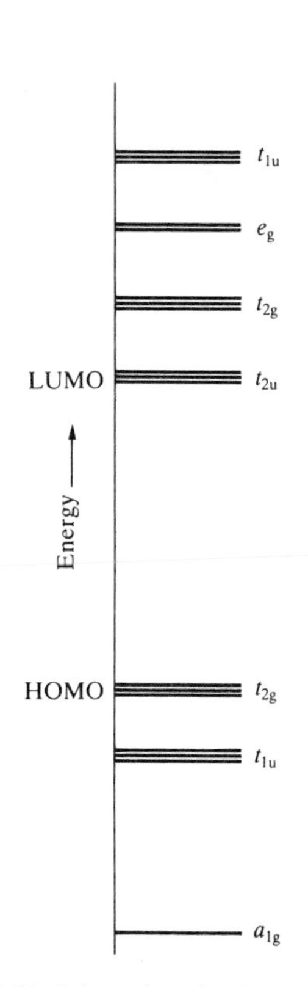

Fig. 11.17 Schematic molecular orbital energy levels of the B atom skeleton of $[B_6H_6]^{2-}$. Orbitals associated with the B—H bonds are not shown.

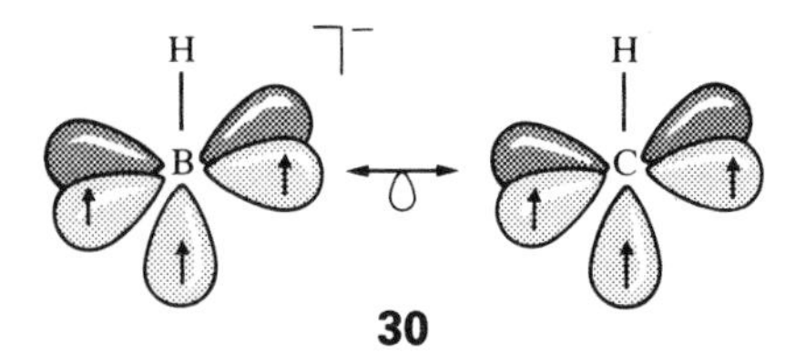

30

Similarly, I_2 leads to the diiodocarborane and NOCl yields $B_{10}H_{10}C_2(NO)_2$.

Although 1,2-$B_{10}C_2H_{12}$ is very stable, the cluster can be partially fragmented in strong base, and then deprotonated with NaH to yield *nido*-$[B_9C_2H_{11}]^{2-}$:

$$B_{10}C_2H_{12} + NaOEt + 2EtOH \rightarrow Na[B_9C_2H_{12}] + B(OEt)_3 + H_2$$
$$Na[B_9C_2H_{12}] + NaH \rightarrow Na_2[B_9C_2H_{11}] + H_2$$

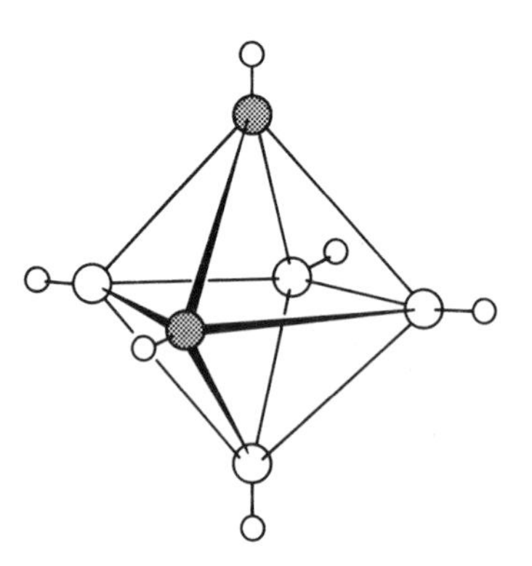

31 *closo*-1,2-$B_4C_2H_6$

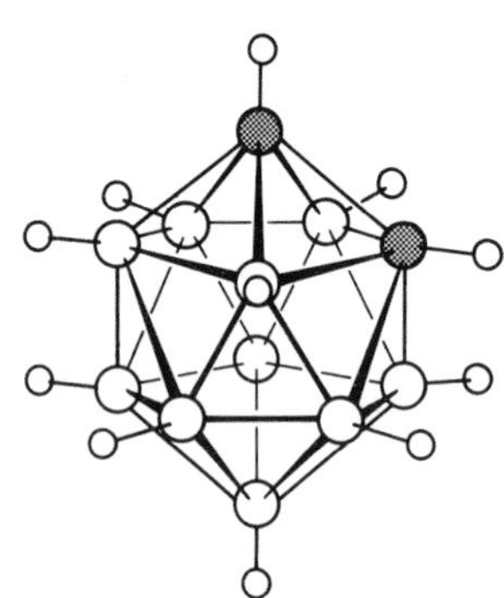

32 *closo*-1,2-$B_{10}C_2H_{12}$

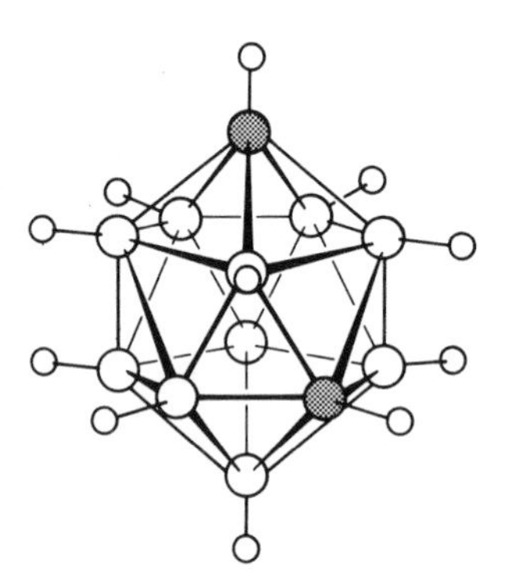

33 *closo*-1,7-$B_{10}C_2H_{12}$

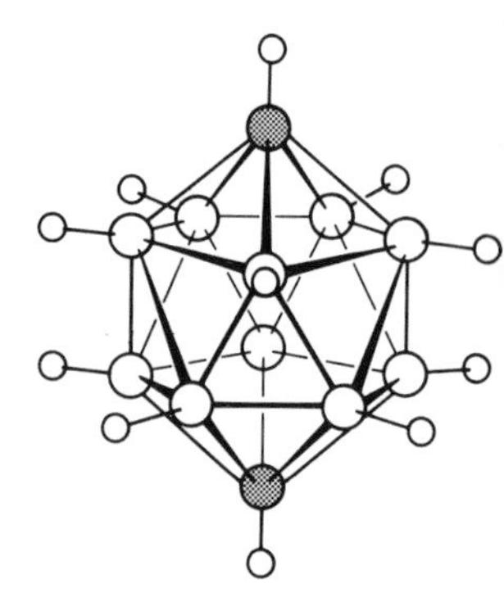

34 *closo*-1,12-$B_{10}C_2H_{12}$

The importance of these reactions is that *nido*-$[B_9C_2H_{11}]^{2-}$ (Fig. 11.18a) is an excellent ligand. In this role it mimics the isolobal cyclopentadienide ligand ($[C_5H_5]^-$, Fig. 11.18b) which is widely employed in organometallic chemistry:

$$2Na_2[B_9C_2H_{11}] + FeCl_2 \xrightarrow{\text{THF}} 2NaCl + Na_2[Fe(B_9C_2H_{11})_2]$$
$$2Na[C_5H_5] + FeCl_2 \xrightarrow{\text{THF}} 2NaCl + Fe(C_5H_5)_2$$

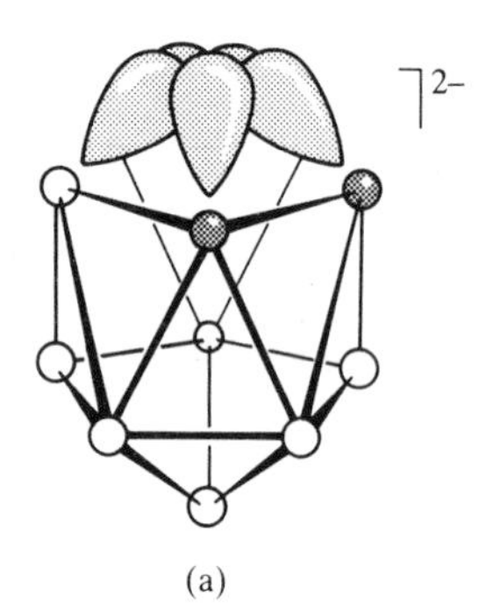

(a)

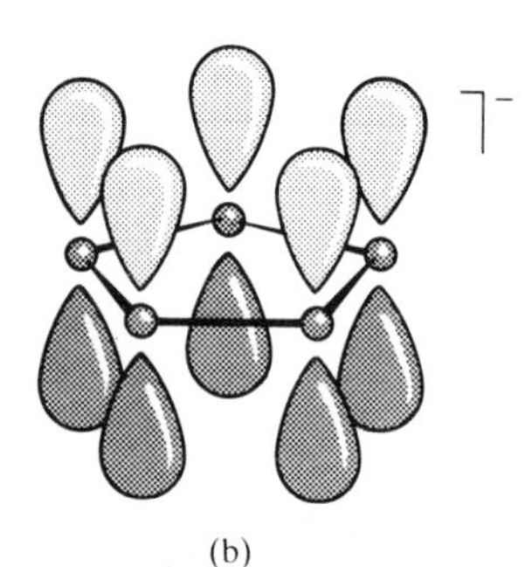

(b)

Fig. 11.18 Isolobal relationship between (a) $[C_2B_9H_{11}]^{2-}$, and (b) $[C_5H_5]^-$. The H atoms have been omitted for clarity.

Example 11.9: *Planning the synthesis of a carborane derivative*

Give balanced chemical equations for the synthesis of 1,2-$B_{10}H_{10}C_2(Si(CH_3)_3)_2$ starting with decaborane (10) and other reagents of your choice.

Answer. The attachment of substitutents to C atoms in 1,2-*closo*-$B_{10}C_2H_{12}$ is most readily carried out using the dilithium derivative $B_{10}H_{10}C_2Li_2$. We first prepare 1,2-$B_{10}C_2H_{12}$ from decaborane:

$$B_{10}H_{14} + 2SR_2 \rightarrow B_{10}H_{12}(SR_2)_2 + H_2$$
$$B_{10}H_{12}(SR_2)_2 + C_2H_2 \rightarrow B_{10}C_2H_{12} + 2SR_2 + H_2$$

This is then lithiated:

$$B_{10}H_{10}C_2H_2 + 2LiC_4H_9 \rightarrow B_{10}H_{10}C_2Li_2 + 2C_4H_{10}$$

The resulting nucleophilic carborane is then reacted with $Si(CH_3)_3Cl$ to yield the desired product:

$$B_{10}H_{10}C_2Li_2 + 2Si(CH_3)_3Cl \rightarrow B_{10}H_{10}C_2(Si(CH_3)_3)_2 + 2LiCl$$

Exercise. Propose a synthesis for the polymer precursor 1,7-$B_{10}H_{10}$-$C_2(Si(CH_3)_2Cl)_2$ from 1,2-$B_{10}H_{10}C_2H_2$ and other reagents of your choice.

Borides, carbides, and silicides

11.15 Borides

The binary compounds of boron with metals exhibit a wide range of compositions and structures. The metal-rich compounds contain isolated B^{3-} ions as well as chains and hexagonal nets of ions. One example of the latter is AlB_2, which has hexagonal graphite-like nets of B atoms with Al atoms between them (Fig. 11.19a). This compound is a golden solid with a high electrical conductivity, presumably owing to delocalization of π electrons in the graphite-like boride net, in much the same way as the alkali metal compounds of graphite are good electrical conductors (Section 11.6). Many of the first-series *d* metals as well as a large number of electropositive metals form similar layered MB_2 compounds.

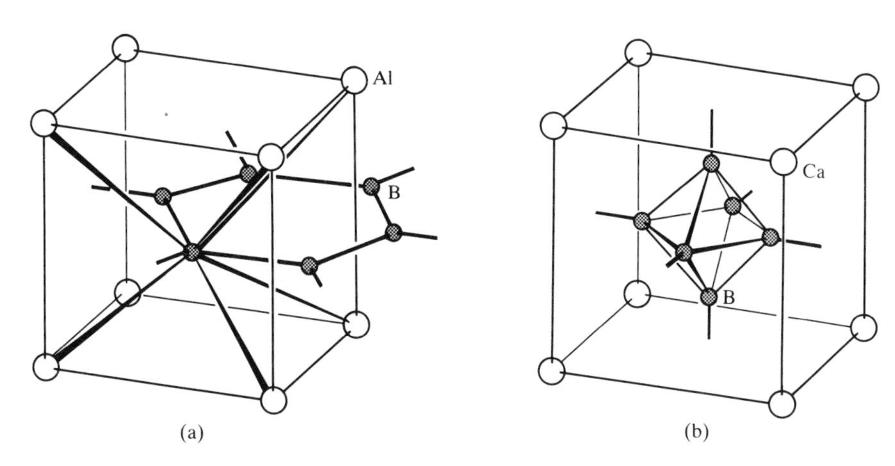

Fig. 11.19 (a) The AlB_2 structure. To give a clear picture of the hexagonal B layer, some B atoms outside the Al cube are displayed. (b) The CaB_6 structure. Note that the B_6 octahedra are connected by a bond between vertices of adjacent B_6 groups. The crystal is a simple cube analog of CsCl. Thus eight Ca surround the central B_6 octahedron. The latter is connected to six B_6 octahedra in adjacent unit cells.

The boron-rich borides, typically MB_6 and MB_{12} where M is an electropositive metal, are of even greater structural interest. In them, the B atoms link to form an intricate network of interconnecting cages. In MB_6 (which are formed by metals such as Na, K, Ca, Ba, Sr, Eu, and Yb) the B_6 octahedra are linked by their vertices to form a cubic framework with a metal ion at its center (Fig. 11.19b). Note the similarity between these anionic linked octahedra and *closo*-$[B_6H_6]^{2-}$ (Section 11.13) although the linked B_6 clusters bear a -1, -2, or -3 charge depending on the cation with which it is associated. In MB_{12} the networks are based on linked cuboctahedra (**35**). This type of compound is formed by some of the heavier electropositive metals, particularly those of the *f* block.

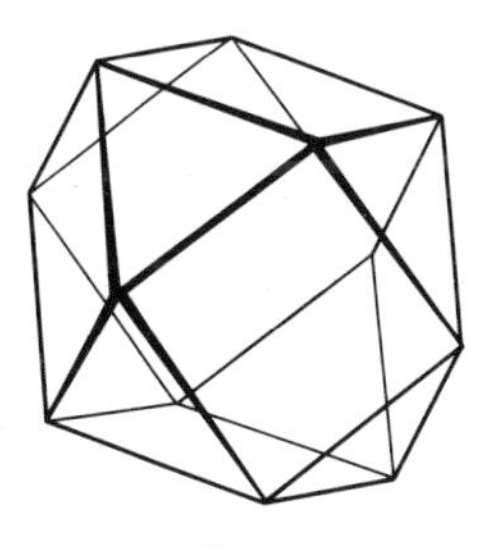

35

11.16 Carbides

Carbon forms binary compounds with most other elements. Many of them are important to daily life, for they include CO_2 and Fe_3C, a major constituent of steel. It proves helpful to classify the binary compounds of carbon with metals and metalloids, the **carbides**, into three main categories:

(1) **Saline carbides**, which are largely ionic solids. These are formed by the elements of Groups 1 and 2 and by aluminum.

(2) **Metallic carbides**, which have a metallic conductivity and luster. These are formed by the *d*-block elements.

(3) **Metalloid carbides**, which are hard covalent solids formed by boron and silicon.

Figure 11.20 summarizes the distribution of the different types in the periodic table. This classification is very useful for the purpose of correlating chemical and physical properties, but (as so often in inorganic chemistry) the borderlines are sometimes indistinct.

Saline carbides

The saline carbides of Group 1 and 2 metals may be divided into three categories: **graphite intercalation compounds**, such as KC_8, **dicarbides** or **acetylides**, which contain the C_2^{2-} anion, and **methides**, which contain formally the C^{4-} anion.

The graphite intercalation compounds are formed by the Group 1 metals. The graphite intercalation compounds are formed by a redox process (Section 11.6). They can be prepared by the reaction of graphite with alkali metal vapor or with metal-ammonia solution. For example, contact between graphite and potassium vapor in a sealed tube at 300°C leads to the formation of KC_8. The alkali metal ions lie in an ordered array (Fig. 11.21). In the case of potassium and most of the other alkali metals, a series of metal–graphite intercalation compounds can be prepared with different metal:carbon ratios, including KC_8 and KC_{16}.

The dicarbides are formed by a broad range of electropositive metals, including those from Groups 1 and 2 and the lanthanides. The C_2^{-2} ion has a very short C—C distance in some dicarbides (for example, 1.19 Å in CaC_2), which is consistent with its being a triply bonded $[C{\equiv}C]^{2-}$ ion isoelectronic with $[C{\equiv}N]^-$ and N≡N. Some dicarbides have a structure related to rock salt, but replacement of the spherical Cl^- ion by the dumb-bell-shaped $[C{\equiv}C]^{2-}$ ion leads to elongation of the crystal along one axis, and a

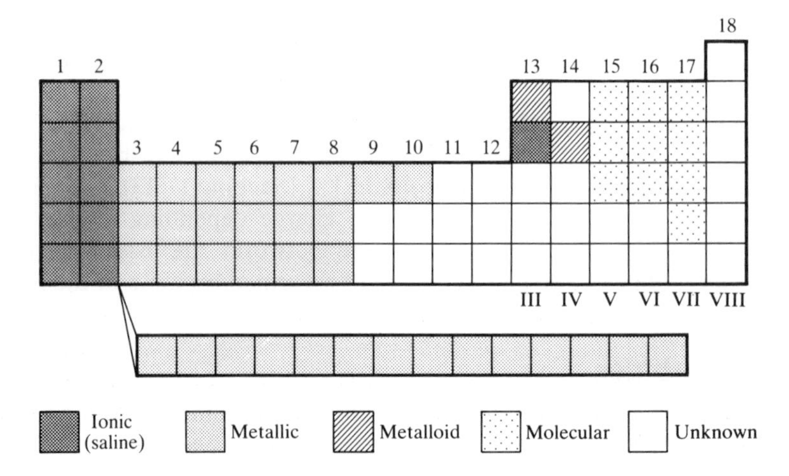

Fig. 11.20 Distribution of the carbides in the periodic table.

resulting tetragonal symmetry (Fig. 11.22). The C—C bond is significantly longer in the lanthanide dicarbides, which suggests that for them the simple triply bonded structure is not a good approximation.

Carbides such as Be_2C and Al_4C_3 are borderline between saline and metalloid, and the isolated C ion is only formally C^{4-}. That the C atom is actually present in some other form is indicated by the crystal structures of methides, which are not those expected for the simple packing of spherical ions.

The principal synthetic routes to the saline carbides and dicarbides of Groups 1 and 2 are very straightforward:

1. *Direct reaction* of the elements at high temperatures:

$$Ca(l) + 2C(s) \xrightarrow{>2000°C} CaC_2(s)$$

The formation of graphite intercalation compounds is another example of a direct reaction, but is carried out at much lower temperatures. The intercalation reaction is more facile because no C—C covalent bonds are broken when an ion slips between the graphite layers.

2. *Reaction of a metal oxide and carbon* at a high temperature:

$$CaO(l) + 3C(s) \xrightarrow{2200°C} CaC_2(l) + CO(g)$$

Crude calcium carbide is prepared in electric arc furnaces by this method. The carbon serves both as a reducing agent to remove the oxygen and as a source of carbon to form the carbide.

3. *Reaction of acetylene with a metal–ammonia solution*:

$$2Na(am) + C_2H_2(g) \rightarrow Na_2C_2(s) + H_2(g)$$

This reaction occurs under mild conditions and leaves the C—C bonds of the starting material intact. As the acetylene molecule is a very weak Brønsted acid, the reaction can be regarded as a redox reaction between a highly active metal and a weak acid to yield H_2 (with H^+ the oxidizing agent) and the metal dicarbide.

The saline carbides are readily oxidized and are susceptible to hydrolysis, the latter leaving the C—C bonds of the anions intact:

$$CaC_2(s) + 2H_2O(l) \rightarrow Ca(OH)_2(s) + HC{\equiv}CH(g)$$

This reaction is readily understood as the transfer of a proton from a Brønsted acid (H_2O) to the conjugate base (C_2^{2-}) of a weaker acid (HC≡CH). Similarly, the controlled hydrolysis or oxidation of KC_8 restores the graphite and produces a hydroxide or oxide of the metal:

$$2KC_8(s) + 2H_2O(g) \rightarrow 16C(\text{graphite}) + 2KOH(s) + H_2(g)$$

Metallic carbides

Most metallic carbides are formed by the *d* metals. They are sometimes referred to as **interstitial carbides** because their structures are often related to those of metals by the insertion of C atoms in octahedral holes. However, this nomenclature gives the erroneous impression that the metallic carbides are not legitimate compounds. In fact the hardness and other properties of metallic carbides demonstrate that strong metal–carbon bonding is present

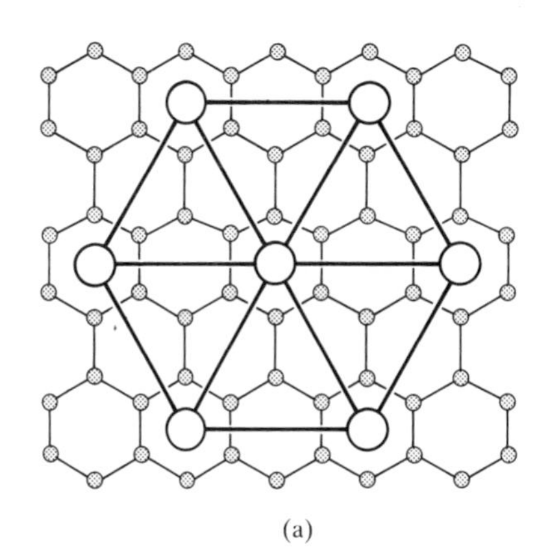

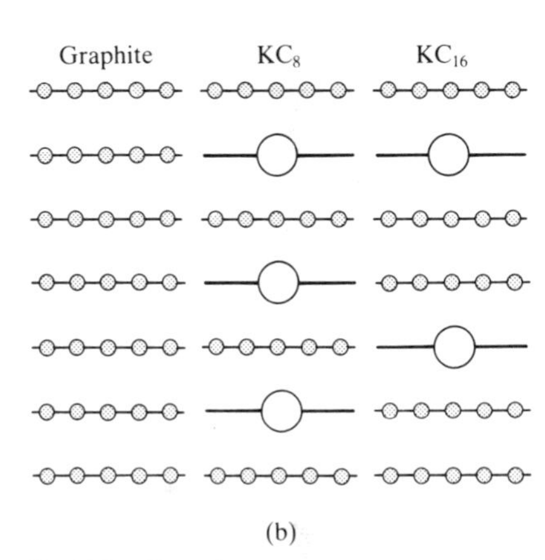

Fig. 11.21 Potassium graphite compounds. (a) Viewed perpendicular to plane of graphite. (b) Viewed parallel with graphite planes (schematic).

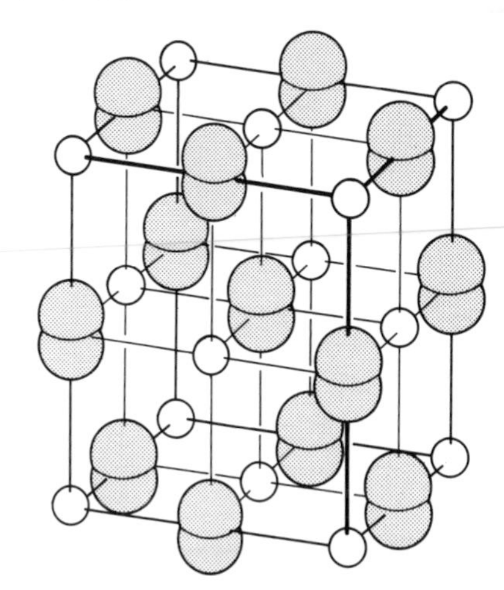

Fig. 11.22 The calcium carbide structure. Note that this structure bears a similarity to that of rock salt. Because C_2^{2-} is not spherical, the cell is elongated along one axis. This crystal is therefore tetragonal rather than cubic.

in them. Some of these carbides are useful materials. Tungsten carbide (WC), for example, is used for cutting tools and high-pressure apparatus such as that used to produce diamond. Cementite (Fe_3C) is a major constituent of steel and cast iron.

Many metallic carbides of composition MC have an fcc or hcp arrangement of metal atoms with the C atoms in the octahedral holes. This results in a rock-salt structure (Fig. 4.11). The C atoms in carbides of composition M_2C occupy only half the octahedral holes between the metal atoms. In both these structures the C atom has only four valence orbitals to bond with six surrounding metal atoms, and so it is in a sense hypervalent. We must therefore adopt a delocalized bonding model using orbitals on C atoms and on the surrounding metal atoms. One such linear combination (**36**) consists of a p_x orbital and two adjacent d orbitals.

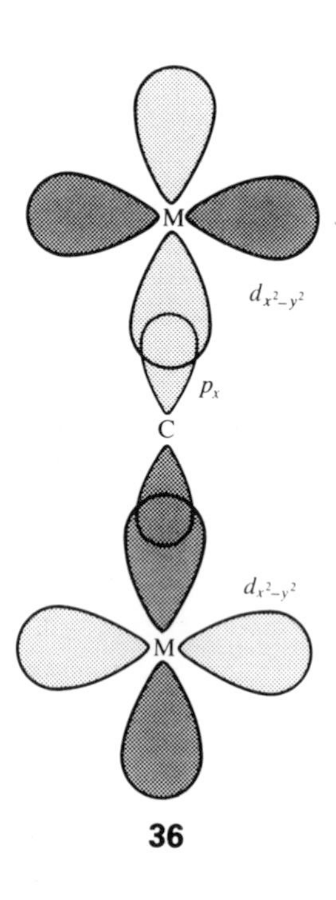

36

It has been found empirically that the formation of simple compounds in which the C atom resides in an octahedral hole of a close-packed structure occurs only if

$$\frac{r_C}{r_M} < 0.59$$

where r_C is the covalent radius of C and r_M the metallic radius of the metal. This relation, known as **Hägg's rule**, also applies to metal compounds containing nitrogen or oxygen. When the ratio is exceeded the compound normally has a low-symmetry structure.

11.17 Silicides

Silicon, like its neighbors boron and carbon, forms a wide variety of binary compounds with metals. Some of them contain isolated Si atoms. The structure of ferrosilicon Fe_3Si, for instance, which plays an important role in steel manufacture, can be viewed as an fcc array of Fe atoms with some replaced by Si. Compounds such as K_4Si_4 contain isolated tetrahedral cluster anions $[Si_4]^{4-}$ that are isoelectronic with P_4. Many of the f-block elements form compounds with the formula MSi_2 that have the layered AlB_2 structure (Fig. 11.19a).

Silicon carbide (SiC) is the most common of the compounds of silicon with nonmetals because it is synthesized on a large scale for use as an abrasive. One polymorph of SiC has the diamond-like sphalerite structure (Fig. 4.13).

Electronic properties

The dependence of modern technology on semiconductor materials began in in the 1950s with the commercial development of germanium transistors. Although silicon is superior to germanium for most semiconductor applications, the latter was used first because it is easier to purify. After methods for purifying silicon were developed in the late 1960s germanium was largely eclipsed. Nevertheless, although silicon is now the semiconductor produced in largest volume, other semiconducting materials are also used and new ones continue to be developed. In this section, we describe some of the desirable properties of semiconducting materials related to the two groups we are considering in this chapter, and see how these properties correlate with composition.

11.18 Band gaps and conductivities

As we saw in Section 3.6, the electrical conductivity of a pure semiconductor can be traced to the thermal population of a conduction band by electrons from the valence band. The electrons in the conduction band and the holes in the valence band are both mobile, and depending on the characteristics of the material one or the other may dominate the conductivity. We also mentioned in Section 3.6 that the introduction of dopants into a pure semiconductor can introduce an imbalance in the number of holes or electrons and hence modify its electrical properties.

The role of the band gap

The thermal population of the conduction band and the accompanying depopulation of the valence band gives rise to an exponential increase of conductivity with temperature:

$$\sigma = \sigma_0 e^{-E_g/2kT}$$

E_g is the energy gap between the valence and conduction bands, k is Boltzmann's constant, and T is the temperature. For present purposes we shall consider σ_0 to be a parameter characteristic of the material. At room temperature, materials with small E_g values (Table 11.9) are generally the best conductors.

It is readily apparent from Table 11.9 that E_g decreases from diamond to tin. Even if gray tin were stable at room temperature its band gap would be too small for most semiconductor applications. Diamond has a large band gap, and accordingly its conductivity is typical of an insulator. Nevertheless it should be possible to incorporate diamond into semiconductor devices, but the problems of synthesis, purification, and selective doping (see below) are formidable.

There are many compound semiconductors. Some of the most widely used are the **III–V compounds**, which are formed from a combination of a boron-group and a nitrogen-group element. (The name III–V comes from the old numbering scheme for the periodic table.) The III–V compounds and the isoelectronic **II–VI compounds** (that is, the 12–16 compounds), such as CdS, have the diamond-like sphalerite structure or the closely related wurtzite structure (Fig. 4.15). As with the elemental semiconductors, E_g decreases as heavier elements are incorporated into the III–V or II–VI material. For example, the band gap of GaN is over twice that of GaAs and the band gap in GaAs is almost a factor of four larger than that of InAs (Table 11.9).

Table 11.9. Band gaps at 25°C of Group 14/IV elements and some III–V compounds

Material	E_g/eV
C	5.47
SiC	3.00
Si	1.12
Ge	0.66
Sn	0
BN	*ca.* 7.5
BP	2.0
GaN	3.36
GaP	2.26
GaAs	1.42
InAs	0.36

Source: S. A. Schwarz, *Kirk-Othmer Encyclopedia of Chemical Technology*, **20**, 601 (1982).

Carrier velocity

An electron passing through a solid may lose its momentum to the solid by collision with an impurity, or with a normal atom that is vibrationally excited away from its normal lattice site. (An analogy is the hurdler who sets a stride to jump regularly spaced hurdles, but trips and falls on one that is out of place.) The collision events between an electron and its lattice thus strongly influence the average velocity of the electron. This **carrier velocity** is an important parameter for high-speed logic chips, where there is a constant quest for greater speed. Gallium arsenide has a much more favorable carrier velocity than silicon, and major efforts are in progress to prepare this material with purity suitable for logic chips.

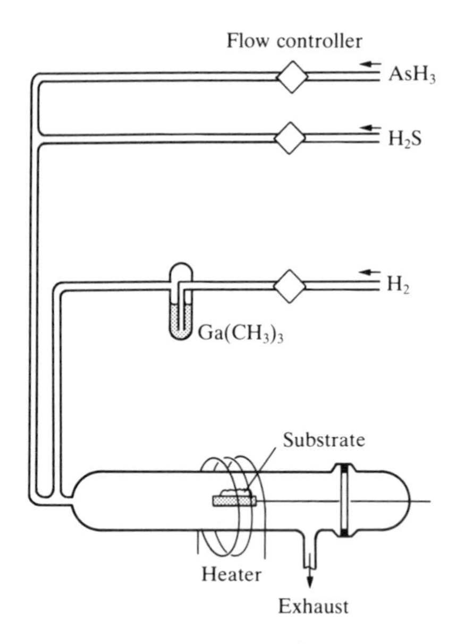

Fig. 11.23 Schematic diagram of the apparatus for vapor phase epitaxial growth of doped gallium arsenide. Arsine gas and trimethylgallium vapor in a hydrogen stream pass over the substrate wafer, which is heated by microwave or radiofrequency irradiation. The dopants, such as H_2S, are introduced as needed to alter the electrical or optical characteristics of the product. The exhaust gases go through chemical or thermal treatment to remove toxic materials.

Gallium arsenide and other compound semiconductors are commonly prepared by a process called **chemical vapor deposition** (CVD, Fig. 11.23). The volatile stream of organometallic or hydride compounds is introduced in a hydrogen atmosphere, and upon passing over a heated substrate crystal, the desired semiconducting compound is deposited, generally as a single crystal film in register with the substrate, the process called **epitaxial growth**. The challenge to the chemist is to produce new volatile compounds suitable for CVD and to understand and control the thermal reaction.

11.19 Optical properties

Just as molecules absorb light with promotion of an electron from a HOMO to a LUMO, photons can excite electrons from the valence band into the conduction band. The band gaps for germanium and silicon are in the infrared region of the spectrum, and so these materials are opaque to visible light since radiation of infrared and higher frequencies can excite electrons across the band gap. In contrast, the band gap for diamond lies in the far ultraviolet region, no lower-frequency visible light is absorbed, and diamond is transparent to visible light.

The detailed shape of the conduction and valence bands strongly influences the electronic transitions in semiconducting materials. For the purposes of describing the excitation of an electron in a solid it is useful to plot the energy of the valence and conduction bands against the wavevector, k (Fig. 11.24), where

$$k = \frac{2\pi}{\lambda}$$

According to the de Broglie relation (Section 1.3), k is proportional to the momentum, so the plot shows how the energy of an electron varies with its linear momentum through the solid.

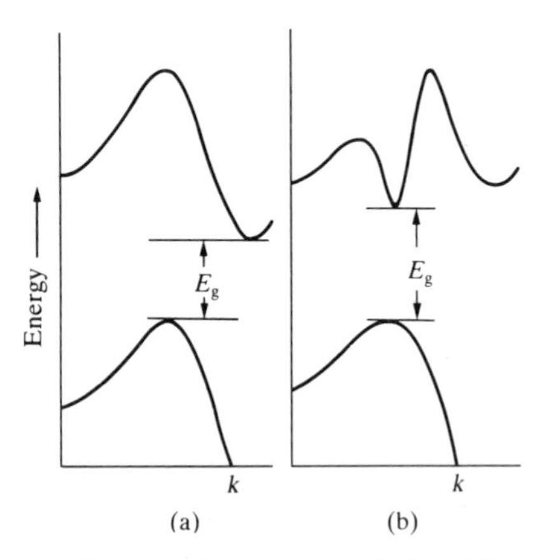

Fig. 11.24 Band structure of (a) indirect-(Si) and (b) direct-(GaAs) gap semiconductor.

We see from Fig. 11.24b that for gallium arsenide the maximum in the valence band occurs immediately below the minimum in the conduction band. This type of material is called a **direct-gap** material. An electronic transition from the top of the valence band to the bottom of the conduction band is allowed in a direct-gap material because the electron does not need to change momentum. The band structure of silicon (Fig. 11.24a) shows it to be an **indirect-gap** material. In an indirect-gap material the top of the valence band and the bottom of the conduction band lie at different wavevectors, and an electron cannot make a transition between the two when the solid is irradiated with photons of energy that match the band gap.

The probability of transition from an excited state to the ground state is governed by the need to match k, since an electron cannot undergo a substantial change in linear momentum when it emits or absorbs a photon. Thus, whereas an electron in the bottom of the conduction band of gallium arsenide has a high probability of making a radiative transition to the top of the valence band, in silicon the probability is very low. The direct-gap property of gallium arsenide accounts for its use in a variety of photoelectronic devices. For example, the red electronic display device known as a **light emitting diode** (LED) is based on gallium arsenide; the semiconductor is doped with P to move the transition into the red. The electrons in silicon cannot revert to the valence band by radiation: they need to couple their linear momentum to that of the nuclei if they are to undergo a change in wavevector. As a result, thermal motion of the atoms is stimulated by the energy of the transition, and the semiconductor becomes hot.

Electronic excitation by photons can also lead to enhanced electrical conductivity. The photoconductor cadmium telluride (CdTe) is one such direct-gap material, which finds use as a red light detector in spectrometers and a variety of position sensors. It is often used in conjunction with a LED as the light source.

Further reading

Boron group

K. Wade, *Electron deficient compounds.* Nelson, London (1971).

R. N. Grimes, *Carboranes.* Academic Press, New York (1970).

E. L. Muetterties (ed.), *Boron hydride chemistry.* Academic Press, New York (1975).

J. F. Liebman, A. Greenberg, and R. E. Williams (ed.), *Advances in boron and boranes.* VCH, New York (1988).

J. D. Kennedy, The polyhedral metalloboranes, in *Prog. Inorg. Chem.*, **33**, 519 (1985); **34**, 211 (1986). These thorough reviews provide a good coverage of this vigorous area of research.

K. Wade and H. J. Banister, *The chemistry of Al, Ga, In, and Tl.* Pergamon, Oxford (1974).

Carbon group

Kirk-Othmer Encyclopedia of Chemical Technology, especially: carbon, semiconductors, silicon.

G. Urry, *Elementary equilibrium chemistry of carbon.* Wiley-Interscience, New York (1989).

Aqueous chemistry and silicate minerals

C. F. Baes and R. E. Mesmer, *The hydrolysis of cations.* Wiley, New York (1976).

W. Stumm and J. J. Morgan, *Aquatic chemistry.* Wiley, New York (1981).

D. W. Breck, *Zeolite molecular sieves.* Wiley, New York (1977).

R. M. Barrer, *Hydrothermal chemistry of zeolites.* Academic Press, New York (1982).

Exercises

11.1 List the elements in Groups 13/III and 14/IV and indicate (a) the metals and nonmetals, (b) those that can crystallize in the diamond structure, (c) those that primarily occur as oxides in nature, (d) those that occur along with metal sulfides in nature.

11.2 Describe the repeating unit in the structures of elemental boron and gallium.

11.3 Sketch the apparatus and give balanced chemical equations for the synthesis of (a) aluminum chloride, (b) tin(II) chloride.

11.4 (a) arrange the following in order of increasing Lewis acidity toward hard Lewis bases: BF_3, BCl_3, SiF_4, $AlCl_3$. (b) In the light of this order, write balanced chemical reactions (or NR) for

(i) $SiF_4N(CH_3)_3 + BF_3 \rightarrow ?$

(ii) $BF_3N(CH_3)_3 + BCl_3 \rightarrow ?$

(iii) $AlCl_3N(CH_3)_3 + BCl_3 \rightarrow ?$

11.5 Preferably without consulting reference material, construct a periodic table and indicate the elements that form saline, metallic, and metalloid carbides.

11.6 Describe the preparation, structure and classification of (a) KC_8, (b) CaC_2.

11.7 Describe the structures of SnO and red PbO and give an indication of how the electronic structure of the ion can be used to rationalize these structures.

11.8 Although solid state scientists frequently prefer to perform fundamental studies on crystalline solids because they are physically and structurally well defined, there are considerable commercial applications for semicrystalline and amorphous solids, many of which are formed by Group 14/IV elements or compounds. List four different examples of amorphous or partially crystalline solids described in this chapter and briefly state their useful properties.

11.9 The lightest *p*-block elements often display different physical and chemical properties from the heavier members. Discuss the similarities and differences by comparison of:

(a) The structures and electrical properties of (i) boron and aluminum and of (ii) carbon and silicon.

(b) The physical properties and structures of the oxides of carbon and silicon.

(c) The Lewis acid–base properties of the tetrahalides of carbon and silicon.

(d) The structures of the halides of boron and aluminum.

11.10 (a) Summarize the trends in relative stabilities of the oxidation states of the elements of Groups 13/III and 14/IV, and indicate the elements that display the inert pair effect.

(b) With this information in mind, write balanced chemical reactions or NR (for no reaction) for the following combinations in acidic solution, and explain how the answer fits the trends.

(i) $Sn^{2+}(aq) + PbO_2(s)$ (excess) $\rightarrow$?
(ii) $Tl^{3+}(aq) + Al(s)$ (excess) $\rightarrow$?
(iii) $In^{+}(aq) \rightarrow$? (air excluded)
(iv) $Sn^{2+}(aq) \rightarrow$? (exposed to air)
(v) $Tl^{+}(aq) \rightarrow$? (exposed to air)

11.11 Verify your answer to Exercise 11.10(b) by the determination of $E^{\ominus}$ for the proposed reaction (Appendix 4). Are your qualitative answer and the indications from $E^{\ominus}$ in agreement?

11.12 By means of balanced chemical equations, contrast the reaction of K_2CO_3 with $HCl(aq)$ and of Na_4SiO_4 with acid.

11.13 Describe in general terms the nature of the $[SiO_3]_n^{2n-}$ ions in jadeite and the silica–alumina framework in kaolinite.

11.14 (a) How many bridging O atoms are in the framework of a single sodalite cage? (b) Describe the (super-cage) polyhedron at the center of the zeolite A structure in Fig. 11.14.

11.15 Contrast the physical properties of pyrophilite and muscovite mica and explain how the properties arise from the composition and structures of these closely related aluminosilicates.

11.16 Give the structural type and describe the structures of B_4H_{10}, B_5H_9, and 1,2-$B_{10}C_2H_{12}$.

11.17 (a) From its formula, classify $B_{10}H_{14}$ as *closo, nido* or *arachno*. (b) Using Wade's rules, determine the number of framework electron pairs for decaborane(14). (c) Verify by detailed counting of valence electrons that the number of cluster valence electrons of $B_{10}H_{14}$ is the same as that determined in (b).

11.18 Starting with $B_{10}H_{14}$ and other reagents of your choice, give the equations for the synthesis of Fe(*nido*-$B_9C_2H_{11})_2^{2-}$, and sketch the structure of this compound.

11.19 (a) What are the similarities and differences in structure of layered BN and graphite? (b) Contrast their reactivity with Na and Br_2.

11.20 Devise a synthesis for the borazines (a) $Ph_3N_3B_3Cl_3$ and (b) $Me_3N_3B_3H_3$, starting with BCl_3 and other reagents of your choice. Draw the structures of the products.

11.21 (a) Describe the trend in E_g for the elements carbon (diamond) through tin (gray), and for cubic BN, AlP, and GaAs. (b) Does the electrical conductivity of silicon increase or decrease when its temperature is changed from 20°C to 40°C? (c) State the functional dependence of conductivity on temperature for a semiconductor and decide whether the conductivity of AlP or GaAs will be more sensitive to temperature.

Problems

11.1 Boron-11 NMR is an excellent spectroscopic tool for inferring the structures of boron compounds. Under conditions in which the $^{11}B—^{11}B$ coupling is absent, it is possible to determine the number of attached H atoms by the multiplicity of a resonance: BH giving a doublet, BH_2 a triplet, and BH_3 a quartet. Also, B atoms on the closed side of *nido* and *arachno* clusters are generally more shielded than those on the open face. Assuming no B—B or B—H—B coupling, predict the general pattern of the ^{11}B—NMR spectra of (a) BH_3CO, (b) $[B_{12}H_{12}]^{2-}$, and (c) B_4H_{10}.

11.2 The anion $[Tl_2Te_2]^{2-}$ (R. B. Burns and J. D. Corbett, *J. Am. chem. Soc.*, **103**, 2627 (1981)) consists of a nearly flat diamond with Te—Tl—Te angles of about 97°. The closest distance across the ring, Tl—Tl, is 3.60 Å and can therefore be considered to be outside bonding distance. (a) Give a Lewis representation of the bonding in this anion. (b) If you were to expose the anion to a Lewis acid, would you expect a complex to form at the Tl atom or at the Te atom? Give your reasoning. (c) Which would be the most promising Lewis acid to use, BF_3, BBr_3 or BH_3?

11.3 Acetylcholine, $[(CH_3)_3N(CH_2)_2OC(O)CH_3]^+$, is an important neurotransmitter, and the physiological properties of the neutral boron analog $(CH_3)_2(BH_3)N(CH_2)_2OC(O)CH_3$ are of interest. (B. F. Spielvogel, F. U. Ahmed, and A. T. McPhail, *Inorg. Chem.*, **25**, 4395 (1986).) Devise a method of synthesis for this analog starting with $[(CH_3)_2HN(CH_2)_2OC(O)CH_3]^+$ and other reagents of your choice.

The nitrogen and oxygen groups

12

As with the rest of the *p* block, we shall see that the elements at the head of these two groups, nitrogen and oxygen, differ significantly from their congeners. Thus, their coordination numbers are generally lower in their compounds and they are the only members of the groups to exist as diatomic molecules under normal conditions. Most of the elements in the two groups can exist in a wide variety of oxidation states, and hence they show a very rich redox chemistry, and to understand their behavior it is important to pay attention to the kinetics of the reactions as well as to the thermodynamics. The formation of covalent molecules and ions such as oxoanions, oxoacids, and chain, ring, and cluster compounds is common in the groups, and we shall see many examples. In agreement with the decrease in metallic character from left to right of the periodic table, there are no true metals in these groups.

The elements

The nitrogen group

12.1 Production and structures of the elements

12.2 Nitrogen activation

12.3 Halides

12.4 Oxides and aqueous redox chemistry

12.5 Compounds of nitrogen with phosphorus

The oxygen group

12.6 Production and structures of the elements

12.7 Halides

12.8 Oxygen and the *p*-block oxides

12.9 Oxides of the *s*-block metals

12.10 Oxides of the *d*-block metals

12.11 Metal–sulfur compounds

12.12 Sulfido complexes of the *d* metals

***p*-Block ring and cluster compounds**

12.13 Anionic clusters

12.14 Polycations

12.15 Neutral heteroatomic rings and clusters

Further reading

Exercises

Problems

14	15	16	17
C	N	O	F
Si	P	S	Cl
Ge	As	Se	Br
Sn	Sb	Te	I
Pb	Bi	Po	At
IV	V	VI	VII

The nitrogen and oxygen groups, specifically Groups 15/V and 16/VI, contain some of the most important elements for geology, life, and industry. However, the heaviest elements of the groups lie at the frontier of nuclear stability, and bismuth ($Z = 83$) is the heaviest element to have stable isotopes. The most common isotope of its neighbor polonium ($Z = 84$), ^{210}Po, is an intense α emitter with a half-life of 138 d.

The elements

All the members of the two groups other than nitrogen and oxygen are solids under normal conditions (Table 12.1), and metallic character generally increases down the groups. However, the trend is not clear-cut because the electrical conductivities of the heavier elements decrease down the groups (Fig. 12.1), which is in contrast to the general trend in the *p* block. The usual trend reflects the closer spacing of the atomic energy levels in heavier elements and hence a smaller separation of the valence and

Table 12.1. Properties of the nitrogen and oxygen group elements

	I/kJ mol^{-1}	χ_P	Radius		Appearance and properties	Common oxidation numbers
			r_{cov}/Å	r_{ion}/Å		
Group 15/V						
N	1410	3.04	0.70		Gas b.p. −196°C	−3, +1, +3, +5
P	1020	2.06	1.10		Polymorphic solid	−3, +3, +5
As	953	2.18	1.21		Dark solid	+3, +5
Sb	840	2.05	1.41		Solid, metallic luster, brittle	+3, +5
Bi	710	2.02	1.51		Solid, metallic luster, brittle	+3, +5
Group 16/VI						
O	1320	3.44	0.66	1.24†	Paramagnetic gas, b.p. −183°C	−2, −1
S	1005	2.44	1.04	1.70†	Yellow polymorphic solid	−2, +4, +6
Se	947	2.55	1.17	1.84†	Polymorphic solid	−2, +4, +6
Te	875	2.10	1.37	2.07†	Solid, silvery, brittle	−2, +4, +6
Po					Solid, all isotopes radioactive	−2, +2, +4, +6

† For oxidation number −2.

Sources: Ionization enthalpies from D. D. Wagman, W. H. Evans, V. B. Parker, R. H. Schumm, I. Halow, S. M. Bailey, K. L. Churney, and R. L. Nuttall, *J. Phys. Chem., Ref. Data* **11** (Suppl. 2) (1982). Electronegativities from A. L. Allred, *J. Inorg. Nucl. Chem.* **17,** 215 (1961). Covalent radii from M. C. Ball and A. H. Norbury, *Physical data for inorganic chemists.* Longman, London (1974). Ionic radii from R. D. Shannon, *Acta Crystallogr.,* **A32,** 1751 (1976).

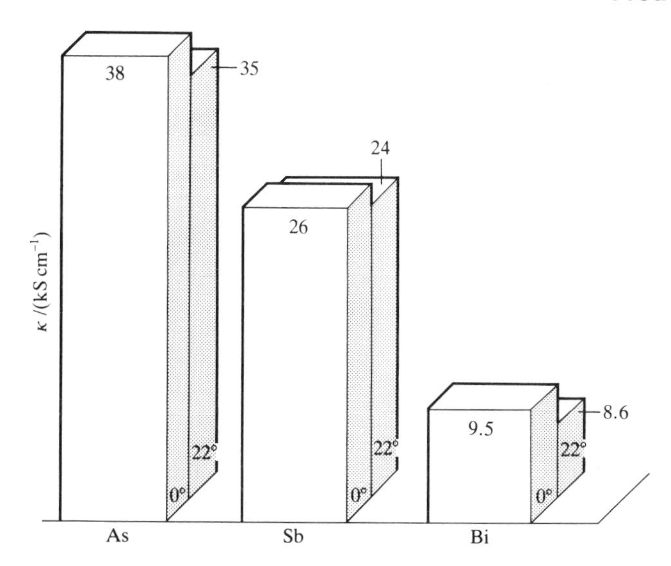

Fig. 12.1 Conductivities of the heavier elements of Group 15/V.

conduction bands. The opposite trends in these two groups may reflect the greater directional character of the bonding between their atoms.

In addition to their distinctive physical properties, nitrogen and oxygen are significantly different chemically from the other members of the groups. For one thing, they are among the most electronegative elements in the periodic table and significantly more electronegative than their congeners. Although oxygen never achieves the group maximum oxidation number of +6, the slightly less electronegative nitrogen does achieve the maximum for its group (+5), but only under much stronger oxidizing conditions than are necessary to achieve this state for its congeners. The small sizes of the N and O atoms also lead to a distinctive character. Thus, nitrogen and oxygen seldom have coordination numbers greater than 4 in simple molecular compounds, but their heavier congeners frequently reach 5 and 6. The heaviest member of Group 15/V, bismuth, requires a very strong oxidizing agent to achieve the group oxidation number, and in most of its compounds it has oxidation number +3. This behavior is a consequence of the inert pair effect, which is also important (as we saw in Chapter 11) for the preceding heavy elements thallium and lead.

The nitrogen group

The members of the nitrogen group are sometimes referred to collectively as the **pnictides** but this name is neither widely used nor officially sanctioned.

12.1 Production and structures of the elements

Nitrogen

Nitrogen is readily available as N_2, for it is the principal constituent of the atmosphere and is obtained from it on a massive scale by the distillation of liquid air. Although the bulk price of N_2 is low, the scale on which it is used is an incentive for the development of less expensive processes than

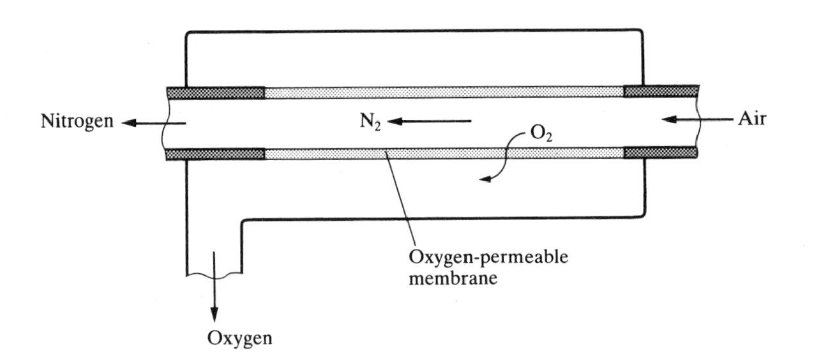

Fig. 12.2 Schematic diagram of a membrane separator for nitrogen and oxygen.

liquefaction and distillation for separating it from oxygen. An active area of current research is the search for practical membrane materials that are more permeable to O_2 than to N_2, which would enable the separation to be performed at about room temperature (Fig. 12.2).

The major nonchemical use of nitrogen gas is as an inert atmosphere in metal processing, petroleum refining, and food processing. We saw in Box 10.1 how nitrogen is used to provide an inert atmosphere in the laboratory, and liquid nitrogen (b.p. −196°C, 77 K) is a convenient refrigerant in both industry and the laboratory. Nitrogen enters the chain of industrial and agricultural chemicals through its conversion into ammonia by the Haber process, which we describe in Chapter 17. Once 'fixed' in this way it can be converted to a wide range of compounds. The destinations of fixed nitrogen are summarized in Chart 12.1, and later in the chapter we shall see something of the chemistry involved in the reactions mentioned there.

Phosphorus

Phosphorus (along with nitrogen and potassium) is an essential plant nutrient. However, as a result of the low solubility of phosphates it is often depleted in soil, and hence is one of the most important elements in balanced fertilizers. Approximately 85 percent of the phosphoric acid produced goes into fertilizer manufacture.

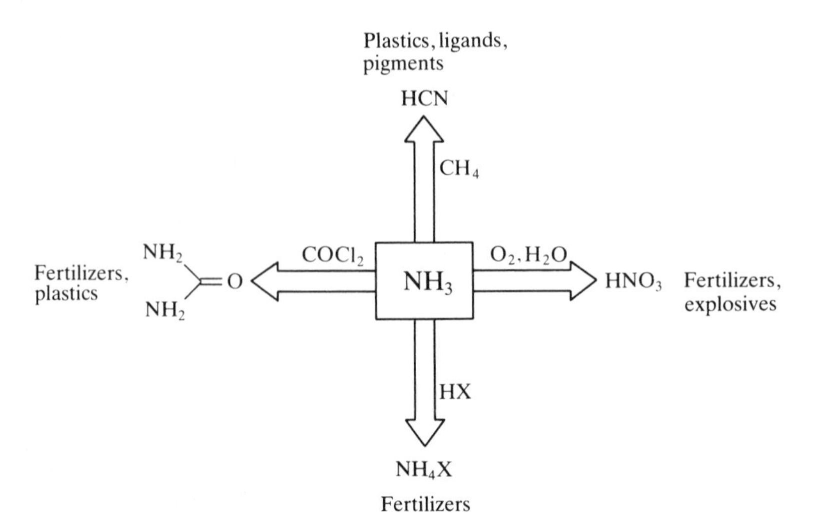

Chart 12.1. The destinations of fixed nitrogen

The principal raw material for the production of elemental phosphorus and phosphoric acid is phosphate rock, which consists primarily of the minerals fluorapatite, $Ca_5(PO_4)_3F$, and hydroxyapatite, $Ca_5(PO_4)_3OH$, and is the insoluble, crushed, and compacted remains of ancient organisms. Phosphoric acid can be produced by a simple acid–base reaction between the rock and concentrated sulfuric acid:

$$Ca_5(PO_4)_3F + 5H_2SO_4(l) \rightarrow 3H_3PO_4(l) + 5CaSO_4(s) + HF(aq)$$

The hydrogen fluoride from the fluorapatite component of the rock is scavenged by reaction with silica to yield the less reactive $[SiF_6]^{2-}$ complex ion.

The product of the treatment of phosphate rock with acid contains *d*-metal contaminants that are difficult to remove, so its use is largely confined to fertilizers and metal treatment. Pure phosphoric acid and most phosphorus compounds are produced via the element because it can be purified by sublimation. The production of the element starts with crude calcium phosphate (as calcined phosphate rock) which is reduced with carbon in an electric arc furnace. (This is another example of the carbon reduction of an highly oxophilic element at very high temperature, Section 8.1.) Silica is added (as sand) to produce calcium silicate:

$$2Ca_3(PO_4)_2 + 6SiO_2 + 10C \xrightarrow{1500°C} 6CaSiO_3 + 10CO + P_4$$

The slag is molten at these high temperatures and so can be easily removed from the furnace. The phosphorus itself vaporizes and is condensed to the solid, which is stored under water to protect it from reaction with air. Most of the element produced in this way is burned to form P_4O_{10}, which is then hydrated to yield pure phosphoric acid.

The solid elements of Group 15/V (as well as those in Group 16/VI) exist as a number of allotropes.[1] **White phosphorus**, for example, is a solid consisting of tetrahedral P_4 molecules (**1**). Despite the small P—P—P angle, the molecules persist in the vapor up to about 800°C, but above that temperature the equilibrium concentration of P_2 becomes appreciable. As with the N_2 molecule, P_2 has a formal triple bond and a short P—P bond length. White phosphorus is thermodynamically less stable than the other solid phases under normal conditions. However, in contrast to the usual practice of choosing the most stable phase of an element as the reference phase for thermodynamic calculations, white phosphorus is adopted since it is better characterized than the other forms.

Red phosphorus is obtained by heating white phosphorus at 300°C in an inert atmosphere for several days. It is normally obtained as an amorphous solid, but crystalline materials can be prepared that have very complex three-dimensional network structures. Unlike white phosphorus, red phosphorus does not spontaneously ignite in air. When phosphorus is heated under high pressure, a series of phases of **black phosphorus** are formed. One of these phases consists of tubes composed of pyramidal three-coordinate P atoms (Fig. 12.3).

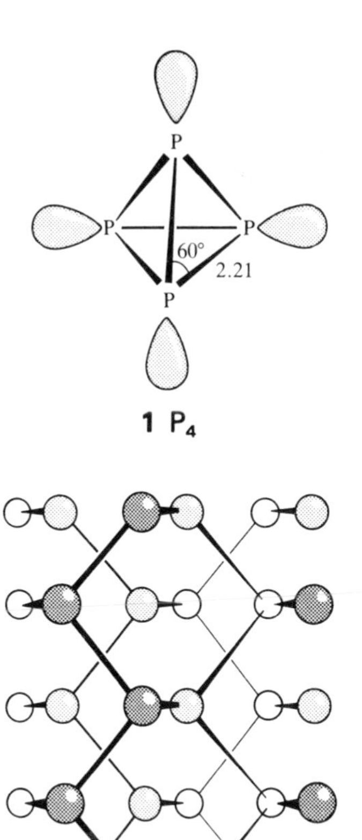

1 P_4

Fig. 12.3 Two of the puckered layers of black phosphorus. Note the trigonal pyramidal coordination of the atoms. Dark atoms are closest to the viewer and open circles are furthest away.

[1] The term *allotrope* refers to different molecular structures of the same element; thus ozone and dioxygen are allotropes of oxygen. The word *polymorph* denotes the same substance in different crystal forms. For example, phosphorus occurs in several solid polymorphs.

Example 12.1: *Electronic structure and chemistry of P_4*

Draw the Lewis structure of P_4, and discuss its possible role as a ligand.

Answer. In the Lewis structure of P_4, there is a lone pair of electrons on each P atom:

This structure, together with the fact that the electronegativity of phosphorus is moderate ($\chi = 2.06$), suggests that P_4 might be a moderately good donor ligand. Indeed, though rare, P_4 complexes are known.

Exercise. Consider the Lewis structure of a segment of bismuth (Fig. 12.4). Is this puckered structure consistent with VSEPR theory?

Arsenic, antimony, and bismuth

The chemically softer elements arsenic, antimony, and bismuth are often found in sulfide ores. Arsenic is usually present in copper and lead sulfide ores, and most of its production is from the flue dust of copper and lead smelters. The solubilities of arsenates, AsO_4^{3-}, are similar to those of phosphates, so traces of these ions are present in phosphate rock.

Arsenic, antimony, and bismuth exist as several allotropes. The most stable structures at room temperature for all three elements are built from puckered hexagonal nets in which each atom has three nearest neighbors. The nets stack in a way that gives three more distant neighbors in the adjacent net (Fig. 12.4). Arsenic vapor resembles phosphorus and consists of tetrahedral As_4 molecules. Bismuth, in common with the α-forms of arsenic and antimony, has a metallic luster and is one of a small class of substances that, like water, expand upon solidification. The electrical conductivity of bismuth is not as high as that of most metals, and its structure is not typical of the isotropic bonding normally found. The band structure of bismuth suggests a low density of conduction electrons and holes, and it is best classified as a semimetal (Section 3.5) rather than as a semiconductor or a true metal.

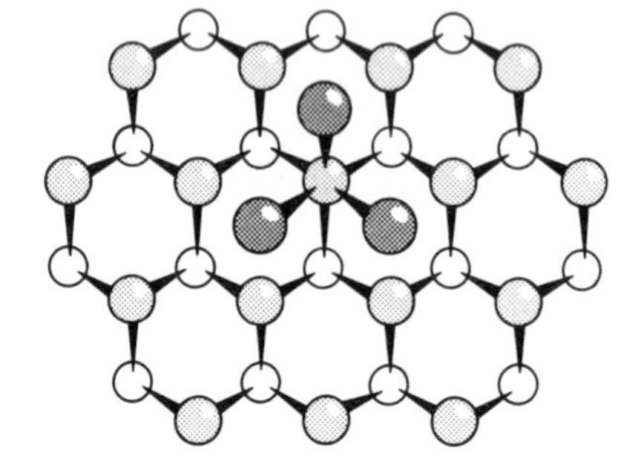

Fig. 12.4. The puckered network structure of bismuth. Each Bi atom has three nearest neighbors; the dark shaded atoms indicate next-nearest neighbors from an adjacent puckered sheet. Unshaded atoms are furthest from the viewer.

12.2 Nitrogen activation

Nitrogen occurs in many compounds, but N_2 itself is strikingly unreactive. A few strong reducing agents can transfer electrons to the molecules at room temperature, leading to scission of the N≡N bond, but usually that needs extreme conditions. The prime example of a successful electron donor is lithium metal, which yields lithium nitride, Li_3N; similarly, when magnesium burns in air it forms the nitride as well as the oxide.

The slowness of the reactions of N_2 appears to be the result of several factors. One is the strength of the N≡N bond and hence the high activation energy for breaking it. Another is the size of the HOMO–LUMO gap in N_2, which makes the molecule resistant to simple electron-transfer redox processes. A third factor is the low polarizability of N_2, which does not encourage the formation of the highly polar transition states that are often involved in electrophilic and nucleophilic displacement reactions.

Cheap methods of nitrogen fixation are highly desirable since they would have a profound effect on the economy. In the Haber process for the production of ammonia (which we discuss in detail in Section 17.7), H_2 and N_2 are combined at high temperatures and pressures over an iron catalyst. Although very successful and used throughout the world, the process requires a costly high-temperature, high-pressure plant; research continues into more economical alternatives.

A clue to the direction nitrogen fixation is likely to take in the future is the observation that bacteria have achieved what inorganic chemists so far have not, namely the room-temperature conversion of atmospheric N_2 to a wide variety of compounds. Dinitrogen complexes of metals were discovered in 1965, and at about the same time the partial elucidation of the structure of the nitrogen-fixing enzyme nitrogenase indicated that the active site contains Fe and Mo atoms (Chapter 19). These developments led to optimism that efficient homogeneous catalysts might be devised in which metal ions will coordinate to N_2 and promote its reduction, and in fact many N_2 complexes have been prepared. In some cases the preparation is as simple as bubbling N_2 through an aqueous solution of a complex;

$$[Ru(NH_3)_5(OH_2)]^{2+}(aq) + N_2(g) \rightarrow [Ru(NH_3)_5(N_2)]^{2+}(aq) + H_2O(l)$$

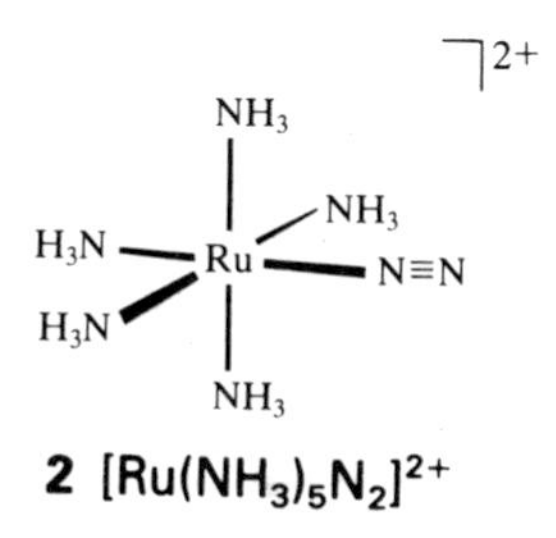

2 $[Ru(NH_3)_5N_2]^{2+}$

As with the isoelectronic CO molecule, end-on bonding (**2**) is typical of N_2 when it acts as a ligand. The N≡N bond is only slightly altered from that in the free molecule both in this end-on complex and in other more elaborate bridging environments. Although new catalysts for N_2 reduction have not yet emerged from this research, there are hopes of it, since it is possible to convert bound N_2 in some of these complexes into NH_4^+:

$$cis\text{-}[W(N_2)_2(P(CH_3)_2(C_6H_5))_4] \xrightarrow{H_2SO_4} N_2 + NH_4^+ + W(VI) \text{ products}$$

12.3 Halides

Halogen compounds of phosphorus, arsenic, and antimony are numerous, and important in synthetic chemistry. This richness stems from the fact that not only do these elements form compounds with all the halogens, but also that they do so with oxidation numbers of +3 and +5. The halogen chemistry of nitrogen and bismuth is less rich because the +5 oxidation state cannot be attained with the less oxidizing halogens. Nitrogen does not reach this high oxidation state in combination with the heavier halogens on account of its high electronegativity and small size; bismuth is big enough and is not electronegative, but the inert pair effect inhibits the formation of Bi(V).

Nitrogen halides

The most extensive series of halogen compounds of nitrogen are the fluorides, and NF_3 is in fact its only exoergic binary halogen compound. This pyramidal molecule is highly unreactive. Thus, unlike NH_3, it is not a Lewis base because the strongly electronegative F atoms make the lone pair of electrons unavailable. Although NF_5 is unknown, NF_3 can be converted into the N(V) species $[NF_4]^+$:

$$NF_3 + 2F_2 + SbF_3 \rightarrow [NF_4^+][SbF_6^-]$$

We see yet again the inability of the small Period-2 *p*-block elements to exceed coordination number 4 in their molecular compounds.

Nitrogen trichloride, NCl_3, is a highly endoergic, explosive, and volatile liquid. It is prepared commercially by the electrolysis of a solution of ammonium chloride, and the resulting gas is used directly as an oxidizing bleach for flour. The extremely unstable nitrogen tribromide, NBr_3, has been prepared, but NI_3 is known only in the form of the highly explosive ammoniate $NI_3{\cdot}NH_3$. In complete contrast to NF_3, X-ray diffraction indicates a polymeric structure for this compound, with linked NI_4 tetrahedra.

Other halides

The trihalides and pentahalides of nitrogen's congeners are used extensively in synthetic chemistry, and their simple empirical formulas conceal an interesting and varied structural chemistry.

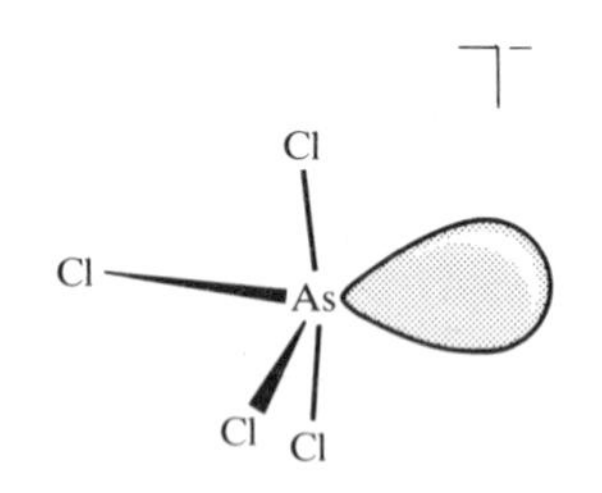

3 $AsCl_4^-$

The trihalides range from volatile liquids and gases, such as PF_3 (b.p. −102°C) and AsF_3 (b.p. −63°C), to solids, such as BiF_3 (m.p. 649°C). A common method of preparation is direct reaction of the element and halogen. For phosphorus, the trifluoride is prepared by metathesis of the trichloride with a fluoride:

$$2PCl_3(l) + 3CaF_2(s) \rightarrow 2PF_3(g) + 3CaCl_2(s)$$

The trichlorides PCl_3, $AsCl_3$, and $SbCl_3$ are useful starting materials for the preparation of a variety of alkyl, aryl, alkoxy, and amino derivatives because they are susceptible to protolysis and metathesis:

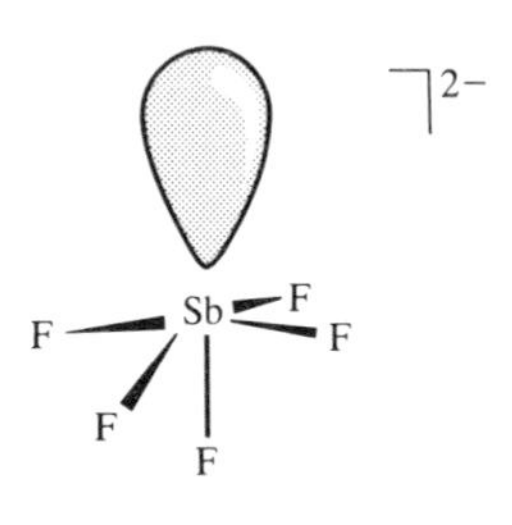

4 SbF_5^{2-}

$$ECl_3 + 3EtOH \rightarrow E(OR)_3 + 3HCl \qquad E = P, As, Sb$$

$$ECl_3 + 6(CH_3)_2NH \rightarrow E(N(CH_3)_2)_3 + 3[(CH_3)_2NH_2][Cl] \qquad E = P, As, Sb$$

Phosphorus trifluoride, PF_3, is an interesting ligand since in some respects it resembles CO. Like that molecule, it is a weak σ-base but a strong π-acid, and complexes of PF_3 exist that are the analogs of carbonyls, such as $Ni(PF_3)_4$ and $Ni(CO)_4$. The acid character is attributed to a P—F antibonding LUMO, which has mainly phosphorus *p*-orbital character when P is bonded to the electronegative F atom. The trihalides also act as mild Lewis acids toward bases such as trialkylamines and halides. Many halide complexes have been isolated, ranging from simple mononuclear species, such as $[AsCl_4]^-$ (**3**) and $[SbF_5]^{2-}$ (**4**), to more complex dinuclear and polynuclear anions linked by halide bridges, such as the polymeric chain $([BiBr_5]^{2-})_n$ in which Bi(III) is surrounded by a distorted octahedron of Br atoms.

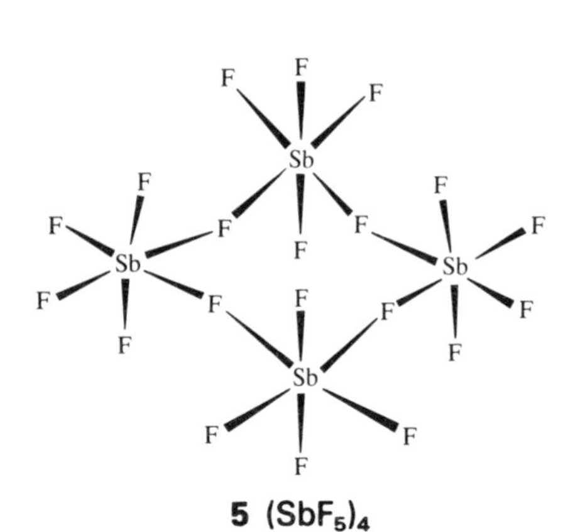

5 $(SbF_5)_4$

The pentahalides vary from highly volatile substances, such as PF_5 (b.p. −85°C) and AsF_5 (b.p. −53°C), to solids, such as PCl_5 (sublimes at 162°C) and BiF_5 (m.p. 154°C). The five-coordinate gas-phase molecules are trigonal bipyramidal, and we saw in Section 2.2 that PF_5 is fluxional. In contrast to the volatility of PF_5 and AsF_5, SbF_5 is a highly viscous liquid in which the molecules are associated through F atom bridges. In solid SbF_5 these bridges result in a cyclic tetramer (**5**) which reflects the tendency of Sb(V) to achieve six-coordination. A related phenomenon occurs with PCl_5, which in the solid state exists as $[PCl_4^+][PCl_6^-]$. In this case the ionic contribution to the lattice energy provides the driving force for the Cl^- ion transfer from one molecule to another.

Unlike the anomalous behavior of the boron trihalides (Section 11.3), the pentafluorides of phosphorus, arsenic, and bismuth are stronger Lewis acids than the other known pentahalides of these elements. As we remarked in Section 6.7, SbF_5 is a very strong Lewis acid; it is much stronger, for example, than the aluminum halides.

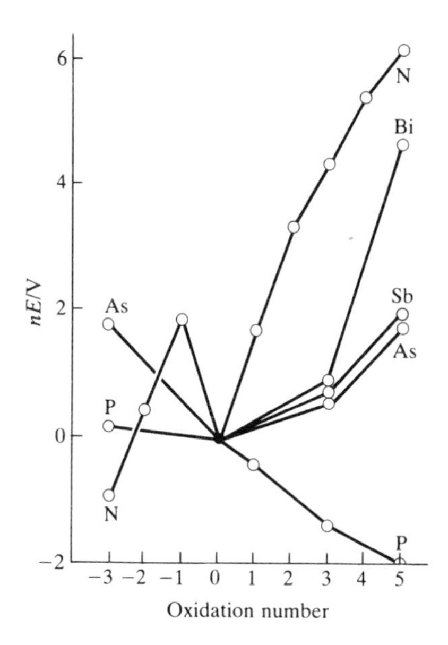

Fig. 12.5 Frost diagrams for the elements of the nitrogen group in acidic solution. The species with oxidation number −3 are NH_3, PH_3, and AsH_3, and those with oxidation numbers −2 and −1 are N_2H_4 and NH_2OH respectively. The positive oxidation states refer to the most stable oxo or hydroxo species in acidic solution, and may be oxides, oxoacids, or oxoanions.

12.4 Oxides and aqueous redox chemistry

We can infer the redox properties of nitrogen-group compounds in acidic aqueous solution from the Frost diagram shown in Fig. 12.5. The slopes of the lines on the right of the diagram show the thermodynamic tendency for reduction of the +5 oxidation states of the elements. They show, for instance, that Bi_2O_5 is potentially a very strong oxidizing agent, which is consistent with the inert pair effect and the tendency of Bi(V) to form Bi(III). The next strongest oxidizing agent is NO_3^-, As(V) and Sb(V) are milder oxidizing agents, and P(V), in the form of phosphoric acid, is extremely weak.

The redox properties of nitrogen are important because of its widespread occurrence in the atmosphere, biosphere, industry, and the laboratory. Nitrogen chemistry also is quite intricate, partly because of the large number of oxidation states attainable, but also because reactions that are thermodynamically favorable are often slow or have rates that depend critically on the identity of the reactants. Carbon chemistry is similarly intricate, as we know from organic and organometallic chemistry, but the mechanisms of the reactions of N_2 and its oxides and oxoanions have turned out to be more difficult to rationalize.

The main points to keep in mind include the fact that, as the N_2 molecule is so inert, redox reactions that consume N_2 are slow. Moreover, for mechanistic reasons that seem to be different in each case, the formation of N_2 is also slow and often sidestepped in aqueous solution (Chart 12.2). As with several other *p*-block elements, the barriers to reactions of high oxidation state oxoanions, such as NO_3^-, are greater than for low oxidation state oxoanions, such as NO_2^-. We should also remember that low pH favors thermodynamically oxidations by oxoanions (Section 8.10). Low pH also often accelerates their reactions by protonation, for this is thought to facilitate N—O bond breaking. Finally, it should be remembered that the reactions of the oxo compounds of nitrogen commonly take place by atom or ion transfer, and outer-sphere electron transfer is rare.

Table 12.2 summarizes some of the properties of the nitrogen oxides, and Table 12.3 does the same for the nitrogen oxoanions. Both tables will help us to pick our way through the details of their properties.

Nitrogen(V) oxoanions

The most common source of N(V) is **nitric acid**, HNO_3, which is a major industrial chemical used in the production of fertilizers, explosives, and a wide variety of nitrogen-containing chemicals. It is produced by modern versions of the **Ostwald process**, which make use of a clever, highly indirect

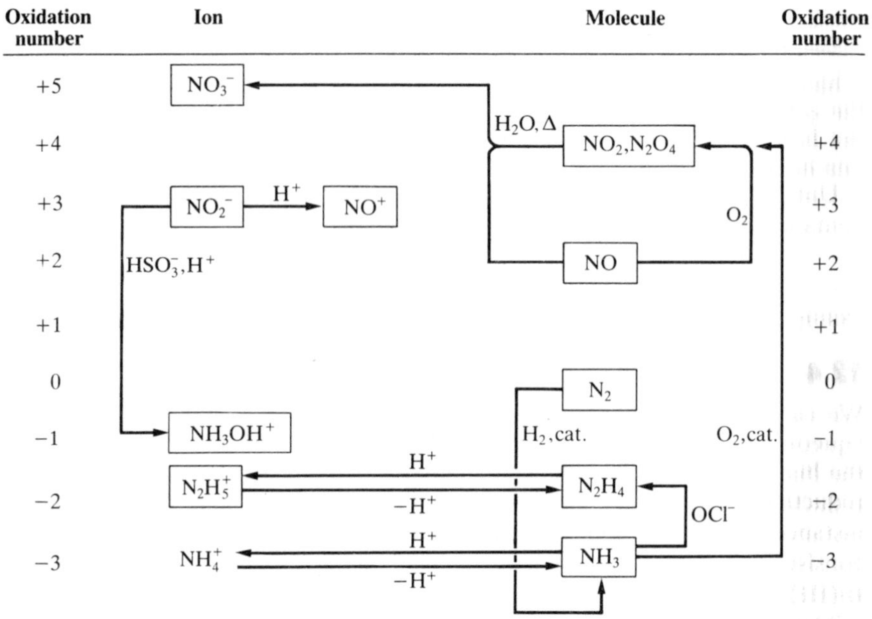

Chart 12.2. Some reactions of common nitrogen-containing molecules and ions

Table 12.2. Oxides of nitrogen

Oxidation number	Formula	Name	Structure (gas phase)	Remarks
+1	N_2O	Nitrous oxide (Dinitrogen oxide)	N—N—O (N—N 1.19) $C_{\infty v}$	Colorless gas, not very reactive
+2	NO	Nitric oxide (Nitrogen monoxide)	N—O (1.15) $C_{\infty v}$	Colorless, paramagnetic gas
+3	N_2O_3	Dinitrogen trioxide	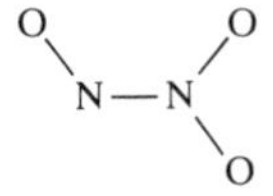 O—N—N(O)—O; Planar, C_s	Forms blue solid (m.p. −101°C) and dissociates into NO and NO_2 in the gas phase
+4	NO_2	Nitrogen dioxide	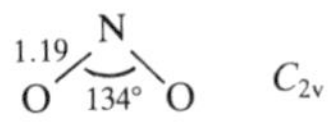 O—N—O (1.19, 134°) C_{2v}	Brown, reactive, paramagnetic gas
	N_2O_4	Dinitrogen tetroxide	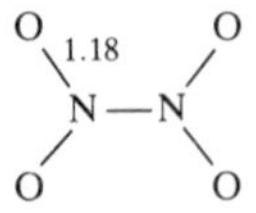 O_2N—NO_2 (1.18); Planar, D_{2h}	Forms colorless liquid (m.p. −11°C); in equilibrium with NO_2 in the gas phase
+5	N_2O_5	Dinitrogen pentoxide	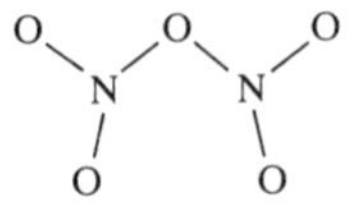 O_2N—O—NO_2; Planar, C_{2v}	Colorless ionic solid $[NO_2][NO_3]$ (m.p. 32°C); unstable in gas phase

Table 12.3. Nitrogen oxoions

Oxidation number	Formula	Name	Structure	Remarks
+1	$N_2O_2^{2-}$	Hyponitrite	$[O-N=N-O]^{2-}$ (trans), C_2	Usually acts as a reducing agent
+3	NO_2^-	Nitrite	$[O-N-O]^-$, N–O 1.24, 115°, C_{2v}	Weak base; as an oxidizing and a reducing agent
+3	NO^+	Nitrosonium	$[N-O]^+$, $C_{\infty v}$	Oxidizing agent and Lewis acid
+5	NO_3^-	Nitrate	$[NO_3]^-$, N–O 1.22, Planar, D_{3h}	Very weak base; an oxidizing agent
+5	NO_2^+	Nitronium	$[O-N-O]^+$, 1.15, Linear, $D_{\infty h}$	Oxidizing agent, nitrating agent, and a Lewis acid

route from N_2 to the highly oxidized compound HNO_3 via the fully reduced compound NH_3. Thus, after nitrogen has been reduced to the -3 state as NH_3 by the Haber process, it is oxidized to the $+4$ state:

$$4NH_3(g) + 7O_2(g) \rightarrow 6H_2O(g) + 4NO_2(g) \qquad \Delta G^{\ominus} = -308.0\ \text{kJ}\ (\text{mol NO}_2)^{-1}$$

The NO_2 then undergoes disproportionation into N(II) and N(V) in water at elevated temperatures:

$$3NO_2(aq) + H_2O(l) \rightarrow 2HNO_3(aq) + NO(g) \qquad \Delta G^{\ominus} = -5.0\ \text{kJ}\ (\text{mol HNO}_3)^{-1}$$

Both steps are thermodynamically favorable. The by-product NO is oxidized with O_2 to NO_2 and recirculated. Such an indirect route is employed because the direct oxidation of N_2 to NO_2 is thermodynamically unfavorable, with $\Delta G_f^{\ominus}(NO_2) = +51\ \text{kJ mol}^{-1}$. In part, this is due to the great strength of the N≡N bond.

Reduction potential data imply that the NO_3^- ion is a fairly strong oxidizing agent; however, the kinetic aspects of its reactions are important, and they are generally slow in dilute acid solution. Since oxygen-coordination promotes N—O bond breaking, concentrated HNO_3 is thermodynamically a more potent oxidizing agent and undergoes more rapid reactions than the dilute acid. A sign of this is the yellow color of the concentrated acid, which indicates its instability with respect to decomposition into NO_2:

$$4HNO_3(aq) \rightarrow 4NO_2(aq) + O_2(g) + 2H_2O(l)$$

This decomposition is accelerated by light and heat.

Reduction of NO_3^- ions rarely yields a single product as so many lower oxidation states of nitrogen are available. For example, a strong reducing agent such as zinc can reduce a substantial proportion of dilute HNO_3 as far as oxidation number −3:

$$HNO_3(aq) + 4Zn(s) + 9H^+(aq) \rightarrow NH_4^+(aq) + 3H_2O(l) + 4Zn^{2+}(aq)$$

A weaker reducing agent, such as copper, proceeds only as far as oxidation number +4 in the concentrated acid:

$$2HNO_3(aq) + Cu(s) + 2H^+(aq) \rightarrow 2NO_2(g) + Cu^{2+}(aq) + 2H_2O(l)$$

With the dilute acid the +2 oxidation state is favored, and NO is formed.

Example 12.2: *The stability of N(V) and Bi(V)*

N(V) and Bi(V) are stronger oxidizing agents than the +5 oxidation states of the three intervening elements. Correlate this observation with trends in the periodic table.

Answer. The light *p*-block elements are more electronegative that the heavier elements and are generally less easily oxidized. Nitrogen is generally a good oxidizing agent in its positive oxidation states. Bismuth is much less electronegative, but favors the +3 oxidation state in preference to the +5 state because of the inert pair effect. The elements phosphorus, arsenic, and antimony are less electronegative than nitrogen and do not show the inert pair effect.

Exercise. Correlate the strength of phosphorus and sulfur as oxidizing agents with periodic trends.

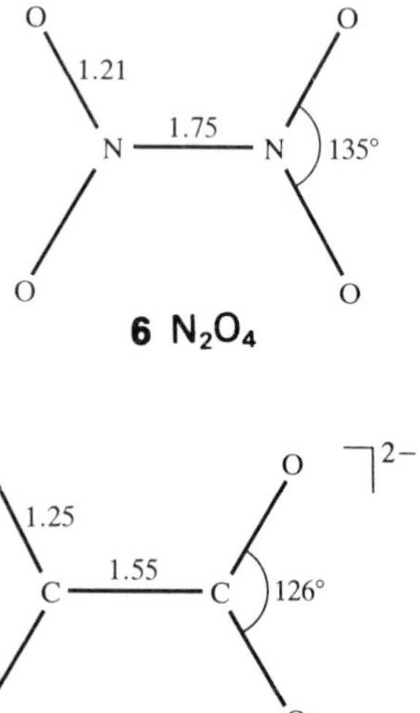

Nitrogen(IV) and nitrogen(III) oxides

Nitrogen(IV) oxide, commonly **nitrogen dioxide**, exists as an equilibrium mixture of the brown NO_2 radical and its colorless dimer N_2O_4:

$$N_2O_4(g) \rightleftharpoons 2NO_2(g) \qquad K_p = 0.115 \text{ at } 25°C$$

O
1.21
1.75
N — N 135°
O O O

6 N_2O_4

O O
1.25
1.55
C — C 126°
O O

7 $C_2O_4^{2-}$

This readiness to dissociate is consistent with the N—N bond in N_2O_4 (**6**) being significantly longer than the C—C bond in the isoelectronic oxalate ion ($C_2O_4^{2-}$, **7**), which does not equilibrate with CO_2^- radicals in the same way. The unpaired electron occupies an antibonding orbital in NO_2 and the transient CO_2^- radical, and so is *less* localized on the more electronegative N atom in NO_2 than it is on C in CO_2^-. Hence the bond in N_2O_4 is relatively weak[2] and N_2O_4 is more likely to dissociate than the oxalate ion.

Nitrogen(IV) oxide is a poisonous oxidizing agent that is present in low concentrations in the atmosphere, especially in photochemical smog. In basic aqueous solution it disproportionates into N(III) and N(V), forming NO_2^- and NO_3^- ions:

$$2NO_2(aq) + 2OH^-(aq) \rightarrow NO_2^-(aq) + NO_3^-(aq) + H_2O(l)$$

In acidic solution (as in the Ostwald process) the reaction product is N(II) in

[2] We remarked in Section 2.7 that bonding electrons are more likely to be found near the more electronegative atom in a bond, and antibonding electrons are more likely to be found near the less electronegative atom.

place of N(III) because nitrous acid readily disproportionates:

$$3HNO_2(aq) \rightarrow NO_3^-(aq) + 2NO(g) + H_3O^+(aq) \qquad E^\ominus = +0.05\ V,\ K = 50$$

Nitrous acid, HNO_2, is a strong oxidizing agent;

$$HNO_2(aq) + H^+(aq) + e^- \rightarrow NO(g) + H_2O(l) \qquad E^\ominus = +1.00\ V$$

and its reactions as an oxidizing agent are often more rapid than its disproportionation. The rate at which it oxidizes another molecule is increased by acid as a result of its conversion to the **nitrosonium ion** NO^+:

$$HNO_2(aq) + H^+(aq) \rightarrow NO^+(aq) + H_2O(l)$$

The nitrosonium ion is a strong Lewis acid and associates rapidly with anions and other nucleophiles. These may not themselves be susceptible to oxidation, as in the case of SO_4^{2-} and F^- ions, which form SO_3ONO^- (**8**) and ONF (**9**) respectively, but the association is often the initial step in an overall redox reaction. Thus the reaction of HNO_2 with I^- ions leads to the rapid formation of ONI:

$$NO^+(aq) + I^-(aq) \rightarrow ONI(aq)$$

followed by the rate-determining second-order reaction between two INO molecules:

$$2ONI(aq) \rightarrow I_2(aq) + 2NO(g)$$

Nitrosonium salts containing poorly coordinating anions, such as $[NO][BF_4]$, are useful reagents in the laboratory as facile oxidizing agents and as a source of NO^+.

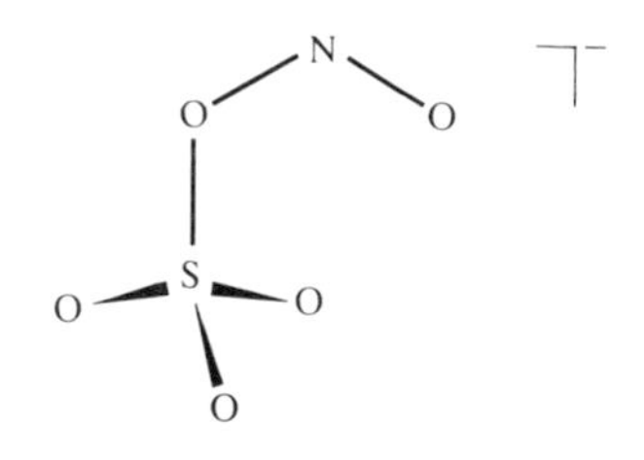

8 $[O_3SONO]^-$

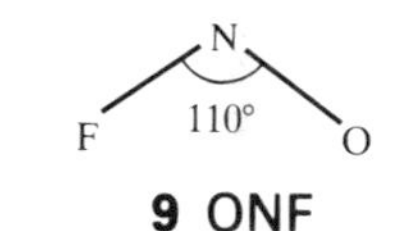

9 ONF

Nitrogen(II) oxide

Nitrogen(II) oxide, more commonly **nitric oxide**, is an odd-electron molecule, but unlike NO_2 it does not form a stable dimer in the gas phase. This difference reflects the greater delocalization of the odd electron in the π^* orbital in NO than in NO_2. Nitric oxide reacts with O_2 to generate NO_2, but in the gas phase the rate law is second order in NO, since a transient dimer $(NO)_2$ is produced that subsequently collides with an O_2 molecule. Because the reaction is second-order, atmospheric NO produced in low concentrations by gasoline and diesel engines is slow to convert to NO_2.

Since NO is endoergic, it should be possible to find a catalyst to convert the pollutant NO to the natural atmospheric gases N_2 and O_2 at its source in exhausts. A practical catalyst of this type is yet to be discovered, and the search for one is an interesting and socially important chemical challenge.[3]

Low oxidation states

The average oxidation number of nitrogen in dinitrogen oxide, N_2O, commonly **nitrous oxide**, is +1; that of the **azide ion**, N_3^-, is $-\frac{1}{3}$. Both species are isoelectronic with CO_2, but their resemblance barely extends beyond their common linear shapes.

Dinitrogen oxide, a colorless unreactive gas, is produced by the comproportionation of molten ammonium nitrate. Care must be taken to avoid

[3] A successful catalyst for NO decomposition under laboratory conditions was found not long ago. It consists of Cu(II) in a zeolite. M. Iwamoto, H. Furukawa, Y. Mire, F. Uemura, S. Mikuriya, and S. Kagawa, *J. chem. Soc. chem. Commun.*, 1272 (1986).

an explosion in this reaction, in which the cation is oxidized by the anion:

$$NH_4NO_3(l) \xrightarrow{250°C} N_2O(g) + 2H_2O(g)$$

Its reduction potentials suggest that N_2O should be a strong oxidizing agent in acidic and basic solutions:

$$N_2O(g) + 2H^+(aq) + 2e^- \rightarrow N_2(g) + H_2O(l) \qquad E^\ominus = +1.77 \text{ V at pH} = 0$$

$$N_2O(g) + H_2O(l) + 2e^- \rightarrow N_2(g) + 2OH^-(aq) \qquad E^\ominus = +0.94 \text{ V at pH} = 14$$

However, kinetic considerations are paramount, and the gas is unreactive toward many reagents at room temperature. One sign of this is that N_2O has been used as the propellant gas for instant whipping cream. Similarly, N_2O was used for many years as a mild anesthetic ('laughing gas'); however, this practice has been discontinued because of some undesirable physiological side-effects.

The N_3^- ion may be synthesized by the oxidation of sodium amide with either NO_3^- ions or N_2O at elevated temperatures:

$$3NH_2^- + NO_3^- \xrightarrow{175°C} N_3^- + 3OH^- + NH_3$$

$$2NH_2^- + N_2O \xrightarrow{190°C} N_3^- + OH^- + NH_3$$

The ion is a reasonably strong Brønsted base, the pK_a of its conjugate acid, **hydrazoic acid** HN_3, being 4.77. It is also a good ligand toward *d*-block metal ions. However, heavy metal complexes or salts, such as $Pb(N_3)_2$ and $Hg(N_3)_2$, are shock-sensitive detonators. Ionic azides such as NaN_3 are thermodynamically unstable but kinetically inert; they can be handled at room temperature, and when heated liberate N_2 smoothly.

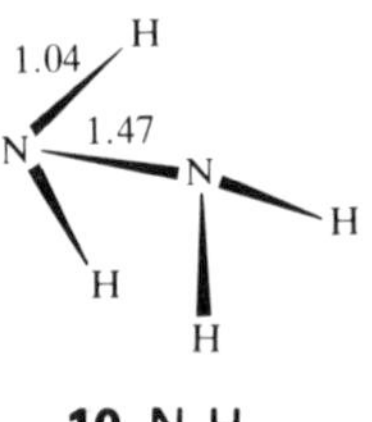

10 N_2H_4

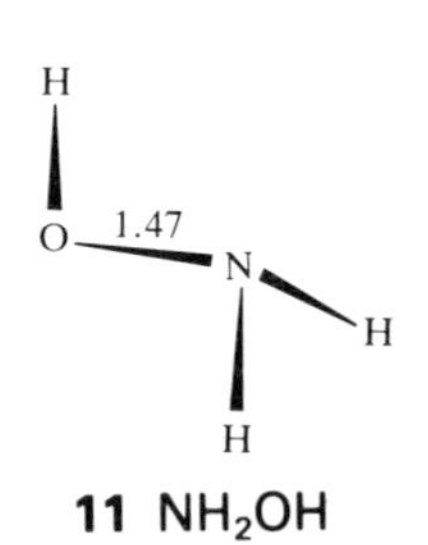

11 NH_2OH

Hydrazine and hydroxylamine

Hydrazine N_2H_4 (**10**), in which the oxidation number of nitrogen is -2, and **hydroxylamine** NH_2OH (**11**), in which it is -1, are isoelectronic molecules that are formally related to NH_3 by the replacement of an H atom by an NH_2 group or an OH group. Both compounds are liquids at room temperature. In each case, the electronegative substituent makes the nitrogen lone pair less readily available and results in weaker Brønsted bases (and hence stronger acidity for the conjugate acids) than NH_3:

	NH_4^+	$N_2H_5^+$	NH_3OH^+
pK_a:	9.26	7.93	5.82

The Lewis base strength is reduced in the same way.

Most commercial hydrazine is prepared by the oxidation of NH_3 by OCl^- ions:

$$2NH_3(aq) + OCl^-(aq) \rightarrow N_2H_4(aq) + Cl^-(aq) + H_2O(l)$$

This is formally a redox reaction. However, the mechanism is more intricate than simple electron transfer because it proceeds through the formation of the intermediate NH_2Cl. Once formed, the NH_2Cl is attacked by the nucleophile NH_3, when Cl^- is displaced and the N—N bond is formed:

$$NH_2Cl(aq) + 2NH_3(aq) \rightarrow H_2NNH_2(aq) + Cl^-(aq) + NH_4^+(aq)$$

Hydrazine (m.p. 2°C, b.p. 113°C) is strongly associated through hydrogen bonding. It is an endoergic compound ($\Delta G_f^\ominus = +149\ \text{kJ mol}^{-1}$), kinetically fairly inert, and widely used as a reducing agent. For example, it is used to reduce dissolved oxygen in boiler water to suppress corrosion. Similarly, alkylhydrazines are used as the reducing component in some rocket fuels. When hydrazine reacts with oxidizing agents, it yields a variety of nitrogen-containing products. One common product is N_2, as in the reactions

$$N_2H_4(aq) + O_2(g) \rightarrow N_2(g) + 2H_2O(l)$$

$$N_2H_4(aq) + 2Cl_2(g) \rightarrow N_2(g) + 4HCl(aq)$$

Hydrazine is a much stronger reducing agent in basic than in acidic solution:

$$N_2(g) + 5H^+(aq) + 4e^- \rightarrow N_2H_5^+(aq) \qquad E^\ominus = -0.23\ \text{V at pH} = 0$$

$$N_2(g) + 4H_2O(l) + 4e^- \rightarrow N_2H_4(aq) + 4OH^-(aq)$$

$$E^\ominus = -1.16\ \text{V at pH} = 14$$

Hydroxylamine is an unstable endoergic compound ($\Delta G_f^\ominus = +23\ \text{kJ mol}^{-1}$) which is prepared by the reduction of NO_2^- ions by HSO_3^- ions in neutral solution, followed by acidification and heating. In the first stage SO_3^{2-} attacks NO_2^- and forms an N—S bond; in the second stage this N—S bond is hydrolyzed:

$$NO_2^-(aq) + 2HSO_3^-(aq) \xrightarrow{0°C} N(OH)(SO_3)_2^{2-}(aq) + OH^-$$

$$\xrightarrow{H^+,\ 50°C} H_3NOH^+(aq) + 2SO_4^{2-}(aq)$$

Although the standard potentials indicate that hydroxylamine can serve as either an oxidizing agent or a reducing agent, the latter reactions generally occur more readily, as in

$$4Fe^{3+}(aq) + 2NH_3OH^+(aq) \rightarrow 4Fe^{2+}(aq) + N_2O(g) + 6H^+(aq) + H_2O(l)$$

This reaction sidesteps N_2, despite its greater stability, and the nitrogen is carried directly from the -1 oxidation state in H_2NOH to the $+1$ state in N_2O by a mechanistically complex reaction.

Example 12.3: *Comparing the redox properties of nitrogen oxoanions and oxo compounds*

Compare the chemistry of (a) NO_3^- and NO_2^- as oxidizing agents; (b) NO_2, NO, and N_2O with respect to their ease of oxidation in air; (c) N_2H_4 and H_2NOH as reducing agents.

Answer. (a) NO_3^- and NO_2^- ions are both strong oxidizing agents. The reactions of the former are often sluggish but are generally faster in strong acid. The reactions of NO_2^- ions are generally faster and become even faster in acidic solution, where the NO^+ ion is a common identifiable intermediate. (b) NO_2 is stable with respect to oxidation in air. NO is thermodynamically susceptible to oxidation but its reaction with oxygen is slow at low NO concentrations because the rate law is second order in NO. (c) Hydrazine and hydroxylamine are both good reducing agents. In basic solution hydrazine becomes a stronger reducing agent.

Exercise. Summarize the reactions that are employed for the synthesis of hydrazine and hydroxylamine. Are these best described as electron transfer processes or nucleophilic displacement reactions?

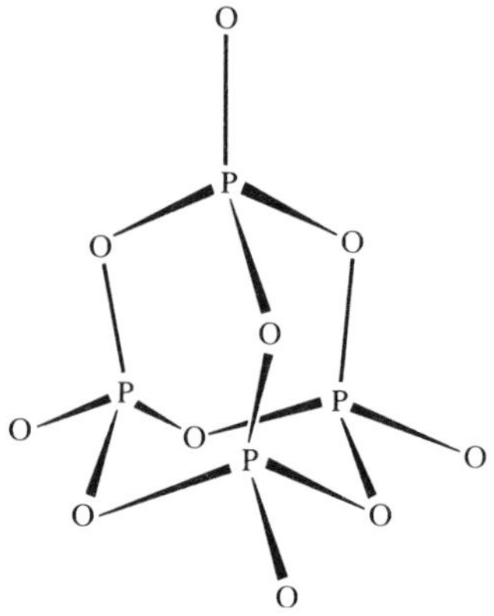

12 P_4O_{10}

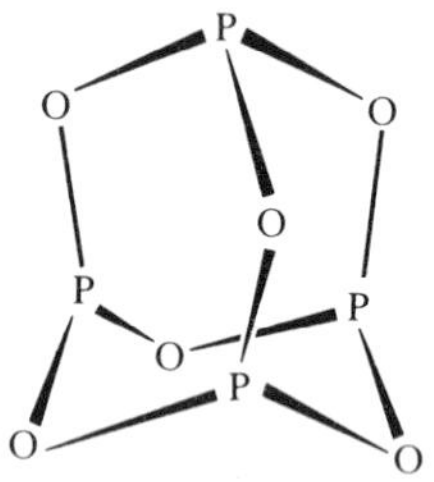

13 P_4O_6

Chart 12.2 is a summary of the reactions of some important nitrogen species. All the reactions shown in solution occur by electrophile–nucleophile interaction rather than simple electron transfer. These include the disproportionation of NO_2 in acidic or basic solution and the disproportionation of NO_2^- in acid. Similarly, the conversion of NO_2^- ions to NH_2OH occurs by nucleophilic attack by HSO_3^- followed by acid hydrolysis, and the oxidation of NH_3 to N_2H_4 by carefully controlled reaction with OCl^- involves a nucleophilic attack of NH_3 on NH_2Cl.

Phosphorus oxides

The complete combustion of phosphorus yields phosphorus(V) oxide, P_4O_{10}. Each P_4O_{10} molecule has a cage structure in which a tetrahedron of P atoms is held together by bridging O atoms, and each P atom has a terminal O atom (**12**). Combustion in a limited supply of oxygen results in the formation of phosphorus(III) oxide, P_4O_6, instead; this molecule has the same O-bridged framework of P_4O_{10}, but lacks the terminal O atoms (**13**). (It is also possible to isolate the intermediate compositions having one, two, or three O atoms terminally attached to the P atoms.) Both oxides can be hydrated to yield the corresponding acids, the P(V) oxide giving phosphoric acid, H_3PO_4, and the P(III) oxide giving phosphorous acid, H_3PO_3. As we remarked in Section 5.6, phosphorous acid has one H atom attached directly to the P atom; it is therefore a diprotic acid and better represented as $(HO)_2PHO$.

Phosphorus oxoanions

We can see from the Latimer diagram in Table 12.4 that elemental phosphorus and most of its compounds other than P(V) are strong reducing agents. White phosphorus disproportionates into phosphine PH_3 (oxidation number $+1$) and hypophosphite ions (oxidation number $+1$) in basic solution:

$$P_4(s) + 3OH^-(aq) + 3H_2O(l) \rightarrow PH_3(g) + 3H_2PO_2^-(aq)$$

Some well-known phosphorus oxoanions are listed in Table 12.5. The approximately tetrahedral environment of the P atom in their structures should be noted, as should the existence of P—H bonds in the hypophosphite and phosphite anions. The synthesis of various P(III) oxoacids and

Table 12.4. Latimer diagrams for phosphorus

Acidic solution

$$H_3PO_4 \xrightarrow{-0.93} H_4P_2O_6 \xrightarrow{0.38} H_3PO_3 \xrightarrow{-0.50} H_3PO_2 \xrightarrow{-0.51} P \xrightarrow{-0.06} PH_3$$

H_3PO_4 to H_3PO_3: -0.28; H_3PO_3 to P: -0.50

Basic solution

$$PO_4^{3-} \xrightarrow{-1.12} HPO_3^{2-} \xrightarrow{-1.57} H_2PO_2^- \xrightarrow{-2.05} P \xrightarrow{-0.89} PH_3$$

HPO_3^{2-} to P: -1.73

Table 12.5. Some phosphorus oxoanions

Oxidation number	Formula	Name	Structure	Remarks
+1	$H_2PO_2^-$	Hypophosphite (dihydrodioxophosphate)	C_{2v}	Facile reducing agent
+3	HPO_3^{2-}	Phosphite (hydrotrioxophosphate)	C_{3v}	Facile reducing agent
+4	$P_2O_6^{4-}$	Hypophosphate		Basic
+5	PO_4^{3-}	Phosphate	T_d	Strongly basic
+5	$P_2O_7^{4-}$	Pyrophosphate		Basic; longer-chain analogs are known

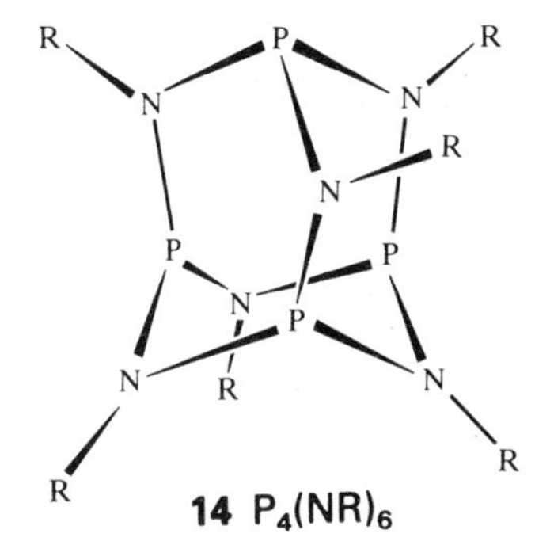

14 $P_4(NR)_6$

oxoanions, including HPO_3^- and alkoxophosphines, is conveniently performed by solvolysis of phosphorus(III) chloride under mild conditions, such as in cold carbon tetrachloride solution:

$$PCl_3(l) + 3H_2O(l) \rightarrow H_3PO_3(aq) + 3HCl(g)$$

$$PCl_3(l) + 3ROH(aq) + 3N(CH_3)_3(aq) \rightarrow P(OR)_3(aq) + 3[HN(CH_3)_3]Cl(aq)$$

Reductions with $H_2PO_2^-$ and HPO_3^{2-} are usually fast. One of the commercial applications of this lability is the use of $H_2PO_2^-$ to reduce $Ni^{2+}(aq)$ ions and to coat surfaces with metallic nickel in the process called 'electrodeless plating'.

15 $[(CH_3)_2PN]_3$

12.5 Compounds of nitrogen with phosphorus

Many analogs of phosphorus–oxygen compounds exist in which the O atom is replaced by the isolobal N—R or N—H group, such as $P_4(NR)_6$ (**14**), the analog of P_4O_6. Other compounds exist in which OH or OR groups are replaced by the isolobal NH_2 or NR_2 groups. An example is $P(NMe_2)_3$, the analog of $P(OMe)_3$. Another indication of the scope of phosphorus–nitrogen chemistry, and a useful point to remember, is that PN is isoelectronic with SiO. For example, various **phosphazenes**, which are chains and rings containing R_2PN units (**15**), are analogous to the siloxanes (Section 10.12) and their R_2SiO units (**16**).

16 $[(CH_3)_2SiO]_3$

17 $[Cl_2PN]_3$

18 $[Cl_2PN]_4$

The cyclic phosphazene dichlorides, which are good starting materials for the preparation of the more elaborate phosphazenes, are easily synthesized:

$$n\mathrm{PCl_5} + n\mathrm{NH_4Cl} \rightarrow (\mathrm{Cl_2PN})_n + 4n\mathrm{HCl}$$

A chlorocarbon solvent and temperatures near 130°C produce the cyclic trimer (**17**) and tetramer (**18**), and when the trimer is heated to about 290°C it changes to polyphosphazene. The P—Cl bonds in this rubbery material make it liable to hydrolysis, but very stable materials can be obtained by substitution:

$$(\mathrm{Cl_2PN})_n + 2n\mathrm{C_2F_5O^-} \rightarrow [(\mathrm{F_5C_2O})_2\mathrm{PN}]_n + 2n\mathrm{Cl^-}$$

Like silicone rubber, the polyphosphazenes remain rubbery at low temperatures since, as with the isoelectronic Si—O—Si group, P—N—P groups are highly flexible.

Example 12.4: *Devising a synthetic route to an alkoxy substituted cyclophosphazene*

Give balanced equations for the preparation of $[\mathrm{NP(OCH_3)_2}]_4$ from PCl_5, NH_4Cl, and $NaOCH_3$.

Answer. The cyclic chlorophosphazene can be synthesized first:

$$4\mathrm{PCl_5} + 4\mathrm{NH_4Cl} \xrightarrow{130°\mathrm{C}} (\mathrm{Cl_2PN})_4 + 16\mathrm{HCl}$$

The Cl atoms are readily replaced by strong nucleophiles such as alkoxides, in the present case to give the desired product:

$$(\mathrm{Cl_2PN})_4 + 8\mathrm{NaOCH_3} \rightarrow [(\mathrm{CH_3O})_2\mathrm{PN}]_4 + 8\mathrm{NaCl}$$

Exercise. Give the equations for the preparation of a high polymer phosphazene containing a PN backbone with two $N(CH_3)_2$ side groups attached to each P atom.

The oxygen group

The elements of the oxygen group are often (and officially) called the **chalcogens**. The name derives from the Greek word for brass, and refers to the association of sulfur and its congeners with copper.

12.6 Production and structures of the elements

Oxygen

Oxygen is readily available as O_2 from the atmosphere and is obtained on a massive scale by the distillation of liquid air. The main commercial motivation is to recover O_2 for steel making, since about 1 ton of oxygen is needed to make 1 ton of steel.

The common allotrope of oxygen, formally **dioxygen** O_2, boils at −183°C and has a faint blue color in the liquid state (which arises from transitions involving pairs of neighboring molecules). The molecular orbital description implies the existence of an O═O double bond; however, as we saw in Section 2.6, the outermost two electrons occupy different antibonding π

orbitals with parallel spins, and the molecule has a triplet ground state with term symbol $^3\Sigma_g$. The other allotrope, **ozone** O_3, boils at −112°C and is an explosive and highly reactive endoergic dark blue gas ($\Delta G_f^\ominus$ = +163 kJ mol^{-1}). The O_3 molecule is angular, in accord with VSEPR theory (**19**), and has bond angle 117°; it is diamagnetic.

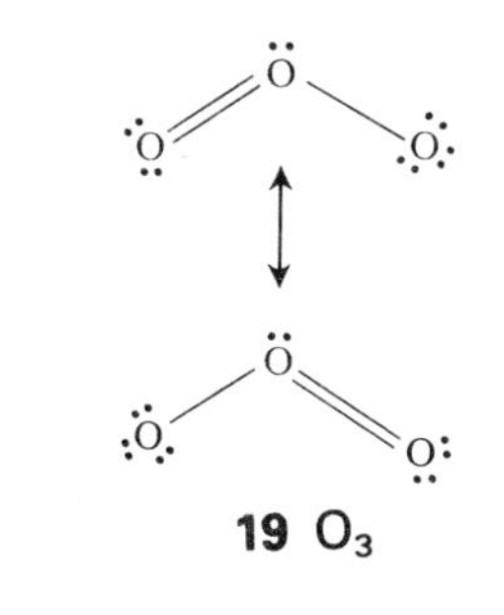

19 O_3

Sulfur

Sulfur is obtained from deposits of the native element, metal sufide ores, and liquid or gaseous hydrocarbons with a high sulfur content (Section 8.2).

Unlike oxygen, all the heavier members of the group favor single bonds over double bonds. As a result, they aggregate into larger molecules or extended structures and hence are solids at room temperature. Sulfur vapor consists partially of paramagnetic disulfur S_2 molecules that resemble O_2 in having a triplet ground state and a formal double bond.

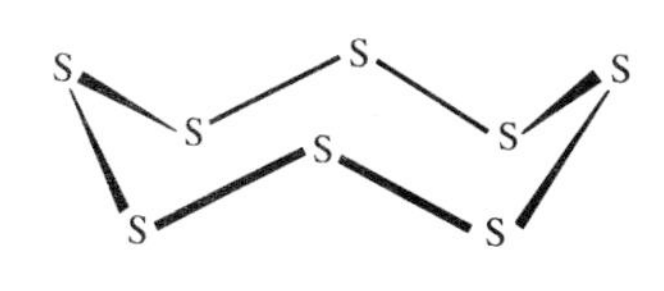

20 S_8

All the crystalline forms of sulfur that can be isolated at room temperature consist of S_n rings. The common orthorhombic polymorph, α-S_8, consists of crown-like eight-membered rings (**20**), but it is possible to synthesize and crystallize rings with from 6 to 20 S atoms. Orthorhombic sulfur melts at 113°C; the yellow liquid darkens above 160°C and becomes more viscous as the sulfur rings break open and polymerize. The resulting helical S_n polymers (**21**) can be drawn from the melt and quenched to form metastable rubber-like materials that slowly revert to α-S_8 at room temperature.

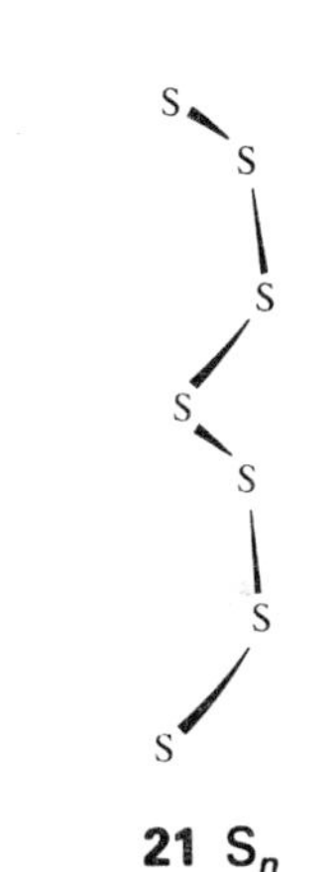

21 S_n

Selenium, tellurium, and polonium

The chemically soft elements selenium and tellurium occur in metal sulfide ores, and their principal source is the electrolytic refining of copper.

Selenium exists as several different polymorphs. As with sulfur, an allotrope of selenium exists that contains Se_8 rings, but the most stable form at room temperature is **gray selenium**, a crystalline material composed of helical chains. The photoconductivity of gray selenium, which arises from the ability of incident light to excite electrons across its reasonably small bandgap (2.6 eV in the crystalline material, 1.8 eV in the amorphous), accounts for its use in photocells. The common commercial form of the element is amorphous **black selenium**. Another amorphous form of selenium, obtained by deposition of the vapor, is used as the photoreceptor in the xerographic photocopying process.

Tellurium crystallizes in a chain structure like that of gray selenium. Polonium crystallizes in a primitive cubic structure and a closely related higher temperature form above 36°C. We remarked in Section 4.2 that a simple cubic array represents inefficient packing of atoms, and polonium is the only element that adopts this structure under normal conditions.

12.7 Halides

Oxygen forms many halogen oxides and oxoanions, and we shall discuss them in Chapter 13. Oxygen forms O_2F_2 and OF_2; the latter is the highest fluoride of oxygen and hence the highest oxidation state (+2) that it reaches.

Sulfur, selenium, tellurium, and polonium have a very rich halogen chemistry, and some of the most common halides are summarized in Table 12.6. The more electronegative elements sulfur and selenium do not form

Table 12.6. Some halides of sulfur, selenium, and tellurium

Oxidation number	Formula	Structure	Remarks
$+\frac{1}{2}$	Te_2X (X = Br, I)	Halide bridges	Silver-gray
+1	S_2F_2	Two isomers:	
		F—S—S—F	Reactive
		S—S(F)F	Reactive
	S_2Cl_2 TeI		
+2	SCl_2	S(Cl)Cl	Reactive
+4	SF_4	S(F)(F)(F)F	Gas
	SeX_4 (X = F, Cl, Br) TeF_4 (X = F, Cl, Br, I)		SeF_4 liquid TeF_4 solid
+5	S_2F_{10} Se_2F_{10}	$F_5S—SF_5$	Reactive
+6	SF_6, SeF_6	SF_6	Colorless gases
	TeF_6		Liquid (b.p. 36°C)

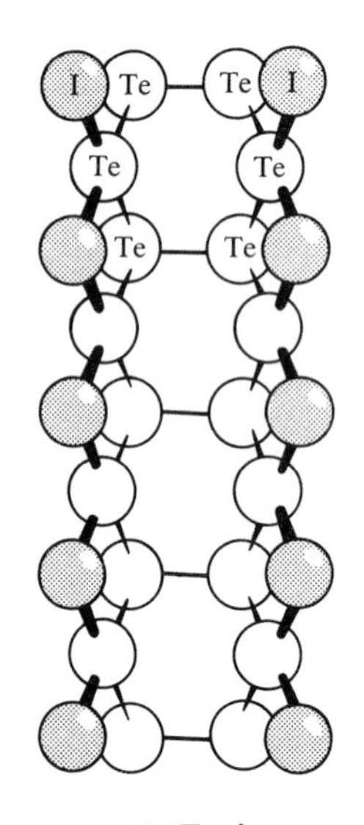

22 Te_2I

simple binary iodides, but the more electropositive tellurium and polonium do. Of the halogens, the small, electronegative F atom alone brings out the maximum group oxidation state of the elements, but it does not form stable binary compounds of selenium, tellurium, and polonium in low oxidation states (+1 and +2). A series of catenated subhalides exist for the heavy members of the group. For example, Te_2I and Te_2Br consists of a ribbon of edge-shared Te hexagons with halogen bridges (**22**). The inability of the less electronegative halogens to bring out the higher oxidation states is understandable from the viewpoint that they are less electronegative than fluorine and their single bond strengths to other elements are generally weaker too. The lack of low oxidation state fluorides may be a consequence of their instability toward disproportionation into the element and a higher oxidation state fluoride.

The structures of the sulfur halides S_2F_2, SF_4, SF_6, and S_2F_{10} (Table 12.6) are all in line with VSEPR theory. Thus, SF_4 has 10 valence electrons around the S atom, two of which form a lone pair in an equatorial position of a trigonal bipyramid. We have already mentioned the theoretical evidence that the molecular orbitals bonding the F atoms to the central atom in SF_6 primarily utilize the sulfur $4s$ and $4p$ orbitals, with the $3d$ orbitals playing a relatively unimportant role (Section 3.2). The same seems to be true of SF_4 and S_2F_{10}.

Sulfur hexafluoride is a gas at room temperature. Its inertness appears to arise because the steric protection of the central S atom suppresses thermodynamically favorable reactions such as the hydrolysis

$$SF_6(g) + 4H_2O(l) \rightarrow 6HF(aq) + H_2SO_4(aq)$$

The sterically less hindered molecule SF_4 is reactive and undergoes rapid partial hydrolysis:

$$SF_4 + H_2O \rightarrow OSF_2 + 2HF$$

Both SF_4 and SeF_4 are very selective fluorinating agents for the conversion of C=O and P=O groups into CF_2 and PF_2 groups:

$$2R{-}\overset{\overset{\displaystyle O}{\|}}{C}{-}R + SF_4 \rightarrow 2RCF_2R + SO_2$$

Other halides of the chalcogens exist, and are summarized in Table 12.6. Sulfur chlorides are commercially important. The reaction of molten sulfur with Cl_2 yields the foul-smelling and toxic substance disulfur dichloride S_2Cl_2, which is a yellow liquid at room temperature (b.p. 138°C). Disulfur dichloride and its further chlorination product SCl_2 (an unstable red liquid) are produced on a large scale for use in the vulcanization of rubber.

12.8 Oxygen and the *p*-block oxides

Oxygen is by no means an inert molecule, and yet many of its reactions are sluggish (a point first made in connection with overpotentials in Section 8.4). For example, a solution of Fe^{2+} is only slowly oxidized by air even though the potential is favorable.

There are several factors that contribute to the appreciable activation energy of many reactions of O_2. One is that, with weak reducing agents, single electron transfer to O_2 is mildly unfavorable thermodynamically:

$$O_2(g) + H^+(aq) + e^- \rightarrow HO_2(g) \qquad E^\ominus = -0.13\ \text{V at pH} = 0$$

$$O_2(g) + e^- \rightarrow O_2^-(aq) \qquad E^\ominus = -0.33\ \text{V at pH} = 14$$

Therefore a single-electron reducing reagent must exceed these potentials to achieve a significant reaction rate. Secondly, the ground state of O_2, with both π^* orbitals singly occupied, is neither an effective Lewis acid nor an effective base, and therefore has little tendency to undergo displacement reactions with *p*-block electrophiles or nucleophiles. Finally, the high bond energy of O_2 (463 kJ mol^{-1}) results in a high activation energy for reactions that depend on its dissociation. We saw in Section 9.3 that radical-chain mechanisms can provide reaction paths that circumvent some of these activation barriers in combustion processes at elevated temperatures, and radical oxidations also occur in solution.

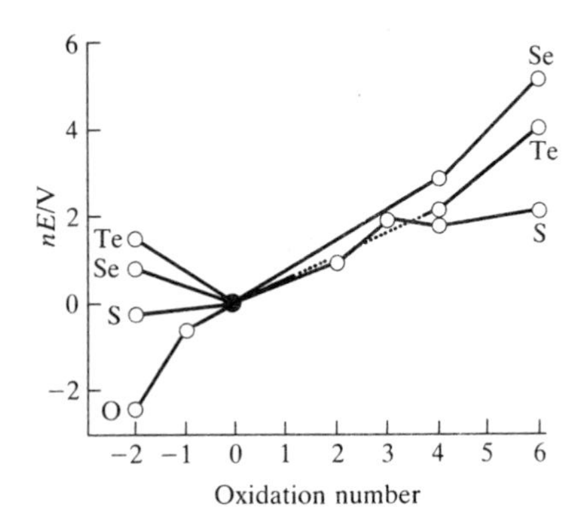

Fig. 12.6 Frost diagram for the elements of the oxygen group in acidic solution. The species with oxidation number −2 are H_2E and for oxidation number −1 the compound is H_2O_2. The positive oxidation states refer to oxides or oxoacids. The species for S(II) is thiosulfate, $S_2O_3^{2-}$, and that for S(III) is hydrogen dithionite, $HO_2SSO_2^-$.

Hydrogen peroxide

The Frost diagram for oxygen (Fig. 12.6) shows that H_2O_2 is unstable with respect to disproportionation. In practice, however, it is not very labile and survives reasonably well at moderate temperatures unless traces of some ions are present, for these act as catalysts. Some insight into the mechanism is obtained by noting that effective catalysts have reduction potentials in the range bounded by +1.76 V, the value for the reduction of H_2O_2 to H_2O, and −0.70 V, the value for the reduction of O_2 to H_2O_2. It is believed that the catalyzing ion shuttles back and forth between two oxidation states as it alternately oxidizes and reduces H_2O_2.

Example 12.5: *Determining whether an ion can catalyze H_2O_2 disproportionation*

Is Fe^{3+} thermodynamically capable of catalyzing the decomposition of H_2O_2?

Answer. The potential for Fe^{3+} reduction to Fe^{2+} is +0.77 V. This falls between the potentials for H_2O_2 reduction to H_2O and for O_2 reduction to H_2O_2, so catalytic decomposition is expected. We can verify that the potentials are favorable. Subtraction of the equations and potentials for reduction of O_2 to H_2O_2 from that for reduction of Fe^{3+} gives:

$$2Fe^{3+}(aq) + H_2O_2(aq) \rightarrow 2Fe^{2+}(aq) + O_2(g) + 2H^+(aq) \qquad E^\ominus = +0.07\ \text{V}$$

Since $E^\ominus > 0$, the reaction is thermodynamically favorable. Next we subtract the equation and potential for the reduction of Fe^{3+} from that for the reduction of H_2O_2 and obtain

$$2Fe^{2+}(aq) + H_2O_2(aq) + 2H^+(aq) \rightarrow 2Fe^{3+}(aq) + 2H_2O(l) \qquad E^\ominus = +0.99\ \text{V}$$

This is also favorable, so catalytic decomposition is thermodynamically favored. In fact, the rates also are high, so Fe^{3+} is an effective catalyst for the decomposition of H_2O_2, and in its manufacture great pains are taken to minimize iron contamination.

Exercise. Determine whether either Br^- or Cl^- is a candidate for the catalytic decomposition of H_2O_2.

S 1.43 120° O O

23 SO_2

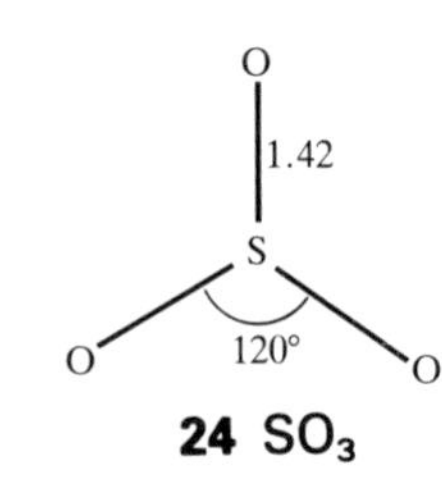

24 SO_3

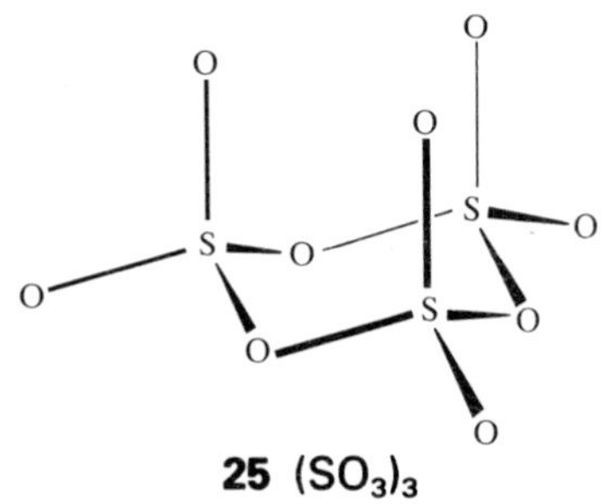

25 $(SO_3)_3$

Sulfur trioxide and sulfur dioxide

The molecules of the two common oxides of sulfur, SO_2 (b.p. −10°C) and SO_3 (b.p. 44.8°C), are respectively nonlinear (**23**) and trigonal planar in the gas phase (**24**). They are both Lewis acids, with the S atom the acceptor site, but SO_3 is the much the stronger and harder acid. The high Lewis acidity of SO_3 accounts for its occurrence as a polymeric O-bridged solid at room temperature and pressure (**25**).

Sulfur dioxide forms weak complexes with simple *p*-block Lewis bases. For example, it does not form a stable complex with H_2O. However, it does form stable complexes with stronger Lewis bases, such as trimethylamine and F^- ions.

Many oxohalides are known for the chalcogens. The most important of these are the **thionyl halides**, OSX_2, and the **sulfuryl dihalides**, O_2SX_2. One laboratory application of thionyl dichloride that we shall describe in Section 13.10 is the dehydration of metal chlorides. The compound $F_5TeOTeF_5$ and its selenium analog are known, and the $[OTeF_5]^-$ ion,

> **Example 12.6:** *Deducing the structures and chemistry of SO_2 complexes*
>
> Suggest the probable structures of SO_2F^- and $(CH_3)_3NSO_2$ and predict their reactions with OH^-.
>
> **Answer.** Although the Lewis structure of SO_2 (**26**) has an electron octet around the S atom, that atom can act as an acceptor (Section 6.1). Both resulting complexes have a lone pair on S, and the four electron pairs result in a trigonal pyramid around S in both complexes (**27**, **28**). The OH^- ion is a stronger Lewis base than either F^- or $N(CH_3)_3$, so exposure of either complex to OH^- will yield the hydrogensulfite ion, HSO_3^-, which has been found to exist in two isomers (**29**) and (**30**).
>
> **Exercise.** Give the Lewis structures and point groups of (a) $SO_3(g)$ and (b) SO_3F^-.

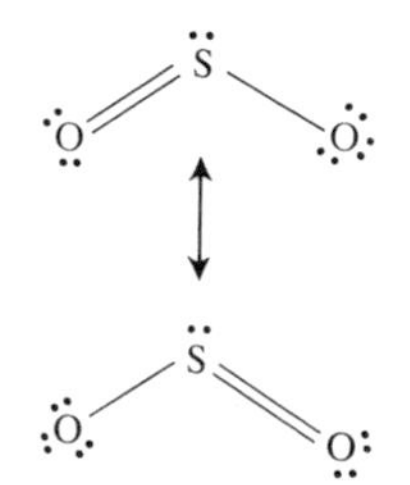

26 SO_2

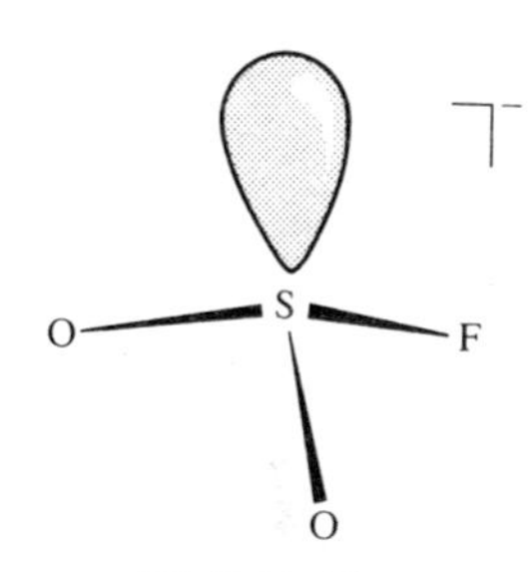

27 $[SO_2F]^-$

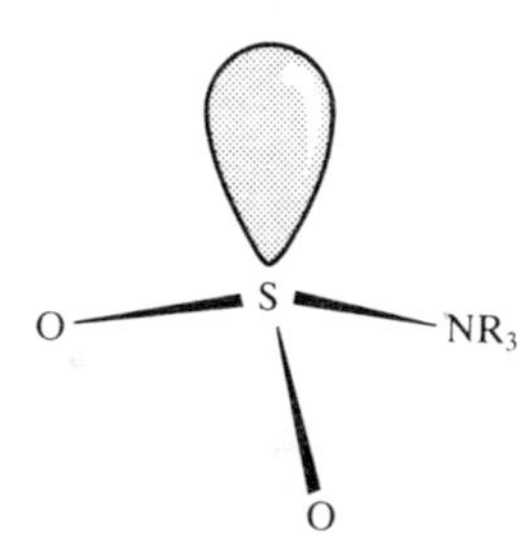

28 NR_3SO_2

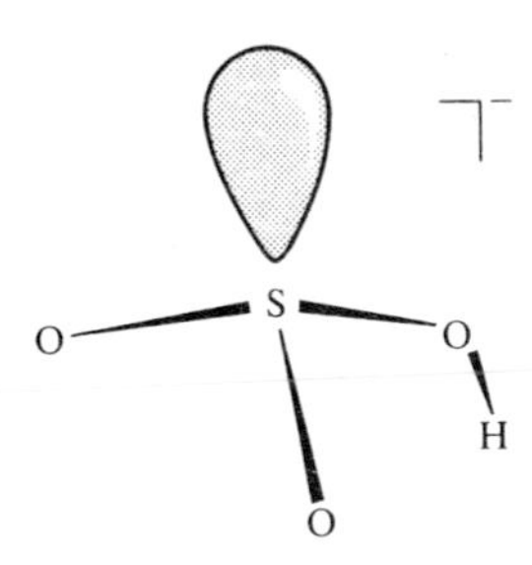

29 HSO_3^-

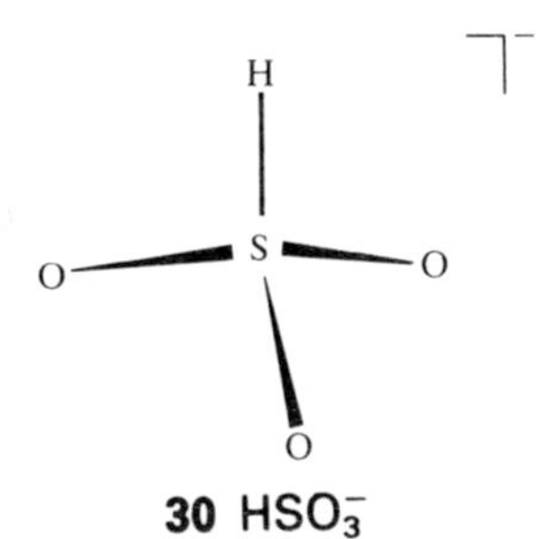

30 HSO_3^-

which is known as 'teflate', is useful when an anion is required that is large, not very basic, and resistant to oxidation.

Redox properties of sulfur oxoanions

A wide range of sulfur oxoanions exist, and many of them are important in the laboratory and in industry. The peroxodisulfate anion, $[O_3SOOSO_3]^{2-}$, for example, is a powerful and useful oxidizing agent:

$$[S_2O_8]^{2-}(aq) + 2e^- \rightarrow 2SO_4^{2-}(aq) \qquad E^\ominus = +1.96\ V$$

Table 12.7 lists some of the most stable examples of the sulfur oxoanions.

Sulfur's common oxidation numbers are -2, 0, $+2$, $+4$, and $+6$, but there are also many S—S bonded compounds that are assigned odd and fractional *average* oxidation numbers. A simple example is the thiosulfate ion, $S_2O_3^{2-}$, in which the average oxidation number of S is $+2$, but in which the environments of the two S atoms are quite different. The thermodynamic relations between the oxidation states are summarized by the Frost diagram (Fig. 12.6). As with many other *p*-block oxoanions, many of the thermodynamically favorable reactions are slow when the element is in its maximum oxidation state ($+6$), as in SO_4^{2-}. Another kinetic factor is suggested by the fact that oxidation numbers of compounds containing a single S atom generally change in steps of 2, which alerts us to look for an O atom transfer path for the mechanism. In some cases a radical mechanism operates, as in the oxidation of thiols and alcohols by peroxodisulfate.

We saw in Section 8.10 that the pH of a solution has a marked effect on the redox properties of oxoanions. This is true for SO_2 and SO_3^{2-}, since the former is easily reduced in acid solution and is therefore a good oxidizing agent whereas the latter in basic solution is not:

$$H_2SO_3(aq) + 4H^+(aq) + 4e^- \rightarrow S(s) + 3H_2O(l) \qquad E^\ominus = +0.50\ V$$

$$SO_3^{2-}(aq) + 3H_2O(l) + 4e^- \rightarrow S(s) + 6OH^-(aq) \qquad E^\ominus = -0.66\ V$$

We have written H_2SO_3 even though (as we saw in Section 5.6) the principal species present is $SO_2(aq)$. The oxidizing character of SO_2 accounts for its use as a mild disinfectant and bleach for foodstuffs, such as dried fruit and wine. In basic solution, SO_3^{2-} ions are good reducing agents:

$$SO_4^{2-}(aq) + H_2O(l) + 2e^- \rightarrow SO_3^{2-}(aq) + 2OH^-(aq) \qquad E^\ominus = -0.94\ V$$

Table 12.7. Some sulfur oxoanions

Oxidation number	Formula	Name	Structure	Remarks
One S atom				
+4	SO_3^{2-}	Sulfite	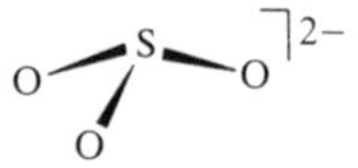	Basic, reducing agent
+6	SO_4^{2-}	Sulfate	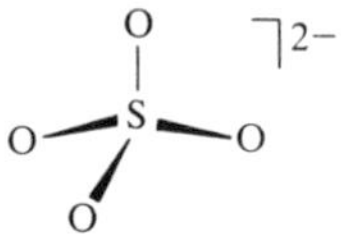	Weakly basic
Two S atoms				
+2	$S_2O_3^{2-}$	Thiosulfate	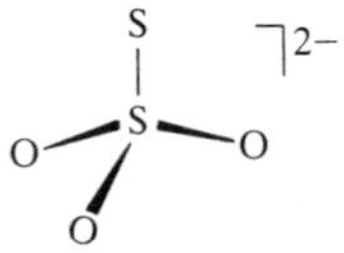	Mild reducing agent
+3	$S_2O_4^{2-}$	Dithionite	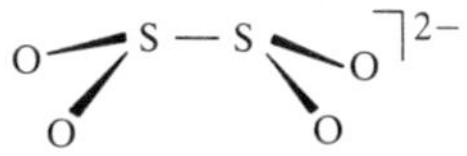	Strong and facile reducing agent
+5	$S_2O_6^{2-}$	Dithionate	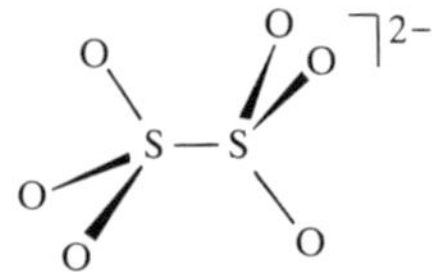	Resists oxidation and reduction
Polysulfur oxoanions				
Variable	$S_nO_6^{2-}$ $3 \leq n \leq 20$	$n = 3$, Trithionate	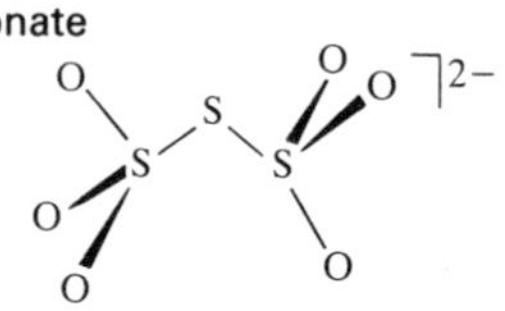	

The known oxoanions of selenium and tellurium are a much less diverse and extensive group. Selenic acid is thermodynamically a strong oxidizing acid:

$$SeO_4^{2-}(aq) + 4H^+(aq) + 2e^- \rightarrow H_2SeO_3(aq) + H_2O(l) \qquad E^\ominus = +1.1 \text{ V}$$

However, like SO_4^{2-} and in common with the behavior of oxoanions of elements in high oxidation states, the reduction of SeO_4^{2-} is generally slow. Telluric acid exists as $Te(OH)_6$ and also as $(HO)_2TeO_2$ in solution. Again, its reduction is thermodynamically favorable but kinetically sluggish.

12.9 Oxides of the *s*-block metals

Since the alkali metals form very stable +1 ions and the alkaline earth metals form +2 ions, we might suppose that their oxides will have the formulas M_2O and MO respectively. These oxides can be prepared, but the oxide chemistry of the *s*-block metals is much more interesting than these formulas suggest. For instance, compounds containing the **superoxide ion** O_2^- and the **peroxide ion** O_2^{2-} are common, and an unusual series of suboxides exists for rubidium and cesium.

Oxides, superoxides, and peroxides

The O^{2-}, O_2^-, and O_2^{2-} ions have already been discussed in several contexts. For example, we dealt with the electronic structure of O_2^- in Example 2.10 and saw there that it is paramagnetic and that its bond order is 1.5. We also saw there that the O_2^{2-} ion is diamagnetic and that its bond order is 1 (like the isoelectronic F_2 molecule).

We saw in Section 4.8 (and specifically in Example 4.7) that the influence of ion size on the stability of polyatomic anions suggests that the rather unstable O_2^- and O_2^{2-} ions will be favored by large cations. These trends are supported by the data in Table 12.8 on the products obtained when the *s*-block metals react with excess oxygen. The data show that the small Li^+ ion and the dipositive lighter Group 2 metal ions are found in combination with the O^{2-} ion. The larger Na^+ and Ba^{2+} ions form stable compounds containing the dinegative O_2^{2-} ion. Finally, the large monopositive K^+, Rb^+, and Cs^+ ions form superoxides.

Table 12.8. Products of the reaction of the Group 1 and 2 metals with excess oxygen

Group 1	Group 2
Li_2O*	BeO*
Na_2O_2†	MgO*
KO_2‡	CaO*
RbO_2‡	SrO*
CsO_2‡	BaO_2†

* Oxide, O^{2-}.
† Peroxide, O_2^{2-}.
‡ Superoxide, O_2^-.

Barium peroxide BaO_2 was at one time used to produce pure O_2 because BaO absorbs O_2 from the atmosphere rapidly at 500°C and releases it at higher temperatures:

$$2BaO(s) + O_2(g) \underset{700°C}{\overset{500°C}{\rightleftharpoons}} 2BaO_2(s)$$

Suboxides

Rubidium and cesium react with a limited supply of oxygen to form a remarkable series of metal-rich oxides, which include Cs_3O, Cs_4O, Cs_7O, and Rb_6O. The compounds are obtained as dark crystals; they are highly reactive and good metallic conductors.

A clue to the explanation of this behavior is that Cs_3O consists of O atoms surrounded by octahedra of six Cs atoms, with neighboring octahedra sharing two opposite faces. This arrangement results in columns of Cs atoms that run through the crystal and form a metallic network (Fig. 12.7). The simple ionic model, which we normally assume can be applied to alkali metal compounds, obviously does not hold for these suboxides.

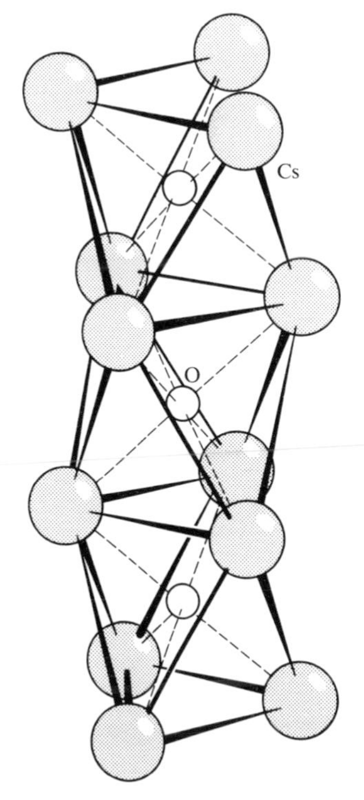

Fig. 12.7 The structure of Cs_3O. Each O atom in this column is surrounded by an octahedron of Cs atoms. The Cs—Cs distances within a column are shorter than those between columns.

12.10 Oxides of the *d*-block metals

The variety of *d*-metal oxidation states that can be attained leads to a very large number of known oxides, ranging from ionic monoxides, such as MnO, to molecular tetroxides, such as OsO_4. Here we shall concentrate on the structures of the monoxides MO and the dioxides MO_2. The solid-state chemistry of metal oxides is developed in more detail in Chapter 18.

Monoxides

The formation of *d*-metal MO compounds is largely confined to members of the first series of the block (Period 4). Monoxides are in fact known for all the first *d* series metals with the exception of scandium and chromium, and

Table 12.9. *d*-Block MO compounds

Group	4	5	6	7	8	9	10
Rock-salt structure	Ti	V	—	Mn	Fe	Co	Ni
	Zr						Pd*

(Rock-salt box encloses Ti, V, —, Mn, Fe, Co, Ni and Zr.)

* PtS structure (square planar four-coordinate metal).

Source: A. F. Wells, *Structural inorganic chemistry.* Oxford University Press (1984), p. 537.

their crystal structures, which are commonly rock-salt, are summarized in Table 12.9.

The monoxides of elements to the left of iron are strong reducing agents. For example, MnO is an excellent scavenger for the removal of oxygen from inert gas streams, and it is claimed that this treatment can reduce the oxygen concentration to parts per billion. One especially interesting monoxide is TiO, for it shows metallic conductivity. This conductivity originates from the overlap of *d* orbitals of neighboring Ti^{2+} ions: *d* orbitals belonging to metal ions on the left of the *d* block, where Z_{eff} is low, are reasonably diffuse and can overlap in this way. Although the monoxide can be prepared with the stoichiometry TiO, the solid actually has about 15 percent Ti^{2+} and O^{2-} ion vacancies. Iron(II) oxide can be prepared with a range of compositions, but it is always iron-poor. Its richest composition is approximately $Fe_{0.93}O$, which is achieved at 570°C. In this case the deviation from the stoichiometry FeO arises from the presence of some Fe^{3+} ions in the lattice, and therefore fewer than one Fe^{2+} ion per O^{2-} ion is needed for electrical neutrality.

Dioxides

Dioxides are known for titanium to manganese in the first series of the *d* block and for a broader range in the second and third series. As indicated in Table 12.10, they often have either the rutile or fluorite structure. The preponderance of dioxides in the second and third series and of monoxides or M_2O_3 compounds in the first correlates with the greater stability of high oxidation states of the heavier *d* metals (Section 8.12).

Table 12.10. *d*-Block MO_2 compounds

Group	4	5	6	7	8	9	10
Rutile (six-coordinate)	Ti	V	Cr	Mn			
	Zr	Nb	Mo	Tc	Ru	Rh	
Fluorite (eight-coordinate metal)	Hf	Ta	W	Re	Os	Ir	Pt

(Rutile box encloses Ti, V, Cr, Mn, Nb, Mo, Tc, Ru, Rh, Ta, W, Re, Os, Ir, Pt; fluorite box encloses Zr and Hf.)

Source: A. F. Wells, *Structural inorganic chemistry.* Oxford University Press (1984), p. 540.

12.11 Metal–sulfur compounds

Sulfur is less electronegative and softer than oxygen; it therefore has a broader span of oxidation states and a stronger affinity for the softer metals on the right of the *d* block. For example, copper(II) sulfide is readily precipitated when hydrogen sulfide is bubbled through an aqueous solution of Cu^{2+}, but $Sc^{3+}(aq)$ does not react. The covalent contribution to the lattice enthalpy of all the softer metal sulfides cannot be overcome in water, and the sulfides generally are very sparingly soluble. Another striking feature of the chemistry of sulfur is its tendency to form catenated sulfide ions. Thus in combination with alkali metals a broad series of compounds containing S_n^{2-} ions can be prepared, such as Na_2S_2 and Na_2S_7.

Monosulfides

As with the *d*-metal monoxides, the monosulfides are most common in the first series (Table 12.11). In contrast to the monoxides, most of the monosulfides have the nickel-arsenide structure (Fig. 4.16). The different structures are consistent with the rock-salt structure of the monoxides being favored by the ionic (harder) cation–anion combinations whereas the nickel-arsenide structure is favored by more covalent (softer) combinations and correspondingly shorter metal–metal separations.

Disulfides

The disulfides of the *d* metals fall into two broad classes (Table 12.12). One class consists of layered compounds with the CdI_2 or MoS_2 structure and the other of compounds containing discrete S_2^{2-} groups.

Table 12.11. Structures of *d*-block MS compounds*

Group	4	5	7	8	9	10
Nickel arsenide	Ti	V	Mn†	Fe	Co	Ni
Rock salt	Zr	Nb				

* Metal monosulfides not shown here for all of Group 6; some of the heavier metals have more complex structures.

† MnS has two polymorphs; one has the rock-salt structure, the other has the wurtzite structure.

Source: A. F. Wells, *Structural inorganic chemistry*. Oxford University Press (1984), p. 752.

Table 12.12. *d*-Block MS_2 compounds*

Group	4	5	6	7	8	9	10	11
	Ti			Mn	Fe	Co	Ni	Cu
Layered	Zr	Nb	Mo		Ru	Rh	Pyrite or marcasite	
	Hf	Ta	W	Re	Os	Ir	Pt	Layered

* Elements not shown either do not form disulfides or have disulfides with more complex structures

Source: A. F. Wells, *Structural inorganic chemistry*. Oxford University Press (1984), p. 757.

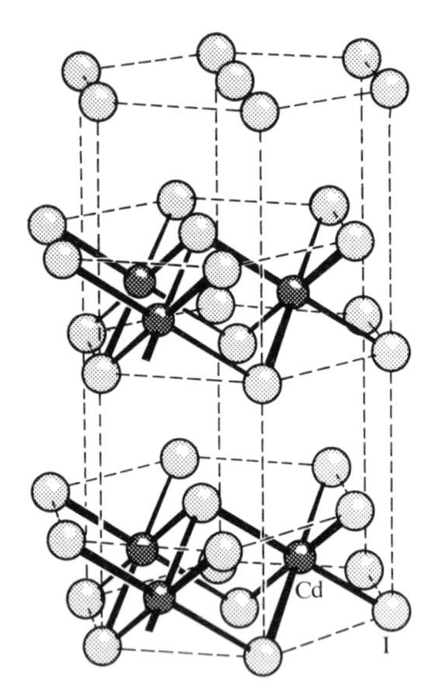

Fig. 12.8 The CdI_2 structure adopted by many disulfides can be viewed as a layered structure. The latter has adjacent sulfide layers without an intervening layer of metal ions. The NiAs structure adopted by many monosulfides is shown in Fig. 4.16.

The layered disulfides are built from sandwiches consisting of a sulfide layer, a metal layer, and then another sulfide layer (Fig. 12.8). These sandwiches stack together in the crystal with sulfide layers in one adjacent to a sulfide layer in the next. Clearly this crystal structure is not in harmony with a simple ionic model, and its formation is a sign of covalency in the bonds between the soft sulfide ion and *d*-metal cations. In these layered structures, the metal ion is surrounded by six S atoms; the coordination environment is octahedral in some cases (PtS_2, for instance) and trigonal prismatic in others (MoS_2). The layered MoS_2 structure is favored by S—S bonding, as indicated by short S—S distances within each of the MoS_2 slabs. The common occurrence of the trigonal prismatic structure in many of these compounds is in striking contrast to isolated metal complexes, where the octahedral arrangement of ligands is by far the most common.

As with graphite, some of the layered sulfides readily undergo intercalation reactions in which ions or molecules penetrate between adjacent sulfide layers, often with accompanying redox reactions:

$$\mathrm{Na}(soln) + \mathrm{TaS_2}(s) \xrightarrow{\mathrm{NH_3}(l)} \mathrm{Na_{0.6}TaS_2}(s)$$

In this reaction, a Na atom gives up an electron to a vacant band in TaS_2, and the Na^+ ion worms its way into positions between the sulfide layers.

Compounds containing discrete S_2^{2-} ions adopt the pyrite or marcasite structure (Fig. 12.9). The stability of the formal S_2^{2-} ion in metal sulfides is much greater than that of the O_2^{2-} ion in peroxides, and there are many more metal sulfides in which the anion is S_2^{2-} than there are peroxides.

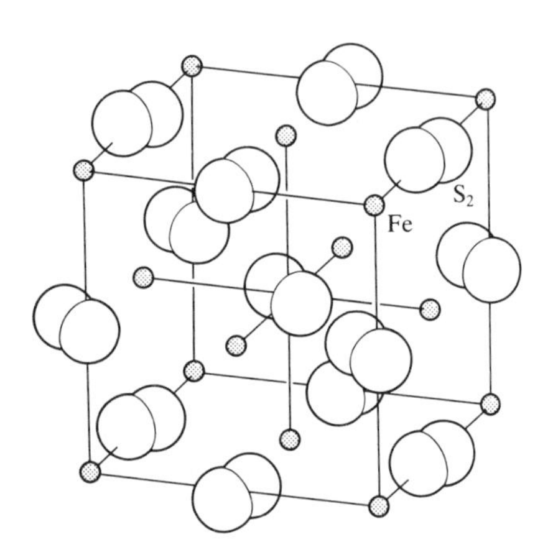

Fig. 12.9 The structure of iron pyrite, FeS_2.

Example 12.7: *Contrasting the structures of two different d-block disulfides*

Contrast the structures of MoS_2 and FeS_2 and justify their existence in terms of the stable oxidation states of the metal ions.

Answer. The point to be decided is whether the metals are likely to be in the +4 oxidation state, in which case the two S atoms would be present as S^{2-} ions. If the +4 oxidation state is relatively inaccessible, the metal might be in the +2 oxidation state, and the S atoms present as the S—S bonded species S_2^{2-}. Since it is a fair reducing agent, S^{2-} will only be found with a metal ion in an oxidation state that is not easily reduced. As with many of the *d* metals in periods 5 and 6, Mo is easily oxidized to Mo(IV); therefore Mo(IV) is stable and can coexist with S^{2-}. Molybdenum(IV) sulfide (MoS_2) has the layered structure typical of metal disulfides. Iron is readily oxidized to Fe(II) but not to Fe(IV). Therefore, Fe(IV) cannot coexist with S^{2-}. The compound is therefore likely to contain Fe(II) and S_2^{2-}. The mineral name for FeS_2 is pyrite; its common name 'fool's gold' indicates its misleading color.

Exercise. MoS_2 is a very effective lubricant. Present a plausible reason for this property.

12.12 Sulfido complexes of the *d* metals

Research on Fe—S cluster complexes has flourished recently.[4] Much of this interest has been stimulated by the discovery that the complexes

[4] R. H. Holm, *Chem. Soc. Rev.*, **10**, 455 (1981); A. R. Butler, C. Glidwell, and M.-H. Li, *Adv. Inorg. Chem.*, **32**, 336 (1988); A. Muller, E. Diemann, R. Jostes, and H. Bogge, *Angew. Chem. Int. Ed. Engl.*, **20**, 934 (1981).

are present in electron-transfer and nitrogen-fixing enzymes. One such model compound, $[Fe_4S_4(SR)_4]^{2-}$, was shown in Fig. 7.4. It is readily prepared from simple starting materials in the absence of air:

$$4FeCl_3 + 4HS^- + 6RS^- + 4CH_3O^- \xrightarrow{\text{Methanol}} [Fe_4S_4(SR)_4]^{2-} + RS\text{—}SR + 12Cl^- + 4CH_3OH$$

The observation that this reaction is successful with many different R groups, and its good yield, indicates that the 4Fe—4S cage is thermodynamically more stable than other possibilities. The HS^- ion provides the sulfide ligands for the cage, the RS^- ion serves both as a ligand and a reducing agent, and the CH_3O^- ion acts as a base. The cubic cluster contains Fe and S atoms at alternate corners, so each S atom bridges three Fe atoms. Each of the thiolate groups, SR^-, occupies a terminal position on an Fe atom. The cluster remains intact upon one-electron reduction to $[Fe_4S_4(SR)_4]^{3-}$, and analogous clusters are implicated in the redox reactions of the enzyme ferredoxin (Chapter 19).

Simple thiometallate complexes such as MoS_4^{2-} can be synthesized easily by passing hydrogen sulfide gas through an aqueous solution of the molybdate or tungstate ion:

$$MoO_4^{2-} + 4H_2S \rightarrow MoS_4^{2-} + 4H_2O$$

These tetrathiometallate anions are building blocks for the synthesis of complexes containing more metal atoms. For example they will coordinate to many dipositive metal ions such as Co^{2+} and Fe^{2+}:

$$Fe^{2+} + 2MoS_4^{2-} \rightarrow [S_2MoS_2FeS_2MoS_2]^{2-}$$

The polysulfides such as S_2^{2-} and S_3^{2-} that are formed by addition of elemental sulfur to a solution of ammonium sulfide can also act as ligands. An example is $[Mo_2(S_2)_6]^{2-}$ (**31**), which is formed from ammonium polysulfide and MoO_4^{2-}; it contains side-bonded S_2^{2-} ligands. The larger polysulfides bond to metal atoms forming chelate rings, as in $[MoS(S_4)_2]^{2-}$ (**32**), which contains chelating S_4 ligands.

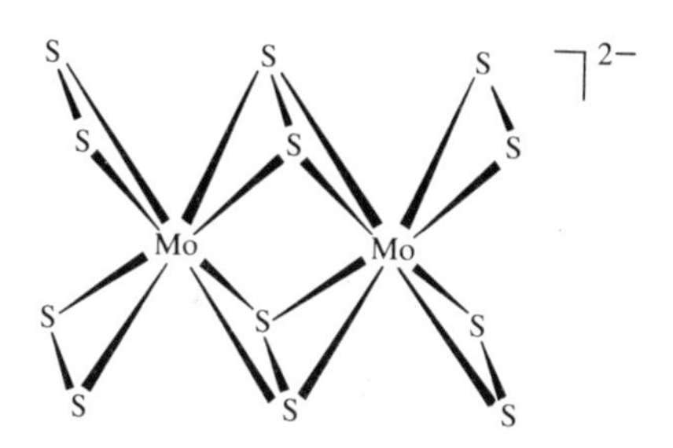

31 $[Mo_2(S_2)_6]^{2-}$

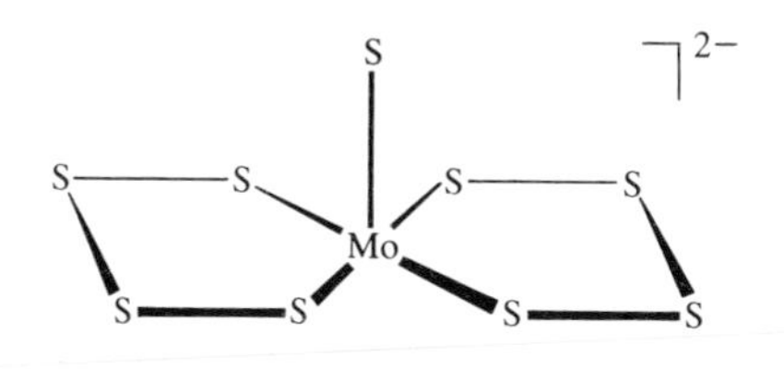

32 $[MoS(S_4)_2]^{2-}$

p-Block ring and cluster compounds

Considerable progress has recently been made in the elucidation of the chemistry of inorganic clusters.[5] We have already discussed boron clusters (the boranes and carboranes), and in Chapter 16 we shall describe clusters formed by *d* metals. In this section we shall concentrate on the clusters formed by the heavier *p*-block elements. These are frequently called **naked clusters** because they generally have no attached groups, such as the H atoms in boron clusters and the CO ligands in *d*-metal clusters. Many of the *p*-block clusters are ions.

12.13 Anionic clusters

Some *p*-block metals and metalloids react with alkali metals to form compounds such as K_2Pb_5. These compounds were investigated by Eduard

[5] R. J. Gillespie, *Chem. Soc. Rev.*, 315 (1979); J. D. Corbett, *Chem. Rev.* **85**, 383 (1985); H. G. von Schnering, *Angew. Chem. Intl. Edn. Engl.*, **20**, 33 (1981). A. H. Cowley (ed.), *Rings, polymers, and clusters of main group elements.* ACS Symposium Series 232, Washington (1983).

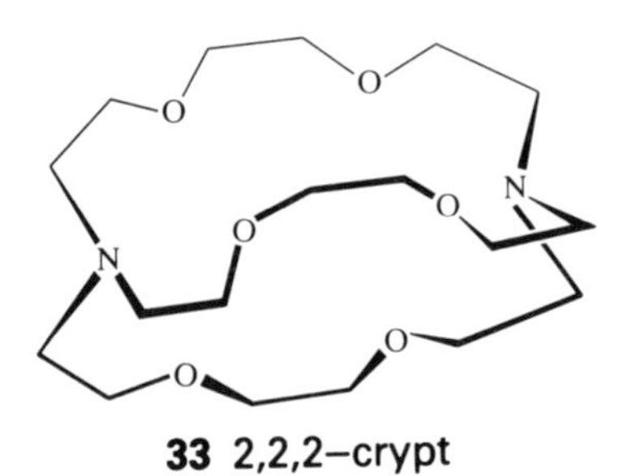

33 2,2,2-crypt

Zintl in Germany in the 1930s, who found that some could be dissolved in liquid ammonia, which is a hospitable solvent to strong reducing agents such as these. A recent breakthrough occurred when John Corbett in the United States discovered that by complexing the alkali metal cation with 2,2,2-crypt (**33**), it was possible to prepare species in ethylenediamine solution that can be crystallized and investigated by X-ray diffraction. The crypt ligand encapsulates the alkali metal ion and thus creates a large cation complex:

$$2KPb_{2.5} + 2\text{crypt} \xrightarrow{\text{en, 50°C}} [K(\text{crypt})]_2[Pb_5]$$

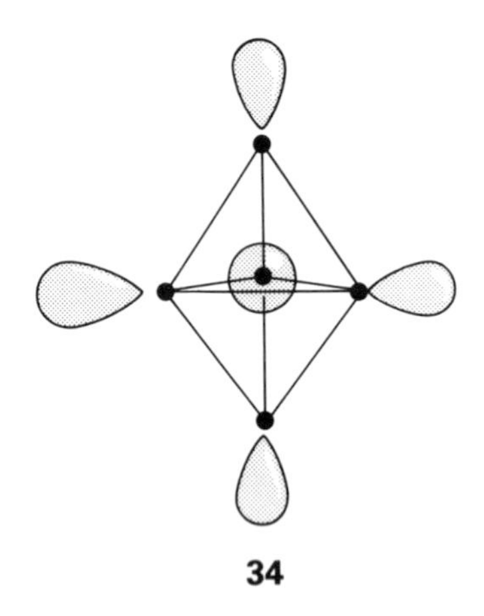
34

The stability of the product stems from the stabilizing effect that large cations have on large anions. Similar techniques have led to a variety of reduced species that have been characterized in the solid state by X-ray diffraction and in some cases in solution by NMR.

The electron-counting correlations (Wade's rules) that we introduced for boranes and carboranes in Section 11.13 are successful with many of these naked *p*-block cluster anions, and we can use Table 11.7 to predict the shapes of deltahedra. As an example, the number of valence electrons available in the Pb_5^{2-} ion is $5 \times 4 = 20$ from the Pb atoms and two more from the charge, giving 22 electrons, or 11 pairs. Of these 11 pairs, one pair on each Pb atom, and so five pairs in all, are unavailable for bonding since, like the electron pair of a B—H unit in boranes, they are assumed to be directed away from the skeleton (**34**). Thus the total number of skeletal electron pairs is $11 - 5 = 6$. This number agrees with the count expected for a 5-atom ($n = 5$) *closo* cluster, since $5 + 1 = 6$.

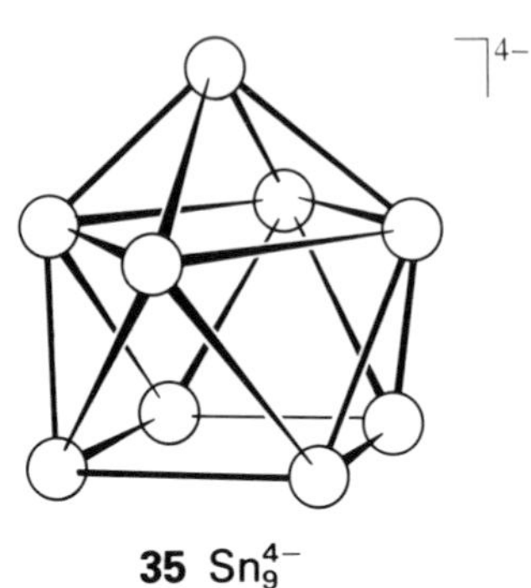

35 Sn_9^{4-}

Example 12.8: *Correlating the electron count and structure for Zintl clusters*

Determine the electron count and the expected structure of (a) the Sn_9^{4-} ion and (b) the diamond shaped Bi_4^{2-} ion.

Answer. (a) The Sn_9^{4-} ion contains $4 \times 9 + 4 = 40$ valence electrons, and hence 20 pairs. Subtracting one nonbonding pair on each atom (a total of 9) leaves 11 pairs. This fits the number expected for a *nido* cluster: $9 + 2 = 11$. In agreement with this conclusion, the structure is a truncated M_{10} deltahedron (**35**) characteristic of a *nido* cluster. (b) The Bi_4^{2-} anion contains $(4 \times 5) + 2 - (4 \times 2) = 14$ skeletal electrons, or 7 pairs. The count expected for an *arachno* cluster, $4 + 3 = 7$, but it is square planar (**36**), rather than the expected butterfly; see the isoelectronic Se_4^{2+} below.

Exercise. Determine the classification from electron count for Ge_9^{2-} and determine whether this is consistent with the structure (**37**).

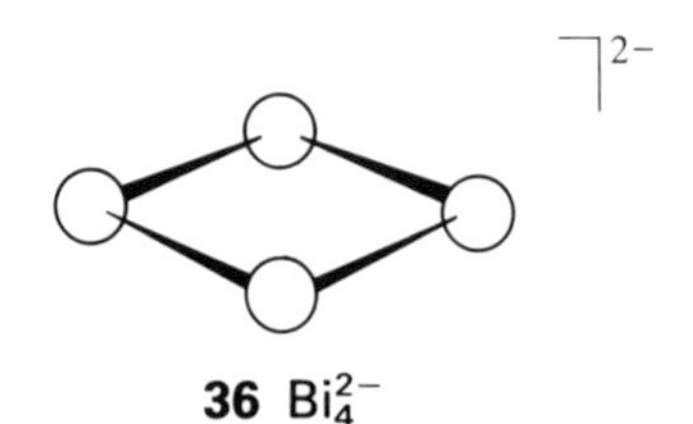

36 Bi_4^{2-}

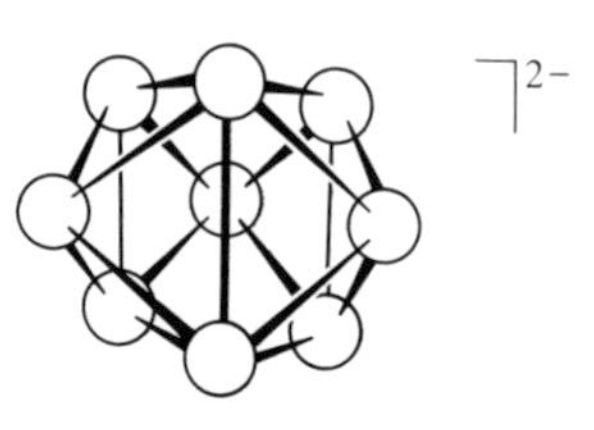

37 Ge_9^{2-}

12.14 Polycations

A large number of cationic chain, ring, and cluster compounds of the *p*-block elements have been prepared. The majority of them contain sulfur, selenium, or tellurium, but they may also contain elements from mercury in Group 12 to the halogens in Group 17/VII. Since these cations are oxidizing agents and Lewis acids, the preparative conditions are quite different from those used to synthesize the highly reducing polyanions. For example, S_8 is oxidized by AsF_5 in liquid SO_2 to yield the S_8^{2+} ion:

$$S_8 + 3AsF_5 \xrightarrow{SO_2} [S_8][AsF_6]_2 + AsF_3$$

A strong acid medium is often used, such as fluorosulfuric acid as the solvent and the peroxide compound FO_2SOOSO_2F as the oxidizing agent:

$$4Se + S_2O_6F_2 \xrightarrow{HSO_3F} [Se_4][SO_3F]_2$$

The cation in the last reaction, Se_4^{2+}, has a square-planar structure (**38**). In the molecular orbital model of the bonding, this square cation has a closed-shell configuration in which the delocalized π bonding a_{1g} and nonbonding e_g orbitals are filled, and a higher-energy nonbonding orbital is vacant. In contrast, most of the larger ring systems can be understood using localized 2c,2e bonds. For these larger rings, the removal of two electrons brings about the formation of an additional 2c,2e bond, thereby preserving the local electron count on each element. This is readily seen for the oxidation of S_8 (**20**) to S_8^{2+} (**39**). An X-ray single crystal structure determination shows that the transannular bonds in (**39**) are long compared with the other bonds. Long transannular bonds are common in these types of compounds.

12.15 Neutral heteroatomic rings and clusters

The P_4 molecule is a good example of a cluster compound that can be described by local 2c,2e bonds. It turns out that several chalcogen derivatives are structurally related to P_4 by the insertion of atoms into P—P bonds to give analogs of the P_4O_n compounds we have already described. Some sulfur compounds in this series are P_4S_3 (**40**), which apparently does not have an oxygen analog, and P_4S_{10}, the analog of P_4O_{10} (**12**).

Sulfur–nitrogen compounds have structures that can be related to the polycations discussed above. The oldest known and perhaps easiest to prepare is the pale yellow to orange tetrasulfurtetranitride (S_4N_4, **41**) which is made by passing ammonia through a solution of S_2Cl_2:

$$6S_2Cl_2 + 16NH_3 \rightarrow S_4N_4 + S_8 + 12NH_4Cl$$

S_4N_4 is endoergic ($\Delta G_f^\ominus = +536\ kJ\ mol^{-1}$) and may decompose explosively. The molecule is an eight-membered ring with the four N atoms in a plane and bridged by S atoms that project above and below the plane. The S—S distance (2.58 Å) suggests that there is a weak interaction between pairs of S atoms. Lewis acids such as BF_3, SbF_5, and SO_3 form 1:1 complexes with

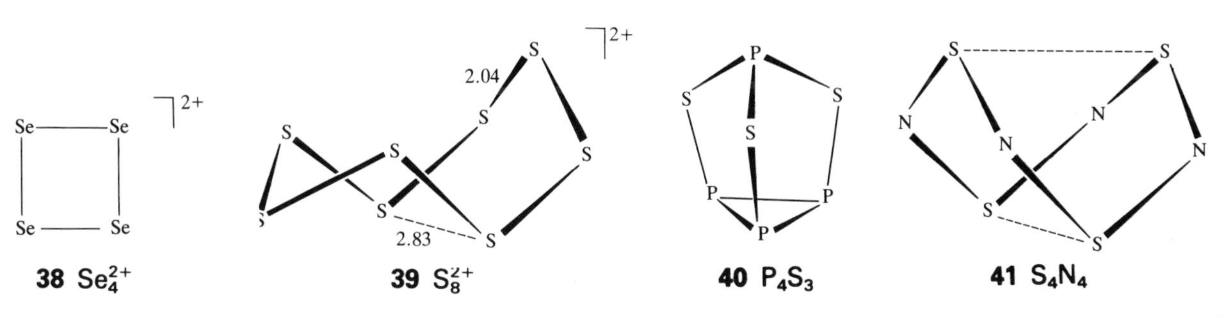

38 Se_4^{2+} **39** S_8^{2+} **40** P_4S_3 **41** S_4N_4

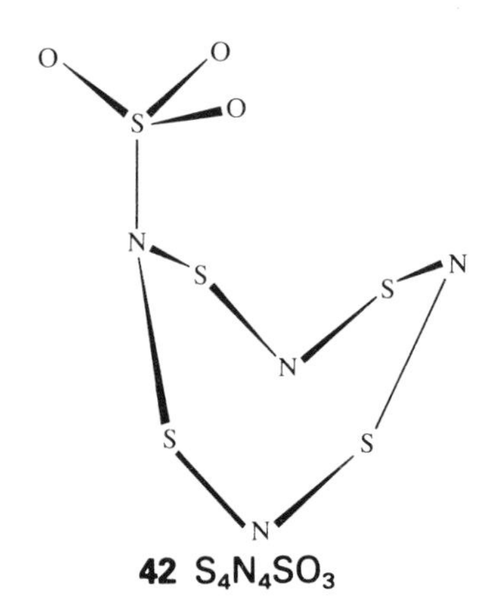

42 $S_4N_4SO_3$

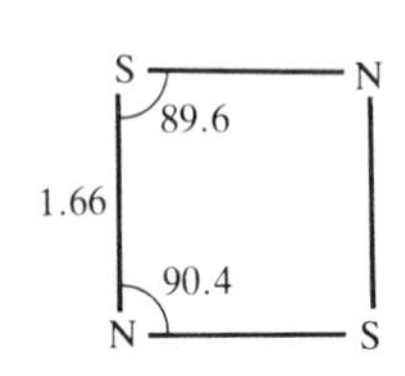

43 S_2N_2

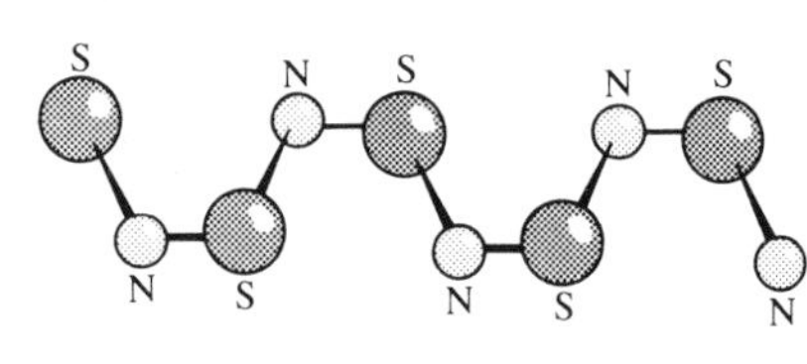

44 $(SN)_x$

one of the nitrogen atoms and in the process the structure of the complex rearranges (**42**).

Disulfurdinitride (S_2N_2, **43**) is formed (together with Ag_2S and N_2) when S_4N_4 vapor is passed over hot silver wool. It is even more touchy than its precursor, for it explodes above room temperature. When allowed to stand at 0°C for several days, it transforms into a bronze-colored polymer $(SN)_x$, (**44**) that has metallic conductivity and is superconducting below 0.3 K. The transition temperature to the superconducting state is very low, but its discovery was important because it was the first example of a superconductor that had no metal constituents. The conductivity of the compound is of interest because it is a zigzag chain and, as we saw in Section 3.5, there are interesting theoretical issues associated with chain conductors.

Further reading

A. J. Bard, R. Parsons, and J. Jordan, *Standard potentials in aqueous solution.* Dekker, New York (1985). See Chapters 4, 6, and 9 particularly. In addition to a discussion of thermodynamic data, these chapters give qualitative information on rates.

A. F. Wells, *Structural inorganic chemistry.* Oxford University Press (1984); especially Chapters 11–12 and 16–20.

A. E. Martell and D. T. Sawyer (eds.), *Oxygen complexes and oxygen activation by transition metals.* Plenum, New York (1988). This volume presents a range of informative review papers on oxygen complexes, and oxygen activation by complexes.

Kirk–Othmer encyclopedia of chemical technology. Wiley, New York (1979). Because of the great industrial importance of N, O, P and S compounds, they receive extensive coverage in these volumes. Volume 15: nitric acid, nitrides, nitrogen, and nitrogen fixation; Volume 16: oxygen and ozone; Volume 17: phosphoric acid, phosphates, phosphorus, phosphides, phosphorus compounds, and peroxides; Volume 22: sulfur, sulfur recovery, sulfuric acid, sulfur trioxide, and sulfur compounds.

Exercises

12.1 List the elements in Groups 15/V and 16/VI and indicate the ones that are (a) diatomic gases, (b) nonmetals, (c) metalloids, (d) true metals. Also indicate those elements that do not achieve the maximum group oxidation number and indicate the elements that display the inert pair effect.

12.2 Contrast the formulas and stability of the oxidation states of the common (a) nitrogen chlorides with phosphorus chlorides, (b) oxygen fluorides with sulfur fluorides.

12.3 Use Lewis structures and VSEPR theory to determine the probable structures of each of (a) SeF_4, (b) $SbF_5(g)$, (c) the $N(CH_3)_3$ complex of PCl_3.

12.4 Select isoelectronic species from C_2^{2-}, O_2, O_2^-, O_2^{2-}, N_2, NO, NO^+, CN^-, and $N_2H_3^-$. Within these isoelectronic groups, describe the probable relative oxidizing strengths and Lewis basicities.

12.5 Give the formula and name of a molecule or molecular ion that is isoelectronic and isostructural with (a) NO_3^-, (b) NO_2^-, (c) N_2O_4, (d) N_2O, (e) N_2, (f) NH_3.

12.6 Contrast the chemistry of each of the isoelectronic pairs in Exercise 12.5 with respect to acidity and strength as an oxidizing agent.

12.7 Starting with $NH_3(g)$ and other reagents of your choice, give the equations and conditions for the synthesis of (a) HNO_3, (b) NO_2^-, (c) NH_2OH, (d) N_3^-.

12.8 Write the balanced chemical equation corresponding to the standard enthalpy of formation of $P_4O_{10}(s)$. Specify the structure and physical state (*s*, *l*, or *g*) of the reactants.

12.9 Without reference to the text, sketch the general form of the Frost diagrams for phosphorus (oxidation states 0 to +5) and bismuth (0 to +5) in acidic solution and discuss the relative stabilities of the +3 and +5 oxidation states of both elements.

12.10 Are reactions of NO_2^- as an oxidizing agent generally faster or slower when pH is reduced? Give a mechanistic explanation for the pH dependence of NO_2^- oxidations.

12.11 Use standard reduction potentials to calculate the standard potential of the reaction of H_3PO_2 with Cu^{2+}. Are HPO_3^{2-} and $H_2PO_2^{2-}$ useful as oxidizing or reducing agents?

12.12 (a) Use standard reduction potentials (Appendix 4) to calculate the standard potential of the disproportionation of H_2O_2 in acid solution. (b) Is Cr^{2+} a likely catalyst for the disproportionation of H_2O_2? (c) Given the Latimer diagram

$$O_2 \xrightarrow{-0.13} HO_2 \xrightarrow{1.51} H_2O_2$$

in acidic solution, calculate $\Delta G^\ominus$ for the disproportionation of hydrogen superoxide (HO_2) into O_2 and H_2O_2, and compare the result with its value for the disproportionation of H_2O_2.

12.13 Which of the solvents ethylenediamine (which is basic and reducing) or SO_2 (which is acidic and oxidizing) might not react with (a) Na_2S_4, (b) K_2Te_3, and (c) $Cd_2(Al_2Cl_7)_2$. Explain your reasons.

12.14 Contrast the common structure for *d*-metal monoxides, with that for the monosulfides. Why are the monoxides much more common for the Period-4 elements than for later periods?

12.15 When equal volumes of nitric oxide (NO) and air are mixed at atmospheric pressure a rapid reaction occurs giving NO_2 and N_2O_4. However, nitric oxide from an automobile exhaust, which is present in the parts per million concentration range, reacts slowly with air. Give an explanation for this observation in terms of the rate law and the probable mechanism.

12.16 Use isoelectronic analogies to infer the probable structures of (a) Sb_4^{2-} and (b) P_7^{3-}. Describe the bonding in these ions.

Problems

12.1 In this chapter the tetrahedral P_4 molecule was described in terms of localized 2c,2e bonds. Determine the number of skeletal valence electrons and from this decide whether P_4 is *closo, nido,* or *arachno*. If it is not *closo,* determine the parent *closo* polyhedron from which the structure of P_4 could be formally derived by the removal of one or more vertices.

12.2 On account of their slowness at electrodes, the potentials of most redox reactions of nitrogen compounds cannot be measured in an electrochemical cell. Instead, the values must be determined from other thermodynamic data. Illustrate such a calculation using $\Delta G_f^\ominus(NH_3, aq) = -26.5\ kJ\ mol^{-1}$ to calculate the standard reduction potential of the N_2/NH_3 couple in basic aqueous solution.

12.3 Summarize the article by J. F. Revelli (*Inorg. Synth.*, **16,** 35 (1979)) by means of a one-page description of the synthesis of TaS_2 and formation of its pyridine intercalation compound. Include sketches of the apparatus and conditions for the reactions.

12.4 A mechanistic study of reaction between chloramine and sulfite has been reported (B. S. Yiin, D. M. Walker and D. W. Margerum, *Inorg. Chem.*, **26,** 3435 (1987)). Summarize the observed rate law and the proposed mechanism. Accepting the proposed mechanism, why should $[SO_2(OH)]^-$ and $[HSO_3]^-$ display different rates of reaction? Explain why it was not possible to distinguish the reactivity of $[SO_2(OH)]^-$ from that of $[HSO_3]^-$.

13 The halogens and the noble gases

The elements
13.1 Dihalogens
13.2 Pseudohalogens
Polyhalogen compounds
13.3 Interhalogens
13.4 Polyhalides
Compounds of halogens and oxygen
13.5 Binary oxides
13.6 Oxoacids and oxoanions
13.7 Redox principles and trends
13.8 Redox reactions of individual species
Metal halides
13.9 Structures of *d*-block halides
13.10 Preparation of anhydrous halides
The noble gases
13.11 The elements
13.12 Compounds
Further reading
Exercises
Problems

15	16	17	18
			He
N	O	F	Ne
P	S	Cl	Ar
As	Se	Br	Kr
Sb	Te	I	Xe
Bi	Po	At	Rn
V	VI	VII	VIII

We now turn to the last two groups in the *p* block, and see that many of the attitudes that were helpful for discussing the two preceding groups will again prove useful. We shall see, for instance, that VSEPR theory successfully predicts the structures of a wide range of the compounds that the halogens form among themselves and with oxygen, and that it is also successful with the few compounds of the noble gases that are now known. As with the oxygen compounds of the nitrogen and oxygen groups, we shall see that the oxoanions of the halogens and of xenon are oxidizing agents that generally act by O atom transfer. Another similarity between the groups, and which we shall emphasize again here, is the useful correlation between the oxidation number of the central atom and the rates of their redox reactions.

We shall consider one important class of halides, those of the *d*-block metals, from the viewpoint of stability and structure. As with the chalcogens, we shall see that high oxidation states are stabilized by the smaller, more electronegative elements, most notably fluorine. We shall also see the converse, the stabilization of low oxidation states by the heavier halogens, most notably iodine. One correlation that we shall emphasize is between the structures of the halides and the size and softness of the halogen.

Although the chemical properties of the noble gases are much less extensive than those of the halogens, we shall see that there are similarities between the two groups in the patterns of their reactions and in the structures of their compounds.

The halogens, the members of Group 17/VII, are among the most reactive nonmetallic elements; the noble gases, their neighbors in Group 18/VIII, are the least. Despite this contrast, there are some resemblances between the two groups, particularly in the structures of their compounds.

Since halogen chemistry is so extensive and has been mentioned many times already, in this chapter we highlight its systematic features, paying particular attention to redox properties and to compounds unique to the halogens. The most important chemical properties of the halogens are the roles of their compounds as intermediates and as end products with unusual properties, such as the polymers poly(vinyl chloride) and polytetrafluoroethylene. The chemical properties of fluorine are especially distinctive, and have extended both our understanding of bonding and the scope of chemistry. One outgrowth of its chemistry, for instance, links the two groups, for it led to the discovery of the noble gas compounds.

The elements

Some physical properties of the elements are listed in Table 13.1. The features to note include their high ionization energies and (for the halogens) their high electronegativities and electron affinities. The halogens have high electron affinities because the incoming electron can occupy an orbital of an incomplete valence shell and experience a strong nuclear attraction; the noble gases have negative electron affinities because their valence shells are full and the incoming electron must enter the orbitals of a new shell.

As in the earlier groups, the element at the head of each group has properties that are distinctively different from its congeners. Perhaps surprisingly in view of its reputation for being very electronegative, the electron affinity of fluorine is lower than that of chlorine. The explanation appears to lie in its smallness and the resulting strong repulsion between the valence electrons. However, fluorine forms ionic compounds with electropositive metals because its low affinity is more than offset by the high lattice enthalpies of compounds containing the small F^- ion. (We have seen that the same is true of O^{2-}.)

Table 13.1. Properties of the halogens and noble gases

Element	Ionization enthalpy/kJ mol^{-1}	Electron affinity/kJ mol^{-1}	χ_P	Ionic radius/Å	Common oxidation numbers
Halogens					
F	1687	328	3.98	1.17	−1
Cl	1257	349	3.16	1.67	−1, 1, 3, 5, 7
Br	1146	325	2.96	1.82	−1, 1, 3, 5
I	1015	295	2.66	2.06	−1, 1, 3, 7
At		270			
Noble gases					
He	2378	−48			0
Ne	2087	−120			0
Ar	1527	−96			0
Kr	1357	−96			0, 2
Xe	1177	−77	2.6		0, 2, 4, 6, 8
Rn	1043				

Sources: As for Table 12.1. Electronegativity of Xe from L. C. Allen and J. E. Huheey, *J. Inorg. Nucl. Chem.*, **42**, 1523 (1980).

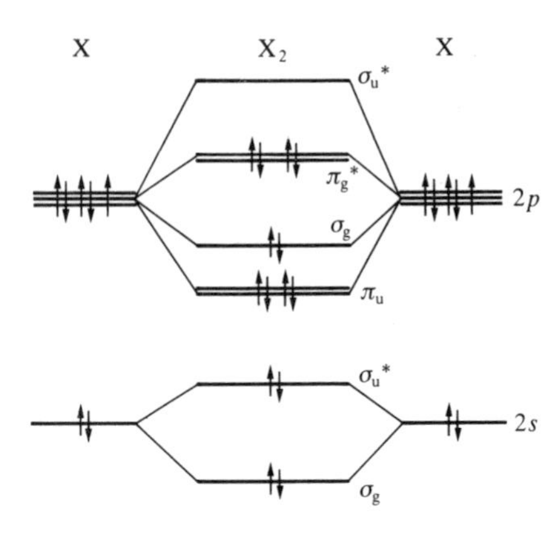

Fig. 13.1. Schematic molecular orbital energy diagram for Cl_2, Br_2, and I_2. For F_2, the order of the upper σ_g and π_u orbitals is interchanged.

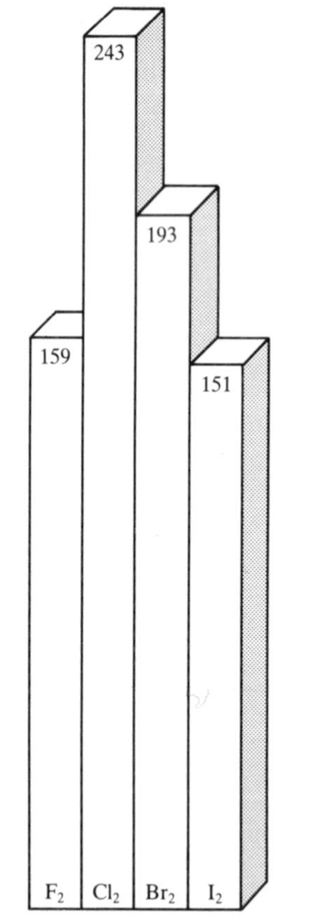

Fig. 13.2 Bond dissociation enthalpies of the halogens (in kJ mol^{-1}).

Since fluorine is the most electronegative of all reactive elements, it never has a positive oxidation number. With the possible exception of astatine, its congeners occur with oxidation numbers ranging from -1 to $+7$. The dearth of chemical information on astatine stems from the lack of any stable isotopes and the relatively short half-life (8.3 h) of the most stable one. Because of this short half-life, astatine solutions are intensely radioactive and may be studied only in high dilution. Astatine appears to exist as the anion At^- and as At(I) and At(III) oxoanions; no evidence for At(VII) has been obtained.

The noble gas with the most extensive chemical properties is xenon. Compounds with Xe—F, Xr—O, and Xe—N bonds are known, and its most important oxidation numbers are +2, +4, and +6. The known chemistry of radon is more limited, for its study is complicated by the high radioactivity of the element.

13.1 Dihalogens

Trends in properties

Among the most striking properties of the halogens are their colors. In the vapor they range from the almost colorless F_2, through yellowish-green Cl_2 and red-brown Br_2, to purple I_2. This progression in color reflects the decrease in the HOMO–LUMO gap on descending the group, for the electronic absorption spectrum arises primarily from transitions in which an electron is promoted from the highest filled σ and π^* orbitals into the vacant antibonding σ^* orbital (Fig. 13.1).

Except for F_2, analysis of the electronic absorption spectra gives precise values for the dihalogen bond dissociation enthalpies (Fig. 13.2). These decrease smoothly down the group below Cl_2. The electronic spectrum of F_2, however, is a broad continuum that lacks structure because the absorption of light is accompanied by dissociation of the already weakly bound molecule. The lack of discrete absorptions make it difficult to estimate the dissociation energy spectroscopically, and thermochemical methods are complicated by the highly corrosive nature of this reactive halogen. When these corrosion problems were solved, the F—F bond enthalpy was found to be less than that of Br_2 and thus out of line with the trend in the group. However, fluorine's low bond enthalpy is consistent with the low single-bond enthalpies of N—N, O—O, and various combinations of these three elements (Fig. 13.3). The simplest explanation (like the explanation of its low electron affinity) is that the strong repulsions within the compact atoms work against the energy advantage of sharing an additional electron when a covalent bond is formed, and hence result in a weak bond.

Chlorine, bromine, and iodine all crystallize in lattices of the same symmetry (Fig. 13.4), so it is possible to make a detailed comparison of distances between bonded and nonbonded neighboring atoms (Table 13.2). The important conclusion is that the latter do not increase as rapidly as the bond lengths. This suggests the presence of secondary weak intermolecular bonding interactions that become stronger on going from Cl_2 to I_2. The interaction also leads to a weakening of the I—I bond within the I_2 molecules, as is demonstrated by the lower I—I stretching frequency and greater I—I bond length in the solid as compared with the gas phase.

Moreover, solid iodine is a semiconductor, and under high pressure becomes metallic.

Occurrence and production

All the halogens except astatine are produced on a significant commercial scale, with chlorine production by far the greatest, followed by fluorine. Chlorine is widely used in industry to make chlorinated hydrocarbons and in applications in which a strong and effective oxidizing agent is needed (including bleaching and water purification).

The elements are found mainly as halides in nature, but the most easily oxidized element, iodine, is also found as iodates. Since many chlorides, bromides, and iodides are soluble, these anions occur in the oceans and in brines. The primary source of fluorine is calcium fluoride, which is highly insoluble and often found in sedimentary deposits.

The principal method of production of the elements is oxidation of the halides (Section 8.2). Although the strongly positive reduction potentials $E^{\ominus}(F_2, F^-) = +2.87$ V and $E^{\ominus}(Cl_2, Cl^-) = +1.36$ V refer to reduction in water, they do suggest that the oxidation of F^- and Cl^- ions requires a strong oxidizing agent, and in practice only electrolytic oxidation is feasible. An aqueous electrolyte cannot be used for fluorine production both because water is oxidized at a much lower potential (1.23 V) and because any fluorine produced reacts rapidly with water.

The isolation of elemental fluorine eluded chemists for most of the nineteenth century until in 1886 the French chemist Henri Moissan prepared it by the electrolysis of a solution of KF in liquid HF using a cell very much

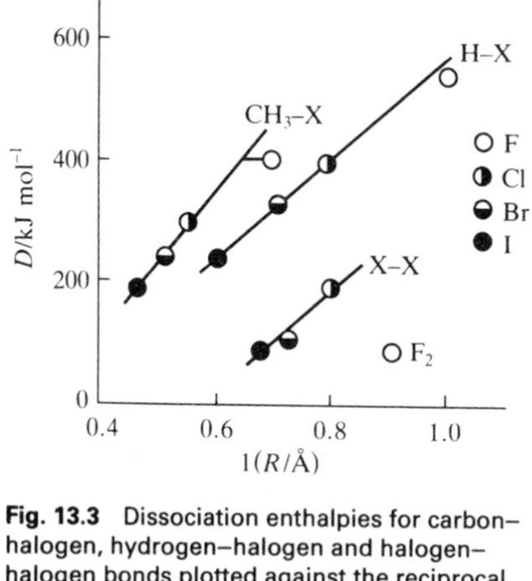

Fig. 13.3 Dissociation enthalpies for carbon–halogen, hydrogen–halogen and halogen–halogen bonds plotted against the reciprocal of the bond length. Note the weakness of the X—F bond. (From P. Politzer, *J. Amer. chem. Soc.*, **91**, 6235 (1969).)

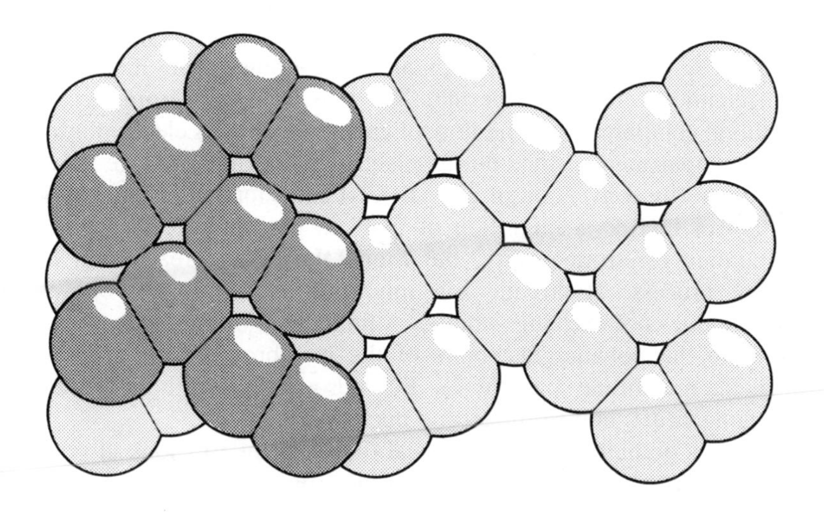

Fig. 13.4 Solid Cl_2, Br_2, and I_2 have similar structures; however, the closest nonbonded interactions are relatively less compressed in Cl_2 and Br_2 than in I_2.

Table 13.2. Bonding and shortest nonbonding distances for solid dihalogens

Element	Temperature/°C	Bond length/Å	Nonbonding distance/Å	Ratio
Cl_2	−160	1.98	3.32	1.68
Br_2	−106	2.27	3.32	1.46
I_2	−163	2.72	3.50	1.29

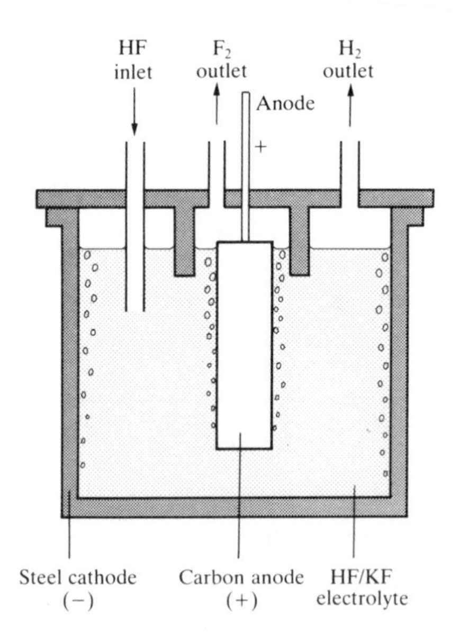

Fig. 13.5 Schematic diagram of an electrolysis cell for the production of fluorine from an HF–KF electrolyte.

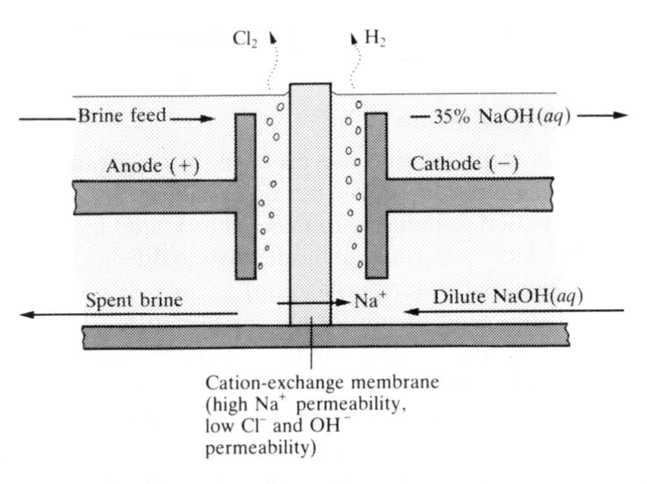

Fig. 13.6 Schematic diagram of a modern chloroalkali cell using a cation-exchange membrane.

like the one still used today (Fig. 13.5). Most commercial Cl_2 production is by the electrolysis of aqueous sodium chloride solution in a **chloroalkali cell** (Fig. 13.6). The reactions at the anode and the cathode are respectively

$$2Cl^-(aq) \rightarrow Cl_2(g) + 2e^-$$

$$2H_2O(l) + 2e^- \rightarrow 2OH^-(aq) + H_2(g)$$

The oxidation of water is suppressed by employing an electrode material that has a higher overpotential for O_2 evolution than for Cl_2 evolution. The best anode material appears to be RuO_2 (Chapter 17).

In modern chloroalkali cells, the anode and cathode compartments are separated by a polymer ion-exchange membrane. The membrane can exchange cations, and hence permits Na^+ ions to migrate from the anode to the cathode compartment. The flow of cations maintains electroneutrality in the two compartments, since during electrolysis negative charge is removed at the anode (by conversion of $2Cl^-$ to Cl_2) and supplied at the cathode (by formation of OH^-). A migration of OH^- in the opposite direction would also maintain electroneutrality, but then OH^- would react with Cl_2 and spoil the process. Hydroxide ion migration is suppressed because the membrane does not exchange anions. A final detail of the process is that the C atoms in the polymer backbone of the membrane are fluorinated: this protects the membrane since the C—F bonds are resistant both to the strong oxidizing agent Cl_2 and to the strong nucleophile OH^-.

Bromine is obtained by the chemical oxidation of Br^- ions in sea water. A similar process is used to recover iodine from certain natural brines that are rich in I^-. The more strongly oxidizing halogen chlorine is used as the oxidizing agent in both processes, and the resulting Br_2 and I_2 are driven from the solution in a stream of air:

$$Cl_2(g) + 2X^-(aq) \xrightarrow{\text{air}} 2Cl^-(aq) + X_2(g) \qquad X = \text{Br or I}$$

13.2 Pseudohalogens

A number of compounds (Table 13.3) have properties so similar to those of the halogens that they are called **pseudohalogens**. For example, like the

Table 13.3. Pseudohalides, pseudohalogens, and corresponding acids

Pseudohalide	Pseudohalogen	$E^\ominus$/V	Acid	pK_a*
CN^- cyanide	NCCN cyanogen	+0.27	HCN hydrogen cyanide	9.2
SCN^- thiocyanate	NCSSCN dithiocyanogen	+0.77	HNCS hydrogen isothiocyanate	−1.9
OCN^- cyanate			HNCO isocyanic acid	3.5
CNO^- fulminate			HCNO fulminic acid	
NNN^- azide			HNNN hydrazoic acid	4.92

* A. Albert and E. R. Serjeant, *The determination of ionization constants*. Chapman and Hall, London (1984).

dihalogens, cyanogen $(CN)_2$ undergoes thermal and photochemical dissociation in the gas phase; the resulting CN radicals are isolobal with halogen atoms and undergo similar reactions. Like the dihalogens, cyanogen takes part in chain reaction with hydrogen:

$$\text{NC—CN} \xrightarrow{\text{heat or light}} 2\text{CN}\cdot$$

$$\text{H}_2 + \text{CN}\cdot \rightarrow \text{HCN} + \text{H}\cdot$$

$$\text{H}\cdot + \text{NC—CN} \rightarrow \text{HCN} + \text{CN}\cdot$$

$$\text{Overall: } \text{H}_2 + \text{C}_2\text{N}_2 \rightarrow 2\text{HCN}$$

Another similarity is the reduction of a pseudohalogen to a pseudohalide:

$$\text{NC—CN}(aq) + 2e^- \rightarrow 2\text{CN}^-(aq)$$

The anionic **pseudohalide** is known in many cases even though the neutral dimer (the pseudohalogen) is not. Covalent pseudohalides similar to the covalent halides of the *p*-block elements are also common (for example, **1**). They are often structurally similar to the corresponding covalent halides (**2**), and undergo similar metathesis reactions. Many of the pseudohalides are ambident bases. The thiocyanate ion, SCN^-, for instance, has a soft base site, S, and a hard base site, N. In the case of CN^-, the C atom is a strong hard base and the N atom is a weak hard base.

As with all analogies, the concepts of pseudohalogen and pseudohalide have many limitations, but they do help to focus attention on certain chemical and structural similarities.

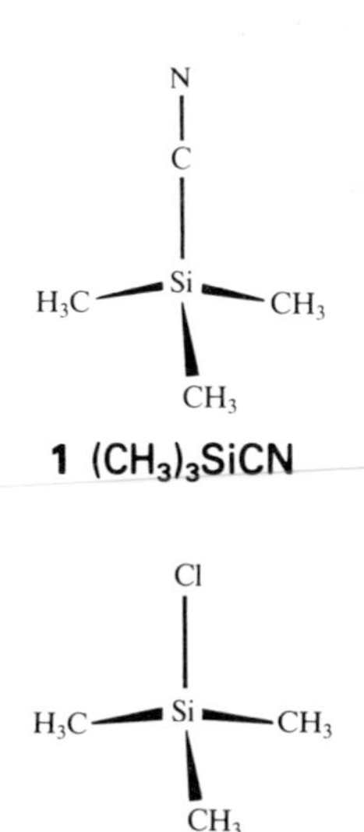

1 $(CH_3)_3SiCN$

2 $(CH_3)_3SiCl$

Polyhalogen compounds

There are many compounds with halogen–halogen bonds. They are of special importance as highly reactive intermediates and for providing useful insights into bonding.

13.3 Interhalogens

The binary molecular **interhalogens** are compounds with the formulas XY, XY_3, XY_5, and XY_7, where the heavier, less electronegative halogen X is the central atom.

Physical properties and structure

The diatomic interhalogens XY have been made for all combinations of the elements, but many of them are thermally unstable. The most stable is ClF, but ICl and IBr can also be obtained in pure crystalline form. Their physical properties are intermediate between those of their component molecules. For example, the deep red ICl (m.p. 27°C, b.p. 97°C) is intermediate between yellowish-green Cl_2 (m.p. −101°C, b.p. −35°C) and black I_2 (m.p. 114°C, b.p. 184°C). An interesting historical note is that ICl was discovered before Br_2 in the early nineteenth century, and this led to the first samples of the dark red-brown Br_2 (m.p. −7°, b.p. 59°C) being mistaken for ICl.

Most of the higher interhalogens are fluorides (Table 13.4). The only neutral interhalogen with the central atom in a +7 oxidation state is IF_7, but the cation ClF_6^+ of Cl(VII) is known. It should not be surprising that there is no neutral ClF_7 molecule because we have seen that Period-3 *p*-block elements do not have coordination numbers greater than 6, probably on account of their size. The omission of BrF_7 can be similarly rationalized. As we shall see later, bromine is reluctant to achieve its maximum oxidation state, a property it shares with some other Period-4 *p*-block elements, notably arsenic and selenium.

The structures of interhalogen molecules are largely as VSEPR theory predicts (Fig. 13.7). The XY_3 compounds (such as ClF_3) adopt a C_{2v} drooping-*T* structure as a consequence of a distorted trigonal bipyramidal array of lone and bonding electron pairs. However, ICl_3 is a Cl-bridged dimer (**3**).

Table 13.4. Representative interhalogens

XY	XY_3	XY_5	XY_7
ClF colorless gas	ClF_3 colorless b.p. 12°C	ClF_5 colorless gas	
BrF* light brown b.p. *ca.* 20°C	BrF_3 yellow liquid	BrF_5 colorless liquid	
IF*	$(IF_3)_n$ yellow solid	IF_5 colorless liquid	IF_7 colorless gas
BrCl*† red-brown gas			
ICl red solid	I_2Cl_6 bright yellow solid		
IBr black solid			

* Very unstable.
† The pure solid is known at low temperatures.

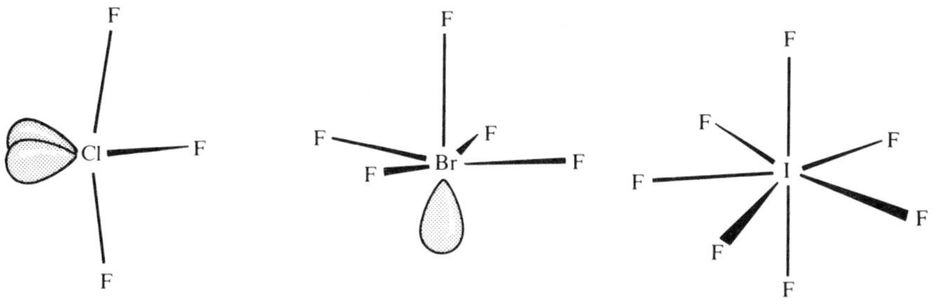

$ClF_3(C_{2v})$ $BrF_5(C_{4v})$ $IF_7(\approx D_{5h})$

Fig. 13.7 The structures of typical interhalogens. The shapes are in accord with the predictions of VSEPR theory.

Example 13.1: *Accounting for the structure of an interhalogen*

Determine the structure of I_2Cl_6 using VSEPR theory and assign the point group.

Answer. There are seven valence electrons on each atom in I_2Cl_6, giving 56 in all, and hence 28 pairs. Given the fact cited above that the molecule has two bridging Cl atoms, a reasonable Lewis structure is (**3**) with six electron pairs on each I atom. According to VSEPR theory, these pairs adopt an octahedral configuration, with the lone pairs in nonadjacent *trans* positions. Thus the bonds to surrounding Cl atoms take on a square-planar configuration and the molecule should be planar. This is verified by X-ray diffraction on single crystals. To assign the point group, we may choose the principal axis as a two-fold axis perpendicular to the plane of the molecule and half-way between the two bridging Cl atoms. There are two additional two-fold axes and a mirror plane perpendicular to this principal axis. These elements define the point group as D_{2h} (Fig. 2.12).

Exercise. Predict the structure and identify the point group of ClO_2F.

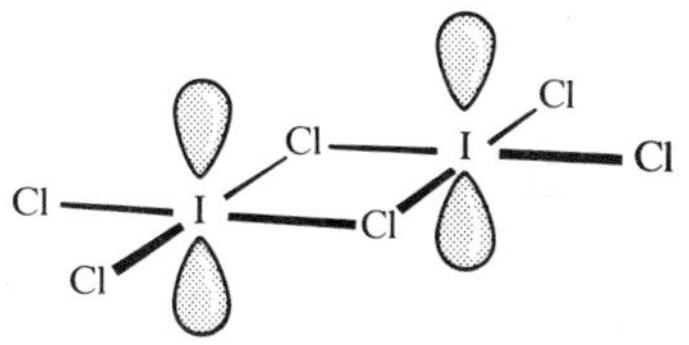

3 I_2Cl_6

The Lewis structure of XF_5 puts five bonding pairs and one lone pair on the central halogen atom, and as expected from VSEPR theory, XeF_5 molecules have square pyramidal structures. As already mentioned, the only known XY_7 compound is IF_7, which we would predict to be a pentagonal bipyramid. The experimental evidence for its actual structure is inconclusive, for it has been both claimed and denied that electron diffraction data support the prediction. As with other hypervalent molecules (Section 3.2), we can account for the bonding in IF_7 without invoking *d*-orbital participation by adopting a delocalized molecular orbital model in which bonding and nonbonding orbitals are occupied but antibonding orbitals are not.

Chemical properties

All the interhalogens are oxidizing agents. As with all the known interhalogen fluorides, the Gibbs energy of formation of ClF_3 is negative, so it is thermodynamically a weaker fluorinating agent than F_2. However, the rate at which it fluorinates substances generally exceeds that of F_2, so it is in fact an aggressive fluorinating agent. In general, the rates of oxidation of interhalogens do not have a simple relation to their thermodynamic stabilities. Thus, ClF_3 and BrF_3 are much more aggressive fluorinating agents than BrF_5, IF_5, and IF_7; iodine pentafluoride, for instance, is a convenient mild fluorinating agent that can be handled in glass apparatus. One use of ClF_3 as a fluorinating agent is in the formation of a passive metal fluoride film on the inside of nickel apparatus used in fluorine chemistry.

Both ClF_3 and BrF_3 react vigorously (often explosively) with organic matter, burn asbestos, and expel oxygen from many metal oxides:

$$2Co_3O_4(s) + 6ClF_3(g) \rightarrow 6CoF_3(s) + 3Cl_2(g) + 4O_2(g)$$

They are produced on a significant scale for the production of UF_6 for ^{235}U enrichment:

$$UF_4(s) + ClF_3(g) \rightarrow UF_6(s) + ClF(g)$$

Bromine trifluoride autoionizes in the liquid state:

$$2BrF_3(l) \rightleftharpoons BrF_2^+ + BrF_4^-$$

This Lewis acid–base behavior is shown in its ability to dissolve a number of fluoride salts:

$$CsF(s) + BrF_3(l) \rightarrow CsBrF_4(soln)$$

Bromine trifluoride is a useful solvent for ionic reactions that must be carried out under highly oxidizing conditions. The Lewis acid character of BrF_3 is shared by other interhalogens, which react with alkali metal fluorides to produce anionic fluoride complexes (Section 13.4).

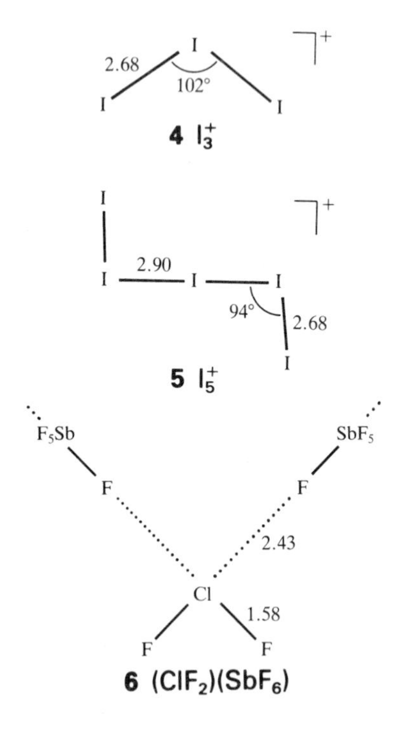

Cationic polyhalogens

Under strongly oxidizing conditions, such as in fuming sulfuric acid, I_2 is oxidized to the blue paramagnetic **diiodonium cation** I_2^+. Similar Br_2^+ species are known. The bond lengths of these cations are shorter than those of the corresponding neutral dihalogens, which is what we expect when an electron is lost from a π^* orbital and the bond order increases from 1 to $1\frac{1}{2}$ (Fig. 13.1). Two higher polyiodonium cations, I_3^+ and I_5^+, are known, and recent X-ray diffraction studies have established the structures shown in (**4**) and (**5**). The shape of I_3^+ is in line with VSEPR theory, since the central I atom has two lone pairs of electrons.

Another class of polyhalogen cations of formula XF_n^+ is obtained when a strong Lewis acid, such as SbF_5, reacts with halogen fluorides (Table 13.5). X-ray diffraction of these solids indicate that the F^- abstraction from the cations is incomplete, so the anions remain weakly associated with them by fluorine bridges (**6**).

Table 13.5. Representative interhalogen cations

$[XF_2]^+$	$[XF_4]^+$	$[XF_6]^+$
$[ClF_2]^+$	$[ClF_4]^+$	$[ClF_6]^+$
$[BrF_2]^+$	$[BrF_4]^+$	$[BrF_6]^+$
	$[IF_4]^+$	$[IF_6]^+$

13.4 Polyhalides

Since the dihalogens (and many interhalogens) have low-lying LUMOs, they can use these empty orbitals to accept an electron pair and hence act as Lewis acids (Section 6.8). In particular, they can form complexes with halide ions acting as Lewis bases and give the range of ions known as **polyhalides**.

Polyiodides

A deep brown color develops when I_2 is added to a solution of I^- ions. This color is characteristic of the homoatomic anions called **polyiodides**, which include **triiodide ions**, I_3^-, and **pentaiodide ions**, I_5^-. These polyiodides are Lewis acid–base complexes in which I^- and I_3^- act as the bases and I_2 acts as the acid.

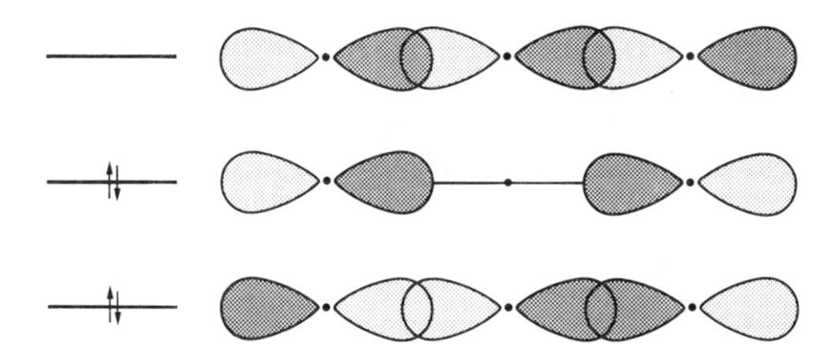

Fig. 13.8 Delocalized molecular orbitals for *p*-block elements with C.N. > 4. This representative fragment is constructed from the *p* orbitals of three atoms in a line.

Many polyiodides have been isolated. Many of them are mononegative ions having an odd number of I atoms and hence a formula that may be symbolized as $[(I_2)_n(I^-)]$. The I_3^- ion is the most stable member of this series. In combination with a large cation, such as $[N(CH_3)_4]^+$, it is symmetrical and linear with a longer I—I bond than in I_2. However, the structure of the triiodide ion, like that of the polyiodides in general, is highly sensitive to the identity of the counterion. For example Cs^+, which is smaller than the tetramethylammonium ion, distorts the I_3^- ion and produces one long and one short I—I bond (**7**). The ease with which the ion responds to its environment is a reflection of the weakness of the delocalized bonds that just manage to hold the atoms together (Fig. 13.8). A more extreme example of sensitivity to the cation is provided by NaI_3, which can be formed in aqueous solution but decomposes when the water is evaporated:

$$Na^+(aq) + I_3^-(aq) \xrightarrow{-H_2O} NaI(s) + I_2(s)$$

This behavior is another example of the instability of large anions in combination with small cations.

The existence and structures of the higher polyiodides are sensitive to the counterion for similar reasons and large cations are necessary to stabilize them in the solid state. In fact, entirely different shapes are observed for polyiodide ions in combination with various large cations, for the structure of the anion is determined in large measure by the manner in which the ions pack together in the crystal. The bond lengths in a polyiodide ion often suggest that it can be regarded as a chain of associated I^-, I_2, or I_3^- units (Fig. 13.9).

Some dinegative polyiodides are known. These contain an even number of I atoms and their general formula is $[I^-(I_2)_nI^-]$. They have the same sensitivity to the cation as their mononegative counterparts.

Other polyhalides

Although polyhalide formation is most pronounced for iodine, other polyhalides are also known. These include Cl_3^-, Br_3^-, and BrI_2^-, which are known in solution and (in partnership with large cations) as solids too. Even F_3^- has been detected spectroscopically at low temperatures in an inert matrix.

In addition to complex formation between dihalogens and halide ions, some interhalogens can act as Lewis acids toward halide ions, thus forming polyhalides. For example, we have already mentioned that BrF_3 will react with CsF to form $CsBrF_4$, which contains the square-planar BrF_4^- anion (**8**).

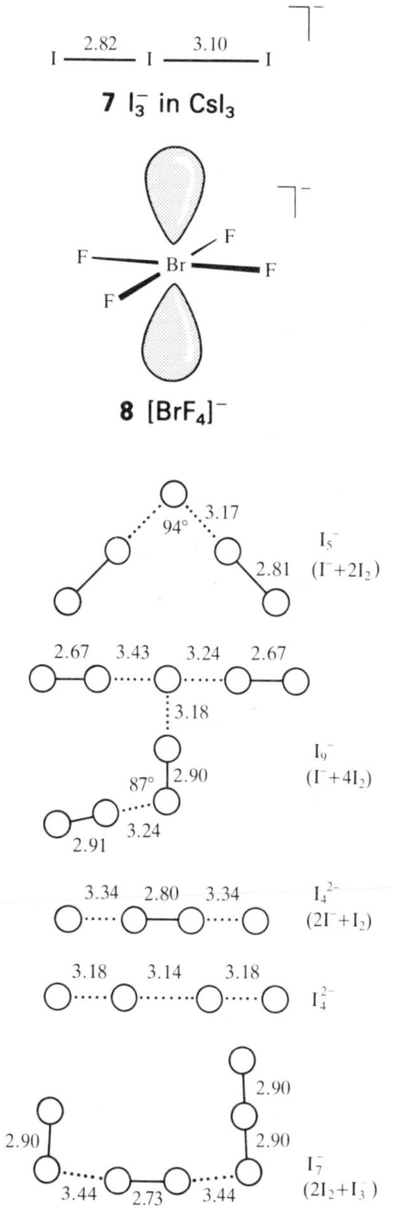

Fig. 13.9 Some representative polyiodide structures and their approximate description in terms of I^-, I_3^-, and I_2 building blocks. The bond lengths and angles vary with the identity of the cation.

Compounds of halogens and oxygen

In contrast to the rather simple formulas and structures of halogen fluorides, the molecular compounds of the halogens with oxygen are a diverse group. The aqueous chemistry of the halogen oxoanions is the most uniform and important, so we shall concentrate on them.

13.5 Binary oxides

Many binary compounds of the halogens and oxygen are known, but most are unstable and not commonly encountered in the laboratory. We shall mention only a few of the most important.

Oxygen difluoride (FOF; m.p. −224°C, b.p. −145°C), the most stable binary compound of oxygen and fluorine, is prepared by passing fluorine through dilute aqueous hydroxide solution:

$$2F_2(g) + 2OH^-(aq) \rightarrow OF_2(g) + 2F^-(aq) + H_2O(l)$$

The pure difluoride is stable in the gas phase above room temperature and does not react with glass. It is a strong fluorinating agent but less so than fluorine itself. As predicted by VSEPR theory (and by analogy with H_2O), the OF_2 molecule is angular.

Dioxygen difluoride (FOOF; m.p. −154°C, b.p. −57°C) can be synthesized by photolysis of a liquid mixture of the two elements. It is unstable in the liquid state and decomposes rapidly above −100°C, but may be transferred (with some decomposition) as a low-pressure gas in a metal vacuum line. Dioxygen difluoride is an even more aggressive fluorinating agent than ClF_3. For example, it oxidizes plutonium metal and its compounds to PuF_6 in a reaction that ClF_3 cannot accomplish:

$$Pu(s) + 3O_2F_2(g) \rightarrow PuF_6(g) + 3O_2(g)$$

The interest in this reaction is in removing plutonium as the volatile hexafluoride from spent nuclear fuel.

Chlorine occurs with many different oxidation numbers in its oxides, for they include

+1	+4	+6	+7
Cl_2O	ClO_2	ClO_3	Cl_2O_7
brown-yellow gas	yellow gas	dark red liquid	colorless liquid

Some of these oxides are radicals, including ClO_2, in which chlorine has the unusual oxidation number of +4, and ClO_3. The latter exists in equilibrium with its dimer, Cl_2O_6, and as $[ClO_2^+][ClO_4^-]$ in the solid.

Chlorine dioxide is the only halogen oxide produced on a large scale. The reaction used is the reduction of ClO_3^- with HCl or SO_2 in strongly acidic solution:

$$2ClO_3^-(aq) + SO_2(g) \xrightarrow{\text{acid}} 2ClO_2(g) + SO_4^{2-}(aq)$$

It is a strongly endoergic compound ($\Delta G_f^\ominus = +121\ \text{kJ mol}^{-1}$) and must be kept dilute to avoid explosive decomposition, and is used at the site of production. Its major uses are for bleaching paper pulp and purifying water.

Table 13.6. Halogen oxoanions

Oxidation	Formula	Name*	Structure	Remarks
+1	ClO^-	Hypochlorite (monoxochlorate)	$[Cl{-}O]^-$ $C_{\infty v}$	Good oxidizing agent
+3	ClO_2^-†	Chlorite (dioxochlorate)	C_{2v}	Strong oxidizing agent, disproportionates
+5	ClO_3^-	Chlorate (trioxochlorate)	C_{3v}	Oxidizing agent
+7	ClO_4^-‡	Perchlorate (tetraoxochlorate)	T_d	Oxidizing agent and weak ligand

* IUPAC names in parentheses.
† Bromite and iodite have not been isolated.
‡ For iodine, IO_4^- and $H_3IO_6^-$ are present in basic solution, and H_5IO_6 is present in acidic solution.

13.6 Oxoacids and oxoanions

The wide range of oxoanions and oxoacids of chlorine and the heavier halogens presents a challenge to those who devise systems of nomenclature. We shall use the most commonly employed names, such as chlorate for ClO_3^-, rather than the systematic names, such as trioxochlorate(V). Table 13.6 is a brief dictionary for converting between the two systems of nomenclature.

Acidities

We saw in Section 5.6 that the strengths of the oxoacids vary systematically with the number of O atoms q and the number of OH groups p on the central atom:

Acid	q/p	Strength	pK_a
HOCl	0	very weak	7.53
HOClO	1	weak	2.00
$HOClO_2$	2	strong	−1.2
$HOClO_3$	3	very strong	−10

At first sight, periodic acid—the I(VII) analog of perchloric acid—appears to be out of line, for it is weak ($pK_{a1} = 3.29$). However, as soon as we note that its formula is $(HO)_5IO$ with $q/p = \frac{1}{5}$, we see it is its structure, not its strength, that is anomalous. The O atoms in the conjugate base $H_4IO_6^-$ are very labile on account of the rapid equilibration

$$H_4IO_6^-(aq) \rightleftharpoons IO_4^-(aq) + 2H_2O(l) \qquad K = 40$$

and in basic solution IO_4^- is the dominant ion. The tendency to have an expanded coordination shell is shared by the oxoacids of the neighboring element tellurium, which in its maximum oxidation state forms the weak acid H_6TeO_6.

Example 13.2: *Calculating the concentration of periodate ions*

Verify that the 4-coordinate form of periodate is in low concentration in solution by calculating the molar concentration of IO_4^- ions in a 0.50 M $H_5IO_6(aq)$ solution with sufficient $HClO_4$ added to achieve $[H^+] = 1.00$ M.

Answer. The acidity constant is

$$K_{a1} = \frac{[H^+][H_4IO_6^-]}{[H_5IO_6]} = 5.1 \times 10^{-4}$$

When $[H^+] = 1.00$ M and $[H_5IO_6] = 0.50$ M,

$$[H_4IO_6^-] = \frac{0.50}{1.00} \times 5.1 \times 10^{-4}\ \text{M} = 2.6 \times 10^{-4}\ \text{M}$$

Since

$$K = \frac{[IO_4^-]}{[H_5IO_6]} = 40$$

we can conclude that

$$[IO_4^-] = 40 \times 2.6 \times 10^{-4}\ \text{M} = 1.0 \times 10^{-2}\ \text{M}$$

Exercise. Calculate the molar concentration of IO_4^- ions in an 0.30 M $H_5IO_6(aq)$ solution that has been acidified to the point that $[H^+] = 0.70$ M.

Oxoanions as ligands

The halogen oxoanions, like many oxoanions, form metal complexes, including the metal perchlorates and periodates we discuss here. In this connection we should note that since $HClO_4$ is a very strong acid and H_5IO_6 is a weak acid, ClO_4^- ions are very weak bases and $H_4IO_6^-$ ions are strong bases.

The hardness, low Brønsted basicity, and single negative charge of ClO_4^- results in its being only weakly basic toward all Lewis acids. Accordingly, perchlorates have often been used to study the properties of hexaqua ions in solution, for they have little tendency to displace a ligand H_2O molecule from the coordination sphere. Similarly, the ClO_4^- ion is used as a weakly coordinating ion that can readily be displaced from a complex by other ligands, or as a medium-sized anion that might stabilize large cationic complexes having easily displaced ligands.

However, the ClO_4^- ion is a treacherous ally. Because it is a powerful oxidizing agent, it should be avoided whenever there are oxidizable ligands or metal ions present (which is commonly the case). In some cases the danger lies in wait, for its redox reactions are generally slow (Section 13.7), and it is possible to prepare many metastable perchlorate complexes or salts that may be handled with deceptive ease. However, once reaction has been initiated by mechanical action, heat, or static electricity, these compounds can detonate with disastrous consequences.[1] Such explosions have injured chemists who may have handled a compound many times before it unexpectedly exploded. Some readily available and more docile weakly basic anions may be used in place of ClO_4^-; they include trifluoromethanesulfonate $[SO_3CF_3]^-$, tetrafluoroborate $[BF_4]^-$, and hexafluorophosphate $[PF_6]^-$.

In contrast to perchlorate, periodate is a facile oxidizing agent (Section 13.7) and is used both to prepare and to stabilize metal ions in high oxidation states. Some of the high oxidation states it can be used to form are very unusual: they include Cu(III) in a salt containing the $[H_3Cu(IO_6)_2]^{4-}$ complex and Ni(IV) in an extended complex containing the $[Ni(IO_6)]^-$ unit. In these complexes the periodate ligand is bidentate.

13.7 Redox principles and trends

The thermodynamic tendency of the halogen oxoanions and oxoacids to participate in redox reactions is well understood. As we shall see, we can summarize their behavior in a Frost diagram that is quite easy to rationalize. It is a very different story with the rates of the reactions, for these vary widely. Their mechanisms are only partly understood despite many years of investigation.

Thermodynamics of redox reactions

We saw in Section 8.9 that one feature of a Frost diagram is that if a species lies above the line joining its two neighbors to each other, then it is unstable with respect to disproportionation into them. From the Frost diagram for the halogen oxoanions and oxoacids in Fig. 13.10 we can see therefore that many of them are susceptible to disproportionation. Chlorous acid, $HClO_2$, for instance, lies above the line joining its two neighbors, and is liable to disproportionation:

$$2HClO_2(aq) \rightarrow ClO_3^-(aq) + HClO(aq) + H^+(aq) \qquad E^\ominus = +0.52\,V$$

The corresponding Br(III) and I(III) oxoanions are not marked in Fig. 13.10 because they are so unstable that in acidic solution they do not exist in solution, except perhaps as transient intermediates.

We also saw in Section 8.9 that the steeper the line joining any two neighbors in a Frost diagram, the stronger the oxidizing power of the couple. A glance at Fig. 13.10 shows that all three diagrams have steeply sloping lines, which immediately shows that all the oxidation states except the lowest (Cl^-, Br^-, and I^-) are strongly oxidizing.

Finally, the decrease in reduction potentials for oxoanions in basic solution as compared with acidic solution (Section 8.10) is evident in their less steep slopes in the Frost diagrams (Fig. 13.10). The numerical comparison for ClO_4^- ions in 1 M acid compared with 1 M base makes this clear:

[1] Two examples are the explosion in a rocket-fuel plant of ammonium perchlorate with the intensity of an earthquake, *Chem. Eng. News*, p. 5, May (1988) and the injury of a chemist by the explosion of the perchlorate salt of a lanthanide complex, *Chem. Eng. News*, p. 4, December (1983).

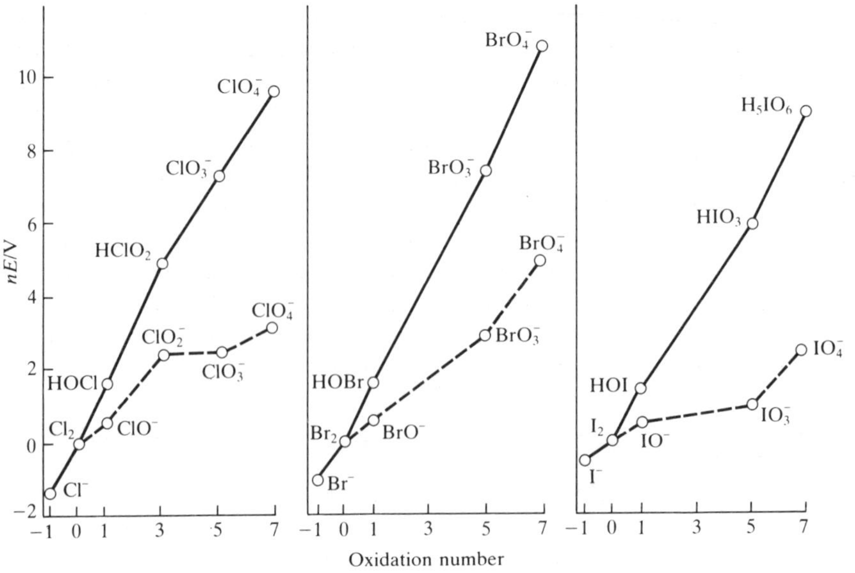

Fig. 13.10 Frost diagrams for chlorine, bromine, and iodine. Full lines are for acidic solution, broken lines for basic solution.

At pH = 0:

$$ClO_4^-(aq) + 2H^+(aq) + 2e^- \rightarrow ClO_3^-(aq) + H_2O(l) \qquad E^\ominus = +1.20\ \text{V}$$

At pH = 14:

$$ClO_4^-(aq) + H_2O(l) + 2e^- \rightarrow ClO_3^-(aq) + 2OH^-(aq) \qquad E_B^\ominus = +0.37\ \text{V}$$

Rates of redox reactions

Although the rates of the halogen oxoanion redox reactions are difficult to rationalize, there are some general patterns that help to correlate the trends and give some clues about the mechanisms that may be involved.

The oxidation of many molecules and ions by halogen oxoanions becomes progressively faster as the oxidation number of the halogen decreases (see Section 8.4). Thus the rates observed are often in the order

$$ClO_4^- < ClO_3^- < ClO_2^- \ll ClO^- \approx Cl_2$$
$$BrO_4^- < BrO_3^- \ll BrO^- \approx Br_2$$
$$IO_4^- < IO_3^- < I_2$$

For example, aqueous solutions containing Fe^{2+} and ClO_4^- are stable for many months in the absence of dissolved oxygen, but an equilibrium mixture of aqueous ClO^- and Cl_2 rapidly oxidizes Fe^{2+}.

Oxoanions of the heavier halogens tend to react most rapidly. This is particularly striking for the elements in their highest oxidation states:

$$ClO_4^- < BrO_4^- < IO_4^-$$

As we have remarked, perchlorates in dilute aqueous solution are usually unreactive, but periodate oxidations are fast enough to be exploited for titrations. The mechanistic details are often complex, but the observed trends in reactivity appear to arise from the rate-controlling halogen–oxygen bond scission. Thus the halogen(VII) oxoanions, which have the shortest and strongest bonds of all the oxoanions, are the least reactive.

The rates of redox reactions also often correlate with the rates of oxygen atom exchange with water. For example, ClO_4^- ions do not exchange their O atoms with water at a measurable rate, but IO_4^- (and ClO^-) undergo rapid exchange. Nucleophilic attack seems to be a feature in the initial stages of these redox reactions.

Oxidations by oxoanions are often accelerated by acid. The oxidation of halides by BrO_3^- ions, for instance, is second-order in H^+:

$$\text{rate} = k[BrO_3^-][X^-][H^+]^2$$

The acid is thought to act by protonating the oxo group in the oxoanion and so aiding oxygen–halogen bond scission. Since oxoanions become stronger oxidizing agents as the acidity is increased, and we now see that their rates are increased too, kinetics and equilibria now unite to bring about otherwise difficult oxidations. An example is the use of a mixture of H_2SO_4 and $HClO_4$ in the final stages of the oxidation of organic matter in certain analytical procedures. It is important to be especially alert to the risk of an explosion when concentrated perchloric acid is mixed with organic compounds or other reducible substances.

Mechanisms of redox reactions

We saw in Section 8.4 that redox reactions of metal complexes may be classified as either outer-sphere electron transfer or inner-sphere atom or electron transfer through a bridging ligand.

Experiments with O-labeled oxoanions have revealed many examples of O-atom transfer, such as

$$2H_2O(l) + Cl\overset{*}{O}_2^-(aq) + 2SO_2(aq) \rightarrow Cl^-(aq) + 4H^+(aq) + 2S\overset{*}{O}O_3^{2-}(aq)$$

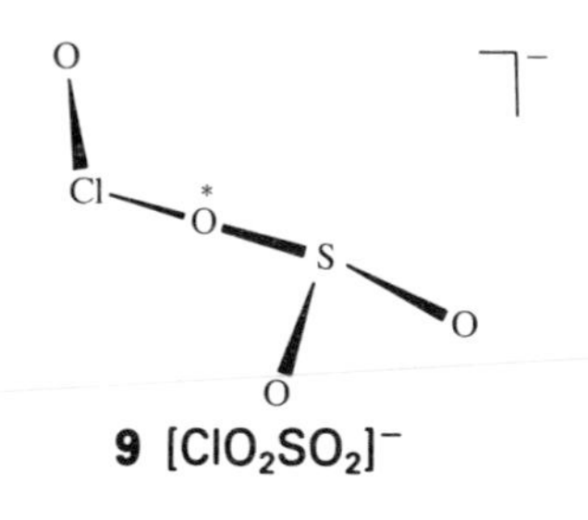

9 $[ClO_2SO_2]^-$

Such reactions appears to take place by an inner-sphere O-atom transfer by a transient oxo-bridged complex (**9**). Electron transfer without atom transfer can also occur across a bridging ligand. The oxidation of $[Fe(CN)_6]^{4-}$ to $[Fe(CN)_6]^{3-}$ by periodate, for instance, appears to proceed by an inner-sphere electron transfer mechanism, presumably involving a CN bridge linking the two complexes and providing a path for the electron transfer.

Outer-sphere electron transfer is also sometimes encountered (or, at least, invoked). The least ambiguous cases involve reaction of the hydrated electron, $e^-(aq)$, a highly transient species produced by γ-radiolysis of water. The hydrated electron reacts very rapidly with BrO_3^-, BrO_4^-, and IO_4^- but not with ClO_3^- and ClO_4^-.

13.8 Redox reactions of individual species

With the general principles of the redox chemistry of the halogens now established, we can consider the characteristic properties and reactions of specific oxidation states.

Halogen(0)

Disproportionation is thermodynamically favored for Cl_2, Br_2, and I_2 in basic solution:

$$X_2(aq) + 2OH^-(aq) \rightleftharpoons XO^-(aq) + X^-(aq) + H_2O(l)$$
$$K = 7.5 \times 10^{15}\ (Cl),\ 2 \times 10^8\ (Br),\ 30\ (I)$$

In writing these and other equilibria in aqueous solution, we conform to the usual convention that the activity of water is 1, so $[H_2O]$ does not appear in the equilibrium constant:

$$K = \frac{[XO^-][X^-]}{[X_2][OH^-]^2}$$

Disproportionation is much less favorable in acid solution:

$$Cl_2(aq) + H_2O(l) \rightleftharpoons HOCl(aq) + H^+(aq) + Cl^-(aq) \qquad K = 3.9 \times 10^{-4}$$

This reaction, a hydrolysis, is thought to proceed by the displacement of X^- by OH^-:

$$H_2O + Cl_2 \rightarrow [H_2O{-}Cl{-}Cl] \rightarrow H^+ + [HO{-}Cl{-}Cl]^-$$
$$\rightarrow H^+ + HOCl + Cl^-$$

It is not yet known if the Lewis complex $H_2O{-}Cl_2$ is actually formed. According to one interpretation, proton transfer is so rapid that the reaction yields $HOCl_2^-$ directly. The $[HO{-}Cl{-}Cl]^-$ complex is at most transitory, for Cl^- is rapidly displaced by OH^-. The overall reaction is so fast that when Cl_2 is used as an oxidizing agent in aqueous solution it is possible that both HOCl and Cl_2 are the actual oxidants.

Since the redox reactions of HOCl and Cl_2 are often fast, Cl_2 in water is widely used as an inexpensive and powerful oxidizing agent. A part of the reason for its usefulness is that the rapid equilibration between ClO^- and Cl_2 gives rise to a variety of possible reaction pathways. For example, the nonlabile complex $[Fe(phen)_3]^{2+}$ reacts more rapidly with Cl_2 than with HOCl because the former can undergo an outer-sphere electron transfer, whereas ClO^- or HOCl appear to require dissociation of phen prior to an inner-sphere redox process. In other situations HOCl is a more facile oxidizing agent than Cl_2. In general, labile aqua complexes such as $[Cr(OH_2)_6]^{2+}$ react faster with ClO^- than with Cl_2.

The equilibrium constants for the hydrolysis of Br_2 and I_2 in acid solution are smaller than for Cl_2, and both elements are unchanged when dissolved in slightly acidified water. Since F_2 is a much stronger oxidizing agent than the other halogens, it produces mainly O_2 and H_2O_2 when in contact with water, so hypofluorous acid, HOF, was discovered only long after its congeners. The first indication of the existence of HOF came from infrared spectral studies of H_2O and F_2 trapped in a matrix of frozen N_2 at about 20 K. With this encouragement, the HOF molecule was finally isolated and chemically characterized in the early 1970s. The work was done by E. H. Appelman in the USA, who carried out the controlled reaction of F_2 with ice at −40°C, and obtained the highly unstable HOF in approximately 50 percent yield:

$$F_2(g) + H_2O(s) \rightarrow HOF(g) + HF(g)$$

Halogen(I)

Hypohalite ions are subject to disproportionation. For instance, the ClO^- ion forms Cl^- and ClO_3^- ions:

$$3ClO^-(aq) \rightleftharpoons 2Cl^-(aq) + ClO_3^-(aq) \qquad K = 1.5 \times 10^{27}$$

This reaction (which is used for the commercial production of chlorates) is slow at or below room temperature for ClO^-, but is much faster for BrO^-. It is so fast for IO^- that this ion has been detected only as a reaction intermediate.

Halogen(III)

Chlorite ions, ClO_2^-, and bromite ions, BrO_2^-, are both susceptible to disproportionation. However, the rate is strongly dependent on pH, and ClO_2^- (and to a lesser extent BrO_2^-) can be handled in basic solution with only slow decomposition. In contrast, chlorous acid, $HClO_2$, and bromous acid, $HBrO_2$, both disproportionate rapidly:

$$2HClO_2(aq) \rightarrow H^+(aq) + HOCl(aq) + ClO_3^-(aq) \qquad E^\ominus = +0.51\ V$$

$$2HBrO_2(aq) \rightarrow H^+(aq) + HOBr(aq) + BrO_3^-(aq) \qquad E^\ominus = +0.50\ V$$

Iodine(III) is even more transitory, and HIO_2 was only recently identified as a transient species in aqueous solution.

These primary reactions are usually accompanied by further reaction. For example, the disproportionation of $HClO_2$ under a broad set of conditions is quite well described by the equation

$$4HClO_2(aq) \rightarrow 2H^+(aq) + 2ClO_2(aq) + ClO_3^-(aq) + Cl^-(aq) + H_2O(l)$$

Since two oxidized products (ClO_2 and ClO_3^-) are produced, the equation can be balanced differently by changing their ratio. The ratio actually observed reflects the relative rates of the alternative reactions.

Halogen(V)

The Frost diagram for chlorine (Fig. 13.10) indicates that chlorate ions, ClO_3^-, are unstable with respect to disproportionation in both acidic and basic solution:

$$4ClO_3^-(aq) \rightarrow 3ClO_4^-(aq) + Cl^-(aq) \qquad E^\ominus = +0.25\ V$$

(This value of $E^\ominus$ corresponds to $K = 1.4 \times 10^{25}$.) However, as $HClO_3$ is a strong acid, and this reaction is slow at both low and high pH, ClO_3^- ions can be handled readily in aqueous solution. Bromates and iodates are thermodynamically stable with respect to disproportionation.

Example 13.3: *Summarizing the redox properties of the halogens*

Summarize the oxidation numbers of the halogens that are thermodynamically susceptible to disproportionation.

Answer. Oxidation states 0, +1, and +3 of the halogens except fluorine are susceptible to disproportionation in either acidic or basic solution.

Exercise. Which dihalogens are thermodynamically capable of oxidizing H_2O to O_2? Which of these reactions are facile?

Halogen(VII)

The perchlorate ion, ClO_4^-, and periodate ion, IO_4^-, are both well known, but the perbromate ion BrO_4^- was not discovered until the late 1960s,[2] which is another example of the instability of the maximum oxidation state of the Period-4 elements arsenic, selenium, and bromine. Although perbromate is not available commercially, it is fairly readily synthesized by the action of fluorine on bromate ions in basic aqueous solution:

$$BrO_3^-(aq) + F_2(g) + 2OH^-(aq) \rightarrow BrO_4^-(aq) + 2F^-(aq) + H_2O(l)$$

Of the three ions, BrO_4^- is the strongest oxidizing agent.

The reduction of BrO_4^- is faster than that of ClO_4^- because perbromate ions undergo electron transfer more rapidly. However, reduction of periodate ions in dilute acid solution is by far the fastest. Periodates are therefore used in analytical chemistry as oxidizing titrants and also in syntheses, such as the oxidative cleavage of diols. As we remarked earlier, the speed of periodate reactions appears to stem from the ease with which the reducing agent can coordinate to the I atom, as in the diol oxidation:

$$HIO_4(aq) + HO\text{–}C(CH_3)_2\text{–}C(CH_3)_2\text{–}OH \longrightarrow \begin{matrix} (CH_3)_2C\text{—}O \\ | \\ (CH_3)_2C\text{—}O \end{matrix} > I(O)(OH)_3$$

$$\longrightarrow 2CH_3COCH_3 + HIO_3 + H_2O$$

Metal halides

The familiar *d*-block halides, such as hydrated iron(II) chloride and cobalt(II) chloride, give an incomplete impression of the overall richness of the chemistry of the *d*-block halides. As well as these conventional compounds, *d*-block halides range from metallic to molecular.

13.9 Structures of *d*-block halides

The *d*-block fluorides have strikingly different properties from the other halides. In particular, the small, electronegative F atom stabilizes metal atoms in high oxidation states. Conversely, it does not stabilize elements in low oxidation states as readily as the heavier halogens do.

Many of the *d*-block halides have structures that can be identified as one of four types. Some have simple structures that are consistent with an ionic model; for instance, they have a fluorite or a rutile structure. Some others have layered structures (like CdI_2) that indicate greater covalency. Halides of metals in low oxidation states often have metal–metal bonded structures. Halides with the metal in a high oxidation state are often molecular.

[2] Appelman's discovery of BrO_4^- and HOF is described in an article titled 'Nonexistent compounds: two case histories', *Acc. chem. Res.*, **4,** 113 (1973). His strategy was to first synthesize, detect and study the important chemical properties of traces of the compound. This information enabled him to devise a practical synthesis for larger quantities.

Table 13.7. Structure of d-metal subhalides

X/M	Compound	Other examples	Type of M–M complex
1.71	Sc_7Cl_{12}	La_7I_{12}	Discrete M_6X_{12} + M
1.60	Sc_5Cl_6	Gd_5Br_8	Single chains (M_6X_{12} type) + M
1.50	Gd_2Cl_3	Y_2Cl_3, Y_2Br_3	Single chains (M_6X_8 type)
1.43	Sc_7Cl_{10}		Double chains (M_6X_8 type) + M
1.00	ZrCl ZrBr	HfCl, ScCl, YCl	Double layer (M_6X_8 type)

Source: A. F. Wells, *Structural inorganic chemistry*. Oxford University Press (1984).

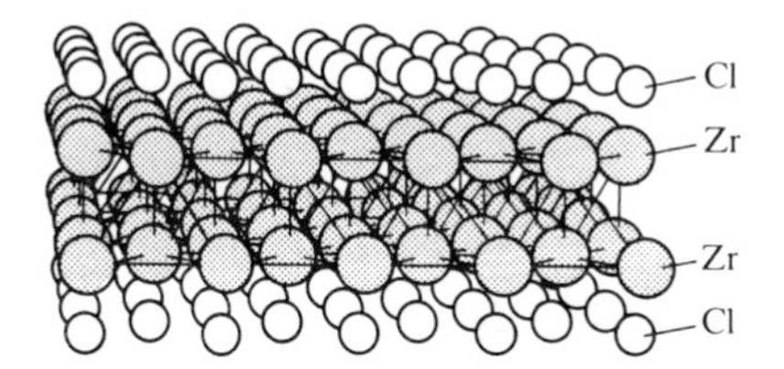

Fig. 13.11 The structures of zirconium chloride consist of layers of metal atoms in graphite-like hexagonal nets.

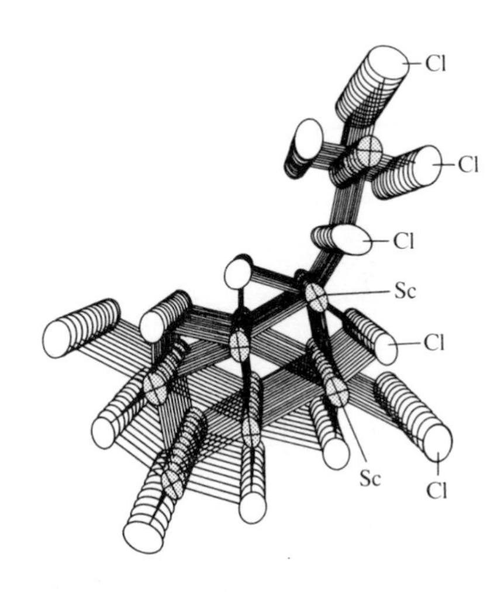

Fig. 13.12 The structures of several scandium subchlorides contain chains of metal atoms. This structure of Sc_7Cl_{10} can be thought of as an infinite collection of $[ScCl_2^+][Sc_6Cl_8^-]$ units. The $ScCl_6$ octahedra are present in the top chain and below this are double chains of distorted octahedra. (From J. D. Corbett, *Acc. chem. Res.*, **14**, 239 (1981).)

Subhalides

As we can see from Table 13.7, many of the early *d*-block metals form halogen-poor subhalides. These compounds have metal–metal bonds; they are deeply colored or metallic in appearance and often have high electrical conductivities. Examples include ScCl and ZrCl (Fig. 13.11) and a series of scandium subchlorides (Fig. 13.12). These subhalides are often synthesized by the reaction of a higher halide and the metal in a sealed inert container at high temperature. For example, ZrCl is prepared in a sealed tantalum reactor:

$$3Zr(s) + ZrCl_4(g) \xrightarrow{600\text{–}800^\circ C} 4ZrCl(s)$$

The occurrence of these metallic halides for elements on the left of the *d* block is generally explained by the large radial extension of their *d* orbitals. The diffuseness of the orbitals allows the metal atoms to achieve metal–metal bonding more readily than the more compact atoms of elements elsewhere in the block. The metallic halides are most likely to be formed by the softer halogens chlorine, bromine, and iodine.

The favorable lattice energies for the small F^- ion in combination with highly charged cations appears to be a prime factor in the disproportionation of fluorides in low oxidation states:

$$2CuF(s) \rightarrow Cu(s) + CuF_2(s)$$

This type of disproportionation conforms with the ionic models discussed in Section 4.8. However, that these simple models have their limitations is shown by the existence of a silver subhalide Ag_2F, which is a metallic solid containing sheets of graphite-like hexagonal nets of Ag atoms.

Dihalides and trihalides of 3*d* metals

As we saw in Section 8.12, the +2 oxidation state is common for 3*d* elements. There are accordingly many halides of this state and they are well described by structures containing isolated M^{2+} ions (Table 13.8). As we see from the table, the fluorides have six-coordinate metal atoms and a rutile structure, which is typical of highly ionic compounds. Most of the chlorides and all the bromides and iodides crystallize in the more covalent layered structures. The metal–metal bonded $ScCl_2$ is distinctly different from the rest of these halides.

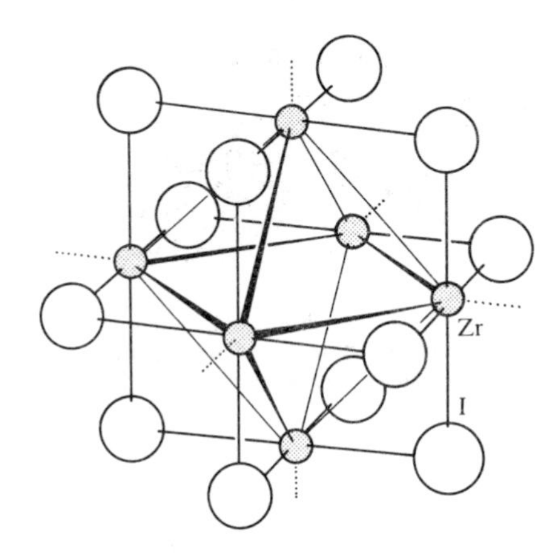

Fig. 13.13 The M_6X_{12} metal cluster present in zirconium iodide. The dotted lines indicate an association with a halogen atom in an adjacent cluster.

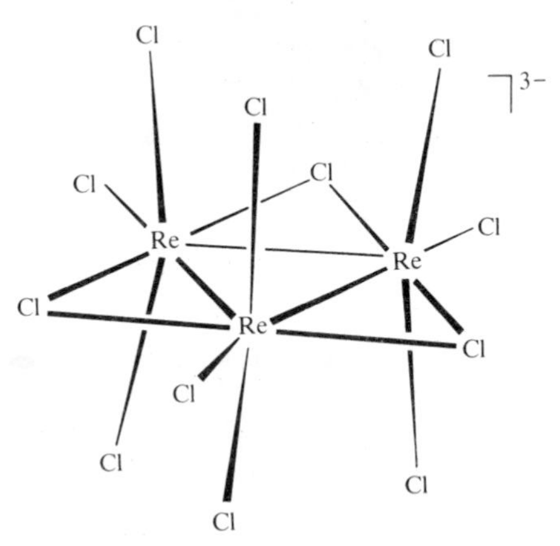

Fig. 13.14 The Re clusters in $[Re_3Cl_{12}]^{3-}$.

Table 13.8. Structures of dihalides*

	Ti	V	Cr	Mn	Fe	Co	Ni	Cu
F	R	R	R	R†	R	R	R	R†
Cl	L	L	R	L†	L	L	L	L†
Br	L	L	L	L†	L	L	L	L†
I	L	L	L	L†	L	L	L	

* R = rutile, L = layered (CdI_2, $CdCl_2$ or related structures).
† The structure is distorted from the ideal type.
Source: Adapted from A. F. Wells, *Structural inorganic chemistry.* Oxford University Press (1984).

A similarly simple situation is found for trihalides of the 3*d* metals. Although their structures are highly varied, they may all be regarded as containing isolated metal ions.

Nominal dihalides and trihalides of 4*d* and 5*d* metals

The Period-4 and Period-5 elements on the left of the *d* block form a fascinating series of metal–metal bonded compounds. For example, ZrI_2 is in fact a solid containing an octahedral Zr_6 unit metal–metal bonded to adjacent clusters (Fig. 13.13). As is typical of complex chain, sheet, or cluster structures, the empirical formulas of these compounds often have metal–halogen ratios that are neither purely MX_2 nor MX_3. An octahedral cluster of metal–metal bonded atoms is common, and in the simple binary halides these clusters are often bridged by the halide ions. Many counterpart metal cluster anions are known that have halide ions in edge-bridging or face-bridging sites. These metal–metal bonded compounds are formed with the softer halide ions (Cl^-, Br^-, I^-) and not by F^- ions, for the latter undergo disproportionation so readily. Silver difluoride (which is a useful fluorinating agent) deserved special mention because it illustrates the ability of fluorine to bring out an oxidation state higher than the usual Ag(I).

One of the most fully studied examples of these compounds is rhenium trichloride, which consists of the Re_3Cl_9 cluster linked to adjoining Re_3Cl_9 clusters by weak halide bridges. The simple complex $[Re_3Cl_{12}]^{3-}$ (Fig. 13.14) is formed in the presence of Cl^- ions because the coordination site formerly occupied by a bridging Cl atom is now occupied by the added Cl^- ion. Neutral ligands can also occupy these coordination sites and result in clusters of general formula $Re_3Cl_9L_3$.

The dihalides and trihalides of elements on the far right of the *d* block, such as $PdCl_2$ and $RhCl_3$, contain isolated cations. The reduced tendency to form metal–metal bonds in this part of the periodic table may be a consequence of the smaller radial extension of the *d* orbitals.

Hexahalides

The +6 oxidation state is easily attained in Group 6 by molybdenum and tungsten (Section 8.12). The hexafluorides are known for both elements (and are highly volatile molecular compounds), but only the more easily oxidized element tungsten forms a hexachloride and a hexabromide. For the other *d*-block elements, only fluorine brings out the +6 state and even then most of the examples are confined to the elements in Periods 4 and 5. Rhenium heptafluoride, ReF_7, can be prepared from the elements under vigorous conditions (400°C and high fluorine pressure). It is a pale yellow volatile solid, with a low melting point (48°C), and resembles the only other known heptafluoride, the interhalogen IF_7.

The hexafluorides of the heavier *d*-block elements have been prepared from Group 6 through Group 10 (as in PtF_6). As expected from the periodic trends in oxidation numbers, WF_6 is not a significant oxidizing agent, but the oxidizing character of the hexafluorides increases to the right, and PtF_6 is so potent that it can oxidize O_2:

$$O_2(g) + PtF_6(s) \rightarrow [O_2]^+[PtF_6]^-(s)$$

Example 13.4: *Deciding on the probable structural type of a metal halide*

List the major structural classes of *d*-metal halides and decide the most likely class for (a) MnF_2, (b) WCl_2, (c) RuF_6, and (d) FeI_2.

Answer. The difluorides of the 3*d* metals (MnF_2) have the simple 'ionic' rutile structure and the heavier halides (FeI_2) are commonly layered. The chlorides, bromides, and iodides of the low oxidation state early 4*d* and 5*d* metals have metal–metal bonds (WCl_2 is in this class; in fact it contains a W_6 cluster). The hexahalides are molecular (RuF_6).

Exercise. Decide the probable general structural type of (a) ReF_6, (b) CoF_2, and (c) WCl_6.

13.10 Preparation of anhydrous halides

Since anhydrous halides are very common starting materials for inorganic syntheses, their preparation and the removal of water from impure commercial halides is of considerable practical importance in the laboratory and in industry. The principal synthetic methods are the direct reaction of a metal and a halogen and the reaction of halogen compounds with metals or metal oxides.[3]

Reaction with halogen

Direct reaction of a halogen and metal may be accomplished in a variety of ways. The halogen vapor may be passed over the metal contained in a tube furnace (like that in Fig. 11.5). This method works well with volatile halides. The direct reaction of the metal with F_2 at elevated temperatures is a common method for the preparation of fluorides. The synthesis of platinum hexafluoride is complicated by the fact that PtF_6 is not stable at high temperatures. One successful synthesis of PtF_6 makes use of the apparatus shown in Fig. 13.15. Alternatively, PtF_6 can be synthesized by heating platinum with fluorine under pressure.

Reaction with a halogen compound

It is sometimes advantageous to use a halogen compound for the synthesis of a metal halide. For example, the reaction of ZrO_2 with CCl_4 vapor in a heated tube is a good method for the preparation of $ZrCl_4$. A significant contribution to the Gibbs free energy of this reaction comes from the

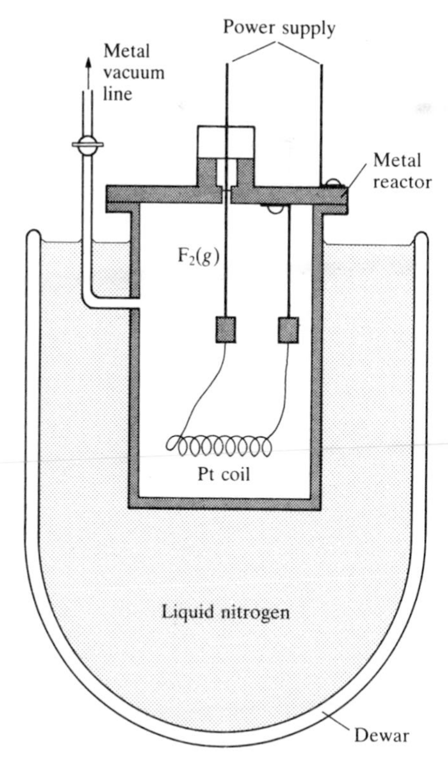

Fig. 13.15 The reactor used for the preparation of platinum hexafluoride. The fluoride is formed on the hot platinum filament, and survives because it immediately condenses on to the walls cooled by liquid nitrogen.

[3] For insight into the procedures and apparatus used in some metal halide syntheses see the articles in *Inorg. Synth.*, **4**, 104 (1953).

production of the strong C=O bond in phosgene:

$$ZrO_2(s) + 2CCl_4(g) \rightarrow ZrCl_4(s) + 2OCCl_2(g)$$

In Sections 13.3 and 13.5 we described the use of some very reactive fluorinating agents ClF_3 and FOOF for the formation of fluorides. Because of their high reactivity, these reagents produce metal fluorides under quite mild conditions.

Dehydration

Anhydrous halides can often be prepared by the removal of water from hydrated metal chlorides. It might be supposed that heating the sample in a dry gas stream would lead to anhydrous halides, but significant hydrolysis generally occurs:

$$2CrCl_3{\cdot}6H_2O(s) \rightarrow Cr_2O_3(s) + 6HCl(g) + 9H_2O(g)$$

The hydrolysis can be suppressed by carrying out the dehydration in a heated tube in a stream of hydrogen halide. Alternatively, a hydrated chloride may be refluxed with thionyl chloride (b.p. 79°C), which produces the desired anhydrous chloride and volatile HCl and SO_2 by-products:

$$FeCl_3{\cdot}6H_2O(s) + 6SOCl_2(l) \rightarrow FeCl_3(s) + 6SO_2(g) + 12HCl(g)$$

Example 13.5: *Choosing a method for the synthesis of a metal halide*

Suggest procedures and give balanced equations for the synthesis of (a) anhydrous $CoCl_2$ from cobalt(II) chloride hexahydrate, (b) $HfCl_4$ from HfO_2, (c) WF_6 from W and other reagents of your choice.

Answer. (a) We can suspect that the method used for $FeCl_3$ (with thionyl chloride) might work. This is in fact the case:

$$CoCl_2{\cdot}6H_2O(s) + 6SOCl_2(l) \rightarrow CoCl_2(s) + 6SO_2(g) + 12HCl(g)$$

In practice, we reflux the mixture and remove excess thionyl chloride while heating under vacuum. (b) Since Hf is very similar chemically to Zr, we can expect to be able to use a synthesis in which CCl_4 vapor is passed over HfO_2 in a heated tube (as for $ZrCl_4$):

$$HfO_2(s) + 2CCl_4(g) \rightarrow HfCl_4(s) + 2OCCl_2(g)$$

(c) The aggressive fluorinating agents F_2 or ClF_3 should be satisfactory:

$$W(s) + 2ClF_3(g) \rightarrow WF_6(s) + Cl_2(g)$$

Exercise. Suggest a conversion of $CoCl_2$ to CoF_3.

The noble gases

The elements in Group 18/VIII of the periodic table have been given and have lost various collective names over the years as different aspects of their properties have been identified and then disproved. Thus they have been called the rare gases and the inert gases, and are currently called the noble gases. The first is inappropriate because argon is by no means rare (it is more abundant than carbon dioxide in the atmosphere). The second has been inappropriate since the discovery of compounds of xenon. The appellation 'noble gases' is now accepted because it gives the sense of low but significant reactivity.

13.11 The elements

Because they are so unreactive and are rarely concentrated in nature, the noble gases eluded recognition until the end of the nineteenth century. Indeed, Mendeleev did not make a place for them in his periodic table because the chemical regularities of the other elements, upon which his table was based, did not suggest their existence. However, in 1868 a new line was observed in the spectrum of the sun that did not correspond to a known element. This line was eventually attributed to a new element, helium, and in due course helium and its congeners were found on earth.

All the noble gases occur in the atmosphere. The atmospheric abundances of argon (0.94 percent by volume) and neon (1.5×10^{-3} percent) make these two elements more plentiful than many familiar elements, such as arsenic and bismuth, in the earth's crust. Xenon and radon are the rarest elements of the group; the latter is a product of radioactive decay and, since it lies beyond bismuth, is itself unstable. The elements neon through xenon are extracted from liquid air by low-temperature distillation. Unlike the atoms of the other noble gases, the light He atom is not retained by the earth's gravitational field, so most of the helium on earth is the product of α-emission in the decay of radioactive elements and has been trapped by rock formations. High helium concentrations are found in certain natural gas deposits (in the USA and eastern Europe) from which it can be recovered by low-temperature distillation.

The low boiling points of the lighter noble gases follow from the weakness of the dispersion forces between the atoms and absence of other forces. Helium has the lowest boiling point (4.2 K) of any substance and is widely used in cryogenics as a very low-temperature refrigerant; it is the coolant for superconducting magnets used for NMR spectroscopy and NMR imaging. When ^4He is cooled below 2.178 K it undergoes a transformation into a second liquid phase known as **helium-II**. This phase is a superfluid, for it flows without viscosity.

On account of its low density and nonflammability, helium is used in balloons and lighter-than-air craft. The greatest use of argon is as an inert atmosphere for the production of air-sensitive compounds and as an inert gas blanket to suppress the oxidation of metals when they are welded. The noble gases are also extensively used in various light sources, including conventional sources (neon signs, fluorescent lamps, xenon flashes) and lasers (He-Ne, Ar-ion, and Kr-ion). In each case, an electric discharge through the gas ionizes some of the atoms and promotes both ions and neutral atoms into excited states which then emit light upon return to a lower state.

Radon, which is a product of nuclear power plants and a product of the radioactive decay of naturally occurring uranium and thorium, is a health hazard because of the ionizing nuclear radiation it produces. It is usually a minor contributor to the background radiation arising from cosmic rays and terrestrial sources. However, in regions where the soil, underlying rocks, or building materials contain significant concentrations of uranium, excessive amounts of the gas have been found in buildings.

13.12 Compounds

The reactivity of the noble gases has been investigated sporadically ever since their discovery, but all attempts to coerce them into reaction were unsuccessful. However, in March 1962, Neil Bartlett, then at the University of British Columbia, observed the reaction of a noble gas.[4] Bartlett's report, and another from Rudolf Hoppe's group in the University of Munster a few weeks later, set off a flurry of activity throughout the world. Within a year, a series of xenon fluorides and oxo compounds had been synthesized and characterized.[5]

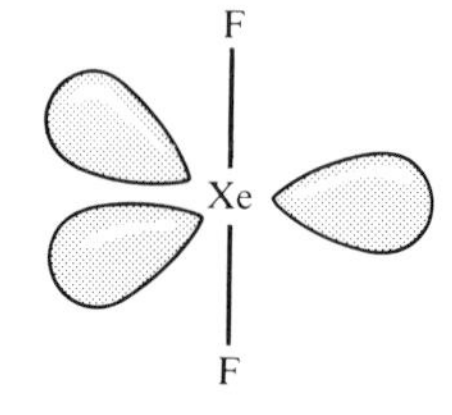

10 XeF_2

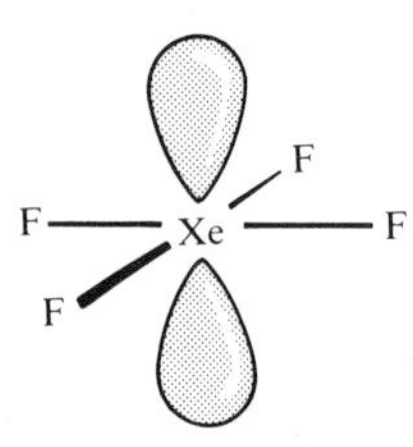

11 XeF_4

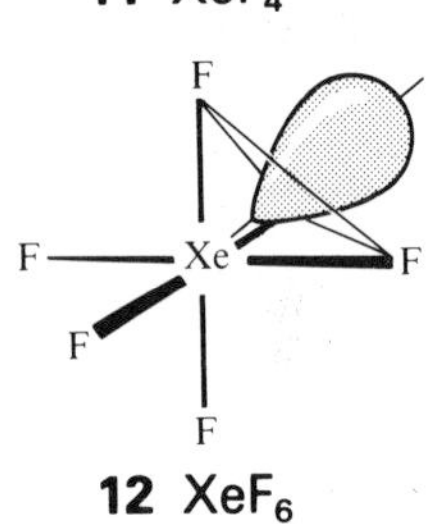

12 XeF_6

Synthesis and structure of xenon fluorides

Bartlett's motivation for studying xenon was based on the observation that PtF_6 can oxidize O_2. It was suspected that the same might be possible with xenon, which has a similar ionization energy. He observed an immediate reaction upon mixing xenon and PtF_6, but the reaction is complex and the formulation of the product (or products) is still in doubt. The direct reaction of xenon and fluorine leads to a series of compounds with oxidation numbers +2 (XeF_2), +4 (XeF_4), and +6 (XeF_6).

The molecules of these compounds are respectively linear (**10**), square-planar (**11**), and distorted octahedral (**12**). The first two structures are consistent with VSEPR theory for a Xe atom with five and six electron pairs respectively and they bear a molecular and electronic structural resemblance to the isoelectronic polyhalide anions I_3^- and ClF_4^- (Section 13.4). The structures of XeF_2 and XeF_4 are well established over a wide timescale by diffraction and spectroscopic methods. Similar measurements on XeF_6 in the gas phase, however, lead to the conclusion that this molecule is fluxional (Section 2.2). Infrared spectra and electron diffraction on XeF_6 show that the instantaneous structure of XeF_6 is not octahedral. A distortion occurs about a threefold axis such that one triangular face of F atoms is opened up to accommodate a lone pair of electrons (**12**). It is thought that the fluxional process is the migration of the distortion from one triangular face to another.

The xenon fluorides are synthesized by direct reaction of the elements, usually in a nickel reaction vessel that has been passivated by exposure to F_2 to form a thin protective NiF_2 coating. This treatment also removes surface oxide, which would react with the xenon fluorides. The synthetic conditions indicated in the following equations show that formation of the higher halides is favored by a higher proportion of fluorine and higher total pressure:

$$Xe(g) + F_2(g) \xrightarrow{400°C,\ 1\ atm} XeF_2(g) \qquad (\text{Xe in excess})$$

$$Xe(g) + 2F_2(g) \xrightarrow{600°C,\ 6\ atm} XeF_4(g) \qquad (Xe:F_2 = 1:5)$$

$$Xe(g) + 3F_2(g) \xrightarrow{300°C,\ 60\ atm} XeF_6(g) \qquad (Xe:F_2 = 1:20)$$

[4] For an enjoyable account of the early failure and final success in the quest for compounds of the noble gases, see P. Lazlo and G. J. Schrobilgen, *Angew. Chem. Int. Ed. Engl.*, **27,** 479 (1988).

[5] John Holloway provides a nice summary of the development of this field in 'Twenty-five years of noble gas chemistry' in *Chem. Brit.*, 658 (1987).

A simple 'windowsill' synthesis also is possible. Xenon and fluorine are sealed in a glass bulb (previously rigorously dried to prevent the formation of HF and the attendant etching of the glass) and the bulb is exposed to sunlight, whereupon beautiful crystals of XeF_2 slowly form in the bulb. It will be recalled that fluorine undergoes photochemical dissociation (Section 13.1), and in this synthesis the photochemically generated F atoms react with Xe atoms.

Reactions of xenon fluorides

The reactions of the xenon fluorides are similar to those of the high-oxidation-number interhalogens, and redox and metathesis reactions dominate. One important reaction of XeF_6 in metathesis with oxides:

$$XeF_6(s) + 3H_2O(l) \rightarrow XeO_3(aq) + 6HF(aq)$$

$$2XeF_6(s) + 3SiO_2(s) \rightarrow 2XeO_3(s) + 3SiF_4(g)$$

Xenon trioxide presents a serious hazard since this endoergic compound is highly explosive. In basic aqueous solution the Xe(VI) oxoanion $HXeO_4^-$ slowly decomposes in a coupled disproportionation and water oxidation to yield a Xe(VIII) perxenate ion, XeO_6^{4-} and xenon. Another important chemical property of the xenon halides is their strong oxidizing power, as illustrated by the following examples:

$$2XeF_2(s) + 2H_2O(l) \rightarrow 2Xe(g) + 4HF(g) + O_2(g)$$

$$XeF_4(s) + Pt(s) \rightarrow Xe(g) + PtF_4(s)$$

As with the interhalogens, the xenon fluorides react with Lewis acids to form cationic xenon fluorides:

$$XeF_2(s) + SbF_5(l) \rightarrow [XeF]^+[SbF_6]^-(s)$$

These cations are associated with the counter-ion by F^- ion bridges. Compounds with Xe—N (**13**) and Xe—C bonds, as in $Xe(CF_3)_2$, have also been synthesized. The latter compound is unstable, with a half-life of 30 min at room temperature.[6] Evidence also exists for the formation of Xe_2^+.

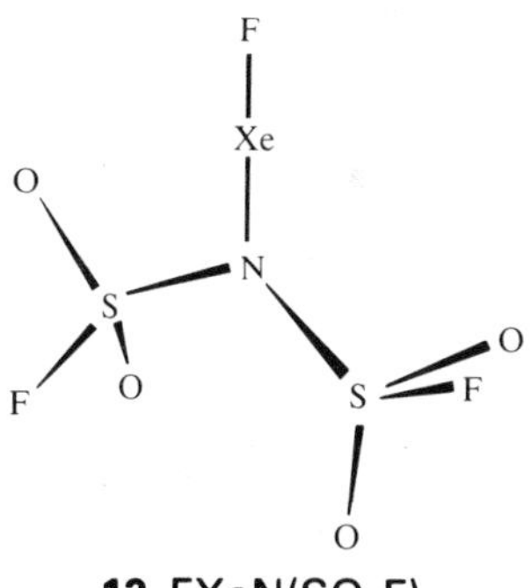

13 $FXeN(SO_2F)_2$

Compounds of other noble gases

Radon has a lower ionization energy than xenon, so we can expect it to form compounds even more readily. Evidence exists for the formation of RnF_2, as well as some other compounds, but their characterization is frustrated by their radioactivity. Krypton has a much higher ionization energy than xenon (Table 13.1) and its ability to form compounds is much more limited. Krypton difluoride is prepared by passing an electric discharge or ionizing radiation through a mixture of fluorine and krypton at low temperatures (−196°C). As with XeF_2, the krypton compound is a colorless volatile solid and the molecule is linear. It is an endoergic and highly reactive compound that must be stored at low temperatures.

[6] Note that in each of these cases, strongly electron withdrawing groups are attached to the C or N, but this is not the case for $HCNXeF^+$. A. A. A. Emara, and G. J. Schrobilgen, *J. chem. Soc. chem. Commun.*, 1644 (1987).

14 XeO_6^{4-}

Example 13.6: *Devising a synthesis and inferring structure of a noble gas compound.*

(a) Describe a procedure for the synthesis of potassium perxenate, starting with xenon and other reagents of your choice. (b) Using VSEPR theory, determine the probable structure of the perxenate ion.

Answer. (a) Xe—O compounds are endoergic, so they cannot be prepared by direct reaction of xenon and O_2. The hydrolysis of XeF_6 yields XeO_3 and as described in the text, the latter in basic solution undergoes redox disproportionation yielding perxenate, $[XeO_6]^{4-}$. Thus the reaction of xenon and F_2 would be carried out at 300°C and 60 atm in a stout nickel container. The resulting XeF_6 might then be converted to the perxenate in one step by exposing it to aqueous KOH solution. The resulting potassium perxenate (which turns out to be a hydrate) could then be crystallized. (b) A Lewis structure of the perxenate ion is shown in (**14**). With six electron pairs around the Xe atom, VSEPR theory predicts an octahedral arrangement of bonding electron pairs and an octahedral overall structure.

Exercise. Write a balanced equation for the decomposition of xenate ions in basic solution for the production of perxenate ions, xenon, and oxygen.

Further reading

A. J. Bard, R. Parsons and J. Jordan, *Standard potentials in aqueous solution.* Dekker, New York (1985). See the chapters on the halogens and the 'inert' gases.

V. Gutmann (ed.), *Halogen chemistry.* Academic Press, New York (1967): Vol. 1, Halogens, interhalogens, polyhalides and noble gas halides; Vol. 2 Nonmetal halides; Vol. 3 Complexes.

Halides of the transition elements: D. Brown, Vol. 1 *Halides of the lanthanides and actinides*; J. H. Canterford and R. Cotton; Vol. 2, *Halides of the first row transition metals*; Vol. 3, *Halides of the second and third transition metals,* Wiley, New York (1968).

H. Selig and J. H. Holloway, Cationic and anionic complexes of the noble gases. In *Top. curr. Chem.*, **124,** 33 (1984).

R. C. Thompson, Reaction mechanisms of the halogens and oxohalogen species in acidic, aqueous solution. In *Advances in inorganic and bioinorganic mechanisms,* Vol. 4 (ed. A. G. Sykes). Academic Press, New York (1986).

J. D. Corbett, Extended metal–metal bonding in halides of the early transition metals. In *Acc. chem. Res.* **14,** 239 (1981).

K. Seppelt and D. Lentz, Novel developments in noble gas chemistry. In *Prog. inorg. Chem.*, **29,** 167 (1982).

P. Hagenmüller (ed.), *Inorganic solid fluorides: physics and chemistry.* Academic Press, New York (1985).

Kirk–Othmer Encyclopedia of chemical technology, Wiley, New York (1980). This series of volumes contains many reviews of the individual halogens and their compounds, as well as of the noble gases.

Exercises

13.1 Preferably without consulting reference material, write out the halogens and noble gases as they appear in the periodic table, and indicate the trends in (a) physical state (*s*, *l*, or *g*) at room temperature and pressure, (b) electronegativity, (c) hardness of the halide ion, and (d) color.

13.2 Describe how the halogens are recovered from their naturally occurring halides. Give balanced chemical equations and conditions where appropriate.

13.3 Sketch the choralkali cell. Show the half-cell reactions and indicate the direction of diffusion of the ions. Give the chemical equation for the unwanted reaction that would occur when OH^- migrates through the membrane and into the anode compartment.

13.4 Sketch the form of the vacant σ^* of a halogen molecule and describe its role in the Lewis acidity of the dihalogens.

13.5 The reduction potential of an oxoanion generally increases with increasing acidity. Demonstrate this phenomenon by calculating the reduction potential of ClO_4^- at pH = 7 and comparing it with the tabulated value at pH = 0.

13.6 With regard to the general influence of pH on the reduction potentials of oxoanions, explain why the redox disproportionation of an oxoanion is often promoted by low pH.

13.7 Based on the analogy between halogens and pseudohalogens: (a) Write the balanced equation for the probable reaction of cyanogen $(CN)_2$ with aqueous sodium hydroxide. (b) Write the equation for the probable reaction of excess thiocyanate with the oxidizing agent $MnO_2(s)$ in acid aqueous solution. (c) Write a reasonable structure for trimethylsilyl cyanide.

13.8 (a) Using VSEPR theory, predict the probable structures of $[IF_6]^+$ and IF_7. (b) Give a plausible chemical equation for the preparation of $[IF_6][SbF_6]$.

13.9 The formation of Br_3^- from a tetraalkylammonium bromide and Br_2 is only slightly exoergic. Write an equation (or NR for no reaction) for the interaction of $[NR_4][Br_3]$ with I_2 in CH_2Cl_2 solution, and give the reasoning behind your answer.

13.10 Explain why $CsI_3(s)$ is stable but $NaI_3(s)$ is not.

13.11 Write plausible Lewis structures for (a) ClO_2 and (b) I_2O_5 and predict their shapes.

13.12 Give the formulas the probable relative acidities of perbromic acid and periodic acid. Which is the more stable?

13.13 Which oxidizing agent reacts more readily in dilute aqueous solution, perchloric acid or periodic acid? Give a mechanistic explanation of the difference.

13.14 For which of the following anions is disproportionation thermodynamically favorable in acid solution: OCl^-, ClO_2^-, ClO_3^- and ClO_4^-? (If you do not know the properties of these ions you should be able to determine them from standard potentials, Appendix 4.) For which of the favorable cases is the reaction very slow at room temperature?

13.15 Which of the following compounds present an explosion hazard? (a) NH_4ClO_4, (b) $Mg(ClO_4)_2$, (c) $NaClO_4$, (d) $[Fe(H_2O)_6][ClO_4]_2$. Explain your reasoning.

13.16 Identify which of the compounds (a) FeF_2, (b) $NiCl_2$, (c) $ScCl_2$, (d) WCl_2 are likely to have simple ionic structures (e.g., rutile structures), layered structures (such as CdI_2 or $CdCl_2$ structures), and metal–metal bonded structures.

13.17 Give balanced chemical equations and conditions for the synthesis of (a) iron(II) chloride tetrahydrate from iron metal and other reagents of your choice, (b) anhydrous iron(II) chloride tetrahydrate from reagents of your choice, (c) platinum hexafluoride from the reagents of your choice, and (d) zirconium monobromide.

13.18 Explain why helium is present in low concentration in the atmosphere even though it is the second most abundant element in the universe.

13.19 Which of the noble gases would you choose as (a) the lowest-temperature liquid refrigerant, (b) an electric discharge light source requiring a safe gas with the lowest ionization potential, (c) the least expensive inert atmosphere.

13.20 By means of balanced chemical equations and a statement of conditions describe a suitable synthesis of (a) xenon difluoride, (b) xenon hexafluoride, (c) xenon trioxide.

13.21 Give the formula and describe the structure of a noble gas species that is isostructural with (a) ICl_4^-, (b) IBr_2^-, (c) BrO_3^-, (d) ClF.

13.22 With what neutral molecules is ClO^- isoelectronic? Are these molecules Lewis acids? Justify the suggestion that nucleophilic attack on ClO^- is the initial step in a disproportionation reaction.

Problems

13.1 Many of the acids and salts corresponding to the positive oxidation numbers of the halogens are not listed in the catalog of a major international chemical supplier: (a) $KClO_4$ and KIO_4 are available but $KBrO_4$ is not; (b) $KClO_3$, $KBrO_3$, and KIO_3 are all available; (c) $NaClO_2$ and $NaBrO_2{\cdot}H_2O$ are available but salts of $[IO_2]^-$ are not; (d) Only ClO^- salts are available but the bromine and iodine analogs are not. Describe the probable reason for the missing salts of oxoanions.

13.2 The reaction of I^- ions is often used to titrate ClO^-, giving deeply colored I_3^- ions, along with Cl^- and H_2O. Although never proved, it was thought that the initial reaction proceed by oxygen transfer from Cl to I, but a recent kinetic study indicates that chlorine transfer is likely to give ICl as the intermediate (K. Kumar, R. A. Day, and D. W. Margerum, *Inorg. Chem.*, **25**, 4344 (1986)). Summarize the evidence for Cl atom transfer.

13.3 Given the bond lengths and angles in I_5^+ (**5**), describe the bonding in terms of 2-center and 3-center σ bonds and account for the structure using VSEPR theory.

13.4 Until the work of K. O. Christe (*Inorg. Chem.*, **25**, 3721 (1986)) F_2 could be prepared only electrochemically. Give chemical equations for Christe's preparation and summarize the reasoning behind it.

13.5 The first compound containing an Xe—N bond was reported by R. D. LeBlond and K. K. DesMarteau (*J. chem. Soc. chem. Commun.*, 554 (1974)). Summarize the method of synthesis and characterization. (The proposed structure was later confirmed in an X-ray crystal structure determination.)

Part 4

d- and *f*-block complexes

We now provide the more thorough treatment of *d*-block complexes that was promised in Chapter 7. Chapter 14 begins with a treatment of many-electron atoms and ions in the gas phase, where it is possible to assess the role of electron–electron repulsions spectroscopically. We then show how to incorporate the effects of these repulsions into the description of complexes by constructing correlation diagrams that relate the energy levels in the two cases, and how to use these diagrams to interpret their spectra. We then extend the theory of bonding to less symmetrical complexes and in particular to the molecular fragments that are used to describe metal–metal bonding. In Chapter 15 we use these concepts to interpret the activation energies and other aspects of the reaction mechanisms of complexes. The primary aim of the latter chapter is to develop an understanding of how experimental evidence is used to infer a mechanism, and how to judge its reliability.

Chapter 16 extends the concepts of coordination chemistry to the description of complexes with metal–carbon bonds. These organometallic compounds have been the subject of intense investigation; partly because they display distinctive characteristics, such as the prevalence of low oxidation states and carbon-centered ligand reactivity. The utility of these topics will become very clear in the treatment of catalysis in Part 5.

Bonding and spectra of complexes

14

The overall aim of this chapter is to see how to analyze the electronic spectra of complexes (particularly those of the *d* block) and hence to enrich our understanding of their bonding. Many of the topics we treat are elaborations of the material we introduced in Chapter 7 on the molecular orbital theory of complexes. The principal new work in the first part of the chapter is the problem of interelectronic repulsions, which we largely ignored in Chapter 7. To see how to take repulsions into account, we must step back from the concepts introduced in that earlier chapter and discuss free atoms, for atomic spectra provide the parameters that we need to take interelectronic repulsion into account quantitatively. We can then go on to see how repulsions compete with the ligand field when the atoms and ions are present as a part of a complex. This chapter also extends the material in Chapter 7 by describing how to apply molecular orbital techniques to systems of lower symmetry than we treated there. In particular, we shall see how to build up a description of the molecular orbitals of complexes from the orbitals of simpler fragments. This procedure will be very useful when we discuss metal cluster compounds and in the treatment of metal–metal bonding.

The electronic spectra of atoms

14.1 Spectroscopic terms

14.2 Terms of a d^2 configuration

The electronic spectra of complexes

14.3 Ligand-field transitions

14.4 Charge transfer bands

14.5 Selection rules and intensities

14.6 Luminescence

14.7 Spectra of *f*-block complexes

14.8 Circular dichroism

14.9 Electron paramagnetic resonance

Lower symmetry and binuclear complexes

14.10 The ML_5 fragment

14.11 Metal–metal bonding and mixed valence complexes

Further information: the addition of angular momenta

Further reading

Exercises

Problems

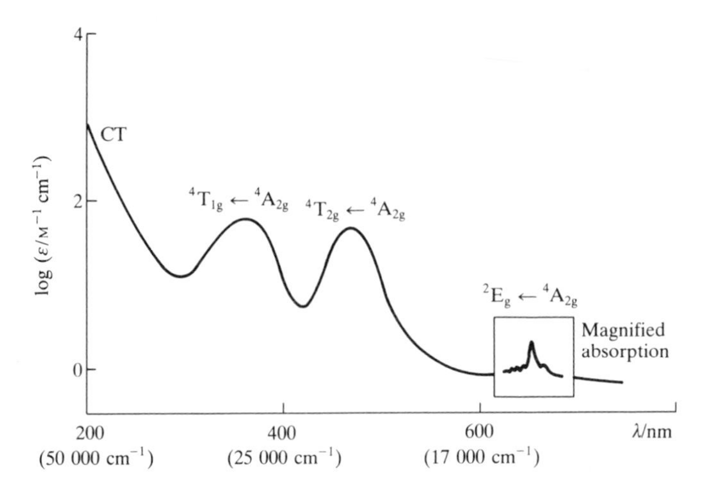

Fig. 14.1 The spectrum of the d^3 complex $[Cr(NH_3)_6]^{3+}$, which illustrates the features studied in this chapter, and the assignments of the transitions as explained in the text.

We shall set the scene for this chapter by examining Fig. 14.1, which shows the electronic absorption spectrum of the d^3 complex $[Cr(NH_3)_6]^{3+}$ in aqueous solution in the range of 50 000 cm^{-1} (200 nm, the ultraviolet) to 10 000 cm^{-1} (1000 nm, the infrared). The band at lowest energy is very weak, and later we shall see that it is an example of a 'spin-forbidden' transition. Next are two bands with intermediate intensities, which will turn out to be transitions between the t_{2g} (the HOMO) and e_g (the LUMO) orbitals of the complex and which are mainly derived from the metal d orbitals. We shall have to explore why there are *two* transitions at different energies even though both correspond to an $e_g \leftarrow t_{2g}$ electron promotion. The splitting is in fact an outcome of the electron–electron repulsions that we have already said are to be one focus of this chapter. The most intense band at high energy (only the tail of it is seen in the illustration) has an entirely different origin from the other three. We shall see that it is an example of a transition in which charge is transferred from the ligands to the central metal atom.

The electronic spectra of atoms

We have been content so far to describe atoms and molecules by giving their electronic **configurations**, the statement of the number of electrons in each orbital (as in $1s^22s^1$ for Li). However, a configuration is an incomplete description of the arrangement of electrons in atoms. In the configuration $2p^2$, for instance, the two electrons might occupy orbitals with different orientations of their orbital angular momenta (that is, with different values of m_l from among the possibilities +1, 0, and −1). Similarly, the designation $2p^2$ tells us nothing about the spin orientations of the two electrons ($m_s = +\frac{1}{2}$ or $-\frac{1}{2}$). The atom may in fact have several different states of total orbital angular momentum, and each one corresponds to a different distribution of electrons. The different ways in which the electrons can occupy the orbitals specified by the configuration are called the **microstates** of the configuration.

14.1 Spectroscopic terms

All the microstates of a given configuration have the same energy only if we ignore interelectronic repulsions. However, interelectron repulsions are in fact strong and cannot always be ignored. One of their effects is to lead to particular distributions of electrons (such as the adoption of specific relative orientations of their orbital angular momenta). They also result in these different distributions having different energies.

Permissible microstates and terms

Each permissible microstate of a configuration is specified by giving the values of m_l and m_s of the electrons of the configuration. For example, if we write the values of m_l for each electron and denote $m_s = +\frac{1}{2}$ and $-\frac{1}{2}$ by the superscripts + and −, the 15 possible microstates of a p^2 configuration are

$$(1^+,1^-) \quad (1^+,0^+) \quad (1^+,0^-) \quad (1^-,0^+) \quad (1^-,0^-) \quad (0^+,0^-)$$

$$(1^+,-1^+) \quad (1^+,-1^-) \quad (1^-,-1^+) \quad (1^-,-1^-)$$

$$(0^+,-1^+) \quad (0^+,-1^-) \quad (0^-,-1^+) \quad (0^-,-1^-) \quad (-1^+,-1^-)$$

The microstates $(1^+,1^-)$ and $(1^-,1^+)$ are identical and are not listed separately because we cannot distinguish individual electrons. Microstates such as $(1^+,1^+)$ are excluded by the Pauli principle for they would correspond to two electrons with the same spin in the same orbital (the orbital with $m_l = +1$).

Interelectronic repulsions result in some microstates having the same energy and some having different energies. For instance, the microstates $(0^+,0^-)$ and $(1^+,1^-)$ both correspond to two electrons in one orbital, with $m_l = 0$ in one case and $m_l = +1$ in the other, and the two arrangements have the same energy. However, they both differ from $(1^-,0^+)$ in which the two electrons occupy different orbitals. If we group together the microstates that have the same energy when electron repulsions are taken into account, we obtain the spectroscopically detectable energy levels called **terms**.

Our first task is to classify all the microstates of a given configuration into terms. The next is to identify transitions between these terms in the spectrum of the atom. The differences in energies of the transitions will then tell us the interelectron repulsion energies within the atom, which is the information we need if we are to make sense of spectra like that in Fig. 14.1.

Russell–Saunders coupling

The most important characteristic of a microstate when assessing the energy of a term is the relative orientation of the spins of the electrons. Next in importance is the relative orientation of the orbital angular momenta of the electrons. Thus we can identify the terms of lighter atoms by noting first their total spin and then their total orbital angular momentum. The process of combining electron angular momenta by summing first the spins, then the orbital momenta, and finally the two resultants, is called **Russell–Saunders coupling**. An alternative procedure, called ***jj*-coupling**, is more appropriate when the spin–orbit coupling of electrons is strong (in heavy atoms).

The total spin $\boldsymbol{S}$ of an atom is the vector sum of the spin angular momenta of the individual electrons:

$$\boldsymbol{S} = \boldsymbol{s}_1 + \boldsymbol{s}_2 + \ldots$$

The magnitude of this total spin is given by the total spin quantum number S:

$$\text{magnitude of } \boldsymbol{S} = \{S(S+1)\}^{1/2}\hbar$$

$S = 1$ if the two spins are parallel (↑↑), and $S = 0$ if they are antiparallel (↑↓). How S is ascribed in atoms with more than two electrons is explained in the *Further information* section at the end of this chapter. Similarly, the total orbital angular momentum $\boldsymbol{L}$ is the vector sum of the individual orbital momenta of the electrons:

$$\boldsymbol{L} = \boldsymbol{l}_1 + \boldsymbol{l}_2 + \ldots$$

Its magnitude is described by the quantum number L:

$$\text{magnitude of } \boldsymbol{L} = \{L(L+1)\}^{1/2}\hbar$$

For two p electrons with $l = 1$, L may be 2 (if the two orbital angular momenta are in the same directions), 0 (if the orbital momenta are in opposite directions), and 1 (if the relative orientation is intermediate).[1] The *Further information* section explains how these values of L are obtained and how to deal with more complicated cases.

By analogy with the notation $s, p, d, \ldots$ for orbitals with $l = 0, 1, 2, \ldots$, the total orbital angular momenta of atomic terms are denoted by the upper-case equivalents:

$L = 0$	1	2	3	4 . . .
S	P	D	F	G then alphabetical (omitting J)

The total spin (S) of an atom is normally reported as the **multiplicity** of the term, the value of $2S + 1$:

$S = 0$	$\frac{1}{2}$	1	$\frac{3}{2}$	2 . . .
$2S + 1 = 1$	2	3	4	5 . . .

The multiplicity is written as a left superscript on the letter representing the value of L, and the entire label of a term is called a **term symbol**. Thus, the term symbol ^{3}P denotes a term (a collection of degenerate states) with $L = 1$ and $S = 1$, and is called a 'triplet' term (more specifically, a 'spin triplet').

Example 14.1: *Deriving term symbols*

Give the term symbols for an atom with the configurations (a) s^1, (b) p^1, and (c) s^1p^1.

Answer. (a) The single s electron has $l = 0$ and $s = \frac{1}{2}$. Since there is only one electron, $L = 0$ (an S term), $S = \frac{1}{2}$, and $2S + 1 = 2$ (a doublet term). The term symbol is therefore ^{2}S. (b) For a single p electron, $l = 1$ so $L = 1$ and the term is ^{2}P. (These terms arise in the spectrum of an alkali metal atom, such as Na.) With one s and one p electron, $L = 0 + 1 = 1$, a P term. The electrons may be paired ($S = 0$) or parallel ($S = 1$). Hence both ^{1}P and ^{3}P terms are possible.

Exercise. What terms arise from an s^1d^1 configuration?

[1] In classical terms we could think of the intermediate case as one electron orbiting round the equator and the other as orbiting over the poles.

14.2 Terms of a d^2 configuration

We have remarked that not all electron assignments are allowed by the Pauli principle. Thus, two electrons cannot both have the same spin and be in an orbital with $m_l = +1$. In other words, an atom in a p^2 configuration cannot simultaneously be a triplet and have $L = 2$. That is, a p^2 configuration does not give rise to a ^{3}D term. On the other hand, the configuration may have a ^{1}D term, for now the two electrons in the same orbital have paired spins, which is permitted. Our first problem is therefore to decide which terms are allowed by the Pauli principle. We shall illustrate the ideas involved by considering a d^2 configuration for then the outcome will be useful in the discussion of complexes later in the chapter. An example of a species with a d^2 configuration is the ground state of a Ti^{2+} ion.

We start the analysis by setting up a table of microstates of the d^2 configuration (Table 14.1). The strategy we then adopt is to identify the values of L and S to which the microstates belong.

Table 14.1 Microstates of the d^2 configuration

M_L	M_S −1	0	+1
4		$(2^+, 2^-)$	
3	$(2^-, 1^-)$	$(2^+, 1^-)(2^-, 1^+)$	$(2^+, 1^+)$
2	$(2^-, 0^-)$	$(2^+, 0^-)(2^-, 0^+)$ $(1^+, 1^-)$	$(2^+, 0^+)$
1	$(2^-, -1^-)(1^-, 0^-)$	$(2^+, -1^-)(2^-, -1^+)$ $(1^+, 0^-)(1^-, 0^+)$	$(2^+, -1^+)(1^+, 0^+)$
0	$(1^-, -1^-)(2^-, -2^-)$	$(1^+, -1^-)(1^-, -1^+)$ $(2^+, -2^-)(2^-, -2^+)$ $(0^+, 0^-)$	$(1^+, -1^+)(2^+, -2^+)$
−1 to −4*			

* The lower half of the diagram is a reflection of the upper half.

The classification of microstates

We need two facts in order to be able to ascribe each microstate to a term. The first is based on the fact that each d electron may have $m_l = +2, +1, 0, -1$, or -2 and $m_s = +\frac{1}{2}$ or $-\frac{1}{2}$. It follows that if one electron has the quantum number m_{l1} and the other has m_{l2}, the total orbital angular momentum will have a z component given by the quantum number

$$M_L = m_{l1} + m_{l2}$$

The same applies to the total spin:

$$M_S = m_{s1} + m_{s2}$$

Thus $(0^+, -1^-)$ is a microstate with $M_L = -1$ and $M_S = 0$ and may contribute to any term for which these two quantum numbers apply.

The second fact we need is that a term with quantum number L has $2L + 1$ states distinguished by the value of M_L. Thus a G term (with $L = 4$) has the nine states with $M_L = +4, +3, \ldots -4$. Similarly, a term with quantum number S (and multiplicity $2S + 1$) has $2S + 1$ states distinguished by different values of M_S. Thus a triplet term, with $S = 1$, has three states with $M_S = +1$, 0, and -1 for each value of M_L. Thus a ^{3}G term has three spin states for each of its nine orbital states, and therefore consists of 27 states in all.

It follows that by picking out the values of M_L that can arise, we can infer the values of L to which the states belong. Similarly, by picking out the values of M_S that can go with a given value of L, we can infer the value of S that is associated with that value of L. The strategy is to identify the state with the largest value of M_L and M_S, and hence to identify the largest values of L and S that may stem from the configuration. We then know that the term accounts for $2L+1$ values of M_L and $2S+1$ values of M_S for each value of M_L, and so can assign $(2L+1)\times(2S+1)$ of the microstates to this term. When that is done, we identify the next largest values of L and S, and continue until all the microstates have been assigned. Keeping track of this process is made easy by crossing out one microstate from each appropriate location as the terms are identified.

First, we note the largest value of M_L, which is +4. This state must belong to a term with $L=4$ (a G term). Table 14.1 shows that the only value of M_S that occurs for this term is $M_S=0$, so the G term must be a singlet. Moreover, since there are nine values of M_L when $L=4$, one of the microstates (or one linear combination of them) in each of the locations in the column below $(2^+, 2^-)$ must belong to this term. We can therefore strike out one microstate for each value of M_L in the central column of Table 14.1,[2] which leaves 36 microstates to classify.

The next largest value is $M_L=+3$, which must stem from $L=3$ and hence belong to an F term. That row contains one microstate in each column (that is, each location contains one unassigned combination of $M_S=+1$, 0, and -1), which signifies a triplet term. Hence the microstates belong to a ^{3}F term. The same is true for one linear combination in each of the rows down to $M_L=-3$, which accounts for a further $3\times7=21$ microstates. If we strike out one state in each of the 21 locations, we are left with 15 to be assigned.

There is one unassigned linear combination in the row with $M_L=+2$ ($L=2$) and the column under $M_S=0$ ($S=0$), which must therefore belong to a ^{1}D term. This term has five values of M_L, which removes one microstate from each row (in the column headed $M_S=0$) down to $M_L=-2$, leaving 10 microstates unassigned. Since these unassigned microstates include one with $M_L=+1$ and $M_S=+1$, nine of them must belong to a ^{3}P term. There now remains only one linear combination of microstates in the central location of the table, with $M_L=0$ and $M_S=0$. This linear combination must be the one and only state of a ^{1}S term.

We can now conclude that the terms of a $3d^2$ configuration are ^{1}G, ^{3}F, ^{1}D, ^{3}P, and ^{1}S, which account for all 45 permitted states:

Term	Number of states
^{1}G	$9\times1=9$
^{3}F	$7\times3=21$
^{1}D	$5\times1=5$
^{3}P	$3\times3=9$
^{1}S	$1\times1=1$
	45

[2] Crossing out one microstate in a group of several microstates signifies that one of their linear combinations has been used. To follow this argument it is a good idea to copy out Table 14.1 and actually cross out microstates.

The energies of the terms

Once we know the values of S and L that can arise from a given configuration, we can identify the ground term using **Hund's rules**. We introduced the first of these rules in Section 1.5, where we expressed it by saying that the lowest energy configuration is achieved if the electrons are parallel. Since more parallel spins implies a higher value of S, we can now express the rule as:

(1) The term with the greatest multiplicity lies lowest in energy.

The rule implies that a triplet term of a configuration (if one is permitted) has a lower energy than a singlet term of the same configuration.

Hund also proposed a second rule, which enables us to predict the relative energies of the terms of a given multiplicity:

(2) For a given multiplicity, the greater the value of L of a term, the lower the energy.

When L is high, each electron can stay clear of its partner and hence have a lower coulombic interaction with it, for (in a classical picture) they are then orbiting in the same direction and can remain diametrically opposite across their orbit. If L is low, the electrons are orbiting in opposite directions and meet more often (and hence repel each other more strongly). The second rule implies that if a configuration can give rise to both a ^{3}F and a ^{3}P term, the ^{3}F term is lower in energy. It follows that the ground term of Ti^{2+} is ^{3}F. For the higher terms, Hund's rules provide some guidance about the energy order but they are not always reliable. Thus, for Ti^{2+} the rules predict the order

$$^3\mathrm{F} < {}^3\mathrm{P} < {}^1\mathrm{G} < {}^1\mathrm{D} < {}^1\mathrm{S}$$

but the observed order is

$$^3\mathrm{F} < {}^1\mathrm{D} < {}^3\mathrm{P} < {}^1\mathrm{G} < {}^1\mathrm{S}$$

The spin multiplicity rule is fairly reliable for predicting the ordering of terms, but the 'greatest L' rule is reliable only for predicting the ground term; there is generally little correlation of L with the order of the higher terms.

If we want to know only the identity of the ground term of an atom or ion, we can identify it immediately. First, we identify the microstate that has the highest value of M_S, which tells us the highest multiplicity of the configuration. Then we identify the highest permitted value of M_L for that multiplicity, for that tells us the highest value of L consistent with the highest multiplicity.

Example 14.2: *Identifying the ground term of a configuration*

What is the ground term of the configurations (a) $3d^5$ of Mn^{2+} and (b) $3d^3$ of Cr^{3+}?

Answer. (a) The d^5 configuration permits occupation of each d orbital singly, so the maximum multiplicity is $2 \times \frac{5}{2} + 1 = 6$, a sextet term. If each of the electrons is to have the same spin quantum number, all must have different m_l values in order to obey the Pauli principle. Thus the m_l values will be $+2$, $+1$, 0, -1, and -2. The sum of these is 0, so $L = 0$ and the term is ^{6}S. (b) For the configuration d^3, the maximum multiplicity corresponds to all three electrons having the same spin quantum number to give $S = 2 \times \frac{3}{2} + 1 = 4$, a quartet. The three m_l values must be different if all spins are the same, which allows a maximum value of $M_L = 2 + 1 + 0 = +3$, indicating $L = 3$, an F term. Hence, the ground term of d^3 is ^{3}F.

Exercise. Identify the ground terms of (a) $2p^2$ and (b) $3d^9$. (Hint: Since d^9 is one electron short of a closed shell with $L = 0$ and $S = 0$, treat it on the same footing as a d^1 configuration.)

Racah parameters

We have remarked several times that the different terms of a configuration have different energies on account of the coulombic repulsion between electrons. These interelectron repulsion energies are given by complicated integrals over the orbitals occupied by the electrons. Mercifully, though, all the integrals for a given configuration can be collected together in three specific combinations, and the repulsion energy of any term of a configuration can be expressed as sums of these three quantities. The three different sums of integrals are called the **Racah parameters** A, B, and C. We do not even need to know the theoretical values of the parameters or the theoretical expressions for them, because it is more reliable to use A, B, and C as empirical quantities obtained from spectroscopy.

Each term stemming from a given configuration has an energy that may be expressed as a linear combination of the three Racah parameters. For instance, for a d^2 configuration a detailed analysis shows that

$$E(^1\text{S}) = A + 14B + 7C$$
$$E(^1\text{G}) = A + 4B + 2C$$
$$E(^1\text{D}) = A - 3B + 2C$$
$$E(^3\text{P}) = A + 7B$$
$$E(^3\text{F}) = A - 8B$$

A, B, and C can be found by fitting these expressions to the observed energies of the terms. Note that A is common to all the terms; hence, if we are interested only in their relative energies, we do not need to know its value. Likewise, if we are interested only in the relative energies of the two triplet terms, we also do not need to know the value of C.

All three Racah parameters are positive (since they represent repulsions). Therefore, provided $C > 5B$, the energies of the terms of the d^2 configuration lie in the order

$$^3\text{F} < {}^3\text{P} < {}^1\text{D} < {}^1\text{G} < {}^1\text{S}$$

This order is nearly the same as we obtained using Hund's rules. However, if $C < 5B$, the advantage of having a high orbital angular momentum is greater than the advantage of having a high multiplicity, and the ^{3}P term lies

Table 14.2. Racah parameters for some *d*-block ions*

	1+	2+	3+	4+
Ti		720(3.7)		
V		765(3.9)	860(4.8)	
Cr		830(4.1)	1030(3.7)	1040(4.1)
Mn		960(3.5)	1130(3.2)	
Fe		1060(4.1)		
Co		1120(3.9)		
Ni		1080(4.5)		
Cu	1220(4.0)	1240(3.8)		

* The table gives the B parameter with the value of C/B in parentheses.

above ^1D (as is in fact the case for the Ti^{2+} ion). Some experimental values of B and C are given in Table 14.2. It should be noted that $C \approx 4B$, so the ions listed there are in the regime where Hund's rules are not valid.

The importance of Racah parameters is that they summarize the energies of all the terms that may arise from a single configuration. The parameters are the quantitative expression of the ideas in Hund's rules, and are more powerful than the rules alone since they also account for deviations from them.

The electronic spectra of complexes

We now return to the spectrum of $[Cr(NH_3)_6]^{3+}$ shown in Fig. 14.1. The two central bands with intermediate intensities are HOMO–LUMO transitions with energies that differ on account of the interelectronic repulsions (as we shall shortly explain). Since both the HOMO and LUMO of an octahedral complex are predominantly metal d orbital in character, with a separation characterized by the strength of the ligand field, these two transitions are called ***d–d* transitions** or **ligand-field transitions**.

The ultraviolet absorption spectrum is very sensitive to ligand substitutions and the polarity of the solvent. This sensitivity suggests that it is a **charge-transfer transition** in which an electron is moved from orbitals that are predominantly ligand in character to orbitals that are predominantly metal in character. The transition is classified as a **ligand-to-metal charge transfer** (LMCT) transition. In some complexes the charge migration occurs in the opposite direction, in which case it is classified as a **metal-to-ligand charge transfer** (MLCT) transition. An example of an MLCT transition is the one responsible for the red color of tris(bipyridine)iron(II), the complex used for the colorimetric analysis of Fe(II). In this case, an electron makes a transition from a d orbital of the central metal into a π^* orbital of the ligand.

14.3 Ligand-field transitions

According to the discussion in Chapter 7, we would expect the octahedral d^3 complex $[Cr(NH_3)_6]^{3+}$ to have the ground configuration t_{2g}^3. We can identify the absorption spectrum in the region near 25 000 cm^{-1} as arising from the excitation $t_{2g}^2e_g^1 \leftarrow t_{2g}^3$ since that wavenumber is typical of ligand field splittings in complexes. However, since there are three t_{2g} orbitals and two e_g orbitals, there are in fact six possible transitions. In the absence of interelectron repulsions, all six lie at the same energy. However, because there are interelectron repulsions, the transition energies depend on which orbitals are specifically involved in the transitions.

Before we embark on the Racah-like analysis of the splitting, it may be helpful to see qualitatively from the viewpoint of molecular orbital theory why there are two bands. The $d_{z^2} \leftarrow d_{xy}$ transition promotes an electron from the xy plane into the already electron-rich z direction (it is electron rich because both d_{yz} and d_{xz} are full). However, the $d_{z^2} \leftarrow d_{xz}$ transition merely relocates an electron that is already largely concentrated along the z axis. In the former case, but not in the latter, there is a distinct increase in electron repulsion. As a result, the two $e_g \leftarrow t_{2g}$ transitions lie at different energies. All the other transitions resemble one or other of these two cases, and three transitions fall into one group and the other three fall into the second group.

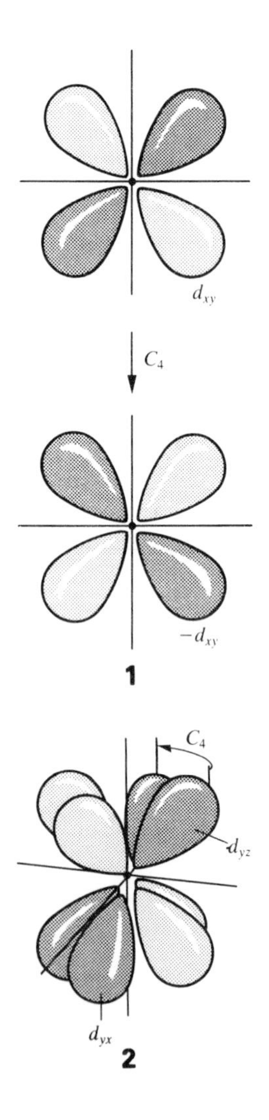

The spectral terms

The two transitions in Fig. 14.1 are labeled ${}^4T_{2g} \leftarrow {}^4A_{2g}$ (at 21 550 cm^{-1}) and ${}^4T_{1g} \leftarrow {}^4A_{2g}$ (at 28 500 cm^{-1}), where in each case the upper state precedes the lower. The labels are **molecular term symbols** and serve a similar purpose to the atomic term symbols we have already encountered. Thus, the left superscript denotes the multiplicity, so the superscript 4 denotes a quartet state with $S = \frac{3}{2}$, as expected when there are three unpaired electrons. However, because L is not a good quantum number in an octahedral environment (because orbital angular momentum is not well defined unless a system has spherical symmetry), the letter denotes the symmetry species of the overall orbital state and not its total orbital angular momentum.

The letters in the term symbols are the upper-case versions of the symmetry species of individual orbitals (just as in atoms upper-case letters are used to denote total angular momenta). Thus, the nearly totally symmetric ground state (with an electron in each of the three t_{2g} orbitals) is denoted A_{2g}. We say only *nearly* totally symmetric, because close inspection of the behavior of the three occupied t_{2g} orbitals shows that all the operations of the O_h point group transform the product $t_{2g} \times t_{2g} \times t_{2g}$ into ± 1 times itself, which identifies it as an A symmetry species (see the character table in Appendix 5). Likewise, since each orbital has even parity (g), the overall parity is also g. However, each C_4 rotation transforms one t_{2g} orbital into the negative of itself (**1**) and the other two t_{2g} orbitals into each other (**2**), so overall there is a change of sign under this operation and its character is -1. Thus the term is A_{2g} rather than the totally symmetrical A_{1g} of the closed shell.

It is more difficult to establish that the term symbols that can arise from the quartet $t_{2g}^2 e_g^1$ excited configuration are ${}^4T_{2g}$ and ${}^4T_{1g}$ and we shall not cover this aspect. The superscript 4 implies that the upper configuration continues to have the same number of unpaired spins as in the ground state, and the subscript g stems from the even parity of all the contributing orbitals. We have argued that the six excitations fall into two groups of three, which is consistent with the symmetry species T in each case.

The energies of the terms

In a free atom we needed to consider only the electron–electron repulsions in order to arrive at the relative ordering of the terms arising from a given

d^n configuration. In an octahedral complex the problem has another aspect because as well as the repulsions we need to take into account the effect of the difference in energy between the t_{2g} and e_g orbitals.

We can illustrate the considerations involved by treating first the simplest case of an atom or ion with a single valence electron. Thus, an s orbital in a free atom becomes a_{1g} in an octahedral field: a totally symmetrical orbital in one environment becomes a totally symmetrical orbital in another environment. As in Chapter 3 in connection with Walsh diagrams, we express the change by saying that the s orbital of the atom **correlates** with the a_{1g} orbital of the complex. Similarly, the totally symmetrical overall S term of a many-electron atom correlates with the totally symmetrical A_{1g} term of an octahedral environment. We have seen that the five d orbitals of a free atom split into the triply degenerate t_{2g} and doubly degenerate e_g sets in an octahedral complex. In the same way, an atomic D term correlates with a T_{2g} term and an E_g term in O_h symmetry. Table 14.3 summarizes the correlations between free ion terms and their terms in an octahedral complex.

Table 14.3. The correlation of spectral terms in O_h complexes

Atomic term	Number of states	Terms in O_h symmetry
S	1	A_{1g}
P	3	T_{1g}
D	5	$T_{2g} + E_g$
F	7	$T_{1g} + T_{2g} + A_{2g}$
G	9	$A_{1g} + E_g + T_{1g} + T_{2g}$

Example 14.3: *Identifying correlations between terms*

What terms in a complex with O_h symmetry correlate with the ^3P term of a bare ion with a d^2 configuration?

Answer. The three p orbitals of a free ion become the triply degenerate t_{1u} orbitals of an octahedral complex. Therefore, if we disregard parity for the moment, a P term of a many-electron atom becomes a T_1 term in the point group O_h. Since d orbitals have g parity, the term overall must be g. The multiplicity is unchanged in the correlation, so the ^3P term becomes the $^3T_{1g}$ term.

Exercise. What terms in a d^2 complex of O_h symmetry correlate with the ^3F and ^1D terms of a free ion?

Weak and strong field limits

Electron–electron repulsions are difficult to take into account, but the discussion is simplified by considering two extreme cases initially. In the **weak field limit** the ligand field is so weak that only electron repulsions are important and the relative energies of the terms are determined by the Racah parameters. The other extreme case is the **strong field limit** in which the ligand field is so strong that the repulsions can be ignored. Then, with the two extremes established, we can consider intermediate cases by drawing a correlation diagram between the two. We shall illustrate what is

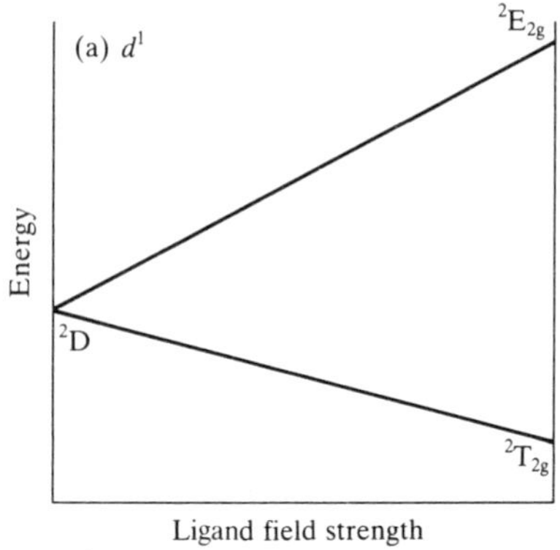

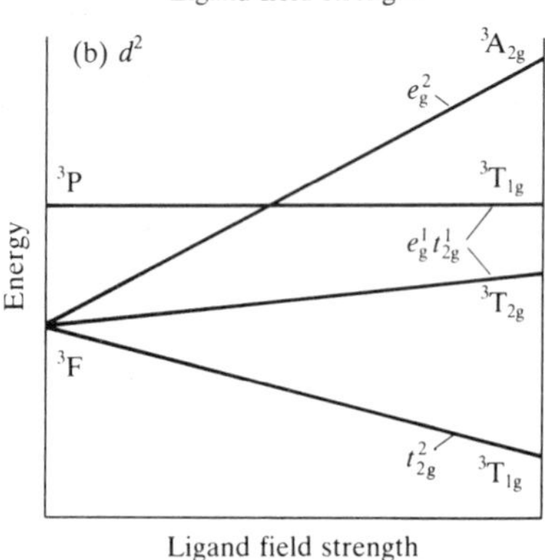

Fig. 14.2 Correlation diagram between a free ion and the strong-field terms of (a) a d^1 configuration and (b) a d^2 configuration.

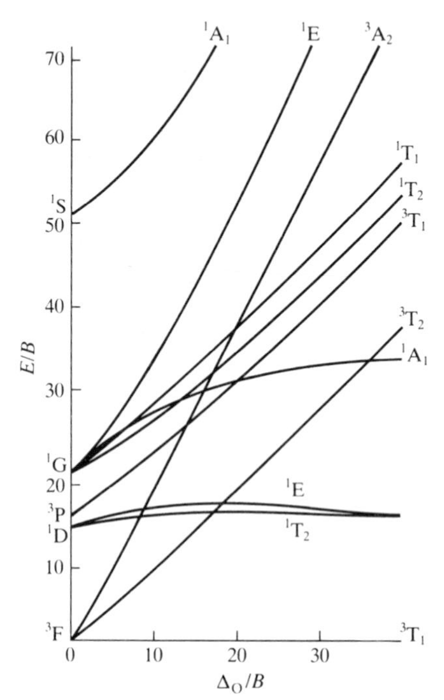

Fig. 14.3 The Tanabe–Sugano diagram for the d^2 configuration. A complete collection of diagrams for d^n configurations is given in Appendix 7 at the back of the book.

involved by considering two simple cases, namely d^1 and d^2. Then we show how the same ideas are used to treat more complicated cases.

The only term of a d^1 configuration in the free atom is ^2D. In an octahedral complex the configuration is either t_{2g}^1, which gives rise to a $^2T_{2g}$ term, or e_g^1, which gives rise to an ^2E_g term. Since there are no electron–electron repulsions to worry about, the separation of the $^2T_{2g}$ and ^2E_g terms is the same as the separation of the t_{2g} and e_g orbitals, which is Δ_O. The correlation diagram for the d^1 configuration will therefore resemble that shown in Fig. 14.2a.

For a d^2 configuration the triplet free atom terms are ^3F and ^3P, as we have seen, and their energies relative to the lower term (^3F) are

$$E(^3F) = 0, \qquad E(^3P) = 15B$$

These two energies are marked on the left of Fig. 14.2b. In the very strong field limit, a d^2 ion has the configurations

$$t_{2g}^2 < t_{2g}^1 e_g^1 < e_g^2$$

The three terms labeled $^3T_{1g}$, $^3T_{2g}$, and A_{2g} are separated in energy by an amount that depends on Δ_O. From the information in Table 7.3 we can write their relative energies as

$$E(t_{2g}^2, T_{1g}) = -0.8\Delta_O \qquad E(t_{2g}^1 e_g^1, T_{2g}) = +0.2\Delta_O \qquad E(e_g^2, A_{2g}) = +1.2\Delta_O$$

Therefore, relative to the energy of the lowest term (t_{2g}^2) their energies are

$$E(t_{2g}^2, T_{1g}) = 0 \qquad E(t_{2g}^1 e_g^1, T_{2g}) = 1.0\Delta_O \qquad E(e_g^2, A_{2g}) = 2.0\Delta_O$$

Since we are ignoring electron repulsions in this limit, the Racah parameters play no role. These configurations give rise to the terms shown on the right in Fig. 14.2b. The $^3T_{1g}$ term of $e_g^1 t_{2g}^1$ is unaffected by the ligand field.

Our problem now is to account for the energies of the intermediate cases, where neither the ligand field nor the electron repulsion terms are dominant. We do this by correlating each free ion term with its counterpart in the complex at the very high field limit. This correlation must be done using group theory, because we need to know how one term becomes another as the spherical symmetry of the free atom is replaced by the lower octahedral symmetry of the complex. However, since tracing the effects of this descent in symmetry is a standard problem in group theory, it has been worked out and tabulated once and for all, and all we need to know are the results (Fig. 14.2).

Tanabe–Sugano diagrams

Once the symmetry analysis has been done, diagrams like those shown in Fig. 14.2 can be constructed for any complex and quantitative calculations carried out for any strength of ligand field. The most widely used version of these diagrams are called **Tanabe–Sugano diagrams** after the Japanese workers who devised them. Tanabe–Sugano diagrams for O_h complexes with the configurations d^2 through d^8 are given in Appendix 7. Figure 14.2 is in fact a simplified version of a Tanabe–Sugano diagram for a d^2 configuration, and the quantitative diagram is shown in Fig. 14.3. The term energies E are expressed as E/B and plotted against Δ_O/B. Since $C \approx 4B$, terms with energies that depend on both B and C can be plotted on the same diagrams. The zero of energy in a Tanabe–Sugano diagram is always taken as that of the lowest term. Hence the diagrams are discontinuous (see that for d^4) when there is a change in the ground term, when the ligand field becomes strong enough to favor electron pairing.

Some lines in a Tanabe–Sugano diagram are curved on account of the mixing of terms of the same symmetry type. States of the same symmetry obey the **noncrossing rule**, which states that if the increasing ligand field causes two weak field terms of the same symmetry to approach, they do not cross (Fig. 14.4) but bend apart from each other.

We shall use the d^3 case that opened this chapter to illustrate how the relevant Tanabe–Sugano diagram (Fig. 14.5) may be used to interpret spectral data even in the absence of a detailed theoretical analysis. We have seen that the two ligand-field transitions occur at 21 550 cm^{-1} ($^4T_{2g} \leftarrow {}^4A_{2g}$) and 28 500 cm^{-1} ($^4T_{1g} \leftarrow {}^4A_{2g}$), a ratio of 1.32. The only point in Fig. 14.5 where this ratio is satisfied is on the far right. Hence, we can read off the values of Δ_O and B from the location of this point.

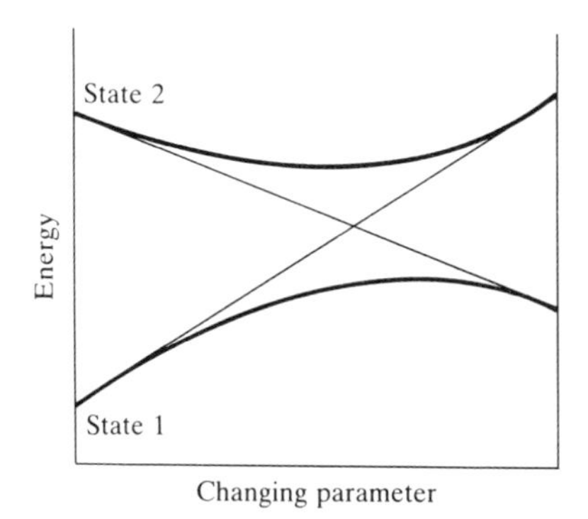

Fig. 14.4 The noncrossing rule states that if two states of the same symmetry are likely to cross as a parameter is changed (as shown by the thin lines), they will in fact mix together and avoid the crossing (as shown by the heavy lines).

Example 14.4: *Calculating Δ_O and B using a Tanabe–Sugano diagram*

Deduce the values of Δ_O and B for $[Cr(NH_3)_6]^{3+}$ from the spectral information in Fig. 14.1 and the Tanabe–Sugano diagram in Fig. 14.5 and Appendix 7.

Answer. The 1.32 transition energy ratio falls at $\Delta_O/B = 32.8$ at the point where the two arrows fit. The tip of the arrow representing the lower energy transition lies at $32.8\,B$ vertically, so equating $32.8\,B$ and 21 550 cm^{-1} gives $B = 657\ cm^{-1}$ and hence $\Delta_O = 21\ 550\ cm^{-1}$.

Exercise. Use the same Tanabe–Sugano diagram to predict the wavenumbers of the first two spin-allowed quartet bands in the spectrum of $[Cr(H_2O)_6]^{3+}$ for which $\Delta_O = 17\ 600\ cm^{-1}$ and $B = 918\ cm^{-1}$.

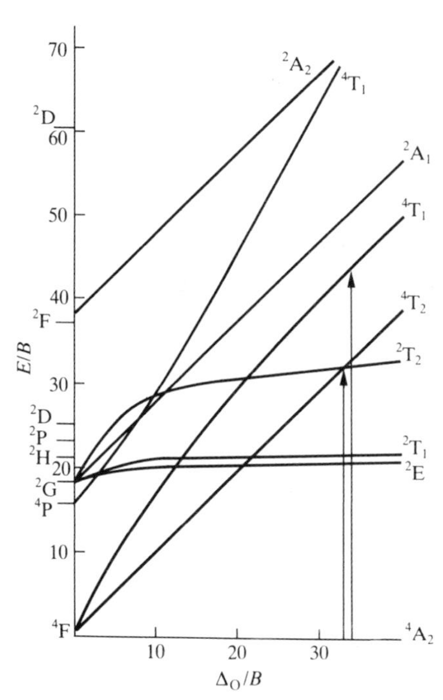

Fig. 14.5 The Tanabe–Sugano diagram for the d^3 configuration.

The nephelauxetic series

In Example 14.4 we found that $B = 657\ cm^{-1}$ for $[Cr(NH_3)_6]^{3+}$. This value is only 64 percent of the value for a free Cr^{3+} ion in the gas phase, and we conclude that electron repulsions are weaker in the complex than in the free ion. This weakening is to be expected, because the occupied molecular orbitals are delocalized over the ligands and away from the metal. The delocalization increases the average separation of the electrons and hence reduces their repulsion.

The reduction of B from its free ion value is normally reported in terms of the **nephelauxetic parameter** β:

$$\beta = \frac{B(\text{complex})}{B(\text{free ion})}$$

(The name is from the Greek for 'cloud expanding'.) The values of β depend on the ligand and vary along the **nephelauxetic series**:

$$F^- > H_2O > NH_3 > CN^-,\ Cl^- > Br^-$$

A small β indicates a large measure of d-electron delocalization on to the ligands and hence a significant covalent character in the complex. Thus the series shows that a Br^- ligand results in a greater reduction in electron repulsions in the ion than an F^- ion, which is consistent with a greater

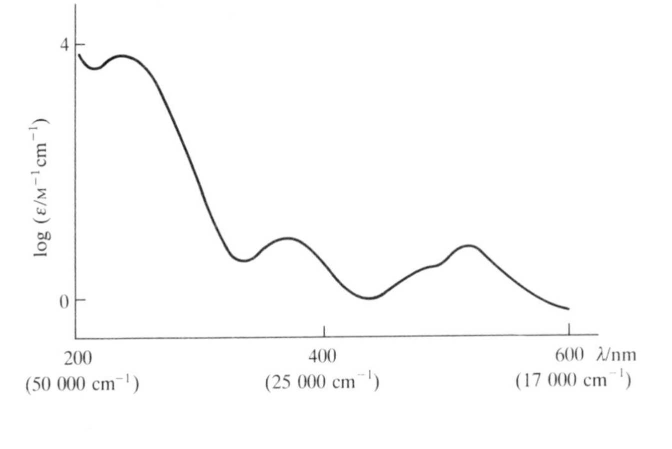

Fig. 14.6 The absorption spectrum of $[CrCl(NH_3)_5]^{2+}$ in water in the visible and ultraviolet regions. The peak corresponding to the transition $^2E \leftarrow {}^4A$ is not visible on this magnification.

covalent character in bromo complexes than in analogous fluoro complexes. Another way of expressing the trend represented by the series is that the softer the ligand, the smaller the nephelauxetic parameter.

The nephelauxetic character of a ligand is different for electrons in t_{2g} and e_g orbitals. Since the σ overlap of e_g is usually larger than the π overlap of t_{2g}, the cloud expansion is larger in the former case. The measured nephelauxetic parameter of an $e \leftarrow t$ transition is an average of the effects on both types of orbital.

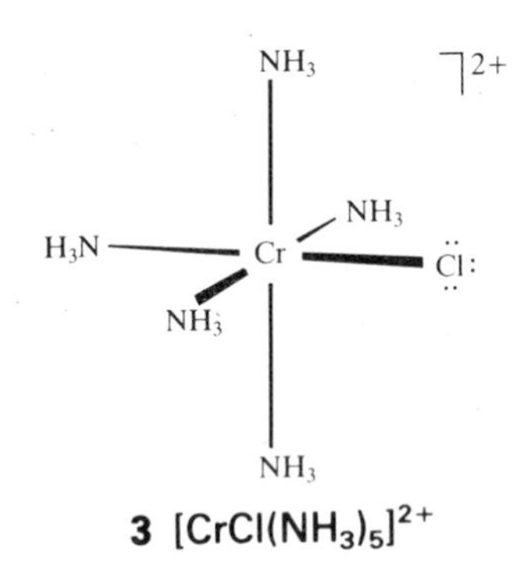

3 $[CrCl(NH_3)_5]^{2+}$

14.4 Charge-transfer bands

Figure 14.6 shows the complete visible and UV spectrum of $[CrCl(NH_3)_5]^{2+}$. It resembles the spectrum shown in Fig. 14.1, and we can recognize the two ligand field bands as the pair in the visible region. The replacement of one NH_3 ligand by one weaker field Cl^- ligand moves the lowest energy band to lower energy than that for $[Cr(NH_3)_6]^{3+}$ in Fig. 14.1, but the reduction in symmetry from O_h to C_{4v} does not produce much additional splitting. The new feature in the spectrum is the strong band in the ultraviolet near 42 000 cm^{-1} (240 nm). This band is at lower energy than the corresponding band in the spectrum of $[Cr(NH_3)_6]^{3+}$ and arises because the Cl^- ligands have π lone pair electrons that are not directly involved in bonding (**3**). The band is an example of an LMCT transition in which a lone-pair electron of Cl^- is promoted into a predominantly metal orbital. Figure 14.7 summarizes the transitions we classify as charge transfer.

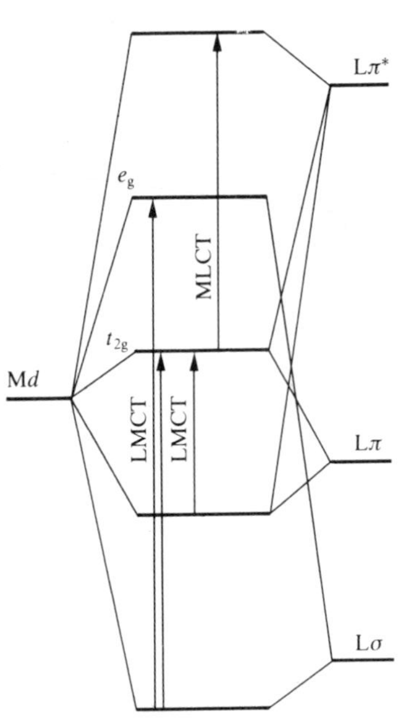

Fig. 14.7 A summary of the charge-transfer transitions in an octahedral complex.

LMCT transitions

As in Fig. 14.6, charge-transfer transitions are generally intense compared with ligand-field transitions. Charge-transfer bands in the visible region of the spectrum (and hence contributing to the intense colors of complexes) may occur if the ligands have lone pairs of relatively high energy (as in sulfur and selenium) or if the metal has low-lying empty orbitals. The color of the artists' pigment 'cadmium yellow', CdS, for instance, is due to the transition $Cd^{2+}(5s) \leftarrow S^{2-}(\pi)$. Similarly, the color of red HgS is a result of the transition $Hg^{2+}(6s) \leftarrow S^{2-}(\pi)$ and red and yellow ochers are iron oxides that are colored on account of the transition $Fe(3de_g^*) \leftarrow O^{2-}(\pi)$.

The tetraoxoanions of metals with high oxidation numbers (such as MnO_4^-) provide what are probably the most familiar examples of LMCT bands. In them, an O lone pair electron is promoted into a low-lying empty e_g metal orbital. High metal oxidation numbers correspond to a low d-orbital population (many are d^0), so the acceptor level is available and low in energy. The trend in LMCT energies is:

Oxidation number	
+7	$MnO_4^- < TcO_4^- < ReO_4^-$
+6	$CrO_4^{2-} < MoO_4^{2-} < WO_4^{2-}$
+5	$VO_4^{3-} < NbO_4^{3-} < TaO_4^{3-}$

The energies of the transitions correlate with the order of the electrochemical series, with the lowest energy transitions taking place to the most easily reduced ions. This correlation is consistent with the transition being the transfer of an electron from the ligands to the metal, corresponding, in effect, to the reduction of the metal ion by the ligands.

Both polymeric and monomeric oxoanions follow the same trends, with the oxidation number of the metal the determining factor. The similarity suggests that these LMCT transitions are localized processes that take place on molecular fragments and are unlike the transitions between delocalized orbitals that are common in semiconductors.

Optical electronegativity

We can express the variation in the position of LMCT bands in terms of molecular orbital energy level diagrams and in particular the **optical electronegativities** of the metal and the ligands. In the simplest case of a complex with an anionic ligand, we focus on the electronegativities χ_L and χ_M of the full ligand π orbitals and of the t_{2g} metal orbitals that are destined to become the π^* orbitals in the complex. The wavenumber of the transition is then written as the difference between the two electronegativities:

$$\tilde{\nu} = C(\chi_L - \chi_M) \quad (1)$$

Optical electronegativities have values comparable to Pauling electronegativities if we take $C = 30\,000\ \text{cm}^{-1}$. If the MLCT transition terminates in an e_g orbital, Δ_O must be added to the energy predicted by this equation.

Table 14.4 lists some optical electronegativities. The values for metals are different in complexes of different symmetry and the ligand values are different if the transition originates from a σ orbital rather than a π orbital.

Table 14.4. Optical electronegativities

Metal	O_h	T_d	Ligand	π	σ
Cr(III)	1.8–1.9		F^-	3.9	4.4
Co(III)†	2.3		Cl^-	3.0	3.4
Ni(II)		2.0–2.1	Br^-	2.8	3.3
Co(II)		1.8–1.9	I^-	2.5	3.0
Rh(III)†	2.3		H_2O	3.5	
Mo(VI)	2.1		NH_3		3.3

† Low-spin complexes

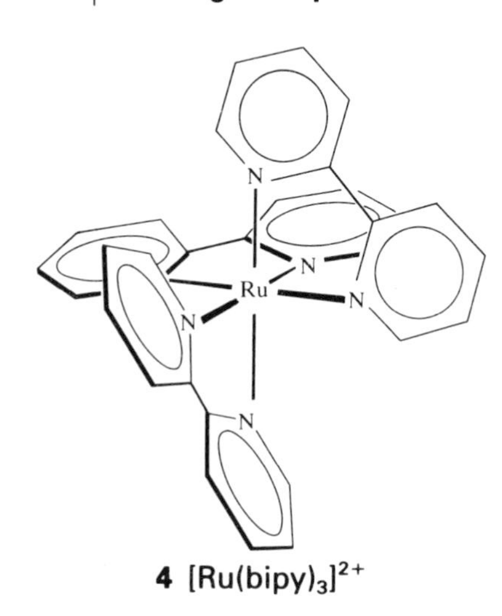

4 $[Ru(bipy)_3]^{2+}$

MLCT transitions

A transfer of charge from metal to ligand is most commonly observed in complexes with an aromatic ligand. The empty π^* orbitals of the ligands lie at sufficiently low energy for promotions of electrons into them to lie at low energy too, particularly if the metal has a low oxidation number. The MLCT excitation of tris(bipyridine)ruthenium(II) (**4**) has been the subject of intense research because the excited state that results from charge transfer is a versatile photochemical redox reagent. MLCT ultraviolet transitions are important in the spectra of complexes of carbonyl and cyano ligands, for which the π^* ligand orbitals are the electron acceptors.

14.5 Selection rules and intensities

The contrast in intensities between typical charge transfer bands and typical ligand field bands raises the question of the factors that control the intensity of absorption bands. In an octahedral, nearly octahedral, or square-planar complex, the maximum molar absorption coefficient ε_{max} (which measures the strength of the absorption)[3] for ligand-field transitions is typically less than or close to $100\,M^{-1}\,cm^{-1}$. In tetrahedral complexes, which have no center of symmetry, ε_{max} might exceed $250\,M^{-1}\,cm^{-1}$. In many cases, charge-transfer bands are allowed by symmetry rules and have intensities governed by the extent of overlap of ground-state and excited-state wavefunctions. Such transitions have ε_{max} of between 1000 and 50 000 $M^{-1}\,cm^{-1}$. Table 14.5 summarizes typical intensities of electronic transitions of complexes of the first series of *d* elements.

Table 14.5. Intensities of spectral bands in 3*d* complexes

Band type	$\varepsilon_{max}/(M^{-1}\,cm^{-1})$
Spin-forbidden	<1
Laporte forbidden *d–d*	20–100
Laporte allowed *d–d*	*ca.* 250
Symmetry allowed (e.g. CT)	1000–50 000

Spin selection rules

The electromagnetic field of the incident radiation cannot change the *relative* orientations of the spins of the electrons in a complex. For example, an initially antiparallel pair ⇅ cannot be converted to ⇈, so a singlet ($S = 0$) cannot undergo a transition to a triplet ($S = 1$). This restriction is summarized by the rule $\Delta S = 0$ for **spin-allowed transitions**.

The coupling of spin and orbital angular momenta can relax the spin selection rule, but such **spin-forbidden** $\Delta S \neq 0$ transitions are generally much weaker than spin-allowed transitions. The intensity of spin-forbidden bands increases as the atomic number increases because the strength of the

[3] The molar absorption coefficient is the constant in the Beer–Lambert law for the transmittance $\tau = I_f/I_i$ when light passes through a length l of solution of molar concentration [X] and is attenuated from an intensity I_i to an intensity I_f:

$$\log_{10}\tau = -\varepsilon l[X]$$

Its earlier name is the 'extinction coefficient'.

spin–orbit coupling is greater for heavy atoms than for light atoms, for it depends on the electric field generated by the nucleus. The breakdown of the spin selection rule by spin–orbit coupling is often called the **heavy-atom effect**. In the first *d* series, in which spin–orbit coupling is weak, spin-forbidden bands have ε_{max} less than about $1\,M^{-1}\,cm^{-1}$; however, spin-forbidden bands are a significant feature in the spectra of later *d*-metal complexes.

The Laporte selection rule

The **Laporte selection rule** is a statement about the change in parity that accompanies a transition:

> In a centrosymmetric molecule or ion, the only allowed transitions are those accompanied by a change in parity.

That is, transitions between g and u terms are permitted, but transitions are forbidden between two g terms and between two u terms:

$$g \leftrightarrow u \qquad g \not\leftrightarrow g \qquad u \not\leftrightarrow u$$

The Laporte selection rule is based on the concept of an **electric dipole transition**, in which the transition gives rise to a transient electric dipole moment. The intensity of such a transition from a state with wavefunction ψ_i to one with wavefunction ψ_f is proportional to the square of the **transition dipole moment**

$$\boldsymbol{\mu}_{fi} = \int \psi_f^* \boldsymbol{\mu} \psi_i \, d\tau$$

where $\boldsymbol{\mu}$ is the electric dipole moment operator $-e\boldsymbol{r}$. The transition dipole moment can be regarded as a measure of the impulse that a transition imparts to the electromagnetic field: a large impulse corresponds to an intense transition; zero impulse corresponds to a forbidden transition. Since $\boldsymbol{r}$ changes sign under inversion (and is therefore u), the entire integral will also change sign under inversion if ψ_i and ψ_f have the same parity, because

$$g \times u \times g = u \quad \text{and} \quad u \times u \times u = u$$

Therefore, because the value of an integral cannot depend on the choice of coordinates,[4] it vanishes if ψ_i and ψ_f have the same parity. However, if they have opposite parity, the integral does not change sign under inversion of the coordinates because

$$g \times u \times u = g$$

and therefore need not vanish.

In a centrosymmetric complex, ligand-field *d–d* transitions are g–g and are therefore forbidden. Their forbidden character accounts for the relative weakness of these transitions in octahedral complexes compared with those in tetrahedral complexes, on which the Laporte rule is silent (as they have no center of symmetry).

The question remains why *d–d* transitions occur at all, albeit weakly. The

[4] An integral is an area, and areas are independent of the coordinates used for their evaluation.

Laporte selection rule may be relaxed in two ways. First, a complex may depart slightly from perfect centrosymmetry, perhaps on account of a distortion imposed by the environment of a complex packed into a crystal or on account of an intrinsic asymmetry in the structure of polyatomic ligands. Alternatively, the complex might undergo an asymmetrical vibration, which also destroys its center of inversion. In either case, a Laporte-forbidden *d–d* band tends to be much stronger than a spin-forbidden transition. As already mentioned, charge-transfer transitions are generally stronger than either kind of *d–d* transitions.

Example 14.5: *Assigning a spectrum using selection rules*

Assign the bands in the spectrum in Fig. 14.6 by considering their intensities.

Answer. Like the hexaammine spectrum in Fig. 14.1, which has *d–d* bands at similar wavenumbers, the complex is moderately strong field ($\Delta_O/B > 20$). If we assume that the complex is approximately octahedral, examination of the Tanabe–Sugano diagram again reveals that the ground state is $^4A_{2g}$. The first excited state is ^2E_g. The next two are $^2T_{1g}$ and $^2T_{2g}$. Transitions to these three will not have ε_{max} greater than $1\ \text{M}^{-1}\ \text{cm}^{-1}$ since they are spin forbidden. Very weak bands at low energy are predicted. They may be too high in energy to be resolved from the more intense bands. The spectrum shows one such weak band which is probably the transition to the lowest energy excited term, ^2E_g. The next two higher terms are $^4T_{2g}$ and $^4T_{1g}$. These are reached by spin allowed but Laporte-forbidden *d–d* transitions, and have ε_{max} near $100\ \text{M}^{-1}\ \text{cm}^{-1}$. In the near UV, the band with ε_{max} near $20\,000\ \text{M}^{-1}\ \text{cm}^{-1}$ corresponds to the LMCT transitions in which the chlorine π lone pair is promoted into a molecular orbital that is principally metal *d* orbital.

Exercise. The spectrum of $[Cr(NCS)_6]^{3-}$ has a very weak band near $16\,000\ \text{cm}^{-1}$, a band at $17\,700\ \text{cm}^{-1}$ with $\varepsilon_{max} = 160\ \text{M}^{-1}\ \text{cm}^{-1}$, a band at $23\,800\ \text{cm}^{-1}$ with $\varepsilon_{max} = 130\ \text{M}^{-1}\ \text{cm}^{-1}$, and a very strong band at $32\,400\ \text{cm}^{-1}$. Assign these transitions using the d^3 Tanabe–Sugano diagram and selection rule considerations. (Hint: NCS^- has low-lying π^* orbitals.)

14.6 Luminescence

For a complex to be **luminescent** and to emit light after it has been excited, the rate of radiative decay by photon emission must compete with thermal degradation of energy as heat in the surroundings. Relatively fast non-radiative decay is quite common at room temperature for *d*-metal complexes, so strongly luminescent systems are comparatively rare. Nevertheless, they do occur, and we can distinguish two types of process. Traditionally, rapidly decaying luminescence was called **fluorescence** and luminescence that persists after the exciting illumination is extinguished was called **phosphorescence**. However, since the lifetime criterion is not reliable, the modern definitions of the two kinds of luminescence are based on the mechanism. Fluorescence occurs when an excited state of the same multiplicity as the ground state decays radiatively into the ground state. The transition is spin-allowed and is commonly fast (a matter of nanoseconds). Phosphorescence is a spin-forbidden process, as we explain below, and hence often slow.

Since light absorption usually populates a state by a spin-allowed transition, the mechanism of phosphorescence involves a nonradiative

intersystem crossing, or conversion from the initial excited state to another excited state of different multiplicity. This second state acts as an energy reservoir because radiative decay to the ground state is spin-forbidden. However, just as spin–orbit coupling allows the intersystem crossing to occur, it also breaks down the spin-selection rule, so the radiative decay can occur, but usually only slowly. A phosphorescent state of a *d*-metal complex may survive for microseconds or longer.

Phosphorescent complexes

An important example of phosphorescence is provided by ruby, which consists of a low concentration of Cr^{3+} ions substituted for Al^{3+} in alumina. Each Cr^{3+} ion is surrounded octahedrally by six O^{2-} ions and the initial excitations are the spin-allowed processes

$$t_{2g}^2e_g^1\ {}^4T_{2g} \leftarrow t_{2g}^3\ {}^4A_{1g} \qquad t_{2g}^2e_g^1\ {}^4T_{1g} \leftarrow t_{2g}^3\ {}^4A_{1g}$$

These absorptions occur in the green and violet regions of the spectrum and are responsible for the red color of the gem (Fig. 14.8). Intersystem crossing to a ^2E term of the t_{2g}^3 configuration occurs in a few picoseconds or less, and red 627 nm phosphorescence occurs as this doublet decays back into the quartet ground state. This red emission adds to the red perceived by the subtraction of green and violet light from white light, and adds luster to the gem's appearance. When the optical arrangement is such that the emission is stimulated by 627 nm photons reflecting back and forth between two mirrors, it grows strongly in intensity. This stimulated emission in a resonant cavity is the process used by Theodore Maiman in the first successful laser (in 1960).

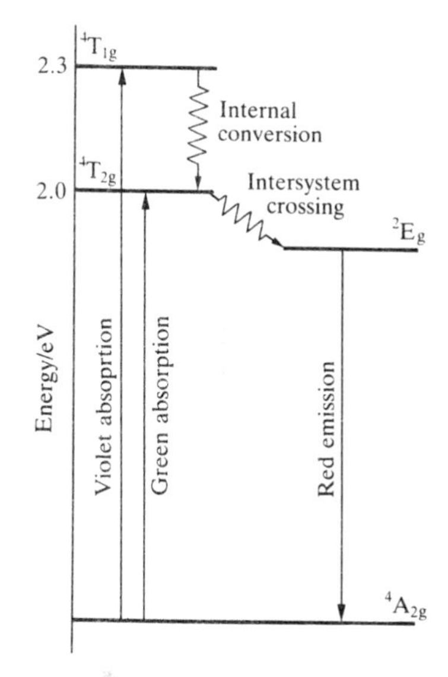

Fig. 14.8 The transitions responsible for the absorption and luminescence of Cr^{3+} ions in ruby.

A similar $^2E \rightarrow {}^4A$ phosphorescence can be observed from a number of essentially octahedral Cr(III) complexes in solution. The emission is always in the red and close to the wavelength of ruby emission. The ^2E term belongs to the t_{2g}^3 configuration, which is the same as the ground state, and the strength of the ligand field is not important. If the ligands are rigid, as in $[Cr(bipy)_3]^{3+}$, the ^2E term may live for several microseconds in solution.

Another interesting example of a phosphorescent state is found in $[Ru(bipy)_3]^{2+}$. The excited singlet term produced by a spin-allowed MLCT transition of this d^6 complex undergoes intersystem crossing to the lower energy triplet term of the same configuration $t_{2g}^5\pi^{*1}$. Bright orange emission then occurs with a lifetime of about 1 μs (Fig. 14.9). The effects of other molecules (quenchers) on the lifetime of the emission may be used to monitor the rate of electron transfer from the excited state.

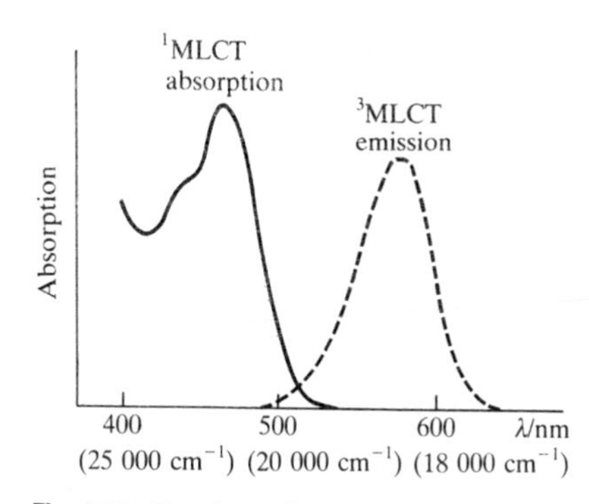

Fig. 14.9 The absorption and phosphorescence spectra of $[Ru(bipy)_3]^{2+}$.

Fluorescent states

Cr(III) d^3 complexes with weak-field ligands may have their $^4T_{1g}$ terms lower than the corresponding phosphorescent ^2E terms because the ligand field splitting is smaller than the electron repulsion energies. Such complexes may be fluorescent. Since the transition populates an antibonding e_g orbital, bond distances change in the excited state. As a result, the emission is considerably shifted to the red from the position of the absorption band.

14.7 Spectra of *f*-block complexes

The partially filled *f* shell can also participate in transitions in the visible region of the spectrum. In one respect, their description is more complex

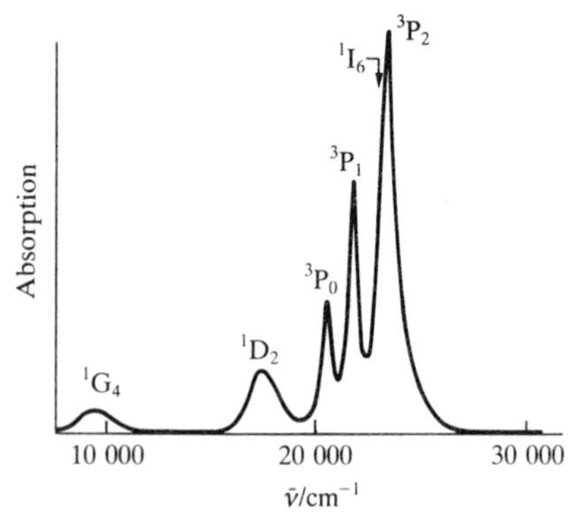

Fig. 14.10 The spectrum of the f^2 $Pr^{3+}(aq)$ ion.

than for the analogous transitions in the d block because there are seven f orbitals. However, it is simplified by the fact that f orbitals are relatively deep inside the atom and overlap only weakly with ligand orbitals. Hence, as a first approximation, their spectra can be discussed in the free-ion limit.

The colors of the aqua ions from $La^{3+}(f^0)$ to $Gd^{3+}(f^7)$ tend to repeat themselves in the reverse sequence $Lu^{3+}(f^{14})$ to $Gd^{3+}(f^7)$:

f electrons:	0	1	2	3	4	5	6	7
	Colorless	green	red	pink	yellow	pink	colorless	
f electrons:	14	13	12	11	10	9	8	

This sequence suggests that the absorption maxima are related in a simple manner to the number of unpaired f electrons. Unfortunately, though, the apparent simplicity is not fully borne out by detailed analysis.

Figure 14.10 shows the spectrum of f^2 $Pr^{3+}(aq)$ in the visible region. The four bands are labeled with the free-ion term symbols, which is appropriate when the interaction with ligands is very weak. The Russell–Saunders coupling scheme remains a good approximation despite the elements having high atomic numbers, because the f electrons penetrate only slightly through the inner shells. As a result, they do not sample the electric field produced by the nucleus very strongly and hence their spin–orbital coupling is weak.

The other noteworthy feature of the spectrum is the narrowness of the bands, which indicates that the electronic transition does not excite much molecular vibration when it occurs. The narrowness implies that there is little difference in the molecular potential energy surface when an electron is excited, which is consistent with an f electron interacting only weakly with the ligands. Since the excited electron interacts only weakly with its environment, the nonradiative lifetime of the excited state is quite long and luminescence can be expected. These complexes tend to be strongly luminescent, and for that reason are used as phosphors on TV screens.

14.8 Circular dichroism

The photon has a spin of 1, and in a state of circular polarization it has a definite **helicity**, or component of angular momentum along its direction of propagation. Left-circularly polarized light consists of photons of one helicity and right-circularly polarized light of photons of the opposite helicity (**5**).

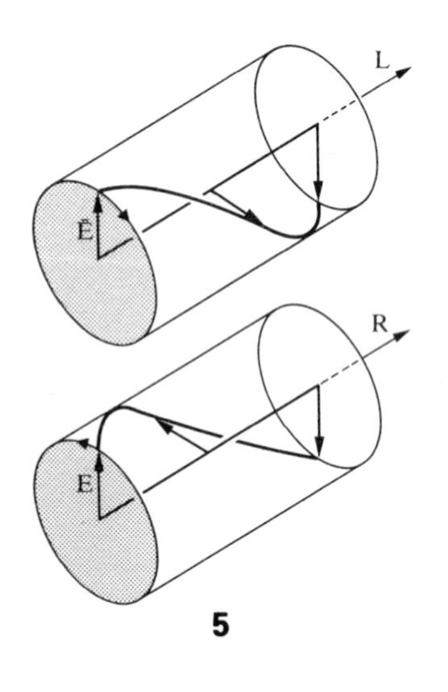

5

A chiral molecule (one lacking an improper-rotation axis, Section 2.4) is **circular dichroic**. That is, the molecule has different absorption coefficients of right- and left-circularly polarized light at any given wavelength. A **circular dichroism spectrum** (a CD spectrum) is a plot of the difference of the molar absorption coefficients for right- and left-circularly polarized light against wavelength. As we see in Fig. 14.11, enantiomers have CD spectra that are mirror-images of each other. The usefulness of CD spectra can be appreciated by comparing the CD spectra of two enantiomers of $[Co(en)_3]^{3+}$ (Fig. 14.11b) with their conventional absorption spectra (Fig. 14.11a). The latter show two ligand field bands, just as though the complex were an octahedral complex, and give no sign that we are dealing with a complex of lower (D_3) symmetry. That difference, however, shows up in the CD spectra where an additional band is seen near 24 000 cm^{-1}.

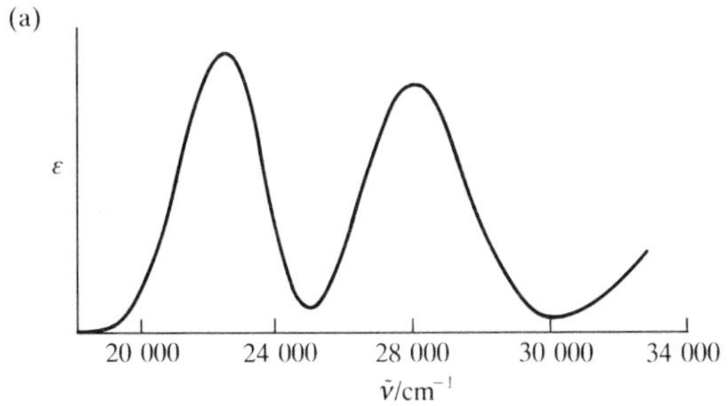

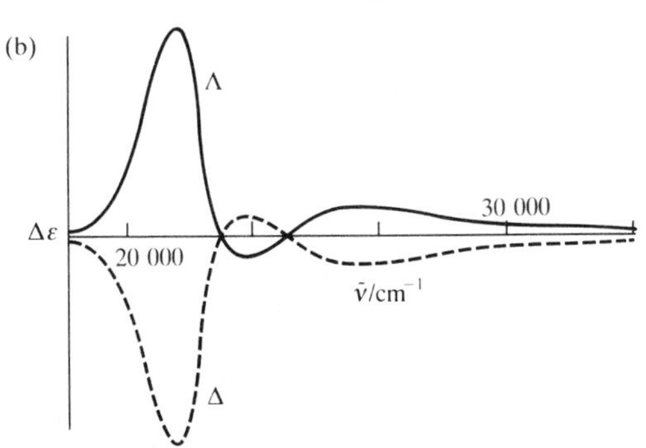

Fig. 14.11 (a) The absorption spectrum of $[Co(en)_3]^{3+}$ and (b) the CD spectra of the two optical isomers.

Circular dichroism spectra are also important because they allow us to identify the absolute configurations of chiral complexes. The same absolute configuration of two complexes with similar electronic configurations give CD spectra of the same sign. Hence CD spectra can be used to relate large families of chiral complexes to the small number of primary cases, including $[Co(en)_3]^{3+}$, for which absolute configurations have been established by X-ray diffraction.

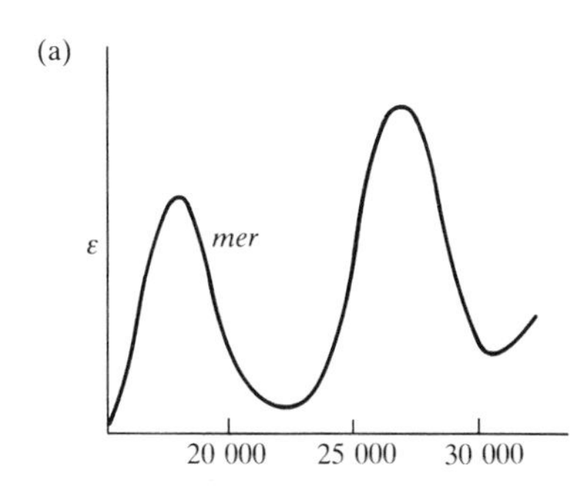

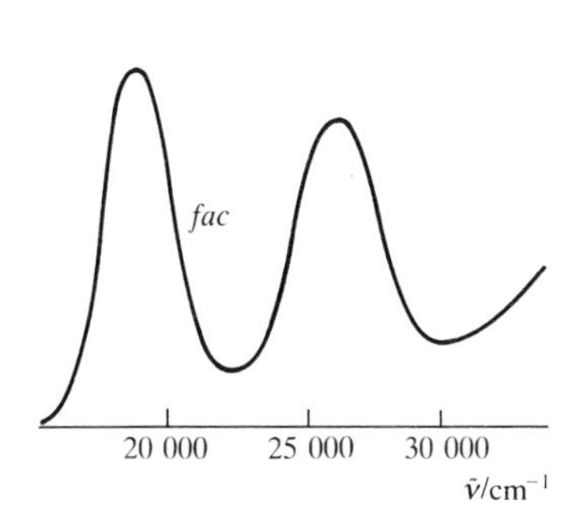

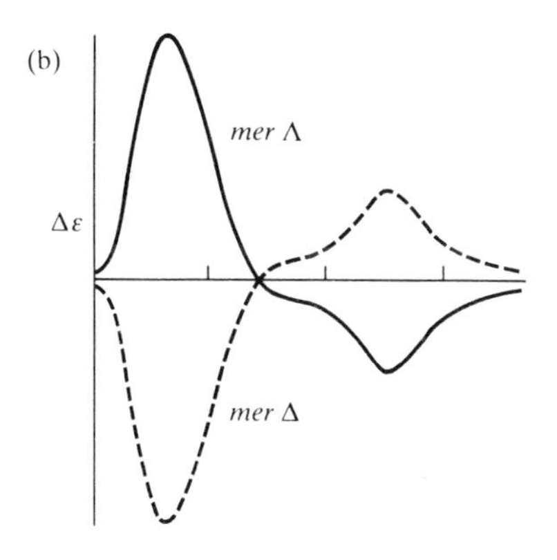

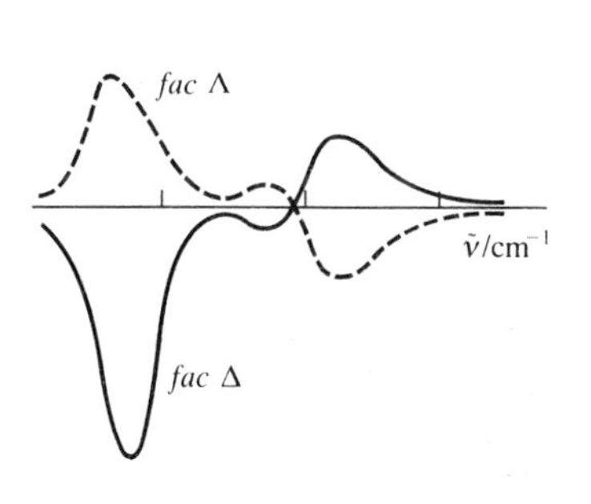

Fig. 14.12 (a) The absorption spectrum of $[Co(ala)_3]$, where ala is alanine and (b) the CD spectra of the isomers and the assignment of their absolute configuration by comparison with the CD spectrum of $[Co(en)_3]^{3+}$.

Circular dichroism spectra are very important in the study of isomerism. For example, the amino acid alanine $CH_3CH(NH_2)COOH$ forms *fac* and *mer* tris complexes with Co(III), and each form can have Λ or Δ isomers. The problem of identifying the arrangement of ligands is first to decide between the *fac* and *mer* isomers, and then between their Λ and Δ optical isomers. The two geometrical isomers have different absorption spectra (Fig. 14.12a) because the immediate symmetry of ligands about the Co(III) center differs. In the *mer* isomer, the principal axis is C_2 along the axis passing through the N and O atoms. In the *fac* isomer, the principal axis is C_3 and it passes through the faces surrounded by N and O respectively. The lower symmetry of the *mer* isomer is likely to lead to a distinctly broader absorption band. Hence a plausible assignment is that the spectrum in Fig. 14.12a is that of the *mer* isomer. The corresponding CD spectra are shown in Fig. 14.12b. The two pairs of curves with peaks of opposite sign clearly indicate which isomers are enantiomers of the same geometrical isomer. Furthermore, we can assign the Δ and Λ absolute configuration labels by comparing these spectra with the CD spectra of $[Co(en)_3]^{3+}$, for its absolute configuration is known from X-ray diffraction.

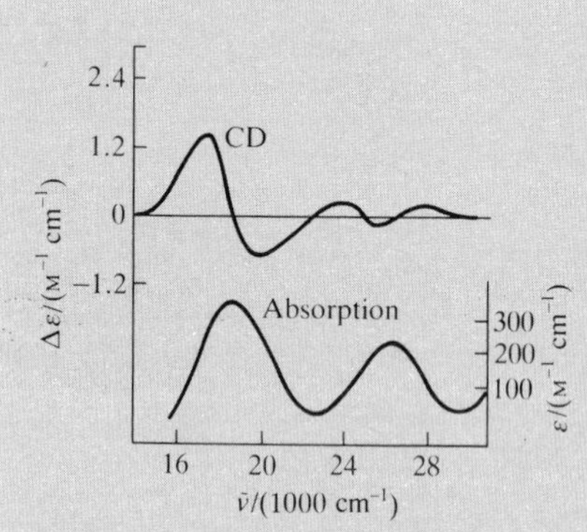

Fig. 14.13 The absorption and CD spectra of (−)546 $[Co(edta)]^-$ referred to in Example 14.6.

Example 14.6: *Using CD spectra to assign a configuration*

The enantiomer commonly labeled (−)546 $[Co(edta)]^-$ because it rotates the plane of linearly polarized 546 nm light (from a Hg spectrum) has the CD spectrum shown in Fig. 14.13. Assign its absolute configuration by comparison with $[Co(en)_3]^{3+}$.

Answer. The complex is similar to the alanine complex with one more O atom bound to Co. Therefore it seems reasonable to continue to correlate the spectra with the two *d–d* transitions of $[Co(en)_3]^{3+}$. The lower energy band has a positive sign followed by a weak negative peak as energy increases. The higher energy absorption band corresponds to weak CD absorption with a larger positive than negative area. These features match the Λ absolute configuration of $[Co(en)_3]^{3+}$.

Exercise. What evidence do you find in Fig. 14.13 that $[Co(edta)]^-$ has lower symmetry at Co than $[Co(en)_3]^{3+}$?

14.9 Electron paramagnetic resonance

The electronic analog of nuclear magnetic resonance is called **electron paramagnetic resonance** (EPR) and is a powerful technique for studying complexes with unpaired electrons (those in other than singlet states). EPR can be used to map the distribution of an unpaired electron in a molecule and, to some extent, decide the extent to which electrons are delocalized over the ligands. EPR can also provide some information about the energy levels of complexes.

The *g*-factor

EPR relies on the fact that an electron spin can adopt two orientations along the direction defined by an applied magnetic field B. The energy difference between the states $m_s = +\frac{1}{2}$ and $-\frac{1}{2}$ is

$$\Delta E = g\mu_B B$$

where μ_B is the Bohr magneton and g is a constant characteristic of the complex. For a free electron, g has the 'free-spin' value $g_e = 2.0023$. If the sample is exposed to electromagnetic radiation of frequency ν, a strong absorption—a resonance—occurs when the magnetic field satisfies the condition

$$h\nu = g\mu_B B \qquad (2)$$

In a typical spectrometer operating near 0.3 T, resonance occurs in the 3 cm band of the microwave region of the electromagnetic spectrum (at about 9 GHz).

The g-factor in a complex differs from g_e by an amount that depends on the ability of the applied field to induce local magnetic fields. We can think of $(g/g_e)B$ as a modification of the applied field that takes into account any locally induced fields. If $g > g_e$, the local field is larger than that applied; if $g < g_e$, it is smaller. The sign and magnitude of the induced local fields depends on the separation of the energy levels of the complex. The closer they are together, the easier it is for the applied field to induce orbital circulation of the electrons and hence produce a local magnetic field. (This is a conclusion from perturbation theory, Section 1.6.)

We can measure g by noting the value of the applied field needed to achieve resonant absorption at a given microwave frequency. If the complex is part of a single crystal, g may be measured along different directions and hence used to infer the symmetry of the complex. For example, the tetragonally distorted D_{4h} $[Cu(H_2O)_6]^{2+}$ complex has $g = 2.08$ in the direction parallel to the C_4 axis and $g = 2.40$ perpendicular to it. The fact that two g factors are measured confirms the existence of the distortion, since an octahedral complex has an isotropic g factor (one that is the same in all directions). The two numerical values of g give information about the energy levels of the complex.

Hyperfine structure

Magnetic nuclei—those with nonzero spin—give rise to an additional magnetic field experienced by an electron in a complex. Since a given magnetic nucleus of spin quantum number I can adopt $2I+1$ different orientations, it can give rise to $2I+1$ different contributions to the local field and hence result in the resonance condition being fulfilled at $2I+1$ different values of the applied field. Hence, EPR spectra are often split into multiplets called their **hyperfine structure**.

Figure 14.14 shows the spectrum of the oxo-Cr(V) cation resulting from the

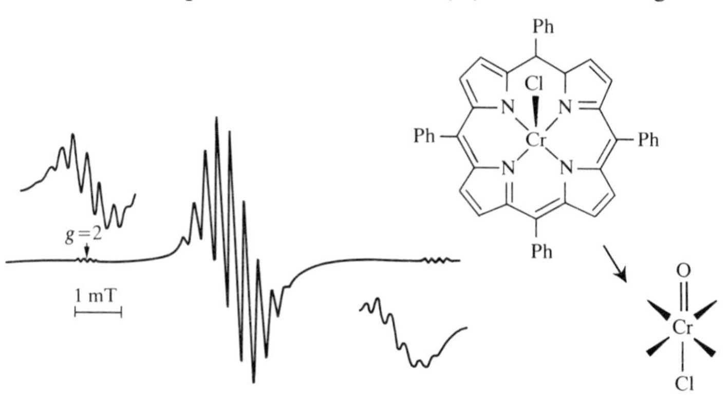

Fig. 14.14 The EPR spectrum of the radical resulting from the oxidation of the Cr(III) porphyrin complex shown to give an oxo-Cr(V) porphyrin with one unpaired electron.

oxidation of a Cr(III) porphyrin complex. The single unpaired electron experiences the local field generated by the four ^{14}N nuclei (for which $I = 1$, so each nucleus produces three lines). One nucleus splits the spectrum into a triplet of lines of equal intensity, another splits each of the three lines into a triplet, and so on, to give the overall pattern of nine lines.[5] Thus, the EPR spectrum is consistent with a complex containing four equivalent N atoms.

The hyperfine splitting constant

We can write the $2I + 1$ resonance conditions responsible for the hyperfine structure as

$$h\nu = g\mu_B(B + Am_I) \tag{3}$$

where A is the **hyperfine splitting constant**. This constant is a characteristic of the nucleus itself (its magnetic moment) and the probability that an unpaired electron will be found near it. By measuring the value of A, we can deduce the probability that an unpaired electron occupies an orbital on that atom and hence build up a description of the molecular orbital it occupies.

The value of A for either the magnetic nuclei of the central nucleus or of the ligands is a very good indication of the degree to which electrons are delocalized from the central metal atom to the surrounding ligands. For instance, for a Cu(II) complex, the hyperfine splitting of the Cu nucleus varies as follows:

Ligand	F^-	H_2O	O^{2-}	S^{2-}
A/mT	9.8	9.8	8.5	6.9

These values show that as the energy of the ligand orbitals approach that of the metal and their interaction increases, the orbital occupied by the unpaired electron becomes more ligand-like and interacts more weakly with the metal nucleus.

Lower symmetry and binuclear complexes

In this section we discuss the molecular orbitals of some complexes that we meet again in later chapters, particularly when we discuss the d-block organometallics (Chapter 16). In particular, we deal here with complexes of lower symmetry than those considered in Chapter 7. We shall show that one very useful technique for constructing molecular orbitals of complicated molecules is to build them up from the valence orbitals of subunits. The technique is especially useful for organometallic molecules because they frequently consist of ML_n subunits (where L is a σ-base, π-acid ligand) bound to an organic molecule.

14.10 The ML_5 fragment

We shall illustrate the technique by deriving the molecular orbitals of the fragment ML_5 from the octahedral molecular orbital diagram. Once we

[5] You should try to show that the intensity distribution is in the ratio 1:4:10:16:19:16:10:4:1.

have established the orbitals of this C_{4v} unit, we can analyze some aspects of the spectra of C_{4v} and D_{4h} complexes. We shall also be able to use the fragment orbitals to construct an orbital diagram of an M_2L_{10} binuclear metal–metal bonded complex.

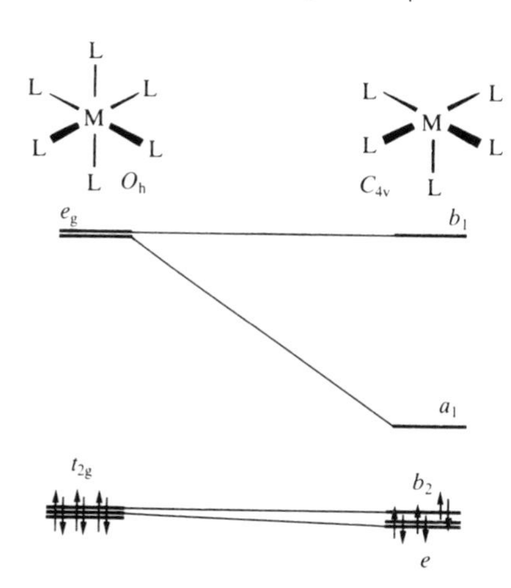

Fig. 14.15 An orbital correlation diagram for the conversion of an ML_6 octahedral complex to an ML_5 fragment. Only the frontier orbitals are shown.

ML_5 molecular orbitals

The HOMO and LUMO of an ML_6 octahedral complex are shown on the left of Fig. 14.15. Suppose we remove one z-axis ligand to give ML_5. There is little effect on the octahedral t_{2g} levels, for they are not σ with respect to the z axis. The $d_{x^2-y^2}$ orbital, which becomes b_1, is largely unaffected by removal of the ligand. The principal change occurs to the d_{z^2} orbital, which becomes of a_1 symmetry, and falls sharply in energy as it loses its antibonding character toward the departing ligand (**6**). As d_{z^2} becomes a_1, it can mix with s and p_z, which are also a_1 in C_{4v} symmetry. We can therefore expect the resulting hybrid orbital to have a large amplitude in the vacant coordination position, which makes it an ideal orbital for acting as a σ acceptor to a new donor ligand.

In C_{4v} symmetry the t_{2g} orbitals are split into a doubly degenerate pair (e) consisting of d_{xz} and d_{yz}, and one other orbital of b_2 symmetry, which is the d_{xy} orbital. The splitting will be smaller than that between the d_{z^2} and $d_{x^2-y^2}$ orbitals because the three orbitals are less directly affected by the ligand that is being removed. However, if the d_{xz} and d_{yz} orbitals are involved in π bonding with the departing ligand there may be an additional effect. Thus, if the ligand is a π base (like Cl^-), the antibonding effect will be reduced and the new e pair will lie below the b_2 orbital (as we have shown in Fig. 14.15). On the other hand, if the ligand is a π acid (like CO), its removal will lead to an increase in the energy of the e orbitals above the b_2 orbital.

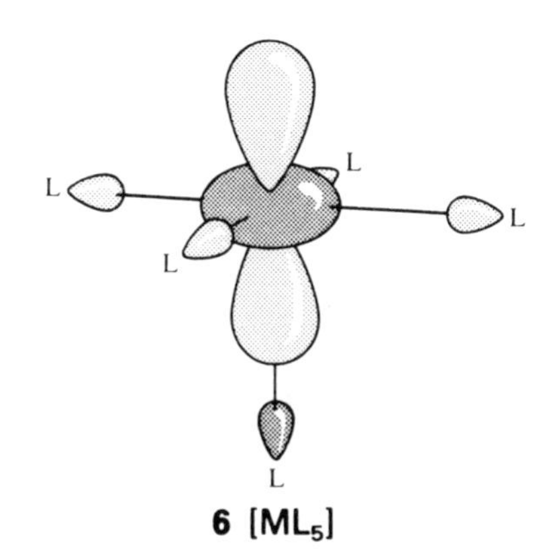

6 $[ML_5]$

Square-pyramidal complexes

We need one more detail before we can use the orbital diagram of the ML_5 fragment to interpret the spectra of five-coordinate complexes. A square-pyramidal five-coordinate complex can have its central M atom above the plane of the four ligands of the pyramid's base. Figure 14.16 shows an example of the energy variations of the HOMO and LUMO as the basal ligands bend away from the metal and the angle ϕ increases from 90° to 120°. This increase in angle reduces the overlap of the ligands with the d_{z^2} and $d_{x^2-y^2}$ orbitals and the net antibonding character is reduced. At the same time, σ overlap with the d_{xz} and d_{yz} orbitals increases and they become more antibonding.

We can use Fig. 14.16 to correlate the structures of square pyramidal complexes with their electron counts. A low-spin d^6 complex should be expected to be a relatively flat pyramid since d_{xz} and d_{yz}, which are filled, rise in energy as ϕ increases from 90°. In contrast, a low-spin d^8 pyramid should have ϕ well above 90° for then d_{z^2} (which is now occupied) is stabilized. In support of this, the d^8 complex $[Ni(CN)_5]^{3-}$ has $\phi = 101°$ whereas the d^6 oxyhemoglobin system has $\phi \approx 90°$.

Trigonal bipyramidal complexes

A trigonal bipyramid is the alternative five-coodinate structure. We show in Fig. 14.17 how the Berry pseudorotation converts a square-planar five-coordinate complex into a trigonal bipyramid and how the natural choice of

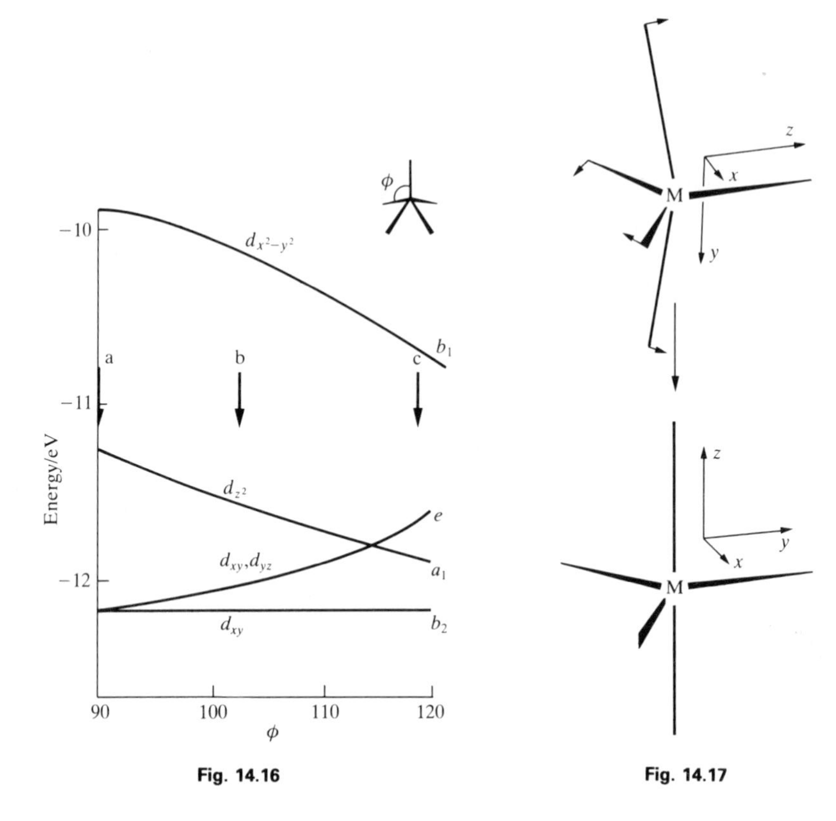

Fig. 14.16 The correlation diagram for the metal *d* orbitals in an ML_5 fragment as the basal ligands bend away from the metal and the angle ϕ increases from 90° to 120°. The arrows show the location of (a) low-spin d^6 oxyhemoglobin, (b) a low-spin d^8 $[Ni(CN)_5]^{3-}$, and (c) a d^{10} macrocyclic complex of Cu(I) with an axial CO ligand.

Fig. 14.17 The Berry pseudorotation of an ML_5 fragment and the corresponding realignment of the natural choice of coordinate axes.

z axis also changes. Because the *z* axis changes, we do not label the orbitals at intermediate stages of the correlation diagram in Fig. 14.18 when the symmetry is lowered to C_{2v}.

Pseudorotation takes the square-planar complex into a trigonal bipyramidal complex by decreasing one L–M–L angle in the *xz* plane and increasing the other in the *yz* plane. Since the *xy* plane is little affected, the d_{xy} orbital becomes, without significant change in energy, a member of e'' in D_{3h}. The other member of this degenerate pair is formed from d_{yz}, for as the L–M–L angle increases, the ligands move into the node of the orbital and stabilize it by reducing its antibonding character. The opposite happens to the d_{xz} orbital of the square-planar complex, and it is destabilized. The fate of the upper two orbitals is harder to describe verbally but can be deduced group theoretically: the d_{z^2} orbital of the square-planar complex becomes an e' orbital of the trigonal bipyramidal complex, and the b_1 orbital becomes a_1'.

Our experience with VSEPR theory and main-group molecules suggests that the trigonal bipyramidal complex is likely to be at least marginally preferred over the square pyramid for d^0 systems. For d^3 and d^4 complexes, the trigonal bipyramidal complex should be favored since one member of the e'' pair of orbitals increases sharply in energy as the complexes depart from this geometry. For low-spin d^5 and d^6 complexes, we can anticipate square-pyramidal geometry since its e orbitals lie below those of the e' orbitals of the trigonal bipyramidal complex.

Promotion of an electron from e to a_1 in the square-pyramidal structure leads to geometrical instability, and in fact irradiation of $W(CO)_4CS$ leads to exchange of the apical and basal ligands without evidence of ligand

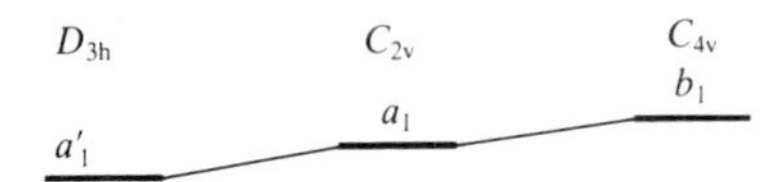

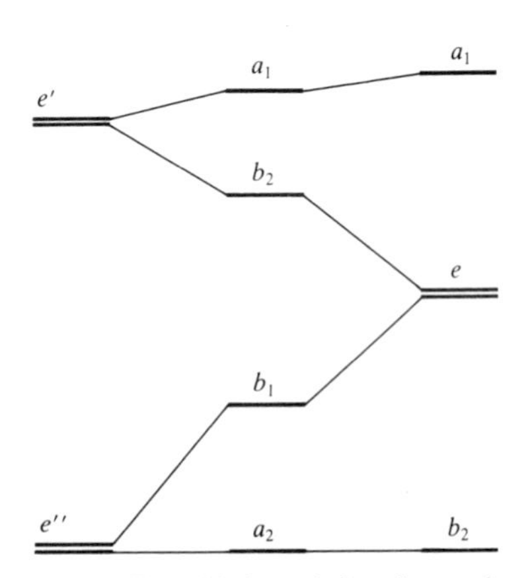

Fig. 14.18 The orbital correlation diagram for the conversion of a trigonal bipyramid (D_{3h}) to a square pyramid (C_{4v}). Only the frontier orbitals that are predominantly metal *d* orbitals are shown.

dissociation.[6] The molecule distorts toward a trigonal bipyramidal complex before relaxing to the square-pyramidal ground state in a photochemical version of a Berry psuedorotation.

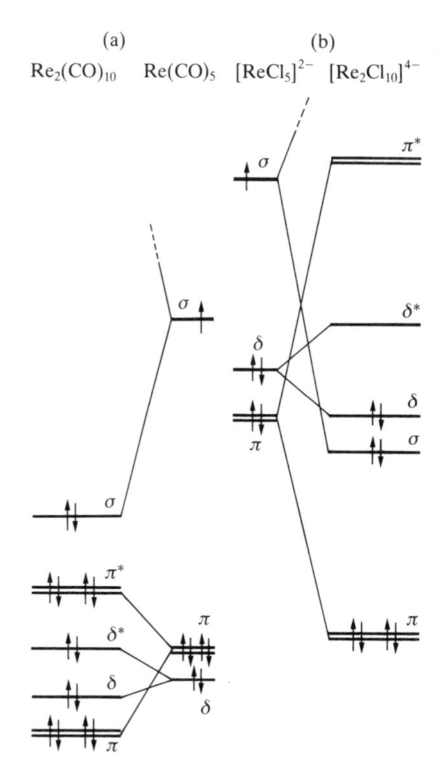

Fig. 14.19 The formation of Re_2L_{10} from two ReL_5 units (a) when the ligands have π acid character (L = CO) and (b) when they have π base character ($L = Cl^-$).

14.11 Metal–metal bonding and mixed valence complexes

The molecular orbitals of polynuclear complexes of the form M_2L_{10}, such as $Mn_2(CO)_{10}$, can be constructed from the orbitals of the two ML_5 fragments. Since a reasonable first approximation is to suppose that the only important interactions are between the frontier orbitals of each complex, we need consider only them.

Metal–metal bonding

Figure 14.19 shows the formation of an M—M bond when the two ML_5 frontier orbitals interact to give σ, π, and δ orbitals. There are two quite different cases. If the metal in ML_5 has a low oxidation number (and hence is electron rich), as in Fig. 14.19a, and the ligands are π acids, the π and δ orbitals lie well below the σ orbitals and the complex is stabilized by back-donation. If the ligands are π bases, as in Fig. 14.19b, the σ orbital is not so far above the π and δ orbitals. Molecular orbital diagrams for the M—M bond in $Re_2(CO)_{10}$ (**7**) and the hypothetical anion $[Re_2Cl_{10}]^{4-}$ (**8**) are shown[7] in Fig. 14.19, which also shows the electron populations for Re(0) and Re(III). The decacarbonyldirhenium complex has a net single σ bond (**7**). In contrast, the analogous chloro complex has a quadruple bond (**8**) consisting of a σ bond, two π bonds and a δ bond; the HOMO is a δ bond.

7 $[Re_2(CO)_{10}]$

8 $[Re_2Cl_{10}]^{4-}$

Figure 14.20 shows the spectrum of $[Bu_4N]_2[Re_2Cl_8]$, where Bu is the butyl group, and its remarkably strong band in the red near 15 000 cm^{-1} (near 650 nm). This band is too intense for a Laporte-forbidden *d–d* transition and has been ascribed to a $\delta^* \leftarrow \delta$ transition of the metal–metal bond.[8] The δ^* excited state has a long lifetime and initiates several very interesting photochemical processes (Section 15.16).

[6] M. Poliakoff, *Inorg. Chem.*, **15,** 2022 and 2892 (1976).

[7] This hypothetical species is related to the well-known quadruply bonded complex $[Re_2Cl_8]^{2-}$ which was illustrated in Section 7.1.

[8] C. D. Cowman and H. B. Gray, *J. Amer. chem. Soc.*, **95,** 8177 (1973). This is probably the only spectroscopic transition to have been featured in a detective novel, Joseph Wambaugh's *The delta star*. Bantum Books, New York (1974).

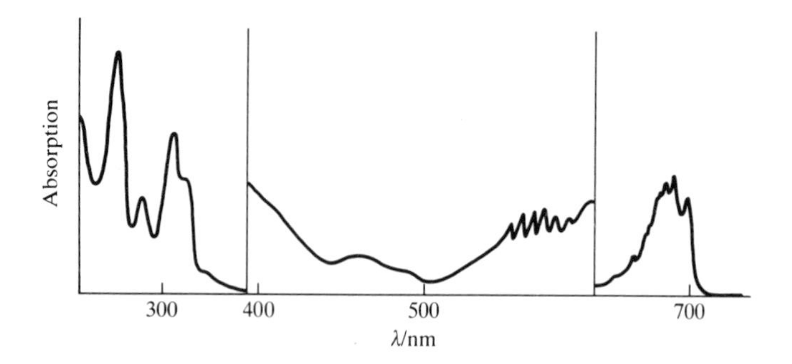

Fig. 14.20 The absorption spectrum of $[Bu_4N]_2[Re_2Cl_8]$ at 5 K (the outer two sections) and at 15 K (the inner section). Note the strong $\delta^* \leftarrow \delta$ band in the red near 15 000 cm^{-1} (670 nm).

Mixed-valence complexes

Another type of binuclear complex with interesting spectroscopic properties is a **mixed-valence complex** with identical metal atoms in different oxidation states. One of the earlier and more important examples is the 'Creutz–Taube ion', $[(NH_3)_5Ru(pyz)Ru(NH_3)_5]^{5+}$ (**9**, pyz is a bridging pyrazine ligand). The question this formula poses is whether the two Ru atoms are a localized Ru(II) and a localized Ru(III) or whether the electrons are sufficiently delocalized to give two equivalent Ru atoms in a +2.5 oxidation state. The problem can be expressed (but not solved) by writing a wavefunction ψ for Ru(II)–Ru(III) and another wavefunction ψ' for Ru(III)–Ru(II) and expressing the overall delocalized ground state by the wavefunction

$$\Psi = c\psi + c'\psi'$$

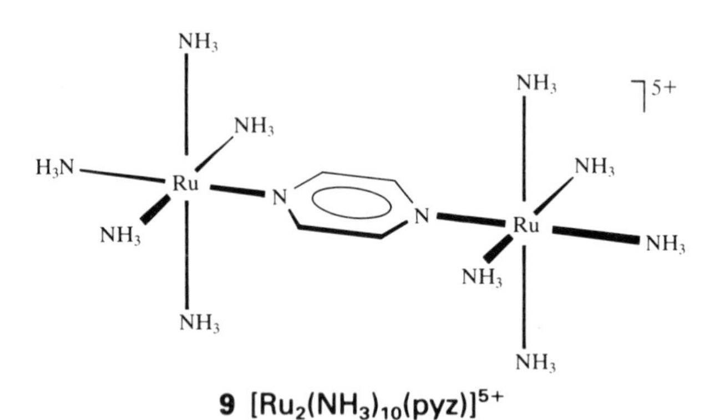

9 $[Ru_2(NH_3)_{10}(pyz)]^{5+}$

$c = 0$ or $c' = 0$ corresponds to complete localization and $c^2 = c'^2$ corresponds to complete delocalization.

A classification of mixed valence compounds that reflects the differences in the strength of the interaction has been proposed by M. B. Robin and Peter Day:

Class I: Fully localized electrons
Class II: Intermediate
Class III: Fully delocalized electrons

Class III is subdivided into IIIA for pairs or clusters of metals and IIIB for indefinitely delocalized solids.

Table 14.6. The Robin–Day classification of IT bands

Class I	Class II	Class III
A and B very different, different ligand fields, etc.	A and B in similar environments, but not equivalent	A and B not distinguishable
Orbitals 'trapped'	Orbitals distinguishable but not fully localized	Orbitals delocalized
IT bands at high energy	IT in visible or near IR	Bands in visible or near IR
Electronic spectra of constituent ions seen	Electronic spectra of constituent ions seen but modified	Electronic spectra of constituent ions not distinguishable
Insulator	Semiconductor	IIIA: Usually insulator IIIB: Usually metallic conductor
Magnetic properties of isolated complex	Magnetic properties of isolated complex except at low temperature	IIIA: Magnetic properties of isolated complex IIIB: Ferromagnetic or paramagnetic

Class I includes a large number of metal oxide and sulfide solids that seem curious from the standpoint of oxidation numbers. For example, the apparent oxidation state of Pb in Pb_3O_4 may be regarded as a mixture of the two common oxidation states Pb(IV) and Pb(II). As outlined in Table 14.6, compounds in this class contain metal ions in ligand fields of different symmetry or strength, and there is definite structural evidence for localized oxidation states (corresponding to $c^2 \gg c'^2$ or vice versa). Spectral and magnetic properties near room temperature are characteristic of the coordination environment of the individual ions.

Class III includes cases where formation of covalent or metallic metal–metal bonds lead to the delocalization of charge ($c^2 \approx c'^2$). The bronze-colored metallic conductor Ag_2F, for example, contains sheets of metal–metal bonded Ag atoms bearing a positive charge, and is a member of class IIIB. In contrast, Ta_6Cl_{15} contains cluster cations of the formula $[Ta_6Cl_{12}]^{3+}$ containing six equivalent Ta atoms. Discrete clusters of this kind belong to class IIIA. The solid is an electrical insulator and does not have a metallic appearance.

Class II includes the compounds for which **intervalence transitions** (IT) are observed in which an electron is excited from the valence shell of one atom into that of the other. As is the case with the Creutz–Taube ion, intervalence transitions are strong if there is an appropriate bridging ligand between the two centers. If the metal ions do not share a bridging ligand, the IT band may be too weak to be observed. The classic example of this type of compound is the pigment Prussian blue. Its structure is a cubic lattice with Fe^{2+} octahedrally coordinated by the C atoms of the CN^- ligands and Fe^{3+} octahedrally coordinated by the N atoms of the ligands. The bridging cyanide provides the conduit for the strong IT transition and the resulting blue color. The magnetic properties of Class II compounds at room temperature give little indication of differences from localized complexes but magnetic interactions between sites become evident at low temperatures.

Example 14.7: *Assigning a compound to a Robin–Day class.*

Magnetite (Fe_3O_4) has half the Fe^{3+} in tetrahedral sites and the rest in half the octahedral sites; the Fe^{2+} ions occupy the rest of the octahedral sites. The solid is black. What is its classification in the Robin–Day scheme?

Answer. The black color suggests the presence of IT bands. The distinct environments exclude Class III. Thus, it is a Class II substance.

Exercise. Is magnetite expected to be an insulator?

Further information: the addition of angular momenta

The total angular momentum quantum numbers L and S are found by working out the various ways that the individual quantized spin and orbital angular momenta of the individual electrons may combine to give a quantized resultant. However, in practice a short cut is provided by the **Clebsch–Gordan series**. This states that the total orbital angular momentum quantum number of an atom with two electrons with individual orbital angular momentum quantum numbers l_1 and l_2 may be one of the values

$$L = l_1 + l_2, l_1 + l_2 - 1, \ldots |l_1 - l_2|$$

The value with $L = l_1 + l_2$ corresponds to a state in which the two electrons have orbital angular momentum in the same direction (like the planets) and hence combine to give a large resultant angular momentum. The smallest value, $|l_1 - l_2|$, arises when the orbital momenta are opposed and their combined momentum is small. For two p electrons, $l_1 = 1$ and $l_2 = 1$, so the possible values of L are

$$L = 2, 1, 0$$

If there are three electrons present, the total orbital angular momentum is obtained by combining l_3 with each value of L obtained by coupling l_1 and l_2. Thus, for each value of L we obtain

$$L' = L + l_3, L + l_3 - 1, \ldots, |L - l_3|$$

and so on for any more electrons that may be present.

As an example, consider the configuration p^2d^1. First, we couple the two p electrons together, with $l_1 = l_2 = 1$:

$$L = 2, 1, 0$$

Then we couple the single d electron ($l_3 = 2$) to each of these values of L:

$$\begin{aligned}&\text{From } L = 2, \ L' = 4, 3, 2, 1, 0\\ &\text{From } L = 1, \ L' = 3, 2, 1\\ &\text{From } L = 0, \ L' = 2\end{aligned}$$

Therefore, the terms that arise are one G, two F, three D, two P, and one S.

A similar Clebsch–Gordan series applies to the spin. When there are two electrons present, the total spin quantum number may be one of the values

$$S = s_1 + s_2, s_1 + s_2 - 1, \ldots |s_1 - s_2|$$

Since s_1 and s_2 are both equal to $\frac{1}{2}$ for electrons, the total spin quantum number S may be 1 (spins parallel) or 0 (spins opposed). If there are three electrons, each value of S is coupled to the spin of the third electron, giving

$$S' = S + s_3, S + s_3 - 1, \ldots, |S - s_3|$$

Further reading

B. N. Figgis, *An introduction to ligand fields.* Wiley, New York (1966). A very readable treatment of the spectra of the d-block metal complexes.

A. B. P. Lever, *Inorganic electronic spectroscopy.* Elsevier, Amsterdam (1984). A comprehensive, up-to-date treatment with a good collection of data.

T. A. Albright, J. K. Burdett, and M. H. Whangbo, *Orbital interactions in chemistry.* Wiley, New York (1985). A good source for the treatment of construction of molecular orbitals by combination of fragments.

Exercises

14.1 Write the Russell–Saunders term symbols for states with the angular momentum quantum numbers (L, S): (a) $(0, \frac{5}{2})$, (b) $(3, \frac{3}{2})$, (c) $(2, \frac{1}{2})$, (d) $(1, 1)$.

14.2 Identify the ground term from each set of terms: (a) ^3F, ^3P, ^1P, ^1G; (b) ^5D, ^3H, ^3P, ^1I, ^1G; (c) ^6S, ^4G, ^4P, ^2I.

14.3 Give the Russell–Saunders terms of the configurations: (a) $4s^1$, (b) $3p^2$. Identify the ground term.

14.4 The free gas-phase ion V^{3+} has a ^3F ground term. The ^1D and ^3P terms lie respectively 10 642 cm^{-1} and 12 920 cm^{-1} above it. The energies of the terms are given in terms of Racah parameters as $E(^3F) = A - 8B$, $E(^3P) = A + 7B$, $E(^1D) = A - 3B + 2C$. Calculate the values of B and C for V^{3+}.

14.5 Write the d-orbital configurations and use the Tanabe–Sugano diagrams to identify the ground term of (a) low-spin $[Rh(NH_3)_6]^{3+}$, (b) $[Ti(OH_2)_6]^{3+}$, (c) high-spin $[Fe(OH_2)_6]^{3+}$.

14.6 Using the Tanabe–Sugano diagrams, estimate Δ_O and B for (a) $[Ni(OH_2)_6]^{2+}$ (absorptions at 8500, 15 400, and 26 000 cm^{-1}) and (b) $[Ni(NH_3)_6]^{2+}$ (absorptions at 10 750, 17 500, and 28 200 cm^{-1}).

14.7 (a) If an Fe(II) complex has a large paramagnetic susceptibility, what is the ground-state label according to the Tanabe–Sugano diagram? What is the description of the states involved in the lowest energy spin-allowed electronic transition?

14.8 The spectrum of $[Co(NH_3)_6]^{3+}$ has a very weak band in the red and two moderate-intensity bands in the visible to near-UV. How should these transitions be assigned?

14.9 Explain why $[FeF_6]^{3-}$ is colorless whereas $[CoF_6]^{3-}$ is colored but exhibits only a single band in the visible.

14.10 The Racah parameter B is 460 cm^{-1} in $[Co(CN)_6]^{3-}$ and 615 cm^{-1} in $[Co(NH_3)_6]^{3+}$. Consider the nature of bonding with the two ligands and explain the difference in nephelauxetic effect.

14.11 An approximately 'octahedral' complex of Co(III) with ammine and chloro ligands gives two bands with ε_{max} between 60 and 80 M^{-1} cm^{-1}, one weak peak with $\varepsilon_{max} =$ 2 M^{-1} cm^{-1} and a strong band at higher energy with $\varepsilon_{max} =$ 2×10^4 M^{-1} cm^{-1}. What do you suggest for the origins of these transitions?

14.12 Ordinary bottle glass appears nearly colorless when viewed through the wall of the bottle but green when viewed from the end (when the light has a long light-path through the glass). The color is associated with the presence of Fe^{3+} in the silicate matrix. Describe the transition.

14.13 $[Cr(OH_2)_6]^{3+}$ ions are pale violet but the chromate ion CrO_4^{2-} is a stronger yellow. Characterize the origins of the transitions and explain the relative intensities.

14.14 Classify the symmetry type of the d_{z^2} orbital in a tetragonal C_{4v} symmetry complex, such as $[CoCl(NH_3)_5]^{2+}$ where the Cl is on the z-axis. (a) Which orbitals will be displaced from their position in the octahedral molecular orbital diagram by π interactions with the lone pairs of the Cl^- ligand? (b) Which orbital will move because the Cl^- ligand is not as strong a σ base as NH_3? (c) Sketch the qualitative molecular orbital diagram for the C_{4v} complex.

14.15 Consider the molecular orbital diagram for a tetrahedral complex (Fig. 7.16) and the relevant d-orbital configuration. Show that the purple color of MnO_4^- ions cannot arise from a ligand field transition.

14.16 Given that the wavenumbers of the two transitions in MnO_4^- are 18 500 cm^{-1} and 32 200 cm^{-1}, explain how to estimate Δ_T from an assignment of the two charge-transfer transitions, even though Δ_T cannot be observed directly.

Problems

14.1 Consider the trigonal prismatic six-coordinate ML_6 complex with D_{3h} symmetry. Divide the *d* orbitals of the metal into sets of defined symmetry type assuming that the ligands are at the same angle relative to the *xy* plane as in a tetrahedral complex. (Hints: 1. Start with the tetrahedral case but assume three ligands above and below the *xy* plane rather than two. 2. The degeneracy of the *d* orbitals is the same in D_{3h} complexes as in trigonal ML_3 and trigonal bipyramidal ML_5 complexes.

14.2 The absorption spectra of the two isomers of $[CoCl_2(en)_2]^+$ are shown in Fig. 14.21. Considering the higher symmetry of the *trans* isomer, suggest which is *cis* and which is *trans*. Supposing that this assignment is correct, predict the CD spectrum of the Λ isomer of *cis*-$[CoCl_2(en)_2]^+$ by comparison with $[Co(en)_3]^{3+}$ shown in Fig. 14.11.

14.3 Figure 14.22 shows the EPR spectra of the anion produced by reduction of the square-planar iron tetraphenylporphyrin $[Fe(TPP)]^-$, of labeled $[^{57}Fe(TPP)]^-$, and of $[Fe(TPP)]^-$ in the presence of excess pyridine. Account for the form of the observed spectra. (G. S. Srivatsa, D. T. Sawyer, N. J. Boldt, and D. F. Bocian, *Inorg. Chem.*, **24**, 2123 (1985).)

14.4 $[Cr(CN)_6]^{3-}$ has absorption bands at 264 nm, 310 nm, and 378 nm. Evaluate Δ_O and *B* using the Tanabe–Sugano diagram. This is not straightforward. First decide how to assign the observed transitions, since the ligand has empty π^* orbitals. The complex also luminesces at 725 nm. Is this luminescence fluorescence or phosphorescence?

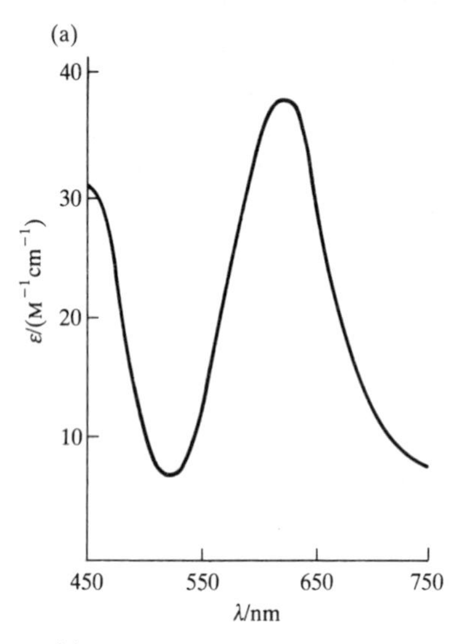

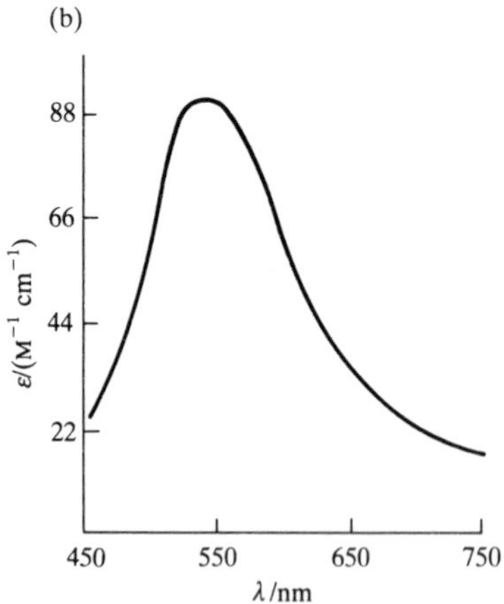

Fig. 14.21 The absorption and CD spectra of two isomers of $[CoCl_2(en)_2]^+$.

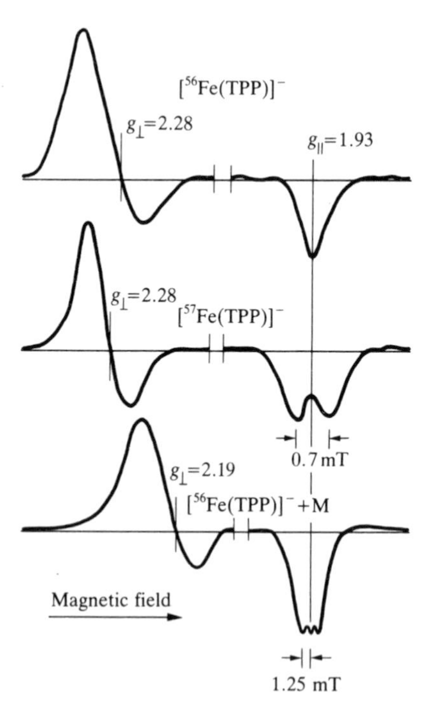

Fig. 14.22 The EPR spectra of the anion produced by reduction of the square-planar iron tetraphenylporphyrin $[Fe(TPP)]^-$, of labeled $[^{57}Fe(TPP)]^-$, and of $[Fe(TPP)]^-$ in the presence of excess pyridine.

Reaction mechanisms of *d*-block complexes

15

We encountered the distinction between labile and inert complexes in Chapter 7, and in Chapter 8 saw some of the factors that influence the rates of electron transfer. Since then many aspects of reactivity have been interpreted in terms of reaction mechanisms. Now we shall look in more detail at the evidence and experiments that are used in the analysis of reaction pathways and so develop a deeper appreciation of the mechanisms of reactions of *d*-block complexes. Since the mechanism of a reaction is rarely known finally and completely, the nature of the evidence for a mechanism should always be kept in mind in order to recognize what other possibilities might also be consistent with it. In the first part of the chapter we describe how reaction mechanisms are classified, and distinguish between the steps in which the reaction takes place and the details of the formation of the activated complex. Then these concepts are used to describe the currently accepted mechanisms for the substitution reactions of complexes, their redox reactions, and the reactions of photoexcited states of the kind mentioned in Section 14.6.

Ligand substitution reactions
15.1 The classification of mechanisms
15.2 Tests of the intimate mechanism
Substitution in square-planar complexes
15.3 Intimate mechanisms
15.4 Stoichiometric mechanisms
Substitution in octahedral complexes
15.5 Rate laws and their interpretation
15.6 Intimate mechanisms
15.7 Stereochemistry
15.8 Base hydrolysis
15.9 Isomerization reactions
15.10 Stoichiometric mechanisms: the *D* intermediate
The mechanisms of redox reactions
15.11 The classification of redox reactions
15.12 The theory of redox reactions
15.13 Oxidative addition
Photochemical reactions
15.14 Prompt and delayed reactions
15.15 *d–d* and charge-transfer reactions
15.16 Metal–metal bonded systems
Further reading
Exercises
Problems

We have seen the central importance of the interplay of equilibrium and kinetics for determining the outcome of inorganic reactions. We need to understand the mechanisms of the reactions in cases where kinetic considerations are dominant, and in this chapter we see how this may be achieved for the reactions of *d*-block complexes. The first two classes of reactions to be fully analyzed were ligand substitution and redox reactions, and we concentrate mainly on them.

Ligand substitution reactions

Ligand substitutions are reactions in which one Lewis base displaces another from a Lewis acid:

$$Y + M—X \rightarrow M—Y + X$$

They include complex formation reactions, in which the **leaving group**, the displaced base X, is a solvent molecule and the **entering group**, the displacing base Y, is some other ligand. An example is

$$[Co(OH_2)_6]^{2+} + Cl^- \rightarrow [CoCl(OH_2)_5]^+ + H_2O$$

Almost all the reactions considered in this chapter take place in water, and we shall not write the phase explicitly in equations.

We saw in Section 6.3 that the equilibrium constants of displacement reactions can be used to rank ligands in order of their strength as Lewis bases. However, a different order may be found if bases are ranked according to the rates at which they react. Therefore, for kinetic considerations the equilibrium concept of basicity is replaced by the kinetic concept of **nucleophilicity** (from the Greek for 'nucleus loving'). By nucleophilicity we mean the ratio of the rate of attack on a complex by a given base Y to the rate of attack by a standard base Y°. The shift from equilibrium to kinetic considerations is emphasized by referring to the displacements as **nucleophilic substitutions**.

We have already seen (in Section 7.8) that the rates of substitution reactions span a very wide range, and that to some extent there is a pattern in their behavior. For instance, aqua complexes of Group I, II, and 12 metal ions, lanthanide ions, and some 3*d*-metal ions in low oxidation states have half-lives as short as nanoseconds. In contrast, half-lives are years long for complexes of heavier *d* metals in high oxidation states, such as those of Ir(III) or Pt(IV). Intermediate half-lives are also observed. Two examples are $[Ni(OH_2)_6]^{2+}$, which has a half-life of the order of milliseconds, and $[Co(NH_3)_5(OH_2)]^{3+}$, in which H_2O survives for several minutes as a ligand before it is replaced by a stronger base.

15.1 The classification of mechanisms

One kinetic aspect of a reaction is its **stoichiometric mechanism**, the sequence of elementary steps by which the reaction takes place. A second kinetic aspect is its **intimate mechanism**, the details of the activation process and the energetics of formation of an activated complex in the rate determining step. The distinction reflects the two types of experiments that are available for studying mechanisms.

Stoichiometric mechanisms

The first stage in the kinetic analysis of a reaction is to study how its rate changes as the concentrations of reactants are varied. This leads to the identification of **rate laws**, the differential equations governing the rate of change of reactant or product concentration. For example, the observation that the rate of formation of $[Ni(OH_2)_5NH_3]^{2+}$ from $[Ni(OH_2)_6]^{2+}$ in the presence of excess ammonia is proportional to the concentration of NH_3 implies that the reaction is first-order in NH_3. The overall rate law is in fact

$$\text{rate} = k[\{Ni(OH_2)_6\}^{2+}][NH_3]$$

It is a fundamental assumption in kinetics that the species appearing in a rate law contribute to the formation or reaction of the activated complex in the rate-determining step, and therefore that an acceptable mechanism must display this contribution. Therefore, together with stereochemical and tracer studies, the establishment and interpretation of the rate law is the route to the elucidation of the stoichiometric mechanism of the reaction.

Three types of stoichiometric mechanism for ligand substitutions have been recognized:

Dissociative *D* Interchange *I* Associative *A*

The first and third are two-step mechanisms in which an intermediate is formed and then transformed into the product.

A **dissociative mechanism** involves a step in which an intermediate of reduced coordination number is formed after the departure of the leaving group. An example is the mechanism proposed for the substitution of hexacarbonyltungsten by a phosphine:

$$W(CO)_6 \rightarrow W(CO)_5 + CO$$

$$W(CO)_5 + PPh_3 \rightarrow W(CO)_5PPh_3$$

The generalized reaction profile is shown in Fig. 15.1a.

An **associative mechanism** (Fig. 15.1c) involves a step in which an intermediate is formed that has a higher coordination number than the original complex. This mechanism is believed to be responsible for many reactions of square planar Pt(II), Pd(II), and Ir(I) complexes. The exchange of $^{14}CN^-$ with the ligands in square-planar $[Ni(CN)_4]^{2-}$ is an example where the proposed intermediate is a compound that can be isolated under other conditions:

$$[Ni(CN)_4]^{2-} + {}^{14}CN^- \rightarrow [Ni(CN)_4({}^{14}CN)]^{3-}$$

$$[Ni(CN)_4({}^{14}CN)]^{3-} \rightarrow [Ni(CN)_3({}^{14}CN)]^{2-} + CN^-$$

An **interchange mechanism**, for which the reaction profile is given in Fig. 15.1b, takes place in one step. In an interchange mechanism, the leaving and entering groups exchange in a single step by forming an activated complex but not a true intermediate. Interchange is common for reactions of $[Ni(OH_2)_6]^{2+}$.

Intimate mechanisms

We shall see that it is useful to distinguish two types of intimate mechanism (the details of the formation of the activated complex), especially when we

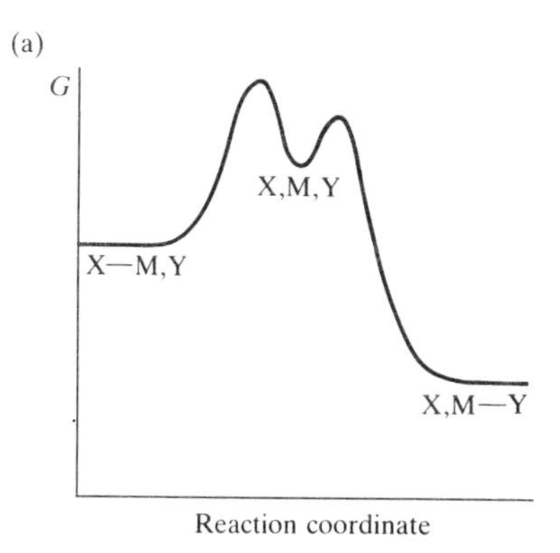

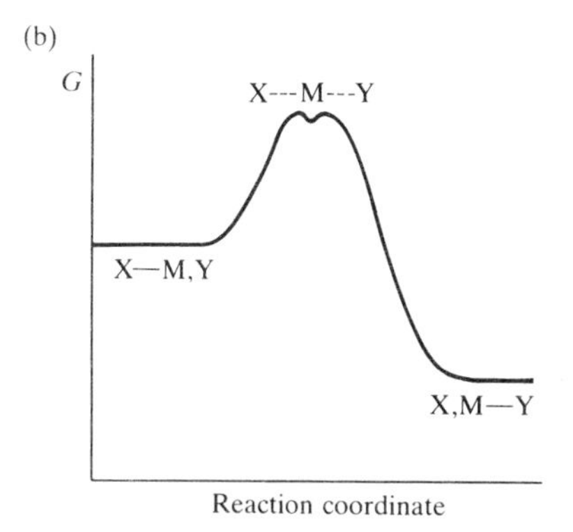

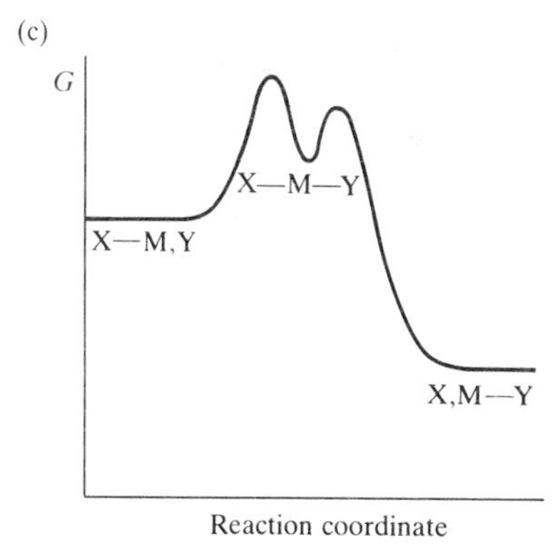

Fig. 15.1 Reaction profiles for (a) dissociative (*D*), (b) interchange (*I*), and (c) associative (*A*) mechanisms. A true intermediate exists in (a) and (c) but not in (b).

are dealing with an interchange mechanism. If the rate constant for the formation of the activated complex depends strongly on the identity of the entering group Y, we can infer that the activated complex must involve significant bonding to it. If the reaction rate is largely independent of Y, it must be controlled largely by the rate at which the bond between the metal and the leaving group X can break. The distinction between reactions with rates that depend strongly on the nature of the entering group and those that do not lead to a classification of reactions according to their intimate mechanisms. Thus, the activation process may be either **dissociative** (d) or **associative** (a).

Table 15.1. Relation between intimate and stoichiometric mechanisms of ligand substitution

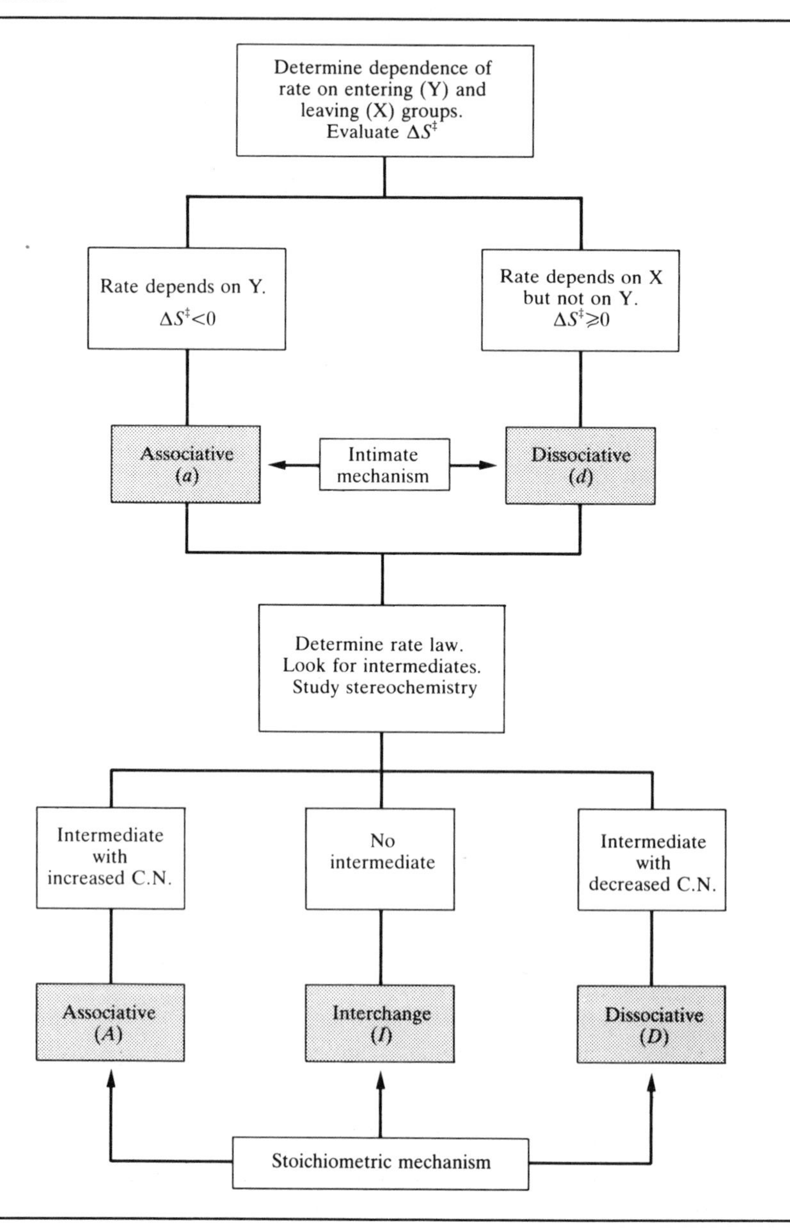

An **associatively activated reaction** is a reaction with a rate that is sensitive to the entering group. Examples are found among reactions of the square-planar complexes of Pt(II), Pd(II), and Au(III) including

$$[PtCl(dien)]^{+} + I^{-} \rightarrow [PtI(dien)]^{+} + Cl^{-}$$

in which replacement of I^- by Br^- changes the rate constant by an order of magnitude.

A **dissociatively activated reaction** (*d*) is a reaction with a rate that is insensitive to changes in the entering group. This category includes some of the classic examples of ligand substitution in octahedral *d*-metal complexes, including

$$[Ni(OH_2)_6]^{2+} + NH_3 \rightarrow [Ni(OH_2)_5NH_3]^{2+} + H_2O$$

Replacement of NH_3 by pyridine in this reaction changes the rate by at most a few percent.

Table 15.1 summarizes the relation between intimate and stoichiometric mechanisms. The dissociative and associative activation mechanisms correspond approximately to the S_N1 and S_N2 mechanisms of nucleophilic substitution in organic chemistry. However, *d* and *a* refer specifically to the intimate mechanism of the activation process.

15.2 Tests of the intimate mechanism

Once the overall stoichiometry of a substitution reaction is known, the determination of its mechanism usually begins with an analysis of the intimate mechanism, and then the stoichiometric mechanism (Table 15.1). We use the composite labels A, I_a, I_d, and D if we know both the stoichiometric and intimate mechanisms. The A and I_a mechanisms are very difficult to distinguish from each other, because the composition of the activated complex is the same in both cases. The distinction hinges on whether or not the intermediate persists long enough to be detectable.

In principle, we can analyze the intimate mechanism by examining the dependence of rate constants on temperature, pressure, solvent permittivity, and solvent viscosity. However, since changes of physical conditions affect many features of a reaction at once, the dependence is often hard to interpret unambiguously. The alternative approach, observing how the rate constant is affected by changes in the structures of the reactants, often turns out to be simpler.

Structural variation

Several structural variables are available for the study of ligand substitution reactions. They include:

- The identities of entering and leaving groups (Y and X respectively).
- The identities of the ligands *cis* or *trans* to the reaction center.
- The steric crowding at the reaction center.
- The overall charge on the complex and solvent properties.
- The metal itself.

It is only rarely that these factors can be studied independently, but each plays a part in the identification of the intimate mechanism. The assumption is normally made that the variation of structure produces only a small perturbation of the pathway of the reaction. However, we should always be alert to the possibility that a small change in structure may open an entirely new reaction pathway. As we saw in Section 7.8, if the rate of reaction is dependent on the identity of the entering ligand (e.g. Cl^- versus Br^-) but independent of the identity of the leaving ligand, the reaction is likely to have an associative (a) intimate mechanism. Conversely, insensitivity to the identity of the entering ligand combined with sensitivity to the identity of the leaving ligand implies a dissociative (d) intimate mechanism.

Variation of physical parameters

Temperature is the most commonly studied physical variable. According to activated complex theory, we can express the temperature dependence of a rate constant k_r as

$$k_r = \kappa \frac{kT}{h} e^{-\Delta G^{\ddagger}/RT} \quad (1a)$$

The factor kT/h (where k is the Boltzmann's constant and h is Planck's constant) describes the rate at which the activated complex attains the transition state (its gateway into products). The factor κ, the transmission coefficient, expresses the probability of the activated complex making a successful transit of the transition state toward product. Attention usually focuses on $\Delta G^{\ddagger}$, the Gibbs free energy of activation, and its decomposition into $\Delta S^{\ddagger}$ and $\Delta H^{\ddagger}$, the entropy and enthalpy of activation:

$$\Delta G^{\ddagger} = \Delta H^{\ddagger} - T\,\Delta S^{\ddagger} \quad (1b)$$

We shall have much to say on these three quantities but there are no simple generalizations about $\Delta H^{\ddagger}$.

An important general factor affecting the entropy of activation is the effect of a change in charge that occurs when an activated complex is formed. If the charge density increases, as happens when two ions of like charge form the activated complex, the solvent molecules will be more ordered around the complex in the process called **electrostriction**. Hence, the entropy decreases, the entropy of activation is negative ($\Delta S^{\ddagger} < 0$), and the rate constant k_r is reduced. If the reaction leads to charge reduction in the activated complex (such as happens when two oppositely charged ions combine), electrostriction is relaxed and the increased disorder of the solvent gives a positive contribution to the entropy of activation. In this case, electrostriction increases the rate constant. If corrected for electrostriction, a large negative value of $\Delta S^{\ddagger}$ is evidence for an associative intimate mechanism, a. For $\Delta S^{\ddagger} \geqslant 0$ a dissociative, d, intimate mechanism is indicated.

The effect of pressure on rate is expressed in terms of the **volume of activation** $\Delta V^{\ddagger}$, the difference in molar volume between the initial reactants and the activated complex:

$$\frac{d \ln k_r}{dp} = \frac{-\Delta V^{\ddagger}}{RT} \qquad (2)$$

In principle, information about the volume change experienced by the complex and entering ligand as the activated complex is formed is a good guide to the mechanism. Unfortunately, the volume of activation is closely related to the extent of electrostriction and a simple relation between $\Delta V^{\ddagger}$ and the geometrical changes in the complex as the activated complex is formed can be expected only for reactions in which the charge does not change. For such 'charge-neutral reactions' dissociatively activated reactions are expected to result in $\Delta V^{\ddagger} > 0$ as a ligand leaves the primary coordination sphere. In contrast, an associatively activated reaction (*a*) is expected to be accompanied by $\Delta V^{\ddagger} < 0$ as an extra ligand is incorporated into the complex. The measurement of $\Delta V^{\ddagger}$ requires special high-pressure equipment, so this determination is less commonly made than $\Delta S^{\ddagger}$, which as we have seen is determined from the temperature variation of the rate. Nevertheless, the mechanistic interpretation of $\Delta V^{\ddagger}$ is generally considered to be the more straightforward of the two.

Substitution in square-planar complexes

We saw in Section 7.8 that nonlabile d^8 complexes of Pt(II), Pd(II) and Au(III) complexes undergo associatively activated substitution. We emphasized in that section only the effect of the entering group and saw that the rates of substitution reactions lie in the order

$$\text{Y: } H_2O < Cl^- < I^- < H^- < PR_3 < CO, CN^-$$

It is quite easy to summarize the dependence of the rate on the leaving group too, because all we need do is to reverse the order:

$$\text{X: } CN^-, CO < PR_3 < H^- < I^- < Cl^- < H_2O$$

Thus, good entering groups (good nucleophiles) are poor (nonlabile) leaving groups.

In this section we shall also consider the role of a **spectator ligand**, a ligand that is present in the complex but not directly involved in the change of coordination number. We shall see that a good nucleophile is a good reaction accelerator if it occupies a *trans* position (T) as a spectator ligand. Thus X, Y, and T play related roles, and knowing how a ligand affects the rate when it is Y allows us to predict how it affects the rate when it is X or T.

15.3 Intimate mechanisms

Experimental observations on the sensitivity of rates to structure variation, as well as pressure and temperature variation, have led to the view that the intimate mechanism of square planar complexes is associative (*a*).

Entering group effects

The characteristic rate law of reactions of the type

$$[PtCl(dien)]^+ + I^- \rightarrow [PtI(dien)]^+ + Cl^-$$

is

$$\text{rate} = (k_1 + k_2[I^-])[\{PtCl(dien)\}^+]$$

where k_1 and k_2 are respectively first- and second-order rate constants. Experimental evidence for this rate law usually comes from a study of substitution in the presence of excess I^- ions, so the rate laws are pseudofirst-order with effective rate constants

$$k_{obs} = k_1 + k_2[I^-]$$

When the observed pseudofirst-order rate constant is plotted against $[I^-]$, the slope of the straight line gives k_2 and the intercept gives k_1.

The joint first- and second-order rate law suggests that there are two reaction pathways, and we shall deal first with the path that leads to second-order kinetics. To do so, we note that we can express the reactivity of the entering group Y (the I^- in the reaction above) in terms of the **nucleophilicity parameter** n_{Pt}:

$$n_{Pt}(Y) = \log \frac{k_2(Y)}{k_2^\circ} \qquad (3a)$$

or, equivalently,

$$\log k_2(Y) = n_{Pt}(Y) + \log k_2^\circ \qquad (3b)$$

where $k_2(Y)$ is the second-order rate constant of the reaction

$$trans\text{-}[PtCl_2(py)_2] + Y \rightarrow trans\text{-}[PtClY(py)_2]^+ + Cl^-$$

and k_2° is the rate constant for the same reaction with a standard base, methanol. We give some value of n_{Pt} in Table 15.2.

One feature of the data in Table 15.2 is that, although the entering groups in the table are all quite simple, the rate constants span nearly nine orders of magnitude. Another feature is that the nucleophilicity of the entering group toward Pt appears to correlate with soft Lewis basicity, thus, $Cl^- < I^-$, $O < S$, and $NH_3 < PPh_3$.

Table 15.2. A selection of n_{Pt} values for a range of nucleophiles

Nucleophile	Donor atom	n_{Pt}
Cl^-	Cl	3.04
C_6H_5SH	S	4.15
CN^-	C	7.00
$(C_6H_5)_3P$	P	8.79
CH_3OH	O	0
I^-	I	5.42
NH_3	N	3.06

The nucleophilicities of the entering groups lie in almost the same order for a range of different Pt complexes. However, there is some variation, and for an arbitrary reaction of the form

$$[L_3PtX] + Y \rightarrow [L_3PtY] + X$$

instead of eqn 3b we write

$$\log k_2(Y) = S n_{Pt}(Y) + C \qquad (4)$$

The parameter S, which is a measure of the sensitivity of the rate constant to the nucleophilicity parameter, is called the **nucleophilic discrimination**

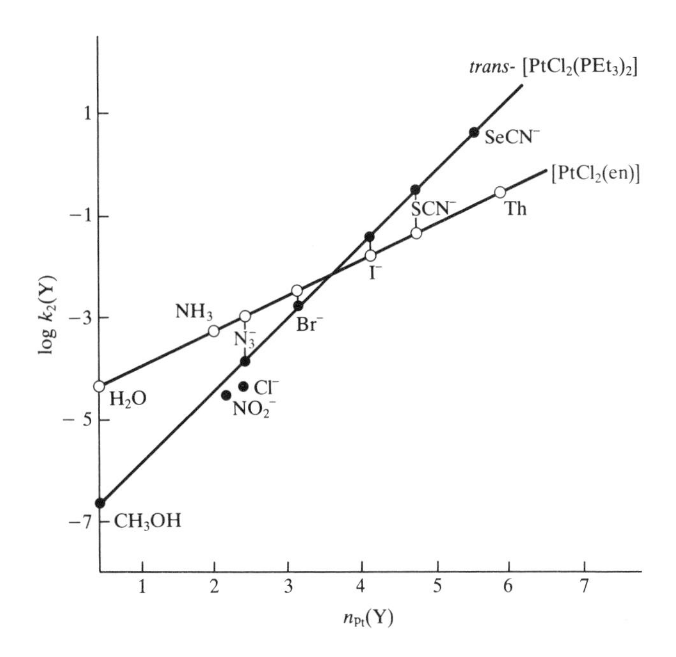

Fig. 15.2 The slope of the straight line obtained by plotting log k_2(Y) against the nucleophilicity parameter n_{Pt}(Y) for a series of ligands is a measure of the responsiveness of the complex to the nucleophilicity of the entering group.

factor. We see from Fig. 15.2 that the straight line obtained by plotting $\log k_2(Y)$ against n_{Pt} for reactions of Y with *trans*-$[PtCl_2(PEt_3)_2]$ is steeper than that for reactions with *trans*-$[PtCl_2(en)]$. Hence, S is larger for the former reaction, which indicates that that reaction's rate is more sensitive to changes in nucleophilicity.

Some values of S are given in Table 15.3. Notice that S is close to 1 in all cases, so all the complexes are quite sensitive to n_{Pt}. This sensitivity is what we expect for reactions with an associative intimate mechanism. Another feature to note is that larger values of S are found for complexes of Pt with softer base ligands.

Table 15.3. Nucleophilic discrimination factors

	S
$[PtCl_2(PEt_3)_2]$	1.43
$[PtCl_2(py)_2]$	1.00
$[PtCl_2(en)]$	0.64
$[PtCl(dien)]^+$	0.65

Example 15.1: *Using the nucleophilicity parameter*

The second-order rate constant for the reaction of I^- with *trans*-$[PtCl(CH_3)(PEt_3)_2]$ in methanol at 30°C is $40\,\text{M}^{-1}\,\text{s}^{-1}$. The corresponding reaction with N_3^- has $k_2 = 7.0\,\text{M}^{-1}\,\text{s}^{-1}$. Estimate S and the first-order rate constant for reaction of this complex with methanol (k_2°) given the n_{Pt} values 5.42 and 3.58 for the two nucleophiles.

Answer. Substituting the two values of n_{Pt} into eqn 3 gives

$$1.60 = 5.42S + C \text{ for } I^-$$

$$0.85 = 3.58S + C \text{ for } N_3^-$$

Solving these two simultaneous equations gives $S = 0.41$ and $C = -0.61$. The value of S is rather small, showing that the discrimination among different nucleophiles is not great. This lack of sensitivity is related to the rather large value of C, which corresponds to the rate constant being large and hence to the complex being reactive. It is commonly found that high reactivity correlates with low selectivity.

Exercise. Calculate the second-order rate constant for the reaction of the same complex with NO_2^-, for which $n_{Pt} = 3.22$.

Table 15.4. The effect of the *trans* ligand in Cl^- replacement reactions of *trans*-$[PtCl(PEt_3)_2L]$

L	k_1/s^{-1}	$k_2/(M^{-1}\,s^{-1})$
CH_3^-	1.7×10^{-4}	6.7×10^{-2}
$C_6H_5^-$	3.3×10^{-5}	1.6×10^{-2}
Cl^-	1.0×10^{-6}	4.0×10^{-4}
H^-	1.8×10^{-2}	4.2
PEt_3	1.7×10^{-2}	3.8

The *trans* effect

The ligands T that are *trans* to the leaving group in square-planar complexes give rise to the spectacular **trans effect**. The *trans* effect is that, if T is a strong σ base or π acid, it greatly accelerates substitution of a ligand that lies *trans* to itself (Table 15.4).

The strength of the *trans* effect on Pt(II) reactions depends on both the σ and π acceptor effects:

For T a σ base:

$$OH^- < NH_3 \approx py < Cl^- < Br^- < CN^-, CO, CH_3^- < I^- < \underline{S}CN^- < PR_3 < H^-$$

For T a π acid:

$$Br^- < I^- < \underline{N}CS^- < \underline{N}O_2^- < CN^- < CO, C_2H_4$$

These orders are roughly the order of increasing overlap of the ligand orbitals with either a σ or a π Pt 6*p* orbital. That is, the greater the overlap, the stronger the *trans* effect. It should be noticed that the same factors contribute to a large ligand-field splitting.

Example 15.2: *Using the trans effect synthetically*

Use the *trans* effect series to suggest synthetic routes to *cis*- and *trans*-$[PtCl_2(NH_3)_2]$ from $[Pt(NH_3)_4]^{2+}$ and $[PtCl_4]^{2-}$.

Answer. Since the *trans* effect of Cl^- is greater than that of NH_3, substitution reactions will occur preferentially *trans* to Cl^-. Reaction of $[Pt(NH_3)_4]^{2+}$ with HCl can lead to $[PtCl(NH_3)_3]^+$. The second step will result in substitution *trans* to Cl so that the action of HCl gives *trans*-$[PtCl_2(NH_3)_2]$. When the starting complex is $[PtCl_4]^{2-}$, reaction with NH_3 can lead first to $[PtCl_3(NH_3)]^-$. A second step should substitute a Cl^- *trans* to a remaining Cl^-. The second reaction step with NH_3 gives *cis*-$[PtCl_2(NH_3)_2]$.

Exercise. Given the reactants triphenylphosphine (PPh_3), NH_3, and $[PtCl_4]^{2-}$, propose efficient routes to *cis*- and *trans*-$[PtCl_2(NH_3)(PPh_3)]$.

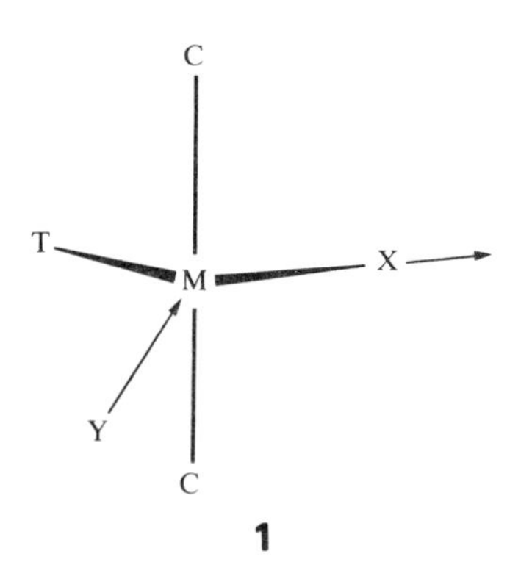

1

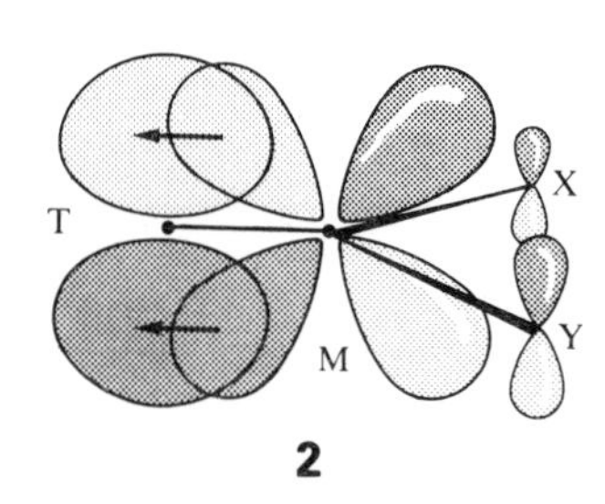

2

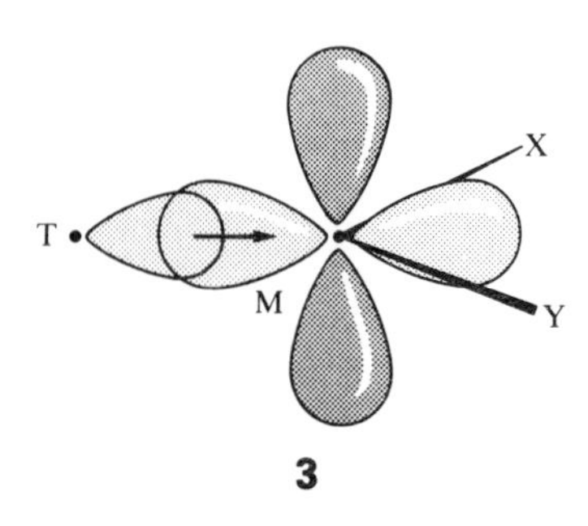

3

The correlation of the strength of the *trans* effect with the nucleophilicity of the entering group and the nonlability of the leaving group provides a strong clue to the form of the activated complex involved in this associatively activated reaction. If the influence of the two *cis* spectator ligands is small, and if the *trans*, entering, and leaving ligands play complementary roles, then we can suspect that a trigonal bipyramidal structure is likely for the activated complex. In this model there are two locations of one kind (the axial locations, occupied by the two *cis* ligands) and three of another kind (the equatorial positions, occupied by X, Y, and T).

The process that might occur when Y attacks from above the plane of the initial complex and X is destined to leave from below it, is shown in (**1**). It is speculated that the π acid can provide a pathway for the removal of electron density from the vicinity of the *trans* ligand and hence facilitate nucleophilic attack on the metal center (**2**). A strong σ-donor T ligand must share a metal orbital with X in a square-planar complex but gains a larger share of the orbital in a trigonal bipyramid (**3**).

Steric effects

Steric crowding at the reaction center is usually assumed to inhibit associatively activated reactions and to facilitate dissociatively activated reactions. Bulky groups can block the approach of attacking nucleophiles, but the reduction of coordination number in a dissociatively activated reaction can relieve the crowding in the activated complex. The rate constants for the hydrolysis of *cis*-$[PtClL(PEt_3)_2]$ complexes (the replacement of Cl^- by H_2O) at 25°C illustrate the point:

L =	pyridine	2-methylpyridine	2,6-dimethylpyridine
k/s^{-1}	8×10^{-1}	2.0×10^{-4}	1.0×10^{-6}

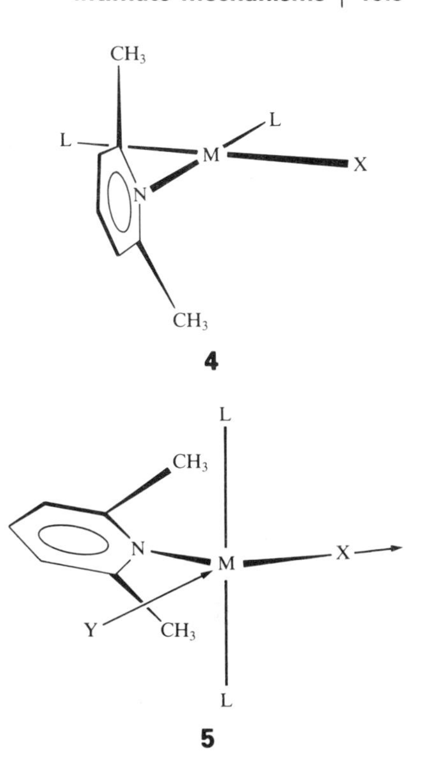

The *ortho*-CH_3 groups of the aryl ligands block positions above or below the plane of the 2-methylpyridine complex and both above and below the plane of the 2,6-dimethylpyridine complex, and so increasingly hinder attack (**4**) by H_2O. The effect is smaller if L is *trans* to Cl^-. This is explained by the CH_3 groups then being further from the entering and leaving groups in the trigonal bipyramidal activated complex if the aryl ligand is in the trigonal plane (**5**).

Stereochemistry

Further insight into the nature of the activated complex is obtained from the observation that substitution of a square-planar complex conserves the original geometry. That is, a *cis* complex gives a *cis* complex product and a *trans* complex gives a *trans* complex product. This is readily explained on the assumption that the activated complex is approximately trigonal bipyramidal (**1**) with the entering (Y), leaving (X), and *trans* (T) groups in the trigonal plane. The steric course of the reaction is shown in Fig. 15.3. We can expect a *cis* ligand to exchange places with the T ligand in the trigonal plane only if the intermediate lives long enough to be stereomobile. That is, it must be a long-lived associative (*A*) intermediate.

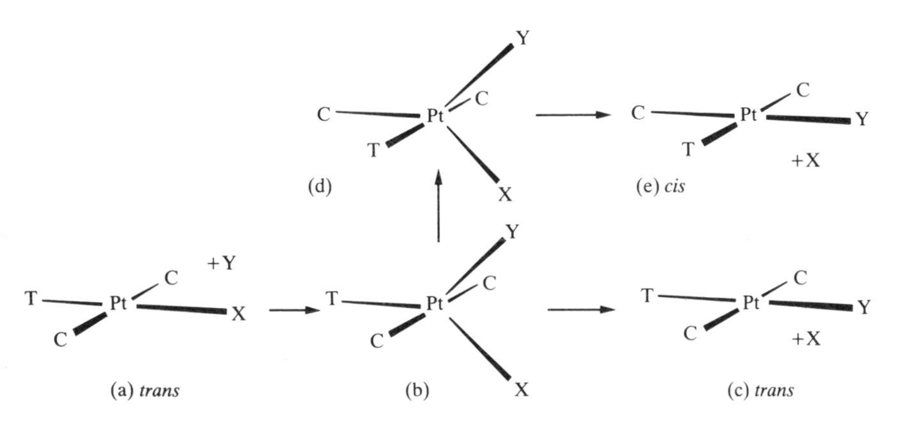

Fig. 15.3 The stereochemistry of substitution in a square planar complex. The normal path (resulting in retention) is from (a) to (c). However, if the intermediate (b) is sufficiently long lived, it can undergo pseudorotation to (d), which leads to the isomer (e).

Temperature and pressure dependence

Table 15.5 summarizes the enthalpies, entropies, and volumes of activation for reactions of Pt(II) and Au(III) complexes. The two striking aspects are the consistently strongly negative values of the entropies and volumes of activation. The simplest explanation is that the entering ligand is being incorporated into the activated complex without release of the leaving group. That is, we can conclude that the intimate mechanism is associative.

Table 15.5. Activation parameters for substitution in square-planar complexes (in methanol as solvent)

Reaction	k_1			k_2		
	$\Delta H^\ddagger$	$\Delta S^\ddagger$	$\Delta V^\ddagger$	$\Delta H^\ddagger$	$\Delta S^\ddagger$	$\Delta V^\ddagger$
trans-$[PtCl(NO_2)(py)_2] + py$				50	−100	−38
trans-$[PtBr(Mes)(PEt_3)_2]^* + SC(NH_2)_2$	71	−84	−46	46	−138	−54
cis-$[PtBr(Mes)(PEt_3)_2]^* + I^-$	84	−59	−67	63	−121	−63
cis-$[PtBr(Mes)(PEt_3)_2]^* + SC(NH_2)_2$	79	−71	−71	59	−121	−54
$[AuCl(dien)]^{2+} + Br^-$				54	−17	

* Mes = 2,4,6-trimethylphenyl. Enthalpy terms in kJ mol^{-1}, entropy in J K^{-1} mol^{-1}, and volumes in cm^3 mol^{-1}

15.4 Stoichiometric mechanisms

The k_1 pathway

The first issue we must address is the first-order pathway in the rate equation; we must decide whether k_1 in the rate law

$$\text{rate} = (k_1 + k_2[\text{Y}])[\text{PtL}_4]$$

does indeed represent the operation of an entirely different reaction mechanism. It turns out that it does not, for we can account for k_1 by an associative reaction that involves the solvent. Thus, the substitution of Cl^- by pyridine in methanol as solvent proceeds in two steps:

$$[\text{PtCl(dien)}]^+ + \text{CH}_3\text{OH} \rightarrow [\text{Pt(dien)(CH}_3\text{OH)}]^{2+} + \text{Cl}^- \qquad \text{(slow)}$$

$$[\text{Pt(dien)(CH}_3\text{OH)}]^{2+} + \text{py} \rightarrow [\text{Pt(dien)(py)}]^{2+} + \text{CH}_3\text{OH} \qquad \text{(fast)}$$

The evidence for this two-step mechanism comes from a correlation of the rates of these reactions with the nucleophilicity parameters of the solvent molecules. It has also been demonstrated explicitly that reactions of entering groups with solvo complexes are rapid compared to the step in which the solvent displaces a ligand. Supporting evidence for the similarity of the k_1 and k_2 pathways is found in Table 15.5, which shows that the entropies and volumes of activation are negative for both.

Associative intermediates

The dip in the Gibbs free energy surface that distinguishes Fig. 15.1b from Fig. 15.1c must be deep if the intermediate is to be detectable chemically and hence justify an *A* rather than an I_a classification. Three types of evidence have been considered.

One type of evidence is the isolation of an intermediate in another related reaction or under different conditions. If an argument by extrapolation to the actual reaction conditions suggests that a moderately long-lived intermediate might exist during the reaction in question, the *A* path is indicated. For example, the synthesis of the first trigonal bipyramidal Pt(II) complex, $[Pt(SnCl_3)_5]^{3-}$, indicated that a five-coordinate Pt complex may be plausible in substitution reactions of square-planar Pt(II) ammine complexes. The fact that $[Ni(CN)_5]^{3-}$ is found in crystals is strong support for its presence when CN^- exchanges with the square planar tetracyanonickelate(II) ion.

A second possible indication of the persistence of an intermediate is the observation of a change in stereochemistry, for this implies that the intermediate has lived long enough to undergo rearrangement (Fig. 15.3). *Cis* → *trans* isomerization is observed in the substitution reactions of certain square-planar phosphine Pt(II) complexes in contrast to the complete retention of original configuration usually observed, which implies that the trigonal bypyramid lives long enough for an exchange between the axial and equatorial ligand positions to occur.

Direct spectroscopic detection of the intermediate may be possible if a sufficient amount accumulates. Such direct evidence, however, requires an unusually stable intermediate with favorable spectroscopic characteristics and has not yet been obtained for Pt(II) complexes.

Substitution in octahedral complexes

The interchange mechanism plays a central role in the discussion of octahedral substitution. The analysis of rate laws for reactions that take place by an interchange mechanism helps us to formulate the precise conditions for distinguishing interchange from dissociative reactions. They can also clarify the conditions required for testing the response of the reaction to changes in nucleophilicity of the entering group.

15.5 Rate laws and their interpretation

The rate law of an octahedral substitution by an interchange mechanism is usually consistent with an **Eigen–Wilkins mechanism**. In this mechanism, an encounter complex is formed in a pre-equilibrium step and then rearranges to the products. The Eigen–Wilkins mechanism can be applied to all bimolecular reactions in solution that occur at rates less than the limit imposed by diffusion.

The Eigen–Wilkins mechanism

The first step in the Eigen–Wilkins mechanism is an encounter in which the complex C and the entering group Y diffuse together. They may also separate at diffusion limited rates, and the equilibrium

$$\mathrm{C} + \mathrm{Y} \rightleftharpoons \{\mathrm{CY}\} \qquad K_{\mathrm{E}} = \frac{[\{\mathrm{CY}\}]}{[\mathrm{C}][\mathrm{Y}]}$$

is established. Since in ordinary solvents the lifetime of a diffusional encounter is approximately 1 ns, the formation of an encounter complex can be treated as a pre-equilibrium in all reactions that take longer than a few nanoseconds. (A special case arises if Y is the solvent. Then the encounter equilibrium is saturated because the complex is always surrounded by solvent.)

The second step is the rate-determining reaction of the encounter complex to give products:

$$\{\mathrm{CY}\} \rightarrow \text{product} \qquad \text{rate} = k[\{\mathrm{CY}\}]$$

When the equilibrium equation is solved for the concentration of {CY} and the result substituted into the rate law for the rate determining step, the overall rate law is found to be

$$\text{rate} = \frac{kK_E[C][Y]}{1 + K_E[Y]}$$

An example is the reaction

$$[Ni(OH_2)_6]^{2+} + NH_3 \rightleftharpoons \{[Ni(OH_2)_6]^{2+}, NH_3\}$$

$$\{[Ni(OH_2)_6]^{2+}, NH_3\} \rightarrow [Ni(OH_2)_5NH_3]^{2+} + H_2O$$

$$\text{rate} = \frac{kK_E[Ni(OH_2)_6][NH_3]}{1 + K_E[NH_3]}$$

It is rarely possible to conduct experiments over a range of concentrations wide enough to test the rate law exhaustively, but this has been done in a few cases. At concentrations of the entering group that are so low that $K_E[Y] \ll 1$, the rate law reduces to the second-order form

$$\text{rate} = k_{obs}[C][Y] \qquad k_{obs} = kK_E$$

Since k_{obs} can be measured and K_E estimated as we describe below, the rate constant k can be found from k_{obs}/K_E. The results for reactions of Ni(II) hexaaqua complexes with various nucleophiles are shown in Table 15.6. The very small variation in k is in sharp contrast to the wide variation typical of associative reactions. This is a model I_d reaction with very slight response to the nucleophilicity of the entering group.

Table 15.6. Complex formation by the $[Ni(OH_2)_6]^{2+}$ ion

Ligand	$k_{obs}/(M^{-1}\,s^{-1})$	K_E/M^{-1}	$(k_{obs}/K_E)/s^{-1}$
$CH_3CO_2^-$	1×10^5	3	3×10^4
F^-	8×10^3	1	8×10^3
HF	3×10^3	0.15	2×10^4
H_2O*			3×10^3
NH_3	5×10^3	0.15	3×10^4
$[NH_2(CH_2)_2NH_3]^+$	4×10^2	0.02	2×10^3
SCN^-	6×10^3	1	6×10^3

* The solvent is always in encounter with the ion so that K_E is undefined and all rates are inherently first-order.

In the special case that Y is the solvent and the encounter equilibrium is saturated, $k_{obs} = k$. Thus, reactions with the solvent can be directly compared to reactions with other entering ligands without needing to estimate a value of K_E.

The Fuoss–Eigen equation

The equilibrium constant for the encounter complex (K_E) can be estimated theoretically using a simple equation proposed independently by R. M. Fuoss of Yale University and M. Eigen of Göttingen. Both sought to take the particle size and charge into account, expecting that larger, oppositely charged particles would meet more frequently than small ions of the same charge. Fuoss used an approach based on statistical thermodynamics and Eigen used one based on kinetics. Their result, which is called the

Fuoss–Eigen equation, is

$$K_E = \frac{4\pi}{3} a^3 N_A e^{-V/kT} \quad (5)$$

a is the distance of closest approach, V the electrostatic potential energy at that distance, and N_A is Avogadro's constant. The equilibrium constant favors the encounter if the reactants are large (so a is large) and oppositely charged (V negative).

15.6 Intimate mechanisms

Many studies of substitution reactions in octahedral complexes support dissociative activation, and we summarize these first. However, octahedral substitutions can acquire some distinct associative activation character in the case of large central ions (as in the second and third series of the *d* block) or where the *d*-electron population at the metal is low (the early members of the *d* block). More room for attack or lower π^* electron density appears to facilitate nucleophilic attack and hence associative activation.

Leaving group effects

We can expect a large effect of the leaving group X in dissociatively activated reactions since their activation energies are determined by the unassisted scission of the M—X bond. When the identity of X can be isolated as the only variable, as in the reaction

$$[CoX(NH_3)_5]^{2+} + H_2O \rightarrow [Co(NH_3)_5(OH_2)]^{3+} + X^-$$

it is found that there is a linear relation between the logarithms of the rate and equilibrium constant of the reaction (Fig. 15.4), and specifically

$$\ln k = \ln K + c$$

Since both logarithms are proportional to Gibbs free energy changes ($\ln k$ proportional to the activation Gibbs energy and $\ln K$ to the standard reaction Gibbs energy) the linear relation between $\ln k$ and $\ln K$ is called a **linear free energy relation** (LFER). In general, a LFER has the form

$$\Delta G^{\ddagger} = a\,\Delta G^{\ominus} + b$$

with a and b constants. In the special case of the reactions described by Fig. 15.4, the experimental fact that $a = 1.00$ shows that changing X has the same effect on the change in free energy for the conversion of Co—X to the activated complex as it has on the change in free energy for the complete elimination of X^- (Fig. 15.5). This observation in turn suggests that in a dissociatively activated reaction the leaving group becomes a solvated ion in the activated complex.

A similar linear free energy relationship (with a slope slightly smaller than 1.0, indicating some *a* character) is observed for the corresponding complexes of Rh(III). However, as we should expect for the softer Rh(III) center, the order of leaving groups is reversed, and the reaction rates are in the order

$$I^- < Br^- < Cl^-$$

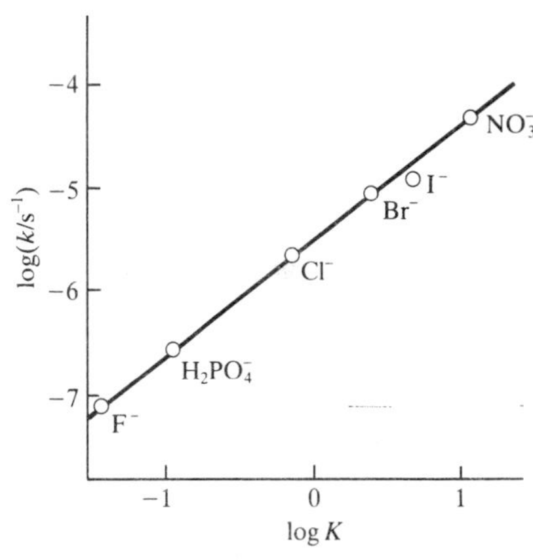

Fig. 15.4 The straight line obtained when the logarithm of a rate constant is plotted against the logarithm of an equilibrium constant shows the existence of a linear free energy relation. This graph is for the reaction

$$[CoX(NH_3)_5]^{2+} + H_2O \rightarrow [Co(NH_3)_5(OH_2)]^{3+} + X^-$$

with different leaving groups X.

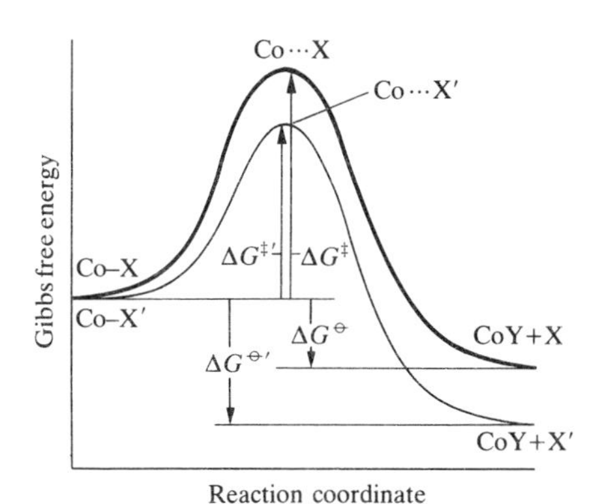

Fig. 15.5 The existence of a linear free energy relation with unit slope shows that changing X has the same effect on $\Delta G^{\ddagger}$ for the conversion of M—X to the activated complex as it has on $\Delta G^{\ominus}$ for the complete elimination of X^-. This reaction profile shows the effect of changing the leaving group from X to X'.

The effects of spectator ligands

Ligands *cis* to a leaving group usually have nonspecific effects on rates. In Co(III), Cr(III) and related octahedral complexes, both *cis* and *trans* ligands affect rates of substitution in proportion to the strength of the bonds they form with the metal. There is no significant *trans* effect. For instance, hydrolysis reactions such as

$$[NiXL_5]^+ + H_2O \rightarrow [NiL_5(OH_2)]^{2+} + X^-$$

are much faster when L is NH_3 than when it is H_2O. This can be explained on the grounds that as NH_3 is a stronger σ base than H_2O, it increases the electron density at the metal and hence facilitates the scission of the M—X bond and the formation of X^-. In the activated complex, the good donor stabilizes the reduced coordination number.

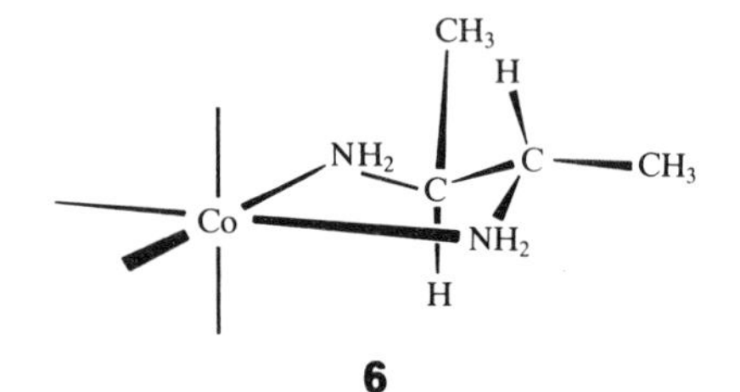

6

7

Steric effects

We can illustrate the steric effects on dissociatively activated reactions by considering the rate of hydrolysis of the first Cl^- ligand in two complexes of the type $[CoCl_2(bn)_2]^+$:

$$[CoCl_2(bn)_2]^+ + H_2O \rightarrow [CoCl(bn)_2(OH_2)]^{2+} + Cl^-$$

The ligand bn is 2,3-butanediamine, and may be either chiral (**6**) or achiral (**7**). The important observation is that the complex formed with the chiral form of the ligand hydrolyzes 30 times more slowly than the complex of the achiral form. The two ligands have very similar electronic effects, but the CH_3 groups are on the opposite sides of the chelate ring in (**6**) and adjacent and crowded in (**7**). The latter is more reactive because the strain is relieved in the dissociatively activated complex with its reduced coordination number. In general, steric crowding favors dissociative activation because the five-coordinate activated complex can relieve strain.

The effect of overall charge

Since nucleophiles are anions or negative ends of dipoles, we can expect greater positive charge or lower negative charge on a complex to favor associatively activated reactions. Conversely, higher overall positive or lower negative charge can be expected to retard the loss of anionic or polar leaving groups, thereby retarding dissociatively activated reactions. It is difficult to isolate the effects of variation in charge from the accompanying electronic effects caused by changes in the ligand. However, observations on a series of Co(III) ammine complexes with net charges +2 and +3 suggest that the increase of charge retards hydrolysis of Cl^- by a factor of 10^2 to 10^3 at 25°C.

Activation energetics

Among the most fruitful attempts to isolate the factors affecting activation parameters is the identification of a correlation between $\Delta H^\ddagger$ and the **ligand field activation energy** (LFAE). The latter is the difference in ligand field stabilization energies of the reactant and a model of the activated complex:

$$\text{LFAE} = (\text{LFSE})^\ddagger - (\text{LFSE})$$

Table 15.7 shows the results of a calculation in which the dissociatively activated complex is assumed to be a five-coordinate square pyramid. The

Table 15.7. Activation parameters for the H_2O exchange reactions $[M(OH_2)_6]^{2+} + H_2{}^{17}O \rightarrow [M(OH_2)_5({}^{17}OH_2)]^{+} + H_2O$

	$\Delta H^{\ddagger}/(\text{kJ mol}^{-1})$	LFSE*/Δ_0	LFSE†/Δ_0	LFAE/Δ_0	$\Delta V^{\ddagger}/(\text{cm}^3\text{ mol}^{-1})$
$Sc^{2+}(d^1)$§		0.4	0.46	−0.06	
$Ti^{2+}(d^2)$		0.8	0.91	−0.11	
$V^{2+}(d^3)$	68.6	1.2	1.0	0.2	−10.1
$Cr^{2+}(d^4$, hs)		0.6	0.91	−0.31	
$Mn^{2+}(d^5$, hs)	33.9	0.0	0.0	0.0	−4.1
$Fe^{2+}(d^6$, hs)	31.2	0.4	0.46	−0.06	+3.8
$Co^{2+}(d^7$, hs)	43.5	0.8	0.91	−0.11	+6.1
$Ni^{2+}(d^8)$	58.1	1.2	1.0	0.2	+7.2

* Octahedral.
† Square-pyramidal.
§ This ion is not stable in water.
hs means high spin.

data show some correlation between large $\Delta H^{\ddagger}$ and large LFAE. Through this approach we can begin to see why Ni^{2+} and V^{2+} complexes are not very labile and why they have higher experimental activation enthalpies. In these two cases there is a significant loss of LFSE on going from the six-coordinate complex to the five-coordinate activated complex.

Associative activation

The last column in Table 15.7 gives $\Delta V^{\ddagger}$ for the H_2O ligand exchange reactions. We see that $\Delta V^{\ddagger}$ increases from $-10\,\text{cm}^3\,\text{mol}^{-1}$ for V^{2+} to $+7.2\,\text{cm}^3\,\text{mol}^{-1}$ for Ni^{2+}. This increase in $\Delta V^{\ddagger}$ follows the increase in the number of nonbonding d electrons from d^3 to d^8 across the first row of the d block. We can interpret the negative volume of activation as the result of the incorporation of the entering H_2O molecule into the activated complex, which suggests that the reactions have some associative character in the activation process. In the earlier part of the d block, associative activation appears to be favored by a low population of d electrons.

Negative volumes of activation are also observed in the second and third rows of the block, such as for Rh(III), and also indicate association of the entering group into the activated complex in the reactions of these complexes. Associative activation begins to dominate when the metal center is more accessible to nucleophilic attack, either because it is large or has a low (nonbonding or π^*) d-electron population, and the mechanism shifts from I_d toward I_a. Table 15.8 shows some data for the formation of Br^-, Cl^-, and NCS^- complexes from $[Cr(NH_3)_5(OH_2)]^{3+}$ and $[Cr(OH_2)_6]^{3+}$. The pentaammine complex shows only a weak nucleophile dependence in contrast to the strong dependence of the hexaaqua complex. The two complexes probably mark the transition from I_d to I_a.

One point of interest in Table 15.8 is that rate constants for aqua complexes are all smaller than those for ammine complexes by a factor of about 10^4. This suggests that the NH_3 ligands, which are stronger σ bases than H_2O, promote dissociation of the sixth ligand more effectively. As we saw above, this is to be expected in dissociately activated processes. The amines can stabilize the reduced coordination number of the activated complex more effectively; the aqua complex needs nucleophilic assistance.

Table 15.8. Kinetic parameters for anion attack on Cr(III) in $[Cr(NH_3)_5L]^{3+}$*

X	$L = H_2O$ $k/(10^8\ M^{-1}\ s^{-1})$	$\Delta H^{\ddagger}$	$\Delta S^{\ddagger}$	$L = NH_3$ $k/(10^4\ M^{-1}\ s^{-1})$
Br^-	1.0	122	8	3.7
Cl^-	2.9	126	38	0.7
NCS^-	180	105	4	4.2

* Enthalpy in $kJ\ mol^{-1}$, entropy in $J\ K^{-1}\ mol^{-1}$

Example 15.3: *Interpreting kinetic data in terms of a mechanism*

The second-order rate constants for formation of $[VX(OH_2)_5]^+$ from $[V(OH_2)_6]^{2+}$ and X^- for $X^- = Cl^-$, NCS^-, and N_3^- are in the ratio 1:2:10. What do the data suggest about the intimate mechanism of substitution?

Answer. Since all three complexes are monoanions of similar size, the encounter equilibrium constants are similar, and second-order rate constants (which are equal to $K_E k_2$) are proportional to first-order rate constants for substitution in the encounter complex, k_2. The greater rate constants for NCS^- than Cl^- and especially the five-fold difference of SCN^- from its close structural analog N_3^-, suggests some contribution from nucleophilic attack and associative activation. (It turns out that there is no such systematic pattern for the same anions reacting with Ni(II), for which the activation is believed to be dissociative.)

Exercise. Using the data in Table 15.7, estimate an appropriate value for K_E and calculate the first-order k_1 for the reaction of $[V(OH_2)_6]^{2+}$ with Cl^- if the observed second-order rate constant is $1.2 \times 10^2\ M^{-1}\ s^{-1}$.

15.7 Stereochemistry

The stereochemical consequences of octahedral substitution are more involved than those of the square-planar complexes. Yet again, Co(III) complexes provide the classic examples, and in Table 15.9 we give some data for the hydrolysis of *cis* and *trans* $[CoAX(en)_2]^{2+}$ where X is the leaving group (either Cl^- or Br^-) and A is OH^-, NCS^-, or Cl^-. The *cis* complexes do not undergo isomerization in conjunction with substitution. The *trans* complexes show a tendency to isomerize to *cis* that increases in the order

$$A = NO_2^- < Cl^- < NCS^- < OH^-$$

Table 15.9. Stereochemical course of reactions of $[CoAX(en)_2]^+$

	A	X	Percentage *cis* in product
cis	OH^-	Cl^-	100
	Cl^-	Cl^-	100
	NCS^-	Cl^-	100
	Cl^-	Br^-	100
trans	NO_2^-	Cl^-	0
	NCS^-	Cl^-	50–70
	Cl^-	Cl^-	35
	OH^-	Cl^-	75

We can understand the data within the framework of an I_d mechanism by recognizing that the five-coordinate metal center in the activated complex may resemble either of the two stable geometries for five-coordination, namely square-pyramidal or trigonal-bipyramidal. As we can see from Fig. 15.6, reaction through the square-pyramidal complex results in retention of the original geometry, but reaction through the trigonal-bipyramidal complex can lead to isomerization. The results suggest that a good π-base ligand *trans* to the leaving group Cl^- favors isomerization, because it can stabilize a trigonal bipyramid (**8**) by π bonding.

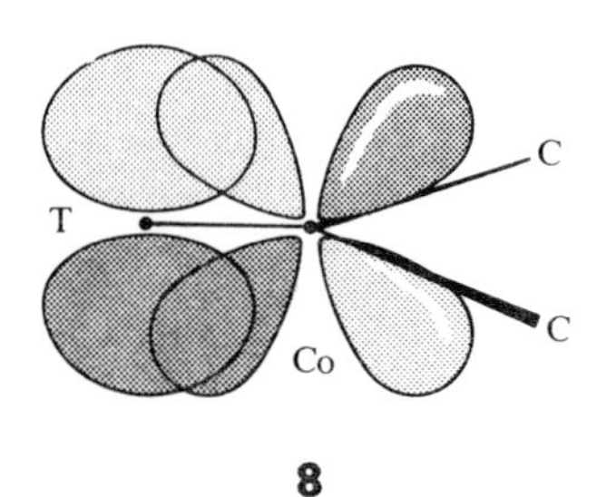

8

15.8 Base hydrolysis

Octahedral substitution is commonly greatly accelerated by OH^- ions. A typical rate law for the consumption of a halo complex is

$$\text{rate} = k[\{CoCl(NH_3)_5\}^{2+}][OH^-]$$

However, an extended series of studies has shown that the mechanism is not a simple bimolecular attack by OH^- on the complex.

The stereochemistry of the reaction indicates that a strong π-base ligand is present and that the steric effects observed are characteristic of a *d* reaction. There is a considerable body of indirect evidence on the problem but one elegant experiment makes the essential point. The conclusive evidence comes from a study of ^{18}O isotope distribution in the product $[Co(OH)(NH_3)_5]^{2+}$. It is found that the $^{18}O/^{16}O$ ratio differs between water and hydroxide ion at equilibrium. The isotope ratio in the product matches that for water, not that for the OH^- ions. Therefore, an H_2O molecule and not an OH^- ion appears to be the entering group.

The mechanism that takes these observations into account supposes that the role of OH^- is to act as a Brønsted base, not an entering group:

$$[CoCl(NH_3)_5]^{2+} + OH^- \rightleftharpoons [CoCl(NH_2)(NH_3)_4]^+ + H_2O$$

$$[CoCl(NH_2)(NH_3)_4]^+ \rightarrow [Co(NH_2)(NH_3)_4]^{2+} + Cl^- \qquad \text{(slow)}$$

$$[Co(NH_2)(NH_3)_4]^{2+} + H_2O \rightarrow [Co(OH)(NH_3)_5]^{2+} \qquad \text{(fast)}$$

In the first step, an NH_3 ligand acts as a Brønsted acid, resulting in the formation of its conjugate base, the NH_2^- ion, as a ligand. Since NH_2^- is a strong π base, it greatly accelerates Cl^- ion loss (see Example 15.4).

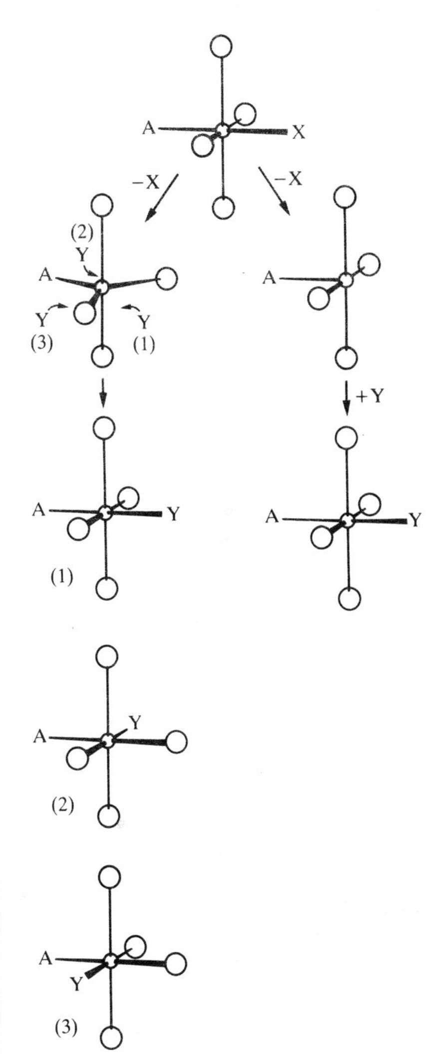

Fig. 15.6 Reaction through a square-pyramidal complex (upper path) results in retention of the original geometry but reaction through a trigonal bipyramidal complex (lower path) can lead to isomerization.

Example 15.4: *Interpreting the stereochemical course of a base hydrolysis reaction*

Substitution of Co(III) complexes of the type $[CoAX(en)_2]^+$ results in *trans* to *cis* isomerization, but only when the reaction is catalyzed by a base. Show that this observation is consistent with the model in Fig. 15.6.

Answer. For base hydrolysis, the conjugate base is the strong π base ligand NHR^-. A trigonal bipyramid of the type shown in the lower path in Fig. 15.6 is possible, which may be attacked in the way shown there.

Exercise. If the directions of attack on the above trigonal bipyramid are random, what *cis–trans* isomer distribution does the analogy with Fig. 15.6 suggest for the products of the type $[CoA(NH_3)_4(OH_2)]^{2+}$?

15.9 Isomerization reactions

Isomerizations are closely related to substitutions; indeed, often a major pathway for isomerization is via substitution. The square-planar Pt(II) and octahedral Co(III) complexes we have discussed can share a five-coordinate trigonal bipyramid activated complex. The interchange of the axial and equatorial ligands in a trigonal bipyramidal complex can be pictured as occurring by a pseudorotation through a square pyramid as shown in (**9**). As we saw above, when a trigonal bipyramid adds a ligand to produce a six-coordinate complex, a new direction of attack of the entering group can lead to an isomerization.

If a chelate is present, isomerization can occur as a consequence of metal–ligand bond breaking without substitution. The exchange of the 'outer' CD_3 group with the 'inner' CH_3 group during the isomerization of tris(acetylacetylacetonato)cobalt(III) (**10** → **11**) shows that ring opening accompanies the isomerization. An octahedral complex can also undergo isomerization by an intramolecular twist without loss of a ligand or breaking a bond. There is evidence, for example, that racemization of $[Ni(en)_3]^{2+}$

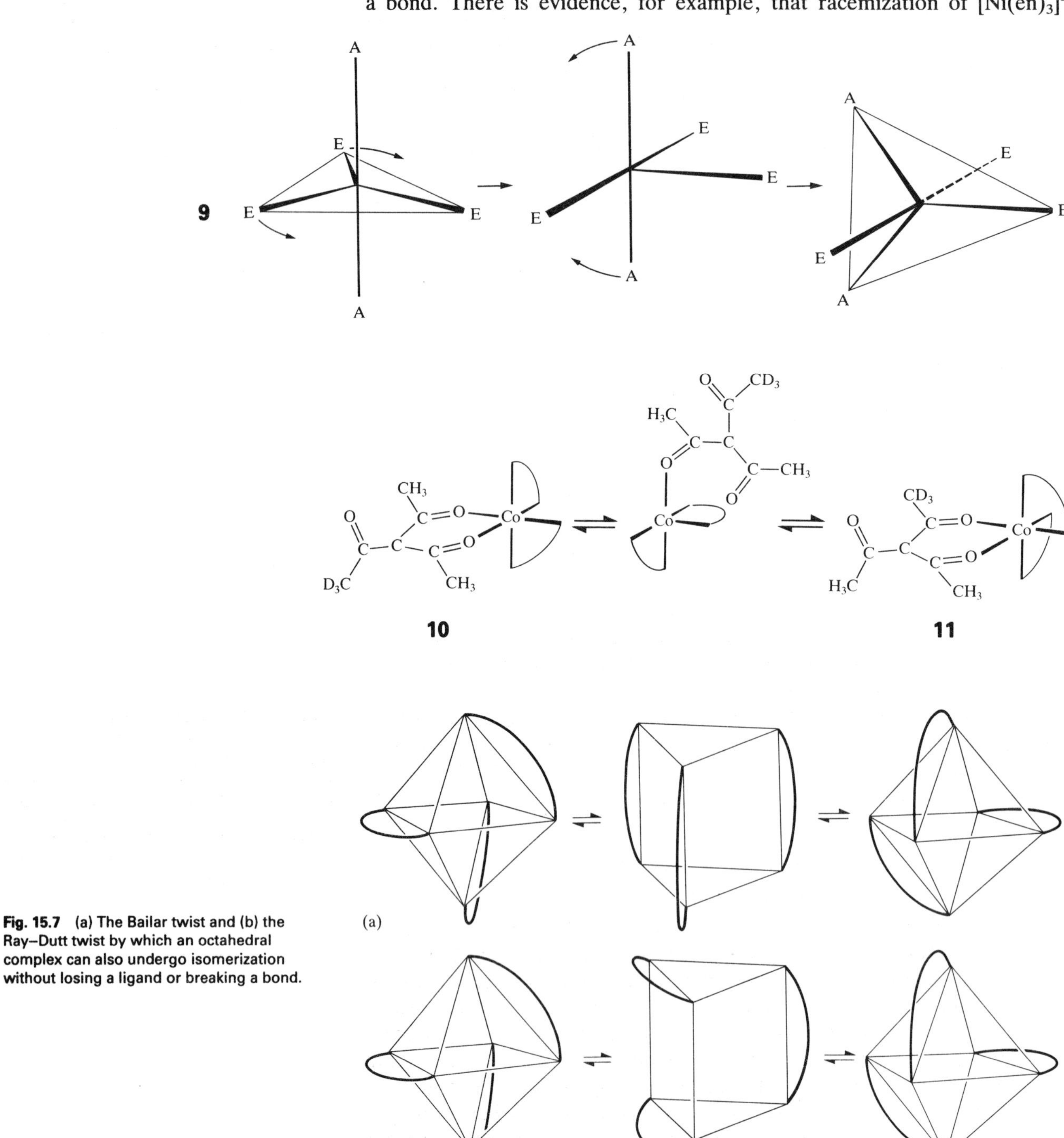

Fig. 15.7 (a) The Bailar twist and (b) the Ray–Dutt twist by which an octahedral complex can also undergo isomerization without losing a ligand or breaking a bond.

occurs by such an internal twist. Two possible paths are the **Bailar twist** and the **Ray–Dutt twist** (Fig. 15.7).

15.10 Stoichiometric mechanisms: the *D* intermediate

We may be able to distinguish a dissociative *D* mechanism from an interchange I_d mechanism by making use of the fact that the intermediate lives long enough for competitive reactions to occur. An example is the substitution reaction of the corrin (**12**) complexes of Co(III), which are related to vitamin B_{12}.

12

The process begins with reversible intermediate formation:

$$\mathrm{Co(corrin)(OH_2)_2} \underset{k'_a}{\overset{k_a}{\rightleftharpoons}} \mathrm{Co(corrin)(OH_2)} + \mathrm{H_2O}$$

The forward reaction is first-order and the reverse reaction is pseudofirst-order since one reactant is the solvent (water). This step is followed by an attack on the intermediate by the entering group:

$$\mathrm{Co(corrin)(OH_2)} + \mathrm{Y} \xrightarrow{k_b} \mathrm{Co(corrin)(OH_2)Y}$$

The overall rate is

$$\text{rate} = k_a k_b \frac{[\mathrm{Co(corrin)(OH_2)_2}][\mathrm{Y}]}{k'_a + k_b[\mathrm{Y}]}$$

If the reaction is studied at low enough entering group concentrations for $k_b[\mathrm{Y}] \ll k'_a$, the rate law reduces to second-order:

$$\text{rate} = k_{obs}[\mathrm{Co(corrin)(OH_2)_2}][\mathrm{Y}] \qquad k_{obs} = \frac{k_a k_b}{k'_a}$$

The result that the rate for any given nucleophile Y depends on the ratio of k_b/k'_a shows that it is the outcome of the competition between Y and H_2O for capture of the five-coordinate intermediate which determines the differences among nucleophiles in a *D* process.

Table 15.10 gives some examples of the variation of k_b with the identity of Y for $[\mathrm{Co(corrin)(OH_2)}]^{2-}$. In contrast to the I_d case, there is a pronounced dependence on the entering group. However, there is a difference from the I_a case in that at high entering group concentrations ($k_b[\mathrm{Y}] \gg k'_a$) the rate law approaches

$$\text{rate} = k_a[\mathrm{Co(corrin)(OH_2)}]$$

so the observed rate constant approaches the common limit k_a (the rate constant for loss of H_2O to form the intermediate) independent of the identity of Y. This limiting behavior is not characteristic of an I_a process. The 20-fold increase in rate of product formation at low concentration associated with the change of Y from HSO_3^- to SCN^- reflects the increase in the rate of the attack on the five-coordinate intermediate. As soon as that attack is fast compared to the rate of formation of the intermediate, the latter becomes limiting and all entering groups react at the same rate.

Probably the first *D* intermediate to be suggested was $[\mathrm{Co(NH_3)_5}]^{3+}$; however, there is now conclusive evidence against its importance. The effect of the leaving group in $[\mathrm{CoX(NH_3)_5}]^{2+}$ on the proportions of $[\mathrm{Co(SCN)(NH_3)_5}]^{2+}$

Table 15.10. Second-order rate constants for reactions of $[\mathrm{Co(corrin)(OH_2)_2}]$

L	SCN^-	I^-	Br^-	HSO_3^-
$k_b/(\mathrm{M^{-1}\,s^{-1}})$	0.23	0.14	0.10	0.017

k_a for H_2O exchange is $14\ \mathrm{s^{-1}}$

and $[Co(NCS)(NH_3)_5]^{2+}$ in solutions containing relatively high concentrations of NCS^- show unambiguously that the leaving group influences the product distribution and therefore that the leaving group is a part of the activated complex. Hence $[Co(NH_3)_5]^{3+}$ does not survive long enough to qualify as a true intermediate and for the reaction to be *D*.[1]

The mechanisms of redox reactions

As we indicated in Chapter 8, redox reactions can occur by the direct transfer of electrons (as in some electrochemical cells). Alternatively, they may occur by the transfer of atoms and ions, as in the transfer of oxygen in reactions of oxoanions. Since redox reactions involve both an oxidizing and a reducing agent, they are always bimolecular in character. (The rare exceptions are reactions in which one molecule has both oxidizing and reducing centers.) The stoichiometric mechanisms depend on whether the centers react without modification of their coordination spheres or whether a ligand substitution must accompany the reaction.

15.11 The classification of redox reactions

In the 1950s, the American chemist Henry Taube[2] identified two stoichiometric mechanisms of redox reactions. One is the **inner-sphere mechanism**, which includes atom transfer processes. In an inner-sphere mechanism, the coordination spheres of the reactants share a ligand transitorily and form a bridged intermediate. The other is an **outer-sphere mechanism**, which includes many simple electron transfers. In an outer-sphere mechanism, the complexes come into contact without sharing a bridging ligand. We encountered both mechanisms briefly in Section 8.4.

Inner-sphere mechanism

The inner-shell mechanism was first confirmed for the reduction of the nonlabile complex $[CoCl(NH_3)_5]^{2+}$ by $Cr^{2+}(aq)$, because the products of the reaction included both Co^{2+} and $[CrCl(OH_2)_5]^{2+}$ (Section 8.4). Moreover, addition of $^{36}Cl^-$ to the solution did not lead to the incorporation of any of the isotope in the Cr(III) product. Furthermore, the reaction is faster than reactions that remove Cl^- from Co(III) or introduce Cl^- into the nonlabile $[Cr(OH_2)_6]^{3+}$ complex. All this must mean that Cl^- has moved directly from the coordination sphere of one complex to that of the other during the reaction.

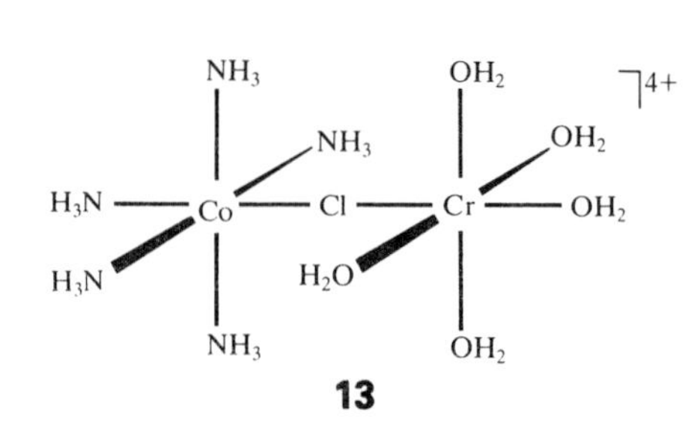

13

The Cl^- attached to Co(III) can easily enter into the labile coordination sphere of $[Cr(OH_2)_6]^{2+}$ to produce a bridged intermediate (**13**). Good bridging ligands are those with more than one pair of electrons available to donate to the two metal centers and include

$$:\ddot{\underset{..}{Cl}}:^- \quad :SCN:^- \quad :N{\equiv}N: \quad :NNN:^- \quad :C{\equiv}N:^-$$

[1] W. G. Jackson, B. C. McGregor, and S. S. Jurisson, *Inorg. Chem.*, **26,** 1286 (1987). These authors argue that $[Co(NH_3)_5]^{3+}$ cannot live longer than a few picoseconds.

[2] Taube has given an excellent retrospective account in his 1983 Nobel Prize lecture reprinted in *Science,* **226,** 1028 (1984).

Outer-sphere mechanism

An example of an outer-shell mechanism is the reduction of $[Fe(phen)_3]^{3+}$ by $[Fe(CN)_6]^{4-}$. Both reactants have nonlabile coordination spheres, and no ligand substitution can occur on the extremely short time scale of the redox reaction. The only possible mechanism is an electron transfer between the two complex ions in outer-sphere contact.

If a redox reaction is faster than ligand substitution, the reaction has an outer-sphere mechanism. Similarly, there is no difficulty in assigning an inner-sphere mechanism when the reaction involves ligand transfer from an initially nonlabile reactant to a nonlabile product. Unfortunately, it is difficult to make unambiguous assignments when the complexes are labile. Much of the study of well-defined examples is directed toward the identification of the parameters that differentiate the two paths in the hope of being able to making correct assignments in more difficult cases.

15.12 The theory of redox reactions

The theory of the intimate mechanisms of redox reactions is well established. This is especially the case for outer-sphere reactions where it is even possible to calculate rate constants in favorable cases.

The concepts needed to discuss outer-sphere processes

The analysis of the outer-sphere mechanism depends on two concepts. One is the **Franck–Condon principle**, which was proposed originally to account for the appearance of the vibrational structure of electronic spectra, where electron transfer occurs within a single molecule. The principle states that electron rearrangements occur so rapidly that nuclei can be considered as stationary until the rearrangement is complete. This separation of the motion of electrons and nuclei is based on the great difference of mass between electrons and nuclei and hence the former's ability to respond almost instantaneously to rearrangements of nuclei and, conversely, for the nuclei to respond only sluggishly to movement of the electrons.

The second concept we need is that electrons are likely to transfer from one site to another only when the nuclei in the two complexes have positions that result in the electron having the same energy on each site. Taken together with the Franck–Condon principle, it follows that the activation energy is governed by the nuclear rearrangements necessary to achieve this matching of energies.

To understand how an outer-sphere reaction occurs, we shall consider an exchange reaction between normal and isotopically labeled (*Fe) centers:

$$[Fe(OH_2)_6]^{2+} + [{}^*Fe(OH_2)_6]^{3+} \rightarrow [Fe(OH_2)_6]^{3+} + [{}^*Fe(OH_2)_6]^{2+}$$

The second-order rate constant is $3.0\,\text{M}^{-1}\,\text{s}^{-1}$ at 25°C and the activation energy is $32\,\text{kJ mol}^{-1}$. With the two theoretical points in mind we must consider the following three factors. First, bond lengths in Fe(II) are longer than those in Fe(III). Hence a part of the activation energy will arise from their adjustment to a common value in both complexes. This adjustment requires a Gibbs energy we call the **inner-sphere rearrangement energy** $\Delta G^{\ddagger}_{IS}$. Secondly, the solvation shell must be reorganized, which requires the **outer-sphere reorganization energy** $\Delta G^{\ddagger}_{OS}$. Thirdly, there is the electrostatic interaction energy between the two reactants ($\Delta G^{\ddagger}_{ES}$). The total

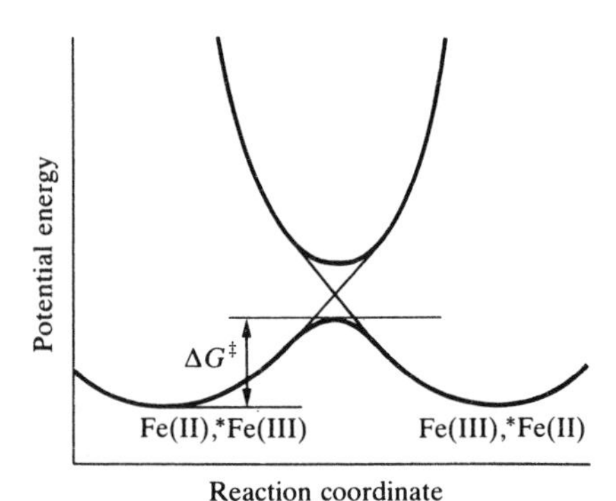

Fig. 15.8 A simplified reaction profile for electron exchange in a symmetrical reaction. On the left of the graph, the nuclear coordinates correspond to Fe(II) and *Fe(III); on the right, the ligands and solvent molecules have adjusted locations and the nuclear coordinates correspond to Fe(III) and *Fe(II) where the * denotes the isotope label.

activation Gibbs energy is therefore

$$\Delta G^{\ddagger} = \Delta G^{\ddagger}_{\mathrm{IS}} + \Delta G^{\ddagger}_{\mathrm{OS}} + \Delta G^{\ddagger}_{\mathrm{ES}}$$

The potential energy curves for reaction

The reactants initially have their normal bond lengths for Fe(II) and *Fe(III) respectively, and the reaction corresponds to motion in which the Fe(II) bonds shorten and the *Fe(III) bonds simultaneously lengthen (Fig. 15.8). The potential energy curve for the products of this symmetrical reaction is the same as for the reactants, the only difference being the interchange of the roles of the two Fe atoms. We have assumed that the metal–ligand deformations resemble a harmonic oscillation and have therefore drawn them as parabolas.

The activated complex is located at the intersection of the two curves. However, the noncrossing rule (Section 14.3 and Fig. 14.4) states that molecular potential energy curves of states of the same symmetry do not cross but instead split into an upper and a lower curve (Fig. 15.8). The noncrossing rule implies that if the reactants in their ground states slowly distort, they follow the path of minimum energy and transform into products in their ground states.

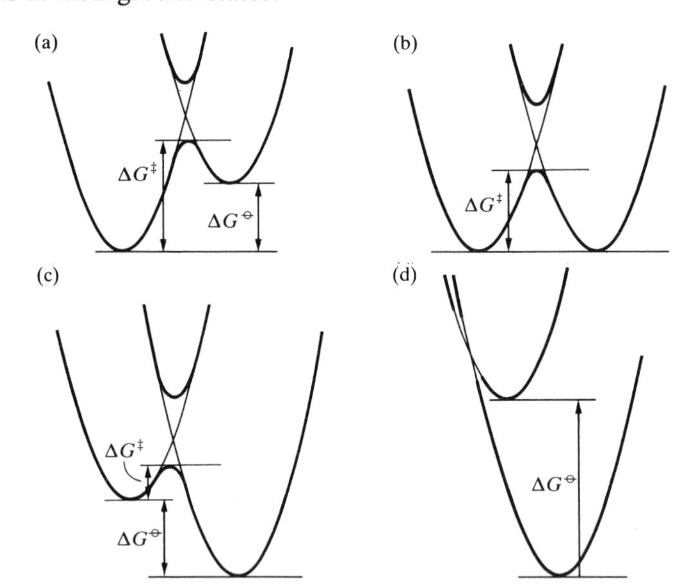

Fig. 15.9 The effect of a change of reaction Gibbs free energy for electron transfer on the activation energy when the shape of the potential surfaces remain constant (corresponding to equal self-exchange rates).

More general redox reactions correspond to a non-zero reaction Gibbs energy, so the parabolas representing the reactants and the products lie at different heights. If the product surface is higher (Fig. 15.9a), the crossing point moves up and the reaction has a higher activation energy. Conversely, moving the product curve down (Fig. 15.9c) leads to lower crossing points and lower activation energy, at least until the crossing begins to occur at the left. At the extreme of exergonic reaction (Fig. 15.9d), the crossing point rises and rates may become slower again.[3]

[3] We will not consider this last point further here. The slowing of rate at high exothermicity is referred to as the 'Marcus inverted region'. In an experimental *tour de force*, appropriate highly energetic oxidizing agents have recently been prepared photochemically and direct evidence for this inverted region has been obtained. See, for example, N. Mataga, in *Photochemical energy conversion* (ed. J. R. Norris and D. Meisel). Elsevier, Amsterdam (1989).

The Marcus equation

The diagrams in Fig. 15.9 suggest that there are two factors that determine the rate of electron transfer. The first is the shape of the potential curves. If the parabolas rise steeply, indicating a rapid increase in energy with distortion, their crossing points will be high and activation energies high too. Shallow potential curves, in contrast, imply low activation energies. Similarly, large changes in the equilibrium internuclear distance mean that the equilibrium points are far apart and the crossing point will not be reached without large distortions. The second factor is the Gibbs reaction energy, and the more favorable it is, the lower the activation energy of the reaction.

These considerations were expressed quantitatively by R. A. Marcus. He derived an equation for predicting the rate constant for an outer-sphere reaction from the exchange rate constants for each of the redox couples involved and the equilibrium constant for the overall reaction. The **Marcus equation** for the rate constant k is

$$k^2 = k_1 k_2 K f \qquad (6)$$

where k_1 and k_2 are the rate constants for the two exchange reactions and K is the equilibrium constant for the overall reaction. The factor f is a complex parameter composed of the rate constants and the encounter rate. It may be taken as near unity for approximate calculations. The idea of a weighted average of the two rates of self-exchange is emphasized by the name **Marcus cross-relation**, which is sometimes used instead.

Example 15.5: *Using the Marcus equation*

Calculate the rate constant at 0°C for the reduction of $[Co(bipy)_3]^{3+}$ by $[Co(terpy)_2]^{2+}$.

Answer. The required exchange reactions are

$$[Co(bipy)_3]^{2+} + [{}^*Co(bipy)_3]^{3+} \xrightarrow{k_1} [Co(bipy)_3]^{3+} + [{}^*Co(bipy)_3]^{2+}$$

$$[Co(terpy)_2]^{2+} + [{}^*Co(terpy)_2]^{3+} \xrightarrow{k_2} [Co(terpy)_2]^{3+} + [{}^*Co(terpy)_2]^{2+}$$

where $k_1 = 9.0\ \text{M}^{-1}\ \text{s}^{-1}$, $k_2 = 48\ \text{M}^{-1}\ \text{s}^{-1}$, and $K = 3.57$, all at 0°C. Setting $f = 1$ gives

$$k = (9.0 \times 48 \times 3.57)^{1/2}\ \text{M}^{-1}\ \text{s}^{-1} = 39\ \text{M}^{-1}\ \text{s}^{-1}$$

This result compares reasonably well with the experimental value, which is $64\ \text{M}^{-1}\ \text{s}^{-1}$.

Exercise. For reactions with k_1 and k_2 as given in the example, what is the rate of electron transfer if the overall reaction has $E^{\ominus} = 1.00\ \text{V}$?

The Marcus equation can be expressed as a linear free energy relation since the logarithm of the rate constants are proportional to the free energies of activation. Thus

$$2 \ln k = \ln k_1 + \ln k_2 + \ln K$$

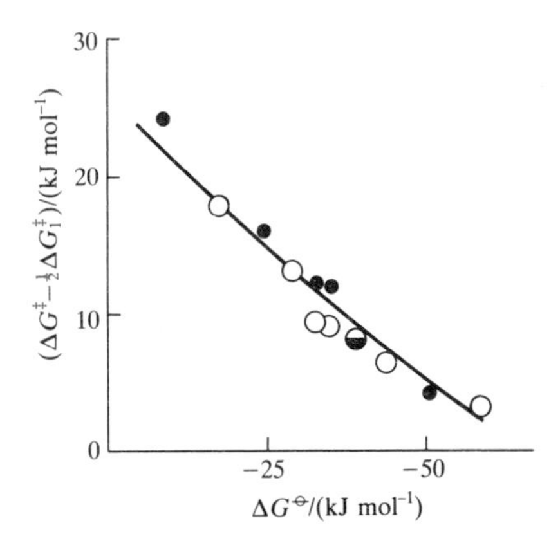

Fig. 15.10 The correlation between rates and overall free energy change for the oxidation of a series of phenanthroline complexes of Fe(II) by Ce(IV) in sulfuric acid (open circles) and $Fe^{2+}(aq)$ in perchloric acid (closed circles).

implies that

$$2\Delta G^{\ddagger} = \Delta G_1^{\ddagger} + \Delta G_2^{\ddagger} + \Delta G^{\ominus} \tag{7}$$

Figure 15.10 shows the correlation between rates and overall free energy change for the oxidation of a series of substituted phenanthroline complexes of Fe(II) by Ce(IV) and $Fe^{2+}(aq)$. The fit to the straight line is a sign of the success of the approximate Marcus equation.[4]

Departures from the Marcus equation are taken as a sign of special features in outer-sphere reactions. These may include the barrier created by change from high to low spin during electron transfer or a change in symmetry, such as from octahedral to distorted tetragonal.

Inner-sphere reactions

Inner-sphere mechanisms are much more difficult to analyze. This is because an outer-sphere reaction must be an electron transfer, but an inner-sphere reaction may be either electron or group transfer. In many cases, it is very difficult to distinguish the two.

Any inner-sphere mechanism must consist of three steps:

(1) Formation of the bridged (μ) complex:

$$M^{II}L_6 + XM'^{III}L'_5 \rightarrow L_5M^{II}—X—M'^{III}L'_5 + L$$

(2) The redox process:

$$L_5M^{II}—X—M'^{III}L'_5 \rightarrow L_5M^{III}—X—M'^{II}L'_5$$

(3) Decomposition of the successor complex to final products:

$$L_5M^{III}—X—M'^{II}L'_5 \rightarrow \text{products}$$

The rate-determining step of the overall redox reaction may be any one of these processes, but most commonly it is the redox step.

The reactions with break-up of the successor complex as the rate-determining step are common only if the configuration of both metals after electron transfer results in substitutional inertness, and then an inert binuclear complex is a characteristic product. A good example is the reduction of $[RuCl(NH_3)_5]^{2+}$ by $[Cr(OH_2)_6]^{2+}$. The rate determining step is the dissociation of $[Ru^{II}(NH_3)_5(\mu\text{-Cl})Cr^{III}(OH_2)_5]^{4+}$.

Reactions in which the formation of the bridged complex is rate-determining tend to be quite similar for a series of partners of a given species. For example, the oxidation of $V^{2+}(aq)$ proceeds at more or less the same rate for a long series of Co(III) oxidants with different bridging ligands. The explanation is that the rate-determining step is the substitution of an H_2O molecule from the coordination sphere of V(II). As is typical for octahedral aqua ions, this reaction has a rate that depends only weakly on the identity of the entering group.

The large class of reactions in which the redox step is rate-determining does not display such simple regularities. Rates vary over a wide range as metal ions and bridging ligands are varied. Because the change of oxidation states requires reorganization of ligands and solvent just as is the case for outer-sphere reactions, the trends indicated by the Marcus relation will also

[4] Tests of the Marcus equation are summarized by D. A. Pennington, in *Coordination chemistry*, Vol. 2, (ed. A. E. Martell), ACS Monograph 174. American Chemical Society, Washington (1978).

find some application to inner-sphere reactions. A very interesting way to study the role of the reorganization barriers is the synthesis of bridged complexes of the type

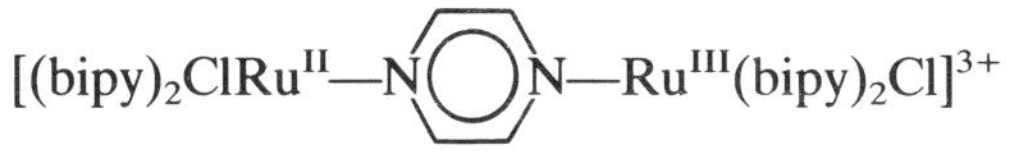

$$[(bipy)_2ClRu^{II}—N\bigcirc N—Ru^{III}(bipy)_2Cl]^{3+}$$

as models for inner-sphere precursor complexes.

All the reactions considered so far result in the change of oxidation number by ±1. Such reactions are still often called **one-equivalent processes**, the name reflecting the largely out-moded term 'chemical equivalent'. Their mechanisms require transfer of electrons or radicals. Similarly, reactions that result in the change in oxidation number by ±2 are often called **two-equivalent processes**. Their mechanisms often involve group transfer and may resemble nucleophilic substitutions. This can be seen by considering the reaction

$$[Pt^{II}Cl_4]^{2-} + [{}^*Pt^{IV}Cl_6]^{2-} \rightarrow [Pt^{IV}Cl_6]^{2-} + [{}^*Pt^{II}Cl_4]^{2-}$$

which occurs through a Cl^- bridge (**14**). The reaction depends on the transfer of a Cl^- ion in the break-up of the successor complex. Note that an extra Cl^- is essential to make the activated complex symmetrical. It enters by adding to the four-coordinate Pt(II) center.

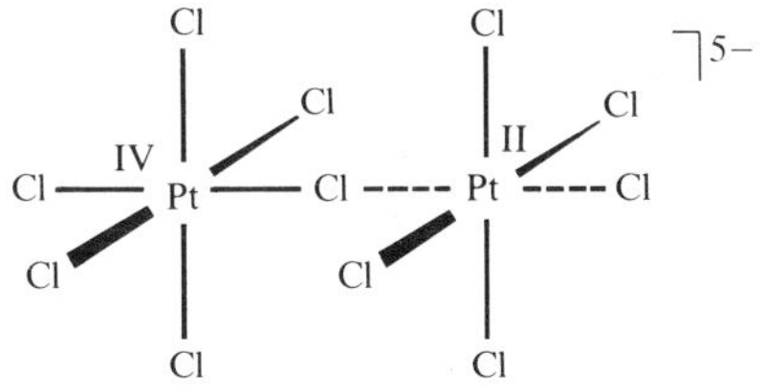

14

15.13 Oxidative addition

Reactions of the type

$$IrCl(CO)L_2 + RI \rightarrow Ir(R)(I)(Cl)(CO)L_2$$

where $L = PPh_3$ and R is an alkyl group are called **oxidative additions**. (H_2 or HX, where X is a halogen, may replace the iodoalkane.) In the reaction, the d^8 square-planar complex uses the nonbonding axial electron pair to bond to the incoming Lewis acid (R^+ or H^+ in this case) and the lone pair of the base X or H coordinates to the metal acting as an acid. The net effect is a metal oxidized to the extent of two electrons, from d^8 to d^6. Since the alkyl group and the anion are both more electronegative than the metal, they formally acquire all four electrons involved in M—R and M—I bonding, so the metal is formally oxidized by +2. Similar reactions are exhibited by the d^{10} species $[Pd(PPh_3)_4]$ and $[Ni(P(OEt)_3)_4]$.

There are three possible mechanisms. Dihydrogen addition (Fig. 15.11a) is a concerted electrophilic attack by the acid and nucleophilic attack by the base. A small isotope effect on replacing H by D suggests that very little breaking of the H—H bond has occured in the activated complex, and recent structural work has yielded stable H_2 side-on complexes that resemble the activated complex (Section 9.3). In the mechanism that involves initiation by nucleophilic attack (Fig. 15.11b) the inversion of configuration at the C atom that occurs during oxidative addition of $(C_6H_5)CH(D)Cl$ to $Pd(PR_3)_4$ strongly suggests attack by Pd lone pair electrons at the C atom. The radical mechanism (Fig. 15.11c) is accelerated by O_2 and light and occurs one electron at a time. Loss of stereochemistry, inhibition by radical scavengers, and sensitivity to O_2 are common indicators of radical pathways.

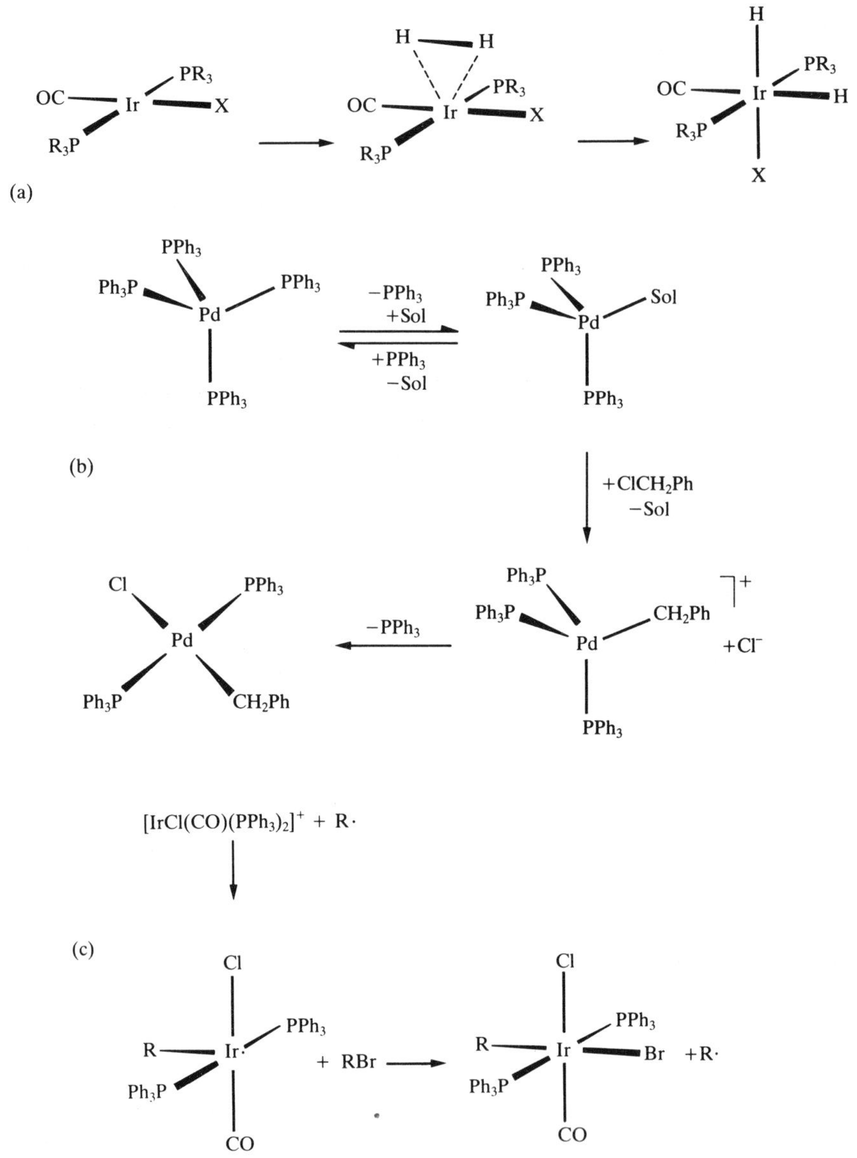

(a)

OC—Ir(PR$_3$)(X)(R$_3$P) → OC—Ir(H—H)(PR$_3$)(X)(R$_3$P) → OC—Ir(H)(H)(PR$_3$)(X)(R$_3$P)

(b)

Pd(PPh$_3$)$_4$ ⇌ (−PPh$_3$, +Sol / +PPh$_3$, −Sol) Pd(PPh$_3$)$_3$(Sol)

↓ +ClCH$_2$Ph, −Sol

[Pd(PPh$_3$)$_3$(CH$_2$Ph)]$^+$ + Cl$^-$

← −PPh$_3$

Pd(Cl)(PPh$_3$)(Ph$_3$P)(CH$_2$Ph)

(c)

[IrCl(CO)(PPh$_3$)$_2$]$^+$ + R·

↓

Ir·(Cl)(R)(PPh$_3$)(Ph$_3$P)(CO) + RBr → Ir(Cl)(R)(PPh$_3$)(Ph$_3$P)(CO)(Br) + R·

Fig. 15.11 The three mechanisms of oxidative addition. (a) Concerted reaction, which is characteristic of (but not limited to) H_2, (b) nucleophilic attack by the metal on the C atom, (c) a radical reaction (in this case, by a chain mechanism).

Photochemical reactions

The absorption of a photon raises the energy of a complex, typically by between 170 and 600 kJ mol^{-1}. Since these energies are larger than typical activation energies it should not be surprising that new reaction channels are opened. However, when the high energy of a photon is used to provide the energy of the primary forward reaction, the back reaction is almost always very favorable, and much of the design of efficient photochemical systems lies in the avoidance of the 'dreaded back reaction'.

15.14 Prompt and delayed reactions

In some cases, the excited states dissociate promptly. Examples include formation of the pentacarbonyl intermediates that initiate ligand substitution in carbonyls

$$Cr(CO)_6 \xrightarrow{h\nu} Cr(CO)_5 + CO$$

and the scission of Co—Cl bonds

$$[Co^{III}(NH_3)_5Cl]^{2+} \xrightarrow{h\nu(\lambda<350\,nm)} [Co^{II}(NH_3)_5]^{2+} + \cdot Cl$$

Both these processes occur in less than 10 ps and hence are called **prompt reactions**.

In the second reaction, the **quantum yield**, the amount of reaction per mole of photons absorbed, increases as the wavelength is decreased and the photon energy increases. The energy in excess of the bond energy is available to the newly formed fragments and increases the probability that they will escape from each other through the solution before they have an opportunity to recombine.

Some excited states have long lifetimes. They may be regarded as energetic isomers of the ground state that can participate in **delayed reactions**. The excited state of $[Ru^{II}(bipy)_3]^{2+}$ created by photon absorption in the metal-to-ligand charge transfer band (Section 14.4) may be regarded as a Ru(III) complexed to a radical anion of the ligand. Its redox reactions can be explained by adding the excitation energy (expressed as a potential using $-FE = \Delta G$ and equating ΔG to the molar excitation energy) to the ground state reduction potential:

$$[Ru(bipy)_3]^{3+} + e^- \rightarrow [Ru(bipy)_3]^{2+} \qquad E^{\ominus} = +1.26\,V\ (-122\,kJ\,mol^{-1})$$

$$[Ru(bipy)_3]^{2+} + h\nu(590\,nm) \rightarrow [^{\ddagger}Ru(bipy)_3]^{2+} \qquad \Delta G = +202\,kJ\,mol^{-1}$$

$$[Ru(bipy)_3]^{3+} + e^- \rightarrow [^{\ddagger}Ru(bipy)_3]^{2+} \qquad E^{\ominus\ddagger} = -0.84\,V\ (+80\,kJ\,mol^{-1})$$

15.15 *d–d* and charge-transfer reactions

We have already seen (Section 14.4) that there are two main types of spectroscopically observable electron promotion in metal complexes, namely *d–d* and charge-transfer transition. A *d–d* transition corresponds to the essentially angular redistribution of electrons within a *d* shell. In octahedral complexes, this redistribution often corresponds to the occupation of M—L antibonding e_g orbitals. An example is the $^4T_{1g} \leftarrow {}^4A_{2g}$ ($t_{2g}^2e_g^1 \leftarrow t_{2g}^3$) transition in $[Cr(NH_3)_6]^{3+}$. The occupation of the antibonding e_g orbital results in a quantum yield close to 1 (specifically 0.6) for the photosubstitution

$$[Cr(NH_3)_6]^{3+} + H_2O \xrightarrow{h\nu} [Cr(NH_3)_5(OH_2)]^{3+} + NH_3$$

This is a prompt reaction, for it can occur in less than 5 ps.

Charge-transfer transitions correspond to the radial redistribution of electron density. They correspond to the promotion of electrons into more predominantly ligand orbitals if the transition is metal-to-ligand or into orbitals of greater metal character if the transition is ligand-to-metal. These excitations commonly initiate photoredox reactions of the kind already mentioned in connection with Co(III) and Ru(II).

Although it is a very useful first approximation to associated photosubstitution and photoisomerization with *d–d* transitions and photoredox

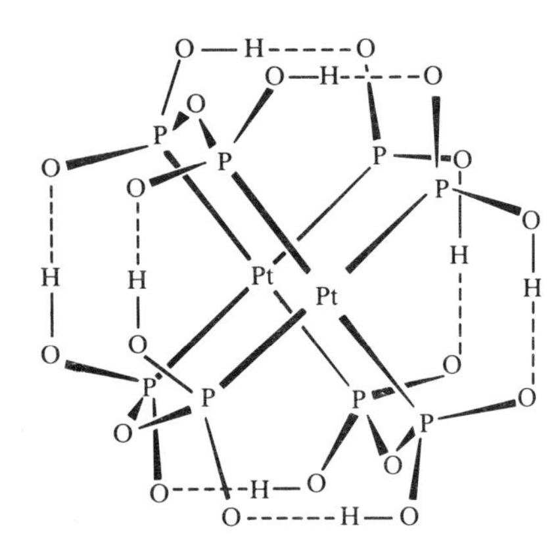

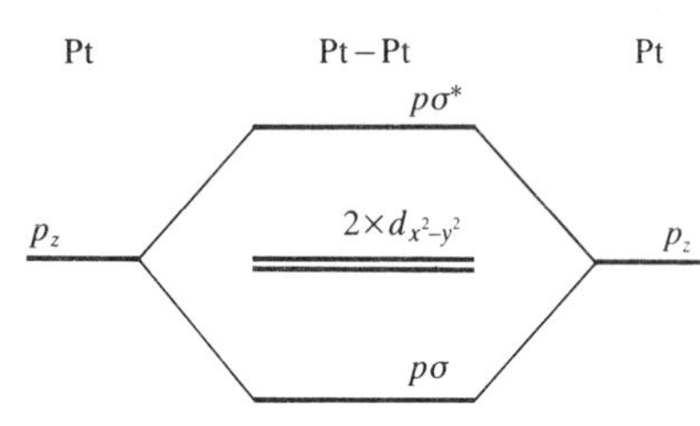

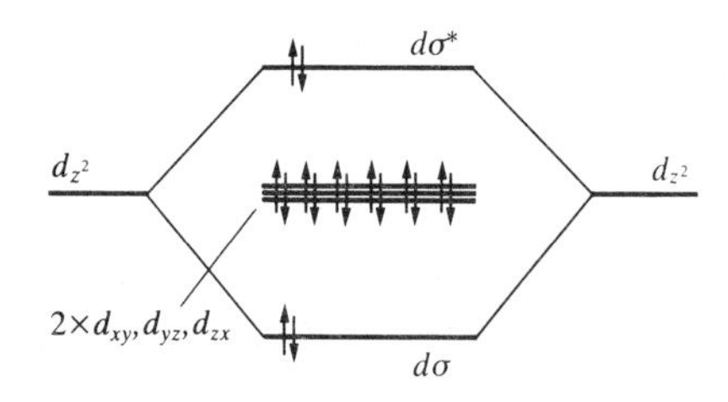

Fig. 15.12 The dinuclear complex $[Pt_2(\mu\text{-}P_2O_5H_2)_4]^{4-}$ called informally 'PtPOP'. The structure consists of two face-to-face square-planar complexes held together by a bridging pyrophosphito ligand. The metal p_z–p_z and d_{z^2}–d_{z^2} orbitals interact along the Pt–Pt axis. The other *p* and *d* obitals are considered to be nonbonding.

with charge transfer transitions, the rule is not absolute. For example, it is not uncommon for a charge transfer band to result in photosubstitution by an indirect path:

$$[Co^{III}Cl(NH_3)_5]^{2+} + H_2O \xrightarrow{h\nu} [Co^{II}(NH_3)_5(OH_2)]^{2+} + \cdot Cl$$

$$[Co^{II}(NH_3)_5(OH_2)]^{2+} + \cdot Cl \longrightarrow [Co^{III}(NH_3)_5(OH_2)]^{3+} + Cl^-$$

In this case, the aqua complex formed after the homolytic fission of the Co—Cl bond is reoxidized by the Cl atom. The net result leaves the Co oxidized and substituted. Conversely, the long-lived ^2E state of $[Cr(bipy)_3]^{3+}$ is a pure *d–d* state which lacks substitutional reactivity in acidic solution. Its lifetime of several microseconds allows the excess energy to enhance its redox reactions. The reduction potential (+1.3 V) calculated by adding the excitation energy to the ground state value accounts for its function as a good oxidizing agent, in which it undergoes reduction to $[Cr(bipy)_3]^{2+}$.

15.16 Metal–metal bonded systems

We encountered the $\delta^* \leftarrow \delta$ transition of multiple metal–metal bonds in Chapter 14, and it should not be surprising that population of an antibonding orbital of the metal–metal system can initiate photodissociation. It is more interesting that such excited states have been shown to initiate multielectron redox photochemistry.

One of the best-characterized systems is the binuclear Pt complex $[Pt_2(\mu\text{-}P_2O_5H_2)_4]^{4-}$ called informally 'PtPOP' (Fig. 15.12)[5]. There is no metal–metal bonding in the ground state of this Pt(II)—Pt(II) d^8–d^8 species. The HOMO–LUMO pattern shown in Fig. 15.12 indicates that excitation populates a bonding orbital between the metal atoms. The lowest-lying excited state has a lifetime of 9 μs. It is a powerful reducing agent, and reacts by both electron and halogen atom transfer. The most interesting oxidation products are Pt(III)—Pt(III) that contain X^- ligands (where X is halogen or pseudohalogen) at both ends and a metal–metal single bond. Irradiation in the presence of $(Bu)_3SnH$ give a dihydride product which can eliminate H_2.

Irradiation of the quadruply bonded dinuclear cluster, $Mo_2(O_2P(OC_6H_5)_2)_4$ (**15**) at 500 nm in the presence of $ClCH_2CH_2Cl$ results in production of ethylene and the addition of two Cl atoms to the two Mo atoms, with a two-electron oxidation.[6] The reaction proceeds in steps of one electron, and requires a complex with the metals blocked by sterically crowding ligands. If smaller ligands are present, the reaction that occurs instead is a photochemical oxidative addition of the organic molecule.

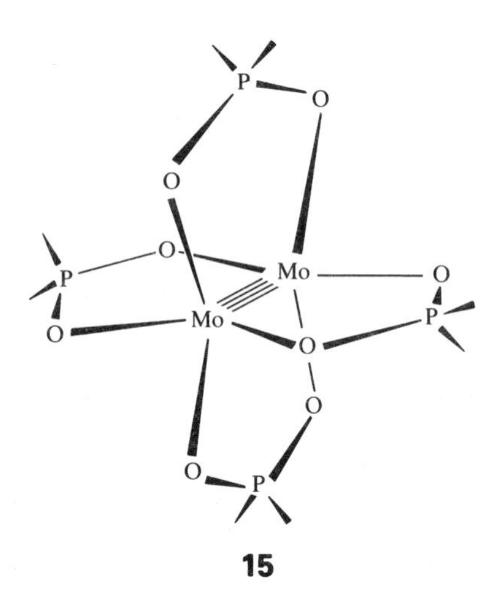

15

Further reading

F. Basolo and R. G. Pearson, *Mechanisms of inorganic reactions*. Wiley, New York (1967). This book is the classic in the field.

[5] The chemistry of this interesting complex is reviewed in D. M. Roundhill, H. B. Gray, and C.-M. Chi, *Acc. chem. Res.*, **22**, 55 (1989).

[6] I. J. Chang and D. C. Nocera, *J. Amer. chem. Soc.*, **109**, 4901 (1987).

Two more recent treatments are:
D. Katakis and G. Gordon, *Inorganic reaction mechanisms.* Wiley, New York (1987).
J. D. Atwood, *Inorganic and organometallic reaction mechanisms.* Brooks-Cole, Monterey (1985).
M. L. Tobe, *Inorganic reaction mechanisms.* Nelson, London (1972). This is a very readable treatment.
C. H. Langford and H. B. Gray, *Ligand substitution processes.* W. A. Benjamin, New York (1966). This book introduced the associative/dissociative notation.

Exercises

15.1 Classify (a) NH_3, (b) Cl^-, (c) Ag^+, (d) S^{2-}, (e) Al^{3+} as nucleophiles or electrophiles.

15.2 The reactions of $Ni(CO)_4$ in which phosphines or phosphites replace CO to give the family $Ni(CO)_3L$ occur at the same rate for different phosphines or phosphites. Is the reaction *d* or *a*?

15.3 Figure 15.1 shows how the Gibbs free energy varies with reaction coordinate for the *D* process in which the rate determining step is the breaking of the bond to the leaving group X. Draw the corresponding diagram for a *D* reaction in which the rate determining step is the addition of the entering group Y to the intermediate.

15.4 In experiment (a), $^{36}Cl^-$ is added to the solution during the study of the *cis* to *trans* isomerism of $[CoCl_2(en)_2]^+$ to determine whether isomerization is accompanied by substitution. In experiment (b), D replaces H in the ammine ligands of $[Cr(NCS)_4(NH_3)_2]^-$ and reduces the rate of replacement of NCS^- by water. Which experiment reveals the stoichiometric mechanism and which reveals the intimate mechanism?

15.5 Write the rate law for formation of $[MnX(OH_2)_5]^+$ from the aqua ion and X^-. How would you undertake to determine if the reaction is *d* or *a*?

15.6 Octahedral complexes of metal centers with high oxidation numbers or of *d* metals of the second and third series are less labile than those of the low oxidation number and *d* metals of the first series of the block. Account for this on the basis of dissociative activation.

15.7 A Pt(II) complex of tetraethyldiethylenetriamine is attacked by Cl^- 10^5 times less rapidly than the diethylenetriamine analog. Explain this in terms of associative activation.

15.8 Does the fact that $[Ni(CN)_5]^{3-}$ can be detected help to explain why substitution reactions of $[Ni(CN)_4]^{2-}$ are very rapid?

15.9 Design two-step syntheses of *cis*- and *trans*-$[PtCl_2(NO_2)(NH_3)]^-$ starting from $[PtCl_4]^{2-}$.

15.10 How does each of the following modifications affect the rate of a Pt, Pd, or Au square-planar complex substitution reaction?
(a) Changing a *trans* ligand from H to Cl.
(b) Changing the leaving group from Cl to I.
(c) Adding a bulky sustituent to a *cis* ligand.
(d) Increasing the positive charge on the complex.

15.11 The rate of attack on Co(III) by an entering group Y is nearly independent of Y with a spectacular exception of the rapid reaction with OH^-. Explain the anomaly. What is the implication of your explanation for the behavior of a complex lacking Brønsted acidity on the ligands?

15.12 Predict the products of the following reactions:
(a) $[Pt(PR_3)_4]^{2+} + 2Cl^-$
(b) $[PtCl_4]^{2-} + 2PR_3$
(c) *cis*-$[Pt(py)_4]^{2+} + 2Cl^-$

15.13 Put in order of rate of substitution by H_2O the complexes (a) $[Co(NH_3)_6]^{3+}$, (b) $[Rh(NH_3)_6]^{3+}$, (c) $[Ir(NH_3)_6]^{3+}$, (d) $[Mn(OH_2)_6]^{2+}$, (e) $[Ni(OH_2)_6]^{2+}$.

15.14 State the effect on the rate of dissociatively activated reactions of Rh(III) complexes of (a) an increase in the positive charge on the complex; (b) changing the leaving group from NO_3^- to Cl^-; (c) changing the entering group from Cl^- to I^-; (d) changing the *cis* ligands from NH_3 to H_2O; (e) changing an ethylenediamine ligand to propylenediamine when the leaving ligand is Cl^-.

15.15 Write out the inner- and outer-sphere pathways for reduction of azidopentaamminecobalt(III) ion with $V^{2+}(aq)$. What experimental data might be used to distinguish between the two?

15.16 The intermediate $[Fe(NCS)(OH_2)_5]^{2+}$ can be detected in the reaction of $[Co(NCS)(NH_3)_5]^{2+}$ with $Fe^{2+}(aq)$ to give $Fe^{3+}(aq)$ and $Co^{2+}(aq)$. What does this suggest about the mechanism?

15.17 The photochemical substitution of $[W(CO)_5(py)]$ with triphenylphosphine gives $[W(CO)_5P(C_6H_5)_3]$. In the presence of excess phosphine, the quantum yield is approximately 0.4. A flash photolysis study reveals a spectrum that can be assigned to the intermediate $[W(CO)_5]$. What product and quantum yield do you predict for substitution of $[W(CO)_5py]$ in the presence of excess triethylamine? Is this reaction expected to be initiated from the ligand field or MLCT excited state of the complex?

15.18 From the spectrum of $[CoCl(NH_3)_5]^+$ which resembles Fig. 14.6, propose a wavelength for photoinitiation of reduction of Co(III) to Co(II) accompanied by oxidation of a ligand.

Problems

15.1 Figure 15.13 (which is from W. R. Muir and C. H. Langford, *Inorg. Chem.*, **7**, 1032 (1968)) shows plots of observed first-order rate constants against $[X^-]$ for the reactions

$$[CoNO_2(dmso)(en)_2]^{2+} + X^- \rightarrow [CoNO_2X(en)_2]^{2+} + dmso$$

Assume that all three reactions have the same mechanism. (a) If the mechanisms were *D*, to what would the limiting rate constants correspond? (b) If the mechanisms were I_d, to what would the limiting rate correspond? (c) What is the significance of the fact that the dmso exchange rate constant is larger than the limits for anion attack? (d) The limiting rate constants are $5\times10^{-4}\,s^{-1}$, $1.2\times10^{-4}\,s^{-1}$, and about $1\times10^{-4}\,s^{-1}$ for Cl^-, NO_2^-, and NCS^- respectively. Does the evidence favor a *D* or an I_d mechanism?

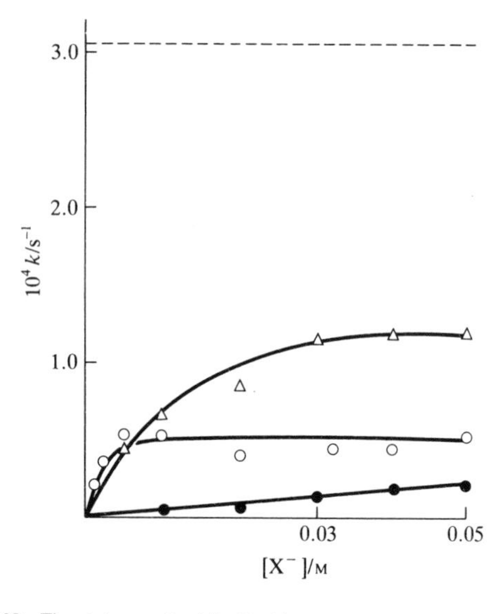

Fig. 15.13 The data required for Problem 15.1.

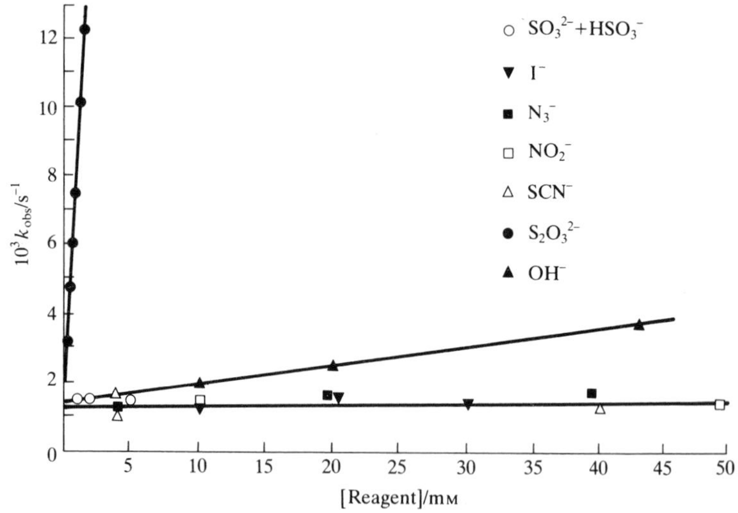

15.2 Given the following mechanism for the formation of a chelate complex:

$$[Ni(OH_2)_6]^{2+} + L\text{—}L \rightleftharpoons \{[Ni(OH_2)_6]^{2+}, L\text{—}L\} \qquad K_E,\ \text{rapid}$$

$$\{[Ni(OH_2)_6]^{2+}, L\text{—}L\} \rightleftharpoons [Ni(OH_2)_5L\text{—}L]^{2+} + H_2O \qquad k_a, k_a'$$

$$[Ni(OH_2)_5L\text{—}L]^{2+} \rightleftharpoons [Ni(OH_2)_4L\text{—}L]^{2+} + H_2O \qquad k_b, k_b'$$

derive the rate law for the formation of the chelate. Discuss the step that is different from that for two monodentate ligands. It is found that the formation of chelates with strongly bound ligands occurs at the rate of formation of the analogous monodentate complex, but that the formation of chelates of weakly bound ligands is often significantly slower. Assuming an I_d mechanism, explain this observation. (See R. G. Wilkins, *Acc. Chem. Res.*, **3**, 408 (1970).)

15.3 The complex $[PtH(PEt_3)_3]$ was studied in deuterated acetone in the presence of excess PEt_3, where Et is CH_3CH_2. In the absence of excess ligand the ^{1}H-NMR spectrum exhibits two triplets. As excess PEt_3 ligand is added the triplets begin to collapse, the line shape depending on the ligand concentration. Give a mechanism to account for the effects of excess PEt_3.

15.4 Figure 15.14 (which is from J. B. Goddard and F. Basolo, *Inorg. Chem.*, **7**, 936 (1968)) shows the observed first-order rate constants for the reaction of $[PdBr(dien)(Et)_4]^+$ with various Y^- to give $[PdY(dien)(Et)_4]^+$. Notice the large slope for $S_2O_3^{2-}$ and zero slopes for X = N_3, I, NO_2, and SCN. Propose a mechanism.

Fig. 15.14 The data required for Problem 15.4.

15.5 The activation enthalpy for the reduction of *cis*-$[CoCl_2(en)_2]^+$ by $Cr^{2+}(aq)$ is $-24\,kJ\,mol^{-1}$. Explain the negative value. (See R. C. Patel, R. E. Ball, J. F. Endicott, and R. G. Hughes, *Inorg. Chem.*, **9**, 23 (1970).)

15.6 The rate of reduction of $[Co(NH_3)_5(OH_2)]^{3+}$ by Cr(II) is seven orders of magnitude slower than reduction of its conjugate base, $[Co(OH)(NH_3)_5]^{2+}$ by Cr(II). For the corresponding reductions with $[Ru(NH_3)_6]^{2+}$ the two differ by less than a factor of 10. What does this suggest about

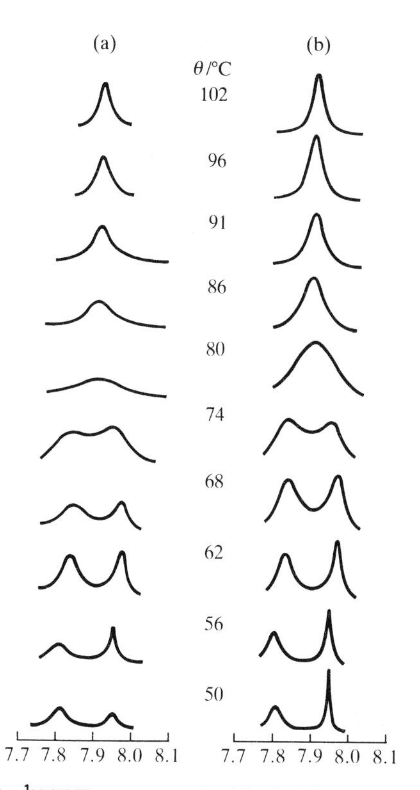

Fig. 15.15 The ^{1}H-NMR spectra required for Problem 15.7.

mechanisms? Comment on water and OH^- as bridging ligands.

15.7 Using the NMR spectra in Fig. 15.15, estimate the rate of exchange between free and coordinated ammonia for the two alkyl Co(III) complexes represented. (See R. J. Guschl, R. Stewart, and T. L. Brown, *Inorg. Chem.*, **13,** 959 (1959).)

15.8 Calculate rate constants for outer-sphere reactions from the following data. Compare your results to the measured values in the last column.

Reaction	$k_1/(\mathrm{M}^{-1}\,\mathrm{s}^{-1})$	$k_2/(\mathrm{M}^{-1}\,\mathrm{s}^{-1})$	$E^{\ominus}/\mathrm{V}$	$k_{obs}/(\mathrm{M}^{-1}\,\mathrm{s}^{-1})$
$Cr^{2+} + Fe^{2+}$	2×10^{-5}	4.0	+1.18	2.3×10^{3}
$[W(CN)_6]^{4-}$ + Ce(IV)	$>4\times10^{4}$	4.4	+0.90	$>10^{8}$
$[Fe(CN)_6]^{4-}$ + MnO_4^-	7.4×10^{2}	3×10^{3}	+0.20	1.7×10^{5}
$[Fe(phen)_3]^{2+}$ + Ce(IV)	$>3\times10^{7}$	4.4	+0.36	1.4×10^{5}

16

d- and *f*-block organometallics

Bonding
16.1 Valence electron count
16.2 Oxidation numbers and formal ligand charges
***d*-Block carbonyls**
16.3 Carbon monoxide as a ligand
16.4 Synthesis
16.5 Structure
16.6 Properties and reactions
Other organometallics
16.7 Hydrogen and open-chain hydrocarbon complexes
16.8 Cyclic polyene complexes
16.9 Early *d*-metal and *f*-metal organometallics
Metal–metal bonding and metal clusters
16.10 Structure
16.11 Syntheses
16.12 Reactions
Further reading
Exercises
Problems

The organometallic chemistry of the *p*-block metals was well understood by the early part of the twentieth century, but that of the *d* and *f* blocks has been developed much more recently. However, the latter has grown into a thriving area that spans interesting new types of reactions, unusual structures, and practical applications in organic synthesis and industrial catalysis. We begin by considering the structures, bonding, and reactions of metal carbonyls, which form the foundation of much of *d*-block organometallic chemistry. Then we consider hydrocarbon ligands, and their variety of structures, bonding modes, and reactions. Finally, we deal with the structures and reactions of metal cluster compounds, which has been one of the most recent areas of the subject to develop. Chapter 17 takes the story further by describing how *d*-block organometallics are used in catalysis.

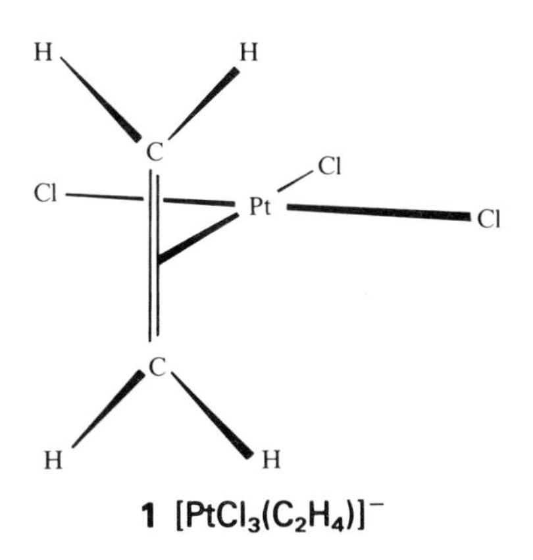

1 $[PtCl_3(C_2H_4)]^-$

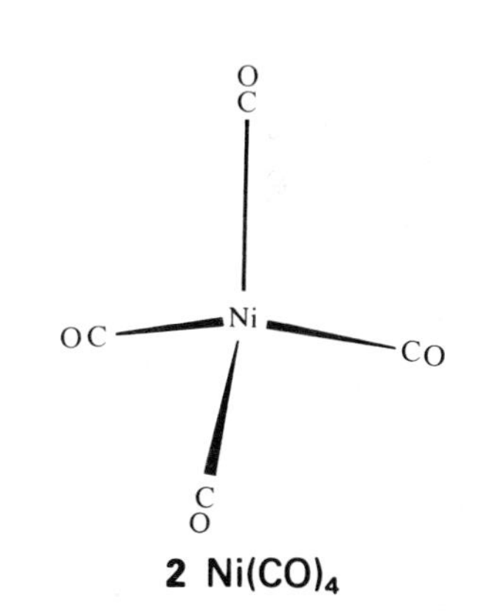

2 $Ni(CO)_4$

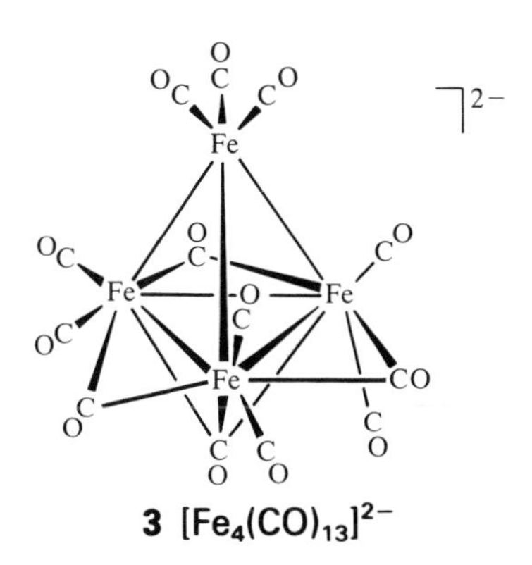

3 $[Fe_4(CO)_{13}]^{2-}$

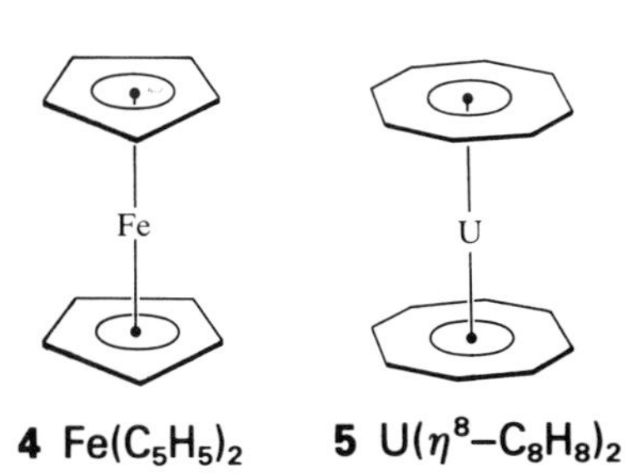

4 $Fe(C_5H_5)_2$ **5** $U(\eta^8\text{–}C_8H_8)_2$

A few *d*-block organometallics were synthesized and partially characterized in the nineteenth century. The first of them (**1**), an ethylene complex of platinum(II), was prepared by W. C. Zeise in 1827. The next major discovery was the first metal carbonyl, tetracarbonylnickel (**2**), which was synthesized by Ludwig Mond, Carl Langer, and Friedrich Quinke in 1890. Beginning in the 1930s, Walter Hieber in Munich synthesized a wide variety of metal carbonyl cluster compounds, many of which are anionic, including $[Fe_4(CO)_{13}]^{2-}$ (**3**). It was clear from this work that metal carbonyl chemistry was potentially a very rich field. However, as the structures of these and other *d*- and *f*-block organometallics are difficult or impossible to deduce by chemical means alone, fundamental advances had to await the development of X-ray diffraction for precise structural data on solid samples and of IR and NMR spectroscopy for structural information about organometallic complexes in solution. The discovery of the remarkably stable organometallic compound ferrocene, $Fe(C_5H_5)_2$, occurred at a time (in 1951) when these techniques were becoming available. The 'sandwich' structure of ferrocene (**4**) was soon correctly inferred from its IR spectrum and then determined in detail by X-ray crystallography.

The stability, structure, and bonding of ferrocene defied conventional bonding description and therefore captured the imagination of chemists. This puzzle in turn set off a train of synthesizing, characterizing, and theorizing that led to rapid development of *d*-block organometallic chemistry. Two highly productive research workers in the formative stage of the subject, Ernst Fischer in Munich and Geoffrey Wilkinson in London, were awarded the Nobel Prize in 1973 for their contributions.[1] Similarly, *f*-block organometallic chemistry blossomed soon after the synthesis and characterization of uranocene (**5**) in Andrew Streitwieser's laboratory at Berkeley in 1968.

We shall adhere to the convention that an **organometallic compound** contains at least one metal–carbon (M—C) bond. Thus compounds (**1**)–(**5**) clearly qualify as organometallics, whereas a complex such as $[Co(en)_3]^{3+}$, which contains carbon but has no C—M bonds, does not. Metal cyanide complexes, such as hexacyanoferrate(II) ions, do have M—C bonds, but their properties are more akin to those of Werner complexes and they are generally not considered to be organometallics. In contrast, complexes of the isoelectronic ligand CO are considered to be organometallic. The justification for this somewhat arbitrary distinction is that many metal carbonyls are significantly different from Werner complexes both chemically and physically.

We shall use the same nomenclature as for Werner complexes (Section 7.2). Ligands are listed in alphabetical order followed by the name of the metal, all of which is written as one word and followed by the oxidation number of the metal in parentheses. The nomenclature employed in research journals, however, does not always obey this rule strictly, and it is common to find the name of the metal buried in the middle of the name of the compound. For example, (**6**) is commonly called benzenemolybdenumtricarbonyl, not benzenetricarbonylmolybdenum(0). All ligands that are radicals when neutral have the suffix -yl, as in methyl, cyclopentadienyl, and allyl.

[1] Wilkinson's personal account, which captures the excitement of these developments, is given in *J. organometal. Chem.*, **100,** 273 (1975).

The IUPAC recommendation for formulas is to write them in the same form as for Werner complexes: metal first followed by formally anionic ligands, listed in alphabetical order. The neutral ligands are then listed in alphabetical order based on their chemical symbol. We shall follow these conventions unless a different order of ligands helps to clarify a particular point.

A further notational point is that organic ligands possess a special kind of versatility: a single ligand may attach to a central metal atom using several of its atoms simultaneously (ferrocene is an example). The **hapticity**[2] η of a ligand is the number of its atoms that are within bonding distance of the metal atom. (The definition is independent of a model of the bonding itself.) For example, a —CH_3 group attached by a single M—C bond is **monohapto**, η^1. The ethylene ligand is **dihapto**, η^2, if its two C atoms are both within bonding distance (**1**). The bonding of a C_5H_5 ligand to the Fe atom in ferrocene (**4**) is **pentahapto**, η^5. The C_5H_5 ligand can also be monohapto (**7**) and trihapto (**8**).

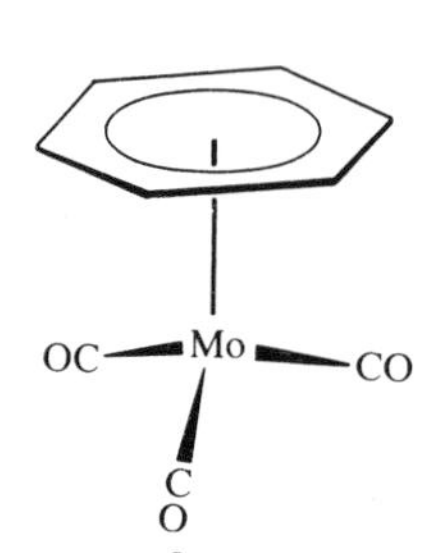

6 $Mo(\eta^6\text{-}C_6H_6)(CO)_3$

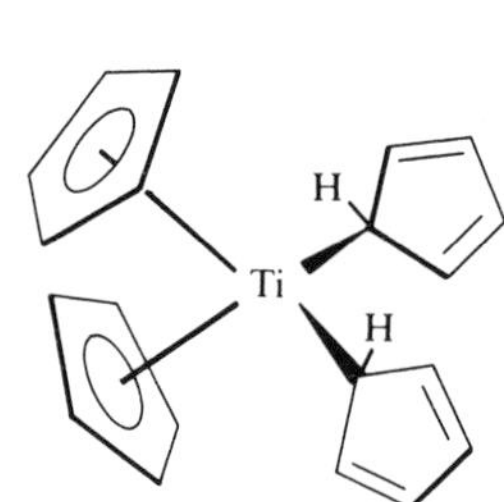

7 $Ti(\eta^1\text{-}C_5H_5)_2(\eta^5\text{-}C_5H_5)_2$

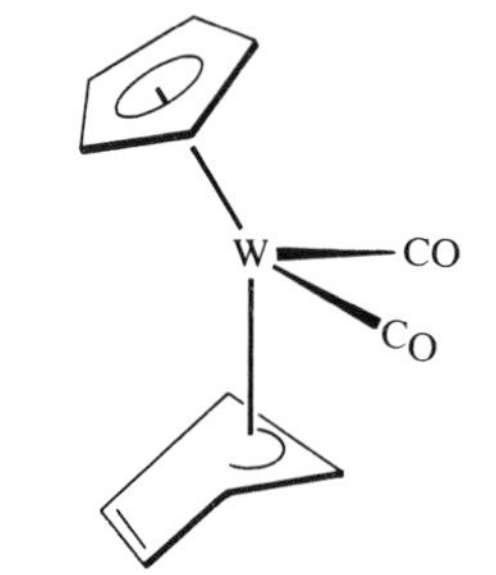

8 $W(\eta^3\text{-}C_5H_5)(\eta^5\text{-}C_5H_5)(CO)_2$

Bonding

In organometallic chemistry, as in most areas of chemistry, the concepts of closed shells and oxidation numbers help to rationalize the structures and reactivity of the compounds.

16.1 Valence electron count

As with *p*-block compounds, the number of valence electrons around the central *d*-metal atom in an organometallic compound gives an indication of its structure and stability. This number may be quite large because the atom can use its *d* orbitals, and most organometallic complexes have either 16 or (more commonly) 18 valence electrons. In contrast to many simple aqua ions and Werner complexes of the 3*d* metals, organometallic complexes are generally closed-shell molecules and ions.

When we count electrons we treat each metal atom and ligand as neutral. If the complex is charged, we simply add or subtract the appropriate number of electrons to the total. We must include in the count all valence electrons of the metal atom and all the electrons donated by the ligands. For example, $Fe(CO)_5$ attains 18 electrons from the eight valence electrons on the Fe atom and the ten electrons donated by the five CO ligands. Table 16.1 lists the maximum number of electrons available for donation to a metal for several common ligands.

[2] This terminology is based on the Greek word *haptein* (ἅπτειν), which means to fasten. It was introduced by F. A. Cotton, *J. Amer. chem. Soc.*, **90**, 6230 (1968).

Eighteen-electron complexes

Most stable organometallic complexes have 18 electrons in the metal valence shell. As we saw in Section 7.4, there are typically nine bonding or slightly antibonding orbitals in a complex, each of which can accommodate two electrons.

Table 16.1. Some organic ligands

Available electrons*	Hapticity	Ligand	Metal–ligand structure
1	η^1	Methyl, alkyl $\cdot CH_3$, $\cdot CH_2R$	M—CH_3
2	η^1	Alkylidene (carbene)	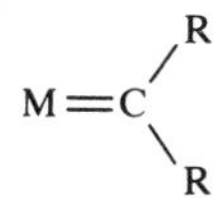 M=C(R)R
2	η^2	Alkene $H_2C{=}CH_2$	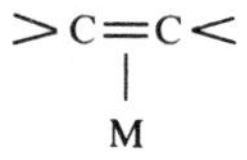 >C=C< with M below
3	η^3	π-Allyl C_3H_5	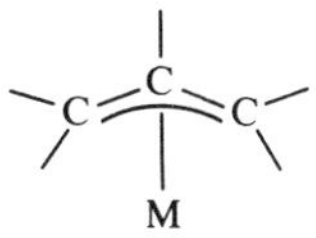 C–C–C with M below
3	η^1	Alkylidyne (carbyne) C—R	M≡C—R
4	η^4	1,3-Butadiene C_4H_6	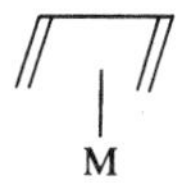 M
4	η^4	Cyclobutadiene C_4H_4	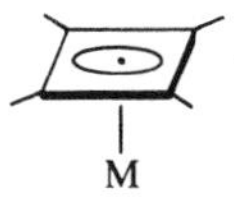 M
5 (3) (1)	η^5 η^3 η^1	Cyclopentadienyl C_5H_5 (Cp)	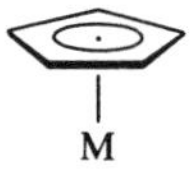 M
6	η^6	Benzene C_6H_6	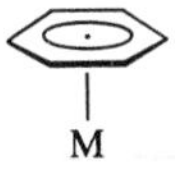 M
6	η^7	Tropylium $C_7H_7^+$	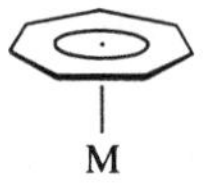 M
6	η^6	Cycloheptatriene C_7H_8	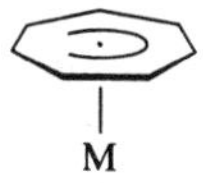 M
8† (6) (4)	η^8 η^6 η^4	Cyclooctatetraene C_8H_8 (cot)	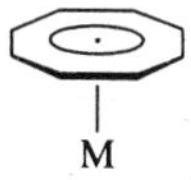 M

* For neutral ligands.
† As with several other polyene ligands in this list, the cot ligand may bond to a metal through only some of the available electron pairs. Variable hapticity is common with cot.

Table 16.2. Formulas and electron count for some Period 4 carbonyls

Group	Formula	Valence electrons	Structure
6	$Cr(CO)_6$	Cr 6 6(CO) 12 18	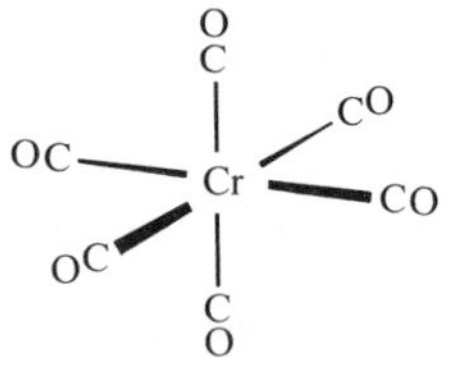
7	$Mn_2(CO)_{10}$	Mn 7 5(CO) 10 M—M 1 18	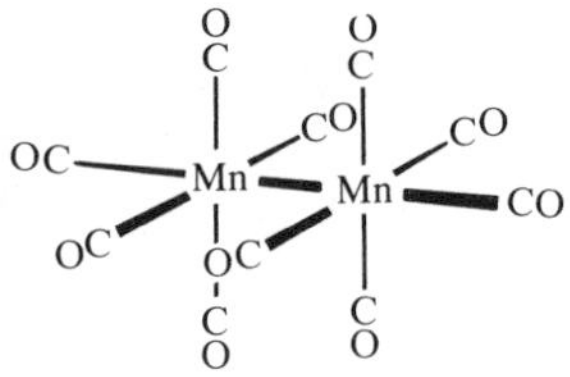
8	$Fe(CO)_5$	Fe 8 5(CO) 10 18	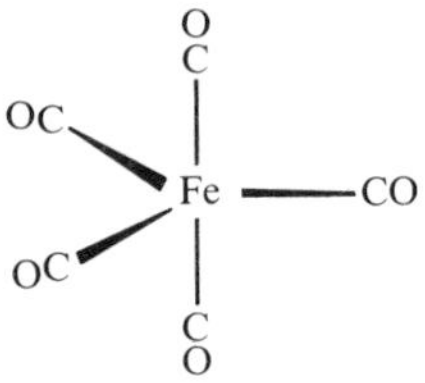
9	$Co_2(CO)_8$	Co 9 4(CO) 8 M—M 1 18	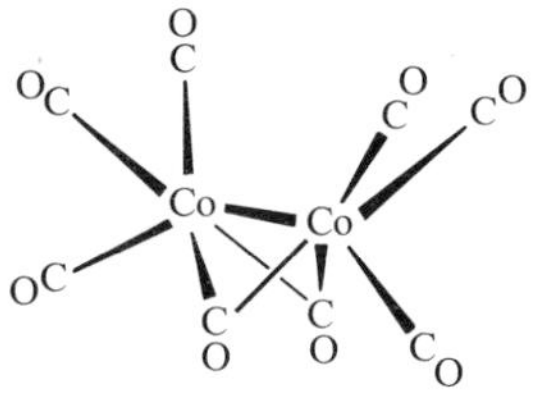
10	$Ni(CO)_4$	Ni 10 4(CO) 8 18	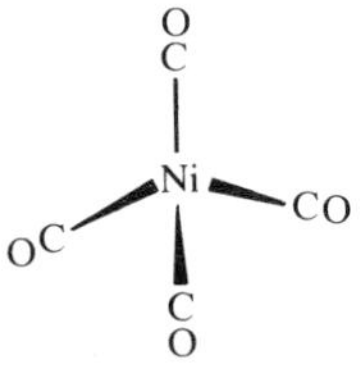

The 18-electron rule nicely systematizes the formulas of metal carbonyls. As shown in Table 16.2, the carbonyls of the Period 4 elements of Groups 6–10 have alternately one and two metal atoms and a decreasing number of CO ligands. The two-metal carbonyls are formed by elements of the odd-numbered groups, which have an odd number of valence electrons and therefore dimerize by forming metal–metal (M—M) bonds. The decrease in the number of ligands across a period matches the need for fewer CO ligands to achieve the 18 valence electrons.

The same principles apply when other soft ligands occupy the metal coordination shell. Two examples are benzene(tricarbonyl)molybdenum(0) (**6**) and bis(cyclopentadienyl)(tetracarbonyl)diiron(1) (**9**), both of which obey the 18-electron rule. The 18-electron rule can be extended to simple polynuclear carbonyls by adding one electron to the valence-electron count of a metal for each M—M bond to that metal atom. However, we shall see later that the 18-electron rule does not apply to M_6 and larger clusters.

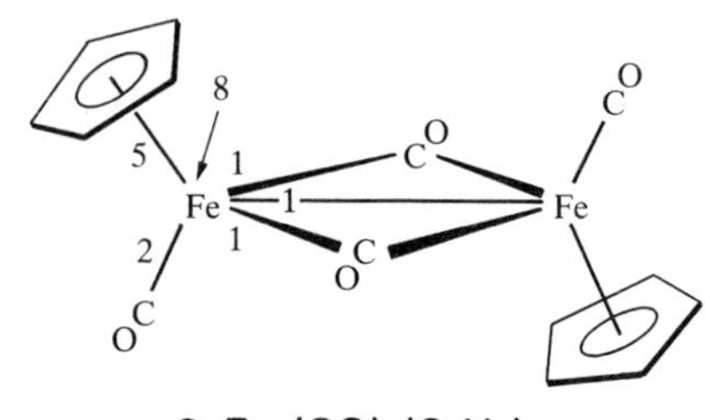

9 $Fe_2(CO)_4(C_5H_5)_2$

Example 16.1: *Counting the valence electrons on a metal atom*

Do (a) $IrBr_2(CH_3)(CO)(PPh_3)$ and (b) $Cr(\eta^5\text{-}C_5H_5)(\eta^6\text{-}C_6H_6)$ obey the 18-electron rule?

Answer. (a) An Ir atom (Group 9) has nine valence electrons; the two Br atoms and the CH_3 radical are one-electron donors; CO and PPh_3 are two-electron donors. Thus the number of valence electrons on the metal atom is $9 + (3 \times 1) + (2 \times 2) = 16$ in accord with the 16–18-electron rule. (b) A Cr atom (Group 6) has six valence electrons, the $\eta^5\text{-}C_5H_5$ ligand has five electrons, and the $\eta^6\text{-}C_6H_6$ ligand has six; so the number of metal valence electrons is $6 + 5 + 6 = 17$. This complex does not obey the 18 electron rule and is not stable. A related but stable 18-electron compound is $Cr(\eta^6\text{-}C_6H_6)_2$.

Exercise. Is $Mo(CO)_7$ likely to exist?

Sixteen-electron complexes

Organometallic complexes with 16 valence electrons are common on the right of the *d* block, particularly in Groups 9–10 (Table 16.3). Examples of such complexes (which are generally square planar) include $[IrCl(CO)(PPh_3)_2]$ (**10**) and the anion of Zeise's salt, $[PtCl_3(C_2H_4)]^-$ (**1**). Square planar 16-electron complexes are particularly common for the d^8 metals of the heavier elements in Groups 9 and 10, especially for Rh(I), Ir(I), Pd(II), and Pt(II). It should be recalled from Section 7.4 that the ligand-field stabilization energy of d^8 complexes favors a low-spin square-planar configuration when Δ is large, as is typical of Period 5 and 6 *d*-metal atoms and ions. The $d_{x^2-y^2}$ orbital is then empty and d_{z^2} is stabilized.

Table 16.3. Scope of the 16/18-electron rule for *d*-block organometallics

Usually less than 18			Usually 18 electrons			16 or 18 electrons	
Sc	Ti	V	Cr	Mn	Fe	Co	Ni
Y	Zr	Nb	Mo	Tc	Ru	Rh	Pd
La	Hf	Ta	W	Re	Os	Ir	Pt

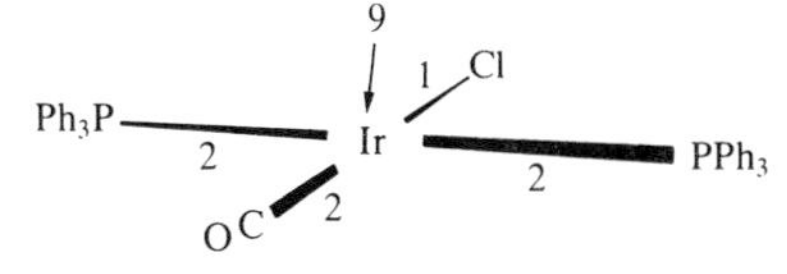

10 *trans*-$[IrCl(CO)(PPh_3)_2]$

Exceptions to the 16/18 electron rule

Exceptions to the 16/18 electron rule are common on the left of the *d* block. Here steric and electronic factors are in competition, and it is not possible to crowd enough ligands around the metal either to satisfy the rule or to permit dimerization. For example, the simplest carbonyl in Group 5, which is $V(CO)_6$, is a 17-electron complex. Other examples of the deviation include $ZrCl_2(\eta^5\text{-}C_5H_5)_2$ with 16 electrons, and $W(CH_3)_6$ which has 12.

Deviations from the 16/18-electron rule are common in neutral biscyclopentadienyl complexes, which are known for most of the *d* metals. Since two η^5-cyclopentadienyl ligands jointly contribute 10 valence electrons, the 18-electron rule can be satisfied only for neutral compounds of the Group 8 metals. The complexes of this kind that do obey the 18-electron rule (such as ferrocene itself) are the most stable. Their stability is suggested by their bond lengths and their redox reactions. For example, the 19-electron complex $Co(C_5H_5)_2$ is readily oxidized to the 18-electron cation $[Co(C_5H_5)_2]^+$. The bonding and electronic structures of the biscyclopentadienyl complexes are described in greater detail in Section 16.8.

16.2 Oxidation numbers and formal ligand charges

The rules for calculating the oxidation number of an element in an organometallic complex are the same as for conventional compounds. Accordingly, nonmetal ligands such as ·H, ·CH_3, and ·C_5H_5 are formally considered to take an electron from the metal atom and hence are assigned oxidation number −1. Thus, ferrocene can be considered to be a compound of Fe(II) and the ferrocinium ion, $[Fe(C_5H_5)_2]^+$, is a compound of Fe(III). It follows that the formal name of ferrocene is bis-η^5-cyclopentadienyliron(II). When planar, the η^8-cyclooctatetraene ligand is considered to extract two electrons and to become (formally) the aromatic dinegative ligand $[C_8H_8]^{2-}$. With two such dinegative ligands in uranocene $U(C_8H_8)_2$ (**5**), the U atom must be assigned an oxidation number of +4, and the formal name of the compound is bis-η^8-cyclooctatetraenyluranium(IV).

Oxidation numbers help us to systematize reactions such as oxidative addition, which we encountered in Section 15.13 and which we shall consider further in Section 16.7. They also bring out analogies between the chemical properties of organometallic complexes and Werner complexes. Thus, the oxidation states U(IV), Fe(II), and Fe(III) are known in both organometallic and Werner complexes. The principal difference between organometallic complexes and Werner complexes with hard ligands is that the soft organic ligands in the former help to stabilize metals in low oxidation states. In organometallic complexes, as in halides and oxides, there is a decreasing ability for the metal to achieve high oxidation numbers on moving from Group 6 to the right of the *d* block.

There are some examples of metal atoms in high formal oxidation states in organometallic compounds. We have already mentioned uranocene, which contains U(IV), and the unstable complex $W(CH_3)_6$, which contains formal W(VI). Another example is $(C_5H_5)ReO_3$ (**11**), which formally contains Re(VII).

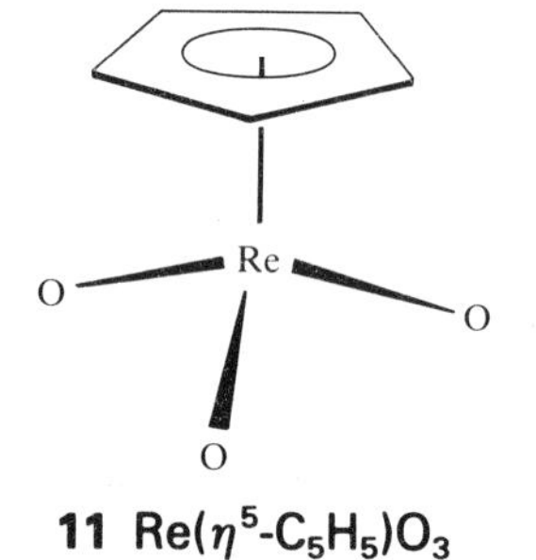

11 $Re(\eta^5\text{-}C_5H_5)O_3$

Example 16.2: *Assigning oxidation numbers*

Determine the oxidation numbers of Ir and Cl in $IrCl(CO)(PPh_3)_2$.

Answer. The ligands CO and PPh_3 are neutral two-electron donors and thus do not influence the oxidation number of Ir. Cl is a one-electron ligand and in combination with metals is assigned oxidation number −1. Since the complex as a whole is neutral (implying that the oxidation numbers sum to zero), the oxidation number of the Ir atoms is +1.

Exercise. Assign the oxidation number of Co in $Co(\eta^5\text{-}C_5H_5)(CO)_2$.

d-Block carbonyls

Simple metal carbonyls[3] can be prepared for most of the *d* metals, but those of palladium and platinum are so unstable that they exist only at low

[3] The simple carbonyls are referred to as *homoleptic* carbonyls, meaning that they are complexes with only one kind of ligand.

temperatures, and no simple carbonyls are known for copper, silver, and gold and for the members of Group 3. The carbonyls have turned out to be useful synthetic precursors for other organometallics and are used in organic syntheses and as industrial catalysts.

16.3 Carbon monoxide as a ligand

Carbon monoxide is the most common π-ligand in organometallic chemistry. Its primary mode of attachment to metal atoms is through the C atom.

The molecular orbitals of CO

The molecular orbital scheme for CO (Fig. 16.1) shows that the HOMO has σ symmetry and is essentially a lobe that projects away from the C atom.

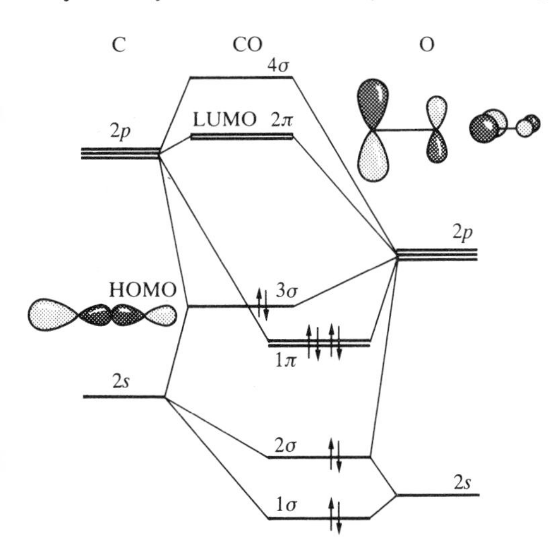

Fig. 16.1 The molecular orbital energy level diagram for CO. The filled 3σ and vacant 2π orbitals are important in metal complex formation.

When CO acts as a ligand, this σ orbital serves as a very weak donor to a metal atom, and forms a σ bond with it (**12**). The LUMOs of CO are the π^* orbitals. These two orbitals play a crucial role because they can overlap metal *d* orbitals that have local π symmetry (such as the t_{2g} orbitals in an O_h complex, **13**). The π interaction leads to the delocalization of electrons from filled *d* orbitals on to the CO ligands, so the ligand also acts as a π acid. The ability of CO to delocalize electron density from the metal accounts for the prevalence of *d*-metal carbonyls with zero or negative oxidation numbers.

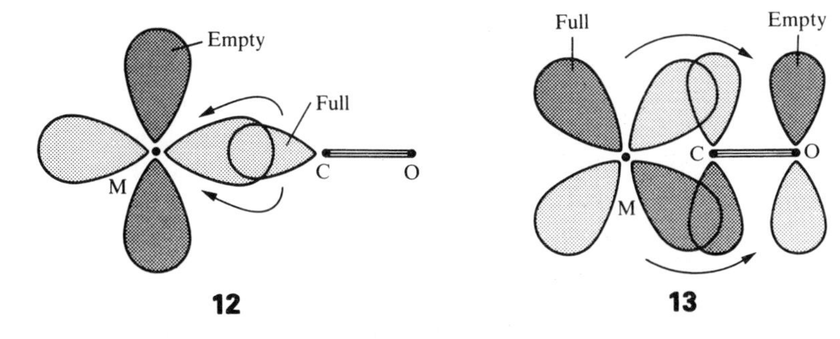

Trends in CO bond lengths and stretching frequencies obtained from IR spectra are in general agreement with the bonding model that we have just described. Thus, both strong σ bases or a formal negative charge on a metal

Table 16.4. The influence of coordination and charge on CO stretching wavenumbers

Compound	$\tilde{\nu}/cm^{-1}$
CO(*g*)	2143
$[Mn(CO)_6]^+$	2090
$Cr(CO)_6$	2000
$[V(CO)_6]^-$	1860
$[Ti(CO)_6]^{2-}$	1750

Source: K. Nakamoto, *Infrared and Raman spectra of inorganic and coordination compounds.* Wiley, New York (1986); Data for $[Ti(CO)_6]^{2-}$ are from S. R. Frerichs, B. K. Stein, and J. E. Ellis, *J. Amer. chem. Soc.*, **109**, 5558 (1987).

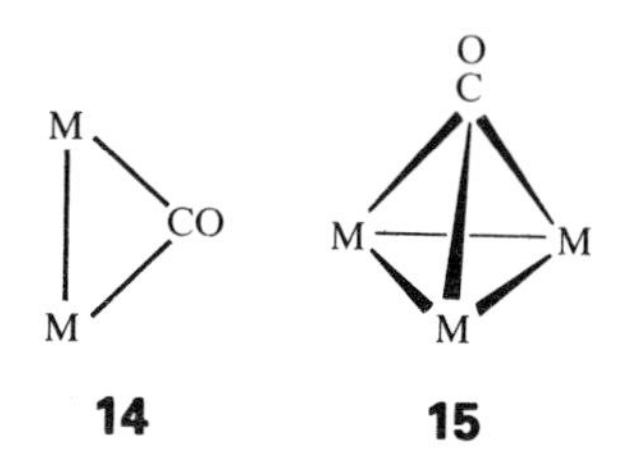

carbonyl anion result in slightly greater CO bond lengths and significantly lower CO stretching frequencies. This is consistent with the bonding model because an increase in the electron density on the metal atom is delocalized over the CO ligands, thereby weakening the CO bond by populating the carbonyl π^* orbital (**13**). A good illustration of this effect is the decrease in stretching frequency with increasing negative charge that is shown in Table 16.4.

Carbon monoxide is versatile as a ligand because it can bridge two (**14**) or three (**15**) metal atoms. Although the description of the bonding is now more complicated, the concepts of σ basicity and π acidity remain useful. The CO stretching frequencies generally follow the order

$$MCO > M_2CO > M_3CO$$

(Fig. 16.2) which suggests an increasing occupation of the π^* orbital as the CO molecule bonds to more metal atoms.

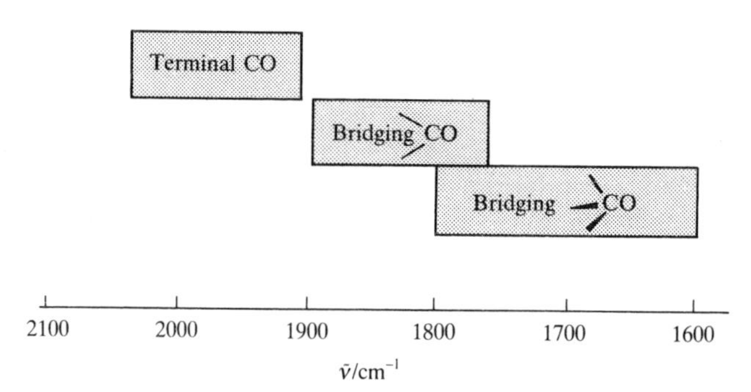

Fig. 16.2 Approximate ranges for the CO stretching wavenumber in neutral metal carbonyls. Note that high wavenumbers (and high frequencies) are on the left, in keeping with the way in which infrared spectrometers generally plot spectra.

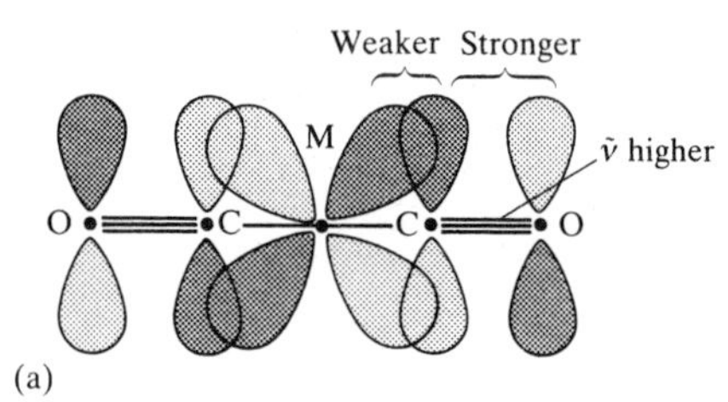

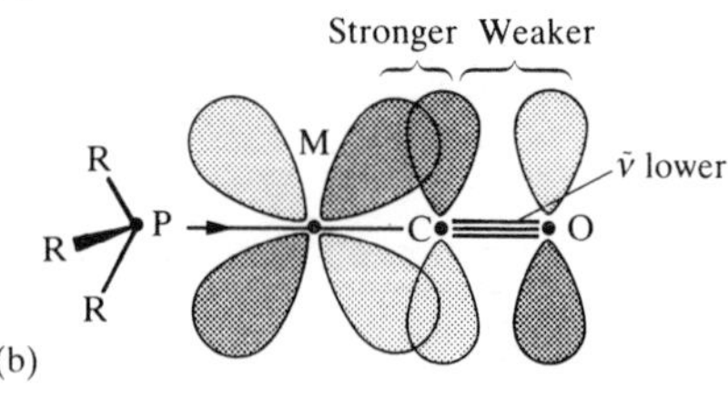

Fig. 16.3 The influence of metal–ligand π bonding on CO bonds in metal carbonyls. (a) An electron-poor metal center that has *d* orbitals lowered in energy because of the positive charge on the metal atom or competing π-bonding ligands. In this case the CO bond length is short and the CO stretching frequency is high. (b) An electron-rich center, which occurs with σ-base ligands or negative charge. Now the CO π-antibonding orbital is more highly populated, so the CO bond is weakened. As a result, the CO bond lengthens slightly and the CO stretching frequency decreases.

Related π-acid ligands

Many other π-acid ligands are found in organometallic complexes, but not all of them contain carbon. Some of them (most importantly, those isoelectronic with CO) can be arranged according to increasing π acidity:

$$C{\equiv}N^- < N{\equiv}N < C{\equiv}NR < C{\equiv}S < N{\equiv}O^+$$

Carbonyl stretching frequencies are often used to determine the order of π-acid strengths for the other ligands present in a complex. The basis of the approach is that the CO stretching frequency is decreased when the CO ligand serves as a π acid. However, as other π acids in the same complex compete for the metal's *d* electrons, they cause the CO frequency to increase (Fig. 16.3). This behavior is opposite to that observed with σ-base ligands. These ligands cause the CO stretching frequency to decrease as they supply electrons to the metal and hence, indirectly, to the CO π^* orbital.

The ligands at the ends of the π-acid series, CN^- and NO^+, differ significantly from CO and are not always good analogs. For example, CN^-, which is electron-rich, is a good σ base and a weak π acid and therefore forms many classical Werner complexes with metal atoms in high oxidation states. At the other extreme, NO^+ is very strongly electron withdrawing. It forms a linearly linked M—NO ligand in complexes in which it is considered to be NO^+ (**16**), but when NO takes on an additional electron to form NO^-, the M—NO link is bent (**17**). For example, in $[IrCl(CO)(PPh_3)_2(NO)]^+$ the IrNO angle is 124°. The other ligands in the series (N_2 through CS) are generally found with metals having low oxidation numbers, and their compounds resemble each other quite closely.

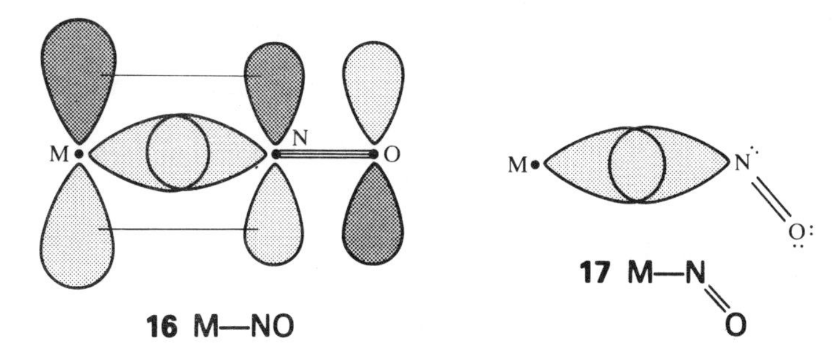

16 M—NO

17 M—N=O

Another useful perspective is obtained by regarding linearly-coordinated NO as a neutral three-electron ligand. Thus two NO ligands can replace three CO ligands and preserve the same electron count on the metal. The tetrahedral nitrosyl complex $Cr(NO)_4$, for instance, has the same electron count as the octahedral carbonyl complex $Cr(CO)_6$.

Two other important π-acid ligands are SO_2 and PF_3. The former can serve as a fairly strong π-acid ligand. Like NO it binds in various ways depending on the number of electrons available. One common case is the coplanar M—SO_2 fragment, in which the bonding is σ donation of an electron pair from the S atom to the metal atom in conjunction with π backbonding from the metal to the SO_2 (**18**). In another extreme, the plane of the SO_2 is tilted with respect to the M—S link (**19**). This type of bonding is generally explained in terms of σ donation to the SO_2 of a lone pair from an electron-rich metal center. In the latter case, SO_2 acts as a simple Lewis acid in its interaction with the metal.

As judged from spectroscopic data, the π-acid character of PF_3 is comparable to that of CO. This was originally explained in terms of the vacant P 3*d* orbitals acting as electron acceptors. Accurate molecular orbital calculations on PF_3, however, indicate that the acceptor orbital is primarily the P—F antibonding σ orbital formed from P 3*p* orbitals (**20**) and that the P 3*d* orbitals play only a minor role. Experimental data on complexes and calculations both indicate that $P(OR)_3$ ligands are somewhat weaker π acids than PF_3 and that both PH_3 and alkylphosphines are very weak π acids but quite good σ bases.

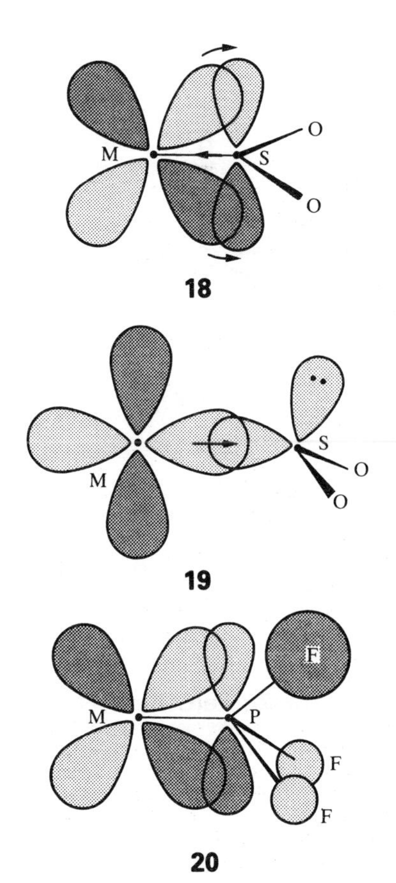

18

19

20

16.4 Synthesis

The two principal methods for the synthesis of binary metal carbonyls are direct combination of carbon monoxide with a finely divided metal and the reduction of a metal salt in the presence of carbon monoxide under pressure. Many polymetallic carbonyls are synthesized from monometallic carbonyls.

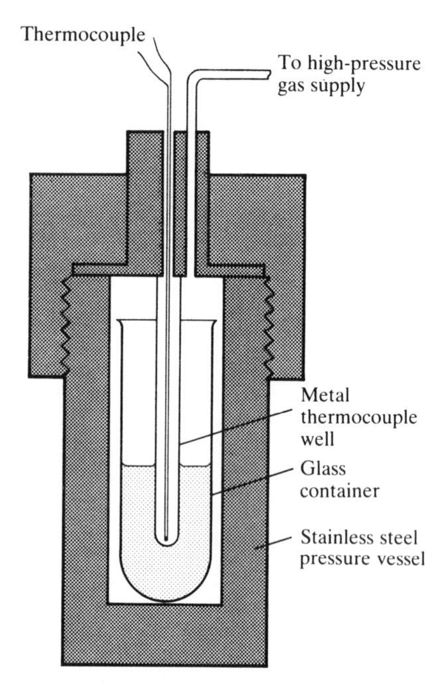

Fig. 16.4 High-pressure reaction vessel. The reaction mixture is in the glass container.

Direct combination

Mond, Langer, and Quinke used the direct combination of nickel and carbon monoxide in their synthesis of the first metal carbonyl, $Ni(CO)_4$:[4]

$$\mathrm{Ni}(s) + 4\mathrm{CO}(g) \xrightarrow[1\ \mathrm{atm}]{30^\circ\mathrm{C}} \mathrm{Ni(CO)_4}(l)$$

Tetracarbonylnickel(0) is in fact the metal carbonyl that is most readily synthesized in this way.

The carbonyls $Cr(CO)_6$, $Fe(CO)_5$, and $Mo(CO)_6$, are thermodynamically more stable than $Ni(CO)_4$, but are formed more slowly. They are therefore synthesized at high pressures and temperatures (Fig. 16.4):

$$\mathrm{Fe}(s) + 5\mathrm{CO}(g) \xrightarrow[200\ \mathrm{atm}]{200^\circ\mathrm{C}} \mathrm{Fe(CO)_5}(l\ \text{at room temperature})$$

$$2\mathrm{Co}(s) + 8\mathrm{CO}(g) \xrightarrow[35\ \mathrm{atm}]{150^\circ\mathrm{C}} \mathrm{Co_2(CO)_8}(s\ \text{at room temperature})$$

Reductive carbonylation

Direct reaction is impractical for most of the remaining *d* metals and **reductive carbonylation**, the reduction of a salt or metal complex in the presence of CO, is normally employed instead. Reducing agents vary from active metals such as aluminum and sodium, to alkylaluminum compounds, H_2, and CO itself:

$$\mathrm{CrCl_3}(s) + \mathrm{Al}(s) + 6\mathrm{CO}(g) \xrightarrow[\text{Benzene}]{\mathrm{AlCl_3}} \mathrm{AlCl_3(soln)} + \mathrm{Cr(CO)_6(soln)}$$

$$3\mathrm{Ru(acac)_3(soln)} + \mathrm{H_2}(g) + 12\mathrm{CO}(g) \xrightarrow[200\ \mathrm{atm},\ \mathrm{CH_3OH}]{150^\circ\mathrm{C}} \mathrm{Ru_3(CO)_{12}(soln)} + \ldots$$

$$\mathrm{Re_2O_7}(s) + 17\mathrm{CO}(g) \xrightarrow[350\ \mathrm{atm}]{250^\circ\mathrm{C}} \mathrm{Re_2(CO)_{10}} + 7\mathrm{CO_2}(g)$$

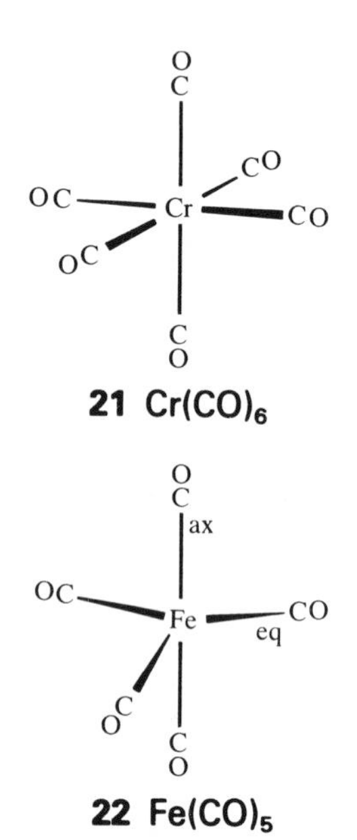

21 $Cr(CO)_6$

22 $Fe(CO)_5$

16.5 Structure

Metal carbonyl molecules generally have well-defined, simple, symmetrical shapes that correspond to the CO ligands taking up the most-distant locations, like electron pairs in VSEPR theory. Thus, the Group 6 hexacarbonyls are octahedral (**21**), pentacarbonyliron(0) is trigonal bipyramidal (**22**), and tetracarbonylnickel(0) is tetrahedral (**2**). Decacarbonyldimanganese(0) consists of two square-pyramidal $Mn(CO)_5$ groups joined by a metal–metal bond. In one isomer of octacarbonyldicobalt(0) the metal–metal bond is bridged by CO.

^{13}C-NMR and IR spectroscopy are widely used to determine the arrangement of atoms in metal carbonyl compounds as separate ^{13}C-NMR signals are observed for inequivalent CO ligands when the molecule is not fluxional (on an NMR timescale), the terminal ligand ^{13}C nucleus being the more shielded. When the molecule is not fluxional, NMR spectra generally contain more detailed structural information than IR spectra. However, IR spectra are often simpler to obtain, and are particularly useful for following

[4] Mond, Langer, and Quinke discovered $Ni(CO)_4$ in the course of studying the corrosion of nickel valves in process gas containing CO. They quickly applied their discovery to develop a new industrial process for the separation of nickel from cobalt. The centennial of this important discovery is celebrated in a special volume of *J. organometal. Chem.* (1990), which is devoted to metal carbonyl chemistry.

Table 16.5. Relation of the number of CO stretching bands in the IR spectrum to structure

Complex	Isomer	Structure*	Point group	Number of bands†
$M(CO)_6$			O_h	1
$M(CO)_5L$			C_{4v}	3‡
$M(CO)_4L_2$	*trans*		D_{4h}	1
	cis		C_{2v}	4††
$M(CO)_3L_3$	*mer*	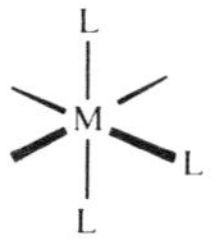	C_{2v}	3††
	fac	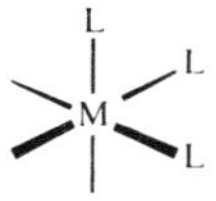	C_{3v}	2
$M(CO)_5$		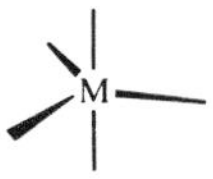	D_{3h}	2
$M(CO)_4L$	*ax*	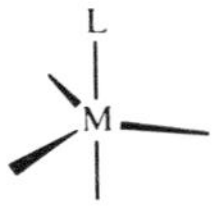	C_{3v}	3**
	eq	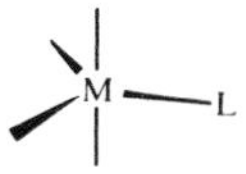	C_{2v}	4
$M(CO)_3L_2$		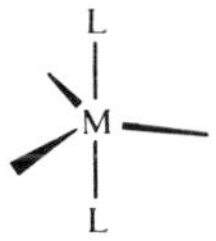	D_{3h}	1
		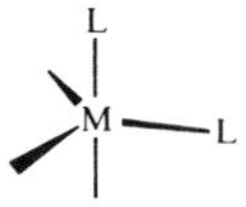	C_s	3
$M(CO)_4$			T_d	1

* Each bonding line that is not terminated by an L is a CO.
† The number of IR bands expected in the CO stretching region is based on formal selection rules. In some cases, fewer bands are observed, as explained in the footnotes.
‡ If the fourfold array of CO ligands lies in the same plane as the metal atom, two bands will be observed.
†† If the *trans* CO ligands are nearly collinear, one fewer band will be observed.
** If the threefold array of CO ligands is nearly planar, only two bands will be observed.

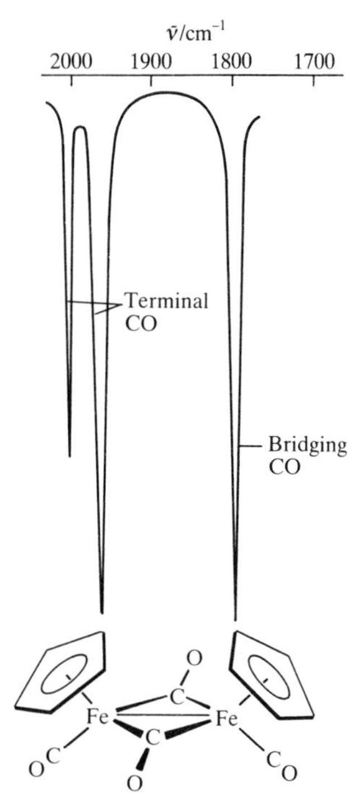

Fig. 16.5 The IR spectrum of $(C_5H_5)_2Fe_2(CO)_4$. Note the two high-frequency terminal CO stretches and the lower frequency absorption of the bridging CO ligands. Although two bridging CO bands would be expected on account of the low symmetry of the complex, a single intense band is observed because the two bridging CO groups are nearly collinear.

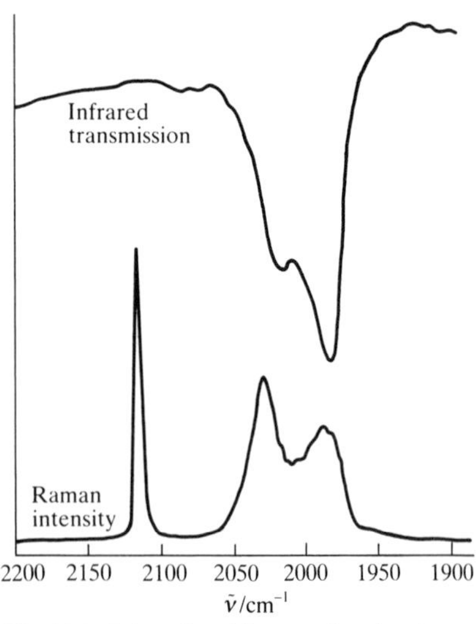

Fig. 16.6 Infrared and Raman vibrational spectra of liquid $Fe(CO)_5$ in the CO stretching region. Note that, in accord with Table 16.5, only two CO stretching bands are observed in the infrared for this D_{3h} complex. The totally symmetric band at 2115 cm^{-1} is absent in the infrared spectrum but strong in the Raman spectrum because the selection rules are different in the two cases.

reactions. Most CO stretching bands occur in the range 2100–1700 cm^{-1}, a region that is generally uncluttered by bands from organic groups. The range of CO stretching frequencies (Fig. 16.2) and the number of CO bands (Table 16.5) are both important for making structural inferences.

If the CO ligands are not related by a center of inversion or a threefold or higher axis of symmetry, the molecule should have one CO stretching absorption for each CO ligand. Thus a bent OC—M—CO group (with only a twofold symmetry axis) will have two infrared absorptions because both the symmetric (**23**) and antisymmetric (**24**) stretches cause the electric

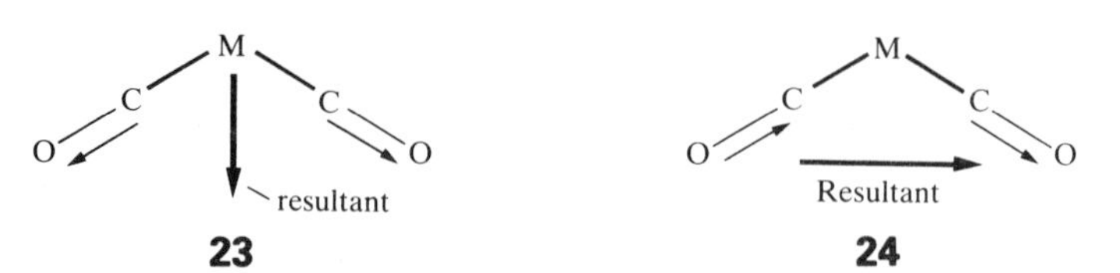

dipole moment to change and are IR active. Highly symmetric molecules have fewer bands than CO ligands. Thus in a linear OC—M—CO group, only one IR band is observed in the CO stretching region because the symmetric stretch leaves the overall electric dipole moment unchanged. As shown in Fig. 16.5, the location of CO ligands may be more symmetrical than the overall symmetry of the complex suggests, and then fewer bands will be observed than are predicted on the basis of the overall point group.

Infrared spectroscopy is also useful for distinguishing terminal CO (MCO) from two-metal bridging CO (M_2CO) and face-bridging CO (μ_3-CO). As we explained earlier, the CO stretches occur at lower frequency in the more highly bridging structures (Figs. 16.2 and 16.5).

Example 16.3: *Determining the structure of a carbonyl from IR data*

The complex $[Cr(CO)_4(P(OPh)_3)_2]$ has one IR absorption band in the CO stretching region. What is the probable structure of this compound?

Answer. A disubstituted hexacarbonyl may exist with either *cis* or *trans* configurations. In the *cis* isomer the four CO ligands are in low symmetry (C_{2v}) and therefore four IR bands should be observed, as indicated in Table 16.5. The *trans* isomer has a square-planar array of four CO ligands (D_{4h}), for which only one band in the CO stretching region is expected (Table 16.5). The *trans* structure is indicated by the data.

Exercise. The IR spectrum of $Ni_2Cp_2(CO)_2$ has a pair of CO stretching bands at 1857 cm^{-1} (strong) and 1897 cm^{-1} (weak). Does this complex contain bridging or terminal CO ligands or both?

We remarked in Section 2.2 that a motionally averaged NMR signal is observed when a molecule undergoes changes in structure more rapidly than the technique can resolve. Although this phenomenon is quite common in the NMR of organometallic compounds, it is not observed in their IR or Raman spectra. An example of this difference is $Fe(CO)_5$, for which the NMR signal shows a single line at $\delta \approx 210$, whereas IR and Raman spectra are consistent with a trigonal bipyramidal structure (Fig. 16.6). We shall meet more NMR evidence for fluxional behavior when we discuss cyclic polyene ligands and metal carbonyl clusters.

16.6 Properties and reactions

Iron and nickel carbonyls are liquids at room temperature and pressure, but all other common carbonyls are solids. All the mononuclear carbonyls are volatile; their vapor pressures at room temperature range from approximately 350 Torr for tetracarbonylnickel(0) to approximately 0.1 Torr for hexacarbonyltungsten(0). The high volatility of $Ni(CO)_4$ coupled with its extremely high toxicity require unusual care in handling it. Although the other carbonyls appear to be less toxic, they too must not be inhaled or allowed to touch the skin.

Since they are nonpolar, all the mononuclear and many of the polynuclear carbonyls are soluble in hydrocarbon solvents. The most striking exception among the common carbonyls is enneacarbonyldiiron(0), $Fe_2(CO)_9$, which has a very low vapor pressure and is insoluble in solvents with which it does not react. In contrast, $Mn_2(CO)_{10}$ and $Co_2(CO)_8$ are soluble in hydrocarbon solvents and readily sublime.

Most of the mononuclear carbonyls are colorless or lightly colored. Polynuclear carbonyls are colored, the intensity of the color increasing with the number of metal atoms. For example, pentacarbonyliron(0) is a light straw-colored liquid, enneacarbonyldiiron(0) forms golden-yellow flakes, and dodecacarbonyltriiron(0) is a deep green compound that looks black in the solid state. The colors of polynuclear carbonyls arise from electronic transitions between orbitals that are largely localized on the metal framework.

The principal reactions of the metal center of simple metal carbonyls are substitution, condensation into clusters, reduction, and oxidation. In certain cases, the CO ligand itself is also subject to attack by nucleophiles and electrophiles.

Substitution

Substitution reactions provide a route to organometallic compounds containing a variety of other ligands. Studies of the rates at which phosphines and other ligands replace CO in $Ni(CO)_4$, $Fe(CO)_5$, and the hexacarbonyls of the Cr group indicate that a dissociative mechanism (*D* or I_d, Section 15.1) is in operation. In the *D* mechanism, an intermediate of reduced coordination number, or more probably a solvated intermediate such as $Cr(CO)_5(THF)$, is produced. This intermediate then combines with the entering group in a bimolecular process:

$$Cr(CO)_6 + Sol \rightleftharpoons Cr(CO)_5Sol + CO$$

$$Cr(CO)_5Sol + PR_3 \rightarrow Cr(CO)_5PR_3 + Sol$$

(Sol denotes a solvent molecule.) In keeping with this interpretation, solvated metal carbonyls have been observed by IR spectroscopy as the photolysis products of metal carbonyls in polar solvents such as THF. The evidence for an I_d mechanism is the rate law for reactions of strong nucleophiles at high concentrations with Group 6 carbonyls, which have a second-order term. Thus the proposed reaction pathways are

$$Cr(CO)_6 \xrightarrow{-CO,\ +Sol} Cr(CO)_5Sol \xrightarrow{+P(CH_3)_3,\ -Sol} Cr(CO)_5P(CH_3)_3 \quad (D)$$

$$Cr(CO)_6 \xrightarrow{+P(CH_3)_3} [Cr(CO)_6P(CH_3)_3]^{\ddagger} \xrightarrow{-CO} Cr(CO)_5P(CH_3)_3 \quad (I_d)$$

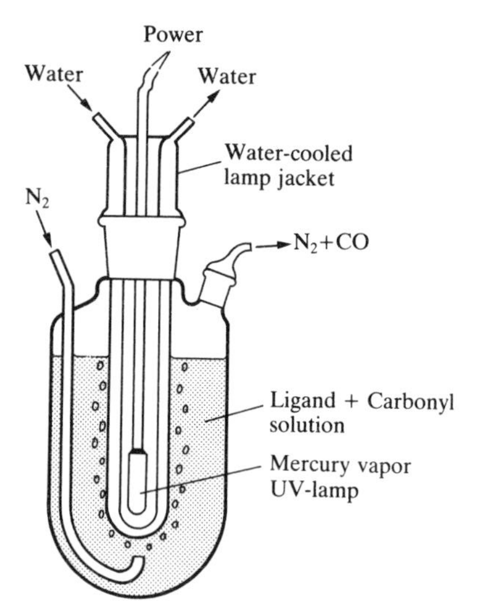

Fig. 16.7 Apparatus for photochemical ligand substitution of metal carbonyls.

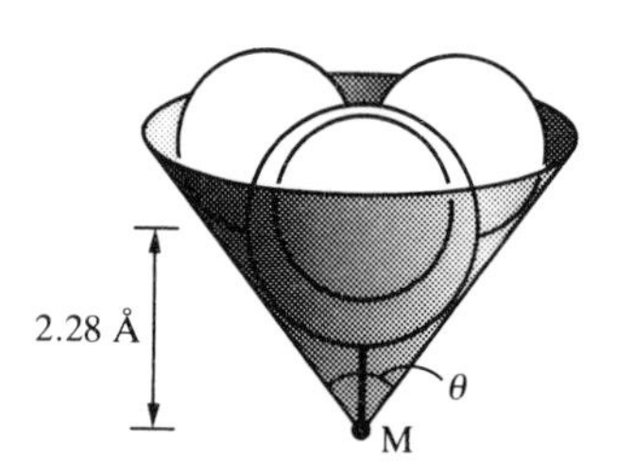

Fig. 16.8 The determination of ligand cone angles from space-filling molecular models of the ligand and an assumed M—P bond length of 2.28 Å.

Loss of the first CO group from $Ni(CO)_4$ occurs easily, and substitution is fast at room temperature. The CO ligands are much more tightly bound in the Group 6 carbonyls, and loss of CO often needs to be promoted thermally or photochemically. For example, the substitution of CH_3CN for CO is carried out in refluxing methyl cyanide using a stream of nitrogen to sweep away the carbon monoxide and thus drive the reaction to completion. To achieve photolysis, mononuclear carbonyls (which do not absorb in the visible) are exposed to near-UV radiation in an apparatus like that shown in Fig. 16.7. As with the thermal process, the photo-assisted substitution reaction leads to the formation of a labile intermediate complex with the solvent (Sol), which is then displaced by the entering group:

$$Cr(CO)_6 \xrightarrow[-CO,\ +Sol]{h\nu} [Cr(CO)_5(Sol)] \xrightarrow[-Sol]{L} Cr(CO)_5L$$

Influence of steric repulsion

As in the reactions of conventional complexes, we can expect steric crowding between ligands to accelerate dissociative processes and to decrease associative processes (Section 15.2). C. A. Tolman (of Dupont) has assessed the extent to which various ligands crowd each other by showing that the volume occupied by a ligand can be approximated by a cone with an angle determined from a space-filling model with, for phosphorus ligands, an M—P internuclear separation of 2.28 Å (Fig. 16.8 and Table 16.6). The ligand CO is 'slim' in the sense of having a small cone angle; $P(t\text{-}Bu)_3$ is 'fat' because it has a large cone angle. Fat ligands have considerable steric repulsion with each other when packed around a metal center.

We can see how the cone angle of a ligand influences the equilibrium constant for ligand binding by examining the dissociation constant of $Ni(PR_3)_4$ complexes (Table 16.7). These complexes are slightly dissociated in solution if the phosphine ligands are slim, such as $P(CH_3)_3$ (cone angle 118°). However, a complex such as $[Ni(P(t\text{-}Bu)_3)_4]$, with fat ligands (cone angle 182°), is highly dissociated. The ligand $P(t\text{-}Bu)_3$ is so bulky that the 14-electron complex $[Pt(P(t\text{-}Bu)_3)_2]$ can be isolated.

Table 16.6. Tolman cone angle for various ligands

Ligand	θ/°	Ligand	θ/°
CH_3	90	$P(OCH_3)_3$	128
CO	95	PEt_3	137
Cl, Et	102	η^5-C_5H_5 (Cp)	136
PF_3	104	PPh_3	145
Br, Ph	105	$C_5(CH_3)_5$ (Cp*)	165
I	107	2,4-$(CH_3)_2C_5H_3$	180
$P(CH_3)_3$	118	$P(t\text{-}Bu)_3$	182
t-Butyl	126	$P(OC_2H_5)_3$	134

Source: C. A. Tolman, *Chem. Rev.*, **77**, 313 (1977); L. Stahl and R. D. Ernst, *J. Amer. chem. Soc.*, **109**, 5673 (1987).

Table 16.7. Cone angle θ and dissociation constant K_d for some Ni complexes*

L	θ/°	K_d/M
$P(CH_3)_3$	118	$<10^{-9}$
$P(C_2H_5)_3$	137	1.2×10^{-5}
$P(CH_3)Ph_2$	136	5.0×10^{-2}
PPh_3	145	Large
$P(t\text{-}Bu)_3$	182	Large

* Data are for $NiL_4 \rightleftharpoons NiL_3 + L$ (in benzene, at 25°C); from C. A. Tolman, *Chem. Rev.*, **77** 313 (1977).

Example 16.4: *Preparing substituted metal carbonyls*

Starting with MoO_3 as a source of Mo, and CO and PPh_3 as the ligand sources, plus other reagents of your choice, give equations and conditions for the synthesis of $Mo(CO)_5PPh_3$.

Answer. A typical procedure is to synthesize $Mo(CO)_6$ first and then to carry out a ligand substitution. Reductive carbonylation of MoO_3 might be performed using $Al(CH_2CH_3)_3$ as a reducing agent in the presence of CO under pressure. The temperature and pressure required for this reaction should be less than those for the direct combination of Mo and CO.

$$MoO_3 + Al(CH_2CH_3)_3 + 6CO \xrightarrow[150°C,\ heptane]{50\ atm} Mo(CO)_6 + \text{oxidation products of } Al(CH_2CH_3)_3$$

The subsequent substitution could be carried out photochemically using the apparatus illustrated in Fig. 16.7.

$$Mo(CO)_6 + PPh_3 \xrightarrow[h\nu]{THF} Mo(CO)_5PPh_3 + CO$$

The progress of the reaction can be followed by IR spectroscopy in the CO stretching region using small samples that are periodically removed from the reaction vessel.

Exercise. If the highly substituted complex $Mo(CO)_3L_3$ is desired, which of the ligands $P(CH_3)_3$ or $P(t\text{-}Bu)_3$ would be preferred? Give reasons for your choice.

Influence of electronic structure on CO substitution

Extensive studies of CO substitution reactions of simple carbonyl complexes have revealed systematic trends in mechanisms and rates.[5]

The 18-electron metal carbonyls generally undergo dissociatively activated substitution reactions. As we saw in Section 15.2, this class of reaction is characterized by insensitivity of the rate to the identity of the entering group. Dissociative activation of an 18-electron complex is usually favored because the alternative, associative activation, would require the formation of an energetically unfavorable 20-electron activated complex.

On the same grounds, we can expect 16-electron organometallic complexes to undergo associatively activated substitution reactions. In that way they can achieve an 18-electron activated complex, which is energetically more favorable than the 14-electron activated complex that would occur in dissociative activation. This is in fact the case: the rates of substitution of 16-electron complexes are sensitive to the identity and concentration of the entering group, which indicates associative activation (Section 15.1). For example, the reactions of $IrCl(CO)(PPh_3)_2$ with triethylphosphine are associatively activated.

Although these generalizations apply to a wide range of reactions, some exceptions are observed, especially if cyclopentadienyl (Cp) or nitrosyl ligands are present. In these cases it is common to find evidence of associatively activated substitution even for 18-electron complexes. The

[5] For an informative article on the development of the major principles see F. Basolo, *Inorg. Chim. Acta*, **50**, 65 (1981).

common explanation is that NO may switch from being a three-electron donor when M—NO is linear to a one-electron donor when it is angular. Similarly, the η^5-Cp five-electron donor can slip relative to the metal and become an η^3 three-electron donor. The resulting electron-deficient central metal is then susceptible to associative substitution.

One final general observation is that the rate of CO substitution in six-coordinate metal carbonyls often decreases as more strongly basic ligands replace CO, and two or three alkylphosphine ligands often represent the limit of substitution. With bulky phosphine ligands, further substitution may be thermodynamically unfavorable on account of ligand crowding. Increased electron density on the metal center, which arises when a π-acid ligand is replaced by a net donor ligand, appears to reduce the rate of CO substitution. The donor character of P(III) ligands as judged by ease of protonation and other data lie in the order[6]

$$PF_3 \ll P(OAr)_3 < P(OR)_3 < PAr_3 < PR_3$$

where Ar is aryl and R is alkyl.

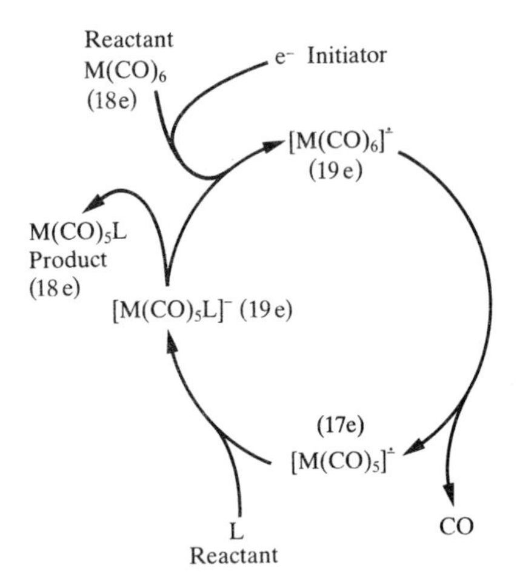

Fig. 16.9 Schematic diagram of an electron transfer catalyzed CO substitution. After addition of a small amount of a reducing initiator, the cycle continues until the limiting reagent $M(CO)_6$ or L has been consumed.

The explanation of the influence of σ-base ligands on CO bonding (Fig. 16.3) is that the increased electron density contributed by the phosphine leads to stronger back π-bonding to the remaining CO ligands and therefore strengthens the M—CO bond. This stronger M—C bond decreases the tendency of CO to leave the metal and therefore decreases the rate of dissociative substitution. In some instances these bonding effects may lead to greater thermodynamic stability for the metal carbonyl than its substitution product.

Recently it has been found that for some metal carbonyls, CO substitution can be catalyzed by electron transfer processes that create anion or cation radicals. A typical process of this type is illustrated in Fig. 16.9, where it can be seen that the key features are the lability of CO in the 19-electron radical anion and the smaller reduction potential of the phosphine-substituted product than for the metal carbonyl starting material.

Reduction to form metal carbonyl anions

Most metal carbonyls can be reduced to metal carbonyl anions. In monometallic carbonyls, two-electron reduction is generally accompanied by loss of the two-electron donor CO ligand:

$$Fe(CO)_5 \xrightarrow{Na,THF} [Fe(CO)_4]^{2-} + CO$$

The metal carbonyl anion contains Fe with oxidation number −2, and it is rapidly oxidized by air. That much of the negative charge is delocalized over the CO ligands is confirmed by the observation of low CO stretching frequencies.

Polynuclear carbonyls containing M—M bonds are generally cleaved by strong reducing agents. Once again the 18-electron rule is obeyed and a mononegative mononuclear carbonyl results:

$$2Na + (OC)_5Mn{-}Mn(CO)_5 \xrightarrow{THF} 2Na^+[Mn(CO)_5]^-$$

Some metal carbonyls disproportionate in the presence of a strongly basic

[6] For a discussion of various measures of phosphine basicity, see R. C. Bush and R. J. Angelici, *Inorg. Chem.*, **27**, 681 (1988).

ligand, producing the ligated cation and the carbonylate anion. Much of the driving force for this reaction is the stability of the metal cation when it is surrounded by strongly basic ligands. Octacarbonyldicobalt(0) is highly susceptible to this type of reaction when exposed to a good Lewis base such as pyridine:

$$\underset{0}{Co_2(CO)_8} + 6py \rightarrow [\underset{+1}{Co}(py)_6][\underset{-1}{Co}(CO)_4] + 4CO$$

Oxidation number: 0 (Co in $Co_2(CO)_8$), +1 (Co in $[Co(py)_6]^+$), −1 (Co in $[Co(CO)_4]^-$)

It is also possible for the CO ligand to be oxidized in the presence of the strong nucleophile OH^-, the net outcome being the reduction of a metal center:

$$3\underset{0}{Fe}(\underset{+2}{C}O)_5 + 4OH^- \rightarrow [\underset{-\frac{2}{3}}{Fe_3}(CO)_{11}]^{2-} + \underset{+4}{C}O_3^{2-} + 2H_2O + 3CO$$

Metal basicity

Metal carbonyl anions can be protonated to produce M—H bonds:

$$[Mn(CO)_5]^-(aq) + H^+(aq) \rightarrow HMn(CO)_5(s)$$

Their affinity for the proton varies widely (Table 16.8). The more negative the anion, the higher its Brønsted basicity and hence the lower the acidity of its conjugate acid, the metal carbonyl hydride.

We commonly refer to *d*-block M—H complexes as hydrides. This reflects the assignment of oxidation number −1 for an H atom attached to a metal atom. Nevertheless, most of the hydrides of metals to the right of the *d* block are weak Brønsted acids. This odd situation simply reflects the formal nature of oxidation numbers. In striking contrast to *p*-block hydrogen compounds, the Brønsted acidity of *d*-block M—H compounds decreases on descending a group.

Neutral metal carbonyls (such as pentacarbonyliron) can be protonated in air-free concentrated acid. Compounds having metal–metal bonds are even more easily protonated. The Brønsted basicity of a metal atom with oxidation number 0 is associated with the presence of nonbonding *d* electrons. Similarly, the tendency to protonate metal–metal bonds can be explained by the transformation of a 2c,2e metal–metal bond into a 3c,2e bond (**25**). A characteristic feature of the Group 6–10 organometallic hydrides is the appearance of a highly shielded ^{1}H-NMR signal ($\delta \approx -8$ to -60); the proton is significantly more shielded than one attached to most elements.

Metal hydrides play an important role in inorganic chemistry and not all of them are generated by protonation. For example, in Section 9.3 we encountered complexes in which an η^2-H_2 is coordinated to a metal center. Even when H_2 adds to a metal center giving conventional H—M—H complexes, it is likely that a transitory η^2-H_2 complex is formed. We discuss the addition of H_2 to metal complexes in Section 16.7 and the role of hydrogen complexes in homogeneous catalysis in Section 17.4.

Metal basicity is turned to good use in the synthesis of a wide variety of organometallic compounds. For example, alkyl and acyl groups can be attached to metal atoms by the reaction of an alkyl or acyl halide with an

Table 16.8. Acidity constant of *d*-metal hydrides in acetonitrile at 25°C

Hydride	pK_a(MH)
$HCo(CO)_4$	8.3
$HCo(CO)_3P(OPh)_3$	11.3
$H_2Fe(CO)_4$	11.4
$CpCr(CO)_3H$	13.3
$CpMo(CO)_3H$	13.9
$HMn(CO)_5$	15.1
$HCo(CO)_3PPh_3$	15.4
$CpW(CO)_3H$	16.1
$Cp^*Mo(CO)_3H$	17.1
$H_2Ru(CO)_4$	18.7
$CpFe(CO)_2H$	19.4
$CpRu(CO)_2H$	20.2
$H_2Os(CO)_4$	20.8
$HRe(CO)_5$	21.1
$Cp^*Fe(CO)_2H$	26.3
$CpW(CO)_2(PMe_3)H$	26.6

Source: E. J. Moore, J. M. Sullivan and J. R. Norton, *J. Amer. chem. Soc.*, **108**, 2257 (1986). Cp* is the pentamethylcyclopentadienyl ligand, $C_5(CH_3)_5$.

$$M—M \xrightarrow{H^+} [M_2(\mu\text{-}H)]^+$$

25

anionic metal carbonyl:

$$[Mn(CO)_5]^- + CH_3I \rightarrow (H_3C)Mn(CO)_5 + I^-$$

$$[Co(CO)_4]^- + CH_3\overset{O}{\overset{\|}{C}}I \rightarrow Co(CO)_4\overset{O}{\overset{\|}{C}}CH_3 + I^-$$

A similar reaction with organometallic halides may be used to form M—M bonds:

$$[Mn(CO)_5]^- + ReBr(CO)_5 \rightarrow (OC)_5Mn\text{—}Re(CO)_5 + Br^-$$

Oxidation to form metal carbonyl halides

Metal carbonyls are susceptible to oxidation by air above room temperature. Although uncontrolled oxidation produces the metal oxide and CO or CO_2, of more interest in organometallic chemistry are the controlled reactions that give rise to organometallic halides. One of the simplest of these is the oxidative cleavage of an M—M bond:

$$\underset{\substack{\uparrow\\0}}{(OC)_5Mn}\text{—}\underset{\substack{\uparrow\\0}}{Mn}(CO)_5 + Br_2 \rightarrow 2\underset{\substack{\uparrow\\+1}}{Mn}Br(CO)_5$$

Oxidation number: 0 0 +1

In keeping with the loss of electron density from the metal when a halogen atom is attached, the CO stretching frequencies of the product are significantly higher than those of $Mn_2(CO)_{10}$.

Reactions of the CO ligand

The C atom of CO is susceptible to attack by nucleophiles if it is attached to a metal atom that is not electron rich. Thus, terminal carbonyls with high CO stretching frequencies are liable to attack by nucleophiles. The *d* electrons in these neutral or cationic metal carbonyls are not extensively delocalized on to the carbonyl C atom, so that atom can be attacked by electron-rich reagents. For example, strong nucleophiles (such as methyllithium, Section 10.6) attack the CO in many neutral metal carbonyl compounds:

$$LiCH_3 + Mo(CO)_6 \rightarrow Li[Mo(\overset{O}{\overset{\|}{C}}\text{—}CH_3)(CO)_5]$$

The resulting acyl compound reacts with carbocation reagents to produce a stable and easily handled neutral product:

$$Li[Mo(COCH_3)(CO)_5] + [(CH_3CH_2)_3O][BF_4]$$
$$\rightarrow Mo(\overset{OCH_2CH_3}{\overset{|}{C}}\text{—}CH_3)(CO)_5 + LiBF_4 + (CH_3CH_2)_2O \quad (1)$$

Compounds of this type, with a direct M=C bond, are often called **Fischer carbenes** because they were discovered in E. O. Fischer's laboratory. The attack of a nucleophile on the C atom is also important for the mechanism of the hydroxide-induced disproportionation of metal carbonyls:

$$(OC)_nM\text{—}C{\equiv}O + OH^-$$
$$\rightarrow (OC)_nM\overset{OH^-}{\overset{|}{C}}{=}O \xrightarrow{3OH^-} [M(CO)_n]^{2-} + CO_3^{2-} + 2H_2O$$

In electron-rich environments the O atom of a CO ligand is susceptible to attack by electrophiles. Once again, IR data give an indication of when this should be expected, since a low CO stretching frequency indicates significant back-donation to the CO ligand, and hence appreciable electron density on the O atom. Thus a bridging carbonyl is particularly susceptible to attack at the O atom:

$$Fe_2Cp_2(CO)_4 + 2Al(CH_2CH_3)_3 \rightarrow Fe_2Cp_2(CO)_4{\cdot}2Al(CH_2CH_3)_3$$
(26)

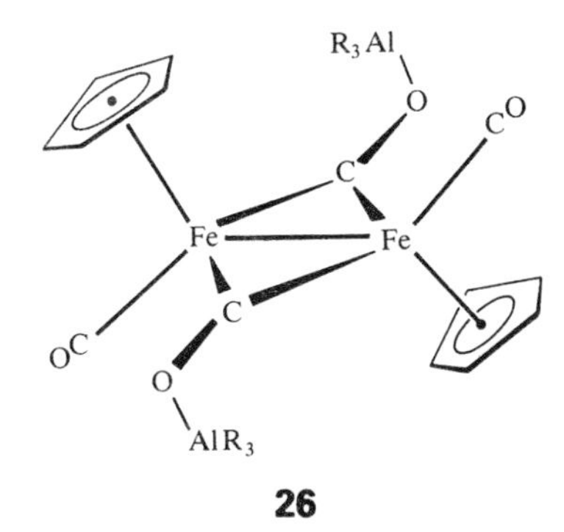

26

Some alkyl-substituted metal carbonyls undergo a **migratory insertion reaction** with the conversion of an alkyl ligand to an acyl ligand, $\overset{\displaystyle O}{\overset{\|}{—C}}R$. There is good evidence that the reaction of $Mn(CH_3)(CO)_5$ occurs by initial migration of the CH_3 group to CO, producing a low concentration of the solvent-coordinated acyl compound. That intermediate then takes on a ligand to produce the stable acyl product:

$$\underset{\displaystyle CH_3}{L_mM{-}CO} \underset{-Sol}{\overset{+Sol}{\rightleftharpoons}} \underset{\displaystyle Sol}{L_mM{-}\overset{\displaystyle O}{\overset{\|}{C}}{-}CH_3} \xrightarrow{L} L_{(m+1)}M{-}\overset{\displaystyle O}{\overset{\|}{C}}{-}CH_3 + Sol$$

We shall see the importance of this reaction in catalysis in Section 17.2.

Example 16.5: *Converting CO to carbene and acyl ligands*

Propose a set of reactions for the formation of $W(CO)_5(\overset{\displaystyle OCH_3}{\overset{|}{C}}{-}Ph)$, starting with hexacarbonyltungsten(0) and other reagents of your choice.

Answer. As with hexacarbonylmolybdenum(0) the CO ligands in hexacarbonyltungsten(0) are susceptible to attack by nucleophiles, and the reaction with phenyllithium should give a C-phenyl intermediate:

$$W(CO)_6 + LiPh \rightarrow Li[W(CO)_5\overset{\displaystyle O}{\overset{\|}{C}}{-}Ph]$$

The anion can then be alkylated on the O atom of the CO ligand by an electrophile:

$$Li[W(CO)_5\overset{\displaystyle O}{\overset{\|}{C}}{-}Ph] + [(CH_3)_3O]BF_4 \rightarrow W(CO)_5\overset{\displaystyle OCH_3}{\overset{|}{C}}{-}Ph + (CH_3)_2O + LiBF_4$$

Exercise. Propose a synthesis for $Mn(CO)_4(PPh_3)(\overset{\displaystyle O}{\overset{\|}{C}}{-}CH_3)$ starting with $Mn_2(CO)_{10}$, PPh_3, Na, and CH_3I.

Other organometallics

A large number of d-block complexes containing hydrocarbon ligands have been investigated from the mid 1950s to the present. We shall discuss the basic structures, bonding, and reactions here and continue the discussion of reactions in Chapter 17 when we describe catalysis. It will be useful to keep an eye on Table 16.1, which summarized the number of electrons that an organic ligand may donate to a metal, as we work through the following examples.

16.7 Hydrogen and open-chain hydrocarbon complexes

Hydrogen

The one-electron ligand H is intimately involved in organometallic chemistry. We have already seen how an M—H bond can be produced by protonation of neutral and anionic metal carbonyls (Section 16.6); in some cases a similar reaction gives a stable product with other organometallics. For example, ferrocene can be protonated in strong acid to produce an Fe—H bond:

$$Cp_2Fe + HBF_4 \rightarrow [Cp_2Fe{-}H]^+[BF_4]^-$$

One of the most interesting and important methods for introducing hydrogen into a metal complex is by oxidative addition (Section 15.13). In this reaction an XY molecule adds to a 16-electron complex ML_4 with cleavage of the X—Y bond to give $M(X)(Y)L_4$ and an increase in the oxidation number of the metal. Thus the addition of either H_2 or HX to a 16-electron complex may produce an 18-electron hydrido complex:

$$IrCl(CO)(PPh_3)_2 + H_2 \longrightarrow IrCl(H)_2(CO)(PPh_3)_2]$$

16e Ir(I) 18e Ir(III)

$$IrCl(CO)(PPh_3)_2 + HCl \longrightarrow IrCl_2(H)(CO)(PPh_3)_2$$

The upper reaction may proceed through a Kubas dihydrogen complex (Section 9.3) in which η^2-H_2 forms transiently.

Alkyl ligands

An alkyl ligand forms an M—C single bond, and in doing so the alkyl group acts as a one-electron monohapto ligand. Many such compounds are known, but they are less common in the *d* block than in the *s* and *p* blocks. This may in part be a result of the modest M—C bond strengths (Table 16.9). Of greater importance are the kinetically facile reactions, such as β-hydrogen elimination, CO insertion, and reductive elimination which lead to the transformation of the alkyl ligand into other groups.

Table 16.9. Bond dissociation enthalpies

	B/(kJ mol^{-1})
$(OC)_5Mn{-}CH_2Ph$	87
$(OC)_5Mn{-}Mn(CO)_5$	94
$(OC)_5Mn{-}CH_3$	153
$(OC)_5Mn{-}H$	213

In a **β-hydrogen elimination reaction**, an H atom on the β-C atom of an alkane is transferred to the metal atom and an alkene is eliminated:

$$L_nM{-}CH_2CH_3 \rightarrow L_nM{-}H + H_2C{=}CH_2$$

We discuss the reverse of this reaction, alkene insertion into the M—H bond, in Chapter 17. Both reactions are thought to proceed through a cyclic intermediate involving a 3c,2e M—H—C interaction:

$$\text{M—CH}_2\text{CH}_3 \longrightarrow \left[\text{M} \begin{matrix} \text{H} & \text{CH}_2 \\ & | \\ & \text{CH}_2 \end{matrix} \right] \longrightarrow \text{M—H} + \text{H}_2\text{C=CH}_2$$

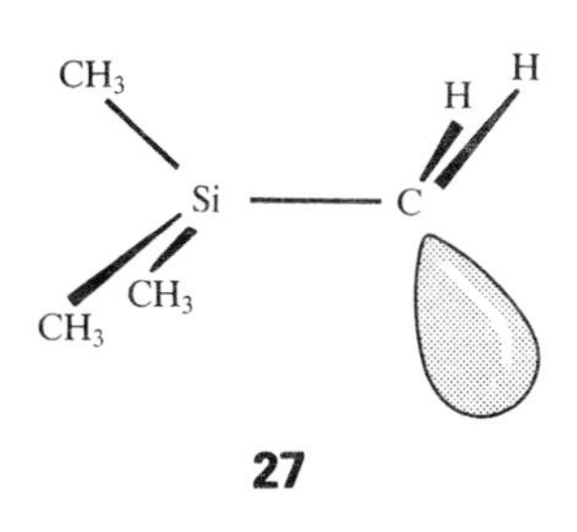

27

Since β-hydrogen elimination is blocked if there are no H atoms on the β-C atom, benzyl $—CH_2(C_6H_5)$ and trimethylsilylmethyl $—CH_2Si(CH_3)_3$ (**27**) ligands are much more robust than ethyl ligands. Similarly, the lack of β-H atoms on the methyl group accounts for the greater stability of complexes containing methyl ligands rather than ethyl ligands. It should be recalled that some *s*- and *p*-block organometallics also undergo β-hydride elimination (Section 10.5).

Alkylidene ligands

The CH_2, CHR, or CR_2 groups are monohapto, two-electron ligands that form the M=C *d–p* double bonds characteristic of Fischer carbenes (**28**, Section 16.6). Because Fischer carbenes are typically formed according to eqn 1 from carbonyl ligands, and therefore contain an electronegative O atom (or an N or S atom), the M=C bond is polarized so as to put a partial positive charge on the C atom (Fig. 16.10a). As a result, Fischer carbenes are attacked at the C atom by nucleophiles. For example, the attack of an amine on the electrophilic C atom of a Fischer carbene results in the displacement of the OR group to yield a new carbene ligand:

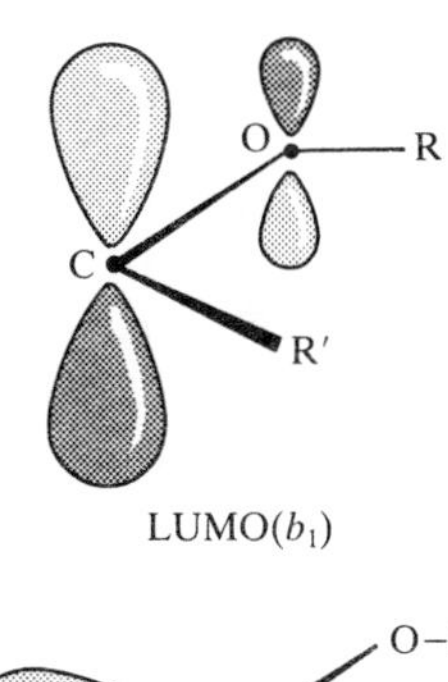

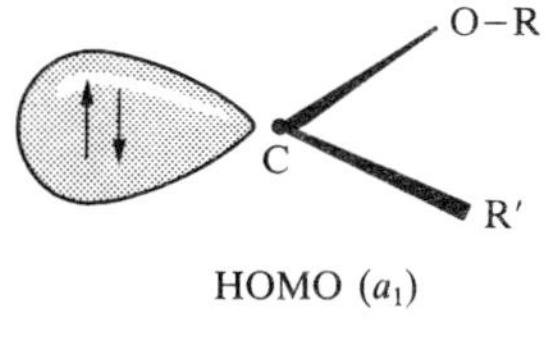

28

$$(\text{OC})_5\text{Cr=C}(\text{OCH}_3)(\text{Ph}) + :\text{NHR}_2 \longrightarrow [(\text{OC})_5\text{Cr—C}(\text{OCH}_3)(\text{Ph})\text{—NHR}_2] \longrightarrow$$

$$(\text{OC})_5\text{Cr=C}(\text{NR}_2)(\text{Ph}) + \text{HOCH}_3$$

A series of purely hydrocarbon alkylidenes has been synthesized for metals on the left of the *d* block. They are informally called **Schrock**

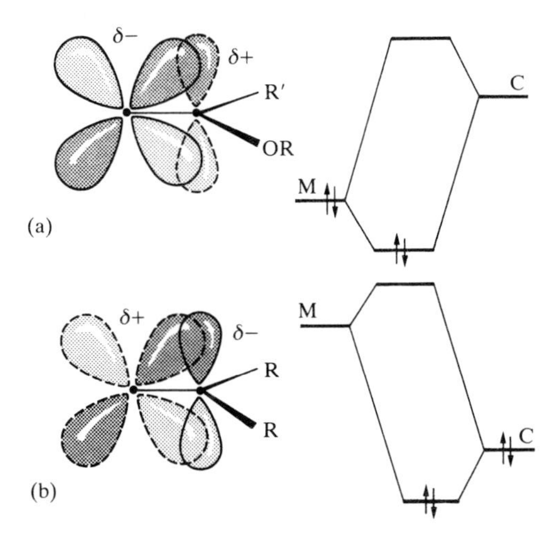

Fig. 16.10 In the Fischer carbene (a), the vacant p_z orbital of CR′(OR) is higher in energy than the metal $d\pi$ orbitals, so the π-electron density is largely concentrated on the metal and the carbene C is electrophilic. For the Schrock carbene (b), the p_z orbital of CR_2 is lower in energy than the $d\pi$ orbitals, so π-electron density is concentrated in the carbon and the carbene C is nucleophilic.

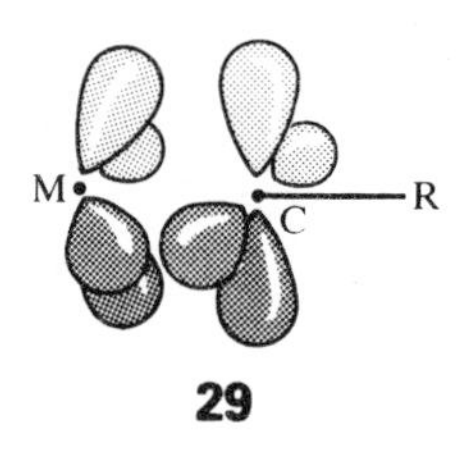

29

carbenes after their discoverer. The M═C bond in these compounds is generally polarized so as to put negative charge on the carbene (Fig. 16.10b). As a result, the Schrock carbenes react with electrophiles such as $(CH_3)_3SiBr$:

$$Cp_2Ti(CH_2)(CH_3) + (CH_3)_3SiBr \rightarrow [Cp_2Ti(CH_2Si(CH_3)_3)(CH_3)]^+ + Br^-$$

Alkylidyne ligands

The alkylidyne ligands have the general formula CH or CR. They are monohapto three-electron ligands and are bound to the metal by a M≡C triple bond involving one σ bond and two d–p π bonds (**29**). The simplest member of this series is methylidyne, CH; the next simplest is ethylidyne, CCH_3.

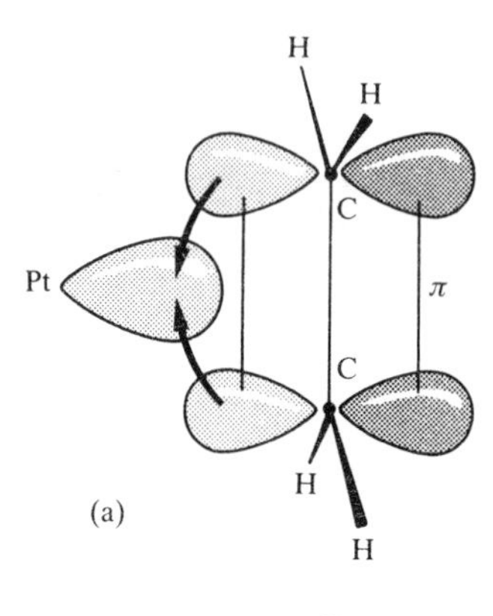

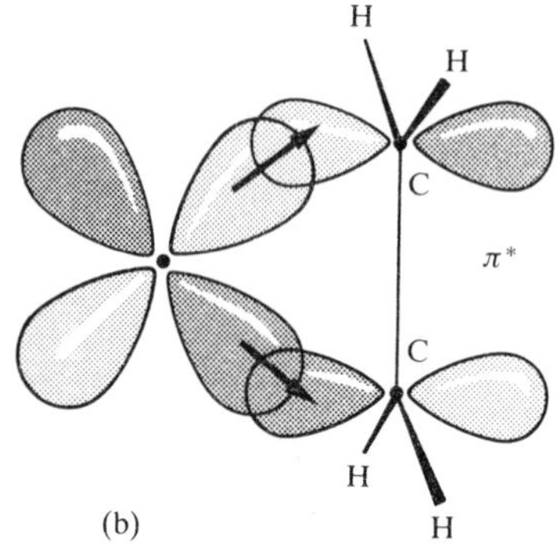

Fig. 16.11 Interaction of ethylene with a metal atom. (a) Donation of electron density from the filled π molecular orbital of ethylene to a vacant metal σ orbital. (b) Acceptance of electron density from a filled $d\pi$ orbital into the vacant π^* orbital of ethylene.

Alkene ligands

The simple alkene ligands are dihapto two-electron donors because the filled π orbital projects toward the metal and donates electrons to suitably oriented metal orbitals (the s, p_z, d_{z^2}, and $d_{x^2-y^2}$ orbitals in Fig. 16.11). In addition, the π^* orbital of the ligand can accept electron density from filled metal orbitals of the appropriate symmetry (the p_x and d_{xz} orbitals in Fig. 16.11). This description of metal–alkene bonding in terms of electron donation from the filled bonding π-orbital of the ligand and simultaneous electron acceptance into the empty antibonding π^*-orbital is called the **Dewar–Chatt model**.

Electron donor and acceptor character appear to be fairly evenly balanced in most ethylene complexes of the d metals, but the degree of donation and back-donation can be altered by substituents. An extreme example is tetracyanoethylene (**30**), which is an abnormally strong electron acceptor ligand on account of its electron-withdrawing nitrile groups. Tetracyanoethylene qualifies as a π-acid ligand because its role as an acceptor dominates its role as a donor. When the degree of donation of electron density from the metal atom to the alkene ligand is small, substituents on the ligand are bent only slightly away from the metal and the C—C bond length is only slightly greater than in the free alkene. With electron-rich metals or electron-withdrawing substituents on the alkene, the back-donation is greater and substituents on the alkene are bent away from the metal, and the C—C bond length approaches that of a single C—C bond. These differences lead some chemists to depict the first group as simple π complexes (**31**) and the second as metallocycles (**32**) with M—C single bonds.

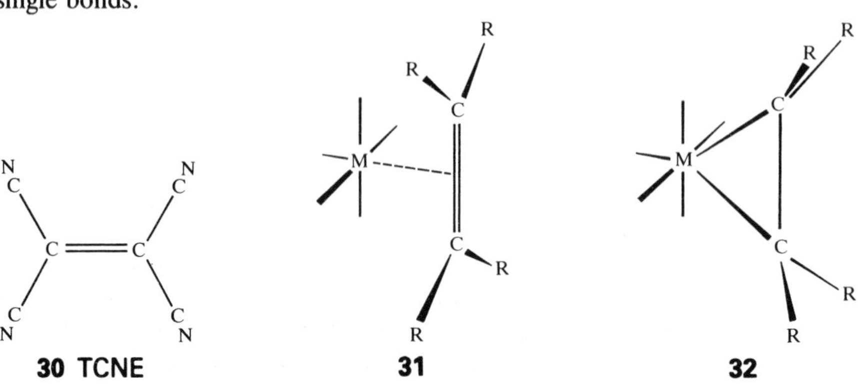

30 TCNE **31** **32**

Diene and polyene ligands

Diene (—C═C—C═C—) and polyene ligands present the possibility of polyhapto bonding. As with the chelate effect in Werner complexes, the resulting polyene complexes are usually more stable than the equivalent complex with individual ligands because the entropy of dissociation of the complex is much smaller than when the liberated ligands can move independently (Section 7.7). For example, bis(η^4-cycloocta-1,5-diene)-nickel(0) (**33**) is more stable than the corresponding complex containing four ethylene ligands. Cycloocta-1,5-diene is a fairly common ligand in organometallic chemistry, where it is referred to engagingly as cod.

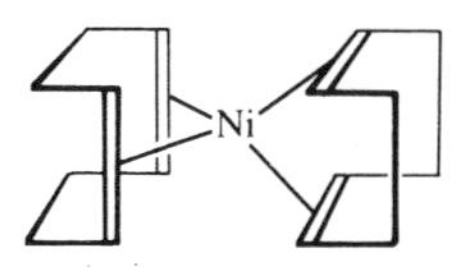

33 Ni(cod)$_2$

34 Fe(η^1-C_3H_5)(CO)$_2$(η^5-Cp)

The π-allyl ligand

The allyl ligand can bind to a metal atom in either of two configurations. As an η^1-ligand (**34**) it is a one-electron donor; as an η^3-ligand (**35**) it is a three-electron donor. In the latter case the terminal substituents are bent slightly out of the plane of the three-carbon backbone and are either *exo* (**36**) or *endo* (**37**) relative to the metal. It is common to observe *exo* and *endo* group exchange, which in some cases is fast on an NMR timescale. A mechanism that involves the transformation $\eta^3 \rightarrow \eta^1 \rightarrow \eta^3$ is often invoked to explain this exchange. Because of this flexibility in type of bonding, η^3-allyl complexes are often highly reactive.

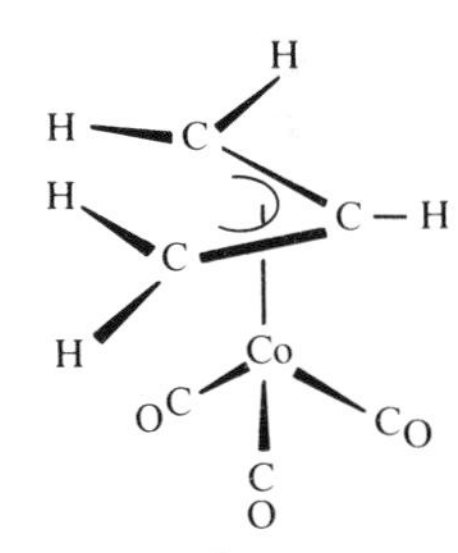

35 Co(η^3-C_3H_5)(CO)$_3$

There are many routes to allyl complexes. One is the nucleophilic attack of an allyl Grignard reagent on a metal halide:

$$2C_3H_5MgBr + NiCl_2 \xrightarrow{\text{Ether}} Ni(\eta^3\text{-}C_3H_5)_2 + 2MgBrCl$$

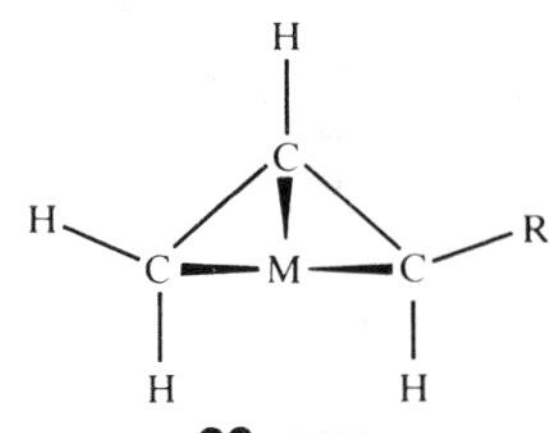

36 *exo*

Conversely, electrophilic attack of a haloalkane on a low-oxidation-state metal center yields allyl complexes:

$$CH_2{=}CHCH_2Cl + [Mn(CO)_5]^- \xrightarrow{-Cl^-} (\eta^1\text{-}CH_2{=}CHCH_2)Mn(CO)_5$$

$$\xrightarrow[-CO]{\Delta} (\eta^3\text{-}CH_2CHCH_2)Mn(CO)_4$$

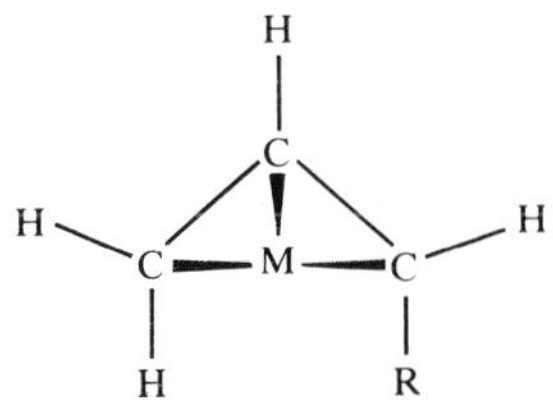

37 *endo*

Butadiene complexes can also be converted to π-allyl complexes by protonation:

Fe(CO)$_3$ + HCl ⟶ CH$_3$ FeCl(CO)$_3$

Alkyne ligands

Acetylene has two π bonds and hence is a potential four-electron donor. However, it is not always clear from experimental information that it acts in this way. Substituted acetylenes form very stable polymetallic complexes in which the acetylene can be regarded as a four-electron donor. An example is η^2-diphenylethynehexacarbonyldicobalt(0) (**38**) where we can view one π bond as donating to one of the Co atom centers and the second π bond as

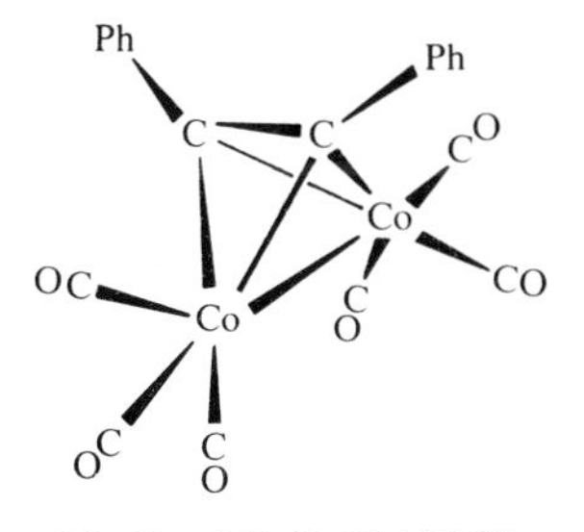

38 Co$_2$(PhC$_2$Ph)(CO)$_6$

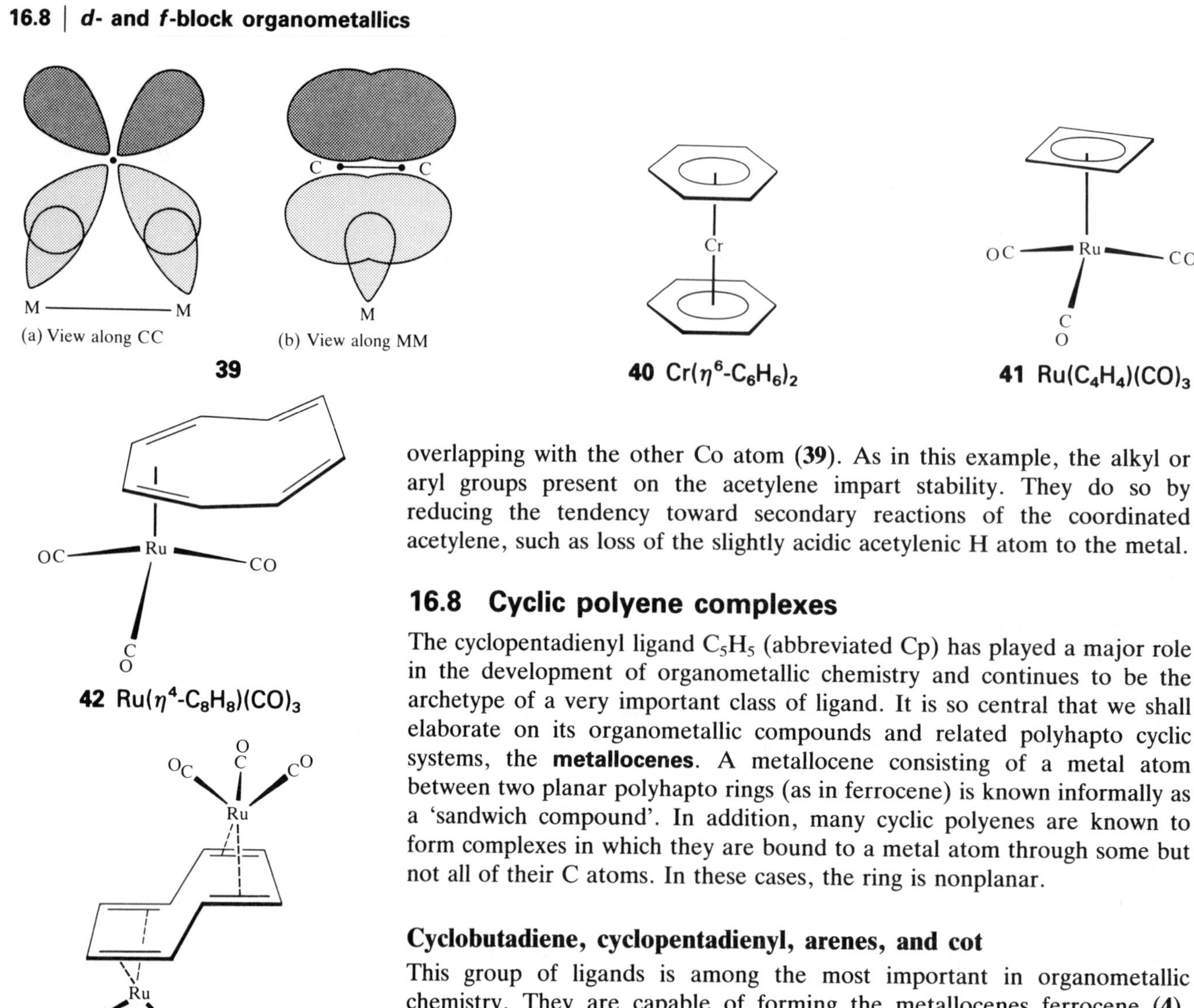

39

40 $Cr(\eta^6\text{-}C_6H_6)_2$

41 $Ru(C_4H_4)(CO)_3$

42 $Ru(\eta^4\text{-}C_8H_8)(CO)_3$

43

44

overlapping with the other Co atom (**39**). As in this example, the alkyl or aryl groups present on the acetylene impart stability. They do so by reducing the tendency toward secondary reactions of the coordinated acetylene, such as loss of the slightly acidic acetylenic H atom to the metal.

16.8 Cyclic polyene complexes

The cyclopentadienyl ligand C_5H_5 (abbreviated Cp) has played a major role in the development of organometallic chemistry and continues to be the archetype of a very important class of ligand. It is so central that we shall elaborate on its organometallic compounds and related polyhapto cyclic systems, the **metallocenes**. A metallocene consisting of a metal atom between two planar polyhapto rings (as in ferrocene) is known informally as a 'sandwich compound'. In addition, many cyclic polyenes are known to form complexes in which they are bound to a metal atom through some but not all of their C atoms. In these cases, the ring is nonplanar.

Cyclobutadiene, cyclopentadienyl, arenes, and cot

This group of ligands is among the most important in organometallic chemistry. They are capable of forming the metallocenes ferrocene (**4**), bisbenzenechromium(0) (**40**), and uranocene (**5**). The simplest ligand of the group, cyclobutadiene, is unstable as the free molecule, but stable complexes are known, including $Ru(CO)_3(C_4H_4)$ (**41**). This is one of many cases in which coordination to a metal atom stabilizes an otherwise unstable molecule.

The cyclopentadienyl radical is a monohapto ligand when it contributes one electron to a σ bond with the metal (**7**), a trihapto ligand when it donates three electrons, but usually it is a pentahapto ligand contributing five electrons. Structure (**8**) showed a compound that contains both η^3 and η^5 ligands. Benzene and its derivatives are most often hexahapto six-electron donors (**40**) in which all six π electrons are shared with the metal. Cyclooctatetraene is a large ligand that is found in a wide variety of bonding arrangements. It uses some of its π electrons for bonding to metals when it is an η^2, an η^4 (**42**), or an η^6 ligand, and in these cases the ring is puckered. Its versatility in bonding is also shown by its ability to act as a bridge between two metal atoms (**43** and **44**). As we have already mentioned, when C_8H_8 is acting as a planar η^8-ligand (**5**), it is best regarded as the planar aromatic $C_8H_8^{2-}$ group.

Bonding in ferrocene

Although details of the bonding in ferrocene are not settled, the molecular orbital energy level diagram shown in Fig. 16.12 accounts for a number of experimental observations. This diagram refers to the eclipsed (D_{5h}) form of the complex, which in the gas phase is about $4\,\mathrm{kJ\,mol^{-1}}$ more stable than the staggered conformation. We shall focus our attention on the frontier orbitals. As shown in Fig. 16.13, the e_1' symmetry-adapted linear combinations of ligand orbitals have the same symmetry as the d_{zx} and d_{yz} orbitals of the metal atom. The lower-energy frontier orbital (a_1') is composed of d_{z^2} and the corresponding symmetry-adapted combination of ligand orbitals (Fig. 16.13), but—by accident—there is little interaction between the ligands and the metal orbitals in this case because the ligand *p* orbitals happen to lie in the conical nodal surface of the metal's d_{z^2} orbital. In ferrocene and the other 18-electron biscyclopentadienyl complexes the a_1' frontier orbital and all lower orbitals are full but the e_1'' frontier orbital and all higher orbitals are empty.

The frontier orbitals are neither strongly bonding nor antibonding. This characteristic permits the possibility of the existence of metallocenes that

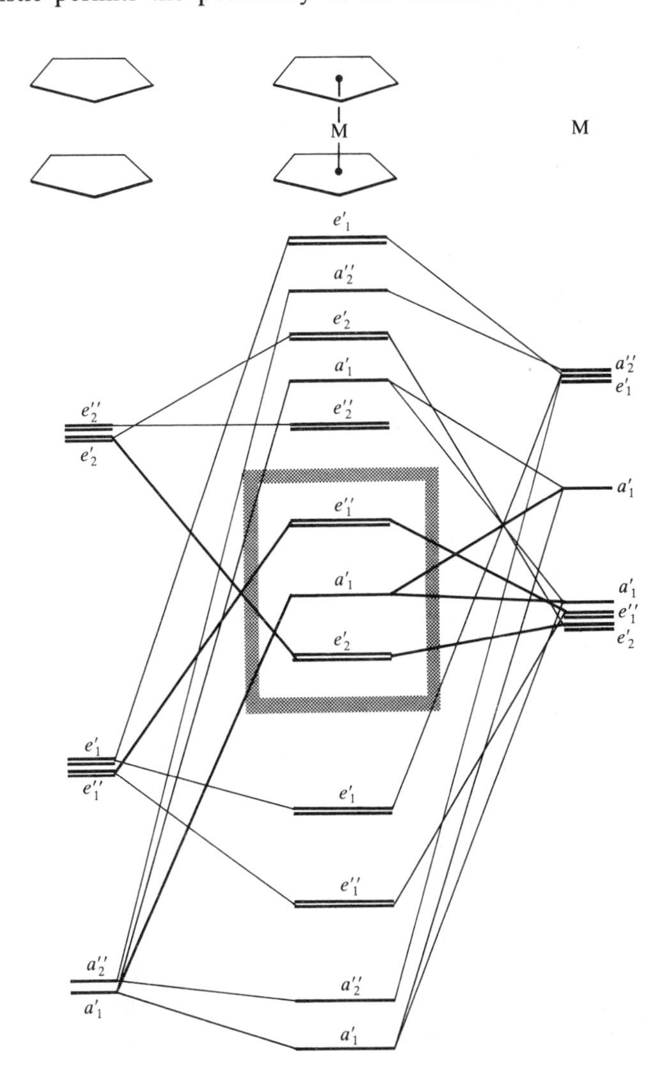

Fig. 16.12 Molecular orbital energy diagram of a metallocene. The energies of the symmetry-adapted π orbitals of the C_5H_5 ligands are shown on the left, relevant *d* orbitals of the metal are on the right and the resulting molecular orbital energies are in the center. 18 electrons can be accommodated by filling the molecular orbitals up to an including the a_1' frontier in the box. Three frontier orbitals are indicated in the box rather than the usual two because the wide range of electron counts for metallocenes involve the filling of these three orbitals.

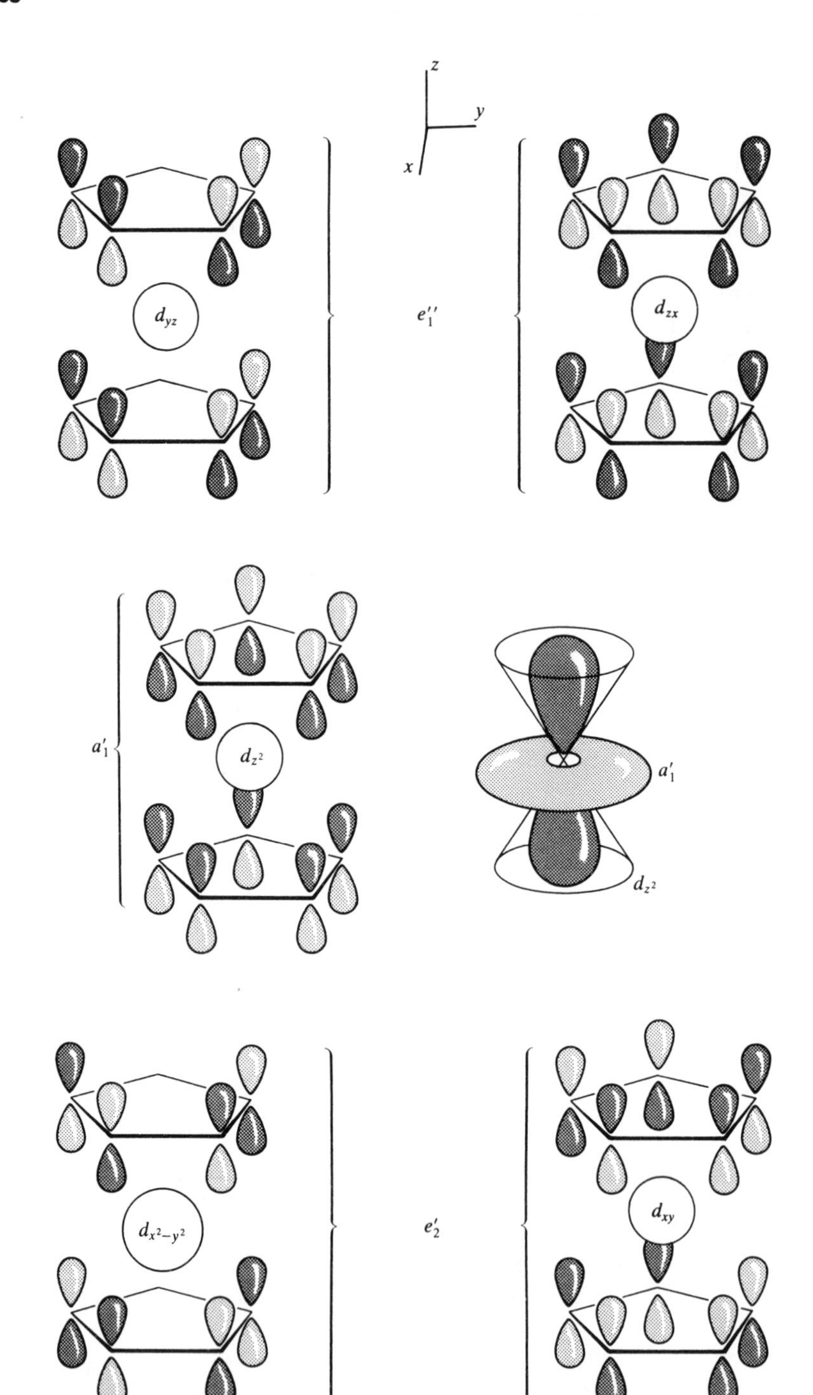

Fig. 16.13 Symmetry-adapted orbital combinations giving rise to the three metallocene frontier orbitals. The accidental coincidence of the nodal surface of d_{z^2} and the C_5H_5 π orbitals results in negligible interaction despite its being symmetry-allowed.

diverge from the 18-electron rule, such as the 17-electron $[Fe(C_5H_5)_2]^+$ and the 20-electron $Ni(C_5H_5)_2$. Deviations from the 18-electron rule, however, do lead to significant changes in M—C bond lengths that correlate quite well with the molecular orbital scheme (Table 16.10). Similarly, the redox properties of the complexes can be understood in terms of the electronic structure. Thus, we have already mentioned that ferrocene is fairly readily oxidized to the ferrocinium ion, $[Fe(C_5H_5)_2]^+$. From an orbital viewpoint,

Table 16.10. Electronic configuration and M—C bond length in $M(\eta^5\text{-}C_5H_5)_2$ complexes

Complex	Valence electrons	Electron configuration	R(M–C)/Å
$V(C_5H_5)_2$	15	$e_2'^2 a_1'^1$	2.28
$Cr(C_5H_5)_2$	16	$e_2'^3 a_1'^1$	2.17
$Mn(C_5H_4CH_3)_2$*	17	$e_2'^3 a_1'^2$	2.11
$Fe(C_5H_5)_2$	18	$e_2'^4 a_1'^2$	2.06
$Co(C_5H_5)_2$	19	$e_2'^4 e_1''^1 a_1'^2$	2.12
$Ni(C_5H_5)_2$	20	$e_2'^4 e_1''^2 a_1'^2$	2.20

* Data are quoted for this complex because $Mn(C_5H_5)_2$ has a high-spin configuration $a_1'^2 e_2'^2 e_1''^2$ and hence an anomalously long M–C bond length (2.38 Å).

this oxidation corresponds to the removal of an electron from the nonbonding a_1' orbital. The 19-electron complex $[Co(C_5H_5)_2]$ is much more readily oxidized than ferrocene because the electron is lost from the antibonding e_1'' orbital to give the 18-electron $[Co(C_5H_5)_2]^+$ ion.

Another useful comparison can be made with octahedral complexes. The e_1'' frontier orbitals of a metallocene are the analog of the e_g orbitals in an octahedral complex, and the a' orbital plus the e_2' pair of orbitals are analogous to the t_{2g} orbitals of an octahedral complex. This formal similarity extends to the existence of high- and low-spin biscyclopentadienyl complexes (Table 16.10). The redox chemistry of metallocenes is permeated by similarities to that of simple octahedral complexes.

Fluxional cyclic polyene complexes

One of the most remarkable aspects of many cyclic polyene complexes is their stereochemical nonrigidity. For example, at room temperature the two rings in ferrocene rotate rapidly relative to each other (**45**). This type of fluxional process is called **internal rotation**, and is similar to the process in ethane.

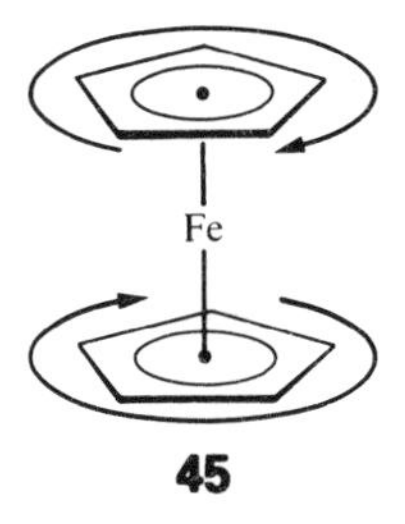

Of greater interest is the stereochemical nonrigidity that is often seen when a conjugated cyclic polyene is attached to a metal atom through some but not all of its C atoms. In such complexes the metal–ligand bonding may hop around the ring, a property called **fluxionality** and, in the informal jargon of organometallic chemists, 'ring whizzing'. A simple example is found in $Ge(\eta^1\text{-}C_5H_5)(CH_3)_3$, where the single site of attachment of the Ge atom to the cyclopentadiene ring hops around the ring in a series of **1,2-shifts**, or motion in which a C—M bond is replaced by a C—M bond to the next C atom around the ring (Fig. 16.14a). The great majority of

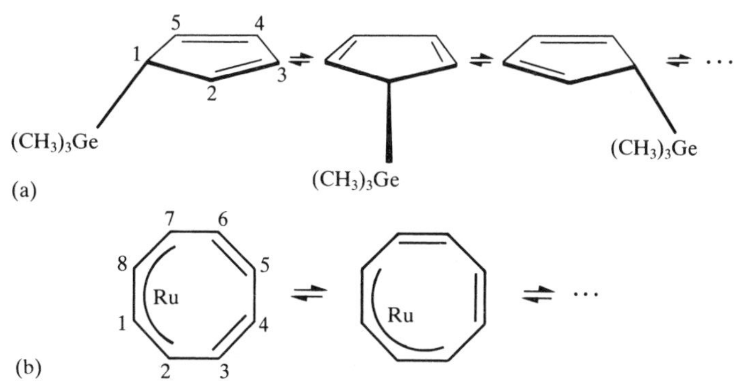

Fig. 16.14 (a) The fluxional process in $Ge(\eta^1\text{-}C_5H_5)(CH_3)_3$ occurs by a series of 1,2-shifts. (b) The fluxionality of $Ru(\eta^4\text{-}C_8H_8)(CO)_3$ can be described similarly. We need to imagine that the Ru atom is out of the plane of the page; the CO ligands have been omitted.

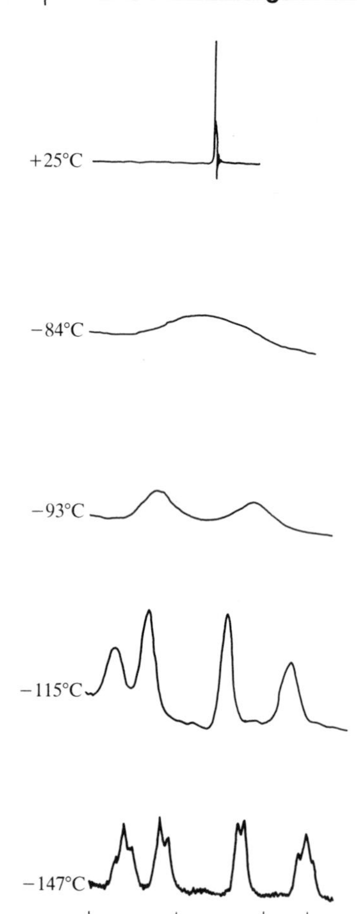

Fig. 16.15 Observed ^{1}H-NMR spectra for (η^4-C_8H_8)Ru(CO)$_3$ at various temperatures. (From: F. A. Cotton, Chapter 10 in *Dynamic nuclear magnetic resonance spectroscopy*, (ed. L. M. Jackson and F. A. Cotton). Academic Press, New York (1975).)

46

fluxional conjugated polyene complexes that have been investigated migrate by 1,2-shifts, but it is not known whether these shifts are controlled by a principle of least motion or by some aspect of orbital symmetry.

NMR provides the primary evidence for the existence and mechanism of these fluxional processes, for when they occur on a time scale of 10^{-2}–10^{-4} s they can be studied by ^{1}H- and ^{13}C-NMR. The complex Ru(η^4-C_8H_8)(CO)$_3$ (**42**) is an illustration of the approach. At room temperature, its ^{1}H-NMR spectrum consists of a single sharp line (Fig. 16.15) that could be interpreted as arising from a symmetrical η^8-C_8H_8 ligand. However, X-ray diffraction studies of single crystals show unambiguously that the ligand is tetrahapto. This conflict is resolved by ^{1}H-NMR spectra at lower temperatures, for as the sample is cooled the signal broadens and then separates into four peaks. These peaks are expected for the four pairs of protons of a η^4-C_8H_8 ligand (**46**). The interpretation is that at room temperature the ring is whizzing around the metal atom rapidly compared with the timescale of the NMR experiment, so an averaged signal is observed (Fig. 16.15). At lower temperatures the motion of the ring is slower, and the distinct conformations exist long enough to be resolved. A detailed analysis of the line shape of the NMR spectra can be used to measure the activation energy of the migration.

Synthesis

Sodium cyclopentadienide is a common starting material for the preparation of cyclopentadienyl compounds. It can be prepared by the action of sodium on cyclopentadiene:

$$2Na + 2C_5H_6 \xrightarrow{\text{THF}} 2Na[C_5H_5] + H_2$$

Sodium cyclopentadienide reacts with *d*-metal halides to produce metallocenes:

$$2Na[C_5H_5] + MnCl_2 \xrightarrow{\text{THF}} Mn(C_5H_5)_2 + 2NaCl$$

The overall result is simple ligand displacement, but the mechanism may be more complicated than this suggests. The biscyclopentadienyl complexes of Fe, Co, and Ni are also readily prepared by this method.

The synthesis of many cyclic polyene metal carbonyls is carried out by photochemical or thermal substitution of metal carbonyls. For example, hexacarbonylchromium(0) can be refluxed with an arene to produce the arene tricarbonyl:

$$Cr(CO)_6 + C_6H_6 \xrightarrow{\Delta} Cr(\eta^6\text{-}C_6H_6)(CO)_3 + 3CO$$

A similar reaction is

$$Fe(CO)_5 + C_8H_8 \rightarrow Fe(\eta^4\text{-}C_8H_8)(CO)_3 + 2CO$$

Reactions

Because of their great stability, the 18-electron Group 8 compounds ferrocene, ruthenocene, and osmocene maintain their ligand–metal bonds under rather harsh conditions, and it is possible to carry out a variety of transformations on the cyclopentadienyl ligands. For example, they undergo reactions similar to those of simple aromatic hydrocarbons, such as

Friedel–Crafts substitution:

$$CH_3\overset{\overset{\displaystyle O}{\|}}{C}Cl + Fe(C_5H_5)_2 \xrightarrow{AlCl_3} Fe(C_5H_5)(C_5H_4\overset{\overset{\displaystyle O}{\|}}{C}CH_3) + HCl$$

It also is possible to replace H on a Cp ring by Li:

$$LiBu + Fe(C_5H_5)_2 \rightarrow Fe(C_5H_5)(C_5H_4Li) + C_4H_{10}$$

As might be imagined, the lithiated product is an excellent starting material for the synthesis of a wide variety of ring-substituted products and in this respect resembles simple organolithium compounds (Section 10.6).

There are many interesting related structures in addition to the simple biscyclopentadiene and bisarene complexes. In the jargon of this area, these are referred to as 'bent sandwich compounds' (**47**), 'half-sandwich' or 'piano stool' compounds (**6**), and, inevitably, 'triple deckers' (**48**).

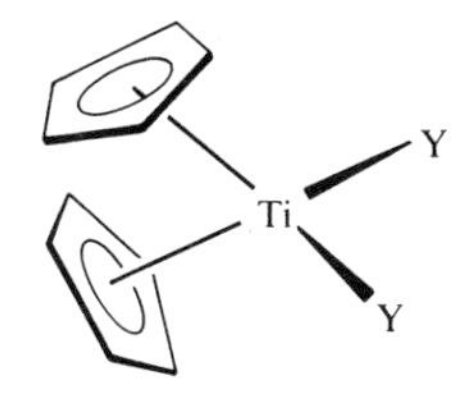

47 $Ti(C_5H_5)_2Y_2$

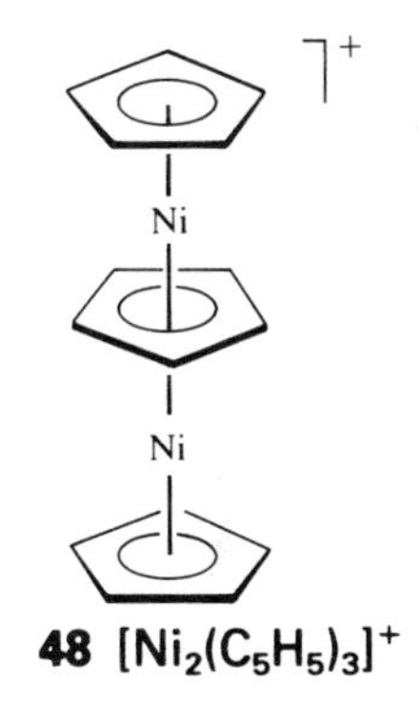

48 $[Ni_2(C_5H_5)_3]^+$

Example 16.6: *Identifying metallocene electronic structure and stability*

By referring to Fig. 16.12, discuss the occupancy and nature of the HOMO in $[Co(\eta^5\text{-}C_5H_5)_2]^+$ and discuss the change in metal–ligand bonding relative to neutral cobaltocene.

Answer. The $[Co(\eta^5\text{-}C_5H_5)_2]^+$ ion contains 18 valence electrons (9 from Co, 10 from the two Cp ligands, less one for the single positive charge). Assuming that the molecular orbital energy level diagram for ferrocene applies, the 18-electron count leads to double occupancy of the orbitals up through a'_{1g}. The 19-electron cobaltocene molecule has an additional electron in the e''_{1g} orbital, which is antibonding with respect to the metal and ligand. Therefore the metal–ligand bond should be stronger and shorter in $[Co(\eta^5\text{-}C_5H_5)_2]^+$ than in $Co(\eta^5\text{-}C_5H_5)_2$. This is borne out by structural data.

Exercise. Using the same molecular orbital diagram, comment on whether the removal of an electron from $Fe(\eta^5\text{-}C_5H_5)_2$ to produce $[Fe(\eta^5\text{-}C_5H_5)_2]^+$ should produce a substantial change in M—C bond length relative to neutral ferrocene.

16.9 Early *d*-metal and *f*-metal organometallics

There are substantial differences in chemical properties between the early *d*-block organometallic compounds (those of Groups 3–5) and those of the late *d* block. The former are similar to *f*-block organometallics and compared with the latter tend to be more oxophilic and halophilic, more active toward C—H bond cleavage, less prone to form unsupported M—M bonds, less likely to form binary carbonyls, and less likely to undergo simple oxidative addition. Many of these properties can be rationalized in terms of the hard, electropositive character, low electron count, and limited number of accessible oxidation states that characterize these early *d*- and *f*-block elements. We shall illustrate oxophilicity and C—H bond cleavage.

Oxophilicity

The hard and electropositive early *d*- and *f*-block elements have an affinity for hard ligands such as O and Cl. The affinity for O is evident in the properties of the carbonyls of these elements. For example, the reaction of

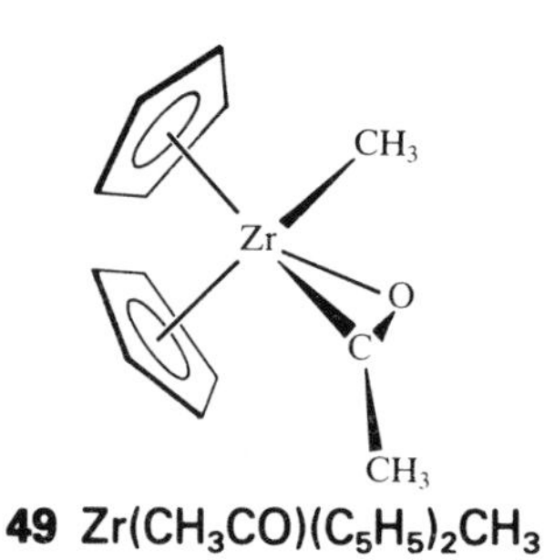

49 $Zr(CH_3CO)(C_5H_5)_2CH_3$

CO with $ZrCp_2(CH_3)_2$ gives a C- and O-bonded η^2-acetyl (instead of a simple C-bonded η^1-acetyl typical of late *d*-metals):

$$ZrCp_2(CH_3)_2 + CO \rightarrow \underset{\textbf{(49)}}{ZrCp_2(CH_3CO)CH_3}$$

Similarly, the reaction between a zirconium dihydride complex and CO in the presence of H_2 results in reduction and migration of the oxygen into the coordination sphere of the Zr atom:

$$Cp_2^*ZrH_2 + CO \rightleftharpoons Cp_2^*Zr(H)_2{-}CO \xrightarrow{H_2} Cp_2^*Zr(OCH_3)(H)$$

(Cp* is pentamethylcyclopentadienyl.) As we explain below, the pentamethylcyclopentadienyl ligand confers greater stability than the unsubstituted cyclopentadiene ligand on early *d*-block organometallics.

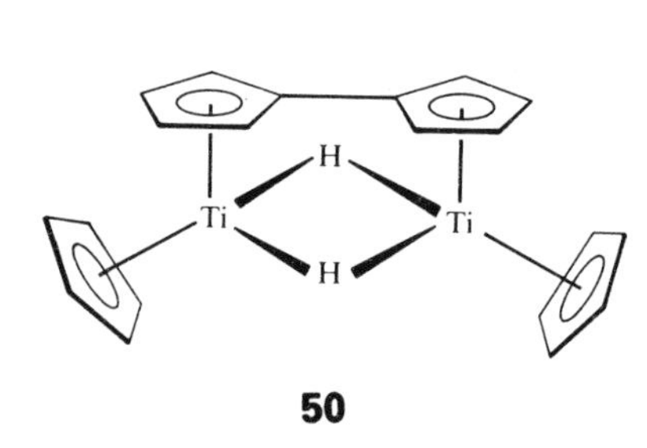

50

C—H cleavage

The early *d* metals have a marked tendency to activate C—H bonds. This tendency is so pronounced, for instance, that attempts to prepare $TiCp_2$ in fact produce the dimer **(50)** in which hydrogen has been transferred to Ti. To suppress this and other unwanted reactions, the pentamethylcyclopentadienyl ligand (Cp*) is often used. This ligand has no H atoms on the ring carbons, so C—H cleavage is avoided. It is also more electron-rich than Cp and therefore forms somewhat stronger metal–ligand bonds. Moreover, the bulk of the Cp* ligand blocks access to the metal and thus hinders associative reactions.

A striking example of a C—H activation, the activation of C—H bonds in methane by a lanthanide organometallic compound, was discovered by Patricia Watson (Dupont). The discovery was based on the observation that $^{13}CH_4$ exchanges ^{13}C with the CH_3 group attached to Lu:

$$LuCp_2^*(CH_3) + {}^{13}CH_4 \rightarrow LuCp_2^*({}^{13}CH_3) + CH_4$$

This reaction can be carried out in deuterated cyclohexane with no evidence for activation of the cyclohexane C—D bonds, presumably because cyclohexane is too bulky to gain access to the metal center. Methane activation has also been observed with actinide organometallics and late *d*-block complexes. Electrophilic metal centers promote this reaction, and a four-center intermediate has been proposed:

$$M{-}R + R'{-}H \rightarrow \left[\begin{matrix} R' & \cdots & H \\ | & & | \\ M & \cdots & R \end{matrix}\right]^{\ddagger} \rightarrow \begin{matrix} R' \\ | \\ M \end{matrix} + \begin{matrix} H \\ | \\ R \end{matrix}$$

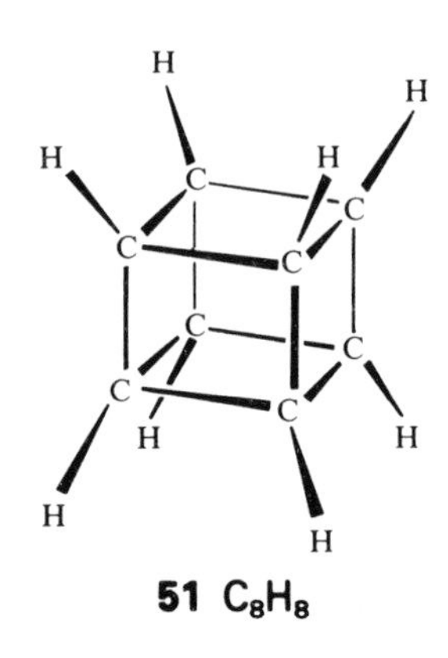

51 C_8H_8

Metal–metal bonding and metal clusters

Organic chemists must make heroic efforts to synthesize their cluster molecules, such as cubane **(51)**. In contrast, one of the distinctive characteristics of inorganic chemistry is the large number of closed polyhedral molecules, such as the tetrahedral P_4 molecule (Section 12.1), the octahedral halide-bridged early d-block clusters (Section 13.9), the polyhedral carboranes (Section 11.13), and the organometallic cluster compounds that we discuss here. We shall include some non-organometallic compounds in this section so as to permit a fuller discussion of metal–metal multiple bonds.

We define **metal clusters** to be molecular complexes with metal–metal bonds that form triangular or larger closed structures (Fig. 16.16). This definition excludes linear M—M—M compounds. It also excludes **cage compounds**, in which several metal atoms are held together exclusively by ligand bridges.[7]

16.10 Structure

Metal–metal bonds[8]

The presence of bridging ligands in a cluster, such as that in Fig. 16.16b, raises the possibility that the atoms are held together by M—L—M interactions rather than M—M bonds. Bond lengths are of some help in resolving this issue. If the M—M distance is much greater than twice the metallic radius, it is reasonable to conclude that the M—M bond is either very weak or absent. However, if the metal atoms are within a reasonable bonding distance, the proportion of the bonding that is attributable to direct M—M interaction is ambiguous. For example, there has been debate in the literature over the extent of Fe—Fe bonding in $Fe_2(CO)_9$ (**52**).

Because of the availability of d orbitals, metal–metal bond orders ranging from 1 to 4 are possible. A large number of two-metal complexes that contain multiple M—M bonds are known, but localized multiple M—M bonds are extremely rare in clusters. As we remarked in Section 13.9, low-oxidation-state metals from the early part of the d block form many halides containing metal–metal bonds. One of these, $[Re_2Cl_8]^{2-}$ (Fig. 7.4), is the first unambiguous example of an M—M quadruple bond of composition $\sigma^2\pi^4\delta^2$. This identification is supported by the stability of the eclipsed conformation (with tetragonal prismatic geometry) and the short Re–Re distance (2.24 Å compared with 2.75 Å in the metal). The orientation of the d_{xy} orbitals between the halide ligands is seen as playing a crucial role for locking-in the observed eclipsed configuration (Fig. 16.17). The separation between the δ and δ^* orbitals is small, and the $\delta^* \leftarrow \delta$ transition gives rise to the blue color of $[Re_2Cl_8]^{2-}$ and the intense red of $[Mo_2Cl_8]^{4-}$.

Triply-bonded M—M systems arise for tetragonal prismatic complexes when both the δ and δ^* orbitals are occupied (Table 16.11). They are more numerous than the quadruply bonded systems. The M≡M bond lengths are often similar to those of quadruply bonded systems because of the weakness of δ bonds. Many of the triply bonded systems also include bridging ligands (Table 16.11). As we see from the table, singly occupied δ or δ^* orbitals lead to formal bond orders of $3\frac{1}{2}$. Once the δ and δ^* orbitals are fully occupied, successive filling of the two higher-lying π^* orbitals leads to further reduction in the bond order from $2\frac{1}{2}$ to 1.

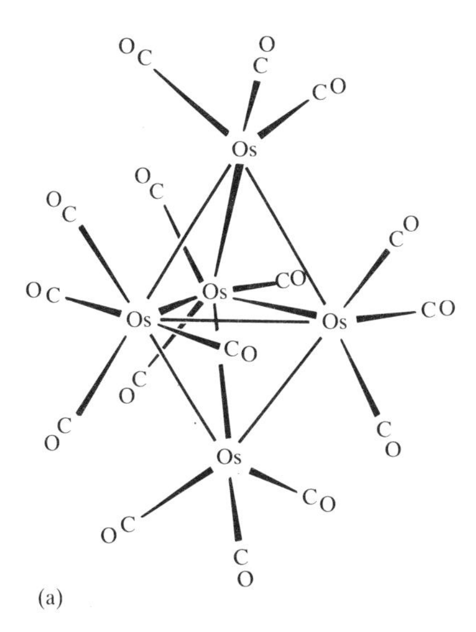

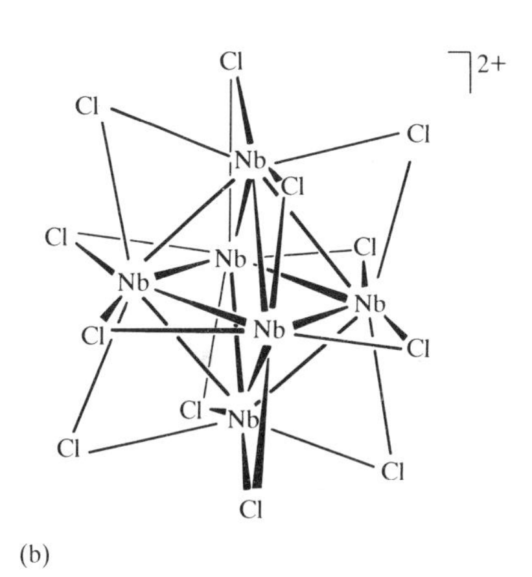

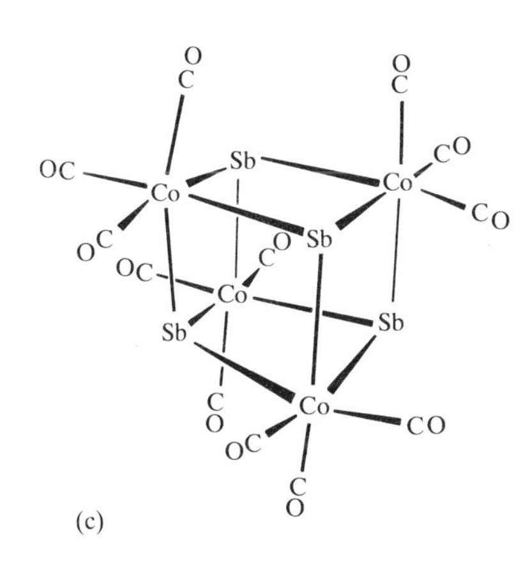

Fig. 16.16 Representative cluster compounds. (a) $Os_5(CO)_{16}$, a typical metal carbonyl cluster, (b) $[Nb_6Cl_{12}]^{2+}$, a typical halide cluster containing a metal from the left of the d block, (c) a mixed Co—Sb cluster, $Sb_4(Co(CO)_3)_4$.

[7] This definition is sometimes relaxed. Occasionally any M—M bonded system is referred to as a cluster, and quite often chemists do not make the distinction between cage and cluster compounds.

[8] F. A. Cotton and R. A. Walton, *Multiple bonds between metal atoms*. Wiley, New York (1983). M. H. Chisholm, *Angew. Chem., Int. Edn. Eng.*, **25**, 21 (1986).

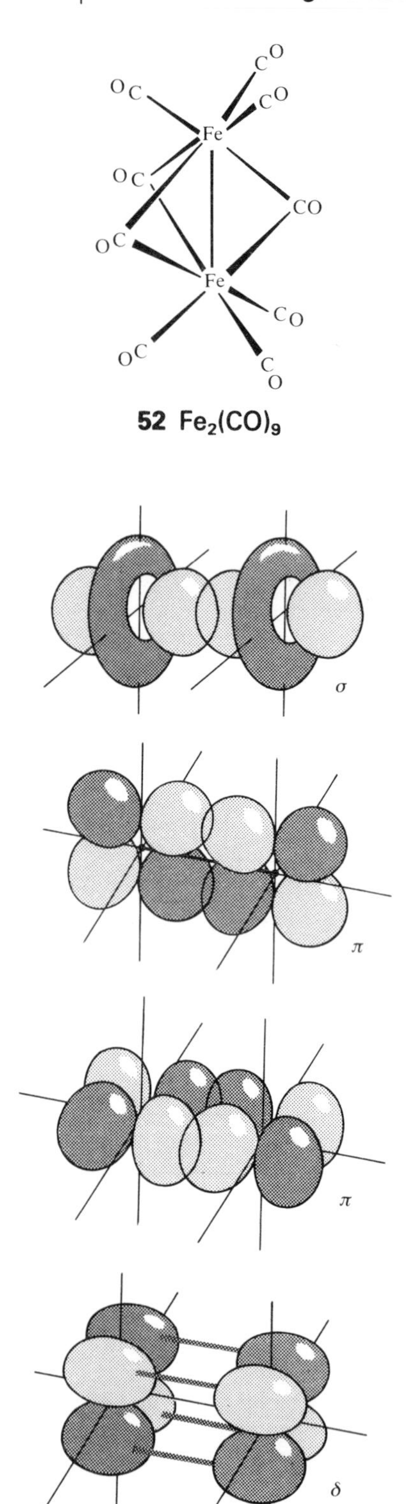

Fig. 16.17 σ, π, and δ interactions between two *d*-block metal atoms situated along the *z*-axis.

Table 16.11. Examples of metal–metal bonded tetragonal prismatic complexes

Complex*	Configuration	Bond order	M–M bond length/Å
$]^{4-}$	$\sigma^2\pi^4\delta^2$	4	2.11
$]^{3-}$	$\sigma^2\pi^4\delta^1$	3.5	2.17
$]^{4-}$	$\sigma^2\pi^4\delta^2\delta^{*2}$	3	2.18
$]^{-}$	$\sigma^2\pi^4\delta^2\delta^*\pi^{*2}$	2.5	2.27
$(CH_3)_2CO$ ... $OC(CH_3)_2$	$\sigma^2\pi^4\delta^2\delta^{*2}\pi^{*2}$	2	2.26
$]^{+}$	$\sigma^2\pi^4\delta^2\delta^{*2}\pi^{*3}$	1.5	2.32

Table 16.11. (contd.)

Complex*	**Configuration**	**Bond order**	**M–M bond length/Å**
Ph, Ph, C, Ph, N, N, N, N, Rh, Rh, N, N, N, N	$\sigma^2\pi^4\delta^2\delta^{*2}\pi^{*4}$	1	2.39

Source: F. A. Cotton and G. Wilkinson, *Advanced inorganic chemistry.* Wiley, New York (1988) and F. A. Cotton. *Chem. Soc. Rev.*, **12**, 35 (1983).
* When multiple bridging ligands are present only one is shown in detail.

Metal–metal bond strengths in metal clusters cannot be determined with great precision, but a variety of evidence—such as the stability of compounds and M—M force constants—indicates that there is an increase in M—M bond strengths down a group in the *d* block. This trend contrasts with the *p* block where element–element bonds are usually weaker for the heavier members of a group. As a consequence of this trend, metal–metal bonded systems are most numerous for the Period 4 and 5 *d* metals.

Cluster structure and electron count

Organometallic cluster compounds are rare for the early *d* metals and unknown for the *f* metals, but a large number of metal carbonyl clusters exist for the elements of Groups 6–10. The bonding in the smaller clusters can be readily explained in terms of a local M—M and M—L electron pair bonding and the 18-electron rule, but octahedral M_6 and larger clusters do not conform to this pattern.

A semiempirical correlation of electron count and structure of the larger organometallic clusters was introduced by K. Wade and refined by D. M. P. Mingos and J. Lauher. These **Wade–Mingos–Lauher rules** are summarized in Table 16.12; they apply most reliably to metal clusters in Groups 6–9. In general (and as in the boron hydrides where similar considerations apply), more open structures (which have fewer metal–metal bonds) occur when there is a higher cluster valence electron (CVE) count.

As an illustration of the influence of correlation between electron count and structure, consider $Rh_4(CO)_{12}$ and $[Re_4(CO)_{16}]^{2-}$:

$Rh_4(CO)_{12}$ tetrahedral		$[Re_4(CO)_{16}]^{2-}$ butterfly	
4Rh:	$4 \times 9 = 36$	4Re:	$4 \times 7 = 28$
12CO:	$12 \times 2 = 24$	16CO:	$16 \times 2 = 32$
		charge:	2
	60		62

Table 16.12. Correlation of cluster valence electron count and structure*

Number of metal atoms		Structure of metal framework	Cluster valence electron count	Example
1	Single metal		18	$Ni(CO)_4$ (**2**)
2	Linear		34	$Mn_2(CO)_{10}$†
3	Closed triangle		48	$Co_3(CO)_9CH$ (**53**)
4	Tetrahedron		60	$Co_4(CO)_{12}$ (**54**)
	Butterfly		62	$[Fe_4(CO)_{12}C]^{2-}$ (**55**)
	Square		64	$Pt_4(O_2CCH_3)_8$
5	Trigonal bipyramid		72	$Os_5(CO)_{16}$
	Square pyramid		74	$Fe_5C(CO)_{15}$
6	Octahedron		86	$Ru_6C(CO)_{17}$
	Trigonal prism		90	$[Rh_6C(CO)_{15}]^{2-}$

* For a more extensive table, see J. W. Lauher, *J. Amer. chem Soc.*, **100**, 5305 (1978).
† Table 16.2.

Example 16.7: *Correlating spectroscopic data, cluster valence electron count, and structure*

The reaction of chloroform with $Co_2(CO)_8$ yields a compound of formula $Co_3(CH)(CO)_9$. NMR and IR data indicate the presence of only terminal CO ligands and the presence of a CH group. Propose a structure consistent with the spectra and the correlation of CVE with structure.

Answer. Electrons available for the cluster are: 27 for 3Co, 18 for 9CO, and 3 for CH (assuming the latter is simply C-bonded: one C electron is used for the C—H bond, so three are available for bonding in the cluster). The resulting total CVE of 48 indicates a triangular cluster. A structure consistent with this and the presence of only terminal CO ligands and a capping CH ligand is illustrated in (**53**).

53 $Co_3(CH)(CO)_9$

Exercise. The compound $Fe_4Cp_4(CO)_4$ is a dark green solid. Its IR spectrum shows a single CO stretch at 1640 cm^{-1}. The ^1H-NMR is a single line even at low temperatures. From this spectral data and the CVE propose a structure for $Fe_4(CO)_4Cp_4$.

Table 16.13. Some isolobal *p*-block and *d*-block fragments*

+ +
Hybrids: H_3C ↔ $(OC)_5Mn$ ↔ $(OC)_4Co$

O
C
Hybrids: C H H ↔ Fe CO CO
C
O

H (CO)3 H(CO)3
Hybrids: C ↔ Co ↔ Fe

H −
Hybrids: B ↔ Fe ↔ Co

H − H
Atomic: B ↔ C

H − H
Hybrids: B ↔ C

−
Atomic: C H H ↔ N H H

−
Hybrids: C H H ↔ N H H

* Showing both atomic orbital and hybrid orbital representations.

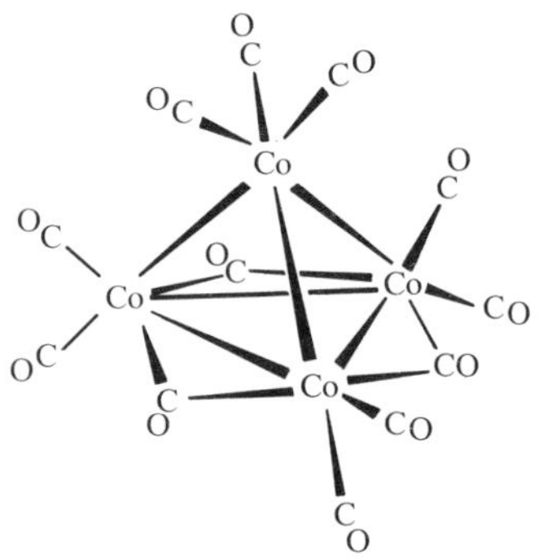

54 $Co_4(CO)_{12}$

Structure and isolobal analogies[9]

A good way of picturing the incorporation of hetero atoms into a framework is to use isolobal analogies. These let us draw a parallel between (**53**) and $Co_4(CO)_{12}$ (**54**), both of which can be regarded as triangular $Co_3(CO)_9$ fragments capped on one side either by $Co(CO)_3$ or by CH. The CH and $Co(CO)_3$ groups are isolobal because each one has three orbitals and three electrons available for participation in framework bonding. A minor complication in this comparison is the occurrence of $Co(CO)_2$ groups along with bridging CO ligands in $Co_4(CO)_{12}$ because bridging and terminal ligands often have similar energies.

Isolobal analogies allow us to draw many parallels between metal–metal bonded systems containing only *d* metals and mixed systems containing *d*-block and *p*-block metals. Additional examples are given in Table 16.13, where it will be noted that a P atom is isolobal with CH; accordingly, a cluster similar to (**54**) is known, but with a capping P atom. Similarly, the ligands CR_2 and $Fe(CO)_4$ are both capable of bonding to two metal atoms in a cluster; CH_3 and $Mn(CO)_5$ can bond to one metal atom.

16.11 Syntheses

One of the oldest methods for the synthesis of metal clusters is the thermal expulsion of CO from a metal carbonyl. For example, when $Co_2(CO)_8$ is heated to about 80°C it forms a tetrahedral cluster:

$$2Co_2(CO)_8 \xrightarrow{\Delta} Co_4(CO)_{12} + 4CO$$

A widely employed and more controllable reaction is based on the condensation of a carbonyl anion and a neutral organometallic complex:

$$\underset{76\ \mathrm{CVE}}{[Ni_5(CO)_{12}]^{2-}} + Ni(CO)_4 \rightarrow \underset{86\ \mathrm{CVE}}{[Ni_6(CO)_{12}]^{2-}} + 4CO$$

The descriptive name **redox condensation** is often given to reactions of this type, which are very useful for the preparation of anionic metal carbonyl clusters. In this example, a trigonal bipyramidal cluster containing Ni with oxidation number $-\frac{2}{5}$ and $Ni(CO)_4$ containing Ni(0) is converted into an octahedral cluster having Ni with oxidation number $-\frac{1}{3}$. The $[Ni_5(CO)_{12}]^{2-}$ cluster, which has four electrons in excess of the 72 expected for a trigonal bipyramid, illustrates a fairly common tendency for the Group 10 metal clusters to have an electron count in excess of that expected from the Wade–Mingos–Lauher rules.

16.12 Reactions

Some of the most common reactions of clusters are ligand substitution, cluster fragmentation, cluster protonation, cluster building, and ligand transformation. We have already discussed cluster building by ligand expulsion and condensation processes; here we shall summarize some reactions of multiply-bonded dimetal complexes and the principal features of cluster reactions.

[9] R. Hoffmann, *Angew. Chem., Int. Ed. Eng.*, **21**, 711 (1982).

Addition to multiple metal–metal bonds[10]

Various molecules can add to a metal–metal triple bond much as they add to multiple carbon–carbon bonds. As with their C analogs, the bond order is decreased in the process:

$$\underset{(\sigma^2\pi^4\delta^2)}{[Re_2Cl_8]^{2-}} + Cl_2 \rightarrow \underset{(\sigma^2\pi^4)}{[Re_2Cl_9]^-} + Cl^-$$

$$Cp(OC)_2Mo{\equiv}Mo(CO)_2Cp + HI \dashrightarrow Cp(OC)_2Mo(\mu\text{-}H)(\mu\text{-}I)Mo(CO)_2Cp \;\; (Mo{=}Mo)$$

$$Cp(OC)_2Mo{\equiv}Mo(CO)_2Cp + Pt(PPh_3)_4 \dashrightarrow Cp(OC)_2Mo{=}Mo(CO)_2Cp\,(\mu\text{-}Pt(PPh_3)_2) + 2PPh_3$$

Substitution versus fragmentation

Because M—M bonds are generally comparable in strength to M—L bonds, there is often a delicate balance between ligand substitution and cluster fragmentation. For example, dodecacarbonyltriiron(0), $Fe_3(CO)_{12}$, reacts with triphenylphosphine under mild conditions to yield simple mono- and disubstituted products as well as some cluster fragmentation products:

$$Fe_3(CO)_{12} + PPh_3 \rightarrow Fe_3(CO)_{11}(PPh_3) + Fe_3(CO)_{10}(PPh_3)_2 + Fe(CO)_5 + Fe(CO)_4(PPh_3) + Fe(CO)_3(PPh_3)_2 + CO$$

However, for somewhat longer reaction times or elevated temperatures, only monoiron cleavage products are obtained. Since the strength of M—M bonds increases down a group, substitution products of the heavier clusters such as $Ru_3(CO)_{11}(PPh_3)$, or $Os_3(CO)_{11}(PPh_3)$ can be prepared without significant fragmentation into mononuclear complexes.

Protonation

The tendency of metal clusters to undergo protonation of the metal framework is even more pronounced than that of simple carbonyls. The Brønsted basicity of clusters is associated with the ready protonation of M—M bonds to produce a formal 3c,2e bond like that in diborane:

$$\underset{2c,2e}{M{-}M} + H^+ \longrightarrow \underset{3c,2e}{[M(\mu\text{-}H)M]^+}$$

For example:

$$[Fe_3(CO)_{11}]^{2-} + H^+ \rightarrow [Fe_3H(CO)_{11}]^-$$

[10] For several reviews see, M. H. Chisholm (ed.), *Reactivity of metal–metal bonds*, ACS Symposium Series 155. American Chemical Society, Washington DC (1981), R. A. Walton, p. 207, M. D. Curtis *et al.*, p. 221, and A. F. Dyke *et al.*, p. 259.

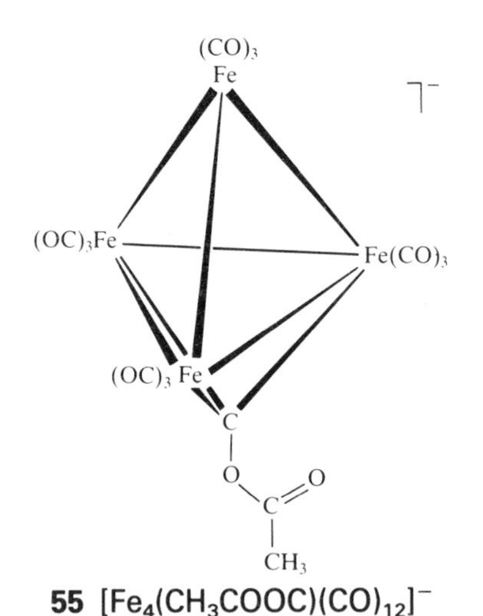

55 $[Fe_4(CH_3COOC)(CO)_{12}]^-$

56 $[Fe_4(C)(CO)_{12}]^{2-}$

As expected, metal cluster anions are much stronger Lewis bases than their neutral analogs.

Cluster-assisted ligand transformations

Transformations of ligands on metal clusters sometimes mimic those on single metal centers. For example, both CO migratory insertion and the attack on CO by nucleophiles have also been observed in clusters. Of greater interest, however, are reactions in which the metal atoms work in concert to bring about ligand transformations. One example is the greater ease of cleaving the CO ligand in a cluster:

$$\underset{(\mathbf{3})}{[Fe_4(\mu_3\text{-}CO)(CO)_{12}]^{2-}} \xrightarrow[-Cl^-]{+CH_3C(O)Cl} \underset{(\mathbf{55})}{[Fe_4(\eta^3\text{-}COAc)(CO)_{12}]^-}$$

where Ac denotes CH_3CO. This acylation of the O atom of CO has been observed only in clusters. The acetate group so produced can be removed to yield a metal carbide cluster with the carbon in an unusual geometric environment:

$$[Fe_4(\eta^3\text{-}COAc)(CO)_{12}]^- \xrightarrow[-NaOAc]{+Na} \underset{(\mathbf{56})}{[Fe_4(\mu_4\text{-}C)(CO)_{12}]^{2-}}$$

Although the C atom in this complex is linked to four Fe atoms, they are not in a tetrahedral array. In penta- and hexairon clusters, the C atom is found in square pyramidal and octahedral coordination environments.

From these examples and other areas of inorganic chemistry—such as carboranes (Section 11.13) and solid metal carbides (Section 11.16)—we see that carbon participates in a wide range of bonding arrangements. We are beginning to see, in fact, that carbon has an even richer chemistry than organic chemistry has taught us to expect.

Further reading

Some excellent introductory texts are available:

C. M. Lukehart, *Fundamental transition metal organometallic chemistry*. Brooks-Cole, Belmont (1985).

A. Yamamoto, *Organotransition metal chemistry*. Wiley-Interscience, New York (1986).

P. Powell, *Principles of organometallic chemistry*. Chapman and Hall, London (1988).

R. H. Crabtree, *The organometallic chemistry of the transition metals*. Wiley-Interscience, New York (1988).

More detailed coverage is available in:

J. P. Collman, L. S. Hegedus, J. R. Norton, and R. G. Finke, *Principles and applications of organotransition metal chemistry*. University Science Books, Mill Valley (1987).

G. Wilkinson, F. G. A. Stone, and E. W. Able (eds.), *Comprehensive organometallic chemistry*. Pergamon Press, Oxford (1982). This multivolume set is a very useful source of information on individual compounds.

D. F. Shriver, H. D. Kaesz, and R. D. Adams (eds.), *The chemistry of metal cluster complexes*. VCH, New York (1990).

Exercises

16.1 Name and draw the structures of (a) $Fe(CO)_5$, (b) $Ni(CO)_4$, (c) $Mo(CO)_6$, (d) $Mn_2(CO)_{10}$, (e) $V(CO)_6$, (f) $[PtCl_3(C_2H_4)]^-$.

16.2 Assign oxidation numbers to the metal atom in (a) $[Fe(\eta^5\text{-}C_5H_5)_2][BF_4]$, (b) $Fe(CO)_5$, (c) $[Fe(CO)_4]^{2-}$, and (d) $Co_2(CO)_8$.

16.3 Count the number of valence electrons for each metal atom in the complexes listed in Exercises 16.1 and 16.2. Do any of them deviate from the 18-electron rule? If so, is this reflected in their structure or chemistry?

16.4 State the two common methods for the preparation of simple metal carbonyls and illustrate your answer with chemical equations. Is the selection of method based on thermodynamic or kinetic differences?

16.5 Suggest a sequence of reactions for the preparation of $Fe(CO)_3$(diphos), given iron metal, CO, diphos (Ph_2P—CH_2—CH_2—PPh_2), and other reagents of your choice.

16.6 Given a series of metal tricarbonyl compounds having the respective symmetries C_{3v}, D_{3h}, and C_s. Without consulting reference material, which of these should display the greatest number of CO stretching bands in the IR spectrum? Check you answer and give the number of expected bands for each by consulting Table 16.5.

16.7 $Ni_3(C_5H_5)_3(CO)_2$ has a single CO stretching absorption at $1761\ cm^{-1}$. The IR data indicate that all C_5H_5 ligands are pentahapto and probably in identical environments. (a) On the basis of these data, propose a structure. (b) Does the electron count for each metal in your structure agree with the 18-electron rule? If not, is nickel in a region of the periodic table where deviations from the 18-electron rule are common?

16.8 Decide which of the two complexes (a) $W(CO)_6$ or (b) $IrCl(PPh_3)_2(CO)$ should undergo the faster exchange with ^{13}CO and justify your answer.

16.9 Which complex, (a) $[Fe(CO)_4]^{2-}$ or $[Co(CO)_4]^-$, (b) $[Mn(CO)_5]^-$ or $[Re(CO)_5]^-$, should be the more basic toward a proton? What are the trends upon which your answer is based?

16.10 What hapticities are possible for the interaction of each of the following ligands with a single *d*-block metal atom such as cobalt? (a) C_2H_4, (b) cyclopentadienyl, (c) C_6H_6, (d) butadiene, (e) cyclooctatetraene.

16.11 Draw plausible structures and give the electron count of (a) $Ni(\eta^3\text{-}C_3H_5)_2$, (b) η^4-cyclobutadienyl-η^5-cyclopentadienylcobalt, (c) $Co(\eta^3\text{-}C_3H_5)(CO)_3$. If the electron count deviates from 18, is the deviation explicable in terms of periodic trends?

16.12 Write out the *d* block of the periodic table. (By this stage, you should not have to consult reference material.) Indicate on this table (a) the elements that form 18-electron neutral Cp_2M compounds, (b) the Period 4 elements for which the simplest carbonyls are dimeric, (c) the Period 4 elements that form neutral carbonyls with six, five, and four carbonyl ligands respectively, and (d) the elements that have the greatest tendency to follow the 18-electron rule.

16.13 Using the 18-electron rule as a guide, indicate the probable number of carbonyl ligands in (a) $W(\eta^6\text{-}C_6H_6)(CO)_n$, (b) $Rh(\eta^5\text{-}Cp)(CO)_n$, and (c) $Ru_3(CO)_n$.

16.14 Give plausible equations that differ from those in the text to demonstrate the utility of metal carbonyl anions in the synthesis of M—C, M—H, and M—M′ bonds.

16.15 Propose a synthesis for $MnH(CO)_5$, starting with $Mn_2(CO)_{10}$ as the source of Mn and other reagents of your choice.

16.16 Give the probable structure of the product obtained when $Mo(CO)_6$ is allowed to react first with LiPh and then with a strong carbocation reagent, $CH_3OSO_2CF_3$.

16.17 Provide a plausible reason for the differences in IR wavenumbers between each of the following pairs: (a) $Mo(PF_3)_3(CO)_3$ 2040, $1991\ cm^{-1}$, versus $Mo(PMe_3)_3(CO)_3$ 1945, $1851\ cm^{-1}$, (b) $MnCp(CO)_3$ 2023, $1939\ cm^{-1}$, versus $MnCp^*(CO)_3$ 2017, $1928\ cm^{-1}$.

16.18 (a) Which compound would you expect to be more stable, $RhCp_2$ or $RuCp_2$? Give a plausible explanation for the difference in terms of simple bonding concepts. (b) Give the equation for a workable reaction that will convert $Fe(\eta^5\text{-}C_5H_5)_2$ to $Fe(\eta^5\text{-}C_5H_5)(\eta^5\text{-}C_5H_4COCH_3)$.

16.19 Give equations for a workable set of reactions that will convert $Fe(\eta^5\text{-}C_5H_5)_2$ to $Fe(\eta^5\text{-}C_5H_5)(\eta^5\text{-}C_5H_4CO_2H)$.

16.20 Sketch the a_1' symmetry-adapted orbitals for the two eclipsed, C_5H_5 ligands stacked together with D_{5h} symmetry. Identify the *s*, *p*, and *d* orbitals of a metal atom lying between the rings that may have nonzero overlap, and state how many a_1' molecular orbitals may be formed.

16.21 $Ni(\eta^5\text{-}C_5H_5)_2$ readily adds one molecule of HF to yield $[Ni(\eta^5\text{-}C_5H_5)(\eta^4\text{-}C_5H_6)]^+$ whereas $Fe(\eta^5\text{-}C_5H_5)_2$ reacts with strong acid to yield $[Fe(\eta^5\text{-}C_5H_5)_2H]^+$. In the latter compound the H is attached to Fe. Provide a reasonable explanation for this difference.

16.22 Write a plausible mechanism, giving your reasoning, for the reactions

(a) $[Mn(CO)_5(CF_2)]^+ + H_2O \rightarrow [Mn(CO)_6]^+ + 2HF$

(b) $Rh(CO)(C_2H_5)(R_3P)_2 \rightarrow RhH(CO)(R_3P)_2 + C_2H_4$

16.23 Contrast the general chemical characteristics of the organometallic complexes of *d*-block elements in Groups 6–8 with those in Groups 3 and 4 with respect to (a) stability of the $\pi\text{-}C_5H_5$ ligand, (b) hydridic or protonic character of M—H bonds, and (c) adherence to the 18-electron rule.

16.24 Give the usual energy ordering for bonding and antibonding σ, π, and δ orbitals in *d*-block dimetal complexes and describe the filling of these levels and M—M bond order in the diamagnetic complexes $Mo_2(py)_2(OAc)_4$ and $Mn_2(CO)_{10}$.

16.25 (a) What cluster valence electron (CVE) count is characteristic of octahedral and trigonal prismatic clusters? (b) Can these CVE values be derived from the 18-electron rule? (c) Determine the probable geometry (octahedral or trigonal prismatic) of $[Fe_6(C)(CO)_{16}]^{2-}$ and $[Co_6(C)(CO)_{16}]$. (The C atom in both cases resides in the center of the cluster and can be considered to be a four-electron donor.)

16.26 Based on isolobal analogies, choose the groups that might replace the group in boldface in
(a) $Co_3(CO)_9$**CH** $\quad OCH_3$, $N(CH_3)_2$, or $SiCH_3$
(b) $(OC)_5Mn$**Mn(CO)$_5$** $\quad$ I, CH_2, or CCH_3

16.27 Ligand substitution reactions on metal clusters are often found to occur by associative mechanisms, and it is postulated that these occur by initial breaking of a M—M bond, thereby providing an open coordination site for the incoming ligand. If the proposed mechanism is applicable, which would you expect to undergo the fastest exchange with added ^{13}CO, $Co_4(CO)_{12}$ or $Ir_4(CO)_{12}$? Suggest an explanation.

Problems

16.1 Propose the structure of the product obtained by the reaction of $[Re(CO)(\eta^5C_5H_5)(PPh_3)(NO)]^+$ with $Li[HBEt_3]$. The latter contains a strongly nucleophilic hydride. (For full details see: W. Tam, G.-Y. Lin, W.-K. Wong, W. A. Kiel, V. Wong, and J. A. Gladysz, *J. Amer. chem. Soc.*, **104,** 141 (1982).)

16.2 Develop a qualitative molecular orbital diagram for $Cr(CO)_6$. Utilize the symmetry orbitals in Appendix 6. First consider the influence of σ bonding and then the effects of π bonding. (For help, see: T. A. Albright, J. K. Burdett, and M. H. Whangbo, *Orbital interactions in chemistry*. Wiley-Interscience, New York (1985).)

16.3 When several CO ligands are present in a metal carbonyl an indication of the individual bond strengths can be determined by means of force constants derived from the experimental IR frequencies. In $Cr(CO)_5(Ph_3P)$ the *cis* CO ligands have the higher force constants, whereas in $Ph_3SnCo(CO)_4$ the force constants are higher for the *trans* CO. Explain which carbonyl C atoms should be susceptible to nucleophiles to attack in these two cases. (For details see D. J. Darensbourg and M. Y. Darensbourg, *Inorg. Chem.*, **9,** 1691 (1970).)

16.4 The complex $Ru_2Cp_2(CO)_4$ is known to exist as three isomers in hydrocarbon solution, one of which contains no CO bridges and presumably has nearly free rotation about the Ru–Ru bond. The two other isomers both have two CO bridges; one has a *cis* disposition of the Cp and terminal CO ligands and the other has a *trans* disposition of these ligands. The IR spectrum of this equilibrium mixture of isomers in solution contains CO stretching bands at 2030, 2024, 2021*, 1993, 1988, 1967*, 1945*, 1834 and 1795 cm^{-1}. The bands marked with * have been assigned to the nonbridged isomer. The variation of the IR spectra with $AlEt_3$ concentration indicate that at an intermediate concentration, a species is produced with the simpler spectrum 2030, 2024, 1933, 1988, 1834, and 1679 cm^{-1}. When the concentration of $AlEt_3$ is increased further the IR spectrum is even simpler: 2046, 2011, 2006, and 1679 cm^{-1}. Propose a chemical basis for the simplification of the IR spectrum at high AlR_3 concentrations. Accompany this answer with equations for the probable reactions and an interpretation of the IR data for the two reaction products. (For further details see: A. A. Alich, N. J. Nelson, D. Strope, and D. F. Shriver, *Inorg. Chem.*, **11,** 2976 (1972).)

16.5 The dinitrogen complex $Zr_2(\eta^5\text{-}Cp^*)_4(N_2)_3$ was isolated and its structure was determined by single-crystal X-ray diffraction. Each Zr is bonded to two Cp* and one terminal N_2. The third N_2 bridges between the Zr atoms in a nearly linear Zr—N═N—Zr array. Before consulting the reference, write a plausible structure for this compound that accounts for the ^1H-NMR spectrum obtained on a sample held at −7°C. This spectrum consists of two lines, indicating that the Cp* rings are in two different environments. At room temperature these rings are equivalent on the NMR time scale and ^{15}N-NMR indicates that N_2 exchange between the terminal ligands and dissolved N_2 is correlated with the process that interconverts the Cp* ligand sites. Propose a way in which this equilibration could interconvert the sites of the Cp* ligands. (For further details see J. M. Manriquez, D. R. McAlister, E. Rosenberg, H. M. Shiller, K. L. Williamson, S. I. Chan, and J. E. Bercaw, *J. Amer. chem. Soc.*, **108,** 3078 (1978).)

16.6 Devise a qualitative molecular orbital scheme for $Ni(\eta^4\text{-}C_4H_4)_2$ using the symmetry-adapted orbitals in Appendix 6.

Part 5

Interdisciplinary topics

Our final set of topics (catalysis, solid state chemistry, and bioinorganic chemistry) span areas that are currently being pursued vigorously in research laboratories and which are the focus of considerable scientific and technological excitement. Research in these fields often combines the talents of inorganic chemists with those of researchers in other disciplines, such as condensed matter physics, biology, and electrical and chemical engineering.

We shall begin by seeing how certain reactions of organometallic compounds can be used in catalysis and, later, how some areas of heterogeneous catalysis can be regarded as examples of surface organometallic chemistry. In many cases, such as in the development of an enantioselective catalyst for the synthesis of a drug to combat Parkinson's disease, the mechanism is subtle. Then we shall extend our discussion of the solid state to include new structural types, non-stoichiometry, and the role of defects. In this way we will be able to discuss high-temperature superconductors. In the final chapter of the book, we shall see how metal ions and metal complexes are involved in some vital biological processes, such as oxygen transport and catalysis.

17 Catalysis

In this chapter we apply the concepts of organometallic chemistry and coordination chemistry to catalysis. We concentrate on the development of general principles, such as the structure of catalytic cycles, in which a catalytic species is regenerated in a reaction, and the delicate balance of reactions required for a successful cycle. We shall see that there are four requirements for a successful catalytic process: the reaction being catalyzed must be thermodynamically favorable, the catalyzed reaction must run at a reasonable rate, the catalyst must have an appropriate selectivity toward the desired product, and it must have a lifetime long enough to be economical. We then survey the five reaction types that are commonly encountered in the homogeneous catalysis of hydrocarbon interconversion and show how these reactions are invoked in proposals about mechanisms. The final part of the chapter develops a similar theme in heterogeneous catalysis, and we shall see that there are many parallels between homogeneous and heterogeneous catalysis lurking beneath differences in terminology. In neither type of catalysis are the mechanisms finally settled, and there is still considerable scope for making new discoveries.

General principles
17.1 Description of catalysts
17.2 Properties of catalysts
Homogeneous catalysis
17.3 Catalytic steps
17.4 Examples
Heterogeneous catalysis
17.5 The nature of heterogeneous catalysts
17.6 Catalytic steps
17.7 Examples
Further reading
Exercises
Problems

A **catalyst** is a substance that increases the rate of a reaction but is not itself consumed. Catalysts are widely used in nature, in industry, and in the laboratory, and it is estimated that they contribute to one-sixth of the value of all manufactured goods in industrialized countries. As shown in Table 17.1, thirteen of the top twenty synthetic chemicals are produced directly or indirectly by catalysis. For example, a key step in the production of the dominant industrial chemical, sulfuric acid, is the catalytic oxidation of SO_2 to SO_3. Ammonia, another chemical essential for industry and agriculture, is produced by the catalytic reduction of N_2 by H_2. Inorganic catalysts are also used for the production of the major organic chemicals and petroleum products, such as fuels, petrochemicals, and polyalkene plastics. Enzymes, a class of elaborate biochemical catalysts, are discussed in Chapter 19.

Table 17.1. The top twenty synthetic chemicals

Synthetic chemical	Rank†	Catalytic process
Sulfuric acid	1	SO_2 oxidation, heterogeneous
Ethylene	2	
Lime	3	
Ammonia	4	$N_2 + H_2$; heterogeneous
Sodium hydroxide	5	
Chlorine	6	Electrocatalysis, heterogeneous
Phosphoric acid	7	
Propylene	8	*a*
Sodium carbonate	9	
1,2-Dichloroethane	10	$C_2H_4 + Cl_2$; homogeneous
Nitric acid	11	$NH_3 + O_2$; heterogeneous
Urea	12	*b*
Ammonium nitrate	13	*b*
Benzene	14	Petroleum refining; heterogeneous
Ethylbenzene	15	Alkylation of benzene; homogeneous
Carbon dioxide	16	
Vinyl chloride	17	Chlorination of C_2H_4; heterogeneous
Styrene	18	Dehydrogenation of ethylbenzene; heterogeneous
Terephthalic acid	19	Oxidation of *p*-xylene; homogeneous
Methanol	20	$CO + H_2$, heterogeneous

† Based on tonnage, from *Chemical and Engineering News* survey of U.S. industrial chemicals (1989).
[a] Primarily used to make polypropylene in a catalytic polymerization.
[b] Synthesis based on starting materials produced by a catalytic process.

In addition to their economic importance and contribution to the quality of life, catalysts are interesting for the subtlety with which they go about their business. Our understanding of the mechanisms of catalytic reactions has improved greatly in recent years because of the availability of isotopically labeled molecules, improved methods for determining reaction rates, and improved spectroscopic and diffraction techniques.

General principles

Catalysts are classified as *homogeneous* if they are present in the same phase as the reagents; this normally means that they are present as solutes in a liquid reaction mixture. Catalysts are *heterogeneous* if they are present in a different phase. Both types of catalysis are discussed in this chapter, and it will be seen that they are fundamentally similar. Of the two, heterogeneous catalysis has had the greater economic impact.

The term 'negative catalyst' is sometimes applied to substances that retard reactions. We shall not use the term because these substances are best considered to be **catalyst poisons** that block one or more elementary steps in a catalytic reaction.

17.1 Description of catalysts

Catalytic efficiency

The **turnover frequency** N (formerly 'turnover number') is often used to express the efficiency of a catalyst. For the conversion of A to B catalyzed by Q and with a rate v,

$$\mathrm{A} \xrightarrow{\mathrm{Q}} \mathrm{B} \qquad v = \frac{\mathrm{d[B]}}{\mathrm{d}t}$$

the turnover frequency is given by

$$N = \frac{v}{[\mathrm{Q}]}$$

if the rate of the uncatalyzed reaction is negligible. A highly active catalyst, one that results in a fast reaction even in low concentrations, has a large turnover frequency.

In heterogeneous catalysis, the reaction rate is expressed in terms of the rate of change in the amount of product (in place of concentration) and the concentration of catalyst is replaced by the amount present. The determination of the number of active sites in a heterogeneous catalyst is particularly challenging, and often the denominator [Q] is simply replaced by the surface area or mass of the catalyst.

Catalytic cycles

The essence of catalysis is a cycle of reactions that consume the reactants, form products, and regenerate the catalytic species. In the example shown in Fig. 17.1, an equilibrium generates a coordinatively unsaturated Rh(I) complex (a). The reactants, hydrogen and ethylene, enter the cycle by addition to the complex (b) and then formation of an alkene complex (c) which rearranges to an alkyl complex (d). In the final step, from (d) to (a), the hydrogenated product leaves the loop with the regeneration of the coordinatively unsaturated Rh(I) complex, and the cycle can continue.

The rhodium in Fig. 17.1 evolves through a series of complexes, each of which facilitates a step in the overall reaction. This multiple role is common in homogeneous catalysis, and strictly speaking there is no single catalyst: a catalytic cycle may contain several complexes, each one of which participates in the overall process. In common parlance the complex introduced into the systems, in this case $RhCl(PPh_3)_3$ (a′), is often referred to as 'the catalyst', but in examples like that illustrated in Fig. 17.1 it is more precise to call it a **catalyst precursor** because it is not within the catalytic cycle.

As with all mechanisms, the cycle has been proposed on the basis of a range of information like that summarized in Table 17.2. Many of the components shown in the table were encountered in Chapter 15 in connection with the determination of mechanisms of substitution reactions. However, the elucidation of catalytic mechanisms is complicated by the occurrence of several delicately balanced reactions, which often cannot be studied in isolation.

Two stringent tests of any proposed mechanism are the determination of rate laws and the elucidation of stereochemistry. If intermediates are postulated, their detection by NMR and IR also provides support. If specific atom transfer steps are proposed, then isotope tracer studies may serve as a test. The influence of different ligands and different substrates are also sometimes informative. Although rate data and the corresponding laws have been determined for many *overall* catalytic cycles, it is also necessary to determine rate laws for the individual steps in order to have reasonable confidence in the mechanism. However, because of experimental complications, it is rare that catalytic cycles are studied in this detail.

17.2 Properties of catalysts

Selectivity

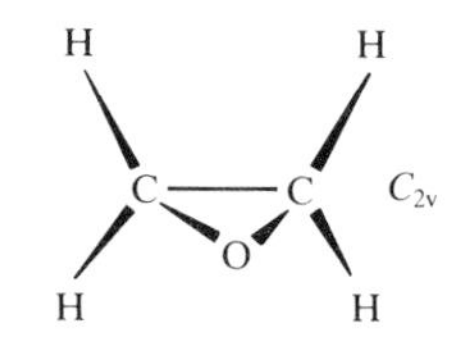

1 Ethylene oxide

A **selective catalyst** yields a high proportion of the desired product with minimum amounts of side products. In industry, there is considerable economic incentive to develop selective catalysts. For example, when metallic silver is used to catalyze the oxidation of ethylene with oxygen to produce ethylene oxide (**1**), the reaction is accompanied by the more thermodynamically favored but undesirable formation of CO_2 and H_2O. This lack of selectivity increases the consumption of ethylene, so chemists are constantly trying to devise a more selective catalyst for ethylene oxide synthesis. Selectivity can be ignored in only a very few simple inorganic reactions, where there is essentially only one thermodynamically favorable product, as in the formation of NH_3 from H_2 and N_2.

Energetics

A catalyst increases the rates of processes by opening up pathways with lower Gibbs free energies of activation ($\Delta G^{\ddagger}$).[1] A catalyst does not affect the Gibbs free energy of the overall reaction ($\Delta G^{\ominus}$) because G is a state function (that is, G depends only on the current state of the system and not on the path that led to the state). The difference is illustrated in Fig. 17.2, where the overall reaction Gibbs free energy is the same in both energy profiles. Thus reactions that are thermodynamically unfavorable cannot be made favorable by a catalyst.

Figure 17.2 also shows that the free energy profile of a catalyzed reaction contains no high peaks and no deep troughs. The lowering of the peaks when a catalyst is present and the new pathway that is thereby opened should be familiar aspects of the action of catalysts. However, an equally important point is that stable catalytic intermediates do not occur in the cycle. Similarly, the product must be released in a thermodynamically favorable step. If, as shown by the dashed line in Fig. 17.2, a stable complex is formed with the catalyst, it would turn out to be the product of the

[1] We discuss reaction rates in terms of the activation Gibbs free energy, not the activation energy E_a alone, because the entropy of activation makes a significant contribution to the barrier to reaction.

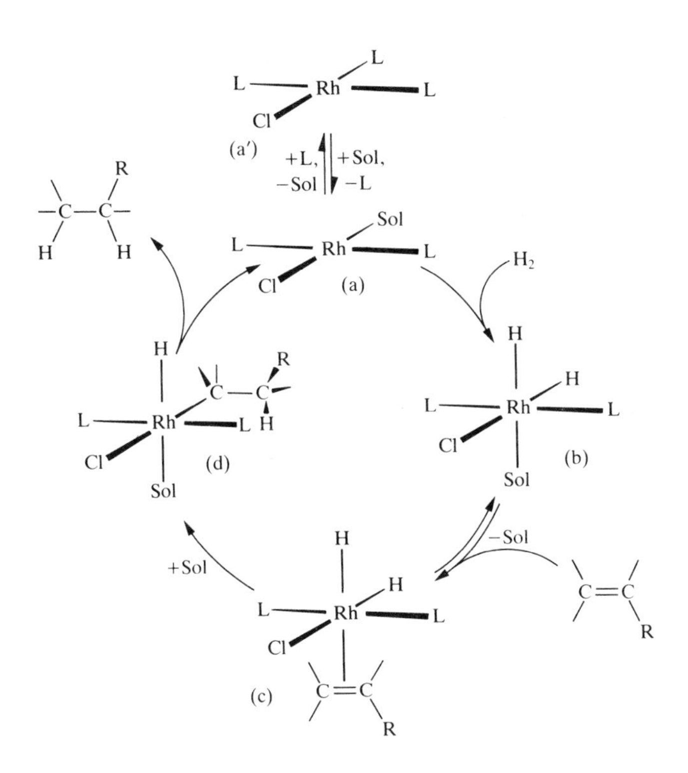

Fig. 17.1 The main catalytic cycle in the homogeneous hydrogenation of an alkene by rhodium–phosphine complexes, $L = PPh_3$.

Table 17.2. Determination of catalytic mechanisms

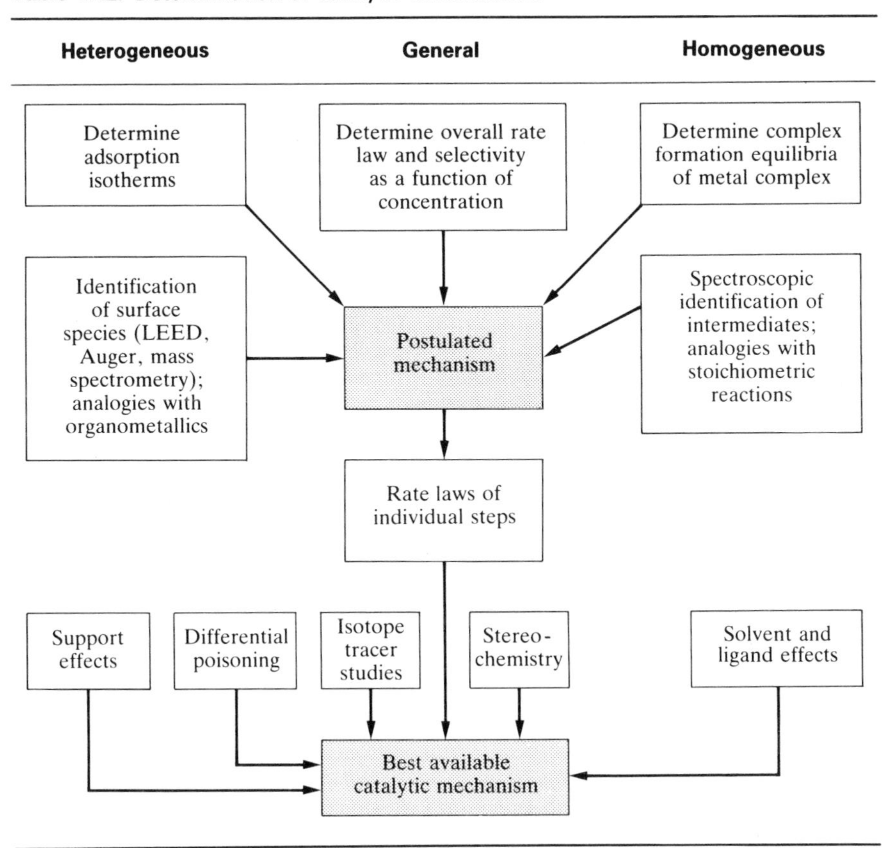

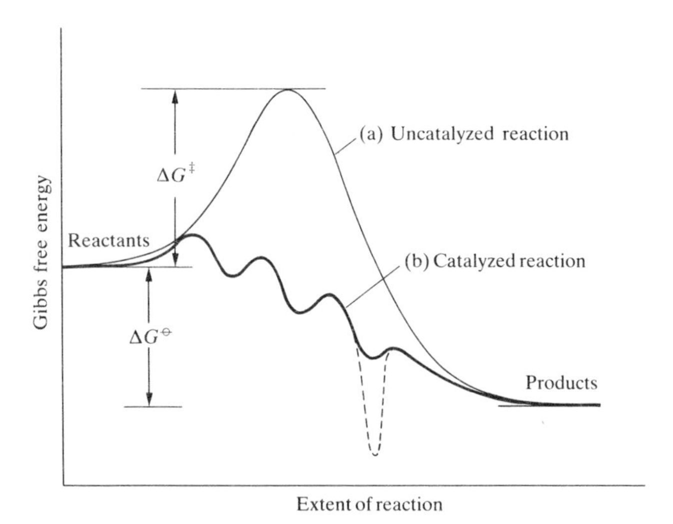

Fig. 17.2 Schematic representation of the energetics of a catalytic cycle. The uncatalyzed reaction (a) has a higher Gibbs energy of activation $\Delta G^{\ddagger}$ than any step in the catalyzed reaction (b). The Gibbs free energy of reaction, $\Delta G^{\ominus}$, for the overall reaction is unchanged from (a) to (b).

reaction and the cycle would not continue. Similarly, impurities may suppress catalysis by coordinating strongly to catalytically active sites. Such impurities act as catalyst poisons.

Homogeneous catalysis

In this section we review some important homogeneous catalytic reactions and describe their currently favored mechanisms. As with nearly all mechanistic chemistry, catalytic mechanisms are inferred from a range of indirect evidence and so a mechanism is subject to refinement or change as more detailed experimental information becomes available. It should also be borne in mind that the mechanism of homogeneous catalysis is more accessible to detailed investigation than that of heterogeneous catalysis because the interpretation of rate data is frequently easier. Moreover, species in solution are often easier to characterize than those on a surface.

From a practical standpoint, homogeneous catalysis is attractive because it is often highly selective toward the formation of a desired product. In large-scale industrial processes homogeneous catalysts have some advantages for exothermic reactions because it is easier to dissipate heat from a solution than from the solid bed of a heterogeneous catalyst.

17.3 Catalytic steps

In this section we review five types of reaction (and in some cases their reverse) which, in combination, account for most of the homogeneous catalytic cycles that have been proposed for hydrocarbon transformations. This list will undoubtedly alter and grow as our knowledge of catalysis deepens through further research.

Ligand coordination and dissociation

The catalysis of molecular transformations generally requires facile coordination of reactants to metal ions and equally facile loss of products from the coordination sphere. Both processes must occur with low activation free energy, so highly labile metal complexes are required. These labile complexes are **coordinatively unsaturated** in the sense that they contain an open coordination site or, at most, a site that is only weakly coordinated.

Square-planar 16-electron complexes are coordinatively unsaturated and are often employed to catalyze the reactions of organic molecules. In Sections 7.8 and 15.3 we met a number of examples of associative reactions between square-planar complexes and entering groups. These facile reactions are quite common for catalytic systems involving ML_4 complexes of Pd(II), Pt(II), and Rh(I), such as the hydrogenation catalyst $RhCl(PPh_3)_3$ featured in Fig. 17.1.

Insertion and elimination

The migration of alkyl and hydride ligands to unsaturated ligands, as in the reaction

$$L + R\text{—M—CO} \longrightarrow L\text{—M—C(=O)—R}$$

is an example of a migratory insertion reaction (or simply 'insertion reaction') discussed in Section 16.5. Another example of the same kind is the migration of an H ligand to a coordinated alkene to produce a coordinated alkyl ligand:

$$H\text{—M}(\eta^2\text{-}CH_2{=}CH_2) \longrightarrow \text{M—}CH_2CH_3$$

The conversion of (c) to (d) in Fig. 17.1 is a hydrogen migration of this type.

The reverse of insertion is elimination. Elimination reactions include the important β-hydride elimination (Section 16.5):

$$\text{—M—}CH_2\text{—}CH_3 \longrightarrow \text{—M(H—}CH_2\text{)—}CH_2 \longrightarrow H\text{—M}(\eta^2\text{-}CH_2{=}CH_2) \xrightarrow[+\text{Sol}]{-C_2H_4} H\text{—M—Sol}$$

Nucleophilic attack on coordinated ligands

The coordination of ligands such as CO and alkenes to metal ions in positive oxidation states results in the activation of the coordinated C atoms toward attack by nucleophiles. Such reactions are useful in catalysis as well as in organometallic chemistry (Chapter 16).

The hydration of ethylene bound to Pd(II) is a good example of catalysis by nucleophilic activation. The stereochemical evidence indicates that the reaction occurs by direct attack on the most highly substituted C atom of the coordinated alkene:

$$[L_3Pd(\eta^2\text{-}HRC{=}CHR)]^{2+} + OH_2 \longrightarrow [L_3Pd\text{—CHR—CHR—OH}]^{+} + H^+$$

It is also possible for the hydroxylation of a coordinated alkene to take place by the prior coordination of H_2O to a metal complex, followed by an insertion reaction:

$$L_4M + H_2O + C_2H_4 \xrightarrow{-2L} L_2M(H_2O)(\eta^2\text{-}CH_2{=}CH_2) \xrightarrow{L} [L_3M{-}CH_2CH_2OH]^- + H^+$$

Similarly, a coordinated CO ligand is attacked by an OH^- ion at the C atom, forming a —CO(OH) ligand, which subsequently loses CO_2:

$$L_5M{-}CO + OH^- \longrightarrow \left[L_5M{-}\overset{\overset{\Large O}{\|}}{C}{-}OH\right]^- \longrightarrow [L_5M{-}H]^- + CO_2$$

We discussed this reaction in Section 16.5 in connection with the formation of metal carbonyl anions. It is thought to be a critical step in the water gas shift reaction catalyzed by metal carbonyl complexes or metal ions on solid surfaces:

$$CO + H_2O \xrightarrow{\text{catalyst}} CO_2 + H_2$$

Oxidation and reduction

Metal complexes are often used for the catalytic oxidation of organic substrates, and in the catalytic cycle the metal atom alternates between two oxidation states. Some common catalytic one-electron couples are Cu^{2+}/Cu^+, Co^{3+}/Co^{2+} and Mn^{3+}/Mn^{2+}; a few examples of catalytic two-electron processes are also known, such as with the couple Pd^{2+}/Pd.

Catalysts containing metal ions are used in large scale processes for the oxidation of hydrocarbons, as in the oxidation of *p*-xylene to terephthalic acid (Table 17.1). The metal ion can play various roles in these radical oxidations, as we can see by considering a mechanism of the following general form:

Initiation (In· = radical initiator):

$$In^{\cdot} + R{-}H \rightarrow In{-}H + R^{\cdot}$$

Propagation:

$$R\cdot + O_2 \rightarrow R{-}O{-}O\cdot$$

$$R{-}O{-}O\cdot + R{-}H \rightarrow R{-}O{-}O{-}H + R\cdot$$

Termination:

$$2R\cdot \rightarrow R_2$$

$$R\cdot + R{-}O{-}O\cdot \rightarrow R{-}O{-}O{-}R$$

$$2R{-}O{-}O\cdot \rightarrow R{-}O{-}O{-}O{-}O{-}R \rightarrow O_2 + \text{nonradicals}$$

The metal ions control the reaction by contributing to the formation of the R—O—O· radicals:

$$Co(II) + R{-}O{-}O{-}H \rightarrow Co(ROOH) \rightarrow Co(III)OH + R{-}O\cdot$$

$$Co(III) + R{-}O{-}O{-}H \rightarrow Co(II) + R{-}O{-}O\cdot + H^+$$

In accord with the remark made above, the metal atom shuttles back and forth between oxidation states in this pair of reactions. A metal ion can also act as an initiator. For example, when arenes are involved it is believed that initiation occurs by a simple redox process:

$$ArCH_3 + Co(III) \rightarrow ArCH_3^{\cdot +} + Co(II)$$

Oxidative addition and reductive elimination

We saw in Sections 15.13 and 16.5 that the oxidative addition of a molecule AX to a complex brings about dissociation of the A—X bond and coordination of the two fragments:

$$L{-}M(L)(L){-}L + AX \longrightarrow L{-}M(A)(L)(L)(L){-}X$$

Reductive elimination is the reverse of oxidative addition, and often follows it in a catalytic cycle.

The mechanisms of oxidative addition reactions vary. Depending upon reaction conditions and the nature of the reactants, there is evidence for oxidative addition by simple concerted reaction, heterolytic (ionic) addition of A^+ and X^-, or radical addition of A· and X·. Despite this diversity of mechanism it is found that the rates of oxidative addition of alkyl halides generally follow the orders

$$\text{primary alkyl} < \text{secondary alkyl} < \text{tertiary alkyl}$$
$$F \ll Cl < Br < I$$

Example 17.1: *Interpreting the reactions in a catalytic cycle*

Figure 17.1 contains two examples of simple association reactions and one of a simple dissociation reaction. Identify them and write balanced chemical equations.

Answer. The catalyst precursor undergoes a ligand dissociation:

$$\underset{(a')}{RhCl(PPh_3)_3} + Sol \rightarrow PPh_3 + \underset{(a)}{RhCl(PPh_3)_2Sol}$$

The product is $RhCl(PPh_3)_2(Sol)$ if a solvent Sol is weakly coordinated. One of the association reactions is the reverse reaction:

$$PPh_3 + RhCl(PPh_3)_2Sol \rightarrow RhCl(PPh_3)_3 + Sol$$

Another association reaction involves ethylene:

$$\underset{(b)}{RhClH_2(PPh_3)_2} + H_2C{=}CHR \rightarrow \underset{(c)}{RhClH_2(PPh_3)_2(H_2C{=}CHR)}$$

Exercise. There is one oxidative addition reaction and one reductive elimination reaction in Fig. 17.1. Give balanced chemical equations for them both and assign oxidation numbers to all the rhodium complexes in the equations.

17.4 Examples

We shall now see how several individual elementary reactions combine forces and contribute to catalytic cycles. The following examples illustrate our current understanding of the mechanisms of some important types of catalytic reactions and give some insight into the manner in which catalytic activity and selectivity might be altered. As we have already remarked, there is usually even more uncertainty associated with catalytic mechanisms than with mechanisms of simpler kinds of reactions. Unlike simple reactions, a catalytic process frequently contains many steps over which the experimentalist has little control. Moreover, highly reactive intermediates are often present in concentrations too low to be detected spectroscopically. The best attitude to adopt toward these catalytic mechanisms is to learn the pattern of transformations and appreciate their implications, but to be prepared to accept new mechanisms that might be indicated by future work.

Table 17.3. Some homogeneous catalytic processes

Hydroformylation of alkenes (Oxo process)

$$RCH{=}CH_2 + CO + H_2 \xrightarrow{\text{Co(I) or Rh(I)}} RCH_2CH_2\overset{\overset{\displaystyle O}{\|}}{C}H$$

Oxidation of alkenes (Wacker process)

$$CH_2{=}CH_2 + O_2 \xrightarrow{\text{Pd(II)+Cu(II)}} CH_3\overset{\overset{\displaystyle O}{\|}}{C}H$$

Carbonylation of methanol to acetic acid (Monsanto process)

$$CH_3OH + CO \xrightarrow{[RhI_2(CO)_2]^-} CH_3\overset{\overset{\displaystyle O}{\|}}{C}OH$$

Hydrocyanation of butadiene to adiponitrile

$$CH_2{=}CHCH{=}CH_2 + 2HCN \xrightarrow{Ni(P(OR)_3)_4} NC{-}CH_2CH_2CH_2CH_2{-}CN$$

Oligomerization of ethylene

$$nCH_2{=}CH_2 \xrightarrow{NiHL} CH_2{=}CH(CH_2CH_2)_{n-2}CH_2CH_3$$

Alkene dismutation (olefin metathesis)

$$2CH_2{=}CHCH_3 \xrightarrow{WOCl_4/AlCl_2Et} CH_2{=}CH_2 + CH_3CH{=}CHCH_3$$

Asymmetric hydrogenation of prochiral alkenes

$$(H)(R)C{=}C(COOR)(NHCOR) + H_2 \xrightarrow{[Rh(DiPAMP)_2]^+} RCH_2C^*(COOR)(NHCOR){-}H$$

90 percent L

Cyclotrimerization of acetylene

$$3CH{\equiv}CH \xrightarrow{Ni(acac)_2} C_6H_6 \text{ (benzene)}$$

Adapted from J. Halpern, *Inorg. Chim. Acta*, **50**, 11 (1981).

The scope of homogeneous catalysis can be appreciated from the examples given in Tables 17.1 and 17.3. The reactions cited there include hydrogenation, oxidation, and a host of other processes. Often the complexes of all metal atoms in a group will exhibit catalytic activity in a particular reaction, but as catalysts the 4*d*-metal complexes are often superior to their lighter and heavier congeners. In some cases the difference may be associated with the greater substitutional lability of 4*d* organometallics in comparison with their 3*d* and 5*d* analogs. It is often the case that the complexes of costly metals must be used on account of their superior performance compared with the complexes of cheaper metals.

Hydrogenation of alkenes

The addition of hydrogen to an alkene to form an alkane is highly favored thermodynamically ($\Delta G^{\ominus} = -101\ \text{kJ mol}^{-1}$ for ethylene to ethane). However, the reaction rate is negligible at ordinary conditions in the absence of a catalyst. Efficient homogeneous and heterogeneous catalysts are known for the hydrogenation of alkenes and are used in such diverse areas as the manufacture of margarine, pharmaceuticals, and petrochemicals.

One of the most studied catalytic systems is the Rh(I) complex $RhCl(PPh_3)_3$ (a′ of Fig. 17.1), which is often referred to as 'Wilkinson's catalyst'. This useful catalyst hydrogenates a wide variety of alkenes at pressures of hydrogen close to 1 atm or less. The cycle shown in Fig. 17.1 is close to the original mechanism proposed by Wilkinson and coworkers but some details have been revised, primarily through the careful mechanistic studies undertaken in Jack Halpern's laboratory at the University of Chicago. This is one of the few catalytic cycles in which each of the postulated steps has been checked kinetically, so we can go through the mechanism in some detail.

The dominant cycle for simple alkenes such as cyclohexene (Fig. 17.3) appears to involve an early step in which Rh(I) oxidatively adds H_2. Kinetic data indicate that in the primary catalytic cycle H_2 adds not to the precursor complex (a′) in Fig. 17.3 but instead to (a), the complex generated by dissociation of the substituted phosphine from the 16-electron precursor. There is also kinetic evidence for an alternative slower but more direct route, which is also shown in Fig. 17.3. In this route, H_2 adds directly to the precursor complex $RhCl(PPh_3)_3$ to yield (b′). The involvement of complex (a), or of its solvent-free version, has been a point of considerable controversy because it violates opinions about the instability of 14-electron complexes and also because (a) has not yet been detected spectroscopically in the catalytic reaction mixture.[2] The kinetic evidence indicates that (a) reacts with H_2 about 10^7 times faster than (a′).

The reaction of the Rh(III) dihydro complex (b) with a large excess of cyclohexene (effectively making the concentration of cyclohexene constant) yields a rate law which is pseudofirst-order in the dihydrido complex:

$$-\frac{d[RhClH_2L_2]}{dt} = k[RhClH_2L_2]$$

More detailed analysis with varying concentrations of the alkene yields a

[2] However, flash photolysis of $RhCl(CO)(PPh_3)_2$ in benzene is known to yield the transient species $RhCl(PPh_3)_2$ (D. Wink and P. C. Ford, *J. Amer. chem. Soc.*, **107**, 1794 (1985)).

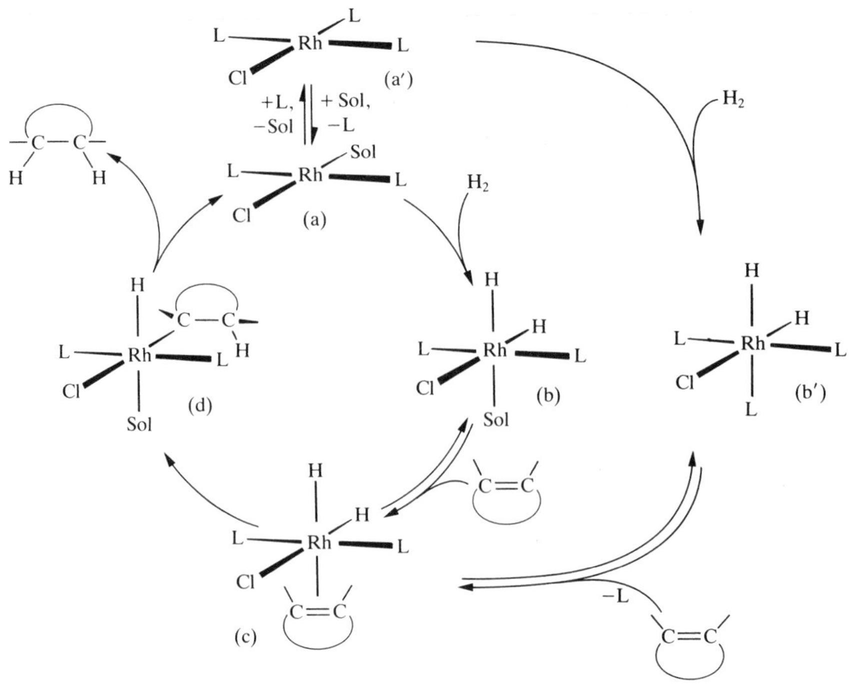

Fig. 17.3 The catalytic cycles implicated in cyclohexene hydrogenation by Wilkinson's catalyst, $RhCl(PPh_3)_3$.

2 $[Rh_2Cl_2L_4]$, D_{2h}

more complex rate expression consistent with a pre-equilibrium reaction between (b) and the 18-electron dihydrido alkene complex (c). This pre-equilibrium is followed by a slower (rate-determining) migratory insertion reaction to produce an alkyl hydrido complex (d). The subsequent reductive elimination of alkane from (d) yields (a) and the cycle is set to repeat. Kinetic data with H_2 and D_2 indicate a faster reaction with H_2, which is consistent with the hydrogen migratory insertion reaction being rate determining.

Experiments with the catalytic addition of D_2 to cyclic alkenes indicate that the *cis* addition of hydrogen to cyclic alkenes is highly stereospecific. Spectroscopic investigation of the reaction mixture shows the presence of several rhodium complexes, but none of the active complexes (a) through (d) has been detected in the reaction mixture. For example, NMR demonstrates the presence of a halogen bridged dimer (**2**) of the unobserved complex (a) in Fig. 17.3. Some of the controversy about this catalytic mechanism stemmed from the inclination of its investigators to include spectroscopically observed complexes within the catalytic loop.

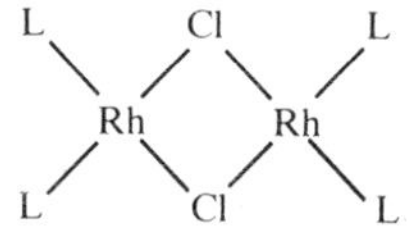

Wilkinson's catalyst is highly sensitive to the nature of the phosphine ligand and the alkene substrate. Analogous complexes with alkyl phosphine ligands are inactive, presumably because they are more strongly bound ligands and do not readily dissociate. Similarly, the alkene must be just the right size: highly hindered alkenes or the sterically unencumbered ethylene are not hydrogenated by the catalyst. It is presumed that the sterically crowded alkenes do not coordinate and that ethylene forms a strong complex which does not react further. These observations emphasize the point made earlier that a catalytic cycle is usually a delicately poised sequence of reactions, and anything that upsets their flow may block catalysis or alter the mechanism.

Wilkinson's catalyst is used in laboratory-scale organic synthesis and in the production of fine chemicals. Related Rh(I) phosphine catalysts that contain a chiral phosphine ligand have been developed to synthesize optically active products in **enantioselective** reactions. The alkene to be hydrogenated must be **prochiral**, which means that it must have a structure that leads to *R* or *S* chirality when complexed to the metal (Fig. 17.4).[3] The resulting complex will thus have two diastereomeric forms depending on the mode of coordination of the alkene. In general, diastereomers have different stabilities and labilities, and in favorable cases one or the other of these effects leads to product enantioselectivity.

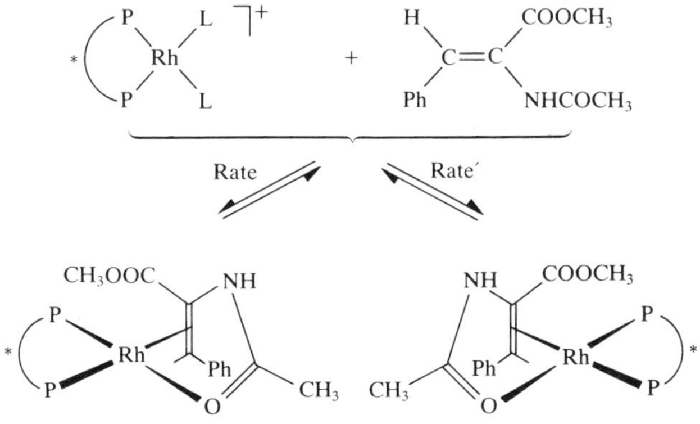

Fig. 17.4 The diastereomeric complexes that may form from a complex with chiral phosphine ligands (*) and a prochiral alkene.

An enantioselective hydrogenation catalyst containing a chiral phosphine ligand referred to as DiPAMP (**3**) is used by Monsanto to synthesize L-dopa (**4**).[4] An interesting detail of the process is that the minor diastereomer in solution leads to the major product. The explanation of the greater turnover frequency of the minor isomer lies in the difference in activation Gibbs free energies (Fig. 17.5).

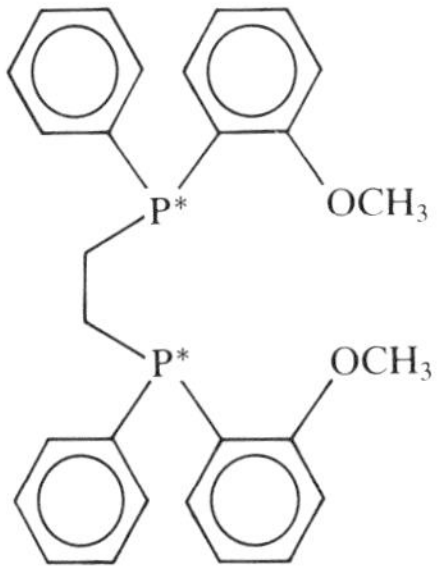

3 DiPAMP

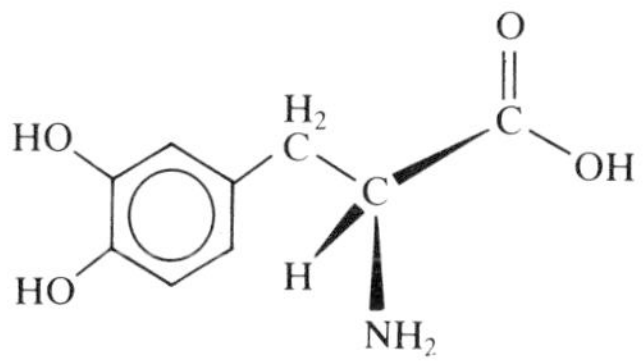

4 L-Dopa

Hydroformylation

In a **hydroformylation reaction** an alkene, CO, and H_2 react to form an aldehyde containing one more C atom than in the original alkene:

$$\mathrm{R{-}CH{=}CH_2 + CO + H_2 \longrightarrow R{-}CH_2{-}CH_2{-}\overset{\displaystyle O}{\overset{\|}{C}}{-}H}$$

Both cobalt and rhodium complexes are employed as catalysts. Aldehydes produced by hydroformylation are normally reduced to alcohols that are used as solvents, plasticizers, and in the synthesis of detergents. The scale is enormous, amounting to millions of tons of product per year.

[3] The designations *R* and *S* for a chiral center are determined as follows. With the element with the lowest atomic number (Z) away from the viewer, the center is *R* if the sequence of Z from highest to lowest for the remaining three atoms decreases in a clockwise manner. The center is *S* if the decrease in Z occurs in a counterclockwise manner. (There are additional rules in the event of atoms having identical atomic numbers.)

[4] L-Dopa is a chiral amino acid used to treat Parkinson's disease. The development of this catalyst and the origin of stereoselectivity in the catalytic synthesis are described by W. S. Knowles, in *Acc. chem. Res.*, **16,** 106 (1983) and by J. Halpern, *Pure Appl. Chem.*, **55,** 99 (1983) respectively.

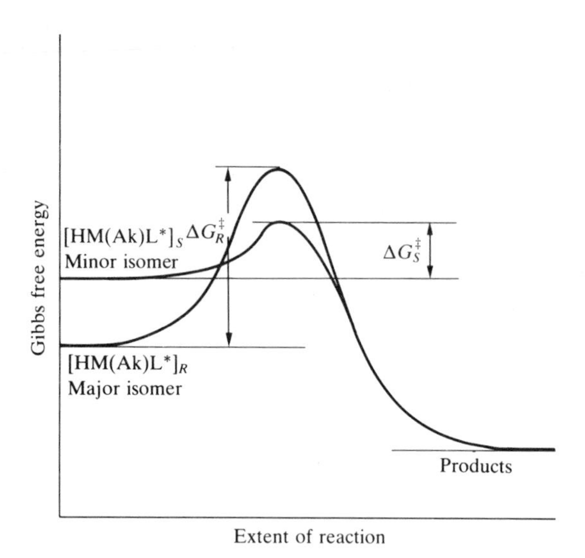

Fig. 17.5 Kinetically controlled stereoselectivity. See Fig. 17.4: L* is a chiral ligand, Ak is the alkene ligand undergoing hydrogenation. Note that $\Delta G_S^\ddagger < \Delta G_R^\ddagger$ so the minor isomer reacts faster than the major isomer.

The term 'hydroformylation' derived from the idea that the product resulted from the addition of formaldehyde to the alkene, and the name has stuck even though experimental data indicate a different mechanism. A less common but more appropriate name is **hydrocarbonylation**. The general mechanism of cobalt carbonyl catalyzed hydroformylation (Fig. 17.6) was proposed in 1961 by Heck and Breslow (of the Hercules Powder Company) by analogy with reactions familiar from organometallic chemistry. Their general mechanism is still invoked, but has proved difficult to verify in detail.

In the proposed mechanism, a pre-equilibrium is established in which octacarbonyldicobalt combines with hydrogen at high pressure to yield the known tetracarbonylhydridocobalt complex:

$$Co_2(CO)_8 + H_2 \rightleftharpoons 2HCo(CO)_4$$

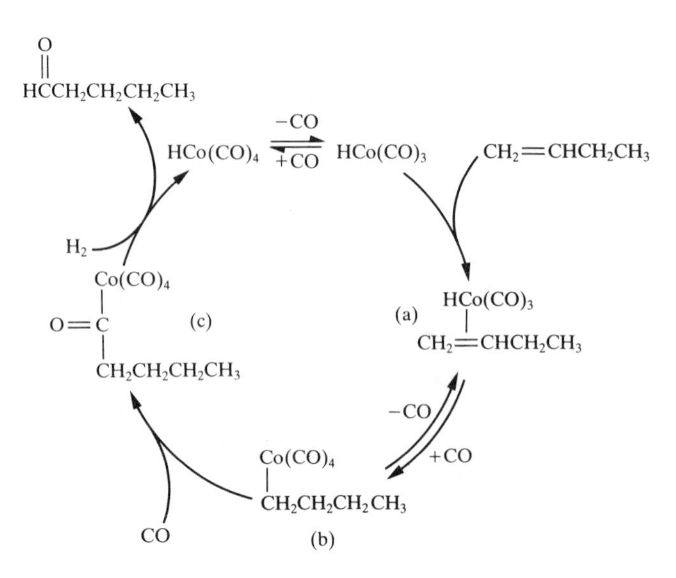

Fig. 17.6 Proposed mechanism for the hydroformylation of 1-butene to pentanal, catalyzed by cobalt carbonyl complexes.

This complex, it is proposed, loses CO to produce the coordinatively unsaturated complex $CoH(CO)_3$:

$$CoH(CO)_4 \rightleftharpoons CoH(CO)_3 + CO$$

It is then thought that $CoH(CO)_3$ coordinates to an alkene, producing (a) in Fig. 17.6, which undergoes an insertion reaction with the coordinated hydrido ligand. The product at this stage is a normal alkyl complex (b). In the presence of CO at high pressure, (b) undergoes migratory insertion yielding the acyl complex (c) which has been observed by infrared spectroscopy under catalytic reaction conditions. The formation of the aldehyde product is thought to occur by attack of either H_2 (as depicted in Fig. 17.6) or by the strongly acidic $CoH(CO)_4$ to yield an aldehyde and regenerate the coordinatively unsaturated $CoH(CO)_3$.

A significant portion of branched aldehyde is also formed in the cobalt catalyzed hydroformylation. This may result from isomerization to a 2-alkylcobalt intermediate followed by the insertion of CO:

$$\underset{\displaystyle CH_3\overset{|}{C}HCH_2CH_3}{Co(CO)_4} \xrightarrow{CO} \begin{array}{c} Co(CO)_4 \\ | \\ O{=}C \\ | \\ CH_3CHCH_2CH_3 \end{array}$$

Hydrogenation then yields a branched aldehyde:

$$\begin{array}{c} Co(CO)_4 \\ | \\ O{=}C \\ | \\ CH_3CHCH_2CH_3 \end{array} + H_2 \longrightarrow \begin{array}{c} O{=}C{-}H \\ | \\ CH_3CHCH_2CH_3 \end{array} + HCo(CO)_4$$

The linear aldehyde is preferred for some applications, such as for the synthesis of biodegradable detergents, and then it is desirable to suppress the isomerization. It is found that addition of an alkylphosphine to the reaction mixture gives much higher selectivity for the linear product. One plausible explanation is that the replacement of CO by a bulky ligand disfavors the formation of complexes of sterically crowded 2-alkenes:

$$\begin{array}{c} Co(CO)_2PBu_3 \\ | \\ CH_2CH_2CH_2CH_3 \end{array} \rightleftharpoons \begin{array}{c} Co(CO)_2PBu_3 \\ | \\ CH_3CHCH_2CH_3 \end{array} \qquad K \ll 1$$

Here again we see an example of the powerful influence of ancillary ligands on catalysis.

In keeping with our comments about the high catalytic activity of $4d$-metal complexes, rhodium phosphine complexes are even more active hydroformylation catalysts than cobalt complexes. One effective catalyst precursor, $RhH(CO)(PPh_3)_3$ (**5**), loses a phosphine ligand to form the coordinatively unsaturated 16-electron complex $RhH(CO)(PPh_3)_2$, which promotes hydroformylation at moderate temperatures and 1 atm (1 atm $\approx 10^5$ Pa). This behavior contrasts with the cobalt carbonyl catalyst, which typically requires 150°C and 250 atm. Because of its effectiveness under convenient conditions, the rhodium complex is useful in the laboratory; because it favors linear aldehyde products, it competes with the phosphine-modified cobalt catalyst in industry.

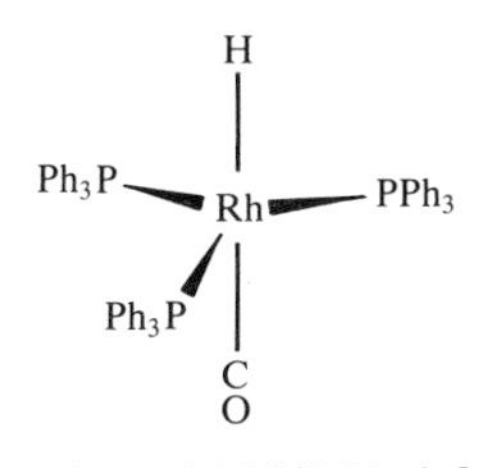

5 $[RhH(CO)(PPh_3)_3]$

Example 17.2: *Interpreting the influence of chemical variables on a catalytic cycle*

An increase in CO partial pressure above a certain threshold decreases the rate of the cobalt-catalyzed hydroformylation of 1-pentene. Suggest an interpretation of this observation.

Answer. The decrease in rate with increasing partial pressure suggests that CO suppresses the concentration of one of the catalytic species. An increase in CO pressure will lower the concentration of $HCo(CO)_3$ in the equilibrium

$$HCo(CO)_4 \rightleftharpoons HCo(CO)_3 + CO$$

This type of evidence was used as the basis for postulating the existence of $HCo(CO)_3$, which is not detected spectroscopically in the reaction mixture.

Exercise. Predict the influence of added phosphine on the rate of hydroformylation catalyzed by $HRh(CO)(PPh_3)_3$.

Monsanto acetic acid synthesis

The time-honored method for synthesizing acetic acid is by aerobic bacterial action on dilute aqueous ethanol, which produces vinegar. However, this process is uneconomical as a source of concentrated acetic acid for industry. A highly successful commercial process is based on the rhodium-catalyzed carbonylation of methanol:

$$CH_3OH + CO \xrightarrow{[RhI_2(CO)_2]^-} CH_3COOH$$

The reaction is catalyzed by all three members of Group 9 (cobalt, rhodium, and iridium), but complexes of the 4*d* metal rhodium are the most active. Originally a cobalt complex was employed, but the rhodium catalyst developed at Monsanto greatly reduced the cost of the process by allowing lower pressures to be used. As a result, the **Monsanto process** is licensed for use throughout the world.

The principal catalytic cycle in the Monsanto process is illustrated in Fig. 17.7. Under normal operating conditions, the rate-determining step is oxidative addition of iodomethane to the four-coordinate, 16-electron

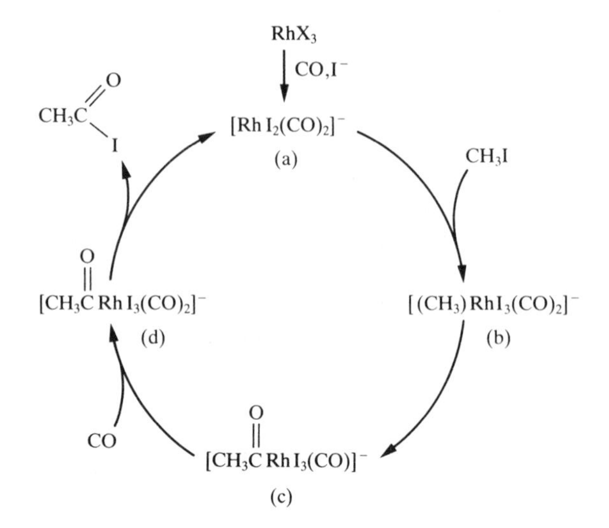

Fig. 17.7 One cycle in the proposed mechanism for the conversion of methanol and CO to acetic acid, catalyzed by rhodium complexes. The acetyl iodide produced by this cycle undergoes hydrolysis to yield acetic acid.

complex $[RhI_2(CO)_2]^-$ (a), producing the six-coordinate 18-electron complex $[(H_3C)RhI_3(CO)_2]^-$ (b). This step is followed by CO migratory insertion, yielding a 16-electron acyl complex (c). Coordination of CO restores an 18-electron complex (d), which is then set to undergo reductive elimination of acetyl iodide with the regeneration of $[RhI_2(CO)_2]^-$. Water then hydrolyzes the acetyl iodide to acetic acid and regenerates HI:

$$CH_3COI + H_2O \rightarrow CH_3COOH + HI$$

No other anion works as well as iodide in this catalytic system, and its special ability arises from several factors. Among them is the greater rate of oxidative addition of iodomethane relative to the other haloalkanes in the rate-determining step. In addition, the soft I^- ion, which is a good ligand for the soft Rh(I), appears to form a five-coordinate complex, $[RhI_3(CO)_2]^{2-}$, which undergoes oxidative addition with iodomethane more rapidly than either $RhI_3(CO)$ or $[RhI_2(CO)_2]^-$. The strong acid HI also is effective in halogenating methanol:

$$CH_3OH + HI \rightarrow CH_3I + H_2O$$

Wacker oxidation of alkenes

The **Wacker process** is primarily used to produce acetaldehyde from ethylene and oxygen:

$$C_2H_4(g) + \tfrac{1}{2}O_2(g) \rightarrow CH_3CHO(g) \qquad \Delta G^{\ominus} = -197\ \mathrm{kJ\ mol^{-1}}$$

Its invention at the Wacker Consortium für Elektrochemische Industrie (West Germany) in the late 1950s marked the beginning of an era of production of chemicals from petroleum feed stock.[5] Although this is no longer a major industrial process, it has some interesting mechanistic features that are worth noting. One step is the palladium(II)-catalyzed hydration of an alkene:

$$C_2H_4 + PdCl_2 + H_2O \rightarrow CH_3CHO + Pd + 2HCl$$

The slow oxidation of Pd(0) back to Pd(II) by oxygen is catalyzed by the addition of Cu(II), which shuttles back and forth to Cu(I):

$$Pd + 2[CuCl_4]^{2-} \rightarrow Pd^{2+} + 2[CuCl_2]^- + 4Cl^-$$

$$2[CuCl_2]^- + \tfrac{1}{2}O_2 + 2H^+ + 4Cl^- \rightarrow 2[CuCl_4]^{2-} + H_2O$$

The overall catalytic cycle is shown in Fig. 17.8. Detailed stereochemical studies on related systems indicate that the hydration of the alkene–Pd(II) complex (a) occurs by the attack of H_2O from the solution on the coordinated ethylene rather than the insertion of coordinated OH. Hydration, to form (b), is followed by two steps that isomerize the coordinated alcohol. First, β-hydride elimination occurs with the formation of (c), and then migratory insertion results in the formation of (d). Elimination of the acetaldehyde and a hydrogen ion then leaves Pd(0). The Pd(0) is converted back to Pd(II) by the auxiliary Cu(II)-catalyzed air oxidation cycle mentioned above.

[5] Ethylene and other alkenes, the starting materials in the Wacker process and many other petrochemical syntheses, are produced by the thermal cracking of saturated hydrocarbons.

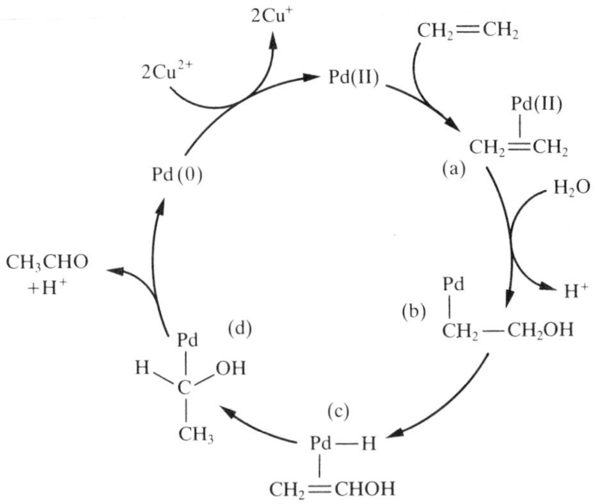

Fig. 17.8 Proposed mechanism for the oxidation of ethylene to acetaldehyde in the Wacker process. Chloride ligands have been omitted. The oxidation number of palladium is +2 at all stages of this cycle except the upper left where reductive elimination of acetaldehyde gives Pd(0), which is oxidized by Cu(II). The complete cycle for the reoxidation of Cu(I) is not shown.

Alkene ligands coordinated to Pt(II) are also susceptible to nucleophilic attack, but only palladium leads to a successful catalytic system. The principal reason appears to be the greater lability of the 4*d* Pd(II) complexes in comparison with their 5*d* Pt(II) counterparts. Furthermore, the potential for the oxidation of Pd(0) to $[PdCl_4]^{2-}$ is more favorable than for the corresponding platinum couple.

Heterogeneous catalysis

Heterogeneous catalysts are used very extensively in industry. One attractive feature is that many of these solid catalysts are robust at high temperatures and therefore make available a wide range of operating conditions. Another reason for their widespread use is that no extra steps are needed to separate the product from the catalyst. Typically, gaseous or liquid reactants enter a tubular reactor at one end and pass over a bed of the catalyst; products are collected at the other end. This same simplicity of design applies to the catalytic converter employed to oxidize CO and hydrocarbons and reduce nitrogen oxides in automobile exhausts (Fig. 17.9).

17.5 The nature of heterogeneous catalysts

Although structural studies are often performed on single crystal surfaces, practical catalysts are generally materials with high surface area, which may contain several different phases. Sometimes the bulk of a high-surface-area material serves as the catalyst, and such a material is called a **uniform catalyst**. One example is the catalytic zeolite ZSM-5, which contains channels through which reacting molecules diffuse. In a **multiphasic catalyst** the high-surface-area material is simply a support on to which an active catalyst is deposited (Fig. 17.10).[6]

[6] The distinctions between uniform and multiphase catalysts are developed by J. M. Thomas, *Angew. Chem., Int. Ed. Eng.*, **12,** 1673 (1988).

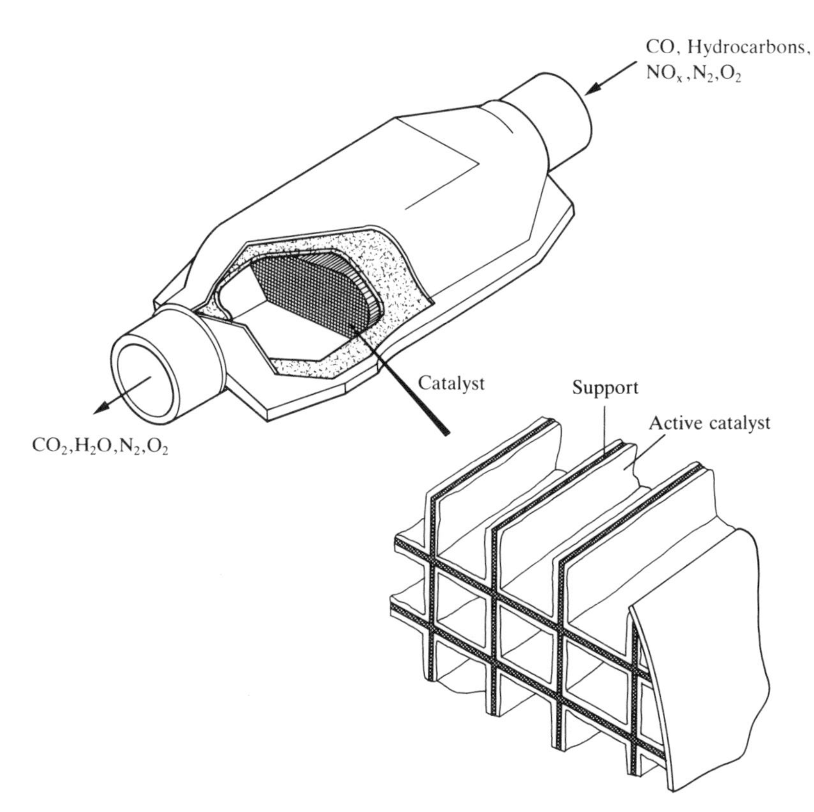

Fig. 17.9 A reactor for heterogeneous catalysis. This automobile catalytic converter oxidizes CO and hydrocarbons and reduces nitrogen oxides. The metal catalyst is supported on a ceramic honeycomb, which is more robust in this application than a bed of loose particles.

Surface area and porosity

In homogeneous catalysis the interaction of reagents and catalysts occurs readily, but special measures must be taken in heterogeneous catalysis to ensure that the reactive molecules achieve contact with catalytic sites. For many applications, an ordinary dense solid is unsuitable as a catalyst because its surface area is quite low. Thus α-alumina, which is a dense material with a low surface area, is used much less as a catalyst support than the microcrystalline solid γ-alumina, which can be prepared with small particle size, and therefore a high surface area. The high surface area results from the many small but connected particles such as those shown in Fig. 17.10. A gram or so of a typical catalyst support has an internal surface area equal to that of a tennis court. Similarly, quartz is not used as a catalyst support but the high-surface-area versions of SiO_2, the silica gels, are widely employed as supports.

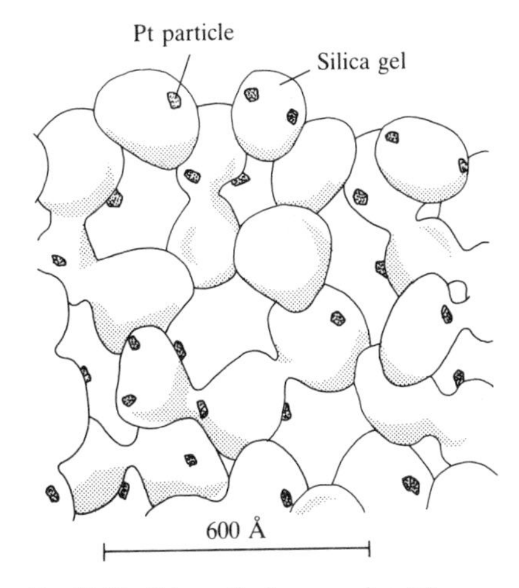

Fig. 17.10 Schematic diagram of metal particles supported on silica gel.

Both γ-alumina and silica gel are metastable materials, but under ordinary conditions they do not convert to their more stable phases (α-alumina and quartz respectively). The preparation of γ-alumina involves the dehydration of an aluminum oxide hydroxide:

$$2AlO(OH) \xrightarrow{\Delta} \gamma\text{-}Al_2O_3 + H_2O$$

Similarly, silica gel is prepared from the acidification of silicates to produce $Si(OH)_4$, which rapidly forms a hydrated silica gel from which much of the adsorbed water can be removed by gentle heating (Section 6.9). When viewed with an electron microscope, the texture of silica gel or alumina appears to be that of a rough gravel bed with irregularly shaped voids between the interconnecting particles (Fig. 17.10).

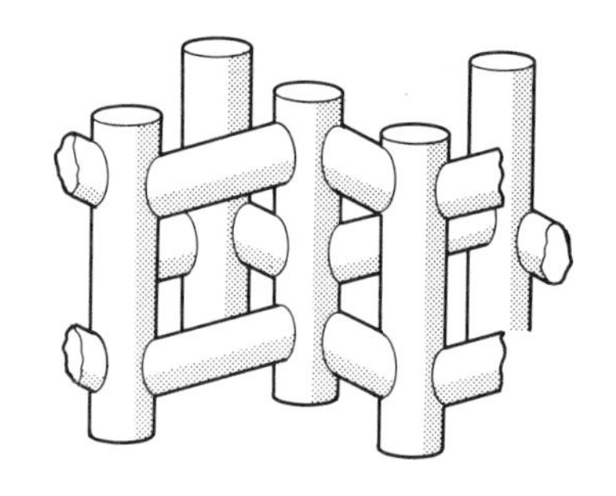

Fig. 17.11 The structure of ZSM-5 has intersecting channels defined by the crystal structure. The tubes in this diagram represent the channels.

Zeolites (Section 11.11) are examples of uniform catalysts. They are prepared as very fine crystals that contain large regular channels and cages defined by the crystal structure (Fig. 17.11). The openings in these channels vary from one crystalline form of the zeolite to the next, but are typically 3 to 10 Å in their smallest diameter. The zeolite absorbs molecules small enough to enter the channels and excludes larger molecules. This selectivity, in combination with catalytic sites inside the cages, provides a degree of control over catalytic reactions that is unattainable with silica gel or γ-alumina. The synthesis of new zeolites and similar shape-selective solids and the introduction of catalytic sites into them is a vigorous area of research (see Section 11.11).

Surface acidic and basic sites

When exposed to atmospheric moisture, the surface of γ-alumina is covered with adsorbed water molecules. As we saw in Section 6.9, dehydration at 100 to 150°C leads to the desorption of water, but surface OH groups remain and act as weak Brønsted acids. At even higher temperatures adjacent OH groups condense to liberate more H_2O and generate exposed Al^{3+} Lewis acid sites as well as O^{2-} Lewis base sites. The rigidity of the surface permits the coexistence of these strong Lewis acid and base sites, which would otherwise immediately form complexes together. Surface acids and bases are highly active for catalytic reactions such as the dehydration of alcohols and isomerization of alkenes (Section 6.9). Similar Brønsted and Lewis acid sites exist on the interior of certain zeolites.

Example 17.3: *Using infrared spectra to probe molecular interaction with surfaces.*

Infrared spectra of hydrogen bonded and Lewis acid complexes of pyridine (py), such as $Cl_3Al(py)$, show that bands near 1540 cm^{-1} can be ascribed to hydrogen bonded pyridine and that bands near 1465 cm^{-1} are due to Al–py Lewis acid–base interactions. A sample of γ-alumina that had been pre-treated by heating to 200°C and then cooled and exposed to pyridine vapor had absorption bands near 1540 cm^{-1} and none near 1465 cm^{-1}. Another sample that was heated to 500°C, cooled, and then exposed to pyridine had bands near 1540 cm^{-1} and 1465 cm^{-1}. Correlate these results with the statements made in the text concerning the effect of heating γ-alumina. (Much of the evidence for the chemical nature of the γ-alumina surface comes from experiments like these.)

Answer. At about 200°C, surface H_2O is lost but OH^- bound to Al^{3+} remains. These groups appear to be mildly acidic as judged by color indicators and the existence of bands around 1540 cm^{-1} that indicate the presence of hydrogen-bonded pyridine generated in the reaction:

```
 OH              O—H···py
 |               |
 Al    + py →    Al
/////           /////
```

When heated to 500°C, much of but not all the OH is lost as H_2O, leaving behind O^{2-} and exposed Al^{3+}. The evidence is the appearance of the 1465 cm^{-1} bands, which are indicative of Al^{3+}—NC_5H_5. However, that not all the OH is lost is indicated by the coexistence of a band at 1540 cm^{-1}.

Exercise. If the statements in Section 6.9 about the effect of heating a γ-alumina are correct, what will be the intensities of the diagnostic infrared bands if the γ-alumina sample is pretreated at 900°C, cooled, exposed to pyridine vapor, and a spectrum determined?

Surface metal sites

Metal particles are often deposited on supports to provide a catalyst. For example, finely divided platinum–rhenium alloys distributed through the pore space of γ-alumina are used to interconvert hydrocarbons, and finely divided platinum–rhodium alloy particles supported on γ-alumina are employed in automobile catalytic converters to promote the combination of O_2 with CO and hydrocarbons to CO_2 and reduction of nitrogen oxides to nitrogen. A supported metal particle about 25 Å in diameter has about 40 percent of its atoms on the surface, and the particles are protected from fusing together into bulk metal by their separation. The high proportion of exposed atoms is a great advantage for these small supported particles, particularly for expensive metals such as platinum.

The metal atoms on the surface of metal clusters are capable of forming bonds such as M—CO, M—CH_2R, M—H, and M—O (Table 17.4). Often the nature of surface ligands is inferred by comparison of infrared spectra with those of organometallic or inorganic complexes. Thus, both terminal and bridging CO can be identified on surfaces by infrared spectroscopy, and the infrared spectra of many hydrocarbons ligands on surfaces are similar to those of discrete organometallic complexes. The case of the N_2 ligand is an interesting contrast, because coordinated N_2 was identified by infrared

Table 17.4. Chemisorbed ligands on surfaces

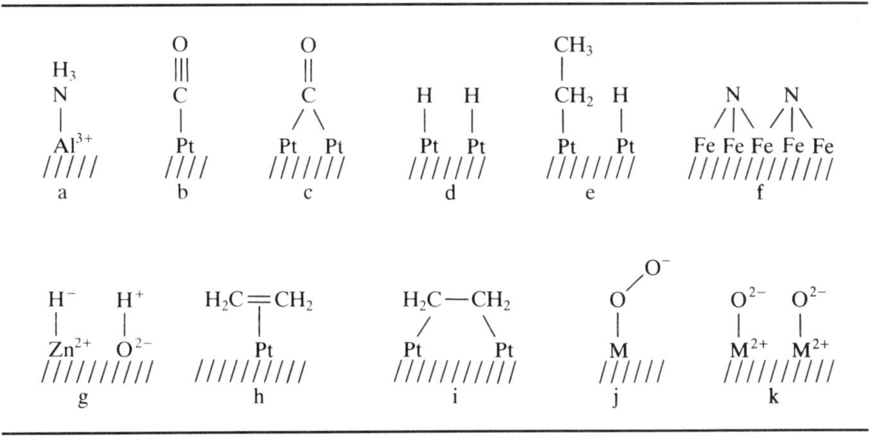

[a] Ammonia adsorbed on the Lewis acid Al^{3+} sites of γ-alumina.
[b,c] CO coordinated to platinum metal.
[d] Hydrogen dissociatively chemisorbed on platinum metal.
[e] Ethane dissociatively chemisorbed on platinum.
[f] Nitrogen dissociatively chemisorbed on iron.
[g] H_2 dissociatively chemisorbed on ZnO.
[h] Ethylene η^2 coordinated to a Pt atom.
[i] Ethylene bonded to two Pt atoms.
[j] O_2 bound as a superoxide to a metal surface.
[k] O_2 dissociatively chemisorbed on a metal surface.
Adapted from: R. L. Burwell, Jr., Heterogeneous catalysis. In *Survey of Progress in Chemistry,* **8,** 2 (1977).

spectroscopy on metal surfaces before dinitrogen complexes were prepared and characterized by inorganic chemists.

The development of new techniques for studying single crystal surfaces has greatly expanded our knowledge of the surface species that may be present in catalysis. For example, the desorption of molecules from surfaces (thermally, or by ion or atom impact) combined with mass spectrometric analysis of the desorbed substance, provides insight into the chemical identity of surface species. Similarly, Auger and X-ray photoelectron spectroscopy (XPS) provide information on the elemental composition of surfaces.

Low-energy electron diffraction (LEED) provides information about the structure of single crystal surfaces and, when adsorbate molecules are present, their arrangement on the surface. One important finding from LEED is that the adsorption of small molecules on a surface may bring about a structural modification of the surface. This surface reconstruction is often observed to reverse when desorption occurs. We are just beginning to see the wide use of scanning tunneling microscopy for the location of adsorbates on surfaces. This striking technique provides a contour map of single crystal surfaces at or close to atomic resolution.[7]

Although most of these modern surface techniques cannot be applied to the study of supported multiphasic catalysts, they are very helpful for revealing the range of probable surface species and circumscribing the structures that may plausibly be invoked in a mechanism of heterogeneous catalysis. The application of these techniques to heterogeneous catalysis is similar to the use of X-ray diffraction and spectroscopy to the characterization of organometallic homogeneous catalyst precursors and model compounds.

17.6 Catalytic steps[8]

There are many parallels between the individual reaction steps encountered in heterogeneous and homogeneous catalysis.

Chemisorption and desorption

The adsorption of molecules on surfaces often activates molecules just as coordination activates molecules in complexes. The desorption of product molecules that is necessary to refresh the active sites in heterogeneous catalysis is analogous to the dissociation of a complex in homogeneous catalysis.

Before a heterogeneous catalyst is used it is usually **activated**. Activation is a catch-all term; in some instances it refers to the desorption of adsorbed molecules such as water from the surface, as in the dehydration of γ-alumina. In other cases it refers to the preparation of the active site by a chemical reaction, such as reduction of metal particles by hydrogen to produce active metal particles.

An activated surface can be characterized by the adsorption of various

[7] These and other modern surface techniques and their help in the elucidation of catalysis are summarized by G. A. Samorjai in *Chemistry in two dimensions*. Cornell University Press (1981).

[8] For an entertaining and informative account of the development of mechanistic concepts in heterogeneous catalysis see R. L. Burwell, Jr., in *Chemtech,* **17,** 586 (1987).

Table 17.5. The tendency of metals to chemisorb simple molecules.

	Gases						
	O_2	C_2H_2	C_2H_4	CO	H_2	CO_2	N_2
Ti, Zr, Hf, V, Nb, Ta, Cr, Mo, W, Fe, Ru, Os	+	+	+	+	+	+	+
Ni, Co	+	+	+	+	+	+	–
Rh, Pd, Pt, Ir	+	+	+	+	+	–	–
Mn, Cu	+	+	+	+	±	–	–
Al, Au	+	+	+	+	–	–	–
Na, K	+	+	–	–	–	–	–
Ag, Zn, Cd, In, Si, Ge, Sn, Pb, As, Sb, Bi	+	–	–	–	–	–	–

+ Strong chemisorption, ± weak, – unobservable.
Adapted from G. C. Bond, *Heterogeneous catalysis*. Oxford University Press (1987), p. 29.

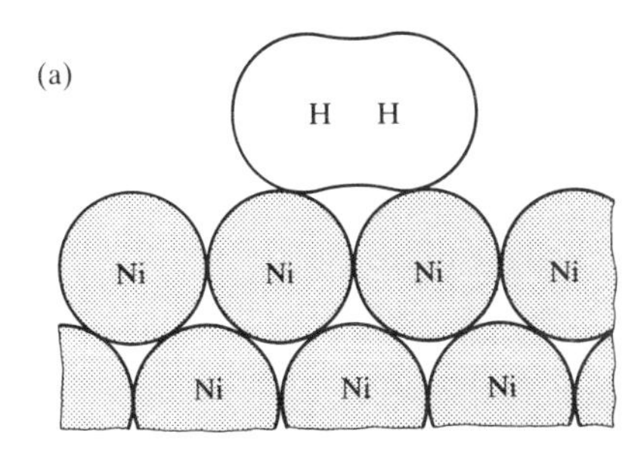

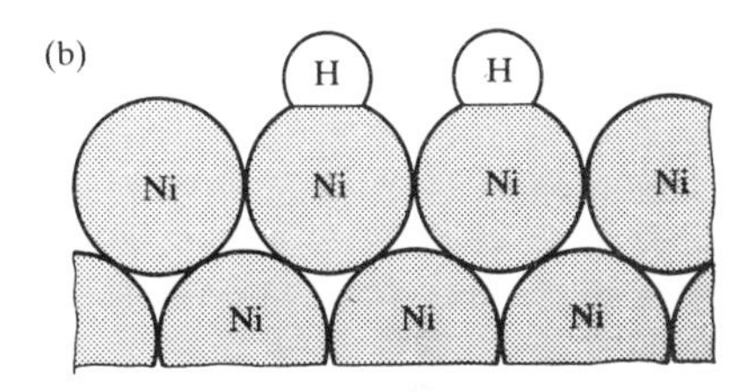

Fig. 17.12 Schematic representation of (a) physisorption and (b) chemisorption of H_2 on a nickel metal surface.

inert and reactive gases. The adsorption may be either **physisorption**, when no new chemical bond is formed, or **chemisorption**, when surface–adsorbate bonds are formed (Fig. 17.12). Low-temperature physisorption of a gas like nitrogen is useful for the determination of the total surface area of a solid, whereas chemisorption is used to determine the number of exposed reactive sites. For example, the dissociative chemisorption of H_2 on supported platinum particles reveals the number of exposed surface Pt atoms.

There are many parallels between the interaction of small molecules with metal surfaces and with low oxidation state metal complexes. Table 17.5 shows that a wide range of metals chemisorb CO, and that many fewer are capable of chemisorbing N_2, just as there is a much wider variety of metals that form carbonyls than form N_2 complexes. Furthermore, just as with metal carbonyl complexes, both bridging and terminal CO surface species have been identified by infrared spectroscopy. The dissociative chemisorption of H_2 (Fig. 17.12) is analogous to the oxidative addition of H_2 to metal complexes.

Even though adsorption is essential for heterogeneous catalysis to occur, it must not be so strong as to block the catalytic sites and prevent further reaction. This factor is in part responsible for the limited number of metals that are effective catalysts. The catalytic decomposition of formic acid on metal surfaces

$$HCOOH \xrightarrow{M} CO + H_2O$$

provides a good example of this balance between adsorption and catalytic activity. It is observed that the reaction is fastest on metals for which the metal formate is of intermediate stability (Fig. 17.13). The plot in Fig. 17.13 is an example of a **volcano diagram**, and is typical of many catalytic reactions. The implication is that the earlier *d*-block metals form very stable surface compounds whereas the later noble metals such as silver and gold form very weak surface compounds, both of which are detrimental to a

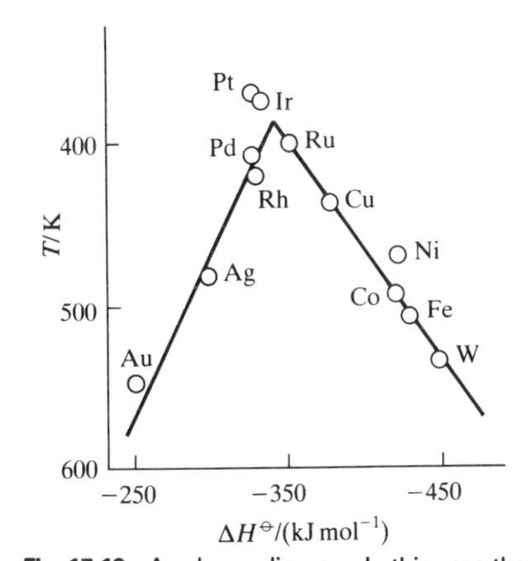

Fig. 17.13 A volcano diagram. In this case the reaction temperature for a set rate of formic acid decomposition is plotted against the stability of the corresponding metal formate as judged by heat of formation (W. J. M. Rootsaert and W. M. H. Sachtler, *Z. Phys. Chem.*, **26**, 16 (1960)).

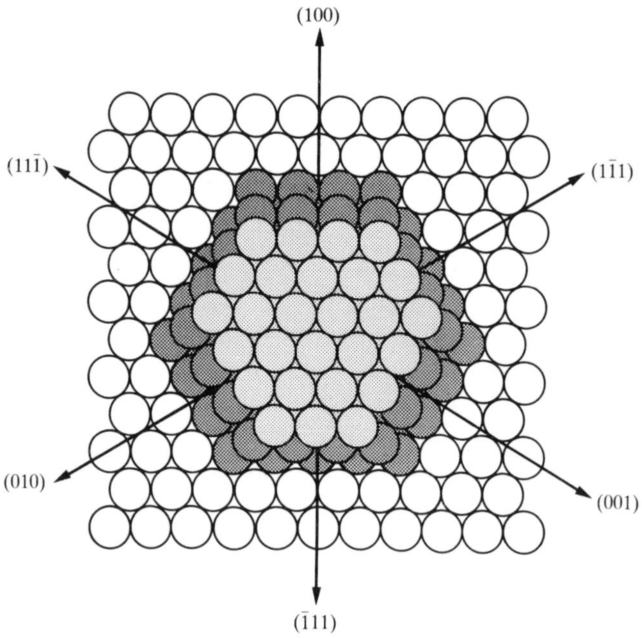

(100)

(11$\bar{1}$)

(1$\bar{1}$1)

(010)

(001)

($\bar{1}$11)

Fig. 17.14 A collection of metal crystal planes that might be exposed to reactive gases. (111), ($\bar{1}$11), etc. are close-packed hexagonal planes. The planes represented by (100), (010), etc. have square arrays of atoms.

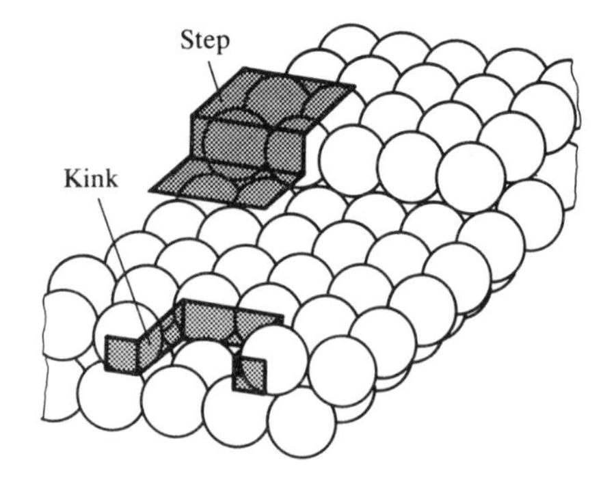

Fig. 17.15 Schematic representation of surface irregularities, steps, and kinks.

catalytic process. Between these extremes the metals in Groups 8 to 10 have high catalytic activity, and this is especially true of the platinum metals. In Section 17.2 we saw a similar high activity of platinum metal complexes in the homogeneous catalysis of hydrocarbon transformations.

The active sites of heterogeneous catalysts are not uniform and many diverse sites are exposed on the surface of a poorly crystalline solid such as γ-alumina or a noncrystalline solid such as silica gel. However, even highly crystalline metal particles are not uniform. A crystalline solid has typically more than one type of exposed plane, each with its characteristic pattern of surface atoms (Fig. 17.14). In addition, single crystal metal surfaces have irregularities such as steps (Fig. 17.15) that expose metal atoms with low coordination numbers. These highly exposed, coordinatively unsaturated sites appear to be particularly reactive. As a result, the different sites on the surface may serve different functions in catalytic reactions. The variety of sites also accounts for the lower selectivity of many heterogeneous catalysts in comparison with homogeneous analogs.

Surface migration

We saw in Chapter 16 that ligands in organometallic clusters are often fluxional. The surface analog of fluxionality in clusters is diffusion, and there is abundant evidence for the diffusion of chemisorbed molecules or atoms on metal surfaces. For example, adsorbed H atoms and adsorbed CO molecules are known to move over the surface of a metal particle. This mobility is important in catalytic reactions, for it allows atoms or molecules to approach each other. Surface diffusion rates are difficult to measure, and reliable data have been obtained only recently.

17.7 Examples

Hydrogenation of alkenes

The hydrogenation of alkenes on supported metal particles is thought to proceed in a manner very similar to that in metal complexes.[9] As pictured in Fig. 17.16, H_2, which is dissociatively chemisorbed on the surface, is thought to migrate to an adsorbed ethylene molecule, giving first a surface alkyl and then the saturated hydrocarbon. When ethylene is hydrogenated with D_2 over platinum, the simple mechanism depicted in Fig. 17.16 indicates that $CH_2D—CH_2D$ should be the product. In fact, a complete range of $C_2H_nD_{6-n}$ ethane molecules is observed. It is for this reason that the third step is written as reversible; the rate of the reverse reaction must be greater than the rate at which the ethane molecule is formed and desorbed in the final step.

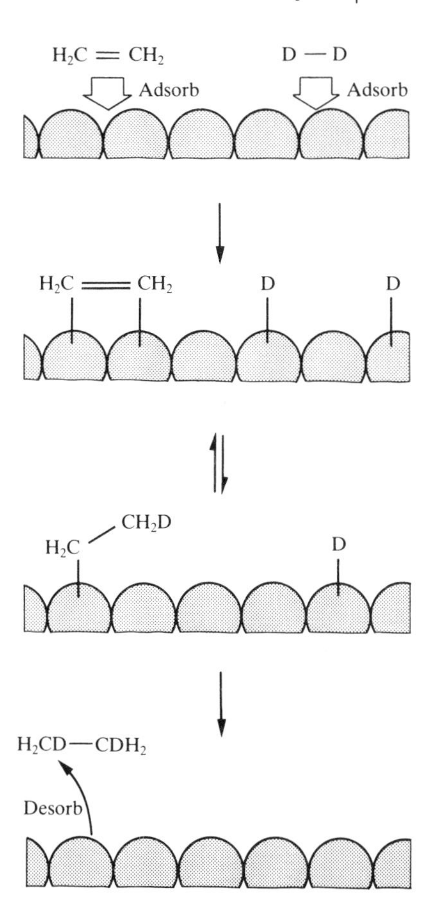

Fig. 17.16 Schematic diagram of the hydrogenation of ethylene by D_2 on a metal surface.

Ammonia synthesis

The synthesis of ammonia

$$N_2 + 3H_2 \rightarrow 2NH_3$$

has already been discussed from several different viewpoints (Sections 9.15 and 12.2). Here we concentrate on details of the catalytic steps. Ammonia is both an exoergic and an exothermic compound at 25°C, the relevant thermodynamic data being $\Delta G_f^{\ominus} = -16.5\ \text{kJ mol}^{-1}$, $\Delta H_f^{\ominus} = -46.1\ \text{kJ mol}^{-1}$, and $\Delta S_f^{\ominus} = -99.4\ \text{J K}^{-1}\ \text{mol}^{-1}$. The negative reaction entropy of formation reflects the fact that two molecules of product gas replace four molecules of reactant gases.

The great inertness of N_2 and to a lesser extent H_2 requires that a catalyst be used to effect the reaction. Iron metal together with small quantities of alumina and potassium salts and other promoters is the catalyst. Extensive studies on the mechanism of ammonia synthesis indicate that the rate determining step under normal operating conditions is the dissociation of N≡N coordinated to the surface. The other reactant, H_2, undergoes much more facile dissociation on the metal surface and a series of insertion reactions between adsorbed species leads to the production of NH_3:

$$\left.\begin{array}{l} N_2(g) \rightarrow \underset{///}{N_2} \rightarrow \underset{////}{2N} \\ H_2(g) \rightarrow \underset{////}{2H} \end{array}\right\} \text{dissociative chemisorption}$$

$$\left.\begin{array}{l} \underset{///}{N} + \underset{///}{H} \rightarrow \underset{////}{NH} \\ \underset{////}{NH} + \underset{///}{H} \rightarrow \underset{////}{NH_2} \\ \underset{////}{NH_2} + \underset{///}{H} \rightarrow \underset{////}{NH_3} \end{array}\right\} \text{insertion of surface atoms}$$

$$\left.\underset{////}{NH_3} \rightarrow NH_3(g)\right\} \text{desorption of product}$$

[9] A milestone in heterogeneous catalysis was Paul Sabatier's observation that nickel catalyzes the hydrogenation of alkenes (1900). He was actually attempting to synthesize $Ni(C_2H_4)_4$, as a follow-up to Mond, Langer, and Quinke's synthesis of $Ni(CO)_4$ (Section 16.4). When he passed ethylene over heated nickel he detected ethane. This led him to include hydrogen with the ethylene, whereupon he observed a good yield of ethane. Major industrial applications soon followed. See P. Sabatier, *Bull. Soc. Chim. Fr.*, **6**, 1261 (1939).

Because of the slowness of the N_2 dissociation it is necessary to run the ammonia synthesis at high temperatures, typically 400°C. However, since the reaction is exothermic, high temperature greatly reduces the equilibrium constant of the reaction, for from the van't Hoff equation

$$\frac{d \ln K}{dT} = \frac{\Delta H^{\ominus}}{RT^2} < 0$$

and K decreases as T is increased. To recover some of this reduced yield, pressures of the order of 100 atm are employed. A catalyst that would operate at room temperature would give good equilibrium yields of NH_3, but such catalysts have not yet been discovered (Section 12.2).

In the course of developing the original ammonia synthesis process, Haber and his coworkers investigated the catalytic activity of most of the metals in the periodic table, and found that the best are iron, ruthenium, and uranium. Cost and toxicity considerations led to the choice of iron (along with potassium oxide and aluminum oxide) for the commercial catalyst. Other metals, such as lithium, are more active for the critical N≡N cleavage step, but in these cases the succeeding steps are not favorable on account of the great stability of the metal nitride. This observation once again emphasizes the detrimental nature of a highly stable intermediate in a catalytic cycle.

SO_2 oxidation

The oxidation of SO_2 to SO_3 is a key step in the production of sulfuric acid (Section 12.8). The reaction of sulfur with oxygen to produce SO_3 gas is exoergic ($\Delta G^{\ominus} = -371\ \mathrm{kJ\ mol^{-1}}$) but very slow, and the principal product of the combustion is SO_2:

$$S(s) + O_2(g) \rightarrow SO_2(g)$$

The combustion is followed by the catalytic oxidation of SO_2:

$$SO_2(g) + \tfrac{1}{2}O_2(g) \rightarrow SO_3(g)$$

This step is also exothermic and, as with ammonia synthesis, has a less favorable equilibrium constant at elevated temperatures. The process is therefore generally run in stages. In the first stage the combustion of sulfur raises the temperature to about 600°C, but by cooling before the last catalytic stage the equilibrium is driven to the right and high conversion of SO_2 to SO_3 is achieved.

Several quite different catalytic systems have been used to catalyze the combination of SO_2 with O_2. The most widely used catalyst at present is potassium vanadate supported on a high-surface-area silica (diatomaceous earth). One interesting aspect of this catalyst is that the vanadate is a molten salt under operating conditions. The current view of the mechanism of the reaction is that the rate-determining step is the oxidation of V^{4+} to V^{5+} by O_2. Schematically the cycle is

$$\tfrac{1}{2}O_2 + 2V^{4+} \rightarrow O^{2-} + 2V^{5+}$$

$$SO_2 + 2V^{5+} + O^{2-} \rightarrow 2V^{4+} + SO_3$$

In the melt the vanadium and oxide ions are part of a polyvanadate complex (Section 5.9), but little is known about the evolution of the oxo species.

Interconversion of aromatics by zeolites

The zeolite-based heterogeneous catalysts play an important role in the interconversion of hydrocarbons and alkylation of aromatics as well as in oxidation and reduction.

The synthetic zeolite ZSM-5 (which was developed in the research laboratories of Mobil Oil) is widely used by the petroleum industry for a variety of hydrocarbon interconversions. ZSM-5 is an aluminosilicate zeolite with a higher silica content and a much lower aluminum content than the zeolites discussed in Section 11.11, and its channels (Fig. 17.11) consist of a three-dimensional maze of intersecting tunnels. As with other aluminosilicate catalysts, the aluminum sites are strongly acidic. The charge imbalance of Al^{3+} in place of tetrahedrally coordinated Si^{4+} requires the presence of an added positive ion. When this ion is H^+ (**6**) the Brønsted acidity of the aluminosilicate can be as great as that of concentrated H_2SO_4.

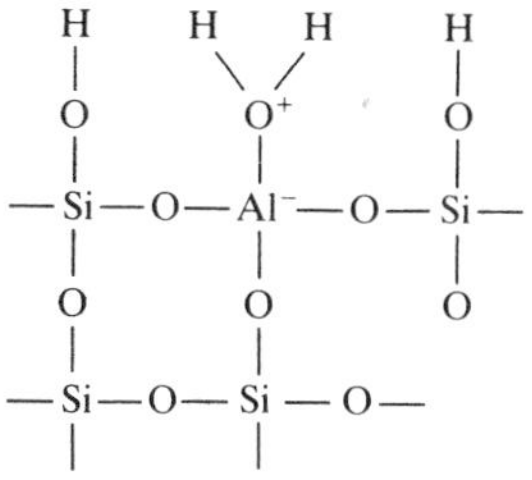

H H H H
O O⁺ O
—Si—O—Al⁻—O—Si—
O O O
—Si—O—Si—O—

6

Reactions such as xylene isomerization and toluene disproportionation are illustrative of the selectivity that can be achieved with acidic zeolite catalysts. The shape selectivity of these catalysts has been attributed to faster diffusion of product molecules that have dimensions compatible with the channels. According to this idea, molecules that do not fit the channels diffuse slowly, and because of their long residence in the zeolite, have ample opportunity to be converted to the more mobile isomers that can rapidly escape. A currently more favored view, however, is that the orientation of reactive intermediates within the zeolite channels favors specific products, such as *para*-dialkylbenzenes.

Aside from their important shape selectivity, acidic zeolite catalysts appear to promote reactions by standard carbocation mechanisms. For example, the isomerization of *m*-xylene to *p*-xylene,

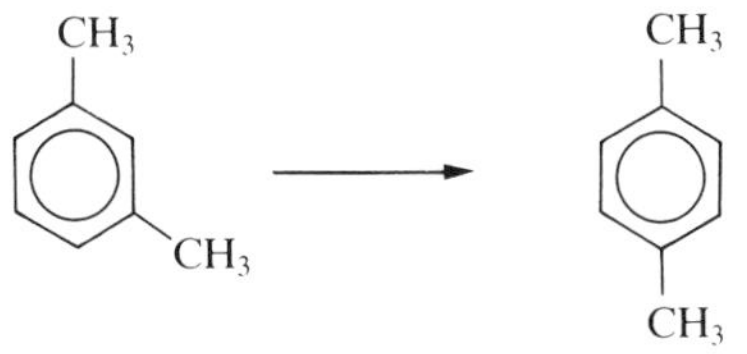

CH₃ CH₃ → CH₃ CH₃

may occur by the following steps:

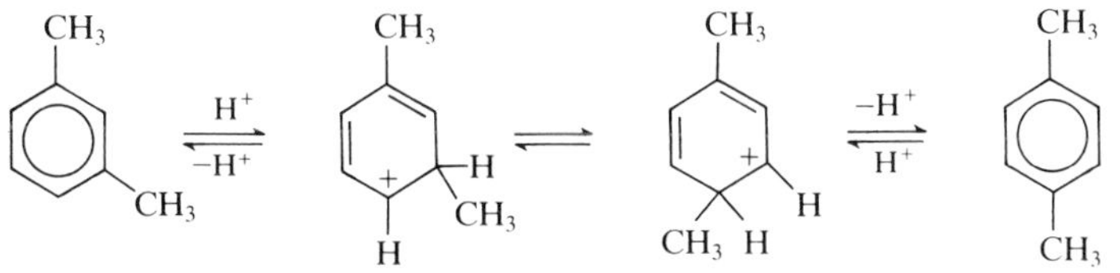

CH₃ CH₃ ⇌ (H⁺ / −H⁺) CH₃ H CH₃ H ⇌ CH₃ CH₃ H H ⇌ (−H⁺ / H⁺) CH₃ CH₃

Another common reaction in zeolites is the alkylation of aromatics with alkenes.

Example 17.4: *Proposing a plausible mechanism for the alkylation of benzene in ZSM-5.*

In its protonated form, ZSM-5 catalyzes the reaction of ethylene with benzene to produce ethylbenzene. Write a plausible mechanism for the reaction.

Answer. The acidic form of ZSM-5 is a strong enough acid to generate carbocations:

$$CH_2{=}CH_2 + H^+ \rightarrow [CH_2{-}CH_3]^+$$

As we saw in Section 6.6, the carbocation can attack benzene. Subsequent deprotonation of the product yields ethylbenzene:

$$[CH_2{-}CH_3]^+ + C_6H_6 \rightarrow C_6H_5{-}CH_2CH_3 + H^+$$

Exercise. A pure silica analog of ZSM-5 can be prepared. Would you expect this to be an active catalyst for benzene alkylation? Explain your reasoning.

Electrocatalysis

The kinetic barrier to electrochemical reactions involving H_2 and O_2 was discussed in Chapter 8. Kinetic barriers are quite common for electrochemical reactions at the interface between a solution and an electrode and, as we saw in Section 8.4, it is common to express these barriers as overpotentials.

The overpotential η of an electrolytic cell is the potential in addition to the zero-current cell potential that must be applied in order to bring about a nonspontaneous reaction within the cell. The overpotential is empirically related to the current density j (the current per unit area of electrode) that passes through the cell by[10]

$$\eta = a + b \log j$$

or equivalently

$$j = ce^{\eta d}$$

where a, b, c, and d are empirical constants. The constant c, the **exchange current density**, is a measure of the rates of the forward and reverse electrode reactions at dynamic equilibrium in the absence of an overpotential. For systems obeying these relations, the reaction rate (as measured by the current density) increases rapidly with increasing applied potential when $\eta d > 1$. If the exchange current density is high, an appreciable reaction rate may be achieved for only a small overpotential. If the exchange current density is low, a high overpotential is necessary. There is therefore considerable interest in controlling the exchange current density. In an industrial process an overpotential in a synthetic step is very costly because it represents wasted energy. However, a low exchange current density may have the beneficial effect of blocking an undesirable side reaction.

A catalytic electrode surface can increase the exchange current density and hence dramatically decrease the overpotential required for sluggish electrochemical reactions, such as H_2, O_2, or Cl_2 evolution and consumption. For example, 'platinum black', a finely divided form of platinum, is very effective for increasing the exchange current density and hence decreasing the overpotential of reactions involving the consumption or evolution of H_2. The role of platinum is to dissociate the strong H—H bond and thereby to reduce the large barrier that this strength imposes on

[10] The exponential relation between the current and the overpotential is accounted for by the 'Butler–Volmer equation,' which is derived by applying activated complex theory to dynamical processes at electrodes. See Chapter 30 of P. W. Atkins, *Physical chemistry*. Oxford University Press and W. H. Freeman & Co. (1990).

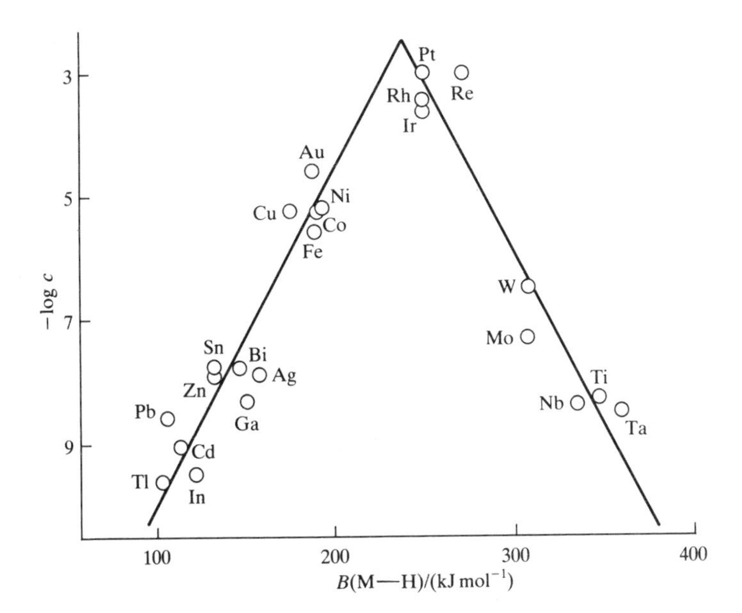

Fig. 17.17 Rate of H_2 evolution expressed as the logarithm of the exchange current density versus M—H bond energies.

reactions involving H_2. Palladium metal also has a high exchange current density (and hence requires only a low overpotential) for H_2 evolution or consumption.

The effectiveness of metals can be judged from Fig. 17.17, which also gives insight into the process. The volcano-like plot of exchange current density against M—H bond enthalpy suggests that M—H bond formation and cleavage are both important in the catalytic process. It appears that an intermediate M—H bond energy leads to the proper balance for the existence of a catalytic cycle and the most effective metals for electrocatalysis are clustered around Group 10.

Ruthenium dioxide is an effective catalyst for both O_2 and Cl_2 evolution and it is also a good electrical conductor. It turns out that at high current densities RuO_2 is more effective for the catalysis of Cl_2 evolution than for O_2 evolution. Therefore RuO_2 is extensively used as an electrode material in commercial chlorine production. The electrode processes that contribute to this subtle catalytic effect do not appear to be well understood.

There is great interest in devising new catalytic electrodes, particularly ones that decrease the O_2 overpotential. Thus, tetraphenylporphyrincobalt(II), Co(TPP), has been deposited on graphite electrodes (on which O_2 reduction requires a high overpotential) and found to promote the electrochemical reduction of O_2. A plausible explanation for this catalysis is that Co^{2+}(TPP) interacts with O_2 to form $Co^{3+}(TPP)(O_2^-)$ on the electrode surface and that subsequent protonation and reduction leads to H_2O_2 and the starting catalyst:

$$Co^{2+}(TPP^{2-}) + O_2 \rightarrow Co^{3+}(TPP^{2-})(O_2^-)$$

$$Co^{3+}(TPP^{2-})(O_2^-) + e^- + H^+ \rightarrow Co^{3+}(TPP^{2-})(O_2H^-)$$

$$Co^{3+}(TPP^{2-})(O_2H^-) + e^- + H^+ \rightarrow Co^{2+}(TPP^{2-}) + H_2O_2$$

The use of tetraphenylporphyrincobalt(II) was motivated by nature's use of metalloporphyrins for oxygen activation, a subject we explore in Chapter 19.

Further reading

Homogeneous catalysis

G. W. Parshall, *Homogeneous catalysis*. Wiley, New York (1980).
C. Masters, *Homogeneous transition-metal catalysis: a gentle art*. Chapman and Hall, London (1981).
B. R. James, *Homogeneous hydrogenation*. Wiley, New York (1973).
R. A. Shelton and J. K. Kochi, *Metal catalyzed oxidations of organic compounds*. Academic Press, New York (1981).
J. P. Collman, L. S. Hegedus, J. R. Norton, and R. G. Finke, *Principles and applications of organotransition metal chemistry*. University Science Books, Mill Valley (1987).

The books by Parshall and Masters provide useful introductions to homogeneous catalysis and the next two are more advanced. For a helpful summary of more recent mechanistic interpretations, the book by Collman *et al.* should be consulted.

Heterogeneous catalysis

R. L. Burwell, Jr. Heterogeneous catalysis. In *Survey of progress in chemistry* **8,** 2 (1977). A brief and instructive account.
G. C. Bond, *Heterogeneous catalysis* (2nd edn). Oxford University Press (1987). Another brief but general introduction to the field.
J. M. Thomas and W. J. Thomas, *Introduction to the principles of heterogeneous catalysis*. Academic Press, New York (1967). A detailed description of the fundamentals and process in heterogeneous catalysis.
C. N. Satterfield, *Heterogeneous catalysis in practice*. McGraw-Hill, New York (1980).
B. C. Gates, J. R. Katzer, and G. C. A. Schuit, *Chemistry of catalytic processes*. McGraw-Hill, New York (1979).

As the titles imply, Satterfield's book focuses in the practice of industrial heterogeneous catalysis, while the book by Gates *et al.* is more chemically oriented.

H. H. Kung. Transition metal oxides. In *Surface chemistry and catalysis*, Elsevier, Amsterdam (1989). This book presents a modern and chemically interesting discussion of oxide catalysts.

Exercises

17.1 Which of the following constitute genuine examples of catalysis and which do not? Present your reasoning. (a) The addition of H_2 to C_2H_4 when the mixture is brought into contact with finely divided platinum. (b) The reaction of an H_2–O_2 gas mixture when an electrical arc is struck. (c) The combination of N_2 gas with lithium metal to produce Li_3N, which then reacts with H_2O to produce NH_3 and LiOH.

17.2 Define the terms (a) turnover frequency, (b) selectivity, (c) catalyst, (d) catalytic cycle, and (e) catalyst support.

17.3 Classify the following as homogeneous or heterogeneous catalysis and present your reasoning. (a) The increased rate in the presence of NO(*g*) of SO_2(*g*) oxidation by O_2(*g*) to SO_3(*g*). (b) The hydrogenation of liquid vegetable oil using a finely divided nickel catalyst. (c) The conversion of an aqueous solution of D-glucose to a D,L mixture catalyzed by HCl(*aq*).

17.4 You are approached by an industrialist with the proposition that you develop catalysts for the following processes at 80°C with no input of electrical energy or electromagnetic radiation.
(a) The splitting of water into H_2 and O_2.
(b) The decomposition of CO_2 into C and O_2.
(c) The combination of N_2 with H_2 to produce NH_3.
(d) The hydrogenation of the double bonds in vegetable oil.
The industrialist's company will build the plant to carry out the process and the two of you will share equally in the

profits. Which of these would be easy to do, which are plausible candidates for investigation, and which are unreasonable? Describe the chemical basis for the decision in each case.

17.5 Addition of PPh_3 to a solution of Wilkinson's catalyst, $RhCl(PPh_3)_3$, reduces the turnover frequency for the hydrogenation of propylene. Give a plausible mechanistic explanation for this observation.

17.6 The rates of H_2 gas absorption (in $L\,mol^{-1}\,s^{-1}$) by alkenes catalyzed by $RhCl(PPh_3)_3$ in benzene at 25°C are hexene: 2910; *cis*-4-methyl-2-pentene: 990; cyclohexene: 3160; 1-methylcyclohexene: 60. Suggest the origin of the trends and identify the affected reaction step(s) in the proposed mechanism (Fig. 17.3).

17.7 Infrared spectroscopic investigation of a mixture of CO, H_2, and 1-butene under conditions that bring about hydroformylation indicate the presence of compound (c) in Fig. 17.6 in the reacting mixture. The same reacting mixture in the presence of added tributylphosphine was studied by infrared spectroscopy and neither (c) nor an analogous phosphine-substituted complex was observed. What does the first observation suggest as the rate limiting reaction in the absence of phosphine? Assuming the sequence of reactions remains unchanged, what are the possible rate limiting reactions in the presence of tributylphosphine?

17.8 (a) Starting with the alkene complex shown in Fig. 17.8a with *trans*-DHC=CHD in place of CrHt, assume dissolved OH^- attacks from the side opposite the metal. Give a stereochemical drawing of the resulting compound of the type 17.8b. (b) Assume attack on the coordinated *trans*-DHC=CHD by an OH^- ligand coordinated to Pd, and draw the stereochemistry of the resulting compound. (c) Does the stereochemistry differentiate these proposed steps in the Wacker process?

17.9 Formic acid is thermodynamically unstable with respect to CO_2 and H_2. Give a plausible mechanism for the catalysis of formic acid decomposition by $IrCl(H)_2(PPh_3)_3$.

17.10 At elevated temperatures, finely divided titanium readily reacts with $N_2(g)$ to form a stable nitride. The rate-determining step in ammonia synthesis over iron is the breaking of the N≡N bond. Why then is titanium ineffective but iron effective as a catalyst for ammonia synthesis?

17.11 Aluminosilicate surfaces were described in the text as strong Brønsted acids, whereas silica gel is a very weak acid. (a) Give an explanation for the enhancement of acidity by the presence of Al^{3+} in a silica lattice. (b) Name three other ions that might enhance the acidity of silica gel.

17.12 Indicate the difference between each of the following:
(a) The Lewis acidity of γ-alumina that has been heated to 900°C versus γ-alumina that has been heated to 100°C.
(b) The Brønsted acidity of silica gel versus γ-alumina.
(c) The structural regularity of the channels and voids in silica gel versus that for ZSM-5.

17.13 Why is the platinum–rhodium in automobile catalytic converters dispersed on the surface of a ceramic rather than used in the form of thin foil?

17.14 Describe the concept of shape-selective catalysts and how this shape selectivity is attained.

17.15 Alkanes are observed to exchange hydrogen atoms with deuterium gas over some platinum metal catalysts. When 3,3-dimethylpentane in the presence of D_2 is exposed to a platinum catalyst and the gases are observed before the reaction has proceeded very far the main product is $CH_3CH_2C(CH_3)_2CD_2CD_3$ plus unreacted 3,3-dimethylpentane. Devise a plausible mechanism to explain this observation.

17.16 The effectiveness of platinum in catalyzing the reaction $2H^+(aq) + 2e^- \rightarrow H_2(g)$ is greatly decreased in the presence of CO. Suggest an explanation.

Problems

17.1 Carbon monoxide is known to undergo dissociative chemisorption on nickel at elevated temperatures, which results in surface carbide and oxide. With this as an initial step, propose a series of plausible reactions for the nickel-catalyzed conversion of CO and H_2 to CH_4 and H_2O.

17.2 At the end of Section 17.7 we saw the reactions that are thought to occur when a polymer film containing tetraphenylporphyrincobalt(II) complexes is deposited on a carbon electrode and exposed to a solution of $[Ru(NH_3)_6]^{3+}$ and O_2 in solution. The large porphyrin complex will diffuse very slowly in the polymer matrix but the smaller ruthenium complex and O_2 should diffuse rapidly. Sketch a cross-section of this electrode and describe the reactions at the electrode surface, the diffusion of species in the polymer and the sequence of reactions that occur in the proposed catalytic cycles.

17.3 When direct evidence for a mechanism is not available, chemists frequently invoke analogies with similar systems. Describe how J. E. Bäckvall, B. Åkermark, and S. O. Ljunggren (*J. Amer. chem. Soc.*, **101**, 2411 (1979)) inferred the attack of uncoordinated water on η^2-C_2H_4 in the Wacker process.

18 Structure and properties of solids

Some general principles
18.1 Defects
18.2 Nonstoichiometric compounds
18.3 Atom and ion diffusion
Prototypical oxides and fluorides
18.4 Monoxides of the 3*d* metals
18.5 Higher oxides
18.6 Glasses
Prototypical sulfides and related compounds
18.7 Layered MS_2 compounds and intercalation
18.8 Chevrel phases
Further information: phase diagrams
Further reading
Exercises
Problems

The physical and chemical properties of solids are a pervasive aspect of inorganic chemistry, and have been discussed throughout the text. However, certain aspects have not yet been covered and are treated here. They include some new principles and the systematic study of the structures and properties of important oxides and sulfides. One new theme, which has a profound impact on modern technology, is the manner in which interruptions in the uniformity of a crystal structure influence properties such as the migration of ions through the solids. In addition, we shall extend the discussion of intercalation compounds, and see how their formation, structures, and properties correlate with the crystal structure and the electronic band structure of the parent compound. We shall see in particular that the appropriate choice of a localized or a delocalized description of the electronic structure of a solid is a great help in correlating their various properties.

Solid state inorganic chemistry is concerned with the synthesis and structures of solids that exhibit new phenomena or possess desirable properties, such as high-temperature superconductivity or ferromagnetism. It is a vigorous and exciting area of inorganic research, partly on account of the technological application of these properties but also because they are challenging to understand. We shall draw on some of the concepts developed in earlier chapters for discussing the solid state, such as lattice energies and band structure, and introduce some additional concepts that are needed if we are to discuss the dynamics of events that occur in the interior of solids. We shall also need to go beyond the view that solids have a well-defined stoichiometry, for interesting phenomena arise from non-stoichiometry and atom mobility.

Some general principles

In Chapter 4 we described the structures of some typical inorganic solids in terms of infinitely repeating regular arrays of atoms. It is now time to modify this idealized picture.

18.1 Defects

All solids contain **defects**, or imperfections of structure or composition. Defects are important because they influence properties such as mechanical strength, electrical conductivity, and chemical reactivity. We need to consider both **intrinsic defects**, which are defects that occur in the pure substance, and **extrinsic defects**, which stem from the presence of impurities. It is also common to distinguish **point defects**, which occur at single sites, from **extended defects**, which are ordered in one, two, and three dimensions. Point defects are random errors in a periodic lattice, such as the absence of an atom from its usual site or the presence of an atom at a site that is not ordinarily occupied. Extended defects involve various irregularities in the stacking of planes.

Why crystals have defects

All solids have a thermodynamic tendency to acquire defects, because defects introduce disorder into an otherwise perfect structure and hence increase its entropy. The Gibbs free energy of a solid with defects has contributions from the enthalpy and the entropy of the sample (*Further information*, Chapter 4):

$$G = H - TS \qquad (1)$$

Entropy is a measure of the disorder of a system, and any solid in which some of the atoms are not in their usual sites has a higher entropy than in a perfect crystal; hence, defects contribute a negative term to the Gibbs energy of the solid. The formation of defects may be endothermic (so H is higher in the presence of defects), but so long as $T > 0$, the Gibbs free energy will have a minimum at a non-zero concentration of defects (Fig. 18.1a), and their formation will be spontaneous. Moreover, as the temperature is raised, the minimum in G shifts to higher concentrations of defects (Fig. 18.1b).

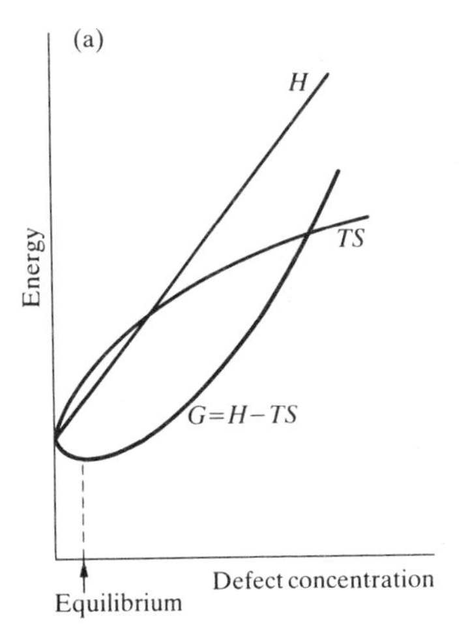

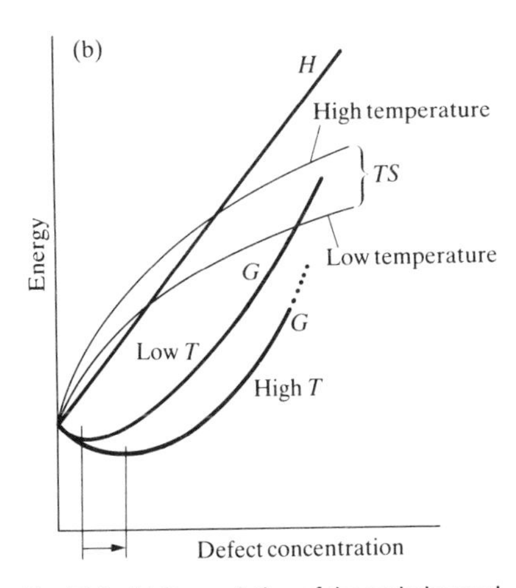

Fig. 18.1 (a) The variation of the enthalpy and entropy of a crystal as the number of defects increases. The resulting Gibbs free energy $G = H - TS$ has a minimum at a non-zero concentration, and hence defect formation is spontaneous. (b) As the temperature is increased, the minimum in the Gibbs free energy moves to higher defect concentrations, and so more defects are present at equilibrium at higher temperatures than at low.

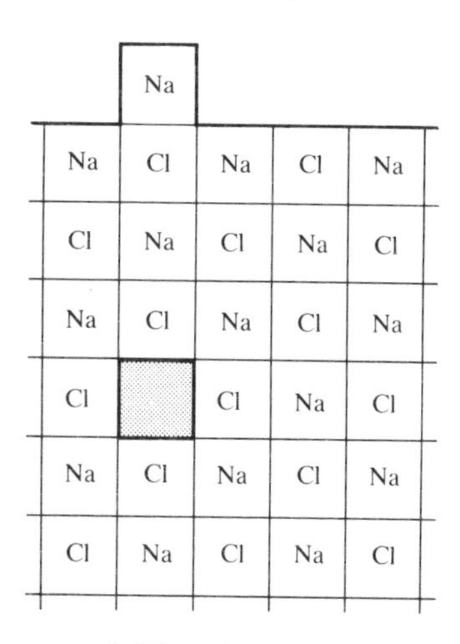

1 Shottky defect

Intrinsic point defects

Point defects are difficult to detect directly. X-ray diffraction, for example, samples the periodic structure of a solid over thousands of angstroms, so small random deviations from periodicity can go unnoticed. Occasionally spectroscopic measurements indicate the presence of an ion in an unusual position or an electron trapped at a particular site in the lattice. The presence of an appreciable number of vacancies or excess atoms may sometimes be inferred from the difference between the measured and calculated densities of a sample (as we illustrate in Example 18.1). The electrical conductivity of a sample, another bulk property, has also been used to infer the presence of defects. Modern electron microscopy has improved our ability to detect defects because its resolution is so high that defects can be observed directly.

In the 1930s, two solid state physicists—Schottky in Germany and Frenkel in Russia—used conductivity and density data to identify specific types of point defect. A **Schottky defect** (**1**) is a vacancy in an otherwise perfect lattice. That is, it is a point defect in which an atom or ion is missing from its normal site in the lattice. The overall stoichiometry of a solid is not affected by the presence of Shottky defects because there are equal numbers of vacancies at M and X sites. A **Frenkel defect** (**2**) is a point defect in which an atom or ion has been displaced into an interstitial site. For example, in silver chloride, which has the rock-salt structure, a small number of Ag^+ ions reside in tetrahedral sites (**3**). The stoichiometry of the compound is unchanged when a Frenkel defect forms.

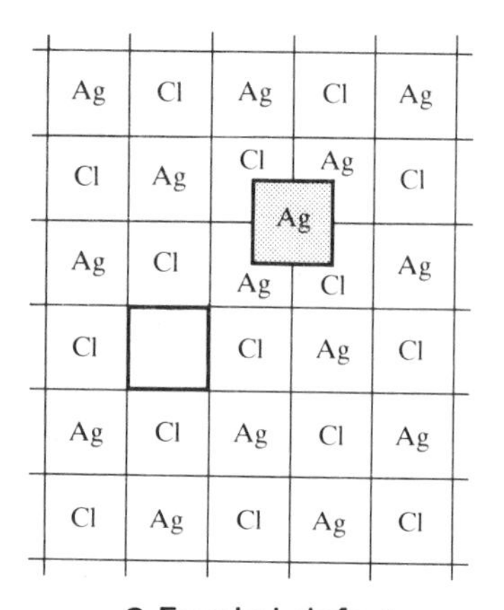

2 Frenkel defect

The concentration of Schottky defects varies considerably from one type of compound to the next. The concentration of vacancies is very low in the alkali metal halides, being of the order of 10^{-12} M at 130°C. That concentration corresponds to about one defect per 10^{14} formula units. On the other hand, some *d*-metal oxides, sulfides, and hydrides have very high concentrations of vacancies. An extreme example is the high-temperature form of TiO, which has vacancies on both the cation and anion sites at a concentration of about 12 M, corresponding to about one defect per 10 formula units. Frenkel defects are most often encountered in the more open structures such as wurtzite and sphalerite, where the coordination numbers are low and the open structure provides sites that can accommodate interstitial atoms.

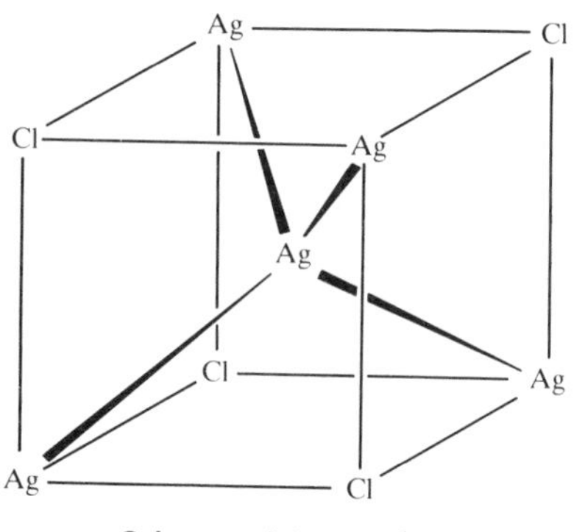

3 Interstitial Ag^+

Example 18.1: *Inferring the type of defect from density measurements*

Titanium monoxide has the rock-salt structure. X-ray diffraction data show that the length of one edge of the cubic unit cell for TiO with a 1:1 ratio of Ti to O is 4.18 Å, and the density determined by volume and mass measurements is 4.92 g cm^{-3}. Do the data indicate that defects are present? If so, are they vacancy or interstitial defects?

Answer. The presence of vacancies (Schottky defects) on the Ti and O sites should be reflected in a lower measured density than that calculated from the size of the unit cell and the assumption that every Ti and O site is occupied. Interstitial (Frenkel) defects would give little if any difference between the measured and theoretical densities. There are four TiO formula units per unit cell (Fig. 4.11). The mass of one formula unit is 63.88 u, so the corresponding theoretical mass in one unit cell is 4×63.88 u $= 255.52$ u, which corresponds to

4.24×10^{-22} g. The corresponding theoretical density is

$$\rho = \frac{4.24 \times 10^{-22}\ \text{g}}{(4.18 \times 10^{-8}\ \text{cm})^3} = 5.81\ \text{g cm}^{-3}$$

which is significantly greater than the measured density. Therefore, the crystal must contain numerous vacancies. Since the overall composition of the solid is TiO, there must be equal numbers of vacancies on the cation and anion sites.

Exercise. The measured density of VO (1:1 stoichiometry) is 5.92 g cm^{-3} and the theoretical density is 6.49 g cm^{-3}. Do the data indicate the presence of vacancies or interstitials?

Schottky and Frenkel defects are only two of the many possible types of defect. Another example is an **atom interchange defect**, which consists of an interchanged pair of M and X atoms or ions. This type of defect is common in metal alloys. For example, a copper–gold alloy of overall composition CuAu has extensive disorder at high temperatures and a significant fraction of Cu and Au atoms are interchanged (**4**).

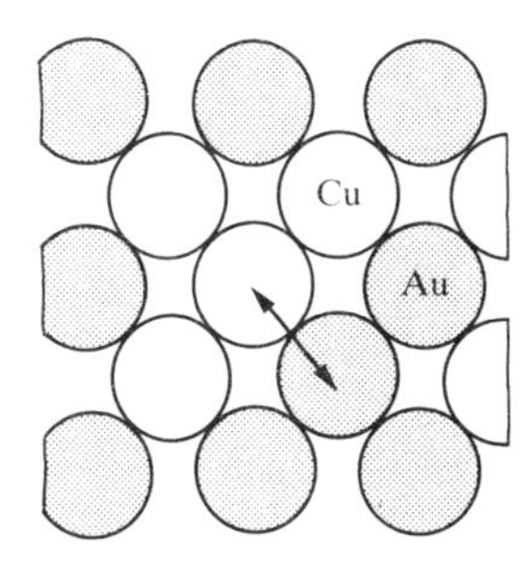

4 Atom interchange

Extrinsic point defects

Extrinsic defects are inevitable because perfect purity is unattainable in crystals of any significant size. In Section 3.6 we discussed impurities that had been introduced intentionally, where a pure substance was deliberately doped with another, such as when As atoms are substituted for Si atoms in a single crystal of silicon in order to modify its semiconducting properties. We can begin to see how the electronic structure of the solid influences the type of defect it is likely to form. Thus, when As replaces Si the added electron from each As atom finds its way into the conduction band. In the more ionic substance ZrO_2, the introduction of Ca^{2+} impurities in place of Zr^{4+} ions is accompanied by the formation of an O^{2-} ion vacancy to maintain charge neutrality (Fig. 18.2). In some cases a change in oxidation state (if one is possible) may be induced by an impurity. For example the introduction of Li_2O into NiO places Li^+ in Ni^{2+} sites, and the system balances charge by oxidation of two Ni^{2+} ions to Ni^{3+} for each Li^+ ion present. As with the doping of silicon by boron, this process introduces holes in the valence band of NiO and the conductivity increases greatly.

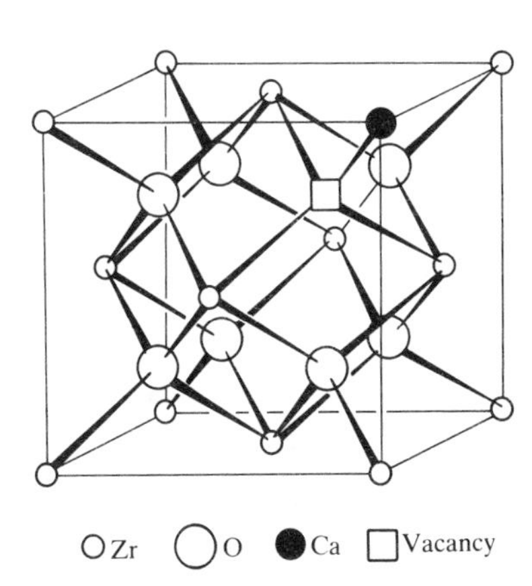

Fig. 18.2 Introduction of a calcium ion into the ZrO_2 lattice produces a vacancy on the oxide sublattice.

Another example of an extrinsic point defect is a **color center**, a generic term for defects responsible for modifications to the infrared, visible, and ultraviolet absorption characteristics of solids that have been irradiated or exposed to chemical treatment. One type of color center is produced by heating an alkali halide crystal in the vapor of the alkali metal. The process results in an alkali metal cation at a normal cation site, but the electron the atom brings into the crystal occupies a halide ion vacancy. A color center consisting of an electron in a halide ion vacancy (**5**), is called an **F-center** (from the German word for color center, *Farbenzenter*). The color results from the excitation of the electron in the localized environment of its surrounding ions, and its quantized energy levels resemble those of an electron in a spherical box (which are like those for particles in rectangular boxes, Section 1.3).

Na	Cl	Na	Cl	Na
Cl	Na	Cl	Na	Cl
Na	e^-	Na	Cl	Na
Cl	Na	Cl	Na	Cl
Na	Cl	Na	Cl	Na
Cl	Na	Cl	Na	Cl

5 F-center

Fig. 18.3 The concept of a crystallographic shear plane. (a) X represents a corner-shared MO_6 octahedron in the ReO_3 structure. The O atoms indicated by small circles are removed and adjacent slabs are translated to give (b) where edge-shared octahedra have been generated.

Extended defects

All the point defects discussed so far entail a significant local distortion of the lattice and in some instances localized charge imbalances too. Therefore it should not be surprising that defects may cluster together and sometimes form lines and planes.

Tungsten oxide illustrates the formation of a plane of defects. As illustrated in Fig. 18.3a, the idealized structure of WO_3 (which is usually referred to as the 'ReO_3 structure'; see below) consists of WO_6 octahedra sharing all vertices. To picture the formation of the defect plane we should imagine the removal of shared O atoms along a diagonal. Then adjacent slabs slip past each other in a motion that results in the completion of the vacant coordination sites around each W atom (Fig. 18.3b). This shearing motion creates edge-shared octahedra along a diagonal direction. The resulting structure was named a **crystallographic shear plane** by the Australian crystallographer A. D. Wadsley, who first devised this way of describing extended planar defects.

Crystallographic shear planes randomly distributed in the solid are called **Wadsley defects**. Such defects lead to a continuous range of compositions, as in tungsten oxide, which ranges from WO_3 to $WO_{2.93}$. If, however, the crystallographic shear planes are distributed in a nonrandom, periodic manner, we should regard the material as a new stoichiometric phase. Thus, when even more O^{2-} ions are removed from tungsten oxide, a series of discrete phases having ordered crystallographic shear planes and compositions W_nO_{3n-2} ($n = 20$, 24, 25, and 40) are observed. Closely spaced shear plane phases are known for oxides of tungsten, molybdenum, titanium, and vanadium. They can be identified by X-ray diffraction, but electron microscopy provides a more general and direct method (Fig. 18.4) because it reveals both ordered and random arrays of shear planes.

18.2 Nonstoichiometric compounds

A **nonstoichiometric compound** is a substance with variable composition but which retains essentially the same basic structure. For example, at 1000°C the composition of wüstite (which is nominally FeO) can vary from $Fe_{0.89}O$ to $Fe_{0.96}O$, and throughout this composition range the major peaks in the X-ray diffraction pattern indicate that the main features of the rock-salt structure are retained. There are gradual changes in the unit cell size as the composition is varied, and some added broad and weak X-ray reflections are observed, but the essential structure remains the same.

Some representative nonstoichiometric hydrides, oxides, and sulfides are listed in Table 18.1. Deviations from stoichiometry are common with *d*-, *f*-, and some *p*-block metals in combination with soft anions, such as S^{2-} and H^- ions, as well as with the harder O^{2-} anion. In contrast, hard anions such as fluorides, chlorides, sulfates, and nitrates form fewer nonstoichiometric compounds.

Nonstoichiometric compounds can be regarded as solid solutions in the sense that, as with liquid solutions, the chemical potentials of the components vary continuously as the composition changes. A sign of this continuous variation in chemical potential is that the partial pressure of O_2 above a nonstoichiometric oxide changes continuously (Fig. 18.5a) because the chemical potential of the oxygen in the gas (and hence its partial

Table 18.1. Representative composition ranges† of nonstoichiometric binary hydrides, oxides, and sulfides

d block		*f* block		
Hydrides				
TiH_x	1 to 2		Fluorite type	Hexagonal
ZrH_x	1.5 to 1.6	GdH_x	1.8–2.3	2.85–3.0
HfH_x	1.7 to 1.8	ErH_x	1.95–2.31	2.82–3.0
NbH_x	0.64 to 1.0	LuH_x	1.85–2.23	1.74–3.0
Oxides				
	Rock-salt type	Rutile type		
TiO_x	0.7–1.25	1.9–2.0		
VO_x	0.9–1.20	1.8–2.0		
NbO_x	0.9–1.04			
Sulfides				
ZrS_x	0.9–1.0			
YS_x	0.9–1.0			

† Expressed as the range of values that x may take.

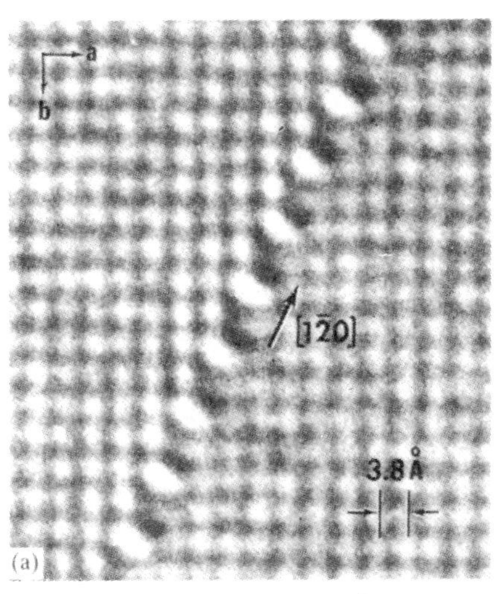

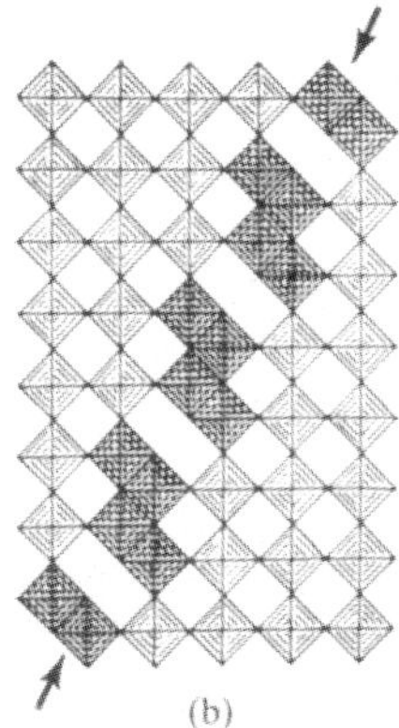

Fig. 18.4 (a) High-resolution electron micrograph lattice image of a crystallographic shear plane (120) in WO_{3-x}. (b) Drawing of the oxygen octahedral polyhedron that surround the tungsten atoms imaged in the electron micrograph. Note the edge-sharing octahedra along the crystallograpic shear plane. (Reproduced by permission from S. Iijima, *J. Solid State Chem.*, **14**, 52 (1975).)

pressure) follows the continuously variable chemical potential of the solid. The variation of pressure with composition agrees with the phase rule (see *Further information* at the end of this chapter) because at constant temperature there are two phases (gas and solid) and two components; hence the variance of the system is $F = 2 - 2 + 1 = 1$, and there is only one degree of freedom, the pressure. If the metal forms only stoichiometric compounds, such as MO_2 and MO, there is an additional phase. Now the variance is $F = 2 - 3 + 1 = 0$, and we expect an invariant, constant partial pressure of O_2 as one solid phase is converted into the other (Fig. 18.5b).

It may be very difficult to distinguish a nonstoichiometric compound from a series of discrete phases if the Gibbs free energies of the individual phases are similar. For example, the partial pressure of oxygen over a series of closely spaced phases may appear to vary continuously rather than in steps (Fig. 18.6), and the presence of separate phases would not be apparent. Discrete stoichiometric phases can also be detected by X-ray diffraction and often by optical or electron microscopy.

18.3 Atom and ion diffusion

A part of the reason why diffusion in solids is much less familiar than diffusion in gases and liquids is that at room temperature it is generally very much slower. However, there are some striking exceptions to this generalization as we shall soon see. Diffusion in solids is in fact very important in many areas of solid state technology, such as semiconductor manufacture, the synthesis of new solids, metallurgy, and heterogeneous catalysis.

General principles of diffusion

The diffusion of particles through any medium is governed by the **diffusion equation**

$$\frac{\partial c}{\partial t} = D\frac{\partial^2 c}{\partial x^2} \qquad (2)$$

where c is the concentration of the diffusing species and D is the **diffusion**

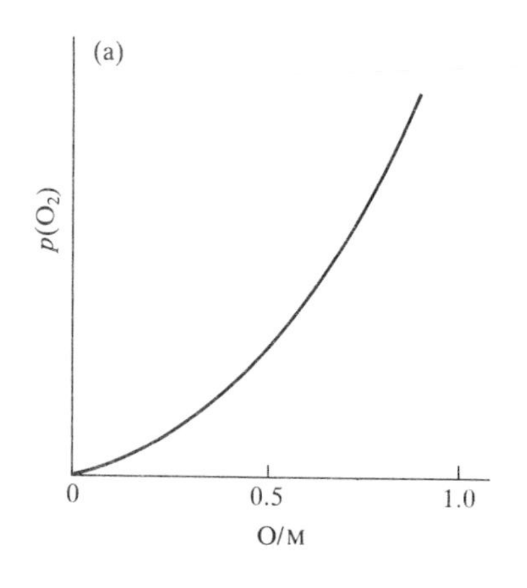

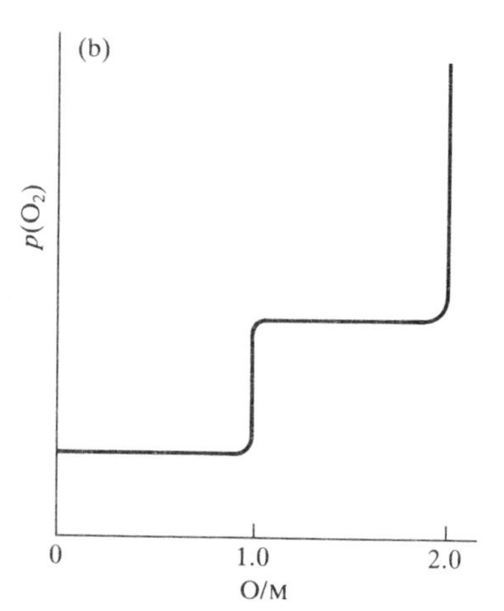

Fig. 18.5 Schematic representation of the variation of the partial pressure of oxygen with composition at constant pressure for (a) a nonstoichiometric oxide; (b) a stoichiometric pair of metal oxides MO and MO_2.

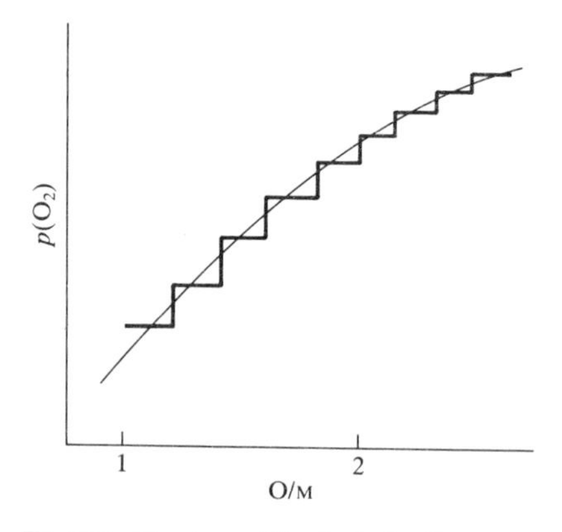

Fig. 18.6 The stepped line indicates the variation in pressure of O_2 in equilibrium with a series of closely spaced discrete phases at constant temperature. In the horizontal regions, two solid phases are present. The smooth curve indicates the observed pressure when the reaction is so slow that equilibrium is not established.

coefficient. The diffusion equation summarizes the fact that the rate of loss of particles in a region is proportional to the spatial inhomogeneity of the concentration there (Fig. 18.7). Where there are more particles than average (in the sense that $\partial^2 c/\partial x^2 < 0$), the particles will tend to disperse and lower the concentration locally ($\partial c/\partial t < 0$). If there is a dip in the concentration locally (and $\partial^2 c/\partial x^2 > 0$), particles will tend to diffuse into the region ($\partial c/\partial t > 0$).[1]

The rate of approach to a uniform distribution of particles is high when D is large. In a solid, D may depend more strongly than in a liquid on the concentration of particles, since the presence of foreign particles affects the local structure of the solid and has a marked effect on the ability of atoms to migrate. The diffusion coefficient increases with temperature, since particle migration is an activated process, and we can write the Arrhenius-like expression

$$D = D_0 e^{-E_a/RT} \tag{3}$$

where E_a is the activation energy for diffusion. Figure 18.8 shows the temperature dependence of the diffusion coefficients for some representative solids at high temperature. According to eqn 3, the slopes of the lines are proportional to the activation energy for atom or ion transport. Thus Na^+ is highly mobile and has a low activation energy for motion in β-alumina, whereas Ca^{2+} in CaO is much less mobile and has a much higher activation energy for diffusion.

Mechanisms of diffusion

The diffusion coefficients for ions can often be understood in terms of the mechanism for their migration and activation barriers the ions encounter as they move. In particular, diffusion coefficients in solids are markedly dependent on the presence of defects. The role defects play is summarized in Fig. 18.9, which shows some commonly postulated mechanisms for atom or ion motion in solids.

As with the mechanisms of chemical reactions, the evidence for the operation of a particular diffusion process is circumstantial. The individual events are never directly observed but are inferred from the influence of experimental conditions on atom or ion diffusion rates. In addition, a detailed analysis of the thermal motion of ions in crystals based on X-ray and neutron diffraction provides strong hints about the ability of ions to move through the crystal, including their most likely paths. Theoretical models, which are often elaborations of the ionic model (Section 4.7), give very useful guidance to the feasibility of these migration mechanisms.

Solid electrolytes

Any electrochemical cell such as a battery, fuel cell, electrochromic display, or electrochemical sensor, requires an electrolyte. Recently there has been considerable interest in the use of solid electrolytes in these applications, and a commercial device that utilizes a solid oxide-ion conductor to measure

[1] The diffusion equation can be regarded as a mathematical form of the remark that 'Nature abhors a wrinkle' (see P. W. Atkins, *Physical chemistry* (4th edn). Oxford University Press and W. H. Freeman and Co. (1990)). Peaks in distributions tend to disperse, and troughs tend to fill.

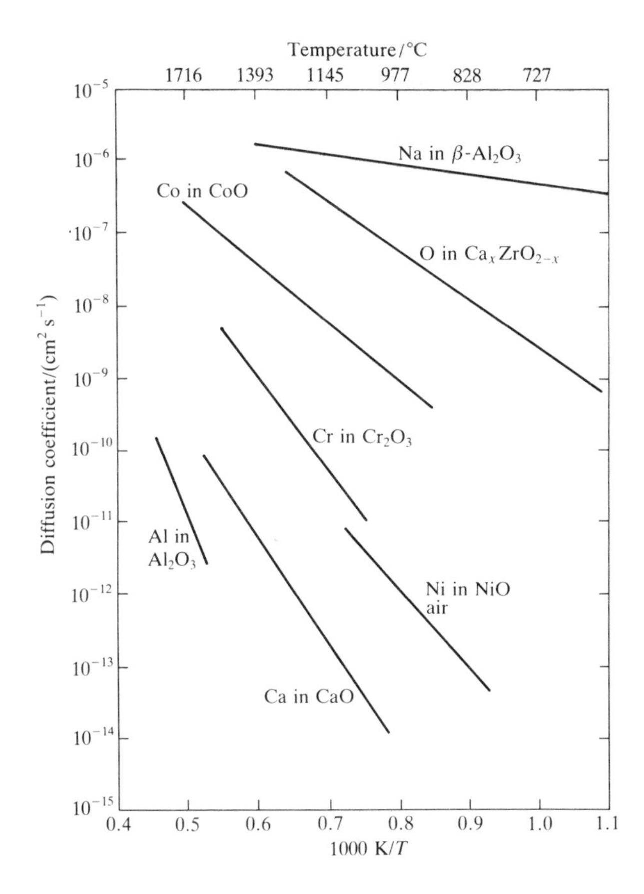

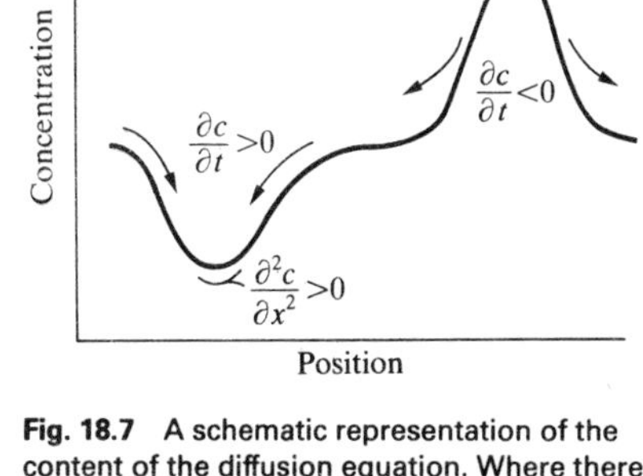

Fig. 18.7 A schematic representation of the content of the diffusion equation. Where there is a trough in the concentration, particles tend to accumulate to remove it; where there is a peak, particles tend to spread away.

Fig. 18.8 Arrhenius plot of the logarithm of the diffusion coefficient of the mobile ion in a series of solids versus $1/T$.

O_2 pressure is described in Section 18.5. Silver tetraiodomercurate(II), Ag_2HgI_4, and β-alumina, a sodium aluminate of composition $Na_{1+x}Al_{11}O_{17+x/2}$, provide examples of two quite different types of solids in which cations are highly mobile. The former is a soft material with a lattice of low rigidity and the latter is a hard refractory solid.

Below 50°C, Ag_2HgI_4 has an ordered crystal structure in which Ag^+ and Hg^{2+} ions are tetrahedrally coordinated by I^- ions (Fig. 18.10a) and there are unoccupied tetrahedral holes. At this temperature its electrical conductivity is low. Above 50°C, the Ag^+ and Hg^{2+} cations of Ag_2HgI_4 are distributed in a disorderly manner over all the tetrahedral sites (Fig. 18.10b). At this temperature the material is a good electrical conductor largely on account of the mobility of the Ag^+ ions. The weak and easily deformed close-packed array of I^- ions gives rise to a low activation energy for ion migration from one lattice site to the next. This example illustrates the general observation that solid inorganic electrolytes have a low-temperature phase in which the ions are ordered on a subset of lattice sites and that at higher temperatures the ions become disordered over the sites and the ionic conductivity increases. There are many similar solid electrolytes having soft lattices, such as AgI and $RbAg_4I_5$, both of which are Ag^+ conductors. The mobility of Ag^+ ions is by no means low, for the conductivity of $RbAg_4I_5$ at room temperature is greater than that of an aqueous sodium chloride solution of moderate concentration.

Sodium β-alumina is an example of a mechanically hard material that is a good ionic conductor. In this case the rigid and dense Al_2O_3 slabs are

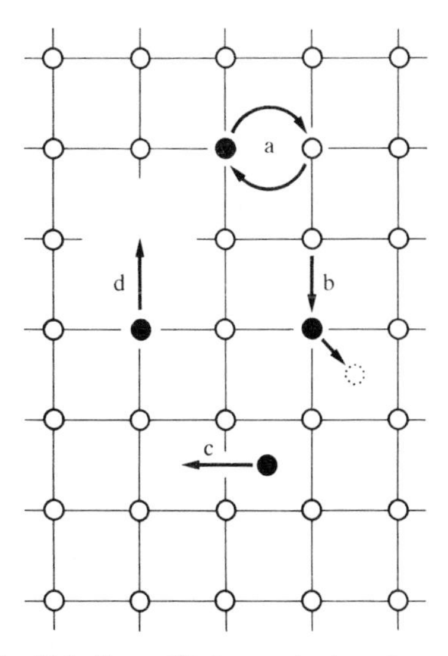

Fig. 18.9 Some diffusion mechanisms for ions or atoms in a solid: (a) exchange, (b) interstitialcy, (c) interstitial, and (d) vacancy.

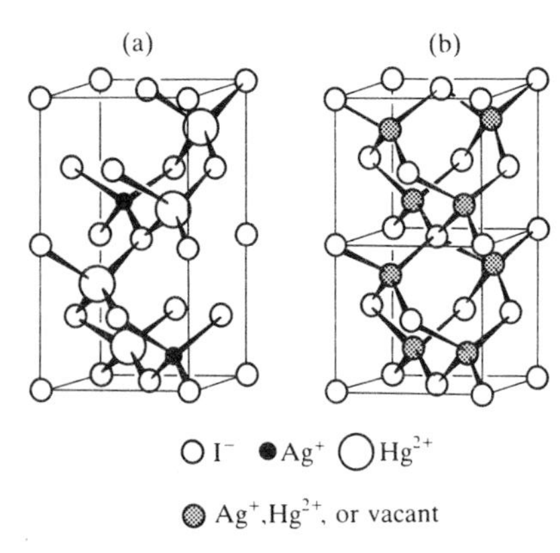

Fig. 18.10 (a) Low-temperature ordered structure of Ag_2HgI_4. (b) High-temperature disordered structure. Ag_2HgI_4 is an Ag^+ ion conductor in the high-temperature form.

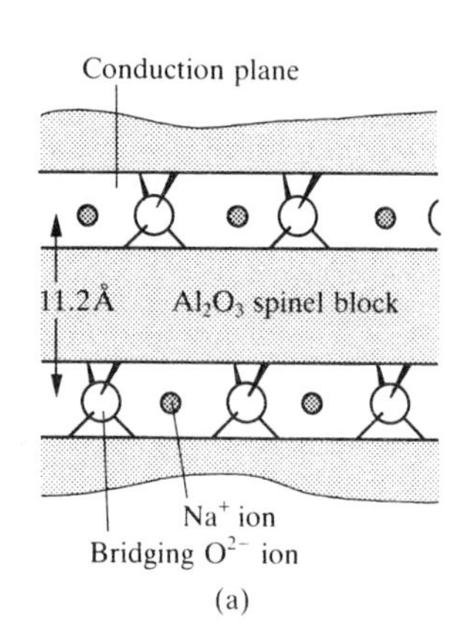

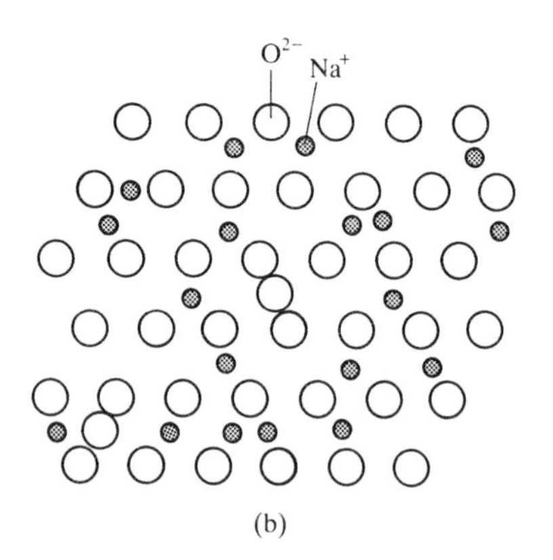

Fig. 18.11 (a) Schematic side view of β-alumina showing the Na_2O conduction planes between Al_2O_3 slabs. The O atoms in this plane bridge the two slabs. (b) A view of the conduction plane. Note the abundance of mobile ions and vacancies in which they can move.

bridged by a sparse array of O^{2-} ions (Fig. 18.11). The plane containing these bridging ions also contains Na^+ ions which can move from site to site because there are no major bottlenecks to hinder their motion. Many similar rigid materials having planes or channels through which ions can move are known, and are called **framework electrolytes**. Another closely related material, sodium β''-alumina, has even less restricted motion of ions than in β-alumina and it has been found possible to substitute dipositive cations such as Mg^{2+} or Ni^{2+} for Na^+. Even the large lanthanide cation Eu^{2+} can be introduced into β'', although the diffusion of such ions is slower than their smaller counterparts. These examples illustrate a fairly common strategy for the formation of new solids by diffusion processes at much lower than the melting point of the host material. Such **low-temperature syntheses** provide routes to a wide variety of solids that are not thermodynamically stable and therefore cannot be prepared from the melt. The rare earth β''-aluminas prepared in this manner are of interest for possible applicaions in lasers.

Example 18.2: *Correlating conductivity and ion size in a framework electrolyte*

Conductivity data on β-alumina containing monopositive ions of various radii show that Ag^+ and Na^+ ions, both of which have $r \approx 1.0$ Å, have activation energies for conductivity close to 17 kJ mol^{-1} whereas that for Tl^+ ($r \approx 1.4$ Å) is about 35 kJ mol^{-1}. Suggest an explanation of the difference.

Answer. In sodium β-alumina and related β-aluminas a fairly rigid framework provides a two-dimensional network of passages that permit ion migration. Judging from the experimental results, the bottlenecks for ion motion appear to be large enough to allow Na^+ or Ag^+ to pass quite readily (with a low activation energy) but too small to let the larger Tl^+ pass through as readily.

Exercise. Why does increased pressure reduce the conductivity of K^+ more than Na^+ in β-alumina?

The common characteristics of solid electrolytes are high concentrations of mobile ions and vacancies and the existence of pathways between the vacancies that provide routes for ion motion with low activation energy. In both Ag_2HgI_4 and sodium β-alumina there are high concentrations of carrier ions (Ag^+ and Na^+ respectively) as well as of vacant sites into which these carriers may diffuse.

Prototypical oxides and fluorides

In this section we explore compounds of the hard oxide and fluoride anions with metals. These compounds provide chemical insight into defects, nonstoichiometry, and ion diffusion, and into the influence of these characteristics on physical properties. This section also describes the structures of solids that are of interest in the synthesis of magnetic and superconducting materials.

18.4 Monoxides of the 3*d* metals

The monoxides of most of the 3*d* metals have the rock-salt structure, so it might at first appear that there would be little more to say about them and

Table 18.2. Monoxides of the Period 4 metals

Compound	**Structure**	**Stoichiometry MO_x**	**Electrical properties**
CaO_x	Rock-salt	1	Insulator
TiO_x	Rock-salt	0.65–1.25	Metallic
VO_x	Rock-salt	0.79–1.29	Metallic
MnO_x	Rock-salt	1–1.15	Semiconductor
FeO_x	Rock-salt	1.04–1.17	Semiconductor
CoO_x	Rock-salt	1–1.01	Semiconductor
NiO_x	Rock-salt	1–1.001	Semiconductor
CuO_x	PtS	1	Semiconductor
ZnO_x	Wurtzite	slight Zn excess	Wide gap semi-conductor

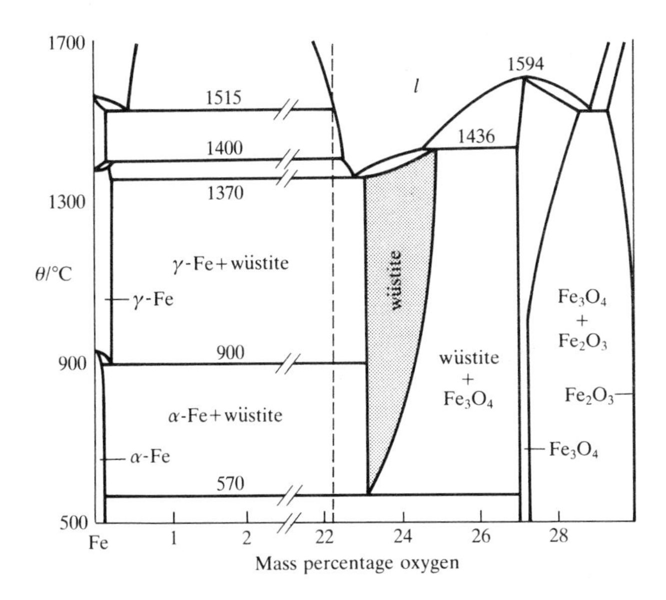

Fig. 18.12 A part of the iron–oxygen phase diagram. The vertical dashed line corresponds to the stoichiometry FeO.

that their properties should be simple. In fact, the structures and properties of these oxides (Table 18.2) are so interesting that they have been repeatedly investigated. In particular, the compounds provide examples of how mixed oxidation states and defects lead to nonstoichiometry, for TiO, VO, FeO, CoO, and NiO can all be prepared with significant deviations from the stoichiometry MO. Their range of chemical character, from mild to aggressive reducing agents, is reflected in the various ways in which they are prepared.

We can see from the phase diagram for iron and oxygen (Fig. 18.12) that the preparation of FeO must be carried out above 570°C.[2] The partial pressure of oxygen must also be carefully controlled. This control is achieved by using an atmosphere of CO and CO_2, for at the high temperature used for the preparation the equilibrium

$$2CO_2(g) \rightleftharpoons 2CO(g) + O_2(g) \qquad K_p = 1.4 \times 10^{-9} \text{ at } 830°C$$

is rapidly established, and the partial pressure of O_2 is governed by the

[2] Phase diagrams like that in Fig. 18.12 play an important role in the discussion of nonstoichiometric compounds. Their significance and interpretation are reviewed in *Further information* at the end of this chapter.

relative proportions of the two carbon oxides:

$$p(O_2)/\text{bar} = K_p \times \frac{p(CO)^2}{p(CO_2)^2}$$

Thus a mixture in which $p(CO_2)$ and $p(CO)$ are equal has an equilibrium partial pressure of O_2 of 1.4 nbar at 830°C. The two carbon oxides form a buffer to maintain the pressure at this constant value: CO_2 can release oxygen if the latter's partial pressure falls, and CO can absorb it if the partial pressure rises.

Defects and nonstoichiometry

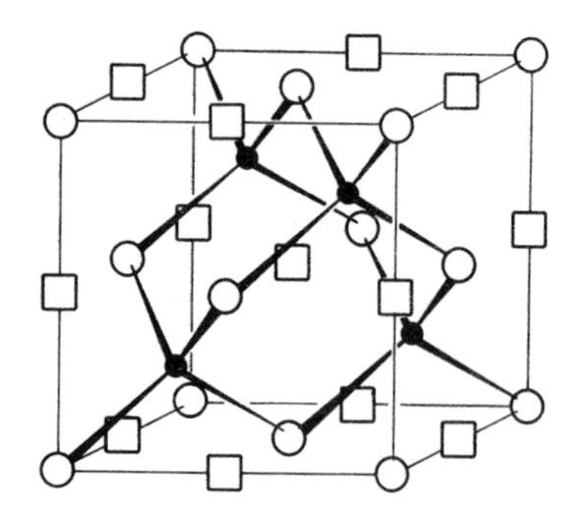

Fig. 18.13 Defect sites proposed for FeO. Note that the tetrahedral Fe^{3+} interstitials and octahedral Fe^{2+} vacancies have clustered together.

The origin of the nonstoichiometry in FeO has been studied in more detail than that in most other compounds. The consensus is that it arises from the creation of vacancies on the Fe^{2+} octahedral sites, and that each vacancy is charge-compensated by the conversion of two Fe^{2+} ions to two Fe^{3+} ions. The relative ease of oxidizing Fe(II) to Fe(III) accounts for the rather broad range of compositions of FeO_x. The Fe^{3+} ions reside in tetrahedral sites of the fcc oxide lattice but the symmetry of the rock-salt structure is not broken because the Fe^{3+} ions are not distributed in an orderly way. At high temperatures the interstitial Fe^{3+} ions associate with the Fe^{2+} vacancies to form clusters distributed throughout the structure (Fig. 18.13). A similar clustering of defects appears to occur with all other 3*d* monoxides, with the possible exception of NiO. It is more difficult to study NiO because its range of nonstoichiometry is extremely narrow, but conductivity and the rate of atom diffusion vary with O_2 partial pressure in a manner that suggests the presence of isolated point defects.

Both CoO and NiO occur in a metal-deficient state, although their range of stoichiometry is not as broad as that of FeO. As indicated by standard potentials in aqueous solution (Section 8.12), Fe(II) is more easily oxidized to Fe(III) than is either Co^{2+} or Ni^{2+}, and this solution redox chemistry correlates well with the much smaller range of oxygen deficiency in NiO and CoO. Chromium oxide does not exist because it is unstable with respect to disproportionation:

$$3CrO(s) \rightleftharpoons Cr_2O_3(s) + Cr(s)$$

However, it is possible to stabilize CrO as a solid solution in CuO. Significant deviation from the stoichiometry MO is out of the question for Group 2 metal oxides such as CaO because for these metals M^{3+} ions are chemically inaccessible.

Electrical conductivity

The 3*d* monoxides MnO, FeO, CoO, and NiO have low electrical conductivities that increase with temperature; hence they are semiconductors. The electron or hole migration in these oxide semiconductors is attributed to a hopping mechanism. In this model the electron or hole hops from one localized metal site to the next. When it lands on a new site it causes the surrounding ions to adjust their locations and the electron or hole is trapped temporarily in the potential well this atomic polarization produces. The electron resides at its new site until it is thermally activated to migrate into another nearby site. Another aspect of this charge hopping mechanism is that the electron or hole tends to associate with local defects so the activation energy for charge transport may also include the energy of freeing the hole from its position next to a defect. The conduction process in MnO, FeO, CoO, and probably NiO occur by such a hopping mechanism, but it is generally not clear whether site-to-site hopping or dissociation from a defect is rate limiting.

Hopping contrasts with the band model for semiconductivity discussed in Section 3.6, where the conduction and valence electrons occupy orbitals that spread through the whole crystal. The difference stems from the less diffuse *d* orbitals in the monoxides of the mid to late 3*d* metals, which are too compact to form the broad bands necessary for metallic conduction. We have already mentioned that when NiO is doped with Li_2O in an O_2 atmosphere a solid solution $Li_xNi^{2+}_{1-x}Ni^{3+}_xO$ is obtained, which has greatly increased conductivity for reasons similar to the increase in conductivity of silicon when doped with boron (Section 3.6). The characteristic exponential increase in electrical conductivity with increasing temperature of metal oxide semiconductors is used in 'thermistors' to measure temperature.

In contrast to the semiconductivity of the monoxides in the center and right of the 3*d* series, TiO and VO have high electric conductivities that decrease with increasing temperature. This metallic conductivity persists over a broad composition range from highly oxygen rich to metal rich. In these compounds a conduction band is formed by the overlap of t_{2g} orbitals of metal ions in neighboring octahedral sites that are oriented toward each other (Fig. 18.14). The radial extension of the *d* orbitals of these early *d*-block elements is greater than for elements later in the period, and a band results from their overlap (Fig. 18.15); this band is in general only partly filled. The widely varying compositions of these monoxides also appears to be associated with the electronic delocalization: the conduction band serves as an accessible source and sink of electrons that can readily compensate for the formation of vacancies.

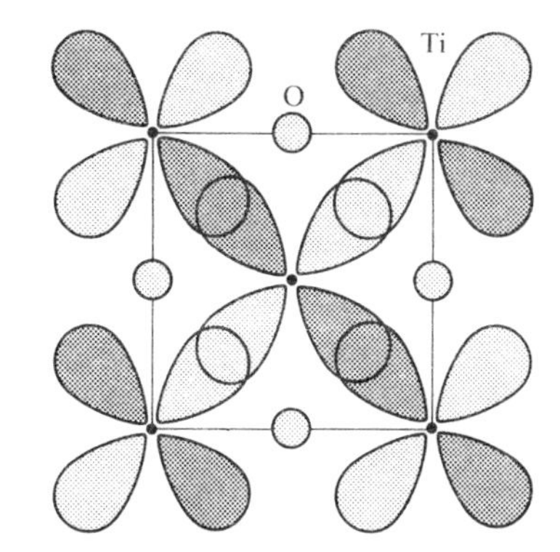

Fig. 18.14 Overlap of d_{xz} orbitals in TiO.

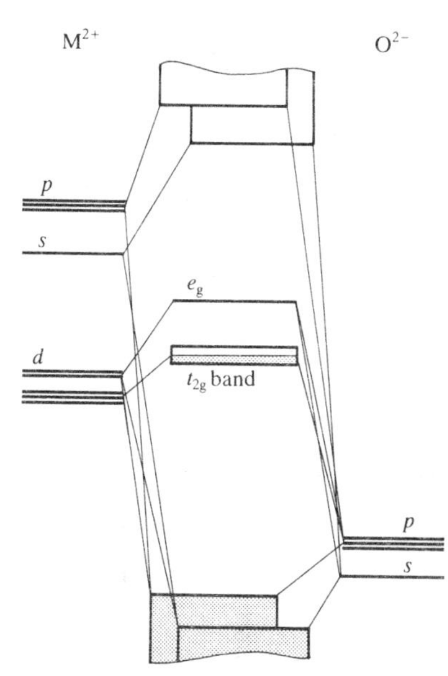

Fig. 18.15 Energy level diagram for early *d*-block metal monoxides. The t_{2g} band is only partly filled and metallic conductivity results.

18.5 Higher oxides

The simple *d*-block oxides Fe_3O_4, Co_3O_4, and Mn_3O_4 and many related mixed-metal compounds have interesting properties. They have structures related to spinel, $MgAl_2O_4$, and may be given the general formula AB_2O_4. The **spinel structure**, which is rather complex (Fig. 18.16), consists of an fcc array of O^{2-} ions in which the A ions reside in one-eighth of the tetrahedral holes and the B ions inhabit half the octahedral holes. It is sometimes useful to emphasize this arrangement of cations by enclosing the octahedral ions in square brackets, and writing $A[B_2]O_4$.

Some compounds with the composition AB_2O_4 possess an **inverse spinel structure**, in which the cation distribution is $M'[MM']O_4$. Lattice energy calculations based on a simple ionic model indicate that for M^{2+} and M'^{3+} the normal spinel structure, $M[M'_2]O_4$, should be the more stable. The observation that many *d*-metal spinels do not conform to this expectation has been traced to the effect of ligand field stabilization energies on the site preferences of the ions.

The **occupation factor** λ of a spinel is the fraction of B atoms in the octahedral sites. Thus $\lambda = 0$ for a normal spinel and 0.5 for an inverse spinel, $M'[MM']O_4$. The distribution of cations in M^{2+}, M^{3+} spinels (Table

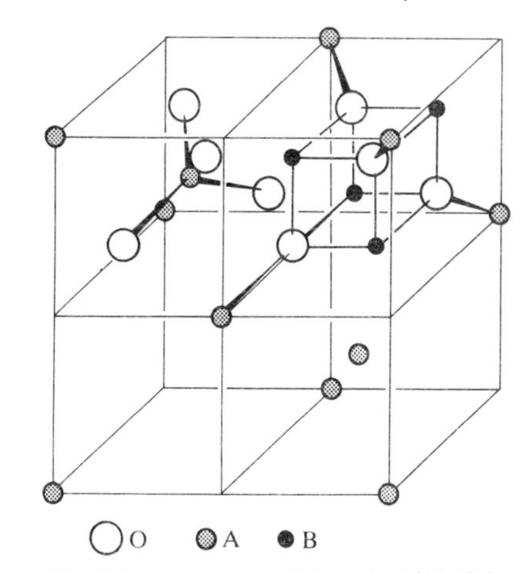

Fig. 18.16 A segment of the spinel (AB_2O_4) unit cell showing the tetrahedral environment of A ions and the octahedral environment of B ions.

Table 18.3. Cation site preference, λ, in some spinels

B \ A		Mg^{2+} d^0	Mn^{2+} d^5	Fe^{2+} d^6	Co^{2+} d^7	Ni^{2+} d^8	Cu^{2+} d^9	Zn^{2+} d^{10}
Al^{3+}	d^0	0	0	0	0	0.38		0
Cr^{3+}	d^3	0	0	0	0	0	0	0
Fe^{3+}	d^5	0.45	0.1	0.5	0.5	0.5	0.5	0
Mn^{3+}	d^4	0					0	
Co^{3+}	d^6				0			0

18.3) illustrates that for the d^0 Al^{3+} and Mg^{2+} ions the normal structure is preferred ($\lambda = 0$) as predicted by electrostatics. Table 18.3 shows that when the M^{2+} is a d^6, d^7, d^8, or d^9 ion and M'^{3+} is Fe^{3+}, the inverse structure is generally favored. This preference can be traced to the lack of ligand field stabilization of the high-spin d^5 Fe^{3+} ion in either the octahedral or the tetrahedral site and the ligand field stabilization of the other d^n ions in the octahedral site. It is important to note that simple ligand field stabilization appears to work over this limited range of cations, but more detailed analysis is necessary when cations of different radius are introduced. Moreover, λ is often found to depend on the temperature.

The inverse spinels of formula AFe_2O_4 are sometimes called **ferrites.** When $RT > J$, where J is the energy of interaction of the spins on different ions, ferrites are paramagnetic. However, when $RT < J$, a ferrite may be either **ferromagnetic** or **antiferromagnetic** (see Box 18.1). The antiparallel alignment of spins characteristic of antiferromagnetism is illustrated by $MgFe_2O_4$, where the cation distribution is nearly $Fe[MgFe]O_4$. In this compound the Fe^{3+} ions (with $S = \frac{5}{2}$) in the tetrahedral and octahedral sites are antiferromagnetically coupled below 9.5 K to give a low net magnetic moment to the solid as a whole.

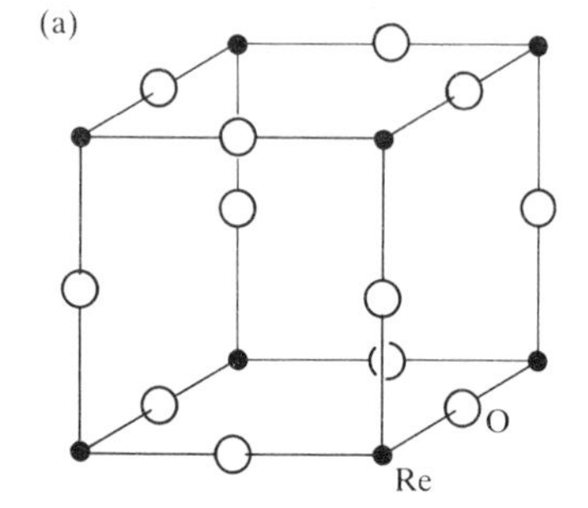

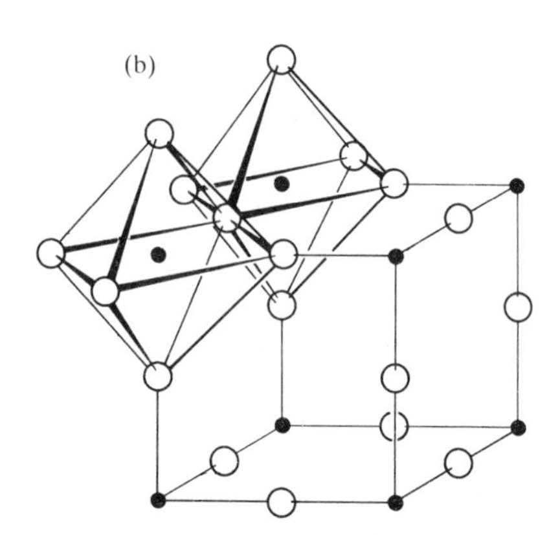

Fig. 18.17 (a) The ReO_3 unit cell; (b) The ReO_3 unit cell plus some oxygen atoms from adjacent unit cells to show the corner shared octahedra.

Example 18.3: *Predicting the magnetic properties of ferrites*

What is the likely magnetic moment per formula unit in the ferrite $MnFe_2O_4$?

Answer. For the inverse spinel structure of the ferrites we expect the Mn^{2+} ion to occupy an octahedral site and the Fe^{3+} ion to be distributed equally over the octahedral and tetrahedral sites. Antiferromagnetic coupling between the tetrahedral and octahedral sites would lead to cancellation of spins on the two Fe^{3+} sites and leave a remaining moment corresponding to five unpaired electrons per d^5 Mn^{2+} ion. Experimental evaluation of the magnetic moment verifies the presence of five spins per formula unit.

Exercise. How many spins per formula unit of $NiFe_2O_4$ are expected at low temperatures?

Perovskites and related phases

To introduce the structure of perovskites we first examine the related but simpler ReO_3 structure, which consists of corner-shared octahedra. Figure 18.17a shows that the unit cell of the ReO_3 structure contains M atoms at the corners of a cube and O atoms on the edges. The presence of the corner-shared octahedra is clarified by extending the structure to more than

Box 18.1 Cooperative magnetism

Diamagnetic and paramagnetic substances were covered in Section 7.5 and are characteristic of individual atoms or complexes. In contrast, properties such as ferromagnetism and antiferromagnetism depend on interactions between electron spins on many atoms and arise from the cooperative behavior of many unit cells in a crystal.

In a **ferromagnetic** substance the spins on different metal centers are coupled into a parallel alignment (Fig. 18B.1a) that is sustained over thousands of atoms in a magnetic **domain**. The net magnetic moment may be very large, since the magnetic moments of individual spins augment each other. Moreover, once established and the temperature maintained below the **Curie temperature** (T_C), the magnetization persists since the spins are locked together. Ferromagnetism is exhibited by materials containing unpaired electrons in *d* or *f* orbitals that couple with unpaired electrons in similar orbitals on surrounding atoms. The key feature is that this interaction is strong enough to align spins but not so strong as to form covalent bonds, where the electrons are paired.

The magnetization *M* of a ferromagnet is not linearly proportional to the applied field *H*. Furthermore, a hysteresis loop (Fig. 18B.2) is observed. For **hard ferromagnets** the loop is broad (Fig. 18B.2a) and *M* is large when the applied field has been reduced to zero. Hard ferromagnets are used for permanent magnets. A **soft ferromagnet** has a narrower hysteresis loop and is therefore much more responsive to a change in the direction of the applied field (Fig. 18.2b). Soft ferromagnets are used in transformers where they must respond to a rapidly oscillating field.

In an **antiferromagnetic** substance, neighboring spins are locked into an antiparallel alignment (Fig. 18B.1b) and the sample has a low magnetic moment. Antiferromagnetism is often observed when a paramagnetic material is cooled to a low temperature and is signaled by a decrease in magnetic susceptibility with decreasing temperature (Fig. 18B.3). The critical temperature for the onset of antiferromagnetism is called the **Néel temperature** (T_N). The spin coupling responsible for antiferromagnetism generally occurs through intervening ligands by a mechanism called **superexchange**. As indicated in Fig. 18B.4, the spin on one metal atom induces a small spin polarization on an occupied orbital of a ligand and this results in alignment of spin on adjacent metal atom that is antiparallel to the first. Many compounds exhibit antiferromagnetic behavior in the appropriate temperature range; for example MnO is antiferromagnetic below −151°C, and Cr_2O_3 below 37°C. Coupling of spins through

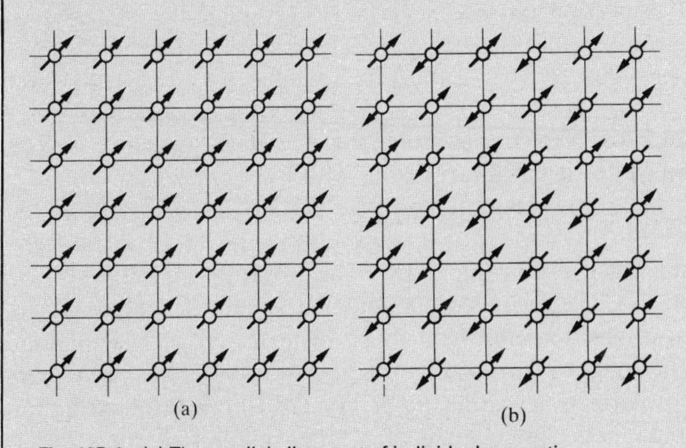

Fig. 18B.1 (a) The parallel alignment of individual magnetic moments in a ferromagnetic material and (b) the antiparallel arrangement of spins in an antiferromagnetic material.

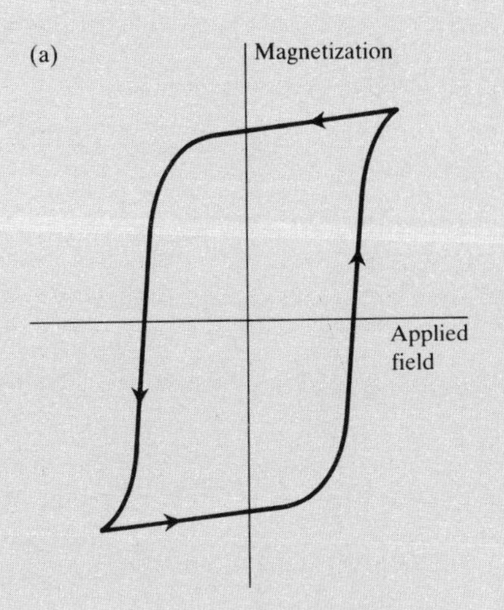

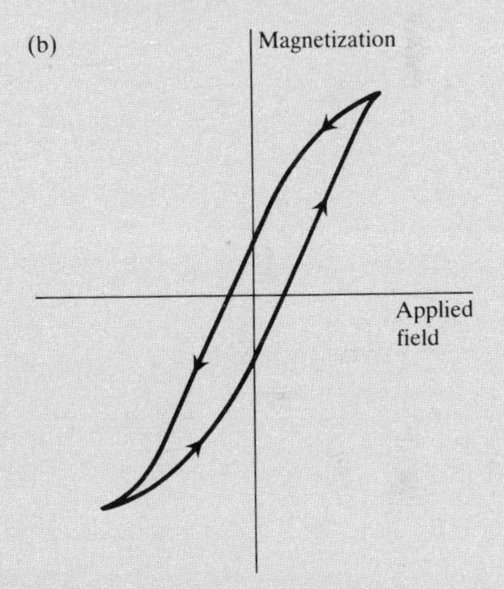

Fig. 18B.2 Magnetization curves for ferromagnetic materials. A hysteresis loop results because the magnetization of the sample with increasing field (→) is not retraced as the field is decreased (←). (a) A hard ferromagnet, (b) a soft ferromagnet.

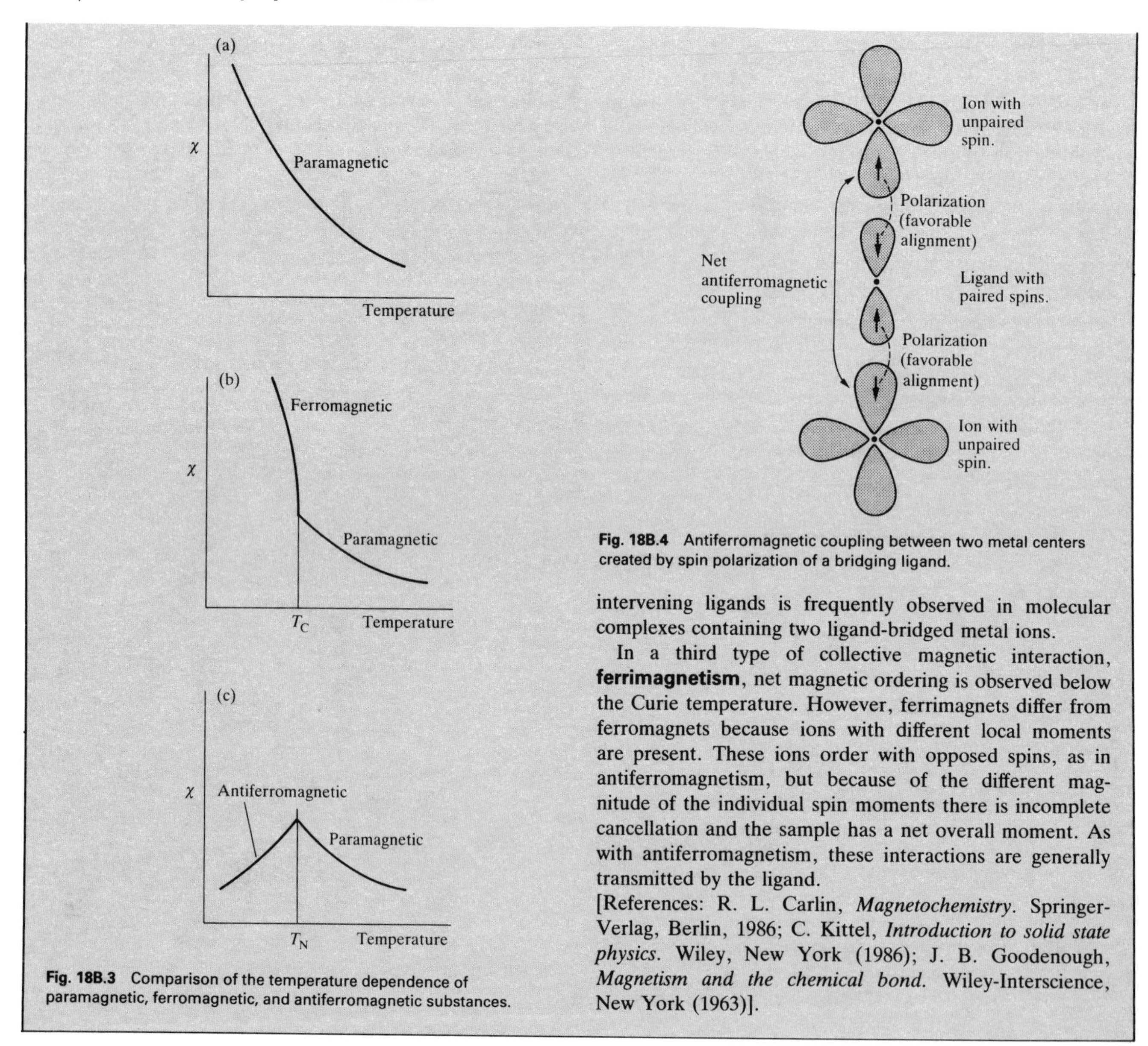

Fig. 18B.3 Comparison of the temperature dependence of paramagnetic, ferromagnetic, and antiferromagnetic substances.

Fig. 18B.4 Antiferromagnetic coupling between two metal centers created by spin polarization of a bridging ligand.

intervening ligands is frequently observed in molecular complexes containing two ligand-bridged metal ions.

In a third type of collective magnetic interaction, **ferrimagnetism**, net magnetic ordering is observed below the Curie temperature. However, ferrimagnets differ from ferromagnets because ions with different local moments are present. These ions order with opposed spins, as in antiferromagnetism, but because of the different magnitude of the individual spin moments there is incomplete cancellation and the sample has a net overall moment. As with antiferromagnetism, these interactions are generally transmitted by the ligand.

[References: R. L. Carlin, *Magnetochemistry*. Springer-Verlag, Berlin, 1986; C. Kittel, *Introduction to solid state physics*. Wiley, New York (1986); J. B. Goodenough, *Magnetism and the chemical bond*. Wiley-Interscience, New York (1963)].

one unit cell (Fig. 18.17b). Note in particular that the ReO_3 structure is very open and that at the center it has a large hole with C.N. = 12.

The perovskites have the general formula ABX_3, in which the 12-coordinate hole of ReO_3 is occupied by a large A ion (Fig. 18.18; a different view of this structure was given in Fig. 4.18). The X ion is generally O^{2-} or F^-; perovskite itself is $CaTiO_3$ and an example of a fluoride perovskite is $NaFeF_3$. The structure is often observed to be distorted in such a manner that the unit cell is no longer centrosymmetric and the crystal has an overall permanent electric polarization. Some polar crystals are **ferroelectric**, for they resemble ferromagnets, but instead of a large number of spins being aligned over a region it is the electric polarization of many unit cells that is aligned.

Another characteristic of many crystals that lack a center of symmetry is **piezoelectricity**, the ability to generate an electrical field when the crystal is under stress or to change dimensions when an electrical field is applied to the crystal. Piezoelectric materials are used for a variety of applications such as pressure transducers, ultramicromanipulators, and sound detectors. Some important examples are $BaTiO_3$, $NaNbO_3$, $NaTaO_3$, and $KTaO_3$, Although a noncentric structure is required for both ferroelectric and piezoelectric behavior, the two phenomena do not necessarily occur for the same crystal. For example, quartz is piezoelectric but not ferroelectric. It is widely used to set the clock rate of microprocessors and watches because a thin sliver oscillates at a specific frequency to produce a small oscillating electrical field.

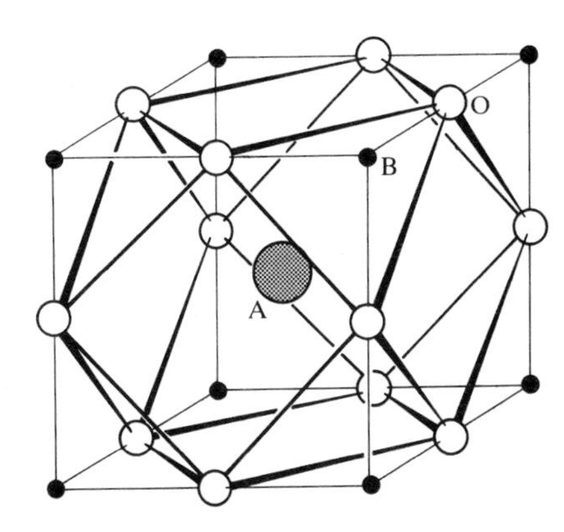

Fig. 18.18 A view of the perovskite (ABO_3) structure emphasizing the dodecahedral coordination of the large A ion. The octahedral environment of the B ion has been shown in Fig. 18.17.

Superconductors

The versatility of the perovskites extends to superconductivity, because some of the high-temperature superconductors first reported in 1986 can be viewed as variants of the perovskite structure. Superconductors have two striking characteristics. Below the critical temperature (T_c) at which the superconducting state is formed, they have zero electrical resistance. They also exhibit the **Meissner effect**, the exclusion of a magnetic field (Fig. 18.19).

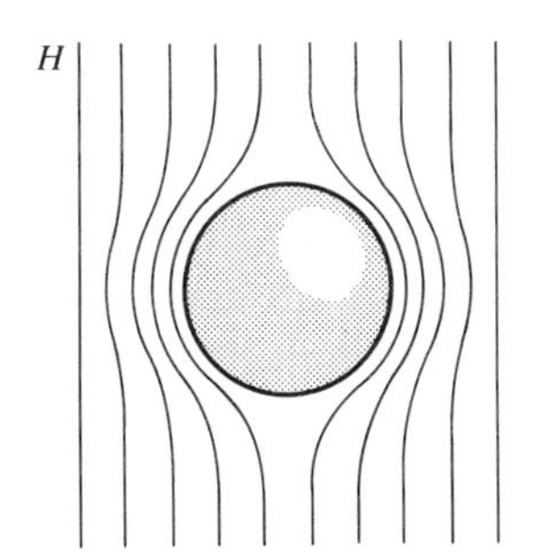

Fig. 18.19 The exclusion of a magnetic field by a superconducting material below T_c (the Meissner effect).

Following the discovery in 1911 that mercury becomes a superconductor below 4.2 K, physicists and chemists made slow but steady progress in the discovery of superconductors with higher values of T_c, at a rate of about 3 K per decade. After 75 years, T_c had been edged up to 23 K. Then, in 1986, **high-temperature superconductors** were discovered.[3] Several materials are now known with T_c well above 77 K, the boiling point of the inexpensive refrigerant liquid nitrogen, and in a few years T_c has increased by more than a factor of five.

Two types of superconductor are known. Those classed as **Type I** show abrupt loss of superconductivity when an applied magnetic field exceeds H_c, a value characteristic of the material. **Type II** superconductors, which include high-temperature materials, show a gradual loss of superconductivity above a field which varies from substance to the next. The Chevrel phases (Section 18.8) have the highest observed values of H_c. Figure 18.20 shows that there is some periodic regularity in the elements that exhibit superconductivity. Note in particular that the ferromagnetic metals iron,

[3] See A. Müller and J. G. Bednorz, The discovery of a class of high-temperature superconductors. *Science,* **217,** 1133 (1987).

Li	Be											B	C	N	O	F	Ne
Na	Mg											Al 1.140 (105)	Si*	P*	S	Cl	Ar
K	Ca	Sc	Ti 0.39 (100)	V 5.38 (1420)	Cr*	Mn	Fe	Co	Ni	Cu	Zn 0.875 (53)	Ga 1.091 (51)	Ge*	As*	Se*	Br	Kr
Rb	Sr	Y*	Zr 0.546 (47)	Nb 9.50 (1980)	Mo 0.92 (95)	Tc 7.77 (1410)	Ru 0.51 (70)	Rh	Pd	Ag	Cd 0.56 (30)	In 3.404 (293)	Sn 3.722 (309)	Sb*	Te*	I	Xe
Cs*	Ba*	Lu 0.1	Hf 0.12	Ta 4.48 (830)	W 0.012 (1.07)	Re 1.4 (198)	Os 0.66 (65)	Ir 0.14 (19)	Pt	Au	Hg 4.15 (412)	Tl 2.39 (171)	Pb 7.19 (803)	Bi*	Po	At	Rn
Fr	Ra	Lr															

La 6.00 (1100)	Ce*	Pr	Nd	Pm	Sm	Eu	Gd	Tb	Dy	Ho	Er	Tm	Yb
Ac	Th 1.37 1.62	Pa 1.4	U*	Np	Pu	Am	Cm	Bk	Cf	Es	Fm	Md	No

Fig. 18.20 Transition temperatures and critical magnetic field, $H_c/(10^{-4}\,\mathrm{T})$. Elements with asterisks have been observed to be superconducting in thin films or at high pressures. (Reproduced with permission from C. Kittel, *Solid state physics*, Wiley, New York (1986).)

cobalt, and nickel do not display superconductivity, nor do the alkali metals or the coinage metals copper, silver, and gold. For simple metals, ferromagnetism and superconductivity never coexist, but in some of the oxide superconductors ferromagnetism and superconductivity appear to coexist on different sublattices of the solid.

One of the most widely studied oxide superconductor materials $YBa_2Cu_3O_7$ (informally called '123', from the proportions of metal atoms in the compound) has a structure similar to perovskite but with missing O atoms (Fig. 18.21). In terms of the structure shown in Fig. 18.18, the A

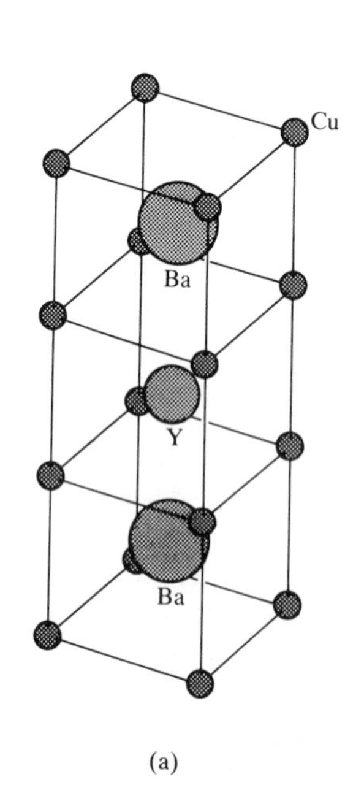

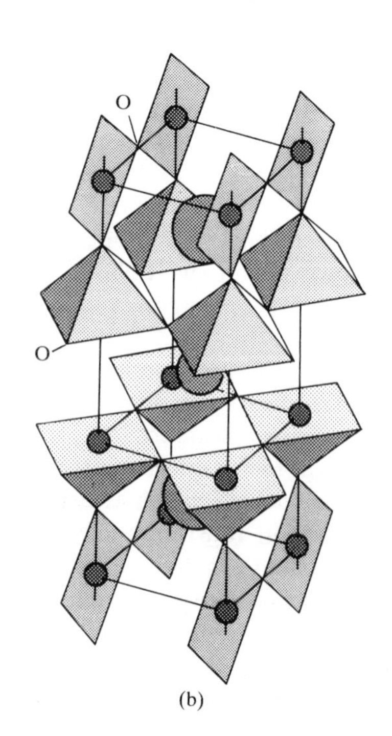

Fig. 18.21 Structure of the $YBa_2Cu_3O_7$('123') superconductor. (a) Metal atom positions. (b) Oxygen polyhedra around the metal atoms, unlike the octahedral environment in perovskite metal ions in 123 are in square-planar and square-pyramidal coordination environments rather than the octahedra in perovskite.

atoms of the original perovskite are yttrium and barium in $YBa_2Cu_3O_7$ and the B atoms are copper. The square planar CuO_4 units are arranged in chains. Sheets and chains of CuO_4 or CuO_5 units can also be detected in other high-temperature superconductors, and it is thought that they are an important component of the mechanism of superconduction. There is as yet no settled explanation of high-temperature superconductivity. It is believed that the Cooper pairs that are responsible for conventional superconductivity (Section 3.7) are important in the high-temperature materials but the mechanism for pairing is hotly debated.[4]

The perovskite-like structure of $YBa_2Cu_3O_7$ is shown in Fig. 18.21. Copper ions occupy the B sites and Y or Ba ions occupy the A sites of the layered lattice. However, unlike in a true perovskite structure, the B sites are not surrounded by an octahedron of O atoms. Some have five O atom neighbors and others have only four (Fig. 18.21). Similarly, Ba^{2+} and Y^{3+} in the A site have less than 12 coordination. Another point to note is that if we assign the usual oxidation numbers (Y + 3, Ba + 2, and O − 2), the average oxidation number of Cu turns out to be +2.33, so it is inferred that $YBa_2Cu_3O_7$ is a mixed oxidation state material that contains Cu^{2+} and Cu^{3+}.

Another prototypical structure, that of potassium tetrafluoronickelate(II) (K_2NiF_4, Fig. 18.22) is related to perovskite and has a bearing on the mechanism of high-temperature superconduction. The compound can be thought of as containing layers from the perovskite structure that share four F atom vertices within the layer and have terminal vertices above and below the layer. These layers are offset, so the terminal vertex in one layer is capped by a K atom in the next layer (Fig. 18.22). Compounds with the K_2NiF_4 structure have come under renewed investigation because some high-temperature superconductors, such as $La_{1.85}Sr_{0.15}CuO_4$, crystallize with this structure. Aside from their importance in superconductivity, compounds with the K_2NiF_4 structure are also of interest because they provide an opportunity to investigate two-dimensional domains of magnetic interactions.

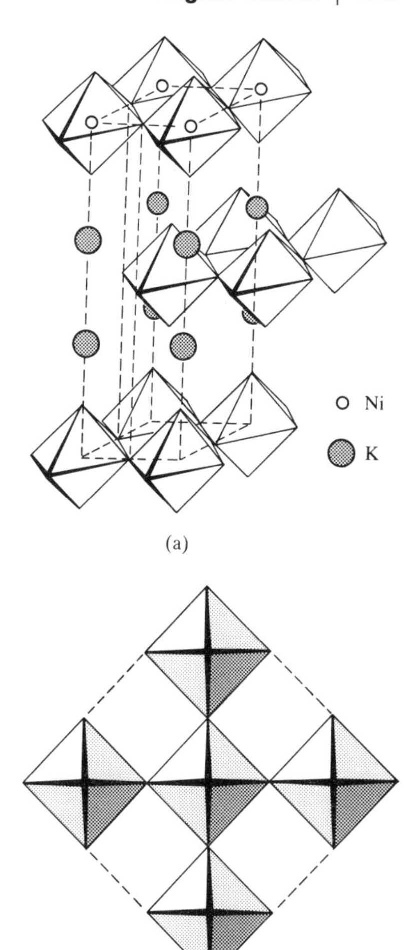

Fig. 18.22 The K_2NiF_4 structure. (a) The displaced layers of NiF_6 octahedra and (b) a view of the NiF_6 octahedra showing the corner sharing of F.

Table 18.4. Some superconductors

Elements	T_c/K	Compounds	T_c/K
Zn	0.88	Nb_3Ge	23.2
Cd	0.56	Nb_3Sn	18.0
Hg	4.15	$LiTiO_4$	13
Pb	7.19	$K_{0.4}Ba_{0.6}BiO_3$	29.8
Nb	9.50	$YBa_2Cu_3O_7$	95
		$Tl_2Ba_2Ca_2Cu_3O_{10}$	122

A few of the new superconductors are listed in Table 18.4 along with other superconducting materials. As illustrated in Fig. 18.23 for $BaPb_{1-x}Bi_xO_3$, superconductivity is often observed for materials that have compositions on the borderline between metallic conducting and semiconducting (or insulating) phases. A possible interpretation is that the types of

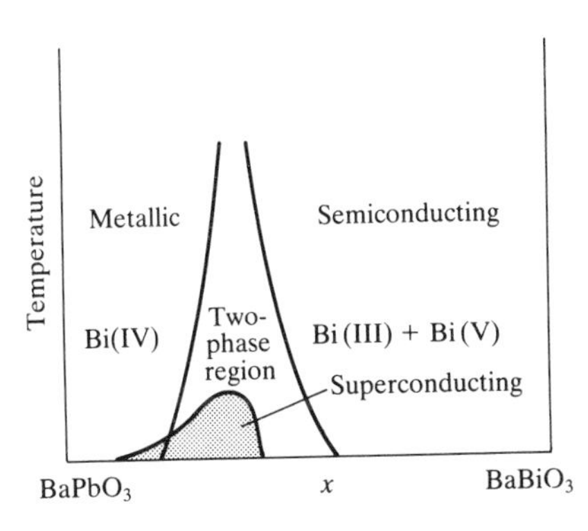

Fig. 18.23 Schematic correlation of phase instability and superconductivity.

[4] For an entertaining account of the people investigating superconductivity see: R. M. Hazen, *The breakthrough. The race for the superconductor*. Summit Books, New York (1988).

processes of instability that lead to changes in the more ordinary conduction process may at low temperatures give rise to superconductivity.

The synthesis of high-temperature superconductors has been guided by a variety of qualitative considerations, such as the demonstrated success of the layered structures and of mixed-oxidation-state copper in combination with heavy *p*-block elements. Additional considerations are the radii of ions and their preference for certain coordination geometries. Some of these materials are prepared simply by heating a mixture of the metal oxides to 800–900°C in an open alumina crucible or a sealed gold tube.

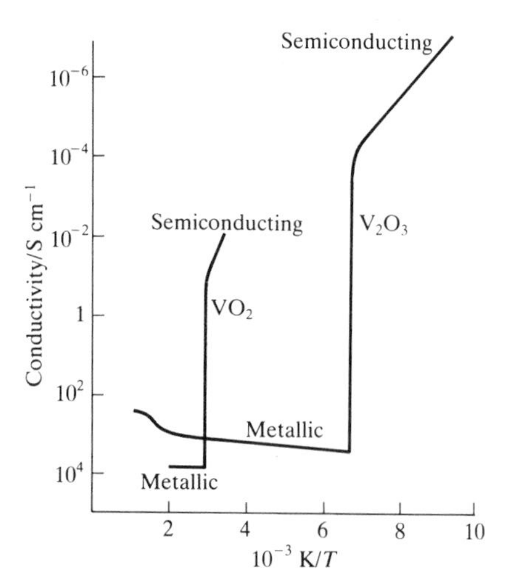

Fig. 18.24 Temperature dependence of the conductivity of VO_2 and V_2O_3 showing the metal-to-semiconductor transition.

M_2O_3 corundum structure

α-Alumina (the mineral corundum) adopts a structure that can be described as a hexagonal close packed array of O^{2-} ions with the cations in two-thirds of the octahedral holes. The details of the structure are complex and we need not go into them. The corundum structure is adopted by the oxides of titanium, vanadium, chromium, iron, and gallium in their +3 oxidation states. Two of these oxides, Ti_2O_3 and V_2O_3, exhibit metallic-to-semiconducting transitions below 410 and 150 K respectively (Fig. 18.24). In V_2O_3 the transition is accompanied by antiferromagnetic ordering of the spins. The two insulators Cr_2O_3 and Fe_2O_3 also display antiferromagnetic ordering.

Another interesting aspect of the M_2O_3 compounds is the formation of solid solutions of dark green Cr_2O_3 and colorless Al_2O_3 to form brilliant red ruby. As we remarked in Section 14.5, this shift in the ligand field transitions of Cr^{3+} stems from the compression of the O^{2-} ions around Cr^{3+} in the Al_2O_3 host lattice (in Al_2O_3 $a = 4.75$ Å and $c = 13.00$ Å; in Cr_2O_3 the lattice constants are 4.93 Å and 13.56 Å respectively). The compression shifts the absorption toward the blue and the solid appears red in white light. The responsiveness of the absorption (and fluorescence) spectrum of Cr^{3+} ions to compression is sometimes used to measure pressure in high-pressure spectroscopic experiments. In this application, a tiny crystal of ruby in one segment of the sample can be interrogated by visible light and the shift in its fluorescence spectrum provides an indication of the pressure inside the cell.

MO_2 and fluorite

The fluorite structure (Fig. 4.14) contains cations that are surrounded by eight anions. As with the cesium chloride structure, this high coordination number is favored by relatively large cations. The structure is found for fluorides such as PbF_2 and oxides such as ZrO_2.

The mobility of F^- in PbF_2 was observed by Michael Faraday, who reported in 1834 that the red-hot solid is a good conductor of electricity. The property of anion conductivity is shared by other crystals having the fluorite structure (Fig. 4.14), and the ion transport is thought to be by an interstitial mechanism in which an F^- ion first migrates from its normal position into an interstitial site and then moves to a vacant F^- site.

Certain oxides that have the fluorite structure conduct by O^{2-} ion migration. The best known of these is ZrO_2 with added CaO to stabilize the fluorite phase (which in its absence is stable only at high temperature). For every Ca^{2+} substituted for Zr^{4+}, a vacancy on an O^{2-} site is created to maintain electroneutrality, and the presence of these vacancies improves the O^{2-} mobility.

A CaO-doped ZrO_2 electrolyte is used in a solid state electrochemical sensor for the oxygen partial pressure in automobile exhaust systems (Fig. 18.25).[5] The platinum electrodes in this cell adsorb O atoms, and if the partial pressure of oxygen is different between the sample and reference side there is a thermodynamic tendency for oxygen to migrate through the electrolyte as the O^{2-} ion. The thermodynamically favored processes are:

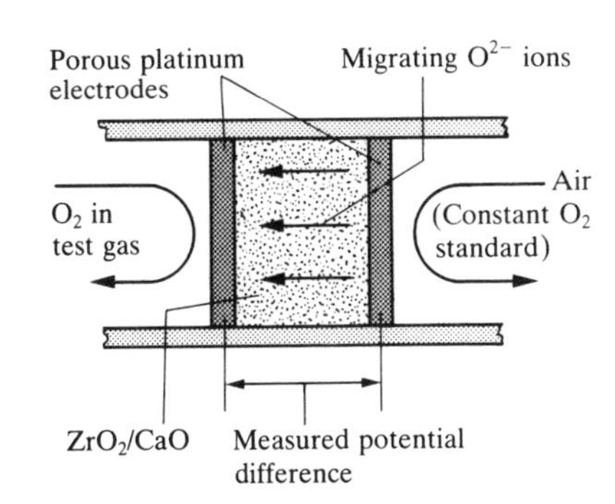

Fig. 18.25 An O_2 sensor based on the solid electrolyte ZrO_2–CaO.

$$\left.\begin{array}{l} O_2(g) + Pt(s) \rightarrow 2O(\text{Pt surface}) \\ O(\text{Pt surface}) + 2e^- \rightarrow O^{2-}(ZrO_2) \end{array}\right\} \text{high } p(O_2) \text{ side}$$

$$\left.\begin{array}{l} O^{2-}(ZrO_2) \rightarrow O(\text{Pt surface}) + 2e^- \\ 2O(\text{Pt surface}) \rightarrow O_2(g) + Pt(s) \end{array}\right\} \text{low } p(O_2) \text{ side}$$

The cell potential is related to the oxygen partial pressures through the Nernst equation (Section 8.3), so a simple measurement of the potential provides a measure of the O_2 partial pressure in the exhaust.

18.6 Glasses

The term **ceramic** is often applied to all inorganic nonmetallic, nonmolecular materials, including both amorphous and crystalline materials. The term **glass** is used in a variety of contexts, but for present purposes it implies an amorphous ceramic with a viscosity so high that it can be considered rigid. A substance in its glassy form is said to be in its **vitreous state**. Although ceramics and glasses have been utilized since antiquity, their development is currently an area of rapid scientific and technological progress. This enthusiasm stems from interest in the scientific basis of their properties and the development of novel synthetic routes to new high performance materials. The most familiar glasses are alkali metal or alkaline earth metal silicates or borosilicates.

A glass is prepared by cooling the melt more quickly than its rate of crystallization. Thus, cooling molten silica gives vitreous quartz. Under these conditions the solid mass has no long-range periodicity as judged by the lack of X-ray diffraction peaks, but spectroscopic and other data indicate that each Si atom is surrounded by a tetrahedral array of O atoms. The lack of long-range order results from variations of the Si—O—Si angles. Figure 18.26 illustrates in two dimensions how a local coordination environment can be preserved but long-range order lost by variation of the bond angles around oxygen. Silicon dioxide readily forms a glass because the three-dimensional network of strong covalent Si—O bonds in the melt does not readily break and reform upon cooling. The lack of strong directional bonds in metals and simple ionic substances makes it much more difficult to form glasses from these materials. Recently, however, techniques have been developed for ultrafast cooling and as a result a wide variety of metals and simple inorganic materials can now be frozen into a vitreous state.

The concept that the local coordination sphere of the glass-forming element is preserved but that bond angles around O are variable was

[5] The signal from this sensor is used to adjust the ratio of exhaust gas to air being fed to the catalytic converter (Fig. 17.9).

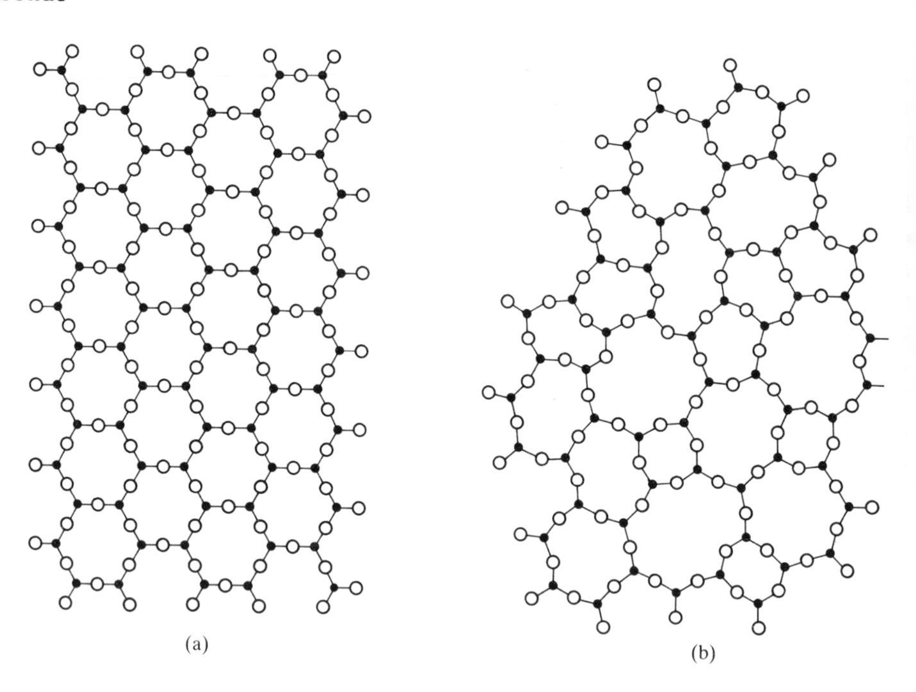

Fig. 18.26 Schematic representation of (a) two-dimensional crystal and (b) two-dimensional glass. (Reproduced with permission from W. H. Zachariasen, *J. Amer. chem. Soc.*, **54**, 3841, (1932).)

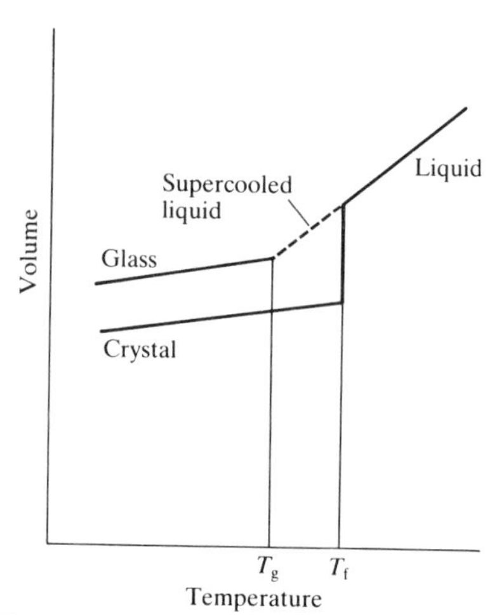

Fig. 18.27 Comparison of the volume change for supercooled liquids and glasses with that for a crystalline material.

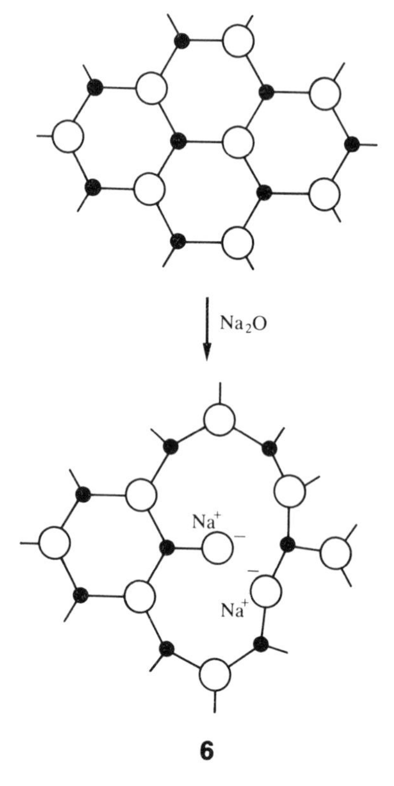

originally proposed by the American crystallographer W. H. Zachariasen in 1932. He reasoned that these conditions would lead to similar free energies and molar volumes for the glass and its crystalline counterpart. Zachariasen also proposed that the vitreous state is favored by corner-shared O atoms, rather than edge- or face-shared, which would enforce greater order. These and other **Zachariasen rules** hold for common glass-forming oxides, but exceptions are known.

An instructive comparison between vitreous and crystalline materials is seen in their change in volume with temperature (Fig. 18.27). When a molten material crystallizes, at T_f, an abrupt change in volume (usually a decrease) occurs. In contrast, a glass-forming material which is cooled sufficiently rapidly persists in the liquid state to form a metastable supercooled liquid. When cooled below the **glass transition temperature**, T_g, the supercooled liquid becomes rigid, and this change is accompanied by only an inflection in the cooling curve.

Although vitreous silica is a strong glass that can withstand rapid cooling or heating without cracking, it has a high glass transition temperature and therefore must be worked at inconveniently high temperatures. Therefore a **modifier**, such as Na_2O or CaO, is generally added to SiO_2. These modifiers disrupt some of the Si—O—Si linkages and replace them by terminal Si—O^- links that associate with the cation (**6**). The consequent partial disruption of the Si—O network leads to glasses that have lower softening points. The common glass used in bottles and windows is called 'sodalime glass' and contains NaO and CaO as modifiers. When B_2O_3 is used as a modifier, the resulting 'borosilicate glasses' have lower thermal expansion coefficients than sodalime glass and thus are less likely to crack when heated. Borosilicate glass is therefore widely used for ovenware and laboratory glassware.

Glass formation is a property of many oxides, and practical glasses have been made from sulfides, fluorides, and other anionic constituents. Some of the best glass formers are the oxides of elements near silicon in the periodic table (B_2O_3, GeO_2, and P_2O_5), but the solubility of most borate and phosphate glasses and the high cost of germanium limit their usefulness.

A current technological revolution is the replacement of electrical signals for voice and data transmission by optical signals. Similarly, optical circuit elements are being developed that may eventually replace electrical integrated circuits. These developments are stimulating considerable research in transparent crystalline and vitreous materials for light transmission and processing. For example, optical fibers for light transmission are currently being produced with a composition gradient from the interior to the surface. This composition gradient modifies the index of refraction and thereby decreases light loss. Fluoride glasses are also being investigated as possible substitutes for oxide glasses, because oxide glasses contain small numbers of OH groups, which absorb near-infrared radiation.

Prototypical sulfides and related compounds

The soft chalcogens sulfur, selenium, and tellurium form binary compounds with metals that often have quite different structures from the corresponding oxides. As we saw in Section 4.6, this difference is consistent with the greater covalency of the compounds of sulfur and its heavier congeners. For example, we noted there that MO compounds generally adopt the rock-salt structure whereas ZnS and CdS adopt the sphalerite or the wurtzite structures. Similarly, the *d*-block monosulfides generally adopt the more characteristically covalent nickel arsenide structure rather than the rock-salt structure of alkaline earth oxides such as MgO. Even more striking are the layered MS_2 compounds of many *d*-block elements in contrast to the fluorite or rutile structures of many *d*-block dioxides (Section 12.10).

Before we discuss these compounds, we should note that there are many metal-rich compounds that disobey simple valence rules and do not conform to an ionic model. Some examples are Ti_2S, Pd_4S, V_2O and Fe_3N and even the alkali metal suboxides, such as Cs_3O (Section 12.9).[6] The occurrence of metal-rich phases is generally associated with the formation of M—M bonds (Section 13.9).

18.7 Layered MS_2 compounds and intercalation

We introduced the layered sulfides and their intercalation compounds in Section 12.11. Here we develop a broader picture of their structures and properties.

Synthesis and crystal growth

The chalcogens react with most metals at or above room temperature, and a wide range of compounds are known. The preparation of crystalline disulfides suitable for chemical and structural studies is often performed by **chemical vapor transport**, which is a generally useful technique in solid state chemistry. It is possible sometimes simply to sublime a compound, but

[6] A discussion of the structure and bonding in metal-rich compounds is given by H. F. Franzen, *Prog. Solid State Chem.*, **12**, 1 (1978).

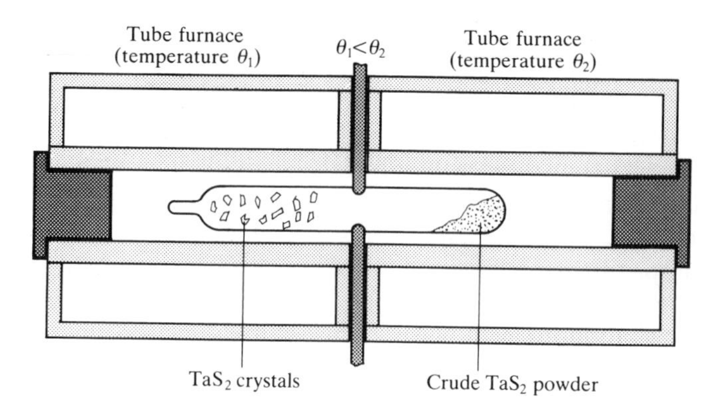

Fig. 18.28 Vapor transport crystal growth and purification of TaS_2. A small quantity of I_2 is present to serve as a transport agent.

the vapor transport technique can also be applied to a wide variety of nonvolatile compounds.

Typically, the crude material is loaded into one end of a borosilicate or fused quartz tube. After evacuation, a small pressure of a chemical transport agent is introduced, the tube is sealed, and placed in a furnace with a temperature gradient. The polycrystalline and possibly impure metal chalcogenide is vaporized at one end and redeposited as pure crystals at the other (Fig. 18.28). The technique is called chemical vapor transport rather than sublimation because the chemical transport agent, which may be a halogen, produces an intermediate volatile species such as a metal halide. Generally only a small amount of transport agent is needed because upon crystal formation it is released and diffuses back to pick up more reactant. For example, TaS_2 can be transported with I_2 in a thermal gradient. At 850°C the reaction with I_2 to produce gaseous products is endothermic:

$$TaS_2(s) + 2I_2(g) \rightarrow TaI_4(g) + S_2(g)$$

and the reaction is favored at this temperature. Therefore, the partial pressure of volatile products is greater at 850°C than at 750°C, so TaS_2 is deposited at the lower temperature. If, as occasionally is the case, the transport reaction is exothermic, the solid is carried from the cooler to the hotter end of the tube.

Structure

As we saw in Section 12.11, the *d*-block disulfides fall into two classes. There are the layered materials formed by metals on the left of the *d*-block, and in the middle and toward the right of the block there are compounds containing formal S_2^{2-} ions, such as iron pyrite. We shall concentrate on the layered materials.

In TaS_2 and many other layered disulfides the *d*-metal ion is located in octahedral holes between close packed AB layers (Fig. 18.29a).[7] The Ta ions form a close-packed-type layer denoted c, and so the metal and adjoining sulfide layers can be portrayed as an AcB sandwich. These AcB sandwich-like slabs form a three dimensional crystal by stacking in sequences such as AcB AcB AcB, etc., where the strongly bound AcB slabs are held to their neighbors by weak dispersion forces. The Mo atoms in MoS_2 reside in the trigonal prismatic holes between sulfide layers that are in

[7] Recall the notation ABC for stacks of close-packed layers in Section 4.2.

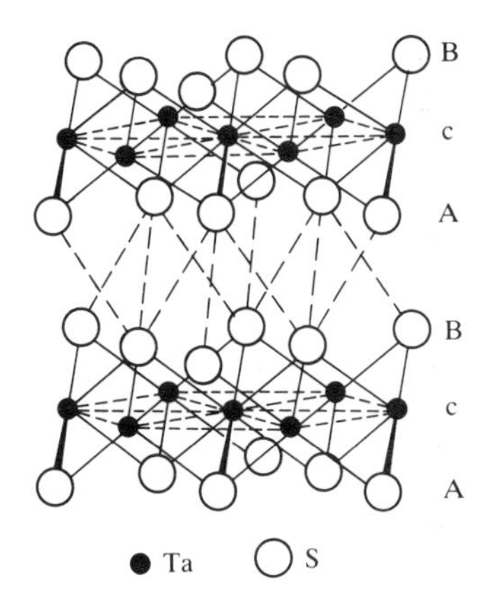

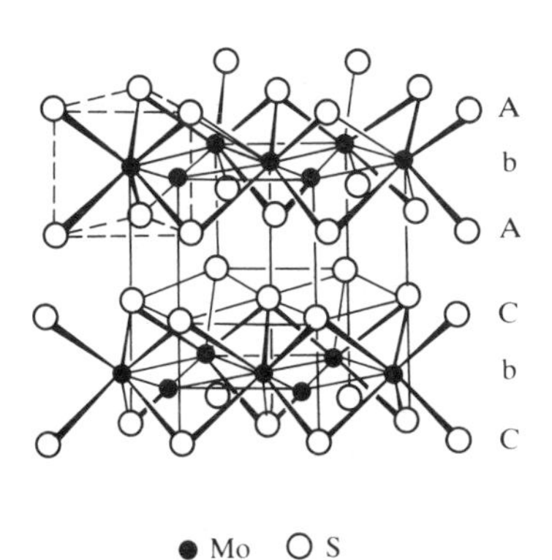

Fig. 18.29 (a) The structure of TaS_2 (CdI_2-type). The Ta atoms reside in octahedral, c, sites between the AB layers of sulfur. (b) The MoS_2 structure; the Mo atoms reside in trigonal prismatic sites between AA or CC sulfide layers.

register with one another (AA, Fig. 18.29b). The Mo atoms, which are strongly bonded to the adjacent sulfide layers, form a close-packed array, and so we can represent each slab as AbA or CbC. These slabs form a three-dimensional crystal by stacking in a pattern AbA CbC AbA CbC etc. Weak dispersion forces are also responsible for holding these AbA and CbC slabs together.

Polytypes (structures that differ only in the stacking arrangement along a direction perpendicular to the plane of the slabs, Section 4.2) also occur. Thus MoS_2 forms several polytypes including one with the sequence CbC AbA and another with the sequence CaC AbA BcB. In addition to variations in stacking sequences, the structures of many *d*-block dichalcogenides show subtle periodic distortions within the individual slabs. These distortions are called **charge density waves** and they will be described after we have established some background information on the band model of metal dichalcogenides.

In the discussion above it was convenient to describe the layered structures in terms of cations and anions. Now we shall use a band model, for then it is easier to account for their significant covalency, electrical conductivities, and chemical properties. Some approximate band structures derived primarily from molecular orbital calculations and photoelectron spectra are shown in Fig. 18.30. They show that the dichalcogenides with

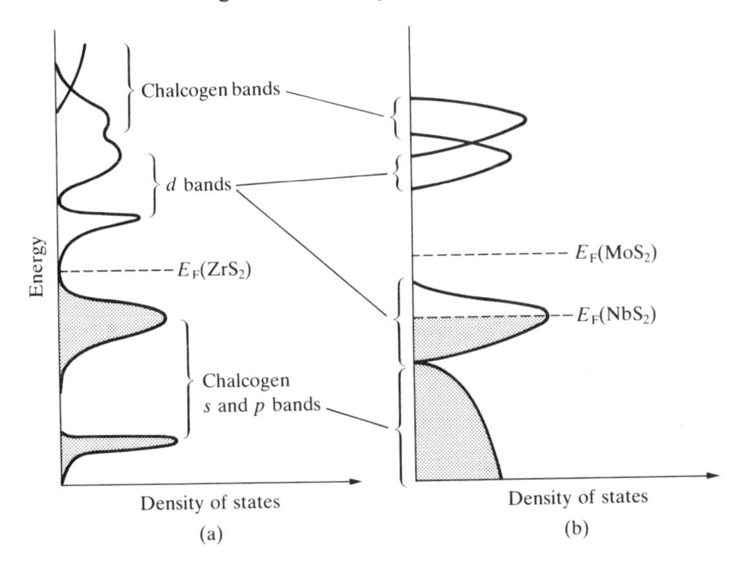

Fig. 18.30 Approximate band structures of dichalcogenides. (a) Octahedral MS_2 compounds; for ZrS_2 (which is d^0 and is illustrated), only the sulfur *s* and *p* bands are full and the compound is a semiconductor. TaS_2 (d^1) has a partially filled *d*-band and it is metallic. (b) Trigonal prismatic MS_2 compounds. NbS_2 (d^1, as illustrated) is metallic while MoS_2 (d^2) is a semiconductor.

Table 18.5. Some alkali metal intercalation compounds

Compound	Δd/Å†
$KZrS_2$	1.60
$NaTiS_2$	1.17
$KTiS_2$	1.92
$RbSnS_2$	2.24
$KSnS_2$	2.67
$Na_{0.6}MoS_2$ (2H)	1.35
$K_{0.4}MoS_2$ (2H)	2.14
$Rb_{0.3}MoS_2$ (2H)	2.45
$Cs_{0.3}MoS_2$ (2H)	3.66

† Change in interlayer spacing in comparison with the parent MS_2.

octahedral and trigonal prismatic metal sites have low lying bands primarily composed of chalcogen *s* and *p* orbitals, higher energy bands derived primarily from metal *d* orbitals, and above these metal and chalcogen bands.

When the metal atom resides in an octahedral site, its *d* orbitals are split into a lower t_{2g} set and a higher e_g set, just as in localized complexes. These atomic orbitals combine to give a t_{2g} band and an e_g band. The broad t_{2g} band may accommodate up to six electrons per metal atom. Thus TaS_2, with one *d* electron per metal, has an only partly filled t_{2g} band, and is a metallic conductor.

A trigonal prismatic ligand field leads to a low energy a_1' level and two doubly degenerate higher energy orbitals designated e' and e'' respectively. Now the lowest band needs only two electrons per atom to fill it. Thus MoS_2, with trigonal prismatic Mo sites and two *d* electrons, has a filled a_1' band and is an insulator (more precisely, a semiconductor).

The formation of charge density waves in some metal dichalcogenides is similar to the onset of the Peierls distortion of one-dimensional structures (Section 3.5). For a partly filled conduction band in one- and two-dimensional systems, a distortion of the lattice will split the band so that the resulting lower component may be filled with electrons and the upper component emptied. Since the energy advantage of this distortion is not large, a static modulation of the structure is usually only evident at low temperatures. Charge density waves are observed at low temperatures in the Group 5 dichalcogenides VS_2, NbS_2, and TaS_2. In these compounds four of the five electrons from each metal are required to fill the *s* and *p* band formed from the sulfur orbitals, and the remaining electron occupies but does not fill the *d*-band. At sufficiently low temperatures the lattice distorts and the *d*-band splits.

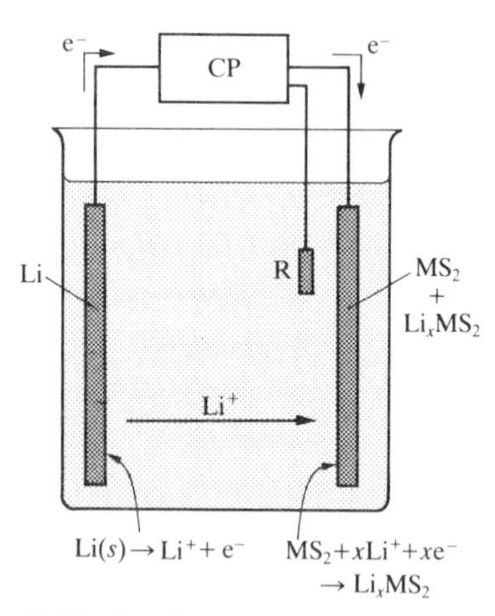

Fig. 18.31 Experimental arrangement for electrointercalation. A polar organic solvent (e.g., propylene carbonate) containing a lithium salt is used as the electrolyte. R is a reference electrode and CP is a coulometer (to measure the charge passed) and a potential controller.

Intercalation and insertion

We have already introduced the idea that alkali metal ions may insert between graphite sheets (Section 11.6) and metal disulfide slabs (Section 12.11) to form intercalation compounds. For a reaction to qualify as an intercalation, or as an **insertion reaction**, the basic structure of the host should not be altered when it occurs.[8] Insertion compound formation is the basis of 'high energy cells', which consist of a lithium anode, a chalcogenide cathode, and an electrolyte dissolved in an inert polar organic solvent. Such a cell can produce well over 2 V.

The π conduction and valence bands of graphite are contiguous in energy (we have seen in fact that graphite is formally a semimetal, Section 3.5) and the favorable free energy for intercalation arises from the transfer of an electron from the alkali metal atom to the graphite conduction band. The insertion of an alkali metal atom into a dichalcogenide involves a similar process. Now, though, the electron is accepted into the *d* band, and the charge-compensating alkali metal ion diffuses to positions between the slabs. The widest range of insertion compounds is found with the dichalcogenides having an empty or nearly empty t_{2g} band, which in principle can

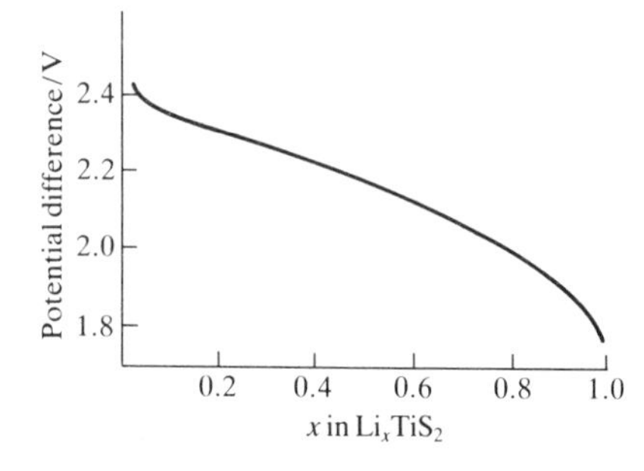

Fig. 18.32 Potential versus composition diagram for the electrointercalation of lithium into titanium disulfide. The composition, x in Li_xTiS_2, is inferred from the charge passed in the course of electrointercalation.

[8] Reactions in which the structure of one of the solid starting materials is not radically altered are called 'topotactic' reactions. They are not limited to the type of insertion chemistry we are discussing here. For example, hydration, and ion exchange reactions may be topotactic.

receive up to six electrons per metal. Insertion is much more limited when the metal is in a trigonal prismatic site: as we have seen, the lowest energy a_1' band can accommodate only two electrons per metal atom. Some representative alkali metal insertion compounds are listed in Table 18.5.

The insertion of alkali metal ions into host lattices can be achieved by direct combination of the alkali metal and the disulfide:

$$TaS_2 + xNa \xrightarrow{800°C} Na_xTaS_2 \qquad x = 0.4 \text{ to } 0.7$$

or by utilizing a highly reducing alkali metal compound or the electrochemical technique of **electrointercalation** (Fig. 18.31). One advantage of electrointercalation is that it is possible to measure the amount of alkali metal incorporated by monitoring the amount of electrons (It/F) passed during the synthesis. It also is possible to distinguish solid solution formation from discrete phase formation. As illustrated in Fig. 18.32, the formation of a solid solution is characterized by a gradual change in potential as intercalation proceeds, whereas the formation of a new discrete phase yields a steady potential over the range in which one solid phase is being converted into the other, followed by an abrupt change in potential when that reaction is complete.

Insertion compounds are examples of mixed ionic and electronic conductors. In general the insertion process can be reversed either chemically or electrochemically. This makes it possible to recharge a lithium cell by removal of lithium from the compound. In a clever synthetic application of these concepts, the previously unknown layered disulfide VS_2 was prepared by first making the known layered compound $LiVS_2$ in a high-temperature process. The lithium was then removed by reaction with I_2 to produce the metastable layered VS_2, which was found to have the TiS_2 structure:

$$2LiVS_2 + I_2 \rightarrow 2LiI + 2VS_2$$

Insertion compounds also can be formed with molecular guests. Perhaps the most interesting are the metallocenes $Co(\eta^5\text{-}C_5H_5)_2$ and $Cr(\eta^5\text{-}C_5H_5)_2$ which are incorporated into a variety of hosts such as TiS_2, $TiSe_2$, and TaS_2 to the extent of about $0.25\,M(\eta^5\text{-}C_5H_5)_2$ per $M'S_2$ or $M'Se_2$. This limit appears to correspond to the space available for forming a complete layer of $M(\eta^5\text{-}C_5H_5)_2^+$. The organometallic compound appears to undergo oxidation upon intercalation, so the favorable free energy in these reactions arises in the same way as in alkali metal intercalation. In agreement with this interpretation, $Fe(\eta^5\text{-}C_5H_5)_2$, which is harder to oxidize than its chromium or cobalt analogs, does not intercalate.

We can imagine the insertion of ions into one-dimensional channels, two-dimensional planes of the type we have been discussing, or channels that intersect to form three-dimensional networks (Fig. 18.33). Aside from the availability of a site for a guest to enter, the host must provide a conduction band of suitable energy to take up electrons reversibly (or, in some cases, be able to donate electrons to the guest). Table 18.6 illustrates that a wide variety of hosts are possible, including metal oxides and various ternary and quaternary compounds. We see that intercalation chemistry is by no means limited to graphite and layered disulfides.

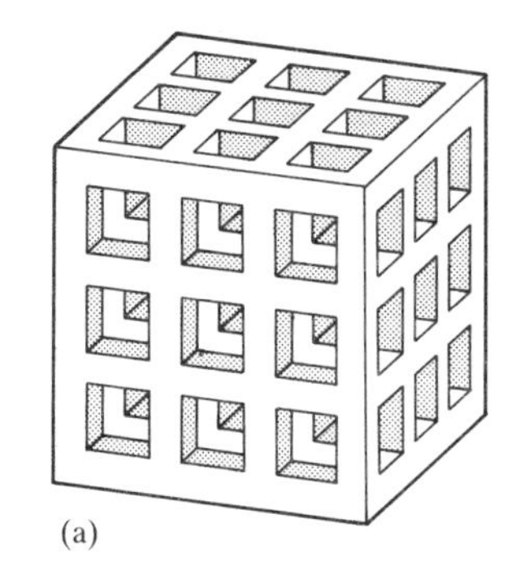

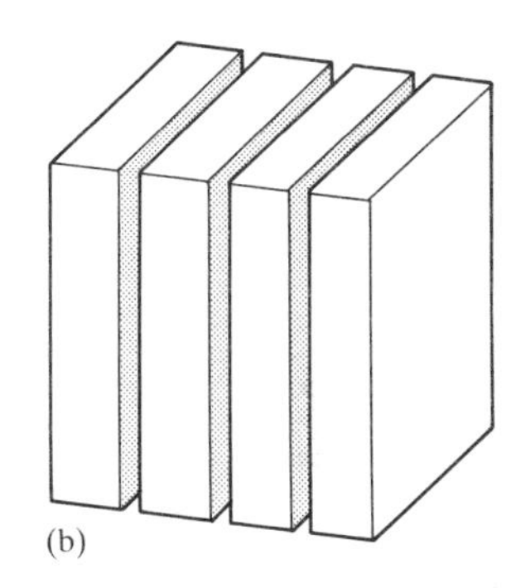

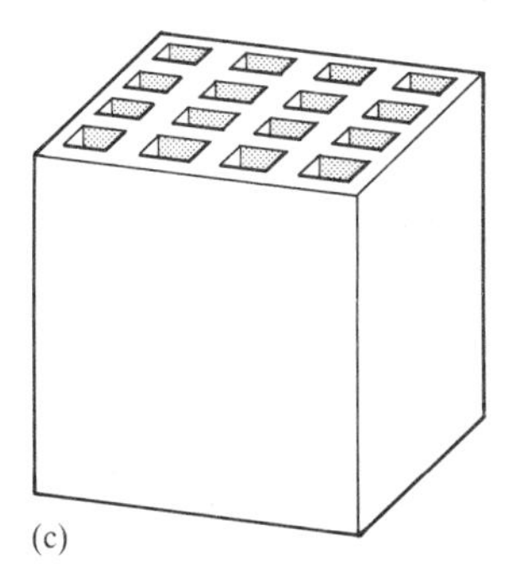

Fig. 18.33 Schematic representation of host materials for intercalation reactions. (a) A three-dimensional host with intersecting channels, (b) a two-dimensional layered compound, and (c) one-dimensional channels.

Table 18.6. Some intercalation compounds

Phase	Composition (x)	Type
$Li_x[Mo_6S_8]$	2.4 to 0.6	3*d*
$Na_x[Mo_6S_8]$	3.6	3*d*
$Ni_x[Mo_6Se_8]$	1.8	3*d*
H_xWO_3	0.6	3*d*
Li_xMoO_3		3*d*
Li_xNiPS_3	0 to 1.5	2*d*

Fig. 18.34 Two views of the Mo_6S_8 unit in a Chevrel compound.

18.8 Chevrel phases[9]

We close this chapter with a discussion of an interesting class of ternary compounds first reported by the French solid state inorganic chemist R. Chevrel in 1971. These compounds, which illustrate three-dimensional intercalation, have formulas such as Mo_6X_8 and $M_xMo_6S_8$; Se or Te may take the place of S and the intercalated M atom may be a variety of metals such as Li, Mn, Fe, Cd, and Pb. The parent Mo_6Se_8 and Mo_6Te_8 are prepared by heating the elements at about 1000°C. A structural unit common to this series is M_6S_8, which may be viewed as an octahedron of M atoms face bridged by S atoms, or alternatively as an octahedron of M atoms in a cube of S atoms (Fig. 18.34). This type of cluster is also observed for some halides of the Period 4 and 5 early *d*-block elements, such as the $[M_6X_8]^{4+}$ cluster found in Mo and W dichlorides, dibromides, and diiodides.

Figure 18.35 shows that in the three-dimensional solid the Mo_6S_8 clusters are canted relative to each other and relative to the sites occupied by intercalated ions. The Mo_6S_8 appear to be canted to allow secondary donor–acceptor interaction between vacant Mo d_{z^2} orbitals (which project outward from the faces of the Mo_6S_8 cube) and a filled donor orbital on S atoms of adjacent clusters (Fig. 18.36).

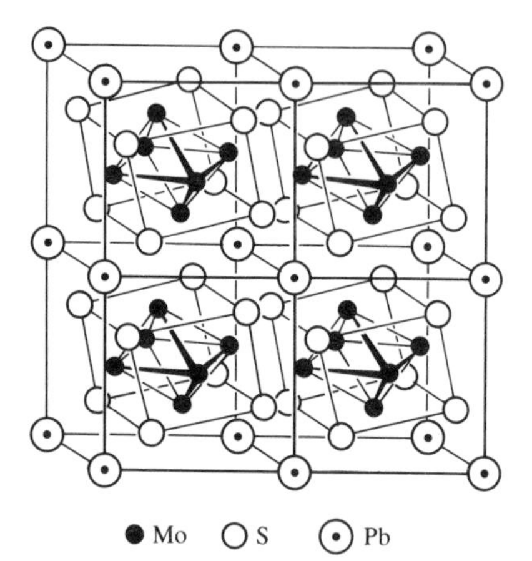

Fig. 18.35 Structure of a Chevrel compound showing the canted Mo_6S_8 unit in the slightly distorted cube of Pb atoms.

A simple molecular orbital treatment of the Mo_6S_8 cluster indicates a capacity for twenty metal cluster valence electrons to fill the bonding *d*-orbitals and another four to fill a nonbonding e_g level. Six Mo atoms have a total of 36 valence electrons (six electrons per Mo atom) and the formation of eight S^{2-} ions will consume 16 of them; hence compounds of composition Mo_6S_8 have 20 cluster electrons, just filling the bonding orbitals. Metal atom guests provide additional electrons that can be accommodated in the nonbonding e_g orbital. In the infinite solid the interaction between these clusters leads to band formation from the orbitals we have just discussed, but the general picture in so far as electron counting is concerned does not change.

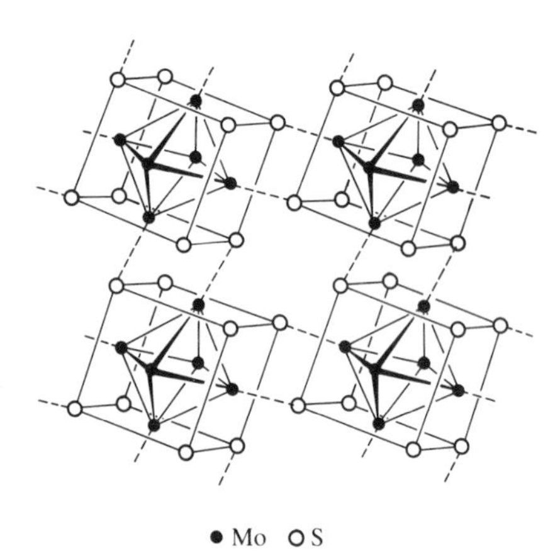

Fig. 18.36 Explanation for the tilt of M_6S_8 units in a Chevrel phase in terms of donor–acceptor interaction between M in one cube with S in the next.

The electrointercalation of lithium into Mo_6S_8 (Fig. 18.37) nicely illustrates the three stages. The two plateaus indicate the formation of discrete compounds, $Li_{0.8}Mo_6S_8$ and $Li_{2.4}Mo_6S_8$, and the broad sloping region indicates a nonstoichiometric phase, $Li_xMo_6S_8$ ($2.4 < x < 3.6$). The explanation of the formation of these individual phases is beyond the scope of simple theory but the upper limit of intercalation ($x = 3.6$) agrees with theory in that it does not exceed the capacity of four electrons in the non-bonding e_g band.

[9] Ø. Fisher, *Appl. Phys. A* **16,** 1 (1978); T. Hughbanks and R. Hoffmann, *J. Amer. chem. Soc.*, **105,** 1150 (1983); R. Schöllhorn, *Angew. Chem. Int. Ed. Eng.*, **19,** 983 (1980).

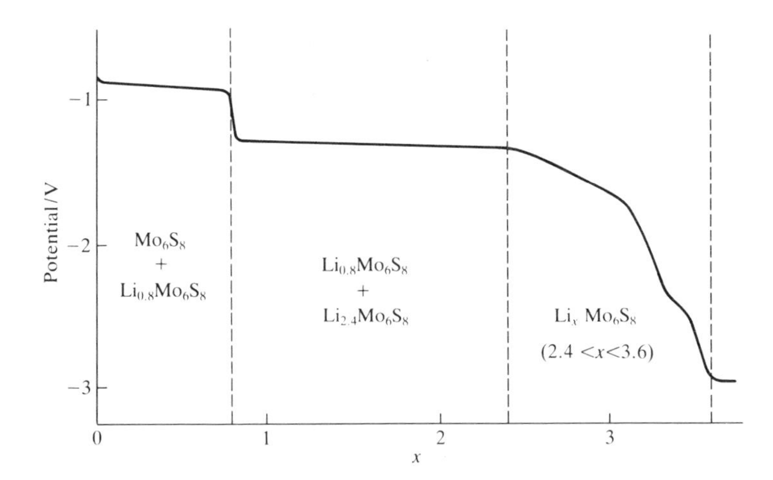

Fig. 18.37 Potential versus amount of Li intercalated into Mo_6S_8 as determined by electrointercalation.

Example 18.4: *Determining the nature of a mixed-metal Chevrel phase*

Chevrel and his coworkers prepared a compound $Mo_4Ru_2Se_8$. Assuming the frontier orbitals are similar to those in Mo_6S_8, determine the capacity of this mixed-metal Chevrel phase to intercalate sodium.

Answer. We may compare the number of metal valence electrons available in the mixed metal system with the maximum of 24 for the Mo_8 Chevrel compounds. Four Mo atoms give 24 electrons and two Ru atoms give 16. Removing 16 for the formation of eight S^{2-} ions yields 24 electrons for the cluster overall, and we predict that it should not intercalate sodium. Thus, if we were simply trying to make another intercalated Chevrel phase, $Mo_4Ru_2Se_8$ would not appear to be a good choice. On the other hand, if we are curious about the outcome, we might try the intercalation reaction because the uncertainty of approximate electronic structure arguments combined with the subtle balance of energies in many reactions leaves the door open for new discoveries.

Exercise. Should $Mo_4Ru_2Se_8$ exhibit metallic conduction or semiconduction?

One of the physical properties that has drawn attention to the Chevrel phases is their superconductivity. Superconductivity persists up to 14 K in $PbMo_6S_8$ and it also persists to very high magnetic fields, which is of considerable practical interest because many applications involve high fields. In this respect the Chevrel phases appear to be significantly superior to the newer oxide-based high-temperature superconductors.

Further information: phase diagrams

A **phase diagram** summarizes the conditions at which phases are in equilibrium with each other or at which only a single phase is present. The number F of intensive variables (such as temperature, pressure, and composition) that may be changed without eliminating the equilibrium among P phases in a system composed of C components is expressed by the **phase rule**:

$$F = C - P + 2$$

In many cases the pressure is regarded as constant, so one degree of freedom is discarded. If in addition we deal only with binary (two-component) systems, $C = 2$. Under these conditions,

$$F = 3 - P$$

When only a single phase of a binary system is present (such as a liquid solution), $F = 2$ and the composition and temperature may be varied at will. In other words, both temperature and composition must be specified to define a point in this one-phase region. If two phases are present in equilibrium, $F = 1$, and there is only one degree of freedom. Thus, if the composition is adjusted, the temperature must be changed to sustain the equilibrium. That is, the equilibrium of two phases is represented by a line in a phase diagram, showing that one variable (temperature) is dependent on the other (composition). If three phases are at equilibrium in a binary system, such as two different solids and a liquid, $F = 0$, and there is no freedom to change the conditions and yet still retain the equilibrium. Thus, three phases meet at a single point in the phase diagram and the location of the point is uniquely determined by the identity of the components (and the pressure, the discarded variable).

As an example, consider the phase diagram in which the components are MgO and CaO (Fig. 18.38). This diagram is fairly typical of systems in which the components do not react to form a compound. The left vertical line corresponds to pure MgO at a series of temperatures and the one on the right to pure CaO. It turns out that MgO and CaO, both of which have the rock-salt structure, have limited solubility in each other. The area labeled MgO *ss* is a solid solution consisting predominantly of MgO and a little CaO. The other solid solution region, CaO *ss*, is mainly CaO with some MgO. We can picture these solid solutions as the usual rock-salt structure in which part of the cation sites are occupied by Mg^{2+} ions and part by Ca^{2+} ions and the anion sites are occupied by O^{2-} ions. In this structure one cation replaces the other without forming a new ordered crystalline phase.

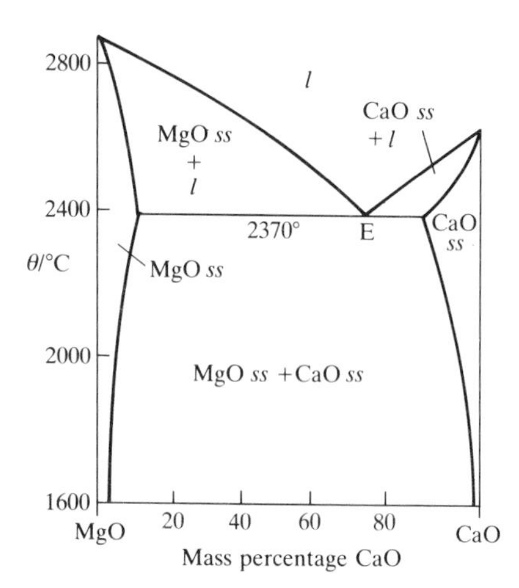

Fig. 18.38 The system MgO–CaO showing simple eutectic behavior and no compound formation between the two components. MgO *ss* is a solid solution of MgO with smaller amounts of CaO. CaO *ss* is a solid containing primarily CaO with lesser amounts of MgO.

At 2000°C the two single-phase solid solution phases are separated by a two-phase region consisting of a physical mixture of the two crystalline solid phases MgO *ss* and CaO *ss*. This illustrates a general thermodynamic requirement that single-phase regions are separated by regions containing two phases.

The diagram has a third single-phase region, labeled *l*, consisting of molten solution of CaO and MgO. Again this single-phase region is separated from the other two single-phase regions by two-phase regions, MgO *ss* + *l*, and CaO *ss* + *l*. All three of the single phases meet at point E. The coexistence of these phases at a single point is in accordance with the principle derived above from the phase rule. E is called a **eutectic point** and its significance can be appreciated by imagining an experiment in which we cool a sample of the liquid phase that has the composition of the eutectic (72 percent CaO by mass). The liquid persists until we reach the eutectic temperature (2370°C) the temperature then remains constant until the whole mass is crystalline. Thus the cooling curve would resemble that for a pure compound, but X-ray diffraction on the solid eutectic would reveal the presence of two distinct crystalline solid solutions, CaO *ss* and MgO *ss*. The phase diagram in Fig. 18.12 has one eutectic point at 1370°C and about 22.7 percent O, and another at about 1500°C and 29 percent O.

In the Fe–O phase diagram (Fig. 18.12) we see that the compound Fe_3O_4 melts sharply at 1594°C. This type of melting, in which the compound does not decompose, is termed **congruent**. Congruently melting compounds are readily prepared from the melt by cooling the molten liquid with the stoichiometry of the compound. Wüstite, on the other hand, does not melt congruently: it melts **incongruently**, meaning that it decomposes as it forms a melt. This is most readily seen for the material at the far right of the wüstite phase field (at about 24.3 percent O). At 1436°C the diagram indicates that wüstite decomposes with the formation of Fe_3O_4 and liquid. In general, incongruently melting compounds cannot be prepared by cooling the liquid of the correct stoichiometry. One strategy for their preparation is to heat the components just below the melting temperature and wait for solid state diffusion to homogenize the mass.

Further reading

The first three books provide good general overviews of inorganic solid state chemistry:

A. R. West, *Solid state chemistry and its applications*. Wiley, New York (1984).

C. N. R. Rao (ed.), *Solid state chemistry*. Dekker, New York (1974).

C. N. R. Rao and J. Gopalakrishnan, *New directions in solid state chemistry*. Cambridge University Press (1986).

A. K. Cheetham and P. Day (eds.), *Solid state chemistry techniques*. Oxford University Press (1987). This book includes an informative chapter on the preparation of solids; otherwise it is devoted to techniques for the characterization of solids.

P. A. Cox, *The electronic structure and chemistry of solids*. Oxford University Press (1987).

J. K. Burdett, *Prog. solid State Chem.* **15,** 173 (1984).

This very accessible book and review article discuss the relationships between the structures and properties of solids.

A. F. Wells, *Structural inorganic chemistry*. Oxford University Press (1984). This large book is an excellent resource for structural information and structural correlations.

B. G. Hyde and S. Andersson, *Inorganic crystal structures*. Wiley, New York (1989). Hyde and Andersson provide a system for understanding complex inorganic crystal structures and apply it to many important structures.

D. M. Adams, *Inorganic solids*. Wiley-Interscience, New York (1984). Adams's more concise book provides a good introduction to the structure of prototypical solids.

The next two books provide good coverage of the topics indicated by their titles.

R. J. D. Tilley, *Defect crystal chemistry and its applications*. Blackie, London and Chapman and Hall, New York (1987).

D. B. Brown (ed.), *Mixed-valence compounds*. Riedel, New York (1981).

R. E. Newnham, *Structure–property relations*, Springer-Verlag, Berlin (1975). This is an accessible volume stressing correlations between structures and properties such as electron transport, ion transport, and ferroelectric and magnetic behavior.

P. W. Atkins, *Physical chemistry*, (4th edn). Oxford University Press and W. H. Freeman & Co, New York (1990). See Chapter 8 for a discussion of the phase rule.

P. G. Bergeron and S. H. Risbud, *Introduction to phase equilibria in ceramics*. The

American Ceramics Society, Columbus (1984). A large and very useful collection of phase diagrams is given in *Phase diagrams for ceramisists*, which is a continuing series published by The American Ceramics Society.

Exercises

18.1 Describe the nature of Frenkel and Schottky defects, and some experimental tests for their occurrence. Does either of these defects by itself give rise to nonstoichiometry?

18.2 From each of the pairs of compounds (a) NaCl and NiO, (b) CaF_2 and PbF_2, and (c) Al_2O_3 and Fe_2O_3, pick the one most likely to have a high concentration of defects. Describe the possible defects and give your reasoning.

18.3 Distinguish intrinsic from extrinsic defects and give an example of each in actual compounds.

18.4 Draw one unit cell in the ReO_3 structure showing the M and O atoms. Does this structure appear to be sufficiently open to undergo Na^+ ion intercalation? If so, where might the Na^+ ions reside?

18.5 Is a crystallographic shear plane a defect, a way of describing a new structure, or both? Explain your answer.

18.6 Sketch the rock-salt crystal structure and locate in the sketch the sites the interstitial sites to which a cation might migrate. What is the nature of the bottleneck between the normal and interstitial sites?

18.7 Contrast the nature of the defects in TiO with those in FeO.

18.8 How might you experimentally distinguish the existence of a solid solution from a series of crystallographic shear plane structures for a material which appears to have variable composition?

18.9 Label the regions in the iron–oxygen phase diagram (Fig. 18.12) in the region 27 to 30 percent oxygen, and state the general principles that you used in assigning the labels.

18.10 Bearing in mind that the redox chemistry of metal oxides can often be correlated with the redox behavior of the ions in solution, explain which of TiO, MnO, and NiO is (a) the most difficult to oxidize with air, (b) the easiest to reduce.

18.11 Describe the electrical conductivity of TiO and NiO and describe the interpretation in terms of the trends in electronic structure across the 3*d* series.

18.12 Above 150°C, AgI is an Ag^+ ion conductor. Suppose a pellet of this compound is sandwiched between two silver electrodes at 165°C and a potential difference of 0.1 V is imposed between the electrodes. Sketch the apparatus and indicate the changes that occur with time at the two electrodes.

18.13 What will be the quantitative changes in the mass of the cathode, the AgI, and the anode after 10^{-3} mol of electrons have been passed through the cell described in Exercise 18.12.

18.14 Constrast the mechanism of interaction between unpaired spins in a ferromagnetic material to that for an antiferromagnet.

18.15 Magnetic measurements on the ferrite $CoFe_2O_4$ indicate 3.4 spins per formula unit. Suggest a distribution of cations between octahedral and tetrahedral sites that would satisfy this observation.

18.16 The superconducting compound $YBa_2Cu_3O_7$ is described as having a perovskite-like structure. Ignoring problems of charge on the ions, describe the difference between this structure and that of perovskite.

Problems

18.1 Given the Δ values that follow, determine the site preference for $A = Ni^{2+}$ and $B = Fe^{3+}$ in normal compared with inverse spinel, assuming that ligand field stabilization is dominant. Δ_O: Fe^{3+}(1400 cm^{-1}), Ni^{2+} (860 cm^{-1}); Δ_T: Fe^{3+} (620 cm^{-1}), Ni^{2+} (380 cm^{-1})

18.2 Magnetite, Fe_3O_4, has the spinel structure and is thus a member of the ferrite group of compounds. In a normal spinel unit cell eight tetrahedral holes would be filled with Fe^{2+} and 16 octahedral holes would contain Fe^{3+}. At room temperature magnetite has high conductivity (about 2×10^2 S cm^{-1}). Are the conductivity data consistent with the normal structure? If not, propose one possible variant of the spinel structure that would lead to high conductivity.

Bioinorganic chemistry

19

Pumps and transport proteins
19.1 Ion pumps
19.2 Oxygen transport
Enzymes exploiting acid catalysis
19.3 Oxaloacetate decarboxylase
19.4 Carboxypeptidases
Redox catalysis
19.5 Cytochromes of the electron transport chain
19.6 Cytochrome P-450 enzymes
19.7 Nitrogen fixation
19.8 Photosynthesis
Further Reading
Exercises
Problems

Organisms must maintain steady states far from equilibrium by a continuous flow of nutrients and energy in a variable environment; to an organism, equilibrium means death. The biological role of inorganic compounds and ions is often associated with the management of these flows and the creation of dynamic spatial structures.

The first section of this chapter illustrates how some elements contribute to dynamical systems. Following that, the discussion turns to the biological catalysts, the enzymes. In most cases, a detailed understanding of their role has been achieved for only a few key steps, and intelligent speculation plays a more prominent role here than in other areas of inorganic chemistry. We organize our discussion of these specific steps according to the reaction types introduced earlier in the text, including Brønsted and Lewis acid–base reactions, redox reactions that depend on electron and group transfer, and photochemically induced electron transfer. Many enzymes exploit a metal complex as the active site. In each case we meet a biochemical step that is reasonably well understood and some inorganic model systems that behave in ways that shed light on the biochemical step.

The flow of ideas between inorganic chemistry and biochemistry happens in both directions. Just as we can bring our knowledge of each type of inorganic reaction to bear on biochemical processes, the elegant solutions that living systems have found suggest new features for inorganic chemistry to imitate and exploit.

Table 19.1 shows the relative concentrations of the elements in the biosphere. The most striking feature of the table is that so many elements appear. What the table does not show, though, is that it is far from true that low concentration means low importance, for we shall see that even the elements that occur in trace quantities play a central role in an organism's fight to avoid equilibrium.

Table 19.1. The distribution of elements in the biosphere

										H							He
Li	Be											B	C	N	O	F	Ne
Na	Mg											Al	Si	P	S	Cl	Ar
K	Ca	Sc	Ti	V	Cr	Mn	Fe	Co	Ni	Cu	Zn	Ga	Ge	As	Se	Br	Kr
Rb	Sr	Y	Zr	Nb	Mo	Tc	Ru	Rh	Pd	Ag	Cd	In	Sn	Sb	Te	I	Xe
Cs	Ba	...															

Less than 0.05 percent; 0.05 to 1 percent; 1 to 60 percent

It proves helpful to classify the elements in a way that suggests their function in the complex dynamic system of a living cell. The simplest classification divides elements according to their occurrence in three different biological environments:

(1) In the extracellular fluids and on the outer wall of the cell membrane.
(2) Inside the cell, starting from the inner wall of the membrane in the aqueous phase known as the cytoplasmic fluid.
(3) In particles called organelles inside the cell, including nuclei and mitochondria (the localized structures in a cell commonly identified with its source of power).

The identification of the location and the host environment of the elements gives the first clue to the role they play in biological systems. Some examples are given in Table 19.2.

Table 19.2. The distribution of elements in the biological cell

Extracellular	Organelles	Cytoplasm
Na, Ca	K, Mg	K, Mg
Cu (Mo)	Fe, Co	Co
	Zn, Ni, Mn	Zn
Cl, Si	P (S)	P (S)
Al	Se	Se

Cell membranes contain channels and special mechanisms for transporting molecules and ions from outside the cell into its interior. In particular, the distributions of the ions Na^+, K^+, and Mg^{2+} are controlled by these pumps and channels in the membranes of cells. Since these three species of ion are major components of the cell, their location establishes a pattern of charge distribution and coulombic fields. That iron is located in specific regions of membranes suggests that it has a highly specific role there, and one of its principal roles is to control the distribution of protons and electrons. Some steps involved in control of electron flow will be described in detail later in the chapter. Electrochemical potential gradients across membranes that stem from the presence of O_2 as an oxidizing agent are controlled by systems that exploit both manganese and iron. The electric fields produced by the ion gradients are of great biological significance, because as well as controlling ATP synthesis, they influence cell volume, render nerve and muscle cells excitable, and drive the transport of sugars and amino acids across the membrane. Some details of the role of ions in this context will also be presented below. The control of physical motion and mechanical stress (e.g. in muscle) are associated with the distribution of zinc, calcium, and magnesium, and we shall touch on a little of calcium's contribution. Zinc is found in the nucleus and is a characteristic metal ion of the genetic apparatus. However, this crucial ability of a biochemical system to transmit information between generations, and the role of metal ions in the functions of the nucleic acids RNA and DNA, is outside the scope of this chapter.

A successful organism must exercise delicate command over a complex pattern of reactions that proceed at high but controlled rates. Indeed, loss of this control is manifest as the organism's disease and death. Since the balanced control of reactions is so crucial to maintaining life, the role of catalysts is critical. In biochemistry the catalysts are proteins called **enzymes**; like all proteins, they are essentially polypeptides, which are macromolecules formed from α-amino acids linked by peptide bonds (**1**). We shall often use the term 'peptide residue', which denotes the component of an amino acid that remains in the chain once the peptide link has formed by elimination of H_2O. Enzymes are named according to the reaction they catalyze. For example, an enzyme that catalyzes ATP hydrolysis is called an 'ATPase'. One that catalyzes ferredoxin reduction is known as a 'ferredoxin reductase'.

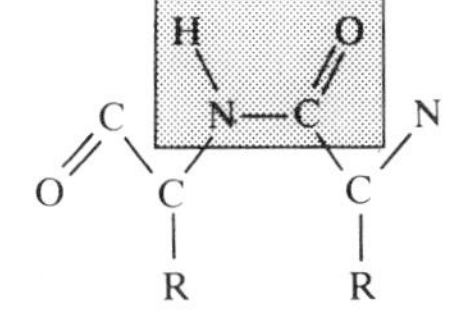

1 Peptide link

Twenty different α-amino acids occur naturally in proteins, and the different side chains that occur along the polypeptide chain possess a range of functional groups, such as alkyl, carboxyl (—COOH), amino (—NH_2), hydroxyl (—OH), and thiol (—SH) groups (Table 19.3). These functional groups confer hydrophobic and hydrophilic character, Brønsted acidity and basicity, and the Lewis basicity necessary for complexation with metal ions. About 30 percent of enzymes are **metalloenzymes**, and are so called because they have a metal atom at the active site. One very important role for the functional groups of the peptide residues is to modify the immediate environment of metal ions in metalloenzymes, and we shall concentrate on this subtle interplay of inorganic and organic chemistry. Metalloproteins are involved in acid-catalyzed hydrolysis (hydrolases), redox reactions (oxidases), and processes that rearrange carbon–carbon bonds (synthases and isomerases). A main theme of this chapter will be the typical mechanisms of metal-containing hydrolases and oxidases.

Metal ions occur in molecules other than enzymes (Table 19.4). Their other roles include the transport and storage functions of proteins and some

Table 19.3. The classification of amino acids $NH_2CHRCOOH$

Type	Name	Abbreviation	R—
Hydrophobic *R*	Glycine	gly	H—
	Alanine	ala	CH_3—
	Valine	val	$(CH_3)_2CH$—
	Leucine	leu	$(CH_3)_2CHCH_2$—
	Isoleucine	ile	$CH_3CH_2CH(CH_3)$—
	Phenylalanine	phe	C_6H_5—CH_2— (phenyl ring)
Inert heteratom in *R*	Tryptophan	trp	indol-3-yl—CH_2— (N–H ring)
Hydroxylic *R*	Serine	ser	$HOCH_2$—
	Threonine	thr	$HOCH(CH_3)$—
	Tyrosine	tyr	HO—C_6H_4—CH_2—
Carboxylic *R*	Aspartic acid	asp	$HOOCCH_2$—
	Glutamic acid	glu	$HOOCCH_2CH_2$—
Amine *R*	Lysine	lys	$H_2NCH_2CH_2CH_2CH_2$—
	Arginine	arg	H_2N—C(=NH)—$NHCH_2CH_2CH_2$—
Amide *R*	Aspargine	asn	$H_2NC(=O)CH_2$—
	Glutamine	gln	$H_2NC(=O)CH_2CH_2$—
Imidazole *R*	Histidine	his	imidazolyl (HN, =N ring)—
Sulfur-containing *R*	Cysteine	cys	$HSCH_2$—
	Methionine	met	$CH_3SCH_2CH_2$—
Others	Proline	pro	pyrrolidine ring (NH) with C(=O)OH

non-protein materials and specialized reactions of non-protein complexes (such as chlorophyll in photosynthesis).

Simple metal complexes can function as Lewis or Brønsted acids, and can participate in redox reactions that involve electron, atom, or group transfer. However, small complexes are insufficiently specific as catalysts to maintain the subtle, multiple dynamic balance of a living system. For subtlety, nature has had to resort to the complexity of macromolecules such as proteins and nucleic acids. The profound contribution of these macromolecules is their ability to control the precise organization of the environment of the reactive site and hence to manage the flow of reactants. In some cases, key steps of the processes are now well known. The analysis of the mechanisms draws on all the ingenuity that can be brought to bear including all our knowledge of inorganic reactions. As a bonus in this stimulating exercise, some entirely novel inorganic pathways have been discovered.

Table 19.4. The classification of biomolecules containing metal ions

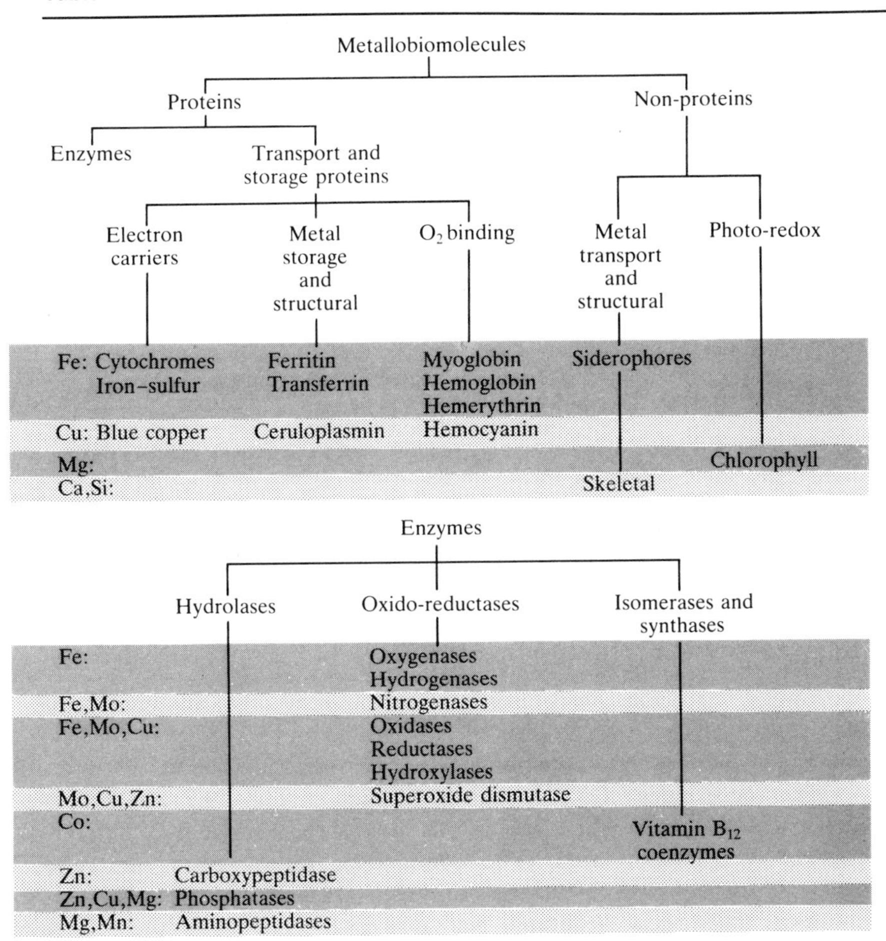

Pumps and transport proteins

As we observed in Section 5.10, the phosphate polymer ATP (adenosine triphosphate) stores energy for the cell in the form of the free energy that can be released on hydrolysis of the —P—O—P— polyphosphate link. Since this method of energy storage is very general, the means by which ATP is synthesized are of central importance to an understanding of the global biochemistry of an organism.

The key features of the conditions favoring condensation of two P—OH units to —P—O—P— with loss of water are control of pH and electrostatic charge. The pH and charge gradients required are created across a cell membrane by the reactions that exploit the electron acceptor properties of the O_2 molecule, and so oxygen is ultimately responsible for driving the build-up of the energy carrier of the cell. This example hints at the intricacy of the network of transport processes that interact to sustain an organism, and we shall begin our discussion of bioinorganic chemistry by considering the operation of one of the pumps that creates a concentration gradient.

19.1 Ion pumps

We deal here with a pump that contributes to **active transport**, the transport of ions through membranes against their natural tendency to diffuse toward lower concentration. The charge distribution established across the membrane by active transport results in an electric potential difference as well as a difference in chemical composition of the fluids inside and outside the cell.

The sodium/potassium pump

Most animal cells have a higher concentration of K^+ ions inside the cell membrane than outside, and a higher concentration of Na^+ ions outside the cell than inside. The immediate source of the energy to sustain this disequilibrium is ATP, and a large amount of ATP is consumed to maintain the concentration gradient.

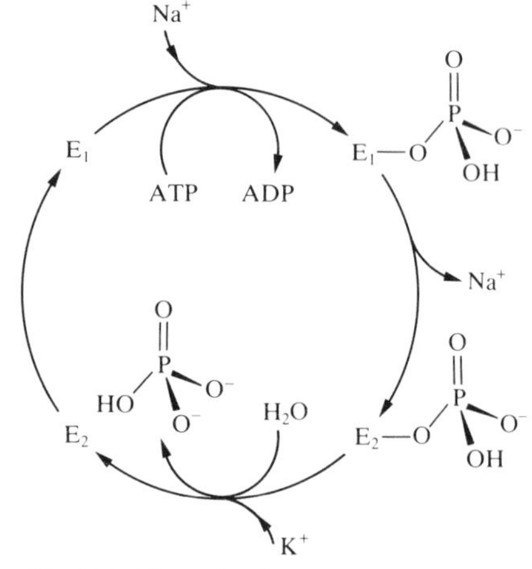

Fig. 19.1 The cycle of enzyme reactions that accomplish sodium pumping. E_1 and E_2 denote the two enzyme conformations.

One clue to the operation of the ion pump came in 1957, when Jens Skou discovered an enzyme of molar mass about 110 kg mol^{-1} that hydrolyzes ATP[1] only if Na^+ and K^+ are present in addition to the Mg^{2+} required by all ATPases (the enzymes that catalyze hydrolysis of ATP). The activity of this enzyme correlates quantitatively with the extent of ion transport. Another clue came from the observation that this ATPase is phosphorylated (having an enzyme–phosphate bond) at an aspartate site (**3**) only in the presence of Na^+ and Mg^{2+} ions. Moreover, the phosphorylated product is hydrolyzed if K^+ ions are present. These observations are summarized by the catalytic cycle shown in Fig. 19.1. It was also observed that the enzyme undergoes a conformational change when it is phosphorylated.

Using these clues, the mechanism outlined in Fig. 19.2 was proposed. First, ATP and three Na^+ ions bind to the inside of the membrane and the enzyme is phosphorylated. The product of the reaction undergoes a conformational change called **eversion**, which is like motion through a revolving door, that brings the bound Na^+ ions to the outside of the cell membrane. There the three Na^+ ions are replaced by two K^+ ions. The attachment of the K^+ ions triggers dephosphorylation, and the loss of the ATP induces a conformational change that carries the two K^+ ions into the interior of the cell, where they are released. The process builds up a charge gradient across the membrane because three Na^+ ions are expelled for two K^+ ions incorporated, and the outer surface becomes relatively positively charged.

An observation that supports the mechanism is that minute (10^{-9} M) concentrations of vanadate ions (VO_4^{3-}) can inhibit the operation of

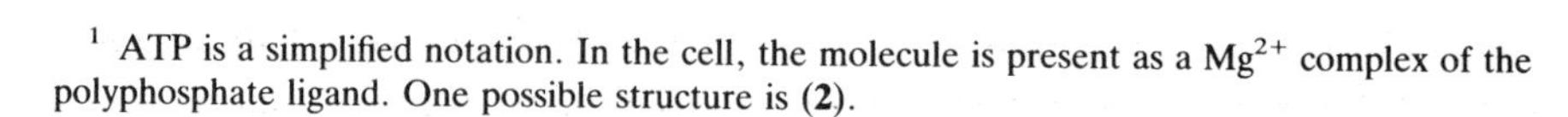

2 MgATP

3 β-Aspartyl phosphate

[1] ATP is a simplified notation. In the cell, the molecule is present as a Mg^{2+} complex of the polyphosphate ligand. One possible structure is (**2**).

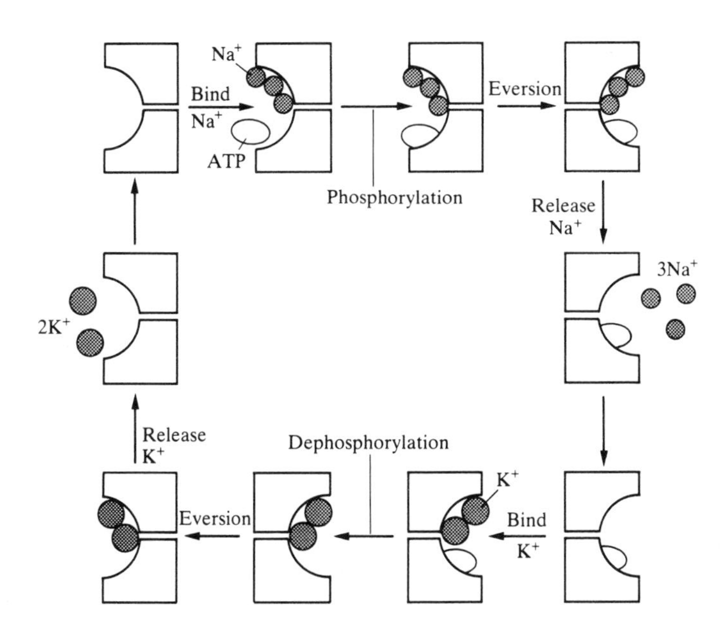

Fig. 19.2 A schematic diagram of the mechanism of sodium pumping showing the role of eversion of the enzyme which transfers ions between the inside and outside of the cell.

ATPase. The difference between phosphate and vanadate ions is that the five-coordinate phosphate intermediate formed when water attacks is unstable and readily breaks down completing hydrolysis. However, the five-coordinate species that vanadate can form does not break down, the conformational change does not ensue, and the bound K^+ ions remain on the outside of the cell membrane.

The selectivity of the process

The novel feature of active transport from an inorganic standpoint is the selectivity of complexation between K^+ and Na^+, which is uncommon in simple systems. We need to examine how the enzyme can be selective first to Na^+ and then to K^+ in its different conformations. We have seen that Group 1 cations are usually not strongly complexed, so selectivity in simple systems is mainly dependent upon coulombic forces and differences in ionic radii. Thus, bases that are stronger than H_2O will bind preferentially to the smaller, hard acid Na^+, and bases that are weaker than H_2O will bind preferentially to K^+ because displacement of H_2O from this softer cation is easier. However, complexes formed with ligands that are weaker bases than H_2O will be very weakly bound indeed, so it is difficult to see how to make a practical ligand selective for K^+ ions.

One possible solution to the problem is revealed by another biochemically important ligand. It is known that crown ethers (**4**) that are antibiotically active increase the permeability of cell membranes to cations. The chelating ligand shown in (**4**) binds both Na^+ and K^+ ions strongly but favors K^+ by a factor of nearly 100. The origin of the selectivity is stereochemical, for a K^+ ions fits neatly into the cavity in the hollow crown. Thus selectivity between Na^+ and K^+ can be achieved by a ligand that is able to adopt conformations providing a chelating site of exactly the right size.

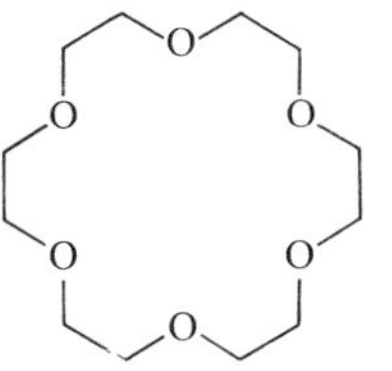

4 Crown ether

Calcium biochemistry

Calcium is the major cation in the structural materials bone and shell, where its role in these hard tissues stems from the insolubility of its carbonates and phosphates, such as calcite, $CaCO_3$, and apatite, $Ca_{10}(PO_4)_6(OH)_2$. However, Ca^{2+} ions also have a very diverse non-structural role, for they function as messengers for hormonal action, in the triggering of muscle contraction and nerve signals, in the initiation of blood clotting, and in the stabilization of protein structures. This diversity is despite the fact that the Ca^{2+} ion is not known for forming robust complexes with simple ligands in solution. It must mean that, as in the case of Na^+ and K^+, ligands are involved that differ significantly from the familiar simple ones.

One muscle protein has a formation constant of approximately 5×10^4 for Ca^{2+} binding. The same site can also bind other metal ions, and the order of formation constants is

$$Ca^{2+} > Cd^{2+} > Sr^{2+} > Mg^{2+}$$

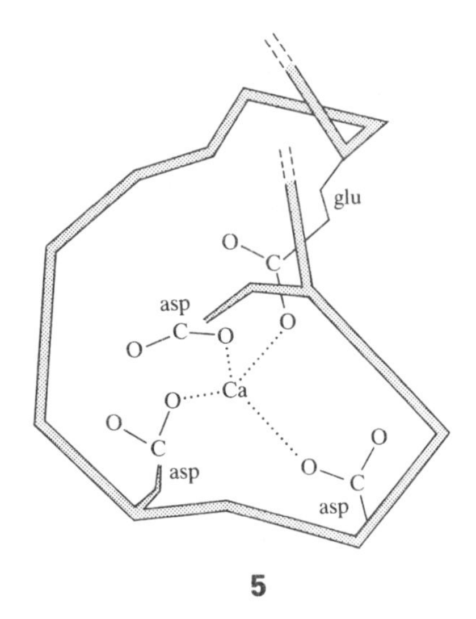

5

We can suspect that, since the site favors Ca^{2+} over Cd^{2+}, it must contain fairly hard ligands. On the other hand, simple electrostatic considerations would lead us to expect that Mg^{2+} should be bound most strongly since it is the smallest of the ions. It is most plausible that a hard site precisely tailored in size to chelate the Ca^{2+} ion exists. Calcium binding proteins are typically rich in aspartate (asp) and glutamate (glu), both of which have carboxylate groups as side chains and hence can act as hard anionic ligands. As an illustration of the kind of structure that occurs, one Ca^{2+} ion binding site that has been characterized by X-ray diffraction consists of four carboxylate O atoms in a distorted tetrahedral array (**5**).

19.2 Oxygen transport

Oxygen transport is a recurring theme in this chapter since its role in respiration and its production in photosynthesis provide examples of a range of biologically important redox reactions, electron transfers, atom transfers, and photochemical processes. We shall set the stage by considering the processes that control the supply of molecular oxygen.

The distribution of oxygen carriers

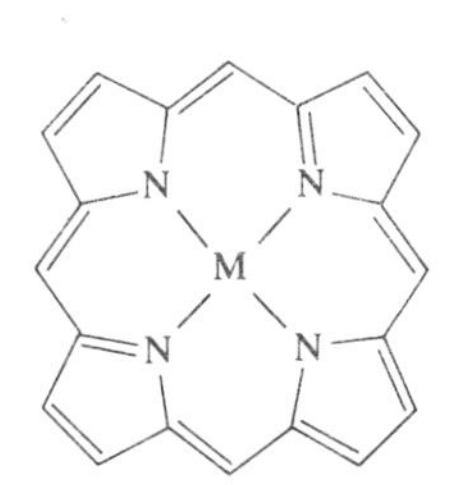

6 Metalloporphyrin

All oxygen carriers that contain iron are found inside cells, whereas carriers that contain copper are also found in extracellular fluids. A part of the reason for the distinction is that the cell interior is a reducing medium that can sustain iron as Fe(II). In addition, the cell supplies a porphyrin ring (**6**) as the ligand that can form a non-labile Fe(II) complex. Since the porphyrin ring ligand is susceptible to oxidative attack it also benefits from the protection of the reductases in the cell. On the other hand, both Cu(I) and Cu(II) can form strong complexes with the imidazole side-chains of proteins and the copper-based carriers may survive free in the bloodstream. The iron and copper carriers can play distinct roles in the same organism since they adopt different strategies for managing oxygen concentration gradients in different compartments.

The three most common oxygen carriers have in common the feature that they are all based on helical proteins. These coiled proteins act like springs that can respond to the strain generated when oxygen binds at one site and can transmit that strain to other sites.[2] Thus, the chemistry of O_2

[2] R. J. P. Williams, *Chimica Scripta*, **28A**, 5 (1988). The argument just quoted is developed in this paper, which surveys the biological reactions of oxygen.

complexation can be a combination of effects of complexation at the metal site and of alteration of the protein environment. The oxygen carriers are not very selective to the identity of the incoming π-acid ligand, except in respect of its size. Thus, they bind NO, CO, CN^-, RNC, N_3^-, and SCN^- as well as O_2.

Structure of the O_2 binding site at Fe(II)

We shall examine the oxygen carrier proteins hemoglobin (Hb) and myoglobin (Mb), both of which are built round Fe(II). The oxidized forms, metmyoglobin and methemoglobin, contain Fe(III) and do not bind O_2.

Naked 'heme', the Fe–porphyrin complex without the accompanying polypeptide, is oxidized irreversibly to Fe(III) by molecular O_2 to give a μ-O_2 bridged dimeric product. This would seem to be a fatal flaw for its biological function. However, this unproductive reaction is prevented by the protein environment in which the heme is imbedded in Hb or Mb.

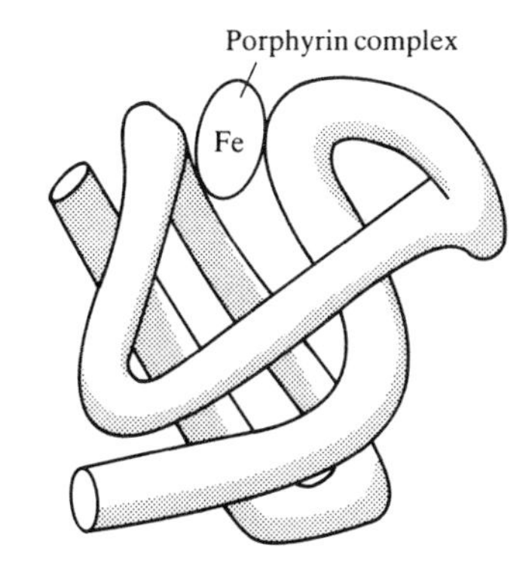

Fig. 19.3 An outline of the structure of myoglobin. The tubular structure represents the polypeptide chain.

The structure of Mb is shown in outline in Fig. 19.3. The porphyrin active site is entwined in the polypeptide chain, which is represented by the tubes in the figure. The protein pocket in which the heme is carried is formed from amino acid residues with nonpolar side groups, and is hydrophobic. The same groups block access of larger molecules to the Fe atom neighborhood and so prevent the formation of the Fe—O_2—Fe bridged species involved in the oxidative reaction of protein-free heme. The hydrophobic groups also deny solvation to ions produced in the oxidation of the heme complex. The result is that the Fe(II) complex can survive long enough to bind and release O_2. This is a typical example of the delicate control of reaction environments that proteins are able to exercise.

The oxygen-free form of Mb is a five-coordinate high-spin Fe(II) complex with four of the coordination positions occupied by four of the porphyrin ring N atoms and the fifth by the N atom of an imidazole ligand of a histidine residue (Fig. 19.4) which couples the heme to the protein. Such five-coordinate heme complexes of Fe(II) are always high spin. The diameter of a high-spin $t_{2g}^4 e_g^2$ Fe(II) ion is larger than that of the central hole in the porphyrin ring, and it lies about 0.40 Å above the plane of the ring (**7**). High-spin Fe(II) porphyrin complexes, including the active site in Hb and

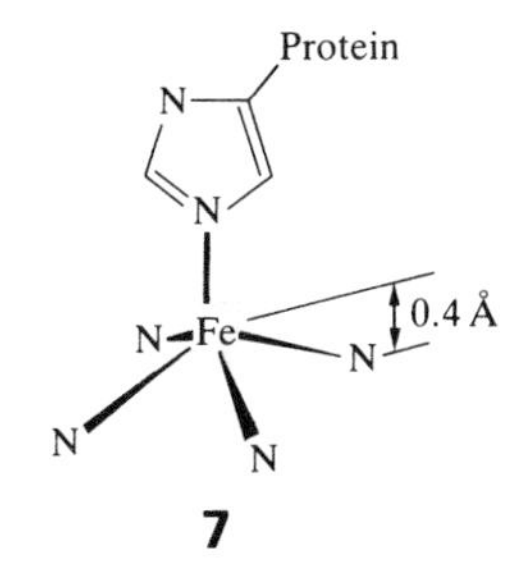

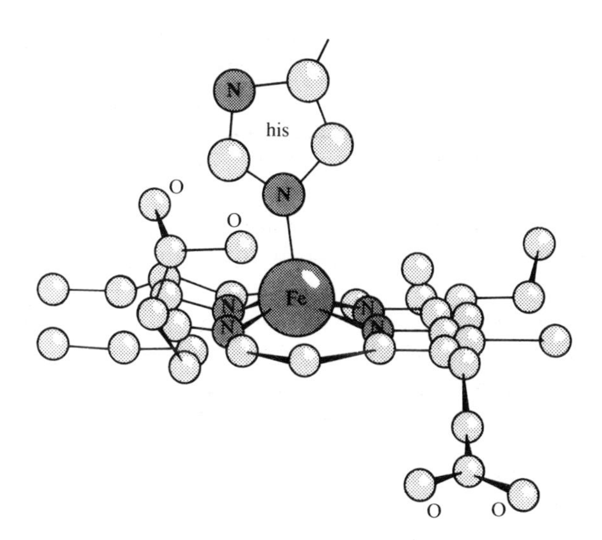

Fig. 19.4 The coordination environment of iron in myoglobin.

8

Mb, involve puckering and twisting of the porphyrin macrocycle, whereas a low-spin Fe(II) ion has a smaller diameter and can fit comfortably into the ring. When O_2 completes the six-coordination environment, the complex converts to low spin (corresponding to the t_{2g}^6 configuration of Fe(II)), the Fe ion shrinks a little, and moves into the plane (**8**). This structural change is verified by X-ray diffraction.

Models of O_2 binding

Several investigations of the interaction of O_2 with metal complexes have been motivated by the hope of understanding the function of oxygen carriers. The results of the work have contributed not only to unraveling the biological problem but also to an understanding of the pathways by which molecular O_2 functions as an oxidizing agent. Among the *d*-block metals that bind O_2 and might provide helpful model systems, it is probably cobalt that has provided the best general picture of the oxidative reactions of oxygen.

Like simple Fe(II) complexes, Co(II) complexes react with O_2 with electron transfer:

$$\mathrm{LCo} + \mathrm{O_2} \rightarrow [\mathrm{LCoO_2^-}]$$

Oxidation number: +2 +3

The product is formally a Co(III) complex of the superoxide ion, O_2^-. It reacts very readily with a second Co(II) complex to give a bridged complex of the peroxide ion O_2^{2-}:

$$[\mathrm{LCoO_2^-}] + \mathrm{LCo} \rightarrow \mathrm{LCo{-}(O_2^{2-}){-}CoL}$$

Oxidation number: +3 +2 +3 +3

9

The structure of the product $[(NH_3)_5Co{-}O_2{-}Co(NH_3)_5]^{4+}$ has been determined (**9**), and the O—O bond distance of 1.47 Å is satisfactorily close to the typical peroxide bond length of 1.49 Å.

The Co(II) complexes of ligands such as salen (the Schiff base chelating ligand in **10**) and a base such as pyridine react with O_2 in a rapid, reversible reaction which is suggestive of Mb or Hb chemistry:

$$[\mathrm{Co(salen)(py)}] + \mathrm{O_2} \rightleftharpoons [\mathrm{Co(O_2)(py)(salen)}]$$

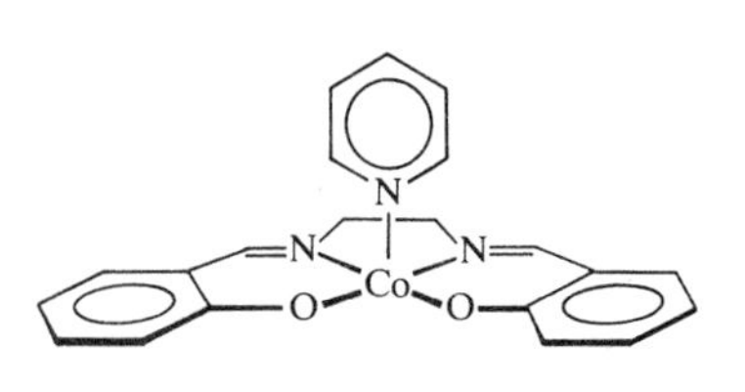

10 [Co(salen)(py)]

An X-ray study of the complex reveals an O—O length of 1.26 Å, which is very close to the bond length in free molecular O_2. However, EPR spectra are consistent with a single unpaired electron interacting with the ^{59}Co nucleus. The latter observation suggests that an O_2^- ligand is coordinated to a low-spin Co(III) center. The single unpaired electron cannot be on the Co atom since a high-spin d^7 complex has three unpaired electrons, and studies of complexes with comparable ligands show that the ligand field is not strong enough to force the complex to have the low-spin configuration (which would have one unpaired electron). A low-spin d^7 formulation is also not supported by an interpretation of the EPR hyperfine coupling constants (Section 14.9). It will be seen to be important for the understanding of Mb and Hb that the cobalt in this model compound is a low-spin d^6 ion coordinated by a superoxide ligand, O_2^-.

One additional proposal about the bonding of O_2 to Co accounts for the long O—O bond lengths that are observed, and takes note of the fact that O_2 is a σ-base and a π-acid. It is suggested that the bonding involves σ

donation of O_2 lone pair electrons and back-donation to the π^* orbitals of O_2 from the d_{xz} or d_{yz} orbitals of Co. Such back-donation weakens and lengthens the O—O bond. The π-acidity of the ligand *trans* to O_2 will exert a substantial influence, for a good π-acid withdraws electron density from Co and hence weakens the O_2 binding.

Example 19.1: *Diagnosing the extent of electron transfer*

Figure 19.5 shows that the logarithms of the equilibrium constant for O_2 binding by cobalt Schiff base complexes with various axial ligands B in the complex [Co(Schiff base)(B)] correlate linearly with reduction potentials for the Co(III)/Co(II) complex couples. What can you conclude from this relation?

Answer. Since log K is proportional to the $\Delta G^\ominus$ of the reaction, and the reduction potential is also proportional to a $\Delta G^\ominus$, the correlation is a linear free-energy relation (Section 15.6). Since the free energies of reduction and O_2 binding increase together, it is reasonable to suggest that electron transfer from Co (and reduction of O_2) is involved in O_2 binding. However, an increase of 0.4 V (40 kJ mol^{-1}) in $E^\ominus$ leads to an increase in log K of only about 2.1 (12 kJ mol^{-1}). The difference suggests that electron transfer to O_2 is incomplete and that Co(II)—O_2 and Co(III)—O_2^- are idealized extremes.

Exercise. The O—O bond lengths in O_2, KO_2, and BaO_2, are 1.21, 1.28, and 1.49 Å respectively. These values provide reference data on the relation between bond length and oxidation state. For the complexes $[Co(CN)_5O_2]^{3-}$, $[Co(bzacen)(py)O_2]$ (bzacen is an acyclic tetradentate ligand suggestive of salen with two N and two O donor atoms), $[(NH_3)_5CoO_2Co(NH_3)_5]^{4+}$, and $[(NH_3)_5Co(O_2)Co(NH_3)_5]^{5+}$, the O—O bond lengths are 1.24, 1.26, 1.47, and 1.30 Å respectively. Comment on the extent of Co to O_2 electron transfer in each complex.

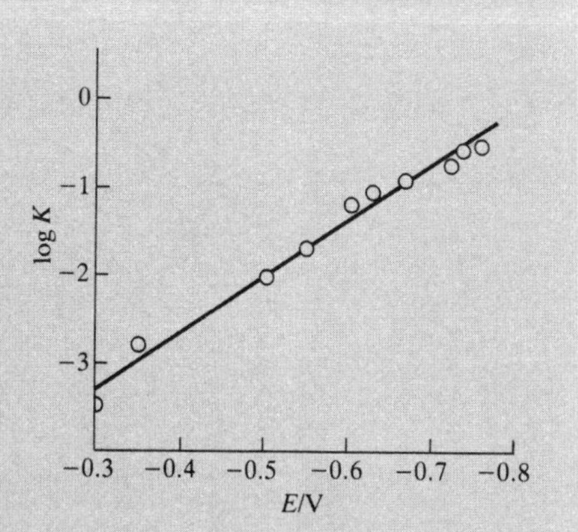

Fig. 19.5 The linear correlation of log K and E for a series of Schiff-base complexes.

Some Fe porphyrin models of reversible O_2 coordination have been synthesized. As we have remarked, the challenge is to prevent binuclear bridged complex formation, and three approaches have been successful. One is the use of steric inhibition in the porphyrin ligand to prevent the approach of a second complex or an additional ligand (as in the biological system itself). A second is the use of low temperatures to slow the dimerization reaction, and the third is to anchor the Fe complex to a surface (e.g. silica gel) so that dimerization is prevented. The first is the one most closely related to the behavior of Mb and Hb.

Steric hindrance of dimer formation is achieved with so-called 'picket fence' porphyrins, where the 'pickets' are a group of blocking substituents projecting from one side of the planar ring. Coordination of a bulky ligand, such as an N-alkylimidazole, can occur only on the unhindered side. An imidazole (Im) is an effective σ-base that favors coordination of a π-acid lying *trans* to itself. It gives the complex an affinity for O_2 similar to that of Mb, and the blocking substituents create a pocket for O_2 and prevent formation of $[Fe(porph)(Im)_2]$. The pickets also prevent reaction with a second Fe center which would give the inactive μ-O_2 species.

One such picket fence complex is shown in Fig. 19.6. It binds O_2 to give a structure very similar to (**8**); the Fe—O—O angle is 136° and the O—O bond length is 1.25 Å. This model system provides the best guidance available about the structure of Mb and Hb oxygen complexes. That the

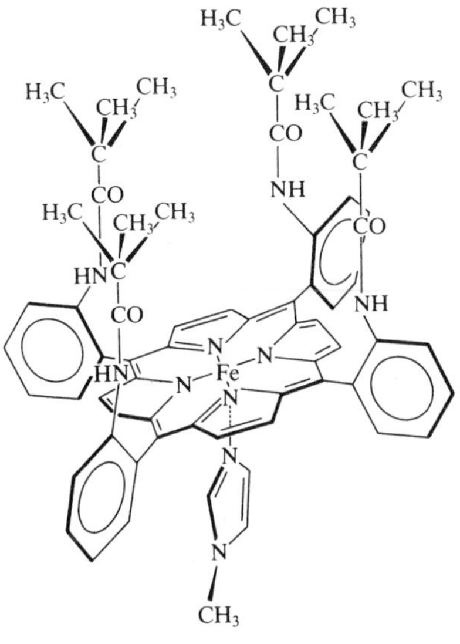

Fig. 19.6 An example of a picket fence porphyrin.

complex is diamagnetic (low spin), might be regarded as evidence of a low-spin d^6 Fe(II) complex of singlet O_2 which would mean that Fe and O_2 reduction is less important than in the Co complexes. However, we must be cautious with this interpretation because other evidence points in a different direction: the O—O stretch lies at 1107 cm^{-1}, which is closer to the O_2^{2-} value of 1145 cm^{-1} than the O_2 value of 1550 cm^{-1}.

The low-spin character could arise as the result of spin pairing of Fe(III) and O_2^-, in which case the complex would be similar to the cobalt model compounds. That such a possibility is very real is underlined by the observation that a d^3 Cr(III) porphyrin complex of oxygen has been synthesized which has only two unpaired electrons. Since d^3 Cr(III) must be high-spin, the only explanation for the spin observed is spin-pairing with an electron from O_2^-.

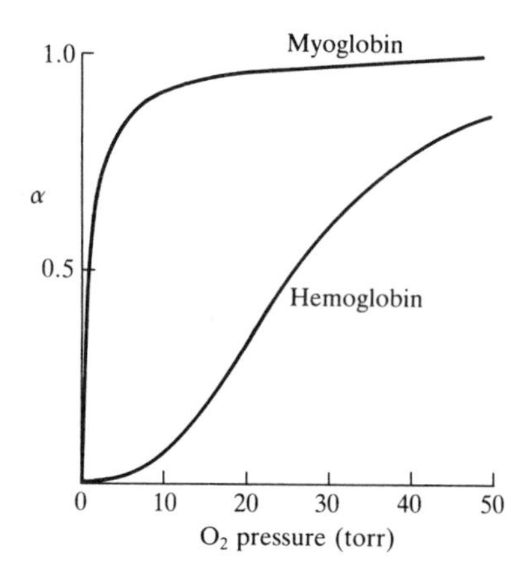

Fig. 19.7 The oxygen saturation curves for Mb and Hb showing the fractional oxygen saturation α as a function of the oxygen partial pressure at pH 7.2.

Hemoglobin and myoglobin functions

The function of hemoglobin (Hb) is to bind O_2 at the high oxygen partial pressures found in lung tissue, to carry it without loss through the blood, and then to release it to myoglobin (Mb). This sequence requires Mb to have a greater oxygen affinity than Hb at low partial pressures.

The oxygen saturation curves for Hb and Mb are shown in Fig. 19.7. The shape of the Mb curve is easily explained in terms of the equilibrium

$$Mb + O_2 \rightarrow MbO_2 \qquad K = \frac{[MbO_2]}{[Mb]p}$$

where p is the partial pressure of oxygen. The **fractional oxygen saturation** α is the ratio of the concentration of the Mb present as MbO_2 to the total concentration of Mb:

$$\alpha = \frac{[MbO_2]}{[Mb] + [MbO_2]}$$

It follows from the equilibrium expression that

$$\alpha = \frac{Kp}{1 + Kp}$$

The Mb curve in the illustration conforms to this equation but the Hb curve is strikingly different.

The Hb curve can be reproduced if the dependence of α on the partial pressure of oxygen is changed to p^n, with n between 2 and 3. Another feature of the Hb binding site is that O_2 binding by Hb is pH dependent despite the absence of an acid group at the heme site, and it is observed that O_2 is released more readily at lower pH. Thus, O_2 is released more readily in cells where the concentration of CO_2 is high. A final detail to accommodate is that the binding of organic phosphates to the protein of Hb far from the heme site also affects O_2 binding.

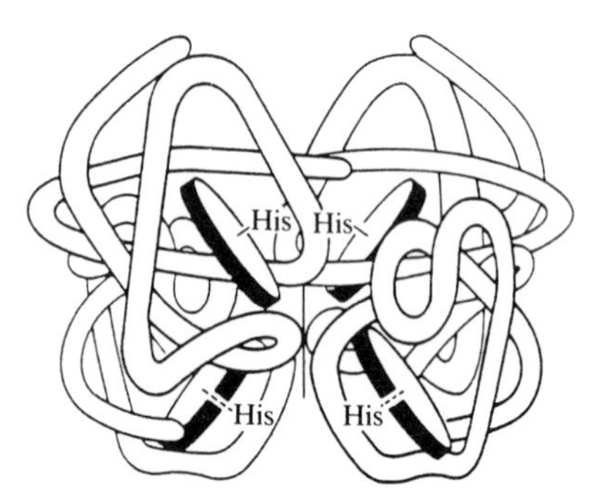

Fig. 19.8 A schematic structure of Hb showing the relationship among the four subunits of the tetramer.

The difference between Mb and Hb is that whereas the former has a single heme group, the latter is essentially a tetramer of Mb, with four heme groups (Fig. 19.8). This difference is crucial, because it allows the four heme units of Hb to take up O_2 cooperatively: once one O_2 is present, the subsequent O_2 molecules can be taken up more readily.

M. F. Perutz, who determined the crystal structure of Hb, argued that one Fe atom in Hb functions as a trigger.[3] When the high-spin Fe atom that lies 0.4 Å above the porphyrin plane changes to low spin and moves into the plane, it pulls the histidine residue of the protein along with it, thus triggering the reorganization of the protein and altering the binding characteristics of the other sites. The pH and phosphate dependence are ascribed to conformational influences stemming from points moderately distant from the metal atom, which underlines yet again the sensitivity of structures to influences at distant sites.

Enzymes exploiting acid catalysis

A metalloenzyme can act as an acid catalyst either on account of the Lewis acidity of the metal ion itself or the Brønsted acidity of a ligand being enhanced by the metal ion. The metal ion in a metalloenzyme may also be able to adopt roles denied to it as a part of a small complex in solution. For instance, the active site of the enzyme may be protected from the bulk aqueous environment in such a way that solvent leveling (Section 5.4) or the effects of solution permittivity are modified. Some of these features have already been seen in the discussion of oxygen transport by Hb.

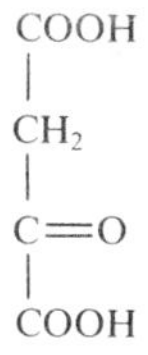

11 Oxaloacetic acid

19.3 Oxaloacetate decarboxylase

We turn now to enzyme catalysts and their models. One of the simplest and most important reactions subject to acid catalysis is CO_2 loss from oxaloacetic acid. This reaction is a step in the widely distributed 'citric acid cycle' of carbohydrate metabolism. The comparison of the enzyme and non-enzyme catalysts provides a good indication of the features that the complexity of proteins can contribute.

The conversion of oxaloacetic acid (**11**, A) to pyruvic acid (**12**) and CO_2

$$HO_2CCOCH_2CO_2H \rightarrow HO_2CCOCH_3 + CO_2$$

CH_3
|
C=O
|
COOH

12 (keto)–Pyruvic acid

is catalyzed by many aqua metal cations. It is accomplished biochemically by the enzyme oxaloacetate decarboxylase which has a metal ion (Mn(II), Zn(II)) at its active site. The rate constant for decarboxylation of the simple complex ZnA is $0.007\ s^{-1}$ at 25°C and that for CuA is $0.17\ s^{-1}$. The uncatalyzed decarboxylation of A^{2-} has a rate constant of only $2 \times 10^{-5}\ s^{-1}$. Notice that the stronger Lewis acid, Cu^{2+}, is only 25 times better as a catalyst than Zn^{2+}. The enzyme-catalyzed pathway, however, is 10^4 times faster than the Cu^{2+}-catalyzed pathway. This is the first glimpse of the striking contribution of protein complexity.

The metal-ion catalyzed reaction is believed to follow the mechanism shown in Fig. 19.9. A ^{13}C kinetic isotope effect of about 6 percent is observed for the metal-ion-catalyzed reaction, which indicates that the cleavage of the C—C bond is rate determining. In contrast, the enzyme-catalyzed reaction shows no ^{13}C isotope effect, but is slower in D_2O than in H_2O. It is clear that C—C bond breaking is no longer the rate-determining step in the enzyme-catalyzed reaction, but that proton transfer is involved.

[3] M. F. Perutz, G. Fermi, B. Luisi, B. Shaanan, and R. C. Liddington give an up-to-date summary of the link between stereochemistry and mechanism in *Acc. chem. Res.*, **20**, 309 (1987).

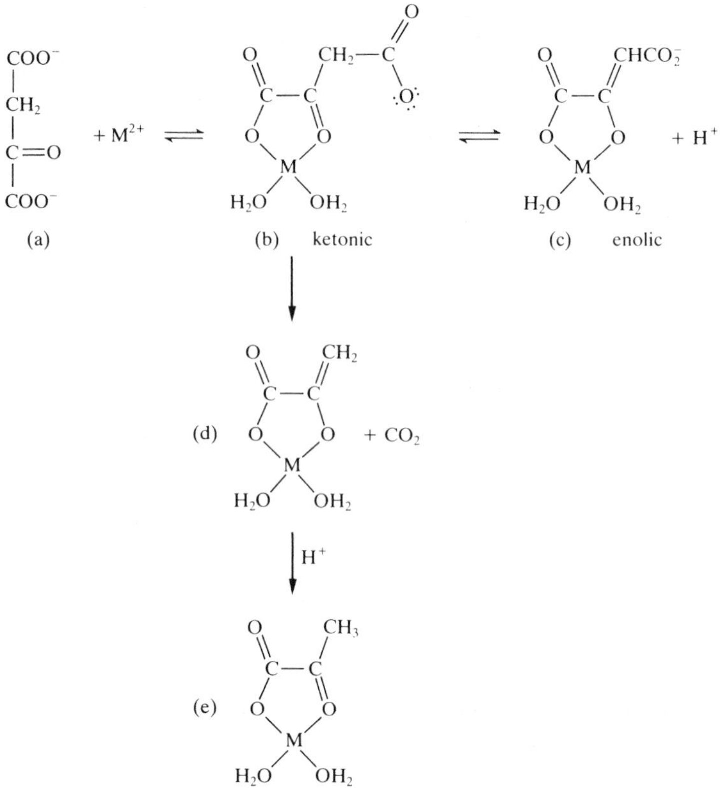

Fig. 19.9 The mechanisms of the conversion of oxaloacetate to pyruvate catalyzed by an M^{2+} Lewis acid catalyst.

(a) (b) ketonic (c) enolic (d) (e)

$$CH_2{=}C(OH){-}COOH$$

13 (enol)–Pyruvic acid

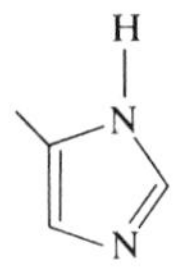

14 Imidazole structure

These observations do not prove that the enzyme-catalyzed reaction follows a different pathway through different intermediates, for if the enzyme can enhance the rate of the C—C cleavage sufficiently, a subsequent step may become rate limiting. The solvent isotope effect does in fact suggest that a proton transfer step has become rate determining. The conversion of the enol form of pyruvate (**13**) to the keto form (**12**), in which the $CH_2{=}$ group has become $CH_3{-}$, is such a step. The rate is determined by the transformation (d)$\rightarrow$(e).

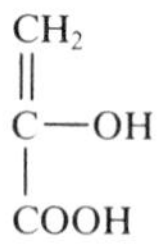

The Lewis acidity of a metal ion can be tailored to a degree by ligand groups that belong to the polypeptide, and the stronger the ligand as a σ-base, the weaker the Lewis acidity of the metal ion. In contrast, the π-acid properties of ligands, such as the conjugated ring ligand imidazole (**14**), may enhance Lewis acidity by accepting electron density from the metal by back donation.

This specific example suggests some general features. One that is not obvious is that in many metal-ion catalyzed reactions the great increase in rate arises primarily from a change in the entropy of activation. For example, the hydrolysis of esters, amides, and peptides involves the formation of a tetrahedral activated complex:

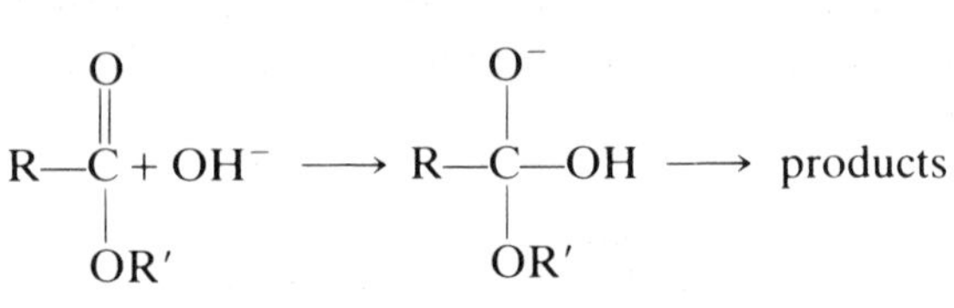

$$R{-}C({=}O)(OR') + OH^- \longrightarrow R{-}C(O^-)(OH)(OR') \longrightarrow \text{products}$$

The negative charge on the carboxyl O atom that develops in the activated complex requires orientation of the solvent around the substrate, and this electrostriction contributes a negative term to the entropy of activation of the order of $-80\,\mathrm{J\,K^{-1}\,mol^{-1}}$. In the metal-ion-catalyzed process, the positive metal center effectively 'solvates' the developing negative charge, and the contribution to the entropy of activation is much less negative.

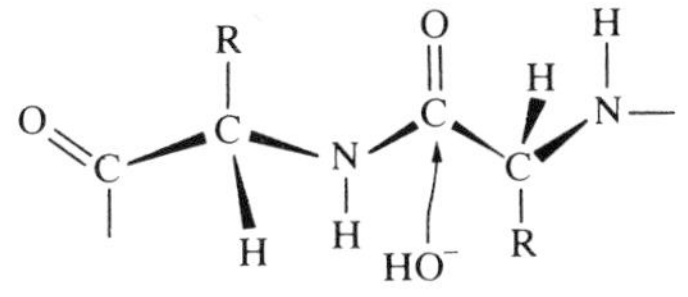

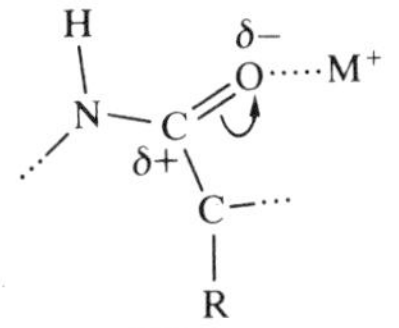

19.4 Carboxypeptidases

We now consider a hydrolytic enzyme. The carboxypeptidases are enzymes containing Zn(II) that catalyze the hydrolysis of the peptide bonds in peptides and proteins:

$$^{-}O_2C{-}CH(R){-}NH{-}C(=O){-}CH(R){-}H \xrightarrow{H_2O} {}^{-}O_2C{-}CH(R){-}NH_3^+ + {}^{-}O_2C{-}CH(R){-}H$$

Bovine carboxypeptidase A is an enzyme composed of 307 amino acids in a single polypeptide chain which binds one Zn^{2+} ion per molecule. The protein has been crystallized, and the hydrogen bonding between groups, the coordination environment of the Zn^{2+} ion, and folding of the protein chain that defines the secondary and tertiary structures have all been identified from X-ray studies. Moreover, comparison of crystals with and without simple substrates show some of the details of the probable reaction precursor.

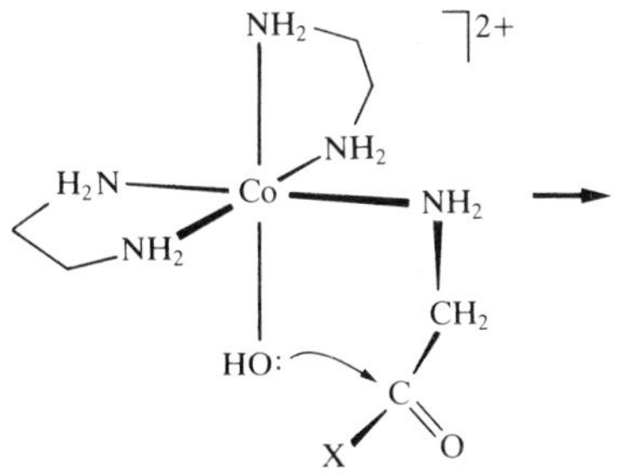

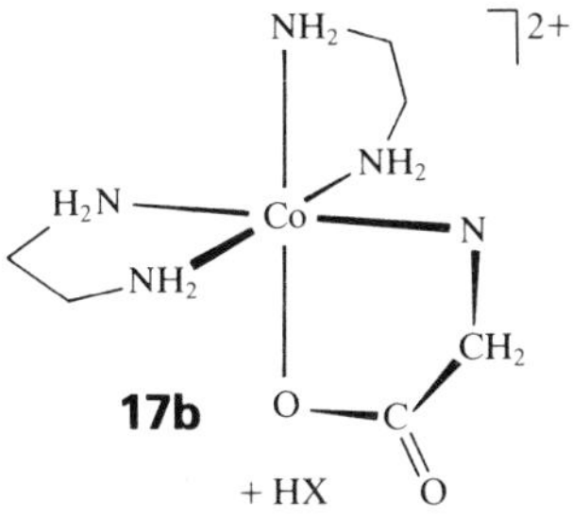

Mechanisms of hydrolysis

The hydrolysis reaction requires the promotion of nucleophilic attack of an O atom on the carbonyl group of the peptide bond (**15**). One way in which a metal ion might function is to promote proton loss from the H_2O molecule to make it into the more nucleophilic OH^- ligand. Alternatively, the metal ion may act as a Lewis acid catalyst by binding the peptide carbonyl group to the metal and so reduce the electron density at its C atom (**16**). In general, either the hydroxo or the Lewis acid mechanism can operate, but to know which path is used by the enzyme requires study of the enzyme structure.

The hydrolysis of peptides and the analogous reaction, ester hydrolysis, has been modeled using octahedral complexes of Co(III), in which the hydroxo mechanism is believed to operate. The hydroxo mechanism is illustrated in (**17**), in which the aqua ligand *cis* to the peptide loses a proton to become a hydroxo ligand (OH^-) that attacks the carbonyl group. The Lewis acid mechanism is illustrated in (**18**) where the carbonyl group is coordinated to Co(III), and the latter is a source of Lewis acid catalysis for the attack of a water molecule or an OH^- ion from the solution. The cobalt(III) model system has been developed into a useful laboratory method for both hydrolytic breakdown and analysis of peptides and the reverse, their synthesis.[4]

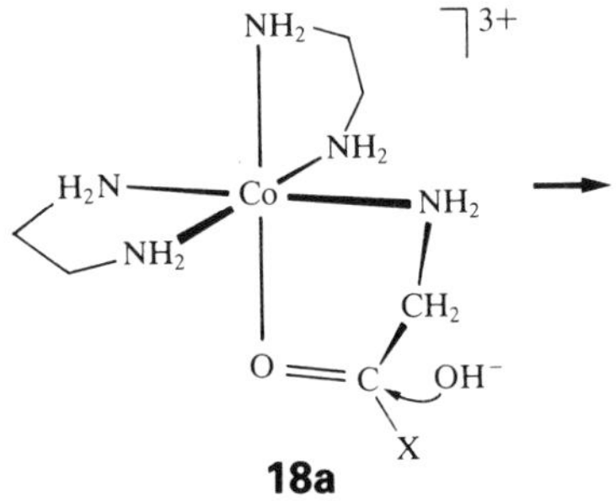

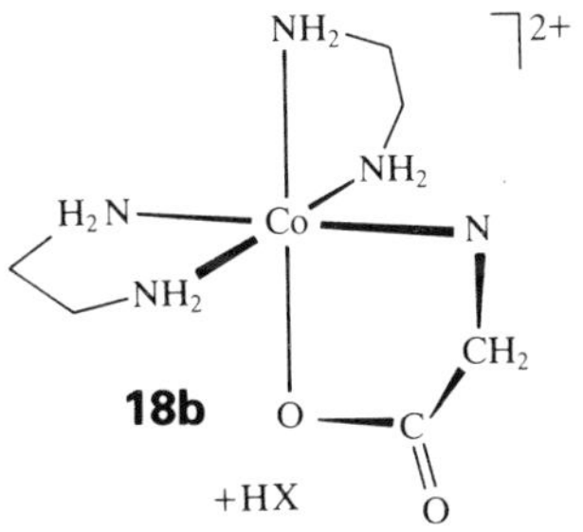

[4] The Co(III) peptide hydrolysis story is well told by P. A. Sutton and D. A. Buckingham in *Acc. chem. Res.*, **20**, 357 (1987).

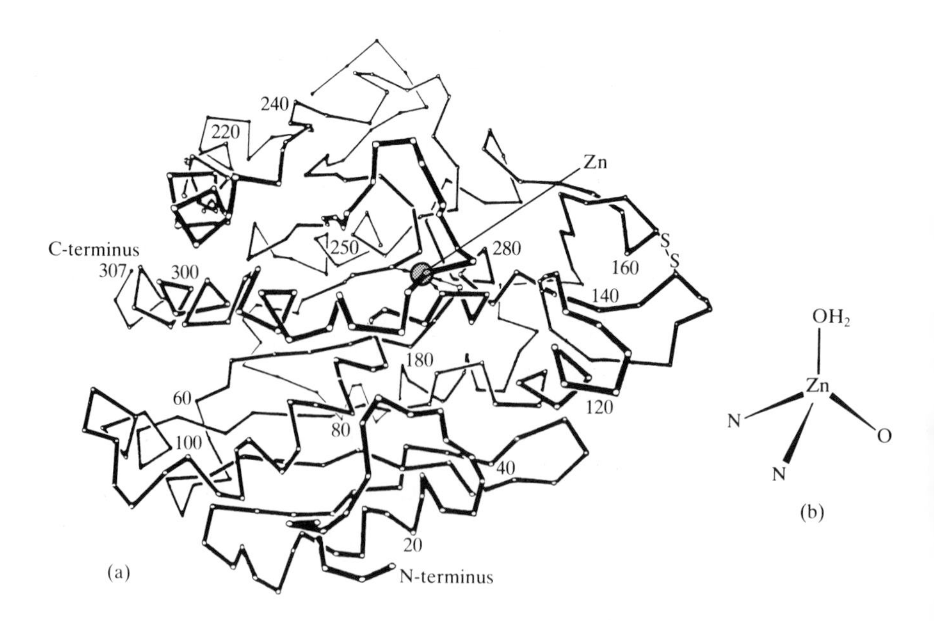

Fig. 19.10 (a) An outline of the structure of carboxypeptidase and (b) the distorted tetrahedral environment of the Zn^{2+} ion. The numbers identify the specific amino acid residues, with the numbering starting at the N-terminus.

Figure 19.10 is a schematic representation of the structure of the actual bovine carboxypeptidase A molecule. The Zn^{2+} ion is situated in a pocket near its center. The four donor atoms of the ligands come from two histidine residues from the protein (69 and 196 in the chain) which are N donors, a glutamic acid residue (which coordinates through an O atom), and an H_2O molecule. The illustration also shows the distorted tetrahedral coordination environment of the Zn^{2+} ion. A second H_2O molecule occupies the enzyme pocket at about 3.50 Å from the Zn^{2+} ion.

As with most enzyme mechanisms, the identification of the correct path is tricky and depends on clever inferences, and more than one possibility must be evaluated. First, we consider the evidence that favors the Lewis acid mechanism of the enzyme's action. Four features are observed when the model peptide substrate glycyltyrosine is bound to the active site. One is that the side chain ($HO—C_6H_4—CH_2—$ in this case) of the terminal amino acid enters a hydrophobic region of the enzyme pocket. Secondly, the terminal carboxylate group interacts coulombically with the positive $—NH_3^+$ group belonging to an arginine residue at site 145. Thirdly, the carboxyl O atom of the peptide that is to be severed displaces the H_2O molecule from the coordination sphere of the Zn^{2+} ion and coordinates in its place. Fourthly, the phenolic OH group of the tyrosine residue at location 248 migrates through about 12 Å to form a hydrogen bond with the NH group of the peptide to be cleaved. The resulting coordination environment is inconsistent with the hydroxo ligand mechanism because no water remains directly coordinated to the Zn^{2+} ion. Therefore, it was proposed that Lewis acid catalysis by Zn^{2+} facilitates nucleophilic attack at the carbonyl C atom.

The details of the Lewis acid mechanism are clarified by noting that the pH of maximum enzyme activity is 7.5. At this slightly basic pH, the carboxylic acid group of the glutamic acid residue at location 270 is present as its conjugate base, the carboxylate group. A hydrogen bond to an H_2O molecule next to the carbonyl group increases the nucleophilicity of the carboxylate and favors attack on the peptide carbonyl. When the carboxylate group attacks, the primary intermediate is a tetrahedral C atom at the site where the peptide is to be broken and an anhydride like structure at the glutamic acid center. That is, glutamic acid is the actual nucleophile which attacks the carbonyl group.

The Lewis acid mechanism just described is based on study of a convenient model of the substrates. It has been criticized on the grounds that the only ligands that can replace an H_2O ligand by an O atom of a carbonyl group at Zn are ligands that form chelates with Zn, and such ligands are not actually substrates susceptible to hydrolysis. That is, the glycyltyrosine model is inappropriate.

A clue to what may be the true mechanism is found in the considerable movement of tyrosine residue 248 toward the active site during substrate binding, which suggests that it is very likely to be involved in the mechanism. Its role has been established using the technique of site-specific mutagenesis[5] to alter the genetic coding for the enzyme from tyrosine at location 248 to phenylalanine. The mutant enzyme is found to have almost the same catalytic activity as in the original enzyme, which suggests that the role of tyrosine is in binding the substrate and not in the cleavage step.

Recent X-ray studies of a model substrate have led to development of an alternative hydroxo mechanism, and the conflicting aspects of the mechanism are now beginning to fall into place.[6] In this picture, the carbonyl group to be attacked is hydrogen bonded to arginine residue 127; an H_2O molecule remains in the coordination sphere of the Zn^{2+} ion and attacks the substrate by the hydroxo mechanism. The primary intermediate in the reaction is a diol structure shown in Fig. 19.11a. This proposal is consistent with the kinetics of the reaction and with observations on binding of analogs of the substrate. Figure 19.11b shows a model compound that has been studied by X-ray diffraction. Notice that the mechanism implies the existence of transient five-coordination at the Zn^{2+} ion.

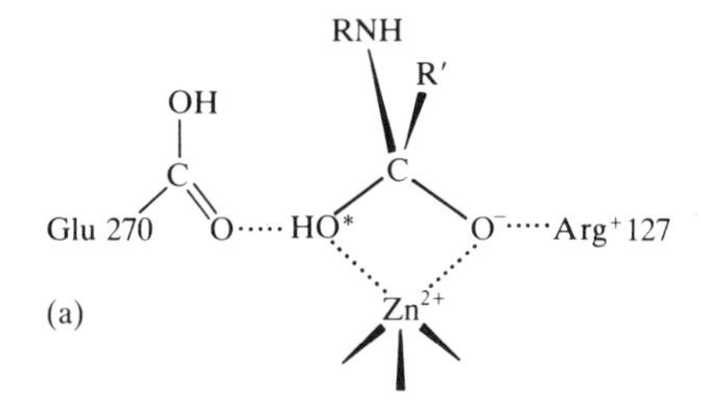

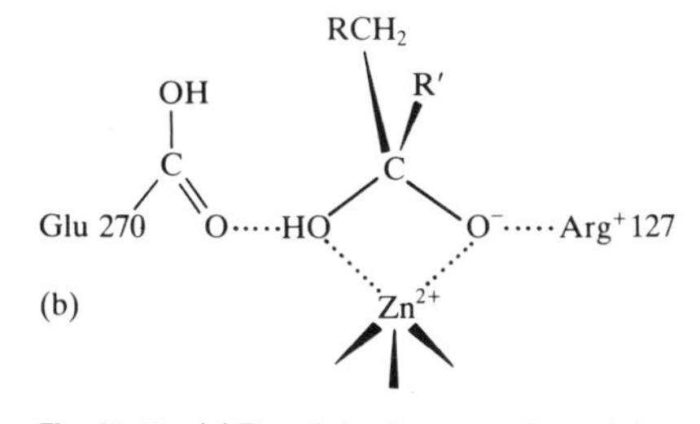

Fig. 19.11 (a) The diol primary product of the reaction of a substrate catalyzed by carboxypeptidase according to the hydroxo mechanism. The starred O atom marks the hydroxo group from the attacking Zn^{2+} ion. (b) An analog of the reaction that has been studied by X-ray diffraction.

Metal substitution

The Zn^{2+} ion of the native enzyme can be replaced by a variety of other M^{2+} ions. The Co^{2+} ion is particularly important because its ligand field spectrum in solution indicates that the ion is in a distorted tetrahedral environment, and hence that the structure determined from the crystal does represent the functioning enzyme in solution.

The order of activity for the hydrolysis of glycyltyrosine by metal-ion substituted carboxypeptidases is

$$\mathrm{Mn(II)} \ll \mathrm{Zn(II)} < \mathrm{Co(II)}$$

Mechanisms that rely on acid catalysis predict a lower activity for Mn(II) because they associate increasing strength of the M—O bond with increasing positive charge at the C atom and stabilization of the activated complex. The Irving–Williams series (Section 7.7) suggests that the Mn—O bond should be the weakest of the three M—O bonds and the Co—O bond the

[5] The technique of site-specific mutagenesis is an extraordinarily powerful tool which has become a central element of the strategy for the study of enzyme mechanisms with the emergence of molecular genetics techniques. This technique changes the genetic coding for protein synthesis to achieve the replacement of a particular amino acid residue in a protein so that its function can be tested.

[6] D. W. Christianson and W. N. Lipscomb, *Acc. chem. Res.*, **22,** 62 (1989) summarize a series of recent studies.

strongest. These conclusions are supported by the IR spectra of model amino acid complexes, for the force constants of the M—O stretch are also in the order

$$\text{Mn(II)} \ll \text{Zn(II)} < \text{Co(II)}$$

Mechanisms of acid catalysis

Carboxypeptidase teaches us that the hydroxo or Lewis acid mechanisms that are available in enzymes are much more elaborate than in simple metal complexes in solution. For instance, the substrate is bound into a favorable place by hydrophobic and ionic interactions. If the attacking nucleophile H_2O is not activated by the metal it may still be activated by a carboxylate group from the protein. The environment in the protein fold may balance aqueous and hydrophobic zones to give the most favorable 'solvent effect', and all these features can operate together.

Example 19.2: *Proposing an enzyme-catalyzed reaction mechanism*

Carbonic anhydrase[7] catalyzes the interconversion of CO_2 to HCO_3^-. The reaction occurs on a millisecond timescale when it is uncatalyzed, and a thousand times faster when catalyzed. Despite the simplicity of CO_2 hydration, the enzyme has many features in common with peptide hydrolysis enzymes. It has a Zn(II) ion coordinated to three histidine N atoms and an H_2O molecule at the active site. Propose a Lewis acid mechanism for attack of water on CO_2 when the latter is bound to Zn at the active site.

Answer. A Lewis acid mechanism would require coordination of CO_2 to the Zn(II) center as Zn—O=C=O. This structure would then be attacked by an H_2O molecule at the C atom to convert it to a coordinated HCO_3^- ion accompanied by proton transfer to the enzyme.

Exercise. The X-ray data on crystallized carbonic anhydrase tend to argue against a Lewis acid mechanism. The active site includes a number of H_2O molecules in a very well-ordered array which suggests that CO_2 is not coordinated, but that the coordinated H_2O is activated for nucleophilic attack by the Zn for nucleophilic attack on a C atom in a hydroxo mechanism. Propose a mechanism for the carbonic anhydrase catalyzed attack of water on CO_2 that is related to the hydroxo mechanism discussed above for carboxypeptidase A.

Redox catalysis

Since photosynthesis and respiration, the two primary energy conversion processes, involve redox reactions, it is clear that a major metabolic role must be played by the enzymes that catalyze oxidation and reduction. As we saw in Chapters 8 and 14, such reactions may occur by electron transfer, atom transfer, and group transfer, and metalloenzymes can exploit all three possibilities. Metalloenzymes are also participants in photochemically induced electron transfer in the steps that initiate photosynthesis.

[7] Carbonic anhydrase is reviewed by D. N. Silverman and S. Lindskog in *Acc. chem. Res.*, **21**, 30 (1988). They discuss a hydroxo mechanism, and alternatives where proton transfer is rate determining.

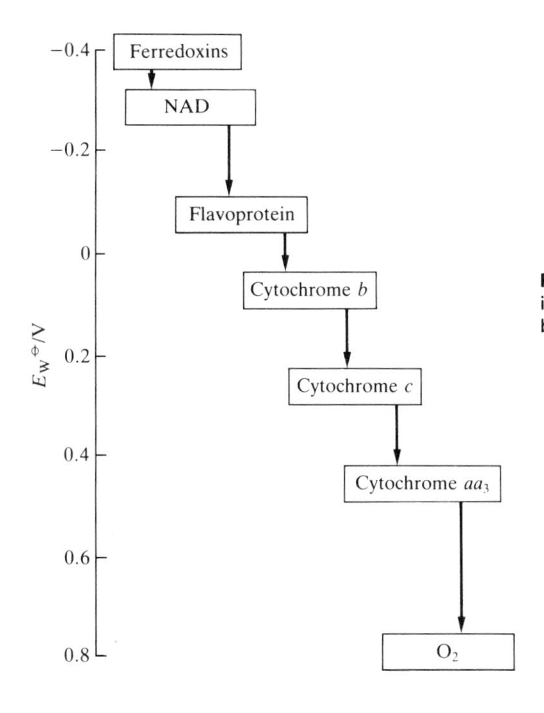

Fig. 19.12 Reduction potentials of some important electron transfer mediators in biological cells at pH = 7.

Reduction and oxidation in a living cell does not occur in a single step but usually involves a series of common compounds, called **mediators**, which act like a series of locks on a canal, allowing oxidation to occur in stages. Once again, control is of paramount importance, for uncontrolled oxidation by oxygen is combustion.

Figure 19.12 illustrates the range of reduction potentials of some mediators found in mitochondria (the organelles in which oxygen utilization is accomplished). They include the heme-containing cytochromes and the flavoproteins, which also mediate the reactions that insert oxygen into organic molecules. Nicotine adenine dinucleotide (**19**, NAD) is a widely distributed substance that participates in reductions by achieving the equivalent of H^- ion transfer.

19 NAD

19.5 Cytochromes of the electron transport chain

Oxygen is a powerful and potentially dangerous oxidizing agent. If the cell is to use it in metabolism, it must keep a safe distance between the sites of oxidation by O_2 and the variety of metabolic reactions being fueled. This isolation is accomplished by the series of proteins of decreasing reduction potential which pass electrons toward oxygen in the mitochondria of the

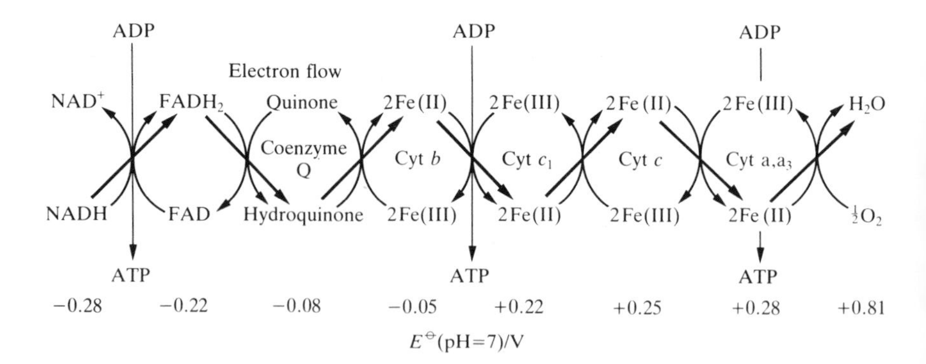

Fig. 19.13 The sequence of reactions in the electron transport chain in the mitochondria of cells where oxygen is utilized.

cell. A schematic version of the mitochondrial electron transport chain is shown in Fig. 19.13.

There are large Gibbs free energy changes at two stages in Fig. 19.13, one at the reaction between the pair of cytochromes denoted cyt b/cyt c_1 and the second where cytochrome a reacts with oxygen (denoted cyt a/O_2). The energy released in these stages can be stored for use elsewhere in the cell by coupling the reaction to the formation of ATP from ADP and HPO_4^{2-}. The chain might seem unnecessarily complex, but it breaks the reaction of O_2 with NAD into discrete steps. This allows coupling to the synthesis of several ATP molecules as well as organizing the process spatially from one side of the membrane (where the cytochromes are located) to the other.

Outer-sphere character

Cytochromes function by shuttling iron between Fe(II) and Fe(III) at the active site and hence are one-electron transfer reagents. The Fe atoms are in porphyrin ring coordination environments buried in the middle of the protein. Despite the complexity of the complexes it is possible to model the encounter complex that is formed when two of these proteins meet to exchange an electron. Figure 19.14 shows a computer model of the encounter between two cytochromes, and it is easy to believe that because the Fe atoms of the two encountering complexes always remain far apart, the reaction cannot be a case of inner-sphere electron transfer. The reaction must take place by long-distance outer-sphere electron transfer.

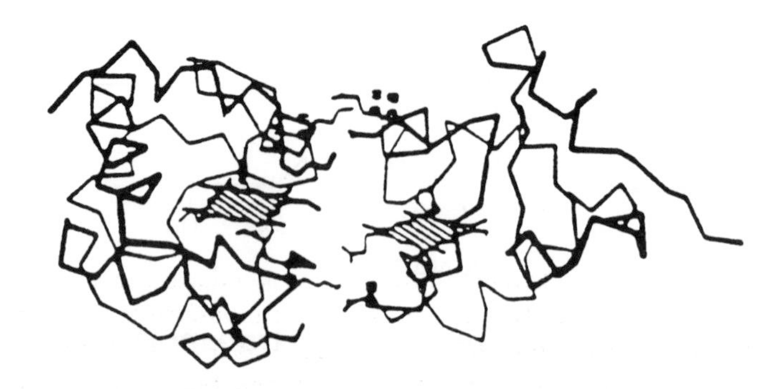

Fig. 19.14 One stage in a computer simulation of the encounter between two electron transfer proteins. The shaded areas are the iron porphyrins.

Table 19.5 shows some rate constants, reduction potentials, and distances of closest approach of the edge of the porphyrin ring containing the Fe atom and its electron transfer partner. It is instructive to compare these protein–protein parameters with similar parameters for reactions of a

Table 19.5. Some rates of electron transfer reactions among redox proteins†

Couple	$E^{\ominus}/\mathrm{V}$	$R/\text{Å}$‡	k/s^{-1}
$Fe^{II}cytb_5/Fe^{III}cytc$	0.2	8	1.5×10^3
$Znapocytc^*/Fe^{III}b_5$	0.8	8	3×10^5
$H_2porfc^*/Fe^{III}b_5$	0.4	8	1×10^4
$Zncytb_5/Fe^{III}cytc$	1.1	8	4×10^3
RuHis33/cyt*c*	0.15	11	40
RuHis/azurin	0.2	10	2.5
RuHis/Mb	0.05	13	0.02
$Fe^{II}ccp/Fe^{III}cytc$	0.4	16	0.25
$Fe^{III}ccp/porfcytc$	1.0	16	180

† Most of the arcane abbreviations are designations of proteins. They are elaborated in biochemical texts. For our purposes we do not need to know the details. All that the different abbreviations mean here is that different members of the family of electron transfer proteins are associated with the different values of $E^{\ominus}$ and RuHis refers to the group $(NH_3)_5Ru$ attached to a histidine (number indicated if necessary) on the enzyme. Distances can be varied by changing the histidine unit to which the Ru is bound. The notation * indicates a reaction using a photoexcited state to make the reaction more exoergic.

‡ Distance between electron donor and acceptor metal centers in Å.

protein with simple metal complexes, and some of these are included in the table too. The distance dependence of the rate constant is usually interpreted in terms of quantum mechanical tunneling, in which a particle may escape through a barrier despite having insufficient energy to surmount it. In simple models of tunneling, the probability of finding the particle outside the barrier declines exponentially with the width of the barrier. The data in Table 19.5 confirm that the distance between the metal centers has the expected effect if comparisons are limited to similar structures.

Applications of Marcus theory

A question posed by the data in Table 19.5 is whether the theories developed to explain outer-sphere electron transfer between small molecules also work for large molecules that make contact only at large distances. For example, we saw in Section 15.12 that the Marcus theory allows us to express the rate constant k_r of any outer-sphere electron transfer as

$$k_r^2 = k_1 k_2 K f \qquad f \approx 1$$

and equivalently

$$\ln k_r = \ln (k_1 k_2)^{1/2} - \frac{\Delta G^{\ominus}}{2RT}$$

The product $k_1 k_2$ of self-exchange rate constants reflects the intrinsic barrier to electron transfer of the two complexes. The equilibrium constant K is a measure of the overall reaction free energy $\Delta G^{\ominus}$. Unfortunately, the intrinsic barrier is not easily estimated for proteins because it is not feasible to study self exchange between two oxidation states of a protein.

The best approach seems to be to consider the outer-sphere reaction of a protein and a small molecule. In such cases the reaction free energy and the self-exchange rate of the small molecule couple can be known. The difficulty lies only in finding the self-exchange rate constant for the protein couple. If we suppose that the Marcus theory does indeed apply to the reaction, we can calculate a self exchange rate constant for the protein couple using the known values of k_r and k_1. Then we can use the value of k_2 so obtained to discuss other reactions of the protein. In general, this approach is quite successful, and supports the view that the Marcus relation is valid.

The variation of k_r with the overall reaction free energy (and, equivalently, with the reduction potentials of the proteins through $\Delta G^{\ominus} = -FE^{\ominus}$) is also as predicted by Marcus theory for proteins with similar structures. However, the structure also plays a role in determining k_r, and different types of protein structures have rate constants that may differ by orders of magnitude. For instance, the first three entries in Table 19.5 form a series with the dependence on $E^{\ominus}$ predicted by the Marcus equation. Similarly, the last two entries may also represent a structurally similar pair. The role of the enzyme structure in determining the intrinsic barrier is clearly indicated by comparing the fourth entry to the first three. The potential is more favorable than any of the first three yet the rate constant is smaller.

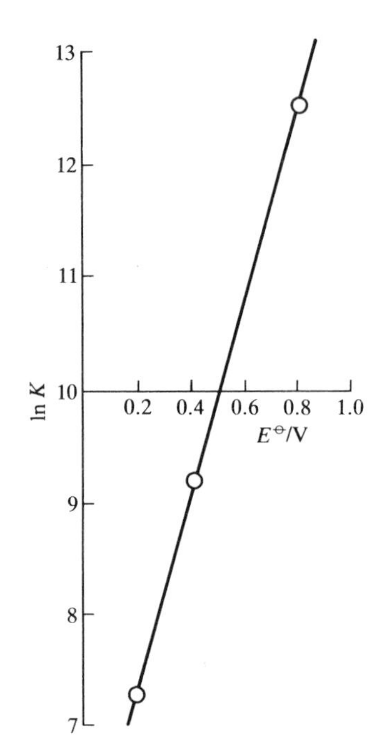

Fig. 19.15 The plot of the data in Example 19.3.

Example 19.3 *Testing the Marcus theory*

Test the asssertion that the first three points in Table 19.5 show the dependence on Gibbs free energy expected from Marcus theory.

Answer. To test the theory, we plot the logarithm of the rate constants against the overall free energy change for the electron transfer. Since $E^{\ominus}$ is proportional to the reaction free energy we may use the reduction potentials. The logarithms of the first three rate constants are 7.31, 12.61, and 9.21. The corresponding potentials are 0.2, 0.8, and 0.4 V respectively. The straight line (Fig. 19.15) confirms the approximately linear relationship predicted by the theory.

Exercise. Compare the first and the sixth, the third and eighth, and the fourth and ninth entries in Table 19.5. The $E^{\ominus}$ values are nearly the same but rates and distances differ. Compare the change in distance to change in $\log k_r$. Do these results confirm exponential distance dependence of the intrinsic barriers to electron transfer or must another factor be invoked?

The general conclusion appears to be that the cytochromes and several other redox proteins are simple outer-sphere one-electron transfer reagents. However, an important aspect is that the protein functions to keep the electron transfer clear of H_2O molecules. In that way the protein eliminates the solvent reorganization energies that contribute to the intrinsic barrier to reaction between the small molecules in solution.

Models of long-range electron transfer

The ability of redox proteins to transfer an electron over long distances is the most fascinating aspect of their behavior, and its recognition has raised a new basic problem for electron transfer theory. Some information on the relevant factors is beginning to emerge from the study of models of long range transfer where an electron donor is attached to an electron acceptor by a rigid and well-defined link. An example of such a link is the rigid three-ring system separating a donor group (D) from an acceptor group (A) which is shown in (**20**).[8] Some evidence is emerging from the study of various structures related to (**20**) that long-range electron transfer occurs along bonds and not across the shortest distances through free space.

D A

20

19.6 Cytochrome P-450 enzymes

'Cytochrome P-450' is the designation of a family of enzymes with iron porphyrin active sites that catalyze the addition of oxygen to a substrate. The most important representative of this class of reactions is

$$\mathrm{R{-}H} + \tfrac{1}{2}\mathrm{O_2} \rightarrow \mathrm{R{-}OH}$$

The insertion of O into an R—H bond—which is just the sort of redox reaction likely to occur by an atom transfer mechanism—is a part of the body's defense against hydrophobic compounds such as drugs, steroid precursors, and pesticides. The hydroxylation of RH to ROH renders the target compounds more water soluble and thereby aids their elimination. The name 'P-450' is taken from the position of the characteristic blue to near-ultraviolet absorption band of the porphyrin called the 'Soret band' which is red-shifted to 450 nm in carbonyl complexes of these molecules.

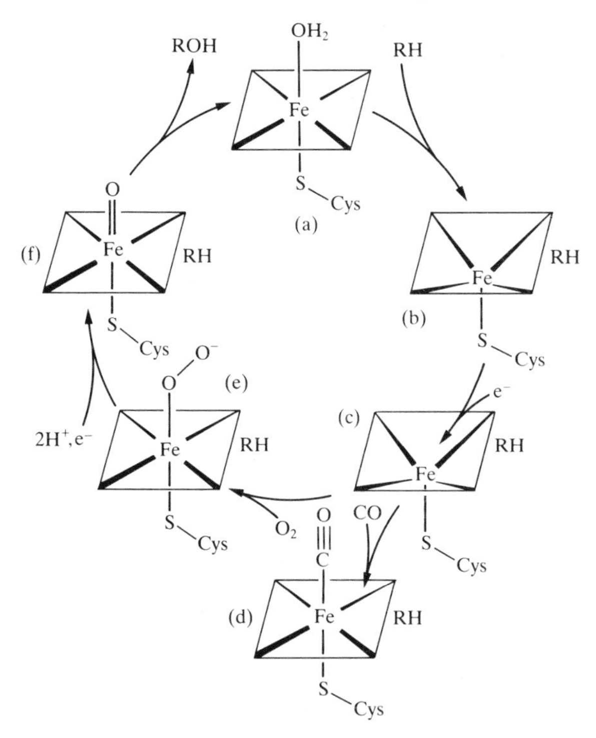

Fig. 19.16 The cycle of reactions of P-450. The resulting state of the enzyme is (a) and the important Fe(IV) oxo species is (f).

The proposed catalytic cycle for P-450 is shown in Fig. 19.16. The sequence begins at (a) in the figure with the enzyme in a resting state with iron present as Fe(III). The hydrocarbon substrate then binds (b), and one electron is transferred (c). The resulting Fe(II) complex with bound substrate proceeds to bind O_2 (e). (At this point in the cycle, a competing

[8] Physical factors controlling long-range electron transfer are analyzed in G. L. Closs and J. R. Miller in *Science*, **240**, 440 (1988). An excellent survey of long-range electron transfer in enzymes is given by G. McLendon in *Acc. chem. Res.*, **21**, 160 (1988).

reaction with CO to give (d) leads to a species which is easily identified and is responsible for the absorption at 450 nm which gives the family its name.) A key reaction is the reduction of the porphyrin ring of the oxygen complex (e) by a second electron, which produces the ring radical anion. Uptake of two H^+ ions then leads to the formation of the Fe(IV) complex (f) which attacks the substrate to insert oxygen. Loss of ROH and uptake of an H_2O molecule at the vacated coordination position brings the cycle back to the resting state.

Structure of the active site

The amino acid sequence of the protein creates a folded structure around the heme site and its Fe atom, so forming an environment shielded from the solution and with a low permittivity. The hydrocarbon substrate is bound nearby, and the C—H bond under attack may be as little as 5 Å from the Fe site. The ligand on the Fe atom lying *trans* to the site of oxygen binding is a thiolate side chain of a cysteine residue (**21**). When O_2 coordinates, the outer end of the O_2 molecule projects into the solution, but the substrate is bound to the enzyme within the hydrophobic pocket.

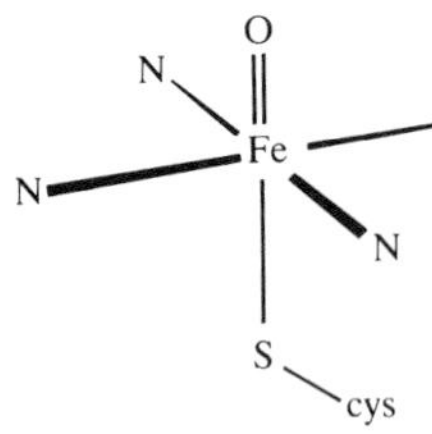

21

Oxygenation mechanism

The precise mechanism of oxygenation remains an intense subject of research. There are thought to be two possibilities, one involving the generation of an oxygen radical species which can attack the C—H bond, and the other the transfer of an O atom to the C—H bond. There are no good inorganic precedents for oxidation of an organic compound by Fe(IV)O and the current debate about the mechanism of the P-450 reaction will almost certainly enlarge the range of possible oxidation mechanisms known to inorganic chemists. The proposal of a radical mechanism entails some unconventional features. Radical oxidations are usually relatively unselective and not stereospecific; P-450 oxidations are quite selective and preserve the optical activity of chiral substrates.

The proposal that oxygenation is accomplished by electrophilic attack of a positive oxygen center on the C—H bond is just one example of a novel mechanism proposed to explain an enzyme reaction. It is supported by evidence of retention of stereochemical configuration and the reduction of reactivity for cases in which the substrate has substituents that block the approach of the O atom to the C—H bond. The transfer of what is effectively an O^+ unit would leave the iron porphyrin reduced from Fe(IV) to Fe(III).

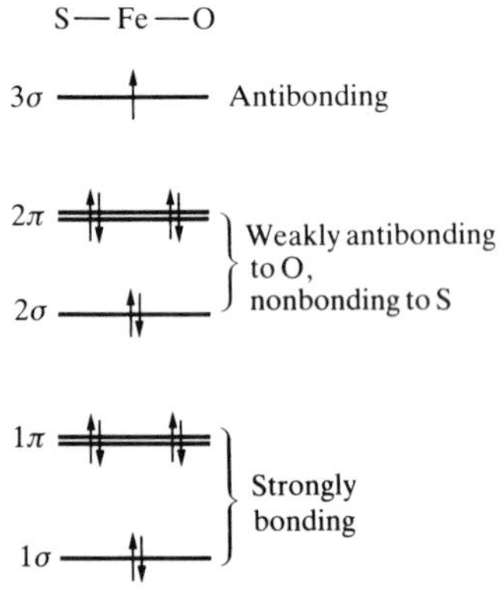

The role of the S atom in this pathway can be expressed in terms of molecular orbitals by focusing attention on the atomic orbitals of the S—Fe—O fragment. If we limit our attention to the S $3p_z$, the Fe $3d_{z^2}$, $3d_{xz}$, $3d_{yz}$, and O $2p$ orbitals, we obtain the orbital scheme shown in Fig. 19.17. The 13 electrons available in the fragment occupy the bonding levels, the weakly antibonding levels, and the last electron occupies the strongly antibonding 3σ orbital. The high formal oxidation number of the iron in the complex can therefore be rationalized in terms of its stabilization by considerable charge donation from the readily polarized S atom into the 2σ and 2π orbitals of the fragment. The special role of the *trans* S atom is to control the degree of positive charge on the O atom.

Fig. 19.17 The molecular orbital scheme of the linear S—Fe—O fragment in a cytochrome P-450 enzyme.

Example 19.4: *Judging the electron configuration of a metalloenzyme*

The Fe=O oxidizing center in P-450 is characterized as an Fe(IV) complex with the porphyrin also oxidized by one electron. Such a picture receives strong support from the ^{57}Fe Mössbauer spectrum[9] of the species. The iron center has magnetic susceptibility indicating three unpaired electrons. To what orbitals are they assigned?

Answer. The Fe(IV) oxidation state corresponds to a d^4 configuration. Assuming it to be approximately octahedral in a strong field environment it should be t_{2g}^4 with two unpaired electrons (**22**). The metal ion is in a 'triplet' state. The oxidation of the porphyrin ring removes one electron from the π HOMO leaving the third unpaired electron there. This configuration could be considered to be a 'doublet' state of the oxidized free ligand.

Exercise. In the resting state of P-450, the complex is effectively five coordinate Fe(III) complexed to four porphyrin N atoms and one cysteine S atom, with overall C_{4v} symmetry. Assign electrons to the Fe d orbitals.

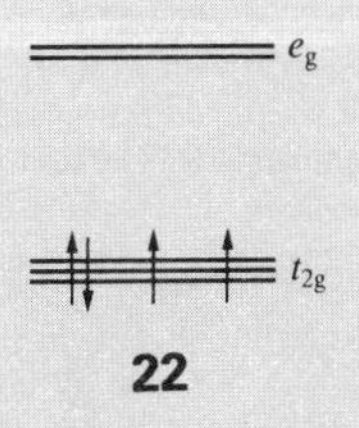

22

19.7 Nitrogen fixation

The *Rhizobium* organisms which live in the root nodules of various legume species (clover, alfalfa, beans, peas, etc.) share with several bacteria and blue-green algae the capacity to 'fix' atmospheric N_2 by converting it to NH_3. The enzyme nitrogenase carries out this redox reaction under essentially anaerobic conditions. The problem the enzyme has solved is how to overcome the great inertness of the N≡N triple bond.

Structure of the active site

Some of the features of the cation of nitrogenase that we must explain are why both molybdenum and iron are essential to its function and why some H_2 is generated when NH_3 is formed. We have to take into account the fact that the enzyme acts in aqueous solution, where there are H^+ ions available, and that it has to achieve the difficult (and unexplained) feat of reducing H^+ ions so as to ensure that most of these atoms attack N_2 in preference to combining together to form H_2.

Some of the structural features that must be accommodated in any mechanism include the fact that two iron–sulfur protein molecules are involved in the nitrogenase complex. The smaller has a molar mass[10] of 60 to 70 kg mol^{-1} (depending on the species) and includes an Fe_4S_4 cluster; it is a strongly reducing reductase. The larger protein, the nitrogenase itself, is a tetramer of four subunits with an overall molar mass of between 220 and 240 kg mol^{-1}. This protein has two Mo atoms, about thirty Fe atoms, and about thirty S atoms that are labile in the sense that they are readily removed from the protein by acid. Extended X-ray absorption fine structure

[9] Mossbauer spectroscopy measures the resonant absorption of a γ-ray by a nucleus held rigidly in a crystal framework so that the emission and absorption of γ-ray photons is recoil-free. The chemical environment of the nucleus affects the absorption frequency and so the spectrum is a probe of the electron distribution. The elements most commonly studied are Fe, Sn, and I. The technique may be used to identify the oxidation state of iron.

[10] Bioinorganic chemists, in common with some biochemists, often prefer to report molar masses in the non-SI unit 'dalton', where 1 dalton = 1 g mol^{-1}. Thus, the two macromolecules to which we now refer have molar masses of between 60 000 and 70 000 dalton and between 220 000 and 240 000 dalton respectively.

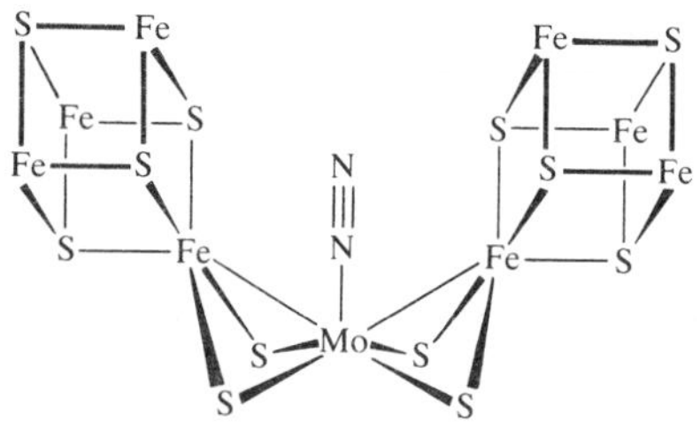

Fig. 19.18 A proposed model of the active site in nitrogenase. The key units are the Mo atom and the iron–sulfur structures.

(EXAFS) measurements, which can identify the atoms bound to a heavy atom center, show that the Mo environment consists of several S atoms at about 2.35 Å and that no Mo=O double bond is present. The data would appear to require an unconventional iron–sulfur cluster since the structure resists conditions (e.g. dilute acid) that react with normal protein-bound iron clusters. All the current proposals for the structure of the active site are speculative. Figure 19.18 shows a structure that is interesting, although it has too many Fe atoms as near neighbors of the Mo and would require that some of the S atoms not bound to Mo be replaced by O to give a correct S/Mo ratio. Several iron–sulfur structures will be discussed below.

Model studies

Since N_2 is isoelectronic with CO it should also function as a σ-base and π-acid ligand. However, N_2 complexes are not as common as carbonyls (Chapter 16). The first example prepared was $[Ru(NH_3)_5N_2]^{2+}$, but this complex does not catalyze nitrogen reduction. Efforts to produce an N_2 complex that did catalyze reduction to ammonia finally met with success when very strong reducing agents were employed. These results also refuted the gloomy expectation that nitrogen reduction is limited to exotic conditions, for it was found that certain molybdenum complexes with phosphine ligands can, after reduction to Mo(0) with a strong reducing agent, promote nitrogen reduction in acid media at room temperature. An example is

$$[MoCl_3(thf)_3] + 3e^- + 2N_2 + \text{excess-dpe} \rightarrow [Mo(N_2)_2(dpe)_2] + 3Cl^- + 3thf$$

$$[Mo(N_2)_2(dpe)_2] + 6H^+ \rightarrow 2NH_3 + N_2 + \text{Mo(VI) products}$$

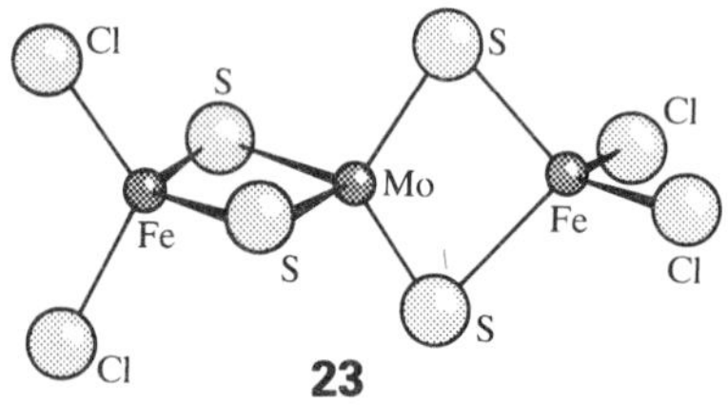

where thf is tetrahydrofuran and dpe is 1,2-bis(diphenylphosphino)-ethane. The strong reducing agent that prepares the system by reducing the initial complex to Mo(0) is a Grignard reagent. It seems likely that the key steps in this reaction, and the enzyme process too, is H atom transfer rather than electron transfer. However, despite the successful reduction of N_2, the above reaction is not catalytic. A synthetic cluster with some of the characteristics of the active site is (**23**), but a good analog has not yet been prepared.

23

Iron–sulfur proteins

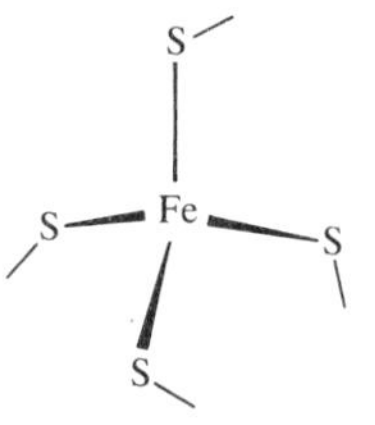

The discussion of nitrogenase draws attention to a class of proteins which we have disregarded up to now despite their role in electron transfer reactions. A recurring theme in bioinorganic chemistry is that complexes in which an Fe atom is tetrahedrally coordinated by S atoms are common as mediators of electron transfer. The S ligands come from cysteine residues in the protein or are bridging atoms that are called 'inorganic' or 'labile' because they can be removed by acid. Proteins that contain a single Fe atom in a coordination environment like (**24**) are called **rubredoxins**. Proteins containing Fe_2 (**25**) and Fe_4 clusters are called **ferredoxins**. We have encountered the ferredoxins already in the mitochondrial electron transport chain and we shall meet them again when we discuss photosynthesis.

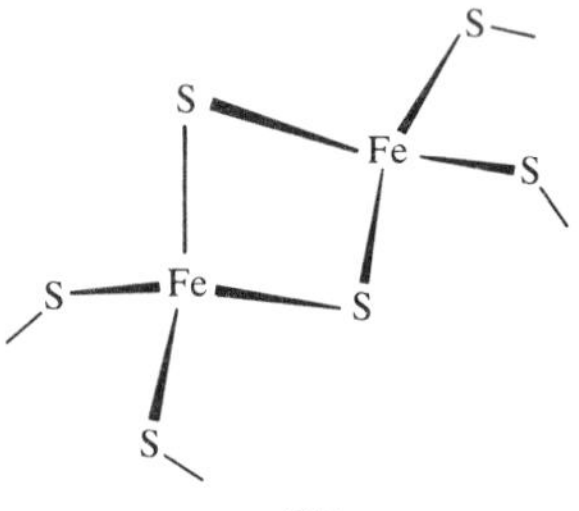

24

25

Modeling these iron–sulfur clusters has opened a significant area of synthetic and structural inorganic chemistry. For instance, it has been found that the reaction of $FeCl_3$, $NaOCH_3$, NaHS, and benzylthiol in methanol gives $[Fe_4S_4(SCH_2Ph)_4]^{2-}$ (**26**). The magnetic susceptibility, electronic spectrum, redox properties, and ^{57}Fe Mössbauer spectrum are analogous to those of ferredoxins. Oxidation number rules suggest two Fe(II) atoms and two Fe(III) atoms, but all spectroscopic methods, including X-ray photoelectron spectroscopy, suggest that all four Fe atoms are equivalent. Consequently, a delocalized electron description is most appropriate and the electron transfer reactions involve orbitals that are delocalized over the cluster with the S and Fe atoms both sharing in the change of oxidation state.

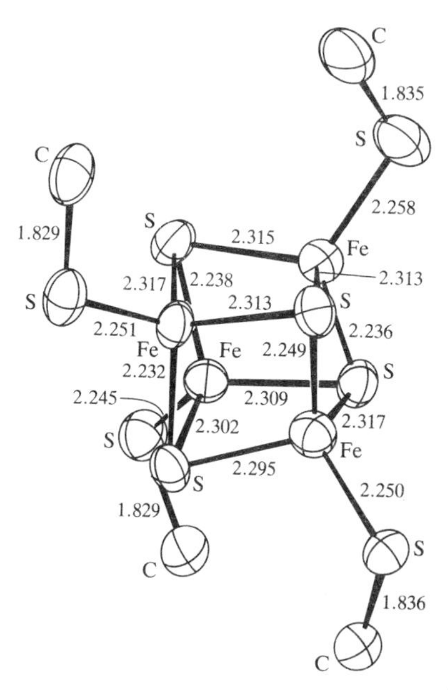

26 $[Fe_4S_4(SCH_2Ph)_4]^{2-}$

Example 19.5 *Writing redox reactions that involve thiolate ligands used in cluster synthesis*

One of the problems that has plagued synthetic chemists in their attempts to prepare model compounds for cysteine complexed metal ions in metalloproteins is the easy oxidation of thiolate anions (RS^-) to RS—SR. Simple complexes with Cu^{2+}—SR and Fe^{3+}—SR bonds which might serve as models for cytochrome P-450 and the ferredoxins (and, in addition, a class known as 'blue copper proteins') are unstable because of this reaction. Write balanced equations for the decomposition of $[Cu(II)L_n(SR)]$ and $[Fe(III)L_n(SR)]$.

Answer. If thiolate ligands are oxidized to disulfides, the metal ion must be reduced to Cu(I) and Fe(II) respectively. The balanced equations, ignoring the possible charges on the complexes, are

$$2[Cu(II)L_n(SR)] \rightarrow 2[Cu(I)L_n] + RSSR$$

$$2[Fe(III)L_n(SR)] \rightarrow 2[Fe(II)L_n] + RSSR$$

Exercise. Early attempts to prepare models of the 2Fe—2S ferredoxins with simple thiolate ligands did not lead to $[Fe_2S_2(RS)_4]^{2-}$ clusters but instead to $[Fe_4S_4(SR)_4]^{2-}$. By considering the average oxidation state of Fe in these clusters, suggest a reason for the difficulty in their preparation.

19.8 Photosynthesis

One of the most remarkable redox reactions is the thermodynamically 'uphill' conversion of water and carbon dioxide into carbohydrates and oxygen using light as an energy source. Formally, the production of carbohydrates involves the reduction of CO_2 and the oxidation of $2H_2O$ to O_2. There are two photochemical reaction centers, **photosystem 1** and **photosystem 2** (PS I and PS II), and the photosynthetic system is organized in the green leaf organelles called **chloroplasts**. A general point to bear in mind in the following is that the primary steps following photoexcitation are outer-sphere electron transfers which have been directly observed by picosecond spectroscopy.

Photosystem I is based on chlorophyll a_1, a magnesium dihydroporphyrin complex (**27**). After PS I has been photoexcited, it can act as a reducing agent for an iron–sulfur complex, and the electrons it transfers to the complex are ultimately used to reduce CO_2. The oxidized form of PS I that remains after the electron transfer step is not strong enough to oxidize water. Its return to the reduced form is used to drive the conversion of two ADP molecules to two ATP molecules through a sequence of mediating species that include several iron-based redox couples and a quinone called plastoquinone.

27 Chlorophyll a_1

The oxidized form of PSII is a strong enough oxidizing agent to oxidize

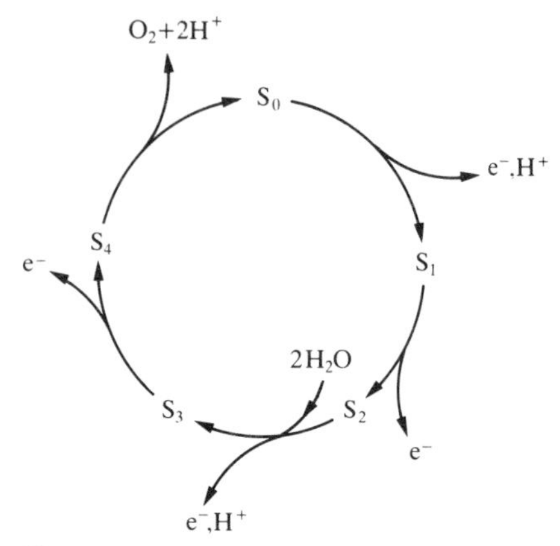

Fig. 19.19 The four steps of oxidation which precede the release of O_2 in the oxygen evolving system of PS II.

water. However, the reaction involves the net transfer of four electrons

$$2H_2O(l) \rightarrow 4H^+(aq) + O_2(g) + 4e^-$$

and is achieved only by a complex series of redox reactions of a manganese based enzyme which transfers electrons to the photochemically active center. Experiments with successive light flashes show that the enzyme system is oxidized in four one-electron steps before accomplishing the four-electron oxidation of water to O_2. If the redox reservoir involves the reduction of Mn(IV) to Mn(II), the enzyme must contain at least two Mn atoms to accommodate the four electrons.

A specific proposal for the cycle of events that take place in the system is shown in Fig. 19.19. The resting state S_0 is oxidized to S_1 by one electron with the loss of one proton. In the second step, S_2 is formed by transfer of a second electron. In the conversion of S_2 to S_3, the third level of oxidation, the electron loss is accompanied by uptake of the two H_2O molecules which supply the oxygen, and a proton is lost. After one more electron has been transfered to give S_4, the critical reaction occurs in which O_2 is released and the Mn enzyme is reduced by the equivalent of four electrons. A proposal for the mechanism of this step is shown in the rearrangement of the adamantane-like cluster shown in Fig. 19.20. The cycle and the proposed O_2 evolution step represent an elegant way to couple one-electron outer-sphere electron transfers into a multiple change in oxidation state by an atom transfer process. Virtually all the oxygen present in our current atmosphere may have been produced in a process like that shown in Fig. 19.19.

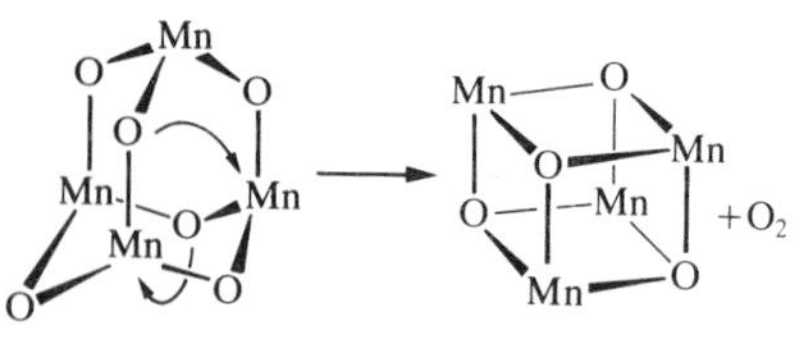

S_4 'Adamantane' form $\rightarrow$ S_0 'Cubane' form $+O_2$

Fig. 19.20 A possible analog of the step S_4 to S_0 in Fig. 19.19 in which an 'adamantane' form of the enzyme converts to a 'cubane' form with loss of O_2.

Chlorophyll

The chromophore in the photosystem is chlorophyll a_1 (**27**). The isolated molecule has the absorption spectrum shown in Fig. 19.21. The strong absorption band in the red (called the Q band) and the band in the blue to near-ultraviolet (the Soret band) are characteristic of porphyrins. Both arise from promotion of electrons from the porphyrin π HOMO to the π^* LUMO. The non-absorbing region between these two bands accounts for the characteristic green color of vegetation. Other pigments—called 'antenna' pigments—are also present in leaves to harvest some light in the absorption gap and transfer its energy to the reaction center.

The initial absorption is from the singlet ground state to a singlet excited state of the chlorophyll Q band (^1Q). Isolated chlorophyll molecules in solution undergo rapid intersystem crossing into the lower-lying spin triplet (^3Q), and this state is responsible for the photochemical electron transfer reactions of the molecule in solution. In contrast, the PS reaction center shortcircuits the intersystem crossing by immediately transferring an electron from ^1Q to electron acceptors that are adjacent to the chlorophyll in the chloroplast. We see yet again that spatial organization is a key aspect of biochemistry. A simplified model of such a reaction is obtained using the porphyrin–amide–quinone molecule (**28**) which exhibits fluorescence properties, suggesting charge transfer from the singlet state of the porphyrin to the quinone that is 10^4 times faster than relaxation to the triplet.

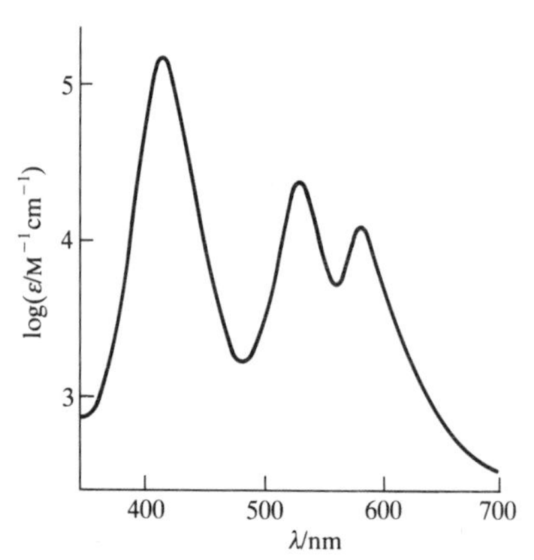

Fig. 19.21 The absorption spectrum of chlorophyll a_1 in the visible region.

Reaction center organization

The spatial organization in the molecule allows the rapid deployment (within 10 ps) of the energy of the singlet excited state. The singlet state is at higher energy than the triplet and its use results in increased energy efficiency: intersystem crossing into a triplet squanders energy as heat. At this stage, spatial organization plays another essential role. The products of reactions that store energy could revert to reactants, so wasting energy, but the system is so organized that electrons are transferred along the reaction sequence faster than the reverse reactions can take place.

A reaction center of a bacteriochlorophyll has recently been crystallized and the structure determined using X-ray diffraction. A 'special pair' of chlorophyll molecules, which has long been recognized to be essential to the rapid initial photoelectron transfer, lies at the site of light absorption. Around this is organized a double chain of the acceptor molecules. One puzzle is that the series of acceptors seems to be duplicated, but the chemical evidence suggests that one component of the structure is dormant and that only one of the chains is active in electron transport.

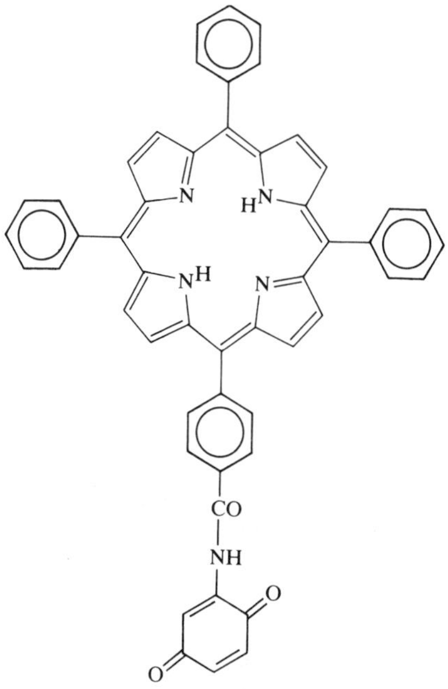

N HN NH N CO NH O O

28

Further reading

L. Stryer, *Biochemistry*. W. H. Freeman & Co., New York (1988). A general textbook of biochemistry which presents mechanisms involving inorganic species in their biochemical context.

E.-I. Ochiai, *Bioinorganic Chemistry*. Allyn and Bacon, Boston (1977). A textbook of bioinorganic chemistry organized according to the elements involved.

R. W. Hay, *Bioinorganic Chemistry*. Ellis Horwood, Chichester and New York (1984). A good brief general introduction.

R. J. P. Williams, *Coord. Chem. Rev.*, **100,** 578 (1990). A view of the implication of considering biological systems as dynamic structures and the implications for the role of the elements.

The November 1985 issue of the *Journal of Chemical Education* is a special issue devoted to bioinorganic chemistry.

Exercises

19.1 In the discussion of O_2 coordination, neutral O_2, O_2^-, and O_2^{2-} were considered as limiting forms. Consider the molecular orbital energy level diagram for O_2 from Chapter 2. What is the implication for bond length and net spin of the choice of these O_2 species as a model for the ligand?

19.2 Does the equilibrium position of the following reaction lie to the left or the right?

$$Hb + Hb(O_2)_4 \rightleftharpoons 2Hb(O_2)_2$$

Use Fig. 19.7 to explain your reasoning. Does the equilibrium position of the reaction

$$Hb(O_2)_4 + 4Mb \rightleftharpoons Hb + 4Mb(O_2)$$

lie to the left or the right? Does your answer depend on the partial pressure of O_2?

19.3 Oxygen is described as a σ-base and a π-acid. Carbon monoxide is an excellent example of this type of ligand. Can you use these facts to propose a mechanism for CO poisoning?

19.4 The use of Co(III) centers for synthesis of peptides depends on the reaction

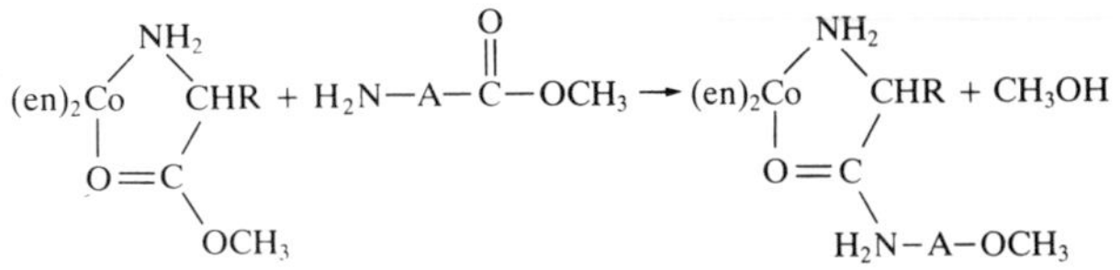

$$(en)_2Co(NH_2CHR)(O{=}C{-}OCH_3) + H_2N{-}A{-}\overset{O}{\overset{\|}{C}}{-}OCH_3 \rightarrow (en)_2Co(NH_2CHR)(O{=}C{-}H_2N{-}A{-}OCH_3) + CH_3OH$$

A stands for a peptide of arbitrary length. What is the nucleophile and what center is subject to catalyzed nucleophilic attack?

19.5 Co(II) substitution for Zn(II) gives a 'spectral probe'. What Co(II) spectral features are exploited? Why does Zn(II) lack them? If Co(II) is to serve as a useful probe for the zinc site, what assumption must be satisfied?

19.6 The diameter of a high-spin Fe(II) is larger that of the 'hole' at the center of the porphyrin ring whereas low-spin Fe(II) is smaller. (a) Draw the electron configurations for the two spin states in an octahedral environment. Why does the high-spin form have the larger radius? (b) Give examples of ligands L that might lead to six-coordinate high and low spin [Fe(porph)L_2] complexes.

19.7 Why are *d* metals such as Mn, Fe, Co, and Cu used in redox enzymes in preference to Zn, Ga, and Ca?

19.8 Considering the specific character of enzyme substrate binding, can you suggest why it is difficult to measure the self-exchange rate constants of redox enzymes?

19.9 Suggest two ways in which the protein parts of the enzymes contribute to making an outer-sphere electron transfer highly specific to one oxidant with one reductant partner.

19.10 Identify one significant role in biological processes for the elements Fe, Mn, Mo, Cu, and Zn.

19.11 What prevents simple iron porphyrins from functioning as O_2 carriers?

19.12 Sketch the steps illustrating a metal complex functioning as (a) a Brønsted acid and (b) a Lewis acid in an enzyme-catalyzed reaction.

19.13 Determine the average oxidation state of iron in structures (**23**), (**24**), and (**25**).

19.14 What is the average change in oxidation number of the Mn atoms in the proposed adamantane to cubane reaction producing O_2 in the S_4 to S_0 step in the PS II cycle?

19.15 Contrast the source of an electron in the reaction of the excited state of chlorophyll to the source in the Fe(II) state of a cytochrome. Could magnesium be used as the metal ion in a cytochrome?

Problems

19.1 It has been observed that the attachment of Fe(TPP), where TPP is tetraphenylporphyrin, to a rigid silica gel support containing a 3-imidazoylpropyl group as a nitrogen donor ligand produced a good reversible oxygen carrier. How does this system prevent irreversible oxidation without a 'picket fence'? (See *Acc. chem. Res.*, **8**, 384 (1975))

19.2 The structures of Hb may be classified as 'relaxed' (R) or 'tense' (T) as alternative terms to oxygenated and deoxygenated. The R and T structures differ in both the relation among the four subunits (the quaternary structure) and the conformation within a subunit (the tertiary structure). Explain how these structural differences relate to the difference in the oxygen binding curve of Hb as compared to Mb.

19.3 Discuss the probable difference in the pockets present in carboxypeptidase and carbonic anhydrase.

19.4 Consider the four features of the binding of glycyltyrosine to carboxypeptidase listed in Section 19.4. Which are related to the spatial organization of the binding of the substrate to the enzyme and which promote the cleavage of the peptide bond? Explain your reasoning.

19.5. The substituted carboxypeptidase enzymes show a reactivity order $Mn < Zn < Ni < Co$. To what extent does this order follow the order of LFSE for tetrahedral complexes and thus support the notion that the determining factor is the bond of the metal to the carbonyl oxygen? What order would you predict if the coordination geometry were octahedral?

19.6 Consider the Frost diagram for oxygen (Fig. 8.6). Using a table of standard potentials, add a point for superoxide. Discuss the biological significance of these potentials. Keep in mind that a biological system is a reduced system that is threatened by oxidizing agents.

19.7 Consider the energy available from a photon at a wavelength of 700 nm. If the photochemical reaction at PS I generates a potential difference of 1 V, what is the efficiency of harvest of energy?

19.8 When photosynthesis uses the combination of PS I and PS II to accomplish the difficult reaction of water oxidation, is this the analog of connecting batteries in parallel or in series? When it organizes the system by stacking chlorophyll in grana, which are in turn stacked in the chloroplasts, is this increasing efficiency by the analog of parallel or series connection?

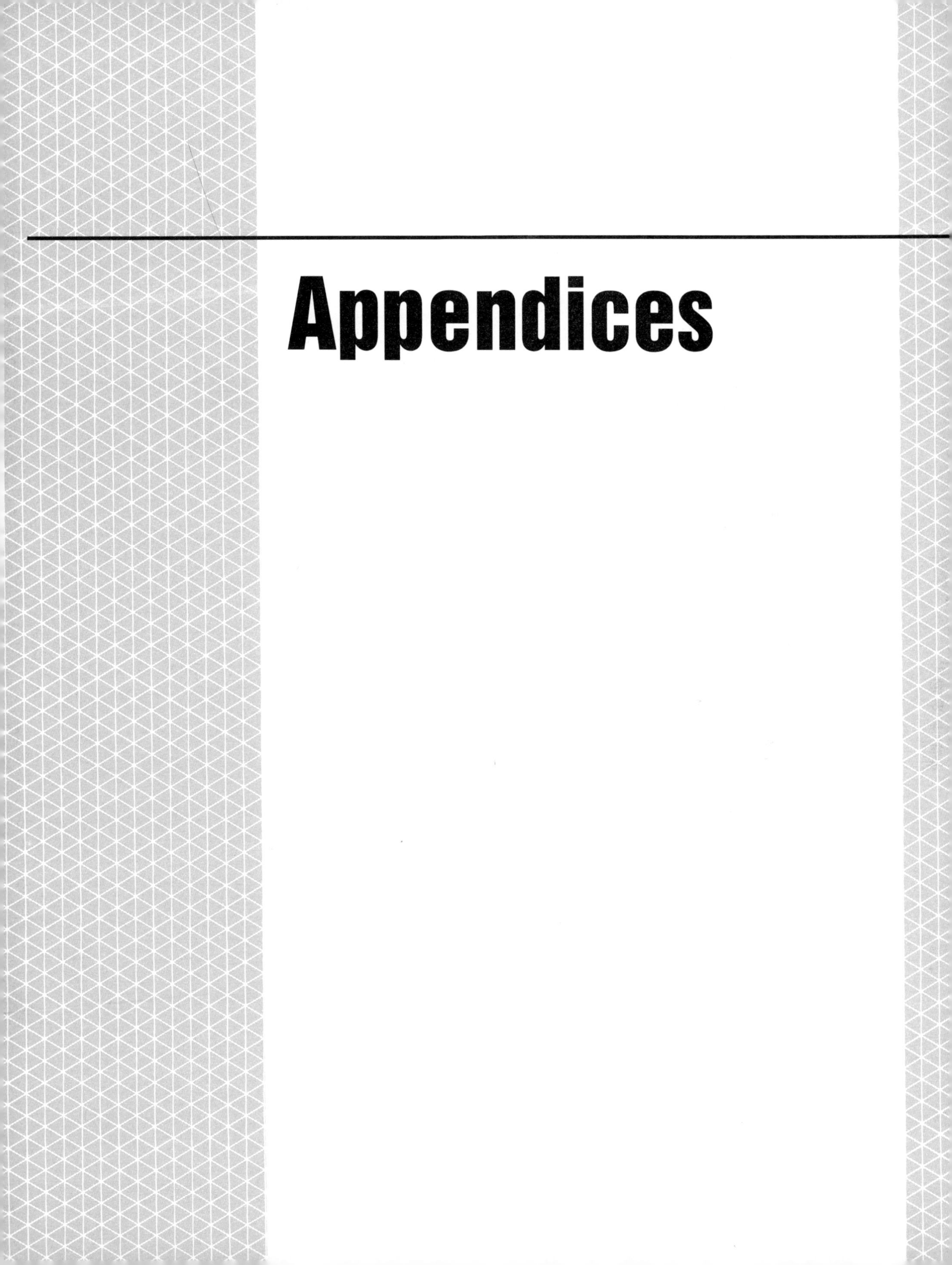

Appendices

Electron configurations of atoms

A1

In the building-up principle, the ground state configurations of neutral atoms are obtained by allowing the available orbitals to be occupied in a specific order with no more than two electrons in any one orbital. Some variation in the order is encountered in the d- and f-block elements in order to accommodate the effects of electron–electron interaction more faithfully. The closed shell configuration $1s^2$, characteristic of helium, is denoted [He]. Lithium can therefore be denoted [He]$2s^1$, showing more clearly that it consists of a single $2s$ electron outside a helium-like core. A similar convention may be used to write the configurations of later elements.

Ground state electron configuration and term symbols are taken from S. Fraga, J. Karwowski, and K. M. S. Saxena, *Handbook of atomic data*. Elsevier, Amsterdam (1976).

Element	Configuration	Term	Element	Configuration	Term
H	$1s^1$	^{2}S	**I**	[Kr]$4d^{10}5s^25p^5$	^{2}P
He	$1s^2$	^{1}S	**Xe**	[Kr]$4d^{10}5s^25p^6$	^{1}S
Li	[He]$2s^1$	^{2}S	**Cs**	[Xe]$6s^1$	^{2}S
Be	[He]$2s^2$	^{1}S	**Ba**	[Xe]$6s^2$	^{1}S
B	[He]$2s^22p^1$	^{2}P	**La**	[Xe]$5d^16s^2$	^{2}D
C	[He]$2s^22p^2$	^{3}P	**Ce**	[Xe]$4f^15d^16s^2$	^{3}H
N	[He]$2s^22p^3$	^{4}S	**Pr**	[Xe]$4f^36s^2$	^{4}I
O	[He]$2s^22p^4$	^{3}P	**Nd**	[Xe]$4f^46s^2$	^{5}I
F	[He]$2s^22p^5$	^{2}P	**Pm**	[Xe]$4f^56s^2$	^{6}H
Ne	[He]$2s^22p^6$	^{1}S	**Sm**	[Xe]$4f^66s^2$	^{7}F
Na	[Ne]$3s^1$	^{2}S	**Eu**	[Xe]$4f^76s^2$	^{8}S
Mg	[Ne]$3s^2$	^{1}S	**Gd**	[Xe]$4f^75d^16s^2$	^{9}D
Al	[Ne]$3s^23p^1$	^{2}P	**Tb**	[Xe]$4f^96s^2$	^{6}H
Si	[Ne]$3s^23p^2$	^{3}P	**Dy**	[Xe]$4f^{10}6s^2$	^{5}I
P	[Ne]$3s^23p^3$	^{4}S	**Ho**	[Xe]$4f^{11}6s^2$	^{4}I
S	[Ne]$3s^23p^4$	^{3}P	**Er**	[Xe]$4f^{12}6s^2$	^{3}H
Cl	[Ne]$3s^23p^5$	^{2}P	**Tm**	[Xe]$4f^{13}6s^2$	^{2}F
Ar	[Ne]$3s^23p^6$	^{1}S	**Yb**	[Xe]$4f^{14}6s^2$	^{1}S
K	[Ar]$4s^1$	^{2}S	**Lu**	[Xe]$4f^{14}5d^16s^2$	^{2}D
Ca	[Ar]$4s^2$	^{1}S	**Hf**	[Xe]$4f^{14}5d^26s^2$	^{3}F
Sc	[Ar]$3d^14s^2$	^{2}D	**Ta**	[Xe]$4f^{14}5d^36s^2$	^{4}F
Ti	[Ar]$3d^24s^2$	^{3}F	**W**	[Xe]$4f^{14}5d^46s^2$	^{5}D
V	[Ar]$3d^34s^2$	^{4}F	**Re**	[Xe]$4f^{14}5d^56s^2$	^{6}S
Cr	[Ar]$3d^54s^1$	^{7}S	**Os**	[Xe]$4f^{14}5d^66s^2$	^{5}D
Mn	[Ar]$3d^54s^2$	^{6}S	**Ir**	[Xe]$4f^{14}5d^76s^2$	^{4}F
Fe	[Ar]$3d^64s^2$	^{5}D	**Pt**	[Xe]$4f^{14}5d^96s^1$	^{3}D
Co	[Ar]$3d^74s^2$	^{4}F	**Au**	[Xe]$4f^{14}5d^{10}6s^1$	^{2}S
Ni	[Ar]$3d^84s^2$	^{3}F	**Hg**	[Xe]$4f^{14}5d^{10}6s^2$	^{1}S
Cu	[Ar]$3d^{10}4s^1$	^{2}S	**Tl**	[Xe]$4f^{14}5d^{10}6s^26p^1$	^{2}P
Zn	[Ar]$3d^{10}4s^2$	^{1}S	**Pb**	[Xe]$4f^{14}5d^{10}6s^26p^2$	^{3}P
Ga	[Ar]$3d^{10}4s^24p^1$	^{2}P	**Bi**	[Xe]$4f^{14}5d^{10}6s^26p^3$	^{4}S
Ge	[Ar]$3d^{10}4s^24p^2$	^{3}P	**Po**	[Xe]$4f^{14}5d^{10}6s^26p^4$	^{3}P
As	[Ar]$3d^{10}4s^24p^3$	^{4}S	**At**	[Xe]$4f^{14}5d^{10}6s^26p^5$	^{2}P
Se	[Ar]$3d^{10}4s^24p^4$	^{3}P	**Rn**	[Xe]$4f^{14}5d^{10}6s^26p^6$	^{1}S
Br	[Ar]$3d^{10}4s^24p^5$	^{2}P	**Fr**	[Rn]$7s^1$	^{2}S
Kr	[Ar]$3d^{10}4s^24p^6$	^{1}S	**Ra**	[Rn]$7s^2$	^{1}S
Rb	[Kr]$5s^1$	^{2}S	**Ac**	[Rn]$6d^17s^2$	^{2}D
Sr	[Kr]$5s^2$	^{1}S	**Th**	[Rn]$6d^27s^2$	^{3}F
Y	[Kr]$4d^15s^2$	^{2}D	**Pa**	[Rn]$5f^26d^17s^2$	^{4}K
Zr	[Kr]$4d^25s^2$	^{3}F	**U**	[Rn]$5f^36d^17s^2$	^{5}L
Nb	[Kr]$4d^45s^1$	^{6}D	**Np**	[Rn]$5f^46d^17s^2$	^{6}L
Mo	[Kr]$4d^55s^1$	^{7}S	**Pu**	[Rn]$5f^67s^2$	^{7}F
Tc	[Kr]$4d^55s^2$	^{6}S	**Am**	[Rn]$5f^77s^2$	^{8}S
Ru	[Kr]$4d^75s^1$	^{5}F	**Cm**	[Rn]$5f^76d^17s^2$	^{9}D
Rh	[Kr]$4d^85s^1$	^{4}F	**Bk**	[Rn]$5f^97s^2$	^{6}H
Pd	[Kr]$4d^{10}$	^{1}S	**Cf**	[Rn]$5f^{10}7s^2$	^{5}I
Ag	[Kr]$4d^{10}5s^1$	^{2}S	**Es**	[Rn]$5f^{11}7s^2$	^{4}I
Cd	[Kr]$4d^{10}5s^2$	^{1}S	**Fm**	[Rn]$5f^{12}7s^2$	^{3}H
In	[Kr]$4d^{10}5s^25p^1$	^{2}P	**Md**	[Rn]$5f^{13}7s^2$	^{2}F
Sn	[Kr]$4d^{10}5s^25p^2$	^{3}P	**No**	[Rn]$5f^{14}7s^2$	^{1}S
Sb	[Kr]$4d^{10}5s^25p^3$	^{4}S	**Lr**	[Rn]$5f^{14}6d^17s^2$	^{2}D
Te	[Kr]$4\mathrm{d}^{10}5s^25p^4$	^{3}P			

Ionization energies

A2

The first ionization energy is the minimum energy required to remove an electron from a neutral atom in the gas phase, the second ionization energy is the minimum energy required to ionize the resulting cation, and so on. Ionization energies are expressed in electronvolts (eV) and may be converted to kJ mol^{-1} and cm^{-1} using

$$1\,eV = 96.485\,kJ\,mol^{-1}$$

$$1\,eV = 8065.5\,cm^{-1}$$

The values given here are taken from various sources, particularly C. E. Moore, *Atomic energy levels*. NBS Circular 467, Washington (1970) and W. C. Martin, L. Hagan, J. Reader, and J. Sugar, *J. phys. Chem. Ref. Data,* **3,** 771 (1974). Values for actinides are taken from J. J. Katz, G.T. Seaborg, and L. R. Morss (eds), *The chemistry of the actinide elements* (2nd edn). Chapman and Hall, London (1986).

Atom	$M \rightarrow M^+ + e^-$	$M^+ \rightarrow M^{2+} + e^-$	$M^{2+} \rightarrow M^{3+} + e^-$
H	13.60		
He	24.59	54.41	
Li	5.320	75.63	122.4
Be	9.321	18.21	153.9
B	8.297	25.15	37.93
C	11.257	24.38	47.88
N	14.53	29.60	47.44
O	13.62	35.11	54.93
F	17.42	34.97	62.70
Ne	21.56	40.96	63.45
Na	5.138	47.28	71.63
Mg	7.642	15.03	80.14
Al	5.984	18.83	28.44
Si	8.151	16.34	33.49
P	10.485	19.72	30.18
S	10.360	23.33	34.83
Cl	12.966	23.80	39.65
Ar	15.76	27.62	40.71
K	4.340	31.62	45.71
Ca	6.111	11.87	50.89
Sc	6.54	12.80	24.76
Ti	6.82	13.58	27.48
V	6.74	14.65	29.31
Cr	6.764	16.50	30.96
Mn	7.435	15.64	33.67
Fe	7.869	16.18	30.65
Co	7.876	17.06	33.50
Ni	7.635	18.17	35.16
Cu	7.725	20.29	36.84
Zn	9.393	17.96	39.72
Ga	5.998	20.51	30.71
Ge	7.898	15.93	34.22
As	9.814	18.63	28.34
Se	9.751	21.18	30.82
Br	11.814	21.80	36.27
Kr	13.998	24.35	36.95
Rb	4.177	27.28	40.42
Sr	5.695	11.03	43.63
Y	6.38	12.24	20.52
Zr	6.84	13.13	22.99
Nb	6.88	14.32	25.04
Mo	7.099	16.15	27.16
Tc	7.28	15.25	29.54
Ru	7.37	16.76	28.47
Rh	7.46	18.07	31.06
Pd	8.34	19.43	32.92
Ag	7.576	21.48	34.83
Cd	8.992	16.90	37.47
In	5.786	18.87	28.02
Sn	7.344	14.63	30.50
Sb	8.640	18.59	25.32
Te	9.008	18.60	27.96
I	10.450	19.13	33.16

Atom	$M \rightarrow M^+ + e^-$	$M^+ \rightarrow M^{2+} + e^-$	$M^{2+} \rightarrow M^{3+} + e^-$
Xe	12.130	21.20	32.10
Cs	3.894	25.08	35.24
Ba	5.211	10.00	37.31
La	5.577	11.06	19.17
Ce	5.466	10.85	20.20
Pr	5.421	10.55	21.62
Nd	5.489	10.73	22.07
Pm	5.554	10.90	22.28
Sm	5.631	11.07	23.42
Eu	5.666	11.24	24.91
Gd	6.140	12.09	20.62
Tb	5.851	11.52	21.91
Dy	5.927	11.67	22.80
Ho	6.018	11.80	22.84
Er	6.101	11.93	22.74
Tm	6.184	12.05	23.68
Yb	6.254	12.19	25.03
Lu	5.425	13.89	20.96
Hf	6.65	14.92	23.32
Ta	7.89	15.55	21.76
W	7.98	17.62	23.84
Re	7.88	13.06	26.01
Os	8.71	16.58	24.87
Ir	9.12	17.41	26.95
Pt	9.02	18.56	29.02
Au	9.22	20.52	30.05
Hg	10.44	18.76	34.20
Tl	6.107	20.43	29.83
Pb	7.415	15.03	31.94
Bi	7.289	16.69	25.56
Po	8.42	18.66	27.98
At	9.64	16.58	30.06
Rn	10.75		
Fr	4.15	21.76	32.13
Ra	5.278	10.15	34.20
Ac	5.17	11.87	19.69
Th	6.08	11.89	20.50
Pa	5.89	11.7	18.8
U	6.19	14.9	19.1
Np	6.27	11.7	19.4
Pu	6.06	11.7	21.8
Am	5.99	12.0	22.4
Cm	6.02	12.4	21.2
Bk	6.23	12.3	22.3
Cf	6.30	12.5	23.6
Es	6.42	12.6	24.1
Fm	6.50	12.7	24.4
Md	6.58	12.8	25.4
No	6.65	13.0	27.0
Lr	4.6	14.8	23.0

A3 Electronegativities

We present several different scales of electronegativity since no one is universally accepted. Linus Pauling, who generated the first scale,[1] argued that the excess energy Δ of an A—B bond over the average energy of A—A and B—B bonds can be attributed to the presence of an ionic component in the molecular wavefunction, and defined the difference in electronegativities as

$$|\chi_A - \chi_B| = 0.208 \times \sqrt{\Delta}$$

where

$$\Delta = E(\text{A—B}) - \tfrac{1}{2}\{E(\text{A—A}) + E(\text{B—B})\}$$

The most recently published[2,3] Pauling electronegativities are labeled χ_P in Table A3.1 and depend on the oxidation state of the atom. For a given element, χ_P increases with increasing oxidation state. The values in the table are for the maximum oxidation state of the element.

Another widely used scale, which was proposed by Allred and Rochow,[4] is based on the view that electronegativity is an atomic property and related to the electrostatic potential (Section 1.6). Allred–Rochow electronegativities (χ_{AR} in Table A3.1) are proportional to the Pauling values and a linear fit permits them to be expressed in 'Pauling units'. Similarly the Mulliken electronegativities described in Section 1.6 can be expressed in Pauling units and are labeled χ_M in the table. A spectroscopic electronegativity scale proposed by Allen[5] is defined in terms of the average energy of the electrons in the *s* and *p* orbitals of an atom:

$$\chi_S = \frac{m\varepsilon_p + n\varepsilon_s}{m + n}$$

The energies ε_p and ε_s are obtained from averaged spectroscopic data. Since spectroscopic data are available for many elements it is possible to calculate electronegativities for the noble gases, for which the thermochemical data needed for the determination of χ_P is lacking (except for krypton and xenon). More detailed discussions of electronegativity scales are available elsewhere.[5,6]

Electronegativity is much less used for the correlation of the properties of compounds of the *d*- and *f*-block elements than of the *s*- and *p*-blocks, but some values of χ_P for the *d* and *f* blocks are given in Table A3.2. Note that the electronegativity order

$$\text{Mn}(1.60) < \text{Fe}(1.64) < \text{Co}(1.70) < \text{Ni}(1.75) = \text{Cu}(1.75) > \text{Zn}(1.66)$$

is similar to the commonly observed order of stability of complexes of these M^{2+} ions and hard donors (Section 7.7).

[1] L. Pauling, *Nature of the chemical bond.* Cornell University Press (1960). In this, the third edition of his book, Pauling switched from the original arithmetic mean (used here) to a geometric mean of the energies.
[2] A. L. Allred, *Inorg. Chem.*, **17,** 215 (1961).
[3] L. C. Allen and J. E. Huheey, *J. Inorg. Nucl. Chem.*, **42,** 1523 (1980).
[4] A. L. Allred and E. G. Rochow, *J. Inorg. Nucl. Chem.*, **5,** 264 (1958).
[5] L. C. Allen, *J. Amer. chem. Soc.*, **111,** 9003 (1989).
[6] S. G. Bratsch, *J. chem. Educ.*, **65,** 34 (1988).

Table A3.1. Electronegativities of the *s*- and *p*-block elements†

Atom	χ_P	χ_{AR}	χ_M	χ_S	Atom	χ_P	χ_{AR}	χ_M	χ_S
H	2.20	2.20	3.06	2.30	Ga	1.81	1.82	1.34	1.76
Li	0.98	0.97	1.28	0.91	Ge	2.01	2.02	1.95	1.99
Be	1.57	1.47	1.99	1.58	As	2.18	2.20	2.26	2.21
B	2.04	2.01	1.83	2.05	Se	2.55	2.48	2.51	2.42
C	2.55	2.50	2.67	2.54	Br	2.96	2.74	3.24	2.69
N	3.04	3.07	3.08	3.07	Kr		2.94	2.98	2.97
O	3.44	3.50	3.21	3.61	Rb	0.82	0.89	0.99	0.71
F	3.98	4.10	4.42	4.19	Sr	0.95	0.99	1.21	0.96
Ne		4.84	4.60	4.79	In	1.78	1.49	1.30	1.66
Na	0.93	1.01	1.21	0.87	Sn	1.96	1.72	1.83	1.82
Mg	1.31	1.23	1.63	1.29	Sb	2.05	1.82	2.06	1.98
Al	1.61	1.47	1.37	1.61	Te	2.10	2.01	2.34	2.16
Si	1.90	1.74	2.03	1.92	I	2.66	2.21	2.88	2.36
P	2.19	2.06	2.39	2.25	Xe	2.60	2.40	2.59	2.58
S	2.58	2.44	2.65	2.59	Cs	0.79	0.86		
Cl	3.16	2.83	3.54	2.87	Ba	0.89	0.97		
Ar		3.20	3.36	3.24	Tl	2.04			
K	0.82	0.91	1.03	0.73	Pb	2.33			
Ca	1.00	1.04	1.30	1.03	Bi	2.02			

† In Pauling units.

Table A3.2. Pauling electronegativities of some *d*- and *f*-block elements

+3	+2	+2	+2	+2	+2	+2	+2	+1	+2
***d* Block (oxidation states as indicated)**									
Sc	**Ti**	**V**	**Cr**	**Mn**	**Fe**	**Co**	**Ni**	**Cu**	**Zn**
1.36	1.54	1.63	1.66	1.55	1.83	1.88	1.91	1.90	1.65
Y	**Zr**	**Nb**	**Mo**	**Tc**	**Ru**	**Rh**	**Pd**	**Ag**	**Cd**
1.22	1.33		2.16			2.28	2.20	1.93	1.69
La	**Hf**	**Ta**	**W**	**Re**	**Os**	**Ir**	**Pt**	**Au**	**Hg**
1.10			2.36			2.20	2.28	2.54	2.00

***f* Block (oxidation state +3)**

Ce	**Pr**	**Nd**	**Pm**	**Sm**	**Eu**	**Gd**	**Tb**	**Dy**	**Ho**	**Er**	**Tm**	**Yb**	**Lu**
1.12	1.13	1.14		1.17		1.20		1.22	1.23	1.24	1.25		1.0
			U	**Np**	**Pu**								
			1.38	1.36	1.28								

Standard potentials

The standard reduction potentials quoted here are presented in the form of Latimer diagrams (Section 8.8) and are arranged according to the blocks in the periodic table in the order *s, p, d, f*. Data in parentheses are uncertain. Most of the data, with occasional corrections, come from *Standard potentials in aqueous solution,* (ed. A. J. Bard, R. Parsons, and J. Jordan). Dekker, New York (1985). Data for the actinides are from L. R. Morss, *The chemistry of the actinide elements,* Vol. 2 (ed J. J. Katz, G. T. Seaborg, and L. R. Morss), p. 1278. Chapman and Hall, London, (1986). The value for $[Ru(bipy)_3]^{3+/2+}$ is from B. Durham, J. L. Walsh, C. L. Carter, and T. J. Meyer, *Inorg. Chem.,* **19,** 860 (1980). Potentials for carbon species in base and some for lead and palladium species are from W. M. Latimer, *Oxidation potentials.* Prentice-Hall, Englewood Cliffs (1952). For further information on standard potentials of unstable radical species see D. M. Stanbury in *Adv. Inorg. Chem.,* **33,** 69 (1989). Potentials are occasionally reported relative to the standard calomel electrode (SCE) and may be converted to the H^+/H_2 scale by adding 0.2412 V. For a detailed discussion of other reference electrodes see D. J. G. Ives and G. J. Janz, *Reference electrodes.* Academic Press, New York (1961).

s Block · Group 1/I

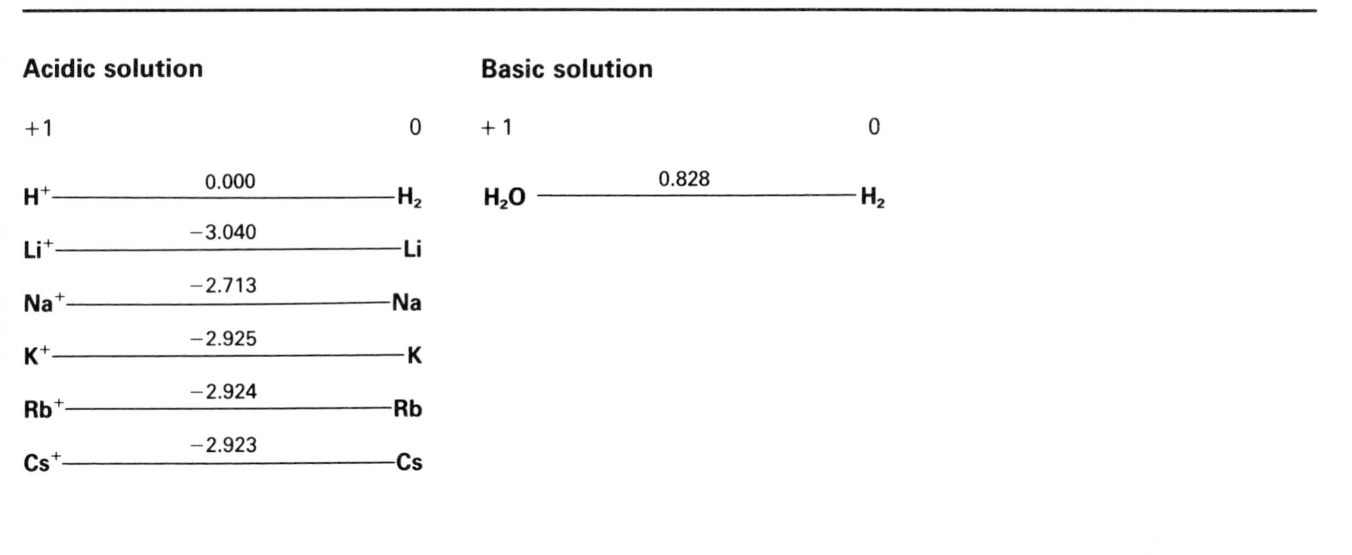

s Block · Group 2/II

Acidic solution

+2		0
Be^{2+}	−1.97	Be
Mg^{2+}	−2.356	Mg
Ca^{2+}	−2.84	Ca
Ba^{2+}	−2.92	Ba
Ra^{2+}	−2.916	Ra

Basic solution

+2		0
$Mg(OH)_2$	−2.687	Mg

p Block · Group 13/III

Acidic solution

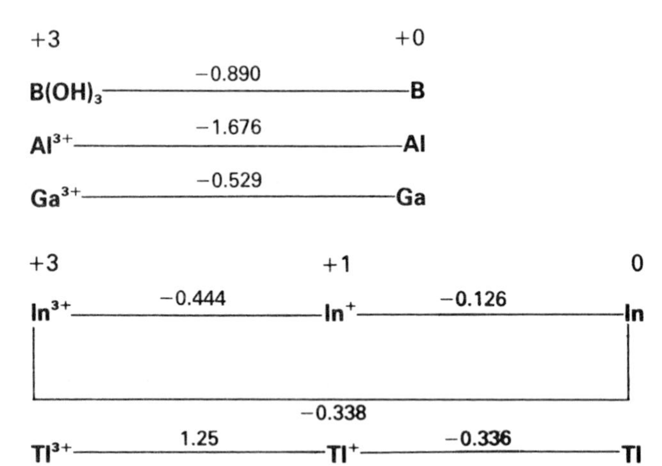

Basic solution

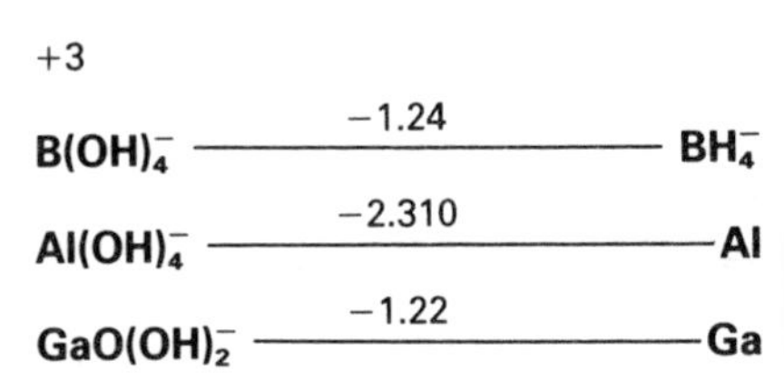

p Block · Group 14/IV

Acidic solution

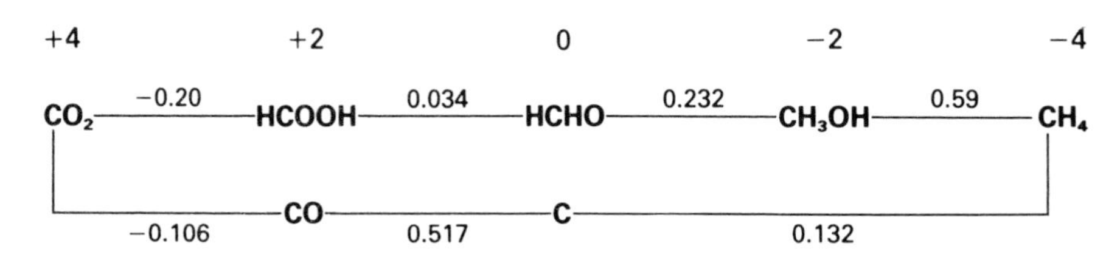

Basic solution

p Block · Group 14/IV (continued)

Acidic solution

+4		+2		0
SiO_2 (quartz)	−0.909			Si
GeO_2 (tetr.)	−0.370	GeO (hyd)	0.225	Ge
Ge^{4+}	0.124			Ge
SnO_2 (white)	−0.088	SnO (black)	−0.104	Sn
Sn^{4+}	0.15	Sn^{2+}	−0.137	Sn
α-PbO_2	1.698	Pb^{2+}	−0.125	Pb
α-PbO_2	1.46	$PbSO_4$	−0.356	Pb

Basic solution

+4		+2		0
SiO_3^{2-}	−1.69			Si
$GeO_2(OH)^-$	−0.89			Ge
$Sn(OH)_6^{2-}$	(−0.93)	$SnOOH^-$	(−0.91)	Sn
PbO_2	2.47	PbO red	−0.54	Pb

p Block · Group 15/V

Acidic solution

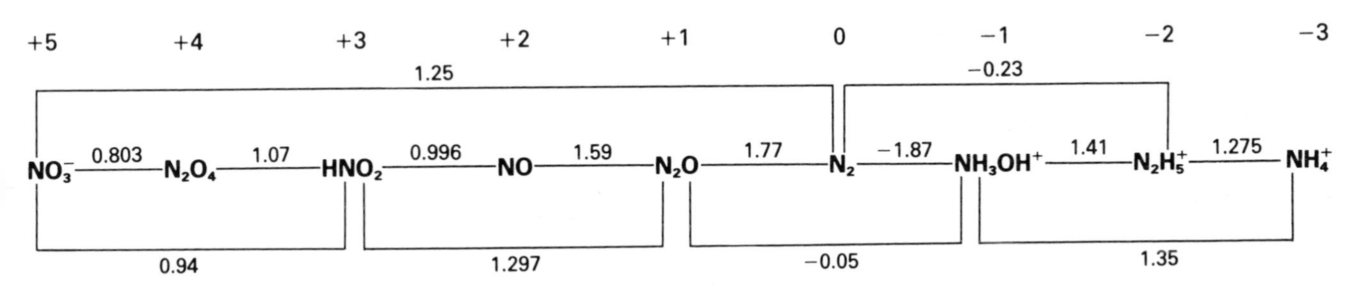

Basic solution

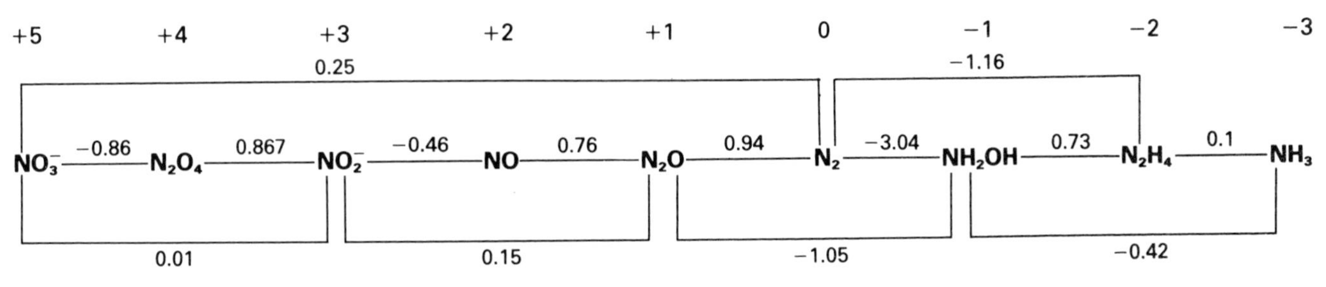

p Block · Group 15/V (continued)

Acidic solution

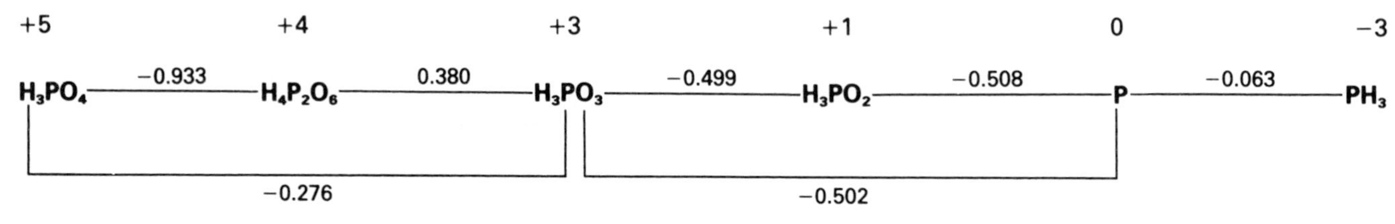

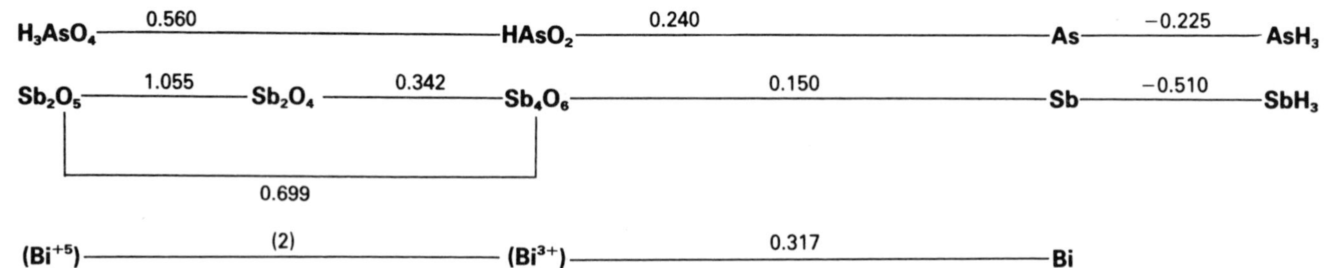

Basic solution

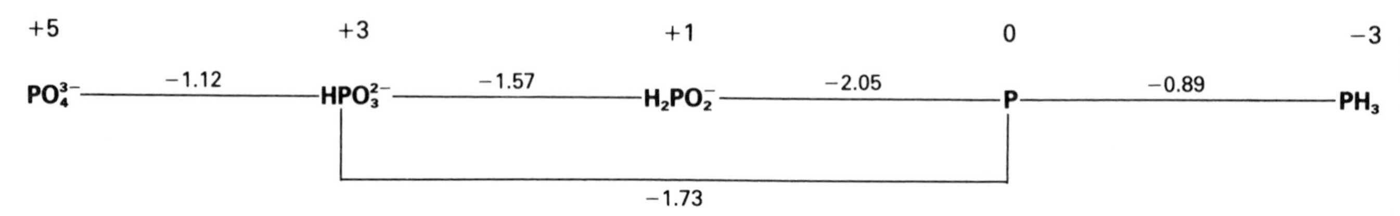

AsO_4^{3-} —(−0.67)— AsO_2^- —(−0.68)— As —(−1.37)— AsH_3

$Sb(OH)_6^-$ —(−0.465)— $Sb(OH)_4^-$ —(−0.639)— Sb —(−1.338)— SbH_3

Bi_2O_3 —(−0.452)— Bi

p Block · Group 16/VI

Acidic solution

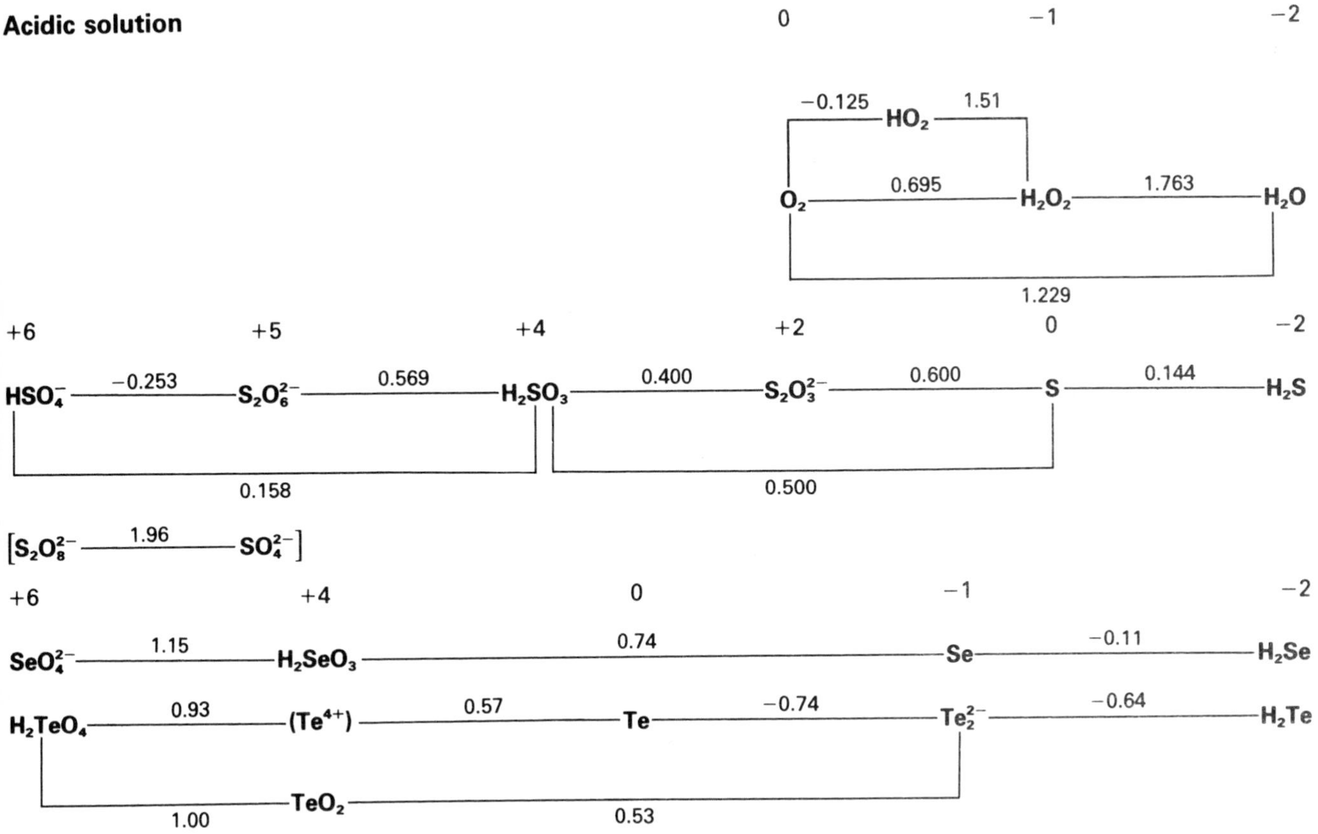

Oxidation states: 0, −1, −2

O_2 —(0.695)— H_2O_2 —(1.763)— H_2O

O_2 —(−0.125)— HO_2 —(1.51)— H_2O_2

O_2 —(1.229)— H_2O

Oxidation states: +6, +5, +4, +2, 0, −2

HSO_4^- —(−0.253)— $S_2O_6^{2-}$ —(0.569)— H_2SO_3 —(0.400)— $S_2O_3^{2-}$ —(0.600)— S —(0.144)— H_2S

HSO_4^- —(0.158)— H_2SO_3

H_2SO_3 —(0.500)— S

[$S_2O_8^{2-}$ —(1.96)— SO_4^{2-}]

Oxidation states: +6, +4, 0, −1, −2

SeO_4^{2-} —(1.15)— H_2SeO_3 —(0.74)— Se —(−0.11)— H_2Se

H_2TeO_4 —(0.93)— (Te^{4+}) —(0.57)— Te —(−0.74)— Te_2^{2-} —(−0.64)— H_2Te

H_2TeO_4 —(1.00)— TeO_2 —(0.53)— Te_2^{2-}

Basic solution

Oxidation states: 0, −1, −2

O_2 —(−0.0649)— HO_2^- —(0.867)— OH^-

O_2 —(−0.33)— O_2^- —(0.20)— HO_2^-

O_2 —(0.401)— OH^-

Oxidation states: +6, +4, +2, 0, −1, −2

SO_4^{2-} —(−0.936)— SO_3^{2-} —(−0.576)— $S_2O_3^{2-}$ —(−0.742)— S —(−0.476)— HS^-

SO_3^{2-} —(−0.659)— S

SeO_4^{2-} —(0.03)— SeO_3^{2-} —(−0.36)— Se —(−0.67)— Se^{2-}

TeO_4^{2-} —(0.07)— TeO_3^{2-} —(−0.42)— Te —(−0.84)— Te_2^{2-} —(−1.445)— Te^{2-}

Te —(−1.143)— Te^{2-}

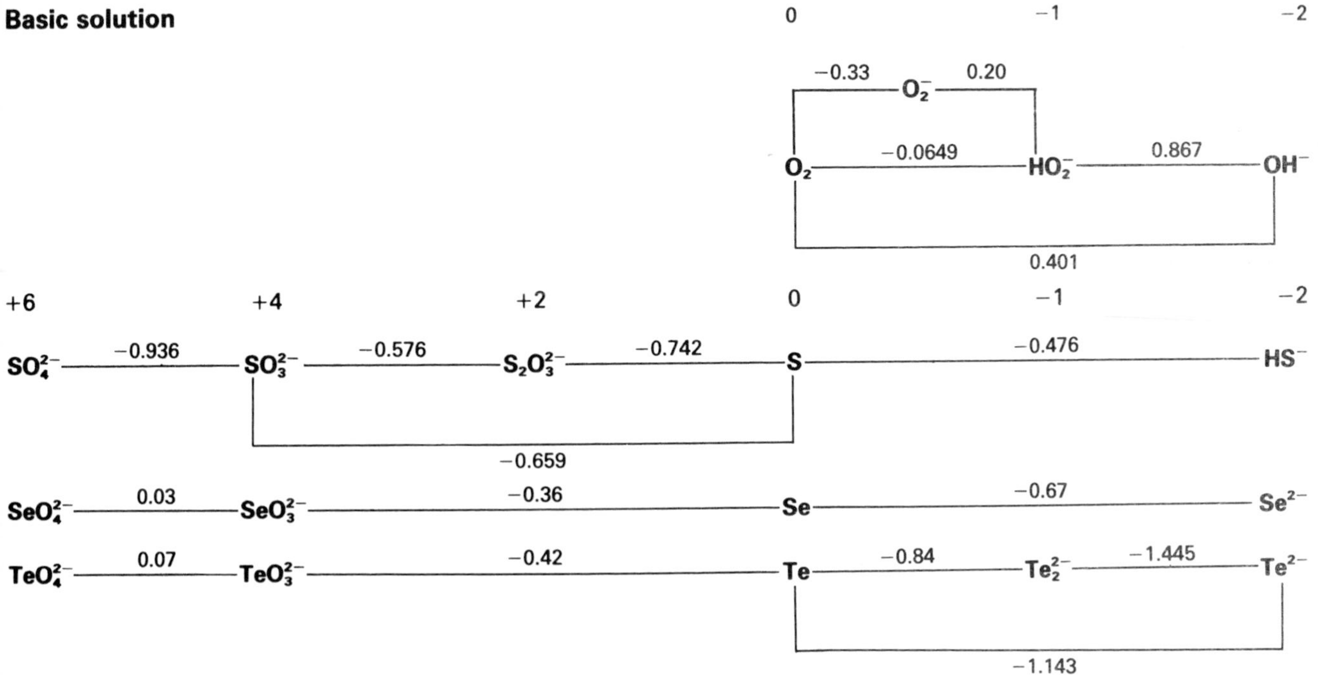

p Block · Group 17/VII

Acidic solution

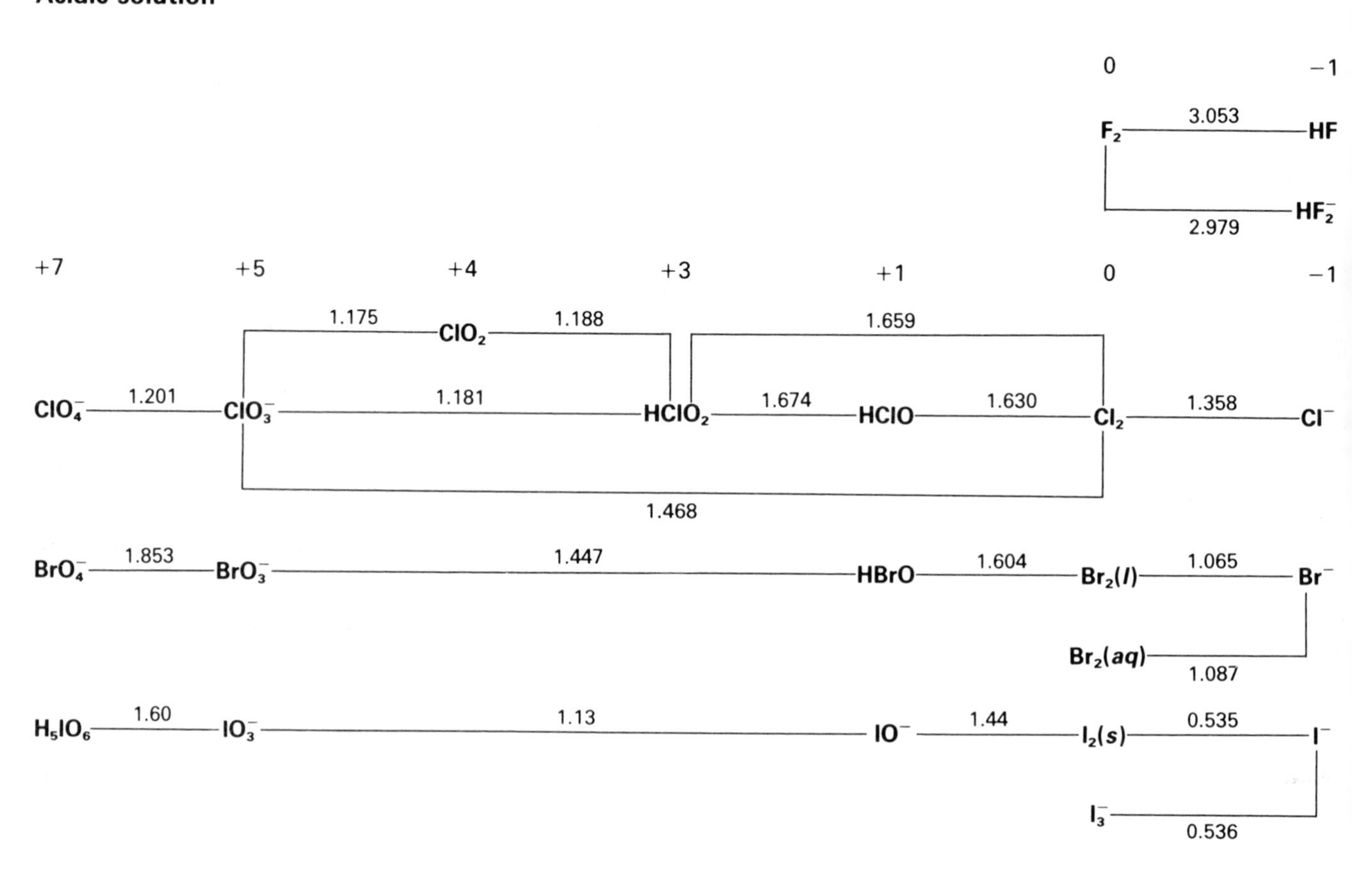

p Block · Group 17/VII (continued)

Basic solution

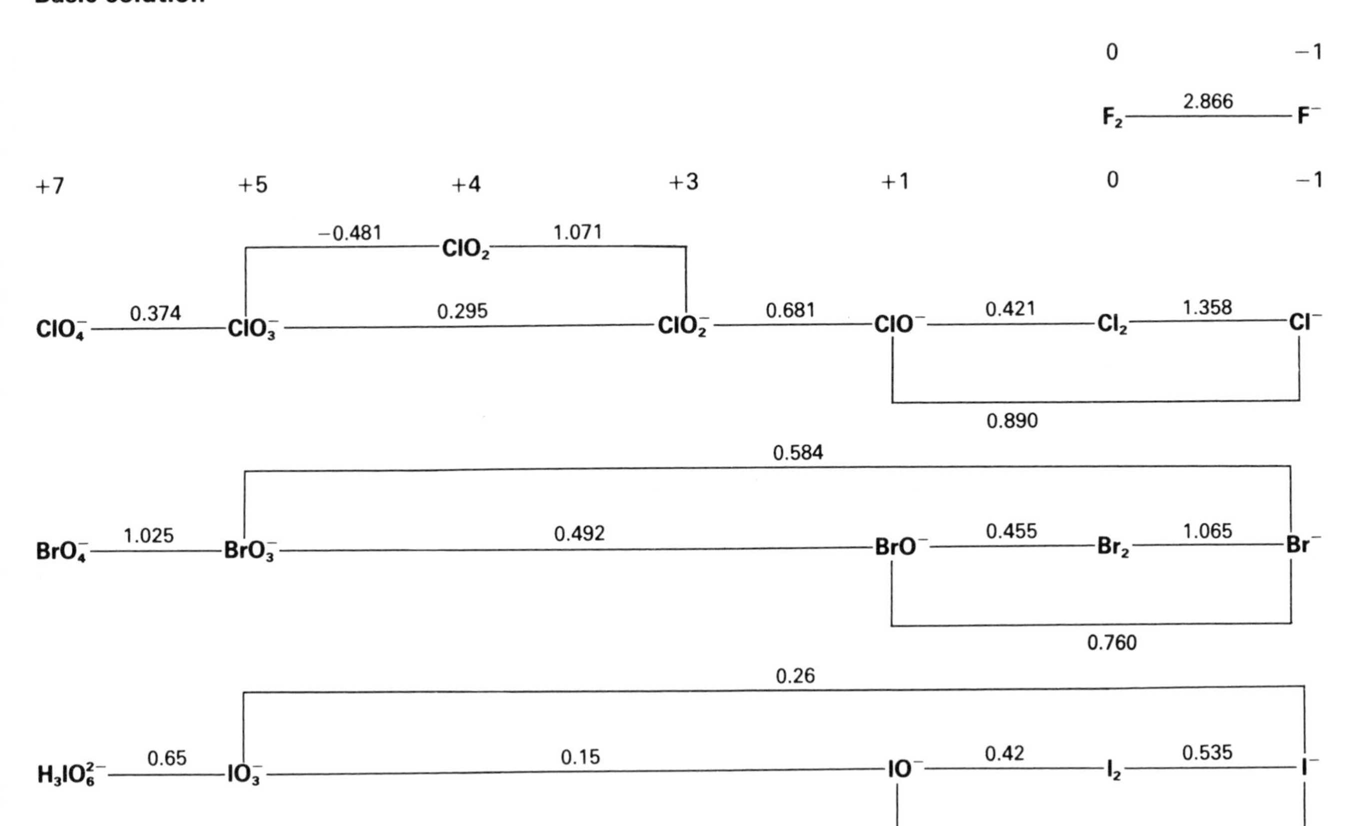

p Block · Group 18/VIII

Acidic solution

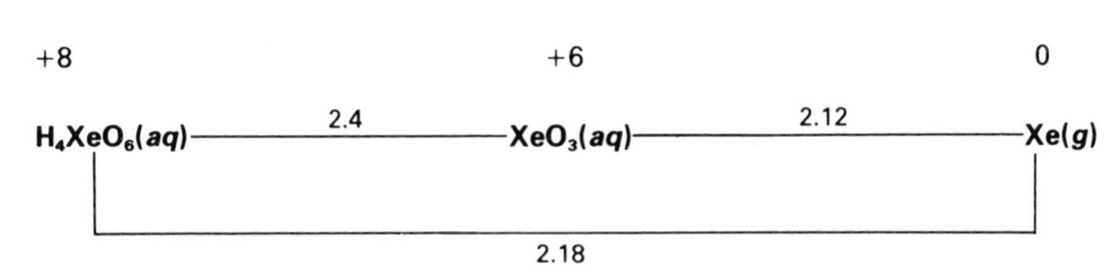

Basic solution

d Block · Group 3

Acidic solution

Basic solution

$Sc(OH)_3$ —— **−2.60** —— Sc

Acidic solution

Y^{3+} —— −2.37 —— Y

La^{3+} —— −2.38 —— La

d Block · Group 4

Acidic solution

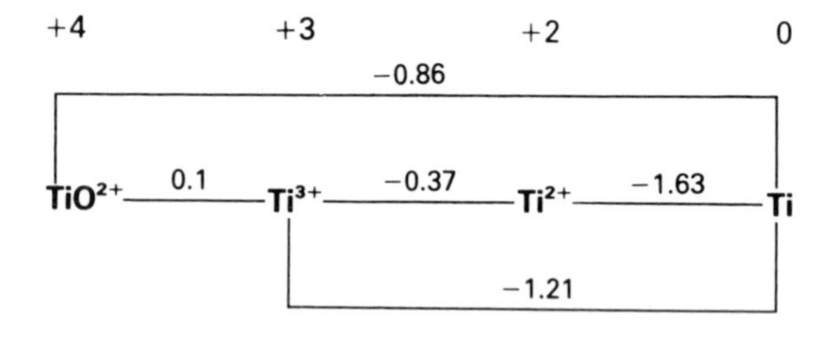

TiO_2 —— −0.56 —— Ti_2O_3 —— −1.23 —— TiO —— −1.31 —— Ti

+4 0

Zr^{4+} —— −1.55 —— Zr

Hf^{4+} —— −1.70 —— Hf

Basic solution

+4 +3 +2 0

TiO_2 —— −1.38 —— Ti_2O_3 —— −1.95 —— TiO —— −2.13 —— Ti

d Block · Group 5

Acidic solution

+5 +4 +3 +2 0

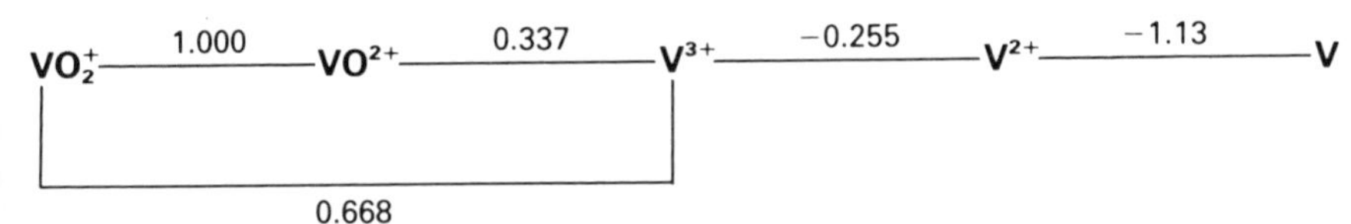

Weakly acidic solution, pH about 3.0–3.5

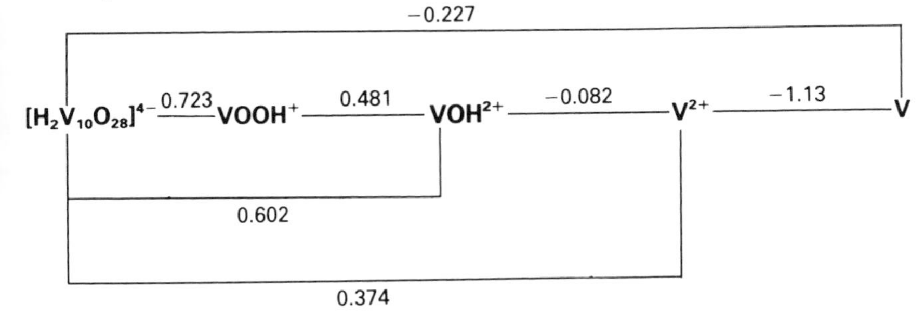

Basic solution

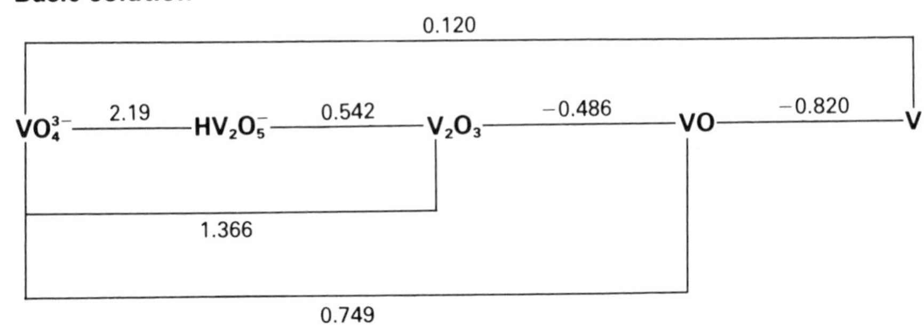

Acidic solution

+5 +3 0

$Nb_2O_5 \xrightarrow{-0.1} Nb^{3+} \xrightarrow{-1.1} Nb$

$Nb_2O_5 \xrightarrow{-0.65} Nb$

+5 0

$Ta_2O_5 \xrightarrow{-0.81} Ta$

$TaF_7^{2-} \xrightarrow{-0.45} Ta$

d Block · Group 6

Acidic solution

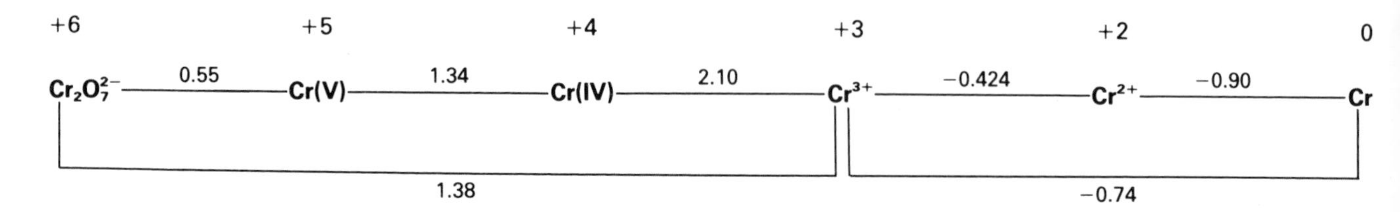

Neutral solution

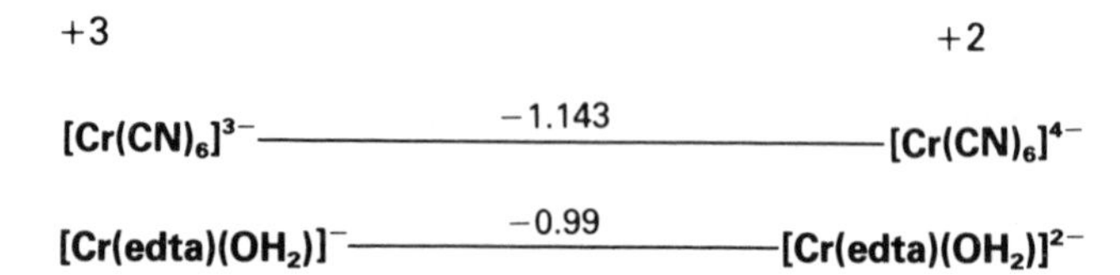

Basic solution

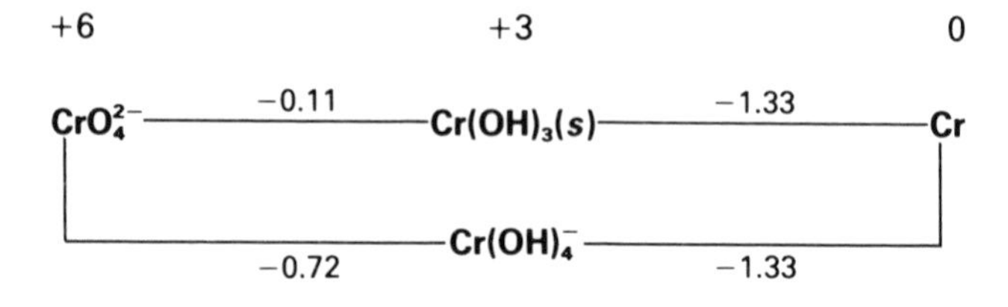

Acidic solution

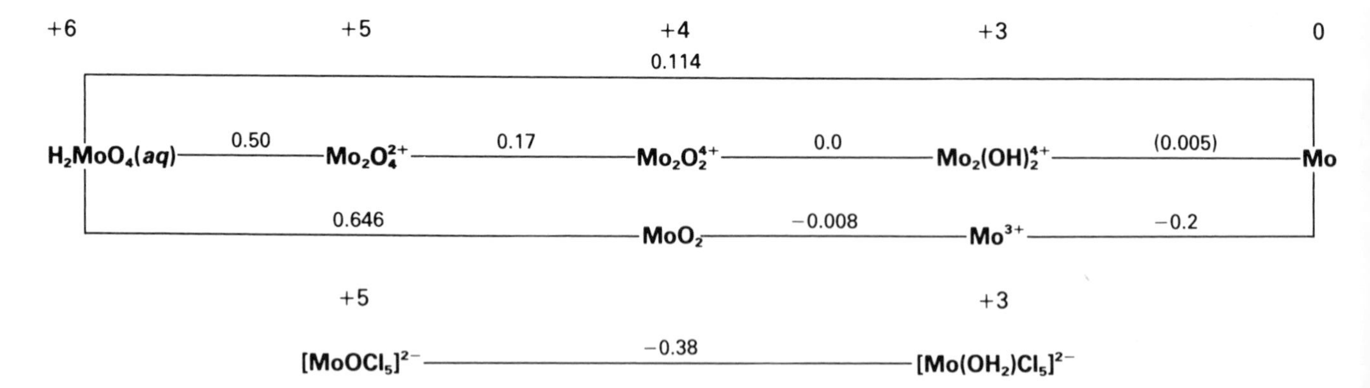

Neutral solution

+5 +4

$[Mo(CN)_8]^{3-}$ —— 0.725 —— $[Mo(CN)_8]^{4-}$

d Block · Group 6 (continued)

Basic solution

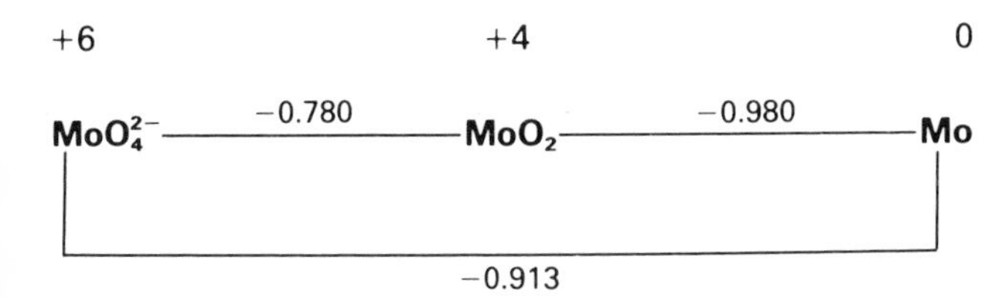

Acidic solution

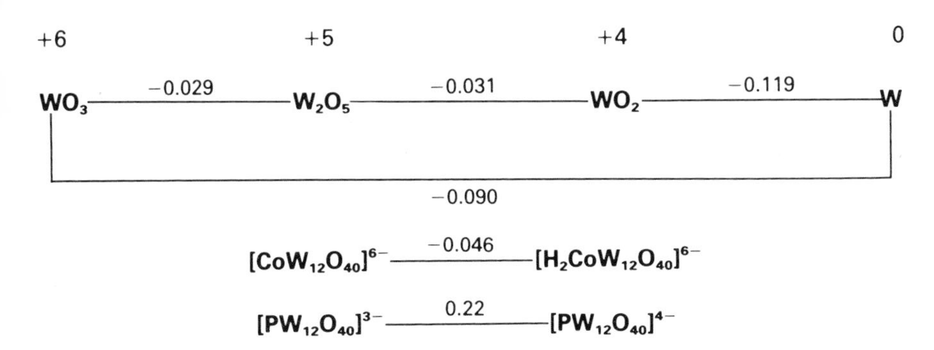

$[CoW_{12}O_{40}]^{6-}$ —(−0.046)— $[H_2CoW_{12}O_{40}]^{6-}$

$[PW_{12}O_{40}]^{3-}$ —(0.22)— $[PW_{12}O_{40}]^{4-}$

Neutral solution

+5 +4

$W(CN)_8^{3-}$ —(0.457)— $W(CN)_8^{4-}$

Basic solution

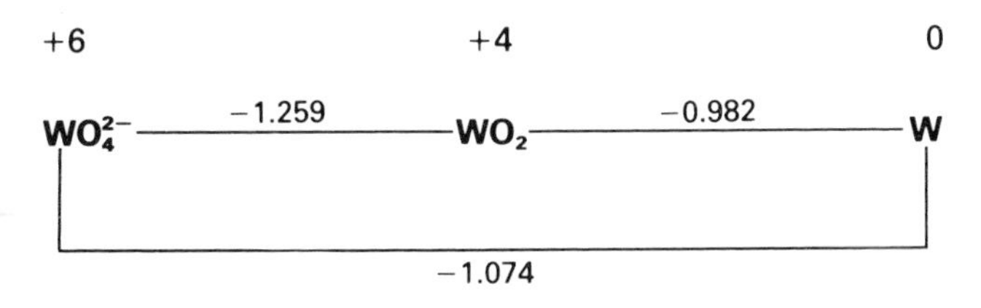

$[W(OH)_4(CN)_4]^{2-}$ —(−0.702)— $[W(OH)_4(CN)_4]^{4-}$

d Block · Group 7

Acidic solution

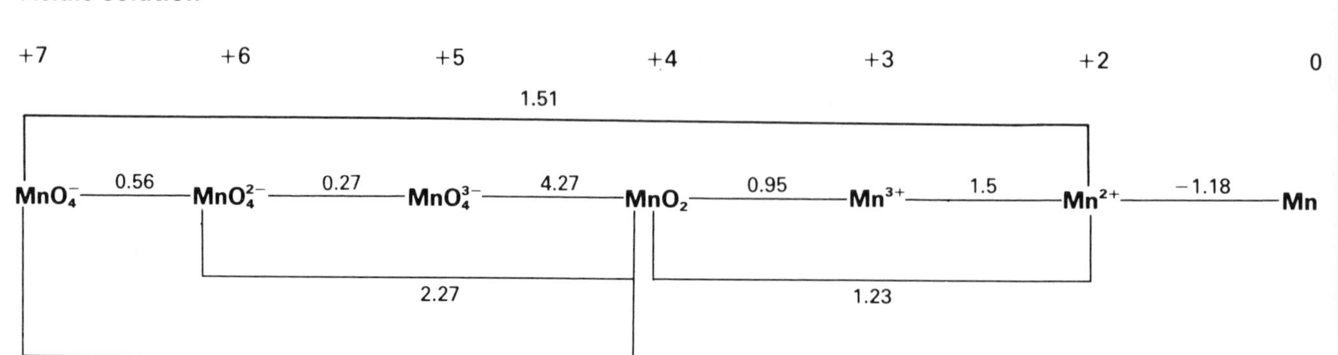

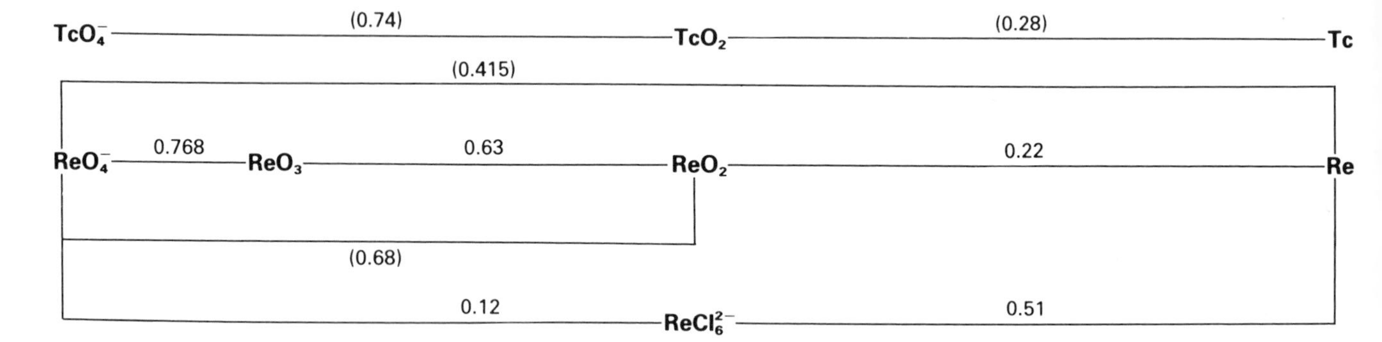

Basic solution

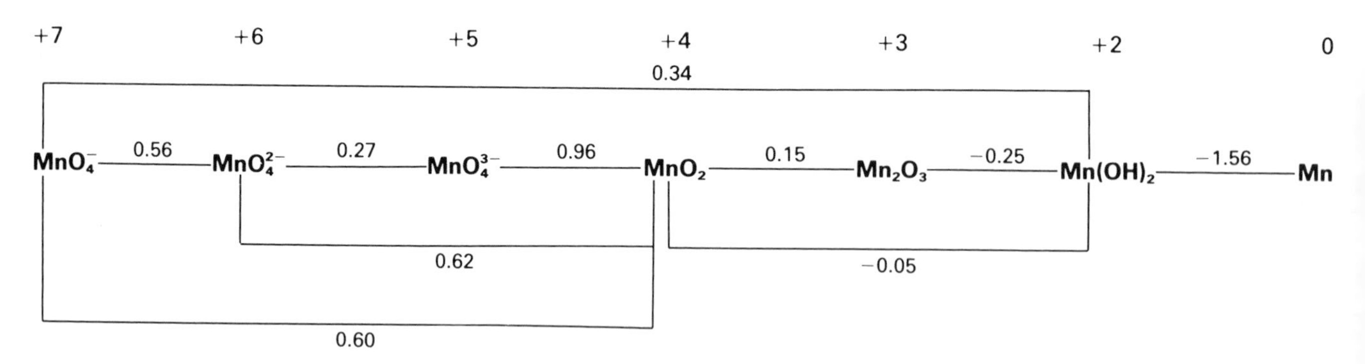

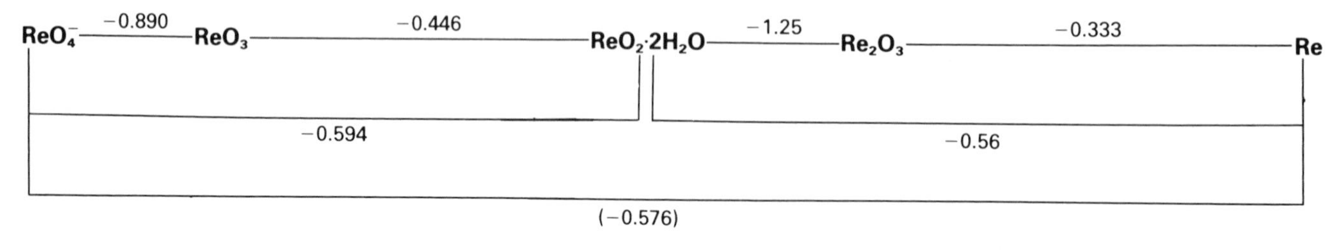

d Block · Group 8

Acidic solution

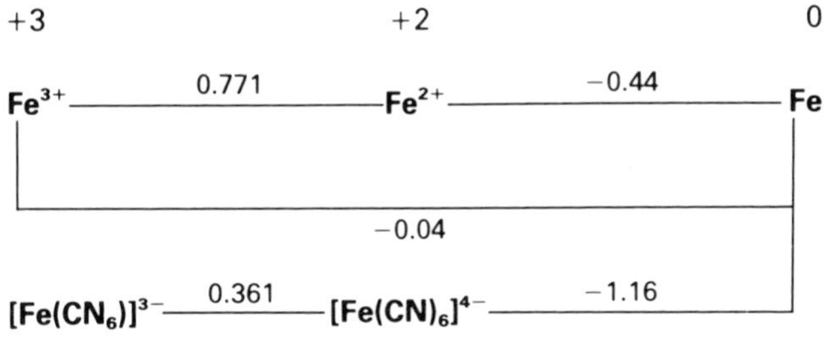

+3 | +2 | 0

Fe^{3+} —0.771— Fe^{2+} —−0.44— Fe

Fe^{3+} —−0.04— Fe

$[Fe(CN_6)]^{3-}$ —0.361— $[Fe(CN)_6]^{4-}$ —−1.16— Fe

Basic solution

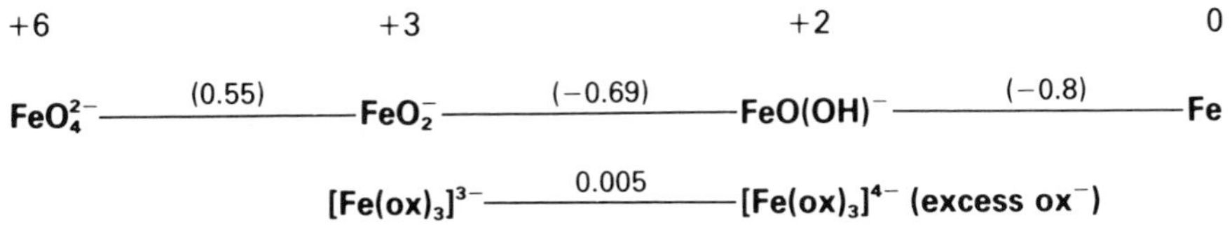

+6 | +3 | +2 | 0

FeO_4^{2-} —(0.55)— FeO_2^- —(−0.69)— $FeO(OH)^-$ —(−0.8)— Fe

$[Fe(ox)_3]^{3-}$ —0.005— $[Fe(ox)_3]^{4-}$ (excess ox^-)

Acidic solution

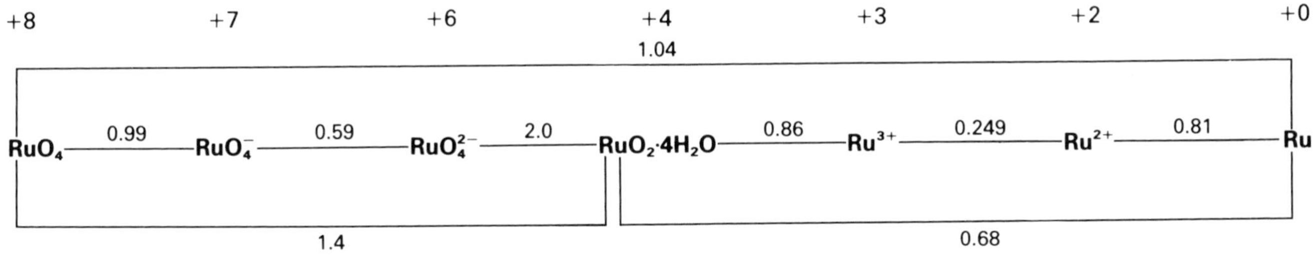

+8 | +7 | +6 | +4 | +3 | +2 | +0

RuO_4 —0.99— RuO_4^- —0.59— RuO_4^{2-} —2.0— $RuO_2{\cdot}4H_2O$ —0.86— Ru^{3+} —0.249— Ru^{2+} —0.81— Ru

RuO_4 —1.04— Ru

RuO_4 —1.4— $RuO_2{\cdot}4H_2O$

$RuO_2{\cdot}4H_2O$ —0.68— Ru

Neutral solution

+3 | +2

$[Ru(NH_3)_6]^{3+}$ —0.10— $[Ru(NH_3)_6]^{2+}$

$[Ru(CN)_6]^{3-}$ —0.86— $[Ru(CN)_6]^{4-}$

$[Ru(bipy)_3]^{3+}$ —1.53— $[Ru(bipy)_3]^{2+}$

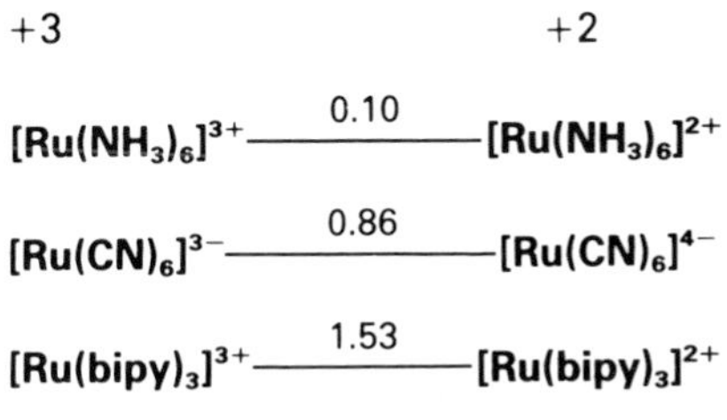

d Block · Group 8 (continued)

Acidic solution

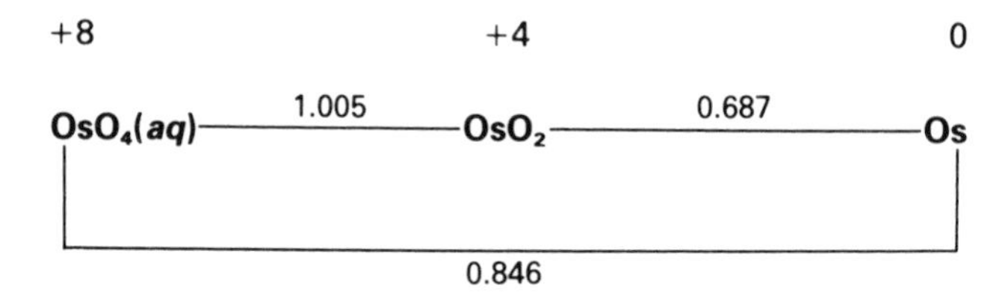

+4 → +3

$[OsCl_6]^{2-}$ —— 0.45 —— $[OsCl_6]^{3-}$

$[OsBr_6]^{2-}$ —— 0.45 —— $[OsBr_6]^{3-}$

+3 → +2

$[Os(CN)_6]^{3-}$ —— 0.634 —— $[Os(CN)_6]^{4-}$

$[Os(bipy)_3]^{3+}$ —— 0.885 —— $[Os(bipy)_3]^{2+}$

d Block · Group 9

Acidic solution

+4 → +3 → +2 → 0

CoO_2 —— 1.416 —— Co^{3+} —— 1.92 —— Co^{2+} —— −0.277 —— Co

Neutral solution

+3 → +2

$[Co(NH_3)_6]^{3+}$ —— 0.058 —— $[Co(NH_3)_6]^{2+}$

$[Co(phen)_3]^{3+}$ —— 0.33 —— $[Co(phen)_3]^{2+}$

$[Co(ox)_3]^{3-}$ —— 0.57 —— $[Co(ox)_3]^{4-}$

Basic solution

+4 → +3 → +2 → 0

CoO_2 —— 0.7 —— $Co(OH)_3$ —— 0.17 —— $Co(OH)_2$ —— −0.733 —— Co

Acidic solution

+3 → 0

Rh_2O_3 —— 0.88 —— Rh

$[RhCl_6]^{3+}$ —— 0.5 —— Rh

Neutral solution

+3 → +2

$[Rh(CN)_6]^{3-}$ —— 0.9 —— $[Rh(CN)_6]^{4-}$

d Block · Group 9 (continued)

Acidic solution

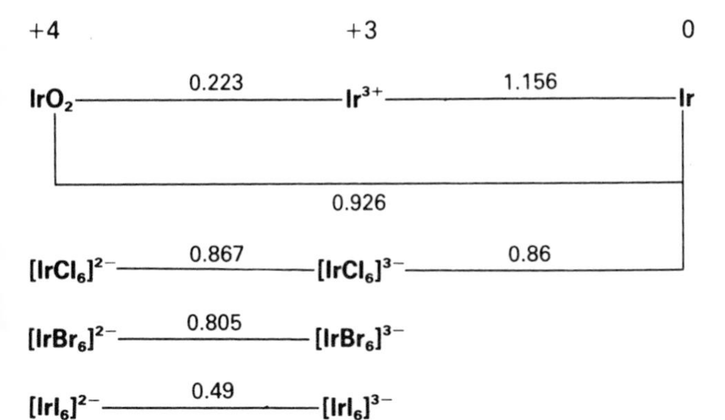

d Block · Group 10

Acidic solution

+6 | +4 | +2 | 0

NiO_4^{2-} —(>1.8)— NiO_2 —(1.593)— Ni^{2+} —(−0.257)— Ni

NiO_4^{2-} to Ni^{2+}: >1.6

Basic solution

NiO_2^{2-} —(>0.4)— NiO_2 —(0.490)— $Ni(OH)_2$ —(−0.72)— Ni

Neutral solution

$[Ni(NH_3)_6]^{2+}$ —(−0.49)— Ni

Acidic solution

+4 | +2 | 0

PdO_2 —(1.194)— Pd^{2+} —(0.915)— Pd

$[PdCl_6]^{2-}$ —(1.47)— $[PdCl_4]^{2-}$ —(0.60)— Pd

$[PdBr_4]^{2-}$ —(0.49)— Pd

d Block · Group 10 (continued)

Basic solution

+4 → +2 → 0

$Pd(OH)_4 \xrightarrow{(0.73)} Pd(OH)_2 \xrightarrow{(0.07)} Pd$

Acidic solution

+4 → +2 → 0

$PtO_2(s) \xrightarrow{1.045} PtO(s) \xrightarrow{0.98} Pt$

$[PtCl_6]^{2-} \xrightarrow{0.726} [PtCl_4]^{2-} \xrightarrow{0.758} Pt$

$[PtBr_6]^{2-} \xrightarrow{0.631} [PtBr_4]^{2-} \xrightarrow{0.698} Pt$

$[PtI_6]^{2-} \xrightarrow{0.329} [PtI_4]^{2-} \xrightarrow{0.40} Pt$

d Block · Group 11

Acidic solution

+2 → +1 → 0

$Cu^{2+} \xrightarrow{0.159} Cu^{+} \xrightarrow{0.520} Cu$

$Cu^{2+} \xrightarrow{0.340} Cu$

+2 → +1 → 0

$[Cu(NH_3)_4]^{2+} \xrightarrow{0.10} [Cu(NH_3)_2]^{+} \xrightarrow{-0.10} Cu$

$Cu^{2+} \xrightarrow{1.12} [Cu(CN)_2]^{-} \xrightarrow{-0.44} Cu$

d Block · Group 11 (continued)

Acidic solution

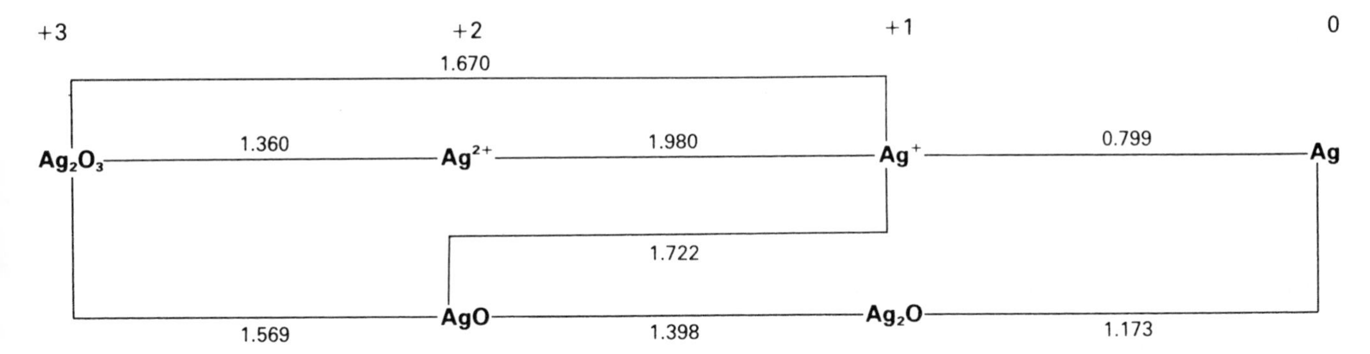

Basic solution

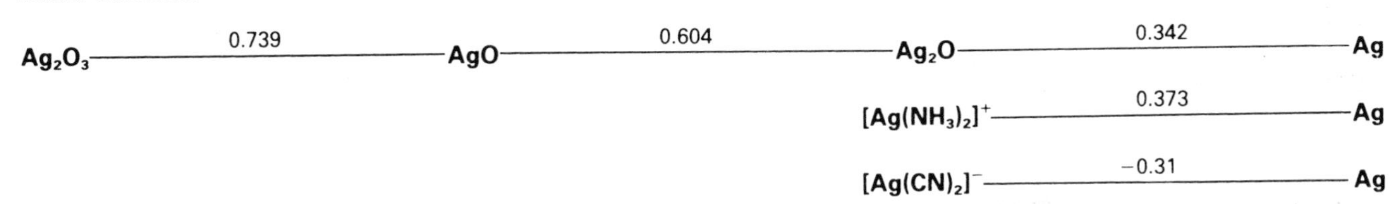

Acidic solution

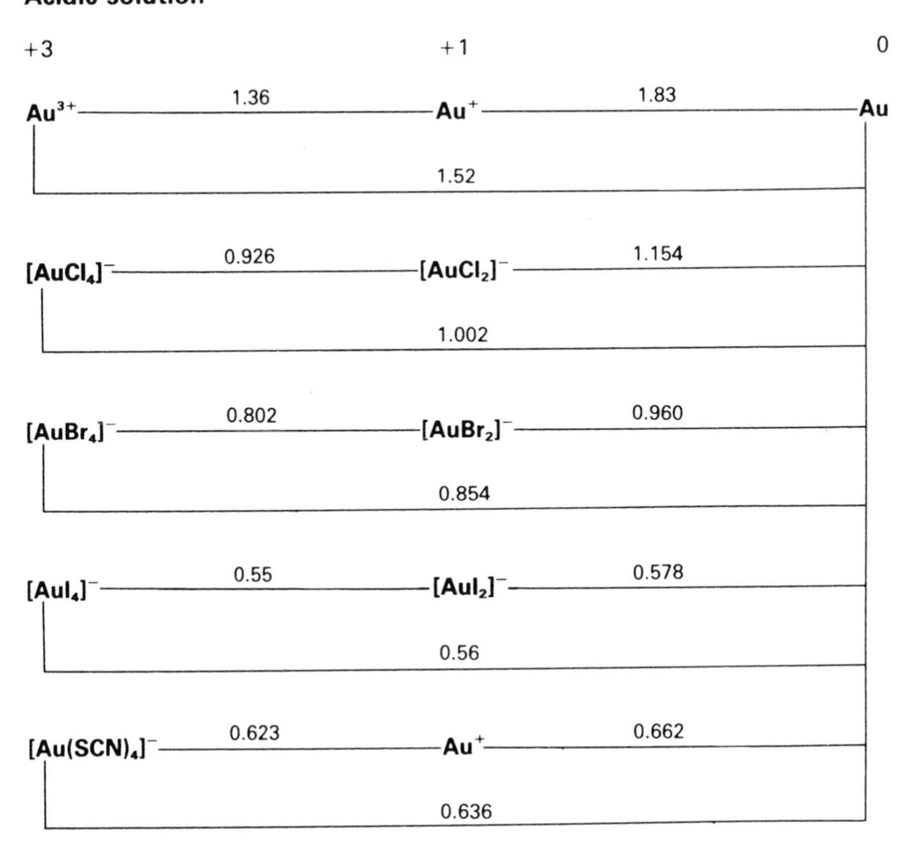

d Block · Group 12

Acidic solution

+2 ———————— 0

Zn^{2+} —— −0.7626 —— Zn

Basic solution

$[Zn(OH)_4]^{2-}$ —— −1.285 —— Zn

$Zn(OH)_2$ —— −1.246 —— Zn

Acidic solution

Cd^{2+} —— −0.4025 —— Cd

Basic solution

$Cd(OH)_2(s)$ —— −0.824 —— Cd

Acidic solution

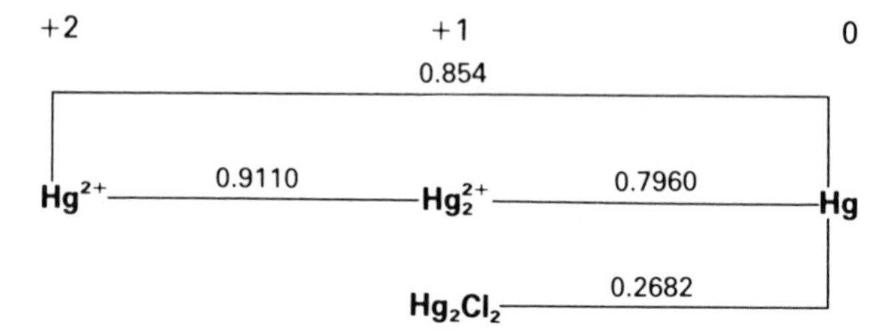

Basic solution

HgO —— 0.0977 —— Hg

f Block · Lanthanides

Acidic solution

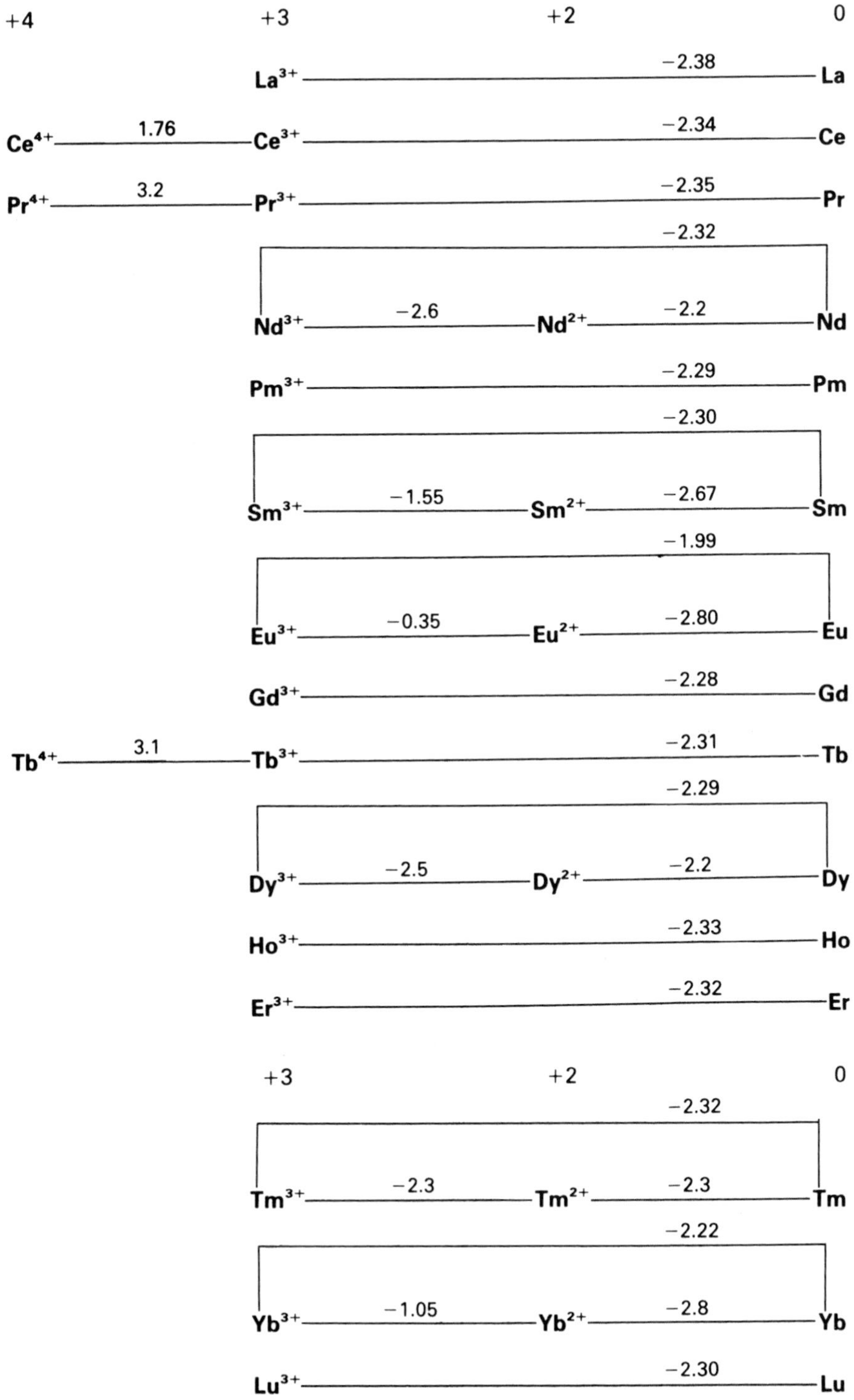

Element	M^{4+} → M^{3+} (+4 → +3)	M^{3+} → M^{2+} (+3 → +2)	M^{2+} → M (+2 → 0)	M^{3+} → M (+3 → 0)
La				−2.38
Ce	1.76			−2.34
Pr	3.2			−2.35
Nd		−2.6	−2.2	−2.32
Pm				−2.29
Sm		−1.55	−2.67	−2.30
Eu		−0.35	−2.80	−1.99
Gd				−2.28
Tb	3.1			−2.31
Dy		−2.5	−2.2	−2.29
Ho				−2.33
Er				−2.32
Tm		−2.3	−2.3	−2.32
Yb		−1.05	−2.8	−2.22
Lu				−2.30

f Block · Actinides

Acidic solution

+6 +5 +4 +3 +2 0

Ac^{3+} —(−4.9)— (Ac^{2+}) —(−0.7)— Ac
Ac^{3+} → Ac: −2.13

Th^{4+} —(−3.8)— (Th^{3+}) —(−4.9)— (Th^{2+}) —(0.7)— Th
Th^{4+} → Th: −1.83

$PaOOH^{2+}$ —(−0.05)— Pa^{4+} —(−1.4)— Pa^{3+} —(−5.0)— Pa^{2+} —(0.3)— Pa
Pa^{4+} → Pa: −1.47

UO_2^{2+} —(0.17)— UO_2^{+} —(0.38)— U^{4+} —(−0.52)— U^{3+} —(−4.7)— U^{2+} —(−0.1)— U
UO_2^{2+} → U^{4+}: 0.27
U^{3+} → U: −1.66
U^{4+} → U: −1.38

NpO_2^{2+} —(1.24)— NpO_2^{+} —(0.64)— Np^{4+} —(0.15)— Np^{3+} —(−4.7)— (Np^{2+}) —(−0.3)— Np
NpO_2^{2+} → Np^{4+}: 0.94
Np^{3+} → Np: −1.79
Np^{4+} → Np: −1.30

PuO_2^{2+} —(1.02)— PuO_2^{+} —(1.04)— Pu^{4+} —(1.01)— Pu^{3+} —(−3.5)— (Pu^{2+}) —(−1.2)— Pu
PuO_2^{2+} → Pu^{4+}: 1.03
Pu^{3+} → Pu: −2.00
Pu^{4+} → Pu: −1.25

AmO_2^{2+} —(1.60)— AmO_2^{+} —(0.82)— Am^{4+} —(2.62)— Am^{3+} —(−2.3)— (Am^{2+}) —(−1.95)— Am
AmO_2^{2+} → Am^{3+}: 1.68
Am^{3+} → Am: −2.07
AmO_2^{2+} → Am^{4+}: 1.21
Am^{4+} → Am: −0.90

f Block · Actinides (continued)

Acidic solution

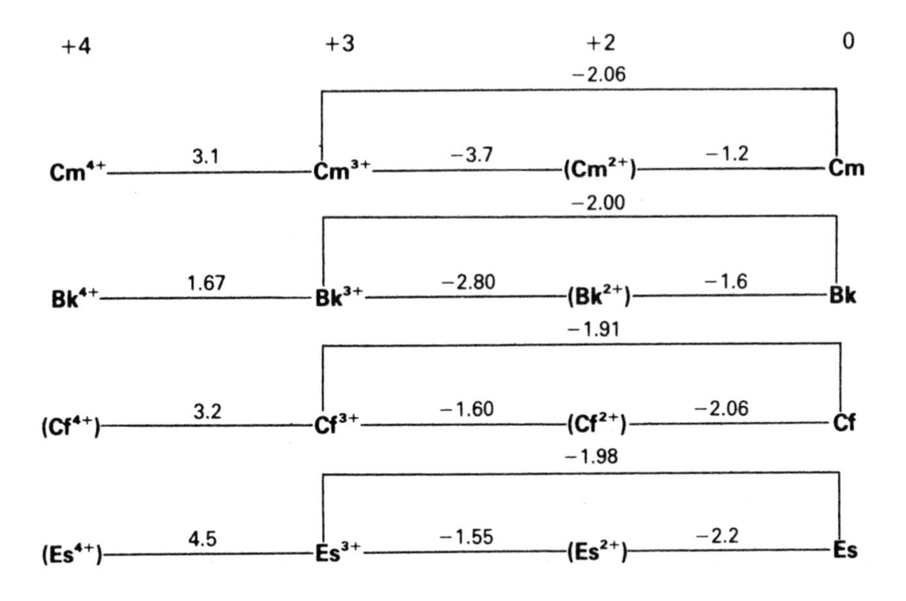

Character tables

The character tables that follow are for the most common point groups encountered in inorganic chemistry. Each one is labeled with the symbol adopted in the *Schoenflies system* of nomenclature (such as C_{3v}). Point groups that qualify as crystallographic point groups (because they are also applicable to unit cells) are also labeled with the symbol adopted in the *International system* (or the 'Hermann–Mauguin system', such as $2/m$). In the latter system, a number n represents an n-fold axis and a letter m represents a mirror plane. A diagonal line indicates that a mirror plane lies perpendicular to the symmetry axis and a bar over the number indicates that the rotation is combined with an inversion.

The symmetry species of the p and d orbitals are shown on the right of the tables. Thus, in C_{2v}, a p_x orbital (which is proportional to x) has B_1 symmetry. The functions x, y, and z also show the transformation properties of translations and of the electric dipole moment. The set of functions that span a degenerate representation (such as x and y that jointly transform as E in C_{3v}) are enclosed within parentheses. The transformation properties of rotations are shown by the letters R on the right of the table.

The groups C_1, C_s, C_i

C_1 (1)	E
A	1

$C_s = C_h$ (m)	E	σ_h		
A'	1	1	x, y, R_z	x^2, y^2, z^2, xy
A''	1	−1	z, R_x, R_y	yz, xz

$C_i = S_2$ ($\bar{1}$)	E	i		
A_g	1	1	R_x, R_y, R_z	$x^2, y^2, z^2, xy, xz, yz$
A_u	1	−1	x, y, z	

The groups C_n

C_2 (2)	E	C_2		
A	1	1	z, R_z	x^2, y^2, z^2, xy
B	1	−1	x, y, R_x, R_y	yz, xz

C_3 (3)	E	C_3	C_3^2	$\varepsilon = \exp(2\pi i/3)$	
A	1	1	1	z, R_z	$x^2 + y^2, z^2$
E	$\begin{Bmatrix} 1 \\ 1 \end{Bmatrix}$	$\begin{matrix} \varepsilon \\ \varepsilon^* \end{matrix}$	$\begin{matrix} \varepsilon^* \\ \varepsilon \end{matrix}$	$(x, y)(R_x, R_y)$	$(x^2 - y^2, xy)(yz, xz)$

C_4 (4)	E	C_4	C_2	C_4^3		
A	1	1	1	1	z, R_z	$x^2 + y^2, z^2$
B	1	−1	1	−1		$x^2 - y^2, xy$
E	$\begin{Bmatrix} 1 \\ 1 \end{Bmatrix}$	$\begin{matrix} i \\ -i \end{matrix}$	$\begin{matrix} -1 \\ -1 \end{matrix}$	$\begin{matrix} -i \\ i \end{matrix}$	$(x, y)(R_x, R_y)$	(yz, xz)

The groups C_{nv}

C_{2v} (2mm)	E	C_2	$\sigma_v(xz)$	$\sigma_v'(yz)$		
A_1	1	1	1	1	z	x^2, y^2, z^2
A_2	1	1	−1	−1	R_z	xy
B_1	1	−1	1	−1	x, R_y	xz
B_2	1	−1	−1	1	y, R_x	yz

C_{3v} (3m)	E	$2C_3$	$3\sigma_v$		
A_1	1	1	1	z	$x^2 + y^2, z^2$
A_2	1	1	−1	R_z	
E	2	−1	0	$(x, y)(R_x, R_y)$	$(x^2 - y^2, xy)(xz, yz)$

C_{4v} (4mm)	E	$2C_4$	C_2	$2\sigma_v$	$2\sigma_d$		
A_1	1	1	1	1	1	z	$x^2 + y^2, z^2$
A_2	1	1	1	−1	−1	R_z	
B_1	1	−1	1	1	−1		$x^2 - y^2$
B_2	1	−1	1	−1	1		xy
E	2	0	−2	0	0	$(x, y)(R_x, R_y)$	(xz, yz)

C_{5v}	E	$2C_5$	$2C_5^2$	$5\sigma_v$		
A_1	1	1	1	1	z	$x^2 + y^2, z^2$
A_2	1	1	1	−1	R_z	
E_1	2	2 cos 72°	2 cos 144°	0	$(x, y)(R_x, R_y)$	(xz, yz)
E_2	2	2 cos 144°	2 cos 72°	0		$(x^2 - y^2, xy)$

C_{6v} (6mm)	E	$2C_6$	$2C_3$	C_2	$3\sigma_v$	$3\sigma_d$		
A_1	1	1	1	1	1	1	z	$x^2 + y^2, z^2$
A_2	1	1	1	1	−1	−1	R_z	
B_1	1	−1	1	−1	1	−1		
B_2	1	−1	1	−1	−1	1		
E_1	2	1	−1	−2	0	0	$(x, y)(R_x, R_y)$	(xz, yz)
E_2	2	−1	−1	2	0	0		$(x^2 - y^2, xy)$

The groups D_n

D_2 (222)	E	$C_2(z)$	$C_2(y)$	$C_2(x)$		
A	1	1	1	1		x^2, y^2, z^2
B_1	1	1	−1	−1	z, R_z	xy
B_2	1	−1	1	−1	y, R_y	xz
B_3	1	−1	−1	1	x, R_x	yz

D_3 (32)	E	$2C_3$	$3C_2$		
A_1	1	1	1		x^2+y^2, z^2
A_2	1	1	−1	z, R_z	
E	2	−1	0	$(x, y)(R_x, R_y)$	$(x^2-y^2, xy)(xz, yz)$

The groups D_{nh}

D_{2h} (mmm)	E	$C_2(z)$	$C_2(y)$	$C_2(x)$	i	$\sigma(xy)$	$\sigma(xz)$	$\sigma(yz)$		
A_g	1	1	1	1	1	1	1	1		x^2, y^2, z^2
B_{1g}	1	1	−1	−1	1	1	−1	−1	R_z	xy
B_{2g}	1	−1	1	−1	1	−1	1	−1	R_y	xz
B_{3g}	1	−1	−1	1	1	−1	−1	1	R_x	yz
A_u	1	1	1	1	−1	−1	−1	−1		
B_{1u}	1	1	−1	−1	−1	−1	1	1	z	
B_{2u}	1	−1	1	−1	−1	1	−1	1	y	
B_{3u}	1	−1	−1	1	−1	1	1	−1	x	

D_{3h} ($\bar{6}m2$)	E	$2C_3$	$3C_2$	σ_h	$2S_3$	$3\sigma_v$		
A_1'	1	1	1	1	1	1		x^2+y^2, z^2
A_2'	1	1	−1	1	1	−1	R_z	
E'	2	−1	0	2	−1	0	(x, y)	(x^2-y^2, xy)
A_1''	1	1	1	−1	−1	−1		
A_2''	1	1	−1	−1	−1	1	z	
E''	2	−1	0	−2	1	0	(R_x, R_y)	(xz, yz)

The groups D_{nh} (continued)

D_{4h} (4/mmm)	E	$2C_4$	C_2	$2C_2'$	$2C_2''$	i	$2S_4$	σ_h	$2\sigma_v$	$2\sigma_d$		
A_{1g}	1	1	1	1	1	1	1	1	1	1		x^2+y^2, z^2
A_{2g}	1	1	1	−1	−1	1	1	1	−1	−1	R_z	
B_{1g}	1	−1	1	1	−1	1	−1	1	1	−1		x^2-y^2
B_{2g}	1	−1	1	−1	1	1	−1	1	−1	1		xy
E_g	2	0	−2	0	0	2	0	−2	0	0	(R_x, R_y)	(xz, yz)
A_{1u}	1	1	1	1	1	−1	−1	−1	−1	−1		
A_{2u}	1	1	1	−1	−1	−1	−1	−1	1	1	z	
B_{1u}	1	−1	1	1	−1	−1	1	−1	−1	1		
B_{2u}	1	−1	1	−1	1	−1	1	−1	1	−1		
E_u	2	0	−2	0	0	−2	0	2	0	0	(x, y)	

D_{5h}	E	$2C_5$	$2C_5^2$	$5C_2$	σ_h	$2S_5$	$2S_5^3$	$5\sigma_v$		
A_1'	1	1	1	1	1	1	1	1		x^2+y^2, z^2
A_2'	1	1	1	−1	1	1	1	−1	R_z	
E_1'	2	2 cos 72°	2 cos 144°	0	2	2 cos 72°	2 cos 144°	0	(x, y)	
E_2'	2	2 cos 144°	2 cos 72°	0	2	2 cos 144°	2 cos 72°	0		(x^2-y^2, xy)
A_1''	1	1	1	1	−1	−1	−1	−1		
A_2''	1	1	1	−1	−1	−1	−1	1	z	
E_1''	2	2 cos 72°	2 cos 144°	0	−2	−2 cos 72°	−2 cos 144°	0	(R_x, R_y)	(xz, yz)
E_2''	2	2 cos 144°	2 cos 72°	0	−2	−2 cos 144°	−2 cos 72°	0		

D_{6h} (6/mmm)	E	$2C_6$	$2C_3$	C_2	$3C_2'$	$3C_2''$	i	$2S_3$	$2S_6$	σ_h	$3\sigma_d$	$3\sigma_v$		
A_{1g}	1	1	1	1	1	1	1	1	1	1	1	1		x^2+y^2, z^2
A_{2g}	1	1	1	1	−1	−1	1	1	1	1	−1	−1	R_z	
B_{1g}	1	−1	1	−1	1	−1	1	−1	1	−1	1	−1		
B_{2g}	1	−1	1	−1	−1	1	1	−1	1	−1	−1	1		
E_{1g}	2	1	−1	−2	0	0	2	1	−1	−2	0	0	(R_x, R_y)	(xz, yz)
E_{2g}	2	−1	−1	2	0	0	2	−1	−1	2	0	0		(x^2-y^2, xy)
A_{1u}	1	1	1	1	1	1	−1	−1	−1	−1	−1	−1		
A_{2u}	1	1	1	1	−1	−1	−1	−1	−1	−1	1	1	z	
B_{1u}	1	−1	1	−1	1	−1	−1	1	−1	1	−1	1		
B_{2u}	1	−1	1	−1	−1	1	−1	1	−1	1	1	−1		
E_{1u}	2	1	−1	−2	0	0	−2	−1	1	2	0	0	(x, y)	
E_{2u}	2	−1	−1	2	0	0	−2	1	1	−2	0	0		

The groups D_{nd}

$D_{2d} = V_d$ ($\bar{4}2m$)	E	$2S_4$	C_2	$2C_2'$	$2\sigma_d$		
A_1	1	1	1	1	1		x^2+y^2, z^2
A_2	1	1	1	−1	−1	R_z	
B_1	1	−1	1	1	−1		x^2-y^2
B_2	1	−1	1	−1	1	z	xy
E	2	0	−2	0	0	(x, y) (R_x, R_y)	(xz, yz)

D_{3d} ($\bar{3}m$)	E	$2C_3$	$3C_2$	i	$2S_6$	$3\sigma_d$		
A_{1g}	1	1	1	1	1	1		x^2+y^2, z^2
A_{2g}	1	1	−1	1	1	−1	R_z	
E_g	2	−1	0	2	−1	0	(R_x, R_y)	(x^2-y^2, xy) (xz, yz)
A_{1u}	1	1	1	−1	−1	−1		
A_{2u}	1	1	−1	−1	−1	1	z	
E_u	2	−1	0	−2	1	0	(x, y)	

D_{4d}	E	$2S_8$	$2C_4$	$2S_8^3$	C_2	$4C_2'$	$4\sigma_d$		
A_1	1	1	1	1	1	1	1		x^2+y^2, z^2
A_2	1	1	1	1	1	−1	−1	R_z	
B_1	1	−1	1	−1	1	1	−1		
B_2	1	−1	1	−1	1	−1	1	z	
E_1	2	$\sqrt{2}$	0	$-\sqrt{2}$	−2	0	0	(x, y)	
E_2	2	0	−2	0	2	0	0		(x^2-y^2, xy)
E_3	2	$-\sqrt{2}$	0	$\sqrt{2}$	−2	0	0	(R_x, R_y)	(xz, yz)

The cubic groups

T_d ($\bar{4}3m$)	E	$8C_3$	$3C_2$	$6S_4$	$6\sigma_d$		
A_1	1	1	1	1	1		$x^2+y^2+z^2$
A_2	1	1	1	−1	−1		
E	2	−1	2	0	0		$(2z^2-x^2-y^2, x^2-y^2)$
T_1	3	0	−1	1	−1	(R_x, R_y, R_z)	
T_2	3	0	−1	−1	1	(x, y, z)	(xy, xz, yz)

The cubic groups (continued)

O_h ($m3m$)	E	$8C_3$	$6C_2$	$6C_4$	$3C_2$ ($=C_4^2$)	i	$6S_4$	$8S_6$	$3\sigma_h$	$6\sigma_d$		
A_{1g}	1	1	1	1	1	1	1	1	1	1		$x^2+y^2+z^2$
A_{2g}	1	1	−1	−1	1	1	−1	1	1	−1		
E_g	2	−1	0	0	2	2	0	−1	2	0		$(2z^2-x^2-y^2, x^2-y^2)$
T_{1g}	3	0	−1	1	−1	3	1	0	−1	−1	(R_x, R_y, R_z)	
T_{2g}	3	0	1	−1	−1	3	−1	0	−1	1		(xz, yz, xy)
A_{1u}	1	1	1	1	1	−1	−1	−1	−1	−1		
A_{2u}	1	1	−1	−1	1	−1	1	−1	−1	1		
E_u	2	−1	0	0	2	−2	0	1	−2	0		
T_{1u}	3	0	−1	1	−1	−3	−1	0	1	1	(x, y, z)	
T_{2u}	3	0	1	−1	−1	−3	1	0	1	−1		

The icosahedral group

I	E	$12C_5$	$12C_5^2$	$20C_3$	$15C_2$		$\eta^{\pm}=\frac{1}{2}(1\pm 5^{1/2})$
A	1	1	1	1	1		$x^2+y^2+z^2$
T_1	3	η^+	η^-	0	−1	(x, y, z) (R_x, R_y, R_z)	
T_2	3	η^-	η^+	0	−1		
G	4	−1	−1	1	0		
H	5	0	0	−1	1		$(2z^2-x^2-y^2, x^2-y^2, xy, yz, zx)$

Further information: P. W. Atkins, M. S. Child, and C. S. G. Phillips, *Tables for group theory*. Oxford University Press (1970).

Symmetry-adapted orbitals

A6

The table gives the symmetry class of the *s*, *p*, and *d* orbitals of the central atom of an AB_n molecule of the specified point group. In most cases, the *z* axis is the principal axis of the molecule; in C_{2v} the *x* axis lies perpendicular to the molecular plane.

The orbital diagrams show the linear combinations of atomic orbitals on the peripheral atoms of AB_n molecules of the specified point groups. Where a view from above is shown, the dot representing the central atom is either in the plane of the paper (for the *D* group) or above the plane (for the corresponding *C* group). Different phases of the atomic orbitals (+ or − amplitudes) are shown by tinting. Where there is a large difference in the magnitudes of the coefficients of orbitals in a particular combination, the atomic orbitals have been drawn large or small to represent their relative contributions to the linear combination. In the case of degenerate linear combinations (those labeled *E* or *T*), any linearly independent combination of the degenerate pair is also of suitable symmetry. In practice, these different linear combinations look like the ones shown here, but their nodes are rotated by an arbitrary angle around the *z* axis.

Molecular orbitals are formed by combining an orbital of the central atom (as in Table A6.1) with a linear combination of the same symmetry.

Table A6.1

	$D_{\infty h}$	D_{2v}	D_{3h}	C_{3v}	D_{4h}	C_{4v}	D_{5h}	C_{5v}	D_{6h}	C_{6v}	T_d	O_h
s	σ	A_1	A_1'	A_1	A_{1g}	A_1	A_1'	A_1	A_{1g}	A_1	A_1	A_{1g}
p_x	π	B_1	E'	E	E_u	E	E_1'	E_1	E_{1u}	E_1	T_2	T_{1u}
p_y	π	B_2	E'	E	E_u	E	E_1'	E_1	E_{1u}	E_1	T_2	T_{1u}
p_z	σ	A_1	A_2''	A_1	A_{2u}	A_1	A_2''	A_1	A_{2u}	A_1	T_2	T_{1u}
d_{z^2}	σ	A_1	A_1'	A_1	A_{1g}	A_1	A_1'	A_1	A_{1g}	A_1	E	E_g
$d_{x^2-y^2}$	δ	A_1	E'	E	B_{1g}	B_1	E_2'	E_2	E_{2g}	E_2	E	E_g
d_{xy}	δ	A_2	E'	E	B_{2g}	B_2	E_2'	E_2	E_{2g}	E_2	T_2	T_{2g}
d_{yz}	π	B_2	E''	E	E_g	E	E_1''	E_1	E_{1g}	E_1	T_2	T_{2g}
d_{zx}	π	B_1	E''	E	E_g	E	E_1''	E_1	E_{1g}	E_1	T_2	T_{2g}

$D_{\infty h}$	C_{2v}
σ_g	A_1 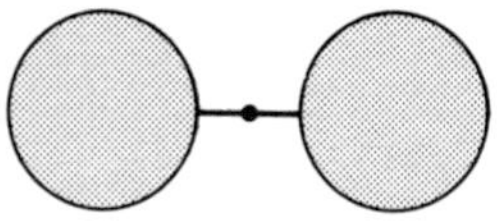
π_g	A_2 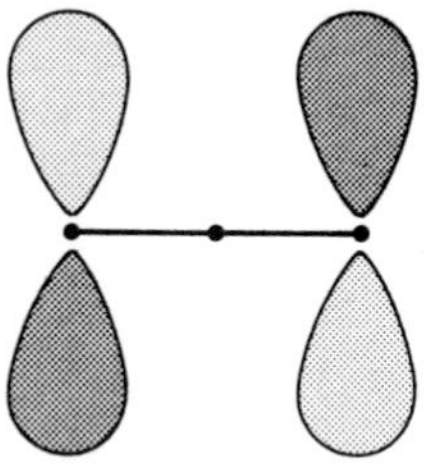
π_u	B_1 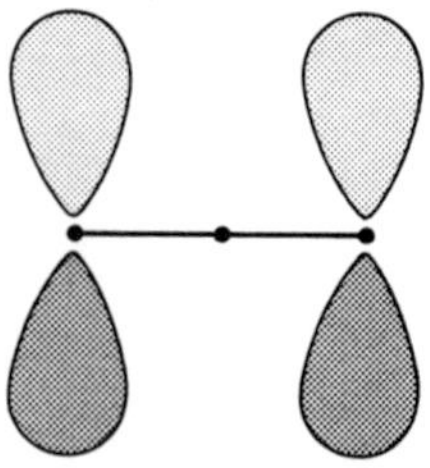
σ_u	B_2 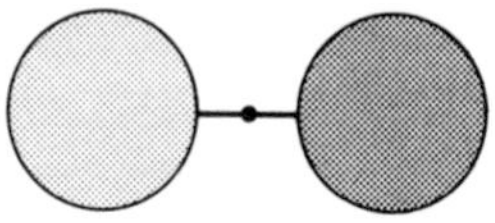
	A_1 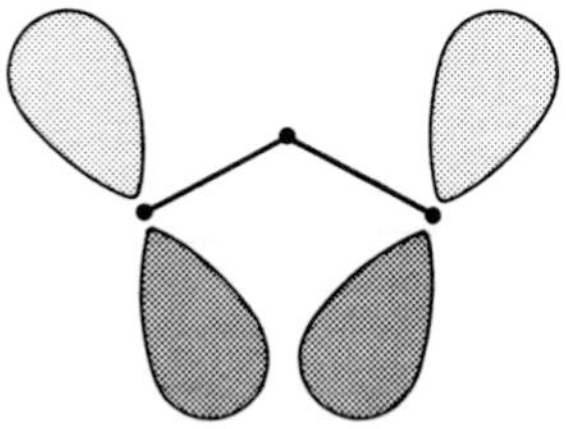
	B_2 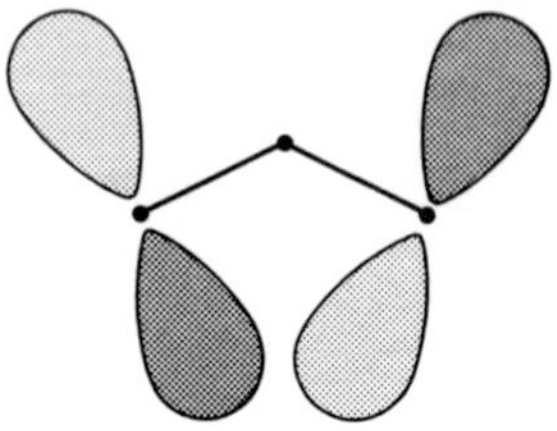

D_{3h}	C_{3v}
A_1'	A_1
A_2'	A_2
E'	E
E'	E
A_1''	A_1

D_{3h}	C_{3v}	(cont.)
E''	E	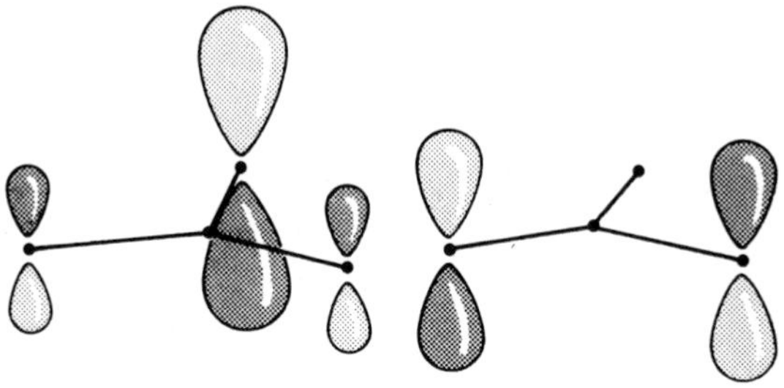

D_{4h}	C_{4v}
A_{1g}	A_1 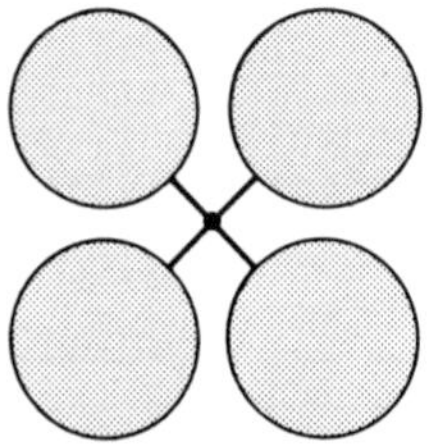
A_{2g}	A_2 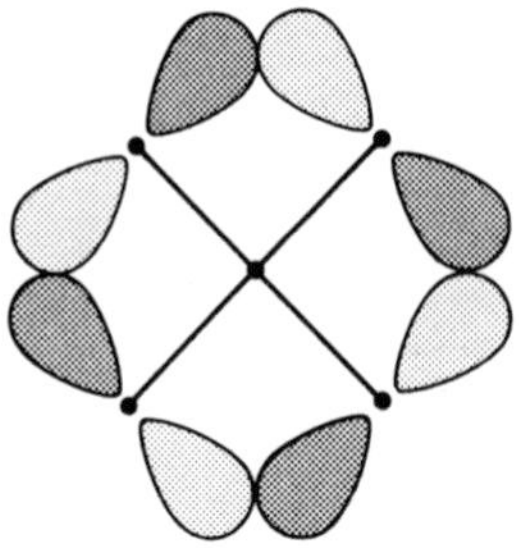
B_{1g}	B_1 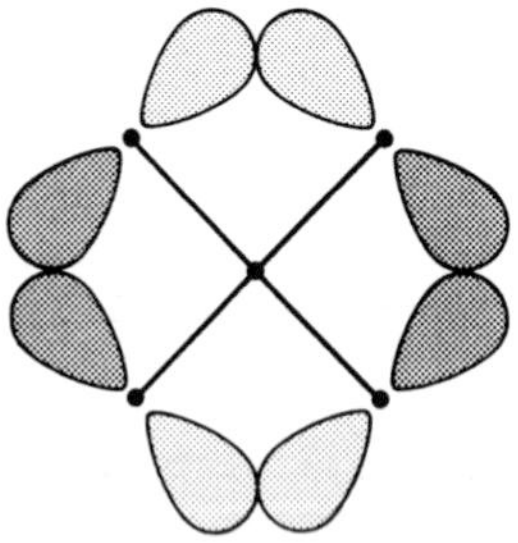
B_{2g}	B_2 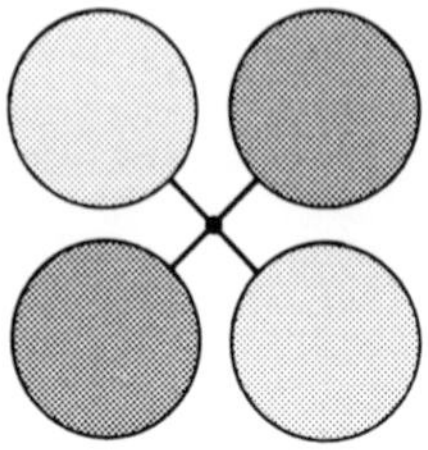

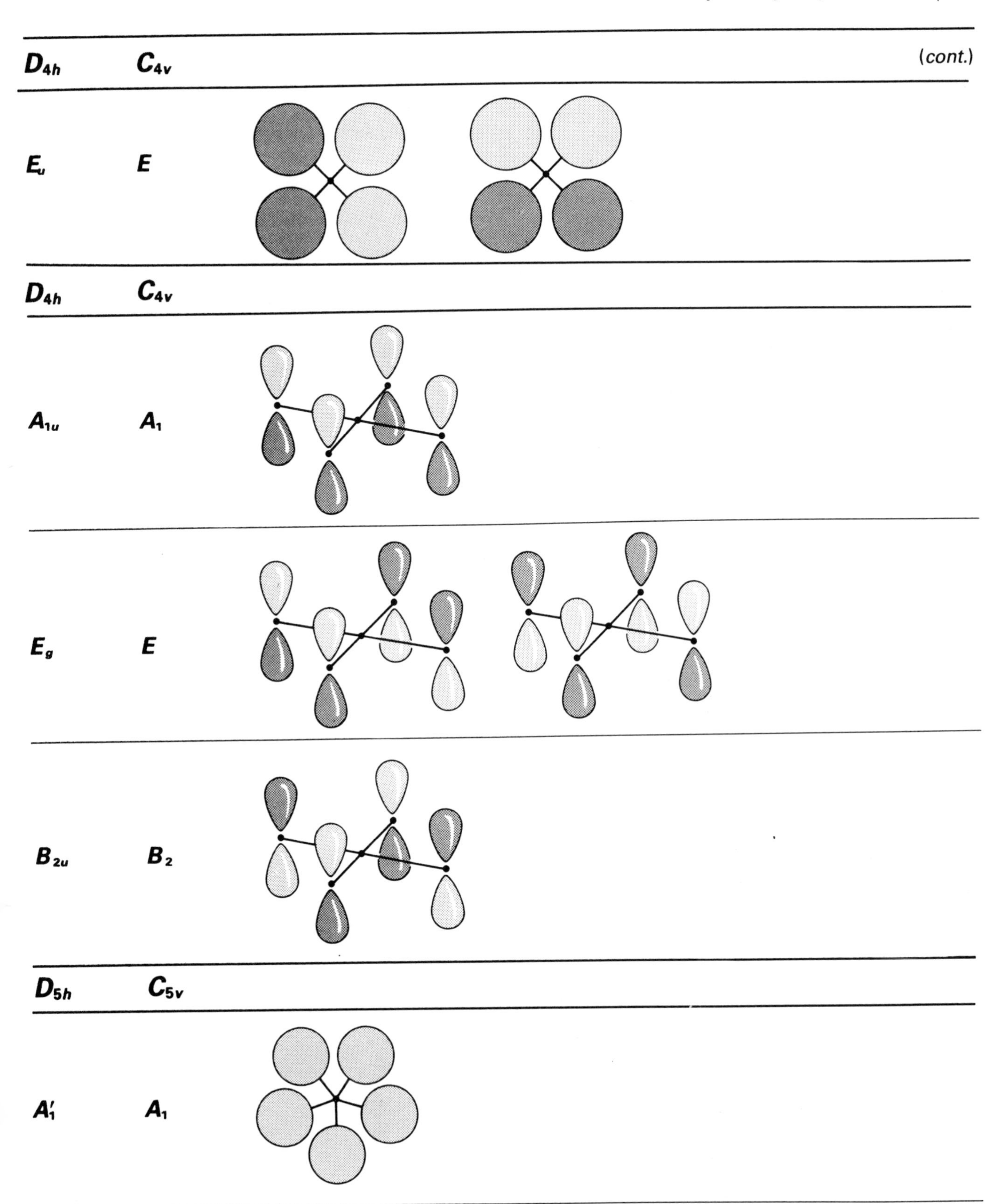

D_{4h}
C_{4v}
(cont.)
E_u
E
D_{4h}
C_{4v}
A_{1u}
A_1
E_g
E
B_{2u}
B_2
D_{5h}
C_{5v}
A_1'
A_1

D_{5h}	C_{5v}	(cont.)
A_2'	A_2	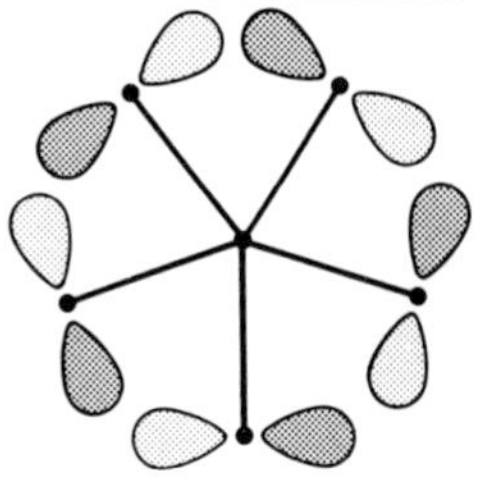
E_1'	E_1	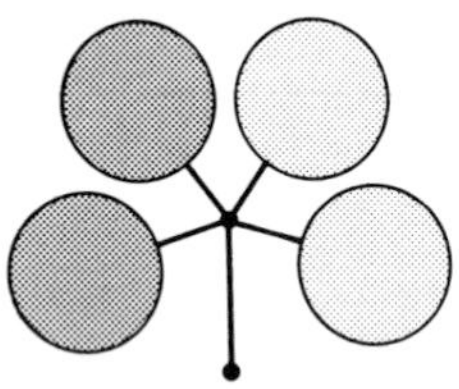 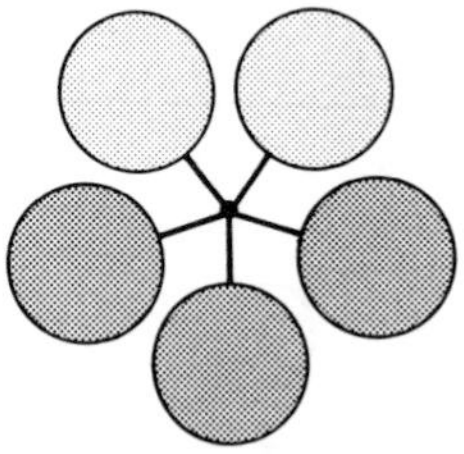
E_2'	E_2	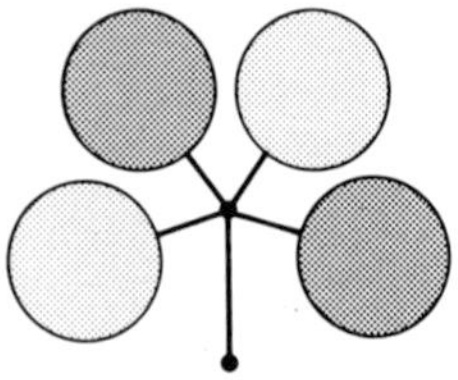 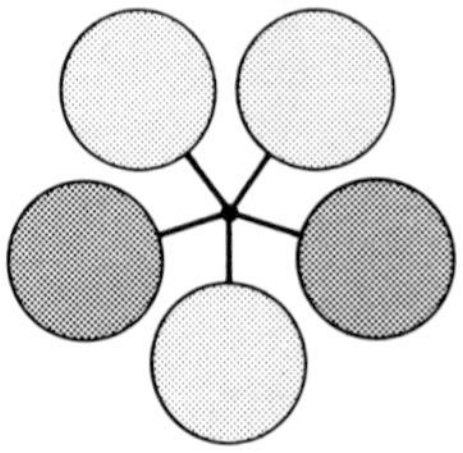
A_2''	A_1	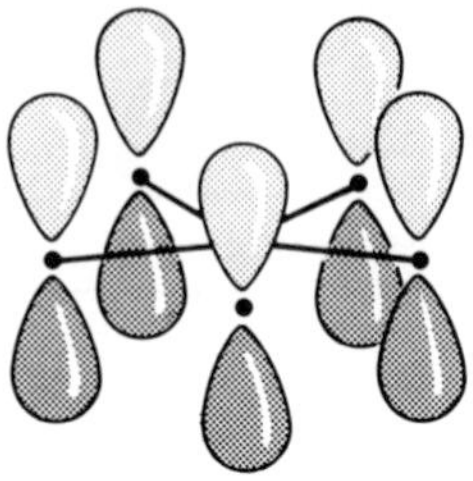
E_1''	E_1	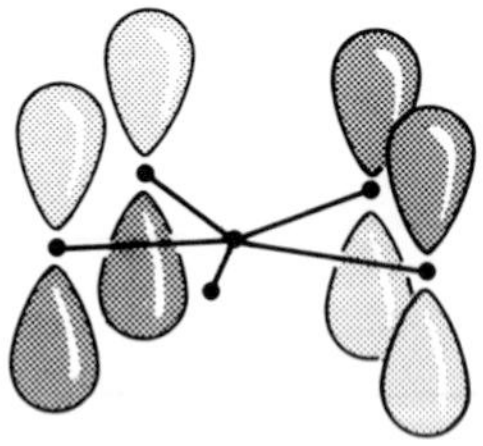 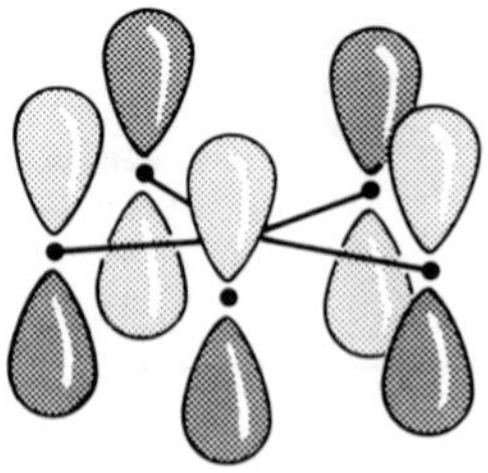
E_2''	E_2	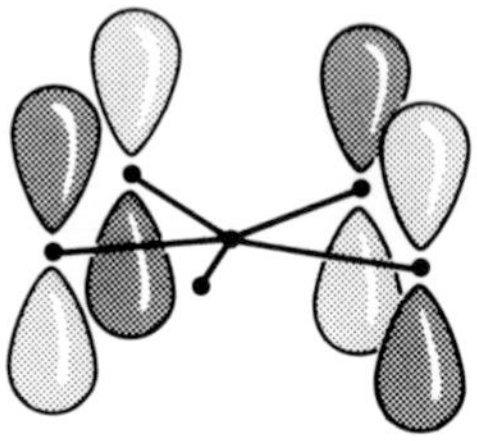 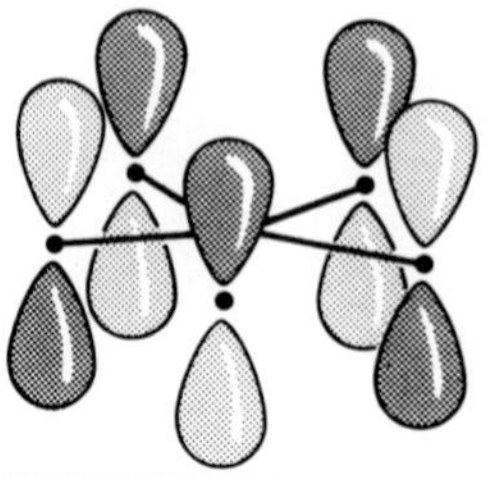

D_{6h}	C_{6v}
A_{1g}	A_1 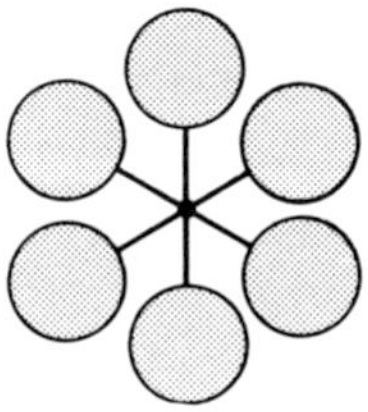
A_{2g}	A_2 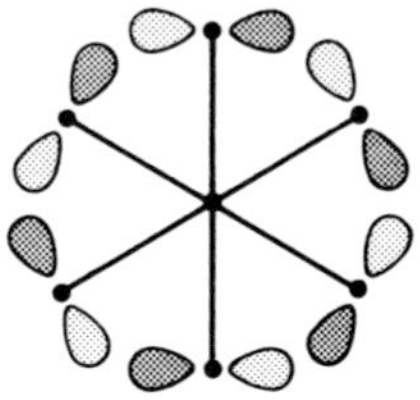
B_{1u}	B_1 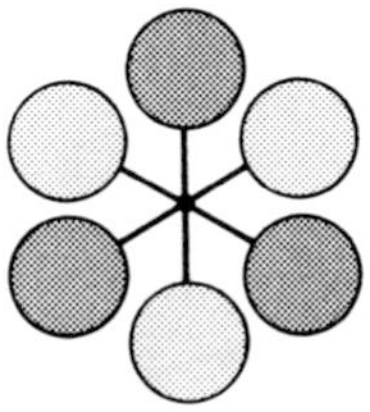
B_{2u}	B_2 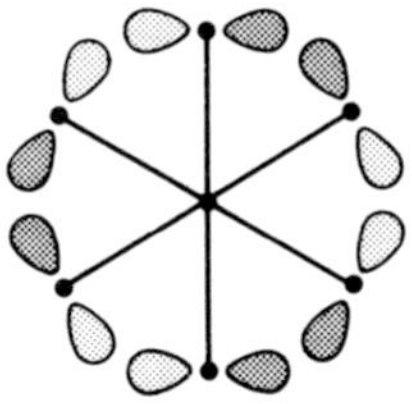
E_{1u}	E_1 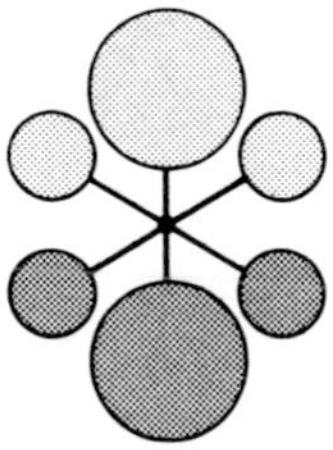 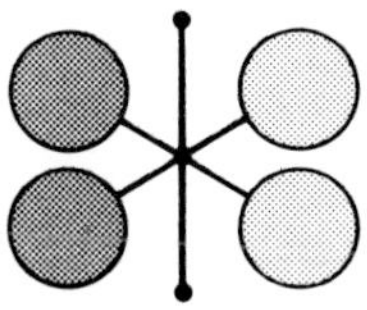
E_{2g}	E_2 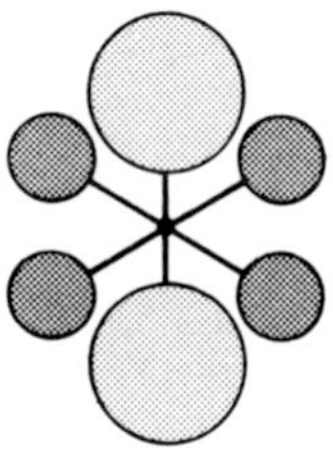 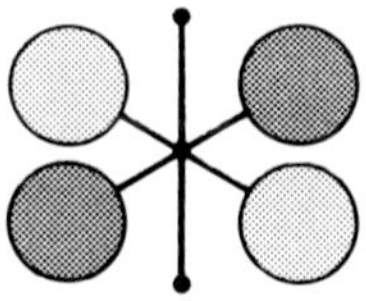

D_{6h}	C_{6v}	(cont.)
A_{2u}	A_1	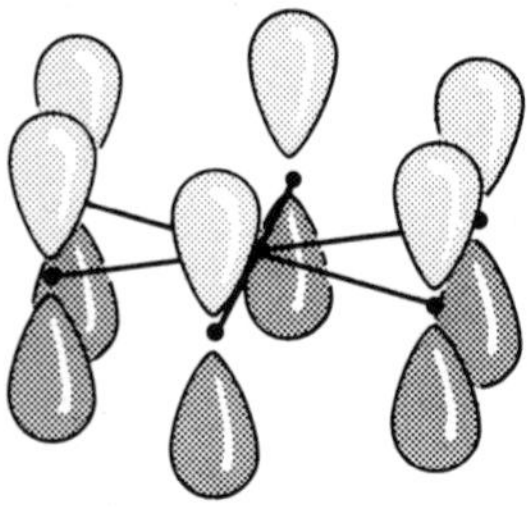
B_{2g}	B_1	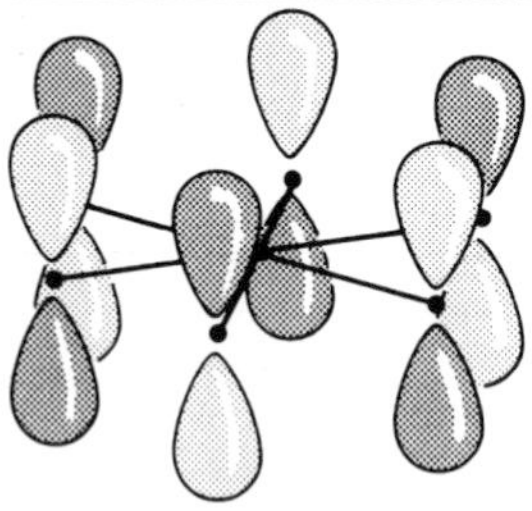
E_{1g}	E_1 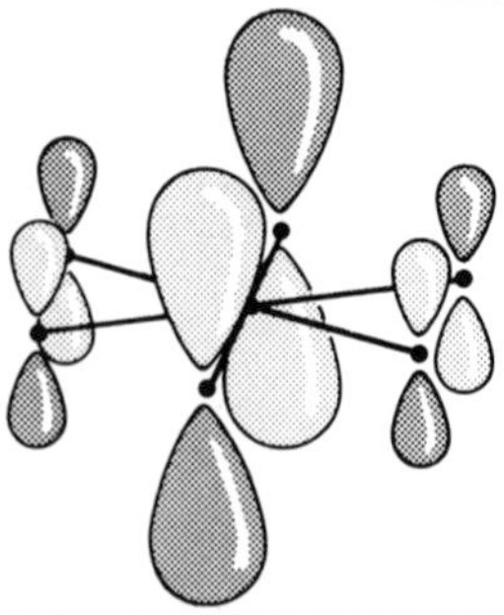 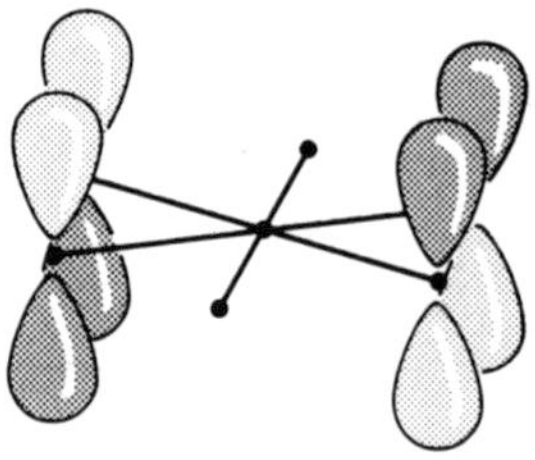	
E_{2u}	E_2 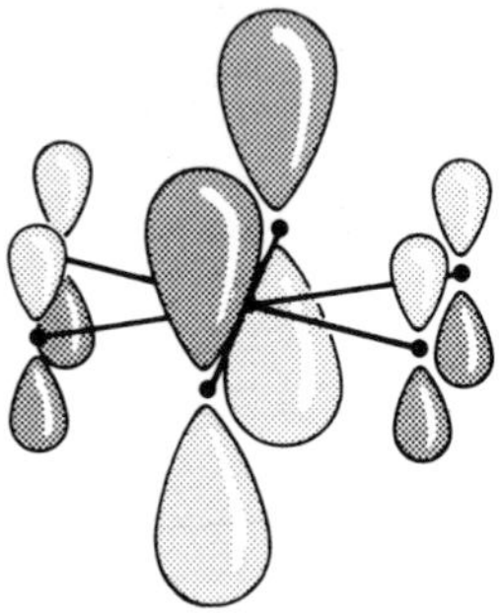 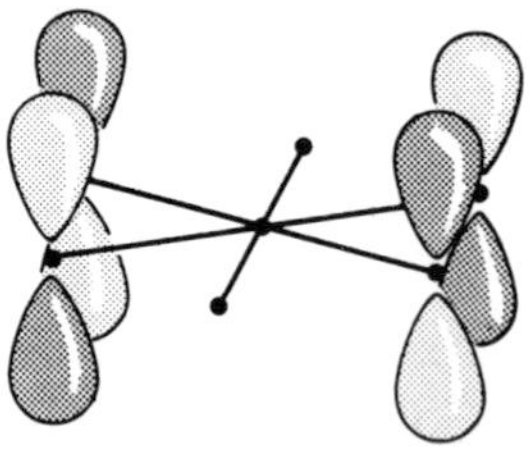	

T_d
A_1 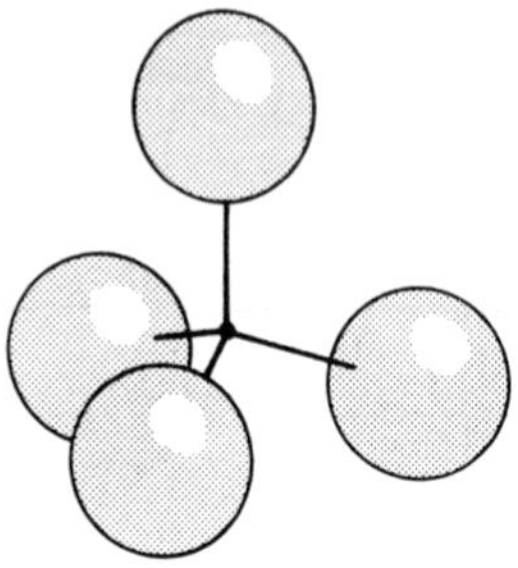

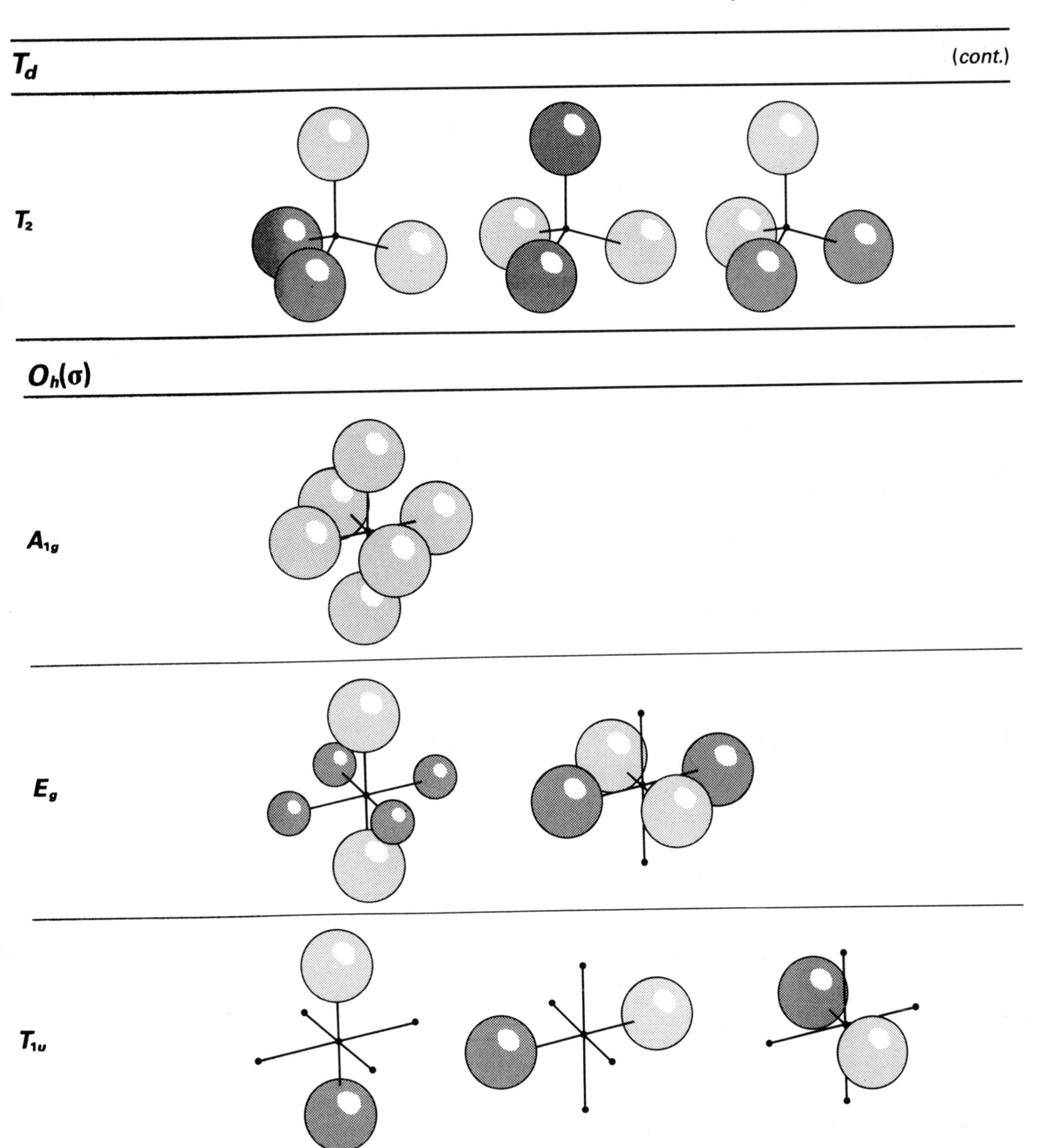

T_d
(cont.)
T_2
$O_h(\sigma)$
A_{1g}
E_g
T_{1u}

$O_h(\pi)$

T_{1u}

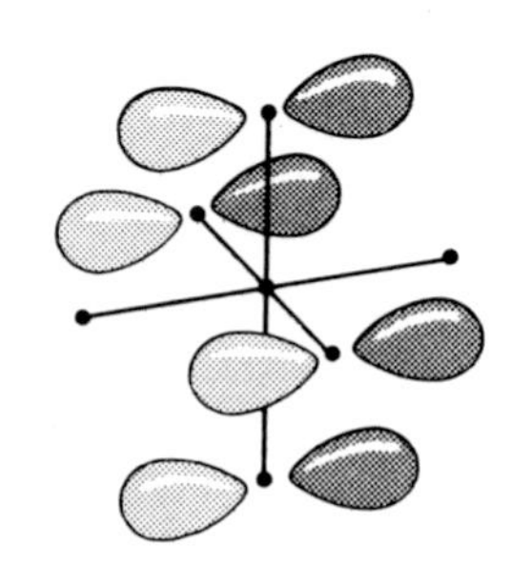
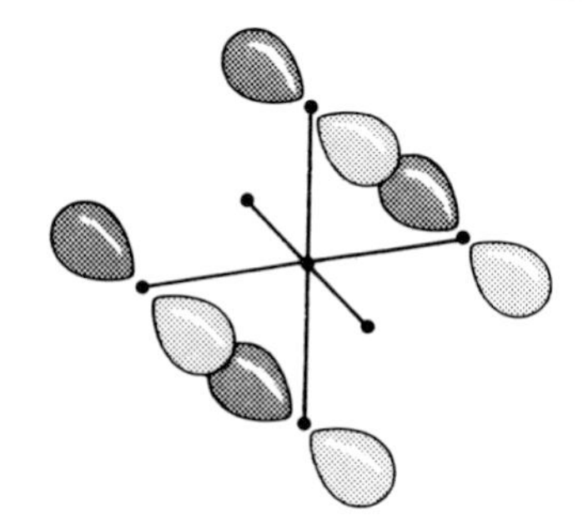
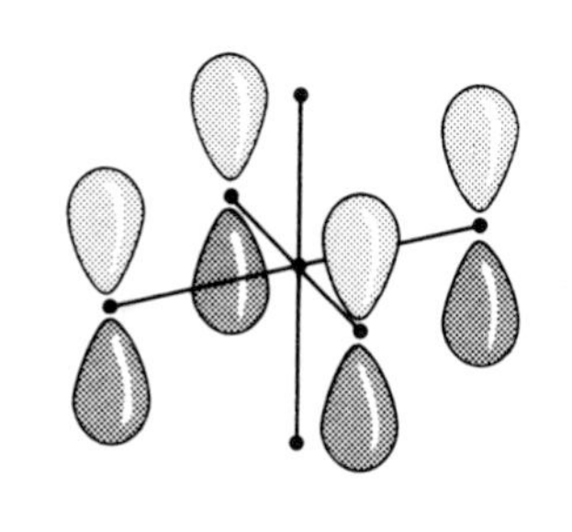

T_{2g}

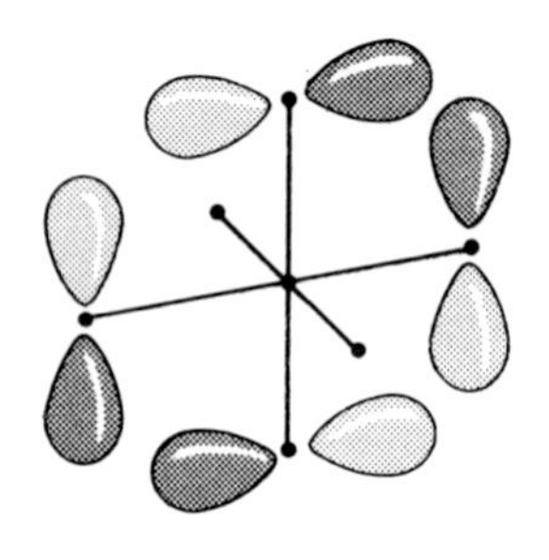
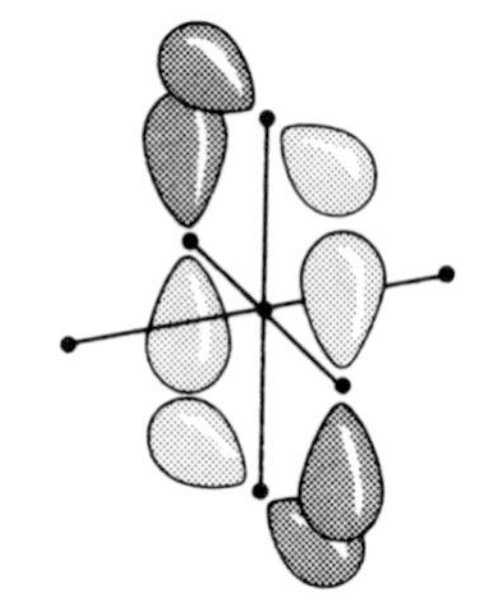
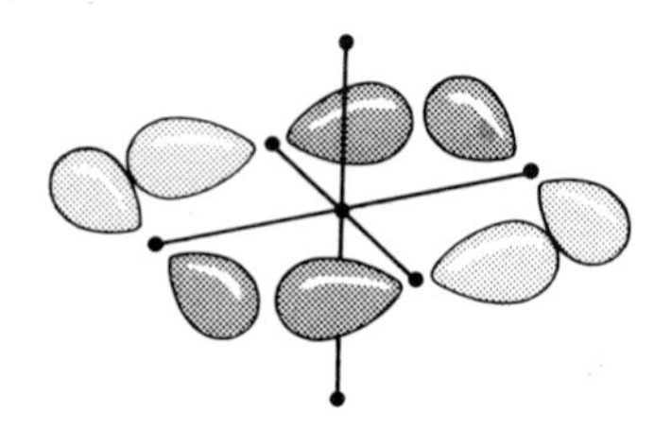

T_{1g}

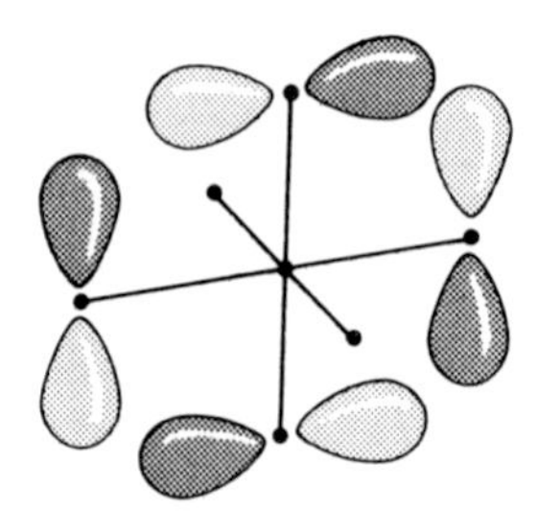
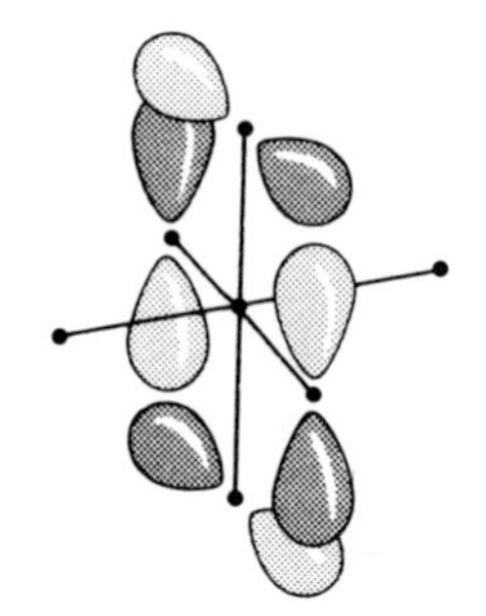
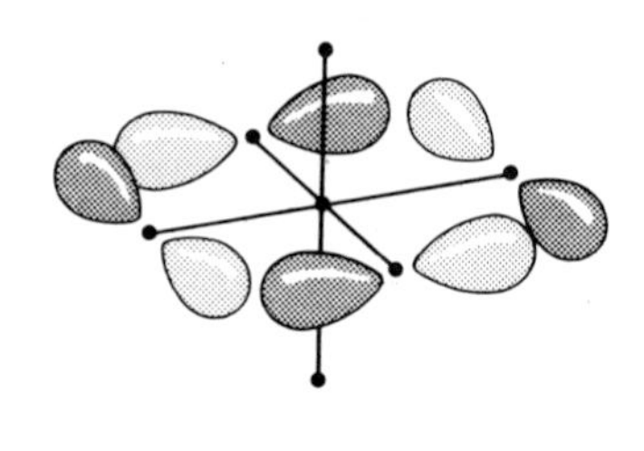

T_{2u}

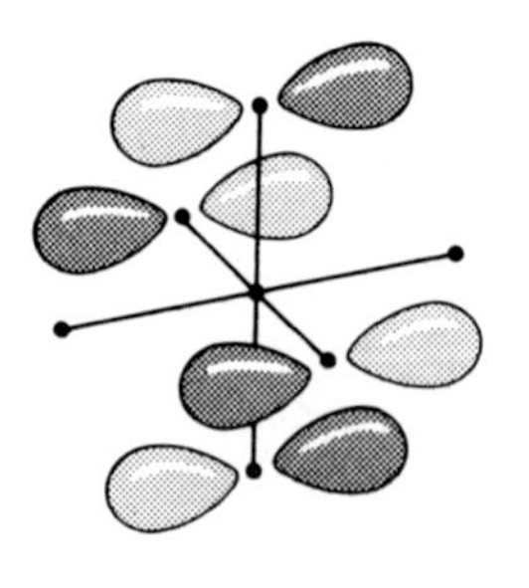
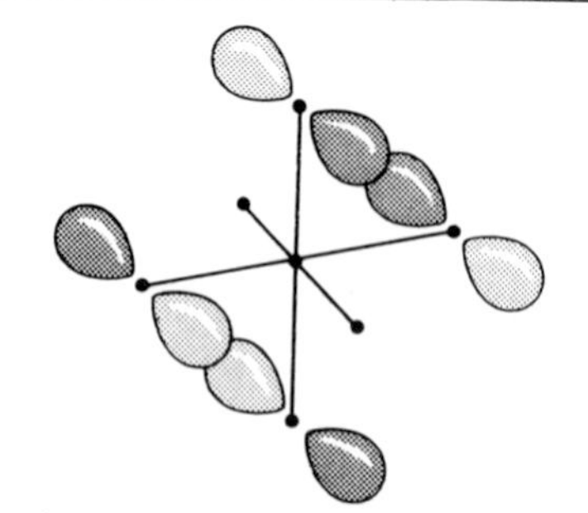
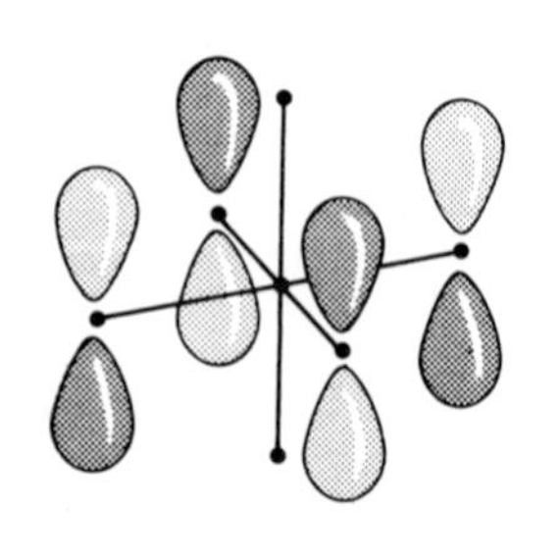

Tanabe–Sugano diagrams

This Appendix collects together the Tanabe–Sugano diagrams for octahedral complexes with electron configurations d^2 to d^9. The diagrams, which were introduced in Section 14.3, show the dependence of the term energies on ligand field strength. The term energies E are expressed as the ratio E/B, where B is a Racah parameter, and the ligand field splitting Δ_O is similarly expressed as Δ_O/B. Terms of different multiplicity are included on the same diagram by making specific, plausible choices about the value of the Racah parameter C, and these are given for each diagram. The term energy is always measured from the ground term, and so there are discontinuities of slope where a low-spin term displaces a high-spin term at sufficiently high ligand-field strengths for d^4 to d^8 configurations. Moreover, the non-crossing rule requires terms of the same symmetry to mix rather than to cross, and this mixing accounts for the curved rather than straight lines in a number of cases. The term labels are those of the point group O_h.

The diagrams were first introduced by Y. Tanabe and S. Sugano, *J. phys. Soc. Japan*, **9,** 753 (1954). They may be used to find the parameters Δ_O and B by fitting the ratios of the energies of observed transitions to the lines. Alternatively, if the ligand field parameters are known, the ligand field spectra may be predicted.

1. d^2 with $C = 4.42B$

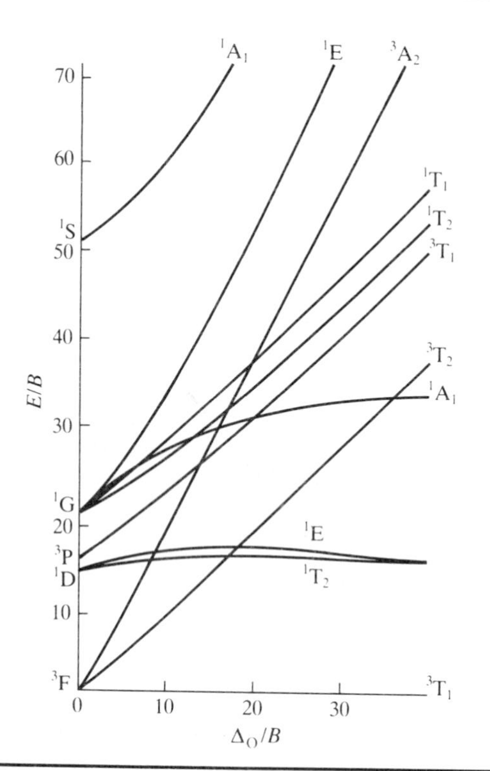

2. d^3 with $C = 4.5B$

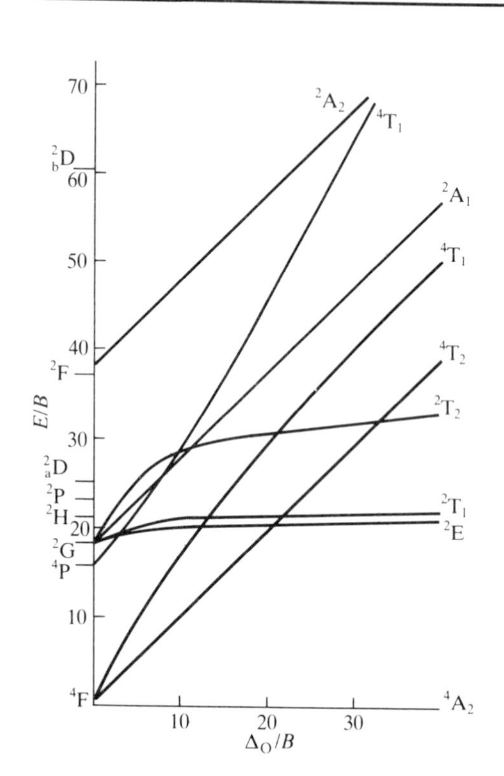

3. d^4 with $C = 4.61B$

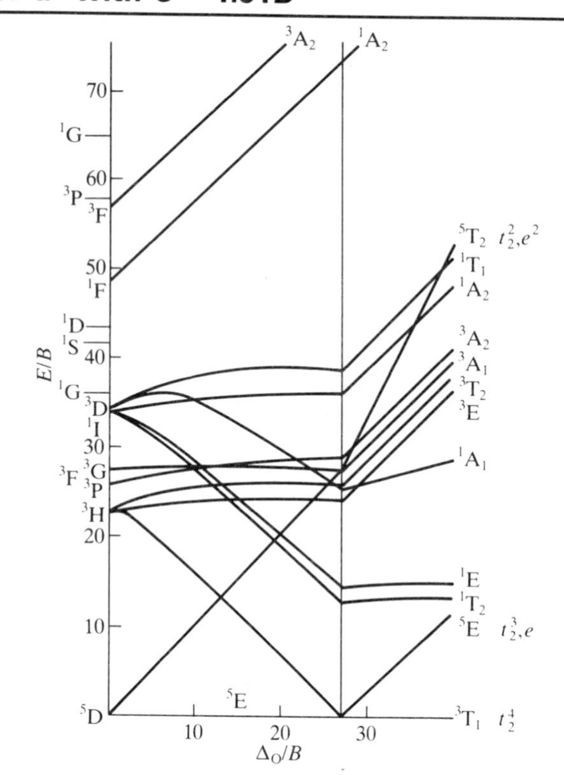

4. d^5 with $C = 4.477B$

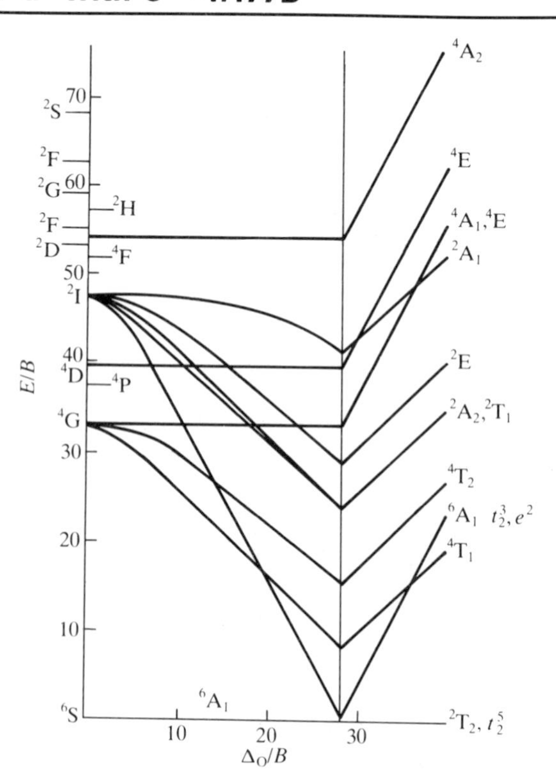

5. d^6 with $C = 4.8B$

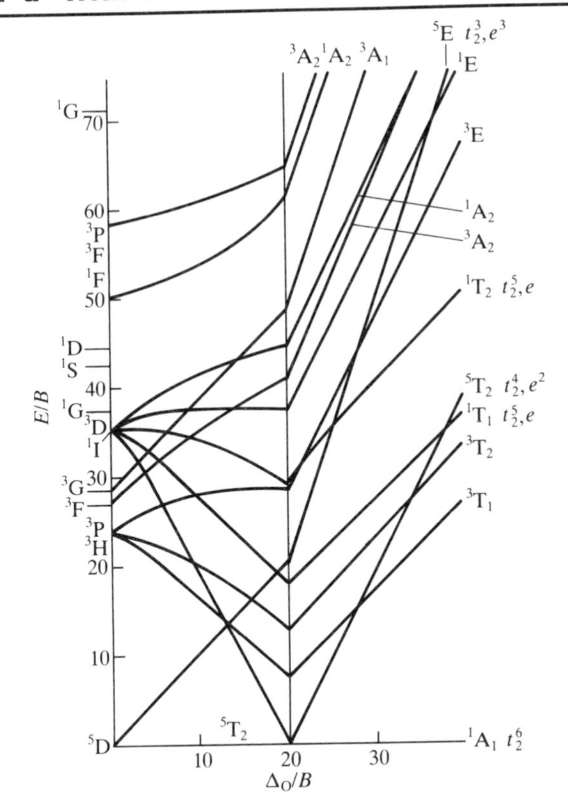

6. d^7 with $C = 4.633B$

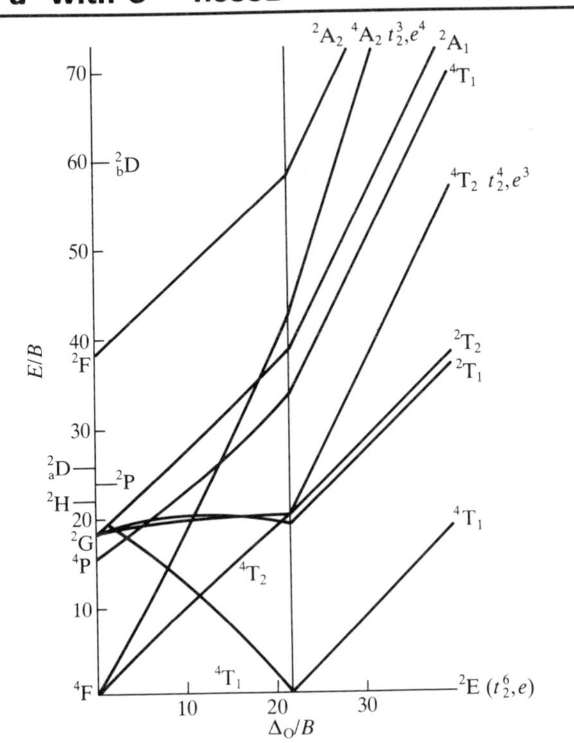

7. d^8 with $C = 4.709B$

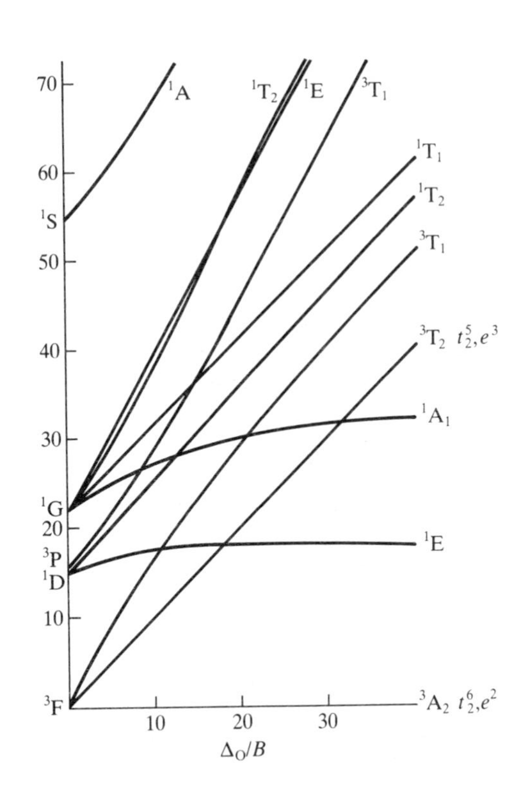

Answers to exercises

The answers below are intended only as a quick check. Complete answers will be found in the accompanying *Guide to solutions for Inorganic Chemistry* by S. H. Strauss.

1

In-chapter exercises

1.1 s, 1; p, 3; d, 5; f, 7; g, 9.
1.2 $r \approx 5.2\, a_0/Z$.
1.3 p electrons are less effective shielders than s electrons in the same shell.
1.4 $[Ar]3d^8$.
1.5 For atoms in the same group, outer electrons in higher n orbitals (heavier atoms) are generally less tightly bound.
1.6 $A_e(C) > A_e(N)$ because the electron adds to a half-filled p orbital in N with added electron–electron repulsion.

End-of-chapter exercises

1.1 See inside front cover.
1.2 Cosmic abundance of H and He is high. Terrestrial abundance is low because of lack of retention by earth's gravity.
1.3 Fe nucleus highly stable and Fe is not volatile; heavier elements have less stable nuclei.
1.4 2.4×10^3 V.
1.5 0.25.
1.6 -13.2 eV.
1.7 **(a)** Identical energies; **(b)** $l = 0, 1, 2, \ldots, n-1$.
1.8 **(a)** Fig. 1.12; **(b)** Fig. 1.13 for s; **(c)** Fig. 1.14 for p.
1.9 See pp. 22–3.
1.10 $I(H, 2s) = 3.4$ eV (calc.) $< I(Li, 2s) = 5.39$ (Table 1.7) $< I(Li^{2+}, 2s) = 30.5$ eV because $Z(Li^{2+}) > Z_{eff}(Li, 2s) > Z(H)$.
1.11 Ca through V, $4s$; Cr, $2d$; Mn, $4s$. General increase in I_2 from left to right due to increase in Z_{eff}.
1.12 C $[He]2s^22p^2$; F $[He]2s^22p^5$; Ca $[Ar]4s^2$; Ga^{3+} $[Ar]3d^{10}$; Bi $[Xe]4f^{14}5d^{10}6s^26p^3$; Pb^{2+} $[Xe]4f^{14}5d^{10}6s^2$; Nd $[Xe]4f^{14}6s^2$; U $[Rn]5f^36d^17s^2$.
1.13 Sc $[Ar]3d^14s^2$; V^{3+} $[Ar]3d^2$; Mn^{2+} $[Ar]3d^5$; Cr^{2+} $[Ar]3d^4$; Co^{3+} $[Ar]3d^6$; Cr^{6+} $[Ar]$; Cu $[Ar]3d^{10}4s^1$; Gd^{3+} $[Xe]4f^7$.
1.14 Generally I_1, A_e, and χ increase toward right because Z_{eff} increases. The trend reverses between Mg and Al for I_1 and between Na and Al for A_e because $3s$ is lower in energy than $3p$. Low values of A_e for Mg and Ar occur because of large energy jump to the next empty subshell or shell. $I_1(P) > I_1(S)$ and $A_e(Si) > A_e(P)$ because the electron adds to a half-filled p subshell in P^- with added electron–electron repulsion.
1.15 Na, $\eta = 2.30$ eV; Na^+, $\eta = 26.22$ eV. Na^+ harder because of high $I_2(Na)$.
1.16 Generally, χ decreases and η decreases with increasing Z down a group or left to right in a period.

2

In-chapter exercises

2.1 :Cl:P:Cl:
:Cl:
2.2 :O::N:O: ↔ :O:N::O: (bent).
2.3 $[XeO_6]^{4-}$ (Xe with six O, one Xe=O shown) + resonance forms
2.4 Square planar.
2.5 2.9×10^{-2} s.
2.6 Along an N—H bond. Four such C_3 axes.
2.7 **(a)** D_{3h}; **(b)** T_d.
2.8 Nonpolar.
2.9 Yes.
2.10 S_2^{2-}: $1\sigma_g^2 2\sigma_u^2 3\sigma_g^2 1\pi_u^4 2\pi_g^4$; Cl_2^-: (as S_2^{2-})$4\sigma_u^1$.
2.11 Similar to Fig. 2.30 with added low energy peak.
2.12 24 kJ mol^{-1}.

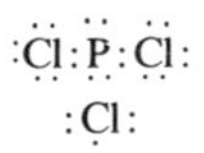

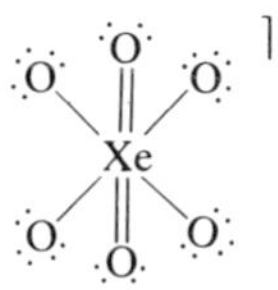

End-of-chapter exercises

2.1 **(a)** 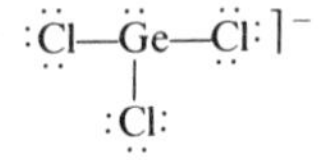 [:Cl—Ge(—:Cl:)—Cl:]⁻ trigonal pyramid, **(b)** 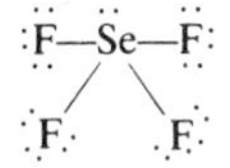 :F—Se(—F:)(F)(F) saw-horse, **(c)** 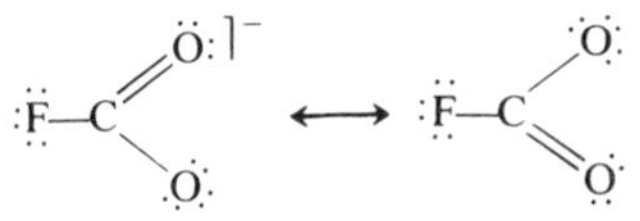 [F—C(=O)—O:]⁻ ⟷ [F—C(—O:)=O]⁻ trigonal planar.
2.2 **(a)** See answer to in-chapter Exercise 2.4. **(b)** 2, with 3 lone pairs on each Cl.
(c) 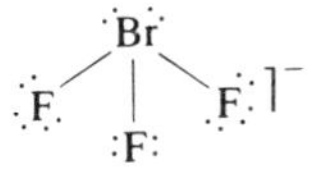 [Br with three F]⁻ drooping T,
(d) See Example 2.5.
(e) [:Cl—I—Cl:]⁻ linear.
2.3 NMR < IR < visible < UV < X-ray < PES.
2.4 **(a)** 1.76 Å; **(b)** 2.17 Å; **(c)** 2.21 Å.
2.5 **(a)** C_3 axis Fig. 2.6, 3σ shown in Example 2.7. **(b)** C_4 and C_2 perpendicular to plane of molecule through Pt; $2C_2$, along each Pt—Cl bond; σ_h, in plane of molecule; $2\sigma_v$, ⊥ to plane of molecule and containing *trans* Cl—Pt—Cl; $2\sigma_d$ ⊥ to plane of molecule and bisecting *cis*-ClPtCl.
2.6 **(a)** Contains i, not S_4; **(b)** contains i, not S_4; **(c)** neither i nor S_4; **(d)** no i; $3S_4$.
2.7 For each of the following the point group is given and for each of these the symmetry elements can be found in Table 2.7 or in Appendix 5.
(a) C_s, **(b)** D_{3h}, **(c)** T_d, **(d)** $C_{\infty v}$, **(e)** C_1, **(f)** D_{4h}.
2.8 **(a)** i plus an infinity of σ and C_n axes ($n = 1$ to ∞); **(b)** C_n ($n = 1$ to ∞), and $\infty\sigma$ along these axes; **(c)** $3C_2$, 3σ. **(d)** i plus one set of C_n ($n = 1$ to ∞), an infinity of σ.
2.9 Nonpolar if C_n axis plus ⊥ C_2 or i,

or multiple noncollinear C_n axes with $n > 2$. **(a)** Polar, **(b)** nonpolar, **(c)** nonpolar, **(d)** polar, **(e)** polar, **(f)** nonpolar.

2.10 Only **(e)** is chiral.

2.11 **(a)** $1\sigma_g^2 2\sigma_u^2$; **(b)** $1\sigma_g^2 2\sigma_u^2 1\pi_u^2$; **(c)** $1\sigma_g^2 2\sigma_u^2 1\pi_u^4 3\sigma_g^1$; **(d)** $1\sigma_g^2 2\sigma_u^2 3\sigma_g^2 1\pi_u^4 2\pi_g^3$.

2.12 **(a)** $1\sigma_g^2 2\sigma_u^2 3\sigma_g^2 1\pi_u^4 2\pi_g^3$; **(b)** $1\sigma_g^2 2\sigma_u^2 3\sigma_g^2 1\pi_u^4 2\pi_g^1$; **(c)** $1\sigma^2 2\sigma^2 1\pi^4$; **(d)** $1\sigma_g^2 2\sigma_u^2 1\pi_u^4 3\sigma_g^2 2\pi_g^2$ ($3\sigma_g$ and $1\pi_u$ might be interchanged).

2.13 **(a)** 2; **(b)** 1; **(c)** 2.

2.14 $1.13 \times 10^{-3}\ s^{-1}$.

2.15 **(a)** Larger b.o., shorter distance; **(b)** smaller b.o., longer distance; **(c)** larger b.o., shorter distance.

2.16 $-483\ kJ\ mol^{-1}$, fortuitously small error $1\ kJ\ mol^{-1}$.

2.17 $2N{\equiv}N \rightarrow N_4$ $\Delta H \approx +912\ kJ\ mol^{-1}$; $2P{\equiv}P \rightarrow P_4$ $\Delta H \approx -238\ kJ\ mol^{-1}$.

3

In-chapter exercises

3.1 See Appendix 6.

3.2 Triple.

3.3 B_{2g}.

3.4 A_1.

3.5 $MgH_2(g)$.

3.6 **(a)** Semiconductor; **(b)** metallic.

3.7 0.706 S.

End-of-chapter exercises

3.1 Four: lowest E ○○○○, ○○●●, ○●●○, ○●○● highest E.

3.2 C_{3v}, double degeneracy (E) maximum, $2p$.

3.3 D_{3h}, double (E), p_x, p_y (where p_z is along C_z).

3.4 H_4: ○○/○○ (A_{1g}); ●○/●○, ●●/○○ (E_u); ○●/●○ (B_{2g})
M: d_{z^2} (A_{1g}); none; d_{xy} (B_{2g}).

3.5 Labeling the H atoms as A and B and making the molecular axis z: $1s_A + 2s_F + 1s_B$; $-1s_A + 2p_{zF} + 1s_B$; p_x, p_y; $1s_A + 2p_{zF} - 1s_B$, $1s_A - 2s_F + 1s_B$.

3.6 **(a)** 1; **(b)** 2/3.

3.7 $1a$ bonding, $1t$ bonding, e nonbonding, $2t$ antibonding, $2a$ antibonding.

3.8 **(a)** NH_3; **(b)** BH_3; **(c)** NH_3.

3.9 Metallic: Fig. 3.28a or 3.29; semiconductor: Fig. 3.26.

3.10 **(a)** n; **(b)** p; **(c)** neither.

3.11 3.35 eV.

3.12 n-type.

3.13 n-type.

3.14 518 nm.

3.15 665.

4

In-chapter exercises

4.1 4 cations, 4 anions, 8 total.

4.2 6:6 coordination.

4.3 $2421\ kJ\ mol^{-1}$.

4.4 V(rock-salt structure)/V(CsCl structure) = 1.01. The rock salt structure is predicted, at odds with the observed CsCl structure, which is difficult to account for theoretically.

4.5 $V = 3.82\ MJ\ mol^{-1}$, $\Delta H_L = 3.89\ MJ\ mol^{-1}$; the ionic description appears appropriate.

4.6 $\Delta H_L = 2.63\ MJ\ mol^{-1}$.

4.7 T(decomp.): $MgSO_4 < CaSO_4 < SrSO_4 < BaSO_4$.

End-of-chapter exercises

4.1 (c), (f).

4.2 Use Fig. 4.3 as a guide.

4.3 The diagonal, d, of a square array of anions of $r = 1$ is the hypotenuse of a right-angled triangle of side 2 units; so $d = 2\sqrt{2} = 2.828$. Subtracting the radii of the anions leaves a hole of 0.828. The radius ratio is $(0.828)/2 = 0.414$.

4.4 Cubic closest packed W sublattice with all octahedral holes filled with C.

4.5 If CuZn has a cesium chloride structure as judged by X-rays the structure must be disordered. Cu and Zn have close-packed structures or nearly close-packed; so the system is neither substitutional nor interstitial.

4.6 See inside front cover. Groups 1, 2, and 17/VII can engage in ionic compound formation. Oxygen should also be included and Be excluded.

4.7 **(a)** 6:6; **(b)** 12.

4.8 **(a)** 8:8; **(b)** 6.

4.9 One Cs^+ and one Cl^-; four Zn^{2+}, four S^{2-}.

4.10 Fluorite structure.

4.11 $r(Se^{2-}) = 1.93$ Å; $r(Ca^{2+}) = 1.03$ Å; $r(Sr^{2+}) = 1.19$; $r(Ba^{2+}) = 1.38$ Å.

4.12 **(a)** 6:6; **(b)** 6:6; **(c)** 6:6; **(d)** borderline but owing to the small size of Be^{2+}, BeO should be covalent; so 4:4 seems likely (4:4 observed).

4.13 Sphalerite 4:4, wurtzite 4:4.

4.14 Using Fig. 4.17 the contribution to unit cell of the central Ti is (1); 8 corner Ti(1); 4 face O (2), 2 interior O (2); therefore, the 2Ti: 4O per unit cell agrees with TiO_2 formula.

4.15 Work as above with Fig. 4.18.

4.16 Estimating $r(K^{2+}) = 1.2$ Å (between Ca^{2+} and Sr^{2+}) $\Delta H_L = 2.4\ MJ\ mol^{-1}$. The instability of KF_2 arises from a very large second ionization enthalpy for potassium.

4.17 $NaCl < LiF < MgO$.

4.18 **(a)** $CaCO_3$; **(b)** $N(CH_3)_4I_3$.

4.19 LiCl has larger difference in cation and anion radii.

4.20 Sr^{2+} or Ba^{2+}.

4.21 Sulfate.

5

In-chapter exercises

5.1 Acid/conjugage base: **(a)** HNO_3/NO_3^-; **(b)** H_2O/OH^-; **(c)** H_2S/HS^-.

5.2 $[Na(OH_2)_6]^+ < [Mn(OH_2)_6]^{2+} < [Ni(OH_2)_6]^{2+} < [Sc(OH_2)_6]^{3+}$.

5.3 pK_a estimated by Pauling's rules: **(a)** 3; **(b)** 8; **(c)** 13.

5.4 Slightly basic solution gives a precipitate, strong base redissolves the precipitate.

End-of-chapter exercises

5.1 Check element symbols inside cover, check assignment of acidity and basicity with Fig. 5.5.

5.2 $[Co(OH)(NH_3)_5]^{2+}$; SO_4^{2-}; CH_3O^-; HPO_4^{2-}; $[Si(OH)_3O]^-$, S^{2-}.

5.3 $C_5H_5NH^+$; $H_2PO_4^-$; OH^-; $CH_3COOH_2^+$; $HCo(CO)_4$; HCN.

5.4 $I^- < F^- < HS^- < HI$.

5.5 $CH_4 < H_2O < HF < HI$.

5.6 $\Delta H = -528\ kJ\ mol^{-1}$; strong hydration of the ions.

5.7 **(i)** O^{2-} too strong, ClO_4^- and NO_3^- too weak; **(ii)** ClO_4^- too weak.

5.8 Electron withdrawing.

5.9 From Pauling's rules, $pK_a = 13$.

5.10 $HSiO_4^{3-} < HPO_4^{2-} < HSO_4^- < HClO_4$.

5.11 **(a)** $[Fe(OH_2)_6]^{3+}$, higher ξ; **(b)** $[Al(OH_2)_6]^{3+}$, higher ξ; **(c)** $Si(OH)_4$, higher ξ; **(d)** $HClO_4$ (higher ox. no. or Pauling's rules); **(e)** $HMnO_4$ (higher ox. no.); **(f)** H_2SO_4 (higher ox. no.).

5.12 $Cl_2O_7 < SO_3 < CO_2 < B_2O_3 < Al_2O_3 < BaO$.

5.13 $NH_3 < CH_3GeH_3 < H_4SiO_4 < HSO_4^- < H_3O^+ < HSO_3F$.

5.14 Ag^+ has more covalent M—O bond.

5.15 Polyanions: As, B, Si, Mo; polyoxocations: Al, Ti, Cu.

5.16 One per M.

5.17 $2PO_4^{3+} + 2H^+ \rightarrow P_2O_7^{4-} + H_2O$, $2[Fe(OH_2)_6]^{3+} \rightarrow [(H_2O)_4Fe(OH)_2Fe(OH_2)_4]^{4+} + 2H^+ + 2H_2O$.

5.18 **(a)** $H_3PO_4 + HPO_4^{2-} \rightarrow 2H_2PO_4^-$; **(b)** $CO_2 + CaCO_3 + H_2O \rightarrow Ca^{2+} + 2HCO_3^-$.

5.19 **14** is O_h; **11** is D_{2h}.

6

In-chapter exercises

6.1 Acid/base; **(a)** $FeCl_3/Cl^-$; **(b)** I_2/I^-; **(c)** $Mn(CO)_5^+/SnCl_3^-$.
6.2 Aluminosilicate: Rb, Sr, Cr; sulfide: Cd, Pb, Pd.
6.3 **(a)** $CaC_2 + 2H_2O \rightarrow Ca(OH)_2 + H_2C_2$; **(b)** $PCl_3 + 3CH_3Li \rightarrow P(CH_3)_3 + 3LiCl$.
6.4 3-Methyltetrahydrofuran.
6.5 **(a)** H_2O; **(b)** benzene.
6.6 **(a)** $-31.5\ kJ\ mol^{-1}$; **(b)** $-75.2\ kJ\ mol^{-1}$.

End-of-chapter exercises

6.1 Elements below Al, Si, N, O, F.
6.2 Acid/base: **(a)** SO_3 and H^+/OH^-, displacement; **(b)** Hg^{2+} and $[B_{12}]/CH_3^-$, displacement; **(c)** $SnCl_2$ and K^+/Cl^-, displacement; **(d)** SbF_5/AsF_3, displacement; **(e)** C_2H_5OH/C_5H_5N, complex (hydrogen bond) formation.
6.3 **(a)** BBr_3, BCl_3, $B(n\text{-}Bu)_3$; **(b)** $(CH_3)_3N$, 4-CH_3py.
6.4 **(a)** <1; **(b)** >1; **(c)** <1; **(d)** >1.
6.5 $F_3B \rightarrow N(CH_3)_2\text{-}F_2P \leftarrow BH_3$.
hard–hard soft–soft
6.6 Steric repulsion.
6.7 **(a)** dmso–BF_3 $\Delta H = -75.3\ kJ\ mol^{-1}$ (most stable); $(CH_3)_2SBF_3$ $\Delta H = -64.8\ kJ\ mol^{-1}$; **(b)** dmso–$I_2$ $\Delta H = -17.6\ kJ\ mol^{-1}$; $(CH_3)_2SI_2$ $\Delta H = -32.6\ kJ\ mol^{-1}$ (most stable).
6.8 **(a)** $4K(C_6H_5) + BCl_3 \rightarrow 3KCl + KB(C_6H_5)_4$; **(b)** $KF + PF_5 \rightarrow KPF_6$; **(c)** $CH_3I + Mg \rightarrow CH_3MgI$, $4CH_3MgI + SnBr_4 \rightarrow (CH_3)_4Sn + 4MgIBr$; **(d)** $2SiH_3I + H_2O \rightarrow H_3Si\text{—}O\text{—}SiH_3 + 2HI$; **(e)** $KF + SO_2 \rightarrow KSO_2F$; **(f)** $AsF_3 + BCl_3 \rightarrow AsCl_3 + BF_3$.
6.9 $SiO_2 + 4HF \rightarrow 2H_2O + SF_4$.
6.10 $Al_2S_3 + 3H_2O \rightarrow Al_2O_3 + 3H_2S$.
hard–soft hard–hard
6.11 **(a)** Hard acid; H_2O; **(b)** hard acid; CH_3OH; **(c)** hard base, H_2O; **(d)** soft base; CH_3CN.
6.12 Cations are susceptible to coordination by bases often with subsequent reaction. Anions are susceptible to protonation and oxidation by protons.
6.13 **(a)** Very weak acid and base properties; **(b)** Lewis base.
6.14 Surface Al^{3+} Lewis acid site coordinates Cl^- and generates CH_3CO^+, which can attack the benzene ring.
6.15 Hg is softer than zinc.
6.16 The higher bp and lower solubility of nonpolar solutes result from a higher degree of hydrogen bonding in H_2O than NH_3. The solubility of salts may be aided by the higher permittivity of H_2O (as a result of the hydrogen-bonding network) and better hydrogen-bonding of anions.
6.17 *f*-block ions are hard acids.
6.18 SiO_2.
6.19 Base: O^{2-} in TiO_2; acid SO_3 in $S_2O_7^{2-}$.
6.20 AsF_5: trigonal bipyramidal (see PCl_5, **2**, in Chapter 2), point group D_{3h}; $X_3BN(CH_3)_3$: C_{3v} (neglecting hydrogens); Al_2Cl_6: D_{2h}.

7

In-chapter exercises

7.1 A: *trans*; B: *cis*.
7.2 Use **10a** and **10b** as a guide.
7.3 **(a)** *cis*-$[PtCl_2(OH_2)_2]$; **(b)** $[Cr(NCS)_4(NH_3)_2]^-$; **(c)** $[Rh(en)_3]^{3+}$.
7.4 **(a)** Chiral; **(b)** achiral; **(c)** achiral.
7.5 Five unpaired electrons, $t_{2g}^3e_g^2$.
7.6 ZnF_2 higher than MnF_2 because of smaller $r(Zn^{2+})$. FeF_2, CoF_2, and especially NiF_2 have added ligand field stabilization energy.
7.7 Added ligand field stabilization energy.

End-of-chapter exercises

7.1 See inside front cover. **(a)** Ti^{2+} d^2, V^{2+} d^3, Cr^{2+} d^4, Mn^{2+} d^5, Fe^{2+} d^6, Co^{2+} d^7, Ni^{2+} d^8, Cu^{2+} d^9, Zn^{2+} d^{10}, similarly for Period 5 and 6 elements in the same group; **(b)** **(i)** tetrahedral MX_4^{2-}: Period 4, **(ii)** Pd^{2+}, Pt^{2+}; **(c)** $Mn^{2+} < Fe^{2+} < Co^{2+} < Ni^{2+} < Cu^{2+} > Zn^{2+}$.
7.2 **(a)** en see **24a,b**; **(b)** ox use **24a,b** as a model; **(c)** hint: both *mer* and *fac* isomers possible; **(d)** see **23a,b**.
7.3 **(a)** *cis*-tetraamminedichlorochromium(III); **(b)** *trans*-diamminetetraisothiocyanatochromate(III); **(c)** bis(ethylenediamine)oxalatocobalt(III), only *cis*.
7.4 **(a)** octahedral $[CoCl(NH_3)_5]^+$ with Cl^- anion; **(b)** octahedral; **(c)** *cis*-$[RuCl_2(en)_2]$; **(d)** +5 cation consisting of two $Cr(NH_3)_5$ units bridged by an OH. There are five uncoordinated Cl^- ions.
7.5 **(a)** *cis* and *trans* isomers possible; **(b)** square planar *cis* and *trans* isomers; **(c)** octahedral, one isomer; **(d)** three isomers: two *trans* one *cis*.
7.6 **(a)** Chiral; **(b)** achiral; **(c)** achiral; **(d)** chiral; **(e)** chiral; **(f)** achiral; **(g)** achiral.
7.7 The first contains the octahedral $[Co(NH_3)_5(OH_2)]^{3+}$ complex ion and is pentaammineaquacobalt(III) chloride; the second contains octahedral $[CoCl(NH_3)_5]^{2+}$ and is pentaamminechlorocobalt(III) chloride.
7.8 The blue complex is $[Cr(OH_2)_6]^{3+}$ and the green complex may be $[CrCl(OH_2)_5]^{2+}$, the formation of which leaves fewer ions in solution, hence lower conductivity.
7.9 Apparently square planar *trans*-$[Pt(NH_3)_2(OH_2)_2]^{2+}$ is produced.
7.10 $[Pt(NH_3)_4]^{2+}[PtCl_4]^{2-}$ with square planar anion and cation; tetraammineplatinum(II) tetrachloroplatinate(II).
7.11 Square planar *trans*-$[PtCl_2(AsR_3)_2]$.
7.12 18e$^-$ only: (b), (e), (f).
7.13 See Appendix 6.
7.14 Unpaired e$^-$/LFSE; **(a)** $0/2.4\Delta_O$; **(b)** $4/0.4\Delta_O$; **(c)** $0/2.4\Delta_O$; **(d)** $3/1.2\Delta_O$; **(e)** $0/2.4\Delta_O$; **(f)** 5/0; **(g)** 0/0.
7.15 $P(C_6H_5)_3$ π acidity; H^- strong σ basicity.
7.16 A plot with a line connecting CuO and MnO (which have no LFSE) extrapolated beyond NiO, indices TiO, VO, FeO, CoO, and NiO may have significant LFSE.
7.17 The electronic configuration $t_{2g}^3e_g^1$ should lead to a Jahn–Teller distortion, probably by elongation of two *trans* O—Cr—O bonds.
7.18 Although t_{2g}^1 is degenerate, a significant Jahn–Teller distortion is not expected by the single occupancy of this nonbonding set of orbitals. The excited state, e_g^1, populates an antibonding orbital and should be distorted. Thus the energy of the excited state will be strongly shifted by ligand vibrations giving rise to a range of transition energies and a broad band.
7.19 **(a)** CH_3NH_2, basicity; **(b)** $NH_2CH(CH_3)CH_2NH_2$, chelate effect; **(c)** NH_3, basicity; **(d)** two $NH_2C_2H_4NHC_2H_4NH_2$, chelate effect.
7.20 Hard ligands, F^- and O^{2-}, prefer the hard, highly positive ions; soft ligands, PR_3 and CO, prefer softer, +1 or neutral metal centers.
7.21 Chelating agents are often used because their complexes have high K_f and therefore give sharp endpoints.
7.22 Cu(II) complexes are labile; Ir(III) complexes are inert.
7.23 Dissociative.
7.24 The rate will depend not only on the identity of the aqua-ion but also on the identity of the incoming ligand.

8

In-chapter exercises

8.1 1700°C.

8.2 −2.1 V at 500°C.

8.3 The potential data indicate that Fe^{2+} can be oxidized; the oxidation of Cl^- is only very slightly favorable at 0.02 V and in practice does not occur at an appreciable rate.

8.4 0.87 V.

8.5 1.94 V.

8.6 −1.18 V, no.

8.7 SO_2 is oxidized to SO_3 which hydrates to form the strong acid H_2SO_4.

8.8 1.43 V.

8.9 The diagram has a dip to Ti^+ ($nE^\ominus = -0.34$ V) and rise to Ti^{3+} ($nE^\ominus = +2.19$ V).

8.10 The slopes for H_2O–H_2O_2 and Mn^{2+}–Mn^{3+} are nearly the same. Frost diagrams are well suited to visualizing general trends in redox potentials but not for detailed calculation.

8.11 Much stronger in acid solution.

8.12 Fe^{2+} is preferred.

8.13 Reduction of $[Ni(NH_3)_6]^{2+}$ should be more favorable (less negative).

8.14 **(a)** Ce^{4+}; Eu^{2+}.

End-of-chapter exercises

8.1 >1400°C.

8.2 **(a)** $4Cr^{2+} + O_2 + 4H^+ \rightarrow 4Cr^{3+} + 2H_2O$; **(b)** $4Fe^{2+} + O_2 + 4H^+ \rightarrow 4Fe^{3+} + 2H_2O$; **(c)** NR: **(d)** NR; **(e)** $2Zn + O_2 + 4H^+ \rightarrow 2Zn^{2+} + 2H_2O$. A competing reaction will be $Zn + 2H^+ \rightarrow Zn^{2+} + H_2$.

8.3 **(a)** $E = 1.23 + (RT/nF)\ln p(O_2) - 9.2\text{pH}$. At pH = 7, $E = 0.20$ V; **(b)** $E = E^\ominus - (RT/nF) \times 13.8$ pH.

8.4 CrO_4^{2-} reduction: $\Delta G^\ominus = +31.8\ \text{kJ mol}^{-1}$, $K = 2.70 \times 10^{-6}$; $[Cu(NH_3)_4]^+$ reduction: $\Delta G^\ominus = +11.6\ \text{kJ mol}^{-1}$, $K = 9.30 \times 10^{-3}$.

8.5 **(a)** $Cl_2 + 2OH^- \rightarrow OCl^- + Cl^- + H_2O$; **(b)** no disproportionation, $Cl_2 + H_2O \rightarrow 2Cl^- + 2H^+ + \frac{1}{2}O_2$, $E^\ominus = 0.13$ V, but overpotential for O_2 evolution is not exceeded and reaction is not kinetically significant; **(c)** thermodynamically yes, but rate is very slow.

8.6 **(a)** $2N_2O \rightarrow N_2 + 2NO$, $3N_2O + 2OH^- \rightarrow 2NO_2^- + 2N_2 + H_2O$, $5N_2O + 2OH^- \rightarrow 2NO_3^- + 4N_2 + H_2O$; all are feasible thermodynamically but N_2O is fairly kinetically inert and these reactions will not occur at a significant rate. **(b)** $Zn + I_3^- \rightarrow Zn^{2+} + 3I^-$, **(c)** $3I_2 + 5ClO_3^- + 3H_2O \rightarrow 6IO_3^- + 5Cl^- + 6H^+$.

8.7 **(a)** Reduced acidity favors reaction; however, in base $Mn(OH)_2$ will precipitate, which will change the equation for the net reaction and alter the thermodynamics; **(b)** favored by acid; **(c)** favored by base; **(d)** no effect.

8.8 **(a)** Atom transfer; **(b)** outer sphere; **(c)** atom transfer, multistep.

8.9 $2ClO_4^- + 16H^+ + 14e^- \rightarrow Cl_2 + 8H_2O$, $E^\ominus = 1.39$ V.

8.10 Slopes of vectors are given by the values of the individual one-electron potentials.

8.11 **(a)** Surface water *ca.* +4.5 to +6.0 V corresponding to Fe_2O_3 (Fig. 8.8); **(b)** Mn^{2+} and possibly Mn_2O_3 (Fig. 8.10); **(c)** SO_4^-; **(d)** Cl^-.

8.12 −0.11 V.

8.13 $E^\ominus$ will be more negative (less favorable).

8.14 Group oxidation number favored: Sc, Ti, V; group oxidation number strong oxidant: CrO_4^{2-}, MnO_4^-; group oxidation number not known: Fe, Co, Ni, Cu, Zn.

8.15 On descending a group in the *d* block, the group oxidation number becomes more stable, e.g. $Cr_2O_7^{2-}/Cr$ + 0.64 V, H_2MoO_4/Mo + 0.11 V, WO_3/W −0.09 V. In the *p* block, the group oxidation number is more stable for a Period 3 element compared with heavier members of its group, e.g. Al^{3+}/Al −1.68 V, Tl^{3+}/Tl +0.72 V.

8.16 For the early part of the actinide series, stable high oxidation states are common: Ac^{3+} (highest possible ox.no.), Th^{4+}, PaO_2^+, UO_2^{2+}. Later actinides show increasing stability of +3 state. For lanthanides, the +3 state is uniformly the most stable.

9

In-chapter exercises

9.1 1301 cm^{-1}.

9.2 Radical–radical termination.

9.3 SrH_2, saline; CuH, borderline; H_2O, molecular, electron rich. H_2O is a good hydrogen bonder.

9.4 Radical chain.

9.5 $CH_2{=}CHCH_3 + BH_3{-}THF \rightarrow CH_3CH_2CH_2{-}BH_2{-}THF$.

End-of-chapter exercises

9.1 See inside front cover and Fig. 9.3.

9.2 **(a)** Barium hydride, saline; **(b)** silane, molecular electron-precise; **(c)** ammonia, molecular electron rich; **(d)** arsine, molecular electron rich; **(e)** palladium hydride, metallic; **(f)** hydrogen iodide, molecular electron rich.

9.3 **(a)** SrH_2; **(b)** HI; **(c)** NH_3.

9.4 Solids: BaH_2, $PdH_{0.9}$; liquids: none; gases: SiH_4, NH_3, AsH_3, HI.

9.5 H_2Se: bent, C_{2v}; P_2H_4: nonplanar, C_2; H_3O^+: pyramidal, C_{3v}.

9.6 **(b)**.

9.7 Similar to Fig. 9.18.

9.8 **(a)** $CH_4 + H_2O \rightarrow CO + 3H_2$, $C + H_2O \rightarrow CO + H_2$, $CO + H_2O \rightarrow CO_2 + H_2$; **(b)** $Mg + 2HCl \rightarrow Mg^{2+} + 2Cl^- + H_2$, $2H_2O \rightarrow 2H_2 + O_2$ by electrolysis.

9.9 $(CH_3)_3SnH$, low Sn—H bond enthalpy.

9.10 **(a)** $H_2O < H_2S < H_2Se$; **(b)** $H_2Se < H_2S < H_2O$.

9.11 **(i)** Combination: $Ca + H_2 \rightarrow CaH_2$; **(ii)** protolysis: $CaF_2 + H_2SO_4 \rightarrow 2HF + CaSO_4$; **(iii)** metathesis: $LiAlH_4 + SiCl_4 \rightarrow SiH_4 + LiAlCl_4$.

9.12 **(a)** $CaSe + 2HCl(aq) \rightarrow CaCl_2 + H_2Se$; **(b)** $SiCl_4 + LiAlD_4 \rightarrow SiD_4 + LiAlCl_4$; **(c)** $2Ge(CH_3)_2Cl_2 + LiAlH_4 \rightarrow 2Ge(CH_3)_2H_2 + LiAlCl_4$; **(d)** $Si + 2HCl \rightarrow SiH_2Cl_2$ (+other chlorosilanes), $2H_2SiCl_2 \rightleftharpoons SiH_4 + SiCl_4$.

9.13 **(i)** No, $B_2H_6 + 3O_2 \rightarrow B_2O_3 + 3H_2O$ or $2B(OH)_3$; **(ii)** see Box 9.1.

9.14 **(i)** $BH_4^- \ll AlH_4^- > GaH_4^-$; **(ii)** AlH_4^-; **(iii)** $GaH_4^- + 4HCl \rightarrow 4H_2 + GaCl_4^-$.

9.15 $Si + 2Cl_2 \rightarrow SiCl_4$ (distill), $SiCl_4 + 4H_2 \rightarrow SiH_4 + 4HCl$.

9.16 Period 2, *p* block: **(i)** highest volatility if not hydrogen bonded. Lowest volatility in the group if hydrogen bonded (NH_3, H_2O, HF); **(ii)** Short strong X—H bonds with H—X—H close to 109°. Period 3, *p* block: **(i)** volatility: AlH_3 nonvolatile, $SiH_4 < CH_4$, $PH_3 > NH_3$, $H_2S > H_2O$, HCl > HF, less hydrogen bonding in Period 3. **(ii)** Weaker longer X—H bonds; **(iii)** ∠H—X—H close to 90°.

9.17 Clathrate hydrates.

9.18 Ignoring nonbonding $p\pi$ electrons on F and designating the F σ symmetry orbital f and the H orbital h: $f - h + f$ (antibonding), $f - f$ (nonbonding), $f + h + f$ (bonding).

9.19 $Cl^- \cdots H_2O$ unsymmetrical double well, FHF^- symmetrical single well.

9.20 ^{19}F doublet due to coupling with ^{31}P ($I = \frac{1}{2}$); ^{31}P quadruplet due to coupling with three ^{19}F ($I = \frac{1}{2}$).

9.21 $HB(CH_3)_3^-$, ^{11}B: 1:1 doublet; $H_2B(CH_3)_2^-$, ^{11}B; 1:2:1 triplet.

9.22 Overlapping multiplets: 1:1:1:1 quartet for $^{11}BH_4^-$, 1:1:1:1:1:1:1 septet for $^{10}BH_4^-$.

10

In-chapter exercises

10.1 $Mg + Hg(CH_3)_2 \rightarrow Mg(CH_3)_2 + Hg$.
10.2 Homolytic Pb—CH_3 bond cleavage.
10.3 Bridging hydrogens and terminal ^tBu groups.
10.4 $(H_3Si)_3N$ planar or only slightly pyramidal; $(H_3C)_3N$ pyramidal.
10.5 $Al_2(CH_3)_6 + 6H_2O \rightarrow 2Al(OH)_3 + 6CH_4$, $Ga(CH_3)_3 + 3H_2O \rightarrow Ga(CH_3)_2(H_2O)_2^+ + OH^- + 3CH_4$.
10.6 C and H electronegativities are similar and the stable compounds are similar, GeR_4, AsR_3 R = H or CH_3.

End-of-chapter exercises

10.1 Sodium acetate and $N(CH_3)_3$ are not organometallic compounds, the rest are.
10.2 See Fig. 10.1, Table 10.1.
10.3 **(a)** Trimethylbismuthine; **(b)** tetraphenylsilicon; **(c)** tetraphenylarsenic(+1); **(d)** potassium tetraphenylboron(−1).
10.4 **(a)** Triethylsilane; electron precise, tetrahedral around Si; **(b)** chlorodiphenylboron, electron deficient, trigonal-planar around B; **(c)** dichlorotetraphenyldialuminum, electron precise; **(d)** ethyllithium (more precisely tetraethyltetralithium), electron deficient; cube with C and Li at alternate corners; **(e)** methylrubidium, saline, NiAs structure.
10.5 $BMe_3 \ll AlMe_3 \gg GaMe_3 \sim InMe_3$. Size of Al greater than B may account for that difference.
10.6 **(a)** See structure **2** and answer to 10.4 **(d)**; **(b)** trigonal planar around B; **(c)** see structure **1**; **(d)** tetrahedral around Si; **(e)** pyramidal around As.
10.7 **(a)** (1), (2), (4); **(b)** (1), (2), (4); **(c)** (1), (2), (4); **(d)** (2); **(e)** (1), (2), (4); **(f)** −; **(g)** (3).
10.8 **(a)** $3LiCH_3 + AsCl_3 \rightarrow As(CH_3)_3 + 3LiCl$, $2LiCH_3 + SiPh_2Cl_2 \rightarrow SiPh_2(CH_3)_2 + 2LiCl$; **(b)** $NH_3 + B(CH_3)_3 \rightarrow H_3NB(CH_3)_3$; **(c)** $Et_4NBr + [HgCH_3][BF_4] \rightarrow BrHgCH_3 + [Et_4N][BF_4]$.
10.9 **(a)** $2Li + CH_3Br \rightarrow LiCH_3 + LiBr$ electropositive metal; **(b)** $2Al + 3Hg(CH_3)_2 \rightarrow Al_2(CH_3)_6 + Hg$ electropositive metal; **(c)** $2Si(CH_3)_3Cl + Al_2(CH_3)_6 \rightarrow 2Si(CH_3)_4 + Al_2Cl_2(CH_3)_4$.
10.10 **(a)** $Na[C_{10}H_8]$ higher LUMO in $C_{10}H_8$; **(b)** $Na_2[C_{10}H_8]$ doubly occupied LUMO of neutral molecule.
10.11 **(a)** $Ca + Hg(CH_3)_2 \rightarrow Ca(CH_3)_2 + Hg$, Ca more electropositive; **(b)** NR, Hg less electropositive; **(c)** $LiCH_3 + SiPh_3Cl \rightarrow LiCl + SiPh_3CH_3$, Li more electropositive; **(d)** NR, Si less electropositive; **(e)** $HSi(CH_3)_3 + C_2H_4 \rightarrow Si(CH_3)_3(C_2H_5)$ hydrosilation conditions.
10.12 **(a)** $Si + 2CH_3Cl \xrightarrow{Cu,\Delta} Si(CH_3)_2Cl_2$ plus other methylchlorosilanes; **(b)** $2Si(CH_3)_2Cl_2 \overset{AlCl_3}{\rightleftharpoons} Si(CH_3)_3Cl + Si(CH_3)Cl_3$.
10.13 Prepare $Si(CH_3)_2Cl_2$ and $Si(CH_3)_3Cl$, see 10.12 **(a)** and **(b)**; $Si(CH_3)_2Cl_2 + H_2O \rightarrow [(CH_3)_2SiO]_4 + 2HCl$, $Si(CH_3)_3Cl + H_2O \rightarrow ((CH_3)_3Si)_2O + 2HCl$, $n[(CH_3)_2SiO]_4 + ((CH_3)_3Si)_2O \xrightarrow{H_2SO_4} (CH_3)_3SiO(Si(CH_3)_2O)_{4n}Si(CH_3)_3$.
10.14 Group 13/III: ox. no. +3 predominates, some Tl(I); Group 14/IV: ox. no. +4 predominates, some Sn(II), Pb(II); Group 15/V: ox. no. +3 predominates, some As(V), Sb(V).
10.15 **(a)** M—C decreases down a group; **(b)** decreases from $Al(CH_3)_3$ to $Tl(CH_3)_3$; **(c)** low for Groups 1 and 2 and AlR_3 high for Group 13/III.
10.16 **(a)** $Sn(CH_3)_4 > Si(CH_3)_4$; **(b)** $B(CH_3)_3 > Li_4(CH_3)_4 > Si(CH_3)_3Cl > Si(CH_3)_4$ toward weak bases, **(c)** $As(CH_3)_3 \gg Si(CH_3)_4$; **(d)** $Li_4(CH_3)_4 \gg Hg(CH_3)_2$.
10.17 **(a)** Schlenk or glove box; **(b)** vacuum line; **(c)** Schlenk or glove box; **(d)** open air; **(e)** open air.
10.18 Low-lying σ^* MO.
10.19 $3R_2SiCl_2 + 6Li[C_{10}H_8] \xrightarrow{\text{low temp.}}$ cyclo-$(SiR_2)_3 + 6LiCl + 6C_{10}H_6$; 2 cyclo-$(SiR_2)_3 \xrightarrow{h\nu} 3R_2Si{=}SiR_2$ (bulky R groups).

11

In-chapter exercises

11.1 **(a)** $BCl_3 + 3EtOH \rightarrow B(OEt)_3 + 3HCl$, easy protolysis of B—Cl; **(b)** $BCl_3 + NC_5H_5 \rightarrow Cl_3BNC_5H_5$, Lewis acidity of BCl_3; **(c)** $BBr_3 + F_3BN(CH_3)_3 \rightarrow Br_3BN(CH_3)_3 + BF_3$, Lewis acidity $BF_3 < BCl_3$.
11.2 $(CH_3)NH_2 + HCl \rightarrow [CH_3NH_3]Cl$, $3[CH_3NH_3]Cl + 3BCl_3 \rightarrow Cl_3B_3N_3(CH_3)_3 + 9HCl$, $Cl_3B_3N_3(CH_3)_3 + 3Li(CH_3) \rightarrow (CH_3)_3B_3N_3(CH_3)_3 + 3LiCl$.
11.3 **(a)** $(CH_3)_3NBF_3 + GaCl_3 \rightarrow CH_3NBCl_3 + GaF_3$, primarily driven by high lattice energy of GaF_3; **(b)** $2TlCl_3 + CH_2O + H_2O \rightarrow 2TlCl + 4HCl + CO_2$.
11.4 **(a)** Electrons added to π^* band; **(b)** electrons removed from π band.
11.5 8^-.
11.6 24.
11.7 16 electrons.
11.8 14 electrons, *nido* structure.
11.9 See synthesis of 1,7-$B_{10}C_2H_{12}$ and $B_{10}C_2H_{10}Li_2$ on p. 360, then react with dimethyldichlorosilane.

End-of-chapter exercises

11.1 **(a)** Metals: Al, Ga, In, Tl, Sn, Pb, possibly Bi; nonmetals: rest; **(b)** C, Si, Ge, Sn; **(c)** oxides in nature: B, Al, Ga, C, Si, Ge, Sn; **(d)** sulfides: Ga, In, Tl, Sn, Pb.
11.2 B_{12} icosahedron; Ga_2.
11.3 $6HCl + 2Al \rightarrow 2AlCl_3 + 3H_2$, see Fig. 11.5; $Sn + 2HCl \rightarrow SnCl_2 + H_2$.
11.4 **(a)** $SiF_4 < BF_3 \leq AlCl_3 < BCl_3$, **(b)** **(i)** $SiF_4N(CH_3)_3 + BF_3 \rightarrow SiF_4 + F_3BN(CH_3)_3$; **(ii)** $F_3BN(CH_3)_3 + BCl_3 \rightarrow BF_3 + Cl_3BN(CH_3)_3$; **(iii)** $Cl_3AlN(CH_3)_3 + BCl_3 \rightarrow Cl_3BN(CH_3)_3 + AlCl_3$.
11.5 See Fig. 11.20.
11.6 **(a)** Graphite + K vapor; **(b)** carbon + Ca at high temperature.
11.7 Fig. 11.10.
11.8 Glassy carbon, soot, silica glass, amorphous silicon.
11.9 **(a)(i)** B hard covalent semiconductor based on linked icosahedra, Al ductile close-packed metal; **(ii)** diamond covalent insulator/semiconductor, graphite layered covalent conductor; Si covalent semiconductor; **(b)** CO and CO_2 molecular gases, SiO_2 hard solid; **(c)** CX_4 not simple Lewis acids, SiX_4 Lewis acids; **(d)** BX_3 molecular, AlX_3 solids.
11.10 B(III), Al(III), Ga(I, III), In(I, III), Tl(*I*, III), C(II and IV), Sn(IV), Ge(II, *IV*), Sn(II, IV), Pb(*II*, IV) (Tl and Pb inert pair effect). **(a)** From Period 5 to Period 6 the elements of Groups III/13 and IV/14 favor the $N-2$ oxidation state. **(b)** **(i)** $Sn^{2+} + PbO_2 + 4H^+ \rightarrow Sn^{4+} + Pb^{2+} + 2H_2O$; **(ii)** $Tl^{3+} + Al \rightarrow Tl + Al^{3+}$; **(iii)** $3In^+ \rightarrow In^{3+} + 2In$; **(iv)** $2Sn^{2+} + O_2 + 4H^+ \rightarrow 2Sn^{4+} + 2H_2O$; **(v)** NR.
11.11 **(i)** 1.85V; **(ii)** 2.04V; **(iii)** 0.318V; **(iv)** 1.09V; **(v)** −0.02V.
11.12 $K_2CO_3(s) + 2HCl(aq) \rightarrow 2K^+ + 2Cl^- + H_2O + CO_2(g)$, $Na_4SiO_4(s) + 4H^+ \rightarrow$

$4Na^+ + (2 - x)H_2O + SiO_2 \cdot xH_2O$ (silica gel).

11.13 Jadeite chains, **19**; kaolinite sheets, Fig. 11.11.

11.14 **(a)** 48 edge-bridging O; **(b)** Fig. 11.13a.

11.15 Fig. 11.12.

11.16 B_4H_{10}: *arachno*, butterfly; B_5H_9: *nido*, square pyramidal; $B_{10}C_2H_{12}$: *closo*, icosahedral.

11.17 **(a)** *nido*, **(b)** $24e^-$.

11.18 $[Fe(nido\text{-}B_9C_2H_{11})_2]^{2-}$, preparation on p. 360.

11.19 **(a)** B and N in register between layers, carbon displaced; **(b)** graphite is reactive, BN is not.

11.20 **(a)** React anilinium chloride with BCl_3, **(b)** react methylammonium chloride with BCl_3, see structure **9**.

11.21 **(a)** E_g: BN $\gg$ AlP $>$ GaAs; **(b)** increase; **(c)** $\sigma = \sigma^0 e^{E_g/2kT}$, AlP more sensitive.

12

In-chapter exercises

12.1 Yes.

12.2 Sulfur stronger.

12.3 Reaction of NH_3 with hypochlorite forms NH_2Cl; then $NH_2Cl + NH_3 \rightarrow N_2H_4 + HCl$; both are nucleophilic displacements.

12.4 $3PCl_5 + 3NH_4Cl \rightarrow (Cl_2PN)_3 + 12HCl$, $n(Cl_2PN)_3 \xrightarrow{\Delta} (Cl_2PN)_{3n}$, $(Cl_2PN)_{3n} + 6nHN(CH_3)_2 \rightarrow [((CH_3)_2N)_2PN]_{3n} + 6nHCl$.

12.5 Both Br^- and Cl^- are thermodynamically capable of decomposing H_2O_2.

12.6 **(a)** SO_3, D_{3h}; **(b)** SO_3F^-, C_{3v}.

12.7 Layered structure with weak S—S interaction between layers.

12.8 Ten cluster skeleton pairs, consistent with *closo* structure.

End-of-chapter exercises

12.1 **(a)** N_2, O_2; **(b)** N through Sb, O through Te; **(c)** Bi (metalloid); **(d)** none; O does not achieve group oxidation number; Bi inert pair effect.

12.2 **(a)** NCl_3 highest, PCl_5 exists; **(b)** OF_2 highest, SF_6 highest.

12.3 **(a)** Saw-horse; **(b)** trigonal bipyramid; **(c)** saw-horse.

12.4 Oxidizing strength order, basicity the inverse: $NO^+ \gg N_2 > CN^- > C_2^{2-}$; $O_2^{2-} > N_2H_5^-$.

12.5 **(a)** Carbonate, CO_3^{2-}; **(b)** ozone, O_3; **(c)** oxalate $C_2O_4^{2-}$; **(d)** azide, N_3^- **(e)** carbon monoxide, CO; **(f)** hydronium, H_3O^+.

12.6 Acidity order (oxidizing strength generally inverse): **(a)** $NO_3^- > CO_3^{2-}$; **(b)** $O_3 > NO_2^-$; **(c)** $N_2O_4 > C_2O_4^{2-}$; **(d)** $N_2O > N_3^-$; **(e)** $N_2 \approx CO$; **(f)** $H_3O^+ > NH_3$.

12.7 **(a)** $4NH_3 + 7O_2 \rightarrow 6H_2O + 4NO_2$, $3NO_2 + H_2O \rightarrow 2HNO_3 + NO$; **(b)** $2NO_2 + 2OH^- \rightarrow NO_2^- + NO_3^- + H_2O$; **(c)** $NO_2^- + 2HSO_3^- + H_2O \rightarrow NH_3OH^+ + 2SO_4^{2-}$; **(d)** $3NaNH_2 + NaNO_3 \rightarrow NaN_3 + 3NaOH + NH_3$.

12.8 $P_4(s) + 5O_2(g) \rightarrow P_4O_{10}(s)$.

12.9 Fig. 12.5.

12.10 Faster, NO^+ a strong electrophile.

12.11 H_3PO_2 reduction of Cu^{2+}, $E^\ominus = 0.84$ V; $H_2PO_2^-$ is a good reducing agent, HPO_3^{2-} is not.

12.12 **(a)** 1.07 V; **(b)** 1.63 V, **(c)** $\Delta G^\ominus(HO_2) = -158$ kJ mol^{-1} and $\Delta G^\ominus(H_2O_2) = -103$ kJ mol^{-1}.

12.13 **(a)** en; **(b)** en; **(c)** SO_2 acid–base considerations.

12.14 Monoxides: rock salt, ionic; sulfides, NiAs, more covalent.

12.15 Rate law second-order in NO.

12.16 **(a)** Sb_4^{2-} butterfly, $P_7^{3-} \approx$ **40**.

13

In-chapter exercises

13.1 Pyramidal, C_s.

13.2 8.8×10^{-3} M.

13.3 F_2, Cl_2, facile with F_2.

13.4 **(a)** Octahedral, possible slight Jahn–Teller distortion; **(b)** rutile; **(c)** octahedral.

13.5 $2CoCl_2 + 3F_2 \rightarrow 2CoF_3 + 3Cl_2$.

13.6 $2XeO_4^{2-} \rightarrow XeO_6^{4-} + Xe + O_2$, $2XeO_4^{2-} + 2H_2O \rightarrow 2Xe + 4OH^- + 3O_2$; two independent reactions (disproportionation of xenate and oxidation of water) so their ratio will depend on reaction conditions.

End-of-chapter exercises

13.1 **(a)** Gases: F_2, Cl_2, noble gases; $Br_2(l)$, $I_2(s)$; **(b)** $F > Cl > Br > I$; **(c)** $F^- > Cl^- > Br^- > I^-$; **(d)** F_2 nearly colorless, Cl_2 yellow, Br_2 red-brown, I_2 black solid, purple gas.

13.2 F_2 nonaqueous electrochemical oxidation, Cl_2 aqueous electrochemical oxidation, Br_2, I_2 oxidation by Cl_2.

13.3 Fig. 13.6; $2OH^- + Cl_2 \rightarrow OCl^- + Cl^- + H_2O$.

13.4 Nodal plane between atoms, positive lobe extends from one end negative from the other, vacant in diatomic, acceptor orbital in complexes.

13.5 For $ClO_4^- + 8H^+ + 8e^- \rightarrow 4H_2O + Cl^-$, +0.20 V at pH = 7.

13.6 Rates of oxoion reactions are increased by H^+.

13.7 **(a)** $(CN)_2 + 2OH^- \rightarrow CNO^- + CN^- + H_2O$; **(b)** $2SCN^- + MnO_2 + 4H^+ \rightarrow Mn^{2+} + (SCN)_2 + 2H_2O$; **(c)** C_{3v}.

13.8 **(a)** Octahedral; **(b)** $IF_7 + SbF_5 \rightarrow [IF_6][SbF_6]$.

13.9 $Br_3^- + I_2 \rightarrow 2IBr + Br^-$, $IBr + Br^- \rightleftharpoons IBr_2^-$.

13.10 Large cations stabilize large anions.

13.11 ClO_2 radical angular; I_2O_5 has oxo bridge between two IO_2.

13.12 Perbromic more acidic and less stable.

13.13 Periodic, labile I—O bond allows atom transfer.

13.14 Disproportionation favored thermodynamically for HClO, $HClO_2$, and ClO_3^-.

13.15 **(a)**, **(d)** Reducible cation with oxidizing perchlorate anion.

13.16 **(a)** Rutile; **(b)** layered; **(c)** M—M bonded; **(d)** M—M bonded.

13.17 **(a)** $Fe(s) + 2HCl(aq) \rightarrow FeCl_2(aq) + H_2(g)$, crystallize; **(b)** $FeCl_2 \cdot 4H_2O + 4SOCl_2 \xrightarrow{\text{reflux}} FeCl_2(s) + 4SO_2(g) + 8HCl(g)$; **(c)** $Pt + 3F_2 \xrightarrow{\Delta,\ \text{high press.}} PtF_6$; **(d)** $ZrBr_4 + 3Zr \xrightarrow{\Delta} 4ZrBr$.

13.18 Helium is not retained by the earth's gravity.

13.19 **(a)** He; **(b)** Xe; **(c)** Ar.

13.20 **(a)** $Xe + F_2 \xrightarrow{\text{sunlight}} XeF_2$; **(b)** $Xe + 3F_2 \xrightarrow{\Delta,\ \text{high press}} XeF_6$; **(c)** $XeF_6 + 3H_2O \rightarrow XeO_3 + 6HF$.

13.21 **(a)** XeF_4; **(b)** XeF_2; **(c)** XeO_3; **(d)** XeF^+.

13.22 ClF is isoelectronic with ClO^-. The dihalogens are Lewis acids. If ClO^- shares this property it may diproportionate *via* $[ClO \cdots ClO]$.

14

In-chapter exercises

14.1 ^3D or ^1D.

14.2 **(a)** ^3P, **(b)** ^2D.

14.3 $^3T_{1g}$, $^3T_{2g}$, $^3A_{2g}$ and $^1T_{2g}$, ^1E_g.

14.4 18 400 cm^{-1}; 27 500 cm^{-1}.

14.5 ^2E_g, $^4T_{1g}$, $^4T_{2g}$ MLCT.

14.6 An extra band.

14.7 No, a good electrical conductor.

End-of-chapter exercises

14.1 ^6S, ^4F, ^2D, ^3P.

14.2 ^3F, ^5D, ^6S.

14.3 **(a)** ^{2}S; **(b)** ^{3}P (ground), ^{1}D, ^{1}S.
14.4 $B = 861\ \text{cm}^{-1}$, $C = 3168\ \text{cm}^{-1}$.
14.5 d^6, $^1A_{1g}$; d^1, $^2T_{2g}$; d^5, $^6A_{1g}$.
14.6 Reading T–S diagrams approximately: **(a)** $B = 770\ \text{cm}^{-1}$, $\Delta_O = 8500$; **(b)** $B = 720\ \text{cm}^{-1}$, $\Delta_O = 10\,750\ \text{cm}^{-1}$.
14.7 $^5T_{2s}$, charge transfer, LMCT.
14.8 Weak: $^3T_{2g} \leftarrow {}^1A_{1g}$; medium $^1T_{1g} \leftarrow {}^1A_{1g}$; strong: $^1T_{2g} \leftarrow {}^1A_{1g}$.
14.9 $[FeF_6]^{3-}$ has only spin-forbidden LF bands, $[CoF_6]^{3-}$ has $t_{2g}^4e_g^2$ and is equivalent to d^1 and has $^5E_g \leftarrow {}^5T_{2g}$.
14.10 CN^- is more covalent.
14.11 Two spin-allowed LF bands, a spin-forbidden LF, and an LMCT band.
14.12 High-spin d^5 has only spin-forbidden LF bands.
14.13 $[Cr(OH_2)_6]^{3+}$, LF bands; CrO_4^{2-} allowed LMCT.
14.14 a_{1g}; **(a)** d_{xz}, d_{yz}; **(b)** d_{z^2}; **(c)** lowest energy xy; (xz, yz); z^2; $x^2 - y^2$ highest energy.
14.15 Mn(VII) is d^0 and has no LF bands.
14.16 The two LMCT bands are to t_{2g} and e_g. The difference, $13\,700\ \text{cm}^{-1}$, is Δ_T.

15

In-chapter exercises

15.1 $5.1\ \text{M}^{-1}\,\text{s}^{-1}$
15.2 $[PtCl_4]^{2-} \xrightarrow{PPh_3} [PtCl_3PPh_3]^- \xrightarrow{NH_3}$ *trans*-$[PtCl_2NH_3PPh_3]$; $[PtCl_4]^{2-} \xrightarrow{NH_3} [PtCl_3NH_3]^- \xrightarrow{PPh_3}$ *cis*-$[PtCl_2NH_3PPh_3]$.
15.3 $K_E \approx 1$, $k = 1.2 \times 10^2\ \text{s}^{-1}$.
15.4 2 *cis* to 1 *trans*.
15.5 $k = 3 \times 10^{18}$, $k = 3.6 \times 10^{10}\ \text{M}^{-1}\,\text{s}^{-1}$.

End-of-chapter exercises

15.1 Nucleophiles; NH_3, Cl^-, S^{2-}; electrophiles; Ag^+, Al^{3+}.
15.2 *d*.
15.3 Maximum after intermediate.
15.4 **(a)** Stoichiometric; **(b)** intimate.
15.5 Rate $= k[Mn(OH_2)_6^{2+}][X^-]/(1 + K_a[X^-])$, vary X^-.
15.6 Stronger covalent bonds to be broken.
15.7 Steric hindrance to nucleophilic attack.
15.8 Yes, a favorable A intermediate.
15.9 $[PtCl_4]^{2-} \xrightarrow{NH_3} [PtCl_3NH_3]^- \xrightarrow{NO_2^-}$ *cis*-$[PtCl_2(NO_2)(NH_3)]$, $[PtCl_4]^{2-} \xrightarrow{NO_2^-} [PtCl_3NO_2]^- \xrightarrow{NH_3}$ *trans*-$[PtCl_2(NO_2)(NH_3)]$.
15.10 **(a)** Decrease rate; **(b)** decrease rate; **(c)** decrease rate; **(d)** increase rate.
15.11 Conjugate base path; Brønsted acidity is required.
15.12 **(a)** *cis*-$[PtCl_2(PR_3)_2]$; **(b)** *trans*-$[PtCl_2(PR_3)_2]$; **(c)** *trans*-$[PtCl_2(py)_2]$.
15.13 $[Ir(NH_3)_6]^{3+} < [Rh(NH_3)_6]^{3+} < [Co(NH_3)_6]^{3+} < [Ni(OH_2)_6]^{2+} < [Mn(OH_2)_6]^{2+}$.
15.14 **(a)** Decreases; **(b)** decreases; **(c)** little effect; **(d)** decreases; **(e)** increases.
15.15 Inner sphere goes through N_3^- bridged complex. Outer sphere has no ligand bridge.
15.16 Inner-sphere mechanism.
15.17 $[W(CO)_5(NEt_3)]$ 0.4; ligand field.
15.18 250 nm.

16

In-chapter exercises

16.1 No (20e$^-$).
16.2 +1.
16.3 Bridging only.
16.4 $P(CH_3)_3$ less steric repulsion.
16.5 Pepare: $[Mn(CO)_5]^-$ (p. 514), $Mn(CO)_5CH_3$ (p. 517) migratory insertion eqn 2 with L = PPh_3.
16.6 Small change.
16.7 CVE = 60; tetrahedral Fe_4 with Cp on each vertex, CO on each face.

End-of-chapter exercises

16.1 **(a)** Pentacarbonyliron(0); **(b)** tetracarbonylnickel(0); **(c)** hexacarbonylmolybdenum(0); **(d)** decacarbonyldimanganese(0); **(e)** hexacarbonylvanadium(0); **(f)** trichloroethyleneplatinate(II).
16.2 **(a)** +3; **(b)** 0; **(c)** −2; **(d)** 0.
16.3 $V(CO)_6$ 17e$^-$, no influence on structure; $[PtCl_3(C_2H_4)]^-$ 16e$^-$, square planar; $[(\eta^5\text{-}C_5H_5)_2Fe][BF_4]$, 17e$^-$ slightly longer Fe—C distances.
16.4 Direct combination, reductive carbonylation.
16.5 $Fe + 5CO \xrightarrow{\text{high press.}} Fe(CO)_5$, diphos + $Fe(CO)_5 \rightarrow Fe(CO)_3(\text{diphos}) + 2CO$.
16.6 C_s (3); C_{3v} (2); D_{3h} (1).
16.7 Ni_3 triangle each Ni capped by Cp, CO above and below nickel triangle and collinear; $18\frac{1}{3}$ e$^-$ per Ni atom, deviations from 18-e$^-$ rule are common in Group 10.
16.8 **(b)** Reacts by an associative process.
16.9 **(a)** $[Fe(CO)_4]^{2-}$ more negative; complex is more basic; **(b)** $[Re(CO)_5]^-$ heavier metal, more basic.
16.10 **(a)** η^2; **(b)** η^1, η^3, η^5; **(c)** η^2, η^4, η^6; in practice the last is observed; **(d)** η^2, η^4, η^2, η^4, η^6, η^8.
16.11 **(a)** 16e$^-$ sandwich; **(b)** 18e$^-$; **(c)** 18e$^-$.
16.12 **(a)** Fe, Ru, Os; **(b)** Mn, Co; **(c)** Cr, Fe, Ni; **(d)** Table 16.3.
16.13 **(a)** 3; **(b)** 2; **(c)** 12.
16.14 See pp. 512 and 513.
16.15 Prepare $Mn(CO)_5^-$ (p. 511), protonate with HCl.
16.16

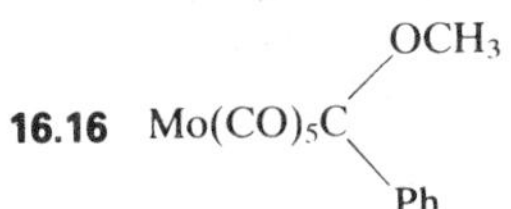

16.17 **(a)** PF_3 π-acid ligand (increases $\nu(CO)$), $P(CH_3)_3$ σ donor ligand (decreases $\nu(CO)$); **(b)** Cp* stronger donor, and therefore lower $\nu(CO)$.
16.18 **(a)** $RuCp_2$ (18e) more stable than $RhCp_2$ (19e); **(b)** $(\eta^5\text{-}C_5H_5)_2Fe + CH_3C(O)Cl \xrightarrow{AlCl_3} (\eta^5\text{-}C_5H_5)(\eta^5\text{-}C_5H_4COCH_3)Fe + HCl$.
16.19 $(\eta^5\text{-}C_5H_5)_2Fe + LiBu \rightarrow (\eta^5\text{-}C_5H_5)(\eta^5\text{-}C_5H_4Li)Fe + BuH$, $(\eta^5\text{-}C_5H_5)(\eta^5\text{-}C_5H_4Li)Fe \xrightarrow{(1)\ CO_2,\ (2)\ H_2O} (\eta^5\text{-}C_5H_5)(\eta^5\text{-}C_5H_4COOH)Fe(C_5H_5)$.
16.20 See Fig. 16.13 and Appendix 6.
16.21 C_5H_6 is a tetrahapto four-electron ligand. Its formation reduces the metal electron count to 18, protonation of Fe in $FeCp_2$ leaves the 18-electron count unchanged.
16.22 **(a)** Attack of H_2O on the CF_2 carbon followed by elimination of HF; **(b)** β-H elimination.
16.23 **(a)** Groups 6–9: stable Cp compounds, Groups 3 and 4: C—H insertion; **(b)** Groups 6 to 8: M—H more acidic than in Groups 3 and 4; **(c)** 18-electron rule violation for many Group 3 and 4 and some Group 9 metals.
16.24 Energy order: $\sigma < \pi < \delta < \delta^* < \pi^* < \sigma^*$; $Mo_2(OAc)_4(py)_2$: $\sigma^2\pi^4\delta^2$, bond order 4; $Mn_2(CO)_{4\text{-}10}$: $\sigma^2\pi^4\delta^2\delta^{*2}\pi^{*2}$, bond order 1.
16.25 **(a)** Octahedral (86), trigonal prismatic, CVE = 90; **(b)** no; **(c)** $[Fe_6C(CO)_{16}]^{2-}$, CVE = 86, octahedral; $[Cu_6C(CO)_{16}]^{2-}$, CVE = 92, trigonal prismatic; **(d)** $[CoC(CO)_{16}]$.
16.26 **(a)** $Si(CH_3)$; **(b)** I.
16.27 $Co_4(CO)_{12}$ has weaker M—M bonds.

17

In-chapter exercises

17.1 Oxidative addition: $Rh^{(I)}ClL_2(Sol) + H_2 \rightarrow Rh^{(III)}ClH_2L_2(Sol)$; reductive elimination: $Rh^{(III)}ClH(C_2H_4R)L_2(Sol) \rightarrow Rh^{(I)}ClL_2(Sol) + C_2H_5R$.
17.2 Decrease rate.
17.3 Single 1465 cm^{-1} band expected.
17.4 No, it lacks acidic sites.

End-of-chapter exercises

17.1 **(a)** Catalytic; **(b)** noncatalytic; **(c)** noncatalytic.
17.2 See **(a)** p. 543; **(b)** p. 545; **(c)** p. 543; **(d)** p. 558.
17.3 **(a)** Homogeneous; **(b)** heterogeneous; **(c)** homogeneous.
17.4 Accomplished with existing technology: **(d)**; thermodynamically prohibited: **(a)** and **(b)**.
17.5 Added PPh_3 suppresses equilibrium concentration of $RhClL_2(Sol)$.
17.6 Steric effects reduce equilibrium formation of **(c)** in Fig. 17.1.
17.7 Formation of (a) or (c) may be rate-limiting.
17.8 Attack of coordinated OH^- gives one enantiomer, attack of solution OH^- gives the other.
17.9 Coordination of formic acid with loss of L, reductive elimination of H_2.
17.10 Titanium nitride is too stable.
17.11 **(a)** An extra proton is required to charge-compensate Al^{3+} substituted for Si^{4+}. **(b)** Ga(III), Sc(III), and Fe(III).
17.12 **(a)** 100°C treatment leaves Al^{3+} four- and six-coordinated; **(b)** 900°C treatment exposes an Al^{3+} coordination site; **(c)** ZSM-5 is highly regular.
17.13 To achieve high surface area.
17.14 See p. 560.
17.15 Initial chemisorption as $C^*H_2C^*HC(CH_3)_2CH_2CH_3$, exchange of H for D at surface attached carbon atoms (*), dissociation (via reductive elimination).
17.16 CO strongly chemisorbs and blocks sites for H_2.

18

In-chapter exercises

18.1 Vacancies on both anion and cation sites.
18.2 Reduction in thickness of conduction plane impedes the larger cation more.
18.3 Two.
18.4 Semiconduction.

End-of-chapter exercises

18.1 Frenkel, interstitial, Schottky, holes; no.
18.2 **(a)** NiO; **(b)** PbF_2; **(c)** Fe_2O_3.
18.3 Intrinsic in pure substance, extrinsic (due to impurities).
18.4 Fig. 18.17; Na^+ in center of cell.
18.5 Both. It is a defect when disordered, and describes a new phase when ordered.
18.6 See Fig. 18.13.
18.7 TiO vacancies on both Ti and O sites; FeO, Fe^{3+} interstitials with vacancies on Fe^{2+} sites.
18.8 Electron microscopy, electron diffraction.
18.9 For example, from 27 to 30 percent O at 1300°C: Fe_3O_4 *ss* (solid solution); Fe_3O_4 *ss* plus Fe_2O_3 *ss*; Fe_2O_3 *ss*.
18.10 **(a)** NiO (MnO is much more easily oxidized than the aqueous potentials suggest); **(b)** NiO.
18.11 TiO metallic, NiO semiconductor.
18.12 Positive Ag electrode: $Ag(s) \rightarrow Ag^+(\text{in AgI}) + e^-(\text{external circuit})$; AgI electrolyte: Ag^+ from + to − electrode; negative Ag electrode: $Ag^+\ (\text{in AgI}) + e^- \rightarrow Ag(s)$.
18.13 Positive electrode mass loss 108 mg, AgI no change, negative electrode mass gain 108 mg.
18.14 Ferromagnet: domains of parallel spins on atoms; antiferromagnet: domains of antiparallel spins on atoms.
18.15 In the inverse spinel structure, Fe[CoFe] with Co^{2+} and Fe^{3+}; three unpaired electrons are expected from the d^7 octahedral Co^{2+}. The structure appears to be close to the inverse spinel formulation.
18.16 See pp. 588 and 589.

19

In-chapter exercises

19.1 1.24 Å is just slightly above O_2, suggesting little electron transfer; 1.26 Å is just less than O_2^- suggesting nearly one electron transferred. 1.47 Å is near O_2^- suggesting transfer of mainly one electron from each Co, and 1.30 Å is just over O_2^- suggesting transfer of just over one electron between the two Co centers.

19.2 Zn–O(H)(H) + O=C=O → Zn–O(H)–C(=O)–O⁻ + H^+

19.3 First and sixth entries have same $E^{\ominus}$ but change of R by 2 Å correlates with greater than 10^2 rate change. The comparison of third and eighth entries is similar. Fourth and ninth show doubling R increases rate by a factor of 10. All imply qualitative agreement with Marcus Theory.
19.4 C_{4v} has the order of d orbitals $xz, yz < xy < z^2 < x^2 - y^2$. The configuration is $(xz)^2(xy)^1(z^2)^1(x^2-y^2)^1$.
19.5 The average oxidation number of Fe in the latter is 2.5. The former is 3. Reduction and disulfide formation can convert the Fe_2 clusters to Fe_4 clusters.

End-of-chapter exercises

19.1 Neutral O_2 with net double bond, triplet; O_2^- longer; $1\frac{1}{2}$ order bond, doublet; O_2^{2-} longer single bond, singlet.
19.2 Left for HbO_2; right for Mb at low $p(O_2)$, left at high $p(O_2)$.
19.3 Binding to hemoglobin blocking O_2.
19.4 Nucleophile is N end of peptide, site of attack is coordinated carbonyl.
19.5 LF spectrum, Zn is d^{10}. Assume same geometry.
19.6 **(a)** High spin $t_{2g}^4e_g^2$, low spin t_{2g}^6; **(b)** high H_2O, low $(C_6H_5)_3P$:.
19.7 Variable oxidation number.
19.8 The two oxidation states are not configured to conform to each other. They bind substrates.
19.9 Specificity in binding to allow close approach, control of inner sphere reorganization energies.
19.10 Consult text.
19.11 Fe—O—O—Fe bridge formation.
19.12 **(a)** $ZnOH_2$ transfers a proton to promote —OH attack; **(b)** Zn—O=C< promotes nucleophilic attack on C.
19.13 +2.5.
19.14 O_2 evolution requires $4e^-$, over four Mn atoms Mn charge changes by −1.
19.15 Chlorophyll–porphyrin π^*, cytochrome Fe(II). Mg could not be used; Mg(III) inaccessible.

Formula index

P: preparation, E: electronic structure, S: structure. Note that in the ordering adopted, ( ) precedes [ ], and then alphabetical. Since there is not complete uniformity in the use of square brackets in formulas, it is important to check the entire entry for each element when seeking a particular compound or complex.

Ag

$[Ag(Cl)_2]^-$ 194
Ag^{2+} 245
Ag_2F 461(S)
Ag_2HgI_4 580(S), 579
AgBr 115(S)
AgCl 115(S), 128
AgI 115(S), 579
$RbAg_4I_5$ 579

Al

$(CH_3)_3Al(py)$ 180
$(CH_3)_3AlN(C_2H_5)_3$ 309(P)
$(CH_3)_3NAlH_3$ 287(P)
$((CH_3)_3N)_2AlH_3$ 287(S)
$[(H_2O)_5AlOHAl(OH_2)_5]^{5+}$ 162
$[Al(OH)_4]^-$ 161, 164
$[Al(OH_2)_6]^{3+}$ 161–3
$[AlH_4]^-$ 193(S), 280, 286
$[AlO_4(Al(OH)_2)_{12}]^{7+}$ 163(S)
$[AlO_4]^{5-}$ 163(S)
$Al(CH_3)_3$ 301, 305–9
$Al(OC_2H_5)_3$ 308(P)
$Al(OH)_3$ 234
$Al_2(C_2H_5)_4H_2$ 287(S)
$Al_2(C_4H_9)_6$ 316
$Al_2(C_6H_5)_6$ 316(S)
$Al_2(CH_2CH_2R)_6$ 317
$Al_2(CH_3)_6$ 301(S), 306, 308, 309, 314, 316(P), 317(P)
Al_2Cl_6 183(S)
$Al_2H_2R_2$ 317(P)
Al_2O_3 161, 186, 234, 242, 340, 559(P), 560, 579, 590
Al_4C_3 363
AlB_2 361(S)
$AlCl_3$ 339(P, S), 340
AlH_3 279, 287
AlN 115(S)
AlO(OH) 559
AlX_3 339(S)
$Cl_3Al(N(CH_3)_3)_2$ 340(P)
$Cl_3Al(py)$ 560
$Cl_3AlN(CH_3)_3$ 340(P)
$Fe_2Cp_2(CO)_4 \cdot 2AlEt_3$ 517(P, S)
Li_3AlH_6 286
$LiAlH_4$ 280, 283, 286(P)
$MgAl_2O_4$ 583(S)
Na_3AlF_6 234
$NaAl(SiO_3)_2$, jadeite 348
R_2AlCl 315(P)

As

$(AsC_6H_5)_6$ 328(P)
$(CH_3)_2AsAs(CH_3)_2$ 328
$[As(C_6H_5)_4]^+$ 328(P)
$[As(CH_3)_4]^+$ 327(P)
$[AsCl_4]^-$ 376(S)
$As(C_6H_5)_5$ 327(S)
$As(CH_3)_3$ 302(S), 305, 327
As_2O_5 161
$As_3(CH_2)_3CCH_3$ 328(S)
As_4 374
AsF_5 399
AsH_3 277(S), 278, 279
AsO_4^{3-} 129, 374
C_5H_5As 327(S)
GaAs 365–6
H_3AsO_4 159(S)
InAs 115(S), 365
NiAs 115(S), 118
R_2AsAsR_2 328
SnAs 115(S)

Au

$[Au(CN)_2]^-$ 235
$[AuCl(dien)]^{2+}$ 476
Cu_3Au 114
CuAu 575

B

$(CH_3)_2B(C_4H_9)$ 316(P)
$(CH_3)_3B(py)$ 177(S)
$(CH_3)_2BOR$ 316(P)
$[(HO)_2B(—OCH_2CH_2O—)]^-$ 336
$[B(C_6H_5)_4]^-$ 309(P)
$[B(OH)_4]^-$ 336
$[B_{12}H_{12}]^{2-}$ 355(S)
$[B_3O_3(OH)_4]^-$ 336(S)
$[B_3O_6]^{3-}$ 336(S)
$[B_5H_5]^{2-}$ 355(S)
$[B_6H_6]^{2-}$ 355, 357(S)
$[B_9C_2H_{11}]^{2-}$ 360(P)
$[B_9C_2H_{12}]^-$ 360(P)
$[BF_4]^-$ 39(E), 182
$[BH_4]^-$ 280, 285–6
$B(C_6H_5)_3$ 309
$B(CH_3)_3$ 170, 177, 302(S), 303(P), 305, 306, 315(P)
$B(OCH_3)_3$ 336(P)
$B(OH)_3$ 335, 336
$B_{10}C_2H_{12}$ 360(P, S)
$B_{10}H_{10}C_2(COOH)_2$ 359(P)
$B_{10}H_{14}$ 283, 296, 356, 358(S)
B_{12} 334
$B_2Br_3Cl_3$ 335(S)
B_2Cl_4 353(P, S)
B_2H_6 81(S, E), 277(S), 279, 283(P)
B_2O_3 334
$B_4C_2H_6$ 360(S)
B_4H_{10} 355(S), 357(S)
B_5H_9 357(S)
BCl 353(P)
$BCl_2CH_2CH_2BCl_2$ 353(P)
BCl_3 182, 315, 335
BF_3 43(S), 53(S), 171, 177(E), 182, 315, 334(P), 335
BF_4^- 129
BH_2 89(S)
BN 337(P, S), 365
B_nH_m 356(S)
BO 284
BP 365
BX_3 182
CaB_6 361(S)
$CH_3CH_2BH_2$ 304(P)
$Cl_3BN_3R_3$ 338(P, S)
$Cl_3BN_3H_3$ 338(P, S)
F_3BNH_3 177(E), 335(P)
H_2NBH_2 338
$H_3BN(CH_3)_3$ 285(P), 337(P)
H_3BNH_3 338(S)
H_3BOR_2 285
H_3NBH_2COOH 338
$LiBH_4$ 283, 285(P)
$N_3B_3H_{12}$ 338(S)
$NaBF_4$ 176
$NaBH_4$ 176
R_2BCl 315(P)
R_2NBCl_2 338(P, S)
R_3BNH_3 177(S)

Ba

$Ba(OH)_2$ 134
$Ba[PbBi]O_3$ 589
$BaCl_2$ 115(S)
$BaCO_3$ 131
BaH_2 274, 279
BaO 393
BaO_2 393
$BaTiO_3$ 115(S)
$YBa_2Cu_3O_7$ 588

Be

$(CH_3)_2Be(TMEDA)$ 313(P)
$(CH_3)Be(OCH_3)$ 313(P)
$[Be(CH_3)_2]_n$ 301(S)
$[BeCl_4]^{2-}$ 194(S)
$Be(CH_3)_2$ 312
$Be(Cp)_2$ 312(P, S)
$Be(^tBu)_2$ 312(S)
Be_2C 363
$Be_3Al_2Si_6O_{18}$ 348(S)
$BeCl_2$ 312
BeF_4^{2-} 129
BeH_2 89(s), 279
BeO 115(S), 161
BeR_2(P) 312

Bi

$Ba[PbBi]O_3$ 589
$Bi(CH_3)_3$ 305, 310
Bi_4^{2-} 398(S)

Br

$(CH_3)_2COBr_2$ 185(S, E)
$[BrF_2]^+$ 172
Br_2 404(E), 405(S), 406(P)
Br_2^+ 410
BrCl 408
BrF 408
BrF_2^+ 410
BrF_3 172, 408, 409
BrF_4^+ 410
BrF_4^- 410, 411(S)
BrF_5 408, 409(S)
BrF_6^+ 410
BrO_3^- 420
BrO_4^- 416, 420
HBr 67(S), 150, 152, 271, 273, 277(S), 279

C

$(CH_3)_2C_6H_4$ 567
$(CN)_2$ 407
$[C_{10}H_8]^-$ 301, 311(P)
Al_4C_3 363
Be_2C 363
$C(CH_3)_4$ 305
C_2^{2-} 362–3
$C_{24}HSO_4$ 343
C_2H_2 53(S), 550
C_2H_4 272(P), 501, 518–20, 547, 550, 557
C_2H_4O 544(S)
$C_2O_4^{2-}$ 380(S)
C_3O_2 344
C_5H_5 500, 501, 522–27(S)
$C_5H_5^-$ 312(P), 526(P)
C_5H_5N, py 171, 201
C_6H_6 501, 522
$C_7H_7^+$ 501
C_7H_8 501
C_8H_8 501, 522, 526(S)
CaC_2 362, 363(P, S)
CH_2 89(S)
CH_3CHO 557–8
$CH_3COCOOH$ 615(P)
CH_3COOH 155–6, 550(P), 556(P)
CH_3OH 542, 550, 556
CH_4 43(S), 53(S), 152, 273, 277(S), 279
CN^- 211, 346, 407, 506
CNO^- 407
CNR 506
CO 66(E), 230, 270, 272, 344–5, 506, 581
CO_2 43(S), 53(S), 161, 230, 345, 581
CO_2^- 380(S)
CO_3^{2-} 43(S), 130, 345
CS 506
CS_2 346
CS_3^{2-} 346
H_2CO_3 159(S), 160
HCN 43(S), 346
HCO_2^- 345
HCOOH 160, 563
$HOOCCH_2COCOOH$ 615
KC_8 362
$Na[C_{10}H_8]$ 301, 311(P)
Na_2C_2 363(P)
$NC(CH_2)_4CN$ 550(P)
NCCN 407
NCSSCN 407
$NH_2CH_2CH_2NH_2$ 200, 202
SCN^- 175–6

Formula index

Ca

$Ca_{10}(PO_4)_6(OH)_2$ 610
CaB_6 361(S)
CaC_2 362, 363(P, S)
$CaCO_3$ 131, 610
CaF_2 115(S), 118(S)
CaH_2 274(S), 279
CaO 115(s), 161, 215, 581, 600
CaS 115(S)
$CaTiO_3$ 115(S), 119(s), 586
$CaZn_2$ 114
$Ca_3(PO_4)_2$ 373
$Ca_5(PO_4)_3F$ 373
$Ca_5(PO_4)_3OH$ 373
$Ca(SiO_3)$ 373

Cd

$[CdBr_3(OH_2)_3]^-$ 219
$[CdBr_4]^{2-}$ 219(P)
Cd 112(S)
$Cd(CH_3)_2$ 302(S), 305
CdI_2 396(S), 420(S)
CdS 115(S), 365, 446, 593
$CdTe$ 367

Ce

Ce^{3+} 258
Ce^{4+} 258
$[Ce(NO_3)_6]^{2-}$ 198(S)
CeH_2 275–6
CeH_3 275–6

Cl

$(ClF_2)(SbF_6)$ 410
$[ClHCl]^-$ 291
$[ClO_2SO_2]^-$ 417(S)
$BrCl$ 408
Cl_2 404(E), 405(S), 406(P)
$Cl_2{\cdot}(H_2O)_{7.25}$ 292(S)
Cl_2O 412
Cl_2O_6 412
Cl_2O_7 412
ClF 408
ClF_2^+ 410
ClF_3 408, 409(S)
ClF_4^+ 410
ClF_5 408
ClF_6^+ 408, 410
ClO^- 246, 413(S), 419
ClO_2 412(P)
ClO_2^- 246, 413(S), 419
ClO_3 412
ClO_3^- 246, 413(S), 419
ClO_4^- 40(E), 129, 246, 413(S), 416
HCl 67(S), 273, 277, 279
$HClO_2$ 159(S), 413, 415, 419
$HClO_3$ 159(S), 413, 419
$HClO_4$ 159(S), 413
$HOCl$ 159(S), 246, 413, 418(P), 419
I_2Cl_6 408
NO_3Cl^{2-} 241
OCl^- 241

Co

Co^{n+} 212, 256, 257
$[(NH_3)_5Co(\mu\text{-}Cl)Cr(OH_2)_5]^{4+}$ 486
$[Co(ala)_3]$ 453
$[Co(bipy)_3]^{2+}$ 489
$[Co(C_2S_2(CN)_2)_2]$ 195(S)
$[Co(Cl)_2(en)_2]$ 204(S)
$[Co(Cl)_2(NH_3)_4]^+$ 196(S)
$[Co(CO_3)(NH_3)_4]^+$ 197(P), 345(S)
$[Co(corrin)(OH_2)_2]$ 485
$[Co(edta)]^-$ 202, 454
$[Co(en)_3]^{3+}$ 205(S), 452–4
$[Co(\eta^3\text{-}C_3H_5)(CO)_3]$ 521(S)
$[Co(NCS)_4]^{2-}$ 215
$[Co(NH_3)_5]^{3+}$ 483
$[Co(NH_3)_5(OH_2)]^{3+}$ 466
$[Co(NH_3)_6]^{2+}$ 197
$[Co(NO_2)_3(NH_3)_3]$ 197(P, S)
$[Co(salen)(py)]$ 612
$[Co(terpy)_2]^{2+}$ 489
$[Co(TPP)]$ 569
$[Co_2(PhC_2Ph)(CO)_6]$ 521(S)
$[CoBr_4]^{2-}$ 193(S), 215
$[CoCl(NH_3)_5]^{2+}$ 241, 483
$[CoCl_2(bn)_2]^+$ 480
$[CoCl_2(en)_2]^+$ 464
$[CoCl_4]^{2-}$ 194(S), 215
$[CoH(CO)_4]$ 273, 515, 554, 556
$[CoL_4]^{2-}$ 215
$[CoX(NH_3)_5]^{2+}$ 479
$[CpCo(\mu^2\text{-}cot)CoCp]$ 522(S)
Co 112(S)
$Co(Cp)_2$ 512, 597
$Co_2(CO)_8$ 502(S), 508(P), 511, 554
$Co_3(CO)_9CH$ 91(S), 532(S)
Co_3O_4 583
$Co_4(CO)_{12}$ 91(S), 532, 534(P, S)
CoF_2 215
CoF_3 410(P)
$CoH(CO)_3P(OPh)_3$ 515
$CoH(CO)_3PPh_3$ 515
CoO 579, 581, 582
CoS 115(S)

Cr

Cr^{n+} 256–7
$[(OC)_5CrCPhNR_2]$ 519
$[Cr(bipy)_3]^{3+}$ 451, 494
$[Cr(CN)_6]^{4-}$ 210(E)
$[Cr(CO)_4(P(OPh)_3)_2]$ 510(S)
$[Cr(CO)_6]$ 493, 502(S), 506, 508(P, S), 511
$[Cr(edta)]^-$ 204(S)
$[Cr(\eta^6\text{-}C_6H_6)(CO)_3]$ 526(P)
$[Cr(\eta^6\text{-}C_6H_6)_2]$ 522(S)
$[Cr(NCS)_6]^{3-}$ 450
$[Cr(NH_3)_6]^{3+}$ 210(E), 434, 441, 445, 493
$[Cr(OH_2)_6]^{2+}$ 210(E), 241, 418
$[Cr(OH_2)_6]^{3+}$ 340
$[CrCl(NH_3)_5]^{2+}$ 446(S)
$[O_3CrOCrO_3]^{2-}$ 162
$Cr(CO)_5$ 493
Cr_2O_3 579, 585, 590
$Cr_2O_7^{2-}$ 243
Cr^{3+} 213, 340
$CrCl_3{\cdot}6H_2O$ 424
$CrCp(CO)_3H$ 515
$CrCp_2$ 597
CrF_6 256
CrO 582
CrO_4^{2-} 129, 162, 447

Cs

$Cs(O_2)$ 393
Cs_3O 393(S), 593
Cs_4O 393
Cs_7O 393
$CsBr$ 117(S)
$CsBrF_4$ 410(P)
$CsCl$ 115(S), 117(S)
$CsCN$ (S) 115(S)
CsF 410
CsH 274, 277, 279
CsI 117(S)
$CsMoS_2$ 596
$CsNb_6I_{11}$ 598

Cu

Cu^{2+} 256
$[Cu(CN)_2]^-$ 194(S)
$[Cu(OH_2)_4(en)]^{2+}$ 220
$[Cu(OH_2)_4(NH_3)_2]^{2+}$ 220
$[Cu(OH_2)_6]^{2+}$ 217(S, E), 220, 455
$[Cu_2(CH_3CO_2)_4(OH_2)_2]$ 199(S)
$[CuH_3(IO_6)_2]^{4-}$ 415
Cu 232(P)
Cu^+ 244, 256
Cu_2O 101
$CuAu$ 575
Cu_3Au 114
$CuCl$ 122, 115(S)
CuF 421
CuH 275
CuH^+ 271
CuI 101
CuO 581
CuS 395
$CuZn$ 114, 115(S)
$YBa_2Cu_3O_7$ 588

e

e^- 417

Er

Er^{3+} 258
ErH_2 577
ErH_3 577

F

$[FHF]^-$ 291
$[H_2F]^+$ 292
ClF 408
F_2 404(E)
HF 65(E), 156, 273, 277(S), 278(S), 279, 290(P), 291
HOF 418(P)
$NaFeF_3$ 586
OF_2 387, 412(P)
O_2F_2 387, 412(P)

Fe

Fe^{2+} 252
Fe^{3+} 252
$[Fe_4(CH_3COOC)(CO)_{12}]^-$ 536(P, S)
$[Fe(CN)_6]^{3-}$ 213, 254, 417
$[Fe(CN)_6]^{4-}$ 254, 417, 487
$[Fe(CO)_4]^{2-}$ 514(P)
$[Fe(\eta^4\text{-}C_8H_8)(CO)_3]$ 526(P)
$[Fe(\eta^5\text{-}Cp)(\eta^1\text{-}C_3H_5)(CO)_2]$ 521(S)
$[Fe(NCS)(OH_2)_5]^{2+}$ 218
$[Fe(OH)(OH_2)_5]^{2+}$ 163(S)
$[Fe(OH)_2(OH_2)_4]^+$ 163(S)
$[Fe(OH_2)_6]^{2+}$ 254, 487
$[Fe(OH_2)_6]^{3+}$ 156(S), 163, 213, 254, 487
$[Fe(phen)_3]^{3+}$ 487
$[Fe_2(CO)_4(Cp)_2]$ 502(S)
$[Fe_2(CO)_9]$ 199(S), 511, 529, 530(S)
$[Fe_2(OH)_2(OH_2)_8]^{4+}$ 163(S)
$[Fe_2Cp_2(CO)_4]{\cdot}2AlEt_3$ 517(S)
$[Fe_3H(CO)_{11}]^-$ 535
$[Fe_4(C)(CO)_{12}]^{2-}$ 532, 536(P, S)
$[Fe_4(CO)_{13}]^{2-}$ 499(S)
$[Fe_4S_4(SCH_2Ph)_4]^{2-}$ 628, 629(S)
$[Fe_4S_4(SR)_4]^{3-}$ 199(S), 397
$[FeCl_4]^-$ 194(S)
$[FeHCp_2]^+$ 518(P)
Fe 113(S), 232
$Fe(CO)_5$ 47(S), 193(S), 502(S), 508(S), 508(P), 510
$Fe_2(CO)_{10}$ 91
Fe_2O_3 101, 232, 252, 581, 590
$Fe_3(CO)_{12}$ 535
Fe_3C 364
Fe_3N 593
Fe_3O_4 232, 462(S), 581, 583
$Fe_4Cp_4(CO)_4$ 532
$Fe_5C(CO)_{15}$ 532
$FeCl_3$ 424
$FeCl_3{\cdot}6H_2O$ 424
$FeCp_2$ 56(S), 252, 499(S), 597
$FeCpH(CO)_2$ 515
FeF_2 215
$FeH_2(CO)_4$ 515
FeO 101, 115(S), 394, 576, 581, 582(S)
FeS 101, 115(S)
FeS_2 396(S)
$FeTiH$ 276
$MgFe_2O_4$ 584
$MnFe_2O_4$ 584
$NaFeF_3$ 586

Ga

Ga^{3+} 257
$(C_2H_5)_2GaOCH_3$ 308(P)
$[(CH_3)_2Ga(en)]^+$ 319(S)
$[Cl_3GaGaCl_3]^{2-}$ 341
$[Ga(C_2H_5)_4]^-$ 318(P)
$[GaH_4]^-$ 286
$Ga(CH_3)_3$ 303(P), 305, 319
$Ga(C_2H_5)_3$ 308, 318(P)
Ga_2O_3 161, 340(S)
$GaAs$ 365
$GaCl_2$ 341
GaH_3 279, 286
GaN 122, 365
GaP 365
GaX 340(P), 341
GaX_2 339(S), 340(P)
GaX_3 339–40
$LiGaH_4$ 286(P)

Gd

Gd^{3+} 258
GdH_2 275, 577
GdH_3 275, 577

Ge

$[Ge(\eta^1\text{-}Cp)(CH_3)_3]$ 525
$[GeCl_6]^{2-}$ 129
$Cl_4GeNCCH_3$ 346
$Ge(CH_3)_4$ 305
Ge_9^{2-} 398(S)
$GeCl_4$ 346
GeH_4 273, 279, 294
GeH_4 277(S), 289(P)
R_2GeCR_2 326
R_2GeGeR_2 326

H

H^- 268, 274
H_2 53(S), 269, 271(P)
H_3 74(E)
H_3^+ 76, 269
H_3O^+ 146(S)
$H_9O_4^+$ 147(S)

Hf

HfH_2 275 577

Hg

$[CH_3Hg]^+$ 314
Ag_2HgI_4 579, 580(S)
$Hg(CH_3)_2$ 302(S), 303, 305, 312, 314
Hg_2Cl_2 199(S)
HgF_2 115(S)
HgR_2 314(P)
HgS 115(S), 446
HgX_2 219

I

$C_6H_6I_2$ 185
$H_4IO_6^-$ 413
HI 273, 277(S), 279
$I(OH)_5O$ 43(S), 159(S)
I_2 184, 404(E), 405(S), 406(P)
I_2^+ 410
I_2Cl_6 408, 409(S)
$I_2O(C_2H_5)_2$ 180
I_3^+ 410(S)
I_3^- 185, 410
I_4^{2-} 411(S)
I_5^- 410, 411(S)
I_7^- 411(S)
I_9^- 411(S)
ICl 408
ICl_3 408
IF 408
IF_3 408
IF_4^+ 410
IF_5 408, 409
IF_6^+ 410
IF_7 408, 409(S)
IO_4^- 129, 413, 416

In

In^{n+} 257
$InAs$ 366
InN 122
InX 341

Ir

$[Ir(Cl)_3(CO)(PPh_3)_2]$ 197(S)
$[Ir(CO)Cl(PPh_3)_2SO_2]$ 184(S)
$[IrCl(CO)(PPh_3)_2(NO)]$ 507
$[IrCl(CO)(PPh_3)_2]$ 270(S), 492, 503(S), 513
$[IrCl_6]^{2-}$ 129
$[IrX(CO)(PR_3)_2]$ 492

K

K_2O 115(S), 118(S)
K_2S 115(S)
KBr 115(S)
KC_8 362
KCl 123
KH 274(S), 279
$KMoS_2$ 596
$KPb_{2.5}$ 398
$KSnS_2$ 596
$KTiS_2$ 596
$KZrS_2$ 596

Kr

KrF_2 427

La

La^{3+} 258
$La(Ni_{0.5}Ir_{0.5})O_3$ 119(S)
$LaNi_5$ 276
$LaNi_5H_6$ 276
LaH_2 275
LaH_3 275

Li

$Li_2(C_6H_5)_2(en)_2$ 311(S)
Li_2O 115(S), 132, 575
Li_3N 280, 374
$Li_4(CH_3)_4$ 301(S), 303, 306
$LiAlH_4$ 280, 283, 286(P)
$LiCH_3$ *see* $Li_4(CH_3)_4$
$LiCl$ 115(S)
LiH 274, 279(P), 281
$LiMo_6S_8$ 598
$LiMoO_3$ 598
$LiNiPS_3$ 598
$LiTiS_2$ 596
$LiZn$ 114

Lu

Lu^{3+} 258
LuH_2 275, 577
LuH_3 275, 577

Mg

$MgBr(C_2H_5)(OEt_2)_2$ 313(S)
$Mg_4Cl_6(C_2H_5)_2(THF)_5$ 313(S)
$[Mg(CH_3)_2]_n$ 301(S)
CH_3MgBr 303(P), 315
$Mg(OH)_2$ 134
Mg_2Sn 114
$MgAl_2O_4$ 583(S)
$MgCO_3$ 131
MgF_2 115(S)
MgH_2 274, 279
MgO 115(S), 242, 600
MgR_2 313(P)
MgS 122
$MgTe$ 122
$MgZn_2$ 114
$RMgX$ 313(P)

Mn

Mn^{n+} 256
$(OC)_5MnOMn(CO)_5$ 91
$[(OC)_5MnRe(CO)_5]$ 516(P)
$[Mn(CO)_5]^-$ 514(P)
$[Mn(CO)_6]^+$ 506
$[Mn(NCS)_6]^{4-}$ 213
$[Mn(OH_2)_6][SO_4]$ 192(S)
$Mn(CO)_5$ 91
$Mn_2(CO)_{10}$ 459, 502(S), 511, 514, 516, 518, 532
Mn_3O_4 583
MnF_2 215
$MnFe_2O_4$ 584
$MnH(CO)_5$ 273, 511, 515(P), 518
MnO 215, 393, 394, 581, 582, 585
MnO_2 115(S)
MnO_4^- 129, 194(S), 243, 446, 447
MnS 115(S)

Mo

Mo^{n+} 257
$(CO)_3Mo(C_5H_5As)$ 328(S)
$[CH_2Mo_4O_{15}H]^{3-}$ 165(S)
$[Mo(CNR)_7]^{2+}$ 198
$[Mo(CO)_5PPh_3]$ 513
$[Mo(\eta^6\text{-}C_6H_6)(CO)_3]$ 500(S)
$[Mo(N_2)_2(dpe)_2]$ 628
$[Mo_2(O_2P(OPh)_2)_4]$ 494(S)
$[Mo_2(S_2)_6]^{2-}$ 397(S)
$[Mo_2Cl_8]^{4-}$ 529
$[MoCl_3(thf)_3]$ 628
$[MoS(S_4)_2]^{2-}$ 397(S)
$[MoS_4]^{2-}$ 397(P)
$[PMo_{12}O_{40}]^{3-}$ 164
$[S_2MoS_2FeS_2MoS_2]^{2-}$ 397(P)
$CsMoS_2$ 596
$KMoS_2$ 596
$LiMo_6S_8$ 598
$LiMoO_3$ 598
$Mo(CO)_6$ 516
Mo_2L_8 530
$Mo_4Ru_2Se_8$ 599
Mo_6Te_8 598
$MoCp(CO)_3H$ 515
MoO_4^{2-} 129, 447
MoS_2 196, 396(S), 595(S)
$NaMo_6S_8$ 598
$NaMoS_2$ 596
$NiMo_6Se_8$ 598
$RbMoS_2$ 596

N

BN 365
CN^- 407, 506
CNO^- 407
CNR 506
H_2NBH_2 338
HCN 43(S), 407
$HCNO$ 407
HN_3 382, 407
$HNCS$ 407
HNO_2 250, 381
HNO_3 159(S), 250, 377, 379(P)
$N(CH_3)_3$ 305
N_2 62(E), 371–2, 374–5, 506
N_2H_4 382(P, S)
N_2O 378, 381, 382(P)
$N_2O_2^{2-}$ 379
N_2O_3 378
N_2O_4 378
N_2O_5 378
N_3^- 382(P), 407
$N_3B_3H_{12}$ 338(S)
$NCCN$ 407
NCS^- 201, 407, 474, 478
$NCSSCN$ 407
NF_3 376
NF_4^+ 376
NH_2 89(S)
NH_2^- 155
$NH_2CH_2CH_2NH_2$ 202
NH_2Cl 382(P)
NH_2OH 249, 382(S), 383(P)
NH_3 40(E), 43(S), 80(E), 150, 155, 273, 277(S), 279, 289(P), 291, 379, 565(P)
NH_4^+ 147, 150, 155
NH_4F 115(S)
NH_4NO_3 249
NO 250, 378, 381, 513
NO^+ 379, 381(P), 506
NO_2 250, 378, 380
NO_2^+ 379
NO_2^- 40(E), 43(S), 201, 241, 379
NO_3^- 41(E), 43(S), 129, 379
NO_3Cl^{2-} 241
NSF_3 43(S)
O_2SNR_3 184
ONF 381(S)
ONI 381(P)
R_2NBCl_2 338(P, S)

Na

$Na[C_{10}H_8]$ 301, 311(P)
Na_2C_2 363(P)
Na_2O 115(S)
Na_2O_2 132
Na_2S 115(S)
Na_2S_2 395
Na_2S_7 395
Na_2Se 115(S)
Na_5Zn_{21} 114
$NaCl$ 98(E), 115(S)
$NaFeF_3$ 586
NaH 274, 279, 281
$NaMo_6S_8$ 598
$NaMoS_2$ 596
$NaTiS_2$ 596
$NaTl$ 114

Nb

$[Nb(NR_2)_5]$ 193(S)
$[Nb_2(Cl)_6(P(CH_3)_3)_4]$ 199(S)
$[Nb_6Cl_{12}]^{2+}$ 529(S)
$NaNbO_3$ 587
NbH 275, 577
NbH_2 275
NbO 577
NbO_4^{3-} 447
NbS_2 595, 596

Nd

NdH_2 275
NdH_3 275
$[Nd(OH_2)_9]^{3+}$ 198

Ni

Ni^{n+} 256
$[Ni(SO_4)(OH_2)_4$ 224
$[Ni(CN)_4]^{2-}$ 216(E), 255, 467
$[Ni(CN)_5]^{3-}$ 195(S), 457, 476
$[Ni(CO)_2(PR_3)_2]$ 345(S)
$[Ni(cod)_2]$ 521(S)
$[Ni(IO_6)]^-$ 415
$[Ni(NH_3)_6]^{2+}$ 219
$[Ni(OH_2)_6]^{2+}$ 219, 466, 478
$[Ni(P(OEt)_3)_4]$ 491
$[Ni(phen)(OH_2)_4]^{2+}$ 224
$[Ni(PR_3)_4]$ 512
$[Ni_2(Cp)_3]^+$ 527(S)
$[Ni_6(CO)_{12}]^{2-}$ 534(P)
$[NiBr_4]^{2-}$ 255
K_2NiF_4 589(S)
$LaNi_5$ 276
$LaNi_5H_6$ 276
$Ni(CO)_4$ 193(S), 194(S), 499(S), 502(S), 508(P), 511, 532

$NiAs$ 115, 118(S)
NiF_2 115(S), 215
$NiMo_6Se_8$ 598
NiO 115(S), 575, 579, 581, 582
NiS 115(S)

O

H_2O 43(S), 88(E), 147, 154–6, 273, 277(S), 279, 290(P), 291, and many other places
H_2O, ice 278(S)
H_2O_2 57(S), 390
O_2 62(E), 386(P), 389
O_2^+ 423(P)
O_2^- 62(E), 393
O_2^{2-} 62(E), 393
O_2F_2 412(P)
O_3 40(E), 43(S), 387(S)
OF_2 412(P)
OH^- 147

Os

$Os_3(CO)_{11}(PPh_3)$ 535
$Os_5(CO)_{16}$ 529(S), 532
$OsH_2(CO)_4$ 515
OsO_4 160, 194(S), 393

P

$((CH_3)_2PN)_3$ 385(S)
$((F_5C_2O)_2PN)_n$ 386
$(C_2H_5)_3PO$ 179
$(Cl_2PN)_3$ 386(S)
$(Cl_2PN)_n$ 386
$[P_4O_{12}]^{4-}$ 165(S)
BP 365
$Ca_5(PO_4)_3F$ 373
GaP 115(S)
$H_2PO_2^-$ 251, 385(S)
H_3PO_3 158(S), 159(S), 384(P)
HPO_3^{2-} 251, 385(S)
P 373(S)
$P(CH_3)_3$ 305
$P(OR)_3$ 385(P)
P_2 373
$P_2O_6^{4-}$ 385(S)
$P_2O_7^{4-}$ 165(S), 385(S)
P_4 373(S)
$P_4(NR)_6$ 385
P_4O_6 384(S)
P_4O_{10} 384(S)
P_4S_3 399(S)
PCl_3 376
PCl_5 43(S), 53(S), 377(S), 386
PCl_6^- 43(S)
PF_3 376(P), 507, 512
PF_5 376
PH_3 251, 273, 277(S), 278, 279, 384(P)
PO_4^{3-} 40(E), 129, 165, 251, 385(S)
$RPPR$ 328

Pb

Pb^{n+} 257
$(CH_3)_3PbH$ 289
$[PbCl_6]^{2-}$ 129
$Ba[PbBi]O_3$ 589
$KPb_{2.5}$ 398
Pb 334
$Pb(CH_3)_4$ 305, 326
$Pb(Cp)_2$ 326
PbH_4 289
Pb_3O_4 347, 461
Pb_5^{2-} 398
$PbCl_2$ 347
$PbCl_4$ 347
PbF_2 590
PbO 333, 347(S)
PbO_2 115(S), 347
PbS 333
$PbSO_4$ 347

Pd

$PdCl_2$ 422
PdH 275, 276
$[Pd(PPh_3)_4]$ 491, 492
Pd_4S 593

Po

Po 112(S)

Pr

Pr^{3+} 258, 452
PrH_2 275
PrH_3 275

Pt

$[(Pt(PCl_3)_2)_3(SnCl)_2]$ 199(S)
$[Pt(C_2O_4)(NH_3)_2]$ 194
$[Pt(Cl)(dien)]^+$ 204(S), 476
$[Pt(Cl)_2(NH_3)_2]$ 194(S)
$[Pt(Cl)_2(PEt_3)_2]$ 295
$[Pt(Cl)_2(py)_2]$ 223
$[Pt(CS_2)(PR_3)_2]$ 346(S)
$[Pt(NH_3)_4]^{2+}$ 193(S)
$[Pt(P(t\text{-}Bu)_3)_2]$ 512
$[Pt(SnCl_3)_5]^{3-}$ 476
$[Pt_2(\mu\text{-}P_2O_5H_2)_4]^{4-}$ 494(S)
$[PtBr(Mes)(PEt_3)_2]$ 476
$[PtCl(NO_2)(py)_2]$ 476
$[PtCl_2(NH_3)_2]$ 474(P)
$[PtCl_2(py)_2]$ 472
$[PtCl_3(C_2H_4)]^-$ 499(S), 503
$[PtCl_4]^{2-}$ 55(S), 194, 474, 491
$[PtCl_6]^{2-}$ 491
$K_2Pt(CN)_4Br_{0.3}\cdot 3H_2O$, KCP 97(S)
$Pt_4(O_2CCH_3)_8$ 532
$PtClH(PEt_3)_3$ 273
PtF_6 423
PtS_2 396(S)
$PtSn$ 115(S)

Pu

Pu^{n+} 259
PuF_6 412(P)
PuH_2 275
PuH_3 275

Rb

$RbAg_4I_5$ 579
RbH 274, 279
RbI 115(S)
$RbMoS_2$ 596
$RbSnS_2$ 596

Re

$[(OC)_5MnRe(CO)_5]$ 516(P)
$[Re(CN)_7]^{4-}$ 116(S)
$[Re(\eta^5\text{-}Cp)O_3]$ 504(S)
$[Re(S_2C_2R_2)_3]$ 196(S)
$[Re_2(CO)_{10}]$ 459(S), 508(P)
$[Re_2Cl_8]^{2-}$ 61(E), 199(S), 529, 535
$[Re_2Cl_{10}]^{4-}$ 459(S)
$[Re_3Cl_{12}]^{3-}$ 422(S)
$[Re_4(CO)_{16}]^{2-}$ 531
$[ReH_9]^{2-}$ 193(S), 198
$[ReO(Cl)_6]^{2-}$ 198
$Re_2(CO)_{10}$ 508(P)
$ReH(CO)_5$ 515
ReO_3 576(S), 584(S)
ReO_4^- 447

Rh

$[Rh(Cl)_2(NH_3)_4]^+$ 193(S)
$[Rh_2Cl_2L_4]$ 552(S)
$[Rh_6C(CO)_{15}]^{2-}$ 532
$[RhCl(CO)(PPh_3)_2]$ 551
$[RhCl(CS)(PR_3)_2]$ 346(S)
$[RhCl(PPh_3)_2]$ 551, 552(S)
$[RhCl(PPh_3)_3]$ 543, 547, 551
$[Rh(I)_2(CO)_2]^-$ 556, 557
$[Rh(I)_3(CO)]^{2-}$ 557
$[RhH(CO)(PPh_3)_3]$ 555
Rh_2L_{10} 530
$Rh_4(CO)_{12}$ 531
$RhCl_3$ 422

Rn

RnF_2 427

Ru

$[(bipy)_2ClRu(\mu\text{-}pyz)Ru(bipy)_2Cl]^{3+}$ 491
$[(NH_3)_5Ru(pyz)Ru(NH_3)_5]^{5+}$ 460(S)
$[(OC)_3Ru(\eta^2\text{-}cot)Ru(CO)_3]$ 522(S)
$[Ru(bipy)_3]^{2+}$ 204(S), 220(S), 448(S), 451, 493
$[Ru(C_4H_4)(CO)_3$ 522(S)
$[Ru(\eta^4\text{-}C_8H_8)(CO)_3]$ 522(S)
$[Ru(\eta^4\text{-}cot)(CO)_3]$ 525
$[Ru(NH_3)_5(OH_2)]^{2+}$ 375
$[Ru(NH_3)_5N_2]^{2+}$ 375(P, S), 628
$[RuCl(NH_3)_5]^{2+}$ 490
$[RuCl_6]^{2-}$ 210(E)
$Mo_4Ru_2Se_8$ 599
Ru_2L_{10} 530
$Ru_3(CO)_{11}(PPh_3)$ 535(S)
$Ru_3(CO)_{12}$ 508(P)
$Ru_6C(CO)_{17}$ 532
$RuCp_2$ 56(S)
$RuCpH(CO)_2$ 515
$RuH_2(CO)_4$ 515
RuO_2 406, 569

S

$(CH_3)_2SO$ 156
$(SN)_n$ 400(P, S)
$[O_3SONO]^-$ 381(S)
$[O_3SOOSO_3]^{2-}$ 391
$[SO_2F]^-$ 391(S)
CH_3SO_3H 158
CS 506
H_2S 235, 273, 277(S), 279, 291
$H_2S_2O_3$ 158(S)
$H_2S_2O_7$ 184
H_2SO_3 160
H_2SO_4 157(S), 184
$HNCS$ 407
HS^- 148
HSO_3^- 391(S)
HSO_3F 183
HSO_4^- 147
Na_2S_2 395
Na_2S_7 395
NCS^- 201
$NCSSCN$ 407
NSF_3 43(S)
$O_2S(NH_2)OH$ 158(S)
$O_2SF(OH)$ 158
O_2SNR_3 184
O_2SX_2 390
O_3SNH_3 180
O_3SNR_3 184
OSX_2 390
R_3NSO_2 391(S)
S 235
S^{2-} 148
S_2Cl_2 388(S), 399
S_2F_2 388(S)
S_2F_{10} 388(S)
S_2N_2 400(S)
$S_2O_3^{2-}$ 392(S)
$S_2O_4^{2-}$ 392(S)
$S_2O_6^{2-}$ 392(S)
S_3O_9 390(S)
S_4N_4 399(P, S)
$S_4N_4SO_3$ 400(S)
S_8 387(S)
S_8^{2+} 399(P, S)
SCl_2 388(S)
SF_4 45(E), 69, 388(S), 389
SF_6 43(S), 69, 81(E), 388(S), 389
SO_2 178, 390(S), 391(S), 507, 566(P)
SO_3 40(E), 43(S), 171, 390(S), 566
SO_3^{2-} 43(S), 391, 392(S)
SO_4^{2-} 40(E), 42(E), 43(S), 129, 147, 391, 392(S)
SOF_4 43(S)

Sb

$(SbF_5)_4$ 376(S)
$[SbF_5]^{2-}$ 376(S)
$[SbF_6]^-$ 172, 183
$RSbSbR$ 328
$Sb(CH_3)_3$ 305
$Sb(Ph)_5$ 43(S)
Sb_2O_5 161
$Sb_4(Co(CO)_3)_4$ 529(S)
$SbCl_5$ 178
SbF_5 172, 184, 376
SbH_3 277(S), 278, 279
SbO_4^{3-} 129
$TlSb$ 115(S)

Sc

Sc^{3+} 256
$ScCl$ 421
$ScCl_2$ 421
$ScCl_3$ 256
Sc_7Cl_{10} 421(S)
ScH_2 275
ScN 115(S)

Se

H_2Se 277(S), 279
H_2SeO_3 159(S), 392
$Mo_4Ru_2Se_8$ 599
$NiMo_6Se_8$ 598
Se_2F_{10} 388(S)
Se_4^{2+} 399(P, S)
Se_8 387
SeF_4 388
SeF_6 388(S)
SeO_4^{2-} 129, 392

Si

$((CH_3)_2HSi)_2N$ 322(P)
$((CH_3)_2SiO)_3$ 385(S)
$((CH_3)_3Si)_2NH$ 322(P)
$(C_6H_5)Si(OC_6H_4O)_2$ 183(S)
$(CH_3)_2SiCH_2$ 324
$(CH_3)_2SiCl_2$ 322
$(CH_3)_3SiCl$ 304(P), 322, 407(S)
$(CH_3)_3SiCN$ 407(S)
$(CH_3)_3SiOH$ 322(P)
$(CH_3)_3SiOSi(CH_3)_3$ 319
$(Mes)_2SiSi(Mes)_2$ 325(S)
$(Si(CH_3)_2)_4$ 323(S)
$(Si(CH_3)_2O)_n$ 322(P, S)
$(Si(Xyl)_2)_3$ 323(P)
$(Xyl)_2SiSi(Xyl)_2$ 324
$[CH_2Si(CH_3)_3]^-$ 320
$[O_3SiOSiO_3]^{6-}$ 348(S)
$[Si_4]^{4-}$ 364
$[Si_6(CH_3)_{12}]^-$ 324(P)
$[Si_6O_{18}]^{12-}$ 348(S)
$[SiF_6]^{2-}$ 129, 346
$[SiO_3^{2-}]_n$ 188
$[SiO_4]^{4-}$ 348(S)
$Al_2(OH)_2Si_4O_{10}$, pyrophilite 349
$Al_2(OH)_4Si_2O_5$, kaolinite 349
$CaAl_2Si_2O_8$, anorthite 7
$CaMgSi_2O_6$, diopside 7
$CaSiO_3$ 373
$CH_3CH_2SiH_3$ 289, 304(P)
K_4Si_4 364
$KAl_2(OH)_2Si_3AlO_{10}$, mica 349(S)
$KAlSi_3O_8$, orthoclase 349(S)
$Mg_3(OH)_2Si_4O_{10}$, talc 349(S)
$NaAlSi_3O_8$, albite 349
R_2SiSiR_2 325(E)
R_nSiX_{4-n} 320
Si 233, 346
$Si(CH_3)_3(C_6H_5)$ 321(P)
$Si(CH_3)_4$ 302(S), 303(P), 305, 320
$Si(OH)_4$ 157(S), 559
$Si(OR)_4$ 288
Si_3H_8 287
Si_4H_{10} 287
SiC 115, 233, 364
$SiCl_3(C_6H_5)$ 321
$SiCl_4$ 304, 346(P)
SiF_4 346
SiH_4 273, 277(S), 279, 280(P), 288
$SiHCl_3$ 288
SiO_2 187, 348, 591
SiX_4 183

Sn

$(SnR_2)_4$ 326(S)
$[(Pt(PCl_3)_2)_3(SnCl)_2]$ 199(S)
$[Pt(SnCl_3)_5]^{3-}$ 476
$[SnCl_4]^{2-}$ 194(S)
$[SnCl_6]^{2-}$ 129
$Cl_2SnN(CH_3)_3$ 180
$KSnS_2$ 596
Mg_2Sn 114
PtSn 115(S)
R_2SnCR_2 326
R_2SnSnR_2 326
R_3SnX 282
$RbSnS_2$ 596
Sn 113(S), 387(S)
$Sn(CH_3)_4$ 305
$Sn(Cp)_2$ 326(S)
Sn_9^{4-} 398(S)
SnAs 115(S)
$SnCl_2$ 171
SnH_4 273, 277(S), 279, 289(P)
SnO_2 161

Sr

$SrCO_3$ 131
SrH_2 274, 279
$SrTiO_3$ 115(S)

Ta

$[Ta(Cl)_4(PR_3)_3]$ 198
$[Ta_6Cl_{12}]^{3+}$ 461
$KTaO_3$ 587
$NaTaO_3$ 587
Ta_6Cl_{15} 461(S)
TaH 275
TaI_4 594(P)
TaO_4^{3-} 447
TaS_2 594, 595(S), 596

Tc

TcO_4^- 447

Te

$[OTeF_5]^-$ 390
$[TeCl_6]^{2-}$ 44
CdTe 367
$F_5TeOTeF_5$ 390
H_2Te 277(S), 279
MgTe 122
Mo_6Te_8 598
$Te(OH)_6$ 159(S)
Te_2Br 388(S)
Te_2I 388(S)
TeF_4 388
TeF_6 388(S)
TeI 388(S)
TeO_4^{2-} 129

Th

Th^{n+} 258–9
$[Th(ox)_4(OH_2)_2]^{4-}$ 198

Ti

Ti^{n+} 256
$[Ti(CO)_6]^{2-}$ 506
$[Ti(\eta^1\text{-}Cp)_2(\eta^5\text{-}Cp)_2]$ 500(S)
$[Ti(OH_2)_6]^{3+}$ 211
$[TiBr_6]^{2-}$ 129
$[TiCl_6]^{2-}$ 129
$BaTiO_3$ 587
$CaTiO_3$ 586
FeTiH 276
$KTiS_2$ 596
$LiTiS_2$ 596
$NaTiS_2$ 596
Ti_2O_3 590
Ti_2S 593
$TiCl_4$ 256
$TiCp_2$ 528
TiH_2 577
TiO 115(S), 215, 394, 574, 577, 581, 583
TiO_2 115(S), 118(S), 161

Tl

Tl^{n+} 257
$[Tl(CH_3)_2]^+$ 318
NaTl 114
$Tl(CH_3)_3$ 305
TlSb 115(S)
TlX 341
TlX_3 341

U

U^{n+} 258–9
$U(\eta^8\text{-cot})_2$ 499(S)
UC 115(S)
UH_3 275
UO_2 115(S)

V

V^{n+} 256
$[V(CO)_6]^-$ 506
$[V(OH_2)_6]^{2+}$ 482
$[VCl_4]^{2-}$ 193(S)
$[V_nO_m]^-$ 164
$[VO(OH_2)_5]^{2+}$ 193(S)
$[VO_2(OH_2)_4]^{2+}$ 164(S)
$[VO_2]^{2+}$ 164(S)
$[VO_4]^{3-}$ 164, 194(S)
$V(CO)_6$ 503
VH 275
VH_2 275
V_2O 593
V_2O_3 590
V_2O_5 161, 164
VCl_4 215
VF_5 256
VO 215, 575, 577, 581
VO_2 577, 590
VO_4^{3-} 447
VS_2 596, 597(P)

W

W^{n+} 257
$[(OC)_6WHW(CO)_6]^-$ 291
$[OWO_4]^{4-}$ 165(S)
$[W(CO)_3(H_2)(P(i\text{-}Pr)_3)_2]$ 270(S)
$[W(\eta^3\text{-}Cp)(\eta^5\text{-}Cp)(CO)_2]$ 500(S)
$[W(N_2)_2(P(CH_3)_2(C_6H_5))_4]$ 375
$[W_6O_{19}]^{2-}$ 164–5(S)
HWO_3 598
$W(CH_3)_6$ 503
$W(CO)_2CpH(PMe_3)$ 515
$W(CO)_4CS$ 458
$W(CO)_6$ 467, 517
WC 364
$WCp(CO)_3H$ 515
WO_2 115(S)
WO_3 576
WO_4^{2-} 447
WS_2 196

Xe

$FXeN(SO_2F)_2$ 427(S)
XeF^+ 427
XeF_2 426(P, S), 427
XeF_4 43(S), 426(P, S)
XeF_6 426(P, S)
XeO_3 427(P)
XeO_6^{4-} 428(S)

Y

$YBa_2Cu_3O_7$ 588
YH_2 275
YH_3 275
YS 577

Zn

$[ZnCl_4]^{2-}$ 194(S)
$CaZn_2$ 114
$CH_3Zn(Cp)$ 314(S)
CuZn 114, 115(S)
LiZn 114(S)
$MgZn_2$ 114
Na_5Zn_{21} 114
Zn 112(S)
$Zn(CH_3)_2$ 302(S), 305, 306, 314(P)
$Zn(CH_3)_2(bipy)$ 314(S)
ZnF_2 215
$ZnFe_2O_4$ 584
ZnH_2 275
ZnO 101, 115(S), 122, 231, 581
ZnS 115(S), 593
ZnS, sphalerite 117(S)
ZnS, wurtzite 118(S)

Zr

$[Zr(CH_3CO)(Cp)_2CH_3]$ 528(P, S)
$[ZrCl_6]^{2-}$ 129
$[ZrF_7]^{3-}$ 198
$KZrS_2$ 596
ZrCl 421(P, S)
$ZrCl_2(\eta^5\text{-}Cp)_2$ 503
$ZrCl_4$ 423(P)
ZrH_2 275, 577
ZrI_2 422(S)
ZrO_2 424, 575, 590
ZrS 577

General index

T denotes that an entry refers to a table of data. For specific substances, also refer to the *Formula index*.

A, Racah parameter 440
absolute configuration 205
CD spectrum 452–4
acceptor
band 101
number 179
acetic acid, acidity window 155
acetylide 362
acid
Brønsted 146
Lewis 170
strength
bond enthalpy 152
Brønsted 147–50
effect of solvation 152
electron affinity 152
electronegativity 152
electrostatic model 157
gas phase 151
hydrogen bonding 152
influencing factors 150–4
ionization energy 152
Lewis 173
oxidation number 161
periodic trends 161
permittivity 154
qualitative analysis 162
acid–base discrimination, of nonaqueous solvents 155
acidic oxides 159, 160
acidity constant 149, 150(T)
actinides, trends in redox stability 258, 259
activation
associative and dissociative 468
enthalpy, octahedral substitution 481(T)
volume, octahedral substitution 481(T)
adduct formation 171
adenosine triphosphate 166, 607
ADP 166, 607
alane (AlH_3) complexes 287
alkali metal
hydrides 274
organometallics 310–12
oxides 392–3
peroxides 393
superoxides 393
alkaline earth organometallics 312–13
alkyl ligands 518, 519
alkylidene ligands 519
allotropes 373
alloys
interstitial 113, 114
substitutional 113, 114
alumina 340, 341, 559–60, 579
surface acidity and basicity 186, 560
aluminosilicates 349–53
surface acidity 186
aluminum 333(T)
halide acidity 182, 340
halides 339
hydrides 286
organometallics 316–18
polycations 162
production 233
ambident base 175
amino acids 605, 606(T)
aminoboranes 338
aminosulfuric acid 158
ammonia 146
boranes 337
molecular orbitals of 79, 80
solvent leveling 154, 155
synthesis 289, 565
amphoteric oxides 161, 162
amphoterism, periodic trends 161
anhydrous metal halides 423
antibonding orbitals 59, 60
octahedral complexes 206
antiferromagnetism 585, 586
antifluorite structure 118
antimony 370(T)
conductivity 371
halides 376
pentafluoride
Lewis acidity 183
structure 374
A_p, proton affinity 151
apatite 373, 610
aqua
acids 156
polymerization 162
trends 157
ion exchange lifetimes 222
arachno boranes 355, 358
argon 403(T), 425
arsenic 370(T)
conductivity 371
halides 376
organometallics 327–8
structure 374
associative mechanism
square planar complexes 476
substitution 223, 467, 468
astatine 403(T)
asymmetric unit 108
atmophile 8
atom
and ion diffusion 577–80
transfer redox 241, 486
atomic
energy levels 11
orbitals 15–22
ATP 165, 607
autoprotolysis constant 149, 154
azide ion 381, 382

B, Racah parameter 440
B–N amino acid analogs 338
β, nephelauxetic parameter 445
β-alumina 579
β-hydrogen elimination 309, 518
Bailar twist 484, 485
band
gap 93
Group 14/IV solids 365
Group III/V compounds 365
structure
gallium arsenide 365
of extrinsic semiconductors 100
intercalation compounds 595
of intrinsic semiconductors 99
of linear chains 92–4
oxides 583
silicon 366
theory of solids 92–101
barium peroxide 393
Berry pseudorotation 195, 457–8
base
Brønsted 146
hydrolysis, octahedral substitution 483
Lewis 170
basic oxides 161–2
basis set 58
minimal 60
bauxite 234
bidentate ligands 200
biomolecules, classification 607
bipyridine ligand 220
bis(cyclopentadienyl)beryllium 312
bismuth 370(T)
cluster anions 398
conductivity 371
inert pair effect 257
structure 374
blast furnace 232, 322–3
body-centered cubic (bcc) 112
Bohr
frequency condition 13
magneton 213
bond
correlations
order/length 69
order/strength 69
strength/length 70
dissociation enthalpy 68
enthalpy
mean 68, 69(T)
p-block hydrogen compounds 280
p-block methyl compounds 305(T)
length 63
of diatomics 67(T)
order 67
bonding orbitals 59
octahedral complexes 206
boranes
see boron, hydrides 283–5, 353–7
borate esters 315, 336
borates 336
borazine 338
boric acid 336
borides 361
Born equation 154
Born–Haber cycle 123, 124
Born–Mayer equation 127
borohydride (tetrahydroborate ion) 285
boron 333(T)
halide acidity 182
halides 334–6
hydrides 283, 353–9
bonding 354–9
classification 354–5
correlation of structural types 358
electron count 354
MO description 357, 359
reactions 284–5
structures 356(T), 358(T)
synthesis 283
nitride 337
organometallics 315–6
production 333
structure 334
subhalides 353
trichloride
hydrolysis 335
Lewis acidity 182, 335
trifluoride
physical properties 335
preparation 334
trihalides
Lewis acidity 182, 335, 340
boron-containing aminoacids 338
boron-group elements
properties 332, 333(T)
boron–nitrogen chemistry 337–9
borosilicate glass 348
bridged complex, redox 490
bromine 403(T)
elemental 403–5
halides 408
oxoanions 416–20
pentafluoride 409
polyhalides 411
production 406
trifluoride, autoionization 410
bromite, disproportionation 419
Brønsted
acid 146
acidity 149
periodic trends 156–7
of surfaces 186
base 146
basicity of surfaces 186
equilibria 148
building-up principle
atoms 21
diatomics 62

C, acid–base parameter 180
C, Racah parameter 440
C—H bond cleavage 528
cadmium telluride, photoconductivity 367
cage compounds 529
calcite 610
calcium
biochemistry 609–10
fluoride, structure of 118
cannula 306
carbanion 308

carbene
Fischer 516, 519
Schrock 519–20
carbides 361–4
classification 362
interstitial 363–4
metallic 363–4
structure 362–4
synthesis 362–3
carbon 333(T)
activated 343–4
black 343–4
diamond 342, 343
dioxide 344–6
in natural waters 253–4
fibres 343–4
graphite 342, 343
monosulfide ligand 506
monoxide 344–6
bridging 506
dipole moment of 66
electronic configuration 66
ligand 505–6
metal bonds 505–6
molecular orbital energies of 66
molecular orbitals 505
reduction of ores 344
partially crystalline 343
reduction of ores 229–33
suboxide 344
carbon-group elements
properties of 332, 333
carbonic acid 158–61
acid strength 159
carbonyl
complexes 504–17
anionic 514–15
basicity 515–16
CO reactions 516–17
IR spectra 509(T), 510
ligand substitution 511–14
migratory insertion 517
oxidation 516
properties 511
reduction 514–5
structure 508–10
synthesis 507, 508
hydrides, p*K* 515(T)
carbonylation, reductive 509
carboranes 357–61
carboxypeptidase 617–20
carrier
scattering 96
velocity 365
catalyst
activation 562–3
definition 542
heterogeneous 558–70
homogeneous 546–58
metal surfaces 561–2
multiphase 558
poisons 543
precursors 543
selectivity 544
shape selective 567
surface acids 560
surface area and porosity 559
surface bases 560
uniform 558
catalytic
cycles 543–4
mechanism
acetic acid synthesis 556–7
alkene hydrogenation 551–3, 565
ammonia synthesis 565, 566
determination 545(T)
enantioselective 553
hydroformylation 553–5
m-xylene isomerization 567
SO_2 oxidation 566
Wacker process 557, 558
catalytic
processes 542(T), 550(T)
steps
heterogeneous 562–4
homogeneous 546–9
catenation 323–4
CD spectrum 452–4
center of inversion 50–2
ceramics 591
cerium(IV) stability in water 243
cesium
chloride structure 117
oxides 393
suboxides 393
chabazite 351(T)
chalocogens 386–400
Frost diagrams 390
chalcophile
definition 7
soft acid character 175
character tables 82–5, 664–70(T)
characters, group theory 83
charge
density waves 595–6
effects, octahedral substitution 480
transfer 185, 441, 446–8
transfer bands 446–8
chelate 202–3
effect 220
chemical
shift 294
typical values 273
vapor
deposition 366
transport 593–4
chemisorption 563
on metals 561
of simple molecules 563(T)
Chevrel phases
intercalation 598
structure 598
superconductivity 599
chiral
complexes 203–5
molecules
and symmetry elements 57
chloralkali cell 406
chlorate
disproportionation 419
chlorine 403(T)
dioxide 412, 413(T)
elemental 404–5
fluoride 408(T), 409
monofluoride 408
oxides 412
oxoacids 413
oxoanions 413–20, 413(T)
production 406
trifluoride
reactions 410
structure 408
chlorite disproportionation 419
chlorophyll a_1 429–31
chloroplasts 629
chromium
complexes
CT spectra 446–8
ligand field stabilization 209
luminescence 448
hexacarbonyl 502, 508(S)
circular dichroism 452–4
class A and B (hard and soft) 173–6
clathrate hydrates 291, 292
Claus process 235
clay 349
cleavage
heterolytic 281–2
homolytic 281–2
symmetrical and
unsymmetrical 285
Clebsch–Gordan series 462–3
close-packed structures 109–12
closo boranes 355–8
clusters
d-block 422, 528–36
p-block 397–400
cobalt complexes
ligand
field splitting 211(T)
stabilization 209(T)
spectrochemical series 212
tetrahedral 215(T)
cobalt(III) stability in water 243
color centers 575
complex 170
formation 171
formation and standard
potentials 254, 255
nomenclature 200–3
square pyramidal 457–9
trigonal bipyramidal 457–9
comproportionation 245
condensation
processes 7
redox 534
cone angle 512
congruent melting 601
conjugate acid and base 148
Cooper pair 101–2
coordination
compounds 192–8, 443–94
see metal complexes 192
equilibria 218–22
number 109
influence on radius 111
metal complexes 192, 194–8,
443–94
sphere 192
coordinative unsaturation 547
copper production 232
copper(I) disproportionation 244–5
correlation
atoms with complexes 443
diagram
D_{3h}–C_{4v} complexes 458
d^1 444
d^2 444
terms 443
corundum structure 590
coupling
jj 435
Russell–Saunders 435–40
covalent radius 68(T)
Creutz–Taube ion 460
cryolite 234
crystal lattice, definition 108
crystallographic shear plane 576
cubic close-packed (ccp) 110
cyanide 346
ligand 506
cyanogen 407
cyclam 203
cycle, thermodynamic 135–6
cyclopentadienyl complexes
reactions 526
synthesis 526–7
cytochrome P-450 625, 626
cytochromes 621, 622
cytoplasm, elements present 604

d isomer 205
d
band 94
block complexes
electronic states 437–41
ligand field transitions 441–46
metal–metal bonded 198–9,
422, 492, 528–36
mixed valence 460–2
metal
clusters 422, 494, 528–36
halides 420–4
hexahalides 422–3
oxides 393, 394
subhalides 421
sulfides 395–6
trends in redox 255–7, 256(T)
orbital 19–21
substitution 217–24, 466–83
d–d transitions 441–6
electron repulsion 442–3
d^8 complexes, structure 195
d^9 complexes, structure 195–6
Δ isomer 205
CD spectrum 454
Δ_O 208, 211(T), 481(T)
Δ_T 215(T)
$\Delta G^‡$ 470
$\Delta H^‡$ 470, 476(*T*), 481(*T*)
square-planar substitution 476(T)
$\Delta S^‡$ 470–1
square-planar substitution 476(T)
$\Delta V^‡$ 470–1, 481(T)
octahedral substitution 481(T)
square planar substitution 475–6,
476(T)
δ
bonds 529
orbital 62
$\delta^* \leftarrow \delta$ transition 459
de Broglie relation 9–10
debye unit, D 66
decomposition temperature, of
salts 130
defects 573–6
atom interchange 575
extended 573, 576
extrinsic 573, 575
intrinsic 573–5
and nonstoichiometry 582
point 573–5
delayed reactions, photochemical 493
deltahedra 354
density of states 96
deuterium 267(T)
abundance 266
isotope effects 266–7
Dewar–Chatt model 520
diamagnetic complexes 212
diamagnetism 212
diamond 342, 343
structure of 113, 342
diastereomers 205
diatomic molecules
heteronuclear 64–6
homonuclear 60–4
diborane
reactions 284–5
synthesis 283
dichromate stability in water 243
dicobalt octacarbonyl 502(T)
dielectric constant 179(T)

diffusion
  coefficient 577–8
  controlled reaction 147
  equation 577
  mechanisms 578
dihydrogen complexes 270
diiodonium cation 410
dimanganese decacarbonyl 502(T)
dimethylsulfoxide, acidity window 156
dimethylberyllium 313
dimethylmercury 314
dioxygen difluoride 412
diphosphate 165
diprotic acid 150
disilenes 325
displacement reaction 172
disproportionation 244
  Latimer diagram 246–8
dissociation
  constant, metal complex 218
  energy 63
dissociative substitution 224, 467
disulfides
  band structure 595
  structure 594
domain, magnetic 585
donor
  band 100
  number 179
Drago–Wayland
  equation 180, 181
  parameters 180(T)

E acid–base parameter 180
effective
  nuclear charge 22
  proton affinity, A_p' 153
Eigen–Wilkins mechanism 477
Eigen–Fuoss equation 479
eight-coordinate complexes 198
eighteen-electron rule 206, 500–4
electric dipole transition 449
electrical conductivity
  solids 98–102
  temperature dependence 92, 95
electrocatalysis 568, 569
electrochemical series 237, 238
electrointercalation 597
electrolytic reduction 233–4
electromagnetic spectrum 13(T)
electron
  affinity 30(T)
  configurations
    atoms 23–6, 635–6(T)
    complexes 209–10
  gain enthalpy 30
  pair donor and acceptor 170
  paramagnetic resonance 454–6
  probability density 14
  properties 4
  spin 16
  transfer 229, 240
    long range 624–5
electron-deficient compound 276
  bonding in 81
electron–electron repulsion 435
electron–precise compound 276
electron–rich compound 276
electronegativity 31–33, 640–1(T)
  Allred–Rochow 32–3, 641(T)
  Mulliken 31, 641(T)
  optical 447(T)
  Pauling 31(T), 641(T)
  spectroscopic 640–1(T)
electronic
  configuration
    complexes 434
    from magnetic moment 213
  spectra of metal complexes 210, 211
  transitions in atoms 12
electrophile 170
electrophilicity 223
electrostatic model, acidity 150
electrostatic parameter, ξ
  definition 130
  hard Lewis acids 174
  ion acidity 157
  and solvation 154
  stabilities of solids 131
electrostriction 470
elements
  abundance 4, 5
  biosphere 604(T)
  cosmic distribution 5
  distribution in cells 604(T)
  extracted by oxidation 234
  Goldschmidt classification 8
  nonvolatile 6
  nucleosynthesis 5–6
  terrestrial abundance 5, 6
  terrestrial distribution 7
  volatile 6
elimination reactions 547
Ellingham diagrams 229–32
enantiomer 57, 203
enantioselectivity 553
encounter complex, octahedral substitution 477
endothermic reactions 135
energetics of catalysis 544
energy levels
  harmonic oscillator 12
  hydrogenic atom 12
  particle in a box 11
entering group 466
  square-planar substitution 471–3
enthalpy 135
  of formation
    p-block methyl compounds 305(T)
entropy 136
  gas evolution 231
  principles of 136
  spontaneous processes 136
  standard molar 136
enzymes 605
  acid catalysis 615
epitaxial growth 366
EPR 49, 324, 454–6
equilibrium constant
  Gibbs free energy 138
ESR, *see* EPR
ethylenediamine complexes, stereochemistry 202
ethylenediaminetetraacetato ligand 202
eutectic point 600
eversion 608
EXAFS 627–8
exchange current density 568
exothermic reactions 135
extended defects 573
extracellular, elements present 604
extraction of the elements 229–35

F center 575
f
  block
    complexes
      coordination number 198
      spectra 451, 452
    metal
      trends in redox stability 255–9
  orbital 20
  organometallics 527
face-centered cubic 110
facial, *fac*, isomers 197
feldspar 349
Fermi
  energy 95, 99
  level 94–6
Fermi–Dirac distribution 94–5, 99
ferredoxins 628
ferrimagnetism 586
ferrites
  antiferromagnetic 584
  ferromagnetic 584
ferrocene 499
  bonding 523–5
ferroelectrics 586
ferromagnetism 585
ferromagnets
  hard 585
  soft 585
ferrous chloride, anhydrous 424
Fischer carbenes 519
five-coordinate complexes 195
fluorescence 450
fluorescent states 451
fluoride ionic conductors 590
fluorine 403(T)
  F—F bond enthalpy 404
  halogen compounds 408–11
  production 405
fluorite 590
  structure of 118
fluorosulfuric acid 158
fluxional molecules 47–9, 525
force constant 12
formal charge 39
formation constant
  and hardness 221
  and ionic parameter, ξ 221
  metal complex 218
  overall, β 218
  stepwise, K_n 218
formation constants
  Irving–Williams series 222
  statistical effect 219
  stepwise 219
four-coordinate complexes 194
fractional oxygen saturation 614
framework
  electrolyte 580
  representation 352
Franck–Condon principle 487
free energy 137, 230
  see Gibbs energy
Frenkel defect 574
frontier orbitals 63
  Lewis acid hardness 174
  octahedral complexes 207
Frost diagram 248–50
  actinides 259
  chromium-group metals 257
  comproportionation 249
  construction 249
  d-block metals 256
  disproportionation 249
  manganese 251
  nitrogen 248
  oxygen 249
  p-block metals 257
  and solute stability 250
  thermodynamic prediction 249
Fuoss–Eigen equation 479

g-factor, EPR 454–6
gallium 333(T)
  arsenide 366
    semiconducting properties 365
  halides 339, 340
  hydrides 286–7
  organometallics 318–9
  structure 334
galvanic cell 236
gap
  band 93, 365–6
  hydride 275
  van der Waals 342
gas-phase acidity 151
  thermodynamic cycle 152
germane 289
germanium 333(T)
  cluster anions 398
  halides 346
  organometallics 326
  semiconducting properties 365
Gibbs energy
  formation 137
    p-block hydrogen compounds 279(T)
  metal oxide reduction 230
  and spontaneous processes 137
  standard 138
  variation with temperature 231
glass 348, 591
  modifiers 592
  transition temperature 592
glove box, inert-atmosphere 306–7
gold production 235
Goldschmidt classification 8
  hard soft interpretation 175
graphite 342
  bisulfates 343
  intercalation compounds 343
greenhouse effect 345
Grignard reagents 313
Grotthus mechanism 147
ground term 439
Group 15/V elements
  Frost diagram 377
  oxides 377–82
  properties 370, 371
Group 16/VI elements
  properties 370, 371

H_3, molecular orbitals 74–7
Haber process 289
Hägg's rule 364
half-reaction 236
halide oxidation 234
Hall–Hérault process 233
halogen oxoanions
  oxo transfer 417–7
  redox mechanisms 417
  redox rates 416
halogen–element bond enthalpies 405
halogen–halogen bond lengths 405
halogens 403(T)
  aqueous redox chemistry 415–20
  atomic properties 403(T)
  bond enthalpies 404
  Frost diagrams 416
  intermolecular interactions 404
  Lewis acidity 184, 185
  occurrence and production 405–6
  polyanions 410–11

halogens 403(T) (*Cont.*)
  polycations 410
  redox disproportionation 418
hapticity 500
hard acids and bases 173–4, 174(T)
hard and soft atoms 33
hardness
  chemical consequences 175
  definition 174
harmonic oscillator
  energy levels 12
  force constant 12
  zero-point energy 12
helicity of light 452
helium 403(T), 425
  cycle 5
helium-II 425
hemoglobin
  O_2 binding 614
Hermann–Mauguin system 664
heterogeneous catalysts 542
heterolytic cleavage 282
heteronuclear diatomic molecules 58
heteropolyanions 164
hexagonal close packed (hcp) 109
hexamethyldisiloxane 319
hexaaquaaluminum(III)
      polymerization 163
hexaaquairon(III) polymerization 163
high
  coordination numbers 198
  spin complexes 210
highest occupied MO 62
holes
  octahedral 110–1
  tetrahedral 111
  in valence band 101
HOMO 62
homogeneous
  catalysis 542, 546–58
  catalytic processes 550(T)
homolytic cleavage 281
homonuclear diatomic molecules 58
Hund's rules 24, 439
hybrid orbitals, shapes of 47(T)
hybridization 45–7, 47(T)
hydration enthalpy 152
  $3d$ ions 214
hydrazine 382, 383
hydrazoic acid 382
hydrides
  metallic 273, 275
  molecular 273
  saline 272, 274
hydridic character 281
hydroboration 285, 304
hydrocarbonylation 554
hydroformylation 553–6
hydrogen 267(T)
  atoms and ions 268
  bonding 278, 291–2
    influence on solvation 153
  chloride, synthesis 290
  compounds
    classification 272, 276
    electron deficient 276
    electron precise 276
    electron rich 276
    NMR spectra 273
    protic character 282
    radial reactions 281, 282
    reaction types 281
    stability 279
    structures of 278
    synthesis 279–81
  cyanide 346
  diffusion in metals 276
  evolution, Nernst equation 239
  fluoride
    autoionization 292
    molecular orbital energies of 65
    synthesis 290
  halides
    acidity 150
  heterolytic dissociation 270
  homolytic dissociation 270
  industrial production 271
  IR of H-containing molecules 267
  isotope effects 267
  isotopes 266
  ligand 518
  mechanism of combustion 290
  molecular properties 269
  peroxide, catalytic
      decomposition 390
  production 271
  purification 276
  radical reactions 270
  see also deuterium and
      tritium 266
  selenide, Se—H stretching
      frequencies 282
  sulfide
    reaction with metal ions 292
    S—H stretching frequencies 282
hydrogenic atom
  energy levels 13
  structure 12
hydrometallurgical extraction 232
hydronium ion 146
hydrosilation 289, 304
hydroxoacids 157
hydroxylamine 382, 383
hyperfine splitting constant 456
hyperfine structure 455
hypervalence 42, 81
hypochlorite, disproportionation 418
hypochlorous acid
  disproportionation 245
  redox reactions 418
hypofluorous acid, preparation 418
hypophosphite 385

I substitution 467
I_a mechanism 469, 481
I_d mechanism 469
improper rotation axis 52
incongruent melting 601
indium 333(T)
  halides 341
  organometallics 319
  properties 333–4
inert
  complexes 222
  gases, see noble gases 424
  pair effect 257
inert-atmosphere techniques 306–7
infrared spectra
  hydrogen compounds 267–8
inner-sphere
  complex 192
  electron transfer 241
  redox mechanism 486, 490
insertion reactions 547, 596
insulators 92
  band model of 98
intensities
  electronic transitions 448
intercalation
  graphite 363
  metal disulfides 395(T), 593–7
interchange substitution 467
interelectron repulsion 435
interhalogen cations 410(T)
interhalogens 408–10, 408(T)
intermetallic compounds 114
International System 664
intersystem crossing 451
intervalence transitions 461
intimate mechanism 466–73
  octahedral substitution 479
  tests 469–71
inverse spinel structure 583
inversion center 51
iodine 403(T)
  elemental 404–5
  heptafluoride 408, 409
  monobromide 408
  monochloride 408
  production 406
  trichloride, structure 408
ion pumps 608, 609
ionic
  model 119
  radius 120(T)
    definition of 119
    and solubility 133
    and stability of solids 131, 132
    trends in 120
  solids 115
ionization
  energy of atom 28, 29(T), 638(T)
  enthalpy of atoms 28
iron
  complexes 209
  ligand field parameters 208
  pentacarbonyl 508
    vibrational spectra 510
  polycations 162
  production 232
  pyrite 396
  sulfur clusters 628, 629
    proteins 628
iron–oxygen phase diagram 581
Irving–Williams series 222
isolobal analogies 90–92, 534
isolobal groups 533(T)
isomerism
  geometric 196–8
  geometrical 192
  optical 203, 204
isomerization
  octahedral complexes 484
  trigonal bipyramid–square
      pyramid 484
isonitrite ligand 506
isopolyanions 164
isopolymetallates 164

jj coupling 435
Jahn–Teller
  distortions 216, 217
  dynamic 217
  theorem 216

K_a 149, 173
K_e, encounter equilibrium 477
K_f 173
K_w 149
kaolinite 349
Kapustinskii equation 129
KCP 97
kinetic factors, redox 239–42
krypton 403(T), 425
  difluoride 427

L isomer 205
l, orbital angular momentum 436
Λ isomer 205
  CD spectrum 454
labile complexes 222
lanthanide
  contraction 28
  organometallics 527–8
  spectra 451–2
  trends in redox stability 258
Laporte selection rule 449
Latimer diagrams 246–8, 643(T)
  basic solution 250
  conditional 250
  disproportionation 247
lattice enthalpy 123–9
  Born–Haber cycle 123
  Born–Mayer equation 127
  Coulomb energy 124
  dispersion forces 126
  Kapustinskii equation 129
  measured and calculated 128(T)
  overlap repulsion 126
  stability of solids 129
  van der Waals forces 126
layered sulfides 593
LCAO 58
lead 333(T)
  cluster anions 397, 398
  halides 347
  inert pair effect 257
  oxides 347
  production 333
  properties 333(T)
  storage battery 347
leaving group 466
  square planar substitution 471
  octahedral substitution 477
leveling effect 155
LED 366
Lewis acid–base
  displacement 172
  interaction 171
  metathesis 172
  MO description 171
Lewis
  acidity
    influence of solvent 178, 179
    of solvents 172
    thermodynamic correlations 180
  acids
    examples 170, 171, 181–3
    hard 173–5
    molten 188
    solid 188
    steric factors 176–7
    strength 172–7
    structural change 176
    surface 186
  bases 170
    hard 173–5
    steric factors 176, 177
    strength 172–7
    surface 186
  basicity
    influence of solvent 178, 179
    of solvents 172
    thermodynamic correlations 180
  structures 38–41
lifetime broadening 47
ligand
  cone angle 512
  field
    activation energy 480
    lattice enthalpy 215
    octahedral complexes 211(T)

octahedral substitution 481(T)
splitting parameter 208
tetrahedral complexes 215
stabilization energies 209(T)
stereochemistry 202
strength
ligands 211
metals 212
strong 443
substitution 224
mechanisms 217–25, 466–86
octahedral complexes 208
tetrahedral complexes 194, 209(T), 215(T)
thermochemical correlation 214
transitions 441
weak 443
ligands
alkene 520
alkyl 518
alkylidene 519
alkylidyne 520
alkyne 521
allyl 521
arene 522
bite angle 202
carbene 519
chelating 202
cyclic polyene 522
cyclobutadiene 522
diene 521
dihydrogen 270–1
hydrogen 518
macrocyclic 202
nomenclature 200–1, 200(T)
organic 501(T)
steric repulsion 512
typical 200(T)
light emitting diodes 366
lime 146
linear
combination of atomic orbitals 58–9, 74–80
free energy relation
octahedral substitution 479
liquid ammonia, acidity window 155
lithophile 8
hard acid character 175
LMCT transitions 446
localized and delocalized bonding 90(T)
low spin complexes 210
low-temperature synthesis of solids 580
lowest unoccupied MO 63
luminescence 450
LUMO 63

m_l, orbital angular momentum 15, 434, 437
m_s, quantum number 437, 438
m-xylene isomerization 567
Madelung constant 125(T)
magnetic
measurements 212
moment
and electronic configuration 213
orbital contributions 213, 214
spin only 213(T)
magnetism, cooperative 585
Marcus
cross-relation 489
equation 489
and enzymes 623, 624
inverted region 488
mechanism
associative 467
dissociative 467
interchange 467
intimate 467
of ligand substitution 222–5
mediator 621
Meissner effect 587
mercury organometallics 314
meridonal, *mer,* isomers 197
metal
basicity 515
borides 361
carbides 361–4
carbonate decomposition 131(T)
carbonyl anions 514–6, 534
carbonyls, electron count 502(T)
clusters 198, 528–536
electron count 531
and structure 532(T)
isolobal analogies 534
ligand transformations 536
synthesis 534
complexes
bonding 205–17
chelate 202
chiral 203–5
coordination number 192–6
nomenclature 200
polymetallic 198, 199
reactions 217–25
structure and symmetry 192–205
structures of 193(T)
halides 420–4
oxides 392–4
silicides 364
sulfides
intercalation 596, 597
layered 593–7
metal–metal
bonded complexes 198–9, 529–36
photochemistry 494
bonding 459, 528–34
bonds
fragmentation 535
protonation 535
multiple bonds 530(T)
addition to 535
metal–sulfur clusters 397
metallocenes 522
electronic structure 523–5
metalloenzymes 605
metalloid 96
metalloporphyrin 610
metals 111–15
close-packed 112
metathesis
defined 172
of organometallics 303
methide 362
methylsulfonic acid 158
mica 349
microstates 434, 435
classification 437
d^2 437
migratory insertion reaction 517
mirror plane 51
mixed-valence complexes 460
ML_5 complexes 456–7
MLCT spectra 448
molecular orbital energies
of the halogens 404
molecular
orbitals
antibonding 60
of $[B_6H_6]^{2+}$ 359
bonding 59
of chains and rings 77–8
of diatomics 58
formation of 79
frontier 63
of heteronuclear diatomics 65
nonbonding 65
octahedral complexes 205–10
Period 2 diatomics 60–1
of polyatomics 74–82
in solids 92–101
symmetry adapted 85–7
shape 42, 43(T)
and hybridization 45–7
timescale of 47, 49(T)
and VSEPR 44–5
and Walsh diagrams 87–9
sieves 350–3
symmetry 50–7
molybdenum disulfide 396
mononuclear acids 158
monoxides
3*d* metals 580–3, 581(T)
defects 582
electrical conductivity 582, 583
Monsanto acetic acid process 556
multiphasic catalysts 558
multiplicity 436
mutagenesis, site specific 619
myoglobin 611
low spin Fe(II) 612
O_2 binding 612–4

Néel temperature 585
neon 403(T), 425
nephalauxetic
parameter 445
series 445
Nernst equation 238, 239
nickel
arsenide structure 118
complexes 203, 222
substitution 224
ligand field parameters 209, 211
tetracarbonyl 502
nicotine adenine dinucleotide 621
nido boranes 355, 358
nine-coordinate complexes 198
nitrate, redox properties 380
nitric
acid
production and use 377
synthesis 379
oxide 381
nitrogen 370(T)
activation 374, 375
complexes 375, 628
fixation 375, 627–9
group, properties 370
halides 375, 376
ligand 506
molecule, electronic configuration 62
N—N bond enthalpy 404
oxides 377–82
oxoanions, structures and properties 379
production 371, 372
redox chemistry 378
trifluoride 375
uses 372
nitrogen(I) oxide 381, 382
nitrogen(II) oxide 381
nitrogen(IV) oxide 380
nitrosonium compounds 381
nitrosyl ligand 506
nitrous
acid, redox properties 381
oxide 381, 382
NMR, chemical shifts 273, 293
further information 292
multinuclear 293
shielding 293, 294
solid state 296
spin–spin coupling 294, 295
noble gases 424–8
atomic properties 403(T)
compounds 426–8
properties 425
nodal plane 60
non-crossing rule 445
nonbonding orbitals 65
octahedral complexes 206
noncomplementary redox reaction 241
nonstoichiometric compounds 576, 577(T)
nuclear
binding energy 6(T)
magnetic resonance 292–6
particles 4
stability 5
synthesis 4
nucleophile 170
nucleophilic
attack 548
of ligands 547
discrimination 472
substitutions 466
nucleophilicity 223, 466
discrimination factor 472, 473(T)
parameter 472(T)
nucleosynthesis 5, 6

octahedral complexes 193, 195
cis–trans isomers 196
distortion of 196
fac–mer isomers 197
octet rule 38–41
one-dimensional solids
bands in 96
Peierls distortion of 97
orbital
angular momentum 16, 434, 436, 462
quenching 212
atomic 15
d 19
distribution function 18
f 19
hydrogenic 17
p 19
penetration 22
quantum numbers 15
s 18
shapes 20
energies of diatomics 61
radial and tangential 357
shapes 17, 671–80(T)
symmetry-adapted 671–80(T)
organelles, elements present 604
organic ligands 501(T)
organoaluminum compounds 316–18
organoarsenic compounds 327–8
organoboron compounds 315
organocadmium compounds 314
organogermanium compounds 326
organolithium compounds 310
organomercury compounds 314

organometallic
  compounds
    alkali metals 310–12
    bonding 500-4
    classification 300
    definition 499
    discovery 499
    early *d* block 527
    f block 527
    gallium, indium, and thallium 318–19
    Lewis acids 309
    ligands 501(T)
    nomenclature 300, 499, 500
    nucleophiles 308
    oxidation 306
    oxidation numbers 504
    radical anion salts 311
    structures 301–2
    stability 305
    synthesis 302–4
  reactions 306, 308–10
organoplumbanes 326
organosilicon compounds 319–25
organostannanes 326
organozinc compounds 314
ORTEP diagram 116
orthophosphate 164–5
outer sphere
  complex 192
  electron transfer 240
  protein 622
  redox mechanism 240, 486, 487
overlap of orbitals 59
overpotential 235, 239, 240, 568
oxaloacetate decarboxylase 615
oxidation 229
  alkenes 557, 558
  number 504
    and stability of solids 133
  radical 548
  of solutes
    by oxygen 245, 246
    by water 242
  states
    of *d*-block metals 256
    of *f*-block metals 258
    of metals, trends 255
oxidative addition 549
  mechanism 491–2
oxide
  acid–base properties 160–1
  d block 393–4, 580–91
  ionic conductors 590, 591
  ores, reduction 229–34
  p block 289–92
  s block 392–3
  transfer 188
oxidizing agent 229
oxoacid 150, 157, 158–60
  structure and acidity 159(T)
  substituted 158
oxoanion
  charge-transfer transitions 446, 447
  reduction, influence of pH 251
oxophilic metal atoms 527
oxygen 370(T)
  carriers
    distribution 610
    models 612
  difluoride 412
  evolution
    catalysis 243
    Nernst equation 244
    overpotential 243
  fluorides 387, 412
  group 370, 371, 386–400
  molecular 386
    electronic configuration 62
  O—O bond enthalpy 404
  redox chemistry 389
  transport 610–15
ozone 387

p
  band 94
  block metal
    trends in redox stability 257
  orbital 19
pK_a 149
pK_f 173
pK_w 149
π-acid ligands 208, 211–12, 506–7
π-base ligands 208, 211–12
π orbitals 61
packing of spheres 108–11
pairing energy 209, 210
palladium complex
  catalysts 550, 557
  substitution 471
paramagnetism 212, 586
passivation of metals 242
Pauli principle 21, 435
Pauling's acidity rules 159
Peierls distortion 97
penetration 22
pentagonal bipyramidal complexes 198
pentaiodide 410
peptide
  hydrolysis 617
  link 605
perbromate, preparation 420
perchlorate
  dangers 414
  weak basicity 414
periodate
  complexes 415
  oxidations 420
periodic
  acid 413, 414
  table, structure 6–9
permanganate stability in water 243
perovskite 584
  structure of 119
peroxide ion 393
peroxydisulfate 391
perturbation theory 33–35
perxenate ion 427
pH 149
phase
  diagrams, 599–601
  rule 599
phenanthroline ligand 220
phosphates 384, 385
phosphazenes 385, 386
phosphine, reaction with metal ions 292
phosphite 384–5
phosphorescence 450
phosphorescent complexes 451
phosphoric acid
  production 373
phosphorous acid, structure and acidity 158
phosphorus 370(T)
  disproportionation 384
  halides 376
  importance 372
  nitrogen compounds 385
  oxides 384
  oxoanions 384–5
  polymorphs 373
  production 373
  trifluoride ligand 507
photochemical
  charge-transfer reactions 493
  d–d reactions 493
  reaction mechanisms 492–4
photoconductors 367
photoelectron spectra 66
  of molecules 63–4
  of solids 96
photon 13
photosynthesis 629–31
photosystem I 629
photosystem II 629
physisorption 563
Pidgeon process 229
piezoelectrics 587
platinum
  chemisorption on 561, 563
  complexes
    cluster 199
    lability 224
    square planar 194, 215–17
    substitution 471–7
  electrocatalysis 568
  hexafluoride 423
  nucleophilicity parameters 472(T)
plumbane 289
pnictides 371
point
  defects 573
  groups 53, 53(T), 54(T)
  of zero charge 164
polar molecules
  and symmetry elements 56
polarizability 33
polonium 370(T)
polyanions 164–5
polyhalides 410, 411
polyhalogen cations 410
polyiodide structures 411
polylithiated 311
polymetallic complexes 198, 199
polymorphism 108
  of metals 113
polyoxo
  anions 164
  cations 163
  ion formation 162–6
polyoxoanion
  structures 164, 165
polyphosphates 165–6
polyprotic acid 150
  successive pK 159
polysulfido complexes 397
polytypes 109
porphyrin 202
  high spin Fe(II) 611
  picket fence 613
potassium ion pumping 608, 609
potential energy
  molecular 63
Pourbaix diagram 252–4
  iron 252
  manganese 254
  natural waters 253, 254
primitive cubic 112
prochiral 553
projection operators 102–4
prompt reactions, photochemical 493
proteins 605
protic
  character 282
  solvent 154
proton
  acceptor 146
  affinity 151
    gas phase 152(T)
    influence of solvation 153
    in water 152(T)
  donor 146
  gain enthalpy 151
  properties 4
  Sponge 220
  transfer 147
    enthalpy 152
    equilibria 147
    Grotthus mechanism 147
    rate constant 147(T)
protonation enthalpy 151
pseudohalides 406, 407(T)
pseudohalogens 406
pseudorotation 195
pumps, biological 604–5, 607–10
pyrazine bridging ligand 460
pyrolysis 288
pyrometallurgical extraction 232
pyrophosphate 166, 385

quantum numbers 15
quantum yield 493
quenching, of orbital angular momentum 212

Racah parameters 440, 441(T)
radial distribution function 18
radius
  atomic 26
  covalent 68(T)
  Hägg's rule 364
  influence of coordination number 111
  ionic 26, 27(T), 129
  metallic 25, 111
  ratio 121
  thermochemical 129(T)
radon 403(T), 425
rare gases 425
  see noble gases
rate law, substitution 467
rates of redox reactions 239–42
Ray–Dutt twist 484, 485
reaction quotient 138, 238
rearrangement energy
  inner-sphere redox 487
red lead 347
redistribution reaction 304
redox
  catalysis 620–7
  condensation 534
  couple 236
  half-reaction 236
  mechanism 486–92
    group transfer 490
    Marcus theory 489
    one-equivalent 491
    potential energy curves 488
    theory 487
    two-equivalent 491
  mediators 621
  protein rates 623(T)
  reaction
    by atom transfer 241
    by electron transfer 240
    noncomplementary 241
    of oxoions 241, 377, 381, 391, 416–7, 426
  redox stability in water 242–5

reducing agent 229
reduction 229
  potentials 235–9, 642(T)
  of solutes by water 243
reductive
  carbonylation 508
  elimination 549
reflection 51
reorganization energy, outer sphere
  redox 487
resonance 40, 41
rhenium
  heptafluoride 422
  trichloride 422
Rhizobium 627
rhodium complex
  catalyst 543–7
Robin–Day classification 460–2
rock-salt structure 115
rotation axis 50
rubredoxins 628
ruby 340, 451, 590
Russell–Saunders coupling 435
rutile, structure of 118–9

S, total spin, 435, 436
s
  band 94
  orbital 17–8
σ
  bonding in complexes 206–7
  orbital 60
sandwich compounds 522, 527
scandium subchlorides 421
Schiff's base 203
Schlenk
  equilibrium 313
  techniques 306–7
Schoenflies system 664
Schottky defect 574
Schrock carbenes 519–20
Schrödinger equation 11
second acidity constant 150
selection rules
  electronic transitions 448–50
  spin 448
selenic acid 392
selenium 370(T)
  elemental 370
  halides 387–9, 388(T)
  polycations 398–9
  polymorphs 387
  tetrafluoride
    reactions 389
semiconductors 92–3, 98–100, 582–3
  acceptor band 101
  direct gap 366
  donor band 100
  doping of 100
  electronic properties 364–7
  extrinsic 100
  indirect gap 366
  intrinsic 99
  n-type 100
  p-type 101
  temperature dependence of
    conductivity 92, 95–6, 98–9
semimetal 96
seven-coordinate complexes 198
shape and timescale 47–9
shielding 22
Si—C bond redistribution 320–2
Si=Si compounds
  electronic structure 325
  reactions 325
  synthesis 324
siderophile 8
silanes 287–9
  reactions 287
  synthesis 289
silazanes 322
silenes 324
silica
  gel
    surface acidity 187
    surface modification 187
  glasses 348
  minerals 161
  surface acidity 187
Silicalite 351
silicates 347–9
silicides 364
silicon 333(T)
  electronic properties 364–6
  halide acidity 183
  halides 346
  hydrides 287–9
    see silanes
  organometallics 319–25
  production 232–3
  semiconducting properties 364–6
  tetrahalides
    Lewis acidity 183
silicon—carbon bond formation 320
silicones 322–3
siloxanes 322
silyl halide protolysis 322
six-coordinate complexes 195–7
sixteen-electron complexes 503
sixteen/eighteen electron rule 503(T)
slag formation 188
smelting 229
sodalite cage 352
sodium
  cyclopentadienide 312
  ion pumping 608, 609
  naphthalide 311
soft acids and bases 173–6, 174(T)
solid electrolytes 578–80
solubility of salts
  correlation with ionic radii 133–5
solvation
  electrostatic theory of 154
  enthalpy
    factors influencing 153–4
  free energy of ions 154
  of acids 153
  parameter 178, 179(T)
solvent leveling 154–5
spectator ligand 471
  octahedral substitution 480
spectra
  d-metal complexes 441–51
  f-block complexes 451–2
spectrochemical series
  ligands 211–2
  metals 212
spectroscopic terms 434–6
sphalerite
  structure of 117
spin
  angular momentum 462–3
  multiplicity 436
  orientations 434
  quantum number 213
spin-allowed transitions 448
spin-forbidden transition 448
spin-only paramagnetism 212
spinel
  occupation factor 583, 584
  structure 583, 584(T)
spontaneous processes 136
square planar complexes 194, 195
  frontier orbitals 216
  molecular orbital energies 215, 216
  substitution 223–4
    activation parameters 476(T)
    entering group 472–3
    leaving group 471
    stoichiometric mechanism 476
square pyramidal complexes 195,
  457–9
stability field of water 243–4
standard
  conditions 237
  electrode potential 236
  enthalpy of formation 135
  potentials 640–63(T)
  reaction enthalpy 135
  reduction potential 237, 238(T),
    642–63(T)
    influence of complex
      formation 254, 255
    pH dependence 250–4
stannane 289
state property 135, 136
stereochemistry
  octahedral activation 482(T)
  square planar substitutions 474–5
steric effect
  on chelate formation 220
  octahedral substitution 480
  square planar substitution 472–3
stoichiometric mechanism 466, 467
  octahedral substitution 485
  square planar substitution 472–3
strong
  acid 149
  field
    case 210
    complexes 212
    limit 443
  ligands 211
structural techniques
  timescale of 47–9
structure
  and acidity 160
  map 122
subatomic particles 4(T)
substitution mechanisms
  determination 468(T)
  octahedral complexes 477–84
  square planar complexes 471–7
substitution reactions
  metal complexes 217
  square planar complexes 223–4
sulfides 395–7
  intercalation 596–8
  layered 593–8
sulfido complexes 396–397
sulfite, redox chemistry 391
sulfur 370(T)
  allotropes 387
  chlorides 389
  containing clusters 399–400
  dioxide 390
    complexes 391, 507
    Lewis acidity and basicity 184
    ligand 507
  halides 387–9, 388(T)
  nitride polymer 400
  nitrogen compounds 399–400
  oxides 390
  oxoanions 391–2, 392(T)
  polycations 399
  production 235
  tetrafluoride, reactions 389
  trioxide 390
sulfur–nitrogen clusters 399–400
sulfur–phosphorus clusters 399
sulfuric acid 146
  production 146, 184, 566
sulfurous acid
  acid strength 160
sulfuryl dihalides 390
superacid 183
superconducting elements 588(T)
superconductors 101, 587–90
  Cooper pair in 101–2
  high temperature 587–90
  Type I 587
  Type II 587
superexchange 585
surface
  acids 186–7
  kinks 564
  migration 564
  steps 564
symmetry
  adapted orbitals 85–7, 102–4,
    671–80(T)
  and chirality 203
  elements 50(T), 50–2
  labels 82–7
  operations 50(T) 50–2
  type 83

talc 349
Tanabe–Sugano diagram 444, 445,
  681–3(T)
tantalum disulfide 396
teflate 390–1
telluric acid 392
tellurium 370(T)
  elemental 370, 387
  halides 388(T)
template synthesis 203
term symbol
  atomic 436
  molecular 442
terms 434–5
  d^2 437
  energy 439, 440
  ground state 439
  molecular 442, 443, 459
  spectroscopic 434
tetradentate ligands 200
tetragonal complexes
  frontier orbitals 216
  molecular orbital energies 215, 216
tetragonal distortion 196
tetrahedral complexes 193, 194
  molecular orbital energies 215
thallium 333(T)
  halides 341
  inert pair effect 257
  organometallics 319
  properties 332–3, 333(T)
thermal
  decomposition of
    carbonates 130(T)
  stability of solids 130–3
thermally stable salts 130
thermochemical radii 129–30, 129(T)
thermodynamics
  of equilibrium 138
  First Law 135–6
  and Gibbs free energy 137
  Second Law 136–8
thionyl halides 390
three-center bonds 81, 82
three-five (III/V) compounds
  electronic properties 364–7

General index

timescale of structure techniques 47–9
tin 333(T)
  cluster anions 397–8
  dihalides 346–7
  gray 342
  organometallics 326
  white 342
Tolman cone angle 512(T)
trans effect 474
transition
  dipole moment 449
  metal
    halides 420–4
    monoxides 393–4, 580–3
    oxides 393–4, 580–91
    sulfides 395–7, 593–7
    see d-block metal
transmetallation 303
tridentate ligands 200
trigonal
  bipyramidal complexes 193, 195, 457–9
  prismatic complexes 196, 484
triiodide 410
tritium
  preparation 266
  radioactivity 266
  see also hydrogen 266
turnover frequency 543
Type I and II superconductors 587

uncertainty principle 10
uniform catalysts 558
unit cell 108
uranocene 499, 504

vacuum line 284
van der Waals
  forces 342
  gap 342
van't Hoff equation 139
vitreous state 591
volcano diagram 563, 569
volume of activation 470, 471
VSEPR 42–5

Wacker process 557, 558
Wade's rules 354–6
Wade–Mingos–Lauher rules 531
Wadsley defects 576
Walsh diagram 87–9
water
  acidity window 154–6
  O—H stretching frequencies 282
  oxidation 235, 243
wave equation 11
wavefunction
  normalization 14
  of a particle 9, 10
  particle in a box 11–12
weak
  acid and base 149
  complexes 212
  field
    case 210
    limit 443
    ligands 211
Werner complexes 192
Wilkinson's catalyst 551
wurtzite
  structure of 118
wüstite 581

X-ray
  diffraction, structure determination 116
  emission spectra
    of solids 95, 96
  photoelectron spectroscopy 562
xenon 403(T), 425–8
  carbon bonds 427
  nitrogen bonds 427
  fluorides 426, 427
  trioxide 427

Zachariasen's rules 592
zeolite 350–3, 560
  catalysts 567
zero-point energy 12, 267
zinc
  blende
    structure of 117
  organometallics 314
  sulfide
    structures 117–8
Zintl
  clusters 397, 398
  phases 114
zirconium
  diiodide 422
  monochloride
    preparation 421
    structure 421
  tetrachloride 423
ZSM-5 351, 560